周邦新

周邦新文选

Selected Works
of Zhou Bangxin

下　卷

上海大学出版社
· 上海 ·

锆合金在 550 ℃/25 MPa 超临界水中的腐蚀行为[*]

摘　要：选用了 Zr‐4，N18(NZ2)，N36(NZ8)和 M5 等 4 种比较典型的锆合金，在 β 相水淬及变形后，经过 580 ℃，5 h 和 650 ℃，2 h 的热处理，用静态高压釜腐蚀试验研究了锆合金样品在 550 ℃，25 MPa 超临界水中的耐腐蚀性能. 结果表明，4 种合金样品的耐腐蚀性能差别明显，Zr‐4 合金会发生疖状腐蚀，而含 Nb 的 N18(NZ2)，N36(NZ8)和 M5 是均匀腐蚀. 获得数量多，分布均匀的纳米尺度的第二相颗粒，对改善锆合金在超临界水中的耐腐蚀性能是有利的，但远不如合金成分的影响巨大. 调整合金成分是改善锆合金耐超临界水腐蚀性能的主要途径.

1　前言

随着核电技术的发展，在 2001 年底国际上提出了第四代核反应堆（Generation Ⅳ nuclear energy systems）的设计概念[1]. 在确定的 6 种反应堆堆型中，超临界水冷反应堆（SCWR）是其中之一. SCWR 是以已商业化的轻水反应堆（LWR）和火力发电站的超临界化石燃烧锅炉（Supercritical Fossil-fired Boilers）技术为基础，将冷却水的工作参数提高到 500～550 ℃ 和 25 MPa 的超临界状态，这样可以采用直接循环的方式推动透平发电机发电. 与 LWR 相比，由于冷却水在整个循环过程中保持超临界的单相状态，省略了蒸汽发生器、稳压器、汽水分离器等大型部件，反应堆的结构大为简化，节省了投资，缩短了建设周期，热效率也由 33％提高到 45％. 在 SCWR 的高温堆芯中，在高温高压超临界水及中子辐照等恶劣的工况条件下，反应堆堆芯中的结构材料，特别是燃料元件的包壳材料的选材和研究开发是一项具有挑战性的工作[2,3].

在 SCWR 工况条件下，现有的锆合金无论在力学性能还是在耐腐蚀性能方面都不能满足要求，但由于锆的热中子吸收截面非常小，开发能满足 SCWR 工况下要求的新锆合金，仍然受到极大的关注[4]. 研究锆合金在超临界水中的腐蚀行为，既可以为 SCWR 工程应用及开发高性能锆合金提供实验及理论依据，也可以促进当前核反应堆锆合金的发展. 美国和韩国已于 2003 年签订了名为"Advanced Corrosion-Resistant Zr Alloys for High Burnup and Generation Ⅳ Applications"的国际合作研究项目[4]，研究锆合金在超临界水中的腐蚀问题是其中的一项工作内容，但还未见相关的论文发表，国内也没有这方面工作的报道. 本工作选取有代表性的几种锆合金，进行归一化的 β 相水淬处理后，再进行变形及不同温度的热处理，研究了它们在 550 ℃/25 MPa 超临界水中的腐蚀行为.

2　实验方法

选用 Zr‐4，N18(NZ2)，N36(NZ8)和 M5 等 4 种比较典型的锆合金，板材由西北有色金

---

* 本文合作者：李强、姚美意、刘文庆、褚于良. 原发表于《稀有金属材料与工程》，2007，36(8)：1358‐1361.

属研究院提供,合金成分见表 1. 将尺寸约为(8×150)mm,厚 1.2 mm 的锆合金片状样品,经酸洗(10%HF+45%HNO₃+ 45%H₂O 体积比的混合酸)与水洗后,真空封装在石英管内,加热到 1 020 ℃保温 20 min 后淬入水中,同时敲碎石英管,使样品迅速冷却,称之为 β 相水淬处理. 将 β 相水淬处理后的样品冷轧至 0.6 mm 厚,裁切成(20×9)mm 的小片,经酸洗及水洗后,分组放置在石英管中,在真空(<5×10⁻³ Pa)中分别进行 580 ℃,5 h 和 650 ℃,2 h 的退火热处理,然后将加热炉推离石英管,用水浇淋石英管进行冷却. 取样经机械磨抛后用双喷电解抛光的方法制备 TEM 样品,所用电解液为 10%HClO₄+90%C₂H₅OH,用 JEM - 200CX 透射电子显微镜观察显微组织.用 JSM - 6700F 扫描电镜观察形貌.

表 1 锆合金的名义化学成分
Table 1 Chemical composition of zirconium alloys($\omega/\%$)

Alloys	Nb	Fe	Cr	Sn	O	Zr
Zr - 4	\	0.22	0.1	1.5	\	Balance
N18(NZ2)	0.35	0.35	0.1	1.0	\	Balance
N36(NZ8)	1.0	0.35	\	1.0	\	Balance
M5	1.0	\	\	\	0.16	Balance

将上述热处理后的样品经过标准方法酸洗和去离子水水洗,在 550 ℃/25 MPa 超临界水中进行静态高压腐蚀试验,腐蚀增重为 4 个试样的平均值.

3 结果和讨论

3.1 不同锆合金样品经 β 淬火、形变及热处理后的显微组织

4 种锆合金样品经过 β 相水淬和 50%变形量的轧制后,分别进行 580 ℃,5 h 和 650 ℃,2 h的热处理,显微组织见图 1. 在这两种温度处理条件下试样都发生了再结晶,形成等轴晶

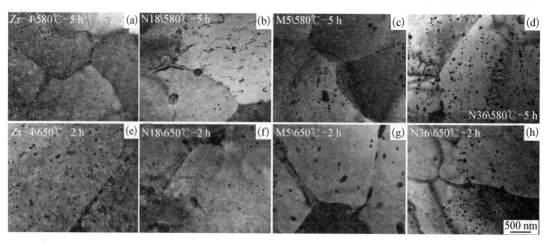

图 1 锆合金样品经 β 淬火、变形及热处理后的显微组织
Fig. 1 TEM micrographs of zirconium alloy specimens treated at 580 ℃, 5 h and 650 ℃, 2 h after β quenching and cold rolling:(a)~(d) Zr - 4, N18, M5, N36 treated at 580 ℃, 5 h;(e)~(h) Zr - 4, N18, M5, N36 treated at 650 ℃, 2 h

粒,但在580 ℃,5 h 处理后的样品中,也观察到有极少部分区域没有再结晶.

580 ℃,5 h 处理的各种锆合金都析出了颗粒细小(<60 nm)的第二相,其数量较多,分布也比较均匀.对于含 Nb 的锆合金,变形后在低于 600 ℃下热处理有利于获得细小并且分布均匀的第二相[5].M5 合金析出的主要是 β-Nb[6],N36(NZ8)的第二相主要是 Zr(NbFe)$_2$ 和 β-Nb[6,7].变形及较低温度的热处理对 Zr-4 合金中第二相的析出也有类似的影响,形成为 10~30 nm 的细小 Zr(Fe,Cr)$_2$ 第二相,有些地方呈条带状分布,这可能与 β 相水淬时合金元素在板条晶界上发生偏聚有关.由于 N18(NZ2)合金 Nb 含量远低于该处理温度下的饱和固溶度,所以析出的主要是 Zr(Fe,Cr)$_2$ 第二相,但其中也会含有一些 Nb[8].与 Zr-4 类似,N18(NZ2)的第二相也呈现出条带状分布的迹象.

与 580 ℃,5 h 处理相比,650 ℃,2 h 处理后的锆合金晶粒尺寸较大,Zr-4 和 M5 样品中的第二相长大并且颗粒数量减少;但 N36(NZ8)和 N18(NZ2)样品的第二相大小并没有明显变化.含 Nb 较多的 N36(NZ8)和 M5 样品在 650 ℃加热时处于 α+β 双相区,因此快速冷却后在晶界上,特别是在三晶交汇处会形成块状的 β-Zr[5].由于海绵锆中有杂质 Fe 的存在,所以 M5 合金同含 Fe 的 N36(NZ8)合金一样,除了 β-Nb 外也会形成 Zr(NbFe)$_2$ 第二相[6],但 M5 析出的颗粒尺度较大(>200 nm).

3.2 锆合金在超临界水中的腐蚀行为

图 2a 为锆合金样品在 550 ℃,25 MPa 超临界水中腐蚀 510 h 的腐蚀增重曲线,图 2b 为 N18(NZ2),N36(NZ8)和 M5 合金最初腐蚀 16 h 时的腐蚀增重曲线.可以看出,在超临界水条件下,成分不同的锆合金腐蚀增重存在极大的差异.对于同种合金样品,580 ℃,5 h 处理的腐蚀增重要比 650 ℃,2 h 处理的低些.Zr-4 的腐蚀增重增加非常迅速,腐蚀时间超过 10 h 后,因样品边缘腐蚀生成的氧化物脱落而无法准确称重.Zr-4 样品腐蚀 1 h 后轧面上就出现了轻微的疖状腐蚀,但片状样品的四周边缘疖状腐蚀非常严重,表现出了强烈的非均匀腐蚀的特性(图 2a).这应与样品经过变形及热处理后形成的织构有关.氧化膜在锆晶体上生长时会呈现明显的各向异性[9],在(0001)面上生长较慢,而在<11$\bar{2}$0>方向生长较快.Zr-4 样品经过轧制变形及热处理后形成了织构,即(0001)面趋向与轧面平行,轧向与横向为耐

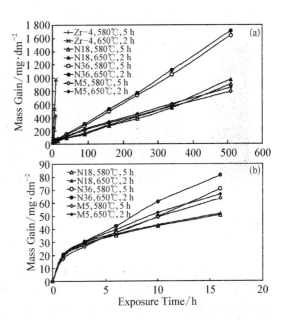

图 2 锆合金在 550 ℃,25 MPa 超临界水中的腐蚀增重变化

Fig. 2 Corrosion behaviors of zirconium alloys in the supercritical-water at 550 ℃, 25 MPa for: (a) 510 h; (b) initial 16 h

腐蚀性能较差的<10$\bar{1}$0>和<11$\bar{2}$0>方向,因而侧面上仍然容易发生疖状腐蚀并发展迅速.

研究不同锆合金腐蚀增重曲线的差异,可以看出合金成分对改善锆合金在超临界水中的耐腐蚀性能有决定性的作用.比较 Zr-4 和含 Nb 的 N18(NZ2),N36(NZ8)和 M5 锆合金的耐腐蚀性能差别,可以看出 Nb 元素的存在对提高锆合金耐超临界水腐蚀的作用非常明显,特别

是能够抑制疖状腐蚀的发生.其作用机制目前还不清楚,有待深入研究.N36(NZ8)与M5合金都含有1%的Nb,但N36(NZ8)的耐腐蚀性能差很多,其中含有Fe并会形成大量Zr(NbFe)$_2$第二相可能是重要的原因.N18(NZ2)具有较好的耐腐蚀性能,与Zr-4相比主要是添加了少量Nb;与N36(NZ8)合金相比,Fe,Sn含量相同,但多了Cr元素,并且Nb含量较低,N18(NZ2)合金中的第二相主要是Zr(Fe,Cr)$_2$.综合比较这几种合金的成分及耐腐蚀性能,还无法归纳出合金元素影响锆合金耐超临界水腐蚀的规律,但至少表明Fe元素添加到Zr-Nb二元合金中会对耐腐蚀性能产生不利的影响.若在含Fe的Zr-Nb系合金中添加适量的Cr元素,使其形成的第二相为Zr(Fe,Cr)$_2$而不是Zr(NbFe)$_2$,则可以改善其耐超临界水腐蚀性能.

N36(NZ8),N18(NZ2)和M5锆合金在550 ℃/25 MPa超临界水中表现为均匀腐蚀(图3a),耐腐蚀性能远优于Zr-4.其中N18(NZ2)合金耐腐蚀性能较好,但M5合金在腐蚀增重超过500 mg/dm^{2}后表现较好,N36(NZ8)合金的耐腐蚀性能与N18(NZ2)和M5合金相比差很多.随着腐蚀时间的延长,均匀腐蚀形成的氧化膜颜色呈现出黑色-黄褐色-深褐色的变化过程,当氧化膜呈现黄褐色时,在放大镜下可以看到裂纹(图3b),在高倍SEM下可观察到大小约为100 nm的等轴晶,在晶界上有微裂纹和孔洞(图3c).腐蚀160 h后,所有样品的氧化膜表面上都出现了肉眼可见的裂纹,但氧化膜并没有脱落.深度腐蚀的样品因两面的氧化膜厚度有差别,产生的应力不均衡而出现翘曲,并且失去韧性易折断,这是因为吸氢量多并在金属基体中生成大量氢化物的结果.

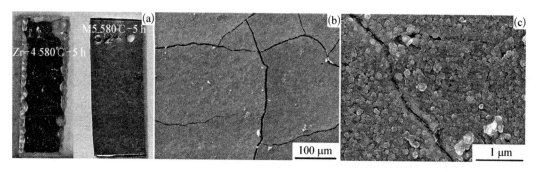

图3 锆合金样品腐蚀后的表面形貌

Fig. 3 The surface morphology of zirconium alloy specimens after autoclave tests:(a) Zr-4 corroded for 16 h and M5 corroded for 160 h;(b),(c):M5 (650 ℃,2 h) corroded for 240 h

在550 ℃/25 MPa超临界水条件下,各种锆合金样品的腐蚀增重都没有出现明显的转折现象.赵文金[8]等研究了Zr-4,N18(NZ2)和N36(NZ8)在500 ℃,10.3 MPa过热水蒸气条件下的腐蚀行为,也没有观测到明显的转折,并且这3种锆合金的腐蚀增重次序类似,表明锆合金在超临界水与高温高压水蒸气条件下的腐蚀机制具有一定的共性.

β相水淬及变形后,再经580 ℃,5 h处理的锆合金样品,其耐腐蚀性能优于经650 ℃,2 h处理的同类合金样品,表明获得数量多并且分布均匀的纳米尺度的第二相颗粒细晶样品,对改善耐超临界水腐蚀性能是有利的,但远不如合金成分的影响巨大.调整合金成分是改善锆合金在超临界水中的耐腐蚀性能的主要途径.

4 结论

(1)获得数量多并且分布均匀的纳米尺度的第二相颗粒细晶样品,对改善耐超临界水腐蚀性能是有利的,但远不如合金成分的影响巨大.

(2) 在 550 ℃/25 MPa 超临界水中,Zr－4 会发生疖状腐蚀,而含 Nb 的 N18(NZ2),N36(NZ8)和 M5 合金是均匀腐蚀.因而合金中添加 Nb 后可以明显提高耐腐蚀性能.

(3) 含 Nb 的 N18(NZ2),N36(NZ8)和 M5 合金由于成分不同,耐腐蚀性能也有明显的差别.调整合金成分是改善锆合金在超临界水中的耐腐蚀性能的主要途径.

参 考 文 献

[1] Bobby Abrams, Douglas Chapin, John Garrick B *et al*. *A Technology Roadmap for Generation IV Nuclear Energy Systems*. U. S. DOE Nuclear Energy Research Advisory Committe,*www. nuclear. gov*, December 2002.

[2] Davit Danielyan. *Supercritical-Water-Cooled ReactorSystem — As One of the Most Promising Type of Generation IV Nuclear Reactor Systems*. *www. nuclear. gov*, November 24, 2003.

[3] Yoshiaki Oka, Seiichi Koshizuka. *Journal of Nuclear Science and Technology*[J], 2001, 38(12): 1081.

[4] Arthur T Motta, Yong Hwan Jeong. *Advanced Corrosion-Resistant Zr Alloys for High Burnup and Generation IV Applications*, Department of Energy International Nuclear Energy Research Initiative DOE/ROK Project Number:. *www. nuclear. gov*, 2003－020－K.

[5] Li Qiang (李强), Liu Wenqing (刘文庆), Zhou Bangxin (周邦新). *Rare Metal Materials and Engineering* (稀有金属材料与工程)[J], 2002, 31(5): 389.

[6] Yong Hwan Jeong, Hyun Gil Kim, To Hoon Kim. *Nucl Mat*[J], 2003, 317: 1.

[7] Lei Ming (雷鸣), Liu Wenqing (刘文庆), Yan Qingsong (严青松) *et al*. *Rare Metal Materials and Engineering* (稀有金属材料与工程)[J], 2007, 36(3): 467.

[8] Zhao Wenjin (赵文金), Miao Zhi (苗志), Jiang Hongman (蒋宏曼) *et al*. *Nuclear Power Engineering* (核动力工程)[J], 2002, 22(2): 124.

[9] Kim H G, Kim T H, Jeong Y H. *Nucl Mat*[J], 2002, 306: 44.

Corrosion Behavior of Zirconium Alloys in Supercritical Water at 550 ℃ and 25 MPa

Abstract: The corrosion behaviors of four zirconium alloys (Zr－4, N18 or NZ2, N36 or NZ8 and M5) have been investigated in the supercritical water (SCW) of 550 ℃ and 25 MPa by autoclave tests. All of the specimens were head-treated at 580 ℃ for 5 h and 650 ℃ for 2 h, respectively, after $\beta-$ quenching and cold rolling. The results showed that the corrosion behaviors of them were distinct. A nodular corrosion appeared on the Zr－4 specimens but uniform corrosions occurred on the others containing Nb alloying element. The nanoscaled second phase particles of uniform distribution and higher density are beneficial to improving the corrosion resistance in SCW. However, the composition of zirconium alloys is a more important factor with a potential possibility for improving the corrosion resistance in SCW by adjusting the composition of zirconium alloys.

温度和脉冲频率对三维原子探针测试结果的影响[*]

摘　要： 在不同的温度和脉冲频率下，用三维原子探针对 Cu - Fe - Ag 合金进行测试. 结果表明，温度和脉冲频率对原子电离的价态有影响；随着温度的升高，Cu 原子以 Cu^+ 电离蒸发的比例增加，以 Cu^{2+} 电离蒸发的比例减少. 随着脉冲频率的提高，Cu 原子以 Cu^+ 电离蒸发的比例减少，以 Cu^{2+} 电离蒸发的比例增加. 温度和脉冲频率对合金成分的测量结果没有影响，对元素同位素丰度的测量结果也没有影响.

三维原子探针（3DAP）是在场离子显微镜（FIM）基础上发展起来的一种分析技术[1-3]. 它是在针状的 FIM 样品尖端叠加脉冲电压使原子电离并蒸发，用飞行时间质谱仪测定离子的质量/电荷比来确定该离子的种类，用位置敏感探头确定原子在原来样品尖端的位置. 通过对不同元素原子的逐个分析，收集数十至百万个离子后，再重新构造出纳米空间中不同元素原子的分布图形. 这种方法能够进行定量分析，是目前最微观、精度较高的一种分析技术. 从逐个分析原子来了解物质微区化学成分的不均匀性，3DAP 是一种不可替代的分析方法.

本文使用的是带能量补偿装置的新一代 3DAP，拥有较高的质量分辨率（$m/\Delta m > 500$），可以将样品冷至 20 K 的低温工作，分析室可以达到 10^{-8} Pa 的超高真空.

由于样品尖端的原子被电离后蒸发，靠质量/电荷比来确定该离子的元素种类，而很多元素可能以几种价态被电离；有的元素有多种同位素；同时由于不同元素的电离能不同，含多种元素的合金，可能存在选择性电离和场蒸发，即电离和场蒸发的难易程度有差异. 本文主要研究温度和脉冲频率的选择对 3DAP 分析结果的影响.

本实验采用含铁的铜合金是一种重要的导电材料，加入铁元素可以通过析出强化提高铜合金的力学性能.

1　实验方案

表 1 为实验用铜合金的化学成分.

表 1　合金的化学成分（原子比）
Tab. 1　Chemical composition of copper alloy

Fe	Ag	Cu
1.13%	0.59%	balance

用电解抛光的方法制备出曲率半径小于 100 nm 的针尖状样品，放入 3DAP 中进行测试. 设定脉冲频率为 5 kHz，在不同的温度下测试，研究温度对测试结果的影响. 设定温度为 50 K，用不同的频率进行测试，研究脉冲频率对测试结果的影响.

* 本文合作者：刘文庆、刘庆冬. 原发表于《真空科学与技术学报》，2007，27（增刊）：53 - 56.

2 实验结果

图1是用 3DAP 在 50 K,脉冲频率为 5 kHz 时得到的质谱图,可以看到$^{63}Cu^{2+}$,
$^{65}Cu^{2+}$,$^{63}Cu^{+}$,$^{65}Cu^{+}$,$^{54}Fe^{2+}$,$^{56}Fe^{2+}$,$^{57}Fe^{2+}$,$^{56}Fe^{+}$,$^{57}Fe^{+}$,$^{107}Ag^{2+}$,$^{109}Ag^{2+}$,$^{107}Ag^{+}$,$^{109}Ag^{+}$
的峰清楚分开,说明该 3DAP 具有很高的质量分辨率. 图2是 Cu 元素的三维空间分布图
(2 nm×3 nm×6 nm),原子层清晰可辨,表明 3DAP 在纵向具有很高的空间位置分辨率.

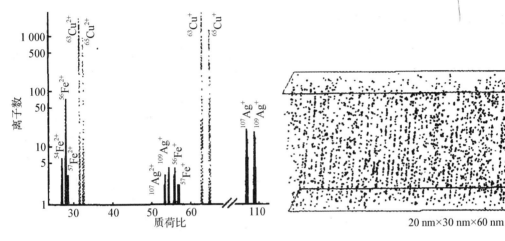

图1 用 3DAP 得到的 Cu 合金质谱图

Fig. 1 Mass spectrum of the copper alloy obtained by 3DAP

图2 Cu 元素的三维空间分布图

Fig. 2 Three dimensional distributions of copper atoms in the alloy

2.1 测试温度对合金成分测量结果的影响

表2是脉冲频率为 5 kHz,在不同温度下测量得到的化学成分. 虽然测试结果与给定的成分有差异,但差别不大,同时结果显示,在不同温度下测量得到的化学成分相当一致,这表明测试温度对化学成分的测量没有影响.

表2 不同温度下测得的化学成分

Tab. 2 Chemical composition of the copper alloy obtained at different temperatures

温度/K	含量/%		
	Fe	Ag	Cu
20	0.87	0.63	balance
30	0.93	0.64	balance
40	0.83	0.64	balance
50	0.78	0.62	balance
60	0.88	0.65	balance

2.2 脉冲频率对合金成分测量结果的影响

表3是试样温度在 50 K,用不同脉冲频率测量得到的化学成分. 结果显示脉冲频率对化学成分的测量没有影响.

表3 不同频率下测得的化学成分

Tab. 3 Chemical composition of the copper alloy obtained at different pulse repetition rates

脉冲频率/kHz	含量/%		
	Fe	Ag	Cu
1	1.55	1.40	balance
2	1.52	0.66	balance
5	1.30	0.65	balance

2.3 测试温度对同位素丰度的影响

表4为Fe,Cu和Ag元素的同位素自然丰度和试样在不同温度下用3DAP测得的同位素丰度,Cu元素同位素测量值和自然丰度符合很好.其中Fe和Ag在合金中含量较低,测试中搜集的原子数少,测量值和自然丰度误差较大.对于Fe元素,丰度很低的$^{54}Fe^{2+}$,$^{57}Fe^{2+}$和$^{58}Fe^{2+}$三种同位素由于受到背低的影响,测量值普遍高于自然丰度值.$^{57}Fe^{2+}$和$^{58}Fe^{2+}$测量值高于自然丰度值可能还有另一个原因,由于制备样品过程中,H元素进入样品,部分$^{56}Fe^{2+}$可能以$(^{56}FeH)^{2+}$和$(^{56}FeH_2)^{2+}$的形式蒸发而被计算成了$^{57}Fe^{2+}$和$^{58}Fe^{2+}$.因此对于Fe元素,丰度的测量值和自然值的误差难以消除.对于Ag元素,由于两种同位素丰度较高且接近,而且Ag和H难以形成AgH_2,只要搜集的原子数足够多,测量值和自然值的误差就会降低到误差范围以内.

试样测试温度对用3DAP得到的Fe,Cu和Ag同位素丰度值没有影响.

表4 不同温度下用3DAP测得的Fe,Cu和Ag同位素丰度

Tab. 4 Isotopic ratios of Fe, Cu and Ag measured by 3DAP at different temperatures

同位素 自然丰度		$^{54}Fe^{2+}$	$^{56}Fe^{2+}$	$^{57}Fe^{2+}$	$^{58}Fe^{2+}$	$^{63}Cu^{2+}$	$^{65}Cu^{2+}$	$^{63}Cu^+$	$^{65}Cu^+$	$^{107}Ag^+$	$^{109}Ag^+$
		5.82	91.66	2.09	0.33	69.1	30.9	69.1	30.9	51.82	48.18
同位素丰度	20 K	6.52	90.59	2.26	0.63	68.56	31.44	68.36	31.64	50.50	49.50
	30 K	7.27	88.63	2.93	1.17	68.13	31.87	68.25	31.75	51.34	48.66
	40 K	7.28	89.01	2.25	1.46	68.62	31.38	67.87	32.13	52.82	47.18
	50 K	6.36	88.87	3.61	1.16	69.11	30.89	68.09	31.91	52.38	47.62
	60 K	7.50	87.64	3.16	1.71	68.44	31.56	68.19	31.81	50.31	49.69

2.4 脉冲频率对同位素丰度的影响

表5为Fe,Cu和Ag元素的同位素自然丰度和在不同脉冲频率下用3DAP得到的同位素丰度,排除误差的因素,结果显示脉冲频率对用3DAP得到的Fe,Cu和Ag同位素丰度值没有影响.

表5 不同脉冲频率下用3DAP测得的Fe,Cu,Ag同位素丰度

Tab. 5 Isotopic ratios for Fe, Cu and Ag measured by 3DAP at different pulse repetition rates

同位素 自然丰度		$^{54}Fe^{2+}$	$^{56}Fe^{2+}$	$^{57}Fe^{2+}$	$^{58}Fe^{2+}$	$^{63}Cu^{2+}$	$^{65}Cu^{2+}$	$^{63}Cu^+$	$^{65}Cu^+$	$^{107}Ag^+$	$^{109}Ag^+$
		5.82	91.66	2.09	0.33	69.1	30.9	69.1	30.9	51.82	48.18
同位素丰度	1 kHz	5.97	90.03	2.98	1.02	68.39	31.61	67.48	32.52	50.31	49.69
	2 kHz	5.62	91.65	2.17	0.56	68.94	31.06	68.13	31.87	52.38	47.62
	5 kHz	6.54	89.18	3.32	0.96	68.53	31.47	68.61	31.39	52.82	47.18

2.5 测试温度对 Cu 元素原子电离价态的影响

设定脉冲频率为 5 kHz,在不同温度下进行测试. 表 6 为在不同温度下得到的各种离子数量. 结果显示 Fe 原子主要以 Fe^{2+} 被电离和蒸发,随着温度的升高,Fe 原子以 Fe^+ 电离蒸发的比例稍有增加,但不明显. 温度对 Cu 元素电离的价态影响十分明显,随着温度的升高,Cu 元素以 Cu^+ 电离蒸发的比例显著增加,以 Cu^{2+} 电离蒸发的比例显著减少(图 3).

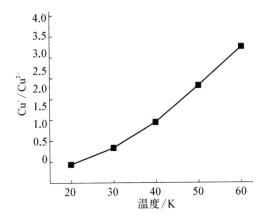

图 3 温度对 Cu 离子价态的影响

Fig. 3 The effect of temperature on copper ion

表 6 不同温度下得到的各种离子数量

Tab. 6 Numbers of all ions obtained at different temperatures

温度 /K	Fe^+	Fe^{2+}	Fe^+/Fe^{2+}	Cu^+	Cu^{2+}	Cu^+/Cu^{2+}
20	19	797	0.024	28 118	64 652	0.43
30	15	853	0.018	41 451	50 212	0.83
40	25	755	0.033	54 436	37 869	1.44
50	45	692	0.065	65 199	27 906	2.33
60	141	1 521	0.093	142 102	43 311	3.36

2.6 脉冲频率对 Cu 元素原子电离价态的影响

设定温度为 50 K,在不同脉冲频率下进行测试. 表 7 为在不同脉冲频率下得到的各种离子数量. 结果显示 Fe 元素主要以 Fe^{2+} 被电离和蒸发,随着脉冲频率的增加,Fe 原子以 Fe^+ 电离蒸发的比例稍有减少,但不明显. 脉冲频率对 Cu 元素电离的价态影响十分明显,随着脉冲频率的提高,Cu 元素以 Cu^+ 电离蒸发的比例减少,以 Cu^{2+} 电离蒸发的比例增加(图 4).

表 7 不同脉冲频率下得到的各种离子数量

Tab. 7 Numbers of all the ions obtained at different pulse repetition rates

脉冲频率 /kHz	Fe^+	Fe^{2+}	Fe^+/Fe^{2+}	Cu^+	Cu^{2+}	Cu^+/Cu^{2+}
1	150	1 274	0.12	79 822	9 628	8.23
2	102	1 245	0.08	76 192	10 681	7.13
5	48	1 146	0.04	72 703	17 592	4.13

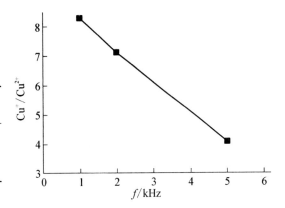

图 4 频率对 Cu 离子价态的影响

Fig. 4 The effect of the pulse repetition rate on copper ion

3 结论

结果表明,温度和脉冲频率对合金成分的测量结果没有影响,对元素同位素丰度的测量结果也没有影响.

当某元素原子可以多种价态蒸发时,温度和脉冲频率对元素原子电离的价态有影响. 随着温度的升高,Cu 元素原子以 Cu^+ 电离蒸发的比例增加,以 Cu^{2+} 电离蒸发的比例减少. 随着脉冲频率的提高,Cu 元素原子以 Cu^+ 电离蒸发的比例减少,以 Cu^{2+} 电离蒸发的比例增加. 温度和脉冲频率对 Fe 元素原子电离价态的影响与 Cu 元素原子相似,只是影响不明显.

在合适的温度和脉冲频率范围内,用 3DAP 可以比较准确地对纳米空间中不同的元素进行定量分析,并给出位置分辨率较高的分布图形.

参 考 文 献

[1] Miller M K, Cerezo A, Hetherington M G, et al. Atom Probe Field Ion Microscopy. Oxford: Oxford Science Publications, Clarendon Press, 1996.

[2] Blavette D, Deconnihout B, Bostel A, et al. Tomographic atom probe. A quantitative three-dimensional nanoanalytical instrument on an atomic scale. Rev Scient Instrum, 1993, 64: 2911.

[3] 周邦新. 三维原子探针——从探测逐个原子来研究材料的分析仪器. 自然杂志,2005,27(3): 125 - 129.

Dependence of Three-Dimensional Atom Probe Analysis on Temperature and Pulse Repetition Rate

Abstract: Ternary alloy of Cu - Fe - Ag was characterized with three-dimensional atom probe at different temperatures and pulse repetition rates. The results show that the temperature and pulse repetition rate strongly affect the ionization mode of the individual composing elements, but little influence the stoichometry of the alloy. For example, as temperature rises, more Cu^+ ions than Cu^{2+} ions are evaporated, whereas as the pulse repetition rate increases, more Cu^{2+} ions than Cu^+ ions are collected. Temperature and pulse repetition rate affect the ionization and field evaporation of Fe in a similar way, though their influence on Fe is weaker then that on Cu.

第二相对 Zr-4 合金在 400 ℃过热蒸汽中腐蚀吸氢行为的影响[*]

摘　要：采用不同的热处理方法制备了 Zr(Fe,Cr)₂ 第二相大小、多少及 Fe/Cr 比不同的样品，并研究了 Zr(Fe,Cr)₂ 第二相对 Zr-4 合金腐蚀时吸氢行为的影响. 结果表明，不同热处理的 Zr-4 样品，经 400 ℃/10.3 MPa 过热蒸汽腐蚀后吸氢量差别很大，与耐腐蚀性能之间并没有直接的对应关系. 经 β 相加热空冷处理的样品第二相较小，也比较少，虽然耐腐蚀性能最差，但吸氢分数最低；经 800 ℃加热处理的样品第二相粗大，数量增多，腐蚀速率略低于 β 相处理的样品，但吸氢分数最高；经 720 ℃和 600 ℃加热处理的样品第二相大小介于前 2 种样品之间，数量也很多，腐蚀速率明显低于 β 相处理的样品，但吸氢分数却高于后者. 认为镶嵌在金属/氧化膜界面处还未被氧化的 Zr(Fe,Cr)₂ 第二相可以作为吸氢的优先通道，因而第二相的大小和数量是影响 Zr-4 合金腐蚀时吸氢行为差别的主要原因.

1　引言

锆合金用作核燃料的包壳，在反应堆运行时，锆与高温高压水反应生成氧化锆的同时放出氢，部分氢被锆吸收，由于氢在 α-Zr 中的固溶度很低（室温时 $<1\ \mu g/g$，400 ℃时为 $200\ \mu g/g^{[1]}$），多余的氢将以氢化锆的形态析出而使锆合金变脆. 过去曾因氢脆导致燃料棒的包壳沿轴向发生开裂的事故[2]. 另外，合金中的氢还会沿应力梯度和温度梯度向张应力大的方向和温度低的方向扩散，产生局域性的氢浓度增高，引起氢致延迟开裂[3,4]. 用透射电镜原位观察的结果证实，氢化锆在裂纹尖端张应力作用下的析出、长大、开裂，以及这种过程的周而复始就构成了氢致延迟开裂[5]. 因而腐蚀和吸氢是锆合金应用中的两个重要问题，涉及到核电站运行时的安全性. 随着燃料组件燃耗的进一步提高，如何降低锆合金腐蚀时的吸氢量越来越受到关注.

一般来说，优化锆合金的成分或热处理制度，提高耐腐蚀性能后可以降低其吸氢量. 但在研究合金成分对锆合金焊接样品吸氢行为影响时，发现样品中的氢含量与耐腐蚀性能间并没有直接的对应关系[6]. 由于 Zr-2(Zr-1.5Sn-0.18Fe-0.1Cr-0.05Ni)、Zr-4(Zr-1.5Sn-0.2Fe-0.1Cr)和 Zr-Sn-Nb 合金（例如 Zr-1.0Sn-0.3Nb-0.35Fe-0.1Cr）中的合金元素 Fe、Cr、Ni 在 α-Zr 中的固溶度很低，大部分将与锆形成颗粒状 Laves 相析出，如 Zr-4 中的 Zr(Fe,Cr)₂，Zr-2 中的 Zr(Fe,Cr)₂ 和 Zr₂(Fe,Ni).

普遍认为锆合金中的第二相对锆合金腐蚀时的吸氢行为有影响，但对第二相如何影响其吸氢行为却没有得到一致的认识. 一种观点认为第二相首先影响氧化膜的特性后（如氧化膜中的微裂纹、孔隙、致密层的厚度等）才引起吸氢行为的差别[7-10]；另一种观点认为氧化膜中暂时未被氧化的第二相为氢提供了短路扩散通道[6,11-13]. 本文旨在进一步研究 Zr(Fe,

* 本文合作者：姚美意、李强、刘文庆、王树安、黄新树. 原发表于《稀有金属材料与工程》，2007,36(11)：1915-1919.

Cr)₂ 第二相大小、多少及 Fe/Cr 比对 Zr - 4 合金在 400 ℃ 过热蒸汽中腐蚀时吸氢行为的影响,并探讨造成这种差别的机理.

2 实验方法

2.1 热处理样品制备

试验用 0.6 mm 厚的 Zr - 4 合金板由中国核动力研究设计院提供,该合金板是按常规工艺制备并经过 600 ℃ 再结晶退火. 为了改变 Zr(Fe,Cr)₂ 第二相的大小、多少及 Fe/Cr 比值,将 Zr - 4 板切成 25 mm×20 mm 的样品,放入真空石英管管式电炉中分别加热至 1 035 ℃ 保温 0.5 h,800 ℃ 保温 36 h,720 ℃ 保温 10 h,然后将电炉推离石英管,在石英管外壁淋水冷却,称为"空冷(AC)".

2.2 腐蚀试验

将上述 4 种不同热处理样品一起放入高压釜中进行腐蚀试验,每种样品各放 10 块以便中间取样进行氢含量测定,腐蚀条件为 400 ℃/10.3 MPa 过热蒸汽. 腐蚀试验前,样品按标准方法用混合酸(体积分数为 10%HF+45%HNO₃+45%H₂O)酸洗和去离子水清洗,腐蚀增重由 3 块试样平均得出.

2.3 氢含量测试及数据处理

腐蚀后样品中的氢含量用 LECO 公司的 RH - 600 型定氢分析仪进行测定,每次分析时取样 0.1~0.15 g,每个样品分析 3 次取其平均值,标准偏差小于 10 μg/g. 分析过程简述如下:将样品称重后放在高纯的石墨坩埚中,感应加热熔化样品,使样品中的氢以氢气的形式释放出来,用惰性气体(如氩气)携带释放出的氢气,通过被加热的氧化铜,氢与氧化铜反应生成水,用红外吸收光谱法测定水的含量,由此确定样品中的氢含量.

样品酸洗后的厚度在 0.56~0.59 mm 范围内,但大多数为 0.57 mm. 当吸氢分数相同时,不同的样品厚度会影响分析得到的氢含量,因此对测得的氢含量数据均按 0.57 mm 这一厚度作了归一化处理

$$C_{H(t=0.57\,mm)} = C_{H(t)} {}^{*} (t/0.57)$$

式中,$C_{H(t=0.57\,mm)}$ 为归一化处理后的氢含量,$C_{H(t)}$ 为厚度为 t 时测得的氢含量,t:样品的实际厚度(mm).

2.4 显微组织观察

用 JEM - 200CX 透射电镜观察合金样品的显微组织及第二相的大小和分布,薄样品用双喷电解抛光制备,电解液为 10%过氯酸乙醇溶液. 用 JEM - 2010F 高分辨透射电镜配置的能谱仪分析第二相的成分,每种样品分析 20 个左右的 Zr(Fe,Cr)₂ 颗粒. 在分析第二相成分时,时常会包含基体部分,不可能得到准确的 Fe、Cr 绝对含量,因此只测定了 Fe/Cr 比值.

从腐蚀后的样品上切下一块,用混合酸溶去金属基体,然后将氧化膜折断,用 JSM - 6700F 高分辨扫描电镜观察氧化膜的断口形貌. 为了提高图像质量,样品表面蒸镀了一层

金.制样方法在文献[14]中已有描述.

3 结果与讨论

3.1 热处理对显微组织及第二相的影响

表1总结了热处理制度与 $Zr(Fe,Cr)_2$ 第二相的大小、多少及 Fe/Cr 比之间的关系. 图1是样品经不同热处理后的显微组织. 1 035 ℃加热时处于 β 相,Zr-4 中的 Fe、Cr 合金元素可以全部溶于 β 相中,冷却时转变成板条状的 α-Zr 晶粒,晶界处析出较小的棒状第二相(宽度<0.1 μm,长度为 0.2~0.4 μm),数量比较少(图 1a). 在 α 相区(600~800 ℃)退火的样品已经完全再结晶,第二相数量较 1 035 ℃/0.5 h(AC)处理的样品多,但随退火温度升高,第二相发生聚集长大,数量逐渐减少,晶粒也发生长大(图 2b~2d). 14#样品中的大部分第二相尺寸≤0.1 μm,只有少数的第二相尺寸>0.2 μm(图 2d),13#样品中的第二相在 0.1 μm 左右,但也有少量的>0.25 μm(图 2c),12#样品中的第二相较大,为 0.15~0.25 μm(图 2b).

表 1 Zr-4 样品中 $Zr(Fe,Cr)_2$ 第二相大小、多少及 Fe/Cr 比与热处理制度的关系

Table 1 Size, amount and Fe/Cr ratio of the second phase $Zr(Fe,Cr)_2$ particles as functions of heat-treatments

Specimen No.	Heat treatment	$Zr(Fe, Cr)_2$ SPPs		
		Size	Number*	Fe/Cr ratio (Mean)
11#	1 035 ℃/0.5 h, AC	<0.1 μm in width and 0.2~0.4 μm in length	+	1.8
12#	800 ℃/36 h, AC	0.1~0.25 μm	++	1.6
13#	720 ℃/10 h, AC	~0.1 μm (most); >0.25 μm (a little)	+++	2.1
14#	As-received (annealing at 600 ℃)	≤0.1 μm (most); >0.2 μm (a little)	++++	1.9

* The amount of $Zr(Fe, Cr)_2$ SPPs increases with the increase of symbol "+"

能谱分析中发现 4 种热处理样品中绝大部分 $Zr(Fe,Cr)_2$ 第二相的 Fe/Cr 比值小于 3.5,而且数据比较分散(1.1~3.4),但也存在个别高 Fe/Cr 比值的第二相(Fe/Cr>50),需说明的是表 1 中列出的 Fe/Cr 平均值只是 Fe/Cr<3.5 的十几颗第二相颗粒的平均结果. 从表 1 可以看出:12#样品中第二相的 Fe/Cr 比值最低(1.6),11#和 14#样品中的中等(1.8~1.9),而 13#样品中的最高(2.1),这与周邦新等以前的研究结果(1 050 ℃/0.5 h 处理样品中第二相的 Fe/Cr 比在 2.1~2.5 之间,800 ℃/3 h 处理样品中第二相的 Fe/Cr 比在1.5 左右,600~700 ℃/3 h 处理样品中第二相的 Fe/Cr 比在 2.0 左右)并不完全一致[14]. 这可能与分析方法不同有关,本实验采用逐个分析第二相的 Fe/Cr 比值再取平均值的方法,而且该平均值不包含个别高 Fe/Cr 比的第二相颗粒,而周邦新等是采用电化学方法将第二相萃取出来,然后分析所有萃取出来的第二相颗粒,其中肯定包括高 Fe/Cr 比的第二相颗粒[14].

3.2 热处理对锆合金腐蚀时吸氢行为的影响

图 2 是样品在 400 ℃/10.3 MPa 过热蒸汽中腐蚀后的增重曲线. 图 3 是腐蚀后样品的吸氢量与腐蚀增重的关系曲线. 从图 2 和图 3 可以看出,11#样品的腐蚀速率最高,耐腐蚀性能最差,但随腐蚀增重增加时吸氢量的增加最少;12#样品的腐蚀速率略低于 11#样品,但吸氢量随腐蚀增重的变化最大;13#和 14#样品的腐蚀速率均明显低于 11#样品,但吸

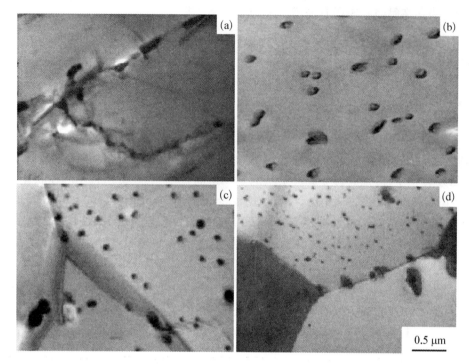

图 1 不同热处理样品的显微组织

Fig. 1 TEM micrographs of Zircaloy‒4 specimens：（a）11♯，（b）12♯，（c）13♯，and（d）14♯

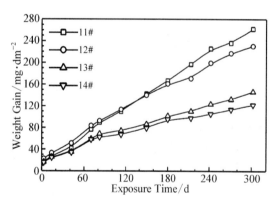

图 2 不同热处理样品在 400 ℃/10.3 MPa 过热蒸汽中 腐蚀时的增重变化

Fig. 2 The weight gain with exposure time obtained from Zircaloy‒4 specimens with different heat-treatments after autoclave test at 400 ℃/10.3 MPa super-heated steam

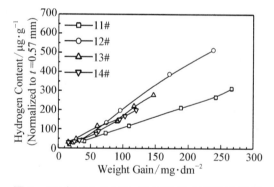

图 3 400 ℃/10.3 MPa 过热蒸汽中腐蚀后的样品吸氢量与腐蚀增重关系曲线

Fig. 3 Hydrogen absorbed content vs weight gain for Zircaloy‒4 specimens with different heat-treatments

氢量随腐蚀增重的变化却明显大于后者. 这与文献[15]报道的变化趋势一致. 一般来说, 提高锆合金的耐腐蚀性能可以降低 Zr 与水蒸气反应放出的氢, 从而减少其吸氢量, 但本实验中发现不同热处理样品腐蚀后的吸氢量与耐腐蚀性能间并没有直接的对应关系, 这一现象在以前研究焊接样品的腐蚀吸氢时也观察到过[6].

　　锆合金腐蚀时的吸氢能力一般用吸氢分数来表示, 它是腐蚀后样品吸入的氢量与腐蚀时理论放氢量之比.表2给出了样品在不同腐蚀增重下计算得到的吸氢分数及其平均值.11♯样品的平均吸氢分数最低(16.3%),数据分散性很小;12♯样品的平均吸氢分数最高(28.8%)且数据分散性很大;13♯样品的平均吸氢分数(26.7%)仅次于12♯样品,数据分散性也很小;14♯样品的平均吸氢分数(20.2%)介于11♯和13♯样品之间,其数据分散性也比较小.4种样品的吸氢能力按12♯>13♯>14♯>11♯的顺序依次减小.特别值得一提的是12♯样品和11♯样品的耐腐蚀性能差不多,但前者的吸氢分数比11♯样品的高出约1倍.这说明热处理对锆合金腐蚀时的吸氢行为有着非常明显的影响.

3.3　第二相影响锆合金腐蚀时吸氢行为机理探讨

基于氢在氧化膜/水介质界面处产生的假定,认为氢在氧化膜中扩散的难易是影响锆合金腐蚀时吸氢性能差异的首要因素.因此,Cox等人认为第二相首先影响氧化膜的特性后(如氧化膜中的微裂纹、孔隙、致密层的厚度等)才引起吸氢行为的差别[7-10].这无法解释11♯样品耐腐蚀性能最差,但吸氢分数最低的实验结果.周邦新等[16]从锆合金在腐蚀过程中氧化膜显微组织和晶体结构演化规律的角度,阐述它们与腐蚀动力学变化之间的密切关系时指出:氧化膜生成时发生体积膨胀而形成的压应力使氧化锆晶体中产生许多缺陷(空位、间隙原子等),稳定了一些亚稳相(立方、四方氧化锆甚至非晶);空位、间隙原子等缺陷在温度、应力和时间的作用下,发生扩散、湮没和凝聚,空位被晶界吸收后形成纳米大小的孔隙簇,弱化了晶粒之间的结合力;孔隙簇进一步发展成为微裂纹,使氧化膜失去了原有良好的保护性,因而发生了腐蚀速率的转折.由于氧化膜生成时产生的压应力是无法避免的,所以氧化膜的显微组织结构在腐蚀过程中发生不断演化也是必然的,促进这种演化过程的各种因素必然会降低锆合金的耐腐蚀性能.

表2　样品在400 ℃/10.3 MPa过热蒸汽中腐蚀的吸氢分数

Table 2　The hydrogen absorbed fraction of specimens corroded in super-heated steam at 400 ℃/10.3 MPa

Specimen No.	Weight gain / mg·dm^{-2}	Hydrogen absorbed fraction/%	Mean value of hydrogen absorbed fraction/%
11♯	73.5	15.4	16.3
	109.5	15.7	
	188.8	16.8	
	262.2	17.4	
12♯	74.5	24.9	28.8
	95.7	28.3	
	171.2	34.1	
	237.9	28.0	
13♯	60.4	27.4	26.7
	116.9	25.8	
	147.0	26.9	
14♯	58.8	18.2	20.2
	95.1	22.6	
	120.3	19.9	

Hatano等[11]在研究第二相大小不同的2种Zr-4样品在高温水蒸气中腐蚀转折前的吸氢行为时,也发现粗大的第二相样品中的吸氢量比细小第二相样品中的多,认为氧化膜中

暂时未被氧化的第二相为氢提供了短路扩散通道,第二相大小是引起吸氢行为差别的主要原因. Tägstrom 等[13]在研究第二相对吸氢行为影响时也提出了类似观点,但认为第二相的数量比其大小对吸氢行为的影响更大. Lelièvre 等[12]用核分析技术研究氢在氧化膜中的分布时,发现在氧化膜中仍与金属接触的未被氧化的第二相附近经常能检测到氢化锆,但在氧化膜中未与金属接触的第二相(不管氧化与否)附近则检测不到氢化锆的存在,由此也提出了与上述类似的观点,即氧化膜中暂时未被氧化的第二相为氢提供了短路扩散通道,但强调指出这些第二相必需仍与金属相接触. 这类观点在解释第二相对氧化初始阶段(如转折前)吸氢行为的影响时是合理的,但对氧化后期,即氧化膜较厚时的吸氢行为影响则无法解释. 因为随着氧化膜厚度的增加,氧化膜中的第二相也会氧化. 如果基于氢在氧化膜/水介质界面处产生的假定,那么氢通过氧化膜传输的难易(即氧化膜特性)又会成为引起锆合金腐蚀时吸氢性能差异的首要因素.

当锆合金与水反应发生腐蚀时,氢是否可能在金属/氧化膜界面处产生呢? 根据 $2H_2O + 2e = H_2 + 2OH^{-1}$ 反应式,除一部分氢在氧化膜/水介质界面处产生外,OH^{-1} 通过氧化膜扩散到金属/氧化膜界面处与锆反应也可以直接生成氧化锆和氢. 根据二次离子质谱仪的分析结果,氧化膜中不仅存在氢,也存在 OH^-,尤其是当锆合金在 LiOH 水溶液中腐蚀时更是如此[17]. 这样,金属/氧化膜界面处的组织特征,如第二相的大小、多少以及种类等就可能成为影响锆合金吸氢行为的主要因素.

$Zr(Fe,Cr)_2$ 本身是一种强烈吸氢的金属间化合物,其吸氢速率比纯 Zr 快[18];另外,$Zr(Fe,Cr)_2$ 第二相氧化比 α-Zr 基体慢[19,20]. 基于以上事实,我们曾提出过一种假设:镶嵌在金属/氧化膜界面处未被氧化的 $Zr(Fe,Cr)_2$ 第二相可以作为吸氢的优先通道,并合理地解释了 Cr 含量不同时,锆合金焊接样品腐蚀后的吸氢量与腐蚀增重不成比例的原因[6],这种观点也可以解释本实验中观察到的现象. 11#样品中的第二相较小,也比较少,因此吸氢分数最低;12#样品中的第二相比较大,且比 11# 样品中的多,因此吸氢分数最高;13# 和 14# 样品中第二相的大小介于前 2 种样品之间,但数量比前 2 种样品的多,其吸氢分数介于 11# 和 12# 之间. 另外,$Zr(Fe,Cr)_2$ 第二相的 Fe/Cr 比对吸氢行为可能也有一定的影响,但从本实验获得的结果还无法进行定量的比较,可以认为第二相的大小和数量是影响 Zr-4 腐蚀时吸氢行为差别的主要原因.

4 结论

(1) 采用不同的热处理方法可以制备出 $Zr(Fe,Cr)_2$ 第二相大小、多少及 Fe/Cr 比不同的 Zr-4 样品,它们经 400 ℃/10.3 MPa 过热蒸汽腐蚀后的吸氢量差别很大,与耐腐蚀性能之间并没有直接的对应关系.

(2) 不同热处理的 Zr-4 样品,它们的腐蚀速率按 600 ℃<720 ℃/10 h<800 ℃/36 h< 1 035 ℃/0.5 h 的顺序依次增加,但腐蚀后的吸氢分数按 800 ℃/36 h>720 ℃/10 h> 600 ℃>1 035 ℃/0.5 h 的顺序依次减小. 第二相的大小和数量是影响 Zr-4 腐蚀时吸氢行为差别的主要原因.

参 考 文 献

[1] Kearns J J. *J Nucl Mat*[J], 1967, 22: 292.

[2] Lemaignan C, Motta A T. In: Kahn R W eds. *Nuclear Materials*, Vol 10B(核科学)[M]. Beijing: Science Press, 1999: 3.

[3] Perryman E C W. *Nucl Energy*[J], 1978, 17: 95.

[4] Simpson C J, Ells C E. *J Nucl Mat*[J], 1974, 52: 289.

[5] Zhou Bangxin (周邦新), Zheng Sikui (郑斯奎), Wang Shunxin (汪顺新). *Acta Meta Sinica* (金属学报)[J], 1989, 25: A190.

[6] Yao M Y, Zhou B X *et al*. *J Nucl Mat*[J], 2006, 350: 195.

[7] Cox B. *J Nucl Mat*[J], 1999, 264: 283.

[8] Khatamian D. *J Alloys and Compounds*[J], 1997, 253-254: 471.

[9] Rudling P, Wikmark G *et al*. In: Sabol G P, Moan G D eds. *Zirconium in Nuclear Industry: Twelfth International Symposium*[C]. Philadelphia: ASTM, 2000: 678.

[10] Lim B H, Hong H S, Lee K S. *J Nucl Mat*[J], 2003, 312: 134.

[11] Hatano Y, Hitaka R *et al*. *J Nucl Mat*[J], 1997, 248: 311.

[12] Lelièvre G, Tessier C *et al*. *J Alloys ande Compounds*[J], 1998, 268: 308.

[13] Tägstrom P, Limbäck M *et al*. In: Sabol G P, Moan G D eds. *Zirconium in Nuclear Industry: Thirteenth International Symposium*[C]. Philadelphia: ASTM, 2002: 96.

[14] Zhou B X, Yang X L. *China Nuclear Science and Technology Report*, CNIC-01073, SINER-0065[R], 1996.

[15] Pan Shufang (潘淑芳), Miao Zhi (苗志) *et al*. In: Zhou Bangxin (周邦新) eds. *Proceedings of Nuclear Materials: Radiation and Corrosion of Fuel Materials for Power Reactor* (核材料: 核反应燃料的辐射和腐蚀)[M]. Beijing: Atomic Energy Press, 1989: 1.

[16] Zhou Bangxin (周邦新), Li Qiang (李强), Yao Meiyi (姚美意) *et al*. *Nuclear Power Engineering* (核动力工程)[J], 2005, 26(4): 364.

[17] Liu Wenqing (刘文庆). *The Effect of Alloying Elements and Water Chemistry on the Corrosion Resistance of Zirconium Alloys* (合金元素及水化学对锆合金耐腐蚀性能影响的研究)[D]. Shanghai: Shanghai University, 2002.

[18] Shaltiel D, Jacob I, Davidov J. *J Less Common Metals*[J], 1977, 53: 117.

[19] Pêcheur D, Lefebvre F, Motta A T *et al*. *Zirconium in the Nuclear Industry: Tenth International Symposium*[C]. Philadel phia: ASTM, 1994: 687.

[20] Pêcheur D. *J Nucl Mat*[J], 2000, 278: 195.

Effect of the Second Phase Particles on the Hydrogen Absorption of Zircaloy-4 Alloy Corroded in Super-Heated Steam of 400 ℃

Abstract: Zircaloy-4 specimens with different amounts, sizes and (Fe/Cr) ratios of the second phase $Zr(Fe, Cr)_2$ particles have been prepared by different heat-treatments in order to further understanding the role of the second phase in hydrogen absorption which is investigated by corrosion tests in the super-heated steam of 400 ℃/10.3MPa. Results show that the amount absorbing hydrogen is quite different with different heat-treatments, not proportional directly to the corrosion resistance. The β-treated specimen, in which the second phase $Zr(Fe,Cr)_2$ particles are finer and less, shows the lowest hydrogen absorption fraction, although it has the highest corrosion rate. The specimens treated at 800 ℃, in which the the second phase particles are coarser and more, shows the highest hydrogen absorption fraction with corrosion rate only a little lower than that of β-treated specimen. The specimen treated at 720 ℃, in which the second phase particles are moderate in size but more in amount, shows a higher hydrogen absorption fraction than the β-treated specimen, but it has much lower corrosion rate than the specimens treated at 800 ℃. So, it is concluded that the amount and size of the second phase $Zr(Fe,Cr)_2$ particles are responsible for different hydrogen absorption behaviors for different heat-treated specimens, as un-oxided second phase $Zr(Fe,Cr)_2$ particles embedded and exposed at the metal/oxide interface could act as a perfect path for hydrogen absorption.

热处理对 Zr-4 合金在 360 ℃ LiOH 水溶液中腐蚀行为的影响[*]

摘　要：研究了热处理对 Zr-4 样品在 360 ℃/18.6 MPa/0.01 mol/L LiOH 水溶液中耐腐蚀性能的影响. 结果表明：经过 β 水淬处理的样品耐腐蚀性能最好，腐蚀 210 d 后的增重与 ZIRLO 和 N18 合金的相当，氧化膜表面黑色光亮；β 水淬后经冷轧再加热到 580～750 ℃ 退火处理的样品，它们的耐腐蚀性能都不如 β 水淬处理的样品，但是在退火的温度范围内，样品的耐腐蚀性能随处理温度的升高而提高. 讨论了热处理对 Zr-4 合金耐腐蚀性能影响的机理，认为不同热处理改变了 Fe+Cr 在 α-Zr 中的过饱和固溶含量是引起耐腐蚀性能差别的主要原因.

1 引言

核电站运行时在一回路水中添加了 H_3BO_3，用 [10]B 作为可燃毒物来控制和调节过剩的核反应性. 为了减少一回路中各种钢构件腐蚀产物的释放及放射性物质的迁移，降低工作人员受辐射剂量水平，需要采用碱性水（pH 7.1～7.2）. 为此，一回路水中在添加 H_3BO_3 的同时，又要用添加 LiOH 来调节 pH 值. 添加 LiOH 后，对燃料包壳锆合金的耐腐蚀性能会产生有害的影响，使发生腐蚀转折的时间缩短，转折后的腐蚀速率增加[1-4].

目前，大多数压水堆仍然用 Zr-4 作燃料的包壳材料. 通过优化加工过程和调节热处理制度，可以进一步提高 Zr-4 合金的耐腐蚀性能，但对其机理并没有得到一致的认识. 由于合金元素 Fe 和 Cr 在 α-Zr 中的平衡固溶度很低，大部分会以 $Zr(Fe,Cr)_2$ 第二相析出，分布在 α-Zr 基体中，第二相的大小和分布与热处理制度有关，所以有人认为 $Zr(Fe,Cr)_2$ 第二相的大小是热处理影响 Zr-4 耐腐蚀性能的主要原因[5-7]；但热处理在改变第二相大小的同时，α-Zr 中固溶的 Fe、Cr 含量也会发生变化，因此，本文作者之一提出过 α-Zr 中 Fe、Cr 的固溶含量或过饱和固溶含量的不同可能是热处理影响 Zr-4 耐腐蚀性能的主要原因[8,9]. 李聪等用电解方法将 Zr-4 样品中 $Zr(FeCr)_2$ 第二相分离后，测定了固溶在 α-Zr 中的 Fe、Cr 含量，结果证明不同热处理样品中固溶在 α-Zr 中的 Fe+Cr 含量差别很大[10]. 用真空电子束焊接成的单片试样研究了合金元素对锆合金耐腐蚀性能影响，发现成分差别不大的区域耐腐蚀性能却存在很大的差别，认为这可能与焊接的复杂热过程引起 α-Zr 基体中的合金元素过饱和固溶含量不同有关[11]. 本文旨在进一步研究固溶在 α-Zr 中的 Fe、Cr 含量不同时，Zr-4 合金在 LiOH 水溶液中腐蚀行为的差别，并探讨影响其腐蚀行为的机理.

2 实验方法

试验用 2 种厚度（0.6 mm 和 2 mm）的 Zr-4 合金板由中国核动力研究设计院提供，其

---

* 本文合作者：姚美意、李强、刘文庆、虞伟均、褚于良. 原发表于《稀有金属材料与工程》，2007，36(11)：1920-1923.

中 0.6 mm 厚的 Zr-4 合金板是按常规工艺制备并经过 600 ℃ 再结晶退火. 为了改变 Fe、Cr 合金元素在 α-Zr 中的固溶含量,把 0.6 mm 厚的 Zr-4 板切成 25 mm×10 mm 的样品,把 2 mm 厚的 Zr-4 板切成 10 mm 宽的小条,将它们分别放入数支石英管中真空密封,再在管式电炉中加热到 1 020 ℃ 保温 10 min,然后淬入水中并快速敲碎石英管,称为"水淬(WQ)". 将水淬后 2 mm 厚的小条冷轧到 0.6 mm,切成 25 mm×10 mm 的样品后放入真空石英管管式电炉中,分别加热至 580 ℃ 保温 2 h、700 ℃ 保温 1 h 和 750 ℃ 保温 1 h,然后将电炉推离石英管,在石英管外壁淋水冷却,称为"空冷(AC)". 图 1 给出了样品的编号、热处理制度和累积退火参数(ΣA_i).

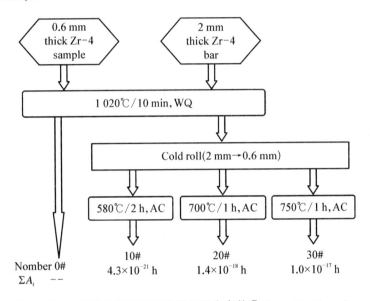

图 1 Zr-4 样品热处理制度及累积退火参数 [$\Sigma A_i = \Sigma i \times \exp(-40\ 000/T_i)$]

Fig. 1 Heat-treatment procedure for Zircaloy-4 specimens and their accumulated annealing parameters

为了对耐腐蚀性能进行比较,将出厂退火态的 0.6 mm 厚样品(编号 14#)和上述热处理样品一起放入高压釜中进行腐蚀试验,腐蚀条件为 360 ℃/18.6 MPa/ 0.01 mol/L LiOH 水溶液. 腐蚀试验前,样品按标准方法用混合酸(体积分数为 10% HF + 45% HNO_3 + 45% H_2O)酸洗和去离子水清洗,腐蚀增重量由 3 块试样平均值得出. 用 JEM-200CX 透射电镜观察样品的显微组织,薄样品用双喷电解抛光制备,电解液为 20% 过氯酸乙醇溶液.

3 结果与讨论

3.1 热处理对腐蚀行为影响

图 2 是几种 Zr-4 样品在 360 ℃/18.6 MPa/0.01 mol/L LiOH 水溶液中腐蚀后的增重变化. β 淬火的 0# 样品耐腐蚀性能最好,腐蚀 210 d 后氧化膜仍然黑亮,腐蚀增重只有 110 mg/dm^2,明显低于出厂退火态的 14# 样品和其他几种样品. 这与 360 ℃ 纯水中的腐蚀规律不一致[9]:在纯水中腐蚀 300 d 时发生了第 2 次转折,转折后腐蚀速率迅速增加. 经过 β

淬火和冷轧变形,重新在 580～750 ℃处理的 10♯、20♯和 30♯样品在腐蚀 70～100 d 时发生了转折,转折后腐蚀速率增加,氧化膜逐渐转变为土黄色;转折后,10♯和 20♯样品的腐蚀速率高于 14♯样品,但 30♯样品的腐蚀速率则明显低于 14♯样品. β 淬火后重新退火处理的温度对 Zr‐4 合金在 360 ℃/0.01 mol/L LiOH 水溶液中的耐腐蚀性能有明显影响,其规律与 Anada[5] 报道的 Zr‐4 合金在 360 ℃纯水中的腐蚀结果一致. 在 α 相区随退火温度升高,腐蚀速率降低,在 700 ℃以上处理时,腐蚀速率明显低于 700 ℃以下处理的样品.

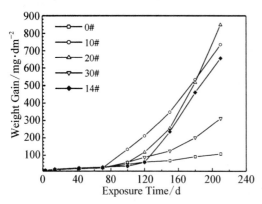

图 2 不同热处理 Zr‐4 样品在 360 ℃/18.6 MPa/0.01 mol/L LiOH 水溶液中腐蚀增重变化

Fig. 2 The weight gains of Zircaloy‐4 specimen with different heat-treatments during the autoclave test in 0.01 mol/L LiOH aqueous solution at 360 ℃/18.6 MPa

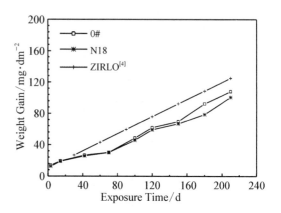

图 3 Zr‐4(0♯样品)与 N18、ZIRLO 合金在 360 ℃/18.6 MPa/0.01 mol/L LiOH 水溶液中腐蚀增重变化

Fig. 3 The weight gains of Zircaloy‐4 (Sample 0♯), N18 and ZIRLO alloys during the autoclave test in 0.01 mol/L LiOH aqueous solution at 360 ℃/18.6 MPa

ZIRLO(Zr‐1Sn‐1Nb‐0.1Fe)合金[4]、E635(Zr‐1.2Sn‐1Nb‐0.4Fe)合金[12] 以及 N18(或称为 NZ2)(Zr‐1Sn‐0.4Nb‐0.3Fe‐0.1Cr)合金[13] 在 360 ℃/LiOH 水溶液中的耐腐蚀性能都明显优于 Zr‐4.本实验中发现 β 淬火的 Zr‐4(0♯)样品也表现出非常优良的耐腐蚀性能.为了与含 Nb 的 N18 和 ZIRLO 合金的耐腐蚀性能比较,将它们在 360 ℃/18.6 MPa/0.01 mol/L LiOH 水溶液中腐蚀的增重曲线一起画在图 3 中(图中 N18 的腐蚀增重数据与文献[13]中的数据相近;ZIRLO 的数据取自文献[4]). 从图 3 可以看出,在腐蚀的 210 d 内,0♯样品的腐蚀增重一直与 N18 和 ZIRLO 合金的相当,表现出非常优良的耐腐蚀性能,这说明过饱和固溶处理可以明显改善 Zr‐4 在 LiOH 水溶液中的耐腐蚀性能.

3.2 热处理影响腐蚀行为机理探讨

应该指出的是成材后的锆管再采用 β 水淬处理在实际生产中是不可行的,但对研究热处理影响 Zr‐4 合金耐腐蚀性能的机理是有用的.至今还未见到关于研究 β 水淬对 Zr‐4 合金在 LiOH 水溶液中耐腐蚀性能影响的报道,但本研究结果表明,Zr‐4 合金经过这样处理后明显提高了耐腐蚀性能,与 Zr‐Sn‐Nb 合金中的 ZIRLO 和 N18 等相当.深入研究这种变化的原因,对于认识锆合金腐蚀机理是有益的.

周邦新等[9] 在研究 Zr‐4 合金耐均匀腐蚀和耐疖状腐蚀性能时,认为固溶在 α‐Zr 中的合金元素含量不同是热处理影响耐腐蚀性能的主要原因.从 Fe、Cr 合金元素过饱和固溶含量不同的角度也可以很好的解释本研究中 4 种不同热处理样品耐腐蚀性能的差别,尤其是 β 水淬的 0# 样品耐腐蚀性能优良的原因. β 水淬的 0# 样品由于冷却速度很快,Fe 和 Cr 可以更多地过饱和固溶在 α-Zr 中,这与 Zr(FeCr)$_2$ 第二相的析出情况也是吻合的,即析出的第二相非常细小且比较少(图 4b 中的箭头所示),而在较低倍数时还分辨不出析出的第二相(图 4a). 当 β 水淬后的样品被重新加热到 α 相区保温冷却后,一方面 α-Zr 中过饱和固溶的 Fe、Cr 含量随 Zr(FeCr)$_2$ 第二相析出而降低,以及随退火温度升高,第二相发生聚集长大,数量逐渐减少(图 4c,4d),但另一方面其平衡固溶度又随温度升高(在 α 相区)而增加. 根据 Zr-(Fe+Cr) 相图可以大致确定[14]:810 ℃时 Fe+Cr 在 α-Zr 中的平衡固溶含量约为 160 $\mu g/g$,750 ℃ 和 700 ℃ 时分别降到 120 $\mu g/g$ 和 100 $\mu g/g$. 李聪等用电解方法将 Zr-4 样品中 Zr(FeCr)$_2$ 第二相分离后,测定了不同热处理样品 α-Zr 中 Fe+Cr 的过饱和固溶含量. 样品在 β 相、α 上限温区加热淬火以及 650 ℃ 加热炉冷,然后经轧制加工和 470 ℃ 去应力退火,Fe+Cr 在 α-Zr 中的固溶含量分别达到了 1 300、780 和 320 $\mu g/g$[10]. 根据以上实验数据和显微组织观察可以确定,经过不同热处理的 10#、20#、30# 和 0# 样品中 Fe+Cr 的固溶(或过饱和固溶)含量将依次增加,这与耐腐蚀性能逐渐变好的趋势是一致的,而耐腐蚀性能与第二相大小之间并无对应的关系.

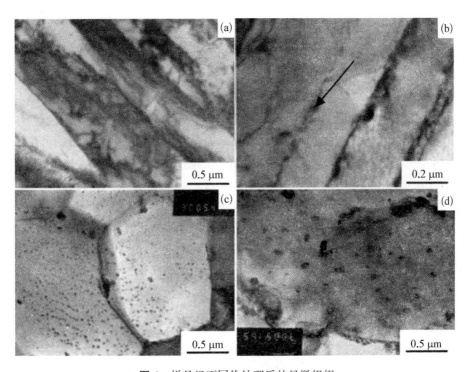

图 4 样品经不同热处理后的显微组织

Fig. 4　TEM micrographs of Zircaloy-4 specimens:(a, b) 0#,(c) 10#, and (d) 20#

周邦新等从锆合金在腐蚀过程中氧化膜显微组织和晶体结构演化规律的角度,阐述了它们与腐蚀动力学变化之间的密切关系[15]:氧化膜生成时发生体积膨胀而形成的压应力使氧化锆晶体中产生许多缺陷(空位、间隙原子等),稳定了一些亚稳相(立方、四方氧化锆甚至非晶);空位、间隙原子等缺陷在温度、应力和时间的作用下,发生扩散、湮没和凝聚,空位被晶界吸收后形成纳米尺寸的孔隙簇,弱化了晶粒之间的结合力;孔隙簇进一步发展成为微裂纹,使氧化膜失去了原有良好的保护性,因而发生了腐蚀速率的转折. 由于氧化膜生成时产生的压应力是无法避免的,所以氧化膜的显微组织结构在腐蚀过程中发生不断演化也是必然的,延缓这种演化过程就可以提高锆合金的耐腐蚀性能. 从这一观点出发,周邦新等[16]进一步研究了合金成分及水化学对锆合金腐蚀时氧化膜显微组织演化的影响,发现在 LiOH 水溶液中腐蚀时,相对 ZIRLO 合金而言,Zr-4 氧化膜中的空位更容易通过扩散凝聚形成孔隙簇和晶界微裂纹,也容易发展成平行于氧化膜/金属界面的裂纹,导致腐蚀转折提早发生,耐腐蚀性能变坏. 这种现象与 Li^+ 和 OH^- 渗入氧化膜后降低了 ZrO_2 表面自由能而促进孔隙的形成有关. 在 ZrO_2 中添加合金元素可以改变它的表面自由能,可以调整这种影响的程度. ZIRLO[4] 和 N18 合金[13] 在 LiOH 水溶液中表现出优良的耐腐蚀性能,这应该与合金中添加了 Nb 有关,Nb 在 $\alpha-Zr$ 中有一定的固溶度. 当锆被氧化时,固溶在 $\alpha-Zr$ 中的合金元素将比第二相中的合金元素更容易固溶到氧化锆中,因而会对 ZrO_2 的表面自由能产生更大的影响. 用这种观点能够比较合理地解释 Zr-4 样品经过不同的热处理后,在 LiOH 水溶液中的耐腐蚀性能存在明显差别的原因,尤其是为什么 β 淬火的样品具有优良的耐腐蚀性能,与 ZIRLO 和 N18 合金相当.

4 结论

(1) 经过 β 水淬处理的 Zr-4 样品,在 360 ℃/18.6 MPa/0.01 mol/L LiOH 水溶液中腐蚀时具有优良的耐腐蚀性能,腐蚀 210 d 后的增重与 ZIRLO 和 N18 合金的相当,氧化膜表面黑色光亮.

(2) β 水淬后经冷轧再加热到 580~750 ℃退火处理的样品,它们的耐腐蚀性能都不如 β 水淬处理的样品,但是在退火的温度范围内,样品的耐腐蚀性能随处理温度的升高而提高.

(3) 不同热处理改变了 Fe+Cr 在 $\alpha-Zr$ 中的过饱和固溶含量,这是引起耐腐蚀性能差别的主要原因.

参 考 文 献

[1] Cox B, Wong Y M. In: Eucken C M, Garde A M eds. *Zirconium in the Nuclear Industry: Ninth International Symposium*[C]. Philadelphia: ASTM, 1991: 643.

[2] Pêcheur D, Godlewski J *et al.* In: Garde A M, Bradley E R eds. *Zirconium in the Nuclear Industry: Twelfth International Symposium*[C]. Philadelphia: ASTM, 2000: 793.

[3] Zhou Bangxin (周邦新), Li Qiang (李强), Huang Qiang (黄强) *et al. Nucl Power Eng* (核动力工程)[J], 2000, 21(5): 439.

[4] Sabol G P, Comstock R J *et al.* In: Garde A M, Bradley E R eds. *Zirconium in the Nuclear Industry: Tenth International Symposium*[C]. Philadelphia: ASTM, 1994: 724.

[5] Anada H, Herb B J *et al.* In: Bradley E R, Sabol G P eds. *Zirconium in the Nuclear Industry: Eleventh International Symposium*[C]. Philadelphia: ASTM, 1996: 74.

[6] Foster P *et al. J Nucl Mat*[J], 1990, 173(2): 164.

[7] Garzarolli G, Steinberg E, Weidinger H G. In: Van Swam L F P, Eucken C M eds. *Zirconium in the Nuclear Industry: Eighth International Symposium*[C]. Philadelphia: ASTM, 1989: 202.

[8] Zhou Bangxin (周邦新) *et al. Chinese Journal of Nuclear Science and Engineering* (核科学与工程)[J], 1995, 15(3): 242.

[9] Zhou Bangxin *et al. China Nuclear Science and Technology Report*, CNIC - 01074, SINRE - 0066 [R], China

Nuclear Information Centre Atomic Energy Press, 1996.

[10] Li C et al. J Nucl Mat[J], 2002, 304: 134.

[11] Yao Meiyi (姚美意), Zhou Bangxin (周邦新), Li Qiang (李强) et al. Rare Metal Materials and Engineering (稀有金属材料与工程)[J], 2006, 35(10): 1651.

[12] Nikulina A V, Markelvo V A et al. In: Bradley E R, Sabol G P eds. Zirconium in the Nuclear Industry: Eleventh International Symposium[C]. Philadelphia: ASTM, 1996: 785.

[13] Zhao Wenjin (赵文金), Miao Zhi (苗志) et al. J Chinese Society for Corrosion and Protection (中国腐蚀与防护学报)[J], 2002, 22(2): 124.

[14] Charquet D, Hahn R et al. In: Van Swam L F P, Eucken C M eds. Zirconium in the Nuclear Industry: Eighth International Symposium[C]. Philadelphia: ASTM, 1989: 405.

[15] Zhou Bangxin (周邦新), Li Qiang (李强), Yao Meiyi (姚美意) et al. Nuclear Power Engineering (核动力工程)[J], 2005, 26 (4): 364.

[16] Zhou Bangxin (周邦新), Li Qiang (李强), Liu Wenqing (刘文庆) et al. Rare Metal Materials and Engineering (稀有金属材料与工程)[J], 2006, 35(7): 1009.

Effect of Heat Treatments on the Corrosion Resistance of Zircaloy - 4 in LiOH Aqueous Solution at 360 ℃/18. 6 MPa

Abstract: The corrosion behavior of Zircaloy - 4 specimens with different heat-treatments was investigated by autoclave tests in 0. 01 mol/L LiOH aqueous solution at 360 ℃/18. 6 MPa. Results showed that the β- water-quench specimens possesses the best corrosion resistance, and the weight gain is comparable to that of ZIRLO and N18 alloys with bright black oxide films after exposure for 210 d. When reheated at 580~750 ℃ after β water quench and cold rolling, the corrosion resistance of the specimens increase with increasing the heat-treatment temperature, but all are worse than that of β- water-quench specimens. The solid solution contents of Fe and Cr in α- Zr can be affected by heat treatment, and will be responsible for the difference of corrosion resistance for different heat treatments.

Zr‑Sn‑Nb 合金耐疖状腐蚀性能的研究[*]

摘　要： 把 N18(NZ2)锆合金样品经过多种不同的热处理后,用高压釜在 500 ℃,10.3 MPa 过热蒸汽中进行腐蚀试验,研究了它们的耐疖状腐蚀性能.结果表明:无论是将样品加热到 β 相,$\alpha+\beta$ 双相还是 α 相后,快冷还是缓冷,它们经过 1 100 h 腐蚀后都没有出现疖状腐蚀.说明在 Zr‑Sn 合金中再添加合金元素 Nb 后,对疖状腐蚀产生了"免疫性".样品在 500 ℃过热蒸汽中的腐蚀增重动力学曲线仍可分为两个阶段,转折发生在氧化膜厚度大约为 3 μm 时.

1　引言

　　锆合金作为核燃料元件包壳材料在反应堆中运行时,与高温水反应生成氧化锆而受到腐蚀.从腐蚀的结果可分为均匀腐蚀与不均匀腐蚀,后者称为疖状腐蚀.一般来说,压水堆一回路水中由于采取了加氢除氧措施,锆合金发生的是均匀腐蚀,而在沸水堆的水介质中含氧量高,锆合金将产生不均匀的疖状腐蚀.

　　发生疖状腐蚀时,先在黑色氧化膜上生成白色斑点,直径 0.1～0.5 mm,截面呈凸透镜状,深度约为直径的 1/5[1].疖状腐蚀是一种成核长大过程,随着疖状斑的增多和长大,最终会连成一片白色氧化膜,由于氧化膜疏松而容易剥落[2,3].以上这些结果都是在 Zr‑Sn 系的 Zr‑4 和 Zr‑2 合金中获得的.本文作者之一在研究新锆合金时,曾观察到数种含 Nb 的锆合金在 500 ℃过热蒸汽中腐蚀 650 h 也未出现疖状腐蚀[4].N18(西北有色金属研究院研制该合金编号 NZ2)是国内研制的含 Nb 新锆合金,尚未制成燃料元件进行堆内辐照考验,究竟这种锆合金是否会产生疖状腐蚀,现在还无法做出结论,文献中也未见到研究热处理和显微组织对含 Nb 锆合金耐疖状腐蚀性能的影响.本工作的目的是研究热处理对 N18(NZ2)锆合金耐疖状腐蚀性能的影响,并了解其机制.

2　实验方法

　　N18(NZ2)锆合金由西北有色金属研究院提供,板材厚度为 1.4 mm,由 500 kg 铸锭加工而成,化学成分列于表 1.先将来料冷轧至 0.68 mm 后切成 20 mm×30 mm 的样品,在不同温度及不同的冷却条件下进行热处理,以便获得不同的显微组织,研究显微组织及第二相对耐疖状腐蚀性能的影响.样品放置在真空石英管中,在 1 000 ℃,0.5 h、820 ℃,1 h、780 ℃,2 h 或 700 ℃,5 h 加热后,采用了两种不同的冷却方式:一种方式是在加热结束后将加热炉推离石英管,在石英管外壁上淋水,让样品在真空石英管中冷却,称为空冷,标为"AC";另一种方式是在加热结束后切断电炉的电源,让样品在真空石英管中随炉冷却,称为

* 本文合作者:姚美意、李强、夏爽、刘文庆、褚于良.原发表于《稀有金属材料与工程》,2007,36(8):1317‑1321.

炉冷,标为"FC".从这样处理的 8 组样品中分出一部分样品再经过 550 ℃,50 h 处理,使可能存在的 β-Zr 分解为 α-Zr 和 β-Nb. 其中还增加了一种处理方式,将 780 ℃,2 h,AC 处理的样品冷轧变形,然后在 550 ℃,50 h 处理,使 β-Zr 分解后得到的 β-Nb 颗粒更细小,这样共获得 17 组不同热处理的样品,可以说几乎包含了所有可能出现的不同显微组织. 样品用混合酸(体积分数为 10%HF+45%HNO₃+45%H₂O)经过标准方法酸洗和去离子水清洗后,用静态高压釜在 500 ℃,10.3 MPa 过热蒸汽中进行腐蚀实验,腐蚀增重由 3 块试样平均得出. 为了与 Zr-4 样品的耐疖状腐蚀性能进行比较,高压釜中同时放入了 Zr-4 样品.

表 1   N18(NZ2)锆合金的化学成分

Table 1   The chemical composition of N18 (NZ2) alloy ($\omega/\%$)

Sn	Nb	Fe	Cr	C	N	O	Zr
1.04	0.33	0.38	0.07	0.01	0.006	0.09	Balance

经过高压釜腐蚀后的样品,用混合酸将局部的金属基体溶去,显露出氧化膜/金属基体界面处氧化膜内表面的形貌,或者将氧化膜折断,用 JSM-6700F 扫描电子显微镜观察,了解氧化膜中显微组织的变化. 为了提高成像质量,观察前将样品表面进行蒸镀金处理.

3   结果和讨论

3.1   热处理对显微组织的影响

图 1 是样品经过不同热处理后得到的几种典型显微组织的透射电镜照片. 根据 Zr-Nb 二元相图可知,样品加热至 1 000 ℃时处于 β 相区,以 AC 处理后得到了板条状的 α-Zr 晶粒,晶界上有一薄层 β-Zr,晶界附近的颗粒状第二相为 Zr(FeCr)₂,过去采用选区电子衍射对这种第二相的晶体结构进行过分析[5],显微组织如图 1a 所示. 样品加热至 780 ℃或 820 ℃时处于 α+β 的两相区,820 ℃加热时的 β 相要比 780 ℃加热时的多,块状的 β-Zr 分布在晶界上或三晶交界处,如图 1b 所示. 这种 β-Zr 并不稳定,在低于 600 ℃重新加热时,会分解为 α-Zr 和 β-Nb(图 1c),并且冷轧变形还能促使其分解,获得更细小的 β-Nb 颗粒,可以明显提高耐腐蚀性能[6,7]. 样品加热至 700 ℃时是 α 单相区,合金中的 Nb 有一部分进入 Zr(FeCr)₂ 第二相中[8,9],形成 ZrNb(FeCr)₂ 或 Zr(FeCrNb)₂,剩下的 Nb 全部固溶在 α-Zr 中,不会形成 β 相,但是在冷却过程中,过饱和固溶在 α-Zr 中的 Nb 会以颗粒状 β-Nb 析出,显微组织如图 1d 所示. 图中较大的第二相是 Zr(FeCr)₂,较小的第二相是 β-Nb. 刘彦章

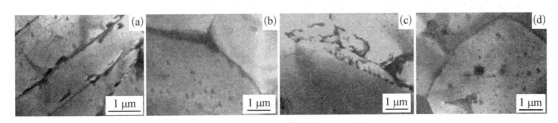

图 1   试样经过不同热处理后几种典型组织的透射电镜照片

Fig. 1   TEM images of some typical microstructures:(a) 1 000 ℃,0.5 h,AC;(b) 820 ℃,1 h,AC;(c) 820 ℃,1 h,AC+550 ℃,50 h;(d) 700 ℃,5 h,AC+550 ℃,50 h

等对这种锆合金的显微组织也进行过仔细的研究[10].

3.2 热处理(显微组织)对腐蚀增重的影响

经过不同热处理的17组样品,在500 ℃过热蒸汽中腐蚀至1 100 h都没有出现疖状腐蚀.经过不同温度加热和采用不同冷却方式处理的样品,再经过550 ℃,50 h处理后,它们的腐蚀增重曲线与未经550 ℃,50 h处理的重合在一起.这大概是因为即使样品加热冷却后存在不稳定的β-Zr,但在500 ℃高温下进行腐蚀时,β-Zr也会很快发生分解,因而看不出预先进行550 ℃,50 h处理对耐腐蚀性能的有利作用.因此,图2和图3中只给出了不同温度加热后采用"AC"和"FC"冷却的两组样品的腐蚀增重曲线.不同温度加热后采用"AC"冷却时,腐蚀增重曲线十分相近,但是,对于加热温度高的样品来说,它的腐蚀增重总是比加热温度低的样品要高一些,这样的规律在采用"FC"冷却时变得十分明显.同时放入高压釜的Zr-4样品,经过7 h腐蚀后已经出现了严重的疖状腐蚀,白色的疖状腐蚀斑几乎连成了一片,腐蚀增重达到了710 mg/dm².

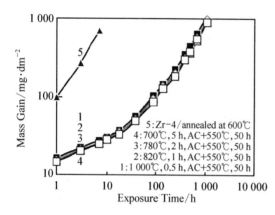 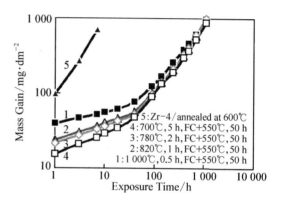

图2 不同温度加热后采用"AC"冷却处理系列的样品 在 500 ℃,10.3 MPa 过热蒸汽中的腐蚀增重
Fig. 2 The mass gains for the specimens heated at different temperatures followed by "AC" after autoclave tests in the super heated steam of 500 ℃, 10.3 MPa

图3 不同温度加热后采用"FC"冷却处理系列的样品在500 ℃, 10.3 MPa 过热蒸汽中的腐蚀增重
Fig. 3 The mass gains for the specimens heated at different temperatures followed by "FC" after autoclave tests in the super heated steam of 500 ℃, 10.3 MPa

从用双对数坐标作出的腐蚀增重随时间变化的曲线中可以看出,这种锆合金样品在500 ℃腐蚀时的动力学曲线仍然可以分为两个阶段,与其他锆合金在400 ℃过热蒸汽中和360 ℃高温高压水中的情况相同.两个腐蚀阶段的腐蚀增重(ΔW_1 和 ΔW_2)随时间变化的关系可以表达为:

$$\Delta W_1 = K_1 t^n \tag{1}$$

$$\Delta W_2 = K_2(t - t_0)^m \tag{2}$$

t_0为第一阶段过渡到第二阶段的转折时间,大约发生在腐蚀40 h时.在这以前,样品表面是黑色光亮的氧化膜,在腐蚀128 h后,氧化膜逐渐变为灰褐色,腐蚀1 100 h后,虽然腐蚀增重达到了1 000 mg/dm²,但氧化膜并未发生剥落.由不同样品的腐蚀增重曲线中得出公式(1)和(2)中各个常数的数值列于表2.在第一阶段腐蚀过程中,除了随炉冷却的样品(但不包

括在 700 ℃ 处理的样品)之外,腐蚀增重与时间之间的变化都符合立方关系,而转折之后成为线性关系. 转折发生在氧化膜厚度为 3 μm 时,比样品在 400 ℃ 或 360 ℃ 腐蚀转折时的氧化膜要厚一些.

由于样品的热处理方式不同,第一阶段中的腐蚀增重也有差别,这应该与显微组织及 α-Zr 中合金元素过饱和固溶含量的不同有关. 在 1 000 ℃ β 相加热,"AC"冷却后,得到的板条状晶粒比"FC"冷却的细小,β-Zr 或 β-Zr 分解后得到的 β-Nb 颗粒以及 $Zr(FeCr)_2$ 第二相颗粒也更细小;另一方面,样品在 β 相加热快冷时,Fe,Cr 合金元素也会过饱和固溶在 α-Zr 中. 用 Zr-4 样品进行实验后的分析结果已经证明了这一点[11]. 并且还有实验结果证明,改变 Fe,Cr 合金元素在 α-Zr 中的过饱和固溶含量,将影响样品的耐腐蚀性能[12]. 因此,采用"FC"冷却时,加热温度高的样品,它的腐蚀增重总是比加热温度低的样品要高,这种差别比采用"AC"冷却时更为明显. 合金中的一部分 Nb 会进入 $Zr(FeCr)_2$ 第二相中[8,9],所以样品在 700 ℃ 加热时是单一的 α 相区,无论加热后是"AC"还是"FC"冷却,也无论是否再经过 550 ℃ 处理,都不会引起显微组织的明显改变,所以在 700 ℃ 不同处理的 4 种样品,它们的腐蚀增重都重叠在一起.

表 2 由腐蚀增重曲线中得出 $\Delta W_1 = K_1 t^n$ 和 $\Delta W_2 = K_2 (t - t_0)^m$ 公式中的几个常数值

Table 2 Some constants in $\Delta W_1 = K_1 t^n$ 和 $\Delta W_2 = K_2 (t - t_0)^m$ obtained from the mass gain curves

Methods of heating treatment	K_1	n	K_2	m
1 000 ℃, 0.5 h, AC or +550 ℃, 50 h	15	0.29	39	0.94
1 000 ℃, 0.5 h, FC or +550 ℃, 50 h	39	0.15	70	0.82
820 ℃, 1 h, AC or +550 ℃, 50 h	15	0.29	52	0.94
820 ℃, 1 h, FC or +550 ℃, 50 h	23	0.23	52	0.94
780 ℃, 2 h, AC or +550 ℃, 50 h, or +rolling before 550 ℃, 50 h	15	0.29	51	0.92
780 ℃, 2 h, FC or +550 ℃, 50 h	20	0.25	51	0.92
700 ℃, 5 h, AC, FC or +550 ℃, 50 h	15	0.29	46	0.96

3.3 氧化膜内表面的形貌及氧化膜断口的形貌特征

图 4 是 700 ℃,5 h,AC+550 ℃,50 h 处理的样品经过 500 ℃,10.3 MPa 过热蒸汽中腐蚀 40 h 和 1 100 h 后氧化膜内表面的形貌. 可以看出,腐蚀转折后氧化膜的内表面仍然比较平整,并不像 Zr-4 样品那样在腐蚀转折后氧化膜内表面变得非常凹凸不平. 这说明这种锆合金在腐蚀转折后氧化膜的生长仍然比较均匀,这可能也是没有发生疖状腐蚀的原因,因为疖状腐蚀是氧化膜生长极不均匀的腐蚀过程. 内表面上的孔洞可能是嵌在氧化膜上还未发生氧化的第二相在溶解金属时被一起溶去后留下的痕迹,尺寸大小为 0.1~0.4 μm,与第二相的尺寸相当.

图 5 是样品经过 40 h 和 1 100 h 腐蚀后氧化膜的断口形貌. 氧化膜中存在孔洞和裂纹. 作者过去曾用高分辨透射电镜观察过 Zr-4 合金在高温高压水中腐蚀时形成的孔洞和裂纹,并讨论了它们形成的机制[13,14]:由于金属锆氧化形成氧化锆时的 P.B. 比为 1.56,氧化膜附着在金属表面不可能自由膨胀,所以其中存在很大的压应力. 在这种情况下生成的氧化锆晶体中会存在许多缺陷,包括空位、间隙原子、线缺陷、面缺陷、甚至体缺陷,在一些局部地区还可以观察到非晶. 缺陷在温度和压应力的作用下会发生扩散和凝聚,在 3 晶交界处或者晶界上形成孔隙,孔隙的集聚和扩展就会形成孔洞和裂纹. 由于这种过程是在氧化膜受到压应力情况下发生的,因此裂纹的扩展方向一定是平行于压应力方向,不可能垂直于压应力方

向,因此裂纹大致与氧化膜/金属的界面平行. N18(NZ2)锆合金在 500 ℃过热蒸汽中腐蚀时,氧化膜中形成的孔隙和裂纹同样也遵循这样的规律. 在断口照片中,可以看到柱状和等轴(接近球形)的氧化锆晶粒,根据以前的研究结果认为,柱状晶中的缺陷通过扩散凝聚而构成新的晶界,发展成等轴晶. 因而,仔细观察数个相连的等轴晶,它们构成的轮廓大致勾画出了原来柱状晶的形貌.

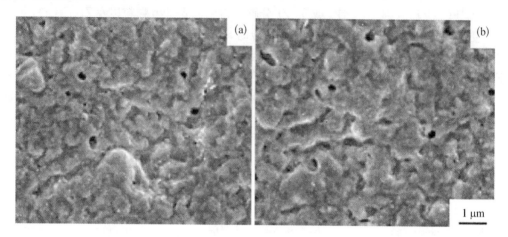

图 4　700 ℃,5 h,AC+500 ℃,50 h 处理的样品在 500 ℃,10.3 MPa 过热蒸汽中腐蚀40 h(a)和 1 100 h(b) 后氧化膜内表面的形貌

Fig. 4 The inner surface morphology of oxide film formed on the specimens treated by 700 ℃, 5 h, AC+500 ℃, 50 h after autoclave tests at 500 ℃, 10.3 MPa super heated steam for 40 h (a) and 1 100 h (b)

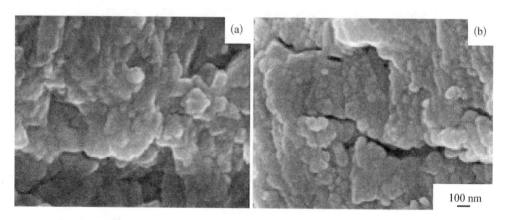

图 5　700 ℃,5 h,AC+500 ℃,50 h 处理的样品在 500 ℃,10.3 MPa 过热蒸汽中腐蚀40 h(a)和 1 100 h(b) 后氧化膜的断口形貌

Fig. 5 The fracture surface morphology of oxide film formed on the specimens treated by 700 ℃, 5 h, AC+500 ℃, 50 h after autoclave tests at 500 ℃, 10.3 MPa super heated steam for 40 h (a) and 1 100 h (b)

3.4　N18(NZ2)锆合金在 500 ℃过热蒸汽腐蚀时不会产生疖状腐蚀的原因

考虑到 N18(NZ2)锆合金中含有合金元素 Fe 和 Cr,如果参照热处理对 Zr-4 合金耐疖状腐蚀性能的影响规律,那么这种锆合金样品在 β 相加热后缓冷,或者在较高的 α 相温度加

热后缓冷,都会产生疖状腐蚀;如果再考虑这种锆合金中含有 Nb,Nb 是稳定 β-Zr 的合金元素,在 β 相或者 $\alpha+\beta$ 双相区加热冷却后,在室温下可以得到含有 β-Zr 的显微组织,这对耐腐蚀性能也会产生有害的影响,这些因素都可能对这种锆合金的耐疖状腐蚀性能产生不利影响.本实验采用的几种不同热处理制度,几乎可以获得各种典型的显微组织,但是在 500 ℃ 过热蒸汽中腐蚀时都没有出现疖状腐蚀.根据 Zr-Nb 二元相图判断,无论将样品如何进行热处理,都会有一定量的合金元素 Nb 固溶在 α-Zr 基体中,这可能是不产生疖状腐蚀的原因.

周邦新从疖状腐蚀的形成过程是"成核长大"的机制出发,提出了一种模型来说明疖状腐蚀的形成过程[3]:由于晶体氧化时氧化膜生长的各向异性,局部区域的氧化膜可以因晶体的不利取向而生长较快,当生长至一定的厚度后,由于形成氧化膜时的 P.B. 比为 1.56,氧化膜的膨胀倾向会对基体金属施加张应力而使其变形,氧离子沿着变形后金属中的位错通道扩散更快,促进了氧化膜的生长而发展成为疖状腐蚀斑.改变 α-Zr 中合金元素的固溶含量或过饱和固溶含量,就可以改变晶体氧化时氧化膜生长的各向异性,也就改变了疖状腐蚀的成核过程,从而影响耐疖状腐蚀性能.如果降低 α-Zr 基体中合金元素的固溶含量,或者使晶粒长大,都会使耐疖状腐蚀性能变坏.实验结果表明,Zr-4 样品的晶粒增大到 0.2～0.5 mm 后,在 400 ℃ 过热蒸汽中腐蚀时就会出现疖状腐蚀[15].如果将 Zr-4 样品加热至 α 相上限温区或者 β 相快冷,可以明显改善耐疖状腐蚀性能,甚至抑制疖状腐蚀产生[16],这是因为在 α-Zr 基体中增加了 Fe,Cr 合金元素的过饱和固溶含量,调整了腐蚀时氧化膜生长的各向异性.Nb 在 α-Zr 中的固溶度比 Fe,Cr 的大,有数量级的差别,利用添加合金元素 Nb 可以获得耐疖状腐蚀非常优良的新锆合金,这也是从以上假设出发推论后得出的一个合理结果.

4 结论

(1) 在不同相区加热处理的 N18(NZ2)锆合金样品,无论是采用快冷还是缓冷,它们经过 500 ℃ 过热蒸汽腐蚀 1 100 h 后都没有出现疖状腐蚀现象.可以认为在 Zr-Sn 合金中添加合金元素 Nb 后,对疖状腐蚀产生了"免疫性".

(2) N18(NZ2)锆合金样品在 500 ℃ 过热蒸汽中的腐蚀增重动力学曲线仍可分为两个不同的阶段,转折发生在氧化膜厚度大约为 3 μm 时.

参 考 文 献

[1] Ogata K, Mishima Y, Okubo T *et al*. *Zirconium in the Nuclear Industry*[C]. Philadelphia:ASTM, 1989:291.

[2] Cheng B, Adamson R B. *Zirconium in the Nuclear Industry*[C]. Philadelphia:ASTM, 1987:387.

[3] Zhou B X. *Zirconium in the Nuclear Industry*[C]. Philadelphia:ASTM, 1989:360.

[4] Zhou Bangxin (周邦新), Zhao Wenjin (赵文金), Miao Zhi (苗志) *et al*. *Proceedings of 1996 Chinese Materials Symposium* (96 中国材料研讨会)[C]. Beijing:Chemical Industry Press, 1997:183.

[5] Zhou Bangxin (周邦新). *Proceedings of 1996 Chinese Materials Symposium* (96 中国材料研讨会)[C]. Beijing:Chemical Industry Press, 1997:187.

[6] Li Qiang (李强), Liu Wenqing (刘文庆), Zhou Bangxin (周邦新). *Rare Metal Materials and Engineering* (稀有金属材料与工程)[J], 2002, 31(5):389.

[7] Liu Wenqing (刘文庆), Li Qiang (李强), Zhou Bangxin (周邦新) *et al*. *Nuclear Power Engineer* (核动力工程)[J], 2003, 24:33.

[8] Caroline T M, Brachet J C, Jago G. *J Nucl Mat*[J], 2002, 305: 224.

[9] Jeong Y H, Kim H G, Kim T H. *J Nucl Mat*[J], 2003, 317: 1.

[10] Liu Yanzhang（刘彦章）, Zhao Wenjin（赵文金）*et al*. *Nuclear Power Engineer*（核动力工程）[J], 2005, 26: 158.

[11] Li Cong, Zhou Bangxin, Zhao Wenjin *et al*. *J Nucl Mat*[J], 2002, 304: 134.

[12] Zhou Bangxin, Zhao Wenjin, Miao Zhi *et al*. *China Nuclear Science and Technology Report*. CNIC - 01074, SINRE - 0066[R], Beijing: Atomic Energy Press, 1996.

[13] Zhou Bangxin（周邦新）, Li Qiang（李强）, Yao Meiyi（姚美意）*et al*. *Nuclear Power Engineer*（核动力工程）[J], 2005, 26: 364.

[14] Zhou Bangxin（周邦新）, Li Qiang（李强）, Liu Wenqing（刘文庆）*et al*. *Rare Metal Materials and Engineering*（稀有金属材料与工程)[J], 2006, 35(7): 1009.

[15] Zhou bangxin（周邦新）, Liqiang（李强）, Miao Zhi（苗志）. *Nuclear Power Engineer*（核动力工程)[J], 2000, 21: 339.

[16] Zhou Bangxin（周邦新）, Li Qiang（李强）, Yao Meiyi（姚美意）*et al*. *Rare Metal Materials and Engineering*（稀有金属材料与工程)[J], 2007, 36(7): 1129.

Nodular Corrosion Resistance of Zr - Sn - Nb Alloy

Abstract: The nodular corrosion resistance of N18（NZ2）zirconium alloy specimens by different heat treatments has been investigated by the autoclave tests in the superheated steam of 500 ℃, 10.3 MPa. The results show that there was no nodular corrosion appeared after the autoclave test for 1100 h in the superheated steam of 500 ℃, 10.3 MPa, regardless the specimens were heat treated in β phase region, in $\alpha +$ β dual phases region or in α phase region followed by fast or slow cooling. It means that the N18（NZ2）zirconium alloy containing niobium alloying element gives an "immunity" against nodular corrosion. From the corrosion dynamic curve of mass gain against exposure time during autoclave tests in the super heated steam of 500 ℃, 10.3 MPa, two stages can be found with a transition at the thickness of 3 μm for the oxide films.

热处理影响 Zr-4 合金耐疖状腐蚀性能的机制*

摘　要：采用 500 ℃，10.3 MPa 过热蒸汽腐蚀方法，研究了热处理对 Zr-4 合金耐疖状腐蚀性能的影响. 试样经过 600 ℃，820 ℃ 和 1 000 ℃ 不同热处理后，耐疖状腐蚀性能明显不同. 提高 Fe，Cr 合金元素在 α-Zr 中过饱和固溶含量，可以明显改善耐疖状腐蚀性能，第二相的大小不是决定的因素. 用高分辨扫描电镜观察了氧化膜的内表面形貌和断口形貌，研究了耐疖状腐蚀性能与氧化膜显微组织之间的关系. 从疖状腐蚀斑的成核与长大，热处理会引起 Fe 和 Cr 合金元素在 α-Zr 中过饱和固溶含量的变化，以及从氧化膜生长的各向异性与 α-Zr 中合金元素过饱和固溶含量的关系出发，讨论了热处理影响耐疖状腐蚀性能的机制.

1　引言

　　锆合金是核动力反应堆中一种重要的结构材料，用作核燃料元件的包壳. 锆合金在高温高压水中或过热蒸汽中服役时，与 H_2O 反应后生成 ZrO_2 和 H_2，锆合金受到腐蚀的同时还会吸收氢. 这种腐蚀过程可分为均匀腐蚀和不均匀的疖状腐蚀，在水质经过加氢除氧的压水堆工况下是均匀腐蚀；在沸水堆工况下，或者是在水质未经加氢除氧处理的压水堆工况下，会出现不均匀的疖状腐蚀. 用高压釜进行腐蚀试验时，在 450~500 ℃ 过热蒸汽中也会产生疖状腐蚀，一般用这种试验方法来研究和评价锆合金的耐疖状腐蚀性能[1].

　　Zr-4 和 Zr-2 合金在加工过程中，通过控制累积退火参数 A 可以改善耐疖状腐蚀性能[2]，需要控制 $A<10^{-18}$[3]. 累积退火参数 A 是一种与加热温度和加热时间相关的参数，在坯料经过 $β$ 相淬火后的后续加工过程中，控制较低的退火温度和较短的退火时间，可以使 A 值 $<10^{-18}$，这时，第二相的颗粒也比较小. 因而认为第二相长大后会使耐疖状腐蚀性能变坏，应该将第二相的大小控制在 $<0.2\ μm$[3,4]. Zr-4 中的第二相是 $Zr(Fe,Cr)_2$，Zr-2 中的是 $Zr(Fe,Cr)_2$ 和 $Zr_2(Fe,Ni)$. 随着退火温度升高或保温时间延长，A 值增大的同时第二相也会长大. 但是，由于 α-Zr 中过饱和固溶的合金元素析出，其含量也会减少. 那么，究竟是第二相大小影响耐疖状腐蚀性能，还是 α-Zr 中合金元素的固溶含量影响耐疖状腐蚀性能，至今还未取得一致的认识. 本研究工作将 Zr-4 试样加热至不同相区后快冷，改变了 α-Zr 中 Fe，Cr 合金元素的过饱和固溶含量，也改变了第二相的大小，然后研究它们耐疖状腐蚀性能的差别，探索热处理影响耐疖状腐蚀性能的机制.

2　实验方法

　　试验试样是工厂生产的 Zr-4 板，厚 0.6 mm，出厂时已经过 600 ℃ 再结晶退火处理. 为了改变 Fe，Cr 合金元素在 α-Zr 中的过饱和固溶含量，重新将切成 25 mm×18 mm 的试样

---

* 本文合作者：李强、姚美意、夏爽、刘文庆、褚于良. 原发表于《稀有金属材料与工程》，2007，36(7)：1129-1134.

放入真空石英管式电炉中,分别加热至 820 ℃保温 1 h 或 1 000 ℃保温 0.5 h,然后将电炉推离石英管,在石英管外壁淋水冷却,称为"空冷". 下面将这样处理的试样标写为 820 ℃,1 h,AC 和 1 000 ℃,0.5 h,AC.

试样在进行 500 ℃,10.3 MP 过热蒸汽腐蚀实验前,按照标准方法用混合酸(体积分数为 10%HF+ 45%HNO₃+45%H₂O)酸洗和去离子水清洗,腐蚀增重由 3 块试样平均得出.经过高压釜腐蚀后的试样,用混合酸将局部的金属基体溶去,得到氧化膜与金属界面处氧化膜的内表面,或者将氧化膜折断得到断口,再用 JSM - 6700F 扫描电子显微镜观察. 为了提高成像质量,将观察的表面进行蒸镀金处理.

3 实验结果

3.1 不同热处理后试样的显微组织与织构

图 1 是试样经过 3 种不同热处理后的显微组织. 600 ℃退火后已经完全再结晶,Zr(FeCr)₂第二相较多;820 ℃退火时处于 α 相上限温区,在晶粒长大的同时,第二相也发生了长大,数量比 600 ℃退火后的明显减少;1 000 ℃退火时处于 β 相,Zr - 4 中的 Fe,Cr 合金元素可以全部溶于 β 相中,冷却时转变成板条状的 α 相晶粒,晶界上析出了细小的 Zr(FeCr)₂第二相. 显微组织随不同热处理的变化过去曾进行过仔细研究[5]. 由于加热温度不同,并且采取了比较快的冷却,除了第二相中的 Fe/Cr 比会发生变化外[6],Fe,Cr 合金元素在 α - Zr 中的过饱和固溶含量也会不同[7].

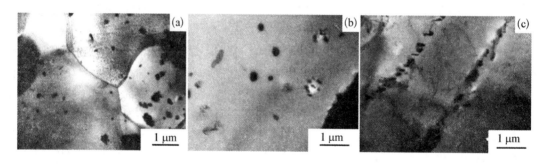

图 1 Zr - 4 试样经过 600 ℃(a)、820 ℃,1 h,AC (b)和 1000 ℃,0.5 h,AC (c) 退火后显微组织的电子显微镜照片

Fig. 1 TEM micrographs of Zircaloy - 4 specimens after annealing:(a) at 600 ℃;(b) at 820 ℃ for 1 h, AC and (c) at 1000 ℃ for 0.5 h, AC

试样经过不同热处理后,用电子背散射菊池线衍射分析方法(EBSD)测定了织构,结果归纳如下: 600 ℃退火后的再结晶织构与文献中的报道完全一致,织构取向为(0001)₂₈° < $10\bar{1}0$ >和(0001)₂₈° < $11\bar{2}0$ >((0001)₂₈° 表示(0001)面沿着轧制方向左右倾斜 28°);820 ℃退火后因为晶粒长大,再结晶织构变得比较集中,(0001)面基本平行于轧面,轧制方向的取向和 600 ℃退火后的一样,没有发生变化;1 000 ℃退火时因为发生了 α→β→α 相变,扰乱了原来的织构,但是由于相变时遵循一定的晶体几何学关系,在被扰乱的织构中,仍然保留了一部分原来 α 相退火后的再结晶织构. 这些结果将在另一篇文章中详细讨论.

3.2 腐蚀增重和试样表面形貌

图 2 是 3 种试样在 500 ℃, 10.3 MPa 过热蒸汽中腐蚀时的增重曲线. 图 3 是腐蚀后试样的表面形貌. 经过 600 ℃ 再结晶退火处理的试样耐疖状腐蚀性能最差, 腐蚀 7 h 后, 试样的一个表面布满了疖状腐蚀斑, 另一个表面上的疖状腐蚀斑几乎连成了一片, 形成了白色氧化膜, 平行试验的 3 个试样都是如此.

试样经过 820 ℃, 1 h, AC 处理后, 耐疖状腐蚀性能明显改善, 在 90 h 腐蚀后, 试样的两面都没有出现疖状腐蚀斑, 但是, 试样周边的耐腐蚀性能并不好, 周边的氧化膜生长较快, 看来在试样的侧面上发生了疖状腐蚀. 经过 90 h 腐

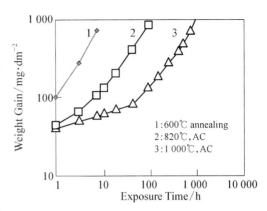

图 2 经过不同热处理的试样在 500 ℃, 10.3 MPa 过热蒸汽中腐蚀时的增重

Fig. 2 The weight gains of specimens by different heat treatments after autoclave tests in the super heat steam of 500 ℃ and 10.3 MPa

蚀后, 由于氧化膜剥落而无法获得准确的腐蚀增重. 这种现象的原因将在本文后面的章节中讨论.

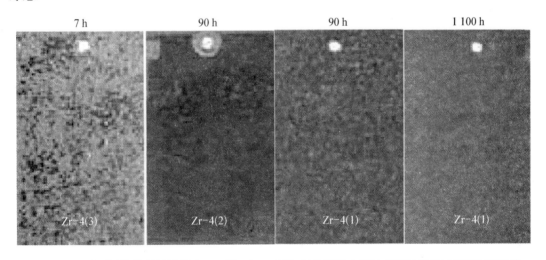

图 3 经过不同热处理的试样在 500 ℃, 10.3 MPa 过热蒸汽中腐蚀不同时间后试样表面的形貌

Fig. 3 The surface morphologies of specimens by different heat treatments after autoclave tests in the super heat steam of 500 ℃ and 10.3 MPa for different exposure times, Zr - 4 (3), Zr - 4 (2) and Zr -4 (1) were annealed at 600 ℃, 820 ℃, AC and 1000 ℃, AC, respectively

试样经过 1 000 ℃, 0.5 h, AC 处理后, 耐疖状腐蚀性能最好, 在 1 100 h 的腐蚀过程中都没有出现疖状腐蚀, 试样周边的氧化膜也没有剥落. 1 100 h 腐蚀后的增重接近 1 200 mg/dm², 氧化膜的厚度达到 80 μm.

3.3 氧化膜内表面和断口的形貌

图 4 是 600 ℃ 再结晶退火的试样腐蚀 7 h 后氧化膜内表面的形貌. 图 4a 中有一个疖状腐蚀斑底部的形貌, 从放大倍率更高的图 4b 中, 可以很清楚地看出, 疖状腐蚀斑的底部凹凸不平形似 "花菜", 这说明疖状腐蚀斑处氧化膜的生长极不均匀. 相对来说, 在疖状腐蚀斑底

部周围的氧化膜比较平整,但是在上面分布着大约 10～20 μm 大小的突起颗粒(由图 4a 中箭头指处),在更高的放大倍率下,它们的形貌如图 4c 所示,这应该是还未发展成疖状腐蚀斑的"核". 通过了解这种"核"的形成原因,对认识疖状腐蚀过程是非常重要的. 图 5a,图 5b 是氧化膜的断口形貌照片. 其中包含了一个疖状腐蚀斑,形状如凸透镜,最厚处达到100 μm,但是旁边黑色氧化膜的厚度只有 3～10 μm. 疖状腐蚀斑的表面有一薄层比较致密的氧化膜,但是下面大部分地区是很疏松的氧化膜,含有裂纹和孔洞,等轴晶的大小≤100 nm. 图 5c 是疖状腐蚀斑旁边黑色氧化膜的断口形貌. 相对于疖状腐蚀斑处的氧化膜来说,黑色氧化膜比较致密,是等轴晶和柱状晶的混合组织.

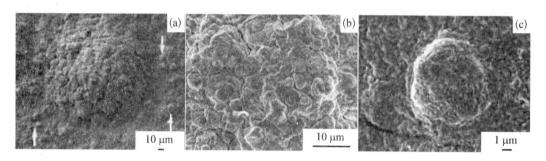

图 4 600 ℃退火的试样经过 7 h 腐蚀后,在氧化膜内表面上疖状腐蚀斑的底部形貌(a,b)及其周围氧化膜的形貌(c)

Fig. 4 The morphology of a nodule corrosive spot on the inner surface of oxide film formed on the specimen annealed at 600 ℃ after autoclave test for 7 h: (a,b) on the bottom and (c) in the surrounding area of the nodule

图 5 600 ℃退火处理的试样经过 7 h 腐蚀后,疖状腐蚀斑断口(a,b)及疖状腐蚀斑周围黑色氧化膜的断口(c)形貌

Fig. 5 The fracture morphology of a nodule corrosive spot on the inner surface of oxide film formed on the specimen annealed at 600 ℃ after autoclave test for 7 h: (a,b) at the nodule corrosive spot and (c) surrounding the spot

　　图 6a,6b 是 820 ℃,1 h,AC 处理的试样在 500 ℃,10.3 MPa 过热蒸汽中腐蚀 90 h 后氧化膜内表面的形貌. 与 600 ℃退火的试样相比有明显的差异. 820 ℃处理的试样腐蚀后,氧化膜内表面形貌有 2 种不同的特征,一种是凹凸不平,另一种是比较平滑,但是上面嵌有不规则的裂纹. 后者应该是非晶氧化锆,当金属基体被酸溶去后,氧化膜中的压应力得到释放,因为变形而产生裂纹,只有非晶体形成裂纹时才会是这样不规则的形貌. 在研究 Zr-4 试样经过 360 ℃高温高压水腐蚀后形成的氧化膜时,也观察到在接近金属基体的氧化膜中有非晶相存在[8,9],李聪等研究 Zr-4 试样上刚形成很薄一层氧化膜的晶体结构时,也观察到非

晶相,还有单斜、四方和立方相等不同的晶体结构,这些不同晶体结构的相区勾画出了原来金属基体的晶粒形貌,因而认为形成不同晶体结构的原因是与金属晶粒表面的晶体取向有关[10]. 820 ℃处理的试样腐蚀后,非晶氧化锆的形成除了与金属晶粒的取向有关外,还会与增加了 α - Zr 中合金元素的过饱和固溶含量有关,当固溶的合金元素转入到氧化锆中后,可能使氧化锆中空位的扩散和凝聚变得比较缓慢,也延缓了非晶向晶态的转变. 图 6c 是氧化膜的断口形貌. 氧化膜中存在孔洞和裂纹,氧化锆的等轴晶≤100 nm.

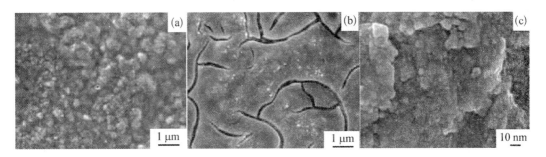

图 6 820 ℃,AC 处理的试样经过 500 ℃,10. 3 MPa 过热蒸汽腐蚀 90 h 后氧化膜的内表面形貌(a, b)及氧化膜的断口形貌(c)

Fig. 6 The morphology of specimen annealed at 820 ℃,AC after autoclave test at 500 ℃ and 10. 3 MPa for 90 h on:(a, b) the inner surface, and (c) the fracture surface of oxide film

图 7a, 7b 是 1 000 ℃,0. 5 h,AC 处理的试样在 500 ℃,10. 3 MPa 过热蒸汽中腐蚀 90 h 后氧化膜内表面的形貌. 与 820 ℃处理的试样相似,也有 2 种不同特征的形貌,1 种是凹凸不平,另 1 种是比较平滑并且上面嵌有不规则的裂纹,这种区域大致勾画出了板条状晶粒的形貌. 在两条分布着裂纹的区域之间,是凹凸不平的区域,这说明非晶氧化锆的形成与金属晶粒表面的晶体取向有关. 在 1 000 ℃处理的试样中,氧化膜内表面上非晶的区域比 820 ℃处理的试样中多,这说明非晶氧化锆的形成与 α - Zr 中合金元素过饱和固溶含量的多少也有关系. 图 7c 是氧化膜的断口形貌. 氧化膜中存在孔洞和裂纹,氧化锆的等轴晶≤100 nm. 在靠近金属基体处,有一层 0. 5 μm 厚的断口比较平滑,没有显露出晶粒组织,这正是非晶的断口特征.

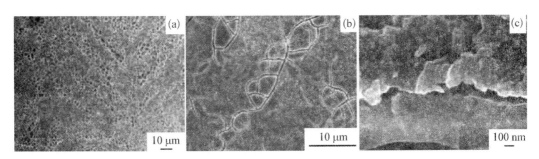

图 7 1 000 ℃,0. 5 h,AC 处理的试样经过 500 ℃,10. 3 MPa 过热蒸汽腐蚀 90 h 后氧化膜内表面的形貌(a, b)及氧化膜的断口形貌(c)

Fig. 7 The morphology of specimen annealed at 1000 ℃ for 0. 5 h,AC after autoclave test at 500 ℃ and 10. 3 MPa for 90 h on:(a, b) the inner surface, and (c) the fracture surface of oxide film

图 8a, 8b 是 1 000 ℃,0. 5 h,AC 处理的试样在 500 ℃,10. 3 MPa 过热蒸汽中腐蚀 1 100 h 后氧化膜的内表面形貌. 氧化膜内表面的凹凸不平勾画出了原来金属基体的板条状

晶粒形貌(图8a),但是没有观察到像图7a,7b中那样的裂纹,说明氧化膜内表面上不存在大面积的非晶区.试样在500℃过热蒸汽中腐蚀时,经过1000 h以上的加热保温,过饱和固溶在α-Zr中的Fe,Cr合金元素会重新析出,降低了合金元素在α-Zr中的固溶含量,不能有效地稳定非晶氧化锆,因而没有出现大面积的非晶裂纹区,氧化膜的内表面又变得凹凸不平(图8b).图8c是氧化膜的断口形貌.氧化膜中存在孔洞和裂纹,是等轴晶和柱状晶的混合组织.

图8 1000℃,0.5 h,AC退火试样经过500℃,10.3 MPa过热蒸汽腐蚀1100 h后的氧化膜内表面形貌(a,b)及断口形貌(c)

Fig. 8 The morphology of specimen annealed at 1 000 ℃ for 0. 5 h, AC after autoclave test at 500 ℃ and 10. 3 MPa for 1 100 h on: (a, b) the inner surface, and (c) the fracture surface of oxide film

4 讨论

Ogata等统计了疖状腐蚀斑的厚度与直径之比,得到了基本是一个恒定值(~0.2)的结果[11].Cheng等定期取出腐蚀试样并对同一地区进行观察,证明疖状腐蚀斑的出现有先有后,是一种成核长大过程[12].晶粒大时形成的疖状腐蚀斑也大[13],当晶粒达到0.2~0.5 mm时,在400℃过热蒸汽中腐蚀时也会出现疖状腐蚀[14],这说明了疖状腐蚀与晶粒组织有关.

要了解热处理影响耐疖状性能的机制,应该从疖状腐蚀斑的成核与长大,热处理引起Fe,Cr合金元素在α-Zr中过饱和固溶含量的变化,以及氧化膜生长的各向异性及其与α-Zr中合金元素过饱和固溶含量的关系等方面来讨论.

4.1 疖状腐蚀斑的成核长大

本文作者之一曾对疖状腐蚀的形成提出过一种解释[13].由于锆合金晶体氧化时氧化膜生长的各向异性,不利的晶面取向会使某些区域中氧化膜生长较快,当发展至一定厚度以后,金属锆氧化形成氧化锆后体积膨胀,对基体金属施加张应力使其变形,氧离子沿着变形后金属中位错通道扩散更快,促进了氧化膜在局部区域中的生长而发展成为疖状腐蚀.这种成核过程可以理解疖状腐蚀斑的出现会有先后,还可以说明疖状腐蚀斑的厚度与直径之间存在一定的关系.在已经出现疖状腐蚀斑的黑色氧化膜内表面上,可以观察到与金属晶粒大小相当的突起状颗粒(图4a,4c),这正是因为基体金属晶粒表面取向不利而引起氧化膜生长较快的结果,这也是疖状腐蚀斑的"核".

4.2 Fe,Cr合金元素在α-Zr中的过饱和固溶含量与热处理的关系

从Zr-Fe和Zr-Cr二元相图可以看出,在820℃和850℃时,Fe和Cr合金元素在α-

Zr 中的固溶度可以分别达到 120 $\mu g/g$ 和 200 $\mu g/g^{[15]}$，在高温 β 相区，Zr - 4 中的 Fe 和 Cr 合金元素可以全部固溶到 β 相中. 因此，当试样在 α 相上限温区或者 β 相加热后快冷，Fe 和 Cr 合金元素会过饱和固溶在 α - Zr 中. 李聪等用电解方法将 Zr - 4 试样中 $Zr(FeCr)_2$ 第二相分离后，测定了 α - Zr 中 Fe 和 Cr 的过饱和固溶含量，当试样在 β 相加热淬火并经轧制加工和 470 ℃ 去应力退火，Fe＋Cr 在 α - Zr 中的过饱和固溶含量达到了 1 300 $\mu g/g^{[7]}$. 将试样加热到 820 ℃ 和 1 000 ℃ 后空冷，Fe 和 Cr 合金元素在 α - Zr 中的过饱和固溶含量一定会比 600 ℃ 退火处理的多，而且 1 000 ℃ 加热空冷的试样又会比 820 ℃ 加热空冷的多.

4.3 氧化膜生长与晶面取向的关系

氧化膜在锆晶体上生长时呈现明显的各向异性特征. Cox[16] 对早期的一些实验结果作了归纳，但是相互之间并不十分一致. 这些实验都是用纯锆的大晶粒试样进行的，不同试验者所用原材料中杂质含量可能不同，这是重要的影响因素. Kim 等[17] 用高纯锆的大晶粒试样经过 360 ℃ 高温水中腐蚀 60 h，观察到 $(11\bar{2}0)$ 面上的氧化膜生长至 20 μm 已开始剥落，但 (0001) 面上的氧化膜厚度只有 2 μm，存在明显的差别. α - Zr 中合金元素固溶含量的多少会影响氧化膜的生长速率，影响耐腐蚀性能[5]. 增加 Fe 和 Cr 合金元素在 α - Zr 中的过饱和固溶含量后，如果可以调整腐蚀时氧化膜生长的各向异性，那么，就可能延缓甚至阻止疖状腐蚀斑的成核，提高耐疖状腐蚀性能. 这样就可以理解为什么试样经过 α 相上限温区或者 β 相加热后快冷，可以提高合金的耐疖状腐蚀性能. 不过还需要更直接的实验数据来证明.

试样在 820 ℃ α 相上限温区加热后快冷，由于织构取向的原因，氧化膜生长较慢的 (0001) 面基本平行于试样的轧面，提高了轧面的耐疖状腐蚀性能，但是试样周边的侧面是耐腐蚀性能较差的 $<10\bar{1}0>$ 和 $<11\bar{2}0>$ 方向，在晶体取向不利的情况下，试样的侧面上仍然会产生疖状腐蚀，只有在 1 000 ℃ β 相加热快冷，进一步提高 α - Zr 中合金元素过饱和固溶含量后，才能完全阻止疖状腐蚀现象.

如果降低 α - Zr 中合金元素的过饱和固溶含量，并同时使晶粒长大，耐疖状腐蚀性能会变得更坏，因为在一个取向不利的大晶粒表面上形成了较厚的氧化膜，这种比较大的"核"更容易发展成白色的疖状腐蚀斑. 这样就可以解释大晶粒的 Zr - 4 试样在 400 ℃ 过热蒸汽中腐蚀时就会出现疖状腐蚀的现象[14].

棒状燃料的端塞与包壳管之间用真空电子束焊接，焊接热影响区的耐疖状腐蚀性能明显优于母材，黑色和白色氧化膜之间存在明显的界线. 观察界线处金属基体的晶粒组织，证明该处并未经过 $\alpha \rightarrow \beta \rightarrow \alpha$ 相变[18]，说明只要加热到 α 相的上限温区后快冷，就可以提高耐疖状腐蚀性能，这与本实验的结果一致. 试样经过 820 ℃ α 相上限温区加热，第二相已发生长大，但是仍然具有较好的耐疖状腐蚀性能，说明第二相的大小并不是影响耐疖状腐蚀性能的决定因素.

5 结论

(1) Zr - 4 合金试样加热到 α 相上限温区或者 β 相快冷后，增加了 Fe 和 Cr 合金元素在 α - Zr 中的过饱和固溶含量，调整了氧化膜生长的各向异性，延缓甚至阻止了黑色氧化膜中疖状腐蚀斑"核"的形成，因而提高了耐疖状腐蚀性能.

(2) 增加 Fe，Cr 合金元素在 α - Zr 中的过饱和固溶含量，促使了非晶氧化锆的形成，延缓了非晶氧化锆向晶态相的演化,降低了氧化膜的生长速率.因而,随着 α - Zr 中 Fe, Cr 合金元素过饱和固溶含量增加,耐均匀腐蚀性能也得到了提高.

参 考 文 献

[1] Johnson A B, Horton R M. *Zirconium in the Nuclear Industry* [C]. Philadelphia: ASTM, 1977: 295.

[2] Thorvaldsson T, Andersson T, Wilson A *et al*. *Zirconium in the Nuclear Industry* [C]. Philadelphia: ASTM, 1989: 128.

[3] Garzarolli F, Steinberg E, Weidinger H G. *Zirconium in the Nuclear Industry* [C]. Philadelphia: ASTM, 1989: 202.

[4] Kruger R M, Adamson R B, Brenner S S. *J Nucl Mat* [J], 1992, 189: 193.

[5] Zhou Bangxin, Zhao Wenjin, Miao Zhi *et al*. *China Nuclear Science and Technology Report*. CNIC - 01074, SINRE - 0066 [R], Beijing: Atomic Energy Press, 1996.

[6] Zhou Bangxin, Yang Xiaolin, *China Nuclear Science and Technology Report*. CNIC - 01073, SINRE - 0065 [R], Beijing: Atomic Energy Press, 1996.

[7] Li C, Zhou B X, Zhao W J *et al*. *J Nucl Mat* [J], 2002, 304: 134.

[8] Zhou Bangxin (周邦新), Li Qiang (李强), Yao Meiyi (姚美意) *et al*. *Nuclear Power Engineering* (核动力工程) [J], 2005, 26: 364.

[9] Zhou Bangxin (周邦新), Li Qiang (李强), Liu Wenqing(刘文庆) *et al*. *Rare Metal Materials and Engineering* (稀有金属材料与工程) [J], 2006, 35(7): 1009.

[10] Li Cong (李聪), Zhou Bangxin (周邦新). *Nuclear Power Engineering* (核动力工程) [J], 1994, 15: 152.

[11] Ogata K, Mishima Y, OkuboT *et al*. *Zirconium in the Nuclear Industry* [C]. Philadelphia: ASTM, 1989: 291.

[12] Cheng B, Adamson R B. *Zirconium in the Nuclear Industry* [C]. Philadelphia: ASTM, 1987: 387.

[13] Zhou B X. *Zirconium in the Nuclear Industry* [C]. Philadelphia: ASTM, 1989: 360.

[14] Zhou Bangxin (周邦新), Li Qiang (李强), Miao Zhi (苗志). *Nuclear Power Engineer* (核动力工程) [J], 2000, 21: 339.

[15] Charquet D, Hahn R, Ortlieb E *et al*. *Zirconium in the Nuclear Industry* [C]. Philadelphia: ASTM, 1989: 405.

[16] Cox B. *J Nucl Mat* [J], 2005, 336: 331.

[17] Kim H G, Kim T H, Jeong Y H. *J Nucl Mat* [J], 2002, 306: 44.

[18] Zhou Bangxin (周邦新), Zheng Sikui (郑斯奎), Wang Shunxin (汪顺新). *J Nuclear Science and Engineering* (核科学与工程) [J], 1988, 8: 130.

Effect of Heat Treatments on Nodular Corrosion Resistance of Zircaloy‐4

Abstract: The effects of heat treatments on the nodular corrosion resistance of Zircaloy‐4 have been investigated in the super-heat steam of 500 ℃ and 10.3 MPa by autoclave tests. It is found that the nodular corrosion resistances are obviously different for the specimens heat treated at 600 ℃, 820 ℃ and 1000 ℃, respectively. The key factors improving the nodular corrosion resistance are the supersaturated solid solution contents of Fe and Cr alloying elements in α - Zr matrix, but not the size of the second phase particles. In order to study the relationship between the nodular corrosion resistance and the microstructure of oxide films, the morphologies of inner surface and fracture surface of oxide films were examined by high resolution scanning electron microscopy. According to the nucleation and growth of nodule corrosive spots, the mechanism concerning the effect of heat treatments on nodular corrosion resistance in Zircaloy‐4 was discussed. It is shown that the supersaturated solid solution contents of alloying elements in α - Zr matrix could be changed by different heat treatments, and the growth of oxide films is anisotropic on the crystallographic plane with different orientations.

三维原子探针及其在材料科学研究中的应用[*]

摘　要： 三维原子探针(3DAP)是一种定量显微分析仪器,通过对不同元素的原子逐个进行分析,可绘出金属样品中不同元素的原子在纳米空间中的分布图形. 从分析逐个原子来了解物质微区化学成分的不均匀性,3DAP 是一种不可替代的分析方法. 本文介绍了 3DAP 的工作原理及样品制备,举例说明了 3DAP 分析技术的应用.

材料的性能与显微组织结构密切相关,而显微组织结构又决定于材料的成分和加工工艺. 观察研究显微组织结构,并分析它们与成分、加工工艺和使用性能之间的关系,已成为开发高性能先进材料过程中一个非常重要的环节,也是材料科学研究领域里的一个重要方面. 在这一环节中,如何正确利用现代分析仪器对显微组织结构进行观察研究,就成为关键的问题,而分析仪器的发展又大大地推动了该领域的研究工作.

三维原子探针(3DAP)大约是在 1995 年才推向市场的新型分析仪器,现在世界上装备了 3DAP 的实验室还为数不多. 它可以给出样品在纳米尺度空间中不同化学元素原子的分布,能够进行定量分析,是目前最微观并且分析精度较高的一种分析技术,这些都是对逐个原子分析后得出的结果. 目前的应用范围还局限于导电物质,在 20～80 K 和超高真空下进行分析.

1　场离子显微镜及原子探针

在了解 3DAP 之前,先描述一下场离子显微镜. 针尖状的样品放置在超高真空室中,并冷却至 20～80 K,在通入微量($\sim 10^{-3}$ Pa)成像气体 He(或者 Ar、Ne)后,在金属样品上加直流电压,由于在样品尖端原子排列凸出的位置处电场强度较强,容易引起该处的成像气体电离,电离后的正离子在电场作用下,沿径向射向荧光屏产生亮斑,得到样品尖端表面的原子像,称为场离子显微镜(FIM)像. 它的放大倍率是观察屏至样品的距离 R(一般是 5～10 cm)与样品尖端的曲率半径 r(一般<100 nm)之比,放大倍数不小于 10^6. 金属样品接正极,在很强的脉冲负电场作用下,样品尖端的原子也会以正离子状态离开表面,称为场蒸发. 在荧光屏的微通导板上加工一个 2 mm 的孔,倾动试样,使 FIM 观察到要研究的原子像对准该孔,然后在样品尖端叠加一脉冲电压(正电压增加),等被研究的原子蒸发飞向荧光屏并穿过孔洞,再配上飞行时间质谱仪测定该离子的质量/电荷比,确定该离子的种类,这就是 Muller 于 1968 年发明的原子探针(一维原子探针 APFIM). 1988 年 Cerezo 等制造出具有"位置敏感探头"的原子探针,但是只能同时探测两种元素的原子. 直到 1993 年,Blavette 等采用 96 通道多阳极探头,同时可以检测多于两种元素的原子,才成为三维原子探针[1].

三维原子探针的关键问题在于探测器的设计和制造. 当触发信号(脉冲电压)施加到样品上,原子从样品尖端表面蒸发,成为离子飞出并击中由微通道板制成的探测器. 微通道板

---

[*] 本文合作者：刘文庆. 原发表于《材料科学与工艺》,2007,15(3)：405–408.

是一种将离子象增强为电子象的"板",由许多<45 μm 玻璃毛细管组成的电子倍增器,入射离子进入毛细管后激发产生二次电子,信号可放大 $10^3 \sim 10^4$ 倍. 微通道板接受离子后可以激发出一束二次电子,由微通道板后面的位置敏感探头探测后确定其位置. 同时通过测定飞行时间可以确定该离子的质荷比 ($m = M/n$),确定是何种离子. 原子是逐层蒸发,逐层探测,数据经过计算机采集处理,再重新构建不同元素的原子在三维空间的分布图形. 深度方向的分辨率估计是 0.06 nm,水平方向的分辨率估计是 0.2 nm,后者主要是由于原子蒸发飞出后飞行轨道失真引起,与原子的热振动有关,因此,样品需要在低温下进行分析.

探测器的效率可达到 60%,每个电脉冲作用下有效的原子蒸发速率为 0.04 ions/pulse,工作电压的脉冲频率是 10 kHz,采用特殊的设计还可达到 200 kHz,这样每秒大约可以检测到 400 ions. 如果要检测 10^6 ions 以便构建纳米空间不同元素原子的分布,采集数据时间需要将近 1h.

英国 Oxford nano Science Ltd 开发的第 2 代 3DAP 中的 PoSAP 系统(optical position sensitive atom probe)是将图像增强像机和阳极阵列光电倍加管组合在一起,可以实时地观察被分析区域的 FIM,并装备了静电反射镜作为能量补偿,提高了质量分辨率,达到 $M/\Delta M = 500$. 在分析时效马氏体钢样品时,钢中所含 Mo、Cu、Ni 合金元素都具有多种同位素,有 ^{58}Ni,^{60}Ni,^{61}Ni,^{62}Ni,^{64}Ni,^{65}Ni,^{63}Cu,^{65}Cu,^{92}Mo,^{94}Mo,^{95}Mo,^{96}Mo,^{97}Mo,^{98}Mo,^{100}Mo,尽管它们之间有的质荷比只相差 1/6,但由于提高了分辨率,质荷比相差这样小的峰都可以分辨. 法国 Cameca 公司开发的经能量补偿 OTAP(optical tomographic atom probe)三维原子探针也具有相似的性能. 最近 Imago 公司推向市场的 LEAP 三维原子探针,采用了局域电极设计,将样品针尖至探头的距离缩短至 80~110 mm,增大了可探测分析的面积,由原有的20×20 nm 增大至 ϕ100 nm(70×70 nm),提高了工作电压的脉冲频率,缩短了采集数据的时间.

2 样品制备

3DAP 分析用的样品要制成针尖状,尖端的曲率半径为 10~100 nm. 通常的办法是先将样品加工成圆截面或方截面的细丝(<ϕ0.5 mm 或<0.5×0.5 mm),再用电解抛光方法制备针尖状的样品. 电解抛光时采用两步法:首先,使一薄层电解液漂浮在密度较大的惰性液体上,丝状样品垂直放入电解液中进行电解抛光,在长度的中部产生细颈直至断开,得到针状样品,但尖端的曲率半径仍不能满足要求,需要进行下一步显微电解抛光;显微电解抛光时用铂丝弯曲成小环作为负极,环中能保留一小滴电解液,将极细的丝状样品插入带有电解液的环中,在样品和 Pt 丝间通上脉冲电源,利用短暂时间的电解抛光来完成针尖样品的制备过程,操作在显微镜下完成,这种方法还可以用来修整已用过的针尖样品,以便重新利用该样品进行分析.

如果采用聚焦离子束加工设备(FIB)来制备针尖状样品,这时可以通过高倍放大后的二次电子象(SEM)观察加工过程,便于对加工过程进行控制. 该方法可以成功地制备出垂直于薄膜膜面的针尖样品,也容易将非常脆的样品加工成针尖状.

3 应用举例

3.1 Cottrell 气团的直接观察

溶质原子对位错的钉扎是了解合金元素影响材料各种力学性能的重要基础. Cotrell 和

Bilby 在 1949 年提出了一个假设：钢中碳原子在位错中偏聚后会围绕着位错形成气团，对位错产生了钉扎作用. 用这种假设可以解释低碳钢中屈服和应变时效现象. 从那时起，溶质原子与位错的弹性交互作用就受到广泛注意. 尽管后来用电阻和内耗测量证实了 0.2wt%C 的碳钢中有 90% 的碳原子在淬火过程中都偏聚在位错和马氏体的界面上，继续在低于 150 ℃时效处理，碳原子还会继续发生偏聚，并没有碳化物析出. 但是，直接观察溶质原子在位错中偏聚后构成的 Cottrell 气团，只有依赖于 3DAP 的分析才能完成，其中关键的问题是要在样品针尖的头部，获得一个布氏矢量垂直于样品轴向的位错. Wilde 等[2] 用 3DAP 研究了三种含碳不同的低碳钢样品（0.10wt%C；0.15wt%C；0.18wt%C）中的 Cottrell 气团，将样品加热到 1 000～1 100 ℃淬火，在室温时效以获得碳原子在位错及界面上的偏聚. 根据 3DAP 分析结果画出 4at%C 等浓度面的区域，给出碳的分布，可获得 Cottrell 气团的三维图像. 气团对于位错中心呈非对称分布，距离位错中心为 7 ± 1 nm，气团中碳的最大浓度为 8 ± 2at%.

　　已获得的钢中碳原子的 Cottrell 气团，不如 FeAl 有序合金中硼原子气团的图像清晰，所以下面给出的是 FeAl 有序合金中的分析结果[3-4]. 含有 400 ppm B(at%) 的 FeAl(40at%

Al)合金，有序化后是 B2 结构，晶体的(100)面是超点阵面，Fe 和 Al 原子相互交替占据该面. 分析时要寻找{001}刃形位错，所以作 3DAP 分析时样品针尖处的晶体是沿[001]方向. 图 1 是 3DAP 的分析结果，图中只给出了一层 Al 原子分布的截面，从图中可以分辨出 Al 原子占据的原子面，面间距约为 0.29 nm，从前面数有 21 排原子面，从后面数有 22 排原子面，这说明图中自上而下存在一个刃型位错，刃型位错的示意图画在图的左上方. B 原子围绕着自上而下的刃型位错成细圆柱状分布，成为 Cottrell 气团，每一个点表示检测到的一个 B 原子. 气团中 B 的最高含量为 3at%，平均含量为 2at%，是 B 添加量的 50 倍. 形成 B 的 Cottrell 气团后，FeAl 合金单晶体的应力应变曲线中也出现了明显的上下屈服点.

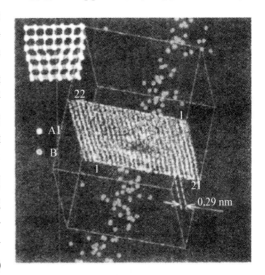

图 1　有序 FeAl 合金中 B 原子在刃型位错附近的 Cottrell 气团，B 原子围绕着刃型位错成细圆柱状分布

3.2　硼在 IF 钢晶界上的偏聚

　　深冲钢板是汽车制造工业中的一种重要原材料. 在超低碳 IF 钢中（interstitial free - IF 钢）再添加微量 Ti 或 Nb 是关键，要求钢板中形成较集中的(111)织构，有利于深冲时的变形，改善深冲性能. 完全去掉 C 和 N 可以提高成型性能，但是在晶界上缺少溶质原子 C 会明显降低界面的结合力. 在低一点的温度下经受冲击性的深冲成型，会引起晶间脆性断裂，称为二次加工脆性，如果为了提高钢的拉伸强度而添加 P、Mn 或 Si 后，问题更为突出. 添加微量 B 是一种补救方法，它并不影响钢的成型性能. 用示踪原子和俄歇电子谱方法研究的结果表明，B 在再结晶初期就会偏聚在晶界上，起到强化晶界的作用. 为了了解 B 在晶界上偏聚的情况，用 3DAP 进行研究是最合适的选择，但在试样的针尖上需要有一条晶界. Seto 等研究了 IF 钢中的晶界偏聚[5]，样品经处理后的晶粒为 14 μm，制成针尖样品后，大约在 1/10

样品的针尖上可以找到晶界. 图 2 是 3DAP 的分析结果,可以看出 B 不仅偏聚在晶界上,还延伸到晶界两侧 1~1.5 nm 处. B 在晶界上的浓度是添加含量的 250 倍,在再结晶的初期,P 不发生明显的晶界偏聚.

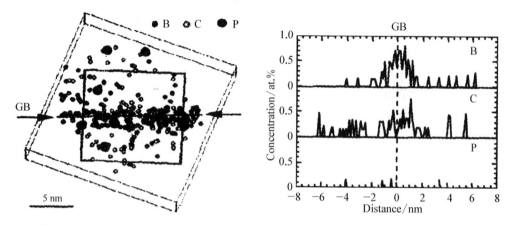

图 2 B、C、P 原子在 IF 钢中晶界(GB)附近的三维分布图及其在晶界附近的浓度分布

3.3 铝合金中弥散相的形成

淬火固溶处理然后时效沉淀强化是增加 Al 合金的强度和硬度的有效方法,Al-Cu 合金是最典型的一种. Al-Cu 合金中添加痕量的 Mg 和 Ag 后,沉淀相中除了 θ(CuAl$_2$)相外,还增加了一种 Ω 相,图 3 是 Al-1.7Cu-0.3Mg-0.2Ag 合金经过固溶时效处理后(190 ℃-8h)的 TEM 明场象,显示出片状的 θ 和 Ω 相[6],与<211>痕迹一致的薄片是在{111}面上的 Ω 相,与[011]痕迹一致的薄片是在(001)面上的 θ 相. Ω 相热稳定性好,不易聚集长大,因此,提高了合金的高温强度. 用 3DAP 研究这种铝合金在淬火时效时 θ 相和 Ω 相的形成过程[7],揭示了 Ω 相热稳定性好的原因. 合金样品(Al-1.9Cu-0.3Mg-0.2Ag)固溶处理后经 180 ℃-15 s 时效,从 3DAP 的分析结果中可以清楚看到形成了 Mg-Ag 富集的 GP 区,其中不含 Cu. 在 180 ℃继续时效时,Cu 向 Mg-Ag 富集的 GP 区中扩散,形成了 20at%Cu-20at%Mg-10at%Ag 的{111}GP 区,再继续时效,Mg 和 Ag 原子又从中心迁移至析出相与 αAl 的界面处,完成了 GP 区转变成 Ω 相的过程,这时 Ω 相的中心部位含 33at%Cu. 样品在 180 ℃时效 10h 后,用 3DAP 分析获得的 θ 相和 Ω 相成分象如图 4 所示.

最近 Larson 等对 3DAP 在分析多层功能薄膜材料的应用方面作了综述和评价[8],由于制备功能薄膜材料是通过多种不同元素交替沉积而成,最薄的层间距只有十多个原子层厚,界面的粗糙程度、原子间的相互扩散以及界面上的偏聚等对薄膜的性能都会有重要的影响. 要从原子尺度上来分析了解显微组织与物理性能之间的关系,3DAP 是唯一可利用的分析手段.

4 结语

利用三维原子探针分析技术,可以获得纳米尺度三维空间内不同元素原子的分布图,分辨率接近原子尺度,在研究纳米材料问题时有着无法替代的作用. 纳米尺度原子团簇的析

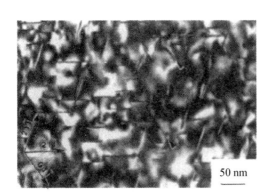

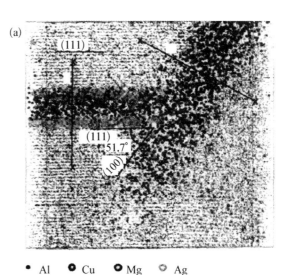

● Al ● Cu ● Mg ○ Ag

图 3 Al－1.7Cu－0.3Mg－0.2Ag 合金经过固
溶时效处理后(190°－8h)的 TEM 明场象

图 4 Al－1.7Cu－0.3Mg－0.2Ag 合金在 180 ℃－10 h
时效后用三维原子探针分析获得的结果

出,以及合金元素在界面上的偏聚等,对材料性能有着重要的影响,研究这些问题正是三维
原子探针分析仪器的特长,它不仅在研究纳米材料方面,在研究传统金属材料方面都可以发
挥重要作用.看来,三维原子探针技术的广泛应用,必将推动材料科学研究工作的发展,人们
将有可能从原子尺度来认识一些目前还不太清楚的问题.由于它不能获得晶体结构方面的
信息,所以有时还需要 TEM 分析来配合.

参 考 文 献

[1] BLAVETTE D, DECONNIHOUT B, BOSTEL A, et al. Tomographic atom probe. A quantitative three-dimensional nanoanalytical instrument on an atomic scale[J]. Rev. Scient. Instrum. , 1993, 64: 2911.

[2] WILDE J, CEREZO A and SMITH G D W. Three-dimensional atomic-scale mapping of a cottrell atmosphere around a dislocation in iron[J]. Scripta Mater. , 2000, 43: 39 – 48.

[3] CADEL E, FRACZHIEWICZ A and BLAVETTE D. Atomic scale observation of cottrell atmosphere in B doped FeAl(B2) by 3D atom probe field ion microscopy[J]. Materials Science and Engineering A, 2001, 309 – 310: 32 –37.

[4] MENAND A, DECONIHOUT B, , CADLE E, et al. Atom probe investigations of fine scale features in intermetallics[J]. Micron. , 2001, 32: 721 – 729.

[5] SETO K, LARSON D J, WARREN P J et al. Grain boundary segregation in boron added interstitial free steels studied by 3-dimensional atom probe[J]. Scripta Mater. ,1999, 40: 1029 – 1034.

[6] HONO K. Nanoscale microstructural analys of metallic materials by atom probe fieldion microscopy[J]. Progress in Materials Science, 2002, 47: 621 – 729.

[7] REICH L, MURAYAMA M and HONO K. Evolution of Ω phase in an Al – Cu – Mg – Ag alloy — a three atom probe study[J]. Acta Mater. , 1998, 46: 6053 – 6062.

[8] LARSON D J, PETFORD-LONG A K, MA Y Q, et al. Information storage materials: nanoscale characterisation by three-dimensional atom probe analysis[J]. Acta Mater. , 2004, 52: 2847 – 2862.

The application of 3DAP in the study of materials science

Abstract: The three-dimensional atom probe (3DAP) is a quantitative micro-analytical instrument, which provides three-dimension maps of the distribution of different element atoms in metallic specimens by chemically analyzing the atoms one-by-one. No other method can take the place of 3DAP to find out the chemical inhomogeneities in nanometer scale of the specimens by the analysis of individual atoms. This review presents the working principal and sample preparation of 3DAP. Unique features of the 3DAP are demonstrated by describing the applications of 3DAP.

A Superior Corrosion Behavior of Zircaloy – 4 in Lithiated Water at 360 ℃/18. 6 MPa by β – Quenching[*]

Abstract: To further understand the mechanism on the heat-treatment effect on corrosion behavior of Zircaloy – 4, β – quenched and recrystallized specimens are corroded in an autoclave with 0. 01 M LiOH aqueous solution at 360 ℃/18. 6 MPa. Results show that the β – quenched specimen possesses excellent corrosion resistance, the weight gain of which is consistently comparable to that of ZIRLO and N18 alloys during 529 days exposure, while the recrystallized specimen exhibits accelerated corrosion rate after 100 days exposure. The increase of the supersaturated solid solution contents of Fe and Cr in α – Zr matrix is responsible for the improvement of corrosion resistance by β – quenching treatment. The fracture surface and inner surface morphology of oxide films indicates that β – quenching treatment significantly slow down the development of pores to cracks in oxide films and reduces the uneven growing tendency of the oxide, which is related to the corrosion behavior.

1 Introduction

It is well known that the presence of LiOH in water at high temperature and pressure is detrimental to the corrosion resistance of Zircaloy – 4 (denoted as Zry – 4), increasing corrosion rate at the post-transition stage dramatically[1-6]. Although the corrosion resistance of Zry – 4 in lithiated water can be further improved by optimizing the thermo-mechanical processing, it is still low compared to that of Zr – Sn – Nb alloys, such as ZIRLO (Zr – 1Sn – 1Nb – 0. 1Fe)[7,8], N18(Zr – 1Sn – 0. 4Nb – 0. 3Fe – 0. 1Cr)[9] and E635 (Zr – 1. 2Sn – 1Nb – 0. 4Fe)[10]. Considering that Nb has a higher solubility limit in α – Zr matrix (about 0. 6% in mass fraction at 610 ℃)[11], the superior corrosion resistance of these Zr – Sn – Nb alloys in lithiated water reminds us whether increase of Fe and Cr contents in α – Zr matrix by supersaturated solid solution treatment, e. g., solution-treated by β – quenching, could further improve the corrosion resistance of Zry – 4 in lithiated water.

So far, there is no general agreement on the reason for the impact of thermo-mechanical processing on the uniform corrosion behavior of Zry – 4. Some researchers [12-14] attribute it to the size and distribution of $Zr(Fe, Cr)_2$ second phase particles (SPPs) since Fe and Cr alloying elements have low solubilities in α – Zr matrix (lower than 200 $\mu g/g$ even at 810 ℃

---

* In collaboration with Yao M Y, Li Q, Liu W Q, Geng X, Lu Y P. Reprinted from Journal of Nuclear Materials, 2008, 374(1 – 2): 197 – 203.

with the maximum solubility) and precipitate as $Zr(Fe, Cr)_2$ SPPs[15]. However, others [16-19] ascribe it to the variation of Fe and Cr contents in α - Zr solid solution because they will also vary with the precipitation of $Zr(Fe, Cr)_2$ SPPs.

Hence, it is necessary to further study the corrosion behavior of Zry - 4 specimens with different heat-treatments, especially solution-treated by β - quenching, in LiOH aqueous solution at 360 ℃/18. 6 MPa. Although it is impractical to employ such treatment at the final stage in manufacturing Zry - 4 tubes, it is helpful for us to understand the corrosion mechanism affected by alloying elements and heat-treatments.

2　Experimental procedure

2. 1　Specimens preparation and microstructure characterization

Specimens with 25 mm×10 mm and 25 mm×20 mm were cut from a Zry - 4 (Zr - 1. 5Sn - 0. 2Fe - 0. 1Cr, mass fraction %) plate with 0. 6 mm thickness, which was fabricated by conventional procedure[20] and underwent a final recrystallization anneal at 600 ℃ for 1 h (accumulated annealing parameter $\sum A_i = 4 \times 10^{-17}$ h, calculating from $\sum A_i = \sum t_i \exp(-40\ 000/T_i)$). The specimens with 25 mm×20 mm were not heat-treated any further (denoted as RA Zry - 4). The specimens with 25 mm×10 mm were sealed in an evacuated quartz capsule at about 5×10^{-3} Pa and solution-treated at 1 020 ℃ (β phase) for 10 min, then quenched in water and broken the capsule simultaneously (denoted as βQ Zry - 4).

The microstructure of the specimens with different heat-treatments was examined by JEM - 200CX transmission electron microscope (TEM). Disk specimens, 3 mm in diameter, were mechanically grinded to about 70 μm in thickness, and then thinned by twin-jet electro-polishing technique at a voltage of 40 - 50 V in a mixture of 10% perchloric acid and 90% ethanol (in volume) at −40 ℃. The contents of alloying elements in α - Zr matrix were measured using EDS fitted in a JEM - 2010F TEM with live time of 1 000 s.

2. 2　Autoclave testing

The corrosion tests for these specimens were performed in an autoclave with 0. 01 M LiOH aqueous solution at 360 ℃/18. 6 MPa. The weight gain was a mean value obtained from three specimens. Prior to the corrosion tests, the specimens were cleaned and pickled in a solution of 10% HF + 45% HNO_3 + 45% H_2O (in volume), sequentially rinsed in cold tap water, boiling deionized water and then blow-dried with warm air.

2. 3　Fracture surface and inner surface characterization techniques

JSM - 6700F high resolution scanning electron microscope (HRSEM) was used to characterize the morphology of fracture surface and inner surface of oxide films formed on specimens after corrosion. The preparation of fracture surface and inner surface samples have been described in Ref. [21] in detail. A schematic diagram of the specimen preparation is shown in Fig. 1. Firstly, the cross-section of the corroded specimen was etched in the same solution as pickling specimens for corrosion tests to dissolve the metal matrix and reveal the metal/oxide interface, and then the upstretched oxide film on the cross-section was broken off. The leaving cross-section sample was used for HRSEM examination of the fracture surface, while the broken oxide flake was used for HRSEM examination of the inner surface. In order to avoid charging by bombardment of electron beams in the microscope, a thin layer of gold was deposited on the oxide samples by sputtering.

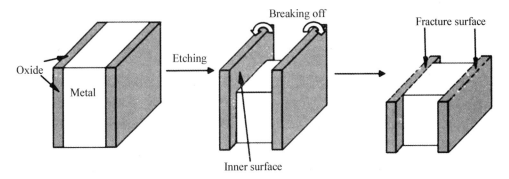

Fig. 1 A schematic diagram of the oxide specimen preparation for the examination of inner and fracture surface

3　Experimental results

3.1　The microstructure of Zry – 4 specimens

Fig. 2 shows the microstructure of βQ Zry – 4 and RA Zry – 4 specimens. The SPPs in βQ Zry – 4 specimen can be hardly distinguished during TEM examination at low magnification (10 000 ×) (Fig. 2(a)), but at higher magnification (50 000 ×), the fine SPPs (<0.05 μm) precipitated along the grain boundaries of lath-like α – Zr grains can be observed (indicated by an arrow in Fig. 2(b)). The SPPs in RA Zry – 4 specimen are much larger in size (0.05 – 0.2 μm) and higher in quantity (Fig. 2(c)). The difference in size and quantity of SPPs indicates that higher Fe and Cr alloying elements should be supersaturated

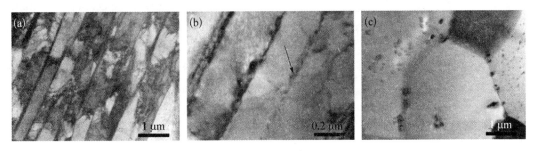

Fig. 2 The TEM micrographs of different specimens: (a, b) βQ Zry – 4 with different magnifications and (c) RA Zry – 4

solid solution in α – Zr after β – quenching.

The results from EDS analysis are summarized in Table 1. From Table 1, the obvious difference in Fe and Cr contents in α – Zr matrix of the two groups of Zry – 4 specimens can be seen. For βQ Zry – 4 specimen, the contents of Fe and Cr in α – Zr matrix amount to about 700 μg/g, respectively, which are much higher than their maximal solubility in α – Zr matrix (120 μg/g for Fe and 200 μg/g for Cr[15]). For RA Zry – 4 specimen, however, the contents of Fe and Cr in α – Zr matrix are too low to be detected. This illustrates that higher contents of Fe and Cr are indeed supersaturated solid solution in α – Zr after β – quenching, which is consistent with the result of SPPs observation. Moreover, the Sn content has no difference in the matrix of two groups of specimens and is close to that of Zry – 4 alloy (1.5%). It shows that 1.5% Sn is completely in the solid solution of α – Zr matrix.

Table 1 The contents of alloying elements in α – Zr matrix in βQ Zry – 4 and RA Zry – 4 specimens

Specimens	Mean content		
	Fe(μg/g)	Cr(μg/g)	Sn (mass fraction %)
βQ Zry – 4	700±400	700±300	1.44±0.09
RA Zry – 4	—	—	1.5±0.03

3.2 Corrosion behavior

Fig. 3 shows the corrosion weight gain as a function of exposure time after the autoclave tests in 0.01 M LiOH aqueous solution at 360 ℃/18.6 MPa for the two groups of specimens. For comparison, the weight gain vs time of ZIRLO[8] and N18 tested in the same condition as this study and RA Zry – 4 tested in deionized water at 360 ℃/ 18.6 MPa is also presented in Fig. 3. Here, the weight gain data of ZIRLO in lithiated water and RA Zry – 4 in pure water was taken from Refs. [8,21], respectively, while the data of N18 was obtained by corroding together with βQ Zry – 4 and RA Zry – 4 specimens in the same autoclave, which is close to Zhao's result[9]. Obviously, a very rapid increase in the post-transition corrosion rate is observed in RA Zry – 4 specimen in lithiated water environment. This is a general phenomenon for Zircaloy corroded in lithiated water. Based on the Li$^+$ incorporated in the oxide layer, some mechanisms were suggested by many researchers, e.g. preferential dissolution of cubic/tetragonal ZrO_2 to form a porous inner layer of oxide[22], the gradual alteration of the barrier layer properties[2,23],

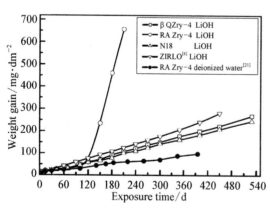

Fig. 3 Corrosion weight gain vs. exposure time of βQ Zry – 4 and RA Zry – 4 specimens tested in 0.01 M LiOH aqueous solution at 360 ℃/18.6 MPa. The weight gain vs. time of ZIRLO[8] and N18 tested in the same condition as this study and RA Zry – 4 tested in deionized water at 360 ℃/18.6 MPa[21] is also presented here

the modification of normal oxide growth by surface LiO groups[24,25], and reduction of the surface free energy of ZrO_2 to promote the formation of pores and cracks[26]. It is not further discussed here.

It is interesting and surprising that βQ Zry - 4 specimen exhibits superior corrosion resistance in lithiated water. Compared to the post-transition corrosion rate of RA Zry - 4 specimen in pure water, the corrosion rate in lithiated water is increased by only about a factor of 3 for βQ Zry - 4 specimen, but about a factor of 33 for RA Zry - 4 specimen. The weight gain of RA Zry - 4 specimen has reached about 700 mg/dm^2 after 210 days exposure, while that of βQ Zry - 4 specimen is only about 268 mg/dm^2 even after 529 days exposure. The weight gain of βQ Zry - 4 specimen is consistently comparable to that of ZIRLO and N18 (Fig. 3), which exhibit superior corrosion resistance in lithiated water. This shows that the corrosion resistance of Zry - 4 in lithiated water can be really further improved by solution-treatment.

3.3 Morphology of the fracture surface and inner surface of oxide films

3.3.1 Fracture surface morphology of oxide films

Fig. 4 shows the fracture surface morphology of the oxide films formed on RA Zry - 4

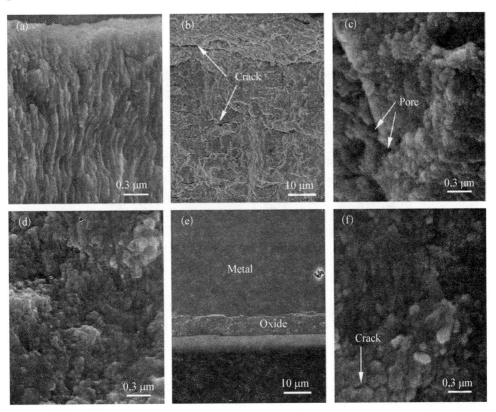

Fig. 4 The fracture surface morphology of the oxide films formed in 360 ℃ lithiated water observed by HRSEM. The upper interface in the figures refers to the metal/oxide interface: (a – c) RA Zry - 4 specimen for 100 days (a) and 210 days (b and c) at different magnifications; (d – f) (βQ Zry - 4 specimen for 100 days (d) and 210 days (e and f) at different magnifications

and βQ Zry - 4 specimens corroded in lithiated water for different time, respectively. At the early stage of oxide growth (100 days exposure), the oxide films on both RA Zry - 4 and βQ Zry - 4 specimens are compact and no large cracks are observed except for a few micro-cracks or pores (Fig. 4(a) and (d)). At the later stage of oxide growth (210 days exposure), the amount of cracks in the oxide film on RA Zry - 4 specimen (indicated by arrows in Fig. 4(b)) is much more than that on βQ Zry - 4 specimen (Fig. 4(e)). At high magnification, more pores can be observed in RA Zry - 4 specimen than in βQ Zry - 4 specimen (Fig. 4(c) and (f)). Here, the cracks are likely an artifact of sample preparation, but the locations of observed cracks must present either as cracks or weak boundaries in the oxide film before sample preparation. Obviously, solution-treatment of the βQ Zry - 4 specimen noticeably slows down the development of pores to cracks in the oxide film.

Moreover, the transformation from columnar grains to equiaxed grains is evident for RA Zry - 4 specimen (Fig. 4(a) and (c)). However, it is not evident for βQ Zry - 4 specimen and equiaxed grains are observed dominantly regardless of corroding time (Fig. 4(d) and (f). The equiaxed grains are less than 100 nm in both specimens for 210 days exposure.

3.3.2 Inner surface morphology of oxide films

Fig. 5 shows the inner surface morphology of oxide films on RA Zry - 4 and βQ Zry - 4 specimens corroded for different time, respectively. Here, the micro-pores observed in oxide films are believed to be artifacts arising from the attack of the solution at SPPs during sample preparation due to their comparative size and slower oxidation of SPPs than the

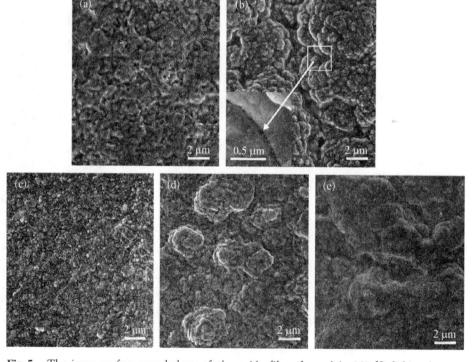

Fig. 5 The inner surface morphology of the oxide films formed in 360 ℃ lithiated water observed by HRSEM: (a and b) RA Zry - 4 specimen for 100 days and 210 days, respectively and (c - e) βQ Zry - 4 specimen for 100 days, 210 days and 300 days, respectively

matrix. At the early stage of oxide growth (100 days exposure), a relatively planar inner surface is evident for both specimens with different heat-treatments (Fig. 5(a) and (c)). However, a typical "cauliflower-like" morphology is evident at the later stage of oxide growth (Fig. 5(b), (d) and (e)), as observed in the post-transition oxide film on Zircaloy - 2 tested in 400 ℃ steam[27] and on Zry - 4 and ZIRLO tested in lithiated water and pure water, suggesting a typical inner surface morphology of the post-transition oxide film[21,26]. Obviously, the growth of oxide film is faster and more uneven for RA Zry - 4 specimen than that for βQ Zry - 4 specimen.

In addition, smooth regions and cracks on the inner surface of oxide films are more easily found in the RA Zry - 4 specimen corroded for 210 days (Fig. 5(b)) than that in βQ Zry - 4 specimen even corroded for 300 days (Fig. 5(e)). Note that the cracks in Fig. 5(b) are also likely artifacts resulting from the stress-relief after etching the metal matrix out during the sample preparation, but there exists a higher inner stress in the oxide film on RA Zry - 4 specimen than βQ Zry - 4 specimen. The smooth regions are likely an amorphous phase (see the bottom left corner in Fig. 5(b), which is a magnification of the rectangle area), which was also observed and indicated by Zhou in the examination of oxide films near the zirconium alloy matrix using TEM, SPM and HRSEM[21,26,28].

4　Discussion

4.1　The reason for improving corrosion resistance by the solution-treatment of β-quenching

Sabol et al. [8] investigated the effect of the β - quenching treatment before tube reduced extrusion (where also denoted as βQ Zry - 4) on corrosion behavior of Zry - 4 tube in lithiated water, and found the corrosion resistance of βQ Zry - 4 tube was inferior to that of conventional Zry - 4 one. It should be pointed out that the βQ Zry - 4 tube in Ref. [8] actually underwent further intermediate recrystallization and a final stress-relief anneal after β - quenching ($\sum A_i = 1.1 \times 10^{-18}$ h). Zhou et al. [17] also investigated the corrosion behavior of βQ Zry - 4 tube (in fact, the tubes were then cold-pilgered and stress-relief annealed at 470 ℃ for 4 h after β - quenching, $\sum A_i = 1.6 \times 10^{-23}$ h) in deionized water at 360 ℃/18.6 MPa, and found that it possessed superior corrosion resistance in deionized water before 300 days exposure, but there was a second transition phenomenon at 300 days and the weight gain increased quickly after second transition. It was proposed that too low $\sum A_i$ value ($<2 \times 10^{-18}$ h) or fine SPPs (<0.05 μm) was not good for uniform corrosion resistance autoclave testing in 350 ℃ water or PWRs[14], and it was reported that the corrosion resistance of Zry - 4 in 350 ℃ water would go worse when the cooling rate exceeded 500 ℃/s from β - Zr region[29]. In this study, however, βQ Zry - 4 specimen with fine SPPs exhibits superior corrosion resistance and the accelerated second transition phenomenon does not appear even during a long-term exposure for 529 days, the weight gain of which is consistently comparable to that of ZIRLO and N18, while RA Zry − 4 specimen with recommended $\sum A_i$ or SPPs size exhibits inferior corrosion resistance (Fig. 3). This illustrates that the SPPs size should not be the reason for improving the corrosion resistance by the solution-treatment of β − quenching.

As mentioned in Section 1, the heat-treatments also vary the contents of alloying elements in α − Zr matrix besides the SPPs size. The results from EDS analysis show that there is no difference in Sn content, but there is obvious difference in Fe and Cr contents in α − Zr matrix of βQ Zry − 4 and RA Zry − 4 specimens; The contents of Fe+Cr in α − Zr matrix reach 1 400 μg/g in βQ Zry − 4 specimen, which are much higher than those in RA Zry − 4 specimen (see Table 1). The results are close to Li's[19], who analyzed the Fe and Cr contents in α − Zr matrix with a combination method of extracting electrochemically Zr(Fe,Cr)$_2$ particles from Zry − 4 and analyzing mass concentration in electrolyte by means of flame atomic absorption spectrometry. Wadman and Andrén[18] also analyzed the matrix composition of Zry − 4 with different heat-treatments using atom probe analysis and studied their corrosion behavior in 400 ℃ steam. Results also showed that Zry − 4 with higher iron concentration in the matrix exhibited the best uniform corrosion resistance. In addition, it was found that β − quenching treatment employed at later period in fabricating Zircaloy tubes was a benefit for improving the nodular corrosion resistance[17,30], which was attributed to an increase in the amount of alloying elements retained in α − Zr matrix. Meanwhile, based on a fact that Nb has a higher solid solubility in α − Zr matrix, Zr − Sn − Nb (e. g. ZIRLO and N18) alloys exhibit superior corrosion resistance in lithiated water (Fig. 2). Thus, it can be concluded that the increase of Fe and Cr contents in α − Zr matrix resulting from β − quenching treatment is responsible for the excellent corrosion resistance of βQ Zry − 4 specimen in lithiated water.

4. 2　The relationship between the corrosion behavior and the microstructure of oxide films

Corrosion behavior is closely associated with the oxide characteristics, such as oxide grain shapes, pores and cracks in the oxide. Generally, the increase in the fraction of the columnar grains is corresponding to the lower weight gain and corrosion rate and the transformation from columnar grains to equiaxed grains is suggested to associate with the corrosion transition[12,31-33]. This seems true for RA Zry − 4 specimen (Fig. 4(a)-(c)), however, the columnar grains are almost not evident in the oxide layer on βQ Zry − 4 specimen with superior corrosion resistance (Fig. 4(d)-(f)). Wadman et al. [34] reported columnar grains were still observed besides equiaxed grains even in a thick oxide layer. This shows that the shapes of the oxide grains may not be correlated consequentially to the corrosion behavior.

The cracks in the oxide film on βQ Zry − 4 specimen is much less than those on RA Zry − 4 specimen, the development of pores to cracks on the former is noticeably slower than that on the latter (Fig. 4), and the growth of oxide film is slower and less uneven for the former than that for the latter (Fig. 5), which is corresponding to their corrosion behavior. In investigating the effect of alloying composition on the microstructure evolution of oxide films on zirconium alloys tested in lithiated water, Zhou et al. [26] also found that the pore clusters and micro-cracks along grain boundaries were more easily formed in the oxide film on Zry – 4 than ZIRLO, and the cracks parallel to the oxide/metal interface were also more easily produced in the former than that in the latter. This illustrates that the increase in solute contents of alloying elements in α – Zr matrix, whether it is obtained by adding alloying elements with higher solid solubility (e. g. adding Nb in Zry – 4) or by heat-treatments (e. g. β – quenching of Zry – 4 in this study), is beneficial for retarding the development of pores to cracks, thereby improve the corrosion resistance. The possible reason is that the alloying elements in α – Zr matrix are easier than those in SPPs to dissolve in ZrO_2 during oxidation. It is known that Sn has much higher solid solubility in α –Zr matrix than Fe, Cr and Nb do, however, higher Sn content is detrimental for the corrosion resistance. Therefore, the effect of alloying elements in α – Zr matrix on the growth of oxide films needs to be further investigated.

5　Conclusion

(1) β – quenching treatment increases the contents of Fe and Cr in α – Zr matrix of Zry –4 alloy to about 700 μg/g, respectively, which are much higher than their maximal solubility in α – Zr matrix.

(2) After treated by β – quenching, Zry – 4 specimens show excellent corrosion resistance in lithiated water, the weight gain of which is consistently comparable to that of ZIRLO and N18 alloys during 529 days exposure. The increase of Fe and Cr contents in α – Zr matrix is responsible for the improvement of corrosion resistance by β – quenching.

(3) The fracture surface and inner surface morphology of oxide films shows that β – quenching treatment significandy slows down the development of pores to cracks in oxide films and reduces the uneven growing tendency of the oxide, which is related to the corrosion behavior.

Acknowledgements

This project is financially supported by the National Nature Science Foundation of China (50371052) and Shanghai Leading Academic Discipline Project (T0101).

References

[1] B. Cox, Y. M. Wong, in: Zirconium in the Nuclear Industry: Ninth International Symposium, ASTM STP, vol. 1132, 1991, p. 643.

[2] D. Pêcheur, J. Godlewski, et al. , in: Zirconium in the Nuclear Industry: Twelfth International Symposium, ASTM STP, vol. 1354, 2000, p. 793.

[3] B. X. Zhou, Q. Li, Q. Huang, et al. , Nucl. Power Eng. 21 (5) (2000) 439 (in Chinese).

[4] N. Ramasubramanian, N. Precoanin, V. C. Ling, in: Zirconium in the Nuclear Industry: Eighth International

symposium, ASTM STP, vol. 1023, 1989, p. 187.

[5] I. L. Bramwell, P. D. Parsons, D. R. Tice, in: Zirconium in the Nuclear Industry: Ninth International Symposium, ASTM STP, vol. 1132, 1991, p. 628.

[6] P. Billot, S. Yagnik, N. Ramasubramanian, et al. , in: Zirconium in the Nuclear Industry: Thirteenth International Symposium, ASTM STP, vol. 1423, 2002, p. 169.

[7] G. P. Sabol, G. R. Kilp, et al. , in: Zirconium in the Nuclear Industry: Eighth International Symposium, ASTM STP, vol. 1023, 1989, p. 227.

[8] G. P. Sabol, R. J. Comstock, et al. , in: Zirconium in the Nuclear Industry: Tenth International Symposium, ASTM STP, vol. 1245, 1994, p. 724.

[9] W. J. Zhao, Z. Miao, et al. , J. Chin. Soc. Corros. Protect. 22 (2) (2002) 124 (in Chinese).

[10] A. V. Nikulina, V. A. Markelvo, et al. , in: Zirconium in the Nuclear Industry: Eleventh International Symposium, ASTM STP, vol. 1295, 1996, p. 785.

[11] Y. H. Jeong, K. O. Lee, H. G. Kim, J. Nucl. Mater. 302 (2002) 9.

[12] H. Anada, B. J. Herb, et al. , in: Zirconium in the Nuclear Industry: Eleventh International Symposium, ASTM STP, vol. 1295, 1996, p. 74.

[13] P. Foster, J. Dougherty, M. G. Burke, et al. , J. Nucl. Mater. 173 (2) (1990) 164.

[14] F. Garzarolli, E. Steinberg, H. G. Weidinger, in: Zirconium in the Nuclear Industry: Eighth International Symposium, ASTM STP, vol. 1023, 1989, p. 202.

[15] D. Charquet, R. Hahn, et al. , in: Zirconium in the Nuclear Industry: Eighth International Symposium, ASTM STP, vol. 1023, 1989, p. 405.

[16] T. Thorvaldsson, T. Andersson, et al. , in: Zirconium in the Nuclear Industry: Eighth International Symposium, ASTM STP, vol. 1023, 1989, p. 128.

[17] B. X. Zhou, W. J. Zhao, Z. Miao, et al. , China Nuclear Science and Technology Report, CNIC – 01074, SINRE – 0066, 1996.

[18] B. Wadman, H. Q. Andrén, in: Zirconium in the Nuclear Industry: Eighth International Symposium, ASTM STP, vol. 1023, 1989, p. 423.

[19] C. Li, B. X. Zhou, W. J. Zhao, et al. , J. Nucl. Mater. 304 (2002) 134.

[20] C. M. Eucken, P. T. Finden, et al. , in: Zirconium in the Nuclear Industry: Eighth International Symposium, ASTM STP, vol. 1023, 1989, p. 113.

[21] B. X. Zhou, Q. Li, M. Y. Yao, et al. , Nucl. Power Eng. 26 (4) (2005) 364 (in Chinese).

[22] B. Cox, M. Ungurela, et al. , in: Zirconium in the Nuclear Industry: Eleventh International Symposium, ASTM STP, vol. 1295, 1996, p. 114.

[23] H. J. Beie, A. Mitwalsky, F. Garazorlli, et al. , in: Tenth International Symposium, ASTM STP, vol. 1245, 1994, p. 615.

[24] N. Ramasubramanian, in: Zirconium in the Nuclear Industry: Ninth International Symposium, ASTM STP, vol. 1132, 1991, p. 613.

[25] N. Ramasubramanian, P. V. Balakrishnan, in: Zirconium in the Nuclear Industry: Tenth International Symposium, ASTM STP, vol. 1245, 1994, p. 378.

[26] B. X. Zhou, Q. Li, W. Q. Liu, et al. , Rare Met. Mater. Eng. 35 (7) (2006) 1009 (in Chinese).

[27] G. Wikmark, P. Ruding, et al. , in: Zirconium in the Nuclear Industry: Eleventh International Symposium, ASTM STP, vol. 1295, 1996, p. 55.

[28] B. X. Zhou, Q. Li, M. Y. Yao, et al. , Rare Met. Mater. Eng. 32 (6) (2003) 417 (in Chinese).

[29] F. Garzarolli, I. Pohlmeyer, et al. , External cladding corrosion in water power reactors, in: Proceeding of a Technical Committee Meeting, International Atomic Energy Agency, Cadarache, France, 1985, p. 66.

[30] T. Kubo, M. Uno, in: Zirconium in the Nuclear Industry: Ninth International Symposium, ASTM STP, vol. 1132, 1991, p. 476.

[31] D. Pêcheur, J. Godlewski, et al. , in: Zirconium in the Nuclear Industry: Eleventh International Symposium,

ASTM STP, vol. 1295, 1996, p. 94.

[32] H. Anada, K. Takeda, in: Zirconium in the Nuclear Industry: Tenth International Symposium, ASTM STP, vol. 1295, 1996, p. 35.

[33] A. Yilmazbayhan, E. Breval, et al. , J. Nucl. Mater. 349 (2006) 265.

[34] B. Wadman, Z. Lei, et al. , in: Zirconium in the Nuclear Industry: Tenth International Symposium, ASTM STP, vol. 1245, 1994, p. 579.

Effect of Single-Step Strain and Annealing on Grain Boundary Character Distribution and Intergranular Corrosion in Alloy 690[*]

Abstract: The effects of single-step thermomechanical treatments on the grain boundary character distribution (GBCD) and intergranular corrosion of Alloy 690 (Ni − 30Cr − 10Fe, wt. %) are investigated. High proportion of low ΣCSL grain boundaries (more than 70% according to Palumbo − Aust criterion) associating with large size grains-cluster microstructure is obtained through one-step thermomechanical treatment of 5% cold rolling followed by annealing at 1, 100 ℃ for 5 min. Nucleation density of recrystallization and multiple twinning are the key factors affecting the GBCD. The grains-cluster is produced by multiple twinning starting from a single recrystallization nucleus. That the mean size of the grains-clusters and proportion of low ΣCSL boundaries decrease with the increasing strain, is caused by the increasing nucleation density of recrystallization with the increase of strain. The specimen with large size grains-cluster microstructure and high proportion of low ΣCSL boundaries exhibits much better resistance to mass loss during intergranular corrosion testing than that with small size grains-cluster microstructure and relatively low proportion of low ΣCSL boundaries.

1 Introduction

Due to its excellent corrosion resistance, nickel based Alloy 690 (Ni − 30Cr − 10Fe, mass %) is used in corrosive and high temperature applications, such as in the petrochemical and the nuclear industries[1, 2]. But with the prolonged service life and improved performance being demanded by industries, the improvement of the resistance to intergranular attack in Alloy 690 should be considered.

Grain boundary structure has long been known to play a critical role in many material properties, such as precipitation[3, 4], corrosion[5, 6], and cracking[7]. Grain boundary engineering (GBE), which was under the light of "grain boundaries design and control" proposed by Watanabe[8], developed a lot, and was fruitful during the last two decades. The grain boundaries related properties of materials can be enhanced by exercising control over the population of low ΣCSL ($\Sigma \leqslant 29$) grain boundaries, as defined by the coincident site lattice (CSL) model[9]. A large body of works proved that the grain boundary character distribution (GBCD) can be altered via proper thermomechanical treatments. Some face-centered cubic (fcc) metal materials with low to medium stacking fault energy,

---

* In Collaboration with Xia Shuang, Chen Wenjue. Reprinted from Journal of Materials Science, 2008, 43: 2990 − 3000.

such as Pb-base alloy[6, 10], Ni-base alloy[11], OFE copper[12], and austenitic stainless steel[13, 14], were successfully applied this concept to promote $\Sigma 3^n$ ($n = 1$, 2, 3 ...) boundaries formation, and their properties were greatly enhanced. This may be an approach to further improve the resistance to intergranular corrosion of Alloy 690, which is also an fcc material with low stacking fault energy.

Thermomechanical treatments that can increase the proportion of low ΣCSL grain boundaries in this class of materials can be categorized as iterative and single-step thermomechanical treatments[15, 16]. The iterative thermomechanical treatments involve several cycles of strain (mostly less than 30%)/high temperature short anneals[11, 17, 18]. The complexity of this kind of processing makes it difficult to fully understand the mechanism of the GBCD manipulation, and unfavorable for industrial applications. Whether each processing cycle has the same mechanism or different mechanisms in contributing to the GBCD is unclear. And how many cycles of the processing are most favorable is also still not agreed upon by different investigations. So it is very difficult to control this kind of processing. The much simpler single-step processing, which involves relatively small strain, less than 10%, followed by low temperature and long time annealing[13, 14], or high temperature and short time annealing[19-21], is also effective for increasing the proportion of low ΣCSL grain boundaries and enhancing the properties. The processing of single-step strain and high temperature short time annealing will be preferable for industrial applications, since it is much simpler than the iterative processing and does not need long time annealing. However, the mechanism of GBCD alteration by single-step processing is unclear, and how the resulting microstructures affect the material property is also ambiguous.

In this article, the effect of single-step strain followed by high temperature short time annealing on the GBCD and the coexistent microstructure of Alloy 690 were reported. The partial annealed state was observed to interpret the relationship between thermomechanical treatments and resulting GBCD. Finally, the intergranular corrosion testing was carried out on the specimens with different GBCD to reveal how this kind of single-step thermomechanical treatment influences material property.

2 Experimental procedures

The composition of Alloy 690 used in this study is listed in Table 1. Strip specimens with 10 mm in width, cut from 2 mm thickness plate were vacuum sealed under 5×10^{-3} Pa in quartz capsules and annealed at 1, 100 ℃ for 15 min, then quenched into water and broken quartz capsules (WQ) simultaneously, for the starting material (solution annealed). The GBCD of the starting material is given in Table 2. There are no textures in the starting material, as shown in Fig. 1. Thermomechanical treatments were performed by cold-rolling with the reduction ratio in thickness varying from 5% to 50%, and subsequently annealed at 1, 100 ℃ for 5 min in vacuum sealed quartz capsules, then quenched into water as described above.

Table 1 Composition of the investigated material (mass %)

Ni	Cr	Fe	C	N	Ti	Al	Si
60.52	28.91	9.45	0.025	0.008	0.4	0.34	0.14

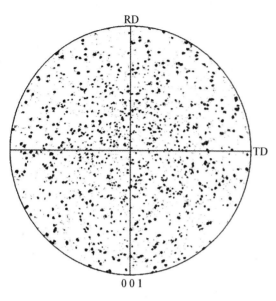

Fig. 1 (001) Pole figure of the starting material (Solution annealed)

Prior to the examination of these specimens, electropolishing was carried out in a solution of 20% $HClO_4$ + 80% CH_3COOH with 30 V direct current. Electron backscatter diffraction (EBSD) was employed for the determination of grain boundary misorientations using the TSL 4.6 laboratories orientation imaging microscopy (OIM) system attached to a Hitachi – S570 scanning electron microscope (SEM) with a tungsten cathode. Operating conditions were: accelerating voltage — 20 kV, working distance — 25 mm, beam incidence angle — 70°. The scans were carried out on a square or rectangular area; with each orientation point being represented as a hexagonal cell, using step sizes about 5 – 7 μm. According to the differences of grain size in each specimen, the areas analyzed by EBSD had dimensions of approximately 1,000×1,000 – 1,500×1,500 μm². Values of the proportion of grain boundaries defined by the CSL model were expressed as a length fraction by dividing the number of hexagonal cells of a particular boundary type with that of the entire grain boundaries. Any adjacent point pair with misorientation exceeding 2° is considered to be a boundary. When analyzing misorientation between crystallites, data points in the middle of grains were used for measurement.

After sensitization treatment at 715 ℃ for 2 h, the specimens were electropolished to get clear finish surfaces for intergranular corrosion testing. The surface areas of specimens were measured carefully. The specimens were weighed to the nearest 0.01 mg, and then immersed in the solution of 65% HNO_3 + 0.4% HF at room temperature for intergranular corrosion testing. Every 24 h, the specimens were taken out and cleaned by water and ethanol, then dried with warm air. Subsequently the specimens were re-weighed to establish mass loss. The surface morphology of the specimens that undergone intergranular corrosion tests was observed by the same SEM used for EBSD experiment.

3 Results and discussion

3.1 Effect of strain and annealing on GBCD

Table 2 shows the GBCD determined according to Palumbo – Aust criterion[22] of the specimens, which were cold rolled with the reduction of 5% - 50% and subsequently annealed at 1,100 ℃ for 5 min and WQ. Alloy 690 is highly propone to twin (Σ3) formation during annealing after strain due to its low stacking fault energy. So Σ3 boundaries made the dominant contribution to the GBCD. When the specimen was annealed at 1,100 ℃ for 5 min after cold rolled 5%, the proportion of low ΣCSL grain boundaries in Alloy 690 was enhanced to the value in excess of 70%, which comprised about 60% Σ3 boundaries and 9% the summation of Σ9 and Σ27 boundaries. This indicated that small strain and subsequent high temperature annealing will markedly promote annealing twinning and multiple twinning in Alloy 690. When the cold rolling reduction increased from 35 to 50%, the low ΣCSL grain boundary has the lowest occurrence among the investigated specimens. In this case the proportions of Σ3 boundaries are about 40%, and the summation of Σ9 and Σ27 boundaries has very low occurrence, only about one seventh of that of the specimen with the maximum proportion of low ΣCSL grain boundaries. This indicated that multiple twinning events had low probability in the case of large strain. The proportion of $\Sigma 3^n$ boundaries decreased with the increase of cold rolling reduction, which resulted in that the proportion of low ΣCSL boundaries also decrease with the increase of strain. Hence the proportion of low ΣCSL boundaries has a very strong dependence on strain.

Table 2 Treatment processes and grain boundary character distribution of specimens (Palumbo - Aust criterion)

Specimen number	Cold rolling (%)	Annealing time (min) at 1,100 ℃	Σ1(%)	Σ3(%)	Σ9+ Σ27(%)	Other low ΣCSL(%)	Total low ΣCSL (Σ≤29) (%)
Starting material	—	15	3	48.5	2.5	0.5	54.9
A	5	5	1.4	60.1	9.3	0.5	71.3
B	10		2.3	54.2	7.2	0.8	64.5
C	20		2.5	48.8	3.7	0.5	55.5
D	35		4.2	40.3	1.2	1.2	46.9
E	50		4	39.8	1.5	1.4	46.7

3.2 The microstructure of the specimens with different GBCD

The OIM map of specimen A, which has the maximum proportion of low ΣCSL grain boundaries is shown in Fig. 2a, and that of specimen C, which has relatively low proportion of low ΣCSL grain boundaries is presented in Fig. 2b. These two OIM maps are given according to Palumbo - Aust criterion. The grains-cluster microstructure is easily identified in Fig. 2a, b. The grains-cluster has the following characteristics: all the boundaries within the cluster can be described by $\Sigma 3^n$ misorientations, whereas the outer boundaries are often crystallographically random with the adjacent grains[23].

For illustrating this kind of microstructure, the misorientations of any two grains within each cluster in Fig. 2a are analyzed. Most of the grain pairs are non-adjacent, and the misorientations of those randomly selected grain pairs are listed in Table 3. It is obvious that the grains within the same cluster have $\Sigma 3^n$ misorientations even when they

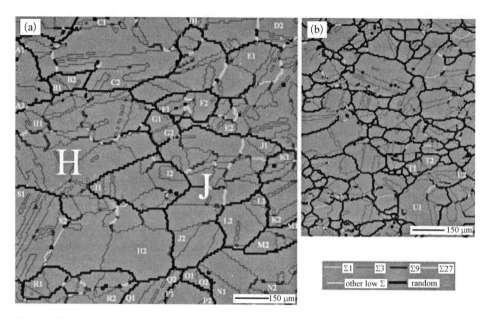

Fig. 2 OIM Maps for specimen A (a) and C (b). Same letter with different number are in the same grains-cluster

Table 3 Misorientations between grains pairs in same grains-cluster
of Fig. 2 and in same twin chain of Fig. 4

Grains pairs	Misorientation	Closest CSL	Deviation ($\Delta\theta$)	$\Delta\theta/\Delta\theta_{\max (B)}$	$\Delta\theta/\Delta\theta_{\max (P-A)}$
A1/A2	59.7°[1 1 1]	Σ3	0.4°	0.05	0.07
B1/B2	59.1°[1 1 −1]	Σ3	0.9°	0.10	0.15
C1/C2	59.4°[1 −1 1]	Σ3	0.9°	0.10	0.15
D1/D2	31.0°[0 1 −1]	Σ27a	0.6°	0.21	0.62
E1/E2	31.9°[0 1 1]	Σ27a	0.5°	0.17	0.52
F1/F2	38.1°[1 1 0]	Σ9	1.1°	0.22	0.45
G1/G2	0.5°[−8 7 12]	Same orientation			
H1/H2	35.0°[−1 2 0]	Σ27b	1.0°	0.35	1.03
I1/I2	35.1°[0 1 2]	Σ27b	0.7°	0.24	0.72
J1/J2	38.7°[−1 0 −1]	Σ9	0.4°	0.08	0.17
K1/K2	35.4°[−1 2 0]	Σ27b	0.1°	0.03	0.10
L1/L2	39.6°[0 1 −1]	Σ9	0.8°	0.16	0.33
M1/M2	0.3°[9 24 −10]	Same orientation			
N1/N2	59.6°[1 1 −1]	Σ3	0.6°	0.07	0.1
O1/O2	59.8°[1 −1 1]	Σ3	0.5°	0.06	0.08
P1/P2	39.8°[0 1 1]	Σ9	1.1°	0.22	0.46
Q1/Q2	59.8°[−1 1 1]	Σ3	0.2°	0.02	0.03
R1/R2	35.8°[0 2 −1]	Σ27b	0.7°	0.24	0.72
S1/S2	59.9°[−1 −1 −1]	Σ3	0.7°	0.08	0.12
T1/T2	59.3°[1 1 1]	Σ3	0.7°	0.08	0.12
U1/U2	39.4°[−1 0 1]	Σ9	0.5°	0.1	0.21
V1/V2	35.4°[0 −2 1]	Σ27b	0.2°	0.07	0.21
W1/W2	59.8°[−1 1 1]	Σ3	0.8°	0.09	0.13
X1/X2	59.5°[1 −1 1]	Σ3	0.5°	0.06	0.08
Average value				0.13	0.30

are at a long distance apart from each other. For example, in Fig. 2a, grains H1 and H2 are in the same cluster H with a size at least of 600 μm, but they also maintain $\Sigma 27b$ misorientation. From grain H1 to grain H2, a line can be drawn only across twin boundaries, thus a long twin chain[24] forms.

The orientation distribution of the grains in the cluster J in Fig. 2a is analyzed in detail, as given in Fig. 3 and Table 4. This cluster contains 29 grains, which are labeled by numbers in Fig. 3a, and their orientations identified on the examination section are shown by color in Fig. 3b. Grains with the same orientation are presented by the same color, as indicated in Fig. 3b, c. There are only 10 different orientations for those 29 grains, i. e. , grains 1, 3, 5, 6, 8, 12, and 16 have the same orientation, which is named orientation α; grain 2 has the orientation β; grains 4 and 14 have the same orientation γ; grains 7, 9, 11, 15, 19, and 20 have the same orientation Φ; grains 10 and 21 have the same orientation θ; grains 13, 17, 24, and 26 have the same orientation ϵ; grains 22, 27, and 29 have the same orientation Ψ; grain 23 has the orientation Ω; grains 18 and 25 have the same orientation δ; grain 28 has the orientation η. The mutual misorientations between these 10 different orientations are shown in Table 4. It is clear that all these orientations have the $\Sigma 3^n$ relationships with each other.

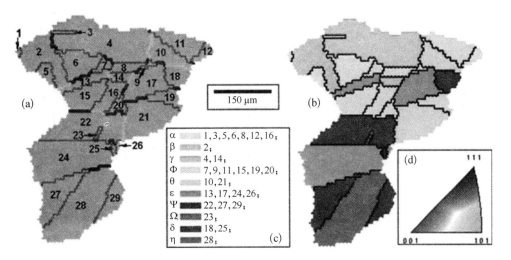

Fig. 3 Orientation distribution of the grains in the cluster J in Fig. 2a. (a) Grains labeled by number (Boundaries colors are as in Fig. 2), (b) Orientation distribution of the grains, (c) Color code for the same orientation grains, (d) Inverse pole figure for color code

This grains-cluster can be divided into two sub-clusters, which contain grains 1 – 21 and grains 22 – 29 respectively. The two sub-clusters are separated by the high order $\Sigma 3^n$ boundaries, i. e. , the $\Sigma 9$ boundaries between grain 15 and grain 22, and between grain 20 and grain 22; the $\Sigma 27$ boundaries between grain 16 and grain 22, and between grain 21 and grain 22. Among the grains in each sub-cluster, a line can be drawn only cross twin boundaries from any grain to any other grain. Thus the twin-chain can link any two grains in each sub-cluster. But between the grains belonging to different sub-cluster, the twin

Table 4 Mutual misorientations between 10 different orientations in the grains-cluster of Fig. 3

	α	β	γ	Φ	θ	ε	Ψ	Ω	δ
β	59.5°[1 −1 −1] /Σ3/0.8°								
γ	59.6°[11 −1] /Σ3/0.6°	38.7°[1 0 1] /Σ9/0.3°							
Φ	59.7°[1 −11] /Σ3/0.7°	38.5°[0 −1 −1] /Σ9/0.6°	39.4°[0 1 1] /Σ9/0.5°						
θ	38.6°[0 1 1] /Σ9/0.5°	35.6°[2 1 0] /Σ27b/0.5°	31.1°[0 −1 −1] /Σ27a/0.6°	59.9°[1 1 1] /Σ3/0.4°					
ε	38.4°[0 −11] /Σ9/0.9°	35.9°[−2 0 1] /Σ27b/0.5°	34.9°[2 −1 0] /Σ27b/0.6°	59.8°[−1 1 1] /Σ3/0.5°	38.6°[1 0 1] /Σ9/0.5°				
Ψ	35.8°[0 −2 1] /Σ27b/0.5°	54.8°[2 3 2] /Σ81b	38.6°[−11 −4] /Σ81a	38.8°[−1 0 1] /Σ9/0.7°	35.1°[0 2 −1] /Σ27b/0.4°	59.6°[−11 −1] /Σ3/0.9°			
Ω	39.0°[−4 1 −1] /Σ81a	49.8°[−3 17 −20] /Σ243b	43.0°[−11 20 −11] /Σ243a	36.1°[1 −2 0] /Σ27b/0.7°	55.1°[−3 −2 −2] /Σ81b	38.7°[0 1 1] /Σ9/0.7°	59.8°[1 −11] /Σ3/0.4°		
δ	36.2°[−1 2 0] /Σ27b/0.9°	39.4°[6 25 −6] /Σ81a	54.6°[−2 −2 −3] /Σ81b	38.7°[0 −11] /Σ9/0.4°	32.6°[−1 0 −1] /Σ27a/1.0°	59.7°[−1 −11] /Σ3/0.5°	39.1°[1 1 0] /Σ9/0.6°	31.6°[21 0 22] /Σ27a/0.6°	
η	39.1°[3 1 5] /Σ81c	34.9°[5 −4 −2] /Σ243d	42.8°[−1 14 −4] /Σ243a	31.7°[1 0 1] /Σ27a/0.3°	38.0°[−1 3 −5] /Σ81c	38.3°[−1 −11 −10] /Σ9/0.7°	59.9°[−1 11] /Σ3/0.7°	39.7°[1 1 0] /Σ9/0.9°	35.0°[0 −1 −2] /Σ27b/0.7°

Note: Orientation relationships are given in the form: Misorientation/Closest CSL/Deviation; the deviations of Σ>27 cannot be accurately measured by the equipment and not given here

chain linking between them is interrupted by one high order $\Sigma 3^n$ boundaries. Nevertheless, some grains located in different sub-clusters have the same orientation. For example, grains 13 and 17 belong to one sub-cluster, and grains 24 and 26 belong to the other one, they also have the same orientation. The OIM map analyzed here represents a twodimensional section of a three dimensional microstructure, it is highly probable that the twin chain in fact connected below or above the examined surface. Since the higher order $\Sigma 3^n$ boundaries are produced by impingement of twins[24].

In Fig. 2b, which is the OIM map of specimen C, the grains-cluster feature of microstructure is also easily identified. For example, the misorientations of grain pairs T1 and T2, and U1 and U2 were also of $\Sigma 3^n$ type, as listed in Table 3. Comparing Fig. 2a and b, it is easy to see that the mean size of grains-cluster in Fig. 2a is much larger than that of Fig. 2b, although the size of the clusters in Fig. 2a is quite inhomogeneous. To study the dependence of the size of grains-cluster on the treatment process, the mean size of clusters in specimens of Table 2 was evaluated by mean distance between random boundaries intersected by line scan of 3,000 μm for 3 times in different areas by EBSD. The result shown in Fig. 4 is that the mean sizes of the grains-clusters decrease with the increase of the cold rolling reduction. Correspondingly, large grains-cluster microstructure has a high proportion of $\Sigma 3^n$ boundaries, while small grains-cluster microstructure has relatively low proportion of $\Sigma 3^n$ boundaries, as shown in Fig. 4 and Table 2.

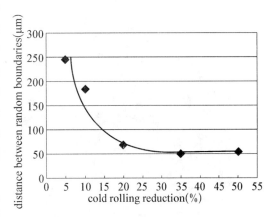

Fig. 4 The relationship between mean dimensions of grains-cluster and the cold rolling reduction

3.3 The microstructure in the partially recrystallized state

All the investigated specimens annealed at 1,100 ℃ for 5 min were fully recrystallized. It is conceivable that the microstructure evolution during annealing after strain should be responsible for the variation of the resulting microstructures and GBCD. So the microstructures before full recrystallization are of great interest to be investigated. Specimens annealed at 1,100 ℃/3 min (specimen A*) and 1 min (specimen C*) for 5% and 20% cold rolled respectively, were analyzed by EBSD and the point-to-point correlations (misorientations) were obtained. The average misorientation of individual grains were automatically obtained and shown by color in Fig. 5. From the color code, it can be known that the blue area denote smaller average misorientation in grains, which indicates the recrystallized area, and the unrecrystallized area, which has larger average misorientation in grains, is presented by green or even warmer hue. The recrystallized and non-recrystallized regions were separated by random boundaries. There are a lot of $\Sigma 1$

boundaries (misorientation across them in the range of $2° - 15°$) and $\Sigma 3$ boundaries in unrecrystallized areas, which are predestined to be wiped off by the growth of recrystallization grains. Therefore, it is meaningless to give statistic of the GBCD for the specimens in the partially recrystallized state.

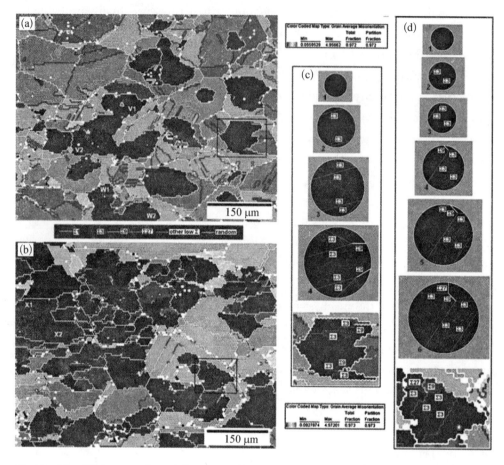

Fig. 5 The OIM maps of partially recrystallized specimens A* (5% cold rolled + 1, 100 ℃/ 3 min) and C* (20% cold rolled + 1, 100 ℃/1 min), (a) and (b) are the grain average misorientation maps of specimens A* and C*, (c) and (d) are the schematic illustration of the possible multiple twinning process of the two recrystallized regions outlined by black framework in (a) and (b) respectively

Most of the recrystallized grains are enwrapped by the non-recrystallized regions in specimen A*, as shown in Fig. 5a. It can be seen that some recrystallized areas contain straight-sided $\Sigma 3$ boundaries, which are annealing twins, while the smaller ones do not contain, and the larger ones contain more. This implies that during the growth of the recrystallized grains by consuming the non-recrystallized regions, continuous chain of twin boundaries can be left behind in the wake of the migrating random boundary, which separated the recrystallized and non-recrystallized regions. We do not touch here, the questions of the mechanisms of twin formation[25-28] and why does the twins continuously form to produce twin chain[24, 29]. We just focus on the question: if twinning takes place continuously to form twin chain, how can it explain the relationship between the thermomechanical treatments and the obtained microstructures.

The twin chains can be easily identified in the partially recrystallized microstructure. For example, in Fig. 5a, the twin chain from grain V1 to V2 involves four twin boundaries; similarly, the twin chain from W1 to W2 involves five twin boundaries. The misorientations of those two grain pairs are listed in Table 3. Most of the areas in specimen C^*, as shown in Fig. 5b are blue, which indicated recrystallized regions. Most of grains-clusters in these regions have already impinged on each other. The twin chains can also be easily identified in the recrystallized regions. For example, the twin chain from grain X1 to X2 involves three twin boundaries, and the misorientations of these two grains are also listed in Table 3. The misorientations of those grain pairs (V1/V2, W1/W2, and X1/X2) are of $\Sigma 3^n$ type, as shown in Table 3. The twin chains in other recrystallized areas of Fig. 5a and b have the similar situation.

Twin chains formed during recrystallization had been studied a lot, and the results were summarized by Berger[29]. Under the light of this knowledge, it is easy to know that each twin chain started from each single orientation, i. e. , single recrystallization nucleus. Concurrently, with the formation the twin chain, $\Sigma 3^n$ boundaries were produced by multiple twinning. To clarify the multiple twinning processes, possible formation processes of two recrystallized areas outlined by black framework in Fig. 5a and b are schematically illustrated in Fig. 5c and d respectively. It is only shown that multiple twinning occurs in two dimensions, and it is conceivable that multiple twinning in three dimensions underwent similar process.

The fully recrystallized microstructures of specimen A and C, as shown in Fig. 2a and b, are developed from the partially recrystallized ones, similar to those of specimen A^* and C^* in Fig 5a and b. Since the starting material is without texture, as shown in Fig. 1, the grains-cluster is extremely unlikely formed by joining two or more twin chains during recrystallization. Rather, the grains-cluster should be developed from a single twin-chain of recrystallization. As analyzed above, the twin-chain is developed from a single recrystallization nucleus. Consequently, the grains-cluster is formed by continuous twinning events starting from a single nucleus of recrystallization, with the associated presence of multiple $\Sigma 3^n$ boundaries. This can explain why the grains in same cluster maintain very close $\Sigma 3^n$ misorentation even at a long distance in Fig. 2.

It is hard to say which grain is the starting nucleus in each twin chains in Fig. 5a and b, but it can be deduced that nucleation density (i. e. the number of nuclei in unit area) of the specimen with 5% strain is much lower than that with 20% strain during annealing. This is according to the common knowledge that smaller strain will result in lower nucleation density during recrystallization[30]. As a consequence, the recrystallization nuclei in the specimen with small strain have much larger area available for growing than those in the specimen with large strain. During growth in larger area, twinning and multiple twinning events have more opportunity to occur, so the size of the grains-cluster and the proportion of $\Sigma 3^n$ boundaries in Fig. 2a are much larger than those in Fig. 2b. Therefore, the proportion of low ΣCSL grain boundaries is high for the specimen with low strain, as shown in Table 2.

In Fig. 2a the size of grains-cluster is quite inhomogeneous. This is caused by that some of these recrystallization nuclei extensively grow to form very large clusters, and others grow to relatively small sizes. In the specimen with small strain, the local orientation will be different with respect to the deformation direction, so stored energy will be inhomogeneous from area to area. As a consequence, nucleation density is inhomogeneous in different areas. Therefore, in Fig. 2a, the size of grains-clusters which developed from each single nucleus is quite inhomogeneous. However, the average nucleation density increases with the increase of strain. So, the mean size of the grains-cluster in the final microstructure and corresponding proportion of low ΣCSL grain boundaries decrease with the increase of strain, as shown in Fig. 4 and Table 2.

3.4 The connectivity of random grain boundaries in the microstructure with high proportion of low ΣCSL grain boundaries

From above analysis, it can be known that the proportion of low ΣCSL grain boundaries is enhanced by the formation of large size grains-clusters, which are the result of low nucleation density and multiple twinning. The grains-cluster is formed by twinning or multiple twinning starting from a single recrystallization nucleus. So all the twins and its variant $\Sigma 3^n$ boundaries intrinsically locate inside the grains-cluster, rather than appear in the network of random boundaries as segments to break up its connectivity.

Some articles reported that when the proportion of low ΣCSL grain boundaries was enhanced, the random boundaries network would be substantially fragmented[20, 31-33], which seemed to be different with the current result. However one point should be noted that the GBCD and OIM maps given above are obtained by applying the Palumbo – Aust criterion. There are two popular criterions in defining the CSL boundaries:

$$\text{Palumbo – Aust criterion}^{[22]}: \Delta\theta_{\max(\text{P-A})} = 15° \Sigma^{-5/6}$$
$$\text{Brandon criterion}^{[34]}: \Delta\theta_{\max(\text{B})} = 15° \Sigma^{-1/2}$$

The deviation of experimentally measured misorientation from exact CSL misorientation is denoted by $\Delta\theta$. The deviation degree from exact CSL can be described by a relative deviation[35] $\Delta\theta/\Delta\theta_{\max}$, where the $\Delta\theta_{\max}$ is the maximum permissible deviation. For the Brandon criterion and Palumbo – Aust criterion, the relative deviations are expressed as: $\Delta\theta/\Delta\theta_{\max(\text{B})}$ and $\Delta\theta/\Delta\theta_{\max(\text{P-A})}$, respectively. Only if the value of these parameters is smaller than 1, the examined misorientation was regarded as CSL in the regime of that criterion.

The OIM map of specimen A, which is given according to the Palumbo – Aust criterion in Fig. 2a, is given again in Fig. 6 according to the Brandon criterion. Comparing Fig. 2a with Fig. 6, it can be seen that the connectivity of the random boundaries, which lie between the grains-clusters, is obviously broken up by segments of low ΣCSL grain boundaries after applying the Brandon criterion. For example, the random boundaries around the two grains-clusters I and J in Fig. 6, are markedly broken up by some low ΣCSL grain boundaries numbered 1 – 10. These 10 boundaries are analyzed in Table 5. For comparing, three boundaries numbered 11 – 13 inside the cluster I are also analyzed in Table 5. The average values $\Delta\theta/\Delta\theta_{max(B)}$ and $\Delta\theta/\Delta\theta_{max(P-A)}$ of boundaries 1 – 10 are 0.86 and 1.97 respectively, while these two values of boundaries 11 – 13 are 0.07 and 0.11, as shown in Table 5. This indicates that the low ΣCSL boundaries that break up the connectivity of the random boundaries have much larger deviation than those located inside the grains-cluster.

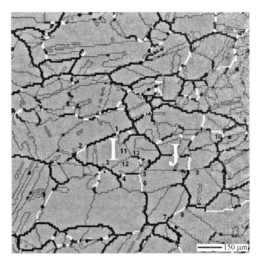

Fig. 6 OIM map for specimen A (Brandon criterion). The connectivity of random boundaries network is substantially interrupted. The interrupter of the random boundaries around two grains-clusters I and J are labeled from 1 to 10. Some boundaries inside cluster I are also labeled (11, 12, 13). (Boundaries colors are as in Fig. 2)

Table 5 Misorientations of the boundaries numbered in Fig. 6

Number	Misorientation	Closest CSL	Deviation ($\Delta\theta$)	$\Delta\theta/\Delta\theta_{max(B)}$	$\Delta\theta/\Delta\theta_{max(P-A)}$
1	46.9°[19 9 21]	Σ29b	2.5°	0.90	2.72
2	46.0°[−6 −7 −13]	Σ21b	2.5°	0.76	2.08
3	41.6°[−19 1 23]	Σ9	4.6°	0.92	1.90
4	47.6°[−20 −21 −1]	Σ11	3.6°	0.80	1.76
5	51.6°[−12 −13 12]	Σ3	8.5°	0.98	1.41
6	47.4°[7 −1 −16]	Σ15	3.6°	0.93	2.28
7	34.8°[−17 −2 −18]	Σ9	4.9°	0.98	2.02
8	38.4°[−1 −1 3]	Σ23	2.1°	0.67	1.89
9	55.1°[8 −9 10]	Σ3	7.1°	0.82	1.18
10	43.9°[0 1 19]	Σ29a	2.3°	0.82	2.50
Average value				0.86	1.97
11	59.8°[1 1 1]	Σ3	0.2°	0.02	0.03
12	59.2°[1 1 1]	Σ3	0.8°	0.09	0.13
13	39.5°[0 1 −1]	Σ9	0.4°	0.08	0.17
Average value				0.07	0.11

To ensure this kind of phenomena, another 60 low ΣCSL boundaries (defined by Brandon criterion) which break up the connectivity of the random boundary network in other area of this OIM map and other OIM maps of specimen A, are also investigated. The results together with the limits of both Brandon and Palumbo – Aust criterion are plotted in Fig. 7. It is obviously that the deviations of most low ΣCSL boundaries interrupt the connectivity of the random boundary network are above the Palumbo – Aust criterion limit.

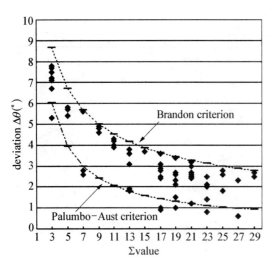

Fig. 7 Deviations of some 60 boundaries (close to low ΣCSL misorientations), which appeared on the random boundaries network in specimen A

The average values of $\Delta\theta/\Delta\theta_{max(B)}$ and $\Delta\theta/\Delta\theta_{max(P-A)}$ for those boundaries are 0. 74 and 1. 75 respectively. This indicates that if a boundary segment that breaks up the connectivity of the random boundary is ranked as a low ΣCSL one according to Brandon criterion, it is highly probable that this boundary will be excluded from the low ΣCSL boundary rank according to Palumbo – Aust criterion. For comparison, the average values of $\Delta\theta/\Delta\theta_{max(B)}$ and $\Delta\theta/\Delta\theta_{max(P-A)}$ for the misorientations between grains within each cluster listed in Table 3 are calculated, 0. 13 and 0. 30 respectively, which are much smaller than those of the boundaries which break up the connectivity of the random boundary network. As analyzed above, the grains-cluster is formed by multiple twinning starting from a single recrystallization nucleus. Since the starting material is without texture, the random boundaries are formed by impingement of unrelated grains-clusters during recrystallization. Therefore, though misorientations of some boundaries between clusters are close to the low ΣCSL (defined by Brandon criterion), they simply happen to be in that situation, and are not geometrically necessary in contrast to the $\Sigma 3^n$ boundaries inside the clusters.

Palumblo, Aust and co-workers[22, 36-38] compared the application of Palumbo – Aust and Brandon criterion in analyzing the experimental results pertaining to interface properties. These results showed that very few boundaries classified as low ΣCSL boundaries according to Palumbo – Aust criterion display attacked, while according to Brandon criterion, more low ΣCSL boundaries display attacked. Therefore, Palumbo – Aust criterion is a better one to define CSL boundaries with special properties. By applying Palumbo – Aust criterion, the connectivity of random boundary network is not substantially broken up in our study as shown in Fig. 2.

The designation of CSL is only based on misorientation. For example Σ3 boundaries are grain boundaries, which have a misorientation within ±8. 6° or ±6° of 60°/⟨111⟩ for the Brandon and Palumbo – Aust criterions respectively. The Σ3 family consists of annealing twins (which are symmetrical tilt boundaries characterized by {111} boundary planes), various tilt and twist boundaries (mostly asymmetric tilts or symmetrical) and grain boundaries which happen to have a Σ3 misorientation but generally have irrational or random boundary planes[39]. According to Randle's study[40], for Σ3 boundaries small relative deviation ($\Delta\theta/\Delta\theta_{max}$), mostly associated with {111} boundary plane, which is of coherent twin boundary, while large relative deviation mostly associated with irrational boundary planes. Grain boundaries which happen to have a Σ3 misorientation by impingement of unrelated grains will generally have irrational or random boundary planes. These kinds of boundaries do not exhibit special properties[40-42]. $\Sigma 3$ boundaries with very large deviations in contrast with those of the twins inside the cluster are also observed in our study, such as boundaries 5 and 9 in Fig. 6. They appear on the network of the random boundaries and break up its connectivity. So this kind of discontinuity of random boundaries is not reliable. Similarly, the deviations of almost all the low ΣCSL boundaries (Brandon criterion) interrupting the random boundary connectivity are comparatively high, as shown in Fig. 7. Hence the marked discontinuity of random boundaries according to Brandon criterion is not reliable.

3.5 Intergranular corrosion phenomena in Alloy 690 with different GBCD

It turns out that the connectivity of random boundary network is very sensitive to the criterion of CSL in the present study. The connectivity of random boundaries network is substantially interrupted when the Brandon criterion is applied, whereas not fragmented obviously when the Palumbo – Aust criterion is used. In this instance, whether material property will be enhanced for the specimen with high proportion of low ΣCSL grain boundaries is valuable to be investigated. Specimen A and C were subjected to anneal at 715 ℃ for 2 h to induce sensitization, which increased the intergranular corrosion susceptibility for intergranular corrosion testing. After sensitization treatment, specimens A and C were applied in intergranular corrosion testing.

Weight losses per area of each specimen as a function of corrosion time are shown in Fig. 8. It is obvious that the weight loss during intergranular corrosion of specimen A is much less than that of specimen C. Surface morphologies of the sensitized specimens A and C undergone 144 h corrosion are shown in Fig. 9a, b and c, d, respectively. In Fig. 9a, which is the surface morphology of specimen A, it is clear that there is a grains-cluster on the point of dropping, and its magnified image is shown in Fig. 9b. In Fig. 9b, the boundaries around the dropping grains-cluster display heavily attacked phenomenon than that of the boundaries inside the cluster. This indicates that the weight loss occurs when individual grains-cluster is removed from the specimen surface, as their surrounding random boundaries are breached by intergranular corrosion. After sensitization treatment, carbide precipitation and associated depletion of Cr are more pronounced in the vicinity of random boundaries than the $\Sigma 3^n$ boundaries inside the cluster. As a consequence, the random boundaries around the grains-cluster are corroded much heavily than the $\Sigma 3^n$ boundaries inside the cluster, as indicated in Fig. 9b.

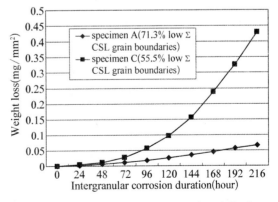

Fig. 8 Weight losses of specimens A and C after sensitization as a function of intergranluar corrosion time

After 144 h corrosion, most surface area of specimen A is integrated, as shown in Fig. 9a, b, while the surface of specimen C has rather bad integrity, where the dropping phenomena are very serious, as shown in Fig. 9c, d. The differences between specimen A and C are the GBCD and the concurrent size of the grains-cluster, as shown in Table 2, Fig. 2, and Fig. 4. These indicate that large grains-cluster is harder to drop from the specimen surface. Based on these observations, the authors attribute the improvement of resistance to mass loss during intergranular corrosion to the large size of the grains-cluster rather than the unreliable discontinuity of the random boundary network as analyzed in section "The connectivity of random ... boundaries".

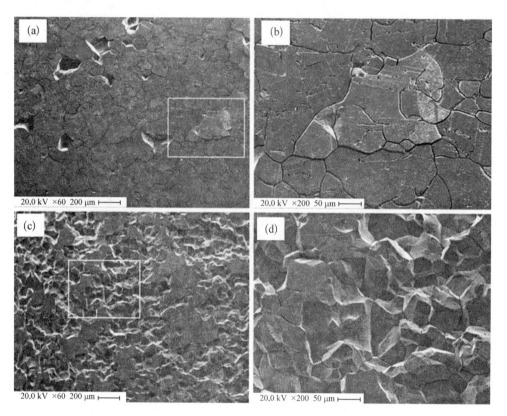

Fig. 9 SEM images of the surfaces of sensitized specimen A (a), C (c), undergone 144 h intergranular corrosion. (b) and (d) are the magnified images of the area outlined by white frameworks in (a) and (c) respectively

Firstly, in the large grains-cluster microstructure with high proportion of low ΣCSL grain boundaries, there are less random boundaries available for being heavily attacked in the specimen surface. Secondly, the intergranular corrosion will go a much longer path and take more time to make the large grains-cluster drop than a smaller one. Thirdly, the large grains-clusters protect the under layer microstructure much better than the smaller ones. Only after the most external clusters drop, the under layer microstructure will suffer as aggressive solution as the external ones do. The small clusters are quicker to drop, and then undersurface clusters are quicker to suffer the corrosive environment. So the weight loss for specimen with small grains-clusters is much severe than that with large ones, as being presented in Fig. 8.

4　Summary and conclusions

(1) In Alloy 690, small strain (5%) and subsequent high temperature (1,100 ℃) annealing for short time (5 min) can produce high proportion of low ΣCSL grain boundaries (more than 70%), which mainly are $\Sigma 3^n$. In this case, the large size twin-induced grains-cluster constitute the microstructure.

(2) Based on the observation of partial and full recrystallization state, it is clear that the nucleation density and multiple twinning are the key factors affecting the GBCD. The grains-cluster is formed by continuous chain of twinning events starting from a single nucleus of recrystallization, and with the associated presence of multiple $\Sigma 3^n$ boundaries. As a consequence, all the grains inside the same cluster have close $\Sigma 3^n$ misorientations even at a long distance. The size of the grains-clusters and the proportion of $\Sigma 3^n$ boundaries decreases with the increase of cold rolling reduction. This is the result of that nucleation density of recrystallization increases with the increase of strain.

(3) In the case of high proportion of low ΣCSL boundaries, connectivity of random boundaries network is very sensitive to the applied criterion of CSL. The connectivity of random boundary network is not fragmented obviously when the Palumbo – Aust criterion is used, whereas substantially interrupted when the Brandon criterion is applied. Almost all those low ΣCSL boundaries (Brandon criterion) appear on the network of the random boundaries have very large deviations. Hence, the high resistance to mass loss during intergranular corrosion testing is attributed to the large size of the grains-cluster accompanied with high proportion of low ΣCSL boundaries, rather than the unreliable discontinuity of the random boundary network.

Acknowledgements

This work was supported by Major State Basic Research Development Program of China (2006CB605001) and Shanghai leading academic discipline project (T0101).

References

[1] Thuvander M, Stiller K (2000) Mat Sci Eng A 281: 96 – 103.

[2] Qiu S, Su X, Wen Y (1995) Nuclear Power Engineering 16: 336 – 340 .

[3] Kurban M, Erb U, Aust KT (2006) Scripta Mater 54: 1053 – 1058.

[4] Bi HY, Kokawa H, Wang ZJ (2003) Scripta Mater 49: 219 – 223.

[5] Bennett BW, Pickering HW (1987) Metall Trans A 18A: 1117 – 1124.

[6] Palumbo G, Erb U (1999) MRS Bull 24: 27 – 32.

[7] Crawford DC, Was GS (1992) Metall Trans A 23A: 1195 – 1206.

[8] Watanabe T (1984) Res Mech 11: 47 – 82.

[9] Kronberg ML, Wilson FH (1949) Trans Am Inst Min Engrs 185: 501.

[10] Lehockey EM, Limoges D, Palumbo G, Sklarchuk J, Tomantschger K, Vincze A (1999) J Power Sources 78: 79 - 83.

[11] Lin P, Palumbo G, Erb U (1995) Scripta Metall Mater 33: 1387 - 1392.

[12] King WE, Schwartz AJ (1998) Scripta Mater 38: 449 - 455.

[13] Shimada M, Kokawa H, Wang ZJ (2002) Acta Mater 50: 2331 - 2341.

[14] Michiuchi M, Kokawa H, Wang ZJ, Sato YS, Sakai K (2006)Acta Mater 554: 5179 - 5184.

[15] Randle V (2004) Acta Mater 52: 4067 - 4081.

[16] Randle V (1999) Acta Mater 47: 4187 - 4196.

[17] Kumar M, Schwartz AJ, King WE (2002) Acta Mater 50: 2599 - 2612.

[18] Alexandreanu B, Capell B, Was G (2001) Mat Sci Eng A 300: 94 - 104.

[19] Lee S- L, Richards NL (2005) Mat Sci Eng A 390: 81 - 87.

[20] Guyot BM, Richards NL (2005) Mat Sci Eng A 395: 87 - 97.

[21] Tan L, Sridharan K, Allen TR (2006) J Nucl Mater 348: 263 - 271.

[22] Palumbo G, Aust KT (1990) Acta Metall Mater 38: 2343 - 2352.

[23] Gertsman VY, Henager CH (2003) Interface Sci 11: 403 - 415.

[24] Gottstein G (1984) Acta Metall 32: 1117 - 1138.

[25] Fullman RL, Fisher JC (1951) J Appl Phys 22: 1350 - 1355.

[26] Gleiter H (1969) Acta Metall 17: 1421 - 1428.

[27] Meyers MA, Murr LE (1978) Acta Metall 26: 951 - 962.

[28] Mahajan S, Pande CS, Imam MA, Rath BB (1997) Acta Mater 45: 2633 - 2638.

[29] Berger A, Wilbrandt P- J, Ernst F (1988) Prog Mater Sci 32: 1 - 95.

[30] Humphreys FJ, Hatherly M (2004) Recrystallization and related annealing phenomena, 2nd ed. Elsevier, Oxford.

[31] Kumar M, King WE, Schwartz AJ (2000) Acta Mater 48: 2081 - 2091.

[32] Schuh CA, Kumar M, King WE (2003) Acta Mater 51: 678 - 700.

[33] Schuh CA, Kumar M, King WE (2003) Z Metallkd 94: 323 - 328.

[34] Brandon DG (1966) Acta Metall 14: 1479 - 1484.

[35] Randle V (1996) The role of the coincidence site lattice in grain boundary engineering. Cambridge University, Cambridge, UK.

[36] Palumbo G, Aust KT, Lehockey EM (1998) Script Mater 38: 1685 - 1690.

[37] Kim SH, Erb U, Aust KT, Palumbo G (2001) Script Mater 44: 835 - 839.

[38] Zhou Y, Aust KT, Erb U, Palumbo G (2001) Script Mater 45: 49 - 54.

[39] Randle V, Davies H (2002) Ultramicroscopy 90: 153 - 162.

[40] Randle V, Davies P (1999) Interface Sci 7: 5 - 13.

[41] Gertsman VY, Bruemmer SM (2001) Acta Mater 49: 1589 - 1598.

[42] Lin H, Pope DP (1993) Acta Metall Mater 41: 553 - 562.

Effect of Water Chemistry and Composition on Microstructural Evolution of Oxide on Zr Alloys*

Abstract: The microstructure of oxide films formed on Zircaloy – 4 and Alloy No. 3, which has a composition similar to ZIRLO™, was investigated by high resolution transmission and scanning electron microscopy, and by scanning probe microscopy after corrosion tests performed at 360 ℃/ 18. 6 MPa in deionized water or lithiated water with 0. 01 M LiOH. The microstructural evolution of the oxide films was analyzed by comparing the microstructure at different depths in the oxide layer. The defects, consisting of vacancies and interstitials, such as points, lines, planes, and volumes, were produced during the oxide growth. Monoclinic, tetragonal, cubic, and amorphous phases were detected and their coherent relationships were identified. The characteristic of oxide with such microstructure had an internal cause, and the temperature and time were the external causes that induced the microstructural evolution during the corrosion process. The diffusion, annihilation, and condensation of vacancies and interstitials under the action of stress, temperature, and time caused stress relaxation and phase transformation. It was observed, in the middle of the oxide layer, that the vacancies absorbed by grain boundaries formed pores to weaken the bonding strength between grains. Pores formed under compressive stress lined up along the direction parallel to the compressive stress. Thus, cracks developed from the pores were parallel to the oxide/metal interface. Li^+ and OH^- incorporated in oxide films were adsorbed on the wall of pores or entered into vacancies to reduce the surface free energy of the zirconium oxide during exposure in lithiated water. As a result, the diffusion of vacancies and the formation of pores were enhanced, inducing the degradation of the corrosion resistance. The relationship between the corrosion resistance of zirconium alloys and the microstructural evolution of oxide films affected by water chemistry and composition is also discussed.

1 Introduction

Zirconium alloys are used as nuclear fuel cladding in water reactors. The reaction between zirconium and water at high temperature forms oxide films on the surfaces. Hydrogen produced during the reaction is partially picked up by the fuel cladding. Studies focused on improving the corrosion resistance of zirconium alloys are important to enable the industry to meet the demand for advanced fuel assemblies that enable extended in-reactor exposure in aggressive environments. The detailed mechanism of the corrosion

* In Collaboration with Li Q, Yao MY, Liu WQ, Chu YL. Reprinted from Journal of ASTM International, 2008, 5(2): 1 – 21. ASTM Symposium on Zirconium in the Nuclear Industry: 15th International Symposium on 24 – 28 June 2007 in Sunriver, OR; M. Limback and B. Kammenzind, Guest Editors.

process is still to be clarified and the development of advanced zirconium alloys continues to rely on the trial and error method.

A new oxide layer is always formed at the oxide/metal interface. The microstructural evolution of the oxide certainly affects the diffusion of oxygen ions and anion vacancies and, consequently, changes the conditions for the continued oxide growth. The volume of the oxide, which is produced by the reaction of Zr and oxygen, to the consumed metal volume, i. e. , the Pilling – Bedworth ratio, is 1. 56 for Zr. A compressive stress parallel to the metal/oxide interface is, consequently, produced in the oxide film due to the restriction of the metal matrix. When the oxide films grow as a new layer of oxide is formed at the oxide/metal interface, the compressive stress in the old layer relaxes. A stress gradient will, consequently, be established through the oxide film. The microstructure of oxide films formed under such conditions is very complicated. It will undergo an evolution as the oxide films grow. What kind of evolution of oxide microstructure will take place under the action of temperature, stress, and time? What are the effects of water chemistry and alloy composition on the microstructural evolution of oxide films? What is the relationship between the corrosion resistance and the microstructural evolution of oxide films? All of these basic questions need to be investigated to enable efficient development of advanced zirconium alloys.

Transmission electron microscopy (TEM) and scanning electron microscopy (SEM) are useful analytical apparatus for investigating the oxide films formed on zirconium alloys. A lot of work has been done in this field[1-10], and a summary is available in Cox's review[11] and in IAEA – TECDOC – 966[12]. Recently, Yilmazbayhan et al. [1] made a detailed TEM study of oxide formed on three different zirconium alloys corroded in pure water and lithiated water. The differences in the morphology of oxide layers, including grain size and shape, oxide phases, textures, and cracks, were investigated and the incorporation of precipitates in the oxide layers was also reported. The investigation of the microstructure at different depths in the oxide films formed on specimens with various chemical compositions corroded in different water chemistries was carried out by TEM. A mechanism for the relationship between the corrosion transition and the formation of pore-clusters through the diffusion and condensation of vacancies under the action of temperature, stress, and time was suggested[10]. There was, however, a lack of direct evidence identified by conventional TEM, due to the limitation of its resolution, which is not sufficient to illustrate the microstructure in detail. In order to understand the microstructural evolution of oxide films and its relations to the corrosion resistance of alloys, further studies of the microstructural differences at different depths in the oxide films are essential.

The development of advanced analytical apparatuses, such as high resolution TEM and SEM equipped with field emission gun (HRTEM and HRSEM) and scanning probe microscopy (SPM), accelerates the development of research work in materials science. If such apparatuses can be employed when investigating the microstructural evolution of oxide films, the mechanism behind the difference in corrosion resistance for alloys with different compositions corroded in different water chemistries could be understood much better. Based on this idea, some primary results, written in Chinese, concerning the microstructural evolution of oxide films on zirconium alloys have been published[13, 14].

Cox paid attention to the pores in oxide films on zirconium alloys and made use of the porosimeter technique to measure the size of pores and cracks in oxide films[15]. He pointed out that the important flaws in post-transition oxide films were pores less than 10 nm in diameter[16]. He wrote in the discussion on a hypothetical transition process that, "If the oxidation rate transition is to be explained, it appears that a process capable of generating a network of fine pores, rather than a mechanical cracking process, is needed." It is concluded that investigations of the nature of pore formation in oxide layers are essential for understanding the corrosion mechanism. The purpose of the present work is to investigate the microstructural evolution of oxide films during the corrosion process, to try to identify the mechanism behind the inevitable formation of pores and micro-cracks in oxide films, and to study the effect of water chemistry and composition on the microstructural evolution of oxide films.

2   Experimental Procedures

2. 1   Materials Preparation and Corrosion Tests

Zircaloy – 4 (Zr – 1. 5Sn – 0. 2Fe – 0. 1Cr) and Alloy No. 3 (Zr – 0. 92Sn – 1. 03Nb – 0. 12Fe), which has a composition similar to ZIRLO^{TM}, were chosen to investigate the effect of alloy composition on the microstructural evolution of oxide films. Commercial Zircaloy – 4 plates, with a thickness of 0. 6 mm, were obtained from a manufacturer in China, and were further annealed at 600 ℃ for 1 h to attain a fully recrystallized structure. A 5 kg ingot of Alloy No. 3 was prepared by consumable-electrode melting twice. The composition of this alloy was analyzed by taking specimens from the bottom, middle, and upper part of the ingot. Through high temperature forging, β – quenching, and multi-stage cold rolling with intermediate annealings at 600 ℃, plates with a thickness of 0. 6 mm were obtained. The final annealing was performed at 580 ℃ for 1 h.

For corrosion tests, 25 by 15 mm^2 specimens were cut from plates. In order to investigate the effect of water chemistry on the microstructural evolution of oxide films, specimens were exposed to deionized water or lithiated water with 0. 01 M LiOH in a static autoclave at 360 ℃/18. 6 MPa. Prior to the corrosion tests, these specimens were pickled in a mixed solution of 10% HF+45% HNO$_3$+45% H$_2$O and 10% HF+30% HNO$_3$+ 30% H$_2$O+30% H$_2$SO$_4$(in volume) for Zircaloy – 4 and Alloy No. 3 respectively, and then rinsed in cold tap water and boiling deionized water. An average value of weight gain was taken from five specimens.

2.2 The Preparation of TEM Specimens

The oxide at different depths of the oxide film represents its formation after different durations of autoclave exposure. The investigation and comparison of the microstructure at different depths of the oxide film will provide information about the microstructural evolution of the oxide film, so thin specimens with a small perforation hole appearing at different depths of the oxide film need to be prepared for TEM examination.

The oxide film on one side of the specimens was removed by grinding after corrosion testing. Then the metal part was dissolved in a mixed acid solution, as is used for pickling specimens before the corrosion tests, until a metal layer with a thickness of ~ 0.1 mm remained for supporting the thin and brittle oxide films. Specimens with a diameter of 3 mm were punched out and protected with adhesive tape on the metal side. A small part of the metal was exposed by cutting a piece of the adhesive tape (less than 0.5 by 0.5 mm^2) at the center of the specimens. Then a drop of mixed acid was used for dissolving the metal to expose the oxide film at the oxide/metal interface. A small penetration hole in the oxide film, at the center of the specimens, was prepared by the ion sputtering thinning method. Rotating specimens were tilted with two parallel argon ion beams by $5°-7°$ during ion sputtering thinning. When two ion beams operate at the same time, a small penetration hole will be obtained at the middle layer of the oxide film. If only one ion beam operates on one side of the specimen, for instance, on the surface of the oxide film, a small penetration hole will be prepared at the bottom layer of the oxide film. On the contrary, a small penetration hole will be prepared at the surface layer of the oxide film. The method of preparing the specimens for TEM examination was described previously[10]. JEM – 2010F TEM was employed for examining the microstructure of the oxide films.

DigitalMicrograph^{TM} was used for processing the high resolution lattice images recorded by a CCD camera. The fast Fourier transforms (FFTs), which are involved in DigitalMicrograph^{TM}, can reveal the periodic content of images and also work on regions of interest with appropriate dimensions placed on an image. They are useful for image analysis and filtering to obtain the information about the crystal structure and the defects in crystals.

2.3 The Preparation of SEM Specimens

The morphology of oxide at the oxide/metal interface, which in this paper is defined as the inner surface of oxide films, can be observed by dissolving the metal part of specimens in the same way as in the preparation of TEM specimens. The specimens for examining the fracture surface of oxide films are easy to prepare by breaking the oxide films on the edge of specimens after dissolving a part of the metal in that area. A very thin layer of gold coating on the specimens, deposited by sputtering, is recommended for improving the image quality. JSM – 6700F SEM was employed for examining the morphology of the fracture surface and of the inner surface of the oxide films.

2.4 The Preparation of SPM Specimens

The grain morphology in oxide films can be examined by SPM due to its very high resolution in the vertical direction, which is not possible with SEM. Specimens for this examination should be mounted in some special materials to protect the oxide films against damage during grinding, polishing, and etching by ion sputtering. Lead-bismuth alloy with the composition at the eutectic point, which has a very low melting point of only about 125 ℃, is suitable for this purpose. The volume expansion during solidification of this alloy will tightly squeeze the specimen mounted in this alloy. The method of preparing the specimens for SPM examination was described previously[17]. The SPM – 6800 apparatus produced by Beijing Zhongke ONAN Tec. Co. Ltd was employed.

3 Experimental Results and Discussion

3.1 Corrosion Behavior

The weight gain versus exposure time for Zircaloy – 4 specimens corroded in 360 ℃/ 18.6 MPa deionized water is shown in Fig. 1(a). The weight gain was close to 100 mg/dm^2 after 395 days exposure and the oxide film was black/shiny. A multi-transition phenomenon is obvious in Fig. 1 (a), as described in the literature[11, 18]. The first transition took place at a weight gain of 35 mg/dm^2 and the second at 70 mg/dm^2; i.e., twice the weight gain observed at the first transition.

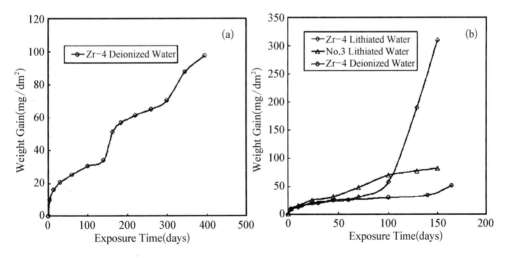

Fig. 1 Weight gains versus exposure time: (a) Zircaloy – 4 specimens corroded in deionized water at 360 ℃/18.6 MPa; (b) Zircaloy – 4 and Alloy No. 3 specimens corroded in lithiated water with 0.01 M LiOH at 360 ℃/18.6 MPa (Zircaloy – 4 specimens corroded in deionized water is also presented)

The weight gain versus exposure time for Zircaloy – 4 and Alloy No. 3 specimens corroded in lithiated water with 0.01 M LiOH at 360 ℃/18.6 MPa is shown in Fig. 1(b).

For comparison, the weight gain for Zircaloy - 4 specimens corroded in deionized water is also depicted in Fig. 1(b). The corrosion resistance of Alloy No. 3 is remarkably superior to that of Zircaloy - 4 described in Ref. [19]. After 150 days exposure, the weight gain reached 310 mg/dm^2 and 82 mg/dm^2 for Zircaloy - 4 and Alloy No. 3, respectively. For Zircaloy - 4 specimens the corrosion transition occurred after 90 days exposure and the corrosion rate increased rapidly after the transition. The oxide films on Zircaloy - 4 specimens became dark brown after 150 days exposure but were black/shiny after 70 days exposure in the pre-transition stage. The oxide films on Alloy No. 3 specimens remained black/shiny until the end of the corrosion tests (150 days). In the pre-transition stage, the weight gain of Zircaloy - 4 specimens corroded in lithiated water was similar to that of specimens corroded in deionized water.

3.2   The Microstructure of Oxide Films at Different Depth Examined by HRTEM

The specimens used for the examination of oxide films by HRTEM, HRSEM, and SPM are listed in Table 1. Zry - 4/W/395d and Zry - 4/Li/150d specimens were chosen for comparing the effect of water chemistry on the microstructural evolution of the oxide. Zry - 4/ Li/70d, Zry - 4/Li/150d, No. 3/Li/70d, and No. 3/Li/150d specimens were chosen for comparing the effect of alloy composition on the microstructural evolution of the oxide. The investigation of oxide microstructure at different depth by HRTEM was used for evaluating the microstructural evolution during the corrosion process.

Table 1   The specimens used for HRTEM, HRSEM, and SPM studies

Specimens designation	Description		
	Alloy composition	Corrosion environment and exposure time	Type of examination
Zry - 4/W/395d	Zircaloy - 4 (Zr - 1.5Sn - 0.2Fe - 0.1Cr)	Deionized water for 395 days	HRTEM, HRSEM, and SPM
Zry - 4/Li/70d	Zircaloy - 4	0.01 M lithiated water for 70 days	HRTEM and HRSEM
Zry - 4/Li/150d	Zircaloy - 4	0.01 M lithiated water for 150 days	
No. 3/Li/70d	Alloy No. 3 (Zr - 0.92Sn - 1.03Nb - 0.12Fe)	0.01 M lithiated water for 70 days	
No. 3/Li/150d	Alloy No. 3	0.01 M lithiated water for 150 days	
No. 3/Li/14d	Alloy No. 3	0.04 M lithiated water for 14 days	SPM
Zry - 4/Li/14d	Zircaloy - 4	0.04 M lithiated water for 14 days	

Since defects exist in the crystals in large quantity, the grain morphology images taken by diffraction contrast are unclear. Crystal defects can be revealed by high resolution lattice images formed by phase contrast. Therefore, the microstructural evolution of oxide can be investigated in detail by taking lattice images, and the crystal structure analysis, combining with the FFT analysis technique, can be carried out in an area of tens of nanometers in scale.

The Microstructure in the Bottom Layer of Oxide Films — The crystal structure in the bottom layer of the oxide films, which is in close vicinity to the metal matrix, was very complicated. Monoclinic, tetragonal, cubic, and amorphous structures can be detected by selected area diffraction, which was described in a previous paper[10]. In order to better understand the microstructural evolution of oxide films, the investigation of the defects in crystals and the dependence of different phases on each other is a key point, which is studied by taking lattice images with HRTEM.

It is difficult to distinguish the differences among the microstructure in the bottom layer of the oxide film on specimens with different chemical composition, exposed in different water chemistries or during different durations. Thus, only the microstructural characteristic observed in the bottom layer of the oxide film on Zry-4/W/395d specimen is described here.

The main volume fraction of the oxide in the bottom layer consisted of monoclinic oxide although the meta-stable tetragonal, cubic and amorphous oxide phases existed simultaneously. The images of oxide grain morphology taken by the method of diffraction contrast are not clear, as shown in Fig. 2(a), due to the existence of a large quantity of defects within the crystals. The moiré pattern, which is formed by two layers of crystal lattice lapped over and twisted with respect to each other by a small angle, can be observed frequently (Fig. 2(b)). It means that a plane defect existed perpendicular to the incidence direction of the electron beam between two layers of crystal lattice. The crystal defects can be clearly revealed in the high resolution lattice images taken by the method of phase contrast. If the lattice images are in good order no matter whether the lines of lattice image are shown in one direction or in two directions, it means that the crystal in that area is perfect. If there are some interruptions or disturbances in the lattice images, besides at the grain boundary areas, some defects must exist in those areas of the crystals.

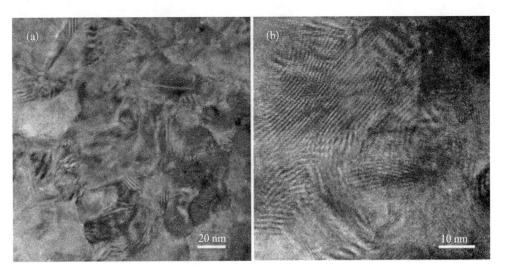

Fig. 2 TEM micrographs in the bottom layer of the oxide films on Zircaloy-4 specimens corroded in deionized water during 395 days

The high resolution lattice images taken in the bottom layer of the oxide films are shown in Figs. 3(a)-3(d). Figures 3(a) and 3(b) show a mixed area of crystalline and amorphous structure. The periodic lines in Fig. 3(a) indicate micro-crystals surrounded by amorphous material, but the main part of the area in Fig. 3(b) is crystalline. By measuring the line spacing and performing FFT analysis (patterns incorporated in Figs. 3(a) and 3(b)), it is concluded that there is cubic crystal structure in both areas, and that the (110) plane is parallel to the specimen surface.

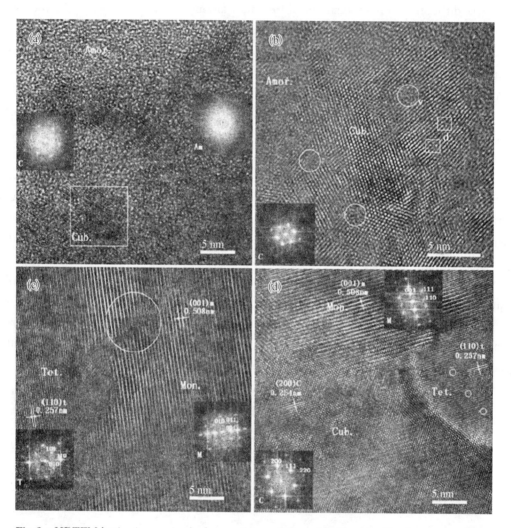

Fig. 3 HRTEM lattice images in the bottom layer of the oxide films on Zircaloy – 4 specimens corroded in deionized water during 395 days: (a, b) mixed area of amorphous and cubic phases; (c) monoclinic and tetragonal phases; (d) monoclinic, tetragonal and cubic phases. Different phases are labeled "Amor," "Cub.," "Tet.," and "Mon." at different areas accordingly in the micrographs. Some edge dislocations and volume defects shown in (b) are labeled by "d" and "v," respectively. Phase transformation from monoclinic into tetragonal induced by stress and strain can be seen within the circle in the upper part of (c). The patterns obtained by FFTs from different phases in different areas are also incorperated in the figures and the lattice spacing measured from lattice images are also given in (c) and (d). Some cross-bandings with "clear" and "blurred" areas indicated by small circles in the right part of (d) show that the distortion of crystal lattice is produced due to the stress induced by the mismatch between monoclinic and tetragonal lattice

The cross-lines of lattice image appear only in the middle part of Fig. 3(b) because of the crystal distortion induced by defects. A lot of linear defects are edge dislocations, some of which are indicated by small square frameworks labeled "d" in Fig. 3(b). Some volume defects in nanometer scale are defined as clusters of atoms in a manner similar to amorphous material, and are embedded in crystalline structure, as indicated by circles labeled "v" in Fig. 3(b). The minimum volume defects are vacancies or interstitials, which are difficult to discern in high resolution lattice images. The co-existence of monoclinic and tetragonal phases with coherent relationship of (001)m ∥ (110)t is shown in Fig. 3(c). There is no distinct phase boundary between them. The FFT patterns indexed for monoclinic and tetragonal phases are also incorporated in Fig. 3(c), and the line spacings of (001)m and (110)t measured from the lattice images are given. In the upper part of Fig. 3(c) outlined by a circle, the lattice image of monoclinic (001) planes is in disorder within a limited area induced by the defects. Thus, a relatively large stress and strain must exist in that area. Some lattice images of tetragonal (110) plane can be found in that area through a careful examination. This is direct evidence of the phase transformation from monoclinic into tetragonal structure induced by stress and strain in zirconium oxide. That is why the tetragonal and other metastable phases can be detected in the bottom layer of oxide films. Figure 3(d) shows a co-existence of monoclinic (in the upper part), cubic (in the lower part), and tetragonal (in the right part) phases. The coherent relations among them with (001)m ∥ (110)t, (001)m ∥ (200)c, and (200)c ∥ (110)t can be identified. The FFT patterns indexed for monoclinic and cubic phases are also incorporated in Fig. 3(d), and the line spacings of (001)m, (110)t, and (200)c measured from the lattice images are given. Through careful examination of the tetragonal crystal area in the right part of Fig. 3(d), some cross-bandings with "clear" and "blurred" areas (indicated by small circles) can be seen. It means that the distortion of the crystal lattice is produced as a result of the stress induced by the mismatch between the monoclinic and the tetragonal lattice. A similar phenomenon can also be observed in other areas with coherent lattice between different phases.

A compressive stress in the oxide films can reach a level higher than 1200 MPa[20]. The diffusion, annihilation, and condensation of defects in the crystal will take place under the action of compressive stress and temperature to induce the microstructural evolution of the oxide film during the corrosion testing. The investigation of the microstructure difference between the bottom layer and the middle layer of oxide films is an approach for understanding the process of microstructural evolution.

The Microstructure in the Middle Layer of Oxide Films — The middle layer of the oxide films has passed through a period of exposure since its formation, and the microstructure in this layer could change gradually. For the Zry – 4/W/395d specimen, it took about 160 days to reach a weight gain of 50 mg/dm^2, which was half of the total weight gain accumulated after completion of the test (Fig. 1(a)). It indicates that the middle layer of the oxide film started to form after 160 days of exposure, and passed through 235 days of continued autoclave exposure after its initial formation. In other words, the microstructural evolution of the middle layer of the oxide film passed through 235 days.

The middle layer of the oxide films on the Zry - 4/Li/150d and No. 3/Li/150d specimens was formed after 120 and 60 days of exposure, respectively (Fig. 1(b)). It indicates that the middle layer of the oxide films on these specimens passed through 30 days (Zry - 4/Li/150d) and 90 days (No. 3/Li/150d) of continued autoclave exposure after their initial formation.

The weight gains were 31 mg/dm^2 and 48 mg/dm^2 for Zry - 4/Li/70d and No. 3/Li/70d specimens, respectively, and the middle layers of the oxide films on these specimens were formed after 10 and 20 days of autoclave exposure, respectively. This means that the microstructural evolution of the middle layer of the oxide films on these specimens continued for 60 days (Zry - 4/Li/70d) and 50 days (No. 3/Li/70d) of autoclave exposure, respectively.

The microstructure in the middle layer of oxide films on Zry - 4/W/395d specimen was examined by HRTEM. When the examination was performed in a relatively thick area on the thin specimens, some interesting features were obtained, as shown in Fig. 4(a). Some pores formed on grain boundaries, particularly on the junctures of three grains, were observed. The pores usually appeared in a clustered manner with the distance between clusters about several tens of micrometers. Grains became spherical with a diameter of about 50 nm at those places where pores formed. These pores of nanometer size can be easily distinguished by their morphology from the pores produced by ion sputtering thinning at thin areas near the perforation hole on thin specimens. This image contrast is obtained mainly by the transparency thickness of specimens, not by the diffraction contrast of grains. Thus, the dark area is thicker than the bright area where the pores formed or were at the point of forming. The high resolution lattice image obtained in the area near pores is better ordered than that obtained in the bottom layer of the oxide films Fig. 4(b). This suggests that the pores are formed by the diffusion and condensation of vacancies under the action of compressive stress and temperature. Vacancies diffused and were absorbed by grain boundaries to form the nuclei of pores. The pores grew through the absorption of the vacancies diffused from the grains, which weakened the bonding strength between grains. Under the action of surface tension, the grains gradually became spherical. The lattice images became better ordered because the defects diffused out of the crystal lattices.

The early stage of the formation of pores on the junctures of grains is shown in Fig. 4(b). A loose area looking like amorphous phase appeared on a juncture of grains by the condensation of vacancies through their diffusion. The pores formed under the compressive stress in the oxide films, which is similar to the precipitation of second phase particles with volume dilation, will line up in the direction parallel to the compressive stress, instead of perpendicular to it. Such a phenomenon was discussed by one of the authors previously[9].

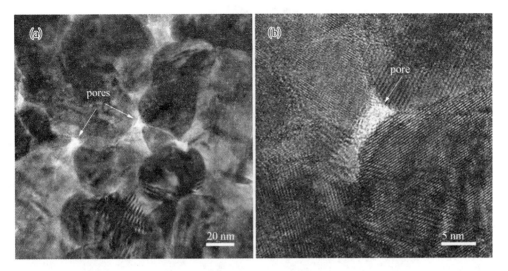

Fig. 4   TEM micrographs in the middle layer of the oxide films on Zircaloy – 4 specimens corroded in deionized water for 395 days. Some pores are indicated by arrows

Thus, the cracks developed through the linking of pores in oxide films are parallel to the direction of compressive stress; i. e. , parallel to the oxide/metal interface.

The microstructure in the middle layer of the oxide film on Zry – 4/Li/150d specimen is shown in Fig. 5(a), which is quite different from Fig. 4(a). In Fig. 5(a), not only pores but also micro-cracks or "loose" areas along grain boundaries are revealed. Also, grains where the micro-cracks or loose areas appeared became spherical. Before the formation of micro-cracks, the loose areas along the grain boundaries had an amorphous appearance as shown in Fig. 5(b). The microstructure of middle layer oxide films on the No. 3/Li/150d specimen is shown in Fig. 5(c). The big hole in Fig. 5(c) could be artificially enlarged by ion sputtering during the preparation of the thin specimen. The micro-cracks or loose areas in a network distribution are not as obvious as in Fig. 5(a). This suggests that the micro-cracks or loose areas in Zircaloy – 4 specimens are much easier to form than those in Alloy No. 3 specimens corroded in lithiated water.

The microstructure in the middle layer of the oxide films on Zry – 4/Li/70d and No. 3/Li/70d specimens, which was in the pre-transition stage, is shown in Fig. 6. As shown in Figs. 5(a) and 5(c), no "loose" area is discernible in the two considered specimens. The defects and different crystal structure are revealed in No. 3/Li/70d specimen as shown in Fig. 6(b). This means that the microstructural evolution of oxide films in the middle layer is not sufficient in the pre-transition stage.

The Microstructure in the Surface Layer of Oxide Films — The pores in the surface layer will be much larger than the pores within the oxide, since the residence time in the autoclave is the longest for this older part of the oxide. However, the larger pores could not be observed by TEM in this layer, because the oxide films near the large ones could easily break into pieces during the preparation of the thin specimens by ion sputtering thinning. The situation of cracks in the surface layer of oxide films is similar to that of the

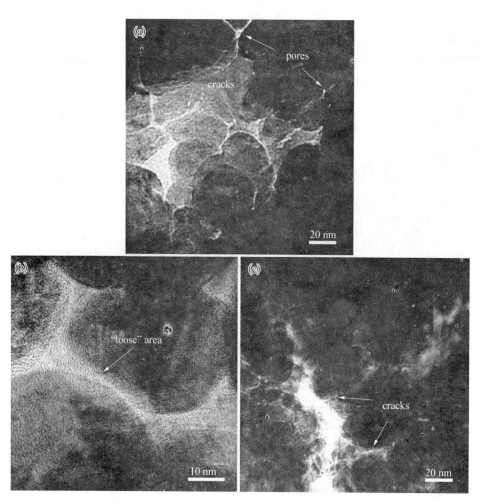

Fig. 5 TEM micrographs in the middle layer of the oxide films on Zircaloy – 4 (a, b) and Alloy No. 3 (c) specimens corroded in lithiated water for 150 days. Some pores and microcracks (or "loose" areas) along grain boundaries are indicated by arrows

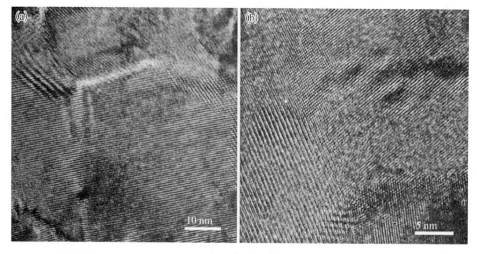

Fig. 6 HRTEM lattice images in the middle layer of the oxide films on Zircaloy – 4 (a) and Alloy No. 3 (b) specimens corroded in lithiated water for 70 days

pores; i. e. , only small cracks can be revealed.

Some typical morphology of pores and cracks revealed in the surface layer of oxide films is shown in Fig. 7. Figure 7(a) shows the pores observed in the surface layer of the oxide film on Zry - 4/W/395d specimen. The pores in the surface layer were much larger compared with those observed in the middle layer of the oxide films (Fig. 4(a), in different magnification). Figure 7(b) shows the pores observed in the surface layer of the oxide film on No. 3/Li/70d specimen. Although no pores were clearly revealed in the middle layer of the oxide film (Fig. 6(b)), the pores in the surface layer were relatively large and micro-cracks or loose structure along grain boundaries could also be seen.

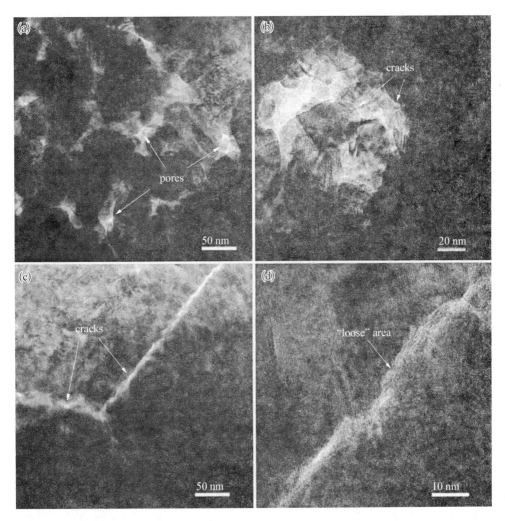

Fig. 7 TEM micrographs in the surface layer of the oxide films on Zircaloy - 4 specimens corroded in deionized water for 395 days (a), on Alloy No. 3 (b), and Zircaloy - 4 specimens (c, d) corroded in lithiated water for 70 days. Some pores, micro-cracks, and "loose" area are indicated by arrows

Some small pores like those observed in the middle layer of the oxide films can also be seen in the surface layer. The high resolution lattice images near the pore areas are better ordered, which is also an indication that the formation of pores is closely related to the diffusion and condensation of vacancies from the grains. Figures 7(c) and 7(d) show the morphology of cracks revealed in the surface layer of the oxide films on Zry – 4/Li/70d specimen. The morphology of cracks viewed at low magnification (Fig. 7(c)) indicates that the formation of the cracks was related to the grain boundary of the original metal grains before oxidation. The morphology of the cracks reviewed at high magnification (Fig. 7(d)) indicates that the formation process of this micro-crack should be linked to the diffusion and condensation of vacancies along grain boundaries. The loose area with 0. 5 – 1. 5 nm in width was formed at first, then developed into discontinuous pores along grain boundaries by the absorption of vacancies from oxide grains. The lattice image, in good agreement with monoclinic structure, shows that the defects were swept out of grains and absorbed by grain boundaries.

3. 3 The Fracture Surface and Inner Surface Morphology of Oxide Films Examined by HRSEM

The Fracture Surface Morphology of Oxide Films — When fracture occurs in polycrystalline materials, the cracks propagate from grain to grain along some crystal planes or grain boundaries. Thus, the fracture surface morphology of oxide films reveals not only the fracture characteristics but also the polycrystalline microstructure. Examination of the fracture surface of oxide films performed by HRSEM provides some valuable information to be used for comparing the microstructure observed by HRTEM.

The fracture surface morphology of oxide films on Zry – 4/W/395d specimen is shown in Fig. 8. The location of the higher magnification pictures is outlined by the white frames in the relatively low magnification. This approach is used also in Figs. 9 and 10. An undulating morphology can be seen on the right side of Fig. 8(a), which is the morphology of the inner surface of the oxide films revealed when the metal matrix was dissolved by acid solution.

A side-step feature, which looks like a crack parallel to the oxide/metal interface, can be seen in Fig. 8(a). In the higher magnification image (Fig. 8(b)), the configuration of the side-step is clear. There are some pores discontinuously distributed in the bottom, and equiaxed grains less than 100 nm in diameter can be seen both in the bottom and in the vertical plane of the side-step. Based on this observation, it is suggested that the side-step features of the fracture surface could be produced when breaking the oxide films due to the weak bonding strength between grains after the formation of pores along the grain boundaries. This is consistent with the observation made by TEM in the middle layer of the oxide film (Fig. 4). Another feature of the fracture surface is shown in Fig. 8(c), in which some cracks are nearly parallel to the oxide/metal interface. The propagation of cracks by joining up with the pores ahead of cracks is shown in a higher magnification picture (Fig. 8(d)). Some small areas with smooth fracture surface near the oxide/metal interface (indicated by an arrow in Fig. 8(f)) could be a fracture surface of amorphous

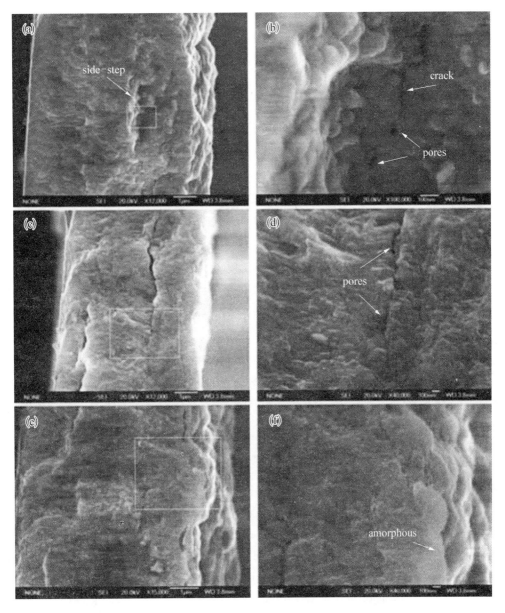

Fig. 8 The fracture surface morphology of the oxide films on Zircaloy - 4 specimens corroded in deionized water for 395 days. Side-step in (a) and pores in (b) are indicated by arrows. The fracture surface morphology in the amorphous phase is indicated by an arrow in (f)

phase. Amorphous phase existing in the bottom layer of oxide films has been verified through HRTEM observations (Fig. 3(a)), and the locations of the amorphous phase, observed by TEM and SEM, coincide with each other.

The fracture surface morphology of oxide films on Zry - 4/Li/150d and Zry - 4/Li/70d specimens is shown in Figs. 9(a)- 9(d). In Fig. 9(a), which is quite different from that of Zry - 4/W/395d specimen (Fig. 8), many cracks nearly parallel to the oxide/metal interface are distributed on the entire fracture surface especially in the area near the oxide/ metal interface. A rough fracture surface, with much higher undulation than that obtained

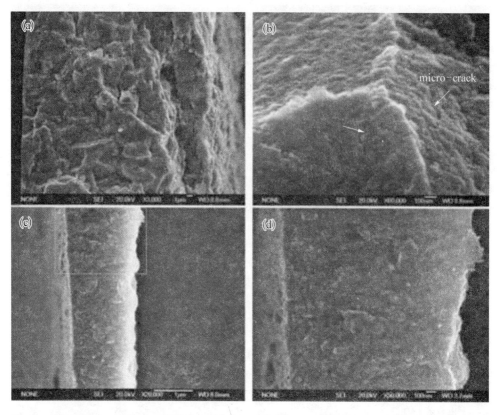

Fig. 9　The fracture surface morphology of the oxide films on Zircaloy – 4 specimens corroded in lithiated water during 150 days (a, b) and 70 days (c, d). Micro-cracks along grain boundaries are indicated by arrows in (b)

from the specimens corroded in deionized water, means that some weak bonding between grains existed in the oxide films before breaking. Equiaxed grains less than 100 nm and micro-cracks along grain boundaries can be revealed in a high magnification micrograph (indicated by arrows in Fig. 9 (b)). The presence of the micro-cracks along grain boundaries, coincident with the observation by TEM (Fig. 5(a)) but different from the specimens corroded in deionized water (Fig. 8), causes the weak bonding between grains.

The fracture surface morphology of oxide films on Zry – 4/Li/70d specimen, which was in the pre-transition stage, is shown in Figs. 9(c) and 9(d). The fracture surface morphology is relatively flat and the oxide films are compact compared with those obtained from Zry – 4/Li/150d specimen (Fig. 9(a)). Some pores but not cracks can be seen in a high magnification micrograph (Fig. 9(d)), and equiaxed grains about 100 nm in diameter constitute the main part of the grain structure. This is a characteristic of the oxide film microstructure in the pre-transition stage. The incorporation of Li^+ and OH^- in oxide films has taken place at this stage[10, 21], but the processes of vacancy diffusion and pore or crack formation need time. Before the formation of pores and cracks, the oxide films possess protective features. Therefore, there is no obvious difference between the pre-transition corrosion rate in lithiated water and deionized water (Fig. 1).

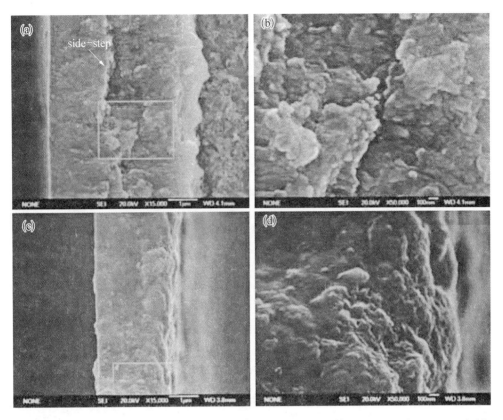

Fig. 10 The fracture surface morphology of the oxide films on Alloy No. 3 specimens corroded in lithiated water during 150 days (a, b) and 70 days (c, d). Side-step is indicated by an arrow in (a)

The fracture surface morphology of oxide films on No. 3/Li/150d specimen is shown in Figs. 10(a) and 10(b). Side-steps which are nearly parallel to the oxide/metal interface can be seen in the Fig. 10(a). Some pores distributed discontinuously can be revealed in a higher magnification micrograph (Fig. 10(b)). The fracture surface morphology of oxide films on No. 3/Li/70d specimen is shown in Figs. 10(c) and 10(d). The oxide films were relatively compact and no cracks on the fracture surface can be seen. Equiaxed grains constitute the main part of the oxide films, which can be identified in a high magnification micrograph (Fig. 10(d)).

The fracture surface morphology is quite different for Zry - 4/Li/150d and No. 3/Li/ 150d specimens. It is illustrated that the tendency towards the formation of pore-clusters and micro-cracks along grain boundaries is also different. The reasons will be discussed in the later part of this paper.

The Morphology of Oxide Inner Surface at the Oxide/Metal Interface — The growth of oxide films towards the metal matrix was inhomogeneous, forming an undulating morphology as revealed on the inner surface of the oxide films when the metal was dissolved. The morphology of the oxide inner surface of Zry - 4/W/395d specimens is shown in Fig. 11. Two types of protuberance can be classified. One was about 0.2 μm in size examined at relatively high magnification (Fig. 11(a)). It is believed that this was caused by the essential process of oxide growth. The other was 5 – 10 μm in size overlapped on the original 0. 2 μm ones to form "cauliflower-like" undulations as shown in Fig. 11(b). Such morphology appeared only after corrosion transition. The undulations can also be observed in the right part of Fig. 8(a), where the metal matrix was dissolved by acid solution on the side face of the oxide film.

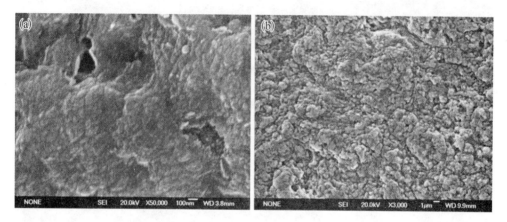

Fig. 11 The inner surface morphology of the oxide films on Zircaloy – 4 specimens corroded in deionized water for 395 days

The inner surface morphology of oxide films on Zry – 4/Li/70d and Zry – 4/Li/150d specimens is shown in Fig. 12. The cauliflower-like morphology on the oxide inner surface can only be observed after 150 days, but not after 70 days exposure. Making the comparison between the morphology of the fracture surface (Fig. 9) and that of the inner surface of the oxide films, it is clear that the cauliflower-like undulations on the oxide inner surface are closely related to the cracks formed in the oxide films and that they appeared only after transition. The inner surface morphology of oxide films on No. 3/Li/150d specimen was not as undulated as that on the Zry – 4/Li/150d specimen (Figs. 12(c) and 12(d)), and there was not a large quantity of cracks on the fracture surface (Fig. 10). The morphology of the oxide inner surface is closely related to the characteristic oxide microstructure, and also represents a different stage of the corrosion process.

It is considered that the pores and cracks in the oxide films are similar places to the oxide film surface, where anion vacancies will be absorbed and ions of oxygen and hydrogen will be produced by the reaction between water molecules and electrons. These places will thus act as resources for providing oxygen ions. This configuration is schematically illustrated in Fig. 13. Taking the pore-clusters or cracks as a center to draw circles with the same radius towards the metal matrix to form an oxide/metal interface, it means that the oxygen ions provided from the pore-clusters and cracks diffused towards the metal matrix to push ahead the growth of the oxide film at the same rate. Thus, the undulation interface between oxide and metal will form. The cauliflower-like morphology on the oxide inner surface appeared after oxidation transition, i. e. , after the formation of

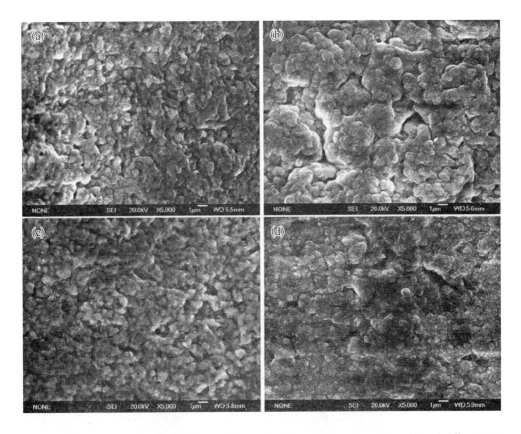

Fig. 12 The inner surface morphology of the oxide films on Zircaloy－4 (a, b) and Alloy No. 3 (c, d) specimens corroded in lithiated water for 70 days (a, c) and 150 days (b, d), respectively

the pore-clusters and cracks in oxide films could be explained.

3. 4 The Grain Morphology of Oxide Films Examined by SPM

The vertical resolution of SPM is about 0. 01 nm, which is much higher than that of SEM. Thus, the morphology of grain undulation and grain size on oxide film surfaces can be clearly observed by SPM[17].

After mechanical polishing and ion sputtering etching, the grain structure on the section of the

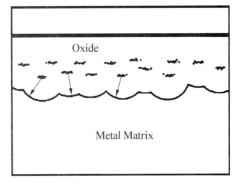

Fig. 13 The schematic diagram showing the formation of cauliflower-like undulation on the oxide/metal interface

oxide film can be revealed. The etching depth given by ion sputtering is shallow. It is not possible to observe by SEM but it is by SPM. The grain morphology on the section of oxide film prepared from Zry－4/W/395d specimen is shown in Fig. 14 (a). Columnar grains constructed by large angle grain boundaries (with thicker boundaries) can be seen, but some equiaxed grains less than 100 nm in diameter constructed by relatively small angle grain boundaries (with thinner boundaries) within the columnar grains can also be seen. Due to the different size of columnar and equiaxed grains, an equiaxed grain morphology examined on the fracture surface can be seen at high magnification, while the examination performed on relatively low magnifications shows only columnar grains such as in Fig. 8(a).

Some ambiguous areas near the oxide/metal interface with a width of about 100 nm could be interpreted as amorphous phase (Fig. 14(b)), because no grain boundaries were revealed by ion sputtering etching. This result is consistent with the observation using HRTEM in the bottom layer of oxide films and using SEM on the fracture surface of oxide films.

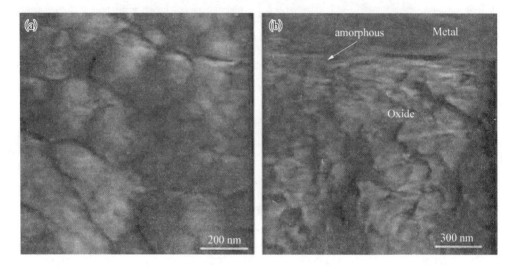

Fig. 14 The grain structure (a) and amorphous layer (b, indicated by an arrow) near the oxide/ metal interface observed on a section of the oxide films on Zircaloy – 4 specimens corroded in deionized water for 395 days observed by SPM after ion sputtering etching

The surface morphology of Zry – 4/Li/14d and No. 3/Li/14d specimens is shown in Fig. 15. The weight gain is 17. 6 mg/dm^2 for Zry – 4/Li/14d specimens and 19. 8 mg/dm^2 for No. 3/Li/14d specimens, respectively. The undulation of grains on the specimen surface is obviously different besides the different size of oxide grains. The expansion of oxide films formed on the specimen surface will be restricted by the metal matrix and the surface free energy of the oxide, the volume of oxide cannot expand freely. Thus, the undulation of the grain morphology on the specimen surface will represent the level of surface free energy. If the surface free energy is smaller, the undulation of grain morphology on the specimen surface will be higher, and vice versa. A section of two grains is schematically shown in Fig. 16. γ_{AB} represents the grain boundary energy that will be affected by the orientation relationship between the two grains, but here it is supposed that the grain boundary energy is a mean value and is not affected by water chemistry. γ_{SA} and γ_{SB} represent the surface free energy of grains A and B, respectively, which will be affected by crystal orientation, water chemistry, temperature, and alloy composition. Here it is also supposed that the surface free energy is a mean value. The angle between γ_{SA} and γ_{SB} is

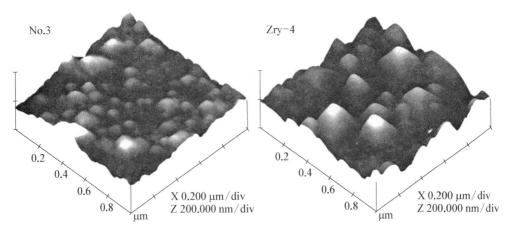

No.3

Zry-4

0.2
0.4
0.6
0.8
μm

X 0.200 μm/div
Z 200.000 nm/div

0.2
0.4
0.6
0.8
μm

X 0.200 μm/div
Z 200.000 nm/div

Fig. 15 Surface morphology of Zircaloy-4 and Alloy No. 3 specimens corroded in lithiated water with 0. 04 M LiOH at 360 ℃/18. 6 MPa for 14 days observed by SPM

α and the junction with γ_{AB} is at "O" point in equilibrium. In the case of smaller γ_{SA} and γ_{SB}, the angle α will also be smaller. Thus, the undulation of grain morphology will be higher, and vice versa. By the analysis of the undulation of grain morphology on the specimen surface as shown in Fig. 15, it is concluded that the surface free energy of oxide formed on Zircaloy-4 specimens corroded in lithiated water is lower than that of oxide formed on Alloy No. 3. Therefore, the corrosion resistance of Zircaloy-4 is worse than that of Alloy No. 3 in lithiated

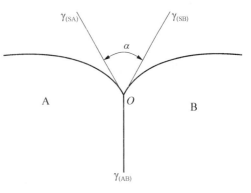

Fig. 16 Schematic diagram showing the equilibrium state between grain boundary energy and surface free energy of grains

water due to the lower surface free energy of the oxide, which can promote the formation of pores and cracks in the oxide film, or in other words, accelerate the microstructural evolution. Besides the water chemistry, the composition of zirconium alloys will also affect the surface free energy of zirconium oxide.

The degradation of corrosion resistance for zirconium alloys corroded in lithiated water has attracted more attention. Some suggestions concerning this mechanism were reviewed[4, 12], but the process has not been clarified so far. The important things are that Li^+ can be detected by secondary ion mass spectroscopy (SIMS) analysis at a certain depth of the oxide film when the specimen is corrode in lithiated water[21-25]. Moreover, OH^- can also be detected by SIMS analysis at the same depth as Li^+ [25]. A major part of lithium in the oxide films is soluble and can be leached out in nitric acid[22]. The results of $^6Li^+/^7Li^+$ exchange experiments indicated that lithium is located mainly in the pores or adsorbed on their walls[21]. Thus, it is indicated that lithium is physically held in the porous oxide.

When Li^+ and OH^- are incorporated in oxide films, they can be adsorbed on the walls of pores or enter into vacancies. Thus, the surface free energy of zirconium oxide will reduce as a universal rule. As a result, the formation of pores by the condensation of vacancies can be enhanced. The results show that the reduction of surface free energy of zirconium oxide for Zircaloy – 4 is more obvious than that for Alloy No. 3 specimens (Fig. 15). Therefore, the corrosion resistance of Zircaloy – 4 was worse than that of Alloy No. 3 in lithiated water. This is a direct result induced by the effect of water chemistry and composition on the microstructural evolution of the oxide. The phenomenon of different corrosion behavior for specimens of different composition tested in lithiated water can be understood in the light of such a mechanism.

3. 5　The Relationship Between Controlling the Microstructural Evolution of Oxide and Improving the Corrosion Resistance of Zirconium Alloys

Based on the above discussion, it is clear that the protective properties of the oxide films will degrade due to the microstructural evolution of the oxide during corrosion tests. The driving forces, from internal factors, are the high compressive stress in the oxide films and high density of defects consisting of vacancies and interstitials in the form of points, lines, planes, and volumes formed during the formation of oxide films. Temperature and corrosion test duration are the external driving forces. The basic process for microstructural evolution of oxide films is that the diffusion, annihilation, and condensation of vacancies and interstitials take place in the oxide under the action of temperature, compressive stress and time. Then the inner stress of oxide crystals will be relaxed and the metastable phases will transform to the stable phase. Vacancies will be absorbed by grain boundaries to form pores on a nanometer scale, which will weaken the bonding strength between grains. In this case, the grains will gradually become spherical due to the action of surface tension. The columnar grains formed in the early stage during the growth of the oxide will change into equiaxed grains. Based on the formation of pore-clusters, the development of cracks in the oxide film will lead to the loss of the protective characteristic; thus, the phenomenon of corrosion transition appears. This is an inevitable result leading from the microstructural evolution of the oxide. The pore-clusters are distributed along the direction of compressive stress. As a result, the cracks developed from pore-clusters in oxide films are approximately parallel to the oxide/metal interface. The variation of corrosion kinetics is induced by the microstructural evolution of the oxide, so the progress of the microstructural evolution is closely related to the corrosion behavior. Therefore, finding and controlling the factors that could retard the microstructural evolution, will improve the corrosion resistance of zirconium alloys.

Besides the temperature and time, the inner stress of the oxide will play an important role in affecting the diffusion and condensation of vacancies. The inner stress of oxide films will increase with the increase of the oxide film thickness at the early stage of oxide growth[20]. Therefore, the microstructural evolution of the oxide will occur till a certain thickness of oxide film is reached. The inner stress of the oxide films relaxes after the formation of pores by the diffusion and condensation of vacancies. Subsequently a repeated accumulation of inner stress will take place as a new layer of oxide grows on the oxide/metal interface, whereupon another cycle of formation of pores and cracks occurs. In the light of this mechanism, the phenomena of the multi-transition in the corrosion kinetic; the stratification of the oxide films observed by transmitted light microscopy[26]; the cyclic variation of protective layer thickness obtained by monitoring in situ impedance spectroscopy[27]; and a compact barrier layer existing on the oxide/metal interface[4] can all be better understood.

The temperature is an essential condition for corrosion tests and the inner stress in oxide films is an inevitable result during the formation of zirconium oxide. Besides the factors of temperature and the inner stress in oxide films, the alloying elements dissolving in zirconium oxide and the surface free energy of zirconium oxide are the important factors which will affect the diffusion of vacancies and the formation of pores.

The solid solution limitation in α - Zr is very low for most alloying elements used for zirconium alloys, except Sn and Nb. These alloying elements, such as Fe, Cr, Ni, and Cu, form metallic compounds with zirconium to precipitate as second-phase particles (SPPs) in the α - Zr matrix. The SPPs in α - Zr will transform into particles embedded in zirconium oxide when the specimens are corroded. It is not easy for the oxides of alloying elements to dissolve into the zirconium oxide. Only the alloying elements, which are already dissolved in the α - Zr matrix, can be dissolved into the zirconium oxide during the corrosion process. In this case, the alloying elements dissolved in zirconium oxide affect the vacancy diffusion and the surface free energy of the zirconium oxide. We have recently presented results that support this suggestion[28]: Zircaloy - 4 specimens with higher contents of Fe and Cr in solid solution in the α - Zr matrix, which was super-saturated solid solution treated by β - quenching, showed excellent corrosion resistance in lithiated water with 0. 01 M LiOH at 360 ℃/18. 6 MPa. The weight gain of this type of Zircaloy - 4 is comparable to that of N18 (Zr - 1. 0Sn - 0. 3Nb -0. 3Fe - 0. 1Cr) and ZIRLO™ after 450 days exposure, while Zircaloy - 4 specimens treated by annealing at 600 ℃ have a transition after 90 days exposure, whereupon the corrosion rate increases rapidly. This means that the increase of Fe and Cr contents in solid solution within zirconium oxide might yield the result of retarding vacancy diffusion or enhancing the surface free energy of the oxide. The corrosion resistance of Zircaloy - 4 specimens corroded in lithiated water is remarkably enhanced therefore (see Fig. 17).

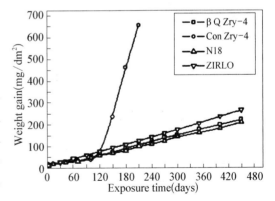

Fig. 17 Weight gain versus exposure time for β-quenched Zircaloy - 4 and conventional Zircaloy - 4 specimens corroded in lithiated water with 0. 01 M LiOH at 360 ℃/18. 6 MPa. The weight gain versus exposure time for ZIRLO[19] and N18 specimens corroded in the same environment are also presented

The nuclei of pores are usually formed at the grain boundaries of an oxide, particularly at the junctures of three grains. If more "boundaries" can be provided in the oxide films, the pores will be in a dispersed distribution after the formation of pore-nuclei by absorbing vacancies on these "boundaries." In this case, the corrosion resistance of the zirconium alloy will be improved due to the retardation of the growth of pores, or in other words, the obstruction of the microstructural evolution of the oxide. We have obtained some results to support this suggestion[29]: Alloy No. 3 specimens with nanometer size particles of β – Nb obtained by β – quenching, deformation and 560 ℃ aging treatments[30] possessed much better corrosion resistance in lithiated water with 0. 04 M LiOH at 350 ℃/16. 8 MPa than Alloy No. 3 specimens with microparticles of β – Nb created through aging treatment without deformation (see Figs. 18 and 19). β – Nb particles with nanometer size will be embedded in the oxide after their oxidation to provide a large quantity of boundaries to act as sinks for vacancies. In this case, the growth of pores will be retarded and the corrosion resistance will be enhanced. The optimum corrosion performance of ZIRLO™ is also achieved by maintaining a uniform distribution of fine second-phase particles[31].

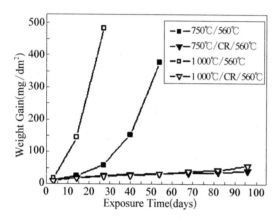

Fig. 18　Weight gain versus exposure time for Alloy No. 3 specimens, fabricated with different thermomechanical processing, corroded in lithiated water with 0. 04 M LiOH at 350 ℃/16. 8 MPa

The variation of the corrosion resistance of zirconium alloys has a certain relation to the microstructural evolution of the oxide. It is necessary to retard the microstructural evolution of the oxide in order to improve the corrosion resistance. For this purpose, it should be taken into account whether the alloying elements in the zirconium alloy can dissolve in α – Zr or form a large quantity of SPPs in the nanometer size range during thermo-mechanical processing. All these subjects are basic issues needing to be further investigated in order to develop advanced zirconium alloys.

4　Conclusions

The microstructure of oxide films formed on Zircaloy – 4 and Alloy No. 3 specimens, which has a composition similar to ZIRLO™, was investigated by HRTEM, HRSEM, and SPM after corrosion tests in deionized water or lithiated water with 0. 01 M LiOH at 360 ℃/18. 6 MPa. The microstructural evolution of the oxide was analyzed by comparing the oxide charateristics at different depths in the oxide film. The relationship between the microstructural evolution of oxide films and the corrosion resistance of zirconium alloys affected by water chemistry and composition can be better understood.

(1) In the bottom layer of the oxide films, in the vicinity of the metal matrix, the

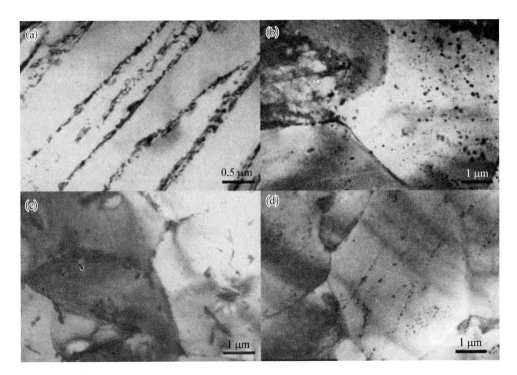

Fig. 19 TEM micrographs of Alloy No. 3 specimens fabricated with different thermo-mechanical processing: (a) 1 000 ℃/0. 5 h annealing and 560 ℃/10 h aging; (b) 1 000 ℃/0. 5 h annealing, 50% cold rolling, and 560 ℃/10 h aging; (c) 750 ℃/0. 5 h annealing and 560 ℃/10 h aging; (d) 750 ℃/0. 5 h annealing, 50% cold rolling, and 560 ℃/10 h aging

defects consisting of vacancies and interstitials, such as points, lines, planes, and volumes, are produced during the oxide growth due to the existence of compressive stress. In addition to the stable monoclinic phase, tetragonal, cubic, and amorphous phases can also be detected. This is a relatively compact layer in the oxide films. The characteristic of the oxide film with such microstructure is the internal cause for the microstructural evolution during the corrosion process.

(2) The diffusion, annihilation, and condensation of vacancies and interstitials under the action of stress, temperature, and time caused stress relaxation in the oxide films and phase transformation of metastable phases into stable phase. Temperature and time are the external causes for the microstructural evolution during the corrosion process.

(3) When analyzing the high resolution lattice images in the bottom layer of the oxide films, it is confirmed that the coherent relationships among monoclinic, tetragonal, and cubic structure with $(001)m \parallel (110)t$, $(001)m \parallel (200)c$, and $(200)c \parallel (110)t$ can be identified.

(4) The vacancies were absorbed by grain boundaries to form the nuclei of pores. The pores grew through the absorption of the vacancies in the grains to weaken the bonding strength between grains. Under the action of surface tension, the grains gradually became spherical. The lattice images became better ordered because the defects diffused out of the grains. The pores formed under the compressive stress in the oxide, and lined up along the

direction parallel to the compressive stress, not perpendicular to it. Consequently, the cracks developed from the pores were parallel to the oxide/metal interface.

(5) Columnar grains constructed by large angle grain boundaries can be observed on sections of the oxide films by SPM after polishing and ion sputtering etching. Equiaxed grains constructed by relatively small angle grain boundaries are, however, also observed within the columnar grains.

(6) Li^+ and OH^- ions incorporated in the oxide films will be adsorbed on the walls of pores or will enter into vacancies to reduce the surface free energy of zirconium oxide when the specimens are corroded in lithiated water. As a result, the diffusion of vacancies and the formation of pores are enhanced, thereby inducing the degradation of corrosion resistance. Through analysis of the undulation of grain morphology on specimen surface, it is concluded that the surface free energy of oxide formed on Zircaloy – 4 specimens corroded in lithiated water is lower than that of oxide formed on Alloy No. 3. As a consequence, the lower corrosion resistance of Zircaloy – 4 as compared to Alloy No. 3 in lithiated water can be understood.

Acknowledgments

The authors wish to express their thanks to the National Natural Science Foundation of China for supporting the research work (50171039, 50371052), and also to Northwest Institute for Nonferrous Metal Research in China for preparing the Alloy No. 3 ingot.

References

[1] Yilmazbayhan, A., Breval, E., Motta, A. T., and Comstock, R. J., "Transmission Electron Microscopy Examination of Oxide Layer Formed on Zr Alloys", *J. Nucl. Mater.*, Vol. 349, 2006, pp. 265 – 281.

[2] Anada, H., and Takeda, K., "Microstructure of Oxides on Zircaloy – 4, 1. 0Nb Zircaloy – 4, and Zircaloy – 2 Formed in 10. 3 – MPa Steam at 673K", *Zirconium in the Nuclear Industry: Eleventh International Symposium*, ASTM STP 1295, E. R. Bradley and G. P. Sabol, Eds., ASTM International, West Conshohocken, PA., 1996, pp. 35 – 54.

[3] Garzarolli, F., Seidel, H., Tricot, R., and Gros, J. P., "Oxide Growth Mechanism on Zirconium Alloys", *Zirconium in the Nuclear Industry: Ninth International Symposium*, ASTM STP 1132, C. M. Eucken and A. M. Garde, Eds., ASTM International, West Conshohocken, PA, 1991, pp. 395 – 415.

[4] Pêcheur, D., Goldlewski, J., Billot, P., and Thomazet, J., "Microstructure of Oxide Films Formed During the Waterside Corrosion of the Zircaloy – 4 Cladding in Lithiated Environment", *Zirconium in the Nuclear Industry: Eleventh International Symposium*, ASTM STP 1295, E. R. Bradley and G. P. Sabol, Eds., ASTM International, West Conshohocken, PA, 1996, pp. 94 – 113.

[5] Wadman, B., Lai, Z., Andren, H. -O., Nystrom, A. -L., Rudling, P., and Pettersson, H., "Microstructure of Oxide Layers Formed During Autoclave Testing of Zirconium Alloys", *Zirconium in the Nuclear Industry: Tenth International Symposium*, ASTM STP 1245, A. M. Garde and E. R. Bradley, Eds., ASTM International, West Conshohocken, PA, 1994, pp. 579 – 598.

[6] Beie, H. -J., Mitwalsky, A., Garzarolli, F., Ruhmann, H., and Sell, H. -J., "Examinations of the Corrosion Mechanism of Zirconium Alloys", *Zirconium in the Nuclear Industry: Tenth International Symposium*, ASTM STP 1245, A. M. Garde and E. R. Bradley, Eds., ASTM International, West Conshohocken, PA, 1994, pp. 615 – 643.

[7] Oskarsson, M. , Ahlberg, E. , Andersson, U. , and Pettersson, K. , "Characterization of Pre-Transition Oxides on Zircaloys", *J. Nucl. Mater.* , Vol. 297, 2001, pp. 77 – 88.

[8] Kim, H. G. , Jeong, Y. H. , and Kim, T. H. , "Effect of Isothermal Annealing on the Corrosion Behavior of Zr-xNb Alloys", *J. Nucl. Mater.* , Vol. 326, 2004, pp. 125 – 131.

[9] Zhou, B. -X. , "Electron Microscopy Study of Oxide Films Formed on Zircaloy – 2 in Superheated Steam", *Zirconium in the Nuclear Industry: Eighth International Symposium*, ASTM STP 1023, L. F. P. Van Swam, and C. M. Eucken Eds. , ASTM International, West Conshohocken, PA, 1989, pp. 360 – 373.

[10] Zhou, B. -X. , Li, Q. , Huang, Q. , Miao, Z. , Zhao, W. -J. , and Li, C. , "The Effect of Water Chemistry on the Corrosion Behavior of Zirconium Alloys", *Nuclear Power Engineering*, Vol. 21, 2000, pp. 439 – 447 (in Chinese).

[11] Cox, B. , "Some Thoughts on the Mechanisms of In-Reactor Corrosion of Zirconium Alloys," *J. Nucl. Mater.* , Vol. 336, 2005, pp. 331 – 368.

[12] IAEA – TECDOC – 996, "Waterside Corrosion of Zirconium Alloys in Nuclear Power Plants", IAEA, Vienna, 1998, ISSN 1011 – 4289, Jan. 1998.

[13] Zhou, B. -X. , Li, Q. , Yao, M. -Y. , and Liu, W. -Q. , "Microstructure Evolution of Oxide Films Formed on Zircaloy – 4 during Autoclave Tests", *Nuclear Power Engineering*, Vol. 26, 2005, pp. 364 – 371 (in Chinese).

[14] Zhou, B. -X. , Li, Q. , Liu, W. -Q. , Yao, M. -Y. , and Zhu, Y. -L. , "The Effect of Water Chemistry and Composition on the Microstructure Evolution of Oxide Films on Zirconium Alloys", *Rare Metal Materials and Engineering*, Vol. 35, 2006, pp. 1009 – 1016. (in Chinese).

[15] Cox, B. , "A Porosimeter for Determining the Size of Flaws in Zirconia or Other Insulating Films", *J. Nucl. Mater.* , Vol. 27, 1968, pp. 1 – 11.

[16] Cox, B. , "Processes Occurring During the Breakdown of Oxide Films on Zirconium Alloys", *J. Nucl. Mater.* , Vol. 29, 1969, pp. 50 – 66.

[17] Zhou, B. -X. , Li, Q. , Yao, M. -Y. et al. , "The Grain Morphology of Oxide Films on Zircaloy – 4", *Rare Metal Materials and Engineering*, Vol. 32, 2003, pp. 417 – 419 (in Chinese).

[18] Bryner, J. S. , "The Cyclic Nature of Corrosion of Zircaloy – 4 in 633K, Water", *J. Nucl. Mater.* , Vol. 82, 1979, pp. 84 – 101.

[19] Sabol, G. P. , Costock, R. J. , Weiner, R. A. , Larouere, P. , and Stanutz, R. N. , "In-Reactor Corrosion Performance of ZIRLO and Zircaloy – 4", *Zirconium in the Nuclear Industry: Tenth International Symposium*, ASTM STP 1245, A. M. Garde and E. R. Bradley, Eds. , ASTM International, West Conshohocken, PA, 1994, pp. 724 – 744.

[20] Zhou, B. -X. , and Jiang, Y. -R. , "Oxidation of Zircaloy – 2 in Air from 500 ℃ to 800 ℃, High Temperature Corrosion and Protection", *Proceedings of the International Symposium on High Temperature Corrosion and Protection*, 26 – 30 June 1990, Shenyang, China, H. -R. Guan, W. -T. Wu, J. -N. Shen, and T. -F. Li, Eds. , Liaoning Science and Technology Publishing House, China, 1991, pp. 121 – 124.

[21] Pêcheur, D. , Godlewski, J. , Peybernes, J. , Faytte, L. , Noe, M. , Frichet, A. , and Kerrec, O. , "Contribution to the Understanding of the Effect of the Water Chemistry on the Oxidation Kinetics of Zircaloy – 4 Cladding", *Zirconium in the Nuclear Industry: Twelfth International Symposium*, ASTM STP 1354, G. P. Sabol and G. D. Moan, Eds. , ASTM International, West Conshohocken, PA, 2000, pp. 793 – 811.

[22] Ramasubramania, N. , Precoanin, N. , and Ling, V. C. , "Lithium Uptake and the Accelerated Corrosion of Zirconium Alloys", *Zirconium in the Nuclear Industry: Eighth International Symposium*, ASTM STP 1023, L. F. P. Van Swam and C. M. Eucken, Eds. , ASTM International, West Conshohocken, PA, 1989, pp. 187 – 201.

[23] Jeong, Y. H. , Kim, K. H. , and Baek, J. H. , "Cation Incorporation into Zirconium Oxide in LiOH, NaOH, and KOH Solutions", *J. Nucl. Mater.* , Vol. 275, 1999, pp. 221 – 224.

[24] Oskarsson, M. , Ahlberg, E. , and Pettersson, K. , "Oxidation of Zircaloy – 2 and Zircaloy – 4 in Water and Lithiated Water at 360 ℃", *J. Nucl. Mater.* , Vol. 295, 2001, pp. 97 – 108.

[25] Zhou, B. -X. , Liu, W. -Q. , Li, Q. , and Yao, M. -Y. , "Mechanism of LiOH Aqueous Solution Accelerating Corrosion Rate of Zircaloy – 4", *Chinese Journal of Materials Research*, Vol. 18, 2004, pp. 225 – 231 (in Chinese).

[26] Yilmazbayhan, A., Motta, A. T., Comstock, R. J., Sabol, G. P., Laiq, B., and Cai, Z. H., "Structure of Zirconium Alloy Oxides Formed in Prre Water Studied with Synchrotron Radiation and Optical Microscopy: Relation to Corrosion Rate", *J. Nucl. Mater.*, Vol. 324, 2004, pp. 6 - 22.

[27] Schefold, J., Lincot, D., Ambard, A., and Kerrec, O., "The Cyclic Nature of Corrosion of Zr and Zr - Sn in High-Temperature Water (633K), A Long-Term in situ Impedance Spectroscopic Study", *J. Electrochem. Soc.*, Vol. 150, 2003, pp. B451 - B461.

[28] Yao, M. -Y., Zhou, B. -X., Li, Q., Chu, Y. -L., Liu, W. -Q., and Lu, Y. -P., "A Superior Corrosion Behavior of Zircaloy - 4 in Lithiated Water at 360 ℃/18. 6 MPa by β - Quenching", *J. Nucl. Mater.* (to be published).

[29] Liu, W. -Q., Li, Q., Zhou, B. -X., and Yao, M. -Y., "The Effect of Microstructure on the Corrosion Behavior for ZIRLO Alloy", *Nuclear Power Engineering*, Vol. 24, 2003, pp. 33 - 36. (in Chinese).

[30] Li, Q., Liu, W. -Q., and Zhou, B. -X., "Effect of Deformation and Heat Treatment on the Decomposition of β - Zr in Zr - Sn - Nb Alloys", *Rare Metal Materials and Engineering*, Vol. 31, 2002, pp. 389 - 392 (in Chinese).

[31] Comstock, R. J., Schoenberger, G., and Sabol, G. P., "Influence of Processing Variables and Alloy Chemistry on the Corrosion Behavior of ZIRLO Nuclear Fuel Cladding", *Zirconium in the Nuclear Industry: Eleventh International Symposium*, ASTM STP 1295, E. R. Bradley and G. P. Sabol, Eds., ASTM International, West Conshohocken, PA, 1996, pp. 710 - 725.

添加合金元素 Cu 和 Mn 对
锆合金中第二相的影响*

摘　要：以 Zr－4 为母合金，分别添加电解纯铜或电解金属锰，用非自耗真空电弧炉熔炼了成分不同的 6 种锆合金．用透射电子显微镜观察了合金中第二相的形貌，用 EDS 分析了第二相的成分，用 SAD 确定了第二相的晶体结构．添加 Cu 元素的合金中有 3 种第二相：$Zr(Fe, Cr)_2$ 粒子、$Zr(Fe, Cr, Cu)_2$ 粒子和含少量 Fe 或不含 Fe 的 Zr_2Cu 粒子；添加 Mn 元素的合金中只有 1 种 $Zr(Fe, Cr, Mn)_2$ 第二相，且随着合金中 Mn 的质量分数从 0.07％增加到 0.35％，$Zr(Fe, Cr, Mn)_2$ 粒子中 Mn 元素的质量分数也升高．

提高核燃料的燃耗，延长燃料组件的换料周期后，对燃料包壳材料锆合金的性能提出了更高的要求，其中改善包壳水侧的耐腐蚀性能又是提高锆合金性能的主要方面．锆合金的耐腐蚀性能受到合金成分、第二相和热处理制度等因素的影响．通过添加合金元素改变和控制显微组织、第二相的种类和氧化膜的显微结构来改善锆合金的耐腐蚀性能，这成为许多科研工作者的研究思路．第二相的大小和分布对合金的力学性能和腐蚀行为都具有重要影响．因此，对在锆合金中添加了不同合金元素后形成的第二相的化学成分、晶体结构、尺寸分布进行全面的了解是十分必要的．

Jeong 等研究出了一种新锆合金，称之为 HANA 系列，其中尤以 HANA－6 合金(Zr－1.1Nb－0.05Cu)的耐腐蚀性能最为优越，可见在 Zr－Nb 合金中添加微量的 Cu 元素，使合金的耐腐蚀性能得到明显提高[1]．这说明 Cu 也是锆合金中一个可添加的重要合金元素．

大部分合金元素在 α－Zr 中的固溶度都比较低，而固溶度较大的 Sn 和 Nb 在添加量合适时能显著提高锆合金的耐腐蚀性能[2-5]．姚美意等[6]研究发现，提高固溶在 α－Zr 基体中的合金元素 Fe 和 Cr 的含量后，能显著提高 Zr－4 合金在 LiOH 水溶液中的耐腐蚀性能．Mn 元素在 α－Zr 中的固溶度也较大，它对锆合金显微组织和耐腐蚀性能的影响在国内外却罕见报道．本工作主要研究在锆合金中添加 Cu 和 Mn 后生成的第二相的种类、结构，为下一步评价合金元素 Cu 和 Mn 对锆合金耐腐蚀性能的影响做准备．

1　实验方法

1.1　合金样品制备

本实验以 Zr－4 为母合金，分别添加电解纯铜或电解金属锰，用非自耗真空电弧炉熔炼成分不同的 6 种锆合金，经过 ICP－AES(电感耦合等离子体原子发射光谱)分析后得到的合金成分列于表 1 中．

---

* 本文合作者：曾奇峰、姚美意、彭剑超、夏爽．原发表于《上海大学学报(自然科学版)》，2008，14(5)：531－536．

表 1 锆合金的化学成分(质量分数)

Table 1 Composition of zirconium alloys (in mass fraction) %

样品编号	Sn	Fe	Cr	Mn	Cu	Zr
B	1.45	0.20	0.10	—	0.046	余量
C	1.40	0.23	0.11	—	0.091	余量
D	1.34	0.22	0.10	—	0.180	余量
E	1.40	0.24	0.10	0.07	—	余量
F	1.39	0.22	0.10	0.14	—	余量
H	1.37	0.23	0.11	0.35	—	余量

所有合金均是经过非自耗真空电弧熔炼炉反复熔炼 5 次,每熔炼 1 次,翻转 1 次,以确保成分均匀.所得的合金锭重约 60 g,将其在 700 ℃预热 0.5 h,然后放在压机(吨位为 15～20 t)上采用三套高度、宽度不同的自制模具中反复热压,得到厚度约 10 mm、宽度约 20 mm 的规整坯料,便于后续的处理.坯料经 700 ℃多次加热和热轧,得到厚度约 3 mm 的板坯,最后经过室温轻微轧制变形,使表面的氧化膜破碎,便于酸洗时去除氧化膜.板坯经过酸洗(酸洗液中 HNO_3,H_2O,HF 体积分数分别为 45%,45%,10%)、自来水冲洗和干燥,在真空中加热到 1 020 ℃进行 1 h 的 β 相均匀化处理,目的是使合金元素充分固溶.处理结束后将炉子推离石英管浇水冷却(本文称之为空冷),再进行多次冷轧和中间退火,确保合金的组织比较均匀,中间退火的参数为 700 ℃/2 h.最终冷轧成 0.65 mm 片材,再进行 600 ℃/3 h 再结晶退火处理,以得到完全再结晶的组织.本工作的加工工艺依据文献[7 - 10].

1.2 透射电镜样品准备

制备透射电镜(TEM)薄试样时,先将样品用化学腐蚀方法减薄到 50 μm 左右,用专用的剪刀剪出 ϕ3 mm 小圆,这种制样方法避免了机械减薄带来的变形和应力.将样品用酒精(90%)和高氯酸(10%)的混合溶液双喷电解抛光制备 TEM 观察用的薄样品,直流电压为 40 V,温度为 -40 ℃.观察所用的透射电镜为 JEM - 200CX 和 JEM - 2010F.

1.3 XRD 样品准备

用真空非自耗电弧熔炼炉熔炼了与 Zr_2Cu 第二相成分相同的合金,经砂纸磨平表面进行 XRD 分析,分析结果用来与 TEM 观察时获得第二相的选区电子衍射(SAD)结果对比.

2 结果和讨论

2.1 显微组织

从图 1 中可以看出,三种冷轧片材在 600 ℃的真空下保温 3 h,均得到等轴的晶粒组织,沉淀相粒子弥散分布在基体晶粒的晶内和晶界上.Zr - 4+0.18% Cu 合金中第二相数目较少,外观上看存在两种类型的第二相:一种较粗大,尺寸大约在 0.25～0.50 μm 之间;另一种较细小,尺寸大都在 0.1 μm 左右.而 Zr - 4+0.14% Mn 合金中第二相数目稍多,尺寸细小的大多为椭圆形,尺寸较大的有些形状比较独特,呈棒状或多边形.Zr - 4+0.35% Mn 合金中第二相数目最多,形状大多为椭圆形或长棒状.

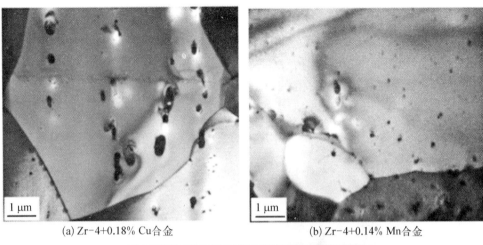

(a) Zr-4+0.18% Cu合金   (b) Zr-4+0.14% Mn合金

(c) Zr-4+0.35% Mn合金

图 1 几种锆合金的显微组织

Fig. 1 TEM micrographs of the zirconium alloys

2.2 XRD 分析结果

图 2 的 XRD 分析结果表明,冶炼的 Zr_2Cu 合金为四方结构,$a=b=0.322\ 0$ nm,$c=1.118\ 3$ nm,与 PDF 卡片 65-7783 一致.

2.3 锆合金中典型的第二相

2.3.1 Zr-4+0.18% Cu 合金中的第二相

针对 Zr-4+0.18% Cu 合金中不同形貌的第二相各 5 个进行能谱分析(EDS)和 SAD 分析,结果表明,合金中存在 3 种第二相:Zr$(Fe,Cr)_2$ 粒子、$Zr(Fe,Cr,Cu)_2$ 粒子和含少量 Fe 或不含 Fe 的 Zr_2Cu 粒子. $Zr(Fe,Cr)_2$ 粒子通常有 C14 六方和 C15 立方两种结构,本

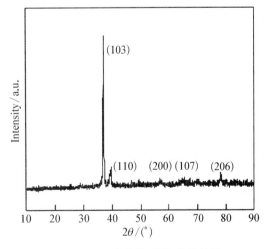

图 2 Zr_2Cu 合金的 XRD 分析结果

Fig. 2 XRD pattern of Zr_2Cu alloy

研究中只观察到 C14 六方结构. Zr(Fe, Cr, Cu)$_2$ 粒子也为六方结构. 图 3 的 SAD 分析表明, Zr$_2$Cu 粒子为四方结构, 这与 XRD 分析得到的结果一致. EDS 分析得到的结果并不完全符合 Zr$_2$Cu 的 Zr/Cu 比, 这是由于微区分析时的电子束有一定穿透深度和扩散范围, 第二相又十分微小, 因此分析时会受到周围 Zr 基体的影响. Hong 等[11]在 Zr-4 的基础上添加 Cu 元素后观察到合金中存在 3 种第二相: 尺寸较小的 Zr(Fe, Cr)$_2$ 粒子、Zr(Fe, Cr, Cu)$_2$ 粒子和尺寸较大的 Zr$_2$Cu 粒子. 这和我们本次的研究结果一致.

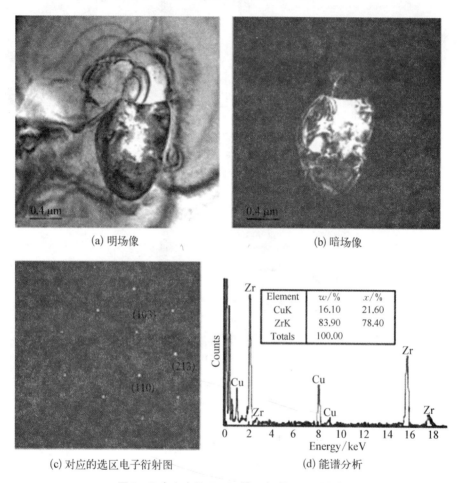

(a) 明场像　　　　　　　　　　(b) 暗场像

(c) 对应的选区电子衍射图　　　　(d) 能谱分析

Element	w/%	x/%
CuK	16.10	21.60
ZrK	83.90	78.40
Totals	100.00	

图 3　D 合金中的 Zr$_2$Cu 第二相的 TEM 图片

Fig. 3　TEM micrographs of Zr$_2$Cu precipitate in D alloy

2.3.2　Zr-4+0.14% Mn 合金中的第二相

针对 Zr-4+0.14% Mn 合金中不同形貌的第二相各 5 个进行 EDS 分析和 SAD 分析, 结果表明合金中只存在一种 Zr(Fe, Cr, Mn)$_2$ 第二相, 没有发现 Zr 和 Mn 元素单独形成的 ZrMn$_2$ 第二相, 且第二相中的 Mn 质量分数均在 10% 左右, 第二相中 Fe/Cr 比接近合金中的 Fe/Cr 比. Mn 元素除了固溶在 α-Zr 基体中, 其余均进入原来 Zr-4 合金中的 Zr(Fe, Cr)$_2$ 粒子中, 并改变了其大小和形状.

TEM 观察发现这些 Zr(Fe, Cr, Mn)$_2$ 第二相粒子的形状比较独特, 为长棒状或多边形, 并且内部含有一些层错. 图 4 的 SAD 分析确定其为面心立方结构, 与在 Zr-4+0.18% Cu 合金中观察到 Zr(Fe, Cr)$_2$ 粒子为六方结构的结果不同. 由此推断 Mn 元素扩散进入

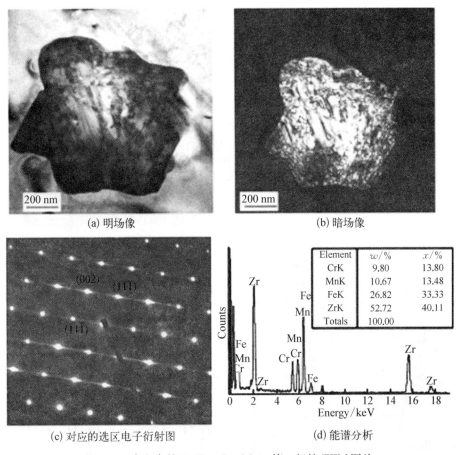

(a) 明场像　　　　　　　　　(b) 暗场像

Element	$w/\%$	$x/\%$
CrK	9.80	13.80
MnK	10.67	13.48
FeK	26.82	33.33
ZrK	52.72	40.11
Totals	100.00	

(c) 对应的选区电子衍射图　　　　　(d) 能谱分析

图 4　F 合金中的 $Zr(Fe, Cr, Mn)_2$ 第二相的 TEM 图片

Fig. 4　TEM micrographs of $Zr(Fe, Cr, Mn)_2$ precipitate in F alloy

$Zr(Fe, Cr)_2$ 粒子中,稳定了 C15 立方型晶体结构,并导致其界面自由能降低,从而使其形状发生比较大的变化. 因为相对于规则的圆形或椭圆形第二相粒子来说,不规则的长棒状或多边形第二相粒子单位体积的界面面积要大,所以一般情况下不易形成,而 Mn 元素进入 $Zr(Fe, Cr)_2$ 粒子后,出现了不规则形状的 $Zr(Fe, Cr, Mn)_2$ 第二相粒子,由此推断是 Mn 元素的进入降低了 $Zr(Fe, Cr, Mn)_2$ 粒子本身的界面自由能.

2.3.3　Zr - 4+0.35% Mn 合金中的第二相

对 Zr - 4+0.35% Mn 合金中不同形貌的第二相进行 EDS 和 SAD 分析,所得结果和 Zr - 4+0.14% Mn 合金中的相似. 只是随着合金中 Mn 元素的增加,$Zr(Fe, Cr, Mn)_2$ 第二相中的 Mn 含量也相应增加,大多在 20% 左右. 它们的形状大多为不规则椭圆形或棒状. 图 5 的 SAD 分析表明,$Zr(Fe, Cr, Mn)_2$ 第二相为面心立方结构,点阵常数也与 F 合金中的一致.

从目前的实验结果来看,在 Zr - 4 合金中添加 Cu 和 Mn 元素所形成的第二相类型不同. 这可能与它们在 α - Zr 中的固溶度不同和扩散速度不同有关. Cu 在 α - Zr 中的固溶度较低且扩散速度较快,加入的 Cu 元素大部分均以第二相的形式析出,因而形成含 Cu 元素较多的 Zr_2Cu 第二相;而 Mn 在 α - Zr 中的溶解度相对较大,加入的 Mn 元素大部分均以固溶的形式存在,而没有形成单独的 $ZrMn_2$ 第二相. 合金中第二相的类型、大小和分布对锆合金

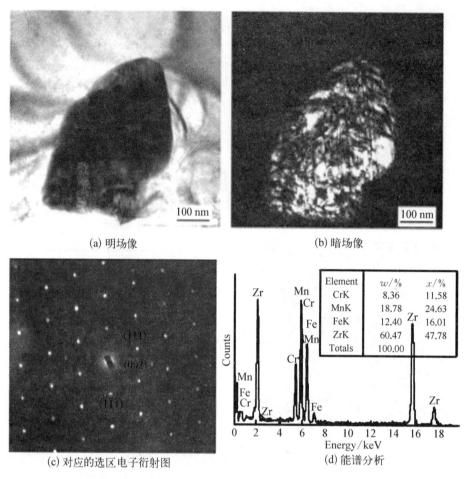

(a) 明场像　　　　　　　　　(b) 暗场像

(c) 对应的选区电子衍射图　　　　(d) 能谱分析

Element	w/%	x/%
CrK	8.36	11.58
MnK	18.78	24.63
FeK	12.40	16.01
ZrK	60.47	47.78
Totals	100.00	

图 5 H 合金中的 $Zr(Fe，Cr，Mn)_2$ 第二相的 TEM 图片

Fig. 5 TEM micrographs of $Zr(Fe, Cr, Mn)_2$ precipitate in H alloy

的耐腐蚀性能起着重要的作用,如形成细小、弥散分布的 β‒Nb 粒子能够明显改善含 Nb 锆合金的耐腐蚀性能[12].而形成这样的第二相,又需要经过正确加工和热处理.第二相的类型对锆合金腐蚀时的吸氢性能也有很大的影响.众所周知,Zr‒2 合金中的 $Zr_2(Fe，Ni)$ 和 Zr‒4 合金中的 $Zr(Fe，Cr)_2$ 本身是一种强烈吸氢的金属间化合物[13],这说明如果添加的合金元素与 Zr 形成的第二相是一种比 Zr 吸氢能力更强的物质,那么含这些强烈吸氢第二相尺寸较大、数量较多的合金吸氢必然多,从减少腐蚀时吸氢的角度来考虑,这种合金元素要尽量少加[14].因此,添加合金元素时必须综合考虑各种因素的影响,并经过正确的加工和热处理,以得到合适的第二相类型、大小和分布.

3　结论

以 Zr‒4 为母合金,添加 Cu 元素形成了 Zr_2Cu 第二相,同时 Cu 元素进入原来 Zr‒4 合金中的部分 $Zr(Fe，Cr)_2$ 粒子中而形成 $Zr(Fe，Cr，Cu)_2$ 粒子.添加 Mn 元素并不形成单独的 $ZrMn_2$ 第二相,而是 Mn 进入原来 Zr‒4 合金中的 $Zr(Fe，Cr)_2$ 粒子中,形成 $Zr(Fe，Cr，$

Mn)₂ 粒子,并稳定了 C15 立方型晶体结构.并且随着合金中 Mn 含量的增加,Zr(Fe,Cr,Mn)₂ 第二相中的 Mn 含量也相应增加.

参 考 文 献

[1] PARK J Y, CHOI B K, SEUNG J Y, et al. Corrosion behavior and oxide properties of Zr – 1. 1 wt% Nb – 0. 05 wt% Cu alloy[J]. Journal of Nuclear Materials, 2006, 359(1/2): 59 – 68.

[2] ETOH Y, SHIMADE S, YASUDA T, et al. Development of new zirconium alloys for a BWR[C]//Zirconium in the Nuclear Industry: Eleventh International Symposium, ASTM STP 1295. Philadelphia: America Society for Testing and Materials, 1996: 825 – 849.

[3] MURAI T, ISOBE T, TAKIZAWA Y, et al. Fundamental study on the corrosion mechanism of Zr – 0. 2Fe, Zr – 0. 2Cr, Zr – 0. 1Fe – 0. 1Cr alloys[C]//Zirconium in the Nuclear Industry: Twelfth International Symposium, ASTM STP 1354. West Conshohocken: America Society for Testing and Materials, 2000: 623 – 640.

[4] SABOL G P. ZIRLO™— An alloy development success[C]//Zirconium in the Nuclear Industry: Fourteenth International Symposium, ASTM STP 1467. Stockholm: America Society for Testing and Materials, 2004: 3 – 24.

[5] YUEH H K, KESTERSON R L, COMSTOCK R, et al. Improved ZIRLO™ cladding performance through chemistry and process modifications[C]//Zirconium in the Nuclear Industry: Fourteenth International Symposium, ASTM STP 1467. Stockholm: America Society for Testing and Materials, 2004: 330 – 346.

[6] YAO M Y, ZHOU B X, LI Q, et al. A superior corrosion behavior of Zircaloy – 4 in lithiated water at 360 ℃/ 18. 6 MPa by β – quenching[J]. Journal of Nuclear Materials, 2008, 374: 197 – 203.

[7] KIM J M, JEONG Y H, JUNG Y H. Correlation of heat treatment and corrosion behavior of Zr – Nb – Sn – Fe – Cu alloys[J]. Journal of Nuclear Materials, 2000, 104: 145 – 149.

[8] BAEK J H, JEONG Y H, KIM I S. Effects of the accumulated annealing parameter on the corrosion characteristics of a Zr – 0. 5Nb – 1. 0Sn – 0. 5Fe – 0. 25Cr alloy[J]. Journal of Nuclear Materials, 2000, 280: 235 – 245.

[9] LIU W Q, LI Q, ZHOU B X, et al. Effect of heat treatment on the microstructure and corrosion resistance of a Zr – Sn – Nb – Fe – Cr alloy[J]. Journal of Nuclear Materials, 2005, 341(2/3): 97 – 102.

[10] KIM J M, JEONG Y H. Influence of thermomechanical treatment on the corrosion behavior of Zr – 1Nb – 0. 2Cu alloys[J]. Journal of Nuclear Materials, 1999, 275: 74 – 80.

[11] HONG H S, MOON J S, KIM S J, et al. Investigation on the oxidation characteristics of copper-added modified Zircaloy – 4 alloys in pressurized water at 360 ℃[J]. Journal of Nuclear Materials, 2001, 297: 113 – 119.

[12] KIM H G, PARK J Y, JEONG Y H. Ex-reactor corrosion and oxide characteristics of Zr – Nb – Fe alloys with the Nb/Fe ratio[J]. Journal of Nuclear Materials, 2005, 345(1): 1 – 10.

[13] SHALTIEL D, JACOB I, DAVIDOV D. Hydrogen absorption and desorption properties of AB₂ laves-phase pseudobinary compounds[J]. J Less-Commom Metals, 1977, 53: 117 – 131.

[14] 姚美意. 合金成分及热处理对锆合金腐蚀和吸氢行为影响的研究[D]. 上海: 上海大学, 2007: 127.

Effect of Alloying Element (Cu and Mn)
Additions on the Precipitates of Zr Alloys

Abstract: A series of zirconium alloys with copper and manganese additions in Zircaloy – 4 were prepared using a vacuum non-consumable arc melting method. TEM observation was carried out to characterize precipitates in alloys. The crystal structure was determined by a selected area electron diffraction (SAD) pattern, and the composition of precipitates was analyzed with energy dispersion spectroscopy (EDS). In the copper-added Zircaloy – 4 specimens, three main precipitates were detected: (1) Zr(Fe, Cr)₂ precipitates, (2) Zr(Fe, Cr, Cu)₂ precipitates, and (3) Zr₂Cu precipitates with a bit of Fe or without Fe. In the manganese-added Zircaloy – 4 specimens, only one kind of Zr(Fe, Cr, Mn)₂ precipitate was observed. Mn content in the precipitates increased as the increase of Mn content from 0. 07% to 0. 35% in the alloy.

Zr－2.5Nb 合金在 550℃/25 MPa 超临界水中腐蚀时的氢致 α/β 相变[*]

摘 要：Zr－2.5Nb 合金经 β 相水淬及冷轧变形后，再经过 580℃/5 h 和 650℃/2 h 的热处理，在静态高压釜中进行 550℃/25 MPa 超临界水腐蚀实验. 用电子显微镜研究了腐蚀前、后合金基体的显微组织. 结果表明在 550℃/25 MPa 超临界水中腐蚀到一定程度后，合金基体内会形成氢稳定的 β－Zr 相，同时合金元素 Nb 扩散进入该相，形成富 Nb/H 的 β－Zr 相. 该相在降温过程中发生分解，形成 ZrH_x、α－Zr 和 Nb 含量不同的 Zr－Nb 相组织.

随着核电技术的发展，在 2001 年底国际上提出了第 4 代核反应堆（Generation IV Nuclear Energy Systems）的设计概念[1]. 在确定的 6 种反应堆堆型中，超临界水冷反应堆（SCWR）是其中之一. 在 SCWR 堆芯的高温（超过 500℃）高压（超过 25 MPa）超临界水及中子辐照等恶劣的工况条件下，反应堆堆芯中的结构材料，特别是燃料元件包壳材料的选材和研究开发是一项具有挑战性的工作[2, 3]. 由于锆的热中子吸收截面非常小，所以锆合金仍然作为 SCWR 燃料包壳的候选材料被予以关注[4]. 目前国际上只有 A. T. Motta 和 Y. H. Jeong 领导的国际合作项目组及本研究小组从事锆合金在超临界水中的腐蚀研究工作[5, 6]. 研究锆合金在超临界水中的腐蚀行为，既可以为 SCWR 工程应用及开发高性能锆合金提供实验及理论依据，也可以促进当前核反应堆用锆合金的研究与发展. 作者对多种不同成分锆合金在超临界水中的腐蚀行为已进行过研究[7]，在这基础上，本工作通过研究 Zr－2.5Nb 合金在 550℃/25 MPa 超临界水（SCW）中腐蚀时合金基体的显微组织变化，以期揭示锆合金在 550℃高温水中腐蚀时发生氢（H）致 α/β 相变的现象与过程.

1 实验方法

1.1 实验样品及热处理

试验用 Zr－2.5Nb 样品来自西北有色金属研究院. 将尺寸约为 8 mm×150 mm，厚 1.2 mm 的锆合金片状样品，经酸洗（10% HF+30% HNO₃+30% H₂SO₄+30% H₂O 体积比的混合酸）与水洗后，真空封装在石英管内，加热到 1 020℃保温 20 min 后淬入水中，同时敲碎石英管使样品迅速冷却，称之为 β 相水淬处理（表示为 Q）. 将 β 相水淬处理后的样品冷轧（表示为 R）至 0.6 mm 厚，切成 20 mm×9 mm 的小片，经酸洗及水洗后，分组放置在真空（<5×10⁻³ Pa）石英管式炉中，分别进行 580℃/5 h 和 650℃/2 h 的热处理，降温过程采取将加热炉推离石英管，用水浇淋石英管进行冷却的方式.

---

[*] 本文合作者：李强、姚美意、刘文庆、褚于良. 原发表于《稀有金属材料与工程》，2008,37(10)：1815－1818.

1.2 腐蚀实验

将热处理后的样品经过标准方法酸洗和去离子水洗,在静态高压釜中进行 550 ℃/25 MPa 超临界水腐蚀试验,定期停釜降温,将样品取出称重,腐蚀增重为 4 个试样的平均值.

1.3 合金基体的显微组织观察

腐蚀前、后的试样经机械磨抛后,用 10% HF+30% HNO_3+30% H_2SO_4+30% H_2O 体积比的混合酸蚀刻,水洗后用洗耳球吹干表面,用 JSM-6700F 扫描电镜观察形貌. 机械磨抛后的样品用酸洗的方法减薄至大约 0.08 mm 后,冲出 φ3 mm 的圆片,用双喷电解抛光的方法制备 TEM 薄样品,所用电解液为 10% $HClO_4$+90% C_2H_5OH,抛光温度约为 −20 ℃. 用配有 INCA 能谱仪(EDS)的 JEM-2010F 透射电子显微镜观察研究显微组织.

2 结果和讨论

2.1 Zr‑2.5Nb 合金在超临界水中的腐蚀增重

图 1 为 Zr‑2.5Nb 合金样品在 550 ℃/25 MPa 超临界水中腐蚀 800 h 的腐蚀增重曲线. 作为对比,图中也给出了 Zr‑4 合金的腐蚀增重曲线. 文献[7]认为,成分不同的锆合金腐蚀增重存在极大的差异,Zr‑2.5Nb 合金样品的耐腐蚀性能远优于 Zr‑4 合金,也优于 M5 合金. Zr‑2.5Nb 合金样品由于 Nb 含量较高,经过 β 相水淬、冷轧变形和热处理后,可以获得分布均匀、数量较多细小的第二相,这可能是其耐腐蚀性能优于 M5 合金的原因之一. 580 ℃/5 h 处理的 Zr‑2.5Nb 合金样品,前期耐腐蚀性能比经过 650 ℃/2 h 处理的样品好,但腐蚀后期变坏,这与合金基体在腐蚀过程中发生了 α/β

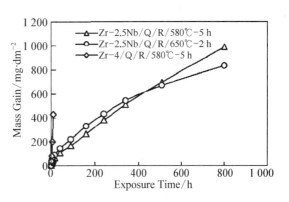

图 1 Zr‑2.5Nb 合金在 550 ℃/25 MPa 超临界水中腐蚀 800 h 的腐蚀增重变化

Fig. 1 Corrosion behaviors of Zr‑2.5Nb alloy in supercritical-water at 550 ℃/25 MPa for 800 h

相变,引起显微组织的变化有关,在后面的章节将对此进行讨论.

2.2 Zr‑2.5Nb 合金样品腐蚀前后合金基体的显微组织及其变化

Zr‑2.5Nb 合金样品经过 β 相水淬和冷轧变形后,分别进行 580 ℃/5 h 和 650 ℃/2 h 热处理的显微组织见图 2. 用 TEM 和 SEM 观察表明,580 ℃/5 h 处理的样品形成等轴晶的 α‑Zr 组织,并且析出大量颗粒细小(<60 nm)、均匀分布的 β‑Nb 第二相(图 2a、2b). 对于 650 ℃/2 h 处理的样品,由于在 $\alpha+\beta$ 双相区加热,因此快速冷却后在晶界上,特别是在三晶交汇处会形成块状的 β‑Zr,在晶粒内除了颗粒状的 β‑Nb 第二相外,还会形成一些棒状 β‑Zr 第二相(图 2c、2d). 在 550 ℃/25 MPa 超临界水中腐蚀实验结果表明(图 1),580 ℃/5 h 热处理的样品有较好的耐腐蚀性能. 上述结果与文献[8]的结果相似,表明无论在 550 ℃/

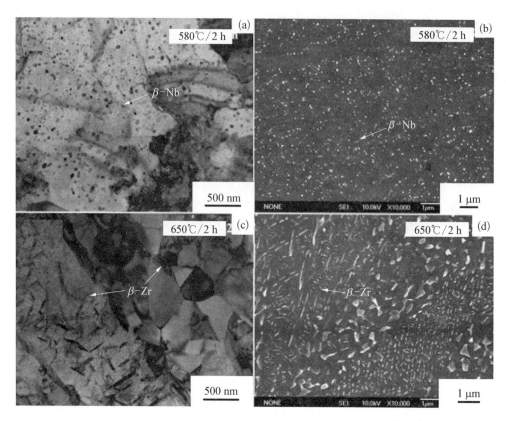

图 2 Zr-2.5Nb 样品经 β 淬火、冷轧变形及热处理后的显微组织

Fig. 2 TEM(a, c) and SEM(b, d) micrographs of Zr-2.5Nb specimens treated at 580 ℃/5 h and 650 ℃/2 h after β quenching and cold rolling

25 MPa 超临界水,或者在 350 ℃/16.8 MPa、0.04 mol/L LiOH 水溶液条件下,含 Nb 锆合金具有大量细小并均匀分布的第二相对改善耐腐蚀性能都是有利的.

图 3、图 4 为 Zr-2.5Nb 合金样品在 550 ℃/25 MPa 超临界水中腐蚀一定时间后,合金基体显微组织的电镜照片. 可以看到与腐蚀前的显微组织相比发生了明显的变化. 650 ℃/2 h 处理的样品经过 16 h 腐蚀后,原来的 Nb 稳定的 β-Zr 第二相在 550 ℃ 腐蚀保温过程中分解成 β-Nb 和 α-Zr(图 3a). 两种条件处理的样品在腐蚀 800 h 后,合金基体显微组织的形貌基本相同(图 3b、图 4a),但是与腐蚀前的显微组织(图 2)相比有很大的差别,在原来的 α-Zr 基体中形成了大量的新相. 这种新相有两种形态,一种是在 α-Zr 晶界处形成的块状相,另一种是在 α-Zr 晶内形成的棒状相. 块状和棒状相的组织中都包含着形态基本相同的并且非常复杂的显微组织,电镜照片中表现为有许多衬度不同的条纹. 选区电子衍射(SAD)和 EDS 分析(图 4b)表明,这些复杂的显微组织包含了 α-Zr、ZrH$_x$ 和 Nb 含量不同的 Zr-Nb 相. 结合 Zr-H 相图进行分析(图 5),可以推断 Zr-2.5Nb 样品在 550 ℃/25 MPa 超临界水中腐蚀时,合金基体发生了如下变化:当锆合金腐蚀时的吸氢量达到一定程度(只要局部区域处超过 6at ％的 H)后,由于样品处于可以发生氢致 β 相变的温度(550 ℃),所以在晶界和晶内部分能量较高的区域形核,发生了 β 相变,形成氢稳定的 β-Zr 相,同时合金元素 Nb 会扩散进入,形成富 Nb/H 的 β-Zr 相;随着吸氢量的增加,该相增多并长大,并在腐蚀实验的降温过程中分解形成了观察到的组织. 580 ℃/5 h 处理的样品在腐蚀后期耐腐蚀

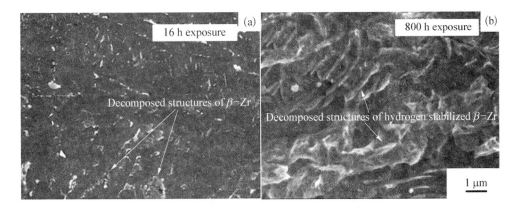

图3 650 ℃/2 h 处理的 Zr–2.5Nb 样品腐蚀 16 h (a)和 800 h (b)后合金基体的 SEM 照片

Fig. 3 SEM micrographs showing the matrix microstructure of Zr–2.5Nb/650 ℃–2 h specimen after corrosion testing for 16 h (a) and 800 h (b)

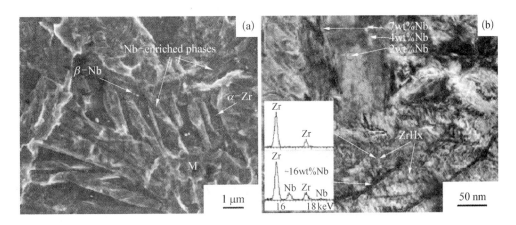

图4 580 ℃/5 h 处理的 Zr–2.5Nb 样品腐蚀 800 h 后的合金基体 SEM(a)和 TEM(b)照片,图 4b 显示的显微组织相当于图 4a 中的 M 区域

Fig. 4 Micrographs of matrix for Zr–2.5Nb/580 ℃/5 h specimen after corrosion testing for 800 h: (a) SEM micrograph, (b) TEM micrograph of the similar region marked as M in Fig. 4a, and the EDS analysis results for different micro areas are also given

性能变差,显然与其发生了这样的相变过程有关,破坏了原来第二相细小并且分布均匀的有利显微组织. 但耐腐蚀性能变得比 650 ℃/2 h 处理的更差,尚难以解释.

在氢致 β 相变及降温时 β–Zr 的分解过程中,合金元素 Nb 的扩散迁移非常活跃. Nb 是稳定 β 相的合金元素,在高温时从 β–Zr 到 Nb 之间可以形成一种完全置换固溶体,所以在 550 ℃腐蚀时形成氢稳定的 β–Zr 后,Nb 元素会向 β–Zr 中扩散富集. 富 Nb/H 的 β–Zr 在降温过程中发生相变,由于 H 的扩散速度较快,并且 H 的固溶度随温度的变化也大,ZrH_x 会先形成,同时形成 α–Zr;Nb 元素在 α–Zr 和 ZrH_x 中固溶度很低,相变时被排斥,在 α–Zr 或 ZrH_x 晶界等处富集,形成富 Nb 的 Zr–Nb 相,在 SEM 二次电子像中显示出白亮的条纹(图 4a),而在 TEM 像中呈现深色条纹(图 4b). 显然,这种相变过程造成显微组织变化的同时,合金元素在基体中的分布也发生了变化,这必定会影响合金的各种宏观性能,作为燃料包壳来说,在服役过程中发生这种情况是不能容忍的. 由于这种氢致 α/β 相变是在

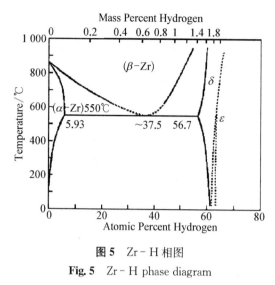

图 5 Zr－H 相图

Fig. 5 Zr－H phase diagram

550 ℃时 Zr－H 之间发生的,对成分不同的锆合金来说,相变的细节可能有一些差别,但是基本过程应该是相似的. 如 Zr－4、N18(NZ2)、N36(NZ8)和 M5 等锆合金,在 550 ℃/25 MPa 超临界水中腐蚀到一定时间后,都观测到发生了氢致 α/β 相变现象,并且实验结果还表明,当合金成分不同时发生氢致 α/β 相变的温度还可能低于 550 ℃.

Motta 和 Jeong 研究了近 10 种体系 30 种不同合金成分的锆合金,在 500 ℃/25.5 MPa 超临界水或 500 ℃过热蒸气中的腐蚀行为,筛选出耐腐蚀性能较好的 Zr－0.4Fe－0.2Cr 和 Zr－1.0Cr-0.2Fe 两种合金,其中前者在 500 ℃/25.5 MPa 超临界水中腐蚀 150 d,腐蚀增重约 150 mg/dm², 显示了开发 SCWR 用锆合金的潜能. 因其腐蚀实验温度只有 500 ℃,所以他们没有观测到氢致 α/β 相变现象.

3 结论

(1) Zr－2.5Nb 合金在 550 ℃/25 MPa 超临界水条件下腐蚀时,由于腐蚀和吸氢过程同时进行,会发生氢致 α/β 相变,形成 H 稳定的 β－Zr,同时合金元素 Nb 扩散进入 β 相,形成富 Nb/H 的 β－Zr. 这种相变过程会造成合金的耐腐蚀性能变坏.

(2) 富 Nb/H 的 β－Zr 在 550 ℃时可以稳定存在,但在温度降低时发生分解,形成 ZrH_x、α－Zr 和 Nb 含量不同的 Zr－Nb 相.

(3) 在 550 ℃水环境条件下,Zr－2.5Nb 合金的耐腐蚀性能尚不能满足核燃料包壳的使用要求,而且从发生氢致 α/β 相变的角度来说,该合金也不能用作超临界核反应堆燃料包壳材料.

参 考 文 献

[1] Abrams Bobby, Chapin Douglas, Garrick B John *et al*. *A Technology Roadmap for Generation IV Nuclear Energy Systems*, US:DOE Nuclear Energy Research Advisory Committee, [2002－12], Http://www.nuclear.gov.

[2] Danielyan Davit. *Supercritical-Water-Cooled Reactor Systemas One of the Most Promising Type of Generation IV Nuclear Reactor Systems*, [2003－12－24], Http://www.nuclear.gov.

[3] Oka Yoshiaki, Koshizuka Seiichi. *Journal of Nuclear Science and Technology*[J], 2001, 38(12):1081.

[4] Motta Arthur T, Jeong Yong Hwan. *Advanced Corrosion-Resistant Zr Alloys for High Burn up and Generation IV Applications*, Department of Energy International Nuclear Energy Research Initiative, DOE/ROK Project Number:[2003－020－K], Http://www.nuclear.gov.

[5] Li Qiang, Zhou Bangxin, Liu Wenqing *et al*. *15th International Symposium on Zirconium in the Nuclear Industry*[C], Sunriver, Oregon, USA. 2007.

[6] Jeong Y H, Park J Y, Gartner E *et al*. *15th International Symposium on Zirconium in the Nuclear Industry*[C], Sunriver, Oregon, USA, 2007.

[7] Li Qiang(李强), Zhou Bangxin(周邦新), Liu Wenqing(刘文庆) *et al*. *Rare Metal Materials and Engineering*(稀有

金属材料与工程)[J]，2007，36(8)：,1358.

［8］Li Qiang(李强)，Liu Wenqing(刘文庆)，Zhou Bangxin(周邦新). *Rare Metal Materials and Engineering*(稀有金属材料与工程)[J]，2002，31(5)：389.

α/β Phase Transformation of Zr – 2. 5Nb Alloy Induced by Hydrogen during Corrosion Testing in SCW at 550 ℃/25 MPa

Abstract: The corrosion behavior for Zr – 2. 5Nb specimens heat-treated at 580 ℃/5 h and 650 ℃/2 h respectively after β- quenching and cold rolling have been investigated in 550 ℃/25 MPa supercritical water (SCW) by autoclave tests. The scanning electron microscopy (SEM) and high resolution transmission electron microscopy (HRTEM) equipped with an energy dispersion spectroscopy (EDS) were employed for examining the matrix microstructure of the specimens before and after corrosion tests. It was noted that the microstructure of the matrix for the alloy specimens were drastically changed after corrosion testing for a certain period. It is considered that H-stabilized β- Zr were formed during corrosion testing at 550 ℃, and at the same time the Nb alloying element diffused into the β – Zr to form Nb/H-enriched β – Zr, which decomposed to ZrH_x, α – Zr and Nb-enriched Zr – Nb phases as the corrosion temperature decreased.

热处理对 N36 锆合金腐蚀与吸氢性能的影响*

摘　要：将 N36 锆合金样品分别进行 1 020 ℃/20 min WQ+C. R. +580 ℃/50 h AC、820 ℃/2 h AC+580 ℃/50 h AC、820 ℃/2 h AC+C. R. +580 ℃/50 h AC、700 ℃/4 h AC+C. R. +580 ℃/50 h AC 4 种不同的热处理. 用透射电镜观察了它们的显微组织，用高压釜腐蚀试验研究了它们在 400 ℃/10.3 MPa 过热水蒸气中的腐蚀与吸氢行为. 结果表明：经 1 020 ℃/20 min WQ+C. R. +580 ℃/50 h AC 处理后样品的耐腐蚀性能最好，其原因在于合金中第二相细小弥散分布在晶界及晶内；而经 820 ℃/2 h AC+580 ℃/50 h AC 处理后样品的耐腐蚀性能最差，第二相主要集中在晶界上，且比较粗大. 经 820 ℃/2 h AC+C. R. +580 ℃/50 h AC、700 ℃/4 h AC+C. R. +580 ℃/50 h AC 处理的样品的腐蚀性能介于两者之间. 热处理对 N36 锆合金腐蚀时的吸氢行为有一定的影响，但不如对 Zr - 4 合金的影响大，这可能是因为 N36 锆合金中的第二相吸氢能力不如 Zr - 4 合金中的 Zr(Fe，Cr)₂ 第二相强的缘故.

锆合金是核反应堆中一种重要的结构材料，用作核燃料包壳，在高温高压水中工作. 锆与高温水反应生成氧化锆的同时放出氢，一部分氢被锆吸收，由于氢在 α - Zr 中的固溶度很低（室温时小于 1 μg/g，400 ℃时为 200 μg/g[1]），多余的氢将以氢化锆的形态析出而使锆合金变脆. 过去曾因氢脆导致燃料棒的包壳沿轴向发生开裂的事故[2]，因而腐蚀和吸氢是锆合金应用中的两个重要问题，涉及到核电站运行时的安全性.

随着燃料组件燃耗的进一步提高，如何提高锆合金的耐腐蚀性能和降低锆合金腐蚀时的吸氢量则越来越受到关注. 一般来说，优化锆合金的成分或热处理制度，可提高耐腐蚀性能，减少其吸氢量，如法国法马通公司开发的 M5 合金的耐水侧腐蚀性能优于优化 Zr - 4 合金，且吸氢量也比优化 Zr - 4 合金少[3]. 但姚美意等人[4]在研究合金成分对锆合金焊接样品吸氢行为影响时，发现样品中的氢含量与耐腐蚀性能间并没有直接的对应关系. 在最近研究热处理对 Zr - 4 合金腐蚀时的吸氢行为影响时也发现了类似的结果，Zr - 4 合金的吸氢与 Zr(Fe，Cr)₂ 第二相的大小、数量密切相关[5].

现在普遍认为锆合金中的第二相对锆合金腐蚀时的吸氢行为有影响，但对第二相如何影响其吸氢行为却没有得到一致的认识. 为了更好地理解锆合金腐蚀时吸氢行为的机理，我们采用不同热处理以获得第二相大小、数量及分布不同的样品，对不同成分锆合金腐蚀时的吸氢行为进行了系统研究. 本文主要介绍我国自主研发的 N36（Zr - 1.02Sn - 1.16Nb - 0.3Fe - 0.12O，mass%）新锆合金在 400 ℃/10.3 MPa 过热蒸气中的腐蚀与吸氢结果，并探讨热处理影响吸氢行为的机理.

1　实验方法和步骤

1.1　样品的制备及显微组织观察

实验用 1 mm 厚的 N36 锆合金板，由西北有色金属研究院提供. 为了改变第二相的大

* 本文合作者：鲁艳萍、姚美意. 原发表于《上海大学学报（自然科学版）》，2008，14(2)：194 - 199.

小、数量及分布,采用 4 种不同的热处理工艺制备样品(表 1).

表 1　样品编号和热处理工艺

Tab. 1　Number of specimens and the heat treatment ways

样品编号	热理工艺
31	1 020 ℃/20 min WQ+C. R. +580 ℃/50 h AC
32	820 ℃/2 h AC+580 ℃/50 h AC
33	820 ℃/2 h AC+C. R. +580 ℃/50 h AC
34	700 ℃/4 h AC+C. R. +580 ℃/50 h AC

31 号样品是先从 1 mm 厚的 N36 锆合金板上切下 10 mm 宽的小条,分别放入数支石英管中真空密封,在管式电炉中加热到 1 020 ℃保温 20 min;然后淬入水中并快速敲碎石英管,称为"水淬(WQ)",将水淬后 1 mm 厚的小条冷轧(C. R.)到 0.6 mm,切成 25 mm×10 mm 的样品;最后将样品放入真空石英管管式电炉中加热至 580 ℃保温 50 h,然后将电炉推离石英管,在石英管外壁淋水冷却,称为"空冷(AC)". 其他 3 种样品是先从 1 mm 厚的 N36 锆合金板上切下 20 mm 宽的小条,其中 32 号样品是先冷轧到 0.6 mm,再进行 820 ℃/2 h,AC 处理;而 33 号和 34 号样品是先分别进行 820 ℃/2 h,AC 处理和 700 ℃/4 h,AC 处理,再冷轧到 0.6 mm,最后这 3 种样品与 31 号样品一样都进行 580 ℃/50 h,AC 处理,样品尺寸为 25 mm×20 mm.

用 H-800 型透射电镜观察合金样品的显微组织及第二相的大小和分布,样品用双喷电解抛光制备,电解液为 20％高氯酸＋80％醋酸溶液.

1.2　腐蚀试验、氢含量测试及氢化锆观察

不同热处理样品经酸洗(酸洗混合酸的体积分数为 10％ HF＋30％ H_2SO_4＋30％ HNO_3)和去离子水冲洗后,在 400 ℃/10.3 MPa 的过热蒸气中进行高压釜腐蚀实验,间隔一定时间取样称重,腐蚀增重是 3 块试样增重的平均值.

腐蚀后样品中的氢含量用 LECO 公司的 RH-600 型定氢分析仪进行测定. 样品酸洗后的厚度在 0.56～0.59 mm 范围内,但大多数为 0.57 mm. 当吸氢分数相同时,不同的样品厚度会影响分析得到的氢含量,因此对测得的氢含量数据均按 0.57 mm 这一厚度作归一化处理,消除厚度影响的具体数据处理方法已在文献[5]中有详细报道.

用 VHX-100 型光学显微镜观察氢化锆,样品先经砂纸抛光,然后用 10％ H_2O_2＋10％ HF＋80％HNO_3 的蚀刻液进行蚀刻.

2　实验结果与讨论

2.1　显微组织

图 1 是不同热处理样品的 TEM 照片. 由于水淬时冷却速度较快,α-Zr 中必然含有过饱和固溶的 Nb,样品重新在 580 ℃保温处理时,发生 α-Zr 中过饱和固溶的 Nb 以 β-Nb 第二相析出,冷轧变形又可以促使 β-Zr 分解时的形核,β-Zr→α-Zr+β-Nb,获得细小的 β-Nb 第二相,所以 31 号样品中第二相非常细小(～0.05 μm),且弥散地分布在晶内与晶界上(图 1(a)). 32 号样品中的第二相主要集中分布在晶界上,相对比较粗大(～0.2 μm)(图

1(b)),这是 820 ℃处理后残留的 β-Zr 在 580 ℃处理时分解的结果. 另外,我们还看到晶粒内有棒状的第二相析出(如图 1(b)中箭头所示). 李强等[6]在研究变形及热处理对 Zr-Sn-Nb 合金中 β-Zr 分解的影响时也观察到过晶粒内这种棒状的第二相,分析认为这些棒状的第二相为 β-Zr. 所以本研究中观察到的棒状第二相可能也是未分解的 β-Zr. 由于冷轧变形促进了 β-Zr 的分解,所以 33 号样品中未见到棒状的第二相,并且第二相比 32 号样品的小(图1(c))(晶粒内~0.05 μm,晶界上 0.1~0.2 μm). 34 号样品中的第二相也是均匀地分布在晶界及晶内(大部分小于 0.1 μm,只有极少数大于 0.2 μm)(图 1(d)),但没有 31 号样品的细小.

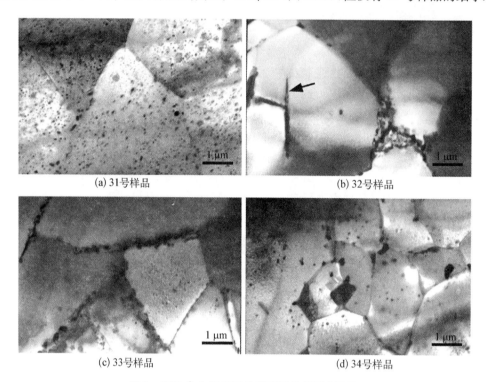

(a) 31号样品   (b) 32号样品

(c) 33号样品   (d) 34号样品

图 1 N36 合金经不同热处理后的 TEM 照片

Fig. 1 TEM micrographs of specimens treated with different heat treatments

2.2 耐腐蚀性能

图 2 是不同热处理样品在 400 ℃/10.3 MPa 过热水蒸气中腐蚀时的增重变化. 结合图 1 和 2 可以看出,第二相细小弥散分布的 31 号样品,腐蚀速率最低,耐腐蚀性能最好;第二相粗大且主要在晶界分布的 32 号样品,腐蚀速率最高,耐腐蚀性能最差;而第二相大小介于 31 和 32 号样品之间并且均匀分布的 33、34 号样品的耐腐蚀性能介于前两种样品之间.

Zr-Nb 合金样品在水或水蒸气介质中腐蚀时,相组织的耐腐蚀性能按照 β-Nb、α-Zr、β-Zr 的顺序依次降低[7-8]. 大量研究结果表明,含 Nb 锆合金基体中固溶的 Nb 含量达到其处理温度下的平衡固溶度,并且获得的 β-Nb 第二相细小弥散分布时,才能表现出优良的抗腐蚀性能[9-12]. 我们的实验结果是与之相符合的.

2.3 吸氢性能

图 3 是腐蚀后样品的氢含量与腐蚀增重的关系曲线. 从图 3 可知,31 号样品的吸氢量随

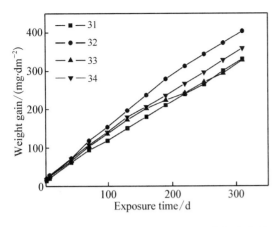

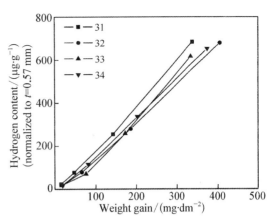

图 2 不同热处理样品在 400 ℃过热水蒸气中腐蚀时的腐蚀增重变化

Fig. 2 Weight gain as a function of the exposure time for specimens with different heat treatments after autoclave test at 400 ℃/10. 3 MPa superheated steam

图 3 氢含量与腐蚀增重的关系曲线

Fig. 3 Plots of hydrogen content vs weight gain for N36 zirconium alloy

腐蚀增重的变化最大,32 号样品的吸氢量随腐蚀增重的变化最小,而 33 号和 34 号样品的吸氢量随腐蚀增重的变化介于 31 号和 32 号样品之间,但 4 种样品在相同腐蚀增重时氢含量的差别并不像 Zr‐4 样品中的那样大.

图 4 是 31 号样品在 400 ℃/10. 3 MPa 过热蒸气中腐蚀不同时间后的金相照片. 从图 4

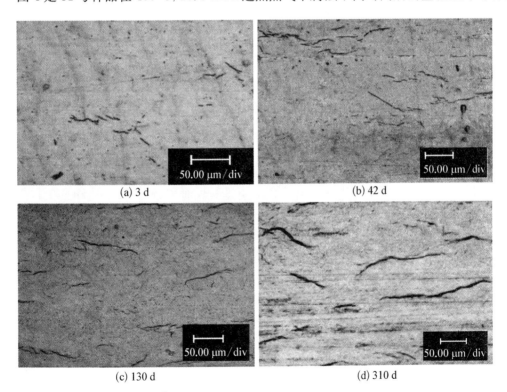

(a) 3 d   (b) 42 d

(c) 130 d   (d) 310 d

图 4 样品 31 经过不同时间腐蚀后氢化锆的形貌

Fig. 4 The morphology of zirconium hydrides in 31 samples exposed different time

可看出随腐蚀时间增加,氢化物逐渐变得粗大,说明氢含量在不断增加,这与氢含量的测试结果是一致的.

锆合金腐蚀时的吸氢能力也用吸氢分数来表示,它是腐蚀后样品吸入的氢量与腐蚀时理论放氢量之比.表2给出了不同热处理N36锆合金样品腐蚀310 d时的吸氢分数.为了与Zr-4合金的吸氢行为相比较,将不同热处理Zr-4合金样品腐蚀302 d时的吸氢分数[5]列在表3中.比较表2和3可知,不同热处理对N36锆合金吸氢的影响(吸氢分数24.9%～26.8%)没有热处理对Zr-4合金吸氢(吸氢分数17.4%～28.0%)的影响大.

表2 N36合金腐蚀310 d时的吸氢分数

Tab. 2 Hydrogen pickup fraction of N36 zirconium alloy with different heat treatments exposed 310 d

样 品 编 号	吸 氢 分 数/%
31	26.8
32	24.9
33	26.3
34	25.1

表3 Zr-4合金腐蚀302 d时的吸氢分数[5]

Tab. 3 Hydrogen pickup fraction of Zr-4 zirconium alloy with different heat treatments exposed 302 d[5]

热 处 理 工 艺	吸 氢 分 数/%
1 035 ℃/0.5 h, AC	17.4
800 ℃/36 h, AC	28.0
720 ℃/10 h, AC	26.9
As-received(600 ℃ an ealling)	19.9

一般来说,提高锆合金的耐腐蚀性能可以降低锆与水蒸气反应放出的氢,从而减少其吸氢量.基于氢在氧化膜/水介质界面处产生的假定,B. Cox[13]认为氧化膜中孔隙和裂纹的几何形状和数量是锆合金氧化过程中吸氢的关键因素,第二相首先影响氧化膜的特性后才引起吸氢行为的差别.这无法解释32样品耐腐蚀性能最差,但吸氢量随腐蚀增重变化最小的实验结果.

Lelievre[14]认为,样品中的第二相是氢扩散的优先通道,但是第二相必须没有被氧化并与金属相接触,这在解释第二相对氧化初始阶段(如转折前)吸氢行为的影响时是合理的,但对氧化后期,即氧化膜较厚时的吸氢行为影响则无法解释.因为随着氧化膜厚度的增加,氧化膜中的第二相也会氧化.如果还是基于氢在氧化膜/水介质界面处产生的假定,那么氢通过氧化膜传输的难易(即氧化膜特性)又会成为引起锆合金腐蚀时吸氢性能差异的首要因素.

姚美意等[5]在研究热处理对Zr-4合金吸氢行为影响时,对氢的产生之处提出了另一假设:认为除一部分氢在氧化膜/水介质界面处产生外,氢也有可能在金属/氧化膜界面处产生.基于以上假设,提出镶嵌在金属/氧化膜界面处未被氧化的第二相可以作为吸氢优先通道的观点,并合理地解释了热处理对Zr-4合金吸氢行为的影响.这也可以解释热处理对N36合金吸氢行为的影响.一方面,31号样品中的第二相最多,它们可提供更多的吸氢通道,但另一方面第二相是最小的,很容易被氧化.所以31号样品的吸氢分数只略大于其他3种样品,32号样品的第二相最少提供的吸氢通道最少,所以吸氢分数相对最低.另外,从平均吸氢分数来看,N36 合金高于 Zr-4 合金,这可能是因为 N36 合金中的第二相比 Zr-4 合金中的第二相多,相界面也可促进氢的吸收,因此平均吸氢分数高.

另外,虽然锆金属本身是一种吸氢材料,但如果添加的合金元素与锆形成的第二相是一种比锆更容易吸氢的物质,这时合金元素对锆合金吸氢性能的影响可能将主要取决于这种吸氢能力强的第二相的大小和数量. 姚美意等[5]最近在研究 Zr-4 合金腐蚀时吸氢行为时,发现 Zr-4 合金的吸氢与 $Zr(Fe, Cr)_2$ 第二相的大小、数量密切相关;如果添加的合金元素与锆形成的第二相是一种并不比锆更容易吸氢的材料,那么合金元素对锆合金吸氢性能的影响主要取决于固溶了一定合金元素的 α-Zr 基体自身,第二相的大小和数量成为影响吸氢的次要因素. N36 锆合金中第二相主要是 β-Nb 和 Zr-Nb-Fe,这些第二相的吸氢能力没有 Zr-4 合金中 $Zr(Fe, Cr)_2$ 第二相的吸氢能力强. 这或许是热处理对 N36 锆合金吸氢的影响没有对 Zr-4 合金影响大的原因.

3　结论

(1) 采用不同的热处理工艺制备第二相大小、数量及分布不同的 N36 锆合金样品,它们经 400 ℃/10.3 MPa 过热蒸气腐蚀后,第二相细小均匀弥散分布的样品耐腐蚀性能最好;而第二相粗大且分布不均匀的样品耐腐蚀性能最差.

(2) 不同热处理样品腐蚀后的吸氢量与腐蚀增重之间并没有直接的对应关系,这可能与合金中第二相的大小和数量有关.

(3) 第二相对 N36 锆合金腐蚀时吸氢行为的影响不如对 Zr-4 合金的大,这可能是因为 N36 锆合金中的第二相吸氢不如 Zr-4 合金中的 $Zr(Fe, Cr)_2$ 第二相吸氢能力强的缘故.

参 考 文 献

[1] KEARNS J J. Terminal solubility and partitioning of hydrogen in the alpha phase of zirconium, zircaloy-2 and zircaloy-4[J]. J Nucl Mat, 1967, 22: 292-303.

[2] LEMAIGNAN C, MOTTA A T. Zirconium alloys in nuclear application[M]. Beijing: Science Press, 1999: 3-48.

[3] MARDON J P, CHARQUET D, JEAN S. Influence of composition and fabrication process on out-of-pile and in-pile properties of M5 alloy[C]//Zirconium in the Nuclear Industry, 12th International Symposium, ASTM STP 1354. West Conshohocken: American Society for Testing and Materials, 2000: 505-524.

[4] 姚美意,周邦新,李强,等. 合金成分对锆合金焊接区腐蚀时吸氢性能的影响[J]. J Nucl Mat, 2004, 33(6): 641-645.

[5] 姚美意,周邦新,李强,等. 第二相对 Zr-4 合金在 400℃过热蒸汽中腐蚀时吸氢行为的影响[J]. 稀有金属材料与工程,2007,36(11): 1915-1919.

[6] 李强,刘文庆,周邦新. 变形及热处理对 Zr-Sn-Nb 合金中 βZr 分解的影响[J]. 稀有金属材料与工程,2002,31(5): 389-392.

[7] LIN Y P, WOO O T. Oxidation of βZr and related phases in Zr-Nb alloy: an electron microscopy investigation[J]. J Nucl Mat, 2000, 277: 11-27.

[8] PECHEUR D. Oxidation of βNb and Zr(Fe, V)₂ precipitates in oxide films formed on advanced Zr-based alloys[J]. J Nucl Mat, 2000, 278: 195-201.

[9] JEONG YH, LEE K O, KIM H G. Correlation between microstructure and corrosion behavior of Zr-Nb binary alloy[J]. J Nucl Mater, 2002, 302: 9-19.

[10] 李中奎,刘建章,周廉,等. 新锆合金耐蚀性能研究[J]. 原子能科学技术,2003,37: 84-87.

[11] 刘文庆,李强,周邦新,等. 显微组织对 ZIRLO 锆合金耐腐蚀性的影响[J]. 核动力工程,2003,24(1)：33－36.

[12] 刘文庆,李强,周邦新,等. 热处理制度对 N18 新锆合金耐腐蚀性能的影响[J]. 核动力工程,2005,26(3)：249－253.

[13] COX B. A mechanism for the hydrogen uptake process in zirconium alloys[J]. Journal of Nuclear Materials, 1999, 264：283－294.

[14] LELIEVRE G, TESSIER C, ILITS X, et al. Impact of intermetallic precipitates on hydrogen distribution in the oxide layers formed on zirconium alloys in a steam atmosphere: a²D (³He, p) α nuclear analysis study in microbeam mode[J]. Journal of Alloys and Compounds, 1998, 268：308－317.

Effect of Heat Treatments on Corrosion and Hydrogen Uptake Behaviors in N36 Zirconium Alloy

Abstract：N36 zirconium alloy specimens were treated in different ways at 1 020 ℃/20 min WQ ＋C. R. (cold rolling) ＋580 ℃/50 h AC, 820 ℃/2 h AC＋580 ℃/50 h AC, 820 ℃/2 h AC＋C. R. ＋580 ℃/50 h AC, and 700 ℃/4 h AC ＋C. R. ＋580 ℃/50 h AC, respectively. Their microstructures were examined by transmission electron microscopy (TEM). The corrosion and hydrogen uptake behaviors of these specimens were investigated after autoclave tests in superheated steam at 400 ℃/10. 3 MPa. Results show that the corrosion resistance of the specimen treated at 1 020 ℃/20 min WQ+C. R. (cold rolling)+580 ℃/50 h AC is the best among them. The reason is that such treatment makes the second phase particles fine and dispersed in α－Zr matrix. The corrosion resistance of the specimen treated at 820 ℃/2 h AC+580 ℃/50 h AC was the worst due to the presence of coarse second phase particles. The corrosion resistance of these specimens treated at 820 ℃/2 h AC+C. R. ＋580 ℃/50 h AC and 700 ℃/4 h AC+C. R. ＋580 ℃/50 h AC is in between. Heat treatments have less influence on hydrogen uptake of N36 than that of Zr－4. This may be the reason that the second phase particles in N36 alloy is less reactive with hydrogen than that of $Zr(Fe, Cr)_2$ SPPs in Zr－4.

小形变量轧制下电工钢中立方织构的形成[*]

摘　要：尝试了小形变量轧制对立方织构电工钢织构形成的影响. 结果表明, 对含碳量较低的 Fe-3.2% Si 铁硅合金的初次再结晶组织施加小形变量形变和退火后, 可以得到具有柱状晶结构的晶粒组织, 并且得到比较集中的立方织构. 柱状晶组织形成过程中伴随着 Σ5 和 Σ11 晶界的减少和最终消失, 这种低 Σ 重位晶界的移动主要是由于 (100) 取向晶粒与基体晶粒之间存在沿 ⟨100⟩ 或 ⟨110⟩ 转动 36.9° 或 50.5° 的取向关系.

立方织构电工钢由于其特殊晶粒取向, 使得其在轧向和横向都可以得到最大的磁感应强度, 因而作为变压器铁芯材料不仅可以简化铁芯制作工艺, 而且可以显著减小变压器的能耗[1]. 这种结构和性能特点自立方织构电工钢问世以来, 一直受到电工钢科研人员的关注[2-4].

电工钢中织构的形成是通过形变与再结晶来实现的. 早先人们在 0.04 mm 厚的样品或纯度较高的铁硅合金中获得了立方织构, 主要是基于降低二次再结晶过程中 (100) 晶面的表面能[5]. 由于厚度和合金纯度的要求, 这种方法在工业生产中难以操作, 因此探索如何在合金纯度较低和较厚板材获得立方织构成为过去几十年研究工作的核心[6-7].

电工钢织构的形成除了与最终退火工艺条件相关外, 材料初次再结晶后的组织至关重要. 在取向硅钢的最终退火过程中, 人们发现最终高斯织构的形成是由于特殊取向晶粒与基体晶粒间存在 Σ9 CSL 晶界[8], 对 {100}⟨021⟩ 织构形成的研究也发现类似的情况[9], 说明最终所期望的织构的获得都与初次再结晶后的组织密切相关. 在立方织构电工钢的织构形成研究中, 初次再结晶组织低 ΣCSL 晶界的变化也引起了重视[10], 但这些变化如何影响织构形成仍然在探索之中. 通过对初次再结晶组织施加小形变量形变, 在不破坏晶粒取向的情况下, 就可以为试样表面层的晶粒施加一定的能量, 使得有利取向的晶粒在合适的退火条件下生长而形成所期望的织构. 基于这样的设想, 本工作通过对特定合金小形变量轧制, 探索了小形变量轧制对立方织构电工钢织构形成的作用, 并初步探讨了特殊晶界对立方织构电工钢织构形成的影响.

1　实验技术

用 10 kg 真空感应炉熔炼含碳量不同的 Fe-3.2% Si 铁硅合金 (化学成分见表 1), 合金铸锭经过热锻成为板坯后, 通过调整控制热轧、冷轧和退火工艺获得一定的初次再结晶组织. 对初次再结晶组织施加小形变量形变 (压下量控制在 10% 以内), 在最终退火过程中通过脱碳控制 γ→α 相变速率, 从而达到利用部分 γ 相的存在或消失来控制基体晶粒和立方取向晶粒的生长, 获得高百分率立方织构的柱状晶组织. 利用光学显微镜观察试样的显微组织, 采用电子背散射衍射 (EBSD) 方法进行织构和晶界的详细分析, 用极图表达织构的集中程

* 本文合作者：王均安、贺英、邱振伟、Frantisek Kovac. 原发表于《上海大学学报(自然科学版)》, 2008, 14(5): 461-466.

度.通过以上分析手段和观察结果,研究小形变量形变对立方织构电工钢织构形成的影响.

表 1　试样的化学成分

Table 1　Chemical composition of the specimens　%

试　样	C	Mn	Si	其他微量元素总量	Fe
A	0.021	1.15	3.15	<0.05	余量
B	0.034	1.17	3.25	<0.05	余量

2　实验结果与讨论

2.1　轧制-退火后试样的织构

图 1 给出了试样经过小形变量冷轧和不同时间退火后从试样表面测得的织构.在所实验的轧制-退火制度下,在两种碳含量不同的合金中都可以相当份额的立方织构.但从(100)极图可以看出,(100)[001]织构比较分散,不仅(100)面不完全平行于轧面,而且[001]方向也与轧向偏离较大.要提高立方织构的集中度和立方晶粒的表面覆盖率,还需要对轧制-退火工艺进一步完善.尽管在含碳量不同的材料中都可以得到立方织构,但材料的化学成分显然影响着再结晶过程.含碳量低的材料经过 1 060 ℃-5 min 退火处理后,基本完成再结晶并且形成较粗大的晶粒组织.而对于含碳量高的材料经过相同条件的处理后,表面层晶粒发生了长大,柱状晶结构还没有形成(见图3).这种结果说明碳在再结晶过程中起到了阻碍基体初次再结晶晶粒的生长作用,同时也减缓了柱状晶组织的形成,但为特殊取向晶粒的长大从动力学上提供了条件.

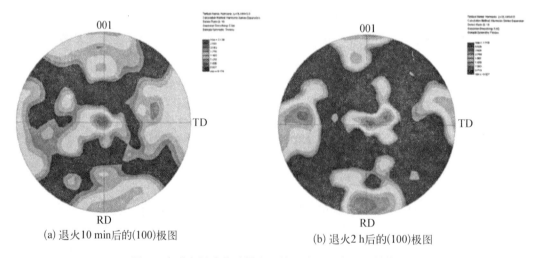

(a) 退火10 min后的(100)极图　　　　　　(b) 退火2 h后的(100)极图

图 1　小形变量冷轧试样在最终退火过程中的织构演化

Fig. 1　Texture evolution of specimens after cold temper rolling and annealing

2.2　小形变量下电工钢的再结晶行为

对经过二次冷轧-初次再结晶退火的试样施加小形变量形变,目的是为试样表面层中有利取向的晶粒施加小的能量,使其在不改变初次再结晶晶粒取向的前提下,在最终退火过程中进一步长大并发展成柱状晶结构组织,获得取向集中度高并且晶粒细小的立方织构电工钢.

　　一般地,材料形变前的组织(尤其是晶粒尺寸)会对后续的再结晶组织及控制产生较大的影响.为了了解这种影响的程度,首先对小形变量轧制前试样的初次再结晶组织进行了考察.从图2可以清楚地看出,适当增加合金中的碳含量,可以明显细化初次再结晶组织,但过高的碳含量并没有进一步细化铁素体组织.

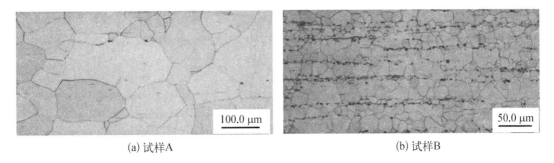

(a) 试样A　　　　　　　　　　　　　　　　(b) 试样B

图2　试样经二次冷轧-初次再结晶退火后的断面组织

Fig. 2　Microstructure of the cross section of specimens after secondary cold rolling and primary recrystallization

　　在小形变量下,通过退火过程获得柱状晶组织的难易程度与合金的化学成分有关(见图3).显然,合金中一定的碳含量对于通过脱碳退火控制并获得柱状晶组织非常重要,在实验选用的含碳量较高的合金中更容易获得长径比较大的柱状晶组织.但碳含量过高会使脱碳

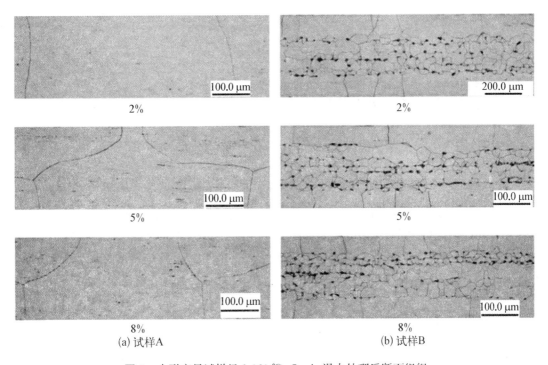

(a) 试样A　　　　　　　　　　　　　　　　(b) 试样B

图3　小形变量试样经1 060 ℃-5 min退火处理后断面组织

Fig. 3　Microstructure of the cross section of specimens after temper cold rolling and annealing at 1 060 ℃-5 min

退火过程变得非常缓慢.

从图 3 不难看出,小形变量对于通过控制脱碳退火获得柱状晶组织是有利的. 对于含碳量低的铁硅合金,形变量越小越有利于贯穿试样厚度的柱状晶组织的形成,但这种柱状晶的

图 4 试样 B 初次再结晶组织经小形变量形变和 1 060 ℃-5 min 退火处理后的纵断面晶粒取向

Fig. 4 Grain orientation on the cross section of specimen B after temper cold rolling and annealing at 1 060 ℃-5 min

长径比较小. 随着合金中碳含量的增加,在小形变范围内,形变量的大小对柱状晶组织的形成影响不十分明显. 研究发现,对于含碳量适中的试样(如试样 B),通过形变量和脱碳退火条件的仔细选择,可以得到长径比较大的柱状晶组织(这种组织中试样轧面上晶粒的直径小,更有利于材料的电磁性能). 在试样断面这些柱状晶除呈现立方取向和旋转立方取向外,较多的晶粒呈现(110)[001]取向(见图 4),这表明轧面将以立方织构为主.

显而易见,碳含量过高不利于贯穿试样厚度的柱状晶组织的形成,漫长的脱碳过程也会使基体晶粒长大. 脱碳退火机理可以用图 5 的模型解释[11]. 在高温退火的初始阶段,试样表面脱碳层(α 相)出现的有利形核点在表面能的作用下首先沿轧面生长,当这些晶粒长大到一定程度并且不能进一步长大时,晶粒将沿试样厚度方向生长. 后一种生长方式主要由脱碳退火过程控制:在试样厚度方向温度梯度的作用下,试样的表面层首先发生 γ→α 相变,相变得到的 α 相为表面层中取向晶粒的向内生长提供了条件,同时 γ→α 相变的可控进行使得基体晶粒的生长得到 γ 相的有效抑制.

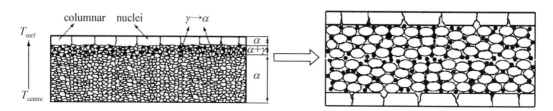

图 5 脱碳退火模型

Fig. 5 Model of decarburization during annealing

2.3 晶界特征以及脱碳退火过程中晶界的移动

比较试样 A 初次再结晶和最终退火一段时间后纵断面上晶界类型及分布(见图 6),不难看出,经过最终退火后在初次再结晶试样中出现频度较多的低 Σ 重位晶界(如 Σ5,Σ7,Σ9 和 Σ11)已消失或明显减弱. 在退火过程中,由于试样中碳含量相对较低,这些晶界和普通的大角晶界易于移动,在较短的退火时间内形成了柱状晶结构. 个别柱状晶内残余的 Σ 重位晶界主要是由于晶粒再结晶不够完善造成的.

从试样 B 初次再结晶和较长时间退火后的晶界分布(见图 7)可以看出,较长时间退火后 Σ5 和 Σ11 晶界所占的百分数显著减少,而 Σ1,Σ7,Σ9 以及其他较高 Σ 重位晶界所占的百分数基本不变,这表明 Σ5 和 Σ11 晶界随着退火时间的延长发生移动并且最终消失. 造成这种结果的主要原因是(100)取向晶粒与基体晶粒之间存在沿⟨100⟩或⟨110⟩转动 36.9°或 50.5°的取向关系. 后者导致了 Σ5 和 Σ11 晶界易于发生移动,并且由于表面能和晶粒尺寸的

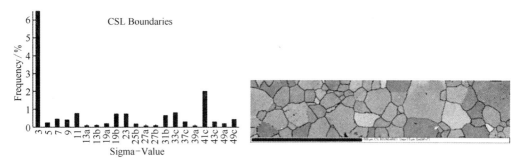

(a) 初次再结晶后

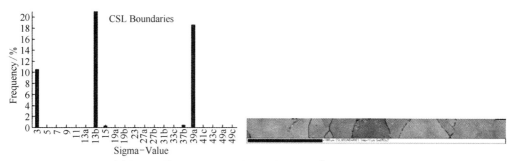

(b) 初次再结晶组织经小形变量形变和1 060℃-5 min退火处理后

图 6 退火处理后试样 A 中的晶界类型及分布

Fig. 6 Grain boundary characterization distribution of the annealed specimens A

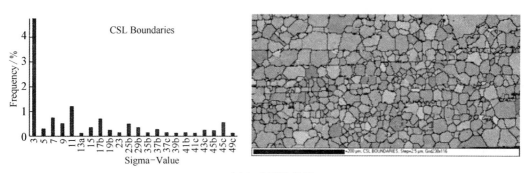

(a) 初次再结晶后

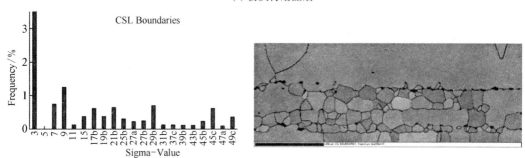

(b) 初次再结晶组织经小形变量形变和1 060℃-5 min退火处理后

图 7 退火处理后试样 B 中的晶界类型及分布

Fig. 7 Grain boundary characterization distribution of the annealed specimens B

双重作用,最终形成立方织构.Σ3 晶界主要是在初次再结晶过程中形成的,随着退火时间的延长,Σ3 晶界数目降低与初次再结晶晶粒数目的减少相关.

　　Harse 等人[10]在立方织构电工钢的研究中发现,在初次再结晶组织中Σ7 晶界的存在对于织构的形成是必要的.但在本研究结果中,对于含碳量低的试样,Σ7 晶界确实在立方织构形成过程中表现出了易于移动并随着退火时间的延长晶界数量显著减少的变化,但对于含碳量较高的试样Σ5 和Σ11 晶界则表现出了更显著的移动性.这种低Σ 重位晶界类型的变化可能与试样的化学成分有关,在脱碳退火过程中不同类型低Σ 重位晶界的变化与立方织构形成的相关性还需要进一步研究.

　　一般认为,铁硅合金中(100)晶粒的生长是由于(100)晶面低表面能的策动,因此获得立方织构的难易在很大程度上取决于最终样品厚度和铁硅合金的纯度.采用小形变量形变与脱碳退火工艺,充分利用了(100)晶面低表面能效应,脱碳退火过程中的 $\gamma \rightarrow \alpha$ 相变过程,以及初次再结晶过程中形成的包含有利低Σ 重位晶界的组织.因此,应该说小形变量对立方织构电工钢织构形成的有利作用是建立在先期完善的形变与再结晶工艺的基础上,只有对合金成分、热轧、冷轧以及退火工艺进行有机的结合,才能实现通过小形变量形变-退火获得立方织构电工钢.

3　结论

　　(1) 选择含碳量适当的铁硅合金,经过小形变量冷轧和控制脱碳退火,可以得到具有柱状晶组织的立方织构电工钢.

　　(2) 柱状晶组织形成过程中伴随着Σ5 和Σ11 晶界的减少和最终消失,这种低Σ 重位晶界的移动主要是由于(100)取向晶粒与基体晶粒之间存在沿⟨100⟩或⟨110⟩转动 36.9°或 50.5°的取向关系.

参 考 文 献

[1] WALTER J L, HIBBARD W R, FIEDLER H C, et al. Magnetic properties of cube textured silicon-iron magnetic sheet[J]. Journal of Applied Physics, 1958, 29(3): 363 – 365.

[2] TOMIDA T, UENOYA S, SANO N. Doubly oriented magnetic steel sheet and method for manufacturing the same: US, 5, 948, 180[P]. 1999 – 09 – 07.

[3] TOMID T, UENOYA S. Cube oriented 3% Si – 1% Mn soft magnetic steel sheets with fine grain structure[J]. IEEE Transactions on Magnetics, 2001, 37(4): 2318 – 2320.

[4] MAO W. Evolution of microstructure and {001}⟨110⟩ texture in electrical steel sheet[C]//Proceedings of the 3rd China – Korea Joint Symposium on Advanced Steel Technology, Shanghai, China. 2003: 109 – 114.

[5] ASSMUS F, BOLL R, GANZ D, et al. Ueber Eisensilizium mit Wuerfeltexture[J]. Zeitschrift fur Metallkunde, 1957, 48(6): 341 –349.

[6] 周邦新. 铁硅合金中形成立方织构的有关问题[J]. 宝钢技术, 2000(5): 53 – 58.

[7] TOMIDA T. Decarburization of 3% Si – 1.1% Mn – 0.05% C steel sheets by silicon dioxide and development of {100}⟨012⟩ texture[J]. Materials Transactions, 2003, 44(6): 1096 – 1105.

[8] SHIMIZU R, HARASE J. Coincidence grain boundary and texture evolution in Fe – 3% Si[J]. Acta Metallurgica, 1989, 37(4): 1241 – 1249.

[9] WANG J, ZHOU B, YAO M, et al. Formation of sharp {100}⟨021⟩ texture in silicon steel sheets[J]. Rare Metal Materials and Engineering, 2004, 33(S1): 187 – 190.

[10] HARASE J, SHIMIZU R, TAKAHASHI N. Coincidence grain boundary and (100)[001] secondary

recrystallization in Fe - 3% Si[J]. Acta Metallurgica et Materialia, 1990, 38(10): 1849 - 1856.

[11] KOVAC F, WANG J, STOYKA V. Investigation of grain boundary motion in non-oriented electrical steels[J].
Acta Metallurgica Slovaca (Metallography 2007), 2007, 13(SI): 176 - 182.

Effect of Temper Rolling on Formation of
Cubic Texture in Electrical Steels

Abstract: Effects of temper rolling on the cubic texture formation in primary recrystallized electrical steel were investigated. The results show that the cubic texture with columnar structure is obtained in the Fe - 3.2% Si electrical steels through temper rolling and final annealing of primary recrystallized sheets. In the formation of columnar structure, $\Sigma 5$ and $\Sigma 11$ grain boundaries decrease and finally disappear. This is beneficial to the growth of (100)[001] grains. The migration of those grain boundaries is due to the special misorientation of 36.9°/⟨100⟩ and 50.5°/⟨110⟩ between (100)[001] grains and primaries and the beneficial annealing environment.

690 合金的晶界特征分布及其对晶间腐蚀的影响[*]

摘　要：运用扫描电子显微镜(SEM)、电子背散射衍射(EBSD)和取向成像显微技术(OIM)研究了 690 合金的晶界特征分布及晶界特征分布对晶间腐蚀性能的影响. 低 ΣCSL (coincidence site lattice, $\Sigma \leqslant 29$)晶界比例高(72.5%)时会出现大尺寸的晶粒团簇, 团簇内部晶粒互有 $\Sigma 3^n$ 取向差关系. 当低 ΣCSL 晶界比例低(46.7%)时, 这种晶粒团簇的大小和内含 $\Sigma 3^n$ 晶界数量都降低. 运用 Palumbo-Aust 标准统计晶界特征分布时, 大尺寸的晶粒团簇之间的随机晶界的连通性几乎不会被打断；但运用 Brandon 标准统计时, 这些随机晶界连通性明显被一些相对偏差较大的低 ΣCSL 晶界打断. 低 ΣCSL 晶界比例高的样品比低 ΣCSL 晶界比例低的样品明显耐晶间腐蚀.

低 ΣCSL(coincidence site lattice, 重位点阵)晶界与随机晶界相比, 有更好的抗晶界偏聚性能[1, 2], 更好的耐腐蚀性能[3]以及抗蠕变性能[4]等. Σ 值代表重位点阵的重合度[5], 低 Σ 指 $\Sigma \leqslant 29$. 如果能提高材料中这种晶界的比例, 材料与晶界相关的性能势必会得到提高. 1984 年 Watanabe 提出了"晶界设计与控制的概念"[6], 随之出现了晶界工程这一研究领域[7]. 通过合适的形变及热处理可以提高材料中的低 ΣCSL 晶界比例, 从而提高材料的整体性能. 其中针对低层错能面心立方金属的研究报道最多, 如：奥氏体不锈钢[8, 9], 铅合金[10, 11], 镍基合金[12, 13], 黄铜[14]等. 通过适当的处理工艺可以提高该类材料中的退火孪晶($\Sigma 3$)比例, 和与退火孪晶相关晶界(如 $\Sigma 9$, $\Sigma 27$)的比例, 这些文献报道中 $\Sigma 3^n$(n 为正整数, 在低 Σ 范畴中只指 $n = 1$, 2, 3 的情况)晶界比例是构成晶界特征分布(grain boundary character distribution, 即各 Σ 值晶界所占的比例)的主体.

690 合金有着良好的耐腐蚀性能, 广泛应用于石油化工工业和核电工业[15-17]. 由于不断要求延长设备及部件的使用寿命, 需要进一步提高材料的耐腐蚀性能. 690 合金也是一种低层错能的面心立方金属, 可通过提高材料中 $\Sigma 3^n$ 晶界比例, 而达到提高低 ΣCSL 晶界比例的目的, 从而进一步提高材料的耐腐蚀性能. 本文分析了含有不同比例低 ΣCSL 晶界 690 合金的显微结构特征及随机晶界连通性, 研究了晶界特征分布对 690 合金晶间腐蚀性能的影响.

1 实验

试验用材料为 690 合金, 其成分列于表 1. 样品固溶处理时, 先将样品封入石英管, 真空度为 5×10^{-3} Pa, 在 1 100 ℃加热 15 min 后淬入水中同时砸破石英管, 使样品淬火. 然后将样品分别冷轧 5%, 20%和 50%后封入石英管, 在 1 100 ℃退火 5 min 后进行淬火. 对处理后的样品进行电解抛光, 电解液为：20% $HClO_4$ + 80% CH_3COOH, 抛光电压为 30 V dc. 电解抛光后获得表面干净的样品, 放入配有 TSL-EBSD 系统的扫描电子显微镜中, 对样品表面逐点逐行进行扫描, 收集背散射电子菊池衍射花样, 经 OIM 系统处理得到一系列晶体学信息.

---

[*] 本文合作者：夏爽、陈文觉. 原发表于《电子显微学报》, 2008, 27(6)：461 - 468.

表 1　材料成分(质量分数%)

表 1　材料成分(质量分数%)

Table 1　Composition of the investigated material（mass%）

Ni	Cr	Fe	C	N	Ti	Al	Si
60.52	28.91	9.45	0.025	0.008	0.4	0.34	0.14

　　所有经过 1 100 ℃退火 5 min 淬火后的样品,再在 715 ℃退火 2 h 进行敏化处理.在此温度的敏化处理不会造成晶界特征分布的变化.将敏化处理后的样品用线切割制成大小基本一致的长方形片状,电解抛光获得干净的表面.测量各样品的表面积,并称重(精确到 0.1 mg).用于晶间腐蚀实验的溶液为 65% HNO$_3$+0.4%HF(体积百分比),选择封闭的聚丙烯塑料盒作为腐蚀实验容器,选择聚四氟乙烯作为样品支架.将各样品悬挂在腐蚀溶液中,进行腐蚀实验.每隔 24 h 将样品取出,在水中清洗 5 次,并在酒精中浸泡 2 次,每次浸泡 10 min,浸泡后用电吹风将样品烘干.然后对各样品再次称重来获得腐蚀失重.这样的腐蚀共进行 216 h,获得了腐蚀失重曲线.腐蚀后材料的表面形貌用扫描电子显微镜观察.

2　结果与讨论

2.1　不同晶界特征分布的显微结构

　　690 合金在固溶处理后分别冷轧 5%,20%和 50%,再在 1 100 ℃再结晶退火 5 min,得到样品 A,B 和 C,按照 Palumbo-Aust 标准统计它们的晶界特征分布示于表 2.690 合金是一种低层错能的面心立方金属,在形变及退火后会产生大量退火孪晶.退火孪晶与母体晶粒有着〈111〉/60°的取向差关系,这正好是 Σ3 重位点阵关系[7]. Σ3 晶界比例随冷轧压下量的增加而下降.当 Σ3 晶界比例高时,会发生明显的多重孪晶(multiple twinning)现象[19],产生较高比例的孪晶相关晶界,如 Σ9 和 Σ27 晶界;当 Σ3 晶界比例相对较低时,Σ9 和 Σ27 晶界比例很低,说明多重孪晶现象不明显.所以 Σ9 和 Σ27 晶界比例之和也随冷轧压下量的增加而下降.其他低 ΣCSL 晶界的比例十分低,不超过随机分布.

表 2　各种样品的处理工艺及晶界特征分布的测量结果(Palumbo-Aust[18]标准)

Table 2　Thermomecanical treatments and grain boundary character distribution statistics of each specimen (Palumbo-Aust criterion[18])

Specimen No.	cold rolling reduction (%)	1 100 ℃annealing time (min)	Σ1 (%)	Σ3 (%)	Σ9+Σ27 (%)	other Low ΣCSL(%)	overall Low ΣCSL (Σ≤29)(%)
A	5		2.4	60.6	9.2	0.3	72.5
B	20	5	2.5	48.8	3.5	0.7	55.5
C	50		4	39.8	1.5	1.4	46.7

　　图 1(a~c)分别是样品 A,B,C 按照 Palumbo-Aust 标准统计的 OIM 图. Gertsman[20] 用晶粒团簇的说法来描述有大量退火孪晶的显微结构.晶粒团簇有以下特点:团簇内所有晶界都有 Σ3^n 的取向差关系,团簇外围基本是随机晶界.但 Gertsman 没有将这种显微结构与材料的处理工艺及材料性能联系起来.为了说明这一显微结构,将含有低 ΣCSL 晶界比例最高的图 1a 中几个晶粒团簇内不相邻晶粒之间的取向差列于表 3.图 1a 中,晶粒 A~F,G~I,J~L 分别位于三个晶粒团簇中.同一个团簇内的晶粒虽然互不相邻,但仍然保持着

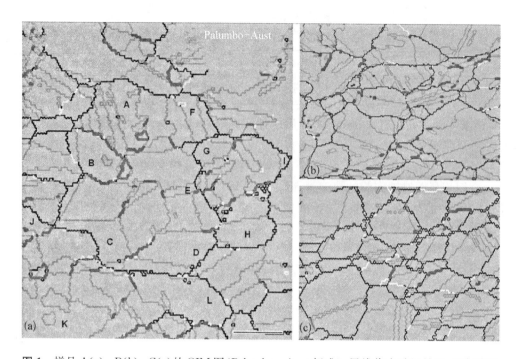

图1 样品 A(a)，B(b)，C(c) 的 OIM 图（Palumbo‑Aust 标准）. 黑线代表随机晶界,细灰线代表 Σ3 晶界,粗灰线代表 Σ9 和 Σ27 晶界,细白线代表 Σ1 和其他低 ΣCSL 晶界. Bar＝100 μm

Fig. 1 OIM maps for specimens A(a)，B(b)，C(c)（Palumbo‑Aust criterion）. Thin gray lines denote Σ3 boundaries，thick gray lines denote Σ9 and Σ27 boundaries，thin white lines denote Σ1 and other ΣCSL boundaries. Bar＝100 μm

$Σ3^n$ 的取向差关系,如表3所示. 这种显微结构是由同一个再结晶晶核在长大过程中不断产生退火孪晶,由于发生多重孪晶而产生了孪晶链,它们之间就会成为 $Σ3^n$ 的取向差关系. 退火孪晶会在不同的(111)面上产生,这种孪晶相遇后会形成 Σ9 晶界;当发生多重孪晶后,这种孪晶之间的相遇又会形成 $Σ3^n$ 晶界,如 Σ27 等晶界,故孪晶比例高的样品中 Σ9 和 Σ27 晶界比例也相对高. 对于再结晶过程中孪晶链及多重孪晶的形成在 20 世纪七八十年代通过高压电子显微镜已有很多研究成果[21],在此不再赘述.

表3 图1(a)中晶粒团簇内不相邻晶粒之间的取向差

Table 3 Misorientations between non-adjacent grains in some grains-clusters of Fig. 1 (a)

non-adjacent grains	misorientation	Closest CSL	deviation($\Delta\theta$)
A/B	$35.2°[\bar{2}\,\bar{1}\,0]$	Σ27b	0.4°
A/C	$35.6°[27\,\overline{1}\,\overline{14}]$	Σ27b	0.8°
A/D	$39.0°[\bar{1}\,21\,\overline{21}]$	Σ9	0.9°
A/E	$38.5°[0\,1\,\bar{1}]$	Σ9	0.8°
A/F	$59.8°[1\,1\,1]$	Σ3	0.5°
B/C	$39.0°[\overline{21}\,21\,1]$	Σ9	0.8°
B/D	$59.2°[\overline{4}\,\overline{11}\,\overline{4}]$	Σ3	0.9°
B/E	$59.7°[1\,1\,\bar{1}]$	Σ3	0.3°
B/F	$39.1°[0\,\bar{1}\,\bar{1}]$	Σ9	0.2°
C/D	$59.1°[1\,\bar{1}\,1]$	Σ3	1.0°
C/E	$59.3°[1\,\bar{1}\,1]$	Σ3	0.9°

non-adjacent grains	misorientation	Closest CSL	deviation($\Delta\theta$)
C/F	39.1°[0 22 21]	$\Sigma 9$	0.7°
D/E	0.5°[19 $\overline{11}$ $\overline{5}$]	$\Sigma 1$	0.5°
D/F	59.7°[$\overline{1}$ 1 1]	$\Sigma 3$	0.5°
E/F	59.7°[$\overline{1}$ 1 1]	$\Sigma 3$	0.3°
G/H	59.8°[$\overline{1}\overline{1}$ 1]	$\Sigma 3$	0.7°
G/I	39.2°[22 21 0]	$\Sigma 9$	0.7°
H/I	59.7°[1 1 1]	$\Sigma 3$	0.3°
J/K	59.9°[17 18 $\overline{17}$]	$\Sigma 3$	0.9°
J/L	35.8°[$\overline{11}$ $\overline{21}$ 0]	$\Sigma 27b$	0.8°
L/K	38.5°[22 $\overline{1}\overline{21}$]	$\Sigma 9$	1.1°

比较图 1 中的 a～c 不难发现,晶粒团簇尺寸有明显差别,随材料的低 ΣCSL 晶界比例降低而递减,并且团簇内 $\Sigma 3^n$ 晶界数量也随之递减. 样品 A 的低 ΣCSL 晶界比例高,所对应的晶粒团簇尺寸就很大,且团簇内有许多 $\Sigma 3^n$ 晶界,构成 $\Sigma 3 - \Sigma 3 - \Sigma 9$ 或 $\Sigma 3 - \Sigma 9 - \Sigma 27$ 的三叉界角(triple junction);而样品 C 的低 ΣCSL 晶界比例低,所对应的晶粒团簇尺寸就小,团簇内很少有 $n > 1$ 的 $\Sigma 3^n$ 晶界(如 $\Sigma 9$ 和 $\Sigma 27$),只出现 $\Sigma 3$ 晶界,这种晶界以单独的直线或直线对的形态存在于团簇内,如图 1c 所示.

2.2　对随机晶界连通性的分析

在考虑低 ΣCSL 晶界比例与材料性能之间关系的大多文献中,一般认为,提高了低 ΣCSL 晶界比例后,低 ΣCSL 晶界会出现在随机晶界的网络上,将其连通性打断,从而使沿晶界的破坏阻断在随机晶界网络上的低 ΣCSL 晶界处[8,12]. 而本研究提高低 ΣCSL 晶界比例的同时会出现大尺寸的晶粒团簇,团簇之间随机晶界基本连续,没有被明显打断,如图 1 所示. 但有一点需要注意,本实验运用的低 ΣCSL 晶界统计标准是 Palumbo - Aust 标准[18]. 另一种常用的统计标准是 Brandon 标准[7]. 这两种统计标准在统计低 ΣCSL 晶界时有一定差别. 这两种标准的表达式分别是:

Palumbo - Aust 标准: $\Delta\theta_{max(P-A)} = 15° \Sigma^{-5/6}$

Brandon 标准: $\Delta\theta_{max(B)} = 15° \Sigma^{-1/2}$

$\Delta\theta$:偏差角,即某一实际重位点阵的取向差与理想重位点阵取向差相差的角度[7].

$\Delta\theta_{max}$:最大偏差角,$\Delta\theta$ 小于 $\Delta\theta_{max}$ 的取向差都定义为该种重位点阵关系[7].

显然运用 Palumbo - Aust 标准统计 CSL 晶界时,除了 $\Sigma 1$ 以外,$\Delta\theta_{max}$ 比运用 Brandon 标准时都小. 也就是 Palumbo - Aust 标准为 CSL 晶界的偏差角限定了更为严格的范围. Palumbo 等研究者在研究了 800 合金的晶间腐蚀,600 合金的晶间应力腐蚀,纯镍的蠕变开裂后,指出运用 Palumbo - Aust 标准统计的低 ΣCSL 晶界被破坏的比例要比运用 Brandon 标准统计的低 ΣCSL 晶界低[22].

为了说明 Palumbo - Aust 标准和 Brandon 标准统计时的具体差别,用 Brandon 标准重新统计图 1a 中的 OIM 图,示于图 2. 比较图 2 和图 1a,用 Brandon 标准统计时,随机晶界明显被一些低 ΣCSL 晶界打断,而运用 Palumbo - Aust 标准统计时随机晶界基本连续. 将图 2 中间位置晶粒团簇外围打断随机晶界的低 ΣCSL 晶界的取向差例于表 4,同时表 4 中还例出了晶粒团簇内部几个 $\Sigma 3^n$ 晶界的取向差. 统计分析了相对偏差,即实际的偏差角与最大偏差

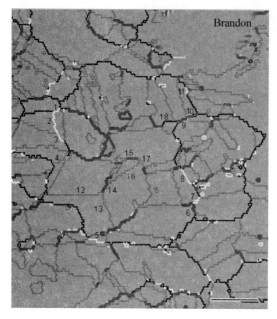

图 2 样品 A 的 OIM 图（Brandon 标准）. 黑线代表随机晶界, 细灰线代表 Σ3 晶界, 粗灰线代表 Σ9 和 Σ27 晶界, 细白线代表 Σ1 和其他低 ΣCSL 晶界.

Bar＝100 μm

Fig. 2 OIM map for specimens A (Brandon criterion). Thin gray lines denote Σ3 boundaries, thick gray lines denote Σ9 and Σ27 boundaries, thin white lines denote Σ1 and other low ΣCSL boundaries. Bar ＝ 100 μm

表 4 对图 2 中标出的各晶界的偏差角的分析

Table 4 Analyzing of the misorientation and deviation from CSL of the boundaries labeled in Fig. 2

No.	misorientation	closest CSL	deviation ($\Delta\theta$)	$\Delta\theta/\Delta\theta_{max(B)}$	$\Delta\theta/\Delta\theta_{max(P-A)}$
1	47.7°[0 $\bar{7}$ 3]	Σ15	2.6°	0.67	1.65
2	47.6°[9 8 9]	Σ19b	2°	0.58	1.54
3	12.8°[9 7 5]	Σ1	12.8°	0.85	0.85
4	51.6°[$\overline{12}$ 13 2]	Σ3	8.5°	0.98	1.41
5	19°[14 $\bar{1}$ 14]	Σ19a	2.7°	0.78	2.07
6	41.2°[1 $\bar{7}$ 7]	Σ9	4.4°	0.88	1.82
7	39.9°[4 11 4]	Σ23	1.0°	0.32	0.90
8	39.7°[$\bar{4}\overline{11}\,\bar{4}$]	Σ23	1.4°	0.45	1.26
9	41.6°[$\overline{19}$ 1 23]	Σ9	4.6°	0.92	1.90
10	34.8°[$\overline{17}\,\bar{2}\,\overline{18}$]	Σ9	4.9°	0.98	2.02
11	25.8°[5 $\bar{6}$ 7]	Σ13b	4.1°	0.99	2.30
			average value	**0.76**	**1.61**
12	59.8°[1 1 1]	Σ3	0.2°	0.02	0.03
13	59.2°[1 1 1]	Σ3	0.8°	0.09	0.13
14	39.5°[0 1 $\bar{1}$]	Σ9	0.4°	0.08	0.17
15	38.9°[0 $\bar{1}$ 1]	Σ9	0.5°	0.10	0.21
16	59.6°[1 1 $\bar{1}$]	Σ3	0.8°	0.09	0.13
17	59.5°[1 1 1]	Σ3	0.6°	0.07	0.10
			average value	**0.08**	**0.13**

角之比：$\Delta\theta/\Delta\theta_{max}$, 对于 Brandon 标准和 Palumbo - Aust 标准来说分别是：$\Delta\theta/\Delta\theta_{max(B)}$ 和 $\Delta\theta/\Delta\theta_{max(P-A)}$. 显然这一值小于 1 时才说明被分析的晶界属于该标准统计的 CSL 晶界.

通过图 2, 图 1a 和表 4 可以看出, 运用 Brandon 标准统计时晶粒团簇之间随机晶界上出现许多低 ΣCSL 晶界, 这些晶界明显阻断了随机晶界, 但这些晶界的偏差角都很大, 运用 Brandon 标准统计的图 2 中, 晶界 1 到晶界 11 的平均相对偏差为 0.76, 大部分偏差角都接近于 Brandon 标准所规定的最大偏差角 $\Delta\theta_{max(B)}$. 而运用 Palumbo - Aust 标准统计这些晶界时,平均相对偏差为1.61,即几乎所有这些出现在随机晶界上的低 ΣCSL 晶界的偏差角都超出了 Palumbo - Aust 标准所规定的最大偏差角,它们被排除在低 ΣCSL 晶界范畴之外,而视作随机晶界. 为进一步说明这一问题,随机统计了样品 A 中 80 个出现在晶粒团簇之间随机晶界上的低 ΣCSL 晶界,将偏差角示于图 3. 同时图 3 中还标出了 Brandon 标准和 Palumbo - Aust 标准各 Σ 值所对应的最大偏差角. 从图中可看出,这些出现在随机晶上的低 ΣCSL 晶界的偏差角大部分都处于两种标准所规定的最大偏差角之间,说明用 Brandon 标准统计时,这些晶界被认为是低 ΣCSL 晶界;而用 Palumbo - Aust 标准统计时它们不是低 ΣCSL 晶界. 所以运用 Palumbo - Aust 这种严格标准来限定 CSL 晶界时,随机晶界几乎是不被打断的,如图 1a 所示.

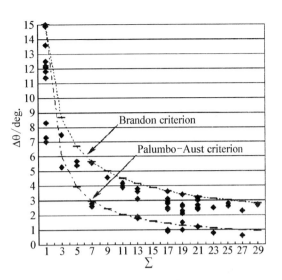

图 3 样品 A 中 80 个出现在晶粒团簇之间随机晶界上的低 ΣCSL 晶界的偏差角

Fig. 3 Deviations of some 80 Low ΣCSL boundaries which appeared on the random boundaries between the grains-cluster in specimen A

分析图 2 晶粒团簇内部的低 ΣCSL 晶界. 这些晶界全部为 $\Sigma 3^n$ 晶界,且大多是 $n=1, 2, 3$,只有晶界 18 的 $n=4$,为 Σ81 晶界,故它不属于低 ΣCSL 晶界,而被划分为随机晶界. 分别用 Brandon 标准和 Palumbo - Aust 标准统计这些晶粒团簇内部的 $\Sigma 3^n (n=1, 2, 3)$ 晶界时,平均相对偏差很小,分别为 0.08 和 0.13. 这些晶界不论用 Brandon 标准还是 Palumbo - Aust 标准统计都属于低 ΣCSL 晶界. 这与用 Brandon 标准统计时,出现在晶粒团簇外围随机晶界上的低 ΣCSL 晶界形成鲜明对比,平均相对偏差相差一个数量级. 表 3 中所列出的同一晶粒团簇内不相邻晶粒之间的相对偏差也较小,分别用 Brandon 标准和 Palumbo - Aust 标准统计时,相对偏差分别为 0.25 和 0.67. 这些相对偏差比相邻晶粒间的相对偏差稍大,这是由于这些晶粒间有较长的距离造成的,但这些相对偏差要比晶粒团簇外围打断随机晶界的低 ΣCSL 晶界的相对偏差小. 通过这些分析,可知统计标准的不同会导致分析随机晶界连通性的结果有很大差别.

2.3 晶界特征分布对晶间腐蚀性能的影响

为了考察低 ΣCSL 晶界比例对 690 合金耐晶间腐蚀能力的作用,采用表 2 中 3 种样品来考验晶间腐蚀能力. 将这 3 种样品在 715 ℃ 退火 2 h 进行敏化处理,浸泡在腐蚀溶液中腐蚀. 图 4 是这 3 种样品单位面积腐蚀失重

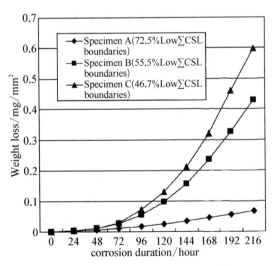

图 4 含有不同比例低 ΣCSL 晶界样品的腐蚀失重与腐蚀时间的关系

Fig. 4 Weight losses of the specimens with different proportions of Low ΣCSL boundaries as a function of intergranular corrosion time

与腐蚀时间的关系. 在腐蚀过程中各样品单位面积失重从小到大,依次为样品 A<B<C. 这说明晶界特征分布对 690 合金样品的腐蚀性能有明显影响,含有的低 ΣCSL 晶界比例越高,材料耐晶间腐蚀性能越好. 图 5 是经过 144 h 腐蚀后,各样品的表面形貌 SEM 像. 比较图 5 中的 a,b,c,可发现样品表面完整性从好到差依次为 A,B,C. 图 5d, e, f,分别是 a, b, c 中方框内的放大图像. 在图 5d 中,有一个正要脱落的晶粒团簇,可看出腐蚀失重是由于样品表面晶粒团簇脱落造成的. 这是因为经敏化处理后晶粒团簇外围的随机晶界处容易析出碳化物,而在晶界附近形成贫铬区[23],从而在腐蚀溶液中容易遭到腐蚀;而晶粒团簇内的 $\Sigma 3^n$ 晶

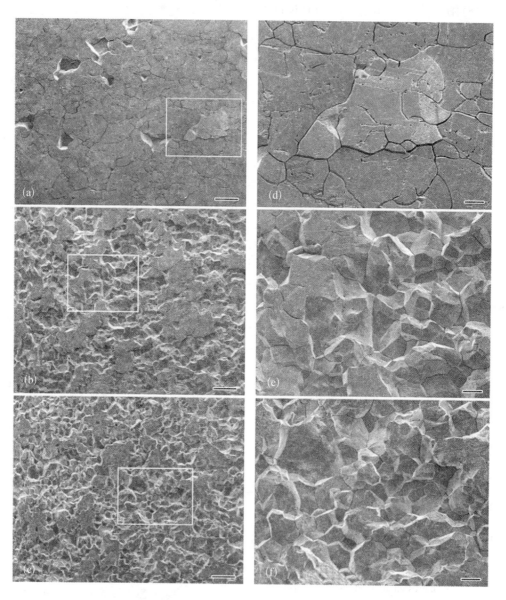

图 5 样品 A(a),B(b),C(c)敏化处理后,经过 144 h 腐蚀表面 SEM 像.
(d~f, Bar=50 μm)是(a~c, Bar=200 μm)中方框内的放大图像

Fig. 5 SEM images of the surfaces of specimen A(a), B(b) and C(c), undergone 144 h intergranular corrosion. (d), (e) and (f) are the magnified images of the area outlined by white frameworks in (a), (b) and (c) respectively

界,在敏化处理后,晶界处析出碳化物的倾向小,晶界附近贫铬现象不如随机晶界处严重[1],从而被腐蚀的程度不严重. 从图 5d 中那个正要脱落的晶粒团簇形貌可证明这一点,团簇外围的晶界腐蚀开裂得很严重,而内部的晶界只有较浅的腐蚀痕迹. 所以腐蚀失重是由于晶粒团簇外围随机晶界遭到腐蚀后,使晶粒团簇从材料表面脱落后造成的.

分析样品 A,B,C 的 OIM 图(图 1),晶粒团簇的大小有明显的差别. 低 ΣCSL 晶界比例越高,晶粒团簇尺寸越大,同时腐蚀失重越小. 首先,低 ΣCSL 晶界比例越高,材料表面的随机晶界就越少,能够被严重腐蚀的晶界就越少. 其次,晶间腐蚀主要沿着随机晶界向材料内部扩展,晶粒团簇尺寸越大,腐蚀沿随机晶界进行使晶粒团簇完全脱离材料所用的时间也就越长. 其三,腐蚀沿随机晶界向材料内部扩展时,随机晶界就是腐蚀介质向材料内部传输的通道. 腐蚀前沿由于化学反应而不断向前推进时,腐蚀介质浓度会下降,这就需要通过扩散使腐蚀前沿与材料表面的腐蚀介质浓度达到平衡,这个扩散通道就是晶粒团簇外围的随机晶界. 如果晶粒团簇越大,这个扩散通道就越长,腐蚀前沿所能达到平衡浓度就越低,腐蚀继续向材料内部扩展的速度也就越慢. 直到整个晶粒团簇脱落下来,下面的组织所接触到的腐蚀介质浓度才会是整个腐蚀溶液的浓度. 这相当于材料表面的晶粒团簇在脱落之前保护了下面的组织,延缓腐蚀向材料内部扩展. 而晶粒团簇尺寸越小,它从材料表面脱落得也越快,这样对内部组织的保护作用也越小.

从前面的分析可知,运用 Brandon 标准统计时,随机晶界连通性会被明显打断,但这些打断随机晶界连通性的低 ΣCSL 晶界的偏差角都较大. 偏差角越大,说明与理想重位点阵取向差的差别越大,这些晶界的特性就越接近随机晶界,而不具有优良性能[23]. 晶粒团簇内部的 $\Sigma 3^n$ 类型晶界的偏差角很小,而具有优良性能. 基于这样的考虑,采用 Palumbo - Aust 标准,得出低 ΣCSL 晶界比例提高后不会使 690 合金随机晶界的网络连通性明显被打断的结论. 所以作者认为材料抗晶间腐蚀能力的提高应归功于形成了大尺寸的晶粒团簇,而不是由于低 ΣCSL 晶界出现在随机晶界的网络上,阻断了随机晶界的连通性,使沿晶界的腐蚀阻止在这些低 ΣCSL 晶界处. 只有当低 ΣCSL 晶界比例明显提高后,晶粒团簇尺寸才会明显变大,抗晶间腐蚀能力才会增强. 这才是提高低 ΣCSL 晶界比例能够提高材料抗晶间腐蚀性能的原因.

3 结论

(1) 690 合金中,低 ΣCSL 晶界比例高时会出现大尺寸的晶粒团簇,团簇内部的晶粒互有 $\Sigma 3^n$ 取向差关系,而晶粒团簇之间的晶界是随机晶界. 当低 ΣCSL 晶界比例低时,这种晶粒团簇的大小和内含 $\Sigma 3^n$ 晶界比例都会降低.

(2) 用严格的 Palumbo - Aust 标准统计时,晶粒团簇之间随机晶界的连通性没有被打断. 而用 Brandon 标准统计时,这种随机晶界的连通性明显被一些相对偏差较大的低 ΣCSL 晶界打断.

(3) 低 ΣCSL 晶界比例高的样品比低 ΣCSL 晶界比例低的样品明显耐晶间腐蚀.

参 考 文 献

[1] Bi H Y,Kokawa H,Wang Z J. Suppression of chromium depletion by grain boundary structural change during twin-induced grain boundary engineering of 304 stainless steel [J]. Scripta Mater,2003,49:219 - 223.

[2] Kurban M, Erb U, Aust K T. A grain boundary characterization study of boron segregation and carbide precipitation in alloy 304 austenitic stainless steel [J]. Scripta Mater, 2006, 54: 1053 - 1058.

[3] Bennett B W, Pickering H W. Effect of grain boundary structure on sensitization and corrosion of stainless steel[J]. Metallurgical Transactions A, 1987, 18: 1117 - 1124.

[4] Don J, Majumdar S. Creep cavitation and grain boundary structure in type 304 stainless steel[J]. Acta Metall Mater, 1986, 34: 961 - 967.

[5] Kronberg M L, Wilson F H. Secondary recrystallization in copper[J]. Trans Am Inst Min Engrs, 1949, 185: 501 - 512.

[6] Watanabe T. Approach to grain boundary design for strong and ductile polycrystals[J]. Res Mech, 1984, 11: 47 - 84.

[7] Randle V. The Role of the Conincidence Site Lattice in Grain Boundary Engineering[M]. UK: Cambridge University Press, 1996.

[8] Michiuchi M, Kokawa H, Wang Z J, Sato Y S, Sakai K. Twin-induced grain boundary engineering for 316 austenitic stainless steel[J]. Acta Mater, 2006, 554: 5179 - 5184.

[9] Shimada M, Kokawa H, Wang Z J. Optimization of grain boundary character distribution for intergranular corrosion resistant 304 stainless steel by twin-induced grain boundary engineering[J]. Acta Mater, 2002, 50: 2331 - 2341.

[10] Lehockey E M, Limoges D, Palumbo G. On improving the corrosion and growth resistance of positive Pb-acid battery grids by grain boundary engineering[J]. Journal of Power Sources, 1999, 78: 79 - 83.

[11] 夏爽,周邦新,陈文觉,王卫国. 高温退火过程中铅合金晶界特征分布的演化[J]. 金属学报,2006,42: 129 - 133.

[12] Lin P, Palumbo G, Erb U. Influence of grain boundary character distribution on sensitization and intergranular corrosion of alloy 600[J]. Scripta Metall Mater, 1995, 33: 1387 - 1391.

[13] Schuh C A, Kumar M, King W E. Analysis of grain boundary networks and their evolution during grain boundary engineering[J]. Acta Mater, 2003, 51: 687 - 700.

[14] Randle V, Davies H. Evolution of microstructure and properties in alpha-brass after iterative processing[J]. Met Mater Trans A, 2002, 33: 1853 - 1857.

[15] Thuvander M, Stiller K. Microstructure of a boron containing high purity nickel-based alloy 690[J]. Mater Sci Eng A, 2000, 281: 96 - 103.

[16] 邱少宇,苏兴万,文燕. 热处理对 690 合金腐蚀性能影响的实验研究[J]. 核动力工程,1995,16: 336 - 340.

[17] Xia S, Zhou B X, Chen W J, Wang W G. Effects of strain and annealing processes on the distribution of $\Sigma 3$ boundaries in a Ni-based superalloy[J]. Scripta Mater, 2006, 54: 2019 - 2022.

[18] Palumbo G, Aust K T. Structure-dependence of intergranular corrosion in high purity nickel[J]. Acta Metall Mater, 1990, 38: 2343 - 2352.

[19] Kopezky C V, Andreeva A V, Sukhomlin G D. Multiple twinning and specific properties of $\Sigma = 3^n$ boundaries in f. c. c. crystals[J]. Acta Metall Mater, 1991, 39: 1603 - 1615.

[20] Gertsman V Y, Henager C H. Grain boundary junctions in microstructure generated by multiple twinning[J]. Interf Sci, 2003, 11: 403 - 415.

[21] Berger A, Wilbrandt P J, Ernst F. On the Generation of new orientations during recrystallization: Recent results on the recrystallization of tensile-deformed fcc single crystals[J]. Progress in Materials Science, 1988, 32: 1 - 95.

[22] Palumbo G, Aust K T, Lehockey E M. On a more restrictive geometric criterion for 'special' CSL grain boundaries [J]. Scri Mater, 1998, 38(11): 1685 - 1690.

[23] Thomson C B, Randle V. Effects of strain annealing on grain boundaries and secure triple junctions in nickel 200[J]. Journal of Materials Science, 1997, 32: 1909 - 1914.

Grain Boundary Character Distribution of Alloy 690 and Its Effect on Intergranular Corrosion

Abstract: Grain boundary character distribution (GBCD), and its effect on the intergranular corrosion in Alloy

690 were investigated by scanning electron microscope (SEM), electron backscatter diffraction (EBSD) and orientation imaging microscopy (OIM). The specimens contained high proportion of low ΣCSL boundaries (72. 5%) and with microstructure of large grains-clusters, exhibited much better resistance to intergranular corrosion than that contained low proportion of low ΣCSL boundaries (46. 7%) and with microstructure of much smaller grainsclusters. In the OIM map with high proportion of low ΣCSL boundaries, the connectivity of the random grain boundary network was almost not broken up, when the Palumbo – Aust criterion was used; whereas that was obviously interrupted by segments of low ΣCSL boundaries with high deviations, when the Brandon criterion was applied in analyzing.

形变及热处理对 690 合金晶界特征分布的影响*

摘　要：应用电子背散射衍射(EBSD)和取向成像显微技术(OIM)研究了材料初始状态、冷轧压下量和 1 100 ℃退火对 690 合金晶界特征分布(GBCD)的影响. 低层错能面心立方金属镍基 690 合金,冷轧 5%后在 1 100 ℃退火 5 min 可使低 ΣCSL(coincidence site lattice,$\Sigma \leqslant 29$)晶界比例提高到 70%以上(Palumbo-Aust 标准),同时形成大尺寸的晶粒团簇. 低 ΣCSL 晶界比例和这种晶粒团簇的尺寸随冷轧压下量的增加而下降. 初始状态的固溶或时效对 690 合金在 1 100 ℃再结晶退火后的晶界特征分布无明显影响.

1984 年,Watanabe 提出了"晶界设计与控制的概念"[1],随之出现了晶界工程(grain boundary engineering,GBE)这一研究领域[2]. 通过合适的形变及热处理工艺,提高材料的低 ΣCSL 晶界比例,就可以改善材料与晶界有关的多种性能,比如：晶间腐蚀[3-5]、断裂[6,7]、合金及杂质元素的偏聚[8,9]、蠕变[10,11]等. CSL 是指重位点阵(coincidence site lattice);Σ 指点阵的重合度,比如 2 个点阵之间每 7 个阵点中就有 1 个重合时,$\Sigma = 7$;CSL 晶界指晶界两侧晶粒之间有重位点阵关系的晶界;低 ΣCSL 晶界指 $\Sigma \leqslant 29$ 的 CSL 晶界.

晶界工程已经成功地运用于许多低层错能的面心立方金属材料,如：奥氏体不锈钢[5],铅合金[3],镍基合金[4,6],铜合金[12]等. 通过适当的处理工艺可以提高该类材料中的退火孪晶($\Sigma 3$)比例和与退火孪晶相关晶界(如 $\Sigma 9$,$\Sigma 27$)的比例,所以这些文献报道中 $\Sigma 3^n$(n 为正整数,在低 Σ 范畴中只指 $n=1, 2, 3$ 的情况)晶界比例是构成晶界特征分布(grain boundary character distribution)的主体. 晶界特征分布是指各种类型晶界所占的比例. 用 CSL 模型来定义晶界,将晶界分为低 ΣCSL 晶界和随机晶界,其中随机晶界指 $\Sigma > 29$ 的晶界和一般大角度晶界.

镍基 690 合金有着优良的耐腐蚀性能,作为压水堆核电站蒸汽发生器的传热管材,是它重要的一个应用方面[13,14]. 由于不断要求延长设备及部件的使用寿命,需要进一步提高材料的耐腐蚀性能. 690 合金也是一种低层错能的面心立方金属材料,所以可以通过提高合金中的退火孪晶比例来提高低 ΣCSL 晶界比例,从而进一步提高材料的耐腐蚀性能. 但对调整 690 合金晶界特征分布的处理工艺还未见报道. 电子背散射衍射 EBSD(electron backscatter diffraction)现在已经成为研究微区晶体取向问题的重要手段[15],在研究晶界特征分布问题中发挥着重要的作用. 本工作借助 EBSD 技术研究了冷轧变形及热处理工艺对 690 合金晶界特征分布的影响.

1　实验

试验用材料为镍基 690 合金,其成分列于表 1. 样品均封入石英管内进行热处理,真空

* 本文合作者：夏爽、陈文觉. 原发表于《稀有金属材料和金属》,2008,37(6)：999-1003.

度为 5×10^{-3} Pa,加热结束后迅速将石英管在水中砸破,使样品淬火(WQ). 冷轧采用辊径为 130 mm 的双辊轧机. 样品进行电解抛光获得干净的表面,电解液成分为:20% $HClO_4$ + 80% CH_3COOH,抛光电压为 30 V. 采用 TSL-EBSD 系统,它安装在 Hitachi-S570 扫描电子显微镜中. 对样品表面逐点逐行进行扫描,收集背散射电子菊池衍射花样,经 OIM 系统处理得到一系列晶体学信息. 测量系统按照 Palumbo-Aust 标准[16]($\Delta\theta_{max} = 15°\Sigma^{-5/6}$)确定 Σ 值,低 ΣCSL 晶界比例以统计晶界长度的百分比计算. 每个样品用 EBSD 扫描 3 处不同区域,每次扫描面积大于 1 000 $\mu m \times$ 1 000 μm. 实验选择了 5%,20% 和 50% 3 种冷轧压下量,对 2 种不同初始状态的 690 合金(固溶状态和时效状态)分别冷轧,再在 1 100 ℃再结晶退火 5 min. 各样品的具体处理工艺列于表 2 中.

表 1 镍基 690 合金成分

Table 1 Composition of Ni based 690 alloy ($\omega/\%$)

Ni	Cr	Fe	C	N	Ti	Al	Si
60.52	28.91	9.45	0.025	0.008	0.4	0.34	0.14

表 2 6 种样品的处理工艺

Table 2 Thermomechanical treatments of six specimen

Specimen No.	S1	S2	S3	S4	S5	S6
Starting stage	Solution annealed (1 100 ℃, 15 min+WQ)			Aged (Solution annealed+715 ℃, 15 h)		
Cold rolling reduction/%	5	20	50	5	20	50
Annealing time (at 1 100 ℃)/min			5			

2 实验结果

图 1 是 690 合金经过 1 100 ℃固溶 15 min 后的 OIM 图和(001)极图. 经固溶处理后样品的低 ΣCSL 晶界的比例为 49.6%,其中 $\Sigma1$ 为 3.5%,$\Sigma3$ 为 43.6%,$\Sigma9 + \Sigma27$ 为 1.5%,其他低 ΣCSL 为 1%. 孪晶界($\Sigma3$)以单独的直线或直线对的形态出现,如图 1a 所示. 从(001)极图来看,材料没有织构的. 在 715 ℃时效 15 h 后,材料的晶界特征分布,OIM 图中晶界的形态,及织构都没有变化,所以不再给出.

对表 2 中样品 S1~S6 进行 EBSD 测定,它们的晶界特征分布如图 2 所示. 690 合金是一种低层错能的面心立方金属,在形变及退火后会产生大量退火孪晶. 退火孪晶与母体晶粒有着 <111>/60°的取向差关系,这正好是 $\Sigma3$ 重位点阵关系[2]. 图 2a 是 $\Sigma3$ 晶界比例与冷轧压下量的关系. 不论初始是固溶状态还是时效状态,经冷轧 5%再在 1 100 ℃退火后,$\Sigma3$ 晶界比例都比较高,超过 60%. 这说明小形变量冷轧后,经再结晶退火可得到高比例的退火孪晶. 随着冷轧压下量的增加,$\Sigma3$ 晶界比例下降. 冷轧 20%后在 1 100 ℃退火样品的 $\Sigma3$ 晶界比例在 45%~50%之间;而冷轧压下量增加到 50%后,$\Sigma3$ 晶界比例则更低. 这说明在同样的退火条件下退火孪晶比例随冷轧压下量增加而下降. 当一个母体晶粒中生成多个退火孪晶时,这些退火孪晶之间也有特定的取向关系. 这些孪晶相遇,就会形成 $\Sigma3^n$ 晶界,这就是多

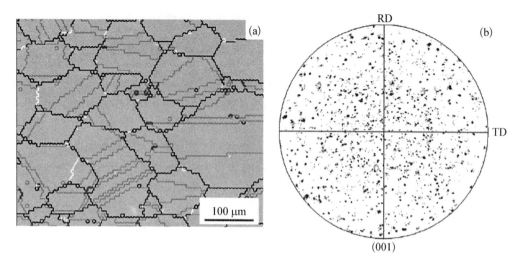

图1 固溶材料的 OIM 图(a)和(001)极图(b)

Fig. 1 OIM map (a) and (001) pole figure (b) of the starting material by (solution annealed). The thin gray lines denote Σ3 boundaries, the thick gray lines denote Σ9 and Σ27 boundaries, the thin white lines denote Σ1 and other low ΣCSL boundaries

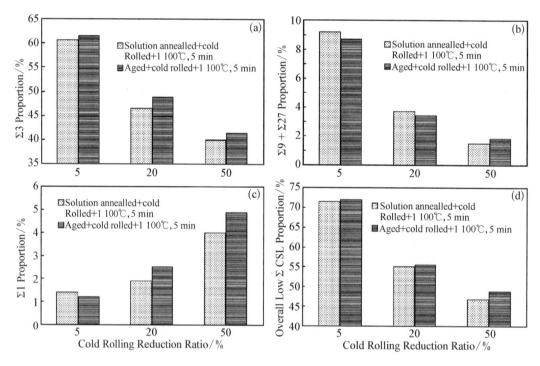

图2 冷轧及热处理对 690 合金晶界特征分布的影响

Fig. 2 Effect of thermomechanical treatment on the grain boundary character distribution: all the x-axis are cold rolling reduction ratio, and the y-axis is the proportion of Σ3(a), Σ9+Σ27(b), Σ1(c), and overall low ΣCSL(d) proportion, respectively

重孪晶(multiple twinning)现象[17]. 所以 Σ9 和 Σ27 晶界比例之和与 Σ3 孪晶界比例有很好的对应关系. 当 Σ3 比例高时, Σ9 和 Σ27 晶界比例之和也相对较高, 如图 2b 所示. 所以 Σ9和 Σ27 晶界比例之和也是随冷轧压下量增加而下降的. Σ1 晶界是指晶粒间取向差小于 15°

的晶界,就是一般意义上的小角晶界.从图 2c 中可知,小角晶界的比例随冷轧压下量的增加而上升.但它们的总体比例比较低,最高不超过 5%.在总体低 ΣCSL 晶界比例中 $\Sigma3^n$ 晶界占绝大多数.这就造成总体低 ΣCSL 晶界比例也随冷轧压下量的增加而下降.原始为固溶或时效不同状态的样品,经过同样的处理后它们的晶界特征分布随冷轧退火工艺变化规律基本一致,这说明固溶或时效不同的初始状态对再结晶后的晶界特征分布影响不大.

本实验中晶界特征分布相近的样品具有相似的 OIM 图,所以将具有不同比例低 ΣCSL 晶界的样品 S1,S2,S3 的 OIM 图示于图 3,而其他样品的 OIM 图不再一一给出.Gertsman[17, 18] 用晶粒团簇来描述有大量退火孪晶的显微结构.晶粒团簇有以下特点:晶粒团簇内所有晶粒之间都有 $\Sigma3^n$ 的取向差关系,晶粒团簇与团簇之间基本是随机晶界.但Gertsman 没有将这种显微结构的特征与材料的处理工艺联系起来.为了说明这种显微结构特征,将含有低 ΣCSL 晶界比例最高的图 3a 中各晶粒团簇内的几个不相邻晶粒之间的取向差列于表 3.分析表 3 中的数据可以看出,同一个团簇内的晶粒虽然互不相邻,但它们之间仍然保持着 $\Sigma3^n$ 的取向差关系,这就是低 ΣCSL 晶界比例高时出现晶粒团簇的内部特征.

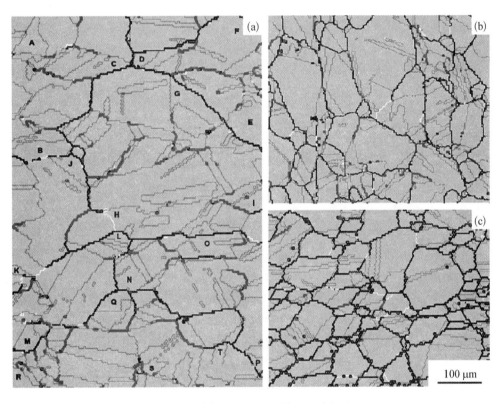

图 3 样品 S1,S2,S3 的 OIM 图

Fig. 3 OIM maps for specimen S1 (a),S2 (b),S3 (c): the thin gray lines denote $\Sigma3$ boundaries, the thick gray lines denote $\Sigma9$ and $\Sigma27$ boundaries, the thin white lines denote $\Sigma1$ and other low ΣCSL boundaries

比较图 3a,3b,3c 不难发现,晶粒团簇尺寸有明显差别,随低 ΣCSL 晶界比例降低而递减,并且团簇内 $\Sigma3^n$ 晶界数量也是递减的.其中样品 S1 的低 ΣCSL 晶界比例最高,所对应的晶粒团簇尺寸最大,并且团簇内有许多 $\Sigma3^n$ 晶界,构成 $\Sigma3$-$\Sigma3$-$\Sigma9$ 或 $\Sigma3$-$\Sigma9$-$\Sigma27$ 的三叉界角(triple junction);而样品 S3 的低 ΣCSL 晶界比例低,所对应的晶粒团簇尺寸小,团簇内

表 3　图 3a 中晶粒团簇内不相邻晶粒之间的取向差

Table 3　Misorientations between non-adjacent grains in some grains-cluster of Fig. 3a

Non-adjacent grains	Misorientation	Closest CSL	Deviation, $\Delta\theta/(°)$
A/B	$59.5°[1\,\bar{1}\,\bar{1}]$	$\Sigma 3$	0.6
A/C	$59.2°[\bar{1}\,\bar{1}\,1]$	$\Sigma 3$	0.9
B/C	$39.1°[\bar{1}\,90\,\bar{20}]$	$\Sigma 9$	0.3
D/E	$38.9°[\bar{1}\,\bar{1}\,0]$	$\Sigma 9$	0.6
D/F	$0.3°[2724]$	$\Sigma 1$	0.3
E/F	$39.1°[\bar{1}\,\bar{1}\,0]$	$\Sigma 9$	0.3
G/H	$59.8°[111]$	$\Sigma 3$	0.5
G/I	$35.8°[\bar{1}\,30\,\bar{2}\,7]$	$\Sigma 27b$	0.5
H/I	$39.1°[0\,1\,\bar{1}]$	$\Sigma 9$	0.3
J/K	$59.8°[\bar{1}\,\bar{1}\,\bar{1}]$	$\Sigma 3$	0.7
L/M	$59.5°[\bar{1}\,\bar{1}\,\bar{1}]$	$\Sigma 3$	0.7
N/O	$59.7°[\bar{1}\,\bar{1}\,1]$	$\Sigma 3$	0.5
N/P	$38.8°[130\,\bar{1}\,4]$	$\Sigma 9$	1.5
O/P	$35.3°[0\,\bar{8}\,15]$	$\Sigma 27b$	1.0
Q/R	$38.7°[\bar{5}\,13\,\bar{2}\,\bar{2}]$	$\Sigma 81a$	0.3
Q/S	$38.9°[22\,\bar{1}\,\bar{2}\,\bar{1}]$	$\Sigma 9$	1.2
Q/T	$31.3°[\bar{1}\,0\,1]$	$\Sigma 27a$	0.3
R/S	$40.0°[22\,\bar{2}\,\bar{1}\,0]$	$\Sigma 9$	1.1
R/T	$59.9°[17\,18\,\bar{1}\,7]$	$\Sigma 3$	0.9
S/T	$59.4°[111]$	$\Sigma 3$	0.8

很少有 $n>1$ 的 $\Sigma 3^n$ 晶界(如 $\Sigma 9$ 和 $\Sigma 27$),而只出现 $\Sigma 3$ 晶界,这种晶界以单独的直线或直线对的形态存在于团簇内,如图 3c 所示.

3　讨论

关于中低层错能面心立方金属形变后退火过程中晶界特征分布的演化机制主要有两种观点.其一是 Randle 等提出的"$\Sigma 3$ 再激发模型"($\Sigma 3$ regenerationmodel)[19].该模型认为再结晶时,2 个含有共格 $\Sigma 3$ 晶界的再结晶晶核长大相遇,相接触部分形成高可动性晶界,该晶界与其中 1 个晶核内的共格 $\Sigma 3$ 晶界相连.该高可动性晶界迁移的同时使得与它相连的共格 $\Sigma 3$ 晶界变长,与另一晶核内的共格 $\Sigma 3$ 晶界相遇产生 $\Sigma 9$ 晶界,从而形成 $\Sigma 3$-$\Sigma 3$-$\Sigma 9$ 这种三叉界角.其中 $\Sigma 9$ 晶界比另外 2 条共格 $\Sigma 3$ 晶界有更高的可动性,它继续迁移与其他共格 $\Sigma 3$ 晶界相遇形成非共格 $\Sigma 3$ 晶界,1 个新的 $\Sigma 3$-$\Sigma 3$-$\Sigma 9$ 三叉界角得以产生.其中那条非共格 $\Sigma 3$ 晶界具有最高可动性,而继续迁移直到耗尽驱动力或再与其他晶界相遇.如此进行下去可实现材料的 $\Sigma 3^n$ 晶界比例提高.但这种模型有一个前提,就是最初 2 个再结晶晶核的取向必须相同或十分相近[20].如果按照定向成核的机制,那么只有在形成了很强的变形织构后,Randle 等提出的机制才可能发生作用.

Kumer 等人通过透射电镜观察 600 合金形变后的退火过程,提出另一种晶界分解的机制[21].材料变形后,形变储能以位错的形式均匀分布在材料中,在随后的退火过程中,Kumer 等认为不产生再结晶晶核,而是形变诱发高 Σ 值的 CSL 晶界迁移(strain-inducedboundary migration).这种晶界迁移将形变组织扫除的同时,该高 Σ 值的 CSL 晶界会分解成 $\Sigma 3$ 晶界和另一可动性更高的 CSL 晶界. 比如 $\Sigma 81$ 晶界分解成 $\Sigma 3$ 晶界和 $\Sigma 27$ 晶界, $\Sigma 51$ 晶界分解成 $\Sigma 3$ 晶界和 $\Sigma 17$ 晶界等, 其中共格 $\Sigma 3$ 晶界能量低, 可动性差, 不易迁移而保留下来, 另一晶界可动性高继续迁移直到再次发生分解, 这样使材料中的低 ΣCSL 晶界比例提高.

本实验所用的材料在变形前没有织构, 如图 1 所示. 并且 5%的形变量也不足以产生形变织构, 那么, 很难想象不同的再结晶晶核会有相同的取向. 在这样的情况下, 同样会得到高比例的 $\Sigma 3^n$ 晶界. 这说明 Randle 等提出的"$\Sigma 3$ 再激发模型"很难解释本实验得到的结果. 一般认为高 Σ 值的 CSL 晶界结构已没有特殊性, 而被视作随机晶界, 固溶后时效处理时, 碳化物会在该类晶界上大量析出. 形变后在 1 100 ℃退火时, 碳化物溶解需要一定时间, 所以在短时间的退火过程中, 这些晶界仍会被碳化物钉扎而很难迁移, 也就不会发生 Kumer 所说的晶界迁移过程中的分解. 本实验采用时效后的样品进行变形和退火处理, 同样得到了高比例的 $\Sigma 3^n$ 晶界, 所以 Kumer 的机制也不能解释本实验得到的结果.

冷轧变形量对 690 合金再结晶后晶界特征分布的影响最大, 这说明再结晶过程在形成晶界特征分布中起了关键作用. 一般认为再结晶包括再结晶晶核的形成、再结晶晶核消耗变形组织而长大并彼此相遇的过程. 再结晶晶核与形变基体之间是一般大角晶界, 它的迁移是"扫除"形变基体的过程[22]. 由于 690 合金的层错能低, 所以这些再结晶晶核外的一般大角晶界, 在迁移吸收位错时很容易发生堆垛层错, 产生退火孪晶[23]. 这种一般大角晶界在靠位错密度差发生迁移时, 不断产生退火孪晶, 由于发生多重孪晶而产生了孪晶链[24], 它们之间就会成为 $\Sigma 3^n$ 的取向差关系. 退火孪晶会在不同的(111)面上产生, 这种孪晶相遇后会形成 $\Sigma 9$ 晶界; 而发生多重孪晶后, 这种孪晶之间的相遇又会形成 $\Sigma 3^n$ 晶界如 $\Sigma 27$, $\Sigma 81$ 等晶界, 所以孪晶比例高的样品中 $\Sigma 9$ 和 $\Sigma 27$ 晶界比例也相对高. 图 3 中同一个团簇内的晶粒虽然互不相邻, 但它们之间仍然保持着 $\Sigma 3^n$ 的取向差关系, 因此, 作者认为图 3 中的晶粒团簇是由同一个再结晶晶核在长大过程中不断产生退火孪晶, 发生多重孪晶形成了孪晶链造成的. 单位面积的形核数量, 即形核密度与形变量有关, 形变量越大, 形核密度就越高, 再结晶晶核在吞并形变基体的过程中所扩张的面积小, 一般大角晶界迁移的距离也就短, 没有更多的机会产生多重孪晶, 致使 $\Sigma 3^n$ 晶界比例低. 所以, 晶粒团簇尺寸和团簇内的 $\Sigma 3^n$ 晶界数量随冷轧压下量的增加而下降, 这样就造成 $\Sigma 3^n$ 晶界比例随冷轧压下量的增加而下降.

4 结论

(1) 冷轧 5%再在 1 100 ℃短时间再结晶退火可使 690 合金低 ΣCSL 晶比例提高到 70%以上(Palumbo‐Aust 标准), 同时形成大尺寸的晶粒团簇. 团簇内的晶粒之间有 $\Sigma 3^n$ 取向差关系.

(2) 晶粒团簇的尺寸和内含 $\Sigma 3^n$ 晶界的数量随冷轧压下量的增加而下降, 从而使低 ΣCSL 晶比例也随冷轧压下量的增加而下降.

(3) 初始状态无论是固溶或时效, 对 690 合金经过冷轧变形再在 1 100 ℃再结晶退火后的晶界特征分布没有明显影响.

参 考 文 献

[1] Watanabe T. *Res Mechanica*[J]. 1984, 11(1): 47.

[2] Randle V. *The Role of the Conincidence Site Lattice in Grain Boundary Engineering*[M]. Cambridge: Cambridge

University Press, 1996: 2.

[3] Palumbo G, Erb U. *MRS Bulletin*[J]. 1999, 24: 27.

[4] Lin P, Palumbo G, Erb U. *Scripta Metallurgica et Materialia*[J], 1995, 33(9): 1387.

[5] Shimada M, Kokama H, Wang Z J. *Acta Materialia*[J]. 2002, 50: 2331.

[6] Crawford D C, Was G S. *Metallurgical Transactions A*[J]. 1992, 23(A): 1195.

[7] Watanabe T, Tsurekawa S. *Acta Mater*[J]. 1999, 47(15): 4171.

[8] Hong Y B, Kokawa H, Zhan J W. *Scripta Materialia*[J]. 2003, 49: 219.

[9] Duh T S, Kai J J, Chen F R *et al*. *Nucl Mater*[J]. 1998, 258: 2064.

[10] Thaveeprungsriporn V, Was G S. *Metallurgical and Materials Transactions A*[J], 1997, 28: 2101.

[11] Lehockey E M, Palumbo G. *Materials Science and Engineering A*[J]. 1997, 237: 168.

[12] Randle V, Owen G. *Acta Mater*[J]. 2006, 54: 1777.

[13] Qiu Shaoyu(邱少宇), Su Xingwan(苏兴万), Wen Yan(文燕). *Nuclear Powered Engineering*(核动力工程)[J]. 1995, 16: 336.

[14] Lemire R J, McRae G A. *Journal of Nuclear Materials*[J]. 2001, 294: 141.

[15] Randle V. *Microtexture Determination and Applications*[M]. London: Maney Publishing, 2003: 3.

[16] Palumbo G. *Scripta Materialia*[J]. 1998, 38(11): 1685.

[17] Gertsman V Y, Henager C H. *Interface Sci*[J]. 2003, 11: 403.

[18] Kopezky C V, Andreeva A V, Sukhomlin G D. *Acta Metall Mater*[J]. 1991, 39: 1603.

[19] Randle V. *Acta Mater*[J]. 1999, 47: 4187.

[20] Randle V. *Acta Mater*[J]. 2004, 52: 4067.

[21] Kumar M, Schwartz A J, King W E. *Acta Mater*[J]. 2002, 50: 2599.

[22] Humphreys F J, Hatherly M. *Recrystallization and Related Annealing Phenomena*, Second ed[M]. Oxford: Elsevier, 2004: 215.

[23] Gottstein G. *Acta Metall*[J]. 1984, 32: 1117.

[24] Berger A, Wilbrandt P J, Ernst F. *Progress in Materials Science*[J]. 1988, 32: 1.

Effect of Deformation and Heat – Treatments on the Grain Boundary Character Distribution for 690 Alloy

Abstract: Effects of initial state, cold rolling reduction ratio, and annealing at 1 100 ℃ on the grain boundary character distribution (GBCD) in 690 alloy have been investigated by electron backscatter diffraction (EBSD) and orientation imaging microscopy (OIM). It can increase the grain boundary ratio of low ΣCSL (coincidence site lattice, $\Sigma \leqslant 29$ by Palumbo – Aust criterion) to be more than 70%, and simultaneously forms the grains-cluster of large size by cold rolling of 5% and subsequent annealing high at 1 100 ℃. The proportion of low ΣCSL grain boundaries and the size of grains-cluster decrease with the increase of strain. The initial state, i. e. solution annealed or aged, does not affect the GBCD obviously.

微量 Nb 的添加对 Zr-4 合金耐疖状腐蚀性能的影响*

摘　要：采用 500 ℃/10.3 MPa 过热蒸汽中腐蚀的方法,研究了添加微量合金元素 Nb 对 Zr-4 合金疖状腐蚀行为的影响. 结果表明：添加 0.05% Nb 后就可以观察到改善 Zr-4 合金耐疖状腐蚀性能的作用,提高到 0.1% 后可进一步改善 Zr-4 合金的耐疖状腐蚀性能,但还不能完全抑制 Zr-4 合金的疖状腐蚀现象.

1　前言

　　锆合金是核反应堆中的一种重要结构材料,用作核燃料的包壳. Zr-2 合金(Zr-1.2～1.7Sn-0.07～0.2Fe-0.05～0.15Cr-0.03～0.08Ni,质量分数%,以下同)是沸水堆中燃料元件使用的包壳材料, Zr-4 合金(Zr-1.2～1.7Sn-0.18～0.24Fe-0.07～0.13Cr)是目前我国大多数压水堆中燃料元件使用的包壳材料. 在沸水堆中或压水堆中一回路冷却水的氧含量较高时, Zr-2 和 Zr-4 合金会发生不均匀的疖状腐蚀. 在堆外一般用高压釜在 450～500 ℃ 过热蒸汽中的腐蚀试验来研究和评价锆合金的耐疖状腐蚀性能. 包壳发生疖状腐蚀时,在黑色氧化膜上出现白色斑点,直径 0.1～0.5 mm,白斑的剖面呈凸透镜状,最厚处可大于 100 μm,比周围的黑色氧化膜厚数十倍,当白斑继续长大,最终会连成白色氧化膜,白色氧化膜比较疏松,在水的冲刷下易于剥落而加速腐蚀[1-3]. 因此,如何提高 Zr-2 和 Zr-4 合金的耐疖状腐蚀性能是一个值得深入研究的问题.

　　通过改变热加工工艺可以改善 Zr-2 和 Zr-4 合金的耐疖状腐蚀性能,文献中一般采用控制累积退火参数 A 值的方法[4],A=$\sum t_i \exp(-40\,000/T_i)$,它表示 β 相淬火后各次加热温度 T_i 和加热时间 t_i 归一化处理后的数值[4]. 研究表明要获得较好的耐疖状腐蚀性能,需要控制 A <10^{-18} h,由于 A 值较小时,获得的第二相也比较小,因此有学者认为合金中较小的第二相(<0.2 μm)对改善耐疖状腐蚀性能是有利的[5,6]. 但我们在以前的研究中却发现,Zr-4 合金经 820 ℃/1 h(A=1.28×10^{-16} h)处理后,样品中的第二相可以达到 0.5 μm,但是耐疖状腐蚀性能明显优于经 600 ℃/1 h 处理的样品;通过 β 相水淬处理提高 Fe、Cr 合金元素在 α-Zr 中的过饱和固溶含量后又进一步改善了 Zr-4 合金的耐疖状腐蚀性能[7]. 因此,我们认为提高 Zr-4 合金耐疖状腐蚀性能的主要原因是提高了 Fe、Cr 合金元素在 α-Zr 中的固溶含量,而不是第二相的大小. 另外,在研究新锆合金时,周邦新等[8,9]发现 Nb 含量高于 0.24% 的几种 Zr-Sn-Nb 锆合金在 500 ℃ 过热蒸汽中腐蚀数百小时也未出现疖状腐蚀现象,这可能与 Nb 在 α-Zr 中的固溶度比 Fe 和 Cr 大有关. 添加微量合金元素 Nb(0.05%～0.1%)对 Zr-4 合金的耐疖状腐蚀性能究竟有多大影响,这是值得研究的问题.

---

* 本文合作者：姚美意、李强、夏爽、刘文庆. 原发表于《上海金属》,2008,30(6)：1-3.

2 实验方法

以 Zr-4 为母合金,分别加入 0.05% 和 0.1% 的 Nb,在实验室规模下用非自耗真空电弧炉熔炼成约 60 g 合金锭,熔炼时通高纯 Ar 气保护. 为保证成分均匀,合金锭每熔炼一次都要翻转,共熔炼 6 次. 为了能在相同的制备工艺条件下比较添加 Nb 后的合金与 Zr-4 合金的疖状腐蚀行为差别,将 Zr-4 母合金也重熔成锭子. 这些合金锭加热至 700 ℃,在压机上经过多次反复热压成规则的条块,再在 700 ℃热轧成 3 mm 的坯料,经过轻微冷轧使表面氧化膜破裂再用酸洗方法去除氧化膜,经过水洗干燥,在 β 相加热(1 020 ℃/0.5 h)均匀化处理,再经过多次冷轧、720 ℃/10 h 中间退火和最终 600 ℃/2 h 再结晶退火等工艺制成腐蚀试验用 0.6 mm 厚的片状试样. 采用 720 ℃/10 h 中间退火工艺是为了获得较大的 A 值($A=3.2\times10^{-17}$ h),在这样的情况下比较添加微量合金元素 Nb 对 Zr-4 合金耐疖状腐蚀行为的影响. 用电感耦合等离子体原子发射光谱(ICPAES)分析得到的合金成分列在表 1 中,结果表明 Nb 的含量与设计值接近,Sn、Fe、Cr 的损耗很少,仍在 Zr-4 合金规定的范围内.

表 1 试验用的几种合金成分

合 金 编 号	合金元素含量(质量%)			
	Sn	Fe	Cr	Nb
Zr-4 重熔	1.44	0.25	0.11	—
Zr-4+0.05Nb	1.41	0.26	0.10	0.054
Zr-4+0.1Nb	1.33	0.22	0.11	0.10

将上述处理的样品按标准方法用混合酸(体积分数为 10% HF+45% HNO$_3$+45% H$_2$O)酸洗和去离子水清洗后,放入高压釜中进行 500 ℃/10.3 MPa 过热蒸汽的腐蚀试验,研究它们的疖状腐蚀行为. 用 10 块试样平均值得出腐蚀增重.

3 实验结果及讨论

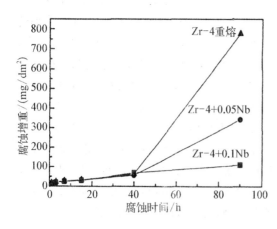

图 1 Zr-4 重熔合金和添加微量 Nb 的 Zr-4 合金在 500 ℃/10.3 MPa 过热蒸汽中的腐蚀结果

图 1 和图 2 分别是 Zr-4 重熔以及在 Zr-4 成分的基础上添加微量合金元素 Nb 的合金在 500 ℃/10.3 MPa 过热蒸汽中的腐蚀增重随时间的变化曲线和腐蚀 90 h 后样品表面的形貌. 从图 1 和图 2 可以看出,Zr-4 重熔样品的腐蚀增重最大,腐蚀 90 h 的增重接近 800 mg/dm^2,腐蚀样品表面出现了许多疖状腐蚀斑,疖状腐蚀最严重;Zr-4 中添加 0.05% Nb 后明显改善了 Zr-4 合金的耐疖状腐蚀性能,腐蚀 90 h 的增重不到 400 mg/dm^2,腐蚀样品表面疖状腐蚀斑的数量显著减少,尺寸也相应减小;添加 0.1% Nb 后又进一步改善了 Zr-4

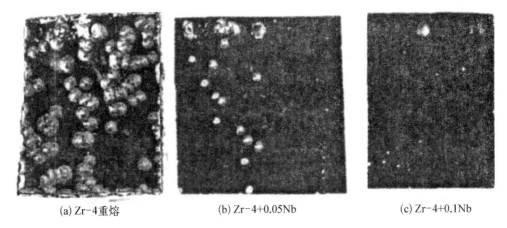

(a) Zr-4重熔　　　　　　(b) Zr-4+0.05Nb　　　　　　(c) Zr-4+0.1Nb

图2　Zr-4重熔合金和添加微量 Nb 的 Zr-4 合金在 500 ℃/10.3 MPa 过热蒸汽中腐蚀90 h后样品表面的形貌

合金的耐疖状腐蚀性能,腐蚀 90 h 的增重只有 110 mg/dm²,样品表面只有一些零星细小的疖状腐蚀斑.这说明在 Zr-4 合金中添加微量合金元素 Nb 后确实可以改善 Zr-4 合金的耐疖状腐蚀性能.

在开发 Zr-Sn-Nb 新锆合金的研究中,周邦新等[8]发现 Nb 含量在 0.24%～1.03%之间的几种锆合金在 500 ℃过热蒸汽中腐蚀 650 h 后都没有出现疖状腐蚀现象,认为 Nb 对锆合金的疖状腐蚀有"免疫性".在进一步研究热处理对我国自主研发的 N18 合金(Zr-1.04Sn-0.33Nb-0.38Fe-0.07Cr)的疖状腐蚀行为影响时也发现,在不同相区加热处理的 N18 锆合金样品,无论是采用快冷还是缓冷,获得的显微组织中不管有无残留的 β-Zr,它们经过 500 ℃过热蒸汽腐蚀 1 100 h 后也都没有出现疖状腐蚀现象[9]. Jeong 等[10]在研究 Nb 含量对 Zr-xNb (x=0～0.6%)合金在 500 ℃过热蒸汽中腐蚀行为影响时,发现当 Nb 含量低于 0.2%时还会出现疖状腐蚀现象.本文的研究结果表明,微量合金元素 Nb 可以明显改善 Zr-4 合金的耐疖状腐蚀性能,但添加 0.1% Nb 后还不能完全抑制疖状腐蚀现象.

4　结论

在 Zr-4 成分的基础上添加 0.05%～0.1% Nb 后可以明显改善 Zr-4 合金的耐疖状腐蚀性能,并随 Nb 含量增加,改善耐疖状腐蚀性能的作用越明显,但添加 0.1% Nb 还不能完全抑制 Zr-4 合金的疖状腐蚀现象.

参 考 文 献

[1] Ogata K，Mishima Y，Okubo T，et al. A systematic survey of the factors affecting Zircaloy nodular corrosion[C]// Zirconium in the Nuclear Industry：Eighth international symposium, ASTM STP 1023. Philadelphia：American Society for Testing and Materials. 1989：291-314.

[2] Cheng B, Adamson R B. Mechanism studies of Zircaloy nodular corrosion[C]//Zirconium in the Nuclear Industry：Seventh international symposium, ASTM STP 939, American Society for Testing and Materials, Philadelphia, 1987, 387.

[3] Zhou B X. Electron microscopy study of oxide films formed on Zircaloy-2 in superheated steam[C]//Zirconium in the Nuclear Industry：Eighth international symposium, ASTM STP 1023. Philadelphia：American Society for

Testing and Materials. 1989: 360 - 373.

［4］Thorvaldsson T，Andersson T，Wilson A，et al. Correlation between 400 ℃ steam corrosion behavior，heat treatment，and microstructure of Zircaloy - 4 tubing［C］//Zirconium in the Nuclear Industry：Eighth international symposium，ASTM STP 1023. Philadelphia：American Society for Testing and Materials. 1989：128 - 140.

［5］Garzarolli F，Steinberg E，Weidinger H G. Microstructure and corrosion studies for optimized PWR and BWR Zircaloy cladding［C］//Zirconium in the Nuclear Industry：Eighth international symposium，ASTM STP 1023. Philadelphia：American Society for Testing and Materials. 1989：202 - 212.

［6］Foster P，Dougherty J，Burke M G，et al. Influence of final recrystallization heat treatment on Zircaloy - 4 strip corrosion［J］. J. Nucl. Mat.，1990，173（2）：164 - 178.

［7］周邦新，李强，姚美意，等. 热处理影响 Zr - 4 合金耐疖状腐蚀性能的机制［J］. 稀有金属材料与工程，2007，36（7）：1129 - 1134.

［8］周邦新，赵文金，等. 新锆合金的研究［C］//96 中国材料研讨会，生物及环境材料，Ⅲ - 2. 北京：化学工业出版社. 1997：183 - 186.

［9］周邦新，姚美意，李强，等. Zr - Sn - Nb 合金耐疖状腐蚀性能的研究［J］. 稀有金属材料与工程，2007，36（8）：1317 - 1319.

［10］Jeong Y H，Kim H G，Kim D J. Influence of Nb concentration in the α - matrix on the corrosion behavior of Zr - xNb binary alloys［J］. J. Nucl. Mater.，2003，323：72 - 80.

Effect of Dilute Nb Addition on Nodular Corrosion
Resistance of Zircaloy - 4 Alloy

Abstract：The effect of dilute alloying element addition of Nb on the nodular corrosion resistance of Zircaloy - 4 alloy was investigated in super heated steam at 500 ℃ and 10. 3 MPa by autoclave tests. Results showed that addition of 0. 05% Nb in Zircaloy - 4 could improve the nodular corrosion resistance obviously. Adding the Nb amount to 0. 1% in Zircaloy - 4 could further greatly enhance the nodular corrosion resistance，but it had not suppressed the nodular corrosion phenomenon completely.

Zr-4 合金薄板的织构与
耐疖状腐蚀性能的关系*

摘　要：采用 500 ℃/10.3 MPa 过热蒸汽中腐蚀的方法,研究热处理及织构取向对 Zr-4 合金薄板耐疖状腐蚀性能的影响. 样品在 820 ℃-1 h 加热快冷后,晶粒发生长大,得到了集中的 $(0001)[11\bar{2}0]$ 和 $(0001)[10\bar{1}0]$ 织构,这时耐腐蚀性能比较好的 (0001) 面基本平行于轧面,而样品四周的侧面是耐腐蚀性能较差的 $(11\bar{2}0)$ 和 $(10\bar{1}0)$ 面. 样品经过这样热处理后,增加了 Fe 和 Cr 合金元素在 α-Zr 中的过饱和固溶含量,提高了样品轧面的耐疖状腐蚀性能,但是样品的侧面上仍然会出现疖状腐蚀现象. 耐疖状腐蚀性能的好坏与织构的晶体取向有关,只有当样品在 1 000 ℃ β 相加热快冷,使 Fe 和 Cr 合金元素在 α-Zr 中的过饱和固溶含量进一步得到提高后,才能完全抑制疖状腐蚀现象.

　　锆合金作为核燃料元件的包壳,在核动力反应堆中是一种重要的结构材料. 锆合金包壳在高温高压水中或过热蒸汽中服役时会发生腐蚀. 这种腐蚀过程可分为均匀腐蚀和不均匀腐蚀(疖状腐蚀),在水质经过加氢除氧的压水堆工况下是均匀腐蚀;在沸水堆工况下,或者是在水质未经加氢除氧处理的压水堆工况下,会出现不均匀的疖状腐蚀. 将锆合金样品放入高压釜中进行 450~500 ℃ 过热蒸汽中腐蚀时,会产生疖状腐蚀,因而可以用这种比较简单的方法来评价锆合金的耐疖状腐蚀性能.

　　Zr-4 合金在加工过程中,通过控制累积退火参数 A 可改善耐疖状腐蚀性能,希望 $A<10^{-18}$ h[1]. 参数 A 是成材过程中坯料经 β 相淬火后,后续加工过程中所有加热温度和加热时间综合参数的总和. 要获得小的 A 值,就是要在 β 相淬火后的后续加工过程中,控制较低的退火温度和较短的退火时间. 这时,β 相淬火后过饱和固溶在 α-Zr 中的合金元素重新析出形成的第二相比较小. 因而,有研究者将耐疖状腐蚀性能与第二相大小联系起来,认为增大第二相的尺寸会使耐疖状腐蚀性能变坏[1-2]. 但 β 相淬火后采用较低的退火温度和较短的退火时间时,过饱和固溶在 α-Zr 中的合金元素重新析出并不充分,一部分 Fe 和 Cr 仍然以过饱和固溶状态存在. 究竟是第二相大小影响疖状腐蚀性能,还是 α-Zr 中固溶的 Fe 和 Cr 合金元素含量影响疖状腐蚀性能,至今还未获得一致的认识.

　　疖状腐蚀是一种成核长大过程,本文作者之一曾对其形成机理提出过一种解释,认为取向不同的晶粒受到腐蚀时氧化膜生长的各向异性是导致疖状腐蚀斑成核的原因[3]. 早期 Bibb 等和 Wilson 用锆单晶在 360 ℃ 高温水或空气中腐蚀氧化后,都观察到氧化膜在锆晶体上生长时的各向异性特征,这些结果在 Cox 最近的文献综述中都有叙述[4]. 近年来,Kim 等[5]用高纯锆的大晶粒样品在 360 ℃ 高温水中腐蚀后,也观察到氧化膜生长速度呈现明显的各向异性特征,在 (0001) 面上氧化膜的生长速度要比在 $(11\bar{2}0)$ 面上的慢很多.

　　我们研究了热处理对 Zr-4 合金耐疖状腐蚀性能的影响[6],观察到增加 Fe 和 Cr 合金元素在 α-Zr 中的过饱和固溶含量后,可以延缓甚至阻止疖状腐蚀斑的成核,因而提高了合

* 本文合作者：姚美意、李强、夏爽、刘文庆. 原发表于《上海大学学报(自然科学版)》,2008,14(5)：441-445.

金的耐疖状腐蚀性能. 如果取向不同的晶粒受到腐蚀时氧化膜生长的各向异性是导致疖状腐蚀斑成核的原因, 那么, 除了 Fe 和 Cr 合金元素在 α-Zr 中的过饱和固溶含量会影响耐疖状腐蚀性能之外, 样品的织构取向也将会影响它的耐疖状腐蚀性能, 本文将叙述这方面的研究结果.

1 实验方法

试验用样品是工厂生产的 Zr-4 薄板, 厚度为 0.6 mm, 出厂时经过 600 ℃再结晶退火处理. 为了改变 Fe 和 Cr 合金元素在 α-Zr 中的过饱和固溶含量, 重新将样品放入真空石英管式电炉中分别加热至 820 ℃保温 1 h 或 1 000 ℃保温 0.5 h, 然后将电炉推离石英管, 在石英管外淋水冷却, 称为"空冷". 下面将上述处理的样品分别标记为 820 ℃-1 h/AC 和 1 000 ℃-0.5 h/AC.

为了了解热处理后织构的变化, 样品经过电解抛光后, 用装备了 EBSD 附件的扫描电子显微镜对样品进行逐点逐行扫描, 通过测定电子背散射衍射菊池线花样来获得微区的晶体取向, 构建 (0001) 极图和样品轧制面法向 (ND)、轧向 (RD) 和横向 (TD) 的反极图, 用这些极图作为分析再结晶织构随热处理变化的依据.

在高压釜中用 500 ℃/10.3 MPa 过热蒸汽的腐蚀试验来研究疖状腐蚀问题. 样品在入釜前按标准方法用混合酸 (体积分数为 10% HF+45% HNO_3+45% H_2O) 进行酸洗和去离子水清洗, 经不同时间腐蚀后停釜冷却, 取样称重, 获得腐蚀增重与腐蚀时间之间的关系, 增重由 3 块试样平均得出.

2 实验结果及讨论

2.1 不同热处理后的显微组织与再结晶织构

样品经过 3 种不同热处理后的显微组织在前一篇文章中已有描述[6], 600 ℃退火后已经完全再结晶, 并有较多的 Zr(Fe, Cr)₂ 第二相; 820 ℃退火时处于 α 相上限温区, 晶粒发生长大, 第二相也发生了长大, 数量比 600 ℃退火后的明显减少; 1 000 ℃退火时处于 β 相温区, 合金元素 Fe 和 Cr 完全固溶在 β-Zr 中, 冷却时 β 相转变成板条状的 α 相晶粒, 晶界上析出了细小的 Zr(Fe, Cr)₂ 第二相. 由于加热温度不同, 并且采取了比较快的冷却, 样品中除了第二相的尺寸、数量及 Fe/Cr 比会发生变化外[7], Fe 和 Cr 合金元素在 α-Zr 中的过饱和固溶含量也会不同[8].

图 1、图 2 和图 3 分别是 3 种热处理样品的 (0001) 极图和 ND, RD 和 TD 的反极图. 样品经 600 ℃退火后, 从统计来看, 再结晶织构取向的 (0001) 面并不严格平行于轧面, 而是沿着轧制方向向着横向左右倾斜 28°, 从 RD 反极图可看出轧制方向的晶体取向分布在 [10$\bar{1}$0] 和 [11$\bar{2}$0] 之间; 820 ℃退火后因为晶粒长大, 再结晶织构变得比较集中, (0001) 面基本平行于轧面, 轧制方向的晶体取向和 600 ℃退火后的一样, 没有发生变化; 1 000 ℃加热冷却后, 因为发生了 $\alpha \rightarrow \beta \rightarrow \alpha$ 相变, 扰乱了原有的织构. 但是在发生相变时遵循着一定的晶体几何学关系: $(110)_\beta$//$(0002)_\alpha$, $\langle 111 \rangle_\beta$//$\langle 11\bar{2}0 \rangle_\alpha$, 所以 (0001)[11$\bar{2}$0] 和 (0001)[10$\bar{1}$0] 六方晶体结构的 α 相织构转变为体心立方晶体结构的 β 相后, 会得到 (110)[$\bar{1}$11] 和 (110)[$\bar{1}$13] 织构, 这

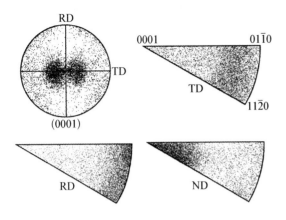

图 1 600 ℃退火后样品的再结晶织构：（0001）正极图；TD，RD，ND 反极图

Fig. 1 Recrystallization textures for specimens annealed at 600 ℃：（0001）pole figure；TD, RD and ND inversed pole figures

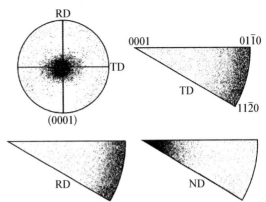

图 2 820 ℃- 1 h/AC 处理后样品的再结晶织构：（0001）正极图；TD，RD，ND 反极图

Fig. 2 Recrystallization textures for 820 ℃- 1 h/AC specimens：（0001）pole figure；TD，RD and ND inversed pole figures

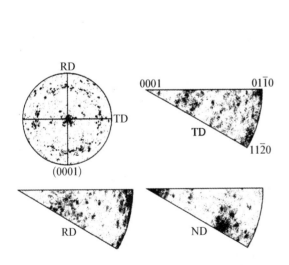

图 3 1 000 ℃- 0. 5 h/AC 处理后样品的再结晶织构：（0001）正极图；TD，RD，ND 反极图

Fig. 3 Recrystallization textures for 1 000 ℃- 0. 5 h/AC specimens：（0001）pole figure；TD，RD and ND inversed pole figures

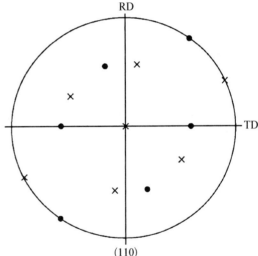

图 4 （110）$[\bar{1}11]$（·）和（110）$[\bar{1}13]$（×）织构的（110）标准极图

Fig. 4 （110）standard pole figure of （110）$[\bar{1}11]$（·）and （110）$[\bar{1}13]$（×）textures

两种织构的（110）标准极图如图 4 所示. 当样品冷却时,这两种 β 相织构取向中所有的{110}面都可能转变为 α 相的（0001）面,这样就扰乱了原来 α 相退火后的再结晶织构;但是平行于轧制面的（110）面转变为 α 相的（0001）面后,仍然会部分保留原来 α 相退火后的再结晶织构;而与轧面相差 60°的那些{110}面转变为 α 相的（0001）面后,就会形成（11$\bar{2}$1）面平行于轧面的"纤维"状织构,这从图 3 中的（0001）极图和 ND 的反极图中得到了证实;与轧面相差 90°的那些{110}面转变为 α 相的（0001）面后,应该会得到（10$\bar{1}$0）或者（11$\bar{2}$0）面平行于轧面的

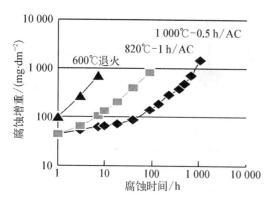

图5 不同热处理的样品在500 ℃/10.3 MPa过热蒸汽中腐蚀时的增重随时间变化曲线

Fig. 5 Weight gains of specimens corroded in the superheated steam at 500 ℃/10.3 MPa

"纤维"状织构,但是从图3的(0001)极图和ND的反极图中并没有观察到这类织构,是否是因为β→α相变时体积发生变化引起应力状态不同而造成的结果,这还有待进一步的研究.

2.2 腐蚀增重和样品表面形貌

图5是3种样品在500 ℃/10.3MPa过热蒸汽中腐蚀时的增重曲线.图6是3种样品经过不同时间腐蚀后表面形貌的实物照片.经过不同热处理的样品,它们的腐蚀增重曲线和腐蚀后的表面形貌有着明显的差别. 600 ℃再结晶退火处理的样品腐蚀增重最快,耐疖状腐蚀性能最差,腐蚀7 h后,样品的表面已经布满了疖状腐蚀斑,如图6(a)所示.

820 ℃-1 h/AC处理的样品腐蚀增重减缓,耐疖状腐蚀性能并没有因为第二相发生长大而变坏,反而得到了明显改善,这应该与增加了Fe和Cr合金元素在α-Zr中的固溶含量有关.在整个腐蚀过程中,片状样品的两个表面上始终没有出现疖状腐蚀斑,但是片状样品周边的耐腐蚀性能并不好,包括悬挂样品的小孔四周,如图6(b)所示.经过10 h腐蚀后,在样品的四周边上出现了白色疖状斑点,随着腐蚀时间增长,白色疖状斑点增多,逐渐连成了白色氧化膜的边框;经过90 h腐蚀后,由于样品周边上白色氧化膜发生严重剥落无法获得准确的腐蚀增重,不得不终止腐蚀试验,但是这时样品的两个表面上仍然没有出现白色疖状斑点.样品在腐蚀时出现这种现象,应该与样品的织构取向分布以及热处理后增加了Fe和Cr合金元素在α-Zr中的固溶含量有关.

1 000 ℃-0.5 h/AC处理的样品腐蚀增重最缓慢,耐疖状腐蚀性能最好,在1 100 h的腐蚀试验过程中,即使在样品的四周边上也没有出现白色疖状斑点,如图6(c)所示.虽然1 100 h腐蚀后的增重接近1 200 mg/dm²,氧化膜的厚度已经达到80 μm(15 mg/dm² = 1 μm),但是氧化膜仍然没有剥落.

(a) 600℃退火样品腐蚀7 h　(b) 850℃-1 h/AC样品腐蚀90 h　(c) 1 000℃-0.5 h/AC样品腐蚀1 100 h

图6 不同热处理的样品在500 ℃/10.3 MPa过热蒸汽中腐蚀不同时间后样品表面的形貌

Fig. 6 Surface morphology of specimens corroded in the superheated steam at 500 ℃/10.3 MPa

从 Zr－Fe 和 Zr－Cr 二元相图可以看出,在 820 ℃和 850 ℃加热时,Fe 和 Cr 合金元素在 α－Zr 中的固溶度可以分别达到 120 μg/g 和 200 μg/g[9],而在高温 β 相区,Zr－4 中的 Fe 和 Cr 合金元素可以全部固溶到 β 相中. 因此,当样品在 α 相上限温区或者 β 相区加热快冷后,Fe 和 Cr 合金元素一定会过饱和固溶在 α－Zr 中. 李聪等[8]用电解方法将 Zr－4 样品中 Zr(FeCr)$_2$ 第二相分离去掉后,测定了 α－Zr 中 Fe 和 Cr 合金元素的过饱和固溶含量,当样品在 β 相加热淬火并经过轧制加工和 470 ℃去应力退火,合金元素 Fe 和 Cr 在 α－Zr 中总的过饱和固溶含量达到了 1 300 μg/g. 本实验将样品加热到 820 ℃和 1 000 ℃后空冷,Fe 和 Cr 合金元素在 α－Zr 中的过饱和固溶含量一定会比 600 ℃退火处理后的高,且 1 000 ℃加热空冷的样品又会比 820 ℃加热空冷后的高.

合金元素固溶在 α－Zr 中后会影响氧化膜的生长速率,从而影响耐腐蚀性能. 增加合金元素 Fe 和 Cr 在 α－Zr 中的过饱和固溶含量后,如果可以减弱样品腐蚀时氧化膜生长的各向异性,那么,根据作者提出的疖状腐蚀成核机理假说[3],就可能延缓甚至阻止疖状腐蚀斑的成核,提高耐疖状腐蚀性能. 这可以说明为什么实验观察到样品经过 α 相上限温区或者 β 相区加热后快冷,样品的耐疖状腐蚀性能得到了明显改善甚至完全抑制了疖状腐蚀现象. 样品在 820 ℃(α 相上限温区)加热后快冷,一定程度上增加了合金元素 Fe 和 Cr 在 α－Zr 中的固溶含量,但由于织构取向的分布特征(见图 2),氧化膜生长较慢的(0001)面基本平行于样品的轧面,提高了轧面的耐疖状腐蚀性能,但是样品的周边是耐腐蚀性能较差的(10$\bar{1}$0)和(11$\bar{2}$0)面,由于 α－Zr 中 Fe 和 Cr 合金元素过饱和固溶含量还不足够高,这时在样品周边的表面上仍然会发生疖状腐蚀,因而 820 ℃－1 h/AC 处理的样品在 500 ℃过热蒸汽中腐蚀时,样品周边的表面上仍然会发生疖状腐蚀,疖状斑增多后形成了白色氧化膜的边框,直到氧化膜发生严重的剥落. 样品只有在 1 000 ℃(β 相区)加热快冷,进一步提高 α－Zr 中 Fe 和 Cr 合金元素过饱和固溶含量后,才能完全抑制疖状腐蚀现象(见图 5 和图 6(c)).

3 结论

(1) Zr－4 薄板样品经 600 ℃再结晶退火后,形成(0001)[11$\bar{2}$0]和(0001)[10$\bar{1}$0]织构,但(0001)面并不严格平行于轧面,而是绕轧向向着横向左右倾斜 28°. 在 α 相区 820 ℃加热时晶粒发生长大,平行于轧面的(0001)织构取向更为集中. 样品加热到 β 相区 1 000 ℃快冷后,虽然原来的织构被扰乱,但由于 α→β→α 相变时遵循一定的晶体学关系,最终仍然保留了一部分原有的再结晶织构取向,还形成了(11$\bar{2}$1)面平行于轧面的"纤维"状织构.

(2) 820 ℃－1 h/AC 热处理增加了 Fe 和 Cr 合金元素在 α－Zr 中的过饱和固溶含量,提高了轧面的耐疖状腐蚀性能. 但是在样品四周的侧面,主要是耐腐蚀性能较差的(10$\bar{1}$0)和(11$\bar{2}$0)面,当腐蚀时间增长后,样品的侧面上仍然会发生疖状腐蚀. 耐疖状腐蚀性能的好坏与织构取向有关.

(3) 样品在 1 000 ℃－0.5 h/AC(β 相区)处理后,Fe 和 Cr 合金元素在 α－Zr 中的过饱和固溶含量进一步得到了提高. 样品在 500 ℃过热蒸汽中经过 1 100 h 的腐蚀过程中都没有出现疖状腐蚀斑,可以完全抑制疖状腐蚀现象.

参 考 文 献

[1] GARZAROLLIF, STEINBERG E, WEIDINGER H G. Microstructure and corrosion studies for optimized PWR and

BWR Zircaloy cladding[C]//VAN SWAM L F P, EUCKEN C M. Zirconium in the Nuclear Industry: Eighth International Symposium, ASTM STP 1023. Philadelphia: American Society for Testing and Materials, 1989: 202 – 212.

[2] FORSTER J P, DOUGHERTY J, BURKE M G, et al. Influence of final recrystallization heat treatment on Zircaloy – 4 strip corrosion[J]. J NuclMat, 1990, 173: 164 – 178.

[3] ZHOU B X. Electron microscopy study of oxide films formed on Zircaloy – 2 in superheated steam[C]//VAN SWAM L F P, EUCKEN CM. Zirconium in the Nuclear Industry: Eighth International Symposium, ASTM STP 1023. Philadelphia: American Society for Testing and Materials, 1989: 360 – 373.

[4] COX B. Some thoughts on the mechanisms of in-reactor corrosion of zirconium alloys[J]. J Nucl Mat, 2005, 336: 331 – 368.

[5] KM H J, KM T H, JEONG Y H. Oxidation characteristics ofbasal (0002) plane and prism (11$\bar{2}$0)plane in HCP Zr [J]. J Nucl Mat, 2002, 306: 44 – 53.

[6] 周邦新,李强,姚美意,等. 热处理影响 Zr – 4 合金耐疖状腐蚀性能的机制[J]. 稀有金属材料与工程, 2007,36(7): 1129 – 1134.

[7] ZHOU B X, YANG X L. The effect of heat treatments on the structure and composition of second phase particles in Zircaloy – 4 [R]. China Nuclear Science and Technology Report, CNIC – 01073, SINRE – 0065. Beijing: China Nuclear Information Centre Atomic Energy Press, 1996.

[8] LIC, ZHOU B X, ZHAO W J, et al. Determination of Fe and Cr content in α – Zr solid solution of Zircaloy – 4 with different heat-treated states[J]. J Nucl Mat, 2002, 304: 134 – 138.

[9] CHARQUETD, HAHN R, ORTLIEB E, et al. Solubility limits and formation of intermetallic precipitates in ZrSnFeCr alloys[C]//VAN SWAM L F P, EUCKEN C M. Zirconium in the Nuclear Industry: Eighth International Symposium, ASTM STP 1023. Philadelphia: American Society for Testing and Materials, 1989: 404 – 422.

Relationship between Nodular Corrosion Resistance and Textures for Zircaloy – 4 Strip

Abstract: The effects of heat treatments and textures on the nodular corrosion resistance of Zircaloy – 4 strip have been investigated in superheated steam of 500 ℃ and 10. 3 MPa by autoclave tests. After heat treatment at 820 ℃– 1 h and fast cooling, concentrated textures with (0001) [11$\bar{2}$0] and (0001) [10$\bar{1}$0] orientations were obtained after grain growth, and the supersaturated solid solution of alloying elements of Fe and Cr in α – Zr matrix was increased simultaneously. (0001) planes with better corrosion resistance were mainly parallel to the rolling plane of the specimens, while (10$\bar{1}$0) and (11$\bar{2}$0) planes with worse corrosion resistance were on the side planes around the specimens. In this case, the nodular corrosion resistance on the rolling plane of the specimens was improved but the nodular corrosion phenomena could also appear on the side planes around the specimens. The nodular corrosion behavior was closely related to the texture orientations. And it could be completely suppressed when the supersaturated solid solution contents of Fe and Cr in α – Zr matrix were further increased after heat treatment in β phase.

用三维原子探针研究压力容器模拟钢中富铜原子团簇的析出[*]

摘　要：用三维原子探针(3DAP)和热时效处理方法研究压力容器模拟钢中富铜原子团簇的析出过程.提高了 Cu 含量的压力容器模拟钢样品经过 880 ℃加热淬火后,在 400 ℃和 500 ℃进行了不同时间的时效处理,显微硬度测试结果表明,在 400 ℃和 500 ℃时效的过程中硬度峰值分别出现在 100 h 和 5 h. 3DAP 分析结果显示,样品在 400 ℃分别时效 100 h,150 h 和 300 h 后,富铜原子团簇的数量密度是递增的,从 1.5×10^{23} m^{-3} 增加到 6.2×10^{23} m^{-3},但富铜原子团簇的长大非常缓慢,团簇的最大等效直径只从 2 nm 增大到了 3.5 nm,团簇中的 Cu 原子数分数 x 为 20%,还含有 Mn 和 Ni,并且观察到 Mn 和 Ni 在团簇和基体金属的界面处发生明显的富集.

目前,绝大多数核电站都采用轻水反应堆,其中又以压水堆为主.压水堆的压力容器采用低合金铁素体钢(A508-Ⅲ)制成,这类材料具有低温脆性,经受中子长期辐照后会使韧脆转变温度升高至室温以上,这影响到核电站的运行安全.由于核反应堆压力容器是不可更换的大型部件,因此,核电站的运行寿命与压力容器钢的辐照脆化过程密切相关,了解压力容器钢的辐照脆化机理,延缓发生脆化的过程,这对于延长核电站的服役期以及降低核电站成本非常重要.

国外大量研究工作已经证实[1-4],压力容器钢在服役的工况下(288 ℃)经过长期中子辐照以后,钢中微量的杂质 Cu 会以细小的富铜原子团簇析出,这是引起韧脆转变温度升高的主要原因.团簇中除 Cu 外,还富集了 Ni 和 Mn,有时还有 Si 和 P 等多种元素[5-9].目前,研究这种只有数纳米大小团簇的析出过程,三维原子探针(3DAP)是无比优越的研究手段[10].

压力容器钢经过中子辐照以后具有放射性,为了简化实验操作,有很多研究者采用了热时效处理的方法来研究模拟钢中富铜原子团簇的析出过程. Miller 等人对 Fe - Cu - Ni(其中 Cu 的质量分数 w 为 1.28%,Ni 的质量分数 w 为 1.43%)模拟钢进行了热时效模拟研究,研究结果表明模拟钢经过热时效处理后析出了富铜原子团簇,Ni 在富铜原子团簇的长大和粗化过程中迁移到了团簇的周围[11].三维原子探针对 Fe - Cu - Ni 和 Fe - Cu - Ni - Mn 合金热时效处理后进行了分析,表明 Ni 和 Mn 出现在 α - Fe 基体/富铜原子团簇界面处,富铜原子团簇的直径在 5～10 nm 之间[12-14].

本研究工作利用 3DAP 研究了压力容器模拟钢(提高了压力容器钢中 Cu 的含量)中富铜原子团簇在热时效时的析出过程,通过热时效处理的方法来模拟中子辐照引起富铜原子团簇的析出,以期了解压力容器钢辐照脆化机理中的一些基础问题.

1　实验方法

为了加速压力容器钢中 Cu 的析出过程,能够用热时效处理的方法研究富铜原子团簇析

---
* 本文合作者：朱娟娟、王伟、林民东、刘文庆、王均安.原发表于《上海大学学报(自然科学版)》,2008,14(5)：525-530.

出的一些规律,本实验所选用的材料是在 A508-Ⅲ 钢成分的基础上提高了 Cu 含量的压力容器模拟钢,钢锭由 50 kg 真空感应炉冶炼,其化学成分如表 1.

表 1　模拟钢的化学成分

Table 1　Composition of the model steel %

	Cu	Ni	Mn	Si	P	C	S	Mo	Fe
原子数分数 x	0.53	0.81	1.60	0.77	0.03	1.00	0.011	0.31	Balance
质量分数 w	0.60	0.85	1.58	0.39	0.016	0.22	0.006	0.54	Balance

将钢锭经过热锻和热轧制成 20 mm 厚的板坯,然后切下一块板坯加热到 1 000 ℃热轧成 4 mm 厚的钢板,最后切成 4 cm×3 cm 和 2 cm×3 cm 的小样品,加热到 880 ℃保温 0.5 h 后水淬;随后在 400 ℃和 500 ℃不同温度下进行等温时效处理,时效处理的保温时间从 0.5~1 000 h 不等.

时效处理后的试样经磨床将两面磨平并用细砂纸抛光,去除脱碳层(0.5 mm)后,在 HV-10 型维氏硬度计上测定硬度,压头所加载荷为 5 kg,加载时间为 15 s,显微硬度是 5 个点测量结果的平均值.

用线切割的方法在试样的中心部位切出边长为 0.5 mm 的方形细棒,长 20 mm,再采用电解抛光的方法将丝状样品制备成曲率半径小于 100 nm 的针尖样品,然后用 3DAP 对针尖样品进行分析,分析时样品冷却至 -203 ℃.

假设团簇为球形,用以下公式得出团簇的等效直径:

$$D = \left(\frac{6n\Omega}{\pi\zeta}\right)^{1/3}, \tag{1}$$

式中,n 为检测到团簇中的总原子数,ζ 为检测参数,值为 0.6,Ω 为原子的体积,对于 bcc 结构的 Fe 为 1.178×10^{-2} nm^{-3}[15]. 由于 Fe 原子数居多,所以以 Fe 原子体积来计算.

计算团簇数量的密度用以下公式:

$$N_v = \frac{N_p\zeta}{n\Omega}, \tag{2}$$

式中,N_p 是分析体积内所检测到的团簇数量,其他参数和式(1)相同.

2　实验结果

2.1　硬度测试

400 ℃和 500 ℃热时效处理后的硬度变化曲线如图 1 所示,从图中可以看出硬度值随着时效时间的增加先下降,后上升,然后又下降. 在 500 ℃时效过程中,硬度值在 5 h 达到峰值,而在 400 ℃时效过程中,硬度峰值出现在 100 h. 这说明时效温度越高,时效硬度峰值出现越早,而且 400 ℃时效时硬度峰值要比 500 ℃时的高.

时效过程中硬度的变化是马氏体的分解和富铜原子团簇析出共同作用的结果. 淬火后模拟钢样品的显微组织为板条状马氏体,碳在马氏体中处于过饱和状态,在时效时会发生分解析出碳化物,硬度下降;同时,在基体中又会析出富铜原子团簇,会使硬度升高,因为 Cu 在

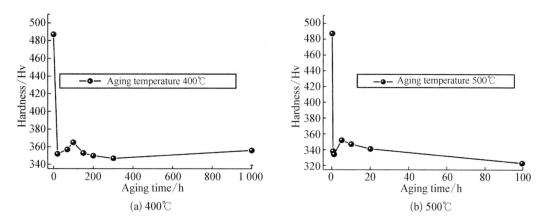

(a) 400 ℃ | (b) 500 ℃

图1 模拟钢样品经880℃加热水淬后分别在400℃和500℃等温时效处理后的硬度变化曲线

Fig. 1 Hardness variation of the model steel specimens aged at 400 ℃ and 500 ℃ after water quench from 880 ℃

400 ℃和500 ℃时平衡固溶度质量分数 w 分别为 0.02% 和 0.08%[11]，都远远低于本次实验模拟钢的 Cu 质量分数 0.6%. 时效初期，淬火后形成的马氏体开始分解，形成回火组织，产生软化作用，故硬度下降；随着时效时间的延长，Cu 原子逐渐偏聚，析出富铜原子团簇，使硬度逐渐上升，一直到达硬度的峰值；随后，富铜原子团簇开始粗化，导致硬度再次降低.

2.2 三维原子探针(3DAP)分析

本次实验分别对 400 ℃在 100 h，150 h 和 300 h 时效处理后的样品进行 3DAP 分析，分析结果如图 2 所示(图中给出了 100 h 和 300 h 的结果)，图中显示出了在纳米尺度范围内被

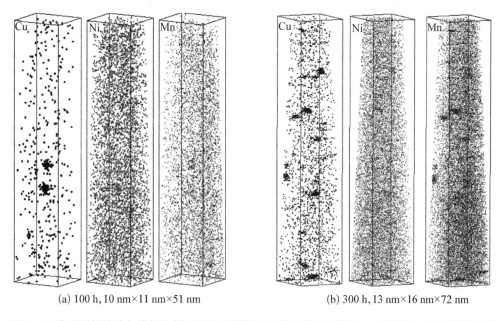

(a) 100 h, 10 nm×11 nm×51 nm | (b) 300 h, 13 nm×16 nm×72 nm

图2 三维原子探针分析模拟钢样品在 400 ℃等温时效处理后得到 Cu，Ni 和 Mn 的原子分布图

Fig. 2 Distribution of Cu, Ni and Mn atoms in model steel specimens after quenching and ageing at 400 ℃ for 100 and 300 h

检测体积中不同元素的原子分布,每一个点表示仪器探测到的一个原子.从图2中可以看出,100～300 h时效处理后样品中的富铜原子团簇的数量是递增的,而且除Cu在团簇中富集外,Mn和Ni元素的原子也在富铜原子团簇处发生明显的偏聚.三维原子探针测量结果用式(2)计算得出富铜原子团簇的数量密度列于表2中.从表中可以看出,富铜原子团簇的数量和数量密度从时效100 h,150 h到300 h都是增加的,数量密度从1.5×10^{23} m^{-3}增加到6.2×10^{23} m^{-3}. Auger等人对法国退役核电站的压力壳钢进行了研究,该压力壳服役了16a之久,其中子辐照注量为$(0.5～16) \times 10^{23}$ nm^{-2},检测到富铜原子团簇的数量密度为$10^{22}～10^{24}$ m^{-3},可见本次实验所得到的结果和真实压力容器钢经中子辐照后的结果相似.

表2　400 ℃时效不同时间后的富铜原子团簇的数量密度

Table 2　Number density of Cu-rich clusters in model steel specimens aged at 400 ℃ for 100, 150 and 300 h

	时效 100 h	时效 150 h	时效 300 h
被检测体积内观察到的团簇数量	2	7	18
团簇数量密度(10^{23} m^{-3})	1.5	2.6	6.2

表3中给出了不同时间时效后富铜原子团簇的等效直径平均值和团簇中Cu的平均含量.从表3中我们可以得出,模拟钢在400 ℃时效100,150和300 h后基体中Cu元素的含量较热时效处理前是明显减小的,但是从100～300 h时效后Cu的含量却是增加的,这可能是因为3DAP探测的体积很小,所以测量误差较大.另外从表3中还可以看出,时效不同时间析出富铜原子团簇中Cu的平均含量没有很大的变化,而且富铜原子团簇的等效直径也变化不大,但是通过分析每个团簇的等效直径(如图3所示),可以看到在时效100 h时,2个团簇的

表3　400 ℃时效100 h, 150 h和300 h时富铜原子团簇等效直径的平均值和团簇中铜的平均含量

Table 3　Average equivalent diameter of Cu-rich a clusters and average content of Cu in Cu-rich clusters

	时效 100 h	时效 150 h	时效 300 h
团簇等效直径平均值/nm	2.0 ± 0.1	2.4 ± 0.4	2.2 ± 0.4
团簇中铜平均原子数分数 $x/\%$	19.71 ± 1.95	20.21 ± 0.78	18.94 ± 0.53
基体中铜的原子数分数 $x/\%$	0.08 ± 0.01	0.12 ± 0.01	0.20 ± 0.01

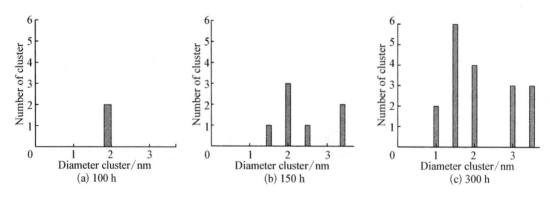

图3　模拟钢样品在400 ℃等温时效处理后的团簇等效直径分布和探测到的团簇数量

Fig. 3　Distribution of equivalent diameter of Cu-rich clusters and the detected number in the model steel specimens aged at 400 ℃ for 100, 150 and 300 h

等效直径是在 1.7～2.0 nm 之间,时效 150 h 时,团簇等效直径分布在 1.0～3.3 nm 之间,在时效 300 h 时,最大团簇等效直径增加到了 3.5 nm,可见在时效过程中先析出的团簇在不断长大,同时又有新的团簇不断析出. 但是富铜原子团簇在 400 ℃时效时的长大是很慢的.

为了更进一步分析富铜原子团簇中各元素的含量,从图 2(b)中截取出一个富铜原子团簇,图 4(a)所示的是该团簇中 Cu, Mn 和 Ni 的原子分布图,所截取的体积为 3 nm×3 nm×7 nm;在分析团簇中各元素含量时, 3DAP 分析软件把原子之间距离大于 0.5 nm 的原子不计算在团簇中[15],因此得到了该富铜原子团簇中 Cu, Mn 和 Ni 的含量分布图(图 4(b)),从图 4(a)中看出,在 Cu 原子富集处,Mn 和 Ni 原子也发生明显的偏聚;从图 4(b)中可以看出,在这个富铜原子团簇中 Cu 和 Mn 的含量最高值原子数分数 x 为 10%,Ni 的含量最高到了 5%,而且 Mn 和 Ni 在铜团簇和基体金属的界面处发生明显的富集,这和文献[9-14]得到的结果一致. 当团簇长大时,必然要排斥 Mn 和 Ni 原子,这可能是富铜原子团簇长大慢的原因.

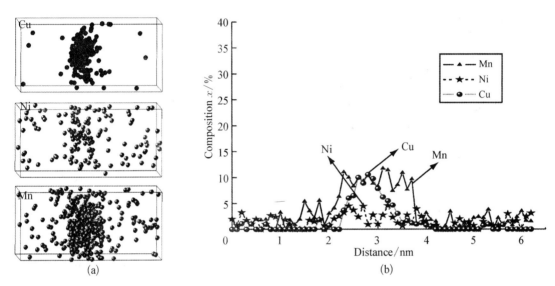

图 4 模拟钢样品在 400 ℃时效 300 h 后所检测到富铜原子团簇中一个团簇的 Cu, Ni 和 Mn 原子分布图(a),分析体积为 3 nm×3 nm×7 nm;该团簇中 Cu, Ni 和 Mn 的成分分布图(b)

Fig. 4 Distribution of Cu, Ni and Mn atoms with in an analyzed volume 3 nm×3 nm×7 nm for one Cu-rich cluster in the specimen aged at 400 ℃ for 300 h (a). The composition profile of the Cu-rich cluster (b)

3 结论

(1) 采用淬火后热时效处理方法,用 3DAP 可以研究提高 Cu 含量后的压力容器模拟钢中富铜原子团簇的析出过程. 模拟钢在 880 ℃加热水淬后,在 400 ℃时效处理 100 h 观察到富铜原子团簇的析出,随着时效时间延长到 300 h,团簇的数量密度从 $1.5×10^{23}$ m^{-3} 增加到 $6.2×10^{23}$ m^{-3}.

(2) 在 400 ℃时效处理时,富铜原子团簇的长大非常缓慢,从 100 h 到 300 h 时效后,最大团簇的等效直径由 2.0 nm 增大到 3.5 nm.

(3) 从 100 h 到 300 h 时效处理后,富铜原子团簇中的成分变化不大,大约含有 Cu, Mn, Ni 的原子数分数 x 分别为 20%,5% 和 4%,Mn 和 Ni 在团簇和基体金属的界面处发生明显的富集,这是阻碍富铜原子团簇长大的主要原因.

参 考 文 献

[1] PAREIGE P J, RUSSELL K F, MILLER M K. APFIM studies of the phase transformations in thermally aged ferritic Fe Cu Ni alloy: comparison with aging under neutron irradiation[J]. Applied Surface Science, 1996, 67: 362 - 369.

[2] PAREIGE P, AUGER P, BAS P, et al. Direct observation of copper precipitation in neutron irradiated Fe - Cu alloys by 3D atomic tomography[J]. Scripta Metall, 1995, 33: 1033 - 1036.

[3] WORRALL GM, BUSWELL J T, ENGLISH C A, et al. A study of the precipitation of copper particles in a ferrite matrix[J]. J Nucl Mater, 1987, 148: 107 - 114.

[4] AUGER P, PAREIGE P, AKAMATSU M. Microstructural characterization of atom clusters in irradiated pressure vessel steels and model alloys[J]. J Nucl Mater, 1995, 225: 192 - 195.

[5] BURKE M G, BRENNER S S. A microstructural investigation of irradiated pressure vessel steel weld metal[J]. J Phys, 1986, 47 - C2: 239 - 244.

[6] MILLER M K, BURKE M G. Characterization of irradiated A533B pressure vessel steel weld[J]. J Phys, 1987, 48 -C6: 429 - 434.

[7] AUGER P, PAREIGE P, WELZEL S, et al. Synthesis of atom probe experiments on irradiation-induced solute segregation in French ferritic pressure vessel steels[J]. J Nucl Mater, 2000, 280: 331 - 344.

[8] BRENNER S S, WAGNER R, SPITZNAGEL J A. Field-ion microscope detection of ultra-fine defects in neutron-irradiated Fe - 0. 34 Pct Cu alloy[J]. Metal Trans, 1978, 9A: 1761 - 1764.

[9] MILLER M K, HOELZER D T, EBRAHIM I F, et al. Characterization of irradiated model pressure steels[J]. J Phys, 1987, 48 - C6: 423 - 428.

[10] 周邦新,刘文庆. 三维原子探针及其材料科学研究中的应用[J]. 材料科学与工艺,2001,15(3): 405 - 411.

[11] PAREIGE P, RUSSELL K F, STOLLER R E, et al. Influence of long-term aging on the microstructual evolution of nuclear reactor pressure vessel material: an atom probe study[J]. J Nucl Mater, 1997, 250: 176 - 183.

[12] FUKUYA K, OHNO K, NAKATA H, et al. Microstructural evolution in medium copper low alloy steels irradiated in a pressurized water reactor and a material test reactor[J]. J Nucl Mater, 2003, 312: 163 - 173.

[13] PAREIGE P, AUGER P, MILOUDI S, et al. Microstructural evolution of the CHOOZ A PWR surveillance program material: small angel neutron scattering and tomographic atom probe studies[J]. Phys, 1997, C2 - 22: 117 -124.

[14] MILLER M K, BURKE M G. An atom probe fieldion microscopy study of neutron-irradiated pressure vessel steels [J]. J Nucl Mater, 1992, 195: 68 - 82.

[15] DIETER I, MICHAEL S, GAGLIANO, et al. Interfacial segregation at Cu-rich precipitates in a high-strength low-carbon steel studied on a sub-nanometer scale[J]. Scripta Metall, 2006, 54: 841 - 849.

3D Atom Probe Characterization of Precipitation of Cu-Rich
Clusters in Pressure Vessel Model Steel

Abstract: Three-dimensional atom probe (3DAP) and thermal aged method were used to characterize precipitation of Cu-rich clusters in pressure vessel model steel. Pressure vessel model steel specimens were quenched at 880 ℃, then aged at 400 ℃ and 500 ℃ for different time. The results of hardness test show that peak hardness reached for 100 h aged at 400 ℃ and 5 h aged at 500 ℃, respectively. Analysis results of 3DAP indicate that the number density of Cu-rich clusters increases from $1. 5 \times 10^{23}$ m^{-3} to $6. 2 \times 10^{23}$ m^{-3}, and equivalent diameter of the biggest clusters increases from 2 nm to 3. 5 nm after aging 100 h and 300 h at 400 ℃. It shows that the growth of Cu-rich clusters is very slow. The content of Cu in Cu-rich clusters is about 20%. Both Mn and Ni atoms are also segregated in Cu-rich clusters and distribution riched at the α - Fematrix/clusters interfaces.

Grain Cluster Microstructure and Grain Boundary Character Distribution in Alloy 690 [*]

Abstract: The effects of thermal-mechanical processing (TMP) on microstructure evolution during recrystallization and grain boundary character distribution (GBCD) in aged Alloy 690 were investigated by the electron backscatter diffraction (EBSD) technique and optical microscopy. The original grain boundaries of the deformed microstructure did not play an important role in the manipulation of the proportion of the $\Sigma 3^n$ ($n = 1$, 2, 3 ...) type boundaries. Instead, the grain cluster formed by multiple twinning starting from a single nucleus during recrystallization was the key microstructural feature affecting the GBCD. All of the grains in this kind of cluster had $\Sigma 3^n$ mutual misorientations regardless of whether they were adjacent. A large grain cluster containing 91 grains was found in the sample after a small-strain (5 pct) and a high-temperature (1 100 ℃) recrystallization anneal, and twin relationships up to the ninth generation ($\Sigma 3^9$) were found in this cluster. The ratio of cluster size over grain size (including all types of boundaries as defining individual grains) dictated the proportion of $\Sigma 3^n$ boundaries.

1 Introduction

ALLOY 690 has been used as a substitute for Alloy 600 as a steam generator tube material in pressurized water reactors because of its excellent corrosion resistance in a broad range of aqueous environments. [1-3] With prolonged service life and improved performance being demanded by the nuclear energy industry, however, the need to improve the resistance to intergranular failure in Alloy 690 should also be considered.

Grain boundary engineering (GBE) has attracted much attention in material science and engineering research[4, 5] since the concept of "grain boundary design" was first proposed by Watanabe. [6] With the help of the automatic indexing electron backscatter diffraction (EBSD) technique, GBE research advanced a great deal. In studies of GBE, grain boundary misorientations are often classified according to the coincident site lattice (CSL) model. [7, 8] The aim of GBE is to enhance the grain-boundary-related properties of materials by increasing the population of low – ΣCSL ($\Sigma \leqslant 29$) grain boundaries. [9, 10] The boundaries that do not have CSL misorientations are often defined as random grain boundaries. However, the CSL grain boundary with a high Σ value cannot be identified automatically by EBSD. Hence, in experimental studies, random grain boundaries include

---

[*] In collaboration with XiaShuang, Chen Wenjue. Reprinted from Metallurgical and Materials Transactions A, 2009, 40A: 3016 – 3030.

general large-angle grain boundaries and CSL grain boundaries with a Σ value large enough that it cannot be identified automatically by EBSD.

The grain boundary character distribution (GBCD) of fcc metal materials with low to medium stacking fault energy (SFE) can be altered through thermal-mechanical processing (TMP) to improve the grain-boundary-related bulk properties. [9-11] Alloy 690 is an fcc material with low SFE; therefore, GBE may increase the resistance of the alloy to intergranular failure. One of the central topics of GBE is how the TMP affects the microstructural evolution that leads to high proportions of low-Σ CSL grain boundaries. A common conclusion on this topic has not been reached. However, there were two common points that can be summarized, as follows: (1) only the proportion of $\Sigma 3^n (n=1, 2, 3 \ldots)$ grain boundaries could be altered in the GBCD, while other low-Σ CSL grain boundaries had very low occurrences in spite of the selection of the TMP; and (2) the magnitude of deformation used in TMP was small, often on the order of 3 to 30 pct.

Thus, the matter turns to how a small deformation plus different annealing treatments enhance the proportions of $\Sigma 3^n$ grain boundaries. Some articles pointed out that conventional recrystallization introduced random grain boundaries into the microstructure. [12, 13] This was thought to be the reason that large amounts of deformation led to relatively low proportions of $\Sigma 3^n$ grain boundaries after recrystallization. In studies in which smaller amounts of deformation were used, the original grain boundaries of the deformed microstructure were thought to play an important role in enhancing the proportion of $\Sigma 3^n$ boundaries during annealing. [12-14] Kumar et al. [12] observed that in the GBE specimens, strain energy was stored throughout the grains, which meant that conventional nucleation of recrystallization did not occur. They attributed the enhancement of the $\Sigma 3^n$ boundaries to the boundary decomposition during strain-induced boundary migration, which is prevalent on annealing of moderately strained microstructures. [34] Shimada et al. [14] attributed the increasing the proportion of $\Sigma 3^n$ boundaries to the introduction of low-energy segments on migrating random grain boundaries during twin emission and boundary-boundary reactions in the grain growth. They thought that a small prestrain accelerated grain growth during TMP, while a large prestrain tended to promote recrystallization, which generated new random grain boundaries. Lee and Richards[13] pointed out that recrystallization does not occur at low levels of strain; rather, the enhancement of $\Sigma 3^n$ boundaries can be understood following the "fine tuning" concept of Thomson and Randle. [15] In contrast to the mechanisms mentioned earlier, some people thought that recrystallization played an important role in enhancing the proportion of $\Sigma 3^n$ boundaries. Randle[5, 16] proposed the "$\Sigma 3$ regeneration model." This model required the impingement of two newly recrystallized grains that have almost the same crystallographic texture and contain twins. [5] The interactions between the twinned portions produce more mobile boundaries. These boundaries migrate to impinge onto other boundaries to regenerate $\Sigma 3$ boundaries. Wang[17] thought that the nuclei of recrystallized grains had incoherent $\Sigma 3$ boundaries with the deformed microstructure. The migration and interactions of these boundaries were believed to enhance the proportion of $\Sigma 3^n$ boundaries.

Our previous work[18, 19] showed that a highly twinned large grain cluster is the typical microstructure produced by GBE and attributed the intergranular corrosion resistance enhancement to the appearance of this kind of large grain cluster. Reed et al.[20] also pointed out the importance of the clusters of twin-related grains, which they called twin-related domains. They used mathematical methods to analyze the multiple twin relationships.[21] However, the relationship between the microstructural evolution during recrystallization and the formation of the grain cluster needs to be further studied.

According to the mechanisms or models mentioned previously, two key questions should be answered to further understand how the TMP manipulates $\Sigma 3^n$ boundaries. One is whether and how the original grain boundaries of the deformed microstructure affect the formation of $\Sigma 3^n$ boundaries. The other is how recrystallization affects the microstructural evolution that results in different proportions of $\Sigma 3^n$ boundaries.

In the current work, the samples were aged after solution annealing prior to the TMP. This pretreatment caused carbide precipitation at the grain boundaries, so that they would be pinned at the beginning of annealing after deformation. In this condition, the observation of partially recrystallized samples facilitated the understanding of what role the original grain boundaries of the deformed microstructure played and how recrystallization affected the GBCD.

2 Experimental procedures

The composition of Alloy 690 used in this study is as follows (wt pct): 28.91Cr, 9.45Fe, 0.4Ti, 0.34Al, 0.14Si, 0.025C, and the balance Ni. Strip samples with 7 – mm widths were cut from a 2 – mm-thick plate, vacuum sealed under 5×10^{-3} Pa in quartz capsules, annealed at 1 100 ℃ for 15 minutes, and then quenched into water and broken quartz capsules (WQ) simultaneously. After that, the samples were aged at 715 ℃ for 15 hours for the starting condition (SC). The SC samples were cold rolled with the reduction ratio in thickness of 5, 20, and 50 pct. They were then annealed at 1 100 ℃ for 1, 2, 3, 4, and 5 minutes in vacuum-sealed quartz capsules and subsequently quenched into water in the same way as described earlier. Those samples were named, and the corresponding TMPs are shown in Table I.

Prior to the examination of these samples, electropolishing was carried out in a solution of 20 pct $HClO_4 + 80$ pct CH_3COOH with a 30 – V direct current. Electroetching for optical microscope observation was performed in the same solution with 5 V. The EBSD was employed for the determination of grain boundary misorientations using the TSL 4.6 Laboratories Orientation Imaging Microscopy (OIM*) system attached to a Hitachi S570 scanning electron microscope (Hitachi Ltd., Tokyo, Japan) with a tungsten filament.

---

* OIM is a trademark of EDAX; TSL and EDAX are a part of AMETEK, Inc., Materials Analysis Division, Mahwah, NJ.

Table I Treatment Processing for Each Sample

Sample Name	a	b	c	d	e	f	g	h	i	j	k	l	m	n	o	p	q	r
SC	1 100 ℃/15 min+WQ+715 ℃/15 h																	
Cold-rolling reduction ratio in thickness (pct)					5						20						50	
Annealing time (min) at 1 100 ℃	0	1	2	3	4	5	0	1	2	3	4	5	0	1	2	3	4	5

Operating conditions were: a $20-kV$ accelerating voltage, $25-mm$ working distance, and 70-deg beam incidence angle. The scans were carried out on a square or rectangular area, with each orientation point being represented as a hexagonal cell, using step sizes on the order of 2 to 7 μm, varied with the grain size. When measuring the grain size, all types of boundaries are included as defining individual grains. The GBCD statistics are automatically calculated as length fractions by the OIM system. When analyzing the misorientation between crystallites that are not interfacing, data points in the middle of grains were manually selected and used for measurement. The CSL relationship identification is automatically carried out for low $-\Sigma$ CSL grain boundaries ($\Sigma \leqslant 29$). The higher-order twin relationships ($\Sigma 3^n$) with $n \geqslant 4$, are identified according to a twinchain analysis as well as the work of Gottstein[22] and Randle.[23]

3　Results and analyzing

3.1　Effects of TMP on GBCD

The samples were aged at 715 ℃ for 15 hours after solution annealing. This treatment caused carbides to precipitate at grain boundaries abundantly in Alloy 690.[2, 24, 25] After this treatment, designated the SC, the sample was tested by EBSD. The OIM map of boundaries with different characters is shown in Figure 1(a) and the (001) pole figure is

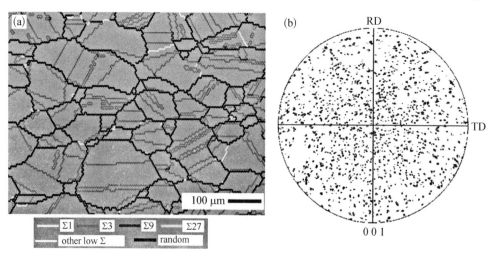

Fig. 1　Microstructure of the starting material (1 100 ℃/15 min+WQ+715 ℃/15 h): (a) OIM map of different types of boundaries (Palumbo－Aust criterion[26]) and (b) (001) pole figure

shown in Figure 1(b). Almost all of the Σ3 boundaries in this map were in the shape of a straight single line or parallel line pairs within grains, which is typical of the morphology of annealing twins in the fcc material with low SFE. There were very few higher-order twin boundaries, which are denoted by the blue (Σ9) and green (Σ27) boundaries. This sample was without texture, as shown in Figure 1(b). The GBCD statistics for this stage are shown in Table II. The samples were subjected to cold rolling for 5, 20, and 50 pct, and subsequently annealed at 1 100 ℃ for 5 minutes. The GBCDs of each sample were obtained by EBSD examination and are shown in Table II. The small strain (cold rolled 5 pct) followed by a short annealing time at a high temperature (1 100 ℃) enhanced the proportion of low − Σ CSL grain boundaries to more than 70 pct, which predominantly are Σ3ⁿ boundaries. Merely increasing the deformation level before annealing decreases the proportion of Σ3ⁿ boundaries dramatically, as shown in Table II.

Table II GBCD Statistics of Each Sample (Palumbo − Aust Criterion[26])

Sample Name	Σ1 (Pct)	Σ3 (Pct)	Σ9+Σ27 (Pct)	Other Low − Σ CSL (Pct)	Overall Low − Σ CSL (Σ≤29) (Pct)
SC	3.5	43.6	1.4	1	49.5
f	1.7	61.3	9.4	0.3	72.7
l	2.5	48.8	3.5	0.7	55.5
r	4.2	40.7	1.6	1.3	47.8

3.2 Optical Observations of Microstructure Changes during TMP

The differences in the GBCDs of each sample in Table II were caused by the differences in the microstructure evolution during TMP. The optical micrographs of the samples listed in Table I are shown in Figure 2, in which the microstructural changes during annealing following different degrees of strain could be clearly seen. Due to the aging treatment after solution annealing, there was a significant amount of carbide precipitation at grain boundaries.[2, 24, 25] After deformation, because of the existence of the carbides, the former grain boundaries could be seen clearly, not only for the low-level strained samples (5 and 20 pct cold rolled) but also for the more heavily strained one (50 pct cold rolled). The electroetching revealed some obvious slip lines inside of the grains. The shape of the original grains did not change very much for the 5 and 20 pct cold-rolled samples, while the grains of the 50 pct coldrolled sample had been elongated.

After deformation, the samples were rapidly heated to the annealing temperature (1 100 ℃). At this temperature, the carbides at the original grain boundaries gradually dissolved, while there were microstructural changes driven by the stored energy resulting from deformation, i.e., recovery and recrystallization. At constant temperature, the dissolving of the carbides was mainly controlled by the diffusion time, while the recrystallization speed was mainly controlled by the stored energy resulting from deformation. Because of the dissolution of the carbides, the traces of former grain boundaries faded as the annealing proceeded at 1 100 ℃. After annealing for 3 to 4

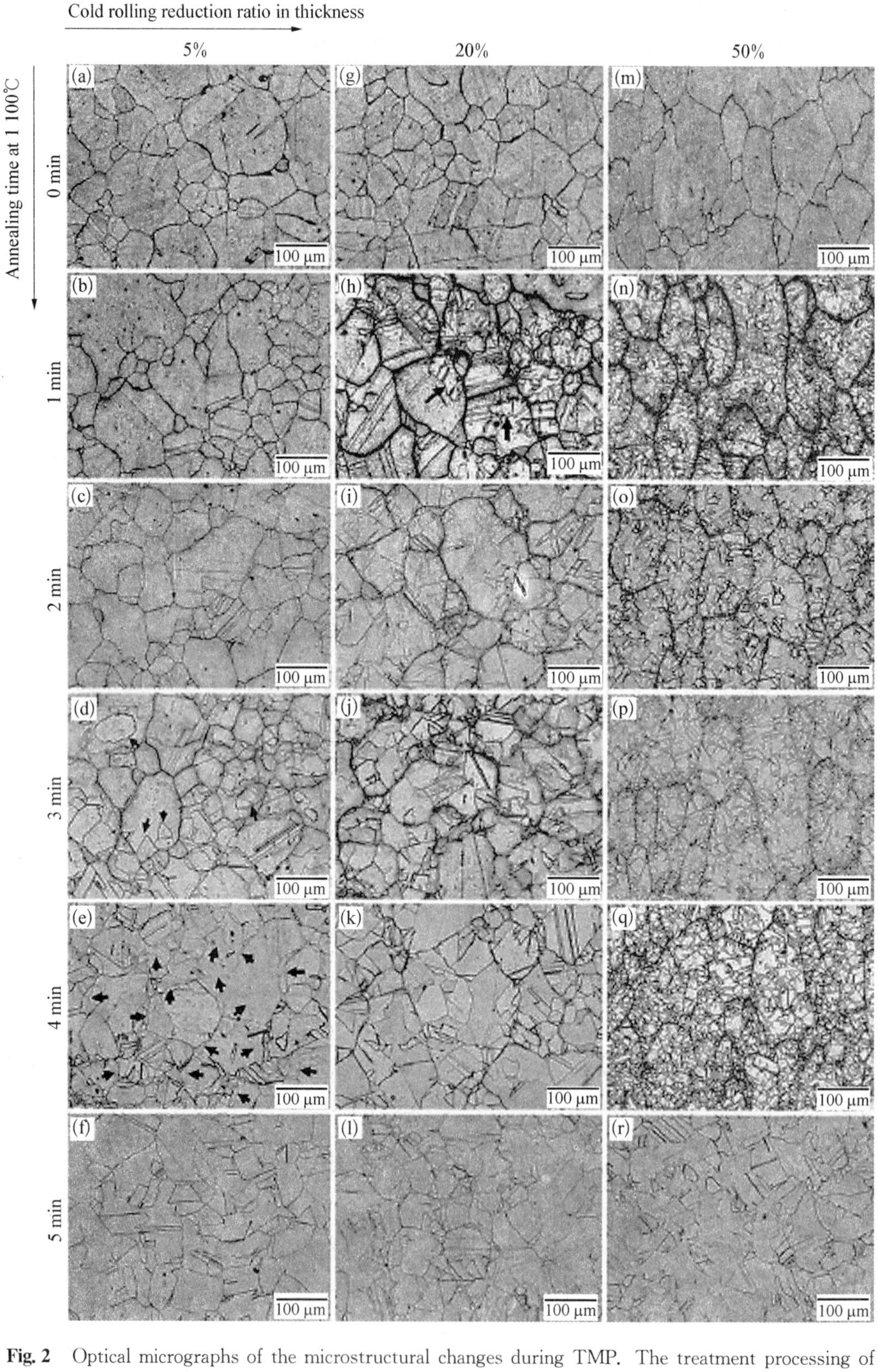

Fig. 2 Optical micrographs of the microstructural changes during TMP. The treatment processing of each sample is given in Table I

minutes, the traces of former grain boundaries became comparatively weak, from which it could be deduced that most of the carbides had dissolved at this time. After etching, the new recrystallizing grains were comparatively more evident, and there were no slip lines inside of them. The recrystallized and nonrecrystallized area could thus be roughly distinguished. The onset time of recrystallization was varied for the samples with different amounts of cold rolling. The new grains in the 5 pct cold-rolled sample could be observed after annealing for 3 minutes, as shown by arrows in Figure 2(d); for the 20 pct coldrolled sample, it was after 1 minute, as shown by arrows in Figure 2(h), while for the 50 pct cold-rolled sample, by 1 minute the recrystallization was complete, as shown in Figure 2(n). Second-phase particles are known to have significant effects on the recrystallization process, because they can exert strong drag effects on the boundaries. At the onset of annealing, the former grain boundaries were pinned by carbides from movement, so strain-induced boundary migration did not occur. Recrystallization had finished before the dissolution of most carbides at the former boundaries, for the 20 and 50 pct cold-rolled samples. Therefore, when the new boundaries between the recrystallizing grains and deformed matrix encountered these carbides, their movements were inhibited. As a consequence, the recrystallization processes took place in areas bounded by the former grain boundaries. The onset of recrystallization and the dissolution of most carbides in the 5 pct cold-rolled sample occurred almost simultaneously. Therefore, it was more likely that the recrystallizing grains could pass across the former boundaries. It can be seen in Figure 2(e) that the traces of former grain boundaries were still visible inside of the recrystallized new grains, as indicated by the arrows.

3.3 EBSD Analysis of Partially Recrystallized Microstructures

The results of the experiment showed that sample "d," which was cold rolled 5 pct and annealed for 3 minutes, and sample "h," which was cold rolled 20 pct and annealed for 1 minute, were partially recrystallized. The recrystallized areas with annealing twins readily inside could be clearly seen, as shown by the arrows in Figures 2(d) and (h). An EBSD technique was used to analyze the crystallographic features of the partially recrystallized microstructures, to understand the GBCD formation.

A part of Figure 2(d) was analyzed using EBSD and is shown in Figure 3(a), with the corresponding second electron microscopy image shown in Figure 3(b). The orientation-contrasted OIM map is shown in Figure 3(c). The orientation gradients along the two dotted lines in Figure 3(c) are shown in Figures 4(a) and (b). Line "1" has a much greater orientation gradient than line "2." This is because line 1 is in the nonrecrystallized area, which contained many geometrically necessary dislocations induced by deformation,[28] while line 2 is in the recrystallized area. Figure 3(d) is an OIM map that designates the grain boundary type. Grain boundaries were here defined as being between any neighboring-orientation point pair with a misorientation exceeding 1 deg in Figure 3(d). There were many $\Sigma 1$ boundaries located in some grains, which were interpreted to be

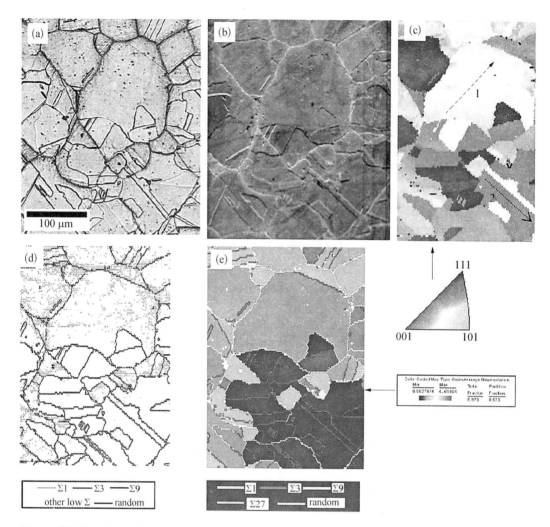

Fig. 3 EBSD analysis of microstructure of the partially recrystallized sample d (SC+5 pct cold rolling/ 3 min annealing at 1 100 ℃): (a) optical micrograph, (b) second electron microscopy image, (c) OIM map contrasted by color of inverse pole figure, (d) OIM map of different types of boundaries (Palumbo - Aust criterion[26]) (any neighboring orientation point pair with misorientation exceeding 1 deg was considered a boundary), and (e) OIM map contrasted by color of grain-average misorientation and overlapped with different types of boundaries (any neighboring orientation point pair with misorientation exceeding 2 deg was considered a boundary)

nonrecrystallized areas of the microstructure.[29] In the other areas, the Σ1 boundaries were sparse, and these areas were believed to be recrystallized.[29] In order to more clearly distinguish the recrystallized and nonrecrystallized areas, the "grain-average misorientation" OIM maps were used; they are shown in Figure 3(e). Grain-average misorientation is the average misorientation angle between the neighboring-orientation point pairs in a grain, and convey short-range orientation variations within a grain.[28] From the color code map for Figure 3(e), it can be seen that the blue areas denote a smaller average misorientation in grains, which corresponds to the recrystallized areas, while the nonrecrystallized areas, which had larger average misorientations in grains, were

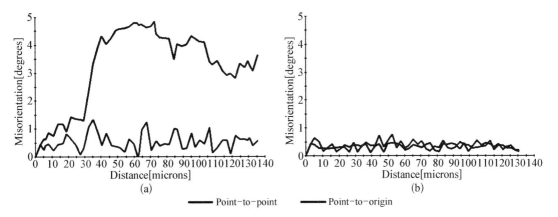

Fig. 4 Orientation gradient along the two lines in Fig. 3(c): (a) for line 1 and (b) for line 2

represented by green or warmer hues. In this map, the boundary types overlapped, but the misorientation definition for a boundary was set to the system default value of 2 deg. Therefore, there were fewer $\Sigma1$ boundaries than are shown in Figure 3(d). The boundaries separating recrystallized and nonrecrystallized regions were random grain boundaries. Inside the recrystallized area, there were many annealing twin – $\Sigma3$ boundaries. In order to investigate the crystallographic characteristics of the recrystallized area, one of the recrystallized areas surrounded by

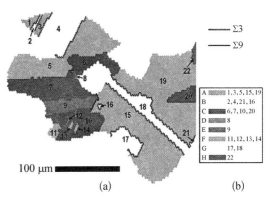

Fig. 5 Twin chain in Fig. 3, which is the microstructure of partially recrystallized sample (SC+5 pct cold rolling/3 min annealing at 1 100 ℃)

random grain boundaries was isolated and was analyzed in Figure 5. This area consisted of 22 grains, which are numbered in Figure 5(a). After analyzing the region using OIM software, these 22 grains could be categorized into eight orientations (an orientation difference less than 1 deg was here recognized as being the same orientation), as shown in Figure 5(b). There were specific $\Sigma3^n$ misorientations between any two orientations among those eight orientations, as listed in Table III. This was the twin-chain phenomenon that results from continuous twinning starting from a single nucleus during recrystallization in fcc materials with low SFE.[30] The twinchain formation was more completely discussed in studying the recrystallization texture evolution.[30-35] Through twinning, new orientations were generated with respect to the original recrystallized grain for obtaining higher mobility boundaries at the front of recrystallization.[30, 31, 34, 36] In the current experiment, the twin-chain formation was the basic characteristic in the partially recrystallized microstructures that could eventually yield high proportions of $\Sigma3^n$ boundaries.

Sample h, which was 20 pct cold rolled and annealed for 1 minute, contained a partially recrystallized microstructure, as determined by EBSD. The orientation-contrast

Table III Mutual Misorientations between Eight Different Orientations in Twin Chain of Figure 5

Orientations	A	B	C	D	E	F	G
B	59.5 deg [−1 1 −1]/ Σ3/0.6 deg	—	—	—	—	—	—
C	59.8 deg [1 1 −1]/ Σ3/0.7 deg	38.9 deg [−1 −1 0]/ Σ9/0.3 deg	—	—	—	—	—
D	38.5 deg [−1 1 0]/ Σ9/0.6 deg	35.5 deg [1 0 2]/ Σ27b/0.2 deg	59.9 deg [1 1 −1]/ Σ3/0.3 deg	—	—	—	—
E	38.6 deg [1 0 1]/ Σ9/0.5 deg	35.8 deg [0 1 −2]/ Σ27b/0.6 deg	59.6 deg [−1 1 −1]/ Σ3/0.7 deg	38.8 deg [−1 0 1]/ Σ9/0.4 deg	—	—	—
F	38.4 deg [0 −1 −1]/ Σ9/0.7 deg	31.5 deg [21 22 0]/ Σ27a/0.5 deg	59.8 deg [1 −1 −1]/ Σ3/0.3 deg	39.4 deg [0 1 −1]/ Σ9/0.8 deg	39.1 deg [0 1 1]/ Σ9/0.4 deg	—	—
G	59.9 deg [−1 −1 −1]/ Σ3/0.4 deg	39.1 deg [0 1 1]/ Σ9/0.5 deg	39.2 deg [1 −1 0]/ Σ9/0.5 deg	32.2 deg [−1 1 0]/ Σ27a/0.6 deg	35.6 deg [0 1 −2]/ Σ27b/0.2 deg	35.1 deg [−1 0 2]/ Σ27b/0.4 deg	—
H	59.5 deg [1 −1 −1]/ Σ3/0.8 deg	39.4 deg [1 0 −1]/ Σ9/0.6 deg	38.7 deg [−1 0 1]/ Σ9/0.3 deg	35.7 deg [0 14 27]/ Σ27b/0.5 deg	32.3 deg [1 1 0]/ Σ27a/0.8 deg	35.4 deg [0 −1 −2]/ Σ27b/0.3 deg	39.5 deg [1 0 1]/ Σ9/0.7 deg

Note: Orientation relationship is given in the following form: misorientation/closest CSL/deviation.

OIM map of this sample is shown in Figure 6(a). The corresponding OIM map based on the grain-average misorientation and overlapped with different types of boundaries was shown in Figure 6(b). The recrystallized area also consisted of twin chains. One twin chain outlined by the rectangular framework in Figure 6(b) was analyzed. All of the grains belonging to this twin chain retained specific $\Sigma 3^n$ mutual misorientations among them, as shown in Table IV. The misorientation definition for a boundary was set to 2 deg, which was the same as that of Figure 3(e). By comparing the nonrecrystallized areas in Figure 3(e) with that in Figure 6(b), it can be seen that there are many more $\Sigma 1$ boundaries in the latter, by the warmer hue color. This was the result of greater deformation. Greater deformation resulted in higher stored energy, which led to a higher nucleation density and an earlier recrystallization onset time for the 20 pct cold-rolled sample. Therefore, smaller-size twin chains had already impinged with each other in the recrystallized area of this sample, as shown in Figures 2(h) and 6(b).

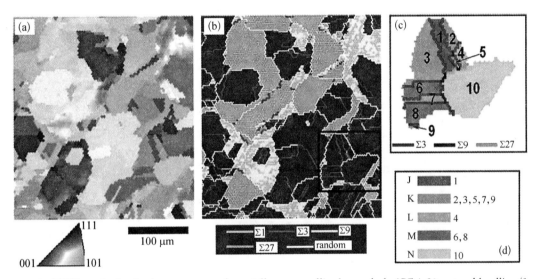

Fig. 6 EBSD analysis of microstructure of partially recrystallized sample h (SC+20 pct cold rolling/1 min annealing at 1 100 ℃): (a) OIM map contrasted by color of inverse pole figure, (b) OIM map contrasted by color of grain-average misorientation, and overlapped with different types of boundaries (Palumbo − Aust criterion[26]) (any neighboring orientation point pair with misorientation exceeding 2 deg was considered a boundary), and (c) twin chain outlined by black framework in (b)

Table IV Mutual Misorientations between Five Different Orientations in Twin Chain of Figure 6(c)

Orientations	J	K	L	M
K	59. 6 deg [1 −1 1]/ $\Sigma 3$/0. 6 deg	—	—	—
L	39. 1 deg [1 0 −1]/ $\Sigma 9$/0. 6 deg	59. 6 deg [−1 1 1]/ $\Sigma 3$/0. 7 deg	—	—
M	39. 3 deg [0 1 1]/ $\Sigma 9$/0. 4 deg	59. 4 deg [−1 −1 1]/ $\Sigma 3$/0. 6 deg	38. 7 deg [−1 0 −1]/ $\Sigma 9$/0. 5 deg	—
N	36. 0 deg [2 −1 0]/ $\Sigma 27b$/0. 6 deg	39. 1 deg [1 0 1]/ $\Sigma 9$/0. 5 deg	31. 9 deg [1 0 1]/ $\Sigma 27a$/0. 6 deg	59. 8 deg [−1 1 1]/ $\Sigma 3$/0. 6 deg

Note: Orientation relationship is given in the form: misorientation/closest CSL/deviation.

The recrystallization process of the sample with 50 pct cold rolling had already finished during a 1 – minute anneal, as shown in Figure 2(n). Therefore, the partially recrystallized microstructure was not obtained. Each area that was bounded by former grain boundaries contained many new grains. One such area in this sample was characterized by EBSD, and the OIM maps are shown in Figure 7. The grain-boundary-type OIM map (Figure 7(b)) showed that inside of the original area bounded by former grain boundaries, there were many smaller size twin chains that had already impinged upon one another. Random boundaries separated these twin chains. This local area in the sample was without texture, as shown in the (001) pole figure.

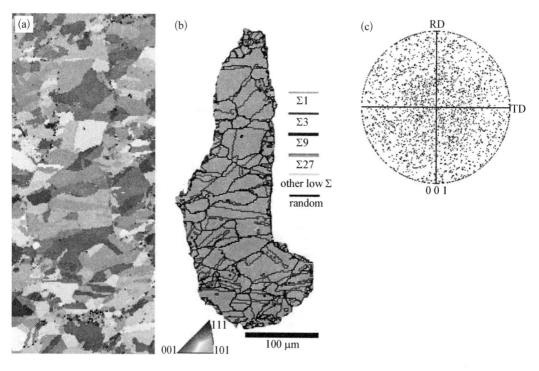

Fig. 7 One area bounded by former grain boundaries in sample "n" (SC + 50 pct cold rolling/1 min annealing at 1 100 ℃): (a) OIM map contrasted by color of inverse pole figure, (b) OIM map of different types of boundaries (Palumbo – Aust criterion[26]), and (c) (001) pole figure of this area

3.4 EBSD Analysis of Fully Recrystallized Samples

The traces of the former grain boundaries in samples annealed for 5 minutes had completely disappeared. This indicated that the carbides had dissolved into the matrix, as shown in Figures 2(f), (l), and (r). At this time, the samples were also in their fully recrystallized state. They were examined by EBSD. The grain-boundary- type OIM maps of the samples with 5, 20, and 50 pct cold rolling and 5 minutes' annealing are shown in Figures 8(a), (b), and (c), respectively. In Figure 8(a), most boundaries were of the $\Sigma 3''$ type, and they connected with each other, forming many triple junctions of $\Sigma 3''$ – type boundaries. In this case, large grain clusters constituted the microstructure. The grain cluster has the following characteristics: All of the boundaries within the cluster can have $\Sigma 3^n$ misorientations, whereas the outer boundaries are often crystallographically random with the adjacent grains. [37] The $\Sigma 3^n$ – type boundaries were included here as defining individual grains. The grain cluster could be recognized as the twin chain after recrystallization.

There was a very large cluster in Figure 8(a); hence, it was analyzed using OIM in detail, and the result is shown in Figure 9. This cluster was composed of 91 grains in the current two-dimensional section. These grains are numbered in Figure 9(a). Among these grains, a continuous line could be drawn across only $\Sigma 3$ boundaries from any grain to any other, with the exception of grain 49 (the reason will be given later in this article). For example, from grains 1 to 91, one route is the following: $1 \rightarrow 2 \rightarrow 3 \rightarrow 23 \rightarrow 28 \rightarrow 36 \rightarrow 53 \rightarrow 62 \rightarrow 76 \rightarrow 78 \rightarrow 80 \rightarrow 88 \rightarrow 89 \rightarrow 91$, which crosses 13 $\Sigma 3$ boundaries. In order to clearly show the twin-chain relationships, all of the grains of this cluster were joined into a twin chain, as shown in Figure 9(b). In this chain, red lines link the interfacing grains with a $\Sigma 3$ twin relationship only if the deviations from exact $\Sigma 3$ CSL are less than 1 deg. However, in Figure 9(b), the higher-order twin relationships ($\Sigma 3^n$) could not be directly identified. For example, from grains 1 to 91, the route mentioned previously crosses 13 $\Sigma 3$ boundaries, but the orientation relationship between grains 1 and 91 was not $\Sigma 3^{13}$ but was $\Sigma 3^5$ (measured misorientation 36.3 deg/[2 4 5], identified as $\Sigma 243d$, according to Reference 23). This was because in this route, some grains had the same orientation. In other words, some orientations turned back to their original orientations after several twinning operations. Thus, from grains 1 to 91, the route mentioned previously could be simplified as $1 \rightarrow 62 \rightarrow 76 \rightarrow 78 \rightarrow 80 \rightarrow 91$ (grains 1 and 53, grains 2 and 23, and grains 88 and 91 had the same orientations, given a 1.5-deg tolerance). Therefore, the misorientation between grains 1 and 91 resulted from five independent twinning operations, $i.e.$, it was $\Sigma 3^5$. The grains in this cluster could be categorized into 28 orientations, as shown in Figure 9(c). A

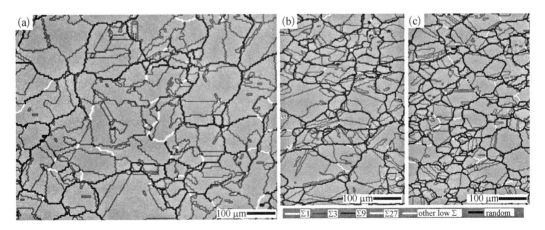

Fig. 8 OIM map of different types of boundaries (Palumbo – Aust criterion[26]) of the fully recrystallized samples. Maps (a), (b), and (c) are of samples f, l, and r, which are cold rolled 5, 20, and 50 pct, respectively, and subsequently annealed at 1 100 ℃ for 5 min

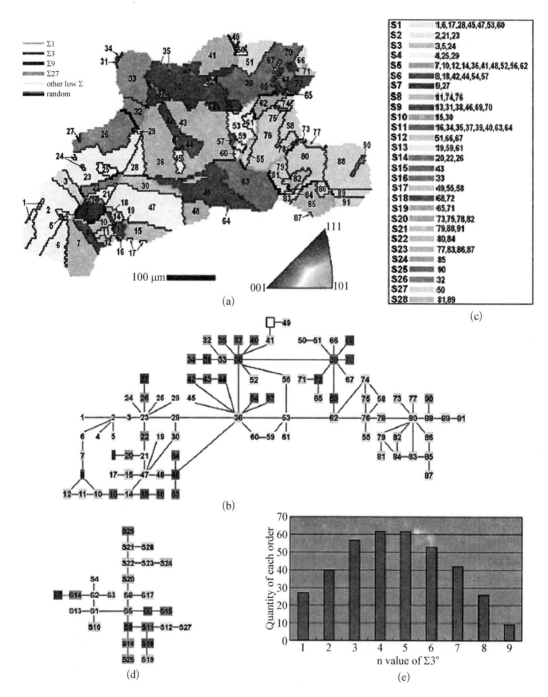

Fig. 9 Twin-chain analysis of large grain cluster in Fig. 8(a): (a) grain cluster containing 91 grains; (b) twin chain of those 91 grains (red lines link the interfacing grains with Σ3 twin relationship); (c) these 91 grains were categorized into 28 orientations; (d) twin chain of these 28 orientations (red lines link the interfacing orientations with Σ3 twin relationship); and (e) quantity of each order twin relationships

simpler form of twin chain including those 28 orientations could be organized, as shown in Figure 9 (d). In this chain, red lines link the interfacing orientations with Σ3 twin relationships. The order of the Σ3^n relationships can be directly identified from this chain.

For example, the misorientation between grains 1 and 91 can be determined by observing from Figure 9(c) that they belong to orientations S1 and S21, respectively. The route from orientation S1 to S21 can be uniquely identified as S1→S5→S8→S20→S22→S21, according to Figure 9(d). The n value of the $\Sigma 3^n$ relationship between grains 1 and 91 is, therefore, 5. Thus, the $\Sigma 3^n$ relationship between any grain pairs inside the cluster can be identified.

Recall that grain 49 (mentioned preciously) has a $\Sigma 9$ boundary interface with the rest of the cluster (via grain 41) in the two-dimensional section. Grain 49 is highly probably interfacing with another grain (denoted by a pane in Figure 9(b)) by a $\Sigma 3$ boundary below or above the presenting two-dimensional section, and that grain interfaces with grain 41 by a $\Sigma 3$ boundary. This is because a $\Sigma 9$ boundary is produced by the intersection of an orientation with its third twin generation. On the other hand, grain 49 belongs to orientation S17, which is a part of the orientation twin chain, as shown in Figures 9(c) and (d). This also verifies that grain 49 belongs to the current cluster.

There were 378 mutual misorientations among those 28 orientations. According to Figure 9(d), every one of those 378 misorientations was associated with a $\Sigma 3^n$ relationship. The distribution of the twin relationship orders is shown in Figure 9(e). The highest twin order is 9. All of these 378 misorientations were also recorded manually in a table in the same format as Table III during OIM analysis. All of these misorientations were of the $\Sigma 3^n$ type.

There were some random grain boundaries inside this cluster, and they are listed in Table V. They actually had higher-order $\Sigma 3^n$ - type ($n > 3$) misorientations, according to the orientation twin-chain analysis shown in Figure 9(d), however. For example, the boundary between grains 9 and 10 is defined as a random grain boundary in Figure 9(a). However, the multiple twin relationship between grains 9 and 10 can be determined by observing from Figure 9(c) that they belong to orientations S7 and S5, respectively. The route from orientation S7 to S5 can be uniquely identified as S7→S14→S2→S1→S5, according to Figure 9(d). Therefore, the relationship between grains 9 and 10 is $\Sigma 3^4$. Without the orientation twin chain as shown in Figure 9(d), it is almost impossible to directly seek out high-order $\Sigma 3^n$ ($n > 3$) misorientations. Therefore, these random grain boundaries inside the cluster were essentially different from the ones formed by the impingement of different grain clusters. The boundary misorientation between grains 63 and 83 is close to a $\Sigma 7$ CSL relationship with the deviation of 2.6 deg, and the boundary misorientation between grains 65 and 74 is close to a $\Sigma 21b$ CSL relationship with the deviation of 2.4 deg. If the Brandon criterion is applied, both boundaries are low - Σ CSL grain boundaries. If the Palumbo - Aust criterion is used, only grain boundary 63/83 is of low - Σ CSL and grain boundary 65/74 is excluded from the low - Σ CSL rank. However, these low - Σ CSL boundaries had higher-order $\Sigma 3^n$ - type relationships, according to the orientation twin chain analysis. This is because there is inevitably some overlap between high-order twin relationships and some low - Σ CSL relationships.[23, 33] Therefore, both the random grain boundaries and the low - Σ CSL grain boundaries inside the cluster were

Table V Misorientations of "Random" Grain Boundary and "Non - $\Sigma3^n$- Type ($n \leqslant 3$)" Low - Σ CSL Boundaries inside Grain Cluster of Figure 9*

Grain Pair-(Orientation Pair)	Boundary Type		Actual $\Sigma3^n$ Relationship	Misorientation
	Brandon Criterion	Palumbo - Aust Criterion		
9/10 -(S7/S5)		random	$\Sigma3^4$	54. 0 deg [2 −2 3]/Σ81b
26/32 -(S14/S26)		random	$\Sigma3^6$	53. 7 deg [−3 −5 8]/Σ729c
49/50 -(S17/S27)		random	$\Sigma3^6$	28. 9 deg [−18 5 −15]/Σ729r
41/50 -(S5/S27)		random	$\Sigma3^4$	60. 9 deg [4 4 −3]/Σ81d
55/59 -(S17/S13)		random	$\Sigma3^4$	38. 6 deg [−1 5 3]/Σ81c
55/63 -(S17/S11)		random	$\Sigma3^4$	53. 9 deg [3 2 −2]/Σ81b
63/79 -(S11/S21)		random	$\Sigma3^6$	54. 3 deg [−7 13 20]/Σ729c
85/89 -(S24/S28)		random	$\Sigma3^4$	55. 0 deg [22 15 15]/Σ81b
8/9 -(S6/S7)		$\Sigma1$	$\Sigma3^5$	11. 8 deg [10 −25 −7]/Σ243c
63/83 -(S11/S23)		$\Sigma7/2. 6$ deg	$\Sigma3^6$	39. 1 deg [7 7 −8]/Σ729a
65/74 -(S19/S8)	$\Sigma21b/2. 4$ deg	random	$\Sigma3^5$	43 deg [11 11 −20]/Σ243a

* They were actually higher-order $\Sigma3^n$- type ($n > 3$) boundaries.

Note: The column "boundary type" is automatically given by the OIM system, in which both the Brandon criterion[27] and the Palumbo - Aust criterion[26] are used. The column "actual $\Sigma3^n$ relationship" is identified by orientation twin-chain analyzing in Fig. 9(d). The column "misorientation" is given in the form: misorientation/closest higher-order twin relationship, in which the higher-order twin relationships are identified according to the work of Gottstein[22] and Randle.[23]

the result of multiple twinning. In addition to the interfacing orientations (boundaries) listed in Table V, there were more noninterfacing orientations that also maintained higher-order $\Sigma3^n$ - type mutual misorientations that overlapped with some low - Σ CSL relationships, as shown in Table VI. Therefore, when analyzing the grain boundary network, care must be taken to identify low - Σ CSL boundaries, especially when the Brandon criterion is used. The Brandon criterion has a high probability of misjudging some higher-order $\Sigma3^n$ - type misorientations as low - Σ CSL boundaries. This is one of the reasons why in most cases the Palumbo - Aust criterion is used in the current study.

Following the definition of the grain cluster, the microstructures of Figures 8(b) and (d) were also constituted of grain clusters. The cluster sizes were much smaller, however, and contained fewer $\Sigma3^n$ - type boundaries. It could be conceived that the microstructures with large clusters containing relatively small twin-related grains would exhibit high proportions of $\Sigma3^n$ boundaries. Therefore, both the cluster size and grain size are important. In order to investigate the dependence of the grain cluster size and grain size on the TMP, the starting material, samples "f," "l," and "r" were analyzed using EBSD in the form of a line scan. For each sample, a 2000 - μm line scan was performed five times in different areas. The result is shown in Figure 10. The mean distance between the random grain boundaries represents the grain cluster size. The mean distance between all kinds of boundaries represents the grain size. The OIM system cannot automatically identify higher-order $\Sigma3^n$ misorientations ($n > 3$). According to a previous analysis, some higher-order $\Sigma3^n$ ($n > 3$) grain boundaries inside the cluster were arbitrarily defined as random grain

Table VI $\Sigma 3^n-$ Type ($n>3$) Noninterfacing Mutual Misorientations Overlapped with Some Low$-\Sigma$ CSL Relationships inside Grain Cluster of Figure 9(a)

Orientation Pair	n of $\Sigma 3^n$	Misorientation	Closest Low$-\Sigma$ CSL/Deviation	
			Brandon Criterion	Palumbo$-$Aust Criterion
S13/S22	5	12.6 deg [-6 19 3]/Σ243c		$\Sigma 1$
S15/S26	5	6.8 deg [-1 23 18]/Σ243i		$\Sigma 1$
S1/S27	5	7.5 deg [-2 16 17]/Σ243i		$\Sigma 1$
S4/S8	4	60.1 deg [9 7 -9]/Σ81d	$\Sigma 3$/6.5 deg	random
S10/S11	4	60.2 deg [4 -3 4]/Σ81d	$\Sigma 3$/7.1 deg	random
S1/S12	4	59.9 deg [17 -17 13]/Σ81d	$\Sigma 3$/6.9 deg	random
S14/S16	4	59.6 deg [-13 10 -13]/Σ81d	$\Sigma 3$/6.6 deg	random
S2/S17	4	60.4 deg [-15 15 11]/Σ81d	$\Sigma 3$/7.8 deg	random
S17/S21	4	59.7 deg [3 -4 -4]/Σ81d	$\Sigma 3$/7.1 deg	random
S13/S26	4	60.1 deg [4 4 -3]/Σ81d	$\Sigma 3$/7.3 deg	random
S2/S27	6	53.6 deg [14 -13 -12]/Σ729p	$\Sigma 3$/7.3 deg	random
S10/S27	6	60.3 deg [-16 20 15]/Σ729s	$\Sigma 3$/7.2 deg	random
S13/S27	6	54.0 deg [-9 8 10]/Σ729p	$\Sigma 3$/7.6 deg	random
S8/S28	4	60.3 deg [-4 -4 3]/Σ81d	$\Sigma 3$/7.5 deg	random
S3/S21	7	35.0 deg [1 -1 -21]/Σ2187ee		$\Sigma 5$/3.1 deg
S7/S23	8	38.8 deg [-4 1 30]	$\Sigma 5$/4.9 deg	random
S19/S20	6	39.3 deg [-15 17 15]/Σ729a		$\Sigma 7$/2.6 deg
S18/S22	6	38.3 deg [10 9 9]/Σ729a		$\Sigma 7$/1.9 deg
S12/S24	8	39.4 deg [18 -19 16]		$\Sigma 7$/2.8 deg
S21/S27	8	34.3 deg [11 -13 10]	$\Sigma 7$/5.6 deg	random
S4/S18	6	49.5 deg [12 1 14]/Σ729q	$\Sigma 11$/4.5 deg	random
S15/S21	6	50.3 deg [-17 15 1]/Σ729q	$\Sigma 11$/3.8 deg	random
S6/S25	6	50.8 deg [-14 1 -13]/Σ729q	$\Sigma 11$/3.1 deg	random
S14/S19	7	31.5 deg [-13 -14 -13]/Σ2187s	$\Sigma 13$b/3.9 deg	random
S11/S25	7	21.2 deg [28 4 3]/Σ2187h	$\Sigma 13$a/4.1 deg	random
S27/S28	9	26.4 deg [15 -11 15]	$\Sigma 13$b/3.9 deg	random
S3/S25	8	48.7 deg [25 -14 -1]	$\Sigma 15$/2.5 deg	random
S4/S25	8	45.7 deg [-27 0 14]	$\Sigma 15$/2.6 deg	random
S14/S25	8	48.8 deg [-9 26 10]	$\Sigma 15$/2.5 deg	random
S3/S28	8	45.8 deg [-26 14 1]	$\Sigma 15$/2.9 deg	random
S3/S15	5	42.9 deg [9 18 10]/Σ243a	$\Sigma 21$b/2.3 deg	random
S14/S16	5	43.1 deg [5 -9 -5]/Σ243a	$\Sigma 21$b/2.6 deg	random
S18/S20	5	43.5 deg [16 -9 9]/Σ243a	$\Sigma 21$b/2.6 deg	random
S18/S21	7	21.5 deg [-8 -10 -9]/Σ2187c	$\Sigma 21$a/2.0 deg	random
S11/S22	5	43.4 deg [-11 11 -20]/Σ243a	$\Sigma 21$b/2.3 deg	random
S19/S22	7	42.8 deg [-20 9 9]/Σ2187a	$\Sigma 21$b/2.7 deg	random
S9/S23	5	42.3 deg [-5 -9 5]/Σ243a	$\Sigma 21$b/2.6 deg	random
S18/S23	7	43.9 deg [-9 9 -20]/Σ2187a	$\Sigma 21$b/2.2 deg	random
S11/S24	7	21.2 deg [8 -7 9]/Σ2187c	$\Sigma 21$a/2.2 deg	random
S24/S27	9	45.0 deg [7 16 8]	$\Sigma 21$b/2.2 deg	random

Note: The column of "n of $\Sigma 3^n$" is identified by orientation twin-chain analyzing in Fig. 9(d). The column of misorientation is given in the form: misorientation/closest higher-order twin relationship, in which the higher-order twin relationships are identified according to the work of Gottstein[22] and Randle,[23] while for the ones with $n>7$, the closest higher-order twin relationships are not given. The column "closest low$-\Sigma$ CSL/deviation" is automatically given by the OIM system, in which both the Brandon criterion[27] and the Palumbo$-$Aust criterion[26] are used.

boundaries by the OIM system. The line scan would intersect with the higher-order $\Sigma 3^n$ ($n>3$) grain boundaries, so the actual mean size of clusters was bigger than the value given in Figure 10. However, both the grain cluster size and the grain size decreased with strain, as shown in Figure 10. Therefore, the ratio of the grain cluster size over the grain size is more important for the GBCD. This ratio can be represented as $R=$ (mean distance between random grain boundaries)/(mean distance between all kinds of boundaries). Figure 11 shows the proportion of $\Sigma 3^n$ boundaries *vs* the value of R. In the current experiment, the value of R was highest for sample f, which underwent 5 pct cold rolling and 5 minutes' annealing at 1 100 ℃, and its proportion of $\Sigma 3^n$ boundaries was highest.

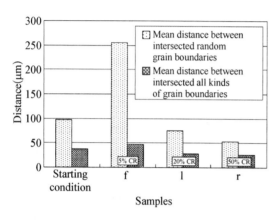

Fig. 10 Grain cluster size and grain size (including $\Sigma 3^n$- type boundaries as defining individual grains) dependence on TMP. The CR denotes cold-rolling reduction ratio before 5 min annealing at 1 100 ℃

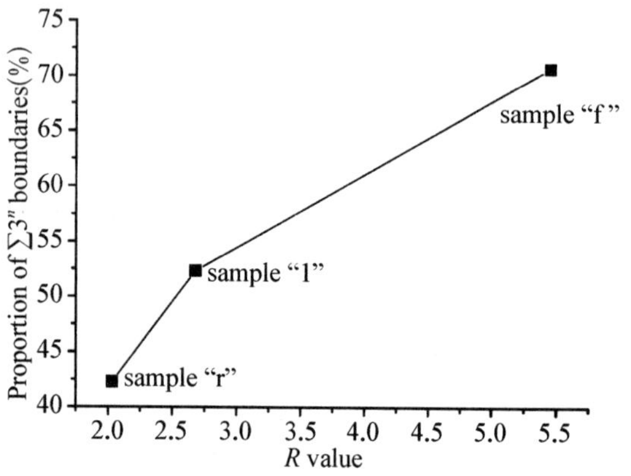

Fig. 11 Relationship between R value and proportion of $\Sigma 3^n$- type boundaries

4　Discussion

The original grain boundaries of the deformed microstructure did not play an important role in the manipulation of the proportion of $\Sigma 3^n$ boundaries. Prior to deformation, there was a significant amount of carbide precipitation at the grain boundaries due to the aging treatment after solution annealing. After deformation, the samples were rapidly heated to the annealing temperature. At the onset of annealing, boundary movement was prohibited by the presence of these carbides; hence, strain-induced boundary migration did not occur. As the annealing progressed at 1 100 ℃, recrystallization took place simultaneously with the dissolving of most carbides in the 5 pct cold-rolled samples (Figures 2(d) and (e)). The boundaries between the recrystallizing grains and the deformed matrix were capable of moving past the original grain boundaries. This could be verified by observing that the mean sizes of clusters of sample f were much larger than that of the SC. Hence, for the samples with 5 pct cold rolling, the original grain boundaries of the deformed microstructure were pinned at the beginning of annealing and disappeared following interaction with migrating grain boundaries during recrystallization. In this way, eventually, the proportions of $\Sigma 3^n$ boundaries were enhanced following recrystallization. In summary, the original grain boundaries of the deformed microstructure in this study did not play an important role in enhancing the proportion of the $\Sigma 3^n$ boundaries.

The size of the grain clusters was controlled by the nucleation density during recrystallization. A small amount of deformation (5 pct cold rolling) led to a lower stored energy for recrystallization. Therefore, fewer recrystallization nuclei formed. Thus, individual nuclei had larger volumes of deformed matrix available to consume. During the new grain growth, grains were prone to twin formation, as shown in Figure 3, because of the low SFE. Because they had a large volume available to occupy, the recrystallizing grains continuously twinned to form a large twin chain, and the associated $\Sigma 3^n$ boundaries were produced by the intersections between different twin generations. [30, 33] After consuming the deformed matrix, the twin chains impinged on each other. The recrystallization process then finished and the large grain cluster had formed. Therefore, small strains led to large grain clusters. It should be pointed out that the size of the grain clusters in sample f was inhomogeneous. Under low-deformation circumstances, the difference in stored energy in different deformed grains is more pronounced. [35] As a consequence, the nucleation density was inhomogeneous. In other words, nucleation occurred in some deformed areas, while other areas were simply consumed. This may be the reason for the inhomogeneous size of the grain clusters.

The grain size also decreased with the increasing of the deformation level (Figure 10). Mahajan's annealing twin formation model[38] pointed out that the higher the velocity of the boundary migration, the narrower the twins. This trend was because the higher velocity favored the nucleation of Shockley partial loops on consecutive {111} planes by growth accidents on propagating {111} steps in the wake of that moving boundary. For the 20 and 50 pct deformed samples in the current experiment, the larger strain provided a greater driving force for the movement of the boundary between the recrystallization front and the deformed matrix and, hence, a higher velocity. Therefore, the widths of the annealing twins in the samples subjected to more strain were narrower than that in the less strained one, as shown in Figures 2, 8, and 10. This was the reason for the grain size decreasing with the increasing level of deformation (Figure 10). On the other hand, for the sample with a large deformation level, there would be many more nuclei during recrystallization. New grain clusters would soon impinge on each other, so there was not much volume available for multiple twinning. Therefore, the proportions of $\Sigma 3^n$ boundaries for the larger strained samples were lower than that of the smaller strained one, although the twins of the larger strained samples were narrower.

Some reports demonstrated that recrystallization would introduce new random grain boundaries into the microstructure and, hence, work against increasing the proportion of $\Sigma 3^n$ boundaries. [12, 13] The current results demonstrated that both high and low proportions of $\Sigma 3^n$ boundaries resulted from recrystallization. It is true that during recrystallization random grain boundaries will be generated, but their movement in large volumes during consumption of the deformed matrix will introduce more $\Sigma 3^n$ boundaries into the microstructure. Therefore, the derivative effect of the random grain boundaries during recrystallization should be considered, *i. e.*, how many twins and multiple twins can they bring to the microstructure, instead of only emphasizing that random grain boundaries are added to the microstructure.

If high proportions of $\Sigma 3^n$ boundaries are desired, it is evident that enlarging the grain cluster size and simultaneously maintaining relatively the small twin-related grain size will be effective, *i. e.*, increase the R value (introduced in the foregoing text). Palumbo *et al.*[39] proposed the "twin limited" microstructure. They thought that as a consequence of both energetic and crystallographic constraints associated with twinning, a CSL distribution consisting entirely of low $-\Sigma$ CSL boundaries was attainable. However, they did not propose how to obtain this kind of microstructure. The large grain cluster in Figure 9 showed us the possibility of obtaining entirely $\Sigma 3^n-$ type boundaries in a certain volume of metal. As discussed previously, the size of the grain cluster was governed by the nucleation density during recrystallization, and the grain size inside of the grain cluster was governed by the driving force for the movement of the boundary between the front of recrystallization and the deformed matrix. If in a piece of deformed metal with a certain volume there is only one nucleus formed, continuous multiple twinning starting from that nucleus occurs during growth to occupy the whole volume. In this way, this piece of metal only contains a single grain cluster, so that all the boundaries in this piece of material would be of the $\Sigma 3^n$ type.

5 Summary

Based on the observation of microstructural changes during recrystallization, it was evident that the original grain boundaries of the deformed microstructure did not play an important role in the manipulation of the proportion of the $\Sigma 3^n-$ type boundaries. This was true not only for the largely deformed samples but also for the samples with small levels of strain (only cold rolled 5 pct), in which the development of twin chains during recrystallization played important roles. Small levels of deformation (5 pct cold rolling) and recrystallization annealing produced microstructures with large twininduced grain clusters that could be recognized as twin chains after recrystallization. In this case, the proportion of $\Sigma 3^n$ boundaries was more than 70 pct. This kind of grain cluster is formed by multiple twinning starting from a single nucleus of recrystallization. A large grain cluster containing 91 grains was analyzed. All of the grains in this cluster have $\Sigma 3^n$ mutual misorientations regardless of whether they are adjacent. Twin relationships up to the ninth generation ($\Sigma 3^9$) were found in this grain cluster. The random grain boundaries and low $-\Sigma$ CSL grain boundaries inside the grain cluster were also the result of multiple twinning, and there is inevitably some overlap between high-order twin relationships and some low $-$

Σ CSL relationships. The ratio of the grain cluster size over the grain size (including all types of boundaries as defining individual grains) governs the proportion of $\Sigma 3^{n}$ boundaries.

Acknowledcments

This work was supported by the Major State Basic Research Development Program of China (Grant No. 2006CB605001), the Innovation Foundation of Shanghai University (Grant No. A10 - 0110 - 08 - 004), and the Shanghai Leading Academic Discipline Project (Grant No. S30107). The authors are grateful to Dr. Qin Bai for assistance during the sample preparation.

References

[1] D. R. Diercks, W. J. Shack, and J. Muscar: *Nucl. Eng. Des.*, 1999, vol. 194, pp. 19 - 30.

[2] M. Thuvander and K. Stiller: *Mater. Sci. Eng.*, A, 2000, vol. 281, pp. 96 - 103.

[3] S. Qiu, X. Su, and Y. Wen: *Nucl. Power Eng.*, 1995, vol. 16, pp. 336 - 40.

[4] M. Kumar and C. A. Schuh: *Scripta Mater.*, 2006, vol. 54, pp. 91 - 92.

[5] V. Randle: *Acta Mater.*, 2004, vol. 52, pp. 4067 - 81.

[6] T. Watanabe: *Res. Mech.*, 1984, vol. 11, pp. 47 - 84.

[7] M. L. Kronberg and F. H. Wilson: *Trans. AIME*, 1949, vol. 185, pp. 501 - 14.

[8] W. Bollmann: *Crystal Defects and Crystal Interfaces*, Springer, Berlin, 1970, pp. 1 - 254.

[9] P. Lin, G. Palumbo, and U. Erb: *Scripta Metall. Mater.*, 1995, vol. 33, pp. 1387 - 92.

[10] E. M. Lehockey, D. Limoges, G. Palumbo, J. Sklarchuk, K. Tomantschger, and A. Vincze: *J. Power Sources*, 1999, vol. 78, pp. 79 - 83.

[11] V. Thaveeprungsriporn and G. S. Was: *Metall. Mater. Trans. A*, 1997, vol. 28A, pp. 2101 - 12.

[12] M. Kumar, A. J. Schwartz, and W. E. King: *Acta Mater.*, 2002, vol. 50, pp. 2599 - 12.

[13] S. L. Lee and N. L. Richards: *Mater. Sci. Eng.*, A, 2005, vol. 390, pp. 81 - 87.

[14] M. Shimada, H. Kokawa, and Z. J. Wang: *Acta Mater.*, 2002, vol. 50, pp. 2331 - 41.

[15] C. B. Thomson and V. Randle: *Acta Mater.*, 1997, vol. 45, pp. 4909 - 16.

[16] V. Randle: *Acta Mater.*, 1999, vol. 47, pp. 4187 - 96.

[17] W. G. Wang and H. Guo: *Mater. Sci. Eng.*, A, 2007, vols. 445 - 446, pp. 155 - 62.

[18] S. Xia, B. X. Zhou, W. J. Chen, and W. G. Wang: *Scripta Mater.*, 2006, vol. 54, pp. 2019 - 22.

[19] S. Xia, B. X. Zhou, and W. J. Chen: *J. Mater. Sci.*, 2008, vol. 43, pp. 2990 - 3000.

[20] B. W. Reed, M. Kumar, R. W. Minich, and R. E. Rudd: *Acta Mater.*, 2008, vol. 56, pp. 3278 - 89.

[21] B. W. Reed and M. Kumar: *Scripta Mater.*, 2006, vol. 54, pp. 1029 - 33.

[22] G. Gottstein: *Acta Metall.*, 1984, vol. 32, pp. 1117 - 38.

[23] V. Randle: *J. Mater. Sci.*, 2006, vol. 41, pp. 653 - 60.

[24] J. J. Kai, G. P. Yu, C. H. Tsai, M. N. Liu, and S. C. Yao: *Metall. Trans. A*, 1989, vol. 20A, pp. 2057 - 67.

[25] H. Li, S. Xia, B. X. Zhou, J. S. Ni, and W. J. Chen: *Acta Metall. Sinica*, 2009, vol. 45, pp. 195 - 98.

[26] G. Palumbo and K. T. Aust: *Acta. Metall. Mater.*, 1990, vol. 38, pp. 2343 - 52.

[27] D. G. Brandon, B. Ralph, S. Ranganathan, and M. S. Wald: *Acta Metall.*, 1964, vol. 12, pp. 813 - 18.

[28] http: //www. edax. com/.

[29] F. J. Humphreys: *J. Mater. Sci.*, 2001, vol. 36, pp. 3833 - 54.

[30] A. Berger, P. J. Wilbrandt, F. Ernst, U. Klement, and P. Haasen: *Prog. Mater. Sci.*, 1988, vol. 32, pp. 1 - 95.

[31] P. Haasen: *Metall. Trans. B*, 1993, vol. 24B, pp. 225 - 39.

[32] P. J. Wilbrandt: *Phys. Status Solidi*, 1980, vol. 61, pp. 411 - 18.

[33] C. V. Kopezky, A. V. Andreeva, and G. D. Sukhomlin: *Acta Metall. Mater.* , 1991, vol. 39, pp. 1603 – 15.

[34] H. Paul, J. H. Driver, C. Maurice, and A. Piatkowski: *Acta Mater.* , 2007, vol. 55, pp. 833 – 47.

[35] F. J. Humphreys and M. Hatherly: *Recrystallization and Related Annealing Phenomena*, 2nd ed. , Elsevier, Oxford, United Kingdom, 2004, pp. 215 – 67.

[36] D. P. Field, L. T. Bradford, M. M. Nowell, and T. M. Lillo: *Acta Mater.* , 2007, vol. 55, pp. 4233 – 41.

[37] V. Y. Gertsman and C. H. Henager: *Interface Sci.* , 2003, vol. 11, pp. 403 – 15.

[38] S. Mahajan, C. S. Pande, M. A. Imam, and B. B. Rath: *Acta Mater.* , 1997, vol. 45, pp. 2633 – 38.

[39] G. Palumbo, K. T. Aust, U. Erb, P. J. King, A. M. Brennenstuhl, and P. C. Lichtenberger: *Phys. Status Solidi* , 1992, vol. 131, pp. 425 – 28.

快淬 NdFeB 磁粉磁性能不均匀性问题的研究[*]

摘　要： 用磁选方法从同一批生产的含合金元素 Zr 和 Co 的快淬 NdFeB 磁粉中选出磁性能明显不同的优劣两种粉磁，研究优劣两种磁粉的成分及显微组织结构差别. 结果表明：优质磁粉的 Zr 含量明显偏低、Fe 含量略高，晶粒结晶完整、晶界衬度清晰，晶界上无其他相存在，晶粒尺寸大都分布在 20～60 nm 范围内，较为均匀；而劣质磁粉的 Zr 含量偏高、Fe 含量稍低，且劣质粉中存在大量的亚稳态组织结构，包括非晶、α-Fe＋非晶以及≤10 nm 衬度不清晰、结构不完整的 $Nd_2Fe_{14}B$ 晶粒. 认为，劣质磁粉中含 Zr 量偏高，提高了发生晶化的开始温度是造成磁性能不均匀的主要原因.

自 1983 年 NdFeB 稀土永磁材料问世以来，随着磁性能的不断提高及成本的降低，应用领域不断扩大，2003 年仅快淬 NdFeB 磁粉制备的粘结 NdFeB 磁体用量就达到 4 476 t，预计 2008 年这一数字将超过 7 000 t. 由于快淬 NdFeB 磁粉的生产过程较为复杂，且对生产设备及工艺依赖性很大，所以容易产生磁性能的波动，特别是用电弧式熔体快淬工艺生产的快淬 NdFeB 磁粉的磁性能波动尤为显著. 根据矫顽力低的快淬 NdFeB 磁粉易磁化，矫顽力高的快淬 NdFeB 磁粉难磁化的特性，用磁选的方法可以将晶化后的快淬 NdFeB 磁粉分选为矫顽力高的优质粉和矫顽力低的劣质粉. 研究优、劣质粉的磁性能、显微组织结构及成分方面的差异，弄清造成这种差异的原因，对制备工艺、尽量减少劣质粉的比例，提高产品的磁性能具有现实意义. 本实验研究了同批次快淬 NdFeB 磁粉的磁性能不均匀性问题.

1　实验方法

用纯度为 99.8％的稀土金属 Nd，工业纯铁，99.85％的 Zr 和电解 Co 以及 FeB 合金制备名义成分为 $Nd_{10.5}Fe_{76}Co_5Zr_{2.5}B_6$ 的五元合金，用自行研制的 ZK-10TCII 型真空电弧快淬炉制备条带，淬速为 18 m/s；用自行研制的真空连续晶化炉以 700 ℃，10 min 的工艺进行晶化处理. 同一批次快淬、晶化的 NdFeB 磁粉经 3 次磁选后，选出 30％左右的劣质粉，余下 70％左右为优质粉. 将选出的优、劣质磁粉分别同时取样，与质量占 3％的环氧树脂均匀混合、模压制备成粘结磁体试样（Φ10 mm×8 mm，ρ＝6.0±0.1 g/cm³），用 DGY-II 磁性测量仪测量磁性能. 将约 30％体积分数的磁粉与 Al 粉混合，经过压形、轧片及离子溅射减薄，制成透射电镜观察用试样[1]，用 JEOL-200CX 透射电子显微镜进行显微组织和晶体结构研究；用 AMRAY-1845FE 扫描电镜对优、劣质快淬 NdFeB 磁粉进行合金成分分析.

* 本文合作者：黄照华、李强、张士岩. 原发表于《稀有金属材料与工程》，2009，38（2）：247-250.

2 结果和讨论

2.1 磁性能的差异

图 1 为晶化后快淬 NdFeB 经磁选优粉和劣粉制备的磁体的退磁曲线. 得到的磁性能列于表 1 中. 由图 1 及表 1 可以看出：优质快淬 NdFeB 磁粉的磁性能明显高于劣质粉. 造成这种差异的因素比较复杂, 同冶炼、快淬和晶化这 3 个主要的工艺环节有密切关系, 如冶炼铸锭、快淬时易引发合金成分的不均匀, 电弧快淬时因快淬薄带厚度的不均匀而引起冷却速度的差异[2, 3], 晶化过程磁粉受热不均匀或晶化工艺不当等因素都会引起显微组织的不均匀. 由此可见, 所有快淬 NdFeB 磁粉都可能存在磁性能的不均匀, 只是不均匀的程度不同而已.

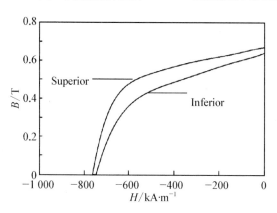

图 1 利用快淬 NdFeB 优、劣质磁粉制备成粘结磁体的退磁曲线

Fig. 1 Demagnetization curves of bonded magnets prepared by the superior and inferior magnetic powders

2.2 合金成分的差异

表 2 为晶化后选出的优、劣质快淬 NdFeB 磁粉的能谱分析结果（由于 AMRAY-1845FE 扫描电镜不能分析超轻元素 B, 故表中只给出了 Nd、Fe、Co 和 Zr 的测定值）. 由表 2 可以看出：优、劣质快淬 NdFeB 磁粉的 Nd、Fe、Zr 含量都存在明显的差异, 其中 Zr 和 Fe 的含量差异最为显著, 优质快淬 NdFeB 磁粉的 Zr 含量明显偏低、Fe 含量略高, 而劣质快淬 NdFeB 磁粉的 Zr 含量则明显偏高、Fe 含量略低. 显然, 优、劣质快淬 NdFeB 磁粉成分的差异是导致其磁性能差异的重要原因. 造成成分差异的因素比较多, 如冶炼工艺、快淬工艺不当都可能引发合金成分的不均匀.

表 1 用优质粉和劣质粉制备的粘结磁体的磁性能对比

Table 1 Magnetic properties of bonded magnets prepared by the superior and inferior magnetic powders

Powders	B_r/T	H_{cj}/kA · m^{-1}	$(BH)_{max}$/kJ · m^{-3}
Saperior	0.67	768	71.2
Inferior	0.64	748	60

表 2 磁选后快淬 NdFeB 优、劣质粉的合金成分(ω/%)

Table 2 Composition of superior and inferior NdFeB powders

Powders	Zr	Nd	Fe	Co
Superior	3.47±0.30	26.70±0.53	65.25±0.67	4.58±0.39
Inferior	4.93±0.35	27.40±0.49	63.47±0.57	4.20±0.37

2.3 显微组织和晶体结构的差异

图 2a、2b 分别为晶化后的优质快淬 NdFeB 磁粉显微组织的 TEM 照片以及选区电子衍射花样. 如图 2a 所示,优质粉呈多边形、晶粒结晶完整、晶界衬度清晰,晶界上无其他相存在,晶粒尺寸大都分布在 20～60 nm 范围内,分布较为均匀,表明快淬 NdFeB 薄带的晶化完全,晶粒尚未长大. 从图 2b 可以看出,这些晶粒是 $Nd_2Fe_{14}B$ 相. 劣质粉中的显微组织有 3 大类型:未晶化组织、晶化不足组织和同一片粉末中存在的晶化程度不均匀组织,以下将详细阐述.

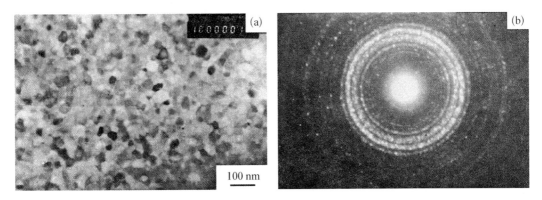

图 2 优质快淬 NdFeB 磁粉显微组织的 TEM 照片(a)和选区电子衍射花样(b)

Fig. 2 TEM micrograph of the superior NdFeB magnetic powder (a) and the selected area diffraction image (b)

图 3a、3b 分别为劣质快淬 NdFeB 磁粉中未晶化粉末显微组织的 TEM 照片和选区电子衍射花样. 如图 3a 所示,未晶化粉末显微组织中未发现有序结构,在其非晶结构的 TEM 照片中有衬度不同的团状结构,可能是成分起伏引起的结果.

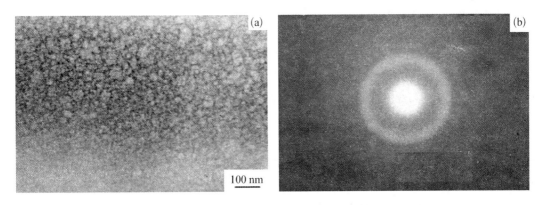

图 3 劣质磁粉中未晶化粉末显微组织的 TEM 照片(a)和选区电子衍射花样(b)

Fig. 3 TEM micrograph of the inferior NdFeB magnetic powder (a) and the selected area diffraction image (b)

图 4 为劣质快淬 NdFeB 磁粉中晶化不充分粉末显微组织的 TEM 照片. 可观察到少量结晶完整的晶粒,但大量存在的还是晶粒大小在 10 nm 以下、衬度不清晰、结晶不完整、但已是 $Nd_2Fe_{14}B$ 相的晶粒组织以及尚处于亚稳态的组织和结构,这些组织具有明显的晶化不足特征. 结合表 2 所示快淬 NdFeB 优、劣质粉的成分分析结果可知:劣质快淬 NdFeB 磁粉中

的 Zr 含量偏高,使合金的晶化温度升高[4],所以在一定的工艺条件下退火处理后,含 Zr 量较高磁粉的显微组织仍存在大量的结晶不完整的组织,有的甚至仍处于亚稳态组织. 这些晶化不足的组织由于矫顽力较低,经磁选时成为劣质粉.

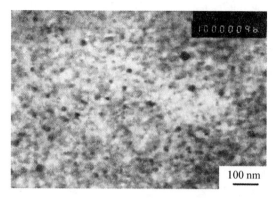

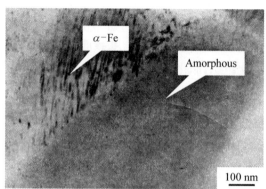

图 4　劣质快淬 NdFeB 磁粉中晶化不充分粉末的 TEM 照片

Fig. 4　TEM micrograph of uncompletely crystallized NdFeB magnetic powder

图 5　同一片粉末中晶化程度不均匀的 TEM 照片

Fig. 5　TEM micrograph of ununiform crystallization in the same section for infeiror powder

另外,值得重视的是在同一片粉末中存在着晶化程度不均匀性问题,一边以非晶结构为主,另一边已形成大量的 α-Fe 初生相,两者间存在明显分界线,如图 5 所示. 甚至还观察到一边是非晶,另一边已形成 $Nd_2Fe_{14}B$ 晶粒的现象. 这种晶化程度的不均匀性显然不能用粉末晶化时受热不均匀来解释,因为在这样一小片粉末中,不可能存在加热不均匀的明显分界线. 结合表 2 给出的快淬 NdFeB 优、劣质粉的成分分析结果来看,只可能是因为同一片粉末中存在 Zr 含量的显著差别而引起晶化温度截然不同的缘故. 故在同一片粉末中存在的晶化程度明显不同的分界线,也是成分差别明显的分界线.

由以上分析可知:无论劣质快淬 NdFeB 磁粉是未晶化组织、晶化不充分组织,还是同一片粉末中存在的晶化程度不均匀的组织,都是因为粉末中存在 Zr 含量的显著偏高而引起晶化温度大幅度升高的结果. 因此适当提高晶化处理温度,对劣质快淬 NdFeB 磁粉重新进行晶化处理,可以使磁性能得到一定的改善(表 3). 劣质粉再晶化后制备的磁体的磁性能得到了改善,但不是很明显. 这主要是由于劣质快淬 NdFeB 磁粉中既含有未晶化非晶组织,又含有晶化不足组织,还有同一片粉末中存在的晶化程度明显不均匀的组织,很难用一种再晶化处理制度来同时满足显微组织明显不同的劣质快淬 NdFeB 粉末的晶化需要,从而限制了磁性能的进一步改善.

表 3　劣质粉再晶化前后的磁性能对比

Table 3　Magnetic properties before and after econd-crystallization for inferior powder

Powders	B_r/T	$H_{cj}/kA \cdot m^{-1}$	$(BH)_{max}/kJ \cdot m^{-3}$
Recrystallized inferior	0.65	756	63.2
Inferior	0.64	748	60

至于为什么会产生 Zr 含量不均匀的现象,是在感应熔铸的合金锭中已经存在成分的不均匀? 还是在电弧重熔快淬时的某种工艺过程造成 Zr 的偏聚? 其中的原因还有待进一步

的分析研究.

3 结论

（1）在本实验条件下，优质快淬 NbFeB 磁粉的磁性能明显高于劣质粉.

（2）劣质快淬 NdFeB 磁粉的 Zr 含量明显高于优质快淬 NdFeB 磁粉. Zr 含量的显著差别是造成快淬 NdFeB 磁粉性能不均匀的主要原因.

（3）优质快淬 NdFeB 磁粉的显微组织是晶粒结晶完整、晶界衬度清晰，晶界上无其他相存在，晶粒尺寸大都分布在 20～60 nm 范围内的较为均匀的 $Nd_2Fe_{14}B$ 晶粒组织；而劣质快淬 NdFeB 磁粉中则存在亚稳态（非晶）组织、α-Fe 与非晶的混合组织以及晶粒尺寸小于 10 nm、衬度不清晰、结晶不完整的 $Nd_2Fe_{14}B$ 晶粒.

（4）劣质快淬 NdFeB 磁粉再晶化后的磁性能得到一定的改善，但改善效果不太明显.

参 考 文 献

［1］Li Qiang(李强), Zhou Bangxin(周邦新). *Rare Metal Materials and Engineering*（稀有金属材料与工程）[J]，2000，29(4)：283.

［2］Zeng Hanming(曾汉民). *Essentials of Advanced Materials for High Technology*（高技术新材料要览）[M]. Beijing：Chinese Science and Technology Press，1993：210.

［3］Chu T Y，Rabenberg L，Mishra R K. *J Appl Phys*[J]，1991，69：6046.

［4］Wu Yisheng（吴义生），Ni Jianseng（倪建森），Xu Hui(徐晖) *et al*. *Journal of The Chinese Rare Earth Society*（中国稀土学报）[J]，2005，23(3)：295.

Ununiformity of Magnetic Properties for Melt-Spun NdFeB Powder

Abstract：Superior and inferior magnetic powders can be separated from the same batch crystallined NdFeB magnetic powders containing Zr and Co elements by magnetic separation method due to their magnetic ununiformity. The compositions and microstructures of the two powders with different magnetic properties were investigated. The results show that the superior powders contain less Zr and a relative more Fe, with integrated grains of about 20～60 nm in size and clear grain boundary without other phases. And the inferior powders contain more Zr and a relative lower Fe and metastable microstructures including the amorphous phases，the α-Fe and amorphous phases，and the un-integrated $Nd_2Fe_{14}B$ phases of $\leqslant$10 nm in size. It is presumed that the ununiform magnetic properties are due to the elevated crystallization temperature induced by the more Zr content for the inferior powders.

镍基 690 合金时效过程中晶界碳化物的形貌演化*

摘　要：用 SEM 研究了含有高比例低 Σ 重位点阵(CSL)晶界的镍基 690 合金在 715 ℃时效过程中不同类型晶界上碳化物的形貌演化. 结果表明：低 ΣCSL 晶界与一般大角晶界上碳化物的形貌演化特征有明显区别. 晶界上的碳化物粒子尺寸随 Σ 值降低而减小, 共格 Σ3 晶界上的小尺寸碳化物在长时间时效后无明显变化；非共格 Σ3 晶界和 Σ9 晶界附近都观察到板条状碳化物, 并随时效时间延长明显长大, 非共格 Σ3 晶界两侧都存在碳化物板条, 而 Σ9 晶界上的碳化物板条只在晶粒一侧生长；Σ27 晶界与一般大角晶界处的碳化物形貌相似, 在晶界附近未观察到板条状碳化物.

统计调查表明, 核反应堆蒸汽发生器传热管损坏的主要原因是材料的晶间应力腐蚀破裂[1], 而产生晶间应力腐蚀的一个主要原因是晶界上碳化物的析出及其引起的晶界附近 Cr 的贫化. 作为压水堆核电站蒸汽发生器新一代传热管材的 690 合金是一种高 Cr 的镍基合金, 固溶后经过时效热处理可控制晶界上碳化物的析出, 调整晶界附近贫 Cr 区的成分, 可以显著提高其耐腐蚀性能[2]. 通过提高低 Σ 重位点阵(coincidence site lattice, CSL)[3] 晶界的比例也可以显著提高 690 合金的抗晶间腐蚀能力[4]. 并且有研究结果表明[5,6], 碳化物形貌对材料的蠕变性能有很大影响. 那么, 低 ΣCSL 晶界对晶界处碳化物的析出及其形貌的影响是值得研究的.

690 合金晶界上的碳化物以富 Cr 的 $M_{23}C_6$ 为主[7-9]. 在奥氏体不锈钢和 690 合金中, 经过固溶后再时效时, 碳化物很容易在晶界上析出[7,10,11]. 根据碳化物的大小和分布, Kai 等[7] 将晶界上析出的碳化物形貌演化按时效时间和温度分为 4 个等级. Trillo 等[11] 通过研究不同 C 含量的 304 不锈钢, 发现在晶界上析出的碳化物形貌受晶界两侧晶粒的取向差影响最大, C 含量很低时, 晶界两侧晶粒取向差较小的晶界上没有碳化物析出, 在晶界两侧晶粒取向差很大的晶界上有细小碳化物析出；当 C 含量升高时, 几乎所有晶界上都有碳化物析出. 在 Σ≤29 的低 ΣCSL 晶界上, 碳化物析出倾向比一般大角晶界小很多[11-13]. 晶界相对于两侧晶粒所处的晶体学面是不同的, 这种差别对晶界上碳化物的形核与长大也有很大影响[8,9,14]. 在奥氏体不锈钢和 690 合金中都观察到在孪晶的非共格界面处有板条状碳化物析出. 而在孪晶的共格界面处却很少观察到有碳化物析出[8,14,15].

通常研究 690 合金晶间碳化物在时效过程中的形貌演化主要集中在一般大角晶界和孪晶晶界上[8,9]. Lim 等[9] 发现低 ΣCSL 晶界和一般大角晶界上碳化物的形貌有很大区别. 但是, 还未见到有关不同类型低 ΣCSL 晶界上碳化物形貌演化的报道. 本工作对 690 合金中不同类型低 ΣCSL 晶界上析出的碳化物进行了观察, 分析了它们在不同时效时间的形貌演化.

1　实验方法

实验所用 690 合金成分(质量分数. %)为：Ni 60.52, Cr 28.91, Fe 9.45, C 0.025, N 0.008,

---

* 本文合作者：李慧、夏爽、伊建森、陈文觉. 原发表于《金属学报》, 2009, 45(2)：198-198.

Ti 0.4，Al 0.34，Si 0.14. 将所有样品真空（真空度约为 $5×10^{-3}$ Pa）密封在石英管中，在 1 100 ℃下保温 15 min. 然后立即淬入水中，并同时砸破石英管进行固溶处理. 为了获得高比例的低 ΣCSL 晶界，利用文献[16]的工艺将固溶处理的样品冷轧 10% 后，真空密封在石英管中 1 100 ℃再结晶退火 5 min，立即淬入水中，并同时砸破石英管. 然后，将样品真空密封在石英管中，在 715 ℃下分别时效热处理 2，5，15，30，50，100 和 200 h 使碳化物析出.

在扫描电子显微镜（SEM）观察之前，先将样品电解抛光与电解蚀刻以显示晶界处的碳化物. 使用的电解液是 20%HClO₄+80%CH₃COOH，在室温下用 30 V 直流抛光 30 s，然后用 5 V 直流蚀刻 3～5 s.

晶界两侧晶粒取向关系利用配备在 Hitachi S570 SEM 中的 TSL 电子背散射衍射（electron backscatter diffraction，EBSD）附件确定，扫描步长为 7 μm. 数据采集后利用取向成像显微系统（orientation imaging microscopy，OIM）自动确定晶界类型，晶界类型的判定采用 Palumbo-Aust 标准[17]，相对于 CSL 模型最大偏差角为 $\Delta\theta_{max} = 15°\Sigma^{-5/6}$. 晶界类型确定后，利用 JSM - 6700F SEM 对不同类型晶界上的碳化物进行观察，并用能谱仪（energy dispersive spectroscopy，EDS）对晶间碳化物的成分进行分析.

2　实验结果与讨论

图 1a 为固溶处理的样品经冷轧 10% 后，在 1 100 ℃再结晶退火 5 min 后的 OIM 图. 由于 690 合金的层错能低，经过这样的加工工艺在 1 100 ℃再结晶退火时，在再结晶晶核的生长过程中，多重孪晶过程得到充分发展，使样品中含有大量的 Σ3，Σ9 和 Σ27 晶界（图 1a），总的低 ΣCSL 晶界比例可超过 70%[4]（图 1b）. 通过 EDS 对碳化物粒子成分的分析表明，所有碳化物都是富 Cr 的，与文献[7,8]中的报道一致.

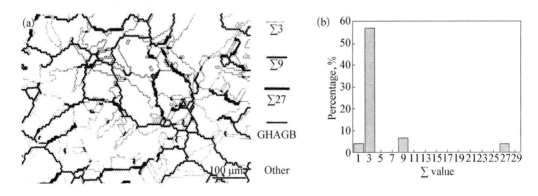

图 1　经过再结晶退火后含有高比例低 ΣCSL 晶界（71%，Palumbo-Aust 标准）样品的 OIM 图与晶界特征分布统计

Fig. 1　OIM map（a）and grain boundary character distribution statistics（b）of the specimens with high proportional（71%. Palurmbo-Aust criterion）low ΣCSL boundaries（the specimens were cold-rollrd 10% and recrystallization annealed at 1 100 ℃ for 5 min；GHAGB — general high angle grain boundary；other — other low ΣCSL boundarics）

在 715 ℃时效不同时间后，不同类型晶界上碳化物的形貌如图 2 所示. 图中的横排是同一种晶界在 715 ℃时效处理 2.30 和 200 h 后的 SEM 二次电子像，图中的纵列是不同类型的晶界在 715 ℃时效处理同一时间后的 SEM 二次电子像. Σ3ᴄ 和 Σ3ᵢ 分别表示退火孪晶的共

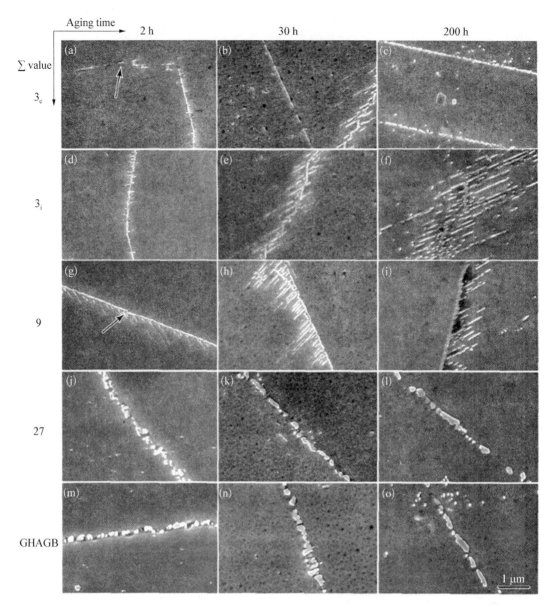

图 2 在 715 ℃时效不同时间后不同类型晶界上碳化物形貌演化

Fig. 2 Morphologies of carbides precipitated on grain boundaries with different characters in the specimens aged at 715 ℃ for different times ($\Sigma 3_c$— coherent twin boundary; $\Sigma 3_i$— incoherent twin boundary; arrow in Fig. 2a shows the $\Sigma 3_c$ boundary, arrow in Fig. 2g shows the plate-like carbide precipitated near $\Sigma 9$ boundary)

格界面和非共格界面. GHAGB 表示一般的大角度晶界. 从图中可以看出,不同类型晶界上的碳化物随着时效时间的延长都在不断长大,但形貌演化的特征却有不同,这种不同的特征如表 1 所示.

不同类型晶界上碳化物的形貌有很大区别,尤其是 $\Sigma 3$ 晶界与其他类型晶界上的碳化物形貌区别更大. 在时效时间很短(2 h)时,$\Sigma 3_c$ 晶界上析出的碳化物很少,如图 2a 中箭头所指的晶界. 即使延长时效时间至 200 h,$\Sigma 3_c$ 晶界上的碳化物粒子也比 $\Sigma 27$ 和一般大角晶界上

表 1　不同类型晶界上碳化物随时效时间的形貌演化

Table 1　Morphology evolution of carbides precipitated at grain boundaries with different characters after different aging times

Grain boundary character	Aging time, h						
	2	5	15	30	50	100	200
$\Sigma3_c$	(······Very fine, discrete······)			(···Very fine, semicontinuous···)			(Fine, continuous)
$\Sigma3_i$	(·················Very fine, discrete·················)						(Fine, semicontinuous)
$\Sigma9$	(Fine, discrete)	(···························Fine, semicontinuous···························)					
$\Sigma27$	(Fine, discrete)	(············Fine, semicontinuous···········)			(Large, discrete)		(Large, semicontinuous)
GHAGB	(Fine, discrete)	(Fine, semicontinuous)	(······Large, semicontinuous······)				(Coarse, discrete)

时效 2 h 时的碳化物粒子要小很多, 如图 2c、j 和 m 所示. $\Sigma3_i$ 晶界上析出的碳化物颗粒随时效时间的延长缓慢长大; 而 $\Sigma3_i$ 晶界附近向两侧晶粒内生长的板条状碳化物随着时效时间的延长明显长大, 如图 2d—f 所示. 通过比较 $\Sigma3_c$ 与 $\Sigma3_i$ 晶界处的碳化物形貌可以看出, 即使晶界两侧的晶粒具有相同取向关系, 但晶界相对于两侧晶粒所处的晶体学面存在差异时, 碳化物的形貌及其演化特征也会有明显的区别.

$\Sigma9$ 晶界上的碳化物形貌也有自身的特点: 当时效时间很短 (2 h) 时, $\Sigma9$ 晶界附近有些碳化物粒子向晶界一侧的晶粒内部生长, 形成细小板条状碳化物, 如图 2g 中箭头所示. 随着时效时间的延长, $\Sigma9$ 晶界上的碳化物粒子缓慢长大 (图 2h 和 i); 而 $\Sigma9$ 晶界附近的板条状碳化物在时效 30 h 后明显变长, 但在时效 200 h 后并没有继续变长, 只是变厚、变宽. $\Sigma9$ 晶界附近的板条碳化物只向一侧的晶粒内生长, 这与 $\Sigma3_i$ 晶界附近的板条碳化物会向晶界两侧的晶内生长存在明显差异. 这是由于晶界上的碳化物一般只与一侧晶粒具有 $\{111\}_\gamma /\!/ \{111\}_{M_{23}C_6}$ 共格取向关系[18], 并且碳化物板条的快速生长方向为 $\langle110\rangle$[8,14]. 由于 $\Sigma3_i$ 晶界两侧晶粒具有绕 $\langle111\rangle$ 轴旋转 $60°$ 的特定取向关系, 使碳化物板条向晶粒两侧生长时仍然保持与奥氏体基体的共格取向关系. 而 $\Sigma9$ 晶界两侧晶粒的取向关系为绕 $\langle110\rangle$ 轴旋转 $38.9°$, 如果碳化物板条向两侧晶粒内部生长, 将有一侧晶粒内部的碳化物板条与奥氏体基体不具有共格取向关系, 那么碳化物板条仅向适合生长的一侧晶粒内部生长.

$\Sigma27$ 晶界和一般大角晶界上的碳化物形貌相似, 碳化物尺寸随时效时间延长稍有不同 (表 1). 但是, $\Sigma27$ 晶界和一般大角晶界处的碳化物与 $\Sigma3$ 和 $\Sigma9$ 晶界处碳化物的形貌特征相比, 界面上的碳化物颗粒尺寸明显大, 并且在晶界附近未出现板条状的碳化物. 这是由于 Σ 值高的晶界和一般大角晶界的界面能高, 碳化物粒子可以充分长大. 界面能的大小可能是影响晶间碳化物形貌的重要因素之一.

3　结论

不同类型晶界上的碳化物随着时效热处理的时间延长都在不断长大, 但碳化物形貌演化的特征却不同. 非共格 $\Sigma3$ 和 $\Sigma9$ 晶界附近都有板条状碳化物形成, 随时效时间的延长, 碳化物板条会明显长大, 两者的区别在于非共格 $\Sigma3$ 晶界两侧都存在碳化物板条, 而 $\Sigma9$ 晶界处只有晶界的一侧生长出碳化物板条. $\Sigma27$ 晶界与一般大角晶界上碳化物的形貌及其演化特征相似, 在晶界两侧均未出现板条状碳化物.

参 考 文 献

[1] Diercks D R. Shack W J, Muscar J. *J Nucl Eng Des*, 1999; 194: 19.

[2] Qiu S Y, Su X W, Wen Y, Yan F G. Yu Y H, He Y C. *Nucl Power Eng*, 1995; 16: 336.
  （邱少宇,苏兴万,文　燕,闫福广,喻应华,何艳春. 核动力工程. 1995; 16: 336）

[3] Kronberg M L, Wilson F H. *Trans AIME*, 1949; 185: 501.

[4] Xia S, Zhou B X, Chen W J. *J Mater Sci*, 2008; 43: 2990.

[5] Min K S, Nam S W. *J Nucl Mater*, 2003; 322: 91.

[6] Spigarelli S, Cabibbo M, Evangelista E, Palumbo G. *Mater Sci Eng*, 2003; A352: 93.

[7] Kai J J, Yu G P, Tsai C H, Liu M N, Yao S C. *Metall Trans*, 1989; 20A: 2057.

[8] Li Q, Zhou B X. *Acta Metall Sin*, 2001; 37: 8.
  （李　强,周邦新. 金属学报,2001; 37: 8）

[9] Lim Y S, Kim J S, Kim H P, Cho H D. *J Nucl Mater*, 2004; 335: 108.

[10] Hall E L, Briant C L. *Metall Trans*. 1984; 15A: 793.

[11] Trillo E A, Murr L E. *Acta Mater*. 1999; 47: 235.

[12] Zhou Y, Aust K T, Erb U. Palumbo G. *Scr Mater*, 2001; 45: 49.

[13] Hong H U, Rho B S, Nam S W. *Mater Sci Eng*, 2001; A318: 285.

[14] Trillo E A, Murr L E. *J Mater Sci*, 1998; 33: 1263.

[15] Sasmal B. *Metall Mater Trans*, 1999; 30A: 2791.

[16] Xia S, Zhou B X, Chen W J, Wang W G. *Scr Mater*, 2006; 54: 2019.

[17] Palumbo G, Aust K T, Lehockey E M. *Scr Mater*, 1998; 38: 1685.

[18] Lewis M H, Hattersley B. *Acta Metall*, 1965; 13: 1159.

Evolution of Carbide Morphology Precipitated at
Grain Boundaries in Ni-based Alloy 690

Abstract: The effects of grain boundary characters on the morphology evolution of the intergranular carbide in Ni-based alloy 690 with high proportional low Σ coincidence site lattice (CSL) boundaries aged at 715 ℃ for 2—200 h were investigated by SEM. The results show that the sizes of intergranular carbides decrease with Σ value decreasing, and only fine carbides at coherent $\Sigma3$ are almost not changed after aging for 200 h. Plate-like carbides precipitated near both incoherent $\Sigma3$ and $\Sigma9$ boundaries, and the carbide plates grow bigger with the aging time prolonging, in which the former precipitated near both sides of incoherent $\Sigma3$ boundary and the latter precipitated near only one side of $\Sigma9$ boundary. The morphology of carbide precipitated at $\Sigma27$ boundary is similar to that precipitated at the general high angle grain boundary, and no plate-like carbides are observed near these boundaries.

高温退火过程中 316 不锈钢晶界特征分布的演化[*]

摘　要： 用扫描电子显微镜(SEM)和电子背散射衍射(EBSD)技术研究了 5％冷变形 316 不锈钢经 1 100 ℃不同时间退火样品的晶界特征分布(GBCD). 结果表明：低 ΣCSL(Σ≤29)晶界比例的提高是在再结晶过程中实现的. 在退火 40 min 时再结晶完成,低 ΣCSL 晶界比例达到 80％,其中 Σ3 晶界比例占总的低 ΣCSL 晶界 80％左右,晶界特征分布得到优化；尺寸较大形状不规则的晶粒团簇(grain-clusters)形成,每个晶粒团簇内部存在大量孪晶界、多重孪晶界和特殊的三叉晶界节点. 部分再结晶状态样品的显微组织特点和晶粒团簇内部孪晶链的分析表明再结晶过程中多重孪晶的发展是提高 316 不锈钢低 ΣCSL 晶界比例的关键.

1　研究晶界特征分布的意义及研究概况

316 不锈钢(316SS)作为快中子反应堆燃料元件的包壳,是堆芯中最重要的结构材料[1]. 在整个寿命期内要保持燃料元件几何尺寸的完整和稳定,要求它必须具有良好的力学性能和辐照性能,随着快堆技术的不断发展,对包壳材料的要求也越来越高,自从 20 世纪 70 年代世界各国核电站发现晶间腐蚀是影响不锈钢构件寿命的主要问题之一,晶间腐蚀成为该领域人们研究的重点.

由于晶界附近贫铬导致的晶间腐蚀[2]问题是影响 316SS 寿命的主要因素,近年来许多人采用晶界工程[3](grain boundary engineering, GBE)技术或者晶界特征分布优化(grain boundary character distribution, GBCD)技术来提高低层错能面心立方金属(镍基合金、铅基合金、铜、奥氏体不锈钢等)材料与晶界相关的性能[4],进行了大量的工作[5-10]. 316SS 是一种低层错能面心立方金属材料,对其进行适当的冷轧热处理后能获得高比例的低 ΣCSL 晶界(亦称特殊晶界),GBCD 得到优化,一般大角晶界(high angle boundary)的网络连通性被有效阻断. 低 ΣCSL 晶界由于具有原子排列有序度高、低的晶界能等特性[8],这种晶界附近不容易发生贫铬现象,因此抗晶间腐蚀的能力较高. 经过优化的试样中这种具有特殊性能的低 ΣCSL 晶界比例得到提高,通过腐蚀实验也证明具有高比例低 ΣCSL 晶界的样品抗腐蚀能力要好于低 ΣCSL 晶界比例低的样品[11-13]. 目前人们通过研究发现 316SS 经小冷变形 (5％)高温短时间退火的机械热处理工艺(Thermal Mechanic Process, TMP)[3, 14]可以获得高比例低 ΣCSL 晶界. 关于提高低 ΣCSL 晶界比例的机制及模型主要有以下几种： (1) Randle提出的"Σ3 孪晶界再激发"模型[15](Σ3 regenerating model)；(2) Kumar 等人提出的"高 ΣCSL 晶界分解"模型[4]；(3) 王卫国等提出的"非共格 Σ3 晶界的迁移与反应"模型[16]；(4) Xia 等[6]通过研究 690 合金提出的"从一个再结晶晶核发展的多重孪晶链"模型. 而目前关于 316SS 晶界特征分布优化机制的研究尚未见报道,本文为了解 316SS 晶界特征分布在形变及退火过程中的演化,试图通过分析经过 5％冷变形,1 100 ℃不同时间退火的部

---

[*]　本文合作者：王坤、陈文觉、夏爽. 原发表于《上海金属》,2009,31(5)：13 - 18.

分再结晶样品的显微组织演化探讨 316SS 中晶界特征分布优化机理，这对提高 316SS 包壳寿命、反应堆工作的安全性和经济性有着重要的现实意义.

2 实验过程

实验用厚 2 mm 热轧 316 不锈钢板，其化学成分如表 1 所示.

表 1 316 不锈钢的化学成分(质量分数，%)

元素	C	Mn	Si	S	P	Ni	Cr	Mo	Nb
含量	0.07	1.53	0.71	0.005	0.01	13.50	17.04	2.40	0.52

将钢板分割成 2 mm×7 mm×70 mm 的样品，真空密封到石英管中，进行真空固溶处理，真空度为 $5×10^{-3}$ Pa，温度 1 100 ℃，保温 30 min 后水淬至室温. 将固溶处理后的试样进行 5% 冷变形，1 100 ℃ 不同时间的再结晶退火处理，退火时间分别为 5 min、10 min、20 min、30 min、40 min、50 min，水淬.

退火试样进行电解抛光，抛光液为 20% 高氯酸与 80% 冰醋酸的混合溶液，抛光电压 40 V，时间 10~15 s，温度 0 ℃. 通过配有 TSL－EBSD 系统的日立 S－570 型扫描电子显微镜(SEM)对抛光样品表面进行逐点逐行扫描，收集由背散射电子 Kikuchi 衍射花样得到的晶体取向信息，再经过软件处理得到不同形式的取向成像显微(orientation image microscope，OIM)图. 本实验中采用 Palumbo-Aust 标准[17]：$\Delta\theta_{max(P-A)} = 15°\Sigma^{-5/6}$ 来确定晶界的类型并计算 Σ 值，通过 OIM 自动地按长度分数统计低 ΣCSL 晶界比例.

3 实验结果与讨论

3.1 晶界特征分布优化

冷轧 5%，1 100 ℃ 退火不同时间样品的不同晶界类型比例的变化趋势如图 1 所示，可见退火 40 min 内，Σ3 晶界比例随着退火时间的延长逐渐上升；退火 50 min 的样品 Σ3 晶界比例下降. Σ3 晶界比例占总的低 ΣCSL 晶界比例的 80% 左右，Σ3 晶界对 GBCD 优化贡献最大. 总的低 ΣCSL 晶界比例的变化趋势与 Σ3 晶界比例变化趋势基本相同：在退火 10~40 min 内总的低 ΣCSL 晶界比例随退火时间的延长而提高，在 40 min 时其比例达到 80%，在退火 50 min 后总的低 ΣCSL 晶界比例下降到 76%. 多重孪晶界 Σ9 和 Σ27 比例只占总的低 ΣCSL 晶界比例的 10% 左右，它们随退火时间的延长变化幅度不如 Σ3 晶界明显，变化趋势与 Σ3 晶界比例变化趋势基本相同. Σ1 晶界(亚晶界)的变化趋势是先下降后上升，尤其是从退火 5 min 到 10 min，Σ1 晶界比例从 26.6% 大幅度下降到 7.7%，在退火 30 min

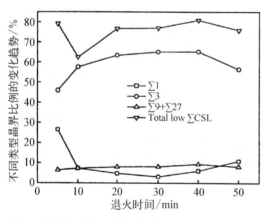

图 1 随退火时间的变化各种类型晶界的变化趋势

后 $\Sigma 1$ 晶界比例又稍微有所上升. 冷轧 5%, 1 100 ℃退火不同时间样品的晶粒平均取向差分布及晶界特征分布, 如图 2 所示. 从图 2 原始的彩色图的色标可以看出, 蓝色表示晶粒平均取向差小, 绿色表示晶粒平均取向差大, 更暖的色调代表晶粒平均取向差更大. 绿色区域的晶粒取向差变化比蓝色区域晶粒取向差变化大得多. 不完全再结晶状态的样品同时包括形变晶粒和再结晶晶粒, 再结晶晶粒内部位错密度低, 相邻点之间的取向差低; 形变晶粒由于经过变形, 内部位错密度较高, 相邻点之间的取向差就较大, 而形变晶粒区域, 各个晶粒的平均取向差也各不相同, 这是因为各个晶粒取向不同, 不同晶粒相对冷轧压下方向形变程度也就不同, 相对冷轧压下方向变形量越大的晶粒中相邻点之间的取向差越大[6], 同时在对样品进行 EBSD 扫描时采用的步长为 3 μm, 晶界的默认门槛值为 2°, 这些细碎的 $\Sigma 1$ 晶界表明在 3 μm 的距离内就有大于 2°的取向差存在, 因此晶粒平均取向差大且有细碎的 $\Sigma 1$ 晶界分布区域代表未发生再结晶的形变基体, 那些晶粒平均取向差小且没有 $\Sigma 1$ 晶界分布(或者 $\Sigma 1$ 晶界分布较少)的区域就是已再结晶状态的显微组织. 如图 2(a)所示, 退火 5 min 样品的"晶粒平均取向差分布"图中, 晶粒取向差大的区域面积较大, 这些区域内部存在大量的 $\Sigma 1$ 晶界和少量的 $\Sigma 3$ 晶界, 这与退火 5 min 样品 $\Sigma 1$ 晶界比例高达 26.6%一致. 极少数晶粒取向差小的再结晶组织内部几乎不存在 $\Sigma 1$ 晶界, 说明由于退火时间很短, 样品大部分变形晶粒还没来得及发生再结晶, 而晶粒取向差大的未再结晶区域内部的 $\Sigma 3$ 晶界应该在退火之前就存在的. 随着退火时间的延长, 如图 2(b)、(c)(退火 10 min、30 min 样品)所示, 晶粒取向差大的区

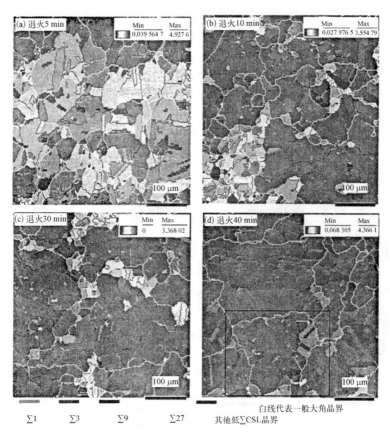

图 2 OIM 软件分析 5%冷变形 1 100 ℃退火不同时间样品的"晶粒平均取向差"分布图和晶界特征分布图

域所占的比例越来越少,并且被再结晶组织所包围,再结晶与未再结晶区域之间是一般大角晶界. 再结晶区域中退火孪晶界 Σ3 晶界逐渐增多,长度在逐渐变大,一般大角晶界在 GBCD 中的密度越来越小,说明在一般大角晶界的迁移的同时,形变晶粒在逐渐被新的再结晶晶粒"吞噬",再结晶区域面积逐渐变大,退火孪晶在不断重复产生,这就是再结晶时的多重孪晶的发展过程[18]. 退火 40 min 的样品,如图 2(d)所示,几乎不存在未再结晶区域,大部分晶粒平均取向差小的再结晶区域相互接触,这些区域之间也是随机晶界,说明形变晶粒几乎全部消除,再结晶基本完成;此时低 ΣCSL 晶界比例得到大幅度提高,一般大角晶界构成了很多尺寸较大形状不规则的封闭环,如图 2(d)的黑线框内,其尺寸达到 200 μm,内部含有很多孪晶界,多重孪晶界(Σ3^n 晶界)和特殊三叉晶界节点(Σ3 - Σ3 - Σ9,Σ3 - Σ9 - Σ27,Σ3 - Σ27 - Σ81).

3.2 孪晶链分析

完成再结晶样品的 GBCD 中存在很多由一般大角晶界构成的封闭环,在一个封闭环内部大部分都是退火孪晶界,存在少部分的 Σ9、Σ27 晶界. Xia 等在研究 690 合金时称这些封闭环为"晶粒团簇(grain-cluster)[6]". 为了解在一个晶粒团簇内部这些晶粒之间的取向关系和低 ΣCSL 晶界的演化,从已完成再结晶的试样的晶界特征分布图中选取一个较大的晶粒团簇(见图 2(d)),该晶粒团簇内部包含 55 个晶粒,如图 3 所示,包含了一个团簇的晶粒取向分布和晶界特征分布. 相邻晶粒之间要么是 Σ3 晶界,要么是 Σ9、Σ27 晶界,一种色泽代表一种取向,与原始的彩色反极图颜色码对照可以了解这些晶粒的晶体取向,从图中看到共有 10 种不同的颜色,晶粒团簇内部有 10 种取向(如晶粒 13、15、16、19、20、23、43、53、54、55 的取向用 α 表示,晶粒 22、36、38 的取向用 β 表示等). 用 OIM 软件分析了这 10 种取向之间的取向差关系,结果表明这 10 种取向之间全部是 Σ3^n 关系,而且偏差角很小,大部分都在 1° 以内,见表 2.

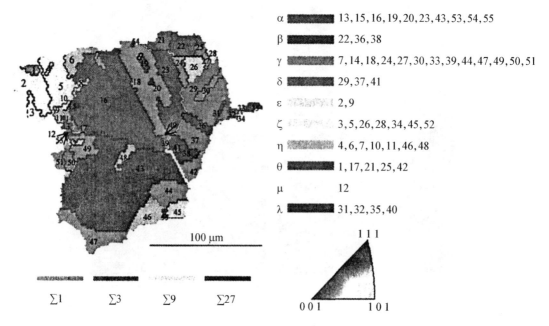

图 3 晶粒团簇内各晶粒的取向分布

表 2　同一晶粒团簇内 10 种取向之间的取向差分析

取向	α	β	γ	δ	ε	ζ	η	θ	μ
β	39.2° [−10−1] Σ9/0.5°								
γ	59.3° [−1−1−1] Σ3/0.9°	31.8° [−10−1] Σ27a/0.6°							
δ	38.7° [101] Σ9/0.7°	59.1° [−14 18 19] Σ3/7.0°	59.1° [−1−1−1] Σ3/0.9°						
ε	35.6° [0−21] Σ27b/0.5°	49.5° [1−67] Σ37o/—	38.5° [−4−4−17] Σ3⁴a/—	42.9° [−13−24−13] Σ21b/2.2°					
ζ	38.9° [21−1−21] Σ9/0.9°	39.4° [4−1−1] Σ37bb/—	35.9° [19 10 0] Σ27b/1.0°	38.6° [101] Σ9/0.3°	59.7° [11−1] Σ3/0.4°				
η	59.7° [1−11] Σ3/0.4°	35.0° [−14−270] Σ27b/0.8°	38.4° [−1−10] Σ9/0.6°	35.1° [14−270] Σ27b/0.5°	38.9° [−110] Σ9/0.6°	60.0° [−111] Σ3/0.3°			
θ	59.9° [1−1−1] Σ3/0.7°	59.7° [−111] Σ3/0.3°	39.2° [−110] Σ9/0.4°	32.4° [−101] Σ27a/0.9°	54.4° [2−32] Σ37s/—	35.4° [0 21 11] Σ27b/0.8°	39.0° [0−11] Σ9/0.2°		
μ	39.3° [−1−10] Σ9/0.8°	38.8° [1−3−5] Σ37z/—	35.3° [−1708] Σ3/7.6°	54.4° [−14 13−20] Σ36p/—	31.7° [1−10] Σ27a/0.4°	39.2° [101] Σ9/0.4°	59.6° [−1−1−1] Σ3/0.7°	32.3° [0−11] Σ27a/1.0°	
λ	34.7° [0 27−13] Σ27b/0.9°	43.0° [−15 21 2] Σ37g/—	39.9° [0 21−22] Σ9/1.1°	59.8° [1−11] Σ3/0.6°	38.7° [13 13 15] Σ7/2.7°	43.1° [−9−5 5] Σ21b/2.6°	39.0° [−87−28] Σ35g/—	38.7° [3−51] Σ37z/—	49.6° [−15−2−13] Σ37g/—

取向差给出的形式:取向差 CSL 的 Σ值/偏差角;当 Σ>49 后,OIM 系统不能显示出偏差角

　　GBCD 的优化与样品的再结晶过程有很大关系.再结晶时单位体积中的形核数量(密度)与形变量有关,冷轧压下量小的样品在再结晶时形核密度低,晶核的生长有较大的空间,产生退火孪晶的几率也高,多重孪晶可以得到充分发展,从而形成很长的孪晶链[6].当退火时间较短时,样品完成再结晶的程度小,还存在大量的形变基体,形核数量较少,一般大角晶界迁移时间很短,迁移距离很小,产生退火孪晶界(Σ3 晶界)的机会很小.随着退火时间的延长,由于新晶核与变形基体存在位错密度差,有畸变晶粒在变形能的驱动下发生再结晶,新晶粒的一般大角晶界不断向畸变晶粒迁移,"扫除"畸变晶粒.退火 40 min 态,形变储存能释放完全,再结晶基本完成.在退火时间超过 40 min 后 ΣCSL 晶界比例也随之下降,如图 1 所示,这同以前的研究结果[13,15]是一致的:再结晶结束后的晶粒长大过程中大角晶界的迁移有"扫除"低 ΣCSL 晶界的作用.

　　小变形量(5%)316SS 在再结晶进行的过程中,一个新的再结晶晶核与形变基体之间的随机晶界有很大的迁移空间,从而退火孪晶不断重复产生,存在特定晶体几何关系的退火孪晶界会形成 Σ3ⁿ 晶界(Σ3、Σ9、Σ27)等具有 Σ3ⁿ 关系的多重孪晶,发展成孪晶链,形成大尺寸

的晶粒团簇. 在同一个晶粒团簇内部,具有不同取向的晶粒相互之间也具备 $\Sigma 3^n$ 关系,说明一个晶粒团簇是由一个再结晶晶核发展而来的. 这样的晶粒团簇尺寸越大,内部包含的低 ΣCSL 晶界数量越多,有利于总的低 ΣCSL 晶界比例提高.

4 结论

(1) 316SS 经 5% 冷变形 1 100 ℃ 退火 40 min 后低 ΣCSL 晶界比例达到了 80%,$\Sigma 3$ 晶界比例占总的低 ΣCSL 晶界的 80% 左右.

(2) 再结晶过程是 316SS 材料 GBCD 优化的关键时期,经过 5% 预冷变形的 316SS 在 1 100 ℃ 退火 40 min 内是再结晶逐步完成的过程,在这个过程中退火孪晶不断形成,使孪晶链得到发展,大尺寸且形状不规则的晶粒团簇形成,从而使总的低 ΣCSL 晶界比例提高. 再结晶完成后的晶粒长大过程不利于 GBCD 的优化.

参 考 文 献

[1] 毛林彬,杨治全,单润华等. 国产快堆燃料元件包壳材料 316 不锈钢的中子辐照效应[M]. 中国核动力研究设计院,2007: 10.

[2] 罗宏,龚敏. 奥氏体不锈钢的晶间腐蚀[J]. 腐蚀科学与防护技术,2006,18(5): 357.

[3] Winning M. Grain Boundary Engineering by Application of Mechanical Stresses[J]. Scr Mater, 2006, 54: 987.

[4] Kumar M, Schwartz A J, King W E. Microstructural Evolution during Grain Boundary Engineering of Low to Medium Stacking Fault Energy fcc Materials[J]. Acta Mater, 2002, 50: 2599.

[5] Lin P, Palumbo G, Erb U, Aust K T. Influence of Grain Boundary Character Distribution on Sensitization and Intergranular Corrosion of Alloy 600[J]. Scr Metall Mater, 1995, 9(33): 1387.

[6] Xia S, Zhou B X, Chen W J. Effect of Single-step Strain and Annealing on Grain Boundary Character Distribution and Intergranular Corrosion in Alloy 690[J]. J Mater Sci, 2008, 43: 2990.

[7] 夏爽,周邦新,陈文觉等. 高温退火过程中铅合金晶界特征分布的演化[J]. 金属学报,2006,42(2): 129.

[8] Randle V. Twinning-related Grain Boundary Engineering[J]. Acta Mater, 2004, 52: 4067.

[9] Song K H, Chun Y B, Hwang S K. Direct Observation of Annealing Twin Formation in a Pb-base Alloy[J]. Mater Sci Eng, 2007, A454 – 455: 629.

[10] Randle V, Davies H. Evolution of Microstructure and Properties in Alpha-brass after Iterative Processing[J]. Metall Mater Trans, 2002, A33(6): 1853.

[11] Michiuchi M, Kokawa H, Wang Z J, et al. Twin-induced Grain Boundary Engineering for 316 Austenitic StainlessSteel[J]. Acta Mater, 2006, 54: 517.

[12] Yang S, Wang Z J, Kokawa H, et al. Reassessment of the Effects of Laser Surface Melting on IGC of SUS 304[J]. Mater Sci Eng, 2008, A474: 112.

[13] 丁霞,陈文觉. 316 不锈钢的晶界分布研究[J]. 上海金属,2006,28(4): 14.

[14] Owen G, Randle V. On the Role of Iterative Processing in Grain Boundary Engineering[J]. Scr Mater, 2006, 55: 959.

[15] Randle V. Mechanism of Twinning-induced Grain Boundary Engineering in Low Stacking-fault Energy Materials[J]. Acta Mater, 1999, 47(15): 4187.

[16] 方晓英,王卫国,周邦新. 金属材料晶界特征分布(GBCD)优化研究进展[J]. 稀有金属材料与工程,2007,36(8): 1500.

[17] Palumbo G, Aust K T, Lehockey E M, et al. On a More Restrictive Geometric Criterion for' Special' CSL Grain Boundaries[J]. Scr Mater, 1998, 38(5): 1685.

[18] Berger A, Wilbrandt P J, Ernst F. On the Generation of New Orientations during Recrystallization: Recent Results on the Recrystallization of Tensile-deformed fcc Single Crystals[J]. Prog Mater Sci, 1988, 32: 1.

Evolution of Grain Boundary Character Distribution in 316 austenitic stainless steel during high temperature annealing

Abstract: Grain boundary character distribution of 316 austenitic stainless steel after cold rolled 5% combined with recrystallization annealed at 1 100 ℃ for different times was studied by scanning electron microscope (SEM) and electron backscatter diffraction (EBSD). The results indicated that the proportion of low-ΣCSL grain boundaries was enhanced during recrystallization process. The recrystallization finished after annealing for 40 minutes, the proportion of low-ΣCSL grain boundaries reached 80%, the amount of $\Sigma 3$ grain boundaries could be up to 80% in all the low-ΣCSL grain boundaries, the grain boundary character distribution was optimized. Large and abnormity shape grain-clusters were formed. There were many annealing twins, multiple twins and special triple junctions in a big grain-cluster. Combining with the study of partly recrystallization samples and annealing twin-chains, it could beconcluded that the development of multiply twins was important to the enhancement of the proportion of low-ΣCSL grain boundaries.

Zr－4合金氧化膜的显微组织研究[*]

摘　要：Zr－4合金样品在高压釜中经过360 ℃/18.6 MPa高温去离子水腐蚀395天后,用扫描电镜观察了氧化膜的断口形貌,用高分辨透射电镜观察了氧化膜不同深度处的显微组织和晶体结构,研究了氧化膜的显微组织在腐蚀过程中的演化过程.氧化膜的晶体结构非常复杂,在靠近金属基体处除了稳定的单斜(m)晶体结构外,还存在非晶、立方(c)和四方(t)等亚稳相,晶体中还生成了很多缺陷.m/t和m/c晶体之间存在共格关系,满足(001)m∥(110)t,(010)m∥(1-10)t;以及(001)m∥(002)c,(010)m∥(020)c的取向关系.氧化膜中的晶体缺陷在应力、温度和时间作用下会发生扩散,并在氧化锆晶界上凝聚生成孔隙,弱化了晶粒之间的结合力并引起显微组织的演化.在内应力的作用下,孔隙之间的扩展和连结发展成微裂纹,氧化膜变得比较疏松,导致了耐腐蚀性能的变化和腐蚀加速.讨论了各种影响显微组织演化因素之间的关系,氧化膜显微组织的演化过程是不可避免的,寻找并控制能够延缓显微组织演化过程的因素是提高锆合金耐腐蚀性能的有效途径.

1　引言

金属锆的热中子吸收截面小,添加合金元素后制成的锆合金具有较好的耐高温水腐蚀性能和力学性能,因而半个世纪来锆合金一直用作水冷反应堆中核燃料元件的包壳材料.上世纪80年代初为了降低核电成本,提出了提高燃耗和延长燃料元件换料周期的需要,因而对锆合金的性能,尤其是耐水侧腐蚀性能提出了更高的要求,许多国家都在研究开发新锆合金.由于对锆合金的腐蚀机理以及合金元素作用的了解还不是非常清楚,在开发新锆合金的过程中仍然沿用着经验的"炒菜"方式.

锆合金在高温高压水(或过热蒸气)中腐蚀时,氧化锆是在氧化膜/金属基体(O/M)的界面上不断生成.氧离子必须从氧化膜外表面通过扩散进入氧化膜,才能在O/M界面处与锆反应,或者氧化锆的阴离子空位从O/M界面处向氧化膜的外表面扩散,所以氧化膜的显微组织及其在腐蚀过程中的演化过程会直接影响合金的耐腐蚀性能.金属锆氧化生成氧化锆时的P.B.比为1.56,形成的氧化膜中将产生很大的压应力,而与氧化膜接触的金属基体会受到张应力.氧化锆在这样大的压应力下生成时,晶体中会生成许多缺陷,这样可以使应力得到部分弛豫.作者用透射电镜(TEM)研究了氧化膜不同深度处的显微组织,观察到在靠近金属基体的氧化膜底层中,氧化锆的相结构非常复杂,含有非晶、立方、四方及单斜结构,并且晶体中存在大量的缺陷[1-4].而这些晶体中的缺陷在应力和腐蚀试验时的温度作用下,会发生扩散和凝聚,在氧化锆晶界上形成孔隙,这就是氧化膜的显微组织在腐蚀过程中必然会不断发生演化的一个重要原因[2-4].要了解这种演化过程,首先应该仔细研究靠近金属基体氧化膜底层中的显微组织,这是刚生成的氧化锆,正是由于这层氧化膜中显微组织的特殊

* 本文合作者：李强、姚美意、刘文庆、褚于良.原发表于《腐蚀与防护》,2009,30(9)：589－594,610.

性,才导致整个氧化膜的显微组织在腐蚀过程中会不断演化的问题.

用同步辐射强 X 射线源研究了氧化膜截面上氧化锆的晶体结构,结果证实在靠近金属基体处的氧化膜中含有较高比例的四方结构的氧化锆[5],而氧化锆在室温下的稳定相是单斜结构.用光学显微镜透射光观察很薄的氧化膜截面可以见到氧化膜呈现致密层/疏松层交替分布的特征[5,6],这应该是氧化膜的显微组织在腐蚀时不断演化后的结果.本工作用高分辨透射电子显微镜(HRTEM)进一步研究了氧化膜底层中的显微组织和晶体结构,这是认识整个氧化膜显微组织演化过程的基础.

2 实验方法

采用静态高压釜对 Zr - 4 合金样品进行 360 ℃/18.6 MPa 去离子水的腐蚀试验. Zr - 4 合金来自工厂提供的 0.6 mm 板材,经 600 ℃—1 h 真空退火得到完全再结晶的组织.

将锆合金板切成 25 mm×15 mm 大小的样品,用标准方法酸洗和去离子水清洗后,在高压釜中进行腐蚀试验.间隔一定时间停釜取样进行称重,腐蚀增重由 5 块试样平均得出.

为观察氧化膜的断口形貌,先用混合酸(10%HF+45%HNO₃+45%H₂O 体积比)将金属基体溶去一部分,然后将高出于金属基体的氧化膜折断.为了提高成像的质量,表面蒸镀一层金,在 JSM - 6700 扫描电镜中观察.

为制备氧化膜的 TEM 薄试样,先将样品一个表面上的氧化膜用砂纸磨掉,然后用混合酸将金属溶去,但是要剩下约 0.1 mm 厚的金属层作为氧化膜的支撑.用专用工具冲出 ϕ3 mm 样品,再用混合酸将样品中心部位的金属溶去,显露出氧化膜,其面积<ϕ0.4 mm,然后用离子溅射减薄方法制备 TEM 观察用的薄样品.为了能观察氧化膜在靠近金属基体处的显微组织和晶体结构,先用单束离子轰击氧化膜的外表面,这时穿孔部位的薄区是在靠近金属基体处的氧化膜底层,最后都要经过双束离子短时间的轰击,以便获得表面清洁的薄样品.如果用双束离子同时轰击,穿孔部位的薄区是在氧化膜的中部,用这种样品观察氧化膜中间层的显微组织.观察所用的 TEM 为 JEM - 2010F.样品制备的方法在前一篇文献中已有描述[1].

采用数字图像方法记录 TEM 的高分辨晶格条纹像,这样便于图像处理.通过快速傅里叶变换(FFT)和反傅里叶变换(IFFT)可以了解图像中的周期结构信息,确定氧化锆的晶体结构及晶体中的缺陷,这种操作可以在相当于晶体空间的数十纳米范围内进行.

3 实验结果和讨论

3.1 样品在 360 ℃/18.6 MPa 去离子水中的腐蚀增重

图 1 是样品经过 395 天腐蚀后的增重曲线,试验结束时的增重将近 100 mg/dm²,氧化膜的厚度接近 7 μm(根据 15 mg/dm²=1 μm 计算).根据增重曲线的变化规律可以看出,样品在 395 天的腐蚀过程中发生了多次转折的现象,这与过去文献中报道过的结果一致[7].

3.2 氧化膜的断口形貌

图 2 是样品腐蚀后氧化膜的断口形貌.照片的左侧是氧化膜的外表面,右侧是氧化膜与

金属基体相连的内表面,这时金属基体已被混合酸溶去,氧化膜内表面显得凹凸不平.断口面高低不平,有许多"台阶",说明氧化膜内存在一些结合力较弱的地方,当氧化膜被折断时,裂纹将沿着这些结合力弱的地方扩展,形成"台阶"状的断口.

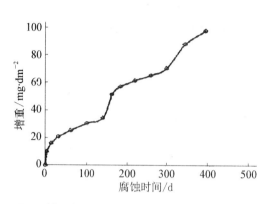

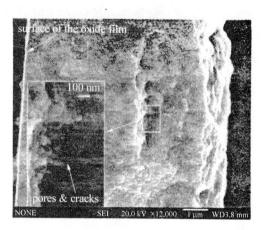

图 1 样品在 360 ℃/18.6 MPa 去离子水中的腐蚀增重曲线

图 2 样品腐蚀后氧化膜断口的 SEM 照片

从断口上还可以观察到氧化膜中的微裂纹,将局部区域进一步放大后镶嵌在图中,能够清晰地显示出其中一些孔隙和微裂纹的形貌,晶粒尺度小于 100 nm.

3.3 靠近金属基体处氧化膜底层中的显微组织

3.3.1 氧化膜底层中的显微组织及晶体中的缺陷

图 3 是用 HRTEM 从氧化膜底层中拍摄到的高分辨晶格条纹像,经过标定后的 FFT 也镶嵌在图中.图 3a 中是非晶和微晶的混合组织,以非晶为主.用 FFT 能够识别的"C"区微晶是立方结构,标定后的图形镶嵌在图 3a 中.图 3b 是一处具有立方晶体结构的晶格条纹像,仔细观察条纹的走向和分布,可以发现该处的晶体中存在许多缺陷,结构并不完整,在 1～2 nm 范围内的高分辨像类似于非晶结构,一部分这种不完整的地区用圆圈标出.经过滤波处

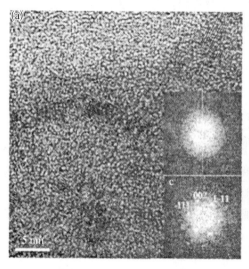

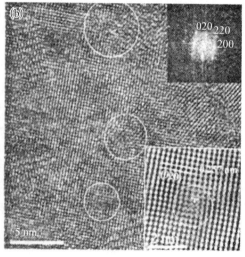

图 3 靠近金属基体氧化膜底层中的 HRTEM 像

理的 IFFT,并经过适当放大后,可以清楚观察到存在位错(圆圈中用"⊥"标出的地方).说明氧化锆在压应力下生成时,可以形成非晶,即使形成晶体,晶体中也会生成许多缺陷,这是因为通过这种方式可以使一部分内应力得到弛豫的缘故.

3.3.2　几种氧化锆晶体结构之间的取向关系

图 4、图 5 和图 6 是几处单斜/立方或者单斜/四方结构毗邻处的 HRTEM 像,图中分别用"A"或"B"标出了晶体结构不同的区域.不同区域的 FFT 并经过晶面指数标定后得到的图形也镶嵌在图中,经过滤波处理的 IFFT 也给出在相应的图中.从 IFFT 图中可以更清楚地观察到晶格条纹像中的细节,测量得到的晶面间距也给出在图中,并标出了晶面指数.根据测量的晶面间距和晶面指数标定结果,可以确定它们的晶体结构.在几张图中的"A"区都是单斜结构(m),而在图 4 中的"B"区是四方结构(t),在图 5 和图 6 中的"B"区是立方结构(c).从标定晶面指数后的图中,以及不同晶体结构的晶格条纹分布关系中都可以看出,m/t 和 m/c 之间存在共格关系,从图 4 中得出 m/t 之间存在(001)m//(110)t;(010)m//(1-10)t 的取向关系;从图 5 中得出 m/c 之间存在(001)m//(002)c;(010)m//(020)c 的取向关系.而从图 6 得出 m/c 之间的关系是(001)m//(002)c;(110)m//(2-20)c,这种取向关系实质上与从图 5 得出的结果是一致的.由于氧化锆的晶体结构比较复杂,数据也有多种报道,我们分析时依据的 PDF 卡片是:37-1484(m);24-1164(t);27-997(c).根据这几张卡片中给出的晶面间距可以计算出共格关系的 m/t 在晶面间距之间存在 1.2%～2.3%的失配,共格关系的 m/c 在晶面间距之间存在 0.25%～2.2%的失配,并且单斜结构的 β 角是 99°,并不是直角.因此,这种共格关系会引起晶格中的应力,这种应力造成晶格的"扭曲",引起局部地区的晶格条纹像"不清晰",或者是晶格条纹像的"不完整",形成一些位错.c、t 亚稳相与 m 相之间存在共格的取向关系,这或许是应力诱发相变时的一种特征,而且在 m/t 和 m/c 之间并没有清晰的相界面.

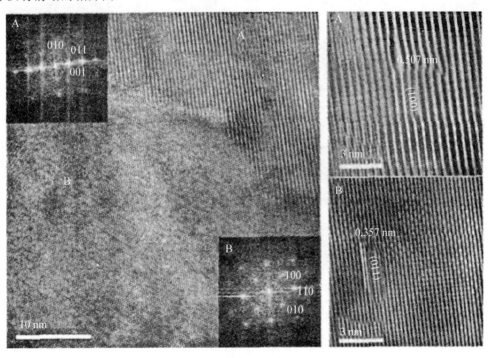

图 4　氧化膜底层中单斜/四方晶体结构处的 HRTEM 像

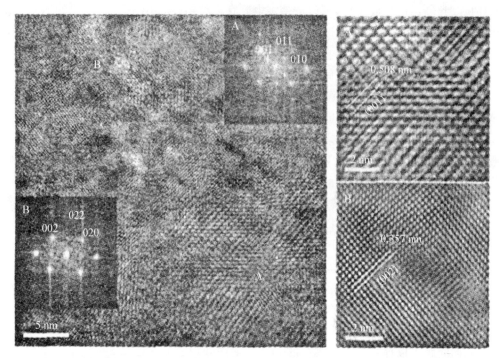

图5 氧化膜底层中单斜/立方晶体结构处的 HRTEM 像

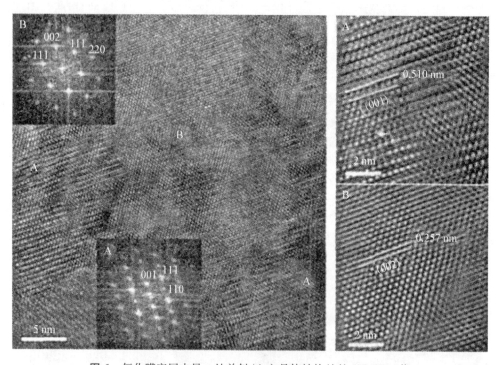

图6 氧化膜底层中另一处单斜/立方晶体结构处的 HRTEM 像

测量得到的晶面间距与卡片中给出的数据有一定的差别,这一方面是由于电镜放大倍率引起的测量误差,另一方面是由于应力引起晶格畸变造成的.即使在 30 nm² 微区范围内的照片中,在不同地方测量晶面间距的结果也有差别,晶格条纹像也不能够清晰一致的显示出来,这都反映了应力引起了晶格畸变后的结果.

3.4 晶体缺陷的扩散凝聚导致显微组织的演化

图 7(a)是从氧化膜中间层拍摄的 TEM 照片,在一些三晶交汇处出现了数纳米大小的孔隙. 出现孔隙处的晶粒,它们的形状逐渐演化成球形,大小约 50 nm. 在放大倍率更高的图 7(b)中可以看到,晶界上存在一些类似非晶的组织,这应该是孔隙形成前的状态. 由此可以推测,氧化膜底层中的晶体缺陷,在应力、温度和时间的作用下,经过扩散后在晶界处凝聚,首先形成非晶状的组织,进一步形成孔隙. 由于氧化膜是在 O/M 界面处不断生成,因此这种演化后的显微组织只有在稍微远离 O/M 界面处的氧化膜中才能观察到. 孔隙在晶界上形成后弱化了晶粒之间的结合力,在内应力的作用下,孔隙之间的扩展和连结,可以发展成微裂纹,氧化膜的显微组织变得比较疏松,导致了耐腐蚀性能的变化,在腐蚀动力学的过程中发生了腐蚀转折和转折后的腐蚀加速.

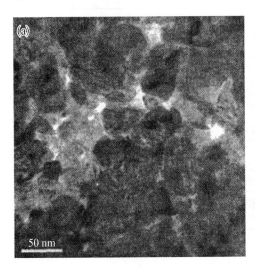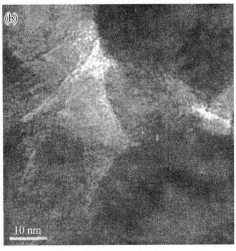

图 7 氧化膜中间层的显微组织在不同放大倍率下的 TEM 照片

3.5 氧化膜显微组织在腐蚀过程中不断演化的原因及其影响因素

锆合金受到腐蚀在表面生成氧化膜时,氧化膜内会产生很大的压应力,在应力下生成的氧化锆除形成了亚稳相外,在晶体中还会形成各种缺陷,这样可以使内应力得到部分弛豫. 这些缺陷在应力、温度和时间的作用下发生扩散、湮没和凝聚,在氧化锆的晶界上形成孔隙,使内应力得到进一步弛豫,亚稳相也转变为稳定的单斜结构. 孔隙在应力作用下可以扩展成微裂纹,使氧化膜变得比较疏松,导致腐蚀过程发生转折. 这样,一方面新的氧化锆在 O/M 界面上不断形成,另一方面已经形成的氧化膜,它的显微组织又会逐渐发生演化,这种过程是不可避免的. 我们在前几篇文章中提出了这样的观点,并进行了比较详细的讨论[2-4]. 氧化膜显微组织的演化对腐蚀性能影响;合金成分及显微组织(第二相的类型、大小和分布等),还有水化学对氧化膜显微组织演化的影响;它们之间的复杂关系可以用图 8 来描述. 图中的中间列表示晶体中的缺陷在应力、温度和时间作用下发生扩散、湮没和凝聚形成孔隙,从而导致显微组织演化而影响耐腐蚀性能的主要过程;两边的两列表示其他因素对显微组织演化的影响,以及显微组织演化后引起的其他相关现象. 晶体中缺陷的扩散过程与缺陷浓度、应力水平以及温度等因素有关;空位通过扩散凝聚形成孔隙时,会增加氧化锆的自由表面,因而这种过程除了与孔隙形核的难易有关外,还与氧化锆的表面自由能大小有关,一些会降低氧化锆表面自由能的因素将会促使孔隙形成. 了解这些因素与合金成分以及显微组织之间的关系,将是发展新锆合金时如何选择合金元素的基础,也是制定材料加工成型时工艺制度的参考依据.

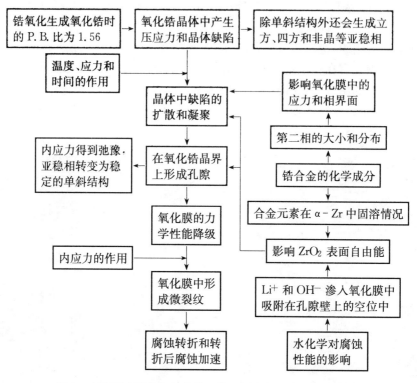

图8 多种影响因素与氧化膜显微组织演化过程之间的关系

大部分合金元素在 α‑Zr 中的固溶度都非常小,虽然锆合金中合金元素的添加量并不多,除了 Sn 和 Nb 之外,其中的大部分将与 Zr 形成金属间化合物以第二相形式析出. 这些不同类型的第二相,其中的大部分都比基体 α‑Zr 更耐腐蚀,所以它们是先被裹进氧化膜中,然后再被氧化形成氧化物. 这种过程会带来两方面的问题:一方面,当第二相氧化时会形成氧化锆和其他金属的氧化物,这些非锆的氧化物是保留在原来的位置还是会向氧化锆基体中扩散? 如果仍然留在原来的位置,那么由于金属氧化后体积发生变化会使氧化膜中产生许多局域性的附加应力,因为大部分金属氧化时的 P.B. 都大于1. 这种应力与氧化膜中原有的压应力相互作用,将会影响晶体中的缺陷扩散而影响氧化膜显微组织的演化. 另外,第二相氧化后与氧化锆基体之间形成的界面,又可以作为空位的尾间,延缓空位通过扩散在氧化锆晶界上凝聚形成孔隙的过程. 另一方面,第二相在压应力的状态下被氧化时,第二相中的其他金属在形成氧化物的同时,有可能会向四周氧化锆的基体中扩散,这样可以使局域性的应力得到弛豫. 当合金元素扩散进入氧化锆后,又可以通过改变氧化锆的表面自由能来影响显微组织的演化过程. 因此,第二相对氧化膜显微组织演化过程的影响是相当复杂的问题,只有仔细研究嵌入到氧化膜中的第二相被氧化的过程后,才能进一步说明这些问题. 不过,无论是第二相中其他金属被氧化后留在原位,还是扩散进入氧化锆晶体中,事先在合金中获得纳米大小的第二相,这对延缓氧化膜显微组织的演化,改善合金的耐腐蚀性能都是有利的.已有的实验结果也表明,在 Zr-Sn-Nb 系的合金中,获得纳米大小分布均匀的第二相后明显改善了合金的耐腐蚀性能[8,9].

核反应堆一回路水中需要添加 H_3BO_3,以[10]B 作为可燃毒物调节反应堆的核反应性.为了控制一回路水的 pH 值以缓解钢构件的腐蚀问题,还要添加 LiOH 使一回路水成为弱碱性.但是添加 LiOH 后对锆合金的耐腐蚀性能产生了有害影响,缩短了腐蚀转折的时间,提高了转折后的腐蚀速率,尤其是对 Zr-4 合金的影响更为显著.过去曾提出过多种机理进行解释,在文献[10,11]中已对这方面的问题进行了总结,但至今还没有得到一致的看法.我们根据锆合金在添加 LiOH 的水溶液中腐蚀时,Li^+ 和 OH^- 会渗入到氧化膜中的实验结果[12-15],用扫描探针显微镜研究了锆合金在不同水化学条件下腐蚀后氧化膜表面的起伏,比较了水化学条件对氧化锆表面自由能的影响,提出了这样的假设:Li^+ 和 OH^- 渗入到氧化膜中吸附在孔隙壁上或者空位中降低了氧化锆的表面自由能,促使孔隙的形成,加速了氧化膜显微组织的演化,因而 LiOH 水溶液对锆合金的耐腐蚀性能产生了有害影响.由此推理,影响氧化锆表面自由能的其他因素也是应该仔细研究的问题.

3.6 从氧化膜显微组织演化过程与锆合金耐腐蚀性能相关性的角度认为需要研究的问题

为了进一步认识氧化膜显微组织的演化过程对锆合金耐腐蚀性能间的影响,分析图 8 中各种因素之间的关系后,认为以下一些问题还值得进行深入研究:

(1) 氧化膜中的应力变化.氧化膜中的应力大小一方面会影响显微组织的演化,另一方面又反映了显微组织演化后的结果.研究合金成分、腐蚀时的温度以及水化学对氧化膜中应力变化的影响,可以从另一个角度来了解氧化膜显微组织的演化过程.

(2) 第二相在氧化膜中被氧化的过程.不同类型第二相中的非锆元素形成什么样的氧化物? 它们是留在原位还是会向周围氧化锆中扩散? 这些问题将会影响氧化膜显微组织的演化过程,也是发展新锆合金时选择合金元素的考虑基础,对于制定材料的加工工艺也有重要的参考价值.

(3) 影响氧化锆表面自由能的因素.在氧化膜显微组织演化过程中,空位通过扩散凝聚形成孔隙是一个重要的环节,形成孔隙的难易直接与氧化锆表面自由能的大小有关.外来离子的渗入以及微量合金成分的变化对氧化锆表面自由能的影响是应该研究的问题.

4 结论

(1) Zr-4 合金样品腐蚀后在靠近金属基体的氧化膜底层中晶体结构非常复杂,除了稳定的单斜(m)结构外,还存在非晶、立方(c)和四方(t)亚稳相,晶体中还生成了很多缺陷.这是因为 P. B. 比为 1.56,氧化锆生成时内部会产生很大压应力的缘故.

(2) m/t 和 m/c 晶体结构之间存在共格关系,满足(001)m//(110)t,(010)m//(1-10)t;以及(001)m//(002)c,(010)m//(020)c 的取向关系.这是由于应力诱发 m→t 和 m→c 相变的结果,两相之间的界面并不清晰.两相之间的晶面间距存在失配,尤其是 m 相结构中的 β 角为 99°,并非直角,因而发生这类相变之后会产生畸变和应力,也会生成一些晶体缺陷.

(3) 在腐蚀过程中,氧化锆晶体中的缺陷在应力、温度和时间作用下会发生扩散,并在氧化锆晶粒晶界上凝聚生成孔隙,弱化了晶粒之间的结合力并引起显微组织的演化.

（4）在内应力的作用下，氧化膜中孔隙之间的扩展和连结会发展成微裂纹，显微组织变得比较疏松，导致了耐腐蚀性能的变化和腐蚀加速．氧化膜显微组织的演化过程是不可避免的，寻找并控制延缓显微组织演化过程的因素，是改善锆合金耐腐蚀性能的途径．

参 考 文 献

［1］周邦新，李强，黄强，等．水化学对锆合金耐腐蚀性能的研究［J］．核动力工程，2000，21(5)：439－447.

［2］周邦新，李强，姚美意，等．锆－4合金在高压釜中腐蚀时氧化膜显微组织的演化［J］．核动力工程，2005，26(4)：364－371.

［3］周邦新，李强，刘文庆，等．水化学及合金成分对锆合金氧化膜显微组织演化的影响［J］．稀有金属材料与工程，2006，35(7)：1009－1016.

［4］Zhou B X, Li Q, Yao M Y, et al. Effect of water chemistry and composition on microstructural evolution of oxide on Zr-Alloys, zirconium in the nuclear industry：15th International Symposium［EB/OL］. Sunriver Oregon US, 2007, June, 24～28. (paper ID JAI 10112, available online at www. astm. org)

［5］Yilmazbayhan A, Motta A T, Comstock R J, et al. Structure of zirconium alloy oxides formed in prre water studied with synchrotron radiation and optical microscopy：relation to corrosion rate［J］. J Nucl Mater ［J］. 2004, 324：6－22.

［6］Yilmazbayhan A, Breval E, Motta A T, et al. Transmission electron microscopy examination of oxide layer formed on Zr alloys［J］. J Nucl Mater, 2006, 349：265－281.

［7］Bryner J S. The cyclic nature of corrosion of Zircaloy-4 in 633K water［J］. J Nucl Mater. , 1979, 82：84－88.

［8］Comstock R J, Schoenberger G, Sabol G P. Influence of processing variables and alloy chemistry on the corrosion behavior of ZIRLO nuclear fuel cladding, Zirconium in the nuclear industry：Eleventh International Symposium［C］//ASTM STP 1295, Bradley E R and Sabol G P, Eds. American Society for Testing and Materials, 1996：710－725.

［9］刘文庆，李　强，周邦新，等．显微组织对ZIRLO锆合金耐腐蚀性的影响［J］．核动力工程，2003，24(1)：33－36.

［10］Cox B. Some thoughts on the mechanisms of in-reactor corrosion of zirconium alloys［J］. J Nucl Mater, 2005, 336：331－368.

［11］IAEA－TECDOC－996. Waterside corrosion of zirconium alloys in nuclear power plants［J］. IAEA, Vienna, ISSN 1011－4289, 1998.

［12］Pecheur D, Godlewski J, Peybernes J, et al. Contribution to the understanding of the effect of the water chemistry on the oxidation kinetics of zircaloy-4 cladding, zirconium in the nuclear industry：Twelfth International Symposium ［C］//ASTM STP 1354, Sabol G P and Moan G D Eds. American Society for Testing and Materials, West Conshohocken PA：2000：793－811.

［13］Ramasubramania N, Precoanin N, Ling V C. Lithium uptake and the accelerated corrosion of zirconium alloys, zirconium in the nuclear industry：eighth international symposium［C］//ASTM STP 1023, Van Swam L F P and Eucken C M Eds. American Society for Testing and Materials, Philadephia, 1989：187－201.

［14］周邦新，刘文庆，李强，等．LiOH水溶液加快Zr－4合金腐蚀速度的机理［J］．材料研究学报，2004，18(3)：225－231.

［15］姚美意，周邦新，李强，等．热处理对Zr－4合金在360 ℃ LiOH水溶液中腐蚀行为的影响［J］．稀有金属材料与工程，2007，36(11)：1920－1923.

Microstructure of Oxide Films Formed on Zircaloy-4

Abstract：The microstructure of oxide films formed on Zircaloy-4 was investigated by HRTEM and HRSEM after autoclave corrosion tests performed at 360 ℃/18. 6 MPa in deionized water for 395 days. The microstructural evolution of the oxide films was analyzed by comparing the microstructure at different depths in the oxide layer. The crystal structure of the oxide films is complicated. Monoclinic(m), tetragonal(t), cubic (c) and amorphous phases were detected in the oxide layer near the metal matrix, and the coherent relationships between m/t and m/c were identified as(001)m//(110)t, (010)m//(1－10)t；and(001)m//

(002)c,(010)m//(020)c. The defects consisting of vacancies and interstitials were produced during the oxide growth. The diffusion, annihilation and condensation of vacancies and interstitials occurred under the action of stress, temperature and time. The vacancies absorbed by grain boundaries formed pores to weaken the bonding strength between grains and to result in the microstructural evolution. Based on the formation of pores along the grain boundaries, the development of micro-cracks in the oxide films will lead to the loss of the protective characteristic, thus the phenomenon of corrosion transition appears and the corrosion rate is accelerated.

The microstructural evolution of the oxide films is an inevitable process. The effects of different factors on the microstructural evolution of the oxide are discussed. Finding and controlling the factors that could retard the microstructural evolution are the effective method to improve the corrosion resistance of zirconium alloys.

晶界类型及时效处理对 690 合金
耐晶间腐蚀性能的影响[*]

摘　要：用 SEM 研究了含有高比例低 Σ 重位点阵(CSL，Σ≤29)晶界的 690 合金在 715 ℃时效不同时间后不同类型晶界的耐晶间腐蚀情况. 结果表明：经过不同时间处理后的样品，腐蚀失重有明显差别. 715 ℃时效 2 h 的样品腐蚀失重最大；时效 2 h 以后，随着时效时间的延长，样品的腐蚀失重减小，失重最小样品失重量为时效 2 h 样品失重量的一半. 在相同腐蚀条件下，晶间腐蚀主要以晶粒团簇外围的随机晶界腐蚀为主，不同类型晶界处的腐蚀程度有很大区别. 孪晶的共格界面腐蚀很轻，孪晶的非共格界面与 Σ9 晶界处可以观察到明显腐蚀痕迹，Σ27 晶界和随机晶界的腐蚀程度相近，都比 Σ3 晶界和 Σ9 晶界腐蚀严重很多.

据报道[1]，美国使用的 690 合金在各种水环境中抗腐蚀性能都很好，因此 690 合金被认为是目前最好的第三代压水堆蒸汽发生器传热管材料. 但随着核电工业的发展，进一步提高蒸汽发生器可靠性是需要研究的问题. 蒸汽发生器传热管道破坏泄漏的主要原因在不同时期有不同的情况，现在与晶界有关的应力腐蚀和晶间腐蚀仍然是蒸汽发生器传热管道破坏的主要原因. 所以如何提高材料抗晶间腐蚀(IGC)和晶间应力腐蚀能力(IGSCC)依然是需要解决的问题[2-3].

众所周知，当富铬的碳化物在晶界析出后，晶界碳化物附近将形成贫铬区[4-6]. 在很多材料中晶间贫铬将大大降低材料的抗晶间腐蚀(IGC)和晶间应力腐蚀(IGSCC)能力. 碳化物的形貌和分布状态对贫铬区的铬含量和材料的抗 IGC 和 IGSCC 能力影响很大[7-10]. 已有研究结果表明，通过提高低 Σ 重位点阵(coincidence site lattice，CSL[11])晶界的比例也可以显著提高 690 合金的抗 IGC 能力[12]，因此，晶界类型是影响材料抗 IGC 和 IGSCC 的另一个重要因素. 作者在研究 690 合金晶界处碳化物形貌随时效时间演化时，发现不同类型晶界处碳化物形貌特征是有很大区别的[13]. 那么，不同类型晶界的抗 IGC 能力也应该是有很大区别的. 本工作利用扫描电子显微镜(SEM)研究了含有高比例低 ΣCSL 晶界的 690 合金经不同时效热处理后，不同类型晶界经晶间腐蚀后的腐蚀形貌的特征.

1　实验方法

实验所用 690 合金成分(质量分数，%)为 Ni 60.52，Cr 28.91，Fe 9.45，C 0.025，N 0.008，Ti 0.4，Al 0.34，Si 0.14. 实验前将所有样品进行真空固溶处理. 为了获得高比例的低 ΣCSL 晶界，利用文献[14]的工艺将固溶处理后的样品进行冷轧与再结晶退火处理. 详细的固溶与再结晶退火处理工艺如文献[13]所示，经过再结晶处理，样品的低 ΣCSL 晶界比例可超过 70%. 然后再将样品真空密封在石英管中，在 715 ℃下分别时效热处理 0，0.5，2，5，15，30，

＊ 本文合作者：李慧、夏爽、陈文觉. 原发表于《中国核科学技术进展报告(第一卷)·核材料分卷》，2009，1：50 - 55.

50,100 h.

　在腐蚀实验之前,将所有时效后样品经过电解抛光获得干净的表面后,测量样品的表面积,并称重(精确到 0.1 mg). 晶界两侧晶粒取向关系利用配备在 Hitachi S570 SEM 中的 TSL 电子背散射衍射(electron backscatter diffraction, EBSD)附件确定,扫描步长为 7 μm. 数据采集后利用取向成像微(orientation imaging microscopy, OIM)系统自动确定晶界类型,晶界类型的判定采用 Palumbo-Aust 标准[15],相对于 CSL 模型最大偏差角为 $\Delta\theta_{max} = 15°\Sigma^{-5/6}$.

　用于晶间腐蚀实验的溶液为 65%HNO_3+0.15%HF+34.85%H_2O,选择封闭的聚丙烯塑料盒作为腐蚀实验容器,选择聚四氟乙烯作为样品支架. 将各样品悬挂在腐蚀溶液中,进行腐蚀实验. 随着实验时间的延长,每隔 2,5 或者 20 h 将样品取出,在水中浸泡 0.5 h,并在酒精中浸泡两次,每次浸泡后用电吹风将样品烘干. 然后对各样品再次称重来获得腐蚀失重. 这样的腐蚀共进行 276 h,获得了腐蚀失重曲线. 所有腐蚀实验都是在室温下进行的. 腐蚀后样品的表面形貌用金相显微镜或者 JSM 6700E SEM 观察.

2　实验结果与讨论

2.1　碳化物析出对腐蚀失重的影响

　　图 1 给出 715 ℃时效不同时间(0～100 h)样品的腐蚀失重的区别. 从图中可以看出,未时效的样品有着非常好的耐腐蚀能力,腐蚀失重很小,这是由于再结晶退火后样品为均一态,晶界处很少有元素的偏聚,不能形成微区原电池,因此晶间腐蚀抗力很好. 从图 1 中可以看出,时效的样品开始随着时效时间的延长腐蚀失重越来越大,时效 2 h 的样品腐蚀失重最大,说明晶界附近的铬浓度已经达到最低点,随着时效时间的延长,贫铬区开始有变平的趋势[4,6,16],样品的失重越来越小. 可以看出时效 2 h 样品腐蚀失重约为时效 100 h 样品失重的 2 倍. 图 1 也说明了,在 715 ℃时效 2 h4 后贫铬区就开始恢复,时效 30 h 后由于贫铬区已经基本变平,腐蚀失重的变化不大,这与邱少宇[17]等报道的结果有些区别,可能是时效前的热处理制度不同和低 ΣCSL 晶界比例不同造成的.

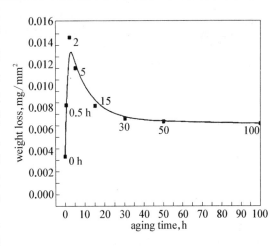

图 1　经 715 ℃时效处理不同时间样品腐蚀 276 h 后的质量损失

2.2　不同类型晶界的晶间腐蚀形貌特征

2.2.1　晶粒团簇对晶间腐蚀性能的影响

　　为研究晶界类型与晶间腐蚀的关系,先用 EBSD 测定样品某一区域各晶界的类型,然后进行晶间腐蚀实验. 经过 7 h 的晶间腐蚀实验后,样品表面已经出现晶间腐蚀的沟痕. 图 2 给出了经过 7 h 腐蚀后各样品的表面形貌. 从图中可以看出,样品表面的腐蚀形貌相似,晶粒团簇外围的随机晶界(如各图中虚线所示网络)腐蚀情况要比晶粒内部的低 ECSL 晶界的腐蚀情况严重很多. 这是因为通过时效热处理后,随机晶界比晶粒团簇内的低 ECSL 晶界更容易形成碳化物且碳化物尺寸更大[13],所以随机晶界附近形成贫铬区更深更宽,因此更容易遭到腐蚀. 从图 2 中也可以看出,样品的腐蚀失重主要是由于晶粒团簇外围的随机晶界晶间腐蚀造成的,所以晶粒团簇大并且低 ECSL 晶界比例高的样品会表现出很好的耐腐蚀效果[12].

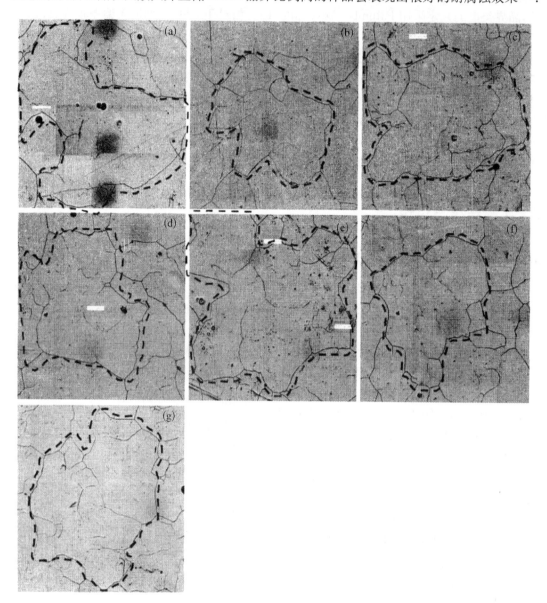

图 2 715 ℃时效处理不同时间的样品经历 7 h 晶间腐蚀后样品的表面形貌的金相照片

时效时间为 a—0.5 h;b—2 h;c—5 h;d—15 h;e—30 h;f—50 h;g—100 h. 图中虚线所示为晶粒团簇外围的随机晶界

2.2.2　晶界类型对晶间腐蚀性能的影响

为了获得不同类型晶界的晶间腐蚀形貌特征,利用 SEM 观察了样品在 715 ℃时效 0～100 h 后,再腐蚀 112 h 后样品的表面形貌(图 3). 从图 3a 中可以看出未经时效处理的样品在此实验条件下基本没有晶间腐蚀迹象,这与图 1 的失重曲线对应很好,通过对不同时效处

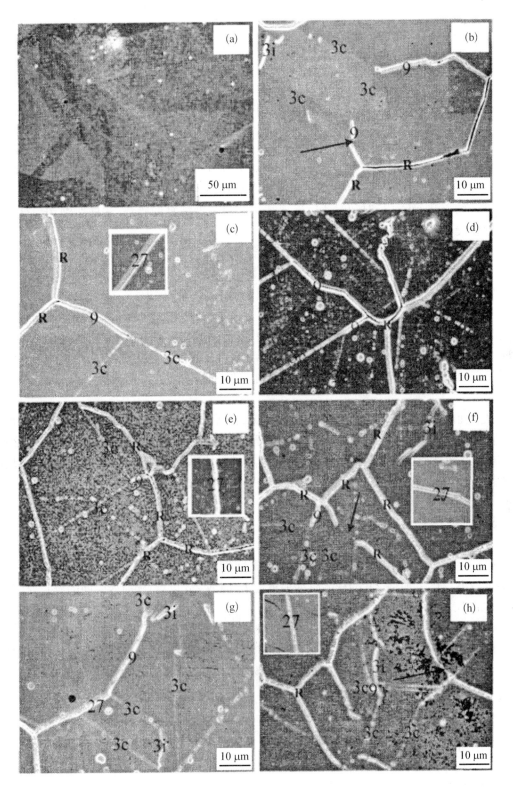

图3 715 ℃时效处理不同时间的样品经历 112 h 晶间腐蚀后样品的表面形貌

时效时间为：a—0 h；b—0.5 h；c—2 h；d—5 h；e—15 h；f—30 h；g—50 h；h—100 h.
晶界上所标注的数字为其 Σ 值 R -随机晶界，Σ3$_i$ -孪晶的非共格界面，Σ3$_c$ -孪晶的共格界面

理后样品的表面腐蚀形貌(图3b~h)的对比可以看出,不同类型晶界的晶间腐蚀形貌特征有很大区别,这是由于不同类型晶界处碳化物形貌的区别及产生贫铬区的差别是引起其耐晶间腐蚀性能变化的主要原因.

在图3中,所有时效处理过的样品(图3b~h)中$\Sigma 3_c$界面(孪晶的共格界面)腐蚀都很轻微,$\Sigma 3_i$界面(孪晶的非共格界面)腐蚀迹象却很明显.$\Sigma 9$晶界的腐蚀程度比$\Sigma 3_i$晶界的腐蚀程度要严重一些.$\Sigma 27$晶界和随机晶界的腐蚀程度相近,但都比$\Sigma 3$晶界和$\Sigma 9$晶界腐蚀严重很多.相同类型的晶界相比,长时间时效处理的样品中晶界腐蚀沟痕较宽较浅(图3e~h),而短时间时效处理的样品腐蚀沟痕较深(图3b~d),显的更为严重,可以与图i中曲线相对应.这是由于开始时贫铬区随着时效时间的增加而加宽加深,一定时间后铬浓度到达一定的最低点又开始上升[4,6,16].长时间时效后出现贫铬区铬浓度恢复的趋势,这样腐蚀沟痕变宽但是腐蚀较慢,沟痕较浅.

从图3中也可以看出,相交的三个晶界腐蚀情况有明显区别.如图3b中的3-3-9三叉晶界,图3c中的3-3-9,9-R-R三叉晶界,图3d中的3-3-9,9-9-R,R-R-R三叉晶界,图3e中的3-3-9,9-R-R三叉晶界,图3f中的3-3-9,9-R-R三叉晶界,图3g中的3-3-9,3-9-27三叉晶界,和图3h中的3-3-9,R-R-R三叉晶界等都说明这一情况.随机晶界与孪晶相接触的晶界片段的腐蚀情况与随机晶界的晶间腐蚀情况有明显差别(如图3b,f和h中箭头所示),这是由于该晶界片段的取向差与它所连接的随机晶界有明显差别,碳化物更容易在该晶界片段附近的随机晶界上形核生长,一定程度上保护了该晶界片段.

从上面的讨论中可以看出,并非所以低ΣCSL晶界都是所谓的"特殊"晶界,只有孪晶的共格界面抗晶间腐蚀性能最好,而孪晶的非共格界面和$\Sigma 9$晶界对晶间腐蚀的抗力较好,$\Sigma 27$晶界与随机晶界性能已经相似.通过晶界工程提高材料抗晶间腐蚀能力的主要原因是大块的晶粒团簇对内部晶粒起保护作用[12],而低ΣCSL晶界并未打断随机晶界的网络(图2).

3 结论

(1) 时效不同时间后的样品,经腐蚀后的腐蚀失重有明显差别.未经时效处理的样品腐蚀失重最小.715 ℃时效2 h的样品腐蚀失重最大.时效2 h以后,随时效时间的延长,样品腐蚀失重减小.

(2) 相同腐蚀条件下,晶间腐蚀主要以晶粒团簇外围的随机晶界腐蚀为主,不同类型晶界处的腐蚀程度有很大区别.$\Sigma 3_c$界面腐蚀很轻,$\Sigma 3_i$界面与$\Sigma 9$晶界处可以观察到明显腐蚀痕迹,$\Sigma 27$晶界和随机晶界的腐蚀程度相近,都比$\Sigma 3$晶界和$\Sigma 9$晶界腐蚀严重很多.

参 考 文 献

[1] Was G. Grain boundary chemistry and intergranular fracture in Nick-based alloys — A review[J]. Corrosion, 1990, 46(4): 319-330.

[2] Diercks D R, Shack W J, Muscar J. Overview of steam generator tube degradation and integrity issues[J]. Nuclear Engineering and Design, 1999, 194: 19-30.

[3] 杨湘,苏兴万,文燕. 国产Inconel690合金管的性能及应用研究[J]. 核动力工程,1997,18(3): 269-272.
(YANG Xiang, SHU Xingwan, WEN Yan. The property and application of Inconel 690 alloy tube[J]. Nuclear Power Engingeering, 1997, 18(3): 269-272)

［4］ Kai J J, Yu G P, Tsai C H, Liu M N, Yao S C. The effects of heat treatment on the chromium depletion, precipitate evolution, and corrosion resistance of Inconel alloy 690[J]. Metallurgical Transactions A, 1989, 20A: 2057－2067.

［5］ Stiller K, Nilsson J O, Kjell N. Structure, chemistry, and stress corrosion cracking of grain boundaries in Alloys 600 and 690[J]. Metallurgical Transactions A, 1996, 27A: 327－341.

［6］ Hall E L, Briant C L. Chromium depletion in the vicinity of carbides in sensitized austenitic stainless steels[J]. Metallurgical Transactions A, 1984, 15A: 793－811.

［7］ Min K S, Nam S W. Correlation between characteristics of grain boundary carbides and creep-fatigue properties in AISI 321 stainless steel[J]. Journal of Nuclear Materials, 2003, 322: 91－97.

［8］ Spigarelli S, Cabibbo M, Evangelista E, PalumboG. Analysis of the creep strength of a low-carbon AISI 304 steel with low-E grain boundaries[J]. Materials Science and Engineering A, 2003, A362: 93－99.

［9］ Alexandreanu B, Capell B, Was G S. Combined effect of special grain boundaries and grain boundary carbides on IGSCC of Ni-16Cr-9Fe-xC alloys[J]. Materials Science and Engineering A, 2001, A300: 94－104.

［10］ Alexandreanu B, Was G S. The role of stress in the efficacy of coincident site lattice boundaries in improving creep and stress corrosion cracking[J]. Scripta Materialia, 2006, 54(6): 1047－1052.

［11］ Kronberg M L, Wilson F H. Secondary recrystallisatian in copper[J]. Trans. Am. Inst. Min. Engrs, 1949, 185: 501－504.

［12］ Xia S, Zhou B X, Chen W J. Effect of single-step strain and annealing on grain boundary character distribution and intergranular corrosion in Alloy 690[J]. Journal of Materials Sciences, 2008, 43(9): 2990－3000.

［13］ 李慧,夏爽,周邦新,倪建森,陈文觉. 镍基690合金时效过程中晶界碳化物的形貌演化. 金属学报,2009,45: 195－198.

(LI Hui, XIA Shuang, ZHOU Bangxin, NI Jiansen, CHEN Wenjue. Evolution of carbide morphology precipitated at grain boundaries in Ni-based Alloy 690. Acta Metallurgica Sinica, 2009, 45: 195－198)

［14］ Xia S, Zhou B X, Chen W J, Wang W G. Effects of strain and annealing processes on the distribution of Σ3 boundaries in a Ni-based superalloy[J]. Scripta Materialia, 2006, 54: 2019－2022.

［15］ Palumbo G, Aust K T, Lehockey E M. On a more restrictive geometric criterion for "Special" CSL grain boundaries [J]. Scripta Materialia, 1998, 38: 1685－1690.

［16］ 李强,周邦新. 690合金的显微组织研究[J]. 金属学报,2001,37: 8－12.

(LI Qiang, ZHOU Bangxin. A study of microstructure of alloy 690 [J]. Acta Metallurgica Sinica, 2001,37: 8－12)

［17］ 邱少宇,苏兴万,文　燕,闫福广,喻应华,何艳春. 热处理对690合金腐蚀性能影响的实验研究核动力工程,1995, 16(4): 336－340.

(QIU Shaoyu, SHU Xingwan, WEN Yan, YAN Fuguang, YU Yinghua, HE Yanchun. Effect of heat treatment on corrosion resistance of Alloy 690 [J]. Nuclear Power Engineering, 1995,16(4): 336－340)

Effect of Grain Boundary Characters and Aging Treatment on Intergranular Corrosion Resistance of Alloy 690

Abstract: The effects of grain boundary characters and aging treatment on the intergranular corrosion resistanceof Alloy 690 with high proportional low E coincidence site lattice (CSL, $\Sigma \leqslant 29$) boundaries aged at 715 ℃ for 0～100 h were investigated by SEM. The results show that, the weight losses of specimens aged for different time are different. After aging treated at 715 ℃ for 2 h, the specimens behave the worst resistance of intergranular corrosion. When prolonging the aging time, the resistance of intergranular corrosion become better. Under the same corrosion condition, intergranular corrosion happened mainly on the outer random grain boundary networks of each grain clusters. Coherent twin boundary was attacked very slightly, incoherent twin boundary and Σ9 boundary were attacked significantly, while Σ27 boundary and random grain boundary attacked seriously.

Features of Highly Twinned Microstructures Produced by GBE in FCC Materials*

Abstract: The proportion of low ΣCSL boundaries of single phase brass, Ni based Alloy 690, 304 stainless steel and Pb alloy were enhanced to about 70% (according to Palumbo-Aust criterion) which mainly were $\Sigma 3^n (n=1, 2, 3\cdots)$ types by small amount of deformation and recrystallization annealing. All the resulting microstructures were constituted of large size grain-cluster, in which all the orientation relationships between any two grains (no matter adjacent or not) could be described by $\Sigma 3^n$ misorientations according to twin chain analyzing. Twin relationships up to 8th generation ($\Sigma 3^8$) were found in one of such clusters. The outer boundaries of grain-clusters were crystallographically random with the adjacent clusters. In the case of high proportion of low ΣCSL boundaries, the connectivity of random boundary network was not fragmented obviously when the Palumbo-Aust criterion was used, whereas substantially interrupted when the Brandon criterion was applied.

1 Introduction

Grain boundary engineering (GBE) provides a good idea to increase the resistance to intergranular failure in materials[1,2]. The grain boundary related properties of materials can be enhanced by exercising control over the population of low ΣCSL ($\Sigma \leqslant 29$) grain boundaries, as defined by the coincident site lattice (CSL) model[3]. In face-centered cubic (FCC) metal materials with low to medium stacking fault energy (SFE), annealing twinning is highly prone to formation during recrystallization[4,5]. Therefore the proportion of $\Sigma 3^n (n=1, 2, 3\cdots)$ boundaries can be promoted by multiple twinning via proper thermal-mechanical processing (TMP), by which means the grain boundary related properties were greatly enhanced in this category of materials, such as Pb-base alloy[6], Ni-base alloy[7], Cu-base alloy[8], and austenitic stainless steel[9]. In the current study, the grain boundary microstructures are analyzed and compared in single phase brass, Ni based Alloy 690, 304 stainless steel and Pb alloy all of which are with high proportion of low ΣCSL boundaries.

2 Experimental Procedures

Strip samples of single phase brass (70Cu-30Zn), Ni based Alloy 690 (Ni-30Cr-

---

* In collaboration with Xia Shuang, Chen Wenjue, Luo Xin, Li Hui. Reprinted from Materials Science Forum, 2010, 638－642: 2870－2875.

10Fe), 304 stainless steel and Pb alloy were subjected to certain types of TMP, as shown in Table 1. After the TMP carried out the samples were mechanically grounded and then electro-polished to get good surface finish for electron backscatter diffraction (EBSD) experiment. EBSD was employed for the determination of grain boundary misorientations using the TSL 4. 6 laboratories orientation imaging microscopy (OIM) system attached to a scanning electron microscope (SEM). Values of the proportion of grain boundaries defined by the CSL model were expressed as a length fraction. When analyzing misorientation between crystallites by the OIM system, data points in the middle of grains were used for measurement.

Table 1 Treatment processes of each material, and the resulting grain boundary character distribution statistics (Palumbo-Aust criterion[10])

Materials	Brass (70Cu - 30Zn)	Alloy 690 (Ni - 30Cr - 10Fe)	304 stainless steel	Pb Alloy
Pre-treatment	50% CR+10 min @ 550 ℃	15 min @ 1100 ℃ +WQ	30 min @ 1100 ℃ + WQ	87% CR+3 min @ 270 ℃
TMP	7% CR + 30 min @ 525 ℃	5% CR + 5 min @ 1100 ℃ + WQ	5% CR + 5 min @ 1100 ℃ + WQ	30% CR + 3 min @ 270 ℃
$\Sigma1(\%)$	4. 2	1. 4	5. 2	3. 7
$\Sigma3(\%)$	60. 6	60. 1	59. 1	51. 7
$\Sigma9+\Sigma27(\%)$	9. 0	9. 3	10. 2	12. 3
Other low ΣCSL (%)	0. 6	0. 5	0. 3	0. 8
Total low ΣCSL ($\Sigma\leqslant29$)(%)	74. 4	71. 3	74. 8	68. 5

Note: CR-Cold Rolling; WQ-Water Quench

3 Results and Discussion

After the samples were pre-treated for homogenization, TMPs were carried out as shown in Table 1. Proportions of low ΣCSL grain boundaries of these materials were effectively enhanced to values about 70% (Palumbo-Aust criterion) by small amount of deformation and different annealing. All these materials are highly propone to twin formation due to their low stacking fault energy. So $\Sigma3^n$ type boundaries make the dominant contribution to the grain boundary character distribution (GBCD), while other type low ΣCSL boundaries have very low occurrences.

Microstructure constituting of large size grain-clusters is a common feature in these highly twinned materials, as shown in Fig. 1. The grain-cluster has the following characteristics: all the boundaries within the cluster can be described by $\Sigma3^n$ misorientations and connecte with each other forming a lot of triple junctions, whereas the outer boundaries are often crystallographically random with the adjacent grains[7,11,12], as outlined by black frameworks in Fig. 1.

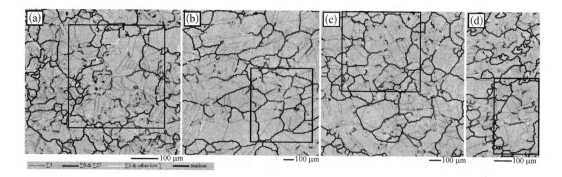

Fig. 1 The OIM maps of different types of grain boundaries in (a) single phase brass, (b) Alloy 690, (c) 304 stainless steel and, (d) Pb alloy, when the proportion of low ΣCSL grain boundaries were enhanced to about 70% (Palumbo-Aust criterion) by the thermal-mechanical processing shown in Table 1. Thin gray lines denote Σ3 boundaries, thick gray lines denote Σ9 and Σ27 boundaries, thin white lines denote Σ1 and other low ΣCSL boundaries, thick black lines denote random boundaries. In each OIM map one grain-cluster is outlined by black framework

A grain-cluster outlined in Fig. 1(c) (304 stainless steel) was analyzed using OIM in detail, and the result is shown in Fig. 2. In the presenting two-dimensional section, this cluster is constituted of 65 grains which are numbered in Fig. 2(a). Among those grains, there is always a line can be drawn only across Σ3 boundaries from any grain to any other one. For example from grain 1 to grain 33, one of the routes can be as follow: 1→4→5→6→44→58→50→19→21→52→33, which crosses 10 boundaries of Σ3 type. In order to clearly show the twin chain relationships, all of the grains of this cluster are joined into the twin-chain, as shown in Fig. 2(c). In this chain, short lines only link the interfacing grains

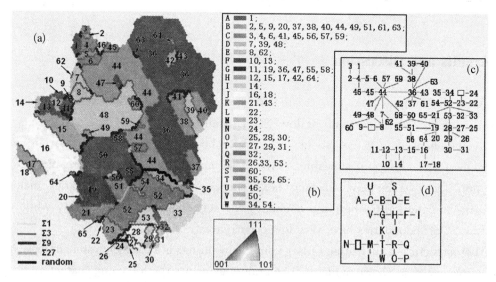

Fig. 2 Twin chain analyzing of the grain-cluster outlined by black framework in Fig. 1(c). Grains are numbered in (a) which is the orientation contrasted OIM map of this grain-cluster; Grains with the same orientation (The orientations with difference smaller than 1.5° are here recognized as same orientation.) are listed in (b); (c) Twin chain analyzing of these 65 grains; (d) Twin chain analyzing of these 23 orientations

with $\Sigma3$ twin relationship, where the deviations from exact $\Sigma3$ CSL are smaller than $1°$ as the threshold. However, in Fig. 2(c), the higher order twin relationships ($\Sigma3^n$) can not be directly identified. For example, from grain 1 to grain 33, the route mentioned above crossing 10 $\Sigma3$ boundaries, however the orientation relationship between grain 1 and 33 is not $\Sigma3^{10}$ but $\Sigma3^6$ (measured misorientation: $[-18 \ -1 \ 16]/50.8°$, identified as: $\Sigma729q$ according to Ref. [13]). This is because of that in this route some grains have the same orientation, in another word, some orientations turn back to their original orientations after several twinning operations. Thus, from grain 1 to grain 33, the route mentioned above can be simplified as $1\rightarrow4\rightarrow44\rightarrow58\rightarrow21\rightarrow52\rightarrow33$ (Grain 4 and grain 6, grain 58 and grain 19 have the same orientations respectively. The orientations with difference smaller than $1.5°$ were here recognized as same orientation.). Therefore, the misorientation between grain 1 and grain 33 results from 6 independent twinning operations, i. e. it is $\Sigma3^6$.

The grains in this cluster can be categorized into 23 orientations, as shown in Fig. 2(b). A simpler form of twin-chain composed of these 23 orientations can be organized, as shown in Fig. 2 (d). In this chain, short lines only link the interfacing orientations with $\Sigma3$ twin relationship. Based on this chain, the order of $\Sigma3^n$ relationships can be directly identified. For example, to identify the misorientation between grain 1 and grain 33, we can look grain 1 and 33 up in Fig. 2 (b) and find that they belong to orientation A and R respectively, and after that the route from orientation A to R is uniquely identified as $A\rightarrow C\rightarrow B\rightarrow G\rightarrow K\rightarrow T\rightarrow R$ composing of 6 independent twinning operations according to Fig. 2 (d), thus the n value of $\Sigma3^n$ relationship between grain 1 and grain 33 is 6. The $\Sigma3^n$ relationship between any grain pairs inside the cluster can be identified in this way. According to the orientation twin-chain, the highest order of twin relationship is $\Sigma3^8$, for example the twin relationships between orientation S and P, orientation I and P are $\Sigma3^8$. All the orientationrelationships between any two grains inside the cluster are of $\Sigma3^n$ type, no matter adjacent or not.

Grain boundary between grain 44 and grain 34 seems a random grain boundary, but it is actually a higher order $\Sigma3^n$ type misorientation. It can be identified as $\Sigma3^4$ according to the orientation twin chain analyzing in Fig. 2 (d). It's measured misorientation is $[19 \ 19 \ 14]/60.3°$, which can also be identified as $\Sigma81d$ according to Ref. [13].

Grain 9 and grain 24 interface with the cluster by $\Sigma9$ boundary rather than directly by $\Sigma3$ boundary in the examining two-dimensional section. They are highly probable interfacing with the cluster directly by $\Sigma3$ boundary below or above the presenting two-dimensional section, because the $\Sigma9$ boundaries are produced by meeting of orientation with its third twin generation[5].

Twin-chain formation was more completely discussed in studying recrystallization texture evolution in FCC metal materials with low SFE by Haasen et al[4,14,15]. Twin-chain is produced by multiple twinning, which is the phenomenon of continuously twinning stating from a single nucleus during recrystallization in this kind of materials. $\Sigma3^n$ type boundaries are produced by the impingement of different twin generations inside the same twin-chain.

Formation of large size grain-cluster is the common feature in the microstructure with high proportion of $\Sigma 3^n$ boundaries. This kind of microstructure is the result of small amount of deformation and annealing as shown in Table 1. Small amount of deformation led to lower stored energy for recrystallization, therefore fewer recrystallization nuclei formed. Thus individual nucleus had larger volume of deformed matrix available to consume. During the new grains growing, continuously twinning occurred to form large size twin-chain, and the associated $\Sigma 3^n$ boundaries were produced by the meeting between different twin generations[5]. After consuming up the deformed matrix the twin-chains impinged with each other, then the recrystallization process finished and the large grain-cluster formed. So, small strain led to big size grain-clusters after recrystallization.

For Pb alloy, the applied cold rolling reduction was 30% which was larger than that for other materials listed in Table 1. However, the microstructure is also constituted of large size grain-cluster. This may be owing to its low melting temperature (T_m). The rolling deformation was conducted at room temperature, which is about 0.5 T_m for Pb alloy. At this temperature, dynamic recover or recrystallization might occur, which would reduce the stored energy for the subsequent static recrystallization annealing. This would reduce the nucleation density of the subsequent static recrystallization, and hence large size grain-cluster also formed.

The connectivity of random boundary network of these highly twined materials are not fragmented obviously when the Palumbo-Aust criterion is used, whereas substantially interrupted when the Brandon criterion[16] is applied, as shown in Fig. 3, where only random boundaries are shown. There are some solid arrows in Fig. 3 (e)-(h), where the connectivity of random boundaries is interrupted by some low ΣCSL grain boundaries according to Brandon criterion. However they are ranked as random boundaries according to Palumbo-Aust criterion, as shown in the same positions in Fig. 3 (a)-(d). This indicates that, when the Brandon criterion is applied, the low ΣCSL grain boundaries breaking up the connectivity of the random grain boundaries are with high deviations from exact CSL relationships. About 100 this kind of low Σ CSL boundaries (defined by Brandon criterion) which break up the connectivity of the random boundary network in the OIM maps (Brandon criterion) of the samples in Table 1 are investigated. The results together with the limits of both Brandon and Palumbo-Aust criterion are plotted in Fig. 4. It is obviously that the deviations of most low ΣCSL boundaries interrupt the connectivity of the random boundarynetwork are above the Palumbo-Aust criterion limit.

As discussed in the foregoing paragraphs, the random boundaries are produced by impingement of different grain-clusters during recrystallization. Moreover, all the twins and its variant $\Sigma 3^n$ boundaries (with small deviations) intrinsically locate inside the grains-cluster rather than appear on the network of random boundaries as segments to break up its connectivity. Therefore, though misorientations of some boundaries between clusters are

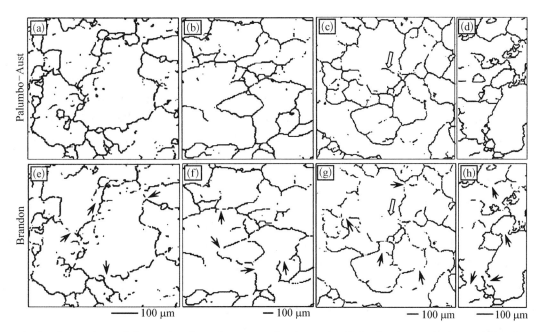

Fig. 3 Comparison of the random boundary network when using Palumbo-Aust criterion and Brandon criterion, in the case of proportions of low Σ CSL grain boundaries were enhanced. (a), (b), (c) and (d) are the random boundary network of single phase brass, Alloy 690, 304 stainless steel and Pb alloy according to Palumbo-Aust criterion respectively, and (e), (f), (g) and (h) are their counterparts according to Brandon criterion

close to the low Σ CSL (defined by Brandon criterion), they simply happen to be in that situation, and are not geometrically necessary in contrast to the $\Sigma 3^n$ boundaries inside the clusters.

The hollow arrow marked boundary in Fig. 3(c) is the grain boundary between grain 44 and grain 34 in Fig. 2(a). It is identified as $\Sigma 3^4$ according to the orientation twin chain analyzing (Fig. 2 (d)) and also as Σ81d according to Ref. [13]. So it is not a low ΣCSL boundary according to Palumbo-Aust criterion as shown in Fig. 3(c), whereas it is misidentified as Σ3 boundary by Brandon criterion. This is because that its misorientation is [19 19 14]/ 60.3° which has 7.2° deviation form the exact

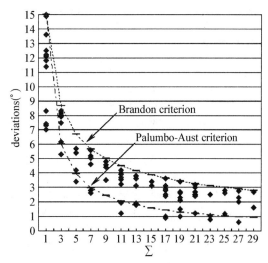

Fig. 4 Deviations of some 100 boundaries (close to low Σ CSL), which appeared on the random boundaries network according to Brandon criterion

Σ3 CSL ([1 1 1]/60°), and lies in the tolerance for Σ3 CSL ([1 1 1]/60°±8.6°) according to Brandon criterion. So, during analyzing the boundary network given by the OIM system, care must be taken when identifying whether a boundary is a low Σ one or a random one.

4 Summary

Grain boundary engineering was carried in single phase brass, Ni based Alloy 690, 304 stainless steel and Pb alloy, all of which are FCC metal materials with low SFE. The proportion of low ΣCSL boundaries can be greatly enhanced by small amount of deformation and recrystallization annealing, and the $\Sigma3^n$ type boundaries make the dominant contribution to the GBCD. There are two common features in the highly twined microstructures:

(1) The microstructure is constituted of large size grain-cluster produced by multiple twinning stating from single nucleus during recrystallization. All of the orientation relationships between any two grains inside the cluster can be described by $\Sigma3^n$ misorientations no matter adjacent or not.

(2) The connectivity of random boundary network is not fragmented obviously when the Palumbo-Aust criterion is used, whereas substantially interrupted when the Brandon criterion is applied.

Acknowledgements

This work was supported by Major State Basic Research Development Program of China (2006CB605001) and Innovation Foundation of Shanghai University (A10 - 0110 - 08 - 004 and A16 - 0110 - 08 - 006).

References

[1] T. Watanabe: Res. Mech. Vol. 11 (1984), p. 47.

[2] V. Randle: Acta Mater. Vol. 52 (2004), p. 4067.

[3] M. L. Kronberg and F. H. Wilson: Trans. AIME Vol. 185 (1949), p. 501.

[4] A. Berger, P. J. Wilbrandt, F. Ernst, U. Klement, and P. Haasen: Prog. Mater. Sci. Vol. 32 (1988), p. 1.

[5] G. Gottstein: Acta Metall. Vol. 32 (1984), p. 1117.

[6] E. M. Lehockey, D. Limoges, G. Palumbo, J. Sklarchuk, K. Tomantschger and A. Vincze: J. Power Sources Vol. 78 (1999), p. 79.

[7] S. Xia, B. X. Zhou and W. J. Chen: J. Mater. Sci. Vol. 43 (2008), p. 2990.

[8] Randle V, Davies H. Met Mater Trans A Vol. 33 (2002), p. 1853.

[9] M. Shimada, H. Kokawa and Z. J. Wang: Acta Mater. Vol. 50 (2002), p. 2331.

[10] G. Palumbo and K. T. Aust: Acta. Metall. Mater. Vol. 38 (1990), p. 2343.

[11] V. Y. Gertsman and C. H. Henager: Interface Sci. Vol. 11 (2003), p. 403.

[12] S. Xia, B. X. Zhou, W. J. Chen and W. G. Wang: Scripta mater. Vol. 54 (2006), p. 2019.

[13] V. Randle: J. Mater. Sci. Vol. 41 (2006), p. 653.

[14] P. Haasen: Metall. Trans. B Vol. 24B (1993), p. 225.

[15] P. J. Wilbrandt: Phys. State. Solidi. Vol. 61 (1980), p. 411.

[16] D. G. Brandon, Acta Metall. Vol. 14 (1966), p. 1479.

利用 EBSD 技术对 690 合金不同类型晶界处碳化物形貌的研究*

摘 要：利用扫描电子显微镜（SEM）、电子背散射衍射（EBSD）技术和取向成像显微（OIM）软件研究了 Ni 基 690 合金中不同类型晶界处碳化物的形貌. 不同类型晶界处析出碳化物的形貌有很大区别,在孪晶的非共格界面（$\Sigma 3_i$）附近,棒状碳化物向两侧晶粒内部生长,而类似的棒状碳化物只向 $\Sigma 9$ 晶界一侧的晶粒内部生长. $\Sigma 3_i$ 与 $\Sigma 9$ 晶界附近的棒状碳化物的生长方向与基体晶粒的 $\{111\}$ 面平行. 晶界上析出的碳化物尺寸随着 Σ 值的升高而明显增大. 在相同的腐蚀条件下,晶间腐蚀的痕迹随着 Σ 值的升高变得严重.

核反应堆蒸汽发生器传热管损坏的重要原因是材料的晶间腐蚀及应力腐蚀破裂[1],而晶界上析出富 Cr 的碳化物引起的晶界附近 Cr 的贫化是晶间腐蚀与晶间应力腐蚀的重要原因之一. 作为压水堆核电站蒸汽发生器传热管材的 690 合金是一种高 Cr 的 Ni 基合金,固溶后经过特殊的时效热处理可控制晶界上碳化物的析出,调整晶界附近贫 Cr 区的成分,可以显著提高其耐腐蚀性能[2]. 研究表明,690 合金也是一种低层错能的面心立方金属,通过提高低 Σ 重位点阵（CSL）晶界的比例也可以显著提高 690 合金的抗晶间腐蚀能力[3]. 并且有研究结果[4,5]表明,碳化物形貌对材料的蠕变性能有很大影响. 因此,低 ΣCSL 晶界对晶界处碳化物的析出形貌的影响是值得研究的.

690 合金晶界上析出的碳化物主要为富 Cr 的 $M_{23}C_6$ [6-8]. 在奥氏体不锈钢和 690 合金中,经过固溶后再时效时,碳化物很容易在晶界上析出[6,9,10]. Trillo 等[10]通过研究不同 C 含量的 304 不锈钢,发现在晶界上析出的碳化物形貌受晶界两侧晶粒的取向差影响最大,晶界两侧晶粒取向差较小的晶界上碳化物析出的倾向明显低于晶界两侧晶粒取向差较大的晶界. 在 $\Sigma \leqslant 29$ 的低 ΣCSL 晶界上,碳化物析出倾向比一般大角晶界小很多[10-12]. 利用透射电镜（TEM）研究的结果发现,晶界相对于两侧晶粒所处的晶体学面是不同的,这种差别对晶界上碳化物的形核与长大也有很大影响[7,8,13]. 在奥氏体不锈钢和 690 合金中都观察到在孪晶的非共格界面处有棒碳化物析出,而在孪晶的共格界面处却很少观察到有碳化物析出[7,13,14].

EBSD 是进行快速而准确的晶体取向测量和相鉴定的强有力的分析工具. 由于它与 SEM 一起工作,使得显微组织如晶粒、相、界面、形变等能与晶体学关系相联系[15]. EBSD 技术可以给出样品表面晶粒的晶体学取向信息,从而判断晶界的取向差. 本文利用 EBSD 技术与 SEM 分析研究了 690 合金中不同类型晶界处碳化物的形貌.

1 实验

实验所用 690 合金成分（质量分数,%）为 Ni 60.52,Cr 28.91,Fe 9.45,C 0.025,

---

* 本文合作者：李慧、夏爽、胡长亮、陈文觉. 原发表于《电子显微学报》,2010,29(1)：730－735.

N 0.008,Ti 0.4,Al 0.34,Si 0.14.将所有样品真空(真空度约为 $5×10^{-3}$ Pa)密封在石英管中,在 1 100 ℃下保温 15 min,然后立即淬入水中,并同时砸破石英管进行固溶处理.为了获得较高比例的低 ΣCSL 晶界,将固溶处理后的样品冷轧 10%,获得厚度约为 0.5 mm 片状样品,再真空密封在石英管中,在 1 100 ℃再结晶退火 5 min 后立即淬入水中,并同时砸破石英管.然后样品在 715 ℃时效,使碳化物析出.对时效后的样品进行晶间腐蚀浸泡实验,以对比不同类型的晶界的抗敏化能力.用于晶间腐蚀实验的溶液为 65% HNO₃+0.15% HF+34.85% H₂O,在室温下腐蚀 112 h.

在 EBSD 测试之前,先将样品电解抛光.使用的电解液是 20% HClO₄ + 80% CH₃COOH,在室温下用 30 V 直流电抛光 30 s.晶界两侧晶粒取向关系利用配备在 Hitachi S570 SEM 中的 TSL - EBSD 附件确定.数据采集后利用 OIM 系统自动确定晶界类型,晶界类型的判定采用 Palumbo-Aust 标准[16],相对于 CSL 模型最大偏差角为 $\Delta\theta_{max} = 15\Sigma^{-5/6}$.为观察晶界处碳化物形貌,将样品在相同电解液中电解蚀刻 3~5 s,蚀刻电压 5 V,温度室温.使用 JSM 6700F 型 SEM 对碳化物形貌进行分析研究.

2 结果与讨论

为了确定 SEM 观察区域与 EBSD 测试区域为相同区域,利用显微硬度计在样品表面标记一个约为 800 μm×800 μm 的区域,如图 1a 中光学显微照片所示.图 1c 给出的 OIM 图为图 1b 中 SEM 像中绿线所画区域的 EBSD 测试结果.在图 1c 中可以看出,样品含有高比例的孪晶相关的低 ΣCSL 晶界.关于获得高比例孪晶相关的低 ΣCSL 晶界的方法与机理已经在文献[3,17,18]中详细论述.由图 1d 可见,在孪晶的非共格界面($\Sigma 3_i$)附近析出棒状碳化物,从晶界向两侧晶粒内部生长.而类似的棒状碳化物只向 Σ9 晶界一侧的晶粒内部生长(图 1e).由于不同类型晶界的界面能有很大区别,晶界的界面能随着 Σ 值的升高而升高[19].因此在相同处理条件下,晶界上析出的碳化物尺寸也随着 Σ 值的升高而明显增大,如图 1d,e,f,g 所示.

从图 1d,e 可看出,$\Sigma 3_i$ 与 Σ9 晶界附近的棒状碳化物具有特殊的生长方向.文献中主要依据 TEM 得到的电子衍射斑来确定这些棒状碳化物的生长方向[6-10,13,14].从图 1d 的 SEM 像可见 $\Sigma 3_i$ 晶界附近的棒状碳化物生长方向与孪晶的共格界面{111}平行,这与理论分析和 TEM 观察结果相同[7,8,14].Σ9 晶界虽然具有特定的取向差,但是晶界所处的晶体学面是不确定的.因此不能单从形貌来判断 Σ9 晶界附近析出棒状碳化物的生长方向.OIM 软件可以给出样品表面晶粒的各晶体学面在样品表面的迹线(Plane Traces),如图 2a 所示.因此也可通过 OIM 软件对 Σ9 晶界附近的棒状碳化物的生长方向进行分析研究.

图 2a 给出了 715 ℃时效 15 h 样品表面区域的不同类型晶界的 OIM 图.在这张 OIM 图中同时给出了晶粒的{001},{011},{111}晶面在样品表面的迹线(分别以不同颜色代表).图 2b 中给出了图 2a 中箭头所示的 Σ9 晶界处碳化物 SEM 像.通过对右下侧晶粒的迹线分析表明棒状碳化物的生长方向与晶粒基体中某一{111}晶面在样品表面的迹线平行.因此,可以判断 Σ9 晶界处析出的棒状碳化物也是在基体晶粒的{111}面上生长的.由于 690 合金中碳化物主要为复杂面心立方结构的 M₂₃C₆,其晶格常数约为 690 合金基体的 3 倍,M₂₃C₆ 碳化物的{111}面与 690 基体的{111}面错配度很低[20].因此碳化物与基体具有这种生长方式时,可以明显降低体系的能量[21].

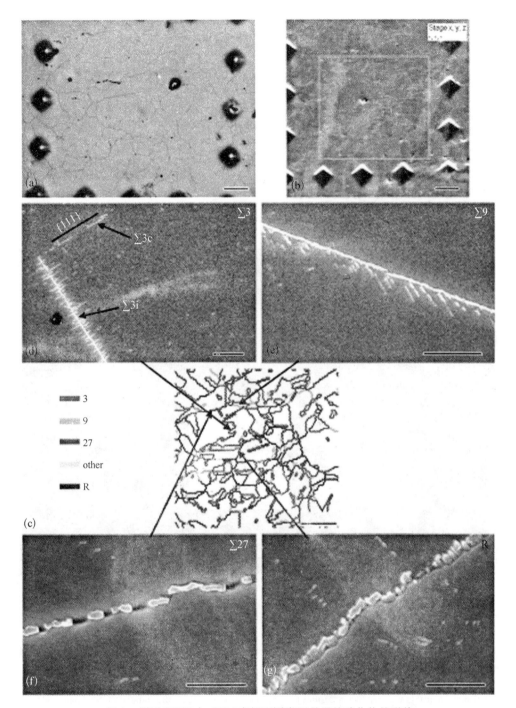

图 1 利用 EBSD 与 SEM 表征不同类型晶界处碳化物的形貌

（a）EBSD 测试区域的光学显微照片；（b）EBSD 测试区域的 SEM 像；（c）EBSD 测试区域的不同类型晶界的 OIM 图；（d）孪晶界处碳化物的 SEM 像（$\Sigma 3_i$：孪晶的非共格界面，$\Sigma 3_c$：孪晶的共格界面）；（e）$\Sigma 9$ 晶界处碳化物的 SEM 像；（f）$\Sigma 27$晶界处碳化物的 SEM 像；（g）随机晶界处碳化物的 SEM 像. （a,b,c）：Bar＝100 μm；（d,e,f,g）：Bar＝1 μm

Fig. 1 Characterization of carb ide morphology by SEM and EBSD technique

(a) Opticalmicrograph；(b) SEM image；(c) OIM map of grain boundaries in the area tested by EBSD. SEM images of carbides precipitated at (d) tw in boundaries，(e) $\Sigma 9$ grain boundary，(f) $\Sigma 27$ grain boundary and (g) Random grain boundary($\Sigma 3_i$: incoherent tw in boundary，$\Sigma 3_c$: coherent tw in boundary). (a, b, c)：Bar＝100 μm；(d, e, f, g)：Bar＝1 μm

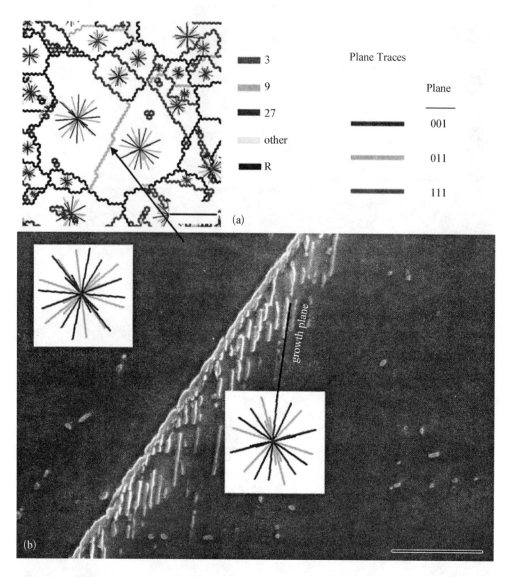

图 2 对 715 ℃时效 15 h 样品,通过晶面迹线分析判断 Σ9 晶界附近棒状碳化物的生长方向

(a) 样品表面的晶界类型 OIM 图,同时给出了{001},{011},{111}面在样品表面的迹线,Bar＝20 μm;(b) 碳化物在 Σ9 晶界处析出的 SEM 像,及通过晶面迹线确定棒状碳化物生长方向,Bar＝1 μm

Fig. 2 Crystal plane traces analys is of the grow th direction of the bar-like carb ide precipitated near Σ9 boundary

(a) OIM map of grain boundaries, traces of {001}, {011}, {111} planes intersected with the present surface are also given in this picture, Bar＝20 μm; (b) SEM in age of carbide precipitated at Σ9 boundary, and identification of the bar-like carbide grow th direction by plane traces analysis, Bar＝1 μm

　　通过对样品的晶间腐蚀实验,可对不同类型的晶界抗晶间腐蚀能力进行对比. 从图 3 可见,不同类型晶界处碳化物形貌的区别引起了晶界敏化程度的差异,从而导致耐晶间腐蚀性能的变化. 利用与前面相同的方法确定晶界类型后,发现所有 Σ3$_c$ 界面腐蚀都很轻微,Σ3$_i$ 界面腐蚀迹象较为明显. Σ9 晶界的腐蚀程度比 Σ3$_i$ 晶界的腐蚀程度要严重一些. Σ27 晶界和随机晶界的腐蚀程度相近,都比 Σ3 晶界和 Σ9 晶界腐蚀严重很多.

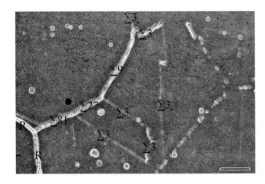

图3 在 715 ℃时效 50 h 后,室温下晶间腐蚀 112 h 后样品的表面形貌. 晶界类型已在相应晶界上标注,$\Sigma 3_i$: 孪晶的非共格界面,$\Sigma 3_c$: 孪晶的共格界面

Bar＝10 μm

Fig. 3 Surface morphology of the specimen underw ent 112 h intergranular corrosion at room temperature. The Σ values of the grain boundaries are indicated on the picture, $\Sigma 3_i$, incoherent twin boundary, $\Sigma 3_c$, coherent twin boundary. Bar＝10μm

3 结论

利用 EBSD 技术与 SEM 可以在确定晶界类型的同时观察对应的碳化物形貌. 应用 OIM 软件分析可以推断碳化物的一些特殊生长方向. 不同类型晶界处析出碳化物的形貌明显不同. 晶界上碳化物的尺寸随着晶界 Σ 值的增加而增大. 孪晶的非共格界面附近可以观察到棒状碳化物向两侧晶粒内部生长,而类似的棒状碳化物只向 $\Sigma 9$ 晶界一侧晶粒生长,这些棒状碳化物在晶粒基体的 $\{111\}$ 面上生长. 晶间腐蚀的程度随着晶界 Σ 值的增加而变严重.

参 考 文 献

[1] Diercks D R, Shack W J, MuscarJ. Overview of steam generator tube degradation and integrity issues[J]. Nuclear Engineering and Design, 1999, 194: 19 – 30.

[2] 邱少宇,苏兴万,文 燕,等. 热处理对 690 合金腐蚀性能影响的实验研究[J]. 核动力工程,1995,16(4):336 – 340.

[3] 夏 爽,周邦新,陈文觉. 690 合金的晶界特征分布及其对晶间腐蚀的影响[J]. 电子显微学报,2008,27(6):461 – 468

[4] Min K S, Nam S W. Correlation between characteristics of grain boundary carb ides and creep-fatigue properties in AISI 321 stainless steel[J]. Journal of Nuclear Materials, 2003, 322: 91 – 97.

[5] Spigarelli S, Cabibbo M, Evangelista E, Palumbo G. Analysis of the creep strength of a low-carbon AISI 304 steel with low-Σ grain boundaries[J]. Materials Science and Engineering A, 2003, A352: 93 – 99.

[6] Kai J J, Yu G P, Tsai C H, et al. Thee ffects of heat treatment on the chromium depletion, precipitate evolution, and corrosion resistance of INCONEL Alloy 690[J]. Metallurgical Transactions A, 1989, 20A: 2057 – 2067.

[7] 李 强,周邦新. 690 合金的显微组织研究[J]. 金属学报,2001,37:8 – 12.

[8] Lim Y S, Kim J S, Kim H P, Cho H D. The effect of grain boundary misorientation on the intergranular $M_{23}C_6$ carbide precipitation in therm ally treated Alloy 690[J]. Journal of Nuclear Materials, 2004, 335: 108 – 114.

[9] Hall E L, Briant C L. Chromium depletion in the vicinity of carbides in sensitized austenitic stainless steels[J]. Metallurgical Transactions A, 1984, 15: 793 – 811.

[10] Trillo E A, Murr L E. Effects of carbon content, deformation, and interfacial energetics on carbide precipitation and corrosion sensitation in 304 stainless steel[J]. Acta Materialia, 1999, 47(1): 235 – 245.

[11] Zhou Y, Aust K T, Erb U, Palumbo G. Effects of grain boundary structure on carbide precipitation in 304L stainless steel[J]. Scripta Materialia, 2001, 45(1): 49 – 54.

[12] Hong H U, Rho B S, Nam S W. Correlation of the $M_{23}C_6$ precipitation morphology with grain boundary characteristics in austenitics tainless steel[J]. Materials Science and Engineering A, 2001, A318: 285 – 292.

[13] Trillo E A, Murr L E. TEM investigation of $M_{23}C_6$ carbide precipitation behavior on varying grain boundary misorientions in 304 stainless steels[J]. Journal of Materials Science, 1998, 33: 1263 – 1271.

[14] Sasmal B. Mechanism of the form ation of lamellar $M_{23}C_6$ at and near twin boundaries in austenitic stainless steels

[J]. Metallurgical Transactions A, 1999, 30A: 2791 – 2801.

[15] 张寿禄. 电子背散射衍射技术及其应用[J]. 电子显微学报,2002,21(5): 703 – 704.

[16] Palumbo G, Aust K T, Lehockey E M, etal. On a more restrictive geometric criterion for "Special" CSL grain boundaries[J]. Scripta Materialia, 1998, 38(11): 1685 – 1690.

[17] Xia S, Zhou B X, Chen W J, Wang W G. Effects of strain and annealing processes on the distribution of $\Sigma3$ boundaries in a Ni-Based superalloy[J]. Scripta Materialia, 2006, 54: 2019 – 2022.

[18] Xia S, Zhou B X, Chen W J. Grain cluster microstructure and grain boundary character distribution in Alloy 690[J]. Metallurgival and Materials Transactions A, 2009, 40A(12): 3016 – 3030.

[19] Bollmann W. Crystal Lattice, Interface, Matrices[M]. Berlin: Springer, 1970.

[20] Lewis M H, Hattersley B. Precipitation of $M_{23}C_6$ in austenitic steels[J]. Acta Metallurgica, 1965, 13: 1159 – 1168.

[21] Wolff U E. Orientation and morphology of $M_{23}C_6$ precipitated in high-nickel austenite[J]. Transactions of the Metallurgical Society of AIME, 1966, 236: 19 – 27.

Study on the morphology of carbide precipitated at grain boundaries with various characters in alloy 690 by EBSD

Abstract: The morphology of carbide precipitated at grain boundaries with various characters in nickel based Alloy 690 were investigated by scanning electron microscopy(SEM), electron backscatter diffraction (EBSD) technical and orientation imaging microscopy (OIM). The morphology of carbides precipitated at grain boundaries with different characters is distinct. The bar-like carbide precipitated near both side of $\Sigma3_i$ boundary, while similar carbides precipitated only near one side of $\Sigma9$ boundary. The barlike carbide near $\Sigma3_i$ and $\Sigma9$ grows on the $\{111\}$ planes of the matrix grain. The size of carbide grainboundary on increases with the increasing of Σ value of the boundary. The intergranular corrosion increases with the increasing of Σ value of the boundary.

变形及热处理影响 Zr - 4 合金显微组织的研究*

摘　要：Zr - 4 合金样品经 1 020 ℃保温 20 min 后水淬处理，形成了一些富集 Fe 和 Cr 的片层状 β - Zr 相（bcc 结构），a＝0.362 nm. 其中 Fe 含量约为 3.65wt％，Cr 含量约为 0.61wt％，Sn 含量约为 1.57wt％. 该 β - Zr 相经变形及 580 ℃热处理后分解，形成 α - Zr 和细小的、带状分布的 Zr(Cr,Fe)₂ 第二相，所以 Zr - 4 合金经 β 相水淬、变形及热处理后也可得到细小且分布较均匀的第二相.

目前大多数压水核反应堆的燃料包壳材料是 Zr - 4 合金，通过优化加工及热处理过程，可以进一步提高其耐腐蚀性能[1-4]. 在生产工艺上通过 β 相淬火后，经过多道次的冷轧及在 800 ℃以下低温再结晶处理，来控制 Zr - 4 合金中 Zr(Cr,Fe)₂ 第二相的分布. 微量 Nb 的添加可以提高锆合金的耐腐蚀性能[5]，对于含 Nb 锆合金，通过变形及热处理则可以获得细小且均匀分布的 β - Nb 第二相，这是由于 Nb 的存在形成了稳定的 β - Zr，β - Nb 是在 610 ℃以下热处理分解形成的[6]；Zr - 4 合金是否也是通过类似的过程得到细小且均匀分布的 Zr(Cr,Fe)₂ 第二相，目前尚未有明确的报道. 因此本文利用高分辨透射电镜等手段对这一问题进行了研究.

1　实验方法

Zr - 4 板材由西北有色金属研究院提供，合金成分见表 1. 将尺寸约为 8 mm×150 mm，厚 1.2 mm 的锆合金片状样品，经酸洗（10％HF＋45％HNO₃＋45％H₂O 体积比的混合酸）和水洗后，真空封装在石英管内，然后将试样加热到 1 020 ℃保温 20 min 后淬入水中，同时敲碎石英管使样品迅速冷却，称之为 β 相水淬处理. 将 β 相水淬处理后的样品冷轧成 0.6 mm 厚的薄片，裁切成 20 mm×9 mm 的小片，经酸洗及水洗后，分组放置在石英管中，在真空（＜5×10⁻³ Pa）中分别进行 580 ℃×5 h，650 ℃×2 h 的退火热处理，然后将加热炉推离石英管，用水浇淋石英管进行冷却. 试样经机械磨抛后用双喷电解抛光的方法制备 TEM 样品，所用电解液为 10％HClO₄＋90％C₂H₅OH，低温抛光（低于－30 ℃）. 用 JEM - 2010F 透射电子显微镜观察显微组织；用 JSM - 6700F 扫描电镜观察形貌，合金表面处理方式为如前所述的酸洗及水洗.

表 1　Zr - 4 合金的名义化学成分（质量分数，％）

元　素	Fe	Cr	Sn	Zr
含　量	0.18～0.24	0.07～0.13	1.2～1.7	余量

＊ 本文合作者：李强、刘仁多、姚美意. 原发表于《上海金属》，2010，32(6)：5-8.

2 实验结果和讨论

2.1 Zr-4 合金经 β 相水淬处理后的显微组织

本实验中,薄片状的 Zr-4 合金样品经 β 相水淬处理后,由于冷却速度较快,发生 β-α' 的马氏体相变,形成晶粒细小的 α-Zr 板条状组织,在 α-Zr 板条晶中形成了厚薄不一的片层状的相(如图 1(a)~(c)),厚的达到 100 nm,薄的约 10 nm,经选区电子衍射(图 1(d))及 EDS(图 2)分析,证实这种片层状的组织为 bcc 结构的 β-Zr 相,a=0.362 nm(图 1(d)),这与文献[3]的报道一致. 该 β-Zr 相富集了合金元素 Fe 和 Cr,其中 Fe 含量约为 3.65wt%, Cr 含量约为 0.61wt%,Sn 含量约为 1.57wt%. 与 Zr-4 合金的名义成分(表 1)相比,该 β-Zr相的 Sn 含量基本符合合金的名义成分,而富集的 Fe 含量则达到合金名义成分的近 18 倍,Cr 达到近 6 倍. 从图 1 可以明显看出,片层状 β-Zr 的两边为相同的 α-Zr 晶粒,β-Zr 的长度方向端头往往终止在 α-Zr 晶内,在同一 α-Zr 晶粒内出现的 β-Zr 相一般有相同的取向关系. 本实验中,虽然 Zr-4 合金中 Fe 含量很低(0.18wt%~0.24wt%),但经 β 相保温后水淬,在发生 β-α' 马氏体相变的过程中,原 β 相中完全固溶的合金元素 Fe 和 Cr 因在 α-Zr 中的固溶度很低,受相变前沿排斥并富集,达到可以稳定 β 相的浓度及温度条件,形成了

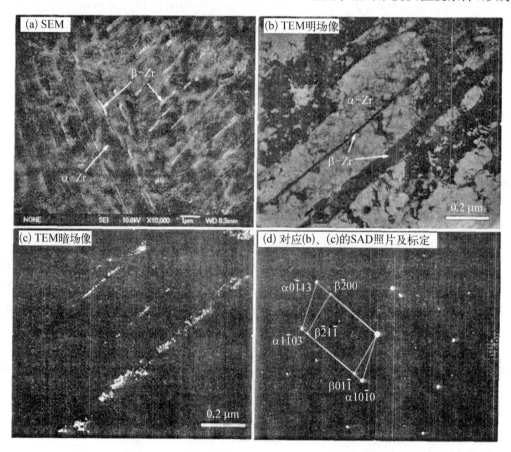

图 1 Zr-4 合金样品经 β 相水淬处理后的显微组织

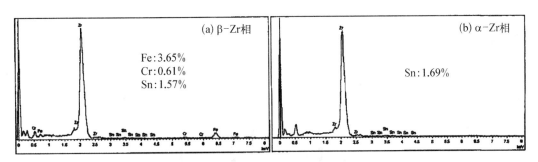

图2 对应图1(b)的EDS谱图(含量为重量百分比)

亚稳态的β-Zr相. 此外,在α-Zr晶粒内几乎见不到其他第二相;在α-Zr和β-Zr晶粒中,用HRTEM可观察到大量因水淬引起的晶体缺陷(图3).

在β相水淬处理的Zr-4合金中观测到β-Zr相,至今鲜有报道;文献[3]中只发现了极少量的残留β-Zr,而在本研究组其他一些相似β相水淬处理的Zr-4合金样品中也没有观测到这种β-Zr相的存在[4],这显然与实验条件密切相关. 一方面Zr-4合金中稳定β相的合金元素Fe、Cr含量很低,另一方面Zr-Fe(Cr)系合金形成β-Zr相的温度(~730 ℃)较 Zr-Nb系(~620 ℃)高,所以在一般的β相处理条件下,Zr-4合金较Zr-Nb系合金难观测到β相.

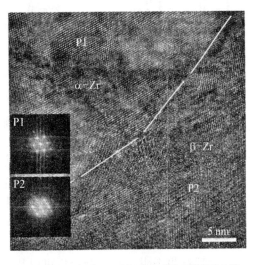

图3 α-Zr与β-Zr界面区域的HRTEM像及相应区域的傅里叶转换图

2.2 Zr-4合金经β相水淬、变形及热处理后的显微组织

Zr-4合金经β相水淬后所形成的合金元素Fe、Cr富集的β相是亚稳相,经过适当的热处理会转变成α-Zr和$Zr(Cr,Fe)_2$第二相. 经50%变形量的冷轧及580 ℃×5 h和650 ℃×2 h两种制度的热处理,Zr-4合金样品的TEM照片见图4,对应第二相的EDS谱图见图5.

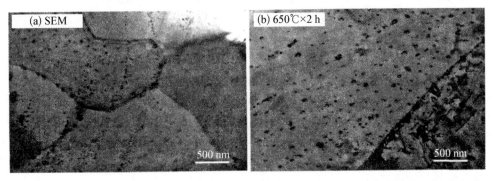

图4 Zr-4合金经β相水淬、变形及热处理后的显微组织

两种温度处理后的显微组织均由等轴的再结晶α-Zr晶粒和$Zr(Cr,Fe)_2$第二相细小颗粒组成. 由于第二相较小,EDS只作为定性分析,且分析时时常会包含基体部分,不可能得

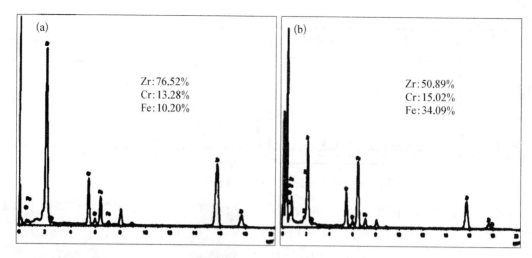

图 5　对应图 4 的第二相 EDS 图谱

(a) 图 4(a)中第二相；(b) 图 4(b)中第二相

到准确的 Fe、Cr 含量，所以 EDS 分析得到的 Zr、Fe、Cr 的成分有时不完全符合 $Zr(Fe,Cr)_2$ 的比例. 随着处理温度的升高，α-Zr 晶粒尺寸增大，$Zr(Cr,Fe)_2$ 第二相尺寸也增大；β 相水淬和冷轧后在 580 ℃加热处理的样品，第二相非常细小，较均匀地分布在 α-Zr 基体中，其中大多数第二相粒径尺寸小于 50 nm，并且可以看出一些第二相粒子呈现明显的条带状分布，这显然是冷轧形变及 β-Zr 分解共同作用的结果. 随着热处理温度的升高(650 ℃×2h 处理)，在 α-Zr 基体晶粒长大的同时，$Zr(Cr,Fe)_2$ 第二相发生溶解、扩散和凝聚，形成较大尺寸的第二相，一般超过 60 nm，分布较均匀，也可看出条带分布的迹象.

上述结果表明，Zr-4 合金也可以与含 Nb 锆合金一样，通过适当的 β 相水淬处理，导致合金元素富集，使形成的 β-Zr 稳定存在，再经过大变形量的冷轧变形及适当的退火热处理，β-Zr 分解形成细小且分布较均匀的第二相. 当然，这种方式只是形成细小且分布较均匀的第二相的一种过程. Zr-4 合金经 β 相水淬处理后，合金元素除了富集在 β-Zr 中，还会有部分过饱和固溶在 α-Zr 基体中[7]，这部分合金元素通过冷轧变形及退火热处理也会析出. 由于冷轧变形使合金基体中存在大量的分布较均匀的晶体缺陷，退火热处理时，这些缺陷会促使过饱和固溶的合金元素形核析出，所以这种形式形成的 $Zr(Cr,Fe)_2$ 第二相也会细小且分布较均匀. 本实验中观测到的第二相应是上述两种过程共同作用的结果，其中呈现明显的条带状分布的 $Zr(Cr,Fe)_2$ 第二相是经 β-Zr 分解过程而形成.

3　结论

Zr-4 合金样品经 1 020 ℃×20 min 保温后水淬处理，会形成一些片层状富集了 Fe 和 Cr 的 β-Zr(bcc结构)相，a=0.362 nm. 其中 Fe 含量约为 3.65wt%，Cr 含量约为 0.61wt%，Sn 含量约为 1.57wt%. 该 β-Zr 相经变形及 580 ℃热处理后会分解，形成 α-Zr 和细小的、带状分布的 $Zr(Cr,Fe)_2$ 第二相，所以 Zr-4 合金经 β 相水淬、变形及热处理后也可以得到细小且分布较均匀的第二相.

参 考 文 献

[1] Steinberg E，Weidinger H G，Schaa A. Zirconium in the Nuclear Industry：Sixth International Symposium[C].

ASTM - STP - 824，1984：106 - 122.

[2] Foster J P, Dougherty J, Burke M G, et al. Influence of final recrystallization heat treatment on Zircaloy-4 strip corrosion[J]. J. Nucl. Mat, 1990, 173(2)：146 - 178.

[3] 周邦新,李　强,苗　志.β相水淬对 Zr - 4 合金耐腐蚀性能的影响[J].核动力工程,2000,21(4)：339 - 343.

[4] 姚美意,周邦新,李　强,等.热处理对 Zr - 4 合金在 360 ℃ LiOH 水溶液中腐蚀行为的影响[J].稀有金属材料与工程,2007,36(11)：1920 - 1923.

[5] 姚美意,周邦新,李强,等.微量铌的添加对 Zr - 4 合金耐疖状腐蚀性能的影响[J].上海金属,2008,30(6)：1 - 3.

[6] 李　强,刘文庆,周邦新.锆锡合金腐蚀转折机理的讨论[J].稀有金属材料与工程,2002,31(5)：389 - 392.

[7] Li C, Zhou B X, Zhao W J, et al. Determination of Fe and Cr content in α - Zr solid solution of Zircaloy - 4 with different heattreated states[J]. J. Nucl. Mat，2002：134 - 138.

Effect of Deformation and Heat Treatment on the Microstructure of Zircaloy - 4

Abstract：Some lamellar β - Zr (bcc，a＝0. 362 nm) richer in Fe and Crwas detected in the quenched Zircaloy - 4 samples after heating at 1 020 ℃ for 20 min. The contents of Fe, Cr and Sn in β - Zr were about 3. 65wt%, 0. 61wt% and 1. 57wt% respectively. This β - Zr would decompose into α - Zr and fine ribbon like second phase Zr(Cr, Fe)₂ particles after deform ation and heat treatment at 580 ℃. So fine and uniformly distributed second phase particles could be obtained in the Zircaloy - 4 alloy through water quenching from β phase, deformation and heat treatment.

核反应堆压力容器模拟钢中富 Cu 原子团簇的析出与嵌入原子势计算[*]

摘　要：利用原子探针层析技术(APT)和热处理时效方法,研究了合金元素 Ni 对核反应堆压力容器模拟钢中富 Cu 原子团簇析出的影响. 实验结果表明,添加合金元素 Ni(0.84wt%)的样品中析出富 Cu 原子团簇的数量密度高于不添加 Ni 的样品,富 Cu 原子团簇内以及团簇和基体界面处都有 Ni 元素的富集现象,这说明合金元素 Ni 会促使富 Cu 原子团簇的析出. 从多体势的角度出发,利用嵌入原子势理论,基于纯金属元素 Fe,Cu,Ni 的多体势参数,建立了 Fe - Cu 二元和 Fe - Cu - Ni 三元体系的嵌入原子多体势. 计算结果表明,当模拟合金中存在 1at% Ni 时有利于富 Cu 原子团簇的析出,这与实验结果相符.

1　引言

　　压水堆核电站的压力容器中装载着核燃料组件(堆芯活性区)及堆内构件,维持着一回路中高温高压冷却水的压力,并对放射性极强的堆芯具有辐射屏蔽作用. 它与核燃料包壳以及反应堆安全壳一起构成了核电站的三道安全屏障. 由于压力容器是不可更换的部件,因而它的寿命决定了整个核电站的寿命[1]. 压力容器采用 Mn-Ni-Mo 低合金铁素体钢制成(A508 - Ⅲ钢),目前的设计寿命是 40 年,在中子辐照以及高温长时间服役时,压力容器钢的韧性降低,韧—脆转变温度升高,是非常大的安全隐患. 研究表明,压力容器钢出现这种辐照脆化的原因是中子辐照产生的晶体缺陷,以及中子辐照和长时间高温时效析出了含有 Mn,Ni,P,Si 等元素的富 Cu 原子团簇[2,3].

　　利用三维原子探针(3DAP)、小角度中子散射仪(SANS)和高分辨透射电镜(HRTEM)等实验仪器,研究了压力容器钢中子辐照监督试样和堆外热模拟实验样品[4-7],Auger 等[4]对中子注量为 12×10^{19} n·cm^{-2} 辐照监督试样进行分析,发现了 2—6 nm 大小的富 Cu 原子团簇,并在富 Cu 原子团簇内以及团簇和基体的界面处出现了多种元素聚集的现象. Cerezo 等[8]利用 Cu 含量为 0.5at%,Ni 含量为 0.3at% 和 1.5at% 的不同合金进行堆外热时效模拟实验,也发现 Ni 出现在富 Cu 原子团簇内以及团簇和基体的界面处,还发现 Ni 含量高的模拟合金中,析出富 Cu 原子团簇的数量密度也较大,说明 Ni 会促使富 Cu 原子团簇的析出. 用计算模拟和物理实验相结合的方法来研究压力容器钢中富 Cu 原子团簇的析出,有利于对问题的深入理解. 这方面的研究主要集中在蒙特卡罗模拟和第一性原理计算上[9-12],而利用嵌入原子势研究富 Cu 原子团簇析出的文章并不多见.

　　自密度泛函理论建立以来,提出了多种形式的多体势,适用于金属材料的多体势可以分为三类,即嵌入原子势(embedded atom method,简称 EAM)[13,14]、F - S 势(Finnis-

＊ 本文合作者：林民东、朱娟娟、王伟、刘文庆、徐刚. 原文发表于《物理学报》,2010,59(2)：1163 - 1168.

Sinclairmodel)[15] 和紧束缚 TB 势(tight binding formalism)[16]. Daw 和 Baskes[13,14] 提出了嵌入原子势模型,但是没有给出具体的基本函数,而且参数要进行复杂的数值拟合才能得出.针对这个缺点,Johnson[17] 提出了分析型嵌入原子势模型,并根据单原子模型中的不变性导出了合金势.基于 Johnson 提出的分析型嵌入原子势,Cai 等[18] 对其考虑了原子间长程作用后,提出了改进的 Cai-Ye 势.

压力容器钢中添加合金元素 Ni 可以提高淬透性并降低服役前的韧—脆转变温度,但是却增加了中子辐照脆化的敏感性[19],其原因至今并不十分清楚.本文利用原子探针层析技术研究了压力容器模拟钢经过淬火和 400 ℃等温时效后,Ni 对富 Cu 原子团簇析出的影响,并基于纯金属元素 Fe,Cu,Ni 的嵌入原子势参数,建立了 Fe-Cu 二元和 Fe-Cu-Ni 三元体系模型合金的分析型嵌入原子多体势,讨论了模拟钢中合金元素 Ni 对富 Cu 原子团簇析出的影响.

2 实验方法和结果

2.1 实验方法

目前,制造核反应堆压力容器采用 Mn-Ni-Mo 低合金铁素体钢(A508-Ⅲ),化学成分如表 1 所示,Cu,P 是其中的杂质元素.

表 1 A508-Ⅲ 和压力容器模拟钢的成分 　　　　　(单位：wt%)

	C	Si	Mn	P	S	Mo	Ni	Cu
A508-Ⅲ	≤0.26	0.15/0.40	1.20/1.50	<0.025	<0.025	0.45/0.55	0.40/1.00	<0.10
样品 1	0.21	0.05	0.15	0.011	0.006	0.01	0.03	0.44
样品 2	0.24	0.01	0.10	0.013	0.006	0.01	0.84	0.48

为了能够用热时效处理的方法研究压力容器钢中富 Cu 原子团簇的析出过程以及合金元素 Ni 的影响,用工业纯铁、电解镍和电解铜配制了 Ni 含量不同的两种合金,并提高了合金中 Cu 的含量,称为压力容器模拟钢.合金的成分如表 1 所示,其中样品 1 为 Fe-Cu 模拟合金,样品 2 为 Fe-Cu-Ni 模拟合金.实验材料由真空感应炉冶炼,铸锭为 40 kg,将钢锭经热锻和热轧制成 4 mm 厚的板材,然后切成 40 mm×30 mm 大小的样品,在 880 ℃,30 min 加热后水淬,然后在 400 ℃等温时效处理 20,100,150,300,1 000 和 2 000 h.热时效后的样品,根据硬度测试结果,选取时效时间为 100,150 和 300 h 的样品利用电火花线切割的方法,切成截面边长为 0.5 mm,长为 20 mm 的方形细棒,再经过电解抛光后制成曲率半径小于 100 nm 的针尖样品,利用 Oxford nanoScience 公司生产的三维原子探针对样品进行分析,数据采集时样品冷却至 −203 ℃,直流脉冲电压频率为 5 kHz,脉冲分数为 20%,所得数据用 PoSAP 软件进行分析.

目前普遍使用 maximum separation envelope method(MSEM)[20] 来界定纳米团簇,其基本原理是纳米团簇由一定数量的溶质原子组成,并且溶质原子的聚集程度要比其在基体中更紧密.一般用最小的原子数目 N_{min} 来表征纳米团簇中某溶质原子的数量,用原子之间的最大距离 d_{max} 来表征纳米团簇中某溶质原子的聚集程度,在这里取 $N_{min} = 10$, $d_{max} = 0.5$ nm.团簇数量的密度用公式 $N_v = N_p \times \zeta / (N \times \Omega)$ 计算,其中 N_v 为团簇数量密度,N_p 为分析体积内所检测到的团簇数量,ζ为检测效率参数取 0.6,N为所搜集的原子总数,Ω为原子的平均体积,对于体心立方结构的铁原子,平均体积 Ω 为 1.178×10^{-2} nm^3,由于所搜集的所有原子中主要为铁原子,所以这里以铁原子的体积来计算.

2.2 实验结果

图 1 为样品 1,2 经过不同时间时效处理后的 Cu 原子空间分布图,图中每一个小点表示仪器检测到的一个 Cu 原子.由于样品针尖的曲率半径不同,所以在数据采集时获得的空间截面大小不同;由于采集数据时会出现针尖断裂现象,所以获得的数据空间在长度方向也不一致.

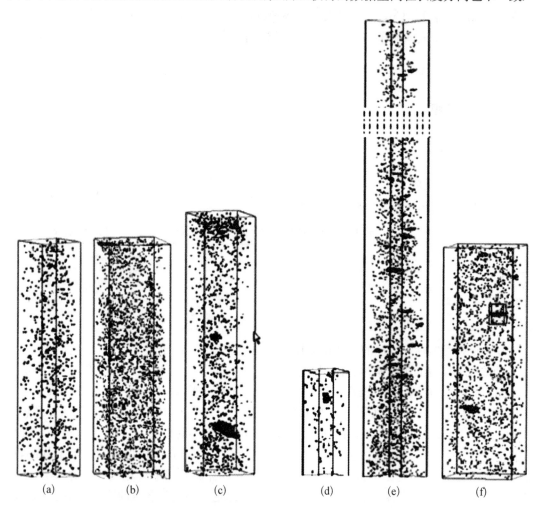

(a) (b) (c) (d) (e) (f)

图 1 样品在 400 ℃时效不同时间 Cu 原子空间分布图 (a) 样品 1 100 h,体积 9.7 nm×10.4 nm×56 nm;(b) 样品 1 150 h,体积 15 nm×15.7 nm×56.7 nm;(c) 样品 1 300 h,空间体积为 13.3 nm×13.7 nm×64.2 nm;(d) 样品 2 100 h,体积 7.5 nm×8.5 nm×25.2 nm;(e) 样品 2 150 h,体积 11.2 nm×11.5 nm×165.7 nm;(f) 样品 2 300 h,体积 14.3 nm×17 nm×56.1 nm

从图 1 可以直观地看出,两种样品经过 400 ℃时效处理后,都会析出纳米大小的富 Cu 原子团簇,但样品 2 中析出的富 Cu 原子团簇的数量总体上多于样品 1,为了进行对比,采用 MSEM[20]方法对团簇进行界定,得到被检测的空间中富 Cu 原子团簇的数量,同时根据团簇数量密度的计算方法,得到富 Cu 原子团簇的数量密度,结果列于表 2 中.从表 2 中可以看出,样品 1 在 300 h 的时效过程中, 富 Cu 原子团簇的数量密度一直在缓慢增加, 而样品 2 在 100 和 150 h 时效后, 不仅富 Cu 原子团簇的数量密度远远大于 1 号样品, 而且在时效 150 h 时达到了极大值, 经过 300 h 时效后富 Cu 原子团簇的数量密度有所下降, 说明这时团簇发生了聚集长大. 这一实验结果清楚表明了合金中添加 Ni(0.84wt%) 会促使富 Cu 原子团簇的析出.

表 2 样品经过不同时效时间富 Cu 原子团簇数量 N_p 以及团簇数量密度 N_v

时效时间/h	N_p		$N_v/10^{23}\ m^{-3}$	
	样品 1	样品 2	样品 1	样品 2
100	1	1	0.6	2.3
150	2	17	0.7	6.1
300	4	5	1.5	1.5

为了了解富 Cu 原子团簇中 Cu, Ni 的分布情况, 选取样品 2 中的一个团簇进行了分析, 得到的结果如图 2 所示. 富 Cu 原子团簇中 Cu 的含量达到 15at% 左右, 而 Ni 的含量也能达到 4at%, 高于合金基体中的平均含量, 并且在富 Cu 原子团簇与基体的界面处 Ni 也有偏聚现象, 这与文献中已经报道过的结果基本一致[4,8,21]. 这一实验结果表明, 添加了 0.84wt% Ni 的样品 2 合金, 虽然 Ni 可以固溶在 $\alpha - Fe$ 基体中, 但可能会出现局域性的成分起伏现象, 在这些高 Ni 区有可能成为富 Cu 原子团簇的形核区, 降低了形核的能量, 提高了形核率, 从而添加合金元素 Ni 会促使富 Cu 原子团簇的析出.

为了进一步理解 Ni 对富 Cu 原子团簇析出的影响, 利用嵌入原子势来计算 Fe-Cu 和 Fe-Cu-Ni 模型合金中富 Cu 原子团簇形成前后能量的变化.

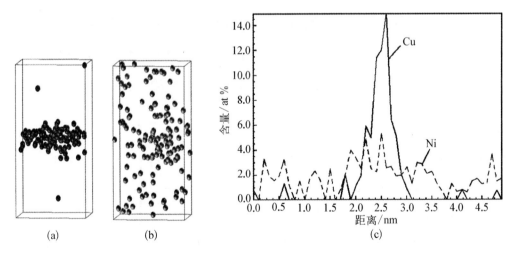

图 2 团簇区域内原子分布图和径向含量分布图 体积 3 nm×3 nm×5 nm (a) Cu, (b) Ni, (c) 径向含量分布图

3 嵌入原子势计算

3.1 嵌入原子势

嵌入原子势模型的基本思想是把系统中的每一个原子都看成是嵌入在由其他原子组成

的基体中的杂质,将系统的能量表示为嵌入能和对势能之和,从而将多原子相互作用归结于嵌入能,所以体系的总能量为

$$E_{\text{total}} = \sum_i F_i(\rho_i) + \frac{1}{2} \sum_{i \neq j} \varphi_{ij}(r_{ij}),\tag{1}$$

其中 E_{total} 为系统总能量,$F_i(\rho_i)$ 为嵌入能,φ_{ij} 是任意两个原子的对势,r_{ij} 是 i 和 j 原子的间距,ρ_i 是所有其他原子在原子 i 处所产生的电子密度,Daw 和 Baskes[13,14]认为是其他原子在 i 处所产生的电子密度的线性之和,即

$$\rho_i = \sum_{j \neq i} f_j(r_{ij}),\tag{2}$$

其中 $f_j(r_{ij})$ 是 j 原子在 i 处产生的电子密度.

文献[20]构建了 Fe,Cu 二元系统多体势的方法,嵌入能函数采用 Banerjea 和 Smith[22]提出的分析型表达形式

$$F(\rho) = -F_0\left[1 - \ln\left(\frac{\rho}{\rho_e}\right)^n\right]\left(\frac{\rho}{\rho_e}\right)^n + F_1\left(\frac{\rho}{\rho_e}\right),\tag{3}$$

其中,$F_0 = E_c - E_v$,E_c 和 E_v 分别表示晶体的结合能和空位形成能,ρ 为的电子密度,ρ_e 为其平衡值,n 是常数,F_1 是需要拟合的势参数,对于 bcc 结构的金属 $F_1 = 0$. 根据文献[20]的建议,电子密度函数[22]均采用如下形式:

$$f(r) = f_e \exp[-\chi(r - r_e)],\tag{4}$$

式中 r_e 是平衡时原子间的最近邻距离,f_e 是控制因子,应用于合金的计算时,该参数不能被约化掉,必须确定 f_e 的值,在本文中 $f_e = \left(\dfrac{E_c}{\Omega}\right)^\gamma$,其中 Ω 为原子体积,χ 是需拟合的参数.

考虑金属晶体结构的不同,针对 bcc 金属的 Fe,同类原子间的对势函数为如下形式[23]:

$$\varphi(r) = k_3\left(\frac{r}{r_e} - 1\right)^3 + k_2\left(\frac{r}{r_e} - 1\right)^2 + k_1\left(\frac{r}{r_e} - 1\right)^1 + k_0,\tag{5}$$

针对 fcc 金属的 Cu,Ni 同类原子间的对势函数采用如下形式[24]:

$$\varphi(r) = -\alpha[1 + \beta(r/r_a - 1)] \times \exp[-\beta(r/r_a - 1)],\tag{6}$$

其中 $k_0,k_1,k_2,k_3,\alpha,\beta,r_a$ 是需要拟合的参数.

Fe,Cu 原子间的交叉势,采用 Gong 等[25]改进的如下线性函数形式:

$$\varphi^{\text{Fe-Cu}}(r) = A[\varphi^{\text{Fe}}(r + B) + \varphi^{\text{Cu}}(r + C)].\tag{7}$$

Fe,Ni 和 Ni,Cu 原子之间的交叉势 $\varphi^{\text{ab}}(r)$ 采取 Johnson 提出的合金势[18],其形式如下:

$$\varphi^{\text{ab}}(r) = \frac{1}{2}\left[\frac{f^{\text{b}}(r)}{f^{\text{a}}(r)}\varphi^{\text{a}}(r) + \frac{f^{\text{a}}(r)}{f^{\text{b}}(r)}\varphi^{\text{b}}(r)\right].\tag{8}$$

3.2 势参数和计算结果

多体势构建的关键是确定多体势模型函数以及确定有效的势参数. 对于 Fe,Cu,Ni 构成

的二元和三元合金系统,势参数则需要通过拟合系统的一些基本物理性质来得到. 金属单质的势参数比较容易得到,因为拟合势参数所需要的物理量一般都能从各种手册中查到,而且这些物理量一般都是比较可靠的实验值. 从文献[20]得到单质 Fe 的势参数 $\chi = 45.091\,44\ \text{nm}^{-1}$,$k_0 = -0.271\,183$,$k_1 = -0.931\,581$,$k_2 = 9.615\,03$,$k_3 = -13.477\,284$,$r_s = 0.29\ \text{nm}$,$r_c = 0.38\ \text{nm}$;单质 Cu 的势参数 $F_1 = 0.676\,073$,$\chi = 111.342\,31\ \text{nm}^{-1}$,$\alpha = 0.725\,977$,$\beta = 3.457\,434$,$r_a = 0.162\,935\,6\ \text{nm}$,$r_s = 0.37\ \text{nm}$,$r_c = 0.44\ \text{nm}$. 从文献[26]得到单质 Ni 的势参数 $F_1 = 1.580$,$\chi = 28.21\ \text{nm}^{-1}$,$\alpha = 0.608$,$\beta = 6.460$,$r_a = 0.219\,7\ nm$,$r_c = 0.39\ nm$. 而异类原子之间的交叉势采用函数关系(7),(8)式的方法处理,涉及 Fe,Cu 原子间的相互交叉作用势参数,从文献[21]得到 $A = 0.453\,151$,$B = 0.008\,982\,7\ \text{nm}$,$C = -0.012\,155\ \text{nm}$.

结合文献[27]的思路,构建了 bcc 结构 Fe 的理想晶体模型,计算体系为 $10 \times 10 \times 10$ 个晶胞,使 Fe 原子布满晶胞的中心和顶点,采用周期性边界条件,原子总数为 2 000 个.

第一种方法是以 20 个 Cu 原子去置换晶胞中的 Fe 原子,一是以平均分散的方式置换晶胞内 Fe 原子,计算总能量为 E_{sca},二是在晶体模型的中心以一个类球形团簇的方式替换 Fe 原子,计算总能量为 E_{clu},两种置换方法的能量差 $E_1 = E_{clu} - E_{sca}$,计算结果为 $E_1 = -0.234\ \text{eV}$. 第二种方法是先用 20 个 Ni 原子以平均分散的方式置换 Fe 原子,表示含 1at%Ni 的 Fe-Ni 合金,然后再按照第一种方法分别用 20 个 Cu 原子以平均分散和类球形团簇两种方式置换 Fe 原子,两种方式置换前后的能量差 $E_2 = E_{clu} - E_{sca}$,计算结果为 $E_2 = -0.517\ \text{eV}$.

嵌入原子势的计算结果显示 $E_1 < 0$,这说明在 Fe-Cu 合金中形成 Cu 原子偏聚后的总能量会降低,Cu 原子有偏聚在一起的倾向. 计算结果显示 $E_1 > E_2$,这表明含有 1at%Ni 的 Fe-Cu-Ni 合金,Cu 原子发生偏聚后系统的总能量会降低更多,从能量的角度说明了 Ni 的存在会促使富 Cu 原子团簇析出.

4 结论

(1)压力容器模拟钢经过 880 ℃加热水淬和 400 ℃时效处理后,会析出纳米大小的富 Cu 原子团簇,添加合金元素 Ni(0.84wt%)的样品中析出富 Cu 原子团簇的数量密度高于未添加 Ni 的样品,说明合金元素 Ni 会促使富 Cu 原子团簇的析出,在富 Cu 原子团簇内以及团簇和基体界面处还有 Ni 元素富集的现象.

(2)嵌入原子势计算结果表明,在 Fe-Cu 和 Fe-Cu-Ni 合金中发生 Cu 原子偏聚时,由于 1at%Ni 存在,使能量降低得更多,从能量变化的角度说明了合金元素 Ni 会促使富 Cu 原子团簇析出.

参 考 文 献

[1] Yang W D 2006 *Reactor Materials Science*（Beijing：Atomic Energy Press）p160（in Chinese）[杨文斗 2006 反应堆材料学(北京：原子能出版社)第 160 页].

[2] Odette G R 1983 *Scripta Metallrgica* **17** 1183.

[3] Williams T J 1996 *ASTM-STP* **1270** 191.

[4] Auger P, Pareige P, Welzel S, van Duysen J C 2000 *J. Nucl. Mat.* **280** 331.

[5] Miller M K, Pareige P, BurKe M G 2000 *Mat. Characterization* **44** 235.

[6] Ulbricht A, Bohmert J 2004 *Physica B* **350** e483.

[7] Lozano-perez S, Sha G, Titchmarsh J M, Jenkins M L, Hirosawa S, Cerezo A, Smith G D W 2006 *J. Mat. Sci.* **41** 2559.

[8] Cerezo A, Hirosawa S, Rozdilsky I, Smith G D W 2003 *Phil. Tras. R. Soc. Lond.* A **361** 463.

[9] Liu C L, Odette G R, Wirth B D, Lucas G E 1997 *Mat. Sci. Eng.* A **238** 202.

[10] Kuriplach J, Melikhova O, Domain C, Becquart C S, Kulikov D, Malerba L, Hou M, Morales A L 2006 *Appl. Surf. Sci.* **252** 3303.

[11] Khrushcheva O, Zhurkin E E, Malerba L, Becquart C S, Domain C, Hou M 2003 *Nuclear Instruments and Methods in Physics Research* B **202** 68.

[12] Vincent E, Domain C, Becquart C S 2006 *J. Nucl. Mat.* **351** 88.

[13] Daw M S, Baskes M I 1984 *Phys. Rev.* B **29** 6443.

[14] Daw M S, Baskes M I 1983 *Phys. Rev. Lett.* **50** 1285.

[15] Finnis M W, Sinclair J E 1984 *Philos. Mag.* A **50** 45.

[16] Tomanek S, Mukherjee S, Bennemann K H 1983 *Phys. Rev.* B **28** 665.

[17] Johnson R A 1989 *Phys. Rev.* B **39** 12554.

[18] Cai J, Ye Y Y 1996 *Phys. Rev.* B **54** 8398.

[19] International Atomic Energy Agency 2005 *Effects of Nickel on Irradiation Embrittlement of Light Water Reactor Pressure Vessel Steels* IAEA - TECDOC - 1441, p1.

[20] Gong H R, Kong L T, Liu B X 2004 *Phys. Rev.* B **69** 054203.

[21] Miller M K 2000 *Atom Prope Tomography: Analysis at the Atomic Level* (New York: Kluwer Academic/Plenum Publishers) p158.

[22] Banerjea A, Smith J R 1988 *Phys. Rev.* B **37** 6632.

[23] Johnson R A, Oh D 1989 *J. Mater. Res.* **4** 1195.

[24] Rose J H, Smith J R, Guinea F, Ferrante J 1984 *Phys. Rev.* B **29** 2963.

[25] Gong H R, Kong L T, Lai W S, Liu B X 2002 *Phys. Rev.* B **66** 104204.

[26] Zhang Q, Lai W S, Liu B X 2000 *J. Phys. : Condens. Mat.* **12** 6991.

[27] Yu S, Wang C Y, Yu T 2007 *Acta Phys. Sin.* **56** 3212 (in Chinese) [于松、王崇愚、于涛. 2007 物理学报 **56** 3212].

Precipitation of Cu-rich clusters in reactor pressure vesselmodel steels and embedded-atom method potentials

Abstract: Atom probe tomography characterization and thermal aging method have been employed to investigate the precipitation of Cu-rich clusters in reactor pressure vessel (RPV) model steels with and without the addition of Ni alloying element. It is observed that the number density of Cu-rich clusters in the PRV model steel with the addition of Ni (0.84wt%) is higher than that without the addition of Ni. The segregation of nickel in the Cu-rich clusters and at the boundaries of clusters/matrix was also observed. It is found that the nickel alloying element can promote the precipitation of Cu-rich clusters in RPV model steel. An analytical embedded atom method model of Fe-Cu and Fe-Cu-Ni system is constructed on the basis of the physical properties of pure constituents Fe, Cu and Ni. The calculation results show that the nickel (1at%) can promote the precipitation of Cu-rich clusters. The calculation results agree with the results of experiments.

晶界特征分布对 304 不锈钢应力腐蚀开裂的影响[*]

摘　要: 通过 5% 的冷轧变形及在 1 100 ℃退火 5 min,使 304 不锈钢中的低 ΣCSL 晶界比例从固溶处理后的 49% 提高到 75%(Palumbo-Aust 标准). 采用 C 型环样品恒定加载方法,在 pH 值为 1.5 的沸腾 25%NaCl 酸化溶液中进行应力腐蚀实验,低 ΣCSL 晶界比例为 75% 的样品在浸泡 120 h 内没有发生应力腐蚀开裂,而低 ΣCSL 晶界比例为 47% 的试样在浸泡 24 h 后就产生了应力腐蚀裂纹. 由断口形貌观察及电子背散射衍射(EBSD)分析表明,开始发生的晶间腐蚀会成为后来应力腐蚀开裂的裂纹源,应力腐蚀开裂由最初的沿晶转变为穿晶形式. 低 ΣCSL 晶界比例提高后的试样因其抗晶间腐蚀性能较好,抑制了在 Cl⁻ 环境下应力腐蚀裂纹的萌生,因而提高了抗应力腐蚀性能.

1984 年,Watanabe[1] 首次提出了晶界设计的概念,并于 20 世纪 90 年代发展为晶界工程研究领域. 该研究通过形变及退火处理,提高低 ΣCSL(coincidence site lattice,CSL[2],即重位点阵,一般认为 $\Sigma \leqslant 29$ 的晶界为低 ΣCSL 晶界)晶界比例,调整晶界特征分布(grain boundary character distribution,GBCD),从而改善材料与晶界有关的多种性能[3-6].

304 不锈钢具有较好的力学性能和优良的抗均匀腐蚀能力,在石油、化工、能源动力等领域中得到了广泛应用,但是,在 Cl⁻、高温水等环境下易发生应力腐蚀开裂[7-8]. 不锈钢在 500~850 ℃之间加热时会在晶界上析出 Cr 的碳化物,在晶界附近产生贫铬区,从而增加不锈钢晶间应力腐蚀的敏感性. 为解决这一问题,可以在奥氏体不锈钢中添加 Ti,Nb 等元素,通过形成它们的碳化物以稳定 C 元素,或者用超低碳不锈钢或高镍合金作为替代材料,但这些方法都提高了成本. 通过晶界工程来改善 304 不锈钢材料的晶界特征分布,既可以控制成本,又可以提高 304 不锈钢材料的使用寿命及安全系数. Shimada 等[6] 研究了 304 不锈钢经不同变形后在 927 ℃退火 72 h 的晶界特征分布,结果表明,将低 ΣCSL 晶界比例提高到 80% 以上时(Brandon 标准[9]),其耐晶间腐蚀性能大幅提高. Jin 等[10] 运用相同的工艺,优化了 304 不锈钢的晶界特征分布,运用三点弯曲加载,在 288 ℃的高温高压水中浸泡腐蚀 500 h,发现未经优化的样品发生了晶间应力腐蚀,并且裂纹尖端终止在低 ΣCSL 晶界处. 但是,Gertsman[8] 分析了 304 不锈钢和 600 合金的裂纹扩展路径,指出并不是所有的低 ΣCSL 晶界都对应力腐蚀开裂免疫,除了共格 Σ3 晶界外,其他类型的晶界都有可能出现应力腐蚀裂纹.

当提高 304 不锈钢低 ΣCSL 晶界比例后,可以使抗应力腐蚀性能得到提高,这些研究结果虽然已有报道,但是腐蚀环境大多为高温、高压水. 在实际工业应用中,含有 Cl⁻ 的环境对 304 不锈钢腐蚀开裂的危害最大,所谓的"氯脆"一直是人们关注的问题. 本研究通过不同的形变量与热处理工艺改变 304 不锈钢的晶界特征分布,考察在含有 Cl⁻ 环境下对材料抗应力腐蚀性能的影响.

1　试验方法

试验所用的 304 不锈钢成分中 C,Si,Mn,Ni,Cr 的质量分数分别为 0.08%,0.58%,

* 本文合作者:罗鑫、夏爽、李慧、陈文觉. 原发表于《上海大学学报(自然科学版)》,2010,16(2):177 – 182.

1.18%,8.75%,18.31%,余量为 Fe. 经 1 100 ℃,30 min 固溶处理,得到初始态材料. 用电火花线切割,将样品加工成片状,以便进行冷轧变形及热处理. 试样编号及处理工艺如表 1 所示. 敏化处理工艺为 675 ℃,2 h.

表 1 试样编号及加工工艺

Table 1 Numbering and treatment processes of each specimen

试 样 编 号	加 工 工 艺
A	固溶处理(处理 A)
B	固溶处理,冷轧 5%,1 100 ℃,5 min,水淬(处理 B)
C	固溶处理,冷轧 40%,1 100 ℃,5 min,水淬(处理 C)
D	固溶处理,冷轧 40%,1 100 ℃,60 min,水淬(处理 D)
E	处理 A,敏化处理
F	处理 B,敏化处理
G	处理 C,敏化处理
H	处理 D,敏化处理

退火处理后的试样用 180♯～1 200♯ 的 SiC 砂纸依次研磨,然后进行电解抛光以制备符合电子背散射衍射(EBSD)测试要求的样品表面. 电解抛光液中 $HClO_4$ 和 CH_3COOH 体积分数分别为 20%,80%,电压为直流 40 V,时间约为 20 s. 金相试样用王水蚀刻,时间约为 10 s. 对于出现应力腐蚀裂纹的样品,用离子溅射减薄仪进行表面抛光,电压为 4 kV,电流为 2 mA,时间约为 2 h.

将试样浸泡在 H_2O_2 和 HF 的水溶液($V(H_2O_2):V(HF):V(H_2O)=3:4:21$)中测试耐晶间腐蚀性能. 腐蚀失重计算公式如下:

$$w_t = (W_t - W_0)/S,$$

式中,W_0 是样品腐蚀前的质量(mg),W_t 是样品腐蚀一定时间 t 后的质量(mg),S 是样品的表面积(mm^2). 应力腐蚀试样采用 C 型环加载方式,试样尺寸按国标 GB15970.5[11]规定加工. 将 C 型环试样嵌入内开口的模具以施加恒定载荷,模具为相同的 304 不锈钢材料,以避免发生电偶腐蚀. 所有的 C 型环样品及模具用电火花线切割,按照编排的程序加工成型,以保证尺寸一致,从而使试样的受力状态一致. 根据 GB15970.5[11]附录 A 的公式得出试样所受载荷约为 400 MPa. 试验介质为磷酸酸化的 25%(质量比)NaCl 溶液[12],溶液的 pH 值为 1.5,加热至沸腾,溶液沸点为 112±1 ℃. 试验在配有冷凝回流装置的锥形瓶中进行.

利用配有 TSL － EBSD(electron backscatter diffraction)系统的 SEM(scanning electron microscope)扫描电镜对抛光后的试样表面进行逐点逐行扫描,得到材料表面各点的取向. 取向成像显微(orientation image microscopy,OIM)系统可以通过晶界两侧晶粒的取向差判定晶界类型,自动统计各种晶界的比例. EBSD 测试步长为 5～10 μm,根据晶粒大小改变,本研究采用 Palumbo-Aust 标准[13]($\Delta\theta_{max} = 15°\Sigma^{-5/6}$)来判定晶界类型.

2 试验结果与讨论

图 1 和图 2 分别为固溶处理后样品的金相显微照片和(001)极图. 图中可见,样品中存在较多的退火孪晶,晶粒取向分布混乱,没有明显的织构. 由 EBSD 测定给出的低 ΣCSL 晶界比例为 49.3%.

图 3 给出了不同试样的晶界特征分布图,各样品的 GBCD 统计结果见表 2. 经 5% 冷轧

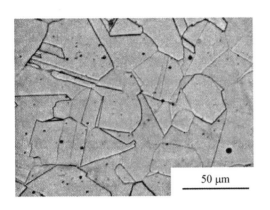

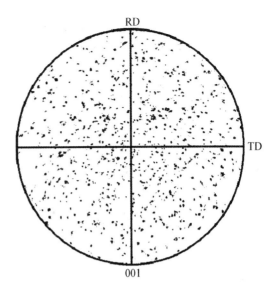

图 1 304 不锈钢经固溶处理后的金相照片

Fig. 1 Opticalm icrograph (solution annealed)

图 2 由 EBSD 测定的(001)极图

Fig. 2 (001) pole figure of the starting material (solution annealed)

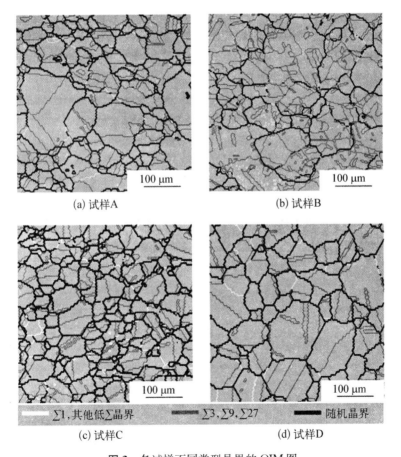

(a) 试样A   (b) 试样B

(c) 试样C   (d) 试样D

图 3 各试样不同类型晶界的 OIM 图

Fig. 3 Grain boundary character distributions are shown by OIM maps for specimens

变形并高温短时退火的试样 B 中,低 ΣCSL 晶界比例提高到 75％ 左右. 而经 40％ 冷轧变形并经相同条件退火的试样 C,其低 ΣCSL 晶界比例为 46％. 延长其退火时间至 60 min,对低 ΣCSL 晶界比例影响不大. 由于敏化处理温度为 675 ℃,不会引起晶界迁移,可以认为敏化处理前后材料的 GBCD 以及晶粒尺寸基本不发生变化. 由 OIM 系统自动统计,图 3(b)～图 3(d)表示的样品 B,C 和 D 的平均晶粒尺寸分别为 32,20 和 35 μm(所有类型晶界均统计在内). 可见试样 B 和 D 的平均晶粒尺寸相差不大,在后续的试验中可以排除晶粒尺寸对试验结果的影响. 低 ΣCSL 晶界比例提高以后,显微组织是由大尺寸的"互有 Σ3^n" 取向关系晶粒的团簇"构成[4,14]. 在晶粒团簇内,Σ3 晶界与其他 Σ3^n 晶界相连构成 Σ3 - Σ3 - Σ9 或 Σ3 - Σ9 - Σ27 的三叉界角,晶粒之间都具有 Σ3^n 的取向关系,而晶粒团簇之间基本由随机晶界构成. 试样 B 形成了较大尺寸的晶粒团簇,其低 ΣCSL 晶界比例得到提高.

表 2 试样的晶界特征分布统计及应力腐蚀裂纹出现时间

Table 2 Grain boundary character distribution statistics and the result of SCC of eachs pecimen

编号	Σ1/％	Σ3/％	Σ9+Σ27/％	总体低 ΣCSL 晶界/％	应力腐蚀裂纹出现时间/h
A	5.8	41.0	1.9	49.3	36
E					48
B	5.2	60.0	10.2	75.2	120 h 内未出现裂纹
F					120 h 内未出现裂纹
C	5.8	38.9	1.0	46.0	48
G					60
D	4.6	39.5	0.1	47.0	36
H					24

图 4 是经过 18 h 晶间腐蚀后,样品 F 及 H 的表面形貌. 从图中可以看出,两种样品表面有明显的差别. 图 4(a)表示的样品 F 表面较完整,显现出一些晶界;图 4(b)表示的低 ΣCSL 晶界比例较低的样品 H,发生了明显的晶间腐蚀,表面已有很多晶粒脱落. 图 5 给出了样品 F 及 H 单位面积腐蚀失重与晶间腐蚀时间的关系. 由图可知,两种样品的腐蚀失重有明显的差别,并且随着晶间腐蚀浸泡时间的延长,腐蚀失重差距越来越大. 由图 4 可知,低 ΣCSL 晶界比例较低的样品由于发生了较严重的晶间腐蚀,会使晶粒脱落,从而造成了明显的失重.

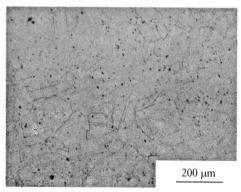 

(a) 样品F,其低ΣCSL晶界比例为75%　　　　(b) 样品H,其低ΣCSL晶界比例为47%

图 4 低 ΣCSL 晶界比例不同的样品,敏化处理后经 18 h 晶间腐蚀的表面形貌

Fig. 4 Optical micrographs of specimen F and H after 18 h intergranular corrosion

由此可见,低 ΣCSL 晶界比例得到提高后,样品的耐晶间腐蚀性能也得到了提高. 这与 Xia 等[4]、Shimada 等[6]的研究结果一致.

样品应力腐蚀测试结果在表 2 中给出. 低 ΣCSL 晶界比例提高到 75% 的样品 B 和 F,在应力腐蚀试验 120 h 内没有出现应力腐蚀裂纹,而其他低 ΣCSL 晶界为 47% 左右的样品,先后在 60 h 内发生了应力腐蚀开裂. 从应力腐蚀裂纹出现时间看,提高低 ΣCSL 晶界比例后,材料的抗应力腐蚀性能也得到提高.

图 6 为试样 G 应力腐蚀开裂后的断口形貌. 图 6(a) 表示了 C 型环试样的整个断口形貌,上方为 C 型环弧顶,裂纹从上向下扩展. 图 6(b) 和图 6(c) 是在断口上下两部分拍摄的 SEM 二次电子像照片. 图 6(b) 为典型的"冰糖状"断口形貌,应力腐蚀类型为晶间应力腐蚀开裂;图 6(c) 为典型的台阶状解理断口形貌,说明此时应力腐蚀裂纹为穿晶裂纹. 由此可见,最初裂纹沿晶界扩展,然后转变为穿晶开裂

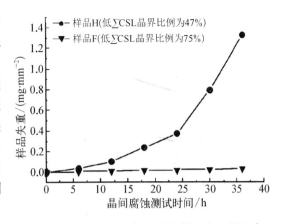

图 5 低 ΣCSL 晶界比例不同的样品由于晶间腐蚀引起的失重与腐蚀浸泡时间的关系

Fig. 5 Relationship between time and weight lost of specimens with varied proportion of low ΣCSL grain boundaries in intergranular corrosion testing

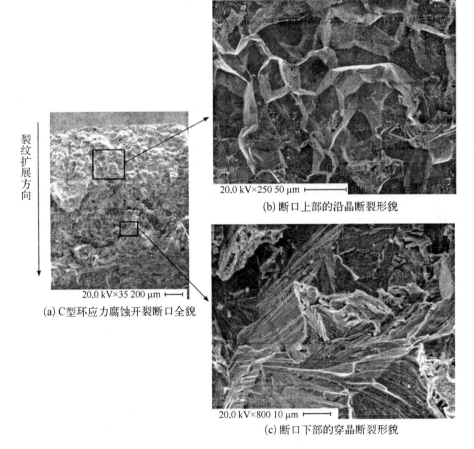

图 6 试样 G 的断口形貌

Fig. 6 Fracture surface morphology

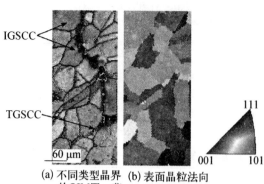

(a) 不同类型晶界
的OIM图，背
底为菊池花样
图像质量图

(b) 表面晶粒法向
取向作为衬度
的OIM图

图 7 试样 G 裂纹附近的 EBSD 测试分析

Fig. 7 Grain boundary OIM map of the cracking region

模式. 图 7 为试样 G 应力腐蚀裂纹附近用 EBSD 测试得到的 OIM 图,可以看到裂纹从沿晶开裂到穿晶开裂的转变. 作为对比试验,将相同尺寸的 C 型环加载试样放入未经磷酸酸化的 25% 沸腾的 NaCl 溶液中,在 168h 内没有任何试样发生应力腐蚀开裂. 试验结束后,样品表面还很光亮,经金相显微镜观察,样品表面没有晶间腐蚀迹象. 可见,在酸化的 Cl^- 环境中,304 不锈钢发生晶间腐蚀后,这种晶间腐蚀会成为应力腐蚀裂纹源. 裂纹先沿晶界扩展(见图 6(b)),当达到一定深度时,产生应力集中,大于应力腐蚀开裂临界应力强度因子 K_{ISCC} 后,裂纹尖端屈服,滑移台阶溶解. 此时应力腐蚀开裂机理可按滑移-溶解理论解释[7,15],材料在 {111} 面上发生滑移,使得钝化膜撕裂,裸露出新鲜表面成为阳极发生电化学反应而溶解,形成穿晶裂纹,出现了典型的解理断口(见图 6(c)). 由晶间腐蚀试验结果可知,低 ΣCSL 晶界比例提高后的样品由于其耐晶间腐蚀性能得到提高,从而抑制了应力腐蚀裂纹的萌生,提高了抗应力腐蚀性能.

3 结论

经 5% 冷轧变形后在 1 100 ℃ 退火 5 min,可以使 304 不锈钢中的低 ΣCSL 晶界比例提高到 75%(Palumbo-Aust 标准),样品的耐晶间腐蚀性能和在含有 Cl^- 环境下的抗应力腐蚀性能都得到了提高. 在低 ΣCSL 晶界比例较低(47%)的样品中,晶间腐蚀会成为应力腐蚀开裂的裂纹源,裂纹扩展由初始的沿晶形式转变为穿晶形式. 低 ΣCSL 晶界比例提高后的试样因其抗晶间腐蚀性能较好,抑制了应力腐蚀裂纹源的萌生,因而提高了抗应力腐蚀性能.

参 考 文 献

[1] KRONBERG M L, WILSON F H. Secondary recrystallization in copper[J]. Trans AIME, 1949, 185: 501-514.

[2] WATANABE T. An approach to grain boundary design of strong and ductile polycrystals[J]. Res Mech, 1984, 11: 47-84.

[3] LIN P, PALUMBO G, ERB U. Influence of grain boundary character distribution on sensitization and intergranular corrosion of alloy 600[J]. Scr Metall Mater, 1995, 33(9): 1387-1392.

[4] XIA S, ZHOU B X, CHEN W J. Effect of single-step strain and annealing on grain boundary character distribution and intergranular corrosion in alloy 690[J]. J Mater Sci, 2008, 43(9): 2990-3000.

[5] PALUMBO G, ERB U. Enhancing the operation life and performance of lead-acid batteries via grain-boundary engineering[J]. MRS Bulletin, 1999, 24: 27-32.

[6] SHIMADA M, KOKAWA H, WANG Z J, et al. Optimization of grain boundary character distribution for intergranular corrosion resistant 304 stainless steel by twin-induced grain boundary engineering[J]. Acta Mater, 2002, 50(9): 2331-2341.

[7] COPSON H R. Physical metallurgy of stress corrosion fracture[M]. NewYork: Interscience, 1959.

[8] GERTSMAN V Y, BRUEMMER S M. Study of grain boundary character along intergranular stress corrosion crack paths in austenitic alloys[J]. Acta Mater, 2001, 49: 1589 – 1598.

[9] BRANDON D G. The structure of high-angle grain boundaries[J]. Acta Mater, 1966, 14: 1479 – 1484.

[10] JIN W Z, YANG S, KOKAWA H, et al. Improvement of intergranular stress corrosion crack susceptibility of austenite stainless steel through grain boundary engineering[J]. J Mater SciTech, 2007, 23(6): 785 – 789.

[11] 中华人民共和国国家质量监督检验检疫总局. GB15970. 5—1998,金属和合金的应力腐蚀试验——第 5 部分 C 型环试样的制备和应用[S]. 北京：中国标准出版社,1998.

[12] ASTM G123—2000, Standard test method for evaluating stress-corrosion cracking of stainless alloys with different nickel content in boiling acidified sodium chloride solution[S]. West Conshohocken, PA: ASTM International, 2000.

[13] LIN P, PALUMBO G, AUST K T. Experimental assessment of the contribution of annealing twins to CSL distributions in FCC materials[J]. Scr Mater, 1997, 36(10): 1145 – 1149.

[14] XIA S, ZHOU B X, CHEN W J. Grain cluster microstructure and grain boundary character distribution in Alloy 690 [J]. Metallurgical and Materials Transactions A, 2009, 40A: 3016 – 3030.

[15] LIU R, NARITA N, ALTSTETTER C, et al. Studies of the orientations of fracture surfaces produced in austenitic stainless steels by stress corrosion cracking and hydrogen embrittlement[J]. Metal Trans A, 1980, 11(9): 1563 – 1574.

Effect of Grain Boundary Character Distribution on Stress Corrosion Cracking in 304 Stainless Steel

Abstract: The proportion of low Σ ($\Sigma \leqslant 29$) coincidence site lattice (CSL) grain boundaries was increased to 75% by 5% cold rolling and subsequent annealing at 1 100 ℃ for 5 min. Stress corrosion cracking (SCC) susceptibility of 304 austenite stainless steel was evaluated through C-ring specimen tests conducted in acidified boiling 25% NaCl solution. The experimental results show that high proportion of low ΣCSL grain boundaries obtained by grain boundary engineering results in obvious improvement of the stress corrosion cracking resistance of 304 stainless steel in chloride solution.

低温退火对铁硅合金中立方织构形成的影响*

摘　要：利用光学显微镜、SEM、TEM 和 EBSD 分析研究了 Fe-3.2%Si 合金在低温初次再结晶退火和较低温度下二次再结晶退火后立方织构的形成. 结果表明：采用低温退火可以在初次再结晶组织中获得(100)[001]织构及(hk0)[001]织构，并且这种初次再结晶组织在较低的二次再结晶温度下可演化成比较集中的立方织构.

作为配电变压器和电动机铁芯材料的冷轧取向及无取向硅钢片是工业生产中一种重要的传统软磁材料. 随着现代通讯和信息技术的发展及能源需求的增长，如何进一步降低硅钢片的铁损和提高其磁感应强度仍然是当前科研中面临的迫切任务[1]. 铁硅合金晶体的[001]方向是易磁化方向，因此，除了提高硅的含量、减小片材厚度及控制磁畴结构等方法可以降低铁损外，通过冷轧和退火的办法获得[001]方向的择优取向是获得高性能硅钢片的首要问题[2-4]. 同高斯织构硅钢片相比，立方织构硅钢片的轧向和横向都是易磁化方向，两个方向都可以得到最大的磁感应强度，因而作为变压器铁芯材料不仅可简化铁芯制作工艺而且可显著减小变压器的能耗[6]. 立方织构硅钢片这种优异的磁性能在过去半个世纪中一直受到关注，并且取得了一定的进展[7].

铁硅合金中立方织构的形成研究总是与具体的制备或生产方法密切关联. 人们先后尝试并提出了不同的获得立方织构的方法：默比乌斯法(以大压下率(75%～80%)冷轧两次，再在 1 250 ℃～1 300 ℃ 高温退火)，阿斯穆斯法(又 50%～-60% 压下率进行多次冷轧，然后在使用吸气剂和催化剂的干氢中于 1 100 ℃～1 300 ℃ 最终退火)，通用电气公司生产法(顺柱状晶体轴线将板坯热轧，接着在高真空或干氢中作长时间高温(1 200 ℃～1 300 ℃)退火，并以 40% 的压下率进行冷轧)，李持曼法(将高斯织构带钢以 60%～90% 压下率冷轧，再在 1 200 ℃ 进行最终退火)，以及交叉轧制法. 并且在形成机制方面基本形成共识：立方织构是在严格控制的保护气氛中通过高温二次再结晶过程形成；为了获得立方织构，坯料(基体金属)必须有一定的取向(在大多数情况下，这种取向是高斯织构{110}〈001〉，同时也可含有其他的伴生织构，其中包括立方织构{100}〈001〉，及过渡性织构{112}〈110〉、{111}〈112〉、{210}〈001〉). 总之，在铁硅合金中获得立方织构的难易程度与最终样品厚度有密切的关系. 对于薄硅钢片(如 0.04 mm 厚)，立方织构的获得已在实验室得到较好的解决. 但对于适合于工业生产的较厚硅钢片，如何在其中获得高立方织构份额还处于探索之中. 尽管人们在控制形核和再结晶方面开展了大量的基础研究工作，并且得到了一些基本认识[7,8]，但对于不同的技术现有的理论认识尚难于指导立方织构硅钢片的工业生产. 因此，研究工作依然聚焦在如何在初次再结晶后获得一定数量(100)[001]取向的晶粒和有利于它们生长的基体织构[9]这一关键问题上.

本文从退火温度对初次再结晶组织和织构的影响入手，研究了铁硅合金中立方织构的

———————————————
* 本文合作者：王均安、王辉. 原发于《电子显微学报》，2010，29(5)：468-474.

形成问题.研究结果表明,通过对合金起始组织和一次冷轧-退火工艺的仔细控制,采用较低温度初次再结晶后再高温再结晶的方法可得到立方织构硅钢片,所采用的加工工艺可用于工业化生产.

1 实验材料及方法

选择含有对形成 γ 相有利的合金元素的 Fe-3.2% Si 铁硅合金成分,用 10 kg 真空感应炉熔炼成 Fe-3.2% Si 铁硅合金.合金铸锭经过热锻成为板坯后,通过已掌握的热轧工艺获得部分(110)[001]织构[10],再经过冷轧、中间退火和二次冷轧(试样最终厚度 0.23 mm)后,通过改变退火温度研究初次再结晶织构变化,发掘既含有(hk0)[001]织构同时又包含一定数量的(100)[001]晶粒的初次再结晶退火条件.在最终高温退火过程中,通过脱碳脱锰过程控制 γ→α 相变速率,利用部分 γ 相的存在来抑制基体晶粒的生长.同时,又利用脱碳退火过程中 γ 相从表层逐渐消失来促使(100)晶粒吞并基体晶粒而发生长大,获得高百分率的集中立方织构.利用光学显微镜和 SEM 观察试样的显微组织,使用 TEM 分析较高温度初次再结晶退火试样的形核机制,用 EBSD 方法表征织构的集中程度.

2 结果与讨论

2.1 初次再结晶织构的获得

采用形变-再结晶工艺在铁硅合金中获得立方织构的过程中,设法得到具有(100)[001]取向晶粒的初次再结晶组织是一个关键步骤.基于这样的考虑,本文在充分调整合金成分的基础上,通过对试样进行热轧(压下量>60%)-退火、一次冷轧(压下量>80%)-退火和二次冷轧(压下量>60%),以及改变二次冷轧试样的初次再结晶退火温度,得到了如图 1 所示的组织和织构变化.显而易见,在初次再结晶组织中存在一定份额的(100)[001]织构,而且初次再结晶晶粒尺寸随退火温度的升高逐渐增大.

二次冷轧试样在 600 ℃~1 050 ℃温度范围内进行初次再结晶后,在低温和高温下都可得到(100)[001]初次再结晶织构及(510)[001]、(410)[001]、(310)[001]取向的晶粒.在低温再结晶处理后晶粒取向主要分布在(100)-(111)连线上,呈带状;高温再结晶处理后晶粒取向主要分布在(100)、(111)周围,比较集中.

图 1 所示的初次再结晶织构对于体心立方结构晶体具有共性.这种织构的获得也得到了相关研究结果的支持(见表 1):具有(111)[11$\bar{2}$]取向的单晶通过形变和退火后可形成弱(hk0)[001]织构;(hk0)[001]取向的铁硅单晶体沿[001]经过>60%冷轧退火后,仍然可得到(hk0)[001]的再结晶织构,并且随着冷轧变形量增加或(hk0)偏离(110)程度的增加,再结晶织构(hk0)[001]会向(100)[001]趋近;具有(110)[001]织构的铁硅多晶体经过 53%~87%变形退火后,可得到(h10)[001]织构,其中 h 值由变形 53%时的 1 变化到变形 87%的 5,已非常接近(100)[001]织构;Mo、W、Nb 单晶的冷轧和再结晶织构也出现同样的变化规律.上述对单晶的研究结果较好地解释了本文对多晶铁硅合金所得到的形变-初次再结晶织构.在热轧-退火板的表面层中本文检测到了弱(110)[001]织构和{111}⟨112⟩基体织构[10].这种特征的织构组织经过一次冷轧和中间退火后,使得织构组分中的⟨112⟩和⟨001⟩方向更

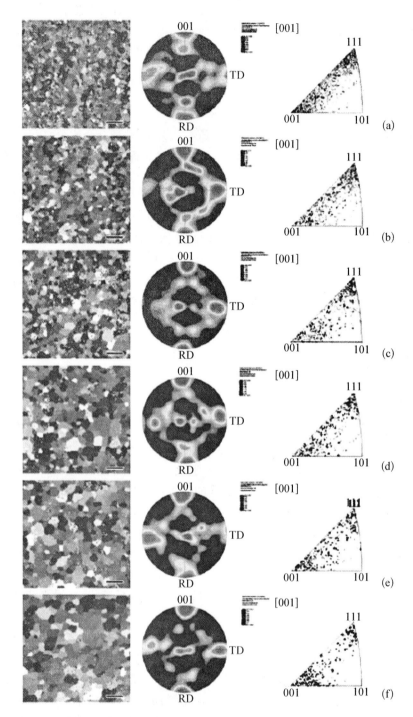

图1 不同温度下初次再结晶织构(依次为 OIM 图,(100)极图和 ND 面反极图). Bar = 100 μm   (a) 600 ℃;(b) 650 ℃;(c) 700 ℃;(d) 800 ℃;(e) 900 ℃;(f) 1 050 ℃

Fig. 1   Orientation image microscopy, (100) pole figure and inverse pole figure on ND plane of the specimens primarily annealed at (a) 600 ℃, (b) 650 ℃,(c) 700 ℃,(d) 800 ℃,(e) 900 ℃ and (f) 1 050 ℃. Bar = 100 μm

表 1　不同取向单晶经过形变和同位再结晶后形成的再结晶晶核及再结晶织构

Table 1　Nuclei and texture of different single crystals after deformation and recrystalliztion[11,12]

单　晶	压下量,%	加 工 织 构	再 结 晶 晶 核	再 结 晶 结 构
(111)[11$\bar{2}$]	70~85	(22$\bar{1}$)[114],(110)[001], (320)[001],(210)[001]	(22$\bar{1}$)[114],(110)[001], (320)[001],(210)[001], (310)[001],(410)[001], (100)[001]	(22$\bar{1}$)[114],(110)[001], (320)[001],(210)[001]
(110)[001]	60~85	{111}⟨112⟩+(110)[001]/ (530)[001]/(310)[001]	(110)[001],(410)[001]	(110)[001],(210)[001], (320)[001]+(210)[001]+ (110)[001]
(320)[001]	85	(210)[001],(310)[001]	(410)[001],(310)[001]	(410)[001],(310)[001]

加平行于轧向,为二次冷轧提供了比较理想的前驱织构. 当然,不同工艺阶段形变量的大小和退火温度对相应阶段的再结晶组织和织构的形成都有显著影响,并最终影响初次再结晶组织和织构.

在初次再结晶过程中(100)[001]取向晶粒的形核可能基于形变带亚晶界上形核,与单晶形变-再结晶的情况一致. 但在高温下,即使在短时间初次再结晶试样中也在形变的{111}晶粒中观察到了立方晶粒(见图 2). 这预示着高温下(100)[001]取向晶粒的形核还可能通过竞争形核的机制进行.

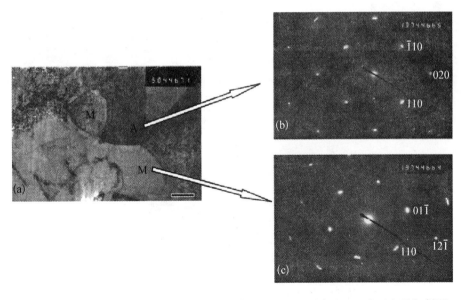

图 2　a:二次冷轧试样经 1 050 ℃~10 s 退火后的 TEM 照片;b:A 选区电子衍射图;
c:A晶粒周围 M 区的电子衍射图. Bar=200 nm

Fig. 2　TEM analysis result of specimens after secondary rolling and annealing at 1 050 ℃ for 10 s. (a) Micrograph, (b) and (c) selected diffraction patterns of grain A and the around grain M respectively. Bar=200 nm

(100)[001]取向晶粒的生长是由于(100)晶面低表面能的策动,这已被 Assmus 等[5]的工作所证实. 在合适的退火气氛条件的协同作用下,在合金的初次再结晶组织中形成了较强的(100)[001]织构和取向接近立方晶粒的弱(hk0)[001]织构(见图 1),这些为高温退火时得到立方织构都提供了比较理想的前驱织构[9].

2.2　最终退火过程中立方织构的形成

高温退火温度对立方织构的形成具有明显的作用.本文将低温初次再结晶退火后的试样再在高温下退火并进行 EBSD 分析(图 3)发现,在 900 ℃～1 000 ℃之间进行退火有利于获得比较集中的立方织构.退火温度过低或过高对于立方织构的获得都是不利的.高温退火温度对织构形成的影响主要归因于对再结晶速度的影响及对脱碳退火过程的影响.在低温下,脱碳退火速率低并且晶粒长大的速度缓慢,即使在有利于(100)[001]取向晶粒生长的气氛环境下,也较难得到集中的立方织构.在高温下,高的脱碳速率使得 γ 相失去了钉扎基体中 α 相的作用,基体中 α 晶粒的长大造成(100)[001]取向晶粒生长困难,最终也难于得到集中的立方织构.

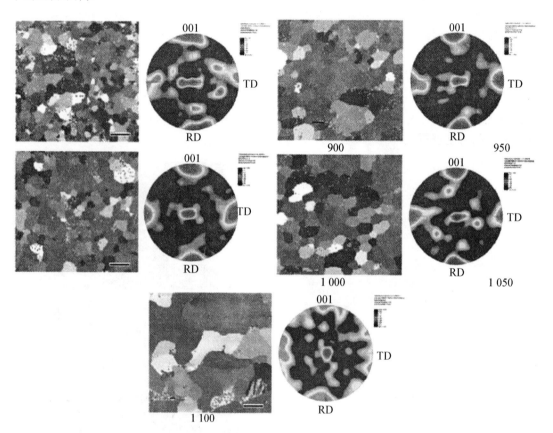

图 3　高温再结晶结构(依次为 OIM 图,(100)极图和 ND 面反极图). Bar＝500 μm

Fig. 3　Orientation image microscopy, (100) pole figure and inverse pole figure on ND plane of the high temperature annealed specimens. Bar＝500 μm

通过形变-退火得到期望的初次再结晶组织后,一般再经过控制二次再结晶过程就可得到具有柱状晶组织的立方织构[13].在这个过程中,既需要采用纯度较高的铁硅合金,利用氧、硫等少量杂质原子或分子在样品表面吸附,来降低(100)晶面的表面能而促使(100)晶粒生长.又需要通过添加某些稳定 γ 相的特定合金元素来控制 γ→α 相变,利用它们得到少量 γ 相可以阻碍基体 α 晶粒生长及在高温退火时还能逐渐去除而发相变的特性.(100)晶粒的生长和(100)[001]织构的形成取决于(100)"晶核"周围基体晶粒的取向、大小及(100)晶粒与基体晶粒之间晶界的迁移性能. 在 α-γ 双相共存的退火条件下, 位于样品表面下的 α 晶粒的生长受到 γ 相的制约, 而处于表面的晶粒由于 γ 相中的碳和锰容易脱掉而率先发生 γ→α 相变, 此时恰当地通过退火气氛、退火温度或其他手段控制 γ→α 相变速率就有可能使表面中的(100)晶粒优先生长. 当然, 这种使(100)晶粒生长的前提是保持(100)晶粒的表面能最低并且(100)晶粒与基体晶粒之间存在低 Σ 重位晶界(表2). 已长大的(100)晶粒随着退火时间的延长会进一步吞并周围较小的基体晶粒. 在这个过程中(100)晶粒与基体晶粒之间不一定必须存在低 Σ 重位晶界, 较小尺寸的基体晶粒被吞并主要是由于其与(100)晶粒的界面能差别造成的. 随着退火时间的延长和脱碳层深度的增大, (100)晶粒从样品表面向中心发展, 最终形成柱状晶体(如图4). 在这样的高温再结晶过程中, 初次再结晶组织和织构起着重要作用.

表 2　(100)晶粒与(hk0)及(111)晶粒之间可能的低 Σ 重位点阵晶界

Table 2　Possible low ΣCSLs boundaries between (100) grain and (hk0) and (111) grains

相 邻 晶 粒	转动轴/转动角度	低 ΣCSL
(110)/(100)	[100]/36.9°	Σ5
	[100]/22.6°	Σ13a
	[100]/28.1°	Σ17a
	[100]/16.3°	Σ25a
(210)/(100)	[100]/36.9°	Σ5
	[100]/28.1°	Σ17a
(320)/(100)	[100]/36.9°	Σ5
(310)/(100)	[100]/36.9°	Σ5
(410)/(100)	[100]/36.9°	Σ5
(111)/(100)	[110]/50.5°	Σ11

在过去的诸多研究中, 人们普遍认为短时间高温退火有利于形成(100)[001]取向晶核[7], 这种看法主要是基于高温退火为(100)[001]取向晶核的竞争形核提供了条件的假设和相应的实验结果. 但是短时间高温退火虽然能得到(100)[001]取向晶核和弱的(100)[001]织构, 但初次再结晶组织和织构却不利于最终立方织构的获得. 这主要是由于在高温下发生初次再结晶时基体晶粒粗化, 后者对二次再结晶退火时得到立方织构是不利的. 相比之下, 本文涉及的低温初次再结晶退火得到弱(100)[001]织构中, 由于(100)[001]取向晶粒的形核是经过充分的回复过程得到, 使得(100)[001]初次晶粒的取向更加集中. 低温退火过程中基体晶粒的生长受到了抑制, 有利于(100)[001]取向晶粒在二次再结晶退火中发展成为立方织构. 从图3中 OIM 图可以看出, 在(100)[001]取向晶粒份额较高的高温再结晶试样中, 依然有个别较大的(111)[112]晶粒存在, 这表明采用低温初次再结晶技术可以得到相对份额较高的立方织构硅钢片. 但获得完全的(100)[001]取向晶粒组织的工艺仍然需要进一步完善.

无疑, 试样初次再结晶组织中(100)[001]取向晶粒的多少和晶粒取向偏离度对最终高温再结晶退火后立方晶粒所占的份额以及立方织构的集中程度有重要影响. 基体晶粒的尺寸和取向也会影响最终立方织构的获得, 但其影响程度相对较弱. 在初次再结晶组织中存在

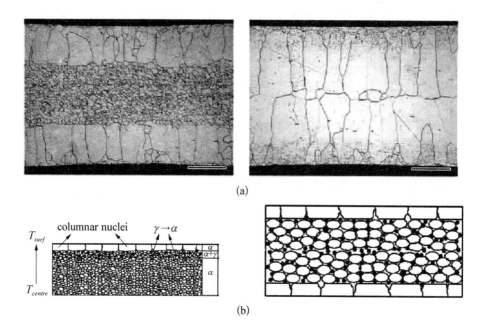

图 4 厚样品在高温退火过程中柱状晶粒结构的形成. Bar＝200 μm （a）高温退火 2 h 和退火结束后材料纵断面的金相组织；（b）高温退火过程中脱碳及柱状晶粒结构形成模型[13]

Fig. 4 Formation of column crystal structure in thick specimens during high temperature annealing. Bar＝200 μm （a）Microstructure of longitudinal section of specimens after 2 h（left）and long-time（right）high-temperature annealing；（b）Model of decarburization and formation of column crystal structure during high-temperature annealing[3]

相当数量的（100）[001]取向晶粒的前提下,无论基体晶粒组织中其他取向晶粒所占比例是否相同,都可以在高温退火后得到立方织构. 这也表明在二次再结晶过程中立方取向晶粒的生长与发展主要归因于退火工艺所造成的（100）[001]取向晶粒表面能的降低,立方织构份额的多少与初次再结晶组织中（100）[001]取向晶粒与其他取向晶粒构成的晶界类型有关.

3 结论

（1）对于经过形变和再结晶退火的多晶铁硅合金,其二次冷轧组织经过低温初次再结晶退火后可以演化成为包含{100}⟨001⟩,{510}⟨001⟩和{410}⟨001⟩织构组分的显微组织.

（2）利用低温初次再结晶可以在 0.23 mm 厚板材上在较低的二次再结晶温度下得到立方织构. 织构的形成是基于形变晶粒的回复和（100）[001]晶粒的定向长大. 这种技术对工业生产立方织构电工钢具有一定的参考价值.

参 考 文 献

［1］Kubot T，Fujikura M，Ushigami Y. Recent progress and future trend on grain-oriented silicon steel[J]. Journal of Magnetism and Magnetic Materials，2000，215－216(SI)：69－73.

［2］Masahashi N，Matuo M，WaLanabe K. Development of preferred orientation in annealing of Fe－3.25％ Si in a high magnetic field[J]. Journal of Materials Research，1998，13(2)：457－461.

［3］Sha Y H，Gao X Y，Dai C L, et al. Effects of electrostatic field on recrystallization texture and grain boundary character distinction in a Fe - 3% Si alloy[J]. Materials Science Forum, 2002, 408 - 412: 1329 - 1334.

[4] Mao W. Evolution of microstructure and {001}〈110〉 texture in electrical steel sheet[C]]//Proceedings of the 3rd China-Korea Joint Symposium on Advanced Steel Technology, Shanghai China: 2003: 109 - 114.

[5] Assmus F, Boll R, Ganz D, et al. Ueber Eisen-Silizium mit Wuerfeltexture[J]. Z Metallkde, 1957, 48(6): 341 - 349.

[6] Walter J L, Hibbard W R, Fiedler H C, et al. Magnetic properties of cube textured silicon-iron magnetic sheet[J]. Journal of Applied Physics, 1958, 29(3): 363 - 365.

[7] Tomid T, Uenoya S. Cube oriented 3% Si-1% Mn soft magnetic steel sheets with fine grain structure[J]. IEEE Tansactions on Magnetics, 2001, 37(4): 2318 - 2320.

[8] Elban W L, Hebbar M A, Kramer J J. Adsorption, surface energy and crystal growth in iron-3 silicon[J]. Metallurgical Transactions A (Physical Metallurgy and Materials Science), 1975, 6A(10): 1929 - 1937.

[9] 周邦新. 铁硅合金中形成立方织构的有关问题[J]. 宝钢技术, 2000, (5): 53 - 58.

[10] 王均安, 周邦新, 贺英, 等. 单道次热轧 Fe-3.2% Si 合金的泉微组织和织构[J]. 钢铁研究学报, 2006, 18(S2): 85 - 89.

[11] 周邦新. 钼单晶体的冷轧及再结晶织构[J]. 物理学报, 1963, 19(5): 297 - 305.

[12] 周邦新, 刘起秀. 钨和铌单晶体的冷轧及再结晶织构[J]. 金属学报, 1965, 8(3): 340 - 345.

[13] 王均安, 贺英, 邱振伟, 等. 小形变量轧制卜电工钢中立方织构的形成[J]. 上海大学学报(自然科学版), 2008, 14(5): 461 - 466.

Effect of low temperature annealing on the formation of cube texture in Fe-3.2% Si alloy

Abstract: Effect of low temperature annealing on the cubic texture formation in Fe-3.2% Si alloy was investigated by means of TEM and EBSD. The results show that the (100)[001] and (hk0)[001] texture components can be obtained in the primarily annealed Fe-3.2% Si alloy through low temperature annealing and the primary microstructure can be involved into relatively sharp cubic texture after subsequent lower temperature secondary recrystallization.

核反应堆压力容器模拟钢中富 Cu 纳米团簇析出早期阶段的研究*

摘　要：采用调质处理后热时效模拟方法，用原子探针层析成像技术研究了核反应堆压力容器模拟钢中富铜纳米团簇的析出过程. 模拟钢经 880 ℃加热水淬，660 ℃高温回火调质处理，并经 400 ℃时效处理 1 000 h 后基体中析出了富铜纳米团簇. 使用 MSEM(maximum separation envelope method)方法重点研究了富铜纳米团簇在析出早期阶段成分变化规律. 结果表明，富铜纳米团簇容易在镍含量较高的位置形核，并随着富铜纳米团簇中铜原子聚集程度的增加，纳米团簇中心处铜含量逐渐增加，镍含量逐渐减少；在纳米团簇与 α-Fe 基体界面处，镍和锰含量逐渐增加，形成了富镍和富锰包裹富铜纳米团簇的结构. 结合实验结果讨论了压力容器钢中合金元素镍及杂质元素磷会增加中子辐照脆化敏感性的原因.

国外的大量研究结果表明，核反应堆压力容器钢在服役工况下经过长期中子辐照后，钢中的杂质元素 Cu 会以富铜纳米团簇析出，而高数量密度的富铜纳米团簇的析出是引起反应堆压力容器钢辐照脆化的主要原因[1-4]. 核反应堆压力容器钢服役时的温度大约在 290 ℃，在此温度下析出的富铜团簇大小约 3 nm[5]，析出早期与 α-Fe 基体共格，为 bcc 晶体结构[6]. 要研究纳米团簇在析出早期的成分变化，原子探针层析成像(atom probe tomography，APT)是非常理想的工具[7-10].

采用中子辐照实验研究压力容器钢中富铜纳米团簇析出的实验费用昂贵，同时中子辐照后样品具有放射性，需要在有防护的条件下进行操作，因而许多研究者采用热时效模拟方法研究压力容器钢中富铜纳米团簇的析出过程，也获得了一些很有价值的实验结果[11-16]. Miller 等[11]对 Fe-1.1% Cu-1.4% Ni(原子分数)模拟钢分别在 300，400，500，550 和 600 ℃进行了长期时效处理；结果表明，模拟钢经 300 ℃/10 000 h、500 ℃/10 h 时效后，模拟钢有富铜原子团簇的析出. 研究还发现，经 500，550 和 600 ℃时效处理后模拟钢中 Cu 的实际溶解度要比 Cu 在 α-Fe 中的理论溶解度低 10%.

尽管很多研究者对核反应堆压力容器钢中富铜纳米团簇的析出过程做了大量的研究，但对析出早期纳米团簇中的成分变化规律仍缺乏了解. 本文利用 APT 研究了压力容器模拟钢热时效时富铜纳米团簇析出的早期阶段，分析了团簇中的成分变化，希望了解其他溶质原子对富铜纳米团簇析出的影响及其相互作用的规律.

1　实验方法

为了能够用热时效处理观察到富铜纳米团簇的析出过程，实验用的模拟钢是在压力容器钢(A508-Ⅲ)成分的基础上提高了 Cu 含量. 实验材料由真空感应炉冶炼，铸锭为 40 kg，其化学成分见表 1.

---

* 本文合作者：王伟、朱娟娟、林民东、刘文庆. 原发表于《北京科学技术大学学报》，2010，32(1)：39-43.

表 1   模拟压力容器钢的化学成分

Table 1   Composition of the model pressure vessel steel

参　数	Cu	Ni	Mn	Si	P	C	S	Mo	Fe
原子分数/%	0.53	0.81	1.60	0.77	0.03	1.00	0.011	0.31	其余
质量分数/%	0.60	0.85	1.58	0.39	0.016	0.22	0.006	0.54	其余

将钢锭热锻和热轧成 4 mm 厚的板坯,然后切成 40 mm×30 mm 的小样品,加热到 880 ℃保温 0.5 h 后水淬;随后在 660 ℃进行 10 h 高温回火即调质处理.最后将调质处理后样品在 400 ℃进行 100,300,1 000 和 2 000 h 等温时效处理.

为避免脱碳层的影响,用电火花线切割的方法在试样的中心部位切出边长为 0.5 mm 的方形细棒,棒长 20 mm,再采用电解抛光的方法将棒状样品制备成曲率半径小于 100 nm 的针尖状样品[17],然后用 APT(Oxford Nano Science 公司制造的三维原子探针)进行分析. APT 采集数据时,样品冷却至−193 ℃,脉冲电压频率为 5 kHz,脉冲分数为 20%,分析成分时所取层间距(grid spacing)为 0.1 nm,所得数据由专门的 Posap 软件分析[18].

目前普遍使用 Maximum Separation Envelope Method(MSEM)[19]来分析 APT 数据并界定纳米团簇的析出.其原理是认为纳米团簇是由一定数量的某溶质原子组成,并且该溶质原子的聚集程度要比在基体中聚集得更紧密.一般用最小的原子数目 N_{min} 来表征纳米团簇中某溶质原子的数量,如果该溶质原子的数量少于设定值,则不认为是纳米团簇的析出;用原子之间的最大距离 d_{max} 来表征纳米团簇中某溶质原子的聚集程度,如果某些溶质原子之间的距离大于设定值,则那些溶质原子将不计入该团簇中.因而,如果从小到大依次设置不同的 d_{max} 值时获得的多个溶质原子团簇(设定 N_{min} 不变),就可以知道这些团簇析出时的先后次序,因为溶质原子从偏聚到成为团簇析出时,团簇中溶质原子之间的距离会越来越接近.同样,如果从小到大依次设置不同的 N_{min} 值时获得的多个溶质原子团簇(设定 d_{max} 不变),也可以做出相似的判断.当然,也可以同时改变 d_{max} 和 N_{min} 的设定值.在对多个团簇之间析出先后次序的判断基础上,研究这些团簇中成分的变化,就可以了解某溶质原子从偏聚到析出过程中与其他溶质原子之间的关系.

纳米团簇的回旋半径(radius of gyration,l_g)可由下式得出:

$$l_g = \sqrt{\frac{\sum\limits_{i=1}^{n}(x_i-\bar{x})^2+(y_i-\bar{y})^2+(z_i-\bar{z})^2}{n}} \tag{1}$$

式中,x_i、y_i 和 z_i 为团簇中的溶质原子在三维空间的坐标,$\bar{x}$、$\bar{y}$ 和 $\bar{y}$ 为团簇质量中心的空间坐标,n 为团簇中的溶质原子总数.以上数据可以由软件分析后直接获得.

计算纳米团簇数量密度 N_v 用下式得出:

$$N_v = \frac{N_p \zeta}{n \Omega} \tag{2}$$

式中,N_p 是分析体积内所检测到的团簇数量;ζ 为检测效率,值为 0.6;n 为所收集的原子总数;Ω 是原子的平均体积,对于 bcc 结构的 Fe 为 $1.178×10^{-2}$ nm³,由于所收集的所有原子中主要为 Fe 原子,所以以 Fe 原子的体积来计算.

2 实验结果与讨论

为了能够研究富铜纳米团簇早期析出时的情况,从 APT 分析得到的结果中,选取了 400 ℃时效 100 h 和 1 000 h 两个样品. 图 1(a)和(b)分别是时效 100 h 的样品(显示的体积为 14 nm×16 nm×78 nm)和时效 1 000 h 的样品(显示的体积为 15 nm×16 nm×74 nm)中 Cu 原子的分布,其中每一个点代表测到的一个 Cu 原子. 当设置 $d_{max}=0.5$ nm、$N_{min}=10$ 时,在 100 h 时效处理后的样品中并未发现析出富铜纳米团簇,而经 1 000 h 时效处理后的样品中出现两个富铜纳米团簇(图 1(b)中箭头所示),这时团簇的数量密度 N_v 为 $6×10^{22}$ m^{-3},团簇的平均回旋半径 l_g 为 1.2±0.3 nm.

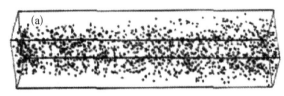

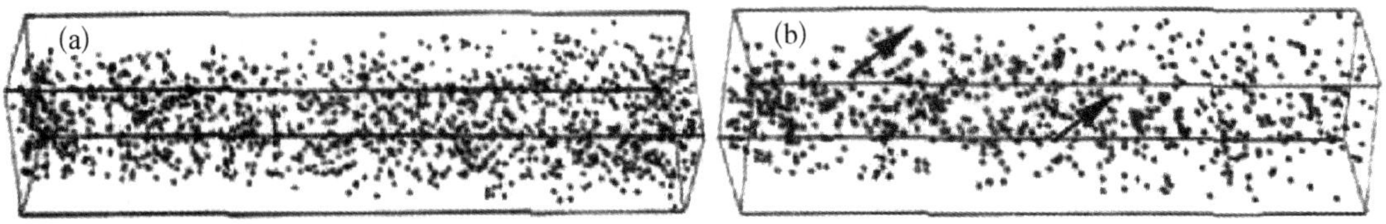

图 1 模拟钢调质处理后经 400 ℃/100 h(a)和 1 000 h(b)时效处理后样品中铜原子的分布图

Fig. 1 Copper atoms distribution of the model steel specimens after aging at 400 ℃ for 100 h (a) and 1 000 h (b)

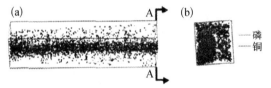

图 2 模拟钢调质后经 400 ℃/1 000 h 时效处理后 P 和 Cu 原子分布图(7 nm×6 nm×73 nm)(a);沿图(a)A—A 方向 P 和 Cu 原子分布图(7 nm×6 nm)(b)

Fig. 2 (a) Copper and phosphorus atoms distribution of the model steel aged at 400 ℃ for 1 000 h (7 nm×6 nm×73 nm);(b) copper and phosphorus atoms distribution in the direction of A-A in the Fig. (a)(7 nm×6 nm)

在经过 1 000 h 时效处理后样品中还观察到 P 原子的偏聚(图 2),文献[2]也报道过相似的现象,并认为 P 偏聚在位错处. 富磷区域的平均 P 的原子分数为 0.8%±0.01%,最高达到 1.5%,为基体 P 含量(0.03%P)的 27~50 倍. 图 3 为沿着磷偏聚方向 Cu 和 P 原子的含量曲线图. 在 P 的偏聚区中也发现了局部地区 Cu 的偏聚,Cu 原子发生偏聚之间的距离平均为 40 nm,这种偏聚可能成为富铜纳米团簇析出时的形核区. 这一结果表明杂质元素 P 的存在会促使富铜原子团簇的析出,而析出高数量密度的富铜纳米团簇是引起韧脆

转变温度升高的主要原因[1-4],因而杂质元素 P 有增大反应堆压力容器钢辐照脆化的倾向.

为研究富铜纳米团簇在析出早期阶段的成分变化,设置 $N_{min}=10$,但 d_{max} 值不同时,对 1 000 h 时效处理后样品中富铜纳米团簇的分布进行了分析,结果如图 4 所示. 当设置 $d_{max}=0.4$ nm 时,只检测到一个团簇;设置 $d_{max}=0.5$ nm 时,有两个团簇;当 d_{max} 增加到 0.7 nm 时,检测到第 3 个团簇;当 $d_{max}=0.8$ nm 时,检测到第 4 个团簇. 把随着设置 d_{max} 增大检测到的团簇依次标记为 4,3,2,1. 在形成富铜纳米团簇的过程中,Cu 原子由原来比较分散的状态逐渐聚集,团簇中原子之间的距离也逐渐减小,因而,从设置 d_{max} 由小到大时分别得到的四个团簇也是它们析出时的先后次序,团簇(4)形成更早一些,在这四个团簇中团簇(1)最晚形成. 表 2 为这四个团簇中检测到的 Cu 原子个数,团簇(2)和(4)中 Cu 原子的个数并没有随设

置 d_{max} 值增加而发生变化;团簇(3)在 $d_{max}=0.5$ 和 0.7 nm 时 Cu 原子个数为 18,当 $d_{max}=0.8$ nm时团簇中 Cu 原子个数为 20,变化也不大.

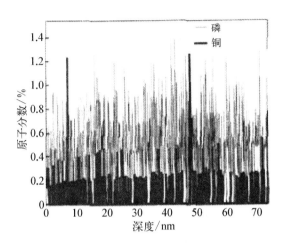

图 3 模拟钢调质后经 400 ℃/1 000 h 时效处理后样品富磷偏聚的 P 和 Cu 原子含量-深度曲线图(7 nm×6 nm×73 nm)

Fig. 3 Composition profile of copper and phosphorus in the model steel after thermal refining and then aging at 400 ℃ for 1 000 h(7 nm×6 nm×73 nm)

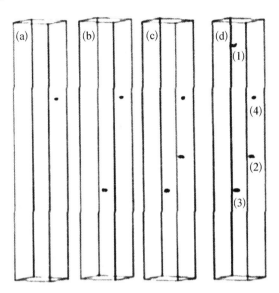

图 4 经 400 ℃/1 000 h 时效处理后设置不同 d_{max} 值时样品中富铜纳米团簇在三维空间分布图(图中基体溶质原子已根据 MSEM 分析被去除)(15 nm×16 nm×74 nm). (a) 0.4 nm;(b) 0.5 nm;(c) 0.7 nm;(d) 0.8 nm

Fig. 4C Three-directional distribution of Cu-rich nano-clusters in the model steel aged at 400 ℃ for 1 000 h for different d_{max} values (the matrix solute atoms have been eliminated based on the MSEM) (15 nm×16 nm×74 nm): (a) 0.4 nm;(b) 0.5 nm; (c) 0.7 nm;(d) 0.8 nm

表 2 不同 d_{max} 对应不同富铜纳米团簇中 Cu 原子个数($N_{min}=10$)

Table 2 Number of Cu atoms in Cu-rich nano-clusters for different values of separation distance d_{max} ($N_{min}=10$)

富铜纳米团簇	Cu 原子个数			
	$d_{max}=0.4$ nm	$d_{max}=0.5$ nm	$d_{max}=0.7$ nm	$d_{max}=0.8$ nm
(1)	—	—	—	13
(2)	—	—	13	13
(3)	—	18	18	20
(4)	24	24	24	24

图 5 为设置 $d_{max}=0.8$ nm 时富铜纳米团簇(1)~(4)中 Cu、Ni 和 Mn 的径向含量—深度曲线图,表 3 为这几个富铜纳米团簇中以及团簇与 α-Fe 基体界面处 Cu、Ni 和 Mn 元素含量. 从图 5 以及表 3 中可以看出,富铜纳米团簇先在 Ni 含量较高的位置形核(团簇 1),随设置 d_{max} 的减小,即随着团簇中 Cu 原子聚集程度的增加,团簇中心处 Cu 含量逐渐增加;同时可以看出,随着团簇中 Cu 原子聚集程度的增加,Ni 原子逐渐从团簇中心处向外围迁移,中心处 Ni 含量逐渐减少,而 Mn 的含量变化不大;在团簇与 α-Fe 基体界面处,Ni 和 Mn 含量

逐渐增加,在团簇中 Cu 原子平均间距达到 0.4 nm 时,在富铜纳米团簇的周围出现 Ni 和 Mn 的富集,形成富 Ni 和富 Mn 层包裹富铜纳米团簇的结构,这个结果与以前文献报道的研究结果相似[2].

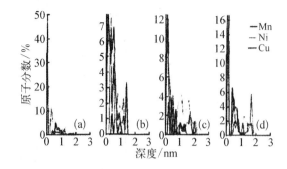

图 5 富铜纳米团簇(1)~(4)中 Cu、Ni 和 Mn 的径向含量-深度曲线图. (a) 团簇(1);(b) 团簇(2);(c) 团簇(3);(d) 团簇(4)

Fig. 5 Radial concentration profiles of Cu, Ni, and Mn elements in Cu-rich nano-clusters: (a) cluster(1);(b) cluster(2);(c) cluster(3); (d) cluster(4)

表 3 设置 d_{max} 值不同时得到的几个富铜纳米团簇中以及团簇与 α-Fe 界面处 Cu,Ni 和 Mn 元素的原子分数

Table 3 Contents of Cu, Ni and Mn elements in different Cu-rich clusters cores and cluster/matrix interfaces　　　　　　　%

富铜团簇	Cu	Ni		Mn	
	团簇中心	团簇中心	团簇/基体界面	团簇中心	团簇/基体界面
(1)	5.0±0.07	11.5±0.08	1.3±0.05	1.3±0.07	2.5±0.09
(2)	7.8±0.03	7.5±0.02	2.3±0.02	1.3±0.02	3.3±0.02
(3)	12.5±0.02	6.5±0.02	3.7±0.02	3.5±0.02	2.5±0.02
(4)	16.5±0.05	5.0±0.04	6.0±0.03	2.0±0.04	7.5±0.04

核反应堆压力容器钢添加 Ni 元素的主要目的是提高钢的淬透性以及降低服役前钢的韧脆转变温度,而 Ni 有增大钢的辐照脆化倾向[2]. 用 APT 分析可以观察到提高镍含量后会促使富铜原子团簇的析出,但是其原因并不十分清楚[20-21]. 本研究发现,富铜原子团簇析出时会在 Ni 含量较高处形核,因而提高钢中的镍含量会促进富铜原子团簇的析出,增大了核反应堆压力容器钢的辐照脆化敏感性.

Ni 和 Mn 在富铜原子团簇与基体界面处的偏聚可以通过不同溶质原子之间的相互作用来解释. 首先,在 α-Fe 基体中存在 Ni 的浓度变化,文献[21]中提出 Ni 和 Mn 存在协同效应,本文认为在这种协同效应的作用下,具有较高扩散系数的 Cu 原子会向富 Ni 位置扩散,在富 Ni 溶质原子周围形成 Cu 原子的聚集;其次,在 400 ℃时 Cu 在 α-Fe 中的平衡固溶度非常小,过饱和固溶的 Cu 趋于析出而形成纳米团簇. 低的平衡固溶度对应高的富铜纳米团簇/α-Fe 基体界面能;相比之下,400 ℃时 Ni 在 α-Fe 中的平衡固溶度要大得多,因此 Ni/α-Fe 基体的界面能要远小于富铜纳米团簇/α-Fe 基体界面能;而 Cu/Ni/α-Fe 的过渡界面具有比 Cu/α-Fe 更低的界面能[22]. 所以,在富铜原子团簇形成时,Ni 容易从团簇中心扩散到富铜纳米团簇/α-Fe 基体界面处形成偏聚,同时由于 Ni 和 Mn 存在协同效应而使得 Mn 与 Ni 之间具有很强的结合力,因此在富集 Ni 的界面处也会出现 Mn 的富集. 最终形成了富 Ni 和富 Mn 包裹富铜纳米团簇的结构.

3 结论

使用 APT 方法研究了核反应堆压力容器模拟钢中富铜纳米团簇析出早期阶段的成分

变化规律.研究结果表明,富铜纳米团簇容易在镍含量较高的位置形核,并随着富铜纳米团簇中 Cu 原子聚集程度的增加,纳米团簇中心处铜含量逐渐增加,Ni 原子逐渐从团簇中心处向外围迁移,中心处 Ni 含量逐渐减少;在纳米团簇与 α-Fe 基体界面处,镍和锰含量逐渐增加,形成了富 Ni 和富 Mn 包裹富铜纳米团簇的结构.

参 考 文 献

[1] Toyama T, Nagai Y, Tang Z, et al. Nanostructural evolution in surveillance test specimens of a commercial nuclear reactor pressure vessel studied by three-dimensional atom probe and positron annihilation. *Acta Mater*, 2007, 55: 6852.

[2] Miller M K, Russell K F, Sokolov M A, et al. APT characterization of irradiated high nickel RPV steels. *J Nucl Mater*, 2007, 361: 248.

[3] Fujii K, Fukuya K, Nakata N, et al. Hardening and microstructural evolution in A533B steels under high-dose electron irradiation. *J Nucl Mater*, 2005, 340: 247.

[4] Pareige P, Radiguet B, Suvorov A, et al. Three-dimensional atom probe study of irradiated, annealed and re-irradiated VVER 440 weldmetals. *Surf Interface Anal*, 2004, 36: 581.

[5] Nagai Y, Hasegawa M, Tang Z, et al. Positron confinement inultrafine embedded particles: Quantum-dot-like state in an Fe-Cu alloy. *Phys Rev B*, 2000, 61(10): 6574.

[6] Golubov S I, Osetsky Yu N, Serra A, et al. The evolution of copper precipitates in binary Fe-Cu alloys during ageing and irradiation. *J Nucl Mater*, 1995, 226: 252.

[7] Zhou B X, Liu W Q. The application of 3DAP in the study of materials science. *Mater Sci Technol*, 2007, 15(3): 405.
(周邦新,刘文庆.三维原子探针及其在材料科学研究中的应用.材料科学与工艺,2007,15(3):405)

[8] Stephenson L T, Moody M P, Liddicoat P V, et al. New techniques for the analysis of fine-scaled clustering phenomena with in atom probetom ography (APT) data. *Microsc Microanal*, 2007, 13: 448.

[9] Karnesky R A, Isheim D, Seidman D N. Directmeasurement of two-dimensional and three-dimensional interprecipitate distance distributions from atom-probe tomographic reconstructions. *Appl Phys Lett*, 2007, 91: Article No. 013111.

[10] Cerezo A, Clifton P H, Lozano-Perez S, et al. Overview: recent progress in three-dimensional atom probe instruments and applications. *Microsc Microanal*, 2007, 13: 408.

[11] Miller M K, Russell K F, Pareige P, et al. Low temperature copper solubilities in Fe-Cu-Ni. *Mater Sci Eng A*, 1998, 250: 49.

[12] Worrall G M, Buswell J T, English C A, et al. A study of the precipitation of copperparticles in a ferrite matrix. *J Nucl Mater*, 1987, 148: 107.

[13] Pareige P J, Russell K F, Miller M K. APFIM studies of the phase transform ationsin thermally ages ferritic FeCuNi alloys: Comparison with aging under neutron irradiation. *Appl Surf Sci*, 1996, 94/95: 362.

[14] Barbu A, Mathon M H, Maury F, et al. A comparison of the effect of electron irradiation and of thermal aging on the hardness of FeCu binary alloys. *J Nucl Mater*, 1998, 257: 206.

[15] Kuri G, Degueldre C, Bertsch J, et al. Local atomic structure in iron copper binary alloys: An extended X-ray absorption fine structure study. *J Nucl Mater*, 2007, 362: 274.

[16] Christien F, Barbu A. Modelling of copper precipitation in iron during therm alaging and irradiation. *J Nucl Mater*, 2004, 324: 90.

[17] Miller M K. *Atom Probe Tomography: Analysis at the Atomic Level*. New York: Kluwer Academic/Plenum Publishers, 2000: 25.

[18] Liu Q D, Chu Y L, Wang Z M, et al. 3D atom probe characterization of alloying elements partitioning in cementite of Nb-V microalloying steel. *Acta Metall Sin*, 2008, 44(11): 1281.
(刘庆冬,褚于良,王泽民,等. Nb-V 微合金钢中渗碳体周围元素分布的三维原子探针表征.金属学报,2008,44(11):

1281)

[19] Miller M K, Kenik E A. Atom probe tomography: a technique for nanoscale characterization. *Microsc Microanal*, 2004, 10: 336.

[20] Cerezo R A, Hirosawa S, Rozdilsky I, et al. Combined atomicscale modeling and experimental studies of nucleation in the solid state. *Philos Trans R Soc London A*, 2003, 361: 463.

[21] IAEA-TECDOC-1441 *Effects of Nickel on Irradiation Embrittlement of Light Water Reactor Pressure Vessel Steels*. Vienna: IAEA, 2005.

[22] Liu C L, Odette G R, Wirth B D, et al. A lattice Monte Carlo simulation of nanophase compositions and structures in irradiated pressure vessel Fe-Cu-Ni-Mn-Si steels. *Mater Sci Eng A*, 1997, 238: 202.

Study on the Early-Stage of Copper-Rich Nano-Clusters Precipitation in Model Nuclear Reactor Pressure Vessel Steel

Abstract: Early-stage for mation of Cu-rich nano-clusters in thermal aging nuclear reactor model pressure vessel steel quenched and tempered was investigated by atom probe tomography (APT). After the initial heat treatment at 880 ℃ for 0.5 h and water quenching, the materials were tempered for 10 h at 660 ℃ and aircooled. Cu-rich nano-clusters were observed in the samples aged at 400 ℃ for 1 000 h. The change in composition of the steel in the early-stage of precipitation of Cu-rich nano-clusters was studied by the maximum separation envelope method (MSEM). Cu nanoclusters are observed to form from high nickel regions. The Cu concentration increases and the Ni concentration decreases at the central cores with increasing Cu atoms congregation. Ni and Mn atoms aggregation on the exterior side of the cluster/matrix interface is also evident. Based on experimental results, the reason that Ni and P can increase the sensitivity to neutron irradiation embrittlement of the nuclear reactor pressure vessel steel is discussed.

304 不锈钢中"晶粒团簇"显微组织的特征与晶界特征分布的关系[*]

摘　要：采用电子背散射衍射(EBSD)及取向成像显微(OIM)技术研究了冷轧变形量对再结晶后 304 不锈钢显微组织及晶界特征分布(GBCD)的影响. 结果表明，经过冷轧变形 5% 及高温(1 100 ℃)短时间(5 min)再结晶退火可将 304 不锈钢的低 ΣCSL($\Sigma \leqslant 29$)晶界比例提高到 75%(Palumbo-Aust 标准)以上，并形成了大尺寸"互有 $\Sigma 3^n$ 取向关系晶粒的团簇". 这是提高低 ΣCSL晶界比例后显微组织的重要特征. 对晶粒团簇中的晶粒进行孪晶链及取向孪晶链的分析，可以确定晶粒团簇内任何两个晶粒之间的多重孪晶阶次. 在一个被分析的团簇内，观察到两个晶粒之间最高为 $\Sigma 3^8$ 取向关系. 随着再结晶退火前冷轧变形量的增加，晶粒团簇的尺寸减小，同时低 ΣCSL 晶界的比例也下降.

1984 年 Watanabe[1]首次提出了"晶界设计"(grain boundary design)的概念，并于上世纪 90 年代发展为晶界工程研究领域. 通过形变及退火处理，提高低 ΣCSL(coincidence site lattice, CSL[2]，即重位点阵，一般认为 $\Sigma \leqslant 29$ 的晶界为低 ΣCSL 晶界)晶界比例，调整晶界特征分布(grain boundary character distribution, GBCD)，从而可以改善材料与晶界有关的多种性能. 镍基合金[3-6]、铅合金[7,8]、奥氏体不锈钢[9-14]等多种低层错能面心立方金属材料，通过合适的形变及退火工艺，能够大幅度提高 $\Sigma 3^n$($n = 1$，2，$3 \cdots \cdots$)晶界比例. Lin 等[6]将 Inconel600 合金的低 ΣCSL 晶界比例提高到 60%～70% 后，腐蚀速率可降低 30%～60%. Palumbo 等[7,8]将晶界工程处理过的铅基合金作为铅酸蓄电池栅板，可使铅酸蓄电池的循环使用寿命提高 2～4 倍.

304 不锈钢是一种低层错能面心立方金属材料，可用晶界工程的方法进一步提高其性能. Lin[9]等将冷轧 10%～67% 的 304 不锈钢在 1 000 ℃退火 5 min，低 Σ-CSL 晶界比例最高达到 62.5%. Thaveeprungsriporn 等[10]将 304 不锈钢变形 3% 后，在 950 ℃退火 10 min，并重复进行 3 次. 3 次处理后得到的低 ΣCSL 晶界比例分别为 47%、33%、57%. Shimada[11]等将冷轧 5% 的 304 不锈钢在 927 ℃退火 72 h 后，低 ΣCSL 晶界比例超过 80%，晶间腐蚀速率下降了约 75%，方晓英等[12]运用类似的工艺使 304 不锈钢的低 ΣCSL 晶界比例得到大幅提高. Kumar[13]研究了大变形量对 304 不锈钢 GBCD 的影响：经 60% 的变形后，样品分别在 700 ℃、900 ℃ 和 1 000 ℃退火 1 h，得到的低 ΣCSL 晶界比例分别为 20%、28.3%、34.5%，可见大变形量对样品退火后低 ΣCSL 晶界比例的提高没有促进作用.

虽然关于提高 304 不锈钢低 ΣCSL 晶界比例的工艺方法已有大量研究报道，但最后的退火温度都与固溶处理的温度不一致. 在实际工业应用中，为了获得满意的力学性能和耐腐蚀性能，通常在成材后需要对 304 不锈钢进行固溶及时效处理. 研究冷轧变形及在固溶处理温度退火如何提高低 ΣCSL 晶界比例的工艺方法，可以使研究成果的推广与目前工业生产的工艺参数衔接. 所以，本文采用 1 100 ℃退火处理，研究提高 304 不锈钢低 ΣCSL 晶界比例

---

* 本文合作者：夏爽、罗鑫、陈文觉、李慧. 原发表于《电子显微学报》，2010，29(1)：678 - 683.

的工艺方法,并分析了显微组织的特征.

1 实验方法

将 304 不锈钢在 1 100 ℃保温 30 min,然后立即淬入水中进行固溶处理,作为研究材料的初始状态. 将固溶处理后的材料用电火花线切割为 30 mm×6 mm×2 mm 的条状样品,然后进行不同压下量的冷轧变形,最后在 1 100 ℃退火.

退火处理后的样品采用砂纸研磨和电解抛光的方法制备符合 EBSD 分析要求的表面. 电解抛光液为:20%HClO$_4$+80%CH$_3$COOH,抛光电压为直流 40 V,时间约 20 s.

利用配有 TSL-EBSD 系统的扫描电镜对电解抛光后的样品表面进行逐点逐行扫描,得到材料表面各点的取向,通过晶界两侧晶粒的取向差判定晶界类型. 由 OIM 系统自动统计不同类型晶界的长度比例. EBSD 测试步长根据晶粒大小不同而改变,在 3~7 μm 之间. 本文采用 Palumbo-Aust 标准[15]:$\Delta\theta_{max} = 15°\sum^{-5/6}$ 判定晶界类型.

2 结果与讨论

对固溶处理的 304 不锈钢进行 EBSD 测试,样品的 OIM 图和(001)极图如图 1 所示. 样品经固溶处理后,大部分 Σ3 晶界(孪晶界)以单独的直线或直线对的形态出现在晶粒内部,如图 1a 示. 该样品的晶界特征分布统计为:Σ1:5.8%,Σ3:41%,Σ9+Σ27:1.9%,其他低 ΣCSL:1%. 低 ΣCSL 晶界比例之和为 49.6%. 从图 1b 极图中(001)极点的分布可见,经固溶处理后的样品没有织构.

固溶处理后的样品分别经 5%、8%、15%、25% 和 40% 的冷轧变形,然后在 1 100 ℃退火

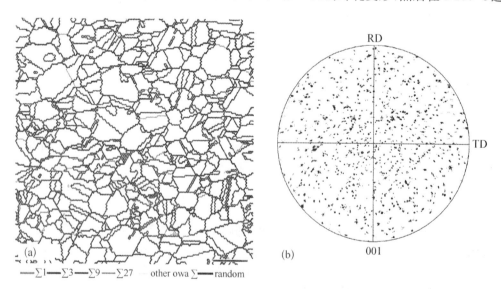

图 1 固溶处理后(1 100 ℃/30 min+水淬)样品的显微组织 (a) 不同类型晶界的 OIM 图; (b) 样品的(001)极图;Bar=100 μm

Fig. 1 Microstructure of the starting material (solution annealed: 1 100 ℃/30 min+water quench) (a) OIM map of different types of boundaries; (b) (001) pole figure; Bar=100 μm

5 min. 图 2 给出了不同冷轧变形量对 1 100 ℃ 再结晶退火后 304 不锈钢 GBCD 的影响. 经 5% 的冷轧变形和 1 100 ℃ 再结晶退火 5 min 处理后,低 ΣCSL 晶界比例提高到 75% 以上. Σ3 晶界(孪晶界)比例接近 60%,Σ9＋Σ27 晶界比例约为 10%. 随着形变量的增大,Σ3、Σ9＋ Σ27 晶界比例均降低,因而使得总体低 ΣCSL 晶界比例降低. 当冷轧变形量大于 15% 后,低 ΣCSL 晶界比例变化不大,都在 45% 左右. Σ1 晶界和其他低 ΣCSL 晶界(非 Σ3ⁿ 类型晶界) 的比例都很低,随冷轧变形量不同并且没有明显的变化规律.

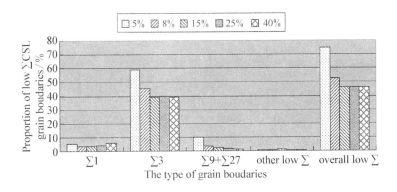

图 2 冷轧变形量与 1 100 ℃ 退火 5 min 后 304 不锈钢 GBCD 的关系

Fig. 2 The relationship between cold rolling reduction and GBCD of 304 stainless steel after recrystallization annealing (1 100 ℃/5 min)

图 3 给出冷轧 5%、15% 和 40% 后,在 1 100 ℃ 退火 5 min 样品的 OIM 图. 从图 3a 可见, Σ3 晶界与其他 Σ3ⁿ 晶界相连构成 Σ3-Σ3-Σ9 或 Σ3-Σ9-Σ27 的三叉界角(triple junction),晶粒之间都具有 Σ3ⁿ 的取向关系,从而构成"互有 Σ3ⁿ 取向关系晶粒的团簇"[4,5,16],以下简称为晶粒团簇. 图 3a 中标出的"A"和"B"是两个典型的晶粒团簇. 晶粒团簇之间是一种基本连续的随机晶界网络(Palumbo-Aust 标准). 下面对晶粒团簇"A"的特征进行详细分析.

在图 4a 所示的二维晶粒组织中,晶粒团簇"A"包含了 65 个晶粒. 在该晶粒团簇中,从任何一个晶粒都可以只穿过孪晶界到达任何另一个晶粒,比如从晶粒 1 到晶粒 33 可以连成如下路线:1 - 4 - 5 - 6 - 44 - 58 - 50 - 19 - 21 - 52 - 33. 图 5a 是对这个晶粒团簇中的晶粒进行孪晶链分析的结果,图中的直线只连接相邻且具有孪晶关系的晶粒(Σ3 取向关系并且偏差角小于 1° 的晶界). 从图可见,晶粒团簇中的每一个晶粒都是整个孪晶链中的一个环节. Haasen[17] 等在上世纪七八十年代关于低层错能面心立方金属再结晶过程中多重孪晶现象已有详尽研究. 大量实验结果[18]表明,孪晶链是再结晶过程中由源自一个再结晶晶核长大时连续产生退火孪晶的多重孪晶过程形成的. 不同阶次孪晶之间长大相遇就形成了 Σ3ⁿ 晶界[18,19]. 经 OIM 软件分析,晶粒团簇"A"截面中观察到的 65 个晶粒可以归纳为 23 种取向(团簇内取向差小于 1° 的晶粒判定为同一取向),如图 4b,4c 所示. 图 5b 给出这 23 种取向的取向孪晶链,图中的直线只连接互有孪晶关系两个取向,并且这两个取向在晶粒团簇中至少有一处相邻. 通过取向孪晶链可知每两种取向间 Σ3ⁿ 关系的 n 值. 比如取向 A 和取向 R 间的孪晶链是:A-C-B-G-K-T-R,它们之间经过了 6 次孪晶关系,就是 Σ3⁶ 关系. 那么前面所说的晶粒 1 属于取向 A,晶粒 33 属于取向 R,这两个晶粒之间就是 Σ3⁶ 关系. 该晶粒团簇内最高存在 Σ3⁸ 取向关系,比如取向 S 和 P 之间,取向 I 和 P 之间都是 Σ3⁸ 关系. 图 3a 中实心箭头指向的晶界片段是 Σ1、Σ3 和 Σ29,虽然它们都是低 ΣCSL 晶界,但是经 OIM 软件分析,它

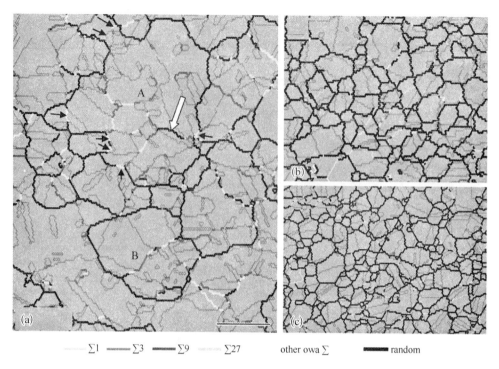

———— Σ1 ———— Σ3 ———— Σ9 ———— Σ27   other owa Σ   ▬▬▬ random

图 3 不同变形量冷轧及在 1 100 ℃退火 5 min 后 304 不锈钢中不同类型晶界的 OIM 图 （a）冷轧 5%；(b) 冷轧 15%；(c) 冷轧 40%；Bar＝200 μm

Fig. 3 OIM maps of different types of boundary in the specimens with different cold rolling and annealing at 1 100 ℃ for 5 min. Cold rolling reduction is (a) 5%, (b) 15% and (c) 40%. Bar＝200 μm

们的相对偏差（实测偏差角与标准规定最大允许偏差角的比值[20]）都接近于 1（Palumbo-Aust 标准）. 而团簇内晶界的相对偏差在 0.2 左右，所以这些出现在晶粒团簇 A 的边界上的低 ΣCSL 晶界，与晶粒团簇内部通过多重孪晶形成的 Σ3^n 晶界的必然性有本质的不同. 它们的出现是随机偶然的，该问题已在文献[4]中详细讨论. 而指向晶粒团簇 A 内部"随机晶界"的空心箭头是晶粒 44 和 34 之间的 Σ3^4 晶界（取向差：60.3°[19 19 14]/Σ81d）. 这虽然属于随机晶界，但仍然是 Σ3^n 晶界中的一种. 图 5a 和 5b 方框中空缺的晶粒或取向连接了两边互有 Σ9 取向关系的晶粒或取向. 出现这种情况很可能是因为在样品观察截面的上方或下方，仍然存在一个没有观察到的晶粒. 该晶粒与"方框"两侧的晶粒或取向仍然保持 Σ3 的取向关系，因为再结晶时的多重孪晶过程是在三维空间中发展的. 通过这样的分析，晶粒团簇 A 中 65 个晶粒之间的多重孪晶链（图 5a），以及 23 个取向之间的孪晶链（图 5b）仍然是完整的.

    图 6 给出不同冷轧变形量与再结晶退火后晶粒团簇大小（由 OIM 软件统计）的关系. 样品在完成再结晶后，冷轧 5%样品的晶粒团簇尺寸最大，团簇内包含的 Σ3^n 晶界也最多. 再结晶时单位体积中的形核数量（形核密度）与形变量有关，形变量越小，形核密度就越低[21]. 晶核有较大的空间生长，一般大角晶界可以迁移较长的距离. 对于低层错能的金属材料来说，在晶界迁移过程中产生退火孪晶的几率高，多重孪晶的过程可以得到充分发展，从而形成很长的孪晶链，这样就能提高 Σ3^n 晶界的比例，形成大尺寸的晶粒团簇. 按照晶粒团簇的定义，图 3b 和 3c 中经过冷轧 15%和 40%退火后的显微组织也是由晶粒团簇构成，不过晶粒团簇的尺寸小，Σ3^n 晶界的比例也低. 随着再结晶退火前形变量的增加，晶粒团簇的尺寸减小，

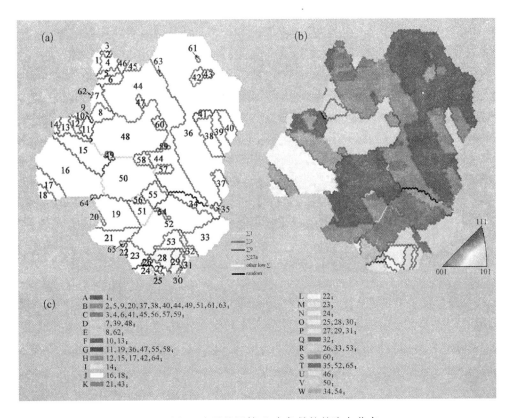

图 4 图 3a 中晶粒团簇 A 内各晶粒的取向分布

Fig. 4 Orientation distribution of grains in grain-cluster A of Fig. 3a Grains are numbered in(a)；(b) Orientation contrasted OIM maps of the grains in this grain-cluster；Grains with the same orientation（The orientations with difference smaller than 1° were here recognized as same orientation）are listed in(c)

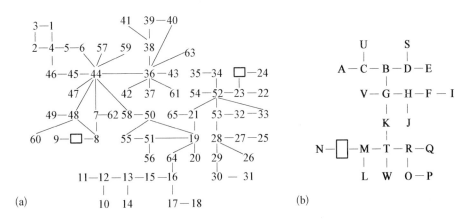

图 5 图 4a 中晶粒团簇 A 的孪晶链分析 （a）65 个晶粒之间的孪晶链；(b) 各种晶粒取向之间的孪晶链

Fig. 5 Twin chain analysis of the large grain-cluster "A" in Fig. 4a （a）Twin chain of the 65 grains；(b) Twin chain of the 23 orientations

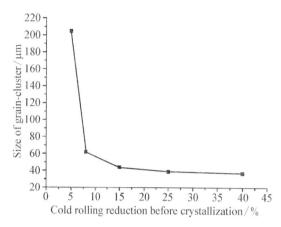

图 6 冷轧变形量对再结晶退火后晶粒团簇尺寸的影响

Fig. 6 Effect of cold rolling reduction on the size of grain-cluster after recrystallization

$\Sigma 3^n$ 晶界比例降低,如图 6,图 3 及图 2 所示. 由于形变量大,再结晶形核密度高,晶核生长的空间小,在多重孪晶过程没有得到充分发展时,晶粒团簇之间相遇,完成了再结晶过程. 因而再结晶退火前的冷轧变形量对控制 304 不锈钢的 GBCD 起了关键作用.

4 结论

在 304 不锈钢中,形成大尺寸的"互有 $\Sigma 3^n$ 取向关系晶粒的团簇"是低 Σ CSL 晶界比例提高后显微组织的重要特征. 晶粒团簇内所有晶粒之间,不论是否相邻都保持 $\Sigma 3^n$ 的取向关系. 这种晶粒团簇是再结晶过程中通过一个再结晶晶核长大时发生多重孪晶形成孪晶链的结果,随着再结晶退火前冷轧变形量的增加,晶粒团簇的尺寸减小,同时低 Σ CSL 晶界的比例也下降.

参 考 文 献

[1] Watanabe T. Approach to grain boundary design for strong and ductile polycrystals[J]. Res Mech, 1984, 11: 47 - 84.

[2] 夏爽,周邦新,陈文觉. 690 合金的晶界特征分布及其对晶间腐蚀的影响[J]. 电子显微学报,2008,27(6): 461 - 468.

[3] Xia S, Zhou B X, Chen W J, Wang W G. Effects of strain and annealing processes on the distribution of $\Sigma 3$ boundaries in a Ni-based superalloy[J]. Scripta Materialia, 2006, 54: 2019 - 2022.

[4] Xia S, Zhou B X, Chen W J. Effect of single-step strain and annealing on grain boundary character distribution and intergranular corrosion in Alloy 690[J]. Journal of Materials Science, 2008, 43: 2990 - 3000.

[5] Xia S, Zhou B X, Chen W J. Grain-cluster microstructure and grain boundary character distribution in Alloy 690[J]. Metall and Mater Trans A, 2009, revised.

[6] Lin P, Palumbo G, Erb U. Influence of grain boundary character distribution on sensitization and intergranular corrosion of alloy 600[J]. Scripta Metall Mater, 1995, 33: 1387 - 1391.

[7] Lehockey E M, Limoges D, Palumbo G, et al. On improving the corrosion and growth resistance of positive Pb-acid battery grids by grain boundary engineering[J]. J Power Sour, 1999, 78: 79 - 83.

[8] Palumbo G, Erb U. Enhancing the operationg life and performance of lead-acid batteries via grain-boundary engineering[J]. MRS Bulletin, 1999, (24): 27 - 32.

[9] Lin P, Palumbo G, Aust K T. Experimental assessment of the contribution of annealing twins to CSL distribution in FCC materials[J]. Scripta Mater, 1997, 36: 1145 - 1149.

[10] Thaveeprungsriporn V, Sinsrok P, Thong-Aram D. Effect of iterative strain annealing on grain boundary network of 304 stainless steel[J]. Scripta Materialia, 2001, 44: 67 - 71.

[11] Shimada M, Kokawa H, Wang Z J. Optimization of grain boundary character distribution for intergranular corrosion resistant 304 stainless steel by twin-induced grain boundary engineering[J]. Acta Mater, 2002, 50: 2331 - 2341.

[12] 方晓英,王卫国,郭红,张欣,周邦新. 金属学报,2007,43(12): 1239.

[13] Kumar B R, Das S K, Mahato B, Arpan D, Chow dhury S G. Effect of large strains on grain boundary character distribution in AISI 304L austenitic stainless steel[J]. Mater Sci Eng, 2007, A454: 239 - 244.

[14] Xia S, Zhou B X, Chen W J, et al. Features of Highly Twined Microstructures Produced by GBE in FCC Materials, Invited Paper for THERMEC' 2009[C]. Materials Science Forum. in press.

[15] Palumbo G, Aus K T, Lehockey E M. On a more restrictive geometric criterion for 'special' CSL grain boundaries [J]. Scripta Materialia, 1998, 38: 1685 – 1690.

[16] Gertsman V Y, Henager C H. Grain boundary junctions in microstructure generated by multiple twinning[J]. Interface Science, 2003, 11: 403 – 415.

[17] Haasen P. How are new orientations generated during primary recrystallization[J]. Metall Tran, 1993, B2: 235 – 239.

[18] Berger A, Wilbrandt P J, Ernst F. On the generation of new orientations during recrystallization: recent results on the recrystallization of tensile-deformed fcc single crystals[J]. Progress in Materials Science, 1988,32: 1 – 95.

[19] Gottstein G. Annealing texture development by multiple twinning in FCC crystals[J]. Acta Metall, 1984, 32: 1117 –1138.

[20] Randle V. The Role of The Conincidence Site Latticein Grain Boundary Engineering[M]. UK: Cambridge University Press, 1996.

[21] Humphreys F J, Hatherly M. Recrystallization and Related Annealing Phenomena[M]. Oxford: Elsevier, 2004.

The Relationship of Grain-Cluster Microstructure and Grain Boundary Characteristic Distribution in 304 Stainless Steel

Abstract: The effect of therm al-mechanical processing on microstructure and grain boundary characteristic distribution(GBCD) of 304 stainless steel was studied by means of electron back-scatter diffraction (EBSD) and orientation imaging microscopy (OIM) technique. The proportion of low $\Sigma(\Sigma \leqslant 29)$ coincidence site lattice (CSL) grain boundaries was increased to 75% by slight cold rolling and annealing at 1 100 ℃ for short time. The large grain-cluster, in which all the grains have $\Sigma 3^n$ ($n = 1, 2, 3 \cdots$) mutualm isorientation, is the key microstructure affecting the GBCD. Twin chain of grains and orientations in a grain-cluster were analyzed, by which means the order of multiple twins between any grains pair could be identified, and the twin relationships up to 8th generation ($\Sigma 3^8$) were found in this grain-cluster. The mean size of the grain-cluster and proportion of low ΣCSL boundaries decreased with the increasing of pre-strain.

金属材料中退火孪晶的控制及利用

——晶界工程研究*

摘　要：退火孪晶是低层错能面心立方金属材料中比较常见的一种显微组织. 斑铜艺人利用"锻打"和"高温烧斑"的工艺，使晶粒发生异常长大后形成退火孪晶，制造了瑰丽斑驳的斑铜艺术品. 虽然他们并不知道出现这些斑纹背后的科学原理，然而他们通过积累的生产实线经验，已经能够控制并利用退火孪晶这种显微组织. 随着现代材料与冶金科学技术的发展，人们对控制材料显微组织的重要性有了更加深刻的认识，也有了更多的方法. 基于退火孪晶的晶界工程研究与实践就是控制与利用退火孪晶这种显微组织的过程.

1　引言

关于"云南斑铜"的文章很多，在百度搜索引擎上输入"斑铜"一词，就能检索到 44 600 篇相关资料. 其中一篇文章中说道："云南斑铜"是云南省会泽地区独有的民间传统手工艺品. 采用铜基合金原料，经过锻打或铸造成型、精工打磨、以及复杂的后工艺处理制作而成，其褐红色的表面呈现出离奇闪烁、瑰丽斑驳的斑纹而独树一帜，堪称金属工艺之冠". 图 1(a)中展示的就是一匹马的斑铜艺术品[1]，在马的胸口有很大的斑纹. 这引起了我们的浓厚兴趣，因为我们正在研究的课题与这种斑纹在本质上有很大关系，这种斑纹其实就是铜合金中粗大的退火孪晶.

在具有低层错能的面心立方金属材料中，比如奥氏体不锈钢，铜合金，铅合金，镍基合金等，退火孪晶是一种常见的显微组织. 图 1(b)是镍基 690 合金显微组织的光学显微镜照片. 图中平行的条状组织就是退火孪晶，比如晶粒 B 就是一个退火孪晶，英文称之"annealing twins"[2]. 由两条平行的共格界面 A/B，和 B/A′组成. 晶粒 A 和 A′具有相同的取向，晶粒 A 和 B 的晶体点阵排列相对于它们之间的界面呈现镜面对称；同样晶粒 B 和 A′的晶体点阵排列相对于它们之间的界面也是这种对称关系. 英文中用复数形式"twins"表示孪晶，就是因为它由两条相同的界面构成. 图 1(b)中的孪晶尺寸比较小，在微米量级；而图 1(a)中"孪斑"的尺寸却很大，肉眼可见. 这是由于"孪斑"是通过二次再结晶得到的[3]. 在初次再结晶后，形成了如图 1(b)中的显微形貌，再进行小量变形及高温退火处理，某些晶粒异常长大，就能形成尺寸在毫米甚至厘米量级的晶粒. 这与"会泽铜辉斑铜工艺制品厂"师傅郭德宏所介绍的工序相符："生斑的制作特别讲究，必须要用 95％的自然铜，一点一点地锻打，再用高温来烧斑. 有的工序、程序要反复多次，有十来道(工序)都是靠我们一锤一锤(敲)打出来的，要烧一火，再打一次. 可以说，一件产品要烧几十火，打几万锤才能最终成型"[4]. 这种捶打及烧制的过程就是促使晶粒异常长大的过程. 在图 1(a)中可以看到斑驳的孪晶斑纹是由于孪晶与

---

*　本文合作者：夏爽、李慧、陈文觉. 原发表于《自然杂志》，2010，32(2)：94-100.

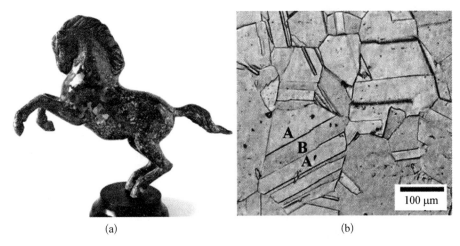

图1 （a）斑铜马艺术品[1]；（b）孪晶的光学显微形貌

母体晶粒的晶体取向不同,工艺品表面经抛光后,对自然光反射能力的差异而造成的.

斑铜艺术品是利用晶粒异常长大后产生了美丽的孪晶斑纹,而在通常的金属材料中减小孪晶的尺寸还能够大幅度提高材料的力学性能. 最近,据中国科学院金属研究所消息,2009 年 1 月 30 日,《科学》报道了中科院金属研究所沈阳材料科学国家(联合)实验室卢磊研究员领导的研究小组与卢柯研究员、丹麦 Risφ 国家实验室的黄晓旭博士合作研究的成果. 卢磊研究员及其合作者采用脉冲电镀沉积技术,在纯铜样品中成功地将孪晶片层平均厚度减小到约 4 nm. 并发现减小孪晶片层厚度可使材料的强度增加,当孪晶片层厚度为 15 nm 时,材料强度达到最大值. 这是利用了共格孪晶界稳定界面结构的特性,在获得了纳米尺寸的晶粒组织后,由于塑性变形机制的变化会导致极值强度的出现,同时表现出一般金属材料所不具备的特殊加工硬化效应.

前面两个例子都说明了控制材料显微组织的重要性. 通过控制显微组织使材料的性能满足使用的要求,这是现代材料科技工作者的一项重要工作内容. 在材料的合成,加工及使用过程中显微组织都有可能发生变化. 人类发展的历史与生产力的发展密切相关,从材料的角度来看,也就是人类在生产过程中利用什么材料和如何利用材料的历史. 在石器时代,人们利用大自然已经"控制"好显微组织的石头;在青铜器时代我们的祖先冶炼和铸造青铜器;在近代钢铁工业中我们除了冶炼和铸造,我们还利用变形和热处理来控制钢铁的显微组织. 现在的信息工业中半导体器件则需要制造纯度极高的单晶硅片.

对于退火孪晶这种显微组织,在 20 世纪初才被人们认知,当时西方的冶金学家在多种面心立方金属中,利用光学显微镜观察到了退火孪晶,但在同样是面心立方金属的铝中却很难观察到退火孪晶[2,5,6]. 后来才发现晶体中的层错能对产生孪晶有很大影响. 金属铝的层错能相当高,所以在晶粒生长过程中很难形成退火孪晶. 自 20 世纪中叶后,随着显微分析手段的发展,特别是电子显微镜的出现,人们对退火孪晶研究后才有了进一步的认识[7-11]. 退火孪晶与母体晶粒有特定的取向关系,用轴角对来表示它们的取向差时,可以写为：<111>/60°. 这表示在立方晶系中,孪晶与母体晶粒之间存在这样的取向关系：绕共同的<111>晶体学方向旋转 60°后就会完全重合起来. 通常利用重合位置点阵(coincidence site lattice,CSL)模型来描述晶界两侧晶粒之间的取向关系[3,12]. 若将两个无限延伸,具有相同点阵结构晶体中的一个,相对于另一个晶体绕某一低指数的晶轴旋转某特定的角度后,这两个晶体点阵中的某些阵点位置会有规则的重合起来. 这些重合位置的阵点在空间将构成三维空间的超点阵,称为重位点阵. 在 CSL 模型中退火孪晶与母体晶粒之间就存在 Σ3 取向关系,所以称之为 Σ3 晶界(Σ 就是重合位置密度,它表示在 CSL 模型中重合的点阵位置数与总共的点阵位置数比值的倒数,Σ 值越小重合点阵位置就越多). 孪晶与母体晶粒之间的共格界面是{111}晶面,共格孪晶界有着十分低的晶界自由体积及自由能,因此具有优良的性能.

关于面心立方金属材料中退火孪晶的形成机制目前有两类观点:① 晶粒长大过程中的生长事故[6,8],即晶粒长大时,一个迁移中的晶界由于某种意外的原因,{111}面的原有堆垛次序发生错误,而产生共格退火孪晶界;② {111}面堆垛层错形核长大[7,10],即在晶界处突出来一个层错小包(stacking fault packets),以它为核心依靠非共格部分界面的迁移使层错小包长大成孪晶. 有一种模型企图统一这两种观点[13]:在一条迁移的晶界后面,由于生长事故,肖克利不全位错环(一种以人名命名的晶体缺陷)在连续的{111}面上形核,然后由于不全位错之间互相排斥使层错长大形成了退火孪晶. 如果这条晶界的迁移速率越高,肖克利不全位错环形核的可能性也就越高,最后产生的退火孪晶密度也就越大. 所以材料的层错能越低就越容易产生退火孪晶. 直到现在还有不少关注于退火孪晶这种显微组织是如何形成的研究报道,但还没有达成统一的认识.

随着现代分析检测仪器的发展,现在已经能够方便的获取及标定晶体样品表面晶粒的取向及自动判定晶粒之间的取向关系,这就是电子背散射衍射(electron backscattered diffraction,EBSD)技术[14,15]. EBSD 设备是安装在扫描电子显微镜(SEM)中的一种分析检测附件. 当电子束入射到样品(晶体材料)表面与材料相互作用会产生多种信号,背散射电子是其中的一种. 背散射电子与晶体样品表面作用之后会产生衍射花样(菊池线花样),如图 2(a)所示. 这种背散射电子衍射花样与晶体的结构和相对于入射电子束晶体的取向位置有关,通过标定这种菊池衍射花样就可以测定微区的晶体取向,鉴别不同的物相结构. 当电子束在多晶体样品表面以栅格状(以一定的步长)扫描时,通过标定菊池花样获得的晶体学取向信息被逐点记录下来,再通过软件分析重构,就能得到以晶体学取向差别为成像依据的取向成像显微图(orientation imaging microscopy,OIM),如图 2(b)所示. 重构的 OIM 图可以反映由晶粒构成的显微形貌,给出晶粒取向分布和晶界结构特征. 所以 EBSD 能够完整地并且定量地确定样品中以晶体学取向信息为基础的显微结构. EBSD 技术为快速方便地确定退火孪晶的晶体学关系提供了保证.

基于人们对退火孪晶的认识和 EBSD 技术的发展,使得控制与利用退火孪晶这种显微组织成为可能. 在 1984 年日本学者 Watanabe[16]提出了晶界设计与控制的概念,在 20 世纪

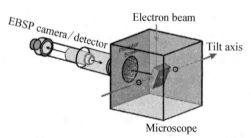

(a) EBSD在SEM中采集背散射衍射花样的示意图

(b) EBSD在晶体样品表面以栅格形式(一定的步长)驻点逐行扫面获得一定区域的晶体取向信息[15]

图 2 电子背散射衍射(EBSD)技术的工作原理简介[15]

90年代发展成为晶界工程(grain boundary engineering，GBE)研究领域. 通过提高材料显微组织中具有优良性能晶界的比例来提高材料的整体性能. 这些具有优良性能的晶界,是由于它们具有有序的结构. 一般认为 Σ 值小于 29 的低 ΣCSL 晶界具有有序的结构.

晶界工程研究的主要内容包括：① 如何通过加工处理工艺提高低 ΣCSL 晶界所占比例；② 低 ΣCSL 晶界在互相连接形成的晶界网络中的分布状态；③ 晶界结构与晶界网络分布状态如何影响材料性能. 通过人们的研究,发现在低层错能面心立方金属材料中,可以通过适当的形变及热处理工艺大幅度提高退火孪晶界及其相关晶界的比例,从而使材料与晶界有关的性能得到大幅度提高. 这里所指的孪晶相关晶界是由形成多重孪晶时,这些孪晶相遇而形成的那些晶界.

面心立方晶体有四组{111}面,发生多重孪晶(multiple twinning)形成孪晶链后,即再结晶过程中晶界迁移消耗形变基体时反复多次产生退火孪晶的过程[11,17,18],发生多重孪晶过程生成的晶粒之间都符合 $\Sigma 3^n (n=1,2,3\cdots)$ 的 CSL 关系,它们之间相遇后形成的晶界就是 $\Sigma 3^n$ 类型晶界,比如 $\Sigma 9,\Sigma 27,\Sigma 81$ 等. 低层错能面心立方金属材料中,通过合适的形变及退火工艺, $\Sigma 3^n$ 类型的低 ΣCSL 晶界比例可以被明显提高,从而与晶界特性相关的多种性能都能得到大幅提高,所以又称为基于退火孪晶的晶界工程[19].

基于退火孪晶的晶界工程中最典型的两个应用案例是：缓解了用于核反应堆蒸汽发生器传热管道 600 合金的应力腐蚀开裂[20]；运用晶界工程的办法处理了铅酸电池电极板从而提高了铅酸电池的使用周期寿命[21]. 在这两个应用案例中,材料性能的明显提高都是通过特殊设计的加工工艺来实现的,并且在 20 世纪 90 年代中期得到了专利保护. Palumbo 等在专利中介绍了这种加工工艺,对 600 合金进行处理可以使低 ΣCSL 晶界由 37% 提高到 71%($\Sigma 3$ 晶界：47%, $\Sigma 9 + \Sigma 27$ 晶界：15%),这使材料的耐腐蚀性能大幅提高,腐蚀速率可降低 30%～60%(以腐蚀失重来衡量),样品腐蚀后的截面照片如图 3 所示[20]. Palumbo 等对铅酸电池的铅合金电极板进行处理,使低 ΣCSL 晶界提高到 70% 以上后($\Sigma 3$ 晶界：56%, $\Sigma 9 + \Sigma 27$ 晶界：12%),耐腐蚀性能明显改善,电池使用寿命提高 1～3 倍[21]. 将通过提高低 ΣCSL 晶界比例处理之后的铅合金电极栅板和未经过这种处理的铅合金电极栅板同样进行实验室腐蚀实验后,得到电极栅板形貌照片如图 4 所示. 未经处理的栅板已经腐蚀破坏,而经过处理的栅板仍然保持原有的栅格形状.

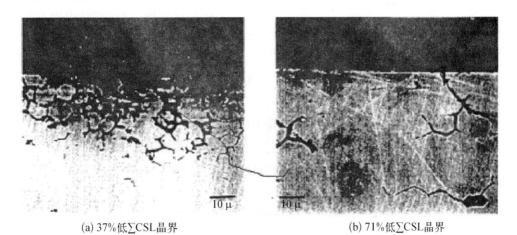

(a) 37%低ΣCSL晶界　　　　　　　(b) 71%低ΣCSL晶界

图 3　低 ΣCSL 晶界比例不同的 600 合金样品腐蚀后截面的 SEM 照片[20]

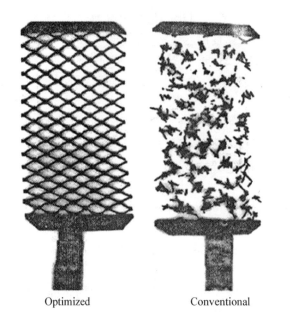
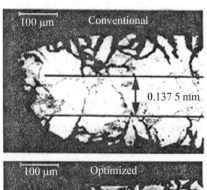
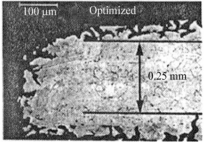

图 4 经过晶界工程处理和未经处理的铅酸电池的铅合金电极板栅板腐蚀比较[21]

当然,晶界工程也可以应用到其他一些低层错能面心立方金属材料中,比如铜合金,奥氏体不锈钢,镍基合金中,也取得了很多进展,在此不一一详细列举.

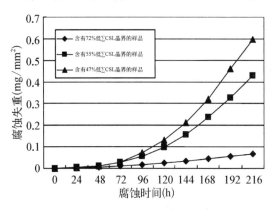

图 5 低 ΣCSL 晶界比例不同的 690 合金样品晶间腐蚀时,由于晶粒脱落造成的失重与腐蚀时间的关系[32]

国内在 690 合金的晶界工程技术上也开展了一些研究工作[22-30].690 合金的一个重要用途是作为压水堆核电站蒸汽发生器的传热管材料.20 世纪 80 年代末,法国首先采用了 690 合金替代 600 合金作为新一代蒸汽发生器的传热管,美国和日本也相继采用.我国大亚湾核电站在国内首先采用了法国提供的 690 合金管,目前国内大批建造的核电站也都是采用 690 合金管.但随着核电工业的发展,进一步提高反应堆运行参数(温度和压力),对 690 合金的性能提出了更高的要求.690 合金也是一种低层错能的面心立方金属材料,通过我们的研究,可以将 690 合金的低 ΣCSL

晶界比例提高到 70% 以上(Σ3 晶界:60%,Σ9+Σ27 晶界:9%,按照较严格的 Palumbo-Aust 标准[31]统计),使 690 合金晶间腐蚀(腐蚀溶液:65%HNO$_3$+0.4%HF)试验 10 天的腐蚀失重减小了近一个数量级,如图 5 所示.这一技术已经获得了中国专利保护[32].

在 690 合金中开展晶界工程研究其实就是控制及利用退火孪晶的过程.下面简要介绍我们在研究过程中得到的几点结果.

(1) 形成大尺寸"互有 Σ3^n 取向关系晶粒的团簇"是提高低 ΣCSL 晶界比例后显微组织的重要特征.

通过研究冷轧变形量、退火温度及退火时间、原始显微组织等条件对获得 690 合金 Σ3^n

类型晶界比例的影响,探明了小量变形(冷轧5%)再在高温(1 100 ℃)短时间(5 min)退火是将690合金低ΣCSL晶界比例提高到70%以上的合理工艺制度[26,27]. 在其他工艺参数不变的情况下,这种晶粒团簇的尺寸和团簇内含Σ3^{n}晶界的数量随再结晶前冷轧压下量的增加而下降. 图6(a)是低ΣCSL晶界比例提高到70%以上后的690合金样品,通过EBSD测定,由OIM系统重构出材料表面的取向衬度OIM图,图中用不同的颜色代表不同取向的晶粒.图6(b)是利用同一数据集重构出的晶界OIM图,图中用不同颜色的线条代表不同类型的晶界.图6(c)是图6(a)或(b)中一个大尺寸晶粒团簇的取向衬度OIM图. 在该图所示的二维晶粒组织中包含了91个晶粒. 在这91个晶粒之间都具有Σ3^{n}的取向关系. 而晶粒团簇外围是随机晶界. 所以我们称这样的显微组织特征为:大尺寸的"互有Σ3^{n}取向关系晶粒的团簇"[23,24].

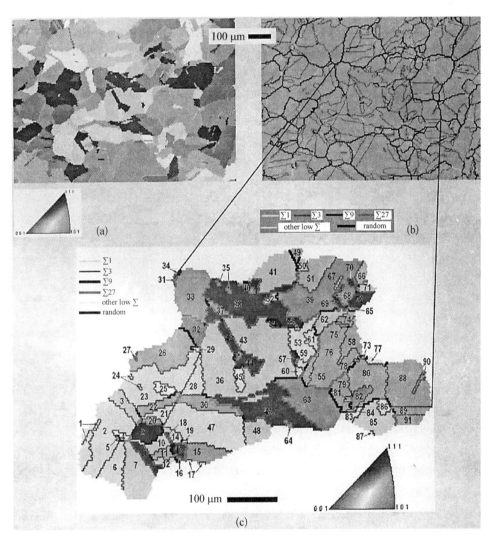

图6 690合金低ΣCSL晶界比例提高后,运用EBSD技术在样品表面获得取向信息,通过OIM软件重构后的显微组织. (a)取向衬度的OIM图;(b)不同类型晶界的OIM图;(c)在二维观察截面包含91个晶粒的大尺寸"互有Σ3^{n}取向关系晶粒的团簇"

这种晶粒团簇是再结晶时,由一个再结晶晶核在长大过程中不断产生退火孪晶,即多重孪晶充分发展的结果. 所以降低再结晶形核密度和促使多重孪晶充分发展是提高低层错能面心立方金属材料孪晶及相关晶界比例的关键因素. 依据这样的规律, 制定了合理的工艺, 在黄铜, 690 合金, 304 不锈钢, 铅合金中均能将低 ΣCSL 晶界比例提高到 70% 以上, 并形成了以大尺寸"互有 $\Sigma 3^n$ 取向关系晶粒的团簇"为特征的显微组织, 如图 7 所示[22,24,30]. 这表明我们针对 690 合金提出的提高低 ΣCSL 晶界比例过程的机理可以普遍适用于多种低层错能面心立方金属材料.

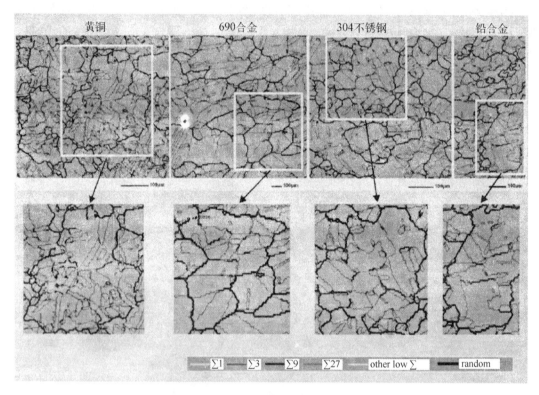

图 7 在黄铜, 690 合金, 304 不锈钢, 铅合金中均能将低 ΣCSL 晶界比例提高到 70% 以上, 并形成了以大尺寸"互有 $\Sigma 3^n$ 取向关系晶粒的团簇"为特征的显微组织

(2) 基于 EBSD 分析技术, 建立了确定高阶多重孪晶关系的方法.

低层错能面立方的金属材料, 经晶界工程处理后会产生大量的退火孪晶界($\Sigma 3$ 晶界)及其派生出来与高阶孪晶相关的晶界, 比如 $\Sigma 9$, $\Sigma 27$ 晶界. 这些晶界的大量出现, 说明了在再结晶过程中源自各再结晶晶核的多重孪晶得到了充分发展, 形成了以大尺寸"互有 $\Sigma 3^n$ 取向关系晶粒的团簇"为特征的显微组织. 然而 EBSD 设备只能自动给出最高为 3 阶的 $\Sigma 3^3$ ($\Sigma 27$)的孪晶取向关系, 无法直接给出更高阶孪晶关系, 比如 $\Sigma 81$, $\Sigma 243$, $\Sigma 729$, $\cdots$.

为此, 我们基于 EBSD 分析技术, 建立了确定高阶多重孪晶关系的方法. 这一方法可以确定大尺寸"互有 $\Sigma 3^n$ 取向关系晶粒的团簇"内部任何两个晶粒之间的多重孪晶关系[22,23]. 以图 6(c)中的晶粒团簇为例, 我们可以将这 91 个晶粒之间的孪晶关系画在图 8(a)中, 从任何一个晶粒都可以只穿过孪晶界到达任何另一个晶粒. 比如, 从晶粒 1 到晶粒 91 可以连成如下的孪晶链: $1 \to 2 \to 3 \to 23 \to 28 \to 36 \to 53 \to 62 \to 76 \to 78 \to 80 \to 88 \to 89 \to 91$, 总共跨过了 13 个孪晶界. 但这并不表示晶粒 1 与晶粒 91 之间具有 $\Sigma 3^{13}$ 的取向关系. 这是由于在这条孪晶链中有的晶粒具有相同的取向. 分析这 91 个晶粒的取向并归类后, 共存在 28 种不同的取向, 如图 8(b)所示. 将这 28 种晶粒取向的孪晶链关系画在图 8(c)中. 这样就可以准确并直

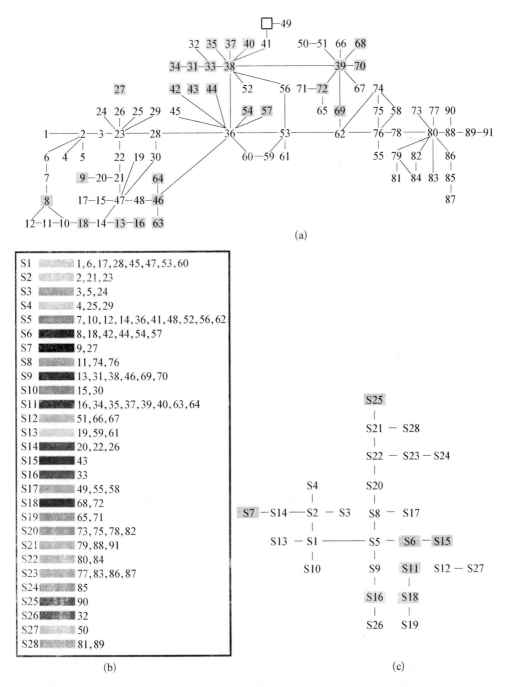

图 8 图 6(c)中晶粒团簇的孪晶链关系.(a) 晶粒孪晶链关系;(b) 团簇内 91 个晶粒具有 28 种取向;(c) 取向孪晶链关系[23]

接得出任何两个晶粒之间的取向关系. 比如:晶粒 1 属于取向 S1,晶粒 91 属于取向 S21,取向 S1 和取向 S21 之间的孪晶链取向关系是:S1→S5→S8→S20→S22→S21,具有 5 次取向不重复的孪晶关系,所以晶粒 1 和晶粒 91 之间是 $\Sigma 3^5$ 的取向关系.

通过这样的分析,可以深入的认识由多重孪晶形成的晶界网络拓扑分布,并且鉴别晶粒团簇内部那些被 EBSD 误判为随机晶界或低 ΣCSL 类型的晶.比如图 6 中,晶粒 8 和 9 之

间的晶界被判定为 $\Sigma 1$ 晶界，其实晶粒 8 和晶粒 9 之间具有 $\Sigma 3^5$ 的取向关系；比如晶粒 55 和 63 之间的晶界被判定为随机晶界，其实它们之间具有 $\Sigma 3^4$ 的取向关系；比如晶粒 63 和 83 之间的晶界被判定为 $\Sigma 7$ 晶界，其实它们之间具有 $\Sigma 3^6$ 的取向关系. 所以，这种确定高阶多重孪晶关系的方法对分析具有大量退火孪晶显微组织中的晶界网络十分有用.

（3）甄别了提高低 ΣCSL 晶界比例后随机晶界的网络连通性会被打断的观点.

有些研究报道指出当材料的低 ΣCSL 晶界比例提高后，随机晶界的网络连通性会被阻断，认为这是材料宏观性能提高的原因[33,34]. 我们的研究结果表明，提高低 ΣCSL 晶界比例后随机晶界的网络依然是连通的，材料性能的提高源于出现以大尺寸"互有 $\Sigma 3^n$ 取向关系晶粒的团簇"为特征的显微组织.

"互有 $\Sigma 3^n$ 取向关系晶粒的团簇"是由再结晶时一个再结晶晶核发展而成，晶粒团簇内部的 $\Sigma 3^n$ 晶界是一般大角度晶界在再结晶时"消耗"形变基体的迁移过程中发生多重孪晶生长后的必然结果. 晶粒团簇外围的晶界是由再结晶时不同再结晶晶核长大经历多重孪晶过程后的晶粒团簇相互接触后形成的，所以它们是随机晶界. 即使有的取向差接近低 ΣCSL 取向关系，被 Brandon 标准[35]判定为低 ΣCSL 晶界，但它们是随机出现的，所以偏差角大. 用更为严格的 Palumbo-Aust 标准[31]判定后大部分就被排除在低 ΣCSL 晶界范畴之外了. 这一现象在前面提到的黄铜，690 合金，304 不锈钢，铅合金的显微组织中都存在. 图 9 中，用 Brandon 标准判定时，随机晶界的网络明显被打断；而用更为严格的 Palumbo-Aust 标准判定后随机晶界的网络基本连通. 所以，严格来说，认为提高低 ΣCSL 晶界比例后随机晶界网络的连通性会被打断的观点是不可靠的[24,25].

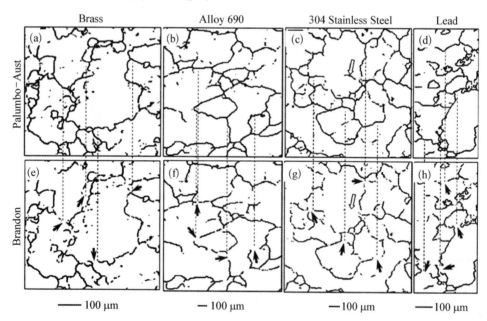

图 9 当低 ΣCSL 晶界比例提高后，分别运用 Palumbo-Aust 标准和 Brandon 标准判断随机晶界网络联通性时的差异.（a），（b），（c）和（d）分别是黄铜，690 合金，304 不锈钢和铅合金运用 Palumbo-Aust 标准时随机晶界的网络，（e），（f），（g）和（h）是运用 Brandon 标准时随机晶界的网络[22]

（4）研究发现在固溶及时效处理后，在孪晶相关晶界附近析出的棒状碳化物实际上是晶界上析出碳化物的二次枝晶.

晶界腐蚀抗力与碳化物在晶界上的析出行为密切相关，而碳化物在晶界上的析出行为又受到晶界类型(晶界结构)的影响[36]. 在研究不同类型晶界上碳化物析出形貌的差异时[28], 发现在 $\Sigma 3_i$(孪晶的非共格界面)及 $\Sigma 9$ 晶界附近都有棒状的碳化物生长, 如图 10 所示, 而在其他类型晶界附近没有这一现象[28].

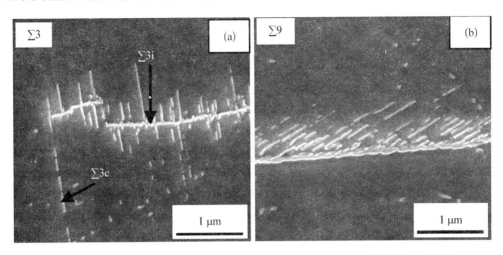

图 10 孪晶界($\Sigma 3_c$ 共格界面, $\Sigma 3_i$ 非共格界面)及 $\Sigma 9$ 晶界处析出的棒状碳化物[28]

通过深度蚀刻的办法使样品产生晶间腐蚀, 当样品表面的晶粒剥落后, 可以看出晶界上析出的碳化物是以枝晶形式生长的, 如图 11(a)所示. 通过深度蚀刻将 $\Sigma 3_i$ 及 $\Sigma 9$ 晶界附近的晶粒基体腐蚀去除之后, 发现这些棒状碳化物实际上是晶界上析出碳化物的二次枝晶[29]. 这提示我们根据二维截面上的观察不能全面反映碳化物的真实形貌.

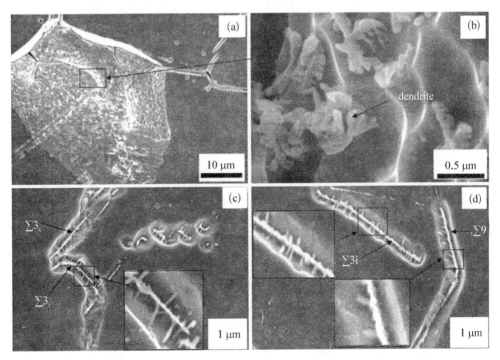

图 11 深度蚀刻后晶界处碳化物的三维形貌[39]: (a)&(b)随机晶界处的碳化物形貌, (c)&(d)孪晶界面及 $\Sigma 9$ 晶界处碳化物的形貌

参 考 文 献

[1] 斑铜[EB/OL]. [2009 - 12 - 21]. http://image. baidu. com/i？ tn＝baiduimage&ct＝201326592&lm=-1&cl=2&fm=ps&word＝%B0%DF%CD%AD.

[2] CARPENTER H C H, TAMURA S. Experiments on the production of large copper crystals[J]. Proceedings of the Royal Society A: Mathematical Physical & Engineering Sciences, 1926, 113: 161.

[3] KRONBERG M L, WILSON F H. Secondary recrystallization in copper[J]. Trans AIME, 1949, 185: 501 - 514.

[4] 中国非物质文化遗产大观：曲靖会译斑铜（图）[EB/OL]. [2009 - 12 - 23]. http://tieba. baidu. com/f？ kz＝585607090.

[5] Carpenter H C H, Tamura S. The formation of twinned metallic crystals[J]. Proceedings of the Royal Society A: Mathematical, Physical & Engineering Science S, 1926, 113: 161 - 182.

[6] FULLMAN R L, FISHER J C. Formation of annealing twins during grain growth[J]. J Appl Phys, 1951, 22: 1350 -1354.

[7] DASH S, BROWN N. An investigation of the origin and growth of annealing twins[J]. Acta Metall, 1963, 11: 1067 - 1075.

[8] GLEITER H. The formation of annealing twins[J]. Acta Metall, 1969, 17: 1421 - 1428.

[9] GINDRAUX G, FORM W J. New concept of annealing-twin formationin face-centered cubic metals[J]. Inst Metals, 1973, 101: 85 - 92.

[10] MEYERS M A, MURR L E. Model for the formation of annealing twins in f. c. c metals and alloys[J]. Acta Metall, 1978, 26: 951 - 962.

[11] BERGER A, WILBRANDT P-J, ERNST F. On the generation of new orientations during recrystallization: recent results on the recrystallization of tensile-deformed fcc single crystals[J]. Progress in Materials Science, 1988, 32: 1 -95.

[12] BOLLMANN W. Crystal defects and crystal interfaces[M]. Berlin: Springer, 1970: 1 - 254.

[13] MAHAJAN S, PANDE C S, IMAM M A, RATH B B. Formation of annealing twins in f. c. c. crystals[J]. Acta Mater, 1997, 45: 2633 - 2638.

[14] Imaging microscopy raising thestandard for electron backscatter diffraction software[EB/OL]. [2007 - 05 - 18]. http://www. edax. com/.

[15] Operator's EBSD training course[EB/OL]. [2006 - 07 - 19]. http://www. oxinst. com/.

[16] WATANABE T. An approach to grain boundary design for strong and ductile polycrystals[J]. Res Mech, 1984, 11: 47 - 84.

[17] GOTTSTEIN G. Annealing texture development by multiple twinning in F. C. C. crystals[J]. Acta Metall, 1984, 32: 1117 - 1138.

[18] CAYRON C. Multiple twinning in cubic crystals: geometric/algebraic study and its application for the identification of the $\Sigma 3^n$ grain boundaries[J]. Acta Cryst, 2007, A63: 11 - 29.

[19] RANDLE V. Twinning-related grain boundary engineering[J]. Acta Mater, 2004, 52: 4067 - 4081.

[20] LIN P, PALUMBO G, ERB U. Influence of grain boundary character distribution on sensitization and intergranular corrosion of alloy 600[J]. Scripta Metall Mater, 1995, 33: 1387 - 1392.

[21] LEHOCKEY E M, LIMOGES D, PALUMBO G, et al. On improving the corrosion and growth resistance of positive Pbacid battery grids by grain boundary engineering[J]. Journal of Power Sources, 1999, 78(1 - 2): 79 - 83.

[22] XIA S, ZHOU B X, CHEN W J, et al. Features of Highly Twinned Microstructures Produced by GBE in FCC Materials[J]. Materials Science Forum, 2010, 638 - 642: 2870 - 2875.

[23] XIA S, ZHOU B X, CHEN W J. Grain-cluster microstructure and grain boundary character distribution in alloy690 [J]. Metallurgical and Materials Transactions A, 2009, 40A: 3016 - 3030.

[24] XIA S, ZHOU B X, CHEN W J. Effect of single-step strain and annealing on grain boundary character distribution and intergranular corrosion in alloy690[J]. Journal of Materials Science, 2008, 43: 2990 - 3000.

[25] 夏爽,周邦新,陈文觉. 690 合金的晶界特征分布及其对晶间腐蚀的影响[J]. 电子显微镜学报,2008,27(6):461 -

468.

[26] 夏爽,周邦新,陈文觉.形变及热处理对 690 合金晶界特征分布的影响[J].稀有金属材料与工程,2008,37(6):999 - 1003.

[27] XIA S, ZHOUB X, CHEN W J, et al. Effects of strain and annealing processes on the distribution of Σ3 boundaries in a Ni-based superalloy[J]. Scripta Materialia, 2006, 54:2019 - 2022.

[28] 李慧,夏爽,周邦新,等.镍基 690 合金时效过程中晶界碳化物的形貌演化[J].金属学报,2009,45:195 - 198.

[29] LI H, XIA S, ZHOU B X, et al. The dependence of carbide morphology on grain boundary character in the highly twinned alloy 690[J]. Journal of Nuclear Materials, 2010, 399:108 - 113. DOI:10.1016/j.jnucmat.2010.01.008.

[30] 夏爽,周邦新,陈文觉,等.铅合金在高温退火过程中晶界特征分布的演化[J].金属学报,2006,42:129 - 133.

[31] PALUMBO G, AUST K T. Structure-dependence of intergranular corrosion in high purity nickel[J]. Acta Metall Mater, 1990, 38:2343 - 2352.

[32] 夏爽,周邦新,陈文觉,等.提高 690 合金材料耐腐蚀性能的工艺方法:中国:ZL200710038731.5[P].

[33] KUMAR M, KING W E, SCHWARTZ A J. Modifications to the microstructural topology in F. C. C. materials through thermomechanical processing[J]. Acta Mater, 2000, 48:2081 - 2091.

[34] SCHUH C A, KUMAR M, KING W E. Analysis of grain boundary networks and their evolution[J]. Acta Mater, 2003, 51:678 - 700.

[36] BRANDON D G, RALPH B, RANGANATHAN S, et al. A field ion microscope study of atomic configuration at grain boundaries[J]. Acta Metall, 1964, 12:813 - 821.

[36] KAI J J, YU G P, TSAI C H, et al. The effects of heat treatment on the chromium depletion, precipitate evolution and corrosion resistance of inconel alloy 690[J]. Metall Trans A, 1989, 20A:2057 - 2067.

Control and Application of Annealing Twins in Metallic Materials: Grain Boundary Engineering

Abstract: Annealing twins is a common microstructure in face centered cubic metallic materials with low stacking fault energy. Craftsman makes speckled copper artworks by forging and high temperature heat treatment. However, they do not know the metallurgical theory behind the processing. Actually, the speckles in the copper artworks are the coarse annealing twins as a result of abnormal grain growth. With the development of modern metallurgy technique, people emphasized more on the microstructure control to obtain desired bulk properties. Twin-induced grain boundary engineering is one ofsuch technique to control and use the annealing twins.

高温高压水环境中锆合金腐蚀的原位阻抗谱特征*

摘　要：利用电化学阻抗谱等技术研究了 360 ℃/18.6 MPa 的 0.01mol/L LiOH 水溶液中退火态 Zr-4 合金在转折前后的腐蚀演化过程. 结果表明，腐蚀转折前，锆合金氧化增重缓慢. 当 Zr 氧化生成 ZrO_2 时，由于体积发生膨胀而在氧化膜中产生压应力，随着氧化膜逐渐增厚，氧化膜外层的压应力因微观缺陷凝聚而得到松弛，使得氧化膜由均匀致密的单层结构演化为外层疏松、内层致密的双层结构，阻抗谱也由单一容抗弧演变为双容抗弧，并出现高频弧减小而低频弧增大的现象. 当氧化膜形成裂纹等宏观缺陷后，腐蚀发生转折，氧化增重加快，阻抗谱的 2 个容抗弧均快速减小. 腐蚀电位和电化学阻抗谱均能原位获得腐蚀转折点的突变信息.

　　锆合金具有低的热中子吸收截面、良好的抗腐蚀性能和力学性能，作为核燃料包壳和堆芯结构材料在核电站中得到了广泛使用[1]. 核电发展的关键是核安全，因此锆合金在高温高压水环境中的抗腐蚀性能对核反应堆安全起着重要作用. 人们通常采用堆外高压釜腐蚀实验方法研究锆合金腐蚀的动力学，并结合 X 射线衍射、扫描电子显微镜、透射电子显微镜等方法分析腐蚀后氧化膜的组织与结构[2-4]. 这些研究方法较为成熟，却难以获得锆合金的原位腐蚀信息.

　　一般认为[5,6]，锆合金的氧化过程为：H_2O 在氧化膜表面吸附并解离为 O^{2-} 和 H^+，阳极反应是 O^{2-} 扩散通过氧化膜与锆基体生成 ZrO_2，阴极反应时电子扩散通过氧化膜并在介质与氧化膜界面上与 H^+ 生成 H_2. 其中阳极反应是腐蚀过程的控制步骤，氧化速率取决于氧化膜内的 O^{2-} 迁移速率. O^{2-} 在氧化膜中的扩散途径是晶界、位错等缺陷，电子在氧化膜中的扩散途径是镶嵌在氧化膜中的金属夹杂物或第二相等. 正因为锆合金的腐蚀过程是一个电化学过程，电化学阻抗谱（EIS）作为一种非破坏性的高灵敏的原位测试技术，在高温高压环境中逐渐得到应用，它可以在不终止腐蚀实验的情况下，通过定期检测金属腐蚀过程中的 EIS 谱，原位获得金属腐蚀的演化信息，深入认识材料的腐蚀机理[7,8]. Gohr 等[9] 用 EIS 研究了不同 Sn 含量的 Zr-4 合金在模拟压水堆（PWR）条件（350 ℃/17 Mpa）下的腐蚀行为，发现 EIS 谱呈现周期性变化特征，这与氧化膜生长的内在性质有关；Sn 含量增加减小了致密层的电荷转移电阻，但对氧化膜厚度没有影响. Ai 等[10] 探索了 Zr 及其合金在 250 ℃下 0.1 mol/L $B(OH)_3$ 和 0.001 mol/L LiOH 水溶液中的腐蚀机制，分析表明，Zr 及其合金的双层氧化膜生长较好地符合点缺陷模型（PDM）. Bojinov 等[11] 报道了 E110 和 Zr-4 合金在模拟 VVER 反应堆环境中的初期腐蚀特征，2 种锆合金的 EIS 值随着溶液中 KOH 浓度升高逐渐下降，腐蚀速率加快.

　　为了深入认识锆合金的腐蚀转折过程，本文利用腐蚀增重曲线、原位电化学阻抗谱和腐蚀电位等方法，分析了高温高压水溶液中锆合金在腐蚀转折前后的腐蚀演化特征.

---

* 本文合作者：杨波、李谋成、姚美意、沈嘉年. 原发表于《金属学报》，2010，46(8)：946-950.

1 实验方法

实验选用 Zr-Sn 系 Zr-4 合金,其合金元素 Sn,Fe 和 Cr 的含量(质量分数,%)分别为 1.5,0.2 和 0.1.采用常规工艺(锻造→β淬火→热轧→多次冷轧及中间退火)制备成板材,并经过 600 ℃/1 h 再结晶退火处理.样品尺寸为 8 mm×50 mm,厚度为 0.5 mm.腐蚀实验前,样品经体积比为 10%HF+45%HNO₃+45%H₂O 混酸酸洗和去离子水清洗.

原位电化学测量的高压釜装置如图 1 所示.工作电极为锆合金实验样品,参比电极和辅助电极均为 Pt 丝.使用预先经过高温氧化绝缘的锆合金样品盘将电极固定,保证各电极之间距离恒定.腐蚀介质为 360 ℃/18.6 MPa 的 0.01 mol/L LiOH 水溶液,通过排水蒸气的方法进行除氧处理.测试仪器为 Solartron 1287/1255B 电化学测试系统,扫描频率范围为 10 mHz—99 kHz,交流激励电压为 20 mV. EIS 谱数据用 ZSmpWin3.10 软件进行拟合.实验结束后用 JSM-6700F 高分辨扫描电镜(SEM)对样品表面进行观察.同时对锆合金样品进行相同腐蚀环境下的传统增重实验.

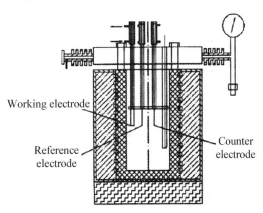

图 1 用于原位电化学测量的高压釜装置示意图
Fig. 1 Schematic diagram of the autoclave for in situ electrochemical measurement

2 实验结果

图 2 示出了 Zr-4 合金在 360 ℃/18.6 MPa 的 0.01 mol/L LiOH 水溶液中的增重曲线.由图可见,约在 3×10³ h 后腐蚀发生转折,转折前增重较为缓慢,而转折后增重迅速增加.图 3

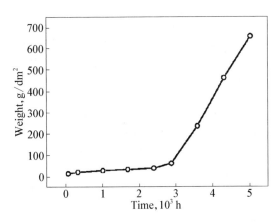

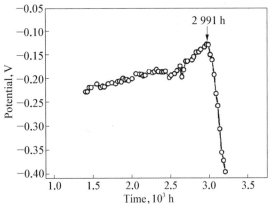

图 2 Zr-4 合金在 360 ℃/18.6 MPa 的 0.01 mol/L LiOH 水溶液中的腐蚀增重曲线
Fig. 2 Weight gain curve of the Zr-4 alloy corrosion in 0.01 mol/L LiOH solution at 360 ℃/18.6 MPa

图 3 Zr-4 合金腐蚀过程中的腐蚀电位曲线
Fig. 3 Corrosion potential curve of the Zr-4 alloy corrossion in 0.01 mol/L LiOH solution at 360 ℃/18.6 MPa

为锆合金腐蚀过程中腐蚀电位随时间的变化曲线. 由图可见,腐蚀电位在 2 991 h 前缓慢升高,而之后则快速降低. 由此得知,锆合金的腐蚀转折点约为 2 991 h,这与增重实验所得结果基本一致.

图 4 示出了锆合金在腐蚀转折前的部分 EIS 谱. 1 597 h 之前,锆合金腐蚀的 EIS 谱由扁平的单一容抗弧组成;随着腐蚀持续进行,EIS 谱逐渐转变为 2 个半圆形容抗弧,并且高频容抗弧逐渐减小,而低频容抗弧逐渐增大(图 4a). 2 557 h 至转折前,高频容抗弧有微小波动,但阻抗值无显著变化,而低频容抗弧略有增大的趋势(图 4b).

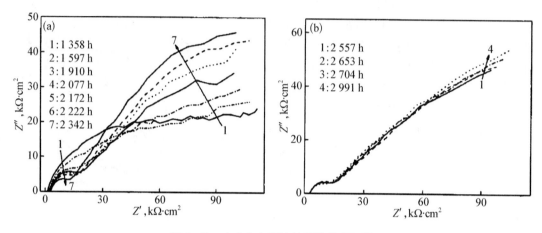

图 4 Zr - 4 合金在腐蚀转折前的 EIS 谱

Fig. 4 EIS spectra of the Zr - 4 alloy before corrosion transition at 1 358—2 342 h (a) and 2 557—2 991 h (b)

图 5 为锆合金在发生腐蚀转折后的 EIS 谱. 由图可见,腐蚀阻抗值随时间快速减小,EIS 谱的高、低频 2 个容抗弧的半径均同时减小. 腐蚀后取出锆合金样品观察发现,氧化膜中分布着许多大小不同的裂纹(图 6).

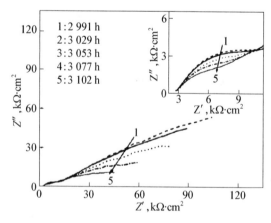

图 5 Zr - 4 合金在腐蚀转折后的 EIS 谱

Fig. 5 EIS spectra of the Zr - 4 alloys after corrosion transition (inset: enlargement of high frequency)

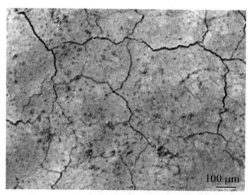

图 6 腐蚀实验后 Zr - 4 合金表面氧化膜的 SEM 像

Fig. 6 SEM morphology of the surface oxide film on Zr - 4 alloy after corrosion test

3 分析讨论

3.1 EIS谱数据解析

由于EIS谱有2个明显的时间常数,结合文献报道[12,13],建立图7所示锆合金腐蚀的等效电路模型.其中,R_1表示溶液电阻;Q_1和R_2分别表示锆合金表面疏松氧化膜层的电容元件和电阻;Q_2和R_3分别表示金属表面致密氧化膜层与溶液界面的电容元件与电阻,这部分阻抗包含了致密膜层和电荷转移过程的信息[14,15].该模型能够较好地拟合分析图4和图5中双容抗弧EIS谱数据.图8以1 597和2 077 h的数据为代表给出了测量点和拟合点的对比图,图中拟合点与测量数据点基本重合.图9给出了腐蚀过程中R_2和R_3的拟合结果,其拟合误差均在5%以内.

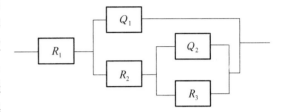

图7 转折前后Zr‐4合金腐蚀的等效电路模型
Fig. 7 Equivalent circuit model of the Zr‐4 alloy corrsion (R_1—solution resistance, Q_1—capacitance of the porous layer, R_2—resistance of the porous layer, Q_2—capacitance of the dense layer, R_3—resistance of the dense layer)

3.2 转折前锆合金的腐蚀机制

在1 358 h附近(图4a),由于锆合金表面氧化膜层比较致密,对基体有较好的保护作用,EIS谱表现为单一容抗弧,同时由于电力线分布不均匀,导致弥散作用较强,EIS谱呈现扁平特征.由于ZrO_2的P. B. 比(Pining-Bedworth的缩写,表示氧化物与形成该氧化物所消耗金属的体积比)为1.56,金属界面上缓慢生成的ZrO_2将使氧化膜层内的压应力不断累积,并导致氧化膜中产生各种微缺陷.随着氧化膜增厚,为了让应力得到松弛,氧化膜外部的微缺陷将发生扩散和凝聚,最终导致氧化膜的外层逐渐形成微孔或微裂纹[16,17],氧化膜也就由整体致密的膜层演化为外层疏松而内层致密的双层结构.因而,约在腐蚀1 910 h后,EIS谱出现2个差别明显的时间常数,高频部分反映疏松膜层的信息,而低频部分反映致密膜层及电荷转移过程的信息.以2 077 h为例,由拟合数据计算可知,高、低频EIS谱的时间常数分别约为9.7×10^{-3}和3.8 s,疏松膜层的时间常数明显小于致密膜层.

由图9可知,在2 077—2 342 h间,内层致密膜的生长使得其电阻R_3逐渐增大,而由于外层膜的缺陷不断增加,其电阻R_2明显降低,在图4a中表现为低频阻抗值增大而高频阻抗值减小.在2 342 h后,由于外层疏松膜对介质阻挡作用的显著降低,导致内层膜出现了一个较为快速的生长.之后直至2 991 h发生腐蚀转折前,内层致密膜处于缓慢生长和向外层疏松膜转化的动态过程中,而外层疏松膜在缓慢增厚的同时微缺陷也进一步增多,因此2个膜层的电阻在此期间变化较小(图4b).此外,演变过程中,内层致密氧化膜始终屏蔽保护着整个金属表面,且其电阻R_3在转折前呈现缓慢增加趋势,故而锆合金腐蚀的腐蚀电位会缓慢升高(图3).

3.3 转折后锆合金的腐蚀机制

2 991 h后,外层氧化膜中微缺陷在生长应力作用下转变为宏观裂纹,使外层氧化膜失

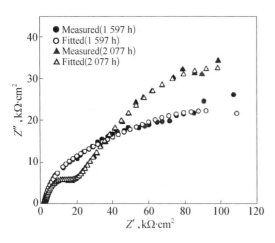

图 8　典型 EIS 谱的拟合结果

Fig. 8　Plots for the fitted results of typical EIS spectra

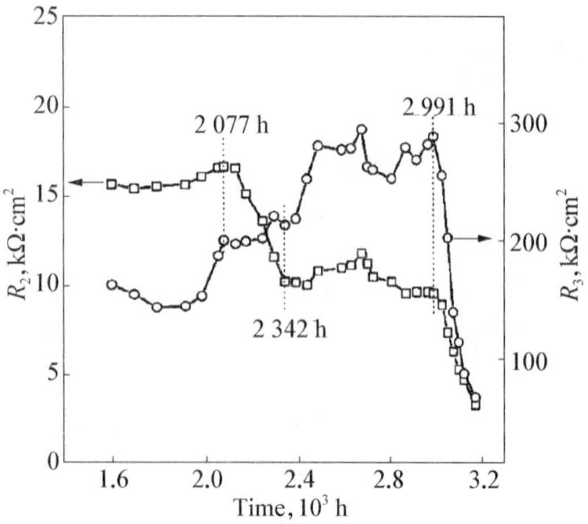

图 9　等效电路模型中 R_2 和 R_3 随时间的变化曲线

Fig. 9　Curves of R_2 and R_3 in Fig. 7 *vs* immersion time

去保护性,同时使腐蚀介质可以更接近金属表面,并且形成局部腐蚀环境,使得局部金属表面难以形成保护性良好的致密氧化膜. 这些裂纹处局部金属的快速腐蚀又会导致外层疏松氧化膜的加速破坏、裂纹增多并扩展. 在图 5 所示 EIS 谱中表现为 2 个容抗弧的快速减小. 与氧化增重快速增大相同,2 个拟合电阻值和腐蚀电位均迅速下降,十分准确地表征了腐蚀转折过程,且说明局部腐蚀环境的恶化使锆合金呈加速腐蚀状态.

4　结论

（1）腐蚀转折前,锆合金腐蚀的电化学 EIS 谱由单一容抗弧转变为双容抗弧,锆合金表面氧化膜出现致密内层和疏松外层的双层结构演化特征.

（2）腐蚀转折后,EIS 谱的 2 个容抗弧均快速减小,裂纹等宏观缺陷处的局部腐蚀环境恶化,2 层氧化膜被破坏而保护作用下降,锆合金腐蚀加快.

参 考 文 献

[1] Huang Q. *Nucl Power Eng*, 1996；17：263.

（黄强. 核动力工程,1996；17：263）

[2] Li Q. *PhD Thesis*, Shanghai University, 2008.

（李强. 上海大学博士学位论文,2008）

[3] Yao M Y. *PhD Thesis*, Shanghai University, 2007.

（姚美意. 上海大学博士学位论文,2007）

[4] Nagy G, Kerner Z, Pajkossy T. *J Nucl Mater*, 2002；300：230.

[5] Pauporte T, Finne J, Lincot D. *Phys Stat Solidi*, 2005；202A：1502.

[6] Pauporte T, Finne J. *J Appl Electrochem*, 2006；36：33.

[7] Indig M E. *Corrsion*, 1990；46：680.

[8] Macdonald D D, Mankowski J, Karaminezhaad-Ranjbar M, Hu Y H. *Corrosion*, 1988；44：186.

[9] Gohr H, Schaller J, Ruhmann H, Garzarolli F. In: Bradley E R, Sabol G P, eds. , *11th International Symposium*. *ASTM STP 1295*, Garmisch-Partenkirchen: ASTM, 1996: 181.

[10] Ai J H, Chen Y Z, Urquidi-Macdonald M, Macdonald D D. *J ELectrochem Soc*, 2007: 154: C43.

[11] Bojinov M, Cai W, Kinnunen P, Saario T. *J Nucl Mater*, 2008: 378: 45.

[12] Wikmark G, Rudling P, Lehtinen B, Hutchinson B, Oscarsson A, Ahlberg E. *11th International Symposium*. *ASTM STP 1295*, Garmisch-Partenkirchen: ASTM, 1996: 55.

[13] Pan J, Leygraf C, Jargelius-Pettersson R F A, Linden J. *Oxid Met*, 1996: 50: 431.

[14] Cao C N. *Principles of Electrochemical Corrosion*. Bei jing: Chemical Industry Press, 2008: 185.

（曹楚南. 腐蚀电化学原理. 北京：化工工业出版社,2008：185）

[15] Cao C N, Zhang J Q. *Introduction to Electrochemical Impedance Spectroscopy*. Beijing: Chemical Industry Press, 2002: 12.

（曹楚南,张鉴清. 电化学阻抗谱导论. 北京：科学出版社,2002：12）

[16] Zhou B X, Li Q, Liu W Q, Yao M Y, Chu Y L. *Rare Met Mater Eng*, 2006: 26: 1009.

（周邦新,李强,刘文庆,姚美意,褚玉良. 稀有金属材料工程,2006：26：1009）

[17] Zhou B X, Li Q, Yao M Y, Liu W Q, Chu Y L. *Nucl Power Eng*, 2005: 26: 364.

（周邦新,李强,姚美意,刘文庆,褚玉良. 核动力工程,2005;26：364）

In Situ Impedance Characteristics of zirconium Alloy Corrosion in High Temperature and Pressure Water Environment

Abstract: The electrochemical impedance spectroscopy (EIS) was used to analyze the corrosion evolution of pre-and post-transition of the annealed Zr – 4 alloy in 0. 01 mol/L LiOH aqueous solution at 360 ℃/ 18. 6 MPa. The results show that oxidation weight gain of Zr – 4 alloy increases slowly before corrosion transition. When Zr is oxidized into ZrO_2, the stress appears in the oxide film due to volume expansion. With gradually thickening of the oxide film, accumulation of micro-defects will relax the stress in outer layer of oxide film. As a result, the uniform and compact oxide layer turns into double layer with a porous outer layer and a dense inner one. Meanwhile the impedance spectra evolve from single arc into double capacitive arcs. When the oxide film forms macro-defects such as cracks, corrosion transition takes place, giving rise to an acceleration of oxidation weight gain and the rapid drop of impedance and corrosion potential. Corrosion potential and electrochemical impedance spectroscopy are able to obtain in situ some information of the transition point.

Zr-4合金表面氧化膜的电化学阻抗谱特征*

摘　要：电化学阻抗谱是分析锆合金表面氧化膜结构及其演化行为的有效方法. 利用 10％HCl 溶液研究了锆合金在 400 ℃, 10.3 MPa 过热蒸汽中腐蚀后的表面氧化膜电化学阻抗行为. 结果表明：锆合金过热蒸汽腐蚀初期表面氧化膜的阻抗谱为单一容抗弧, 随着腐蚀进行而演变为双容抗弧. 氧化膜表现为双层膜结构特征. 氧化膜阻挡作用的降低是锆合金过热蒸汽腐蚀发生转折的一个原因. 锆合金中第二相粒子对氧化膜阻抗谱及合金耐蚀性有较大影响.

锆合金通常作为燃料元件包壳和堆芯结构材料, 在反应堆中会受到高温高压水或者过热蒸汽的腐蚀而在合金表面生成氧化膜. 用透射电镜等方法研究发现, 腐蚀转折前后氧化膜的显微组织结构存在较大差异[1,2]. 但这些方法的制样难度较大, 且制样过程中还可能使原有孔隙扩大而破坏原有的显微组织.

电化学阻抗谱(EIS)技术是一种对样品非破坏性的测试技术, 它可以得到很多电极界面的结构信息, 是研究氧化膜结构的有效方法. 一般, 致密氧化膜的阻抗谱中仅存在一个时间常数, 且氧化膜的厚度与阻抗值大小存在着对应关系；如果氧化膜具有多层结构, 则其阻抗谱中有多个时间常数, 每个时间常数则对应着不同膜层的结构特性[3-6]. 目前氧化膜的 EIS 研究多在中性介质中进行, 仅有少量文献报道了酸性介质中氧化膜的阻抗行为[3,4].

考虑到锆合金氧化膜的开路电位在中性溶液介质中不易达到稳态, 而在酸性介质中(氢氟酸除外)不发生溶解, 本实验选择在 10％HCl 溶液中对 400 ℃, 10.3 MPa 过热蒸汽腐蚀不同时间后的 Zr-4 合金样品进行研究, 并结合腐蚀增重和合金显微组织, 分析锆合金氧化膜的结构特性.

1　实验

先将 0.6 mm 厚的 Zr-4 合金板切成 25 mm×20 mm 的样品, 然后, 经过 600 ℃再结晶退火 1 h, 空冷(记为 Z600), 或在真空管式电炉中加热至 800 ℃保温 36 h, 空冷(记为 Z800). 用 JEM-200CX 透射电镜观察合金样品的显微组织及第二相的大小和分布.

腐蚀实验在静态高压釜内进行, 条件为 400 ℃, 10.3 MPa 过热蒸汽. 实验前, 样品用混合酸(体积分数为 10％HF+45％HNO$_3$+45％H$_2$O)酸洗和去离子水清洗, 每种样品采用 3 块平行试样.

采用三电极系统测试电化学阻抗谱. 锆合金为工作电极, 铂片为辅助电极, 饱和甘汞电极为参比电极. 溶液介质为 10％HCl 溶液, 样品测试前在室温浸泡 24 h. 测试仪器为 PAR M273A 恒电位仪和 5210 锁相放大器. 扫描频率范围为 10 MHz～99 kHz. 交流激励电压为 5 mV. 阻抗谱数据用 ZSimpWin3.10 软件进行拟合.

---

* 本文合作者：钟祥玉、杨波、李谋成、姚美意、沈嘉年. 原发表于《稀有金属材料与工程》, 2010, 39(12)：2165-2168.

2 实验结果

2.1 显微组织

图 1 给出了两种热处理 Zr-4 合金样品的 TEM 照片[7]. 在 α 相区(600,800 ℃)退火的两种样品已经完全再结晶,600 ℃退火样品中多数第二相粒子直径较小(<0.1 μm),且分布比较弥散. 与 600 ℃退火样品相比,800 ℃退火样品中第二相粒子因发生聚集长大而直径较大(0.15~0.25 μm),数量明显减少.

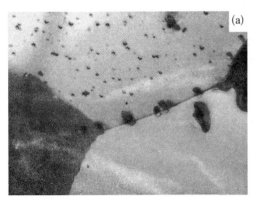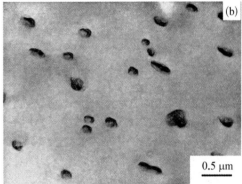

图 1 两种样品的 TEM 照片
Fig. 1 TEM images of Zircaloy-4 specimens: (a) Z600 and (b) Z800

2.2 腐蚀增重曲线

图 2 为两种不同热处理 Zr-4 合金的腐蚀动力学曲线[7]. 由图可知,两种样品腐蚀增重的转折点都在 42 d 附近,转折之后腐蚀速度明显加快. 但是,在任一腐蚀时间下,600 ℃退火态样品比 800 ℃退火处理样品的增重都小、耐蚀性更高,尤其是转折之后,两种样品的增重差别不断增大.

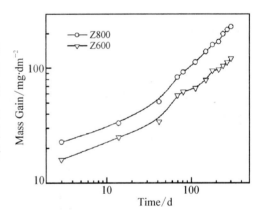

图 2 两种样品在 400 ℃,10.3 MPa 过热蒸汽中的腐蚀增重曲线
Fig. 2 Mass gain vs exposure time for Zircaloy-4 specimens in 400 ℃, 10.3 MPa super-heated steam

2.3 电化学阻抗谱

2.3.1 同种样品在过热蒸汽中腐蚀不同时间后的阻抗谱

Z600 样品在 400 ℃,10.3 MPa 过热蒸汽中腐蚀 14、42 和 302 d 后,其氧化膜在室温下的阻抗谱如图 3 所示. 图中 Z' 和 Z'' 分别为阻抗的实部和虚部. 由图可知,随着过热蒸汽腐蚀时间延长,高频部分逐渐形成了一个完整的容抗弧(图 3 中插图),而低频容抗值从 14 d 到 42 d 略有增大,但到 302 d 时显著降低. 这说明氧化膜微观组织结构在过热蒸汽腐蚀过程中发生了明显变化.

2.3.2　不同热处理的两种样品过热蒸汽腐蚀相同时间后的阻抗谱

Z600 样品和 Z800 样品在 400 ℃,10.3 MPa 过热蒸汽中腐蚀 42 d 后,在室温下测得的阻抗谱如图 4 所示. 与 Z600 样品相比,Z800 样品在 10％HCl 溶液中浸泡 24 h 以后低频容抗值明显较小,但其高频容抗弧更加完整、值也较大(图 4 中插图). 这说明热处理制度不同的两种样品经 42 d 过热蒸汽腐蚀后表面形成的氧化膜微观组织结构存在较大的差异.

图3　Z600 样品腐蚀不同时间后在 10％HCl 溶液中的阻抗谱

Fig. 3　EIS plots in 10％HCl solution for Z600 specimen for different oxidation time (Inset is the magnification at high frequency)

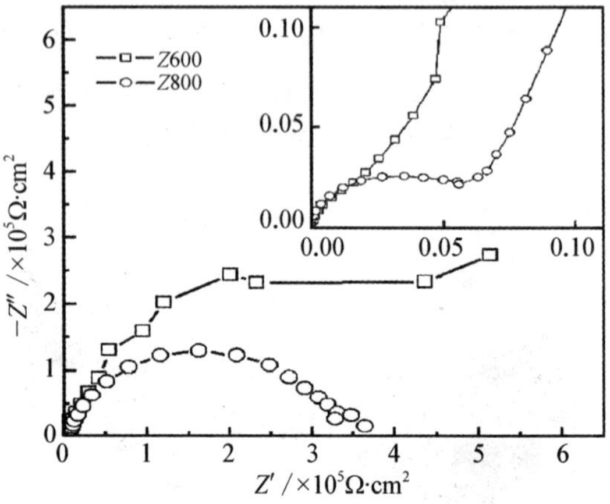

图4　不同热处理的两种样品在 10％HCl 溶液中浸泡 24 h 后的阻抗谱

Fig. 4　EIS plots of Z600 and Z800 specimens after immersion for 24 h in 10％ HCl solution (Inset is the magnification at high frequency)

3　讨论

锆合金高温腐蚀形成的表面氧化膜通常由疏松的外层和致密的内层组成[6],可用如图 5 所示的等效电路来表征[5]. 图中:R_s 为溶液电阻,R_1 和 CPE_1 分别为样品表面外层疏松氧化物膜的电阻和电容元件,R_2 和 CPE_2 分别为样品表面内层致密氧化物膜的电阻和电容元件. 常相位电容元件的阻抗可表示为:

$$Z_{(CPE)} = 1/[Y_0(j\omega)^n]$$

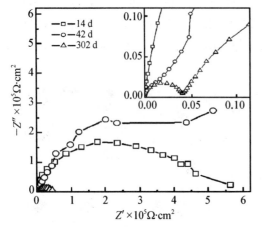

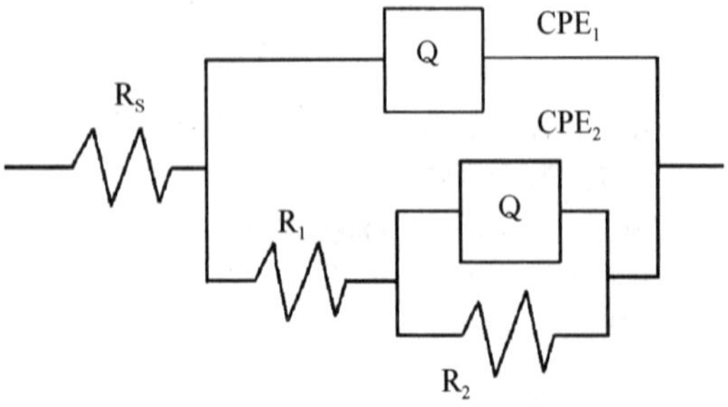

图5　锆合金表面氧化膜的等效电路图

Fig. 5　Equivalent circuit diagram for the oxide film on Zr alloys

其中 $j = \sqrt{-1}$,为虚数单位,ω 为交流信号的角频率,Y_0 为容抗导纳的模,n 为弥散系数.

运用该模型对图 3 和图 4 中阻抗谱进行解析,表 1 给出了主要参数的拟合结果. Z600 样品腐蚀 14 d 后表面形成的氧化物层较薄,且致密性较好,其阻抗谱呈现单一时间常数特征,难以用该模型进行解析. 随着在过热蒸汽中腐蚀时间延长到 42 d,氧化膜不断生长而使阻抗值增大,并开始出现双层结构特征. 最初形成的致密氧化膜在生长过程中,内应力随着氧化膜不断增厚而增大,为了释放应力,氧化膜中形成新的微裂纹或者使原有的裂纹扩展,从而导致外层疏松氧化膜逐渐形成. 另一方面,在金属与氧化膜界面处又不断有新的致密的氧化膜生成,于是阻抗谱呈现出两个时间常数特征. 当腐蚀进行到 302 d 时,Y_{01} 比 42 d 时减小了约两个数量级(见表 1),说明疏松的外层氧化膜明显增厚,并形成一个完整的高频容抗弧. 但该层膜缺陷不断增多、变大,其阻抗值 R_1 却并未增大,而是更小;与腐蚀 42 d 的样品相比,腐蚀 302 d 的样品,Y_{02} 变大、R_2 变小,表明内层氧化膜相对较薄、保护性降低,这可能是图 2 中在 42 d 左右发生转折后腐蚀速率变快的一个主要原因. 由此可知,锆合金表面外层疏松氧化膜在发生转折前就开始形成,并且其保护性较差、也不会随着厚度增加而增强;转折后内层氧化膜的保护作用降低、锆合金腐蚀更快.

表 1 Zr - 4 合金腐蚀不同时间后阻抗谱的拟合结果

Table 1 Fitted results for EIS plots of corroded Zr - 4 alloys with different exposure time

Sample	Time/d	$Y_{01}/\times 10^{-8}$	R_1	$Y_{02}/\times 10^{-6}$	$R_2/\times 10^5$
Z600	42	41.4	4 402	2.81	7.69
Z600	302	0.488	3 919	5.78	1.69
Z800	42	1.39	6 500	1.06	3.37

热处理制度不同的 Z600 样品和 Z800 样品,在腐蚀 42 d 后,它们的阻抗谱的差别较大. 与 Z600 样品相比,Z800 样品表面外层氧化膜较厚,使得 Y_{01} 更低、R_1 更大;Y_{02} 更低,表明内层氧化膜也较厚;但 R_2 却较小,说明 Z800 样品内层膜缺陷更多或者缺陷更大,阻挡过热蒸汽腐蚀的作用更差,这与图 2 中腐蚀增重更大的结果是一致的. 其原因可能是:$Zr(Fe,Cr)_2$ 第二相粒子比合金基体的电位要正,腐蚀时第二相粒子将作为阴极而残存在氧化膜中,但这些未氧化的第二相随着腐蚀时间延长也会逐渐被氧化,从而导致氧化膜产生局部附加应力,形成裂纹等缺陷,或者使已有裂纹扩展[8]. Z800 样品中第二相粒子的尺寸明显比 Z600 样品中的要大(图 1),使得内层氧化膜的缺陷更多或者缺陷更大,从而其阻抗值较小、耐蚀性更低.

4 结论

(1) Zr - 4 合金在 400 ℃,10.3 MPa 过热蒸汽中腐蚀初期形成的氧化膜较为致密、阻抗谱由单一容抗弧组成;随着腐蚀进行,氧化膜逐渐演变为双层结构,内层致密、外层多缺陷,阻抗谱呈现双容抗弧特征;双层结构演变始于腐蚀发生转折之前. 转折后氧化膜阻挡过热蒸汽腐蚀的作用降低,使得锆合金腐蚀速率比转折前更快.

(2) 在 α 相区(600,800 ℃)退火,升高合金退火温度使第二相粒子长大,将导致锆合金氧化膜的阻抗值减小、保护性降低,从而降低锆合金的耐蚀性.

参 考 文 献

[1] Zhou Bangxin(周邦新),Li Qiang(李强),Yao Meiyi(姚美意)*et al*. *Nuclear Power Engineering*(核动力工程)[J],2005,26(4):364.

[2] Garcia E A, Beranger G. *J Nucl Mater*[J],1999,273:221.

[3] Barberis P, Frichet A. *J Nucl Mater*[J],1999,273:182.

[4] Gebhardt O, Hermann A. *Electrochimica Acta*[J],1996,41:1181.

［5］Pan J，Leygraf C，Jargelius-Pettersson R F A *et al*. *Oxidation of Metals*［J］，1999，50：441.

［6］Oskarsson M，Ahlberg E，Pettersson K. *J Nucl Mater*［J］，2001，295：97.

［7］Yao Meiyi(姚美意)，Zhou Bangxin(周邦新)，Li Qiang(李强) *et al*. *Rare Metal Materials and Engineering*(稀有金属材料与工程)［J］，2007，36(11)：1915.

［8］Li Cong(李聪)，Zhon Bangxin(周邦新)，Miao Zhi(苗志)*et al*. *Corrosion Science and Protection Technology*(腐蚀科学与防护技术)［J］，1996，8(3)：242.

EIS Characterization of the Oxide Film on Zr‑4 Alloy Surface in Super-Heated Steam

Abstract：Electrochemical impedance spectroscopy (EIS) is a useful approach to characterize the oxide film structure and its evolution behavior for zirconium alloys. Oxide films formed on Zr‑4 alloys with different heat treatment in super-heated steam at 400 ℃, 10. 3 MPa were characterized by EIS in 10% HCl solution. Results show that the EIS plot is one capacitive arc for the initial oxide film on zircaloy surface in super-heated steam，and then it evolves into two capacitive arcs with the corrosion proceeding. The oxide film is of dual-film structure. The corrosion transition of zirconium alloys in super-heated steam is partly related to the decreased protection of the oxide film. The second phase particles in zirconium alloys play an important role in impedance response of oxide films and corrosion resistance of alloys.

加工工艺对 N18 锆合金在 360 ℃/18.6 MPa LiOH 水溶液中腐蚀行为的影响*

摘 要：将经过不同加工工艺处理的 N18 锆合金样品放入高压釜中，在 360 ℃/18.6 MPa/0.01 mol/L LiOH 水溶液中进行 310 d 的长期腐蚀，用 TEM 和 SEM 观察样品的显微组织，研究加工工艺对 N18 合金腐蚀行为的影响. 结果表明，样品在冷轧退火处理之前进行 β 相水淬处理，得到尺寸在几十纳米均匀弥散分布的第二相，其耐腐蚀性能最好；提高加工过程中的中间退火温度至 740 ℃，样品由于第二相长大到数百纳米使耐腐蚀性能明显变坏，腐蚀到 150 d 时发生转折，转折后的腐蚀速率急剧增加；样品经 780 ℃ 保温 2 h 和 800 ℃ 保温 2h 处理后，其显微组织中存在 β-Zr，这对 N18 锆合金的耐腐蚀性能是有害的，但只要在后续的加工热处理过程中能使 β-Zr 分解，获得细小的第二相，则合金的耐腐蚀性能可恢复到较好的水平.

随着核电站燃料燃耗的加深和燃料元件换料周期的延长，对燃料包壳材料锆合金的性能提出了更高要求. 为此，许多国家都研究开发了新型锆合金[1]. 有代表性的是美国的 ZIRLO 合金和俄罗斯的 E635 合金等. 我国在上世纪 90 年代也开发研究了 N18 和 N36 新型锆合金. 这些合金同属 Zr-Sn-Nb 系列，它们在 LiOH 水溶液中的耐腐蚀性能明显优于 Zr-4 合金[2-4]. 除了合金成分的影响之外，通过优化热处理制度，改变合金的显微组织和第二相分布，也能有效地提高锆合金的耐腐蚀性能[5-9]. 在之前研究 Zr-4 合金的耐腐蚀性能时，有研究[5]认为 Zr(Fe, Cr)₂ 第二相的大小和分布是热处理影响 Zr-4 耐腐蚀性能的主要原因；也有研究[6,7]认为改变热处理制度获得不同大小第二相的同时，α-Zr 中过饱和固溶的 Fe 和 Cr 含量也在发生变化，后者才是影响 Zr-4 合金耐腐蚀性能的重要因素. Comstock 和 Sabol[8]以及本课题组[9]研究了热处理制度对 ZIRLO 锆合金显微组织和耐腐蚀性能的影响，认为得到均匀细小的 β-Nb 第二相是提高 ZIRLO 锆合金耐腐蚀性能的关键. N18 合金是一种低 Nb 的 Zr-Sn-Nb 合金，当加热温度达到 780 ℃ 或更高时，就会进入 $\alpha+\beta$ 双相区[10]，由于 β-Zr 中可以固溶更多的 Nb 以及其他的合金元素，使 β-Zr 在冷却过程中被稳定下来，但在 600 ℃ 长期加热又会发生分解，造成第二相的不均匀分布. 即使样品加热时未进入双相区，在 α 相区加热时第二相也会发生长大，这种显微组织的变化会对 N18 的耐腐蚀性能产生怎样的影响是应该研究的问题. 本文旨在观察不同热处理制度对 N18 锆合金在 LiOH 水溶液中腐蚀行为的影响，并探讨其机理，为优化加工工艺提供依据.

1 实验方法

本实验选用了 3 mm 厚的 N18(Zr-1Sn-0.35Nb-0.35Fe-0.1Cr，质量分数，%)锆合金板材，按照图 1 所示的工艺流程进行轧制加工和热处理，获得 A—G 共 7 类样品. A 和 B 类样品的加工工艺是为了观察显微组织中存在 β-Zr 相时对合金耐腐蚀性能的影响；其余

* 本文合作者：张欣、姚美意、李士炉. 原发表于《金属学报》，2011，47(9)：1112-1116.

的加工工艺是为了研究第二相的大小和分布对合金耐腐蚀性能的影响,为了排除合金元素在 α‑Zr 中含量不同带来的影响,C—G 类样品最终均在 580 ℃进行长时间的退火处理. 样品按标准方法用混合酸(10%(体积分数)HF＋45% HNO₃＋45% H₂O)酸洗和去离子水清洗,经过称重和测量尺寸后,放入高压釜中进行腐蚀,腐蚀条件为 360 ℃/18.6 MPa/0.01 mol/L LiOH 的水溶液. 腐蚀增重由 3 块试样平均值得出. 未经腐蚀的样品用砂纸研磨和机械抛光,在腐蚀剂(甘油：氢氟酸：硝酸＝6：3：1,体积比)中腐蚀 20 s 后,立刻依次用浓硝酸和去离子水清洗,制备扫描电镜(SEM)观察用样品;制备透射电镜(TEM)观察的薄样品时,先在酸洗液中进行化学抛光,减薄至约 0.1 mm 后用冲模获得直径 3 mm 的圆片,再采用双喷电解抛光,电解液为 10%高氯酸乙醇溶液. 用 JSM‑6700F SEM 和 JEM‑200CX TEM 观察样品的显微组织. 用 JEM‑2010F TEM 上的能谱仪(EDS)分析样品中第二相的成分.

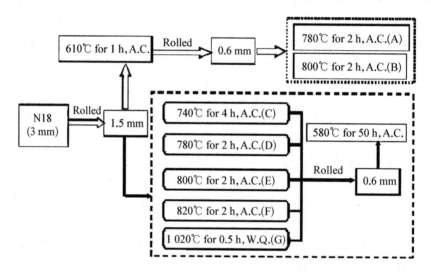

图 1　N18 锆合金样品的加工及热处理工艺图

Fig. 1　Procedures of cold rolling and heat treatments for N18 zirconium alloy specimens A—G (specimens A and B are used in examining the effect of β‑Zr on corrosion resistance and other specimens are used in examining the effect of size and distribution of the second phase particles (SPPs) on corrosion resistance)

2　实验结果与分析

2.1　合金的显微组织

图 2 给出了 B,C,F,G 4 种样品显微组织的 SEM 像. 图 2a 显示样品 B 晶界处存在 β‑Zr,同时晶内还有粒状析出相. N18 锆合金在 780 ℃加热时进入了 α＋β 双相区,随着加热温度的升高,β‑Zr 含量增多[10]. 样品 B 在 800 ℃保温 2 h 后空冷,由于冷却速率较快,β‑Zr 来不及分解而残留在晶界处,这些残留的 β‑Zr 固溶了较多的 Fe,Cr 和 Nb 等合金元素. EDS 分析表明,晶内粒状析出相是含少量 Nb 的 Zr(Fe, Cr)₂,这与本课题组[9]之前报道的结果一致. 图 2b 显示样品 C 中第二相粒子较大,最大尺寸约为 1 μm,这与样品在 740 ℃保温较长时间有关. 样品 D,E,F 经 780~820 ℃保温 2 h 处理后再进行冷轧和随后的 580 ℃保温 50 h

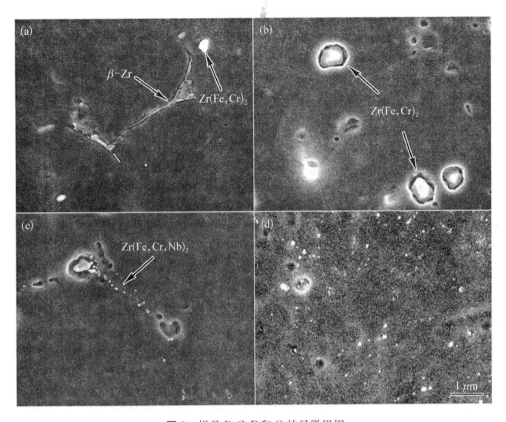

图 2　样品 B,C,F 和 G 的显微组织

Fig. 2　SEM images of specimens B (a), C (b), F (c) and G (d)

退火处理,β-Zr 发生分解,沿着原晶界析出了细小的第二相,图 2c 给出了样品 F 的显微组织照片. EDS 分析表明(图 3),样品 F 中较大和较小的第二相粒子都含 Zr,Nb,Fe 和 Cr,较小的第二相粒子含 Nb 较多,但 Fe/Cr 比基本不变. 这种含有 Zr,Fe,Cr 和 Nb 的第二相应该是 Zr(Fe, Cr, Nb)$_2$,与文献[11]报道结果一致. 图 2d 是经 β 相水淬后再经过冷轧和 580 ℃保温 50 h 退火的样品 G 的显微组织,第二相粒子分布较均匀,尺寸细小,只有几十纳米.

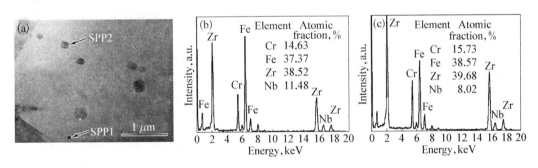

图 3　样品 F 的显微组织及第二相的 EDS 结果

Fig. 3　TEM image of specimen F (a) and EDS analysis of SPP1 (b) and SPP2 (c) in Fig. 3a

2.2　腐蚀增重

图 4 是经过不同加工工艺处理的 N18 样品和用作对比的 Zr-4 样品在 360 ℃/18.6 MPa/

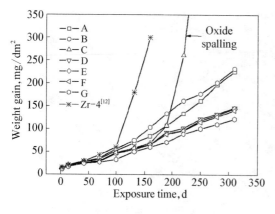

图 4 不同加工工艺处理的 N18 锆合金和 Zr–4 样品在 360 ℃/18.6 MPa/0.01 mol/L LiOH 水溶液中的腐蚀增重曲线

Fig. 4 Curves of weight gains *vs* exposure time of specimens A—G and Zr–4 during corroding in 0.01 mol/L LiOH aqueous solution at 360 ℃/8.6 MPa

0.01 mol/L LiOH 水溶液中腐蚀后的增重曲线, 其中 Zr–4 合金的腐蚀增重数据取自文献 [12]. 可以看出, 在冷轧退火处理之前进行 β 相水淬的样品 G 具有最好的耐腐蚀性能, 腐蚀 310 d 后氧化膜仍然比较黑亮, 腐蚀增重只有 120 mg/dm², 明显低于其他样品; 经 780 ℃ 保温 2 h 和 800 ℃ 保温 2 h 处理的样品 A 和 B 中均有 β–Zr 存在, 它们的耐腐蚀性能很差, 腐蚀 310 d 后氧化膜变为土黄色; 但在 780,800 和 820 ℃ 处理后再进行冷轧和随后 580 ℃ 保温 50 h 退火处理的样品 D, E, F, 由于 β–Zr 发生分解, 获得了细小分布的第二相后又明显改善了合金的耐腐蚀性能; 在 740 ℃ 保温 4 h 处理后经冷轧和随后的 580 ℃ 保温 50 h 退火处理的样品 C, 其耐腐蚀性能最差, 在腐蚀 150 d 后发生转折, 转折后腐蚀速率急剧增加, 腐蚀 220 d 后氧化膜开始剥落 (此时增重达到 260 mg/dm²), 但其耐腐蚀性仍优于 Zr–4 合金.

3 讨论

文献 [8,9] 在研究显微组织对 Zr-Sn-Nb 合金耐腐蚀性能的影响时指出: 得到细小均匀分布的第二相粒子和降低固溶在 α–Zr 中的 Nb 含量是提高 Zr-Sn-Nb 合金耐腐蚀性能的关键. 从这两个角度可以较好地解释本文的实验结果. 经 β 相水淬后再经过冷轧和退火处理的样品 G 中, 第二相粒子尺寸只有几十纳米且分布均匀, 其耐腐蚀性能最佳; 如果提高样品加工过程中的中间退火温度 (低于 α+β 双相区温度), 第二相粒子将会长大到数百纳米, 除了少量的 Nb 存在于 Zr(Fe, Cr)₂ 中, 其余都固溶于 α–Zr 中, 在冷轧和后续的 580 ℃ 保温 50 h 时, Nb 也较难析出, α–Zr 中固溶较高的 Nb, 这使得样品 C 的耐腐蚀性能明显变坏, 在腐蚀 150 d 后发生转折, 转折后腐蚀速率急剧增加; 当样品在 800 ℃ 保温 2 h 并快速冷却后, 其显微组织中有 β–Zr 残留在晶界处; β–Zr 的存在对锆合金的耐腐蚀性能是有害的 [13-15], 但在后续的冷轧和 580 ℃ 保温 50 h 退火后, β–Zr 分解为 α–Zr 和细小的沿原晶粒晶界分布的 Zr(Fe, Cr, Nb)₂, 虽然第二相粒子分布并不均匀但是合金的耐腐蚀性能可恢复到较好的水平.

Zr 在 LiOH 水溶液中的腐蚀过程是: O²⁻ 或 OH⁻ 通过氧化膜中的晶界或阴离子空位扩散到金属/氧化膜界面处, 在金属/氧化膜界面处生长 [16-18]. 因此, 氧化膜的性质以及氧化膜显微组织在腐蚀过程中的演化与锆合金的腐蚀行为密切相关. 本课题组 [12,19] 从锆合金在腐蚀过程中氧化膜显微组织和晶体结构演化规律的角度, 阐述了它们与腐蚀动力学变化之间的密切关系: 由于 Zr 生成 ZrO₂ 时的 P. B. 比 (金属氧化物与金属体积之比) 为 1.56, 氧化膜体积发生膨胀, 同时又受到金属基体的约束, 因此, 氧化膜内部会形成很大的压应力, ZrO₂ 在这种条件下生成时, 晶体中会产生许多缺陷, 稳定了一些亚稳相; 空位、间隙原子等缺陷在温度和应力的作用下,发生扩散、湮没和凝聚,空位被晶界吸收后形成纳米大小的孔隙簇,弱化了晶粒之间的结合力;孔隙簇进一步发展成为微裂纹,使氧化膜失去了原有良好的保护性,因而,发生了腐蚀速率的转折. 图 5 是样品 B,F,G 在 360 ℃/18.6 MPa/0.01 mol/L LiOH 水溶液中腐蚀 190 d 后氧化膜的断口形貌照片. 可以看出,样品 B 在腐蚀 190 d 后氧化膜厚度达到 9 μm,氧化膜断口起伏不平,并有近似平行金属/氧化膜界面的横向微裂纹,ZrO_2 晶粒多为等轴晶,这表明氧化膜在被折断前,内部存在一些弱的结合面;样品 F 氧化膜断口起伏程度明显比样品 B 轻,没有明显的横向微裂纹,只观察到一些孔隙,部分柱状晶正在向等轴晶转变;样品 G 氧化膜断口起伏不大,晶粒组织较为致密,基本以柱状晶为主. 等轴晶是柱状晶中的缺陷通过扩散凝聚构成新的晶界而形成的,这些孔隙不断增加就会发展成为微裂纹,使氧化膜失去了原有良好的保护性. 由于氧化膜生成时产生的压应力是无法避免的,所以氧化膜的显微组织在腐蚀过程中发生不断演化也是必然的. 如能延缓这种显微组织的演化过程,就可以提高锆合金的耐腐蚀性能.

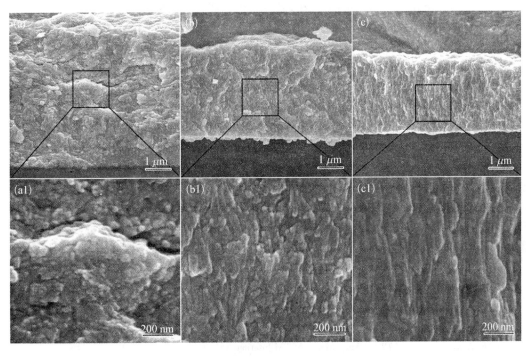

图 5 样品 B, F, G 腐蚀 190 d 的氧化膜断口形貌照片

Fig. 5 Low (a—c) and locally high magnified (a1—c1) fractographs of cross section of oxide films on specimens B (a, a1), F(b, b1) and G(c, c1) corroded for 190 d

目前,在锆合金中添加的合金元素除了 Sn 和 Nb 在 α-Zr 中的固溶度稍大之外,大部分合金元素在 α-Zr 中的固溶度都非常小,它们将与 Zr 形成金属间化合物以第二相形式析出. 这些不同类型的第二相(其中的大部分都比基体 α-Zr 更耐腐蚀)先被裹进氧化膜中,然后再被氧化形成氧化物. 当合金中的第二相尺寸为几十纳米且均匀弥散分布时,细小弥散分布的第二相粒子在氧化膜刚刚形成时会与氧化膜形成相界面,这些界面会作为空位的尾闾,延缓空位通过扩散在 ZrO_2 晶界上凝聚形成孔隙的过程,从而提高锆合金的耐腐蚀性能. 当合金中的第二相分布不均匀时,在第二相集中分布的区域发生氧化后会使氧化膜中产生局部附加应力;同时,当第二相粗大时,其发生氧化后在氧化膜中产生的局部附加应力更加集中,

这都会加速 ZrO_2 晶体中的缺陷扩散凝聚形成孔隙和发展成为微裂纹的过程,从而使合金的耐腐蚀性能变坏.样品 C 的氧化膜发黄脱落也证实了这一点.从这个角度也很好地解释了细小弥散分布的第二相粒子能够提高锆合金的耐腐蚀性,而第二相粗大的样品 C 的耐腐蚀性能最差的原因.

4 结论

（1）样品在冷轧退火处理之前进行 β 相水淬处理,得到尺寸在几十纳米、均匀弥散分布的第二相,其耐腐蚀性能最好;提高加工过程中的中间退火温度,第二相长大到数百纳米后,耐腐蚀性能明显变坏.说明获得细小均匀分布的 $Zr(Fe, Cr, Nb)_2$ 第二相是提高 N18 锆合金耐腐蚀性能的必要条件.

（2）$\beta-Zr$ 的存在对 N18 锆合金的耐腐蚀性能是有害的,但只要在后续的加工热处理过程中能使 $\beta-Zr$ 分解,获得细小的第二相,则合金的耐腐蚀性能可恢复到较好的水平.

参 考 文 献

[1] Zhao W J, Zhou B X, Miao Z, Peng Q, Jiang Y R, Jiang H M, Pang H. *Atom Energ Sci Technol*, 2005；39 (suppl.)：1.

（赵文金,周邦新,苗志,彭倩,蒋有荣,蒋宏曼,庞华. 原子能科学技术,2005；39(增刊)：1)

[2] Sabol G P, Comstock R J. In：Garde A M, Bradley E R eds., *Zirconium in the Nuclear Industry*：*10th International Symposium*, *ASTM STP 1245*, Baltimore, MD：ASTM International, 1994：724.

[3] Nikulina A V, Markelov V A. In：Bradley E R, Sabol G P eds., *Zirconium in the Nuclear Industry: 11th International Symposium*, *ASTM STP 1295*, Garmisch-Partenkirchen Germany：ASTM International, 1996：785.

[4] Zhao W J, Miao Z, Jiang H M. *J Chin Soc Corros Protect*, 2002；22：124.

（赵文金,苗志,蒋宏曼. 中国腐蚀与防护学报,2002；22：124)

[5] Anada H, Herb B J. In：Bradley E R, Sabol G P eds., *Zirconium in the Nuclear Industry*：*11th International Symposium*, *ASTM STP 1295*, Garmisch-Partenkirchen Germany：ASTM International, 1996：74.

[6] Zhou B X, Zhao W J, Miao Z, Pan S F, Li C, Jiang Y R. *Chin J Nucl Sci Eng*, 1995；15：242.

（周邦新,赵文金,苗志,潘淑芳,李聪,蒋有荣. 核科学与工程,1995；15：242)

[7] Yao M Y, Zhou B X, Li Q, Liu W Q, Yu W J, Chu Y L. *Rare Met Mater Eng*, 2007；36：1920.

（姚美意,周邦新,李强,刘文庆,虞伟均,褚于良. 稀有金属材料与工程,2007；36：1920)

[8] Comstock R J, Sabol G P. In：Sabol G P, Bradley E R eds., *Zirconium in the Nuclear Industry: 11th International Symposium*, *ASTM STP 1295*, Ann Arbor：ASTM International, 1996：710.

[9] Liu W Q, Li Q, Zhou B X. *Nucl Power Eng*, 2003；24：33.

（刘文庆,李强,周邦新. 核动力工程,2003；24：33)

[10] Liu Y Z, Zhao W J, Peng Q, Sun C L. *Nucl Power Eng*, 2005；26：158.

（刘彦章,赵文金,彭倩,孙长龙. 核动力工程,2005；26：158)

[11] Isobe T, Matsuo Y. In：Eucken C M, Garde A M eds., *Zircorium in the Nuclear Industry: 9th International Symposium*, *ASTM STP 1132*, Kobe, Japan：ASTM International, 1991；346.

[12] Zhou B X, Li Q, Liu W Q, Yao M Y, Chu Y L. *Rare Met Mater Eng*, 2006；35；1009.

（周邦新,李强,刘文庆,姚美意,褚于良. 稀有金属材料与工程,2006；35；1009)

[13] Choo K N, Kang Y H, Pyum S I. *J Nucl Mater*, 1994；209：226.

[14] Jeong Y H, Kim H G, Kim T H. *J Nucl Mater*, 2003；317：1.

[15] Shen Y F, Yao M Y, Zhang X, Li Q, Zhou B X, Zhao W J. *Acta Metall Sin*, 2011；47：899.

（沈月锋,姚美意,张欣,李强,周邦新,赵文金. 金属学报,2011；47：899)

[16] Liu J Z. *Structure Nuclear Materials*. Beijing：Chemical Industry Press, 2007：111.

（刘建章. 核结构材料. 北京：化学工业出版社，2007：111）

[17] Yao M Y, Wang J H, Peng J C, Zhou B X, Li Q. *J ASTM Int*, 2011；8：13.

[18] Liu W Q, Li Q, Zhou B X, Yao M Y. *Rare Met Mater Eng*, 2004；33：728.

（刘文庆，李强，周邦新，姚美意. 稀有金属材料与工程，2004；33：728）

[19] Zhou B X, Li Q, Yao M Y, Liu W Q, Chu Y L. In: Kammenzind B, Limbäck M eds. , *Zirconium in the Nuclear Industry: 15th International Symposium*, ASTM STP 1505, Sunriver, Oregon, USA：ASTM International, 2009：360.

Effect of Thermal Processing on the Corrosion Resistance of Zirconium Alloy N18 in LiOH Aqueous solution at 360 ℃/18. 6 MPa

Abstract：The effect of thermal processing on corrosion behavior of N18 alloy has been investigated by autoclave tests in 0. 01 mol/L LiOH aqueous solution at 360 ℃ and 18. 6 MPa, and the microstructures of these specimens were examined by TEM and SEM. The results show that the corrosion resistance of specimens is improved obviously by β phase quenching before cold rolling and annealing due to nano second phase particles (SPPs) precipitated dispersively, but when the intermediate annealing temperature increased to 740 ℃ before cold rolling and annealing, the corrosion resistance is lowered greatly due to SPPs coarsened to several hundred nanometers, where the corrosion rate accelerates markedly after the corrosion transition at 150 d exposure. The existence of β-Zr phase in the microstructures of specimens treated at 780 and 800 ℃ for 2 h in dual phase region is harmful to the corrosion resistance, but the corrosion resistance returns to a better level after the decomposition of β-Zr phase during the following processes of cold rolling and annealing at 580 ℃ to obtain fine SPPs.

Study of the Initial Stage and Anisotropic Growth of Oxide Layers Formed on Zircaloy – 4[*]

Abstract: An in situ investigation of the epitaxial oxide layer formed on a thin specimen heated in a transmission electron microscope was carried out. Some dot-like grains about 10 nm in size were formed on the surface of a relatively thin area. The dot-like grains are monoclinic zirconium oxide and have an orientation relationship of $(001)m/\!/(0\bar{1}11)\alpha - Zr$, $(001)m/\!/(1\bar{1}01)\alpha - Zr$, with the $\alpha - Zr$ matrix. Some long strip-like grains, probably a new kind of zirconium suboxide, were formed on the surface of a relatively thick area. The strip-like grains have a bcc structure with a lattice parameter a = 0. 66 nm and have an orientation relationship of $(110)_{bcc}/\!/(10110)_{\alpha\text{-}Zr}$, $[1110]_{bcc}/\!/[0001]_{\alpha\text{-}Zr}$ with the $\alpha - Zr$ matrix. The relationship between the thickness of oxide layers and the grain orientations of the $\alpha - Zr$ matrix was studied with coarse-grained Zircaloy – 4 specimens through autoclave corrosion tests at 500 and 400 ℃ in superheated steam, and at 360 ℃ in both lithiated and deionized water for long time exposure. The results show that the anisotropic growth of oxide layers on the grain surface with different orientations is considerable. However, the relationship between the thickness of oxide layers and the grain orientations of the $\alpha - Zr$ matrix varies with corrosion temperature and water chemistry. The largest variation of oxide thickness developed during corrosion tests at 500 ℃. The thickest oxide layers were formed on those grains whose surface orientations were distributed around the planes from $(01\bar{1}0)$ to $(\bar{1}2\bar{1}0)$. The thicker oxide layers on these grains were further developed into nodular corrosion. When the specimens were corroded at 360 ℃ in lithiated water, the thickest oxide layers formed on those grains, whose surface orientations tilted from 15 to 30° away from the (0001) plane. When the specimens were corroded at 400 ℃ in superheated steam and at 360 ℃ in deionized water, the difference between the thickness of oxide layers on different grain surfaces was less prominent.

1 Introduction

Zirconium alloys of a hexagonal close-packed crystal structure have a prominent anisotropic characteristic in comparison with metals of a cubic structure. The anisotropic characteristic is bound to be reflected in the corrosion behavior of zirconium alloys since strong textures are produced in sheet or tubular materials during the fabrication process. It is worthwhile to make further investigations of the anisotropic growth of the oxide layer because this is one of the essential issues for better understanding of the corrosion

---

[*] In collaboration with Peng J C, Yao M Y, Li Q, Xia S, Du C X. Reprinted from Journal of ASTM International, 2011, 8(1): 620 – 648.

mechanism of zirconium alloys.

It is hard, if not impossible, to prepare single crystals of zirconium alloys, even specimens with large grains, because the mobility of the grain boundaries is obstructed by the formation of many second phase particles (SPPs) distributed in the α – Zr matrix and the decrease of the temperature of the α/β phase transformation after the addition of alloying elements. Most existing research on the relationship between the growth of an epitaxial oxide layer and the crystal orientation of a metal matrix was carried out with small single crystals or coarse grains of pure zirconium, and the testing time was relatively short in different conditions and at different temperatures, such as 1 h in oxygen at 415 ℃[1], 3 days in water (18. 6 MPa) at 360 ℃[2], 40 min in steam (0. 1 MPa) at 500 ℃[3], and 1 h in dry air at 350 – 400 ℃[4]. The thickness of epitaxial oxide layers was very small, only 25 nm to～1 μm. A summary of oxidation anisotropy data was given by Ploc[5] and Cox[6] and was described in more detail by Kim et al. [7]. The results were inconsistent with each other. The reasons for the inconsistency might be attributed to the fact that the corrosion tests were carried out in different conditions, at different temperatures, and for different exposure times. Kim et al. [7] discussed the oxidation characteristics at the initial stage, which are controlled by oxygen diffusion, and the preferred growth of oxide. At this stage, the growth of oxide is related to the crystallographic relationship between the oxide layer and the metal matrix. But at the later stage, the microstructure difference between oxide layers, including the microstructural evolution during the corrosion process, should be taken into account for further growth of the oxide as discussed in a previous paper[8], because the oxygen ions pass the barrier of the oxide layer and reach the oxide/metal interface to form zirconium oxide.

Kim et al. [7] investigated the oxidation characteristics of the basal (0001) plane and the prism (11$\bar{2}$0) plane with small single crystals of pure zirconium. Oxidation tests were carried out in water (18. 9 MPa) at 360 ℃ for 200 h exposure. The results showed that the oxidation rate of the (11$\bar{2}$0) prismatic plane was faster than that of the (0001) basal plane. When the oxide layer on the prism (11$\bar{2}$0) plane was spalling off at～20 μm after 60 h exposure, the oxide layer on the basal (0001) plane was only～2 μm. Zhou et al. [9] pointed out that the nodular corrosion resistance of Zircaloy – 4 specimens was not only related to the temperature of heat treatments, but also the textures of the specimen. This illustrates that the anisotropic character of oxide growth is involved in the corrosion process. In order to clarify the inconsistent results obtained in the past, corrosion tests at different temperatures and in different water chemistry using Zircaloy – 4 coarse-grained specimens are considered in the present work. In situ investigation of the epitaxial oxide layer formed on a thin specimen heated in a transmission electron microscope (TEM) was also performed. The goal of this work is to highlight the role of anisotropic growth and the microstructural evolution of the oxide layer in corrosion behavior at different temperatures and in different water chemistry for long-term exposure.

2　Experimental Procedures

2.1　The Preparation of Coarse-Grained Specimens

In order to investigate the relationship between the oxide layer thickness and the grain orientation of the metal matrix, the effect of grain boundaries on the growth of the oxide layer must be taken into account. In order to minimize this effect, specimens with coarse grains should be prepared.

Flat Zircaloy-4 samples with a composition of 1.44% Sn, 0.24% Fe, and 0.11% Cr (in mass fraction), 20×8 mm^2 in size, were cut from a 1 mm thick plate, vacuum sealed under 5×10^{-3} Pa in quartz capsules, annealed at 1 020 ℃ for 20 min, and then quenched into water while simultaneously breaking the quartz capsules. They were pickled in a mixed solution of 10% HF + 45% HNO$_3$ + 45% H$_2$O (in volume) to remove the blue thin oxide on the surface caused by water quenching and then rinsed in cold tap water. The specimens were vacuum annealed again at 800 ℃ for 10 h to grow the coarse grains. Since a microstructure with fine martensite and a high density of dislocations was produced by water quenching from the high temperature β phase region, coarse grains could be grown during annealing at 800 ℃ in the high temperature alpha phase region. Annealing at 700 ℃ in a vacuum for 100 h, after the growth of coarse grains at 800 ℃, was adopted to reduce the supersaturated solid solution of alloying elements in the α-Zr matrix. Finally, specimens with coarse grains of 0.2-0.8 mm in diameter were obtained. Some SPPs in the α-Zr matrix grew up to the range between 0.5 and 0.8 μm since specimens were annealed at 800 ℃ for 10 h. But some SPPs in less than 0.1 μm can be also detected, which might be precipitated during annealing at 700 ℃.

2.2　Corrosion Tests and the Determination of the Relationship Between the Oxide Layer Thickness and the Grain Orientations of the Metal Matrix

The specimens divided into four groups were first pickled in a mixed solution of 10% HF + 45% HNO$_3$ + 45% H$_2$O, then rinsed in cold tap water, and finally rinsed in boiling deionized water. Corrosion tests were carried out in static autoclaves at 500 ℃/10.3 MPa or 400 ℃/10.3 MPa superheated steam and at 360 ℃/18.6 MPa deionized water or lithiated water of 0.01M LiOH.

After the corrosion tests, the specimens were mounted into a Bi-Pb alloy to prepare specimen cross-sections. Then, the specimen sections were ground with sand paper, mechanically polished, and pickled in a mixed acid solution as mentioned above to remove the deformation layer on the section surface. Electron backscatter diffraction (EBSD) was employed for the determination of grain orientations of the metal matrix on the specimen sections using a HKL™ EBSD equipment attached to a CamScan Apollo-300 scanning electron microscope (SEM), and the thickness of the oxide layers was also measured from

SEM micrographs at the magnification of 500 – 5 000 times.

2. 3 The Preparation of Transmission Electron Microscope Specimens for In Situ Investigation of the Epitaxial Oxide Layer

The double-jet electropolishing method was adopted to prepare thin specimens for TEM examination. Electropolishing was carried out at about − 40 ℃ with 50 V dc in an electrolyte of $10\%HClO_3+90\%$ ethanol (in volume). Thin specimens placed in a specimen heating holder were heated in the specimen chamber of a TEM (Hitachi H – 800). After making a general examination, the specimens were cooled down and transferred to a JEM – 2010F TEM for recording high resolution lattice images.

A DigitalMicrograph^{TM} was used for processing the high resolution lattice images recorded by charge coupled device (CCD) camera. The fast Fourier transforms (FFTs), which are involved in DigitalMicrograph^{TM}, can reveal the periodic contents of images and also work on regions of interest with appropriate dimensions placed on an image. They are useful for image analysis and filtering to obtain information about the crystal structure and the defects in crystals.

2. 4 The Preparation of the Specimen Surface for Scanning Electron Microscope Examination

A JSM – 6700F SEM was employed for examining the grain morphology of the oxide layer on the specimen surface after corrosion tests. A thin layer of iridium coating on the specimens applied by sputtering is required for improving the quality of images.

3 Experimental Results and Discussion

3. 1 Initial Stage of the Oxidation

Even in the high vacuum condition ($\sim10^{-3}$ Pa) of the TEM, a thin oxide film formed with different interference colours that outline the original metal grain structure after the specimen was heated for 2 days at 200 ℃, as shown in Fig. 1. This indicates that the in situ investigation of epitaxial oxide layer growth on zirconium alloys by TEM is possible.

When a thin specimen was heated at 400 ℃ for 2 h, different morphologies and different crystal structures of the epitaxial oxide layers were observed around the penetrated hole of the thin specimen at different distances from the edge of the hole, as shown in Fig. 2. At the

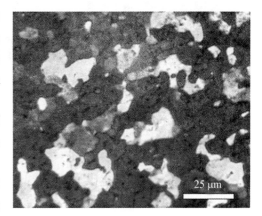

Fig. 1 A thin oxide film formed with different interference colours to outline the original grain structure of a Zircaloy – 4 specimen heated at 200 ℃ in TEM for 2 days of exposure (optical miscroscopy)

Fig. 2 Different morphologies and different crystal structures of the epitaxial oxide layers formed around the hole of a thin specimen at different distances

relatively thin area around the hole, some dot-like grains about 10 – 20 nm in size were detected by a selected area diffraction (SAD) pattern and the bright/dark field images, as shown in Fig. 3. Figure 4(a) shows the high resolution lattice image for both the dot-like grains and the α – Zr matrix, and the FFT patterns indexed for both areas are also incorporated in this figure. Figure 4(b) shows the high resolution lattice image for the dot-like grains only. By analyzing the indexed FFT patterns and measuring the line spacing, it is indicated that the dot-like grains are monoclinic zirconium oxide (m – ZrO_2), and have an orientation relationship of $(001)_m /\!/ (0111)_{\alpha\text{-}Zr}$, $(111)_m /\!/ (1101)_{\alpha\text{-}Zr}$ with the α – Zr matrix. This relationship is inconsistent with the data obtained through experiments or by theoretical prediction in the past, which were discussed by Ploc[5]. Perhaps this relationship is different on different α – Zr crystal planes on which the m – ZrO_2 grows.

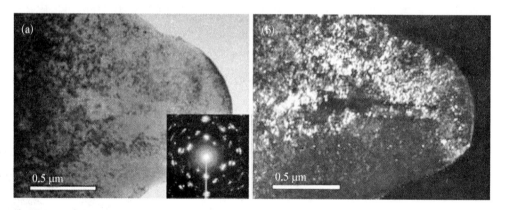

Fig. 3 (a) The bright field image, (b) dark field image, and SAD pattern of the dot-like grains formed on the relatively thinner area near the hole of a thin specimen

As the lattice parameter of the α – Zr is known, the spacing of crystal planes for m – ZrO_2 in the thin layer oxide can be determined accurately by calibration of the values obtained from measuring the lattice spacing on the high resolution lattice image with the inverse FFT operation. The d values of (001), ($\bar{1}$11) and (111) planes for m – ZrO_2 of the dot-like grains are 0.482, 0.273, and 0.240 nm, respectively, and the d values of these planes for the standard m – ZrO_2 (No. 37 – 1484) are 0.508 70, 0.316 47, and 0.284 06 nm. It is clear that the lattice parameter of m – ZrO_2 formed on the thin specimens at the initial stage is smaller than that of standard m – ZrO_2 owing to the constraint of the α – Zr matrix due to a high Pilling-Bedworth (PB) ratio of 1. 56. The difference of the spacing for different crystal planes compared to the standard $m - ZrO_2$ is about $5\% - 15\%$ smaller. There must be a high compressive stress produced during the formation of the oxide layer. When looking at the high resolution lattice image in Fig. 4(b) carefully, the distortion of the lattice image by the existence of defects, including some dislocations, and the formation of a "cell structure" of about 10 nm with a little difference in orientation between each cell are revealed. Some parts of the image containing defects are further enlarged to see them in detail and are also incorporated in Fig. 4(b). This is the reason that the SAD pattern shown in Fig. 3(a) is like a diffraction pattern of a strong fiber texture. The nanocrystals of $m - ZrO_2$ in the thin oxide layer have detached from the restriction of the matrix and adjusted their orientation as a unit of about $10 - 20$ nm to form Moiré patterns as shown in Fig. 5 after further growth of the oxide layer.

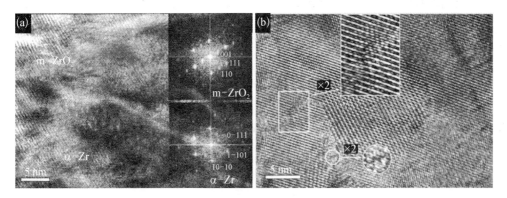

Fig. 4 High resolution lattice images of the dot-like grains showing the orientation relationship with (a) α - Zr matrix and (b) some defects in the lattice

Many long strip-like grains grown at relatively thick areas on the thin specimen with a certain distance from the edge of the hole were observed. This distance in Fig. 2 is between 6 and 10 μm. These strip-like grains formed on the (0001) plane of the α - Zr matrix crossed each other to form a 60° angle as shown in Fig. 6 and preferentially grew up along the $\langle 11\bar{2}0 \rangle_{\alpha\text{-Zr}}$ direction at a rate of $\sim$100 nm/min. A SAD pattern including the α - Zr matrix and the long strip-like grains was indexed and is incorporated in Fig. 6. Through the analysis of the SAD pattern, a bcc structure with a lattice parameter $a = 0. 66$ nm and the orientation relationship of $(110)_{bcc} // (10\bar{1}0)_{\alpha\text{-Zr}}$, $[110]_{bcc} // [0001]_{\alpha\text{-Zr}}$ with the α - Zr matrix was identified. This long strip-like phase is likely a new kind of zirconium suboxide that has not been documented yet. Oxygen was adsorbed on the surface of the specimen and diffused into the α - Zr matrix during specimen heating at 400 ℃ in vacuum; consequently, the oxygen content of the matrix at the relatively thick area will be lower than at the relatively thin area. This might explain the fact that the oxide structure formed on the surface at different thickness areas around the hole is different. An attempt was made to investigate whether this kind of phase could transform to other structures after a long time exposure at 400 ℃. But this failed due to contamination after heating for a long time.

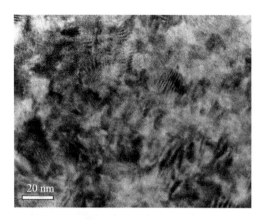

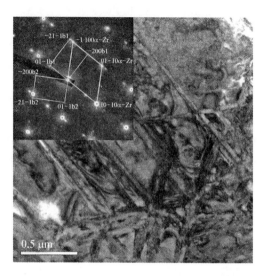

Fig. 5 The Moiré patterns in the areas of the dot-like grains formed

Fig. 6 The morphology and SAD pattern of the long strip-like grains formed on the relatively thicker area at a certain distance from the edge of the hole

3. 2　Corrosion Behavior of the Coarse-Grained Specimens During Autoclave Testing

The weight gains versus exposure time for the specimens corroded at 360 ℃ in deionized water and in lithiated water and at 400 and 500 ℃ in superheated steam are shown in Fig. 7(a)-7(d), respectively. An average value of weight gain was taken from four to six specimens. The variation in weight gains for parallel coarse-grained specimens was higher than the spread for fine-grained specimens. This means that the grain number on the surface of coarse-grained specimens is not sufficient to give reasonable statistical figures of weight gain. In other words, big specimens with large surface areas are needed, since the thickness of oxide layers on metal grains with different orientations is different.

There is no obvious transition phenomenon, particularly for the coarse-grained specimens corroded at 360 ℃ in lithiated water. In general, the corrosion transition appears after 100 days exposure of Zircaloy-4 fine-grained specimens corroded at 360 ℃ in lithiated water, and the corrosion rate is accelerated markedly after the transition. The difference of the transition phenomena between coarse-grained and fine-grained specimens is related to the anisotropic growth of the oxide layer and the textures of the specimens to be discussed in the latter part of this paper.

3. 3　Surface Morphology of the Specimens After Corrosion Tests

The surface morphology of the specimens after corrosion tests is shown in Fig. 8. The undulation of the oxide layer on the specimen surface is obvious due to the variation of the oxide thickness on different grains, but it was developed to a different extent for specimens

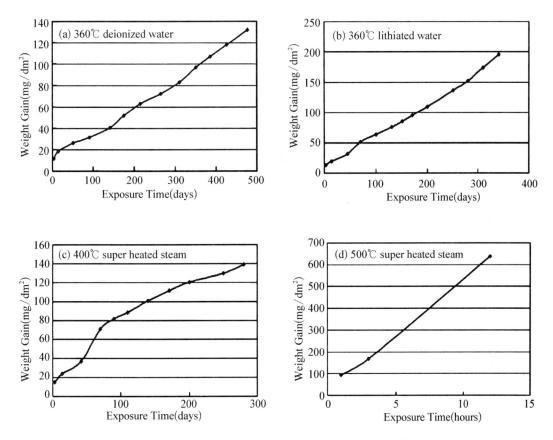

Fig. 7 The weight gain versus exposure time of the specimens corroded at different temperatures and in different water chemistry: (a) 360 ℃/18. 6 MPa deionized water; (b) 360 ℃/18. 6 MPa lithiated water with 0. 01M LiOH; (c) 400 ℃/10. 3 MPa superheated steam; and (d) 500 ℃/10. 3 MPa superheated steam

corroded at different temperatures and in different water chemistry. The most serious one was developed during the corrosion test performed at 500 ℃ in superheated steam. The thicker oxide layers on those grains were developed in a lenticular shape when the exposure time lasted from 3 to 12 h as shown in Fig. 9. It shows that nodular corrosion took place. A lot of cracks on the surface of nodules look like a flower as shown in Fig. 10(a). When the corrosion tests were carried out at 360 ℃ in lithiated water, thicker oxide layers on some grains were developed after 340 days exposure, but they were not in a lenticular shape when examined on the sections and not many cracks appeared on the surface of the thickest oxide layers as shown in Fig. 10(b). The difference of oxide layer thickness on the different grains for the specimens corroded at 400 ℃ in superheated steam after 280 days exposure and at 360 ℃ in deionized water after 476 days exposure was also detected, but it was not as serious as the specimens corroded at 500 ℃ in superheated steam or at 360 ℃ in lithiated water. Due to the similarity of the surface morphology of specimens corroded at 360 ℃ in deionized water to those corroded at 400 ℃ in superheated steam, only the surface morphologies of specimens corroded at 400 ℃ in superheated steam are given in Fig. 8.

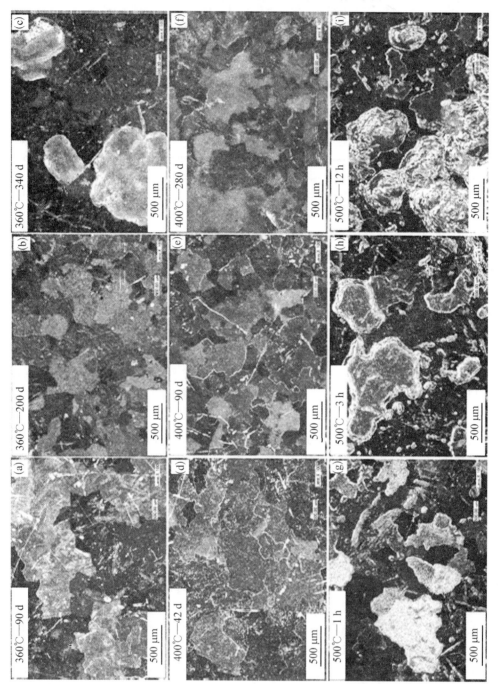

Fig. 8 Optical micrographs of the surface morphology of the specimens corroded at different temperatures, in different water chemistry, and for different time exposures: [(a)-(c)] Corroded at 360℃ in lithiated water; [(d)-(f)] corroded at 400℃ in superheated steam; and [(g)-(i)] corroded at 500℃ in superheated steam

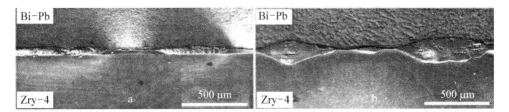

Fig. 9 The morphology of the specimen sections corroded at 500 ℃ in superheated steam for (a) 3 h exposure and (b) 12 h exposure showing the development of nodular corrosion (SEM)

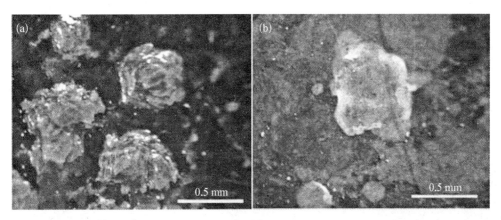

Fig. 10 The surface morphology of (a) the nodules corroded at 500 ℃ in superheated steam for 12 h exposure and (b) the thickest oxide layers corroded at 360 ℃ in lithiated water for 340 days exposure

3. 4 Grain Morphology of Oxide on the Specimen Surface

In order to get more information about the characteristics of anisotropic growth of oxides, the grain morphology of oxides on the surface of the specimens corroded at different temperatures in different water chemistry and for different exposure times was examined using a SEM. Particularly, the morphology difference of oxide grains on the oxide layers with different thicknesses for a same specimen was compared. Cracks on the oxide surface, micro-cracks along grain boundaries, and pores on triple junctions of grain boundaries were revealed on the relatively thick oxide layers after long time exposure. The difference of oxide grain morphology on the oxide layers of different thicknesses is more obvious when the specimens were corroded at 500 ℃ in superheated steam and at 360 ℃ in lithiated water than at 400 ℃ in superheated steam and at 360 ℃ in deionized water. Some typical morphologies of oxide grains are shown in Fig. 11, and the corrosion conditions, weight gains, and the average thickness of oxide layers for these specimens used for the examination of the grain morphology of oxide are given in Table 1. The average thickness of an oxide layer is calculated according to its weight gain (1 μm = 15 mg/dm^2). The average thickness of oxide layers for these four specimens is comparable. The morphology of grains formed on the thicker oxide layer areas of the specimen corroded at 500 ℃ after

1 h exposure is shown in Fig. 11(a). Some lumps protruded on the oxide surface to make it relatively rough, and grains less than 30 nm in diameter were revealed much more clearly by the pores and micro-cracks formed along grain boundaries. But a relatively smooth surface with the micro-cracks along the grain boundaries was revealed on thinner oxide layers as shown in Fig. 11(b). The morphology of grains formed on thicker or thinner oxide layers for the specimen corroded at 400 ℃ after 90 days exposure is shown in Fig. 11(c), 11(d). Except for a little difference in grain size, pores and microcracks along grain boundaries were revealed in both cases. The different morphologies of grains formed on thicker or thinner oxide layers for the specimen corroded at 360 ℃ in lithiated water after 90 days exposure are shown in Fig. 11(e), 11(f). It is interesting to note that grains with some orientation on the surface could grow up into a strange shape like whiskers, about 50 nm in diameter and up to several hundred nanometres in length. The mechanism of this phenomenon is not clear, but it must be related to the decrease in the free surface energy of zirconium oxide formed in lithiated water as compared to the discussion of the growth mechanism of metal whiskers of Sn in the middle of the last century[10]. The decrease of

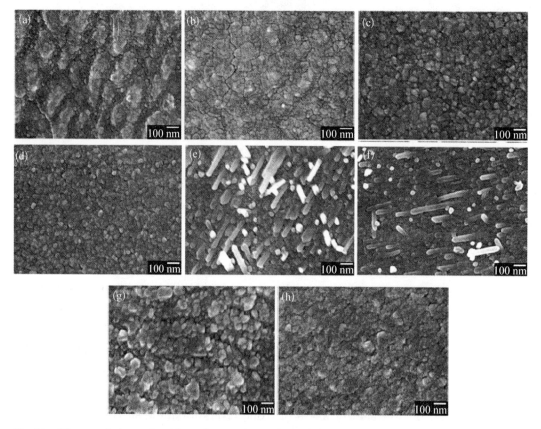

Fig. 11　The morphology of oxide grains on the surface of the specimens corroded at [(a) and (b)] 500 ℃ in steam for 1 h exposure, [(c) and (d)] 400 ℃ in steam for 90 days exposure, [(e) and (f)] 360 ℃ in lithiated water for 90 days exposure, and [(g) and (h)] 360 ℃ in deionized water for 265 days exposure. (a), (c), (e), and (g) show the grain morphology on the thicker oxide layers, and (b), (d), (f), and (h) show the grain morphology on the thinner oxide layers

Table 1 Specimens used for the examination of the grain morphology of oxide

	Exposure Time	Weight Gain (mg/dm²)	Average Thickness of Oxide (μm)
500 ℃ in superheated steam	1 h	94.5	6.3
400 ℃ in superheated steam	90 days	81.6	5.4
360 ℃ in lithiated water	90 days	61.5	4.1
360 ℃ in deionized water	265 days	65.8	4.4

the free surface energy of zirconium oxide formed in lithiated water was discussed in a previous paper[8]. More whisker-like grains on the relatively thick oxide layers had grown up towards different directions from the surface in comparison with those on the thinner oxide layers. It seems that a certain crystal direction is parallel to the axis of whisker-like grains. This means that grain orientations of oxide in thicker oxide layers varied considerably compared with those in thinner oxide layers; therefore, this difference must be related to the orientations of the original metal grains. The microstructure and the orientation of oxide layers formed on the metal grains with different orientations must be different. In order to clarify this relationship, some more detailed investigations are needed. The same results were observed on the specimens corroded at 360 ℃ in lithiated water for 200 days exposure, but the whisker-like grains were more on the thicker oxide layers than those for 90 days exposure. The different morphologies of grains formed on thicker or thinner oxide layers for the specimen corroded at 360 ℃ in deionized water after 265 days exposure are shown in Fig. 11(g), 11(h). There were no whisker-like grains grown on the surface of oxide layers, but the pores and micro-cracks were obviously developed on thicker oxide layers than on thinner oxide layers.

3.5 The Relationship Between the Thickness of the Oxide Layer and the Surface Orientation of Metal Grains

Figure 12 shows the relationship between the oxide layer thickness and the surface orientation of metal grains obtained from the specimen corroded at 500 ℃ in superheated steam for 3 h exposure. The weight gain of this specimen is 161 mg/dm² and the average thickness of the oxide layer is 10.7 μm. The poles of the surface orientation of 97 metal grains on which the oxide layer thickness was measured are plotted in an inverse pole figure as shown in Fig. 12(a). There is no texture in the specimen as the distribution of poles is random. The difference of oxide layer thickness is divided into three ranges: 3 – 6, 10 – 20, and 30 – 60 μm; then three lines of different thickness are used to link the poles having the same range of oxide layer thickness as shown in Fig. 12(b). The anisotropic growth of the oxide layer is prominent, and the thickness difference of the oxide layers on different grains of the α – Zr matrix is more than 20 times. The oxide layers are the thickest on the surface orientation of metal grains distributed in the range from $(01\bar{1}0)$ to $(12\bar{1}0)$, and the maximum thickness reached 60 μm. A mean thickness of oxide layers from 10 to 20 μm is observed on the surface orientation of metal grains distributed in the range from 40 to 60°

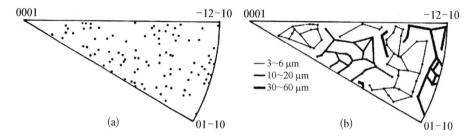

Fig. 12 (a) Inverse pole figure showing the orientations of the metal grain surface on which the thickness of the oxide layer was measured. (b) The relationship between the oxide thickness and the grain surface orientations of the metal matrix for the specimens corroded at 500 ℃ in superheated steam after 3 h exposure

away from (0001). On other surface orientations of metal grains, the thickness of oxide layers is only 3 - 6 μm.

Figure 13 shows the relationship between the oxide layer thickness and the surface orientation of metal grains obtained from the specimen corroded at 400 ℃ in superheated steam for 280 days exposure. The weight gain of this specimen is 140 mg/dm^2 and the average thickness of the oxide layer is 9.3 μm. The poles of the surface orientation of 86 metal grains on which the oxide layer thickness was measured are plotted in an inverse pole figure as shown in Fig. 13(a). The oxide layer thickness is divided into two ranges: 5 - 8 and 9 - 12 μm; then two lines of different thickness are used to link the poles having the same range of oxide layer thickness as shown in Fig. 13(b). Two poles having an oxide layer thickness of 14 - 15 μm are incorporated in the range of 9 - 12 μm, and denoted with an underline in Fig. 13(b). In general, the thickness of oxide layers is somewhat different on different surface orientations of metal grains after the specimen is corroded at 400 ℃ in superheated steam, but it is not as prominent as for the specimen corroded at 500 ℃ in superheated steam. Some relatively thick oxide layers from 9 to 12 μm were observed on the surface orientation of metal grains distributed within 50° around the [0001] direction as shown in Fig. 13(b).

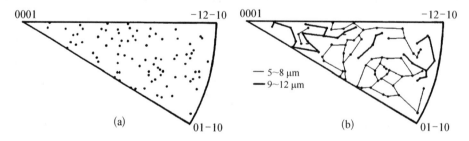

Fig. 13 (a) Inverse pole figure showing the orientations of the metal grain surface on which the thickness of the oxide layer was measured. (b) The relationship between the oxide thickness and the grain surface orientations of the metal matrix for the specimens corroded at 400 ℃ in superheated steam after 280 days exposure

Figure 14 shows the relationship between the oxide layer thickness and the surface orientation of metal grains obtained from the specimen corroded at 360 ℃ in lithiated water for 340 days exposure. The weight gain of this specimen is 210 mg/dm^2 and the average thickness of the oxide layer is 14 μm. The poles of the surface orientation of 92 metal grains on which the oxide layer thickness was measured are plotted in an inverse pole figure as shown in Fig. 14(a). The oxide layer thickness is divided into three ranges: 8 – 11, 12 – 20, and 25 – 65 μm; then three lines of different thickness are used to link the poles having the same range of oxide layer thickness as shown in Fig. 14(b). The thickest oxide layers were detected on the surface orientation of metal grains distributed in the range from 15 to 30° away from the (0001) plane, and the maximum thickness of the oxide layer reached 65 μm. The mean thickness of oxide layers, from 12 to 20 μm, was observed on the surface orientation of metal grains distributed about 15° around the area where the thickest oxide layer was detected. In other areas, the thickness of oxide layers was only 8 – 11 μm, except for three poles that had an oxide layer thickness of 15 – 20 μm.

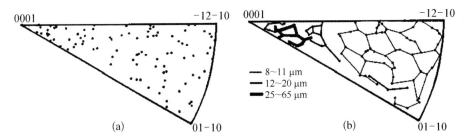

Fig. 14 (a) Inverse pole figure showing the orientations of the metal grain surface on which the thickness of the oxide layer was measured. (b) The relationship between the oxide thickness and the grain orientations of the metal matrix for the specimens corroded at 360 ℃ in lithiated water after 340 days exposure

Figure 15 shows the relationship between the oxide layer thickness and the surface orientation of metal grains obtained from the specimen corroded at 360 ℃ in deionized water for 476 days exposure. The weight gain of this specimen is 134 mg/dm^2 and the average thickness of the oxide layer is 8.9 μm. The poles of the surface orientation of 119 metal grains on which the oxide layer thickness was measured are plotted in an inverse pole figure as shown in Fig. 15(a). The oxide layer thickness is divided into two ranges: 5.5 – 8.5 and 9 – 11 μm; then two lines of different thickness are used to link the poles having the same range of oxide layer thickness as shown in Fig. 15(b). Some relatively thick oxide layers, from 9 to 11 μm, were detected on the surface orientation of metal grains distributed within 45° around the [0001] direction. This relationship between the oxide layer thickness and the surface orientation of metal grains is different in comparison with that obtained from specimens corroded at 360 ℃ in lithiated water, but it is similar to specimens corroded at 400 ℃ in superheated steam.

The summary of weight gain, average thickness of the oxide layer, and the range of the oxide thickness distributed on different metal grains of four specimens used for analyzing the relationship between the oxide thickness and the grain orientation of the metal matrix is given in Table 2. The average thickness of the oxide layer for these four

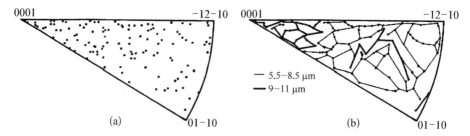

Fig. 15 (a) Inverse pole figure showing the orientations of the metal grain surface on which the thickness of the oxide layer was measured. (b) The relationship between the oxide thickness and the grain orientations of the specimens corroded at 360 ℃ in deionized water for 476 days exposure

specimens is comparable, from 9 to 14 μm, but the range of the oxide thickness distributed on different metal grains is obviously different. The anisotropic growth of the oxide layer is more prominent for specimens corroded at 500 ℃ in superheated steam and at 360 ℃ in lithiated water than those corroded at 400 ℃ in superheated steam and at 360 ℃ in deionized water.

Table 2 Specimens used for analyzing the relationship between the oxide thickness and the grain orientations of the metal matrix

	Exposure Time	Weight Gain (mg/dm²)	Average Thickness of Oxide (μm)	The Range of the Oxide Thickness Distributed on Different Metal Grains (μm)
500 ℃ in superheated steam	3 h	161	10. 7	3 – 60
400 ℃ in superheated steam	280 days	140	9. 3	5 – 12
360 ℃ in lithiated water	260 days	210	14	8 – 65
360 ℃ in deionized water	476 days	134	8. 9	5. 5 – 11

3. 6 The Variation of the Relationship Between the Anisotropic Growth of the Oxide Layer and the Surface Orientation of Metal Grains

It is clear from Figs. 12 – 15 that the relationship between the anisotropic growth of the oxide layer and the surface orientation of metal grains varies with the corrosion temperature and water chemistry for long time exposure. This means that the thickness of oxide layers does not only depend on the surface orientation of metal grains. There must be another factor playing an important role in affecting the growth of the oxide layers at the later stage of the oxidation.

A new oxide layer is always formed on the oxide/metal interface. The microstructural evolution of the oxide certainly affects the diffusion of oxygen ions and anion vacancies, and, consequently, changes the conditions for continued oxide growth. Zhou et al. [8] investigated the effect of water chemistry and composition on the microstructural evolution of oxide on zirconium alloys, and pointed out that the defects consisting of vacancies and interstitials were produced during oxide growth due to the high compressive stress caused by a high PB ratio of 1.56. The characteristics of oxides with such a microstructure have an internal cause, and the temperature and the time are the external causes that induced the microstructural evolution through the diffusion, annihilation, and condensation of vacancies and interstitials under the action of stress, temperature, and time during the corrosion process. It was observed in the middle of the oxide layer that the vacancies absorbed by grain boundaries formed the pores to weaken the bonding strength between grains, and then the microcracks formed through the development of the pores. The corrosion temperatures, as well as the water chemistry, certainly affect the microstructural evolution and, consequently, the behavior of oxide growth. The results obtained by in situ investigation of epitaxial oxide layer growth using TEM also show that the defects were produced during oxide growth (Fig. 4).

Li$^+$ and OH$^-$ incorporated in oxide films were adsorbed on the wall of pores or entered into vacancies[11-13] to reduce the free surface energy of zirconium oxide during exposure in lithiated water[8]. As a result, the diffusion of vacancies and the formation of pores were enhanced, including the acceleration of the microstructural evolution and the degradation of the corrosion resistance. In other words, the growth rate of oxide layers was accelerated. The relationship among the corrosion temperature and water chemistry, the surface orientation of metal grains, and the anisotropic growth of oxide layers is schematically shown in Fig. 16. The most important factor affecting the characteristic of oxide layer growth is the microstructural evolution of the oxide layers at the later stage of oxidation, which is put into a thicker frame in Fig. 16. In the early stage of the oxide layer growth, the diffusion of oxygen into the α - Zr matrix and the oxide crystal matched with the metal crystal orientation play an important role. Therefore, the epitaxial oxide layer grown on metal grains with different interference colours can outline the original grain structure of the α - Zr matrix (Fig. 1). But at a later stage of the oxidation, the oxide layer growth mostly depends on the microstructural evolution of the oxide layers. As a result, the variation of the relationship between the anisotropic growth of the oxide layer and the surface orientation of the metal grains with the corrosion temperature and water chemistry

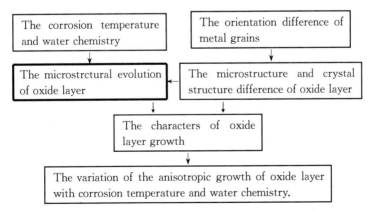

Fig. 16 Schematic diagram showing the relationship among the corrosion temperature and water chemistry, crystal orientation of metal grains, and anisotropic growth of oxide layer

could be better understood. But the mechanism of this phenomenon is not clear so far. Further investigation is needed.

3.7 The Relationship Between the Corrosion Resistance and the Textures of the Specimens

The thickness of the oxide layers on the metal grains with different orientations is markedly different after the coarse-grained specimens were corroded for a long time exposure. The relationship between the anisotropic growth of the oxide layers and the surface orientation of the metal grains varies with the temperature and water chemistry of the corrosion tests. It is deduced that there must be a relationship between the corrosion resistance and the textures of the specimens, which could also vary with corrosion temperature and water chemistry.

A typical (0001) pole figure and inverse pole figures of normal direction (ND), rolling direction (RD), and transverse direction (TD) for a Zircaloy - 4 sheet specimen after annealing at 600 ℃ for 2 h are shown in Fig. 17. The textures of tubular specimens are very similar to those of sheet specimens. Such textures can be described as that the majority of the grains have their c-axis distributed within 20° around an axis tilted ±28° away from the ND of the sheet towards the TD. The textures become concentrated by grain growth after annealing at 820 ℃ for 1 h as shown in Fig. 18. In this case, the majority of the grains have their c-axis distributed within 20° around an axis parallel to the ND of the sheet specimen, and the planes distributed in the range from ($01\bar{1}0$) to ($\bar{1}2\,\bar{1}0$) are parallel to the side planes around the sheet specimen as shown by RD and TD inverse pole figures in Fig. 18. Comparing the inverse pole figures of the textures (Figs. 17 and 18) to the inverse pole figures showing the relationship between the thickness of oxide layers and the metal grain orientations (Figs. 12 and 14), it is clear that the oxide layer on the normal plane of the sheet specimens corroded at 360 ℃ in lithiated water becomes thicker than that on the side planes around the sheet specimens after a long time exposure. And the development of the nodular corrosion on the side planes of the specimens corroded at 500 ℃ in superheated steam is easier than that on the normal plane of the specimens. This is a principle for the discussion on the relationship between the corrosion resistance and the textures of the specimens.

The weight gain versus exposure time for fine-grained and coarse-grained specimens corroded at 360 ℃ in lithiated water is plotted together in Fig. 19. The data of fine-grained specimens are taken from Zhou et al.[8]. The weight gain for the fine-grained specimens shows a transition phenomenon after 100 days exposure, and the corrosion rate is accelerated markedly after the transition in comparison with the specimens corroded at 360 ℃ in deionized water or at 400 ℃ steam[8,14]. But this is not the case for coarse-grained specimens with no texture corroded in lithiated water at 360 ℃ (Fig. 7). The reason for this is that the texture brings the crystal planes of the majority of the grains, on which the oxide layer grows faster after the transition, to be parallel to the normal plane of the specimens. Although the oxide layer on some grain surfaces is thicker than on others in

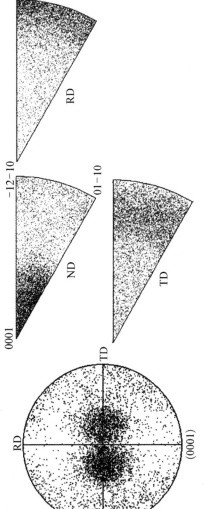

Fig. 17 Recrystallization textures of the specimen annealed at 600°C for 2 h: (0001) pole figure; ND, RD, and TD inverse pole figures

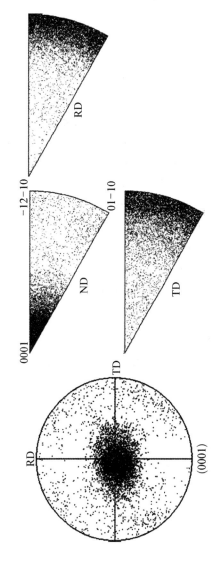

Fig. 18 Recrystallization textures of the specimen annealed at 820°C for 1 h: (0001) pole figure; ND, RD, and TD inverse pole figures

coarse-grained specimens corroded in lithiated water, there is no transition phenomenon in the curve of weight gain versus exposure time since there is no texture.

The nodular corrosion resistance is improved by the increase of iron and chromium solid solution in the α-Zr matrix after increasing the annealing temperature in the alpha phase region[15] or irradiated in a reactor[16] to enhance the dissolution of small SPPs. Figure 20 shows the weight gain versus exposure time for Zircaloy-4 specimens corroded at 500 ℃ in superheated steam after heat treatment at 600 ℃ for 2 h or at 820 ℃ for 1 h[9]. The nodular corrosion resistance is obviously improved by the increase of the annealing temperature in the alpha phase region, since the solid solution of iron and chromium alloying elements in the α-Zr matrix has been increased. The surface morphology of the specimens corroded at 500 ℃ in superheated steam is also shown in Fig. 21[9]. For the specimens annealed at 600 ℃, many nodules distribute on the surface of the specimen after 7 h exposure. For the specimens annealed at 820 ℃, there are no nodules on the normal plane but a very thick oxide layer developed from the nodules around the edges of the specimen after 90 h exposure. The nodules were completely eliminated on the normal plane of the specimens, but not on the edges of the specimens. This phenomenon is attributed to the fact that the texture brings the different crystal planes on the different surface of the specimens. By comparing the texture of the specimens annealed at 820 ℃ (Fig. 18) to the relationship between the oxide thickness and the grain surface orientations of the α-Zr matrix for the specimens corroded at 500 ℃ (Fig. 12(b)), it can be seen that the normal plane of the specimens annealed at 820 ℃ have a good resistance against nodular corrosion, but this is not the case for the side planes around the sheet specimens.

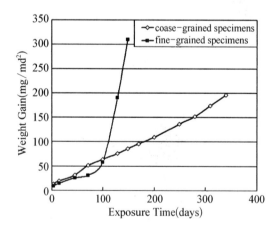

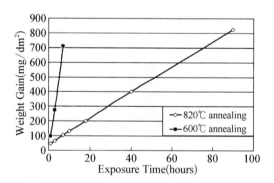

Fig. 19 Weight gain versus exposure time for Zircaloy-4 fine-grained and coarsegrained specimens corroded at 360 ℃ in lithiated water

Fig. 20 Weight gain versus exposure time during the corrosion tests at 500 ℃ in superheated steam for Zircaloy-4 specimens annealed at 600 ℃ for 2 h and at 820 ℃ for 1 h

The corrosion behavior of irradiated Zircaloy was investigated by Cheng et al. [16]. The results showed that nodular corrosion as indicated by 520 ℃ steam testing was completely eliminated after the tubular specimens had been irradiated in BWRs at a temperature near

288 ℃. But post-irradiation annealing at 650 ℃ for 2 h caused a return of nodules to the specimen only on the edges, not on the tubular surface. This phenomenon is also attributed to the textures of the specimens because the textures of tubular specimens are very similar to that of sheet specimens.

4 Conclusions

Fig. 21 The morphology of Zircaloy – 4 specimens corroded at 500 ℃ in superheated steam: (a) Specimen annealed at 600 ℃ for 2 h after 7 h exposure; (b) specimen annealed at 820 ℃ for 1 h after 90 h exposure

The investigation of the relationship between the thickness of oxide layers and the grain orientations of the α – Zr matrix was performed using Zircaloy – 4 coarse-grained specimens through corrosion tests at 500 ℃ and at 400 ℃ in superheated steam, and at 360 ℃ in lithiated water and in deionized water for long time exposure. In situ investigation of the epitaxial oxide layer formed on a thin specimen heated at 400 ℃ in TEM was also carried out. The following conclusions have been obtained.

(1) Dot-like grains about 10 nm in size were formed on the surface of a relatively thin area around the hole of a thin specimen. The dot-like grains are monoclinic zirconium oxide (m – ZrO_2), and have an orientation relationship of $(001)_m$ ∥ $(0111)_{\alpha\text{-}Zr}$, $(111)_m$ ∥ $(1101)_{\alpha\text{-}Zr}$ with the α – Zr matrix. Long strip-like grains, probably a new kind of zirconium suboxide, were formed on the surface of a relatively thick area at a certain distance from the edge of the hole of a thin specimen. The strip-like grains have a bcc structure with a lattice parameter $a=0.66$ nm, and have an orientation relationship of $(110)_{bcc}$ ∥ $(1010)_{\alpha\text{-}Zr}$, $[110]_{bcc}$ ∥ $[0001]_{\alpha\text{-}Zr}$ with the α – Zr matrix.

(2) The anisotropic growth of oxide layers was most prominent when the corrosion tests were carried out at 500 ℃ in superheated steam in comparison with the corrosion tests at 400 ℃ in superheated steam and at 360 ℃ in deionized water. The thickest oxide layers were formed on those grains whose surface orientations were distributed around the planes from $(01\bar{1}0)$ to $(\bar{1}2\bar{1}0)$. The thicker oxide layers on those grains were further developed into nodular corrosion to form a lenticular shape observed on the sections as the exposure time lasted from 3 to 12 h.

(3) When the specimens were corroded in lithiated water at 360 ℃, the thickest oxide layers formed on those grains, whose surface orientations tilted from 15 to 30° away from the (0001) plane, but they were not in a lenticular shape.

(4) When the specimens were corroded in superheated steam at 400 ℃ or in deionized water at 360 ℃, the difference of oxide layer thickness on different metal grain surfaces was less prominent. Some relatively thick oxide layers were observed on the surface

orientation of metal grains distributed within 45° around the [0001] direction.

(5) The relationship between the anisotropic growth of oxide layers and the surface orientation of metal grains varies with the corrosion temperature and water chemistry for long time exposure, and it does not only depend on the surface orientation of the metal grains. In order to explain this phenomenon, it is suggested that the microstructural evolution of the oxide layers plays an important role in the growth of oxide layers at the later stage of oxidation.

(6) There is a relationship between the corrosion resistance and the textures of the specimens. This relationship varies with corrosion temperature and water chemistry, since the relationship between the anisotropic growth of oxide layers and the surface orientation of metal grains is also varied with corrosion temperature and water chemistry.

Acknowledgments

This project was financially supported by the National Nature Science Foundation of China (Grant Nos. 50871064 and 50971084), the National High Technology Research and Development Program of China (Grant No. 2008AA031701), and the Shanghai Leading Academic Discipline Project (Grant No. S30107). The writers wish to acknowledge the Instrumental Analysis and Research Center of Shanghai University for help in the microstructural analysis.

References

[1] Pemsler, J. P., "Diffusion of Oxygen in Zirconium and Its Relation to Oxide and Corrosion," *J. Electrochem. Soc.*, Vol. 105, 1958, pp. 315–322.

[2] Bibb, A. E. and Fascia, J. R., "Aqueous Corrosion of Zr Single Crystals," *Trans. Metall. Soc. AIME*, Vol. 230, 1964, pp. 415–419.

[3] Wanklyn, J. N., "Corrosion of Zirconium Alloys," *ASTM Spec. Tech. Publ.*, Vol. 368, 1964, pp. 66.

[4] Wilson, J. C., "Anisotropy of Oxidation in Zr (Abstract of Paper Presented at the 15th Annual USAEC Corrosion Symposium)," ORNL–3970, 1966.

[5] Ploc, R. A., "Transmission Electron Microscopy of Thin (<200 nm) Thermally Formed ZrO_2 Films," *J. Nucl. Mater.*, Vol. 28, 1968, pp. 48–60.

[6] Cox, B., "Some Thoughts on the Mechanisms of In-Reactor Corrosion of Zirconium Alloys," *J. Nucl. Mater.*, Vol. 336, 2005, pp. 331–368.

[7] Kim, H. G., Kim, T. H., and Jeong, Y. H., "Oxidation Characteristics of Basal (0002) Plane and Prism (11–20) Plane in HCP Zr," *J. Nucl. Mater.*, Vol. 306, 2002, pp. 44–53.

[8] Zhou, B. X., Li, Q., Yao, M. Y., Liu, W. Q., and Chu, Y. L., "Effect of Water Chemistry and Composition on Microstructural Evolution of Oxide on Zr Alloys," *Zirconium in the Nuclear Industry: 15th International Symposium*, ASTM–STP–1505, Sunriver, OR, 2009, B. Kammenzind and M. Limback, Eds., ASTM International, West Conshohocken, PA, pp. 360–383.

[9] Zhou, B. X., Yao, M. Y., Li, Q., Xia, S., and Liu, W. Q., "Relationship Between Nodular Corrosion Resistance and Textures for Zircaloy–4 Strips," *J. Shanghai Univ.*, Vol. 14, 2008, pp. 441–445 (Natural Science Edition).

[10] Ge, T. S., "Metal Whiskers and the Theoretical Strength of Metals," *Crystal Defects and Metal Strength*, Vol. 2, Science Publishing House, Beijing, 1964, pp. 191–246, (in Chinese).

[11] Ramasubramanian, N. , Precoanin, N. , and Ling, V. C. , "Lithium Uptake and the Accelerated Corrosion of Zirconium Alloys," *Zirconium in the Nuclear Industry: Eighth International Symposium*, ASTM-STP-1023, San Diego, CA, 1989, L. F. P. Van Swam and C. M. Eucken, Eds. , ASTM International, West Conshohocken, PA, pp. 187-201.

[12] Pêcheur, D. , Godlewski, J. , Peybernes, J. , Fayette, L. , Noe, M. , Frichet, A. , and Kerrec, O. , "Contribution to the Understanding of the Effect of the Water Chemistry on the Oxidation Kinetics of Zircaloy-4 Cladding," *Zirconium in the Nuclear Industry: 12th International Symposium*, ASTM-STP-1354, G. P. Sabol and G. D. Moan, Eds. , Toronto, Canada, 2000, ASTM International, West Conshohocken, PA, pp. 793-811.

[13] Oskarsson, M. , Ahlberg, E. , and Pettersson, K. , "Oxidation of Zircaloy-2 and Zircaloy-4 in Water and Lithiated Water at 360 ℃ ," *J. Nucl. Mater.* , Vol. 295, 2001, pp. 97-108.

[14] Nikulina, A. V. , Markelov, V. A. , Peregud, M. M. , Bibilashvili, Y. K. , Kotrekhov, V. A. , Lositsky, A. F. , Kuzmenko, N. V. , Shevnin, Y. P. , Shamardin, V. K. , Kobylyansky, G. P. , and Novoselov, A. E. , "Zirconium Alloy E635 as a Material for Fuel Rod Cladding and Other Components of VVER and RBMK Cores," *Zirconium in the Nuclear Industry: 11th International Symposium*, ASTM-STP-1295, Garmisch-Partenkirchen, Germany, 1996, E. R. Bradley and G. P. Sable, Eds. , ASTM International, West Conshohocken, PA, pp. 785-804.

[15] Zhou, B. X. , Zhao, W. J. , Miao, Z. , Pan, S. F. , Li, C. , and Jiang, Y. R. , "The Effect of Heat Treatments on the Corrosion Behavior of Zircaloy," *China Nuclear Science and Technology Report Nos.* CNIC-01074 and SINRE-0066, China Nuclear Information Centre Atomic Energy Press, Beijing, China, 1996.

[16] Cheng, B. C. , Kruger, R. M. , and Adamson, R. B. , "Corrosion Behavior of Irradiated Zircaloy," *Zirconium in the Nuclear Industry: Tenth International Symposium*, ASTM-STP-1245, Baltimore, MD, 1994, A. M. Garde and E. R. Bradley, Eds. , ASTM International, West Conshohocken, PA, pp. 400-418.

Study on the Role of Second Phase Particles in Hydrogen Uptake Behavior of Zirconium Alloys[*]

Abstract: In an effort to better understand the role of second phase particles (SPPs) in the hydrogen uptake of zirconium alloys, four alloys and four heat treatments for each alloy were chosen to prepare specimens with different SPPs size distributions and area fractions. The hydrogen uptake performance of these specimens was investigated after autoclave testing in 400 ℃/10. 3 MPa steam. Results show that the hydrogen uptake is not always in a strict corresponding relationship with the corrosion resistance among the specimens, but it is closely related to the size, area fraction, and compositions of the SPPs. In the case of Zry - 2 and Zry - 4, the hydrogen uptake fraction (HUF) increased with increasing size and area fraction of the SPPs. The dependence was more notable for the Zry - 2 than the Zry - 4. In the case of N36 and N18, the HUF had only a slight variation with the size and area fraction of the SPPs. No matter which heat treatment was employed, the corrosion resistance of the N18 specimens was superior to the N36 specimens, but the HUF of the former was larger than that of the latter. These results clearly demonstrate that the effect of the size and area fraction of SPPs on the hydrogen uptake depends on the SPP compositions. Pressure-composition-temperature and kinetics of absorbing and desorbing hydrogen tests were conducted on $Zr(Fe, Cr)_2$, $Zr_2(Fe, Ni)$, $Zr(Nb, Fe)_2$, and $\beta - Nb$ alloys (which may be found as SPPs in the four zirconium alloys tested) as well as on pure zirconium. Results show that $Zr(Nb, Fe)_2$, $Zr_2(Fe, Ni)$, and $Zr(Fe, Cr)_2$ alloys have a stronger reversible ability for hydrogen absorption and desorption than $\beta - Nb$ alloy and pure zirconium. Based on the testing results, a model correlating the hydrogen uptake performance to the reversible ability of the SPPs to absorb and desorb hydrogen is proposed. The model can successfully explain the results.

1 Introduction

Zirconium alloys have been widely used as the cladding material for fuel rods in pressurized water reactors. They are exposed to water at high temperature and high pressure during service. The corrosion reaction with water to form the oxide has to be matched by the evolution of hydrogen. A fraction of the hydrogen produced from the corrosion is taken up by the alloy. Since the solid solubility of hydrogen in $\alpha - Zr$ is very low and reduces with decreasing temperature (e. g. , it is less than 1 $\mu g/g$ at ambient temperature and it is 200 $\mu g/g$ at 400 ℃ [1]), the remainder of the hydrogen is precipitated

---

* In collaboration with Yao M Y, Wang J H, Peng J C, Li Q. Reprinted from Journal of ASTM International, 2011, 8(2): 466 - 495.

as zirconium hydride, causing hydrogen embrittlement of the alloy. An in situ transmission electron microscope (TEM) observation has confirmed that the zirconium hydride precipitates, grows, and cracks at the tip of cracks induced by tensile stresses, which is the essential process of hydrogen-induced delayed cracking (HIDC)[2,3]. Studies in the past 40 years have indicated that corrosion, hydrogen uptake, and the concomitant HIDC are mechanisms for the failure of zirconium alloys used as nuclear fuel cladding as well as other structural parts in light water and heavy water reactors [4-7]. With the requirement of extending fuel assembly burn-up, hydrogen ingress is receiving more attention.

The general understanding is that the higher the weight gain, the higher the hydrogen uptake for a given alloy, as the hydrogen produced as a result of the corrosion reaction has also increased. So, generally speaking, an improvement in the corrosion resistance by adjusting the heat treatment procedure or alloying composition can also reduce the amount of hydrogen uptake. For example, the hydrogen uptake of M5[8] and ZIRLO[9] alloys are significantly lower than that of Zircaloy - 4 (denoted as Zry - 4). In a previous paper[10], however, it was found that a lower weight gain did not show a lower amount of hydrogen uptake in the molten zone of zirconium-alloy welding specimens. That is to say, the relative hydrogen uptake from specimen to specimen does not always correspond to the relative corrosion weight gain from specimen to specimen. It was obvious that the hydrogen uptake increased with the increase of Cr content in the molten zone. This illustrates that hydrogen uptake is also a function of alloy composition.

Recently, more and more investigators have realized that the second phase particles (SPPs) play an important role in the hydrogen uptake of zirconium alloys, but there is no general agreement on how the SPPs affect the hydrogen uptake. The first thought is that the SPPs affect the oxide layer characteristics (such as cracks, pores, and dense barrier layer thickness), thereby influencing the hydrogen uptake[11-16]. The second thought is that the unoxidized SPPs in the oxide film can act as a preferred path for hydrogen transport through the oxide film[7,17-21]. The third thought is that the SPPs in the metallic state acts as trapping sites while the microcracks resulting from the transition of tetragonal-to-monoclinic ZrO_2 provide the fast transport route of hydrogen through the oxide[22,23].

In order to better understand the role of SPPs in the hydrogen uptake of zirconium alloys, four zirconium alloys and four heat treatments for each alloy were chosen to prepare specimens with different SPPs compositions, amounts, and sizes. In this paper, the hydrogen uptake behavior of these specimens corroded in 400 ℃/10. 3 MPa steam is investigated and the involved mechanism is discussed.

2　Experimental Procedures

2. 1　Materials

The four alloys used for the present investigation were Zry - 4, Zircaloy - 2 (Zry - 2),

N36, and N18. Their chemical compositions are listed in Table 1. The Zry - 4 and Zry - 2 are well known Zr - Sn alloys. The N18 alloy is an advanced Zr - Sn - Nb alloy with a low Nb content that was developed by China. Compared to the N18 alloy, the N36 alloy does not contain Cr but contains a higher Nb content. In fact, its composition is close to the ZIRLO alloy. From previous investigations, it is known that the SPPs are $Zr(Fe, Cr)_2$ in Zry - 4[19], $Zr_2(Fe, Ni)$ and $Zr(Fe, Cr)_2$ in Zry - 2[7,20,24], β - Nb and $Zr(Nb, Fe)_2$ in N36[25-27], and $Zr(Fe, Cr)_2$ with Nb in N18[26]. The as-received Zry - 4 and Zry - 2 plates at 0. 6 mm thickness were fabricated by conventional procedures[28] and underwent a final recrystallization anneal at 600 ℃ for 1 h. The as-received N36 and N18 plates at 1 mm thickness were also fabricated by conventional procedures for Zr - Sn - Nb alloys[29] and underwent a final recrystallization anneal at 580 ℃ for 2 h.

Table 1 The composition of zirconium alloys (in mass fraction%)

Alloy	Sn	Fe	Cr	Ni	Nb	O	C	Si	N	H
Zircaloy - 4	1. 5	0. 2	0. 1	/	/	0. 11	0. 010	<0. 004	0. 007	0. 003
Zircaloy - 2	1. 5	0. 2	0. 1	0. 05	/	0. 12	0. 012	<0. 004	0. 007	0. 002
N36	1. 0	0. 3	0. 1	/	0. 35	0. 10	0. 010	<0. 004	0. 006	0. 003
N18	1. 0	0. 3	/	/	1. 0	0. 13	0. 010	<0. 004	0. 005	0. 002

2. 2 Heat Treatments

Zry - 4 and Zry - 2 specimens — The heat treatment procedures for the Zry - 4 and Zry - 2 alloys are shown in Fig. 1(a). First, 25×20 mm specimens were cut from the as-received plates. Then the specimens were heated to 1 020 ℃ for 0. 5 h, 800 ℃ for 36 h, and 720 ℃ for 10 h in a vacuum furnace followed by pushing the furnace away from the quartz tube and cooling by spraying water (denoted as AC.) The specimens treated in the above three ways and the as-received specimens annealed at 600 ℃ will be described hereafter as Specimens 11, 12, 13, and 14 for the Zry - 4, and Specimens 21, 22, 23, and 24 for the Zry - 2, respectively.

N36 and N18 Specimens — The heat treatment procedures for the N36 and N18 alloys are shown in Fig. 1(b). In the case of Specimen 31 for the N36 alloy and Specimen 41 for the N18 alloy, the heat treatment procedures are as follows: (1) Strips at 8 mm width were cut from the as-received plates. (2) The strips were sealed in evacuated quartz capsules at about 5×10^{-3} Pa and heated to 1 020 ℃ (β phase) for 20 min. (3) The capsules were then quenched into water and broken simultaneously (denoted as **WQ**). (4) The strips were cold rolled from 1 mm to 0. 6 mm and were cut into specimens with 25 mm in length. (5) The specimens were heated to 580 ℃ for 50 h to decompose the residual β - Zr from the former β - quenching. In the case of Specimens 33 and 34 for the N36 alloy and Specimens 43 and 44 for the N18 alloy, they were first heated to 820 ℃ for 2 h and 700 ℃ for 4 h, respectively; then they were prepared in the same way as the procedures (4) and (5) for Specimen 31. In the case of Specimens 32 and 42, they were first cold rolled from

1 mm to 0.6 mm; then they were heated to 820 ℃ for 2 h; finally, they were prepared in the same way as procedure (5) for Specimen 31. Prior to each heat treatment, the specimens were cleaned and pickled in a solution of 10% HF+45% HNO$_3$+45% H$_2$O (in volume) for the Zry-4 and Zry-2, and in a solution of 10% HF+30% HNO$_3$+30% H$_2$SO$_4$+30% H$_2$O (in volume) for the N36 and N18 alloys.

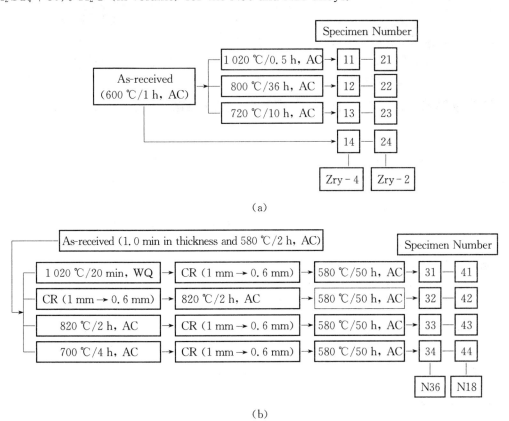

Fig. 1　The heat treatment procedures and corresponding designation of (a) Zry-4 and Zry-2 and (b) N36 and N18 alloys

2.3　Microstructural Characterization

The SPPs distribution was examined using a JSM-6700F scanning electron microscope (SEM). The specimens for SEM examination were prepared by pickling in a solution of 30% HF+10% HNO$_3$+60% glycerol (in volume). The SPPs analysis was carried out at an acceleration voltage of 10 kV and at a magnification of between ×5 000 and ×20 000 depending on the specimen (Fig. 2). The size distribution and the area fraction of the SPPs were measured using German SIS image analysis system (Version 5.0). More than eight fields and 400 particles per specimen were analyzed to determine the size distribution and area fraction.

In the case of the Zr-Sn-Nb alloys, it was found that the SPPs were more finely and uniformly distributed when the specimens were heat-treated by β-quenching, cold rolling, and annealing at 580 ℃, which is lower than the monotectic reaction temperature

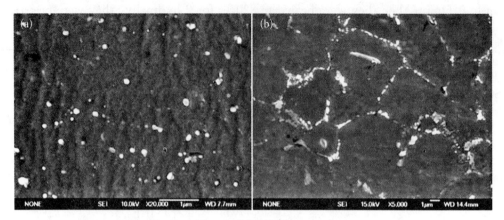

Fig. 2 SEM micrographs of the SPPs in Specimens (a) 31 and (b) 32

(610 ℃), as shown in Fig. 2 (a); while the SPPs distribute mainly along the grain boundaries when the specimens were heat-treated in the α+β two-phase region before final annealing at 580 ℃, as shown in Fig. 2(b). It is well known that β–quenching puts more alloying elements into a supersaturated solid solution in the α matrix. Meanwhile, subsequent cold rolling and annealing make the alloying elements supersaturated in solid solution in the α matrix to precipitate as SPPs and the residual β–Zr to decompose into α–Zr and β–Nb. Therefore, the SPPs are more finely and uniformly distributed in the α–Zr when the specimens were heat-treated in such a way. When heat-treated in the α+β two-phase region, the residual β–Zr is mainly distributed along the grain boundaries, so after the subsequent annealing at 580 ℃, the SPPs resulting from the decomposition of the residual β–Zr are mainly distributed along the grain boundaries.

2. 4　Corrosion Tests

The corrosion tests were performed in H_2O steam at 400 ℃/10. 3 MPa in a static autoclave. The corrosion behavior was evaluated by weight gain as a function of exposure time. The reported weight gain is a mean value obtained from three specimens. Prior to the corrosion tests, the specimens were cleaned and pickled in the same solution as was used to prepare them for heat treatment, sequentially rinsed in cold tap water, boiling deionized water, and then blow-dried with warm air.

2. 5　Hydrogen Analysis

The hydrogen content in all the corroded specimens after each exposure was analyzed by the hot vacuum extraction method with LECO Model RH – 600 equipment. The reported hydrogen content is also a mean value obtained from three measurements for each corroded specimen and the standard deviation was less than 10 $\mu g/g$. For each measurement, a corroded specimen with a weight between 0. 1~0. 2 g was used. All tested specimens were ultrasonically cleaned in acetone before measurement.

The hydrogen uptake fraction (HUF), which is defined as a ratio of the measured hydrogen content in the specimen to the theoretical value calculated from the corrosion reaction, was used to characterize the hydrogen performance.

2. 6 Second Phase Particle Alloys Preparation and Hydrogen Absorption and Desorption Characteristic Tests

As mentioned in the introduction, the SPPs may play an important role in the hydrogen uptake of zirconium alloys. Hence, a separate study of the hydrogen absorption and desorption characteristics of the SPP alloys is carried out to further understand the hydrogen uptake mechanism affected by the SPPs. Many studies have been performed on the hydrogen absorption and desorption characteristics of the $Zr(Fe, Cr)_2$ alloy as a hydrogen storage material[30-33], but they mainly have been concerned on its characteristics at near ambient temperature. So, it is necessary to investigate the hydrogen absorption and desorption characteristics of SPP alloys at high temperature.

According to the SPP compositions in the four alloys, five alloys, including two $Zr(Fe, Cr)_2$ alloys with Fe/Cr atomic ratios of 2 and 0. 5 (denoted as $Zr(Fe, Cr)_2$ R2 and $Zr(Fe, Cr)_2$ R0. 5, respectively), one $Zr_2(Fe, Ni)$ alloy with an Fe/Ni atomic ratio of 1, one $Zr(Nb, Fe)_2$ alloy with an Nb/Fe atomic ratio of 1, and one $\beta-Nb$ alloy (10Zr-90Nb, mass fraction percent), were chosen for this study. Each SPP alloy was arc melted six times under argon in a non-consumable arc melting furnace. X-ray diffraction analysis showed that the five SPP alloys were of the same structure as the SPPs in the four investigated zirconium alloys. The pressure-composition-temperature (P - C - T) and kinetic characteristics for the five SPP alloys and pure zirconium in a powder state (38. 5~55 μm in size) were measured at 4 MPa H_2 pressure, 360 ℃ using a Suzuki-Shokan multi-channel Seivert-type P - C - T system. Before the P - C - T and kinetics tests, each sample was activated three times with a 1 h hydrogen absorbing cycle at 4 MPa H_2 pressure followed by a 1 h hydrogen desorbing cycle at 360 ℃.

3 Results

3. 1 Microstructural Characteristics

Zry - 4 Specimens — The results from the SEM analysis of the specimens are summarized in Table 2. In the case of the Zry - 4, the mean diameter of the SPPs in Specimen 11 is very small, only 35 nm, whereas Specimen 12 has the largest mean diameter of the SPPs, 280 nm. Specimens 13 and 14 have similar mean SPPs diameters of 216 and 180 nm, respectively. The area fraction of the SPPs in Specimen 11 is also very small, only 0. 1%, whereas Specimens 12 and 13 have a larger and similar SPPs area fraction, 1. 9% and 2. 0%, respectively. Specimen 14 has a moderate SPPs area fraction, 1. 2%. The SPPs size distributions of the specimens are illustrated in Fig. 3(a). It is clear that Specimen 11, which was heat-treated at 1 020 ℃/0. 5 h, by far has the smallest SPPs size distribution with 85% of the SPPs being less than 50 nm. Specimens 12, 13, and 14, which were heat-treated in the α - phase temperature range (800～600 ℃), all have a higher fraction of SPPs in sizes between 50 and 300 nm, 70%～80%; the fraction of SPPs larger than 300 nm decreases with decreasing heat-treatment temperature.

Table 2　The mean size and area fraction of the SPPs in the four zirconium alloys with different heat treatments

Alloy	Number	SPPs mean size(nm)	SPPs area fraction(%)
Zry - 4	11	35	0. 1
	12	280	1. 9
	13	216	2. 0
	14	180	1. 2
Zry - 2	21	61	0. 6
	22	494	1. 9
	23	323	1. 3
	24	136	0. 7
N36	31	77	5. 9
	33	320	5. 9
	33	209	6. 0
	34	197	5. 4
N18	41	58	2. 3
	42	350	1. 6
	43	243	2
	44	242	2. 9

Zry - 2 Specimens — In the case of the Zry - 2, the mean diameter of the SPPs in Specimen 21 is very small, only 61 nm, whereas Specimen 22 has the largest mean diameter of the SPPs, 494 nm, followed by Specimens 23 (323 nm) and 24 (136 nm). The area fractions of the SPPs in Specimens 21 and 24 are smaller and similar, only 0. 6% and 0. 7%, respectively, while Specimen 22 has the largest SPPs area fraction, 1. 9%. Specimen 23 has a moderate SPPs area fraction, 1. 3%. The SPPs size distributions of the specimens are illustrated in Fig. 3(b). It is clear that Specimen 21, which was heat-treated at 1 020 ℃/0. 5 h, by far has the smallest SPPs size distribution with 50% of the SPPs less than 50 nm. Specimens 22, 23, and 24, which were heat-treated in the α - phase temperature range (800～600 ℃), all have a higher fraction of SPPs in sizes between 50 and 300 nm, 53%～74%, and the fraction of SPPs larger than 300 nm decreases with decreasing heat-treatment temperature. The trend for Zry - 2 specimens is to have larger SPPs than Zry - 4 specimens.

N36 Specimens — In the case of the N36 alloy, the mean diameter of the SPPs in Specimen 31 is very small, only 77 nm, whereas Specimen 32 has the largest mean diameter of the SPPs, 320 nm. Specimens 33 and 34 have a moderate and similar mean diameter of the SPPs of 209 and 197 nm, respectively. The area fractions of the SPPs among the four specimens are close to each other, 5. 4%～6. 0%. The SPPs size distributions of the specimens are illustrated in Fig. 3(c). It is clear that Specimen 31, which was heat-treated

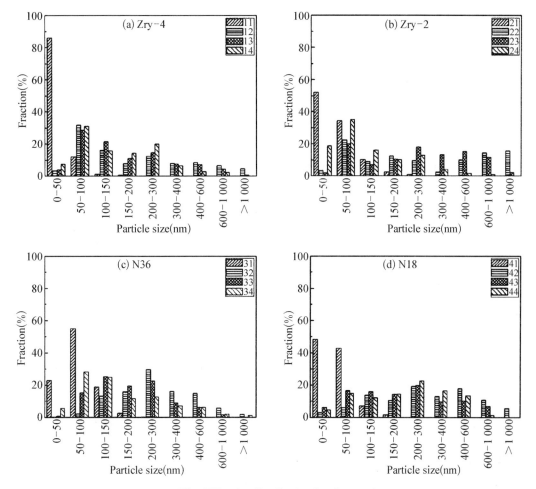

Fig. 3 The SPPs size distribution for the specimens

by the "1 020 ℃/20 min，WQ+CR+580 ℃/50 h，AC" process，by far has the smallest SPPs size distribution with 80% of the SPPs less than 100 nm in diameter. Specimens 32，33，and 34 all have a higher fraction of the SPPs between 100 and 300 nm，59%，67%，and 50%，respectively. The cold rolling before annealing at 580 ℃/50 h decreases the SPPs size.

N18 Specimens — In the case of the N18 alloy，the mean diameter of the SPPs in Specimen 41 is very small，only 58 nm，whereas Specimen 42 has the largest mean diameter of the SPPs，350 nm. Specimens 43 and 44 have a moderate and similar mean diameter of the SPPs of 243 and 242 nm，respectively. The area fractions of the SPPs in the four specimens are 1. 6%~2. 9% and decrease in the sequence of Specimen 44>41>43>42. The SPP size distributions of the specimens are illustrated in Fig. 3(d). It is clear that Specimen 41，which was heat-treated by the "1 020 ℃/20 min，WQ+CR+580 ℃/50 h，AC" process，by far has the smallest SPPs size distribution with 90% of the SPPs less than 100 nm in diameter. Specimens 42，43，and 44 all have a similar size distribution of SPPs with less than 400 nm diameter and the SPPs size fractions with less than 400 nm diameter decrease in the sequence of Specimen 44>43>42.

Comparing the effect of heat treatments on the size distribution and area fraction of the SPPs in the four alloys (Table 2 and Fig. 3), it is clear that the heat treatment effect decreases in the sequence of Zry‒2>Zry‒4>N18>N36 for the SPPs size distribution while it decreases in the sequence of Zry‒4>Zry‒2>N18>N36 for the SPPs area fraction. Moreover, the SPPs area fraction is the largest for the N36 alloy due to a higher content of alloying elements, followed by the N18 alloy.

3.2　Corrosion and Hydrogen Uptake Behavior

Zry‒4 Specimens — The hydrogen uptake content ΔW_H of the Zry‒4 specimens is plotted against the weight gain ΔW_O in Fig. 4(a). In the early stage of corrosion, the HUF, i. e. , $d\Delta W_H/d\Delta W_O$, is similar among the four types of Zry‒4 specimens. As the corrosion proceeds, the $d\Delta W_H/d\Delta W_O$ shows differences among the four types of Zry‒4 specimens and increases in the sequence of Specimens 11<14<13<12. It is well known that at the same weight gain, the theoretical hydrogen content from corrosion is the same. So, if the hydrogen uptake of these specimens is the same, these curves should overlap. However, it is obvious that the hydrogen contents are different among the four specimens at the same weight gain. This illustrates that the heat treatments have a significant effect on the hydrogen uptake of the Zry‒4 from corrosion.

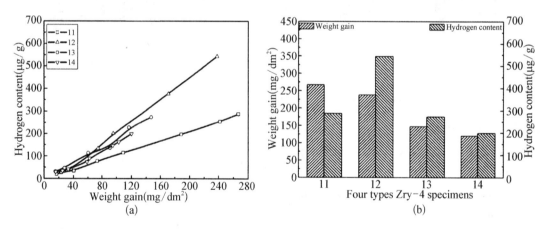

Fig. 4　The corrosion and hydrogen uptake behavior of the Zry‒4 specimens tested in 400 ℃/10. 3 MPa steam: (a) The curve of the hydrogen uptake versus weight gain. (b) Corrosion weight gain and hydrogen uptake content of the specimens corroded for 302 days

Figure 4(b) shows the weight gain and hydrogen uptake content of the Zry‒4 specimens corroded in 10. 3 MPa steam at 400 ℃ for 302 days. Specimen 11 exhibits the largest weight gain (267 mg/dm²) but a lower hydrogen uptake content (288 μg/g). The weight gain of Specimen 12 (238 mg/dm²) is only a little lower than that of Specimen 11; however, the hydrogen uptake content of the former (543 μg/g) is much higher than that of the latter. The weight gain of Specimen 13 (147 mg/dm²) is about half of Specimen 11 but the hydrogen uptake content of the former (273 μg/g) is similar to the latter. In a word, the weight gain increases in the sequence of Specimens $14<13<12<11$ but the hydrogen uptake content increases in the sequence of Specimens $14<13<11<12$. This illustrates that the hydrogen uptake does not have a strict corresponding relationship with the corrosion resistance. That is to say, it is not always true that the higher the weight gain of the specimen, the higher the hydrogen content in the specimen for different heat-treated specimens.

Zry-2 Specimens — Figure 5 shows the corrosion and hydrogen uptake behavior of the Zry - 2 specimens corroded in 10.3 MPa steam at 400 ℃. In the early stage of corrosion, the $d\Delta W_H/d\Delta W_O$ is similar among the four types of Zry-2 specimens. As the corrosion proceeds, the $d\Delta W_H/d\Delta W_O$ shows differences among the four types of Zry - 2 specimens and increases in the sequence of Specimens $21\approx24<23<22$ (Fig. 5(a)). This difference illustrates that the heat treatments also have a significant effect on the hydrogen uptake of Zry - 2 from corrosion.

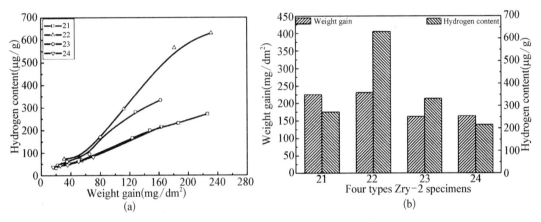

Fig. 5 The corrosion and hydrogen uptake behavior of the Zry - 2 specimens tested in 400 ℃/10.3 MPa steam: (a) The curve of the hydrogen uptake versus weight gain. (b) Corrosion weight gain and hydrogen uptake content of the specimens corroded for 302 days

From Fig. 5(b), it can be seen that Specimens 21 and 22 exhibit similar weight gains (226 and 231 mg/dm²) but the former has a much lower hydrogen uptake content than the latter (272 and 630 µg/g, respectively). Specimens 23 and 24 also exhibit similar weight gain (162 and 164 mg/dm²); however, the former has a higher hydrogen uptake content than the latter (333 and 216 µg/g). In a word, the weight gain changes in the sequence of $24\approx23<21\approx22$ but the hydrogen uptake content changes in the sequence of $24<21<23<22$. This further illustrates that the hydrogen uptake does not have a strict corresponding relationship to the corrosion resistance.

N36 Specimens — Figure 6 shows the corrosion and hydrogen uptake behavior of the N36 specimens corroded in 10.3 MPa steam at 400 ℃. It is clear that the heat treatments have little effect on the hydrogen uptake of the N36 specimens but the hydrogen uptake does not always have the same corresponding relationship to the weight gain. For example, Specimen 32 exhibits the largest weight gain; however, the hydrogen uptake content is not the largest.

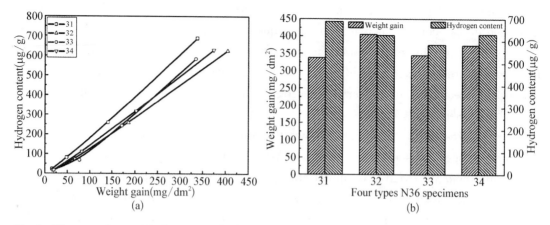

Fig. 6 The corrosion and hydrogen uptake behavior of the N36 specimens tested in 400 ℃/10. 3 MPa steam: (a) The curve of the hydrogen uptake versus weight gain. (b) Corrosion weight gain and hydrogen uptake content of the specimens corroded for 302 days

N18 Specimens — Figure 7 shows the corrosion and hydrogen uptake behavior of the N18 specimens corroded in 10. 3 MPa steam at 400 ℃. It is clear that the heat treatments also have little effect on the hydrogen uptake of the N18 specimens and again, the hydrogen uptake does not always have the same corresponding relationship to the corrosion resistance.

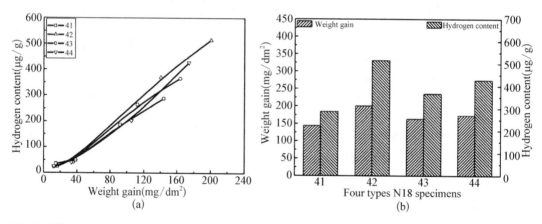

Fig. 7 The corrosion and hydrogen uptake behavior of the N18 specimens tested in 400 ℃/10. 3 MPa steam: (a) The curve of the hydrogen uptake versus weight gain. (b) Corrosion weight gain and hydrogen uptake content of the specimens corroded for 302 days

3.3 Relationship Between Hydrogen Uptake and Second Phase Particles

It is clear from Figs. 4 – 7 that the phenomenon of no strict corresponding relationship between hydrogen uptake and corrosion resistance generally exists among the different heat-treated specimens with the same composition as well as across the alloys with different compositions in this study. For example, the weight gain of the N36 specimens is much larger than that of the N18 specimens but the HUF of the former is lower than the latter (as shown in Figs. 6 and 7). This indicates that the hydrogen uptake behavior is indeed markedly influenced by the chemical composition and the heat treatments.

It is well known that the chemical composition of zirconium alloys will affect the composition of the SPPs that form and the heat treatments will affect the size and amount of the SPPs. Figure 8 shows the relationship between the HUF and the size and area fraction of the SPPs for the specimens tested in 10. 3 MPa steam at 400 ℃ for 302 days. In the case of the Zry - 4 and Zry - 2, the HUF increases with the increase in the size and area fraction of the SPPs (Fig. 8(a) and 8(b)). In the case of the N36 or N18 alloys, it is not a simple relationship between the HUF and the size and area fraction of the SPPs. The size and area fraction of the SPPs show a less monotonic effect on the hydrogen uptake behavior. Moreover, it is obvious that the difference in hydrogen uptake among the four specimens for the four alloys decreases in the sequence of Zry - 2>Zry - 4>N18>N36 (Fig. 8).

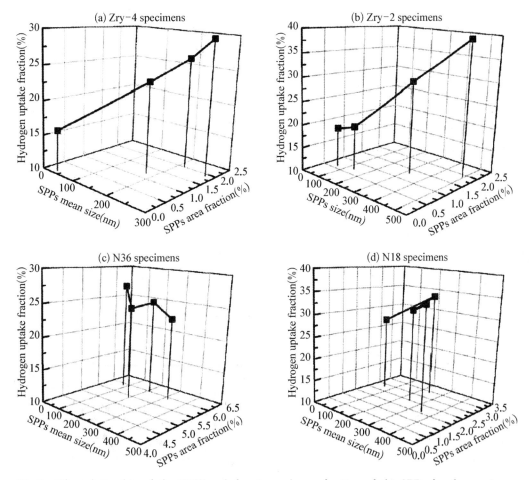

Fig. 8 The relationship of the HUF and the size and area fraction of the SPPs for the specimens tested in 10. 3 MPa steam at 400 ℃ for 302 days

3. 4 Hydrogen Absorption and Desorption Characteristics of the Second Phase Particles Alloys and Pure Zr

Figure 9 shows the hydrogen absorption and desorption characteristic curves of the five SPPs alloys and pure Zr tested in 4 MPa H_2 pressure at 360 ℃. The kinetics data of the absorbing and desorbing hydrogen are summarized in Table 3. From the activation curves (Fig. 9(a)), it can be seen that for the $Zr(Fe, Cr)_2$, $Zr(Nb, Fe)_2$, and β - Nb alloys, the amount of hydrogen absorption are similar among the three activation times for each alloy. This indicates that the alloys can absorb and desorb hydrogen freely and have been activated completely after three activation cycles in 4 MPa H_2 pressure at 360 ℃. For the $Zr_2(Fe, Ni)$ alloy and pure Zr, however, the amount of hydrogen absorption is greatly different between the first and the following two activation cycles. The amount of hydrogen absorption decreases from 3.1 to 1.5 H/M for $Zr_2(Fe, Ni)$ alloy and from 1.94 to 0.1 H/M for pure Zr. This illustrates that a fraction of the hydrogen is transformed into the irreversible hydride for the $Zr_2(Fe, Ni)$ alloy and most of the hydrogen is transformed into the irreversible hydrides for pure Zr. In the P - C - T curves (Fig. 9(b)), which were determined after three activation cycles, it is obvious that the contents of the reversible hydrogen absorption and desorption (H_{rev}, which is defined as the difference between the maximum content of hydrogen absorption and the hydrogen content on the P - C - T curve at 0.01 MPa H_2 pressure, as shown in Fig. 9(b)) decrease in the sequence of $Zr(Nb, Fe)_2 > Zr_2$ $(Fe, Ni) > Zr(Fe, Cr)_2$ R0.5 $> Zr(Fe, Cr)_2$ R2 $>$ β - Nb $>$ Zr, and their H_{rev} values are 1.93, 1.19, 0.89, 0.42, 0.28, and 0.14 H/M, respectively (Table 3). The H_{rev} values of the $Zr(Fe, Cr)_2$ alloys increase with decreasing Fe/Cr ratio, which agrees with the results reported in the literature[30-34]. Kinetic tests of hydrogen absorption and desorption (Table 3), which are measured after determining the P - C - T curves, show that $Zr(Nb, Fe)_2$, Zr_2 (Fe, Ni), and $Zr(Fe, Cr)_2$ alloys absorb and desorb hydrogen with a higher speed and larger content. That is to say, these materials have a strong reversible ability to absorb and desorb hydrogen, which decreased in the above sequence. However, β - Nb alloy and pure Zr can only absorb hydrogen easily but desorb hydrogen difficultly. That is to say, they have a weak reversible ability to absorb and desorb hydrogen. It should be pointed out that the hydrogen absorption and desorption kinetics of pure Zr does not represent its intrinsic

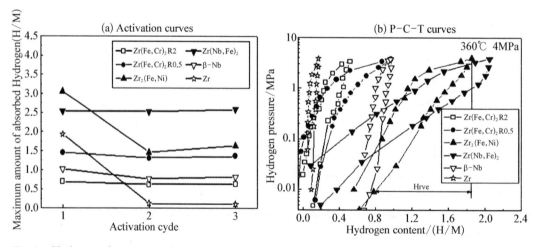

Fig. 9 Hydrogen absorption and desorption characteristic curves of the SPPs alloys and pure Zr tested in hydrogen atmosphere at 360 ℃ and 4 MPa

characteristic due to its difficulty in hydrogen desorption but it illustrates that hydrogen is difficult to be further absorbed when a hydride layer is formed on the Zr matrix. Based on the above results, it can be concluded that the reversible ability for hydrogen absorption and desorption of Zr(Nb, Fe)$_2$, Zr$_2$(Fe, Ni), and Zr(Fe, Cr)$_2$ alloys are larger than that of β - Nb alloy and pure Zr. This indicates that hydrogen can be more easily and quickly captured and transported by these alloys.

Table 3   The H_{rev} and kinetics of absorbing and desorbing of the SPPs alloys and pure Zr tested in hydrogen atmosphere at 360 ℃ and 4 MPa

Materials Number	Kinetics [(H/M)/s]		H_{rev}(H/M)
	Absorbing Hydrogen	Desorbing Hydrogen	
Zr(Fe, Cr)$_2$ R2	0.020	0.012	0.42
Zr(Fe, Cr)$_2$ R0.5	0.032	0.013	0.89
Zr$_2$(Fe, Ni)	0.083	0.051	1.19
Zr(Nb, Fe)$_2$	0.220	0.060	1.93
β - Nb	0.059	0.006	0.28
Zr	0.044	0.007	0.14

4   Discussion

It is generally believed that hydrogen is generated at the oxide/environment interface through the discharge of a proton by an electron migrating through the oxide film. Thus, the oxide characteristics (e. g. , pores, small cracks, and the dense barrier layer thickness) are key factors affecting the hydrogen uptake behavior. Based on such an assumption, some investigators believe that SPPs affect the hydrogen uptake behavior by affecting the oxide characteristics[11-16]. According to this thought, the hydrogen uptake of zirconium alloys from corrosion should have a direct corresponding relationship with the corrosion resistance. That is to say, the hydrogen content in the specimen with superior corrosion resistance should be less than that in the specimen with inferior corrosion resistance because on one hand, the specimen with superior corrosion resistance generates less hydrogen and on the other hand, the specimen with superior corrosion resistance possesses less pores and cracks in the oxide. Thus, it is more difficult for hydrogen to transport through the oxide. high resolution scanning electron microscope observation of the fracture surface of the oxide film on Zry - 4 specimens corroded for 302 days also proves that the pores and cracks in the oxide films on Specimens 13 and 14 with superior corrosion resistance are indeed less than that of Specimens 11 and 12 with inferior corrosion resistance (Fig. 10). However, under the same exposure time of 302 days, the weight gain of Specimen 11 is 267 mg/dm^2 and its HUF is 16% while the weight gain of Specimen 12 is a little less than that of Specimen 11, 238 mg/dm^2, but its HUF is 29.6%, which is about twice as much as Specimen 11. That is to say, the hydrogen uptake is not always in a strict corresponding relationship with the corrosion resistance among the four Zry - 4 specimens.

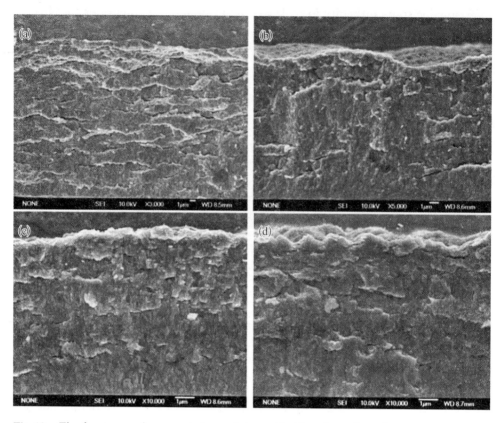

Fig. 10　The fracture surface morphology of the oxide films formed on the Zry – 4 specimens (a) 11, (b) 12, (c) 13, and (d) 14 in a steam at 400 ℃/10. 3 MPa for 302 days

This phenomenon of no strict corresponding relationship between hydrogen uptake and corrosion resistance generally exists in the different heat-treated specimens with the same composition as well as across the alloys with different compositions in this study (Figs. 4 – 7) and was also observed in many other studies[10,19,20]. Therefore, the thought that SPPs only affect the hydrogen uptake behavior by affecting the oxide characteristics cannot reasonably explain the results that the hydrogen uptake is not always in a strict corresponding relationship with the corrosion resistance.

Hatano et al. [19,20] studied the hydrogen uptake behavior of two types of Zry – 4 and Zry – 2 specimens containing fine and coarse SPPs in the pre-transition period of oxidation in steam and also found that the amount of hydrogen uptake of the specimen containing the fine SPPs was smaller than that of the specimen containing the coarse ones, which they explained with the model that the Zr(Fe, Cr)$_2$ SPPs remaining unoxidized in the oxide film act as the short-circuiting route in the hydrogen transport through the oxide film. Tägstrom et al. [7] also put forward a similar thought as Hatano's, but believed that the amount of the SPPs had a greater influence on the hydrogen uptake behavior than the size of the SPPs. Lelièvre et al. [18] studied the deuterium distribution in the oxide layers formed on Zry – 4 and Zry – 2 specimens in 400 ℃ steam by nuclear analysis. The results showed that zirconium hydrides were very often detected under the metal/oxide interface in the vicinity of the SPPs in a 0.7 μm thick oxide, while this correlation was not observed in a 1.4 μm thick oxide. This was interpreted as an indication of the role of SPPs as hydrogen diffusion short-circuits through the oxide layer, but only when these SPPs are metallic and still in contact with the metal, i.e., at the very early stage of oxidation. This thought can reasonably explain the effect of SPPs on the hydrogen uptake behavior of zirconium alloys at the early stage of oxidation but not for the case at the later stage of oxidation, because the SPPs in the oxide film will be oxidized with the process of oxidation. If still based on the assumption that the hydrogen is generated at the oxide/environment interface, the hydrogen transporting through the oxide film, i.e., the oxide characteristic, will again become the key factor affecting the hydrogen uptake behavior.

Is hydrogen generated at the metal/ oxide interface when zirconium alloys react with water or H_2O steam? Based on the hydrogen defect theory in oxides[35-38], when metals or alloys are oxidized in atmospheres containing H_2O steam, the proton easily combines with the oxygen ion to form new ions, such as OH^-. The charge and size of OH^- are smaller than those of the oxygen ion. The activation energy for migration of OH^- through the oxide is lower than that of oxygen ion. So, it can be reasonably concluded that the diffusion of OH^- is faster than the oxygen ion. Secondary ion

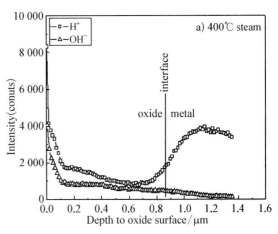

Fig. 11 The profiles of H^+ and OH^- ions in the oxide layer formed on Specimen 14 in 10.3 MPa steam at 400 ℃ (Its weight gain is 13.2 mg/dm²)

mass spectroscopy (SIMS) analysis was used to detect the H^+ and OH^- ion profiles in the oxide layer on Specimen 14, which was formed by oxidation in 10.3 MPa H_2O steam at 400 ℃ to a weight gain of about 13 mg/dm². The results show that both H^+ and OH^- ions exist in the oxide layer (Fig. 11). Since the oxidizing reaction of zirconium alloys in water or H_2O stream always takes place at the metal/oxide interface, it is possible that there are half-cell reactions of $2H_2O+2e \rightarrow H_2+2OH^-$ (cathodic reaction) and $2OH^-+Zr \rightarrow ZrO_2+H_2+2e$ (anodic reaction). Based on the hydrogen defect theory and the SIMS results, it can be hypothesized that besides hydrogen being generated at the oxide/environment interface, OH^- can also directly react with zirconium to produce hydrogen at the metal/ oxide interface after transporting through the oxide layer. Moreover, a fraction of the hydrogen generated at the oxide/environment interface will also transport through the oxide to the metal/oxide interface. Thus, the microstructural characteristics at the metal/ oxide interface, e.g., the size, amount, and composition of the SPPs, will likely become the key factors affecting the hydrogen uptake from corrosion.

From Fig. 8, it can be seen that the hydrogen uptake indeed does have a good corresponding relation-ship with the size or area fraction of the SPPs. In the case of the

Zr – Sn alloys (Zry – 2 or Zry – 4), the HUF of the four specimens increased significantly with increases in the SPPs size and SPPs area fraction. The dependence was more notable for the Zry – 2 alloy than for the Zry – 4 alloy. In the case of the Zr – Sn – Nb alloys (N36 or N18), the HUF of the four specimens had only a slight variation with the size and area fraction of the SPPs. No matter which heat treatment was employed, the corrosion resistance of the N18 specimens was obviously superior to the N36 specimens but the HUF of the N18 specimens was larger than that of the N36 specimens. These results clearly demonstrate that the effect of the size and area fraction of the SPPs on the hydrogen uptake is different for the zirconium alloys with different types of SPPs.

P – C – T and kinetics tests on the SPP alloys in a hydrogen atmosphere have revealed that the $Zr(Nb, Fe)_2$, $Zr_2(Fe, Ni)$, and $Zr(Fe, Cr)_2$ alloys possess a stronger reversible ability for hydrogen absorption and desorption than pure Zr and β – Nb (Fig. 9). Moreover, the SPPs oxidation is slower than that of α – Zr[39-41]. An in situ investigation of the hydride decomposition and reprecipitation by TEM showed that when irradiated by the convergent electron beam in the TEM, the hydride in the Zry – 4 decomposed; when ending the irradiation, fine hydrides reprecipitated around the $Zr(Fe, Cr)_2$ SPPs during cooling (Fig. 12)[42]. This illustrates that the hydrogen could be captured by $Zr(Fe, Cr)_2$ SPPs faster and in greater quantity than by the α – Zr matrix and then released to form hydrides around the SPPs during cooling. It provides a direct proof that the $Zr(Fe, Cr)_2$ SPPs possess a stronger reversible ability for hydrogen absorption and desorption than the α – Zr matrix.

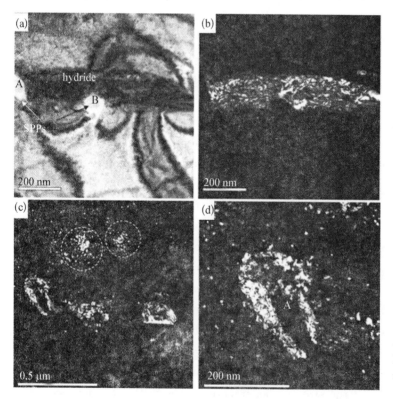

Fig. 12 In situ investigation results of the hydride decomposition and reprecipitation in the Zry – 4 by TEM: The bright filed image (a) and dark field image (b) of the hydride and SPPs around it before electron irradiation; the lower magnification image (c) and higher magnification image (d) of fine hydrides reprecipitated around the Zr (Fe, Cr)₂ SPPs during cooling after the original hydrided irradiated by the convergent electron beam

Based on the above analysis, a model for the effect of SPPs on the hydrogen uptake is proposed as follows (as shown in Fig. 13): The water molecules in the corrosion medium get electrons diffusing from the metal/oxide interface to the outer surface of the oxide film and the reaction of $H_2O+e\rightarrow H+OH^-$ takes place; OH^- ions go through the oxide film to the metal/oxide interface and react with zirconium ($2OH^-+Zr\rightarrow ZrO_2+2H+2e$) to generate hydrogen; moreover,

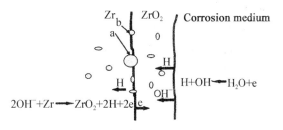

Fig. 13 Schematic diagram of the hydrogen uptake model of zirconium alloys from corrosion

a fraction of the hydrogen generated at the oxide/environment interface will also transport through the oxide to the metal/oxide interface. Thus, hydrogen migrating through to the metal/oxide interface will be a key step for hydrogen uptake. In this case, the SPPs' reversible ability for absorbing and desorbing hydrogen will affect the hydrogen uptake performance. When the SPPs possess a stronger reversible ability to absorb and desorb hydrogen, much more hydrogen can be captured and the hydrogen can be transported faster into the matrix by them embedded at the metal/oxide interface. So, the hydrogen uptake is higher. In this case, the size and amount of the SPPs will have a great effect on the hydrogen uptake. When the SPPs possess a weaker reversible ability for absorbing and desorbing hydrogen, the role in transporting the hydrogen into the matrix is not obvious. So, the hydrogen uptake is lower. In this case, the size and amount of the SPPs have little effect on the hydrogen uptake.

The hydrogen uptake model can successfully explain the results in this study. The SPPs are $Zr(Fe, Cr)_2$ in $Zry-4$[19], $Zr_2(Fe, Ni)$ and $Zr(Fe, Cr)_2$ in $Zry-2$[7,20,24], $\beta-Nb$ and $Zr(Nb, Fe)_2$ in N36[25-27], and $Zr(Fe, Cr, Nb)_2$ in N18[26]. The reversible ability for hydrogen absorption and desorption for the $Zr_2(Fe, Ni)$ alloy is stronger than for the $Zr(Fe, Cr)_2$ alloy, so the HUF of $Zry-2$ specimens is larger than that of $Zry-4$ specimens. The HUF difference among the four $Zry-2$ specimens is also larger than that among the four $Zry-4$ specimens. Specimens 12 and 22 containing the coarsest and higher area fraction SPPs possess the highest HUF; the SPPs in Specimens 11 and 21 containing the finest and lowest area fraction SPPs possess the lowest HUF. Because $Zr(Nb, Fe)_2$ and $Zr(Fe, Cr)_2$ alloys possess a stronger reversible ability for hydrogen absorption and desorption, N18 specimens with $Zr(Fe, Cr, Nb)_2$ SPPs show a higher HUF among the four alloys. Most of the SPPs in the N36 alloy are $\beta-Nb$, so the HUF of N36 specimens is relatively low.

5 Conclusions

Zry-4, Zry-2, N36, and N18 alloys and four different heat treatments for each alloy were chosen to prepare the specimens with different SPPs compositions, amounts, and sizes. The hydrogen uptake performance of these specimens was investigated after autoclave testing in 400 ℃/10. 3 MPa steam. Meanwhile, P - C - T and kinetics of absorbing and desorbing hydrogen tests were carried out on $Zr(Fe, Cr)_2$, $Zr_2(Fe, Ni)$, $(Zr, Nb)_2Fe$, $Zr(Nb, Fe)_2$, and β - Nb alloys (which may be found as SPPs in the above four zirconium alloys) as well as pure Zr. The following conclusions have been obtained.

(1) The hydrogen uptake shows a significant difference and is not always in a strict corresponding relationship with the corrosion resistance among the specimens. It is closely related to the size, area fraction, and composition of the SPPs, and the effect of the size and area fraction of the SPPs on the hydrogen uptake depends on the SPP compositions.

(2) In the case of the Zr - Sn alloys (Zry - 2 or Zry - 4), the HUF of the four specimens increased significantly with an increase in size and area fraction of the SPPs. The dependence was more notable for the Zry - 2 than for the Zry - 4.

(3) In the case of the Zr - Sn - Nb alloys (N36 or N18), the HUF of the four specimens had only a slight variation with the SPPs size and SPPs area fraction. No matter which heat treatment was employed, the corrosion resistance of the N18 specimens was obviously superior to the N36 specimens, but the HUF of the N18 specimens was larger than that of the N36 specimens.

(4) $Zr(Nb, Fe)_2$, $Zr_2(Fe, Ni)$, and $Zr(Fe, Cr)_2$ alloys possess a stronger reversible ability for hydrogen absorption and desorption than β - Nb alloy and pure Zr, and their reversible ability for hydrogen absorption and desorption decreases in the above sequence. The $Zr(Fe, Cr)_2$ alloys' reversible ability for hydrogen absorption and desorption increases with decreasing Fe/Cr ratio.

(5) A model for the effect of SPPs on the hydrogen uptake is proposed: Getting the hydrogen through to the metal/oxide interface is the key step for hydrogen uptake. Thus, the hydrogen uptake performance is closely related to the SPPs' reversible ability for absorbing and desorbing hydrogen. When the SPPs possess a stronger reversible ability for absorbing and desorbing hydrogen, the size and amount of the SPPs embedded at the metal/oxide interface will have a great effect on the hydrogen uptake. When the SPPs possess a weaker reversible ability for absorbing and desorbing hydrogen, the size and amount of the SPPs embedded at the metal/oxide interface have little effect on the hydrogen uptake.

Acknowledgments

The writers would like to express their thanks to Mr. Xinshu Huang and Shu'an Wang of the Key Laboratory for Nuclear Fuel and Materials, Nuclear Power Institute of China for help in the analysis of the hydrogen contents. The writers also wish to acknowledge Mr. Yuliang Chu and Weijun Yu of the Instrumental Analysis and Research Center of Shanghai University for help in the microstructural analysis. This study is partly supported by grants from the National High Technology Research and Development Program of China (863 Program) (Grant No. 2008AA031701) and the Shanghai Leading

Academic Discipline Project (Grant No. S30107).

References

[1] Kearns, J. J., "Terminal Solubility and Partitioning of Hydrogen in the Alpha Phase of Zirconium, Zircaloy – 2 and Zircaloy – 4," *J. Nucl. Mater.*, Vol. 22, 1967, pp. 292 – 303.

[2] Cann, C. D. and Sexton, E. E., "An Electron Optical Study of Hydride Precipitation and Growth at Crack Tips in Zirconium," *Acta Metall.*, Vol. 28, 1980, pp. 1215 – 1221.

[3] Zhou, B. X., Zeng, S. K., and Wang, S. X., "In Situ Electron Microscopy Study on Precipitation of Zirconium Hydrides Induced by Strain and Stress in Zircaloy – 2," *Acta Metall. Sin.*, Vol. 25, 1989, pp. A190 – 195.

[4] Perryman, E. C. W., "Pickering Pressure Tube Cracking Experience," *Nucl. Energy*, Vol. 17, 1978, pp. 95 – 105.

[5] Simpson, C. J. and Ells, C. E., "Delayed Hydrogen Embrittlement in Zr – 2. 5 wt % Nb," *J. Nucl. Mater.*, Vol. 52, 1974, pp. 289 – 295.

[6] Ploc, R. A., "The Effect of Minor Alloying Elements on Oxidation and Hydrogen Pickup in Zr – 2. 5Nb," *Zirconium in the Nuclear Industry: 13th International Symposium*, ASTM – STP – 1423, Annecy, France, 2002, ASTM International, West Conshohocken, PA, pp. 297 – 310.

[7] Tägstrom, P., Limbäck, M., Dahlbäck, M., Andersson, T., and Pettersson, H., "Effects of Hydrogen Pickup and Second-Phase Particle Dissolution on the In-Reactor Corrosion Performance of BWR Claddings," *Zirconium in the Nuclear Industry: 13th International Symposium*, ASTM – STP – 1423, Annecy, France, 2002, ASTM International, West Conshohocken, PA, pp. 96 – 118.

[8] Mardon, J. P., Charquet, D., and Senevat, J., "Influence of Composition and Fabrication Process on Out-of-Pile and In-Pile Properties of M5 Alloy," *Zirconium in the Nuclear Industry: 12th International Symposium*, ASTM – STP – 1354, Toronto, Canada, 2000, ASTM International, West Conshohocken, PA, pp. 505 – 524.

[9] Sabol, G. P., Kilp, G. R., Balfour, M. G., and Roberts, E., "Development of a Cladding Alloy for Higher Burnup," *Zirconium in the Nuclear Industry: Eighth International Symposium*, ASTM – STP – 1023, San Diego, CA, 1989, ASTM International, West Conshohocken, PA, pp. 227 – 244.

[10] Yao, M. Y., Zhou, B. X., Li, Q., Liu, W. Q., and Chu, Y. L., "The Effect of Alloying Modifications on Hydrogen Uptake of Zirconium-Alloy Welding Specimens During Corrosion Tests," *J. Nucl. Mater.*, Vol. 350, 2006, pp. 195 – 201.

[11] Cox, B., "A Mechanism for the Hydrogen Uptake Process in Zirconium Alloys," *J. Nucl. Mater.*, Vol. 264, 1999, pp. 283 – 294.

[12] Cox, B. and Wong, Y. -M., "A Hydrogen Uptake Micro-Mechanism for Zr Alloys," *J. Nucl. Mater.*, Vol. 270, 1999, pp. 134 – 146.

[13] Cox, B., "Hydrogen Uptake During Oxidation of Zirconium Alloys," *J. Alloys Compd.*, Vol. 256, 1997, pp. 244 – 246.

[14] Rudling, P. and Wikmark, G., "A Unified Model of Zircaloy BWR Corrosion and Hydriding Mechanisms," *J. Nucl. Mater.*, Vol. 265, 1999, pp. 44 – 59.

[15] Elmoselhi, M. B., "Hydrogen Uptake by Oxidized Zirconium Alloys," *J. Alloys Compd.*, Vol. 231, 1995, pp. 716 – 721.

[16] Khatamian, D., "Hydrogen Diffusion in Oxides Formed on Surfaces of Zirconium Alloys," *J. Alloys Compd.*, Vols. 253 – 254, 1997, pp. 471 – 474.

[17] Barberis, P., Ahlberg, E., Simic, E., Charquet, D., Lemaignan, C., Wikmark, G., Dahlbäck, M., Tägtström, P., and Lehtinen, B., "Role of the Second-Phase Particles in Zirconium Binary Alloys," *Zirconium in the Nuclear Industry: 13th International Symposium*, ASTM – STP – 1423, Annecy, France, 2002, ASTM International, West Conshohocken, PA, pp. 33 – 55.

[18] Lelièvre, G., Tessier, C., Iltis, X., Berthier, B., and Lefebvre, F., "Impact of Intermetallic Precipitates on Hydrogen Distribution in the Oxide Layers Formed on Zirconium Alloys in a Steam Atmosphere: A²D (³He, p)∝

Nuclear Analysis Study in Microbeam Mode," *J. Alloys Compd.* , Vol. 268, 1998, pp. 308 – 317.

[19] Hatano, H. , Hitaka, R. , Sugisaki, M. , and Hayashi, M. , "Influence of Size Distribution of Zr(Fe, Cr)$_2$ Precipitates on Hydrogen Transport through Oxide Film of Zircaloy – 4," *J. Nucl. Mater.* , Vol. 248, 1997, pp. 311 –314.

[20] Hatano, Y. , Isobe, K. , Hitaka, R. , and Sugisaki, K. , "Role of Intermetallic Precipitates in Hydrogen Uptake of Zircaloy – 2," *J. Nucl. Sci. Technol.* , Vol. 33, No. 12, 1996, pp. 944 – 949.

[21] Ramasubramanian, N. , Billot, P. , and Yagnik, S. , "Hydrogen Evolution and Pickup During the Corrosion of Zirconium Alloys: A Critical Evolution of the Solid State and Porous Evolution Oxide Electrochemistry," *Zirconium in Nuclear Industry: 13th International Symposium*, ASTM – STP – 1423, Annecy, France, 2002, ASTM International, West Conshohocken, PA, pp. 222 – 244.

[22] Lim, B. H. , Hong, H. S. , and Lee, K. S. , "Measurement of Hydrogen Permeation and Absorption in Zirconium Oxide Scales," *J. Nucl. Mater.* , Vol. 312, 2003, pp. 134 – 140.

[23] Große, M. , Lehmann, E. , Steinbrück, M. , Kühne, G. , and Stuckert, J. , "Influence of Oxide Layer Morphology on Hydrogen Concentration in Tin and Niobium Containing Zirconium Alloys After High Temperature Steam Oxidation," *J. Nucl. Mater.* , Vol. 385, 2009, pp. 339 – 345.

[24] Ruding, P. , Lundblad-Vannesjö, K. , Vesterlund, G. , and Massih, A. R. "Influence of Second-Phase Particles of Zircaloy Corrosion in BWR Evironments," *Zirconium in the Nuclear Industry: Seventh International Symposium*, ASTM – STP – 939, Strasbourg, France, 1987, ASTM International, West Conshohocken, PA, pp. 292 – 306.

[25] Comstock, R. J. , Schoenberger, G. , and Sable, G. P. , "Influence of Processing Variables and Alloy Chemistry on the Corrosion Behavior of ZIRLO Nuclear Fuel Cladding," *Zirconium in the Nuclear Industry: 11th International Symposium*, ASTM – STP – 1295, Garmisch-Partenkirchen, Germany, 1996, ASTM International, West Conshohocken, PA, pp. 710 – 725.

[26] Li, Z. K. , Zhou, L. , and Zhao, W. J. , "Effect of Intermediate Annealing on Out-of Pile Corrosion Resistance of Zirconium-Based Alloy," *Rare Met. Mater. Eng.* , Vol. 30, No. 1, 2001, pp. 52 – 54.

[27] Toffolon-Masclet, C. , Barberis, P. , Brachet, J. -C. , Mardon, J. -C. , and Legras, L. , "Study of Nb and Fe Precipitation in α – phase Temperature Range (400 to 500 ℃) in Zr – Nb –(Fe – Sn) Alloys," *Zirconium in the Nuclear Industry: 14th International Symposium*, ASTM – STP – 1467, Stockholm, Sweden, June 13 – 17, 2004, ASTM International, West Conshohocken, PA, pp. 81 – 101.

[28] Eucken, C. M. , Finden, P. T. , Trapp-Pritsching, S. , and Weidinger, H. G. , "Influence of Chemical Composition on Uniform Corrosion of Zirconium-Base Alloys in Autoclave Tests," *Zirconium in the Nuclear Industry: Eighth International Symposium*, ASTM – STP – 1023, San Diego, CA, 1989, ASTM International, West Conshohocken, PA, pp. 113 – 127.

[29] Sabol, G. P. , Comstock, R. J. , Weiner, R. A. , Larouere, P. , and Stanutz, R. N. , "In-Reactor Corrosion Performance of ZIRLO™ and Zircaloy – 4," *Zirconium in the Nuclear Industry: Tenth International Symposium*, ASTM – STP – 1245, Baltimore, MD, 1994, ASTM International, West Conshohocken, PA, pp. 724 – 744.

[30] Shaltiel, D. , Jacob, I. , and Davidov, D. "Hydrogen Absorption and Desprption Properties of AB$_2$ Laves-Phase Pseudobinary Alloys," *J. Less-Common Met.* , Vol. 53, 1977, pp. 117 – 131.

[31] Qian, S. H. , 1989, "Hysteresis and Sloping Plateau Pressures in Zr(Fe$_x$, Cr$_{1-x}$)$_2$ – H Systems," Ph. D. dissertation, Windsor University, Windsor, ON, Canada.

[32] Douglas, G. I. , 1985, "Hydrogen Storage Characteristics of Zr(Fe$_x$, Cr$_{1-x}$)$_2$," Ph. D. dissertation, Windsor University, Windsor, ON, Canada.

[33] Esayed, A. , 1993, "Hysteresis in Metal-Hydrogen Systems," Ph. D. dissertation, Windsor University, Windsor, ON, Canada.

[34] Kakiuchi, K. , Itagaki, N. , Furuya, T. , Miyazaki, A. , Ishii, Y. , Suzuki, S. , Terai, T. , and Yamawaki, M. , "Role of Iron for Hydrogen Absorption Mechanism in Zirconium Alloys," *Zirconium in the Nuclear Industry: 14th International Symposium*, ASTM – STP – 1467, Stockholm, Sweden, 2004, ASTM International, West Conshohocken, PA, pp. 349 – 365.

[35] Norby, T., Dyrlie, O., and Kofstad, P., "Ptotonic Conduction in Acceptor-Doped Cubic Rare-Earth Sesquioxides," *J. Am. Ceram. Soc.*, Vol. 75, No. 5, 1992, pp. 1176 – 1181.

[36] Norby, T. and Kofstad, P., "Electrical Conductivity of Y_2O_3 as a Function of Oxygen Parti Pressure in Wet and Dry Atmospheres," *J. Am. Ceram. Soc.*, Vol. 69, No. 11, 1986, pp. 784 – 789.

[37] Norby, T. and Kofstad, P., "Proton and Native-Ion Conductivites in Y_2O_3 at High Temperatures," *Solid State Ionics*, Vol. 20, 1986, pp. 169 – 184.

[38] Thomas, D. G. and Lander, J. J., "Hydrogen as Donor in Zinic Oxide," *J. Chem. Phys.*, Vol. 25, No. 6, 1956, pp. 1136 – 1142.

[39] Pêcheur, D., Lefebvre, F., Motta, A. T., Lemaignan, C., and Charquet, D., "Oxidation of Intermetallic Precipitates in Zircaloy − 4: Impact of Irradiation," *Zirconium in the Nuclear Industry: Tenth International Symposium*, *ASTM − STP − 1245*, Baltimore, MD, 1994, ASTM International, West Conshohocken, PA, pp. 687 –705.

[40] Pêcheur, D., "Oxidation of β − Nb and $Zr(Fe, V)_2$ Precipitates in Oxide Films Formed on Advanced Zr-Based Alloys," *J. Nucl. Mater.*, Vol. 278, 2000, pp. 195 – 201.

[41] Weidinger, H. G., Ruhmann, H., Cheliotis, G., Maguire, M., and Yau, T.-L., "Corrosion-Electrochemical Properties of Zirconium Intermetallics," *Zirconium in the Nuclear Industry: Ninth International Symposium*, *ASTM−STP−1132*, Kobe, Japan, 1991, ASTM International, West Conshohocken, PA, pp. 499 – 534.

[42] Peng, J. C., 2009, "In-Situ Investigation of Oxide Formation and Hydride Precipitation in Zircaloy − 4 by Transmission Electron Microscope," M. S. dissertation, Shanghai University, Shanghai, China, pp. 34 – 38.

$Zr(Fe_x, Cr_{1-x})_2$ 合金在 400 ℃ 过热蒸汽中的腐蚀行为[*]

摘 要：用真空非自耗电弧炉熔炼了 3 种与锆合金中常见第二相粒子成分相同的合金 $Zr(Fe_x, Cr_{1-x})_2$（$x=1,2/3,1/3$），研究锆合金中第二相粒子的腐蚀行为，第二相合金的粉末在高压釜中经 400 ℃/10.3 MPa 过热蒸汽腐蚀不同时间后，利用 XRD 和能量过滤 TEM 对腐蚀产物进行了物相分析，结果表明：Cr 对第二相合金的氧化速率有很大的影响，增加 Cr 含量可以提高第二相合金的耐腐蚀性能；由于 Fe 和 Cr 在 ZrO_2 中的固溶度极低，第二相合金被腐蚀形成 ZrO_2 时，Fe 和 Cr 被排出并形成 α - Fe(Cr) 和 γ - Fe(Cr)，最终腐蚀生成 $(Fe, Cr)_3O_4$. 不同成分第二相合金的腐蚀行为不同，会对锆合金氧化膜的显微组织演化产生不同的影响，因而也对锆合金的耐腐蚀性能产生不同的影响.

为了提高核电的经济性，需要加深核燃料的燃耗，这样就延长了核燃料元件的换料周期，因而对燃料元件包壳材料锆合金的性能提出了更高的要求，包括耐腐蚀性能、吸氢性能、力学性能及辐照尺寸稳定性等，其中耐水侧腐蚀性能尤为重要[1].

改善锆合金耐腐蚀性能的途径有多种，主要是添加合金元素和优化热加工制度[2]. 大多数合金元素在 α - Zr 中的固溶度低，通常与 Zr 形成金属间化合物，以第二相粒子的形式析出，通过优化热加工制度可以改变第二相粒子的大小和分布，同时也改变了合金元素在基体中的固溶含量[3]. 目前尚不清楚到底是前者还是后者影响了锆合金的耐腐蚀性能、或是两者兼而有之.

锆合金在腐蚀过程中，α - Zr 基体首先氧化生成 ZrO_2，第二相粒子被镶嵌到氧化膜中. 随着时间推移，第二相粒子在氧化膜的包裹下发生氧化. α - Zr 基体和第二相粒子独自氧化的同时又相互影响，各自的稳定腐蚀产物构成了锆合金氧化膜的最终状态[4]. 因此，第二相粒子的腐蚀行为对锆合金氧化膜显微组织的演化过程有重要影响. 人们采用了多种思路研究上述问题，透射电子显微镜（TEM）观察是常用的方法[5-7]. 但含 Cu，Fe 的第二相在制样过程中极易脱落，而且第二相粒子尺寸小、含量低，这都给观察和分析工作带来不利影响.

锆合金中第二相粒子的尺寸一般为十至数百纳米，直接研究嵌在锆合金氧化膜中如此细小的第二相的氧化过程是十分困难的. 为了排除基体影响，有学者[8,9]采用电化学阳极溶解的方法从锆合金中提取出第二相粒子，但是按照这种方法收集到的第二相粒子的数量非常有限. 因此，本文直接熔炼了与第二相成分相同的合金，研究这种合金粉末的腐蚀过程. 这样既获得了大量的第二相粒子，又排除了基体 Zr 的影响，可以为了解锆合金氧化膜中第二相粒子的腐蚀过程提供依据.

1 实验方法

根据 Zr - 4 合金中常见的第二相成分，用真空非自耗电弧炉熔炼了 $Zr(Fe_x, Cr_{1-x})_2$

* 本文合作者：曹潇潇、姚美意、彭剑超. 原发表于《金属学报》，2011,47(7)：882 – 886.

($x=1$,2/3,1/3)合金,本文将它们称为第二相合金.钮扣锭的重量约 60 g,为了获得成分均匀的钮扣锭,每熔炼一次将钮扣锭翻转一次,共熔炼 5 次.由于这些金属间化合物都属于 Laves 相,硬度高、韧性低,其钮扣锭在冷却过程中由于相变引起的内应力而裂开,可以用玛瑙研钵予以粉碎,用不锈钢筛网将粉末粒径控制在 38.5—55 μm 范围内.

将 3 种第二相合金的粉末分别置于不锈钢片制成的小舟中,悬挂于高压釜里,在400 ℃/10.3 MPa 过热蒸汽下进行 30 和 150 h 腐蚀后,经真空干燥箱除湿后再对其进行 X 射线衍射(XRD)分析.

腐蚀后的粉末样品和纯度为 99.9% 的 Al 粉按照体积比 2∶8 混合之后,经模具压制成直径 10 mm,厚度约 2 mm 的块体,经过冷轧成厚度约 0.3 mm 的薄片后,再从中冲出直径 3 mm 的圆片.圆片经砂纸研磨至 80 μm 以下,再用离子减薄方法制备符合 TEM 观察要求的薄样品.在离子减薄过程中,Al 的减薄速率比第二相合金粒子要快,多数第二相合金的颗粒会由于失去 Al 粉的包裹而脱落,形成筛网状的穿孔,但是在样品的薄区中仍然可以找到可供 TEM 观察的第二相合金粒子.由于第二相合金粒子还没有被完全腐蚀氧化,因此,观察粒子的边缘区和中心区就恰好能揭示第二相合金的腐蚀演变过程,这种制样方法在以往的研究[10]中已有详细描述.减薄后的样品在配有矫正式 Ω 能量过滤器的 Zeiss Libra 200EF-TEM 分析型 TEM 上观察,加速电压为 200 kV,采集电子能量损失谱(EELS),谱仪入口光阑直径为 100 μm.明场和暗场像均以能量宽度为 5 eV 的狭缝选择零损失峰成像,以消除由非弹性散射电子引入色差的影响[11].用选区电子衍射(SAD)和快速 Fourier 变换(FFT)分析技术来确定腐蚀产物的晶体结构.

2 实验结果与讨论

2.1 第二相合金腐蚀前后的 XRD 分析

腐蚀前对各种第二相合金进行了 XRD 分析.当 $x=1$ 时,$ZrFe_2$ 第二相合金的谱线与标准谱线(PDF 卡片 65-3034)对应较好;当 $x=2/3$ 和 1/3 时,Fe/Cr 比分别为 2 和 0.5 的第二相合金与标准谱线(PDF 卡片 42-1289 $ZrFe_{1.5}Cr_{0.5}$)相比,分别整体向左偏移了 0.2° 和 0.7°.这是因为第二相合金中的 Fe/Cr 比值与标准谱线所用的样品相比偏小,晶格中将有更多的 Fe 原子被 Cr 原子取代,Cr 原子半径大于 Fe,从而扩大了晶格间距,致使谱线向左偏移.

3 种第二相合金在腐蚀前及经过 30,150 h 腐蚀后的 XRD 谱如图 1 所示.$ZrFe_2$ 第二相合金腐蚀 30 h 后,$ZrFe_2$ 的特征谱线已完全消失,说明这时 $ZrFe_2$ 已被完全氧化.谱线中观察到了明锐的立方 ZrO_2 特征谱线(PDF 卡片 49-1642),弱的单斜 ZrO_2 特征谱线(PDF 卡片 37-1484).α-Fe 的 110 特征峰(PDF 卡片 06-0696)也非常明显,这是因为 Fe 在 ZrO_2 中的固溶度非常低,合金腐蚀时形成 ZrO_2 后 Fe 被排出凝聚形成 α-Fe 晶体.还观察到立方 Fe_3O_4 的特征谱线(PDF 卡片 65-3107),这是 α-Fe 的氧化产物.在 150 h 腐蚀产物的谱线中,α-Fe 的特征峰变得很弱,而立方 Fe_3O_4 特征峰增强.同时,观察到单斜 ZrO_2 的特征峰随腐蚀时间延长逐渐增强,立方 ZrO_2 的特征峰逐渐减弱,这是亚稳态立方 ZrO_2 向稳态单斜 ZrO_2 演变的结果.$ZrFe_2$ 长时间腐蚀后的稳定产物是单斜 ZrO_2 和 Fe_3O_4,这与文献[12]中报道的结果吻合.

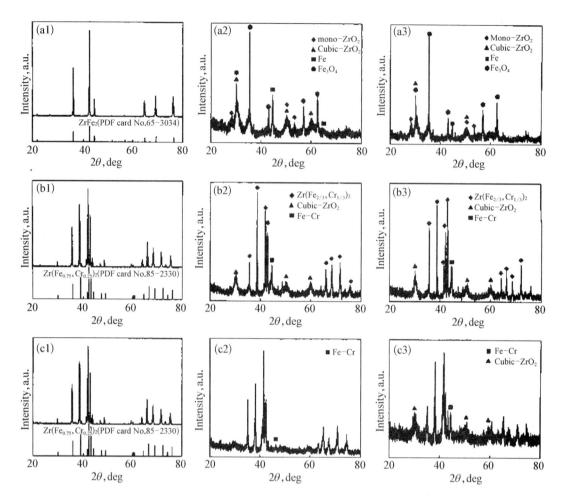

图 1 3 种第二相合金在腐蚀前及腐蚀 30，150 h 后的 XRD 谱

Fig. 1 XRD spectra of the alloys $Zr(Fe_x，Cr_{1-x})_2$ with $x=1(a1-a3)$，$x=2/3(b1-b3)$ and $x=1/3$（c1—c3）before（a1，b1，c1）and after corrosion tests for 30 h（a2，b2，c2）and 150 h（a3，b3，c3）

与 $ZrFe_2$ 相比，Fe/Cr 比为 2 的 $Zr(Fe_{2/3}，Cr_{1/3})_2$ 第二相合金经 30 h 腐蚀后，$Zr(Fe_{2/3}，Cr_{1/3})_2$ 的特征峰强度降低并且宽化，但仍十分明显，说明还有大量 $Zr(Fe_{2/3}，Cr_{1/3})_2$ 未被氧化. 立方 ZrO_2 的谱线十分弱，强度明显弱于 $ZrFe_2$ 第二相合金腐蚀 30 h 后腐蚀产物中立方 ZrO_2 谱线的强度. 观察不到单斜 ZrO_2 的特征谱线，但是可以观察到 bcc 结构 Fe - Cr 的特征谱线（PDF 卡片 34 - 0396）. 在腐蚀 150 h 后的腐蚀产物中，$Zr(Fe_{2/3}，Cr_{1/3})_2$ 特征峰的强度明显降低，立方 ZrO_2 的谱线增强，但仍然未发现单斜 ZrO_2 的谱线. 同时，bcc 结构 Fe - Cr 的特征谱线变得更明显.

相对于上述 2 种第二相合金而言，Fe/Cr 比为 0.5 的 $Zr(Fe_{1/3}，Cr_{2/3})_2$ 第二相合金经过 30 和 150 h 腐蚀后，$Zr(Fe_{1/3}，Cr_{2/3})_2$ 基体的特征谱线仍然非常明显，说明这种成分的合金腐蚀速率很慢. 经过 150 h 腐蚀后，$Zr(Fe，Cr)_2$ 的谱线发生宽化，并出现了弱的 bcc 结构 Fe - Cr 的特征谱线（PDF 卡片 34 - 0396）. 只是在 150 h 腐蚀产物中，基体谱线变得更为低矮、宽化，bcc 结构 Fe - Cr 的特征谱线略有增强，并出现弱的立方 ZrO_2 谱线. 从以上结果可以看出，3 种第二相合金的氧化速率由快到慢的序列为：$ZrFe_2 > Zr(Fe_{2/3}，Cr_{1/3})_2 >$

$Zr(Fe_{1/3}, Cr_{2/3})_2$,增加 Cr 含量可以明显提高第二相合金的耐腐蚀性能.

2.2 第二相合金腐蚀后的 TEM 分析

第二相合金粉末的表面先被腐蚀,随时间延长腐蚀逐渐深入到粉末内部,因此,氧化膜厚度方向上组织形貌和晶体结构的变化就反映了第二相合金腐蚀时的演化过程. 氧化膜沿深度方向可分为:外层氧化膜、内层氧化膜和未腐蚀的第二相合金. 在 3 种第二相合金中,$ZrFe_2$ 第二相合金腐蚀 30 h 后就形成了包括单斜和立方结构的 ZrO_2,同时还观察到 α - Fe 和 Fe_3O_4 的特征谱线,表明腐蚀后生成的氧化膜中包含了比较丰富的信息,因此,选用 $ZrFe_2$ 第二相合金腐蚀 30 h 后的样品进行 TEM 观察分析.

图 2a 是 $ZrFe_2$ 第二相合金颗粒外层氧化膜的 TEM 像及 SAD 图. 可观察到尺寸为 10—30 nm 的晶粒,在三晶交汇处可以观察到孔隙. 由于是纳米量级结构,在样品厚度内可能重叠着 2 个取向不同的晶粒,也可能在一个晶粒内存在面缺陷,造成两层之间取向的微小差别,因而在图 2a 中还可以观察到 moiré 条纹. 由 SAD 图的标定结果可知,这是单斜结构的 ZrO_2,这种晶粒组织的形貌与锆合金腐蚀后表面氧化膜的形貌非常相似.

在距离表面一定深度的内层氧化膜中,晶粒的形貌变得更不清晰,这种衍射衬度不清楚的图像应该与晶体中存在大量缺陷有关. 从高分辨像(图 2b)中可以看出,即使在几十纳米范围内,条纹也显得杂乱,说明晶体中存在许多缺陷,包括了位错和一些缺陷团簇构成的局域非晶,这些缺陷放大后的形貌也嵌镶在图 2b 中. Zr 氧化生成 ZrO_2 时体积发生膨胀,P. B. 比为 1.56,由于金属基体的约束,氧化膜内会产生很大的压应力,这时生成的 ZrO_2 中会产生许多晶体缺陷,包括空位、间隙原子、位错等,使应力得到部分弛豫,同时也形成了一些亚稳相,包括四方和立方结构的 ZrO_2,甚至非晶相[13-15]. 第二相合金在表面生成氧化膜时,与

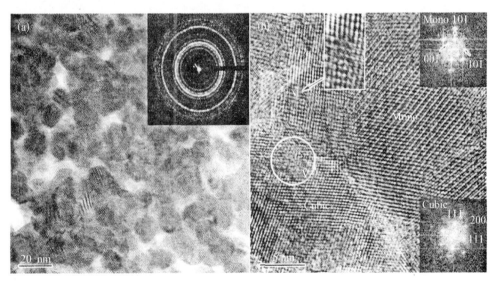

图 2 $ZrFe_2$ 第二相合金颗粒外层氧化膜的 TEM 像和 SAD 图,以及内层氧化膜的高分辨像和 FFT 得到的晶体结构信息

Fig. 2 TEM image of outer oxide layer on $ZrFe_2$ alloy powder oxidized for 30 h and SAD pattern of monoclinic ZrO_2 (a), and HRTEM image of inner oxide layer containing monoclinic and cubic ZrO_2 and their FFT patterns (b) (some dislocations and volume defects shown by letters d and v, respectively)

锆合金也会有相似之处,因而在 ZrO_2 晶体中生成了许多缺陷以及立方结构的亚稳相.用 FFT 分析局部区域内的晶体结构后可以看出,一些很小的立方结构晶体镶嵌在单斜结构的晶体中.

在内层氧化膜中还可以观察到一些尺寸约为 100 nm 的晶粒,它们的衍射衬度清晰,晶粒内部还发生了很大的畸变,其明场和暗场像如图 3a 和 b 所示. SAD 图如图 3c 所示,其中 2 套斑点属于 fcc 结构,1 套斑点属于 bcc 结构,晶面间距与 γ - Fe 和 α - Fe 完全吻合.应用 EELS 对该区域进行分析,在 705 eV 处有明锐的峰,经 EELS Atlas 核对,这是 Fe 的 $L_{2,3}$ 边对应的能量损失谱,如图 3d 所示.图 3 中的晶粒是第二相合金腐蚀生成 ZrO_2 时排出的 Fe 凝聚成的 Fe 晶体,由于 Fe 晶体在 ZrO_2 中析出时受到很大的压应力,使致密的 γ 相更加稳定,同时,也受到严重的挤压变形. γ - Fe 的(111)和 α - Fe 的(110)晶面间距分别为 0.207 5 和 0.203 9 nm,十分接近,在晶体受到挤压变形谱线发生宽化后,在 XRD 谱中无法区分,因而在图 1 中只标出了 α - Fe 的 110 峰.在研究 Zr(Fe, Cr)$_2$ 合金腐蚀时,在 ZrO_2 中也曾观察到 fcc 结构的 γ - Fe 存在[12].

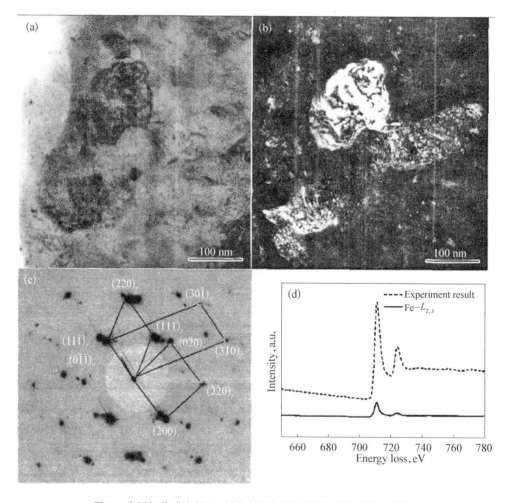

图 3　内层氧化膜中的 Fe 晶粒的明场和暗场像,SAD 图与 EELS

Fig. 3　Bright field image (a), dark field image (b) of iron grain in inner oxide layer, corresponding SAD pattern (c) and EELS (d)

2.3 讨论

锆合金受到腐蚀时,在氧化膜和金属的界面上不断发生氧化过程,阳离子或阴离子空位需要通过在已经生成的氧化膜中扩散,因此,氧化膜的显微组织以及这种显微组织在腐蚀过程中的演化都会对锆合金的耐腐蚀性能产生影响[13-15].大多数合金元素在锆合金中的固溶度都很低,会与 Zr 形成金属间化合物以第二相形式析出,正如前面已经提到,大多数第二相的耐腐蚀性能比 α-Zr 基体好,在锆合金表面形成氧化膜时,第二相先被裹入氧化膜中,然后在氧化膜中发生氧化.一方面,氧化膜中第二相与 ZrO_2 之间的界面可以作为吸收空位的尾闾,起到延迟氧化膜中微裂纹的形成和显微组织演化的作用;另一方面,在第二相发生氧化后,体积又会膨胀,在氧化膜中产生了附加应力,这种应力又会促使氧化膜中的空位扩散,加速显微组织的演化[12].无论从发挥第二相的有利作用或克服有害作用来考虑,得到纳米尺度并且均匀分布的第二相对改善锆合金的耐腐蚀性能都是有利的.如果能够提高第二相合金的耐腐蚀性能,那么它们在进入氧化膜后可以延缓氧化过程,这样就可以发挥有利的作用而克服有害的影响.

3 结论

(1) Cr 含量对 $Zr(Fe_x, Cr_{1-x})_2$ 第二相合金的氧化速率有很大的影响,增加 Cr 含量可以提高第二相合金的耐腐蚀性能.3 种第二相合金的氧化速率由快到慢的顺序为:$ZrFe_2 >$ $Zr(Fe_{2/3}, Cr_{1/3})_2 > Zr(Fe_{1/3}, Cr_{2/3})_2$.

(2) 由于 Fe 和 Cr 在 ZrO_2 中的固溶度很低,在第二相合金被腐蚀形成 ZrO_2 时,Fe 和 Cr 被排出并形成 bcc 或 fcc 结构的 Fe(Cr) 晶体,随腐蚀时间延长,最终再形成 $(Fe, Cr)_3O_4$ 氧化物.

(3) 不同成分第二相合金的腐蚀行为不同,当它们作为锆合金中的第二相时,对锆合金氧化膜的显微组织演化也会产生不同的影响.

参 考 文 献

[1] Zhao W J, Zhou B X, Miao Z, Peng Q, Jiang Y R, Jiang H M, Pang H. *At Energy Sci Technol*, 2005；39(suppl)：1.
(赵文金,周邦新,苗志,彭情,蒋有荣,蒋宏曼,庞华. 原子能科学技术,2005;39(增刊);1)

[2] Zhou B X. *J Met Heat Treat*, 1997；18(3)：8.
(周邦新. 金属热处理学报,1997;18(3);8)

[3] Zhou B X, Yang X L. *Nucl Power Eng*, 1997；18：511.
(周邦新,杨晓林. 核动力工程,1997;18;511)

[4] Pêcheur D, Lefebvre F, Motta A T, Lemaignan C, Wadier J F. *J Nucl Mater*, 1992；189：318.

[5] Li C, Zhou B X, Miao Z. *Corros Prot*, 1996；8：242.
(李聪,周邦新,苗志. 腐蚀科学与防护,1996;8;242)

[6] Pêcheur D. *J Nucl Mater*, 2000；278：195.

[7] Baek J H, Jeong Y H. *J Nucl Mater*, 2002；304：107.

[8] Yang X L, Zhou B X, Jiang Y R, Li C. *Nucl Power Eng*, 1994；15：79.
(杨晓林,周邦新,蒋有荣,李聪. 核动力工程,1994;15;79)

[9] Toffolon-Masclet C, Brachet J C, Jago G. *J Nucl Mater*, 2002；305：224.

[10] Li Q, Zhou B X. *Rare Met Mater Eng*, 2000；29：283.

（李强,周邦新. 稀有金属材料与工程,2000；29：283）

[11] Engerton R F. *Electron Energy Loss Spectroscopy*. 2nd Ed. , New York：Plenum Press，1996：323.

[12] Zhou B X，Miao Z，Li C. *Nucl Power Eng*，1997；18：53.

（周邦新,苗志,李聪. 核动力工程,1997；18：53）

[13] Zhou B X，Li Q，Yao M Y，Liu W Q，Chu Y L. *Nucl Power Eng*，2005；26：364.

（周邦新,李强,姚美意,刘文庆,褚于良. 核动力工程,2005；26：364）

[14] Zhou B X，Li Q，Liu W Q，Yao M Y，Chu Y L. *Rare Met Mater Eng*，2006；35：1009.

（周邦新,李强,刘文庆,姚美意,褚于良. 稀有金属材料工程,2006；35：1009）

[15] Zhou B X，Li Q，Yao M Y，Liu W Q，Chu Y L. In：Kammenzind B，Limbäck M，eds，*Zirconium in the Nuclear Industry: 15th Int Symp Zirconium in the Nuclear Industry*，*ASTM STP 1505*，West Conshohochen：American Society for Testing and Materials，2009：360.

Corrosion Behavior of Zr(Fe$_x$, Cr$_{1-x}$)$_2$ Alloys in 400 ℃ Superheated Steam

Abstract：To study the corrosion behavior of second phase particles in zirconium alloys, Zr(Fe$_x$,Cr$_{1-x}$)$_2$($x=$ 1，2/3，1/3) metallic compounds which have the same composition as the second phase particles in Zr-4 alloy were prepared by vacuum non-consumable arc melting. XRD and energy filtered TEM were employed for analyzing the corrosion products, the element distribution and grain morphology after corrosion tests of Zr (Fe$_x$, Cr$_{1-x}$)$_2$ metallic compounds powder at 400 ℃ and 10.3 MPa superheated steam with different exposure times. The results show that Cr has a very strong effect on the corrosion resistance of Zr(Fe$_x$, Cr$_{1-x}$)$_2$ metallic compounds, increasing Cr content can improve the corrosion resistance. When Zr(Fe$_x$, Cr$_{1-x}$)$_2$ oxidation starts, zirconium oxide is formed while elements Fe and Cr are expelled from the zirconium oxide due to their low solid solubility in the oxide. α-Fe(Cr) and γ-Fe(Cr) are formed and then oxidized to the stable corrosion product (Fe, Cr)$_3$O$_4$. The different corrosion behaviors of metallic compounds will affect the microstructure evolution of zirconium oxide layer differently during the corrosion process, and hence affect the corrosion resistance of zirconium alloys.

APT 和萃取复型研究压力容器模拟钢中富 Cu 团簇的析出*

摘　要：用原子探针层析技术（APT）和萃取复型方法研究了核反应堆压力容器模拟钢中富 Cu 原子团簇的析出. 提高了 Cu 含量的压力容器模拟钢经过 880 ℃水淬，再经过 660 ℃/10 h 调质处理，随后在 370 ℃进行不同时间的时效处理，利用 APT 对时效 4 500 h 的样品分析结果显示，样品中富 Cu 原子团簇的数量密度达到 3.1×10^{23} m^{-3}. 富 Cu 团簇大小在 1～5 nm 范围内时，随着团簇长大含 Cu 量也迅速增加. Ni 和 Mn 除了在 Cu 团簇中发生偏聚外，还会在团簇的周围发生富集. 用 4%硝酸酒精溶液作腐蚀液可以将小至 5 nm 的富 Cu 团簇从 α‐Fe 中萃取出来，HRTEM 和 EDS 分析显示，富 Cu 团簇中存在孪晶结构，虽然大多数团簇是含 Cu 10%～80%（原子分数）的 Cu‐Fe(Ni，Mn)合金，但都是 9R 或 fcc 结构的均匀固溶体.

目前服役中的核电站以压水堆为主，压力容器是压水堆核电站的大型重要部件，装载着核燃料组件、堆内构件和一回路高温高压的冷却水，并担负着屏蔽辐射和防止燃料元件破损后裂变产物外逸等功能. 压力容器一般采用低合金铁素体钢（A508‐III）并在内壁堆焊一层不锈钢制成，此材料经受中子长期辐照后韧脆转变温度会升高至室温以上，严重影响核电站的安全运行. 由于压力容器是不可更换的部件，因而核电站有一定的设计寿命，一般为 40 a，目前希望能延长至 60 a. 国外大量研究工作[1‐6]已证实，压力容器钢在服役温度下（288 ℃）经受中子长期辐照后，除了材料受到中子辐照损伤之外，钢中的杂质元素 Cu 会以富 Cu 原子团簇形式析出，这是引起韧脆转变温度升高的主要原因[6]. 团簇中除 Cu 外，还富集了 Ni，Mn，P，Si 等元素[5,6]. Monzen 等[7]用透射电子显微镜（TEM）研究了 Fe‐Cu 二元合金（Fe‐1.5%Cu，质量分数）中富 Cu 原子团簇从 bcc 到 fcc 转变过程中的 2 种亚稳态 9R 和一种扭曲的 3R 结构. 9R 结构是一种每三层密排面就会发生一次堆垛层错的单斜结构，在高倍电镜下其形貌类似鲱鱼鱼骨状. 3R 结构是扭曲的 fct 结构. Miller 和 Brenner[8]在 1981 年第一次用原子探针（AP）研究了辐照监督件 A302B 试样，辐照后固溶在基体中 Cu 和 P 的含量明显减少，并检测到大量的富 Cu 原子团簇，在团簇中还富集着 P，Al 等元素. Toyama 等[9]利用原子探针层析技术（APT）研究了随堆辐照的压力容器钢监督试样，观察到辐照脆性变化过程中纳米富 Cu 团簇的析出及演化过程中的信息，并观察到在富 Cu 团簇周围还富集着其他元素等情况.

要研究富 Cu 原子团簇，APT[10]和高分辨透射电子显微镜（HRTEM）是非常重要的分析方法. APT 方法可以获得纳米团簇的大小和成分信息，用 HRTEM 研究薄膜样品可以得到富 Cu 团簇的晶体学信息，但是由于纳米大小的团簇是镶嵌在 α‐Fe 的基体中，分析这种团簇的晶体结构时容易受到基体的干扰. 因此，用萃取复型的方法把第二相从基体中分离出来进行研究，可以避免基体的干扰.

本文采用热时效方法研究了提高 Cu 含量后压力容器模拟钢中富 Cu 原子团簇的析出，

---

* 本文合作者：楚大锋、徐刚、王伟、彭剑超、王均安. 原发表于《金属学报》，2011，47(3)：269‐274.

在 APT 分析纳米富 Cu 原子团簇析出及其成分变化的基础上,再采用碳膜萃取复型的方法借助 HRTEM 研究了富 Cu 原子团簇的晶体学信息.

1 实验方法

为了容易观察到富 Cu 原子团簇的析出过程,本实验所选用的材料是在 A508 - III 钢成分的基础上提高了 Cu 含量的压力容器模拟钢,钢锭由 50 kg 真空感应炉冶炼,其化学成分如表 1 所示.将钢锭经过热锻和热轧制成 4 mm 厚的板材,最后切成 40 mm×30 mm 的小样品,加热到 880 ℃ 保温 0.5 h 后水淬,再加热到 660 ℃ 保温 10 h 进行调质处理;随后在 370 ℃下进行 1 000～6 000 h 不同的时效处理,研究富 Cu 原子团簇的析出过程.

表 1　实验用压力容器模拟钢的化学成分

Table 1　Chemical composition of the pressure vessel model steel

Composition	Cu	Ni	Mn	Si	P	C	S	Mo	Fe
Atomic fraction,%	0.53	0.81	1.60	0.77	0.03	1.00	0.011	0.31	Bal.
Mass fraction,%	0.60	0.85	1.58	0.39	0.016	0.22	0.006	0.54	Bal.

为了避免脱碳层的影响,时效处理后的试样用电火花线切割方法从试样中心部位切出长 20 mm,截面边长为 0.5 mm 的方形细棒,分别用 25% (体积分数,下同)的高氯酸乙酸和 2% 的高氯酸 2 - 丁氧基乙醇作电解液,分两步电解抛光将棒状样品制备成曲率半径小于 100 nm 的针尖状样品,APT 的制样及工作原理在文献[11,12]中已有详细描述.用 Oxford nanoScience 公司生产的三维原子探针对针尖样品进行分析,工作温度为 50 K[13],脉冲电压频率为 5 kHz,脉冲分数为 20%.

取另一块时效处理后的试样制备金相样品,试样经过砂纸研磨和机械抛光,用 4% 硝酸酒精溶液腐蚀 9～10 s,然后用 DM - 250 型真空镀膜机在试样表面喷镀一层碳膜.用刀片将碳膜划成大约 3 mm×3 mm 的小方块,在 4% 硝酸酒精溶液中再次腐蚀 65～70 s,样品用无水酒精清洗干净后,将试样缓慢浸入去离子水中,试样与液面夹角为 45°.利用水的表面张力将碳膜从样品表面上剥离下来漂浮在水面上,用 Mo 网捞出晾干,用 JEM - 2010F 型 HRTEM 进行观察.

虽然单质 Cu 在 4% 硝酸酒精溶液中会被腐蚀溶解,但析出了富 Cu 原子团簇的 α - Fe 在 4% 硝酸酒精溶液中腐蚀时,Fe 和 Cu 之间将形成原电池,由于 Fe 的电极电位比 Cu 低,所以在腐蚀时 α - Fe 作为阳极会被腐蚀掉,而富 Cu 团簇作为阴极则受到保护.据此原理,本实验采用 4% 硝酸酒精溶液作腐蚀剂,可以将富 Cu 原子团簇从 α - Fe 中萃取出来.

用 Gatan 公司的 Digital Micrograph 软件对获得的富 Cu 原子团簇晶格条纹像进行 Fourier 变换(FFT)分析以及晶面间距的测量,进而确定团簇的晶体结构.

2 实验结果及分析

2.1 APT 实验结果及分析

用 APT 分析经过 370 ℃ 时效 4 500 h 后样品的结果如图 1 所示.图 1 中给出了 Cu,

Ni, Mn 元素的原子分布, 每一个点表示仪器探测到的一个原子. 从图 1 中可以明显看到这时已经析出了富 Cu 的原子团簇. 使用 maximum separation envelope method（MSEM）来分析 APT 数据[14], 该方法需要设置团簇中溶质原子的一些限定条件, 如团簇中溶质原子之间的最大距离不超过某一值, 以及满足该限定值的原子数量, 以此为限定条件来探测是否存在该种溶质原子的团簇. 现设溶质原子 Cu 之间的最大距离 $d=0.5$ nm, 最小 Cu 原子数目为 $N=10$, 由此得出在 13 nm×14 nm×56 nm 大小的分析空间内有 5 个富 Cu 原子团簇, 依次用序号 1～5 标出（图 1）. 通过计算得到这时团簇的数量密度为 3.1×10^{23} m^{-3}. 团簇中除了 Cu 元素外, Mn 和 Ni 也在团簇中发生明显的偏聚. 这种现象和真实压力容器钢经中子辐照后的结果相似[5,6,12,15]. 图 1 中有 2 个大小约为 5 nm 的富 Cu 原子团簇, 还有几个尺寸更小, 只有 1—2 nm 大小的 Cu 原子团簇, 这些小的团簇应该是 Cu 原子团簇析出时的早期阶段.

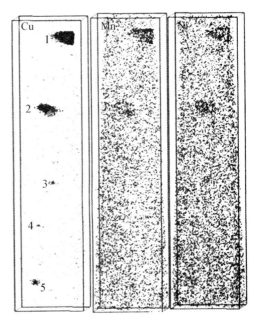

图 1 APT 分析样品在 370 ℃时效 4 500 h 后得到的 Cu, Mn 和 Ni 的原子分布图

Fig. 1 Distributions of Cu, Ni and Mn atoms within an analyzed volume 13 nm×14 nm×56 nm in the specimen aged at 370 ℃ for 4 500 h, Cu-rich clusters labeled 1, 2, 3, 4 and 5 from top to bottom

分析这几个团簇的成分可以看出, 团簇从小长大, Cu 的含量从 5%（原子分数, 下同）增加至 80% 以上. 团簇中 Ni 和 Mn 的含量变化不大, 但 Ni 和 Mn 往往会在 Cu 团簇与 α-Fe 基体的界面处发生偏聚.

图 2 和 3 分别给出了图 1 中 Cu 团簇 5 和团簇 1 的成分分析结果. 团簇 5 中 Cu, Mn 和 Ni 的原子分布图如图 2a 所示, 成分分布图如图 2b 所示. 该团簇中 Cu 含量为 5%～10%, Ni 和 Mn 含量大约分别为 2% 和 3%. 团簇 1 中 Cu, Mn 和 Ni 的原子分布图如图 3a 所示, 成分分布图如图 3b 所示. 该团簇中 Cu 含量并不均匀, 在 70%～99% 之间, Ni 和 Mn 在 Cu 团簇内部及 Cu 团簇与 α-Fe 基体的界面处的含量分别都达到 8% 和 15% 左右. Ni 和 Mn 除了在 Cu 团簇中发生偏聚外, 还会在团簇的周围发生富集. 这样, 随着富 Cu 团簇的长大, 必然要排斥 Mn 和 Ni 原子, 使其向基体四周扩散, 这可能就是富 Cu 原子团簇长大慢的原因.

2.2 碳膜萃取复型实验结果及分析

从 370 ℃经过 1 000, 3 000, 4 500 和 6 000 h 时效处理的 4 个样品中, 用萃取复型方法分别获得了 15, 20, 17 和 21 个 Cu 团簇. 统计结果显示, 随着时效时间延长, 用萃取复型得到的 Cu 团簇平均直径从 11 nm 增大到 20 nm, 团簇略有长大, 从萃取复型样品上观察到的团簇数量也有所增加. 绝大多数的团簇都是球形, 在时效 6 000 h 的样品中也曾观察有短棒状的 Cu 团簇, 这与文献[16]中研究 Fe-1.23%Cu（质量分数）二元合金的结果相似, 一般来说, 富 Cu 原子团簇长大超过 30 nm, 团簇就会逐渐生长成为短棒状. 这可能是因为当团簇充分长大后破坏了原有的共格关系, 改变了 Cu 团簇和 α-Fe 之间界面能的缘故.

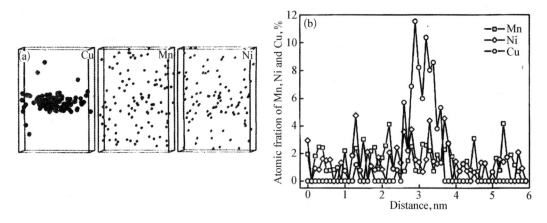

图 2 APT分析图1中5号富Cu原子团簇的Cu，Mn和Ni原子分布图及成分分布图

Fig. 2 Distributions of Cu, Ni and Mn atoms within an analyzed volume 3 nm×4 nm×6 nm for one Cu-rich cluster (No. 5 in Fig. 1) in the specimen aged at 370 ℃ for 4 500 h (a) and the composition profile (b)

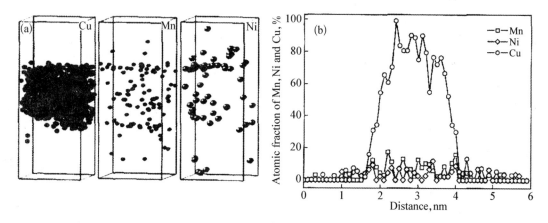

图 3 APT分析图1中1号富Cu原子团簇的Cu，Mn和Ni原子分布图及成分分布图

Fig. 3 Distributions of Cu, Ni and Mn atoms within an analyzed volume 2 nm×2 nm×6 nm for one Cu-rich cluster (No. 1 in Fig. 1) in the specimen aged at 370 ℃ for 4 500 h (a) and the composition profile (b)

图4a 是从 370 ℃时效 1 000 h后样品中得到的萃取复型 TEM 像. 经能谱(EDS)分析，证明球形的萃取物都是富 Cu 团簇，大小为 9～16 nm，一些团簇中存在孪晶结构. 其中一个富 Cu 团簇(图 4a 中箭头所示)的 EDS 分析结果如图 4b 所示. 团簇中含 Cu 将近 87%，其余主要是 Fe 和少量的 P，图中可以清晰地看到 Mo 的 L 系谱线，这是因为碳膜是由 Mo 网支撑引起的，计算团簇成分时 Mo 元素不考虑在内. 该富 Cu 团簇中有 P 存在，众所周知，钢中的 P 在时效过程中会在位错等晶体缺陷处发生偏聚，而位错等缺陷同样可以作为富 Cu 原子团簇析出时的成核位置[12]，因而在富 Cu 原子团簇中会存在 P.

用萃取复型方法也观察到了尺寸在 5 nm 左右的 Cu 原子团簇. 图 5 是从 370 ℃时效 3 000 h样品中观察到 3 nm 大小的富 Cu 团簇. 虽然 APT 分析结果证明了存在 1～2 nm 大小的 Cu 团簇(图 1)，但是这样小的 Cu 团簇即使被萃取黏附在 100 nm 左右厚的碳膜上也不易分辨. 因而用萃取复型方法统计 Cu 团簇的平均大小比过去用 APT 分析得到的尺寸要大[17]，但是结果仍然说明了 Cu 团簇的长大速度很慢.

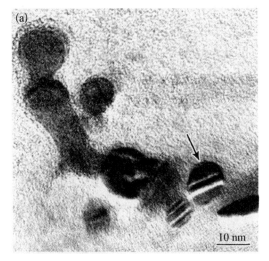
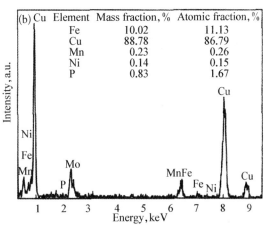

(b)	Element	Mass fraction, %	Atomic fraction, %
	Fe	10.02	11.13
	Cu	88.78	86.79
	Mn	0.23	0.26
	Ni	0.14	0.15
	P	0.83	1.67

图 4　370 ℃时效 1 000 h 后得到的萃取复型 TEM 像及一个富 Cu 团簇(箭头所指)的 EDS 分析结果

Fig. 4　TEM micrograph of an extraction replica from the specimen aged at 370 ℃ for 1 000 h (a) and EDS analysis results of one cluster indicated by arrow in Fig. 4a (b)

图 6a 为时效 3 000 h 样品中一个富 Cu 原子团簇的 HRTEM 晶格条纹像,大小为 12 nm. 可以看出其中存在孪晶结构,图中标出了 2 条孪晶界的位置. 标定晶面指数后的 FFT 图镶嵌在图 6a 中,图中得到了 2 套互为孪晶取向关系的斑点. 分析得出此团簇为 fcc 结构,(110)晶面平行于图片. 根据测量晶面间距计算得出晶格常数 $a=$ 0.37 nm,与纯 Cu 的晶格常数($a=0.361\,5$ nm)相近. 该团簇的 EDS 分析结果如图 6b 所示. 团簇中含 Cu 将近 78%. 其余主要是 Fe(将近 17%). 图 7 为时效 6 000 h 样品中一个富 Cu 原子团簇的 HRTEM 晶格条纹像,大小为 15 nm. 根据文献[7]给出的 bcc 转变为 9R 结构时的晶体取向关系. 分析得出此团簇为 9R 单斜结构,标定晶面指数后的 FFT 图镶嵌在图 7 中.

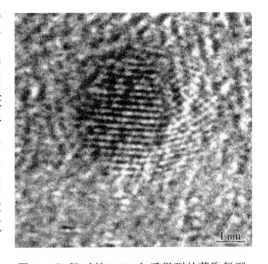

图 5　370 ℃时效 3 000 h 后得到的萃取复型中一个 3 nm 大小的富 Cu 团簇

Fig. 5　A Cu-rich cluster on an extraction replica from the specimen aged at 370 ℃ for 3 000 h

从 HRTEM 晶格条纹像中可以看出,这种纳米大小含 Fe 的富 Cu 相是固溶状态的单相,并不像块体材料的 Cu Fe 合金会形成多相组织. 这与 Cu 和 Fe 的粉末机械合金化的过程相似,当粉体的晶粒通过高能球磨达到纳米尺度后,会得到单一的 fcc 或 bcc 结构[18]. 这大概是晶体纳米尺度效应的结果.

370 ℃经过 1 000,3 000,4 500 和 6 000 h 时效处理的 4 个样品中共萃取获得了 73 个富 Cu 团簇,其中 32 个富 Cu 团簇拍摄到了清晰 HRTEM 像,通过 FFT 分析得出其中有 6 个是 fcc 结构. 26 个为 9R 或扭曲的 3R 结构. 从富 Cu 团簇的大小与晶体结构的关系上,看不出 9R 结构向 fcc 结构转变与团簇大小之间的明显规律,并不像文献[19]中指出的那样.

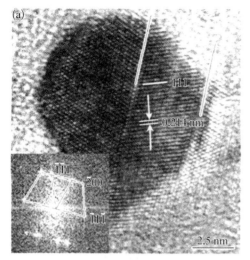

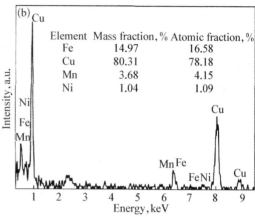

Element	Mass fraction, %	Atomic fraction, %
Fe	14.97	16.58
Cu	80.31	78.18
Mn	3.68	4.15
Ni	1.04	1.09

图 6 370 ℃时效 3 000 h 后得到萃取复型上一个 Cu 团簇的 HRTEM 像和 FFT 图以及该团簇的 EDS 分析结果

Fig. 6 HRTEM micrograph (a) and EDS analysis results (b) of one cluster on an extraction replica from the specimen aged at 370 ℃ for 3 000 h, the FFT pattern indexed for this cluster is inseted in Fig. 6a

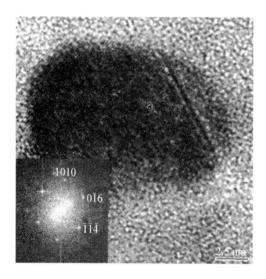

图 7 370 ℃时效 6 000 h 后得到萃取复型上一个 9R 结构 Cu 团簇的 HRTEM 像和标定后的 FFT 图

Fig. 7 HRTEM micrograph and indexed FFT pattern (inset) of one cluster with 9R crystal structure on an extraction replica of the specimen aged at 370 ℃ for 6 000 h

3 结论

（1）在 370 ℃时效时，富 Cu 原子团簇析出后的长大速度很慢. Ni 和 Mn 除了在 Cu 团簇中发生偏聚外，还会在团簇的周围发生富集，这可能是阻碍团簇长大的原因.

（2）当富 Cu 原子团簇在 5 nm 以下时，团簇中的 Cu 含量随团簇长大而迅速增加，但 Ni 和 Mn 的含量变化不大.

（3）萃取复型得到大小为 5～28 nm 的富 Cu 原子团簇，通过 HRTEM 和 FFT 分析表明，大多数是单斜结构，一部分是 fcc 结构，点阵常数比纯 Cu 略大. 虽然大多数团簇是含 Cu 为 10%～80% 的 Cu‑Fe(Ni，Mn)合金，但都是固溶体，并不是多相组织.

感谢上海大学分析测试中心在样品分析中给予的大力支持,感谢朱晓勇,杜晨曦等同志在论文完成过程中给予的帮助!

参 考 文 献

[1] Pareige P, Russell K F, Miller M K. *Appl Surf Sci*, 1996; 94: 362.

[2] Worall G M, Buswell J T, English C A, Hetherington M G, Smith G D. *J Nucl Mater*, 1987; 148: 107.

[3] Akarnatsu M, Van Duysen J C, Pareige P, Auger P. *J Nucl Mater*, 1995; 225: 192.

[4] Preige P, Van Duysen J C, Auger P. *Appl Surf Sci*, 1993; 67: 342.

[5] Auger P, Pareige P, Welzel S. Van duysen J C. *J Nucl Mater*, 2000; 280: 331.

[6] Miller M K, Burke M G. *J Phys*, 1987; 48C: 429.

[7] Monzen R, Iguchi M, Jenkins M L. *Philos Mag Lett*, 2000; 80: 137.

[8] Miller M K, Brenner S S. In: Swanson L W, Bell A, eds., *Proc 28th Int Field Emission Symposium*, The Oregon Graduate Center, Beaverton, OR, 1981; 6: 27.

[9] Toyama T, Nagai Y, Tang Z, Hasegawa M, Almazouzi A, Van Walle E, Gerard R. *Acta Mater*, 2007; 55: 6852.

[10] Zhou B X, Liu W Q. *Mater Sci Technol*, 2007; 15: 405.

   (周邦新,刘文庆. 材料科学与工艺,2007; 15: 405)

[11] Miller M K. *Atom Probe Tomography: Analysis at the Atomic Level*. New York: Kluwer Academic/Plenum Publishers, 2000: 25.

[12] Miller M K, Russell K F. *J Nucl Mater*, 2007; 371: 145.

[13] Zhu C, Xiong X Y, Cerezo A, Hardwicke R, Krauss G, Smith G D. *Ultramicroscopy*, 2007; 107: 808.

[14] Wang W, Zhu J J, Lin M D, Zhou B X, Liu W Q. *J Univ Sci Technol Beijing*, 2010; 1: 39.

   (王伟,朱娟娟,林民东,周邦新,刘文庆. 北京科技大学学报,2010; 1: 39)

[15] Pareige P, Auger P, Moloudi S, Van Duysen J C, Akamatsu M. *Ann Phys*, 1997; 22C: 117.

[16] Hornbogen E, Glenn R C. *Trans Metall Soc AIME*, 1960; 218: 1064.

[17] Zhu J J, Wang W, Lin M D, Liu W Q, Wang J A, Zhou B X. *J Shanghai Univ*, 2008; 14: 525.

   (朱娟娟,王伟,林民东,刘文庆,王均安,周邦新. 上海大学学报(自然科学版),2008; 14: 525)

[18] Li B L, Zhu M, Li L, Luo K C, Li Z X. *Acta Metall Sin*,1997; 33: 420.

   (李柏林,朱敏,李隆,罗堪昌,李祖鑫. 金属学报,1997; 33: 420)

[19] Othen P J, Jenkins M L, Smith G D W. *Philos Mag*, 1994; 70A: 1.

APT and Extraction Replica Characterization of Cu-Rich Clusters Precipitated in Pressure Vessel Model Steels

Abstract: The precipitation of Cu-rich clusters in reactor pressure vessel (RPV) model steel was investigated by means of atom probe tomography (APT), extraction replica (ER) and HRTEM. RPV model steel was prepared by vacuum induction furnace melting with higher content of Cu (0.6%, mass fraction). The ingot (about 50 kg of weight) was forged and hot-rolled to 4 mm in thickness and then cut to specimens of 40 mm× 30 mm. Those specimens were further heat treated by 880 ℃/0.5 h water quenching and 660 ℃/10 h tempering, and finally aged at 370 ℃ for different time from 1 000 to 6 000 h. Four percent nitric acid alcohol solution was used as an etchant to extract the precipitates of Cu-rich clusters from α-Fe matrix. The results obtained by APT analysis show that the number density of Cu-rich clusters reaches 3.1×10^{23} m^{-3} in the specimen aged at 370 ℃ for 4 500 h, and the Cu content in the clusters increases rapidly during their growth from 1 nm to 5 nm. The segregation of Ni and Mn elements within and around the Cu-rich clusters was detected. The results obtained by ER, EDS and HRTEM analyses show that the majority of Cu-rich clusters are Cu-Fe (Ni, Mn) alloys with 10%~80% Cu (atomic fraction), but they are a single phase with 9R or fcc crystal structure.

Zr - 4 合金腐蚀初期氧化膜的显微组织研究*

摘 要: 用 Zr - 4 合金的大晶粒样品在 360 ℃/18.6 MPa 的 0.01 mol/L LiOH 水溶液中进行 5 h 的腐蚀实验,借助 SEM,EBSD 和 HRTEM 等分析方法,研究了腐蚀初期氧化膜生长的各向异性、显微组织和氧化物结构,氧化膜的厚度因基体表面晶粒取向不同而有差异(376—455 nm), 在基面(0001)和棱柱面(01$\bar{1}$0)附近晶面上生长的氧化膜最厚;氧化膜由单斜、立方和四方结构的氧化物组成;氧化膜的结构和晶粒取向分布也随基体晶粒取向不同而存在差异,在(0001)附近晶面上生成的氧化膜中氧化物结构种类较多、晶粒取向差别较大,这种氧化膜的显微组织在腐蚀过程中可促使氧化膜更快生长.

Zr 的热中子吸收截面低,在高温高压水中具有良好的耐腐蚀性能和力学性能,因而锆合金被用作压水堆核电站中核燃料元件的包壳材料,是一种重要的核反应堆结构材料. 为了降低核电成本,需要提高核燃料的燃耗,延长换料周期,这就需要发展耐水侧腐蚀性能更加优良的新型锆合金. 开展锆合金耐腐蚀性能的研究,加深对锆合金腐蚀机理的认识,有助于指导新型锆合金的研究和开发. 锆合金具有 hcp 结构,其氧化膜生长的各向异性是腐蚀过程中的一个基础问题.

Zhou 等[1]用 Zr - 4 合金的大晶粒样品在不同温度和不同水化学条件的高压釜中进行了长期的腐蚀实验,观察到氧化膜生长与金属晶粒取向之间存在明显的各向异性特征,尤其在 360 ℃ LiOH 水溶液和 500 ℃过热蒸汽中腐蚀时更为显著,并且这种各向异性特征还会因腐蚀温度和水化学条件的不同而变化. 为了解释这种现象,提出了样品在长期腐蚀时,氧化膜的生长除了与金属晶粒的表面取向有关之外,还会受到氧化膜自身显微组织和晶体结构演化过程的影响,而这种演化过程除了会受到腐蚀时的温度和水化学条件的影响之外,还与氧化膜初始的显微组织和晶体结构的差别有关. 因此,研究锆合金腐蚀初期氧化膜的显微组织和晶体结构特征与金属晶粒取向之间的关系是非常重要的.

Ploc[2-4]发现氧化膜在 α - Zr 的(0001),(10$\bar{1}$0)和(11$\bar{2}$0)面上生成时与金属基体具有一定的晶体学关系. Wadman 等[5,6]发现 Zr - 4 合金氧化时最早期形成氧化膜的结构主要为非晶和 5—10 nm 的微晶. Kim 等[7]研究了小尺寸的纯 Zr 单晶在 360 ℃/18.6 MPa 水中的腐蚀行为,结果表明氧化膜在(11$\bar{2}$0)晶面上生长的速率明显比(0001)晶面上的快. Bakradze 等[8]发现在纯 O_2 中纯 Zr 单晶(10$\bar{1}$0)面较(0001)面更容易氧化. Charquet 等[9]认为 Zr - 4 合金(0001)面在 500 ℃过热蒸汽中具有良好的耐腐蚀性能. Li 等[10]研究了纯 Zr 单晶的(0001)和(10$\bar{1}$0)面上氧化初期氧扩散速率的差别. 李聪和周邦新[11]发现 Zr - 4 合金多晶样品在 350 ℃空气中氧化时,初期形成氧化膜(厚度小于 100 nm)的结构也因金属晶体的取向不同而有差别. 这些研究结果虽然因腐蚀条件及合金成分的不同而不完全一致,但都说明了氧化膜生长的各向异性与基体表面晶粒取向之间存在密切的关系. 因此,要了解其中的详细

---

* 本文合作者:杜晨曦、彭剑超、李慧. 原发表于《金属学报》,2011,47(7):887 - 892.

关系和规律还需要进行深入细致的研究工作. 本文采用 Zr-4 合金的大晶粒样品,研究了样品在 360 ℃/18.6 MPa 的 0.01 mol/L LiOH 水溶液中腐蚀 5 h 后的氧化膜,观察了氧化膜的厚度、显微组织和晶体结构的变化与基体表面晶粒取向的关系,讨论了长期腐蚀时氧化膜生长过程中的问题.

1 实验方法

在研究氧化膜各向异性生长时,为了避免晶界的干扰,需要制备大晶粒样品. 由于添加合金元素后降低了锆合金 α/β 的相变温度,并形成了许多细小弥散分布的第二相,因而晶界迁移十分困难,通过退火获得大晶粒样品并不容易. 采用 β 相区加热淬火和 α 相区高温退火的方法获得了晶粒大于 0.5 mm 的粗晶样品[1]. 本研究工作应用同样的方法,将 1 mm 厚的 Zr-4 板材切割成 30 mm×8 mm 大小的片状样品,经 10%HF+45%HNO₃+45%H₂O(体积分数)酸洗,去离子水清洗并干燥后真空封装在石英管中,在 1 020 ℃加热 30 min 后进行 β 相水淬,在石英管淬入水中时,同时敲碎石英管. 用细砂纸磨去淬火时样品表面生成的蓝色氧化膜,对样品清洗后重新真空封装在石英管中,在 800 ℃进行 10 h 真空退火处理,这样就可以得到晶粒尺寸为 0.2—0.8 mm 的大晶粒样品. 然后在 700 ℃进行 100 h 退火处理以降低合金元素在 α-Zr 中的过饱和固溶量.

样品进行高压釜腐蚀之前,先进行电解抛光以获得平整的表面,抛光液为 20%HClO₄+80%CH₃COOH,抛光电压为直流 30 V. 样品抛光后用去离子水清洗,然后放入装有 0.01 mol/L LiOH 水溶液的高压釜中,在 360 ℃/18.6 MPa 条件下腐蚀 5 h. 腐蚀后的样品用 VHX-100 型光学显微镜(OM)和 JSM-6700F 型扫描电子显微镜(SEM)观察氧化膜的表面晶粒形貌;另取一些样品从一面磨薄后用酸洗液在 Zr-4 合金基体上腐蚀出一条缝,直到显出氧化膜时把样品掰断,在断面上喷金后用 SEM 从截面上测定氧化膜的厚度;另外,用专用工具从磨薄的样品上冲出直径 3 mm 的圆片,并用装配有背散射电子衍射(EBSD)的 SEM(Apollo300)测定试样表面晶粒的取向;在测定过表面取向的晶粒上腐蚀一个小凹坑直到显出氧化膜,氧化膜外表面经过离子减薄后用 JEM-2010F 型高分辨透射电子显微镜(HRTEM)研究氧化膜的显微组织和晶体结构.

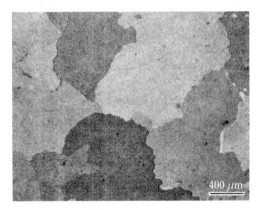

2 实验结果与讨论

2.1 氧化膜颜色、厚度与基体表面晶粒取向的关系

样品在 360 ℃/18.6 MPa 的 0.01 mol/L LiOH 水溶液中腐蚀 5 h 后表面生成的氧化膜呈彩色,如图 1 所示. 氧化膜的不同颜色勾画出了金属晶粒的轮廓,晶粒尺寸为 0.2—0.8 mm. Zr 的

图 1 Zr-4 样品在 360 ℃/18.6 MPa 的 0.01 mol/L LiOH 水溶液中腐蚀 5 h 后氧化膜的表面颜色

Fig. 1 Colors of the oxide layer on the large grained (0.2-0.8 nm) Zr-4 specimen corroded in 0.01 mol/L LiOH aqueous at 360 ℃/18.6 MPa for 5 h (different colors corresponding to different thicknesses)

氧化膜很薄时是透明的,研究[12]表明,光的干涉造成了氧化膜的颜色不同,这反映了氧化膜厚度的差别,并与合金晶粒的表面取向有关.为叙述方便,在测量了氧化膜颜色与厚度的关系后,本文用氧化膜的颜色表示氧化膜的不同厚度.

利用 SEM 观察氧化膜的截面,测量氧化膜的厚度(图 2a).把测得的氧化膜颜色分为绿色、黄色和红色三类,每种颜色氧化膜的厚度测量 3 个数据,同一类颜色的氧化膜厚度大致相同,氧化膜颜色与厚度平均值的关系如图 2b 所示.不同颜色氧化膜的厚度在 376—455 nm 范围内,并按照绿色、黄色和红色的顺序依次增加.

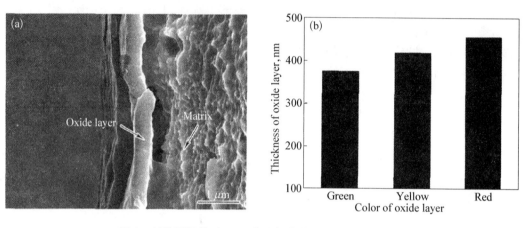

图 2 氧化膜的截面 SEM 像及氧化膜厚度与颜色的关系

Fig. 2 Cross sectional SEM image of oxide layer (a) and the relationship between the thickness and the color of oxide layer (b)

用 EBSD 测定 Zr-4 合金表面的晶粒取向,图 3 是样品腐蚀 5 h 后氧化膜颜色与金属晶粒表面取向的关系.图 3a 是 29 个晶粒表面取向的倒极图.将测量得到的氧化膜颜色划分为黄色、绿色和红色三类,然后用 3 种粗细不同的线段将氧化膜颜色相近的晶面极点连在一起,如图 3b 所示.可以看出,红色氧化膜对应的晶面极点除了在基面(0001)附近有两点外,其他都分布在棱柱面($\bar{1}2\bar{1}0$)和($01\bar{1}0$)之间;黄色氧化膜对应的晶面极点分布在倒极图的中央位置;绿色氧化膜对应的晶面极点位于棱柱面和黄色氧化膜对应的晶面极点之间.结果表明,Zr-4 合金早期氧化膜的生长就存在各向异性特征,但是氧化膜厚度与金属晶体取向之间的关系又与样品在 360 ℃/18.6 MPa 的 0.01 mol/L LiOH 水溶液中长期腐蚀 340 d 后的

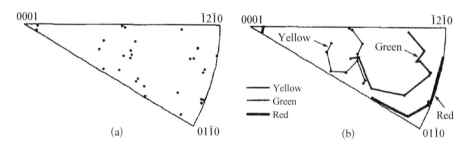

图 3 EBSD 测定的 29 个晶粒取向在倒极图中的位置以及氧化膜颜色与晶粒取向之间的关系

Fig. 3 Positions of the 29 grain orientations in the hcp Zr inverse pole figure measured by EBSD (a) and the distribution of the colors (thicknesses) of the oxides on the 29 grains after corrosion (b)

结果[1]不同,产生这种差别的原因一定与氧化膜显微组织在腐蚀过程中的演化有关,因为锆合金的氧化反应是在氧化膜和金属的界面上发生的,O^{2-}需要扩散通过已经形成的氧化膜后才能到达界面上与 Zr 发生反应,这时氧化膜的显微组织将会影响 O^{2-} 的扩散.因此,在氧化膜生长后期,表面晶粒取向对氧化膜生长的影响将成为次要的因素,而氧化膜的显微组织差别,也就是腐蚀过程中氧化膜显微组织的演化过程将成为影响氧化膜生长的重要因素.本文的研究结果与文献[7-9,13]用成分不同的合金或者在不同腐蚀条件下得到的结果也不完全相同.这说明氧化膜生长的各向异性不仅与腐蚀的时间有关,还会因腐蚀温度、腐蚀介质和合金成分的不同而发生变化.

2.2 氧化膜表面晶粒的形貌

用 SEM 观察了不同颜色氧化膜表面的晶粒形貌,晶粒尺寸为 30—60 nm.不同颜色的氧化膜厚度虽然有差别,但晶粒形貌没有显著的不同.图 4 是黄色和红色氧化膜表面的晶粒形貌,尽管这时氧化膜的厚度还很薄,但是晶粒之间已经出现了微裂纹(箭头指示处).

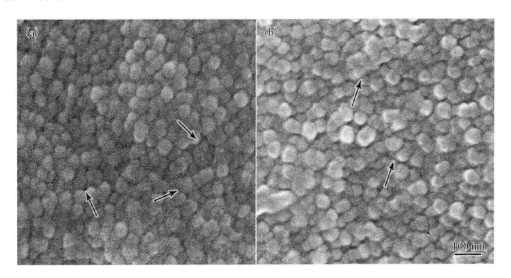

图 4　黄色和红色氧化膜表面晶粒的 SEM 像

Fig. 4　Grain morphologies on the yellow (a) and red (b) oxide surface observed by SEM (microcracks indicated by arrows)

2.3 氧化膜的显微组织和晶体结构

许多研究结果[14-19]表明,锆合金腐蚀时形成的氧化膜中除了有稳定的单斜结构外,还存在非稳定的四方、立方结构以及非晶结构,且氧化膜的显微组织在腐蚀过程中的演化对氧化膜的生长和锆合金的耐腐蚀性能具有重要的影响.因而,研究锆合金表面不同晶粒取向上最初形成的氧化膜显微组织和相组成的差别就显得尤为重要.实验选取了 4 个取向不同的晶粒,它们的取向分布如图 5 所示.这 4 个晶粒表面上生成氧化膜的颜色分别为红色

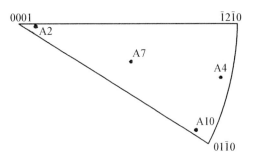

图 5　4 个晶粒的取向

Fig. 5　Orientations of four grains which were selected for the investigation of oxide microstructure

（A2 和 A10）、黄色（A7）和绿色（A4）. 图 6 是氧化膜的选区电子衍射（SAD）图. 改变选区光阑的大小或者移动选区的位置，除了衍射强度有变化之外，衍射花样并没有太大的改变，说明在同一个基体晶粒表面上生成的氧化膜具有相同的显微组织和相组成，但是在不同取向晶粒表面上的氧化膜却有明显的差别. 分析 A2，A7 和 A10 的 SAD 图可知，氧化膜中主要是单斜结构，但是晶粒之间的取向分布却明显不同，A7 的 SAD 图类似于单晶的衍射花样，说明氧化膜中小晶粒之间的取向差别比较小，而 A2 的 SAD 图几乎形成了连续的衍射环，说明氧化膜中小晶粒之间的取向差别比较大，A10 氧化膜中的情况介于前两者之间. 虽然 A2 和 A10 都是红色较厚的氧化膜，但是氧化膜中小晶粒取向分布差别的程度不同. 从 TEM 暗场像（图 7）中也可以看出氧化膜中晶粒取向分布的差别，在 A2 氧化膜中可以观察到单个的晶粒，这是因为晶粒之间存在较大的取向差别，而 A7 氧化膜中只能观察到"成片"的晶粒，"成片"晶粒的取向之间差别比较小. 在 A2，A7 和 A10 氧化膜中除了单斜结构之外，还有少量的四方和立方结构，由于大部分四方和立方晶体的衍射峰都与单斜晶体的衍射峰重叠，从少数几条可区分的衍射峰可以确定四方和立方结构的存在，并标注在图 6 的 SAD 图中. A4 氧化膜主要为立方和单斜结构，SAD 图类似于单晶的衍射花样，说明小晶粒之间的取向差别比较小，取向趋于一致.

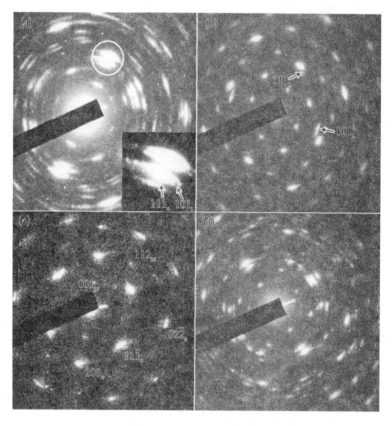

图 6 4 个晶粒上形成的氧化膜的 SAD 图

Fig. 6 SAD patterns of oxide layers on grains A2 (a), A7 (b), A4 (c) and A10 (d) (oxide layers on grains A2, A7 and A10 consisted mainly of monoclinic (m) structural oxide, and that on A4 consisted of cubic (c) and monoclinic oxides, some weak diffraction lines produced by tetragonal (t) structure oxides are indicated by arrows in Figs. 6a and b)

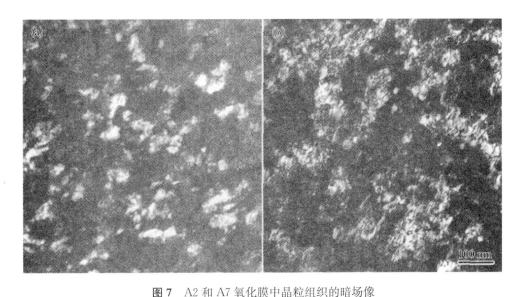

图 7　A2 和 A7 氧化膜中晶粒组织的暗场像

Fig. 7 Dark field images of oxide layers formed on grains A2 (a) and A7 (b)

图 8 是 A2 和 A10 氧化膜的 HRTEM 像. 从晶格条纹的形貌和分布可以看出氧化膜中氧化物结构不止一种. A2 氧化膜的显微组织和相组成比 A10 复杂, 从图 6 的 SAD 图也可以看到这种差别. 晶格条纹清晰的区域可以采用 Fourier 变换(FFT)分析方法确定氧化物结构种类, 分别用 m, t 和 c 标志单斜、四方和立方结构的区域, 并选取了一些经过标定的 FFT 图镶嵌在相应的图中. 由于 Zr 氧化生成 ZrO_2 时体积发生膨胀, P. B. 比为 1.56, 在氧化膜中会产生很大的压应力, 为了使应力得到部分弛豫, 形成 ZrO_2 时会生成许多缺陷, 包括空位、间隙原子、位错等, 同时也稳定了一些亚稳相[16], HRTEM 像中晶格条纹的"杂乱"程度充分反映了这种过程, 不同相之间的界面并不很清晰, 包含了大量的晶体缺陷. 在有的地方还可以观察到非晶相的存在. 图 9 是 A7 氧化膜中非晶和微晶的混合区域, FFT 图镶嵌在图中, 经

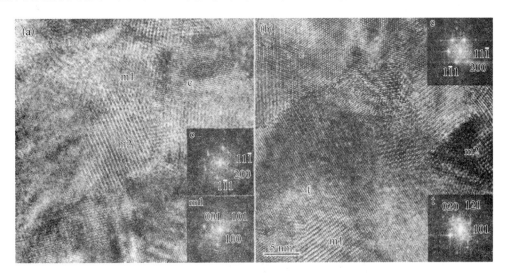

图 8　A2 和 A10 氧化膜的 HRTEM 像以及不同区域的 FFT 图

Fig. 8 HRTEM images of oxide layers formed on grains A2 (a) and A10 (b) and FFT patterns of different areas with monoclinic (m), cubic (c), and tetragonal (t) structures

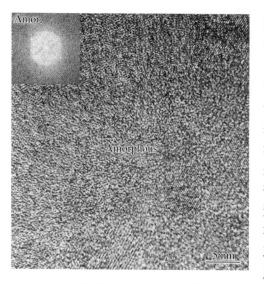

图 9 A7 氧化膜中非晶和微晶的混合区域

Fig. 9 Mixed area of amorphous structure and microcrystals in the oxide layer formed on grain A7

测量得到微晶的一组面间距为 0.244 nm,这可能是立方结构的(200)面或四方结构的(110)面.

从以上分析可以看出,在取向不同的晶粒表面上,腐蚀初期形成的氧化膜不仅厚度不同,氧化物结构的种类和晶粒取向分布特征也存在差异,这为氧化膜在腐蚀过程中显微组织发生进一步的演化奠定了相应的基础.氧化物结构种类多并且晶粒取向差别大的氧化膜,在腐蚀过程中显微组织更容易发生演化.缺陷经过扩散凝聚容易形成孔隙,进而发展成微裂纹,这样的显微组织有利于 O 的扩散,也促使了氧化膜的继续生长.取向在(0001)和(01$\bar{1}$0)附近的 A2 和 A10,虽然腐蚀 5 h 后都形成了红色较厚的氧化膜,但是氧化物结构种类和晶粒取向分布特征存在明显的差别,A2 氧化膜中的晶粒取向差别比 A10 氧化膜更大,在腐蚀过程中氧化膜的显微组织更容易发生演化,这可能就是样品在 360 ℃ LiOH 水溶液中长期腐蚀后,取向在(0001)附近晶面上生长的氧化膜的厚度比在(01$\bar{1}$0)附近晶面上的氧化膜厚得多的原因[1].

3 结论

(1) 对 Zr-4 合金大晶粒样品在 360 ℃/18.6 MPa 的 0.01 mol/L LiOH 水溶液中腐蚀 5 h 后氧化膜生长的各向异性和氧化膜显微组织特征进行了研究,氧化膜的厚度因金属晶粒表面取向不同而有差异,在 376—455 nm 范围内,并按照绿色、黄色和红色的顺序依次增加.红色氧化膜对应的晶面极点分布在基面(0001)和棱柱面(01$\bar{1}$0)取向附近,黄色氧化膜对应的晶面极点分布在倒极图的中央位置,绿色氧化膜对应的晶面极点位于棱柱面和黄色氧化膜对应的晶面极点之间.氧化膜的晶粒尺寸为 30—60 nm.

(2) 在取向不同的晶粒表面上生成的氧化膜不仅厚度不同,氧化膜的晶体结构和晶粒取向分布特征也存在差异.在(0001)取向附近晶面上生成的氧化膜具有更复杂的结构和晶粒取向分布特征,也包含了更多的晶体缺陷.在继续腐蚀的过程中,这种氧化膜的显微组织和晶体结构更容易发生演化,从而促使了 O 的扩散和氧化膜的生长.

参 考 文 献

[1] Zhou B X, Peng J C, Yao M Y, Li Q, Xia S, Du C X, Xu G. *J ASTM Int*, 8(1), DOI: 10.1520/JAI 102951.

[2] Ploc R A. *J Nucl Mater*, 1982; 110: 59.

[3] Ploc R A. *J Nucl Mater*, 1983; 115: 110.

[4] Ploc R A. *J Nucl Mater*, 1983; 113: 75.

[5] Wadman B, Andren H O, Falk L K L. *Colloque De Physique*, 1989; C8: 303.

[6] Wadman B, Andren H O. *Zirconium in the Nuclear Industry: 9th International Symposium*, ASTM STP 1132, West Conshohocken PA: ASTM International, 1991: 461.

[7] Kim H G, Kim T H, Jeong Y H. *J Nucl Mater*, 2002; 306: 44.

[8] Bakradze G, Jeurgens P H, Mittemeijer J. *Surf Interface Anal*, 2010; 42: 588.

[9] Charquet D, Tricot R, Wadier J F. *Zirconium in the Nuclear Industry: 8th International Symposium*, *ASTM STP 1023*, West Conshohocken PA: ASTM International, 1989: 374.

[10] Li B, Allnatt A R, Zhang C S, Norton P R. *Surf Sci*, 1995; 330: 2.

[11] Li C, Zhou B X. *Nucl Power Eng*, 1994; 15: 152.

（李聪,周邦新. 核动力工程,1994; 15: 152）

[12] Peachey L D. *J Biophysic Biohem Cytol*, 1958; 4: 233.

[13] Pemsler J P. *J Electrochem Soc*, 1958; 105: 315.

[14] Zhou B X, Li Q, Liu W Q, Yao M Y, Chu Y L. *Rare Met Mater Eng*, 2006; 35: 1009.

（周邦新,李强,刘文庆,姚美意,褚于良. 稀有金属材料与工程,2006; 35: 1009）

[15] Ploc R A. *J Nucl Mater*, 1968; 28: 48.

[16] Zhou B X, Li Q, Yao M Y, Liu W Q, Chu Y L. *Zirconium in the Nuclear Industry: 15th International Symposium*, *ASTM STP 1505*, West Conshohocken PA: ASTM International, 2009: 360.

[17] Zhou B X, Li Q, Yao M Y, Liu W Q, Chu Y L. *Corros Prot*, 2009; 30: 589.

（周邦新,李强,姚美意,刘文庆,褚于良. 腐蚀与防护,2009; 30: 589）

[18] Park J Y, Kim H G, Jeong Y H, Jeong Y H. *J Nucl Mater*, 2004; 335: 433.

[19] Yilmazbayhan A, Motta A T, Comstock R J, Sabol G P, Lai B, Cai Z H. *J Nucl Mater*, 2004; 324: 6.

Investigation of Microstructure of Oxide Layers Formed Initially on Zr‑4 Alloy

Abstract: Zr‑4 specimens with coarse grain of 0.2–0.8 mm were prepared to investigate the anisotropic growth of oxide layers formed initially on the grain surface with different orientations during corrosion tests in autoclave at 360 ℃/18.6 MPa in 0.01 mol/L LiOH aqueous after 5 h exposure. SEM, EBSD and HRTEM were adopt to measure the thickness of oxide layers, to determine the grain orientation of the matrix surface and to investigate the microstructure of oxide layers. The thicknesses of oxide layers formed on different grains varied in the range of 376–455 nm. The thickest oxide layers were detected on the grains with the orientations nearby basal plane (0001) and prismatic plane $(01\bar{1}0)$. The oxide layers have monoclinic, cubic, tetragonal crystal structures. Besides the thickness difference of oxide layers, the crystal structure and misorientation of nano-grains in oxide layers formed on different grains were also significantly different, and the most complicated oxide layer was formed on the grain with orientation nearby (0001) plane. Such kind of microstructure has more crystal defects, and larger ability for promoting the diffusion of oxygen ions and the growth of oxide layer.

水化学和腐蚀温度对锆合金氧化膜中压应力的影响[*]

摘　要：在不同水化学条件下，对 Zr－4 和 N18 管状样品进行腐蚀实验，然后用氧化膜卷曲法测量腐蚀样品氧化膜中的压应力，研究腐蚀温度、水化学对氧化膜中压应力随厚度变化的影响规律. 实验结果表明，Zr－4 和 N18 样品氧化膜中的压应力均按 360 ℃去离子水＞400 ℃过热蒸汽＞360 ℃ LiOH 水溶液的顺序依次减小. 在 360 ℃ LiOH 水溶液中腐蚀时，氧化膜中的压应力最低，这与 Li[+] 和 OH[-] 会渗入氧化膜，降低氧化锆表面自由能，从而加速氧化膜中空位的扩散凝聚、孔隙的形成和微裂纹发展的过程有关. 高温使空位的扩散加快，促进了氧化膜中压应力的弛豫过程.

锆合金由于其热中子吸收截面小，具有足够的热强度，以及在高温、高压水和过热蒸汽中具有良好的耐腐蚀性等特点，使其成为压水堆核电站燃料元件包壳材料的最佳选择[1]. 为了进一步降低核电成本，需要加深核燃料的燃耗，延长燃料组件的换料周期，因此，对燃料元件包壳的耐水侧腐蚀性能提出了更高的要求，研发高性能锆合金成为解决这一问题的关键.

锆合金氧化过程是在氧化膜/金属基体的界面上不断发生的，氧离子需要穿过氧化膜后才能到达氧化膜/金属基体的界面，或者说阴离子空位需要从氧化膜/金属基体的界面向氧化膜表面扩散. 所以，氧化膜的显微组织及其在腐蚀过程中的演化，又会影响氧离子或阴离子空位在氧化膜中的扩散，从而直接影响合金的耐腐蚀性能. 近几年对氧化膜显微组织的研究结果[2-6]表明，氧化膜的显微组织在腐蚀过程中的演化过程与腐蚀动力学有着密切的关系，如果能对这种演化过程进行控制并使其延缓，就可以提高合金的耐腐蚀性能.

金属锆氧化生成氧化锆时体积会发生膨胀，二者体积比为 1：1.56[7]. 由于金属基体约束氧化锆体积的膨胀，氧化膜生成后在其内部会产生很大的压应力. 氧化膜中的压应力对氧化膜显微组织的演化过程和晶体结构都有重要的影响，这是导致氧化膜的显微组织不断演化的一个重要原因，互相之间也存在因果关系[8-9]. 因此，研究氧化膜厚度、合金成分、温度及腐蚀时的水化学条件对锆合金氧化膜中压应力的影响，可以从另一角度来研究腐蚀机理和氧化膜的显微组织演化过程.

1　实验方法

1.1　腐蚀样品制备

样品为退火态的 Zr－4 和 N18 管材，成分如表 1 所示. 用电火花线切割方法将其切割为 15 mm 长的小段，经过酸洗($V(HNO_3)：V(H_2O)：V(HF)＝45\%：45\%：10\%$)、自来水

* 本文合作者：耿建桥、姚美意、王锦红、张欣、李士炉、杜晨曦. 原发表于《上海大学学报(自然科学版)》，2011，17(3)：293－296.

冲洗和干燥后,在真空中进行 580 ℃—2 h 退火处理.

表 1　实验用锆合金的名义成分

Table 1　Nominal composition of the zirconium alloy　　　　　　　　%

	Sn	Fe	Cr	Nb	Zr
Zr-4	1.5	0.2	0.1	—	Balance
N18	1.0	0.3	0.1	0.35	Balance

退火后的样品按照标准方法酸洗和去离子水清洗[10]后,放入高压釜中进行 360 ℃/18.6 MPa 去离子水和浓度为 0.01 mol/dm³ 的 LiOH 水溶液以及 400 ℃/10.3 MPa 过热蒸汽的腐蚀试验.通过测量样品的腐蚀增重来评定耐腐蚀性能.

1.2　应力测量方法及样品制备

由于锆合金的氧化过程是在氧化膜/金属界面处发生,随着氧化膜的生长,氧化膜外层中的压应力会逐步弛豫,结果在氧化膜厚度方向形成了应力梯度.当氧化膜在管状样品的外表面形成时,由于氧化膜沿管子周向的压应力可以通过氧化膜生成后管子直径的增加而得到部分释放,因而氧化膜沿轴向的压应力将大于周向的.在酸液中腐蚀掉金属基体后,脱离基体束缚的氧化膜就在应力梯度作用下沿着管状样品的轴向卷曲[2].假设氧化膜外表面应力为 0,沿厚度方向应力梯度保持不变,通过测量氧化膜的厚度 t、杨氏模量 E 和氧化膜卷曲后的曲率半径 r,可计算出氧化膜/金属界面处的压应力为[7]

$$\sigma = E\,\frac{t}{r}. \tag{1}$$

实际上,上述假设并非严格成立.只有当氧化膜较厚时,其外表面应力才为 0.随着氧化过程的进行,氧化膜的显微组织在不断发生演变[2-6],导致氧化膜不同厚度处显微组织中孔隙和微裂纹的分布存在差别.故沿氧化膜厚度方向的应力梯度分布也会发生变化,氧化膜的杨氏模量在不同厚度处也存在差异.为了简化计算过程,本研究采用同一杨氏模量来计算应力值.用式(1)计算出的压应力与氧化膜中的实际值有一定差别,但这是一种比例关系[11],不影响对氧化膜内应力随腐蚀过程变化趋势的分析.

用砂纸磨掉腐蚀后样品端面上的氧化膜,然后将其浸在酸洗液中.待基体完全溶解后,管状样品外壁氧化膜在应力梯度作用下卷曲.用 VHX-100 型光学显微镜测量氧化膜卷的直径,根据腐蚀增重计算氧化膜厚度(15 mg/dm² = 1 μm),杨氏模量为 206 GPa[2],根据式(1)计算氧化膜的压应力.

2　实验结果及讨论

图 1 为 Zr-4 和 N18 样品在 3 种不同水化学条件下的腐蚀增重曲线,它们之间都有一个共同点,在腐蚀转折以前,Zr-4 和 N18 合金的耐腐蚀性能非常相近,直到腐蚀转折后,它们之间的差别才逐渐显示出来.在 360 ℃/18.6 MPa/0.01 mol/dm³ LiOH 水溶液中腐蚀时,Zr-4 合金腐蚀 100 d 时发生转折(见图 1(a)),这与文献[1,5]的结果吻合.N18 合金在腐蚀 115 d 时也发生转折,但转折后 N18 的腐蚀速率仍然低于 Zr-4.在 360 ℃去离子水中腐蚀

250 d后,N18的耐腐蚀性能优于Zr-4的特征逐渐显示出来(见图1(c)).在400℃过热蒸汽中腐蚀90 d后,N18的耐腐蚀性能逐渐不如Zr-4(见图1(b)),与它们在360℃去离子水及LiOH水溶液中腐蚀时的变化规律相反,但是它们之间的差别不如在LiOH水溶液中腐蚀时那样显著.这说明锆合金在不同的水化学条件下腐蚀时,影响氧化膜显微组织演化过程的因素将有所不同.要彻底了解这方面的问题,还需要进行深入的研究.

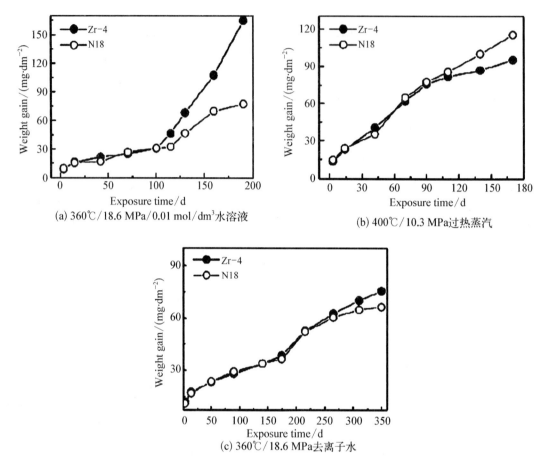

(a) 360℃/18.6 MPa/0.01 mol/dm³水溶液

(b) 400℃/10.3 MPa过热蒸汽

(c) 360℃/18.6 MPa去离子水

图1 不同腐蚀条件下Zr-4和N18合金样品的腐蚀增重曲线

Fig. 1 Weight gain *vs.* exposure time of Zr-4 and N18 specimens corroded in different water Chemistry

图2为Zr-4和N18样品在不同条件下腐蚀后氧化膜/金属界面处压应力与氧化膜厚度的变化曲线.从图2可以看出,无论是在Zr-4还是N18样品上形成的氧化膜,其应力在一定厚度范围内都是随着氧化膜厚度的增加而减小.在360℃去离子水中腐蚀的样品氧化膜中的压应力最大,当氧化膜达到一定厚度后,应力减小的速度变缓;400℃过热蒸汽中腐蚀次之;360℃/LiOH水溶液中腐蚀最低.同时还可以观察到,腐蚀条件对氧化膜中压应力的影响比合金成分的影响大,如在氧化膜厚度为0.8 μm左右时,360℃去离子水中腐蚀的Zr-4样品氧化膜中的压应力为960 MPa,N18的压应力为953 MPa;400℃过热蒸汽中腐蚀的Zr-4样品氧化膜中的压应力为773 MPa,N18的压应力为865 MPa;360℃/18.6 MPa/0.01 mol/dm³ LiOH水溶液中腐蚀的Zr-4样品氧化膜中的压应力仅为380 MPa,N18的压应力为371 MPa.

从图2可明显看出,氧化膜中压应力随厚度增加而减小,这主要是由于氧化膜显微结构在腐蚀过程中发生演化的结果. 氧化锆在压应力作用下时,会生成许多空位和间隙原子构成的点、线、面和体等不同形式的缺陷[3,6]. 晶体中的缺陷在应力、温度和时间的作用下发生扩散、凝聚或湮灭,会被氧化锆晶界吸收形成孔隙. 而氧化膜中孔隙的扩展、连通形成微裂纹,导致氧化膜中压应力弛豫. 因而,氧化膜中的压应力会随厚度的增加而减小,这也是锆合金腐蚀过程中发生转折的主要原因.

样品在 360 ℃/18.6 MPa/0.01 mol/dm³ LiOH 水溶液和去离子水中腐蚀时,氧化膜中压应力差距如此之大,这是 Li⁺ 和 OH⁻ 会渗入氧化膜的缘故[11-12]. 离子渗入到氧化膜后,会被吸附在空位和孔隙壁上,降低了氧化锆的表面自由能[13]. 用扫描探针显微镜观察样品

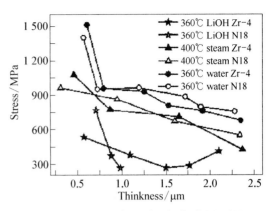

图 2 Zr-4 和 N18 腐蚀样品氧化膜/金属界面应力与氧化膜厚度变化之间的关系

Fig. 2 Relationship between the compressive stress at oxide/metal interface and the thickness of the oxide films on Zr-4 and N18 corroded specimens

腐蚀后氧化锆晶粒在表面上起伏程度的差别已经证实了这样的结果[5-6]. 这样使得形成孔隙和微裂纹所需的能量也相应降低,加速了空位通过扩散凝聚形成孔隙,以及微裂纹发展的过程[6],使压应力得到弛豫,因此,在 LiOH 水溶液中腐蚀时样品氧化膜中的压应力比在去离子水中腐蚀时的低.

400 ℃ 过热蒸汽中腐蚀样品氧化膜中的压应力始终小于 360 ℃ 去离子水中腐蚀的样品,这是由于温度的作用可以使扩散系数成指数增加[14],使得氧化膜中空位的扩散和凝聚更加容易. 400 ℃ 时形成孔隙和微裂纹的过程比 360 ℃ 去离子水中时更快,所以加快了氧化膜显微结构的演化过程,也导致氧化膜中的压应力更容易得到弛豫. 所以,400 ℃ 过热蒸汽中腐蚀样品氧化膜中的压应力低于 360 ℃ 去离子水中腐蚀时的样品.

3 结论

(1) 在 3 种不同的水化学条件下,Zr-4 和 N18 样品在腐蚀转折前的耐腐蚀性能非常相近. 转折后随着腐蚀时间的延长,2 种样品的耐腐蚀性能差别增大,但其变化规律并不完全一致. 在 360 ℃ 去离子水和 LiOH 水溶液中腐蚀时,N18 的耐腐蚀性能优于 Zr-4;但是,在 400 ℃ 过热蒸汽中腐蚀时,N18 的耐腐蚀性能不如 Zr-4.

(2) 氧化膜中的压应力随厚度的增加而减小,腐蚀温度和水化学条件对氧化膜中压应力的影响十分明显. 在 3 种水化学条件下,Zr-4 和 N18 氧化膜中的应力水平均按 360 ℃ 去离子水>400 ℃ 过热蒸汽>360 ℃ LiOH 水溶液的顺序依次减小.

(3) 样品在 LiOH 水溶液中腐蚀时氧化膜中的应力水平最低,这与 Li⁺ 和 OH⁻ 会渗入氧化膜中降低氧化锆的表面自由能有关. 氧化锆的表面自由能降低后,可以促使空位凝聚形成孔隙以及微裂纹的发展,加速了氧化膜显微组织结构的演化过程,使氧化膜中的压应力得到弛豫,这也是锆合金在 LiOH 水溶液中腐蚀时耐腐蚀性能差的原因.

参 考 文 献

[1] 杨文斗. 反应堆材料学[M]. 北京:原子能出版社,2006:17-28.

［2］ZHOU B X. Electron microscopy study of oxide films formed on zircaloy－2 in superheated steam［C］∥Zirconium in the Nuclear Industry: Eighth International Symposium, Philadelphia, ASTM STP 1023. 1989: 360－373.

［3］周邦新,李强,黄强. 水化学对锆合金耐腐蚀性能影响的研究［J］. 核动力工程,2000,21(5): 439－472

［4］周邦新,李强,姚美意. 锆－4 合金在高压釜中腐蚀时氧化膜显微组织的演化［J］. 核动力工程,2005,26(4): 364－371.

［5］周邦新,李强,刘文庆. 水化学及合金成分对锆合金腐蚀时氧化膜显微组织演化的影响［J］. 稀有金属材料与工程,2006,36(7): 1009－1016.

［6］ZHOU B X, LI Q, YAO M Y. Effect of water chemistry and composition on microstructural evolution of oxide on Zr-alloys［C］∥ Zirconium in the Nuclear Industry: Fifteenth International Symposium, West Conshohocken, ASTM STP 1505. 2009: 360－383.

［7］周邦新,蒋有荣. 锆－2 合金在 500～800 ℃空气中氧化过程的研究［J］. 核动力工程,1990,11(3): 233－239.

［8］QIN W, NAM C, LI H L. Effect of local stress on the stability of tetragonal phase in ZrO_2 film［J］. Journal of Alloys and Compounds,2007, 437: 280－284.

［9］BENALI B, HERBST G M, GALLET I. Stres driven phase transformation in ZrO_2 film［J］. Applied Surface Science, 2006, 253: 1222－1226.

［10］姚美意. 合金成分及热处理对锆合金腐蚀和吸氢行为影响的研究［D］. 上海:上海大学,2007.

［11］刘文庆,李强,周邦新. 水化学对 Zr－4 合金氧化膜/基体界面处压应力的影响［J］. 稀有金属材料与工程,2004,33(7): 728－730.

［12］JEONG Y H, BAEK J H, KIM S J. Cation incorporation into zirconium oxide in LiOH NaOH, and KOH solutions［J］. Journal of Nuclear Materials, 1999, 270: 322－333.

［13］傅献彩,沈文霞. 物理化学［M］. 北京:高等教育出版社,2006: 356－375.

［14］胡庚祥,蔡珣. 材料科学基础［M］. 上海:上海交通大学出版社,2000: 138－140.

Effect of Water Chemistry and Corrosion Temperature on Compressive Stress of Oxide Films on Zirconium Alloys

Abstract: To investigate the effect of corrosion temperature and water chemistry on compressive stress with the increase of oxide film thickness, Zr－4 and N18 tubular specimens were corroded in three kinds of water chemistry. Compressive stress of oxide films formed on the outside of the tubular specimens was measured with a curling method of oxide film. Experimental results indicate that compressive stress in oxide films decreases in the sequence of 360 ℃ deionized water＞400 ℃ super-heated steam＞360 ℃ lithiated water. Li^+ and OH^- ions incorporated into the oxide films decrease the surface free energy of oxide. As a result, the diffusion of vacancies, formation of pores and development of micro-cracks are enhanced. This is the reason why compressive stress of oxide films formed in 360 ℃ lithiated water is the lowest. Diffusion of vacancies is promoted at higher temperature, resulting in relaxation of compressive stress in oxide films. Therefore compressive stress of oxide films formed in 400 ℃ super-heated steam is lower than that formed in 360 ℃ deionized water.

晶界网络特征对 304 不锈钢晶间
应力腐蚀开裂的影响[*]

摘　要：通过晶界工程（GBE）处理，可使 304 不锈钢样品中的低 ΣCSL 晶界比例提高到 70%（Palumbo Aust 标准）以上，同时形成了大尺寸的"互有 $\Sigma3^n$ 取向关系晶粒的团簇"显微组织. 采用 C 型环样品恒定加载方法，在 pH 值为 2.0 的沸腾 20% NaCl 酸化溶液中进行应力腐蚀实验. GBE 样品在平均浸泡 472 h 后出现应力腐蚀裂纹. SEM，EBSD 和 OM 分析表明，应力腐蚀开裂（SCC）为沿晶开裂（IGSCC）和穿晶开裂（TGSCC）的混合型. 而未经 GBE 处理的样品在平均浸泡 192 h 后出现多条应力腐蚀主裂纹，且多为沿晶界裂纹. 经过 GBE 处理的样品中大尺寸的晶粒团簇及大量相互连接的 $\Sigma3-\Sigma3-\Sigma9$ 和 $\Sigma3-\Sigma9-\Sigma27$ 等 $\Sigma3^n$ 类型的三叉界角，阻碍了 IGSCC 裂纹的扩展，从而提高了 304 不锈钢样品的抗 IGSCC 性能.

　　304 不锈钢具有良好的综合性能，是核电反应堆中重要的结构材料，常用于各种堆内构件和回路管道等的制备[1]. 但晶间腐蚀（IGC）和晶间应力腐蚀开裂（IGSCC）一直是 304 不锈钢在服役过程中的重要失效形式，对核反应堆的安全可靠运行带来极大危害. 晶界敏化是导致这些问题的主要原因之一[1-3]. 304 不锈钢在 500—800 ℃ 之间加热时会在晶界上析出富 Cr 的碳化物，从而在晶界附近产生贫 Cr 区，增加了 304 不锈钢 IGC 和 IGSCC 敏感性. 为解决这一问题. 可以在奥氏体不锈钢中添加 Ti，Nb 等元素，通过形成它们的碳化物以稳定 C 元素. 或者用超低碳不锈钢如 304 L 和 316 L 代替 304 不锈钢，但这些方法不仅提高成本，而且即使在这些超低碳不锈钢中也会发生沿晶应力腐蚀开裂[4,5].

　　1984 年，Watanabe[6]首次提出了晶界设计（grain boundary design）的概念，并于上世纪 90 年代发展为晶界工程（grain boundary engineering，GBE）这一研究领域，旨在通过大幅度增加低 ΣCSL（coincidence site lattice[7]，重位点阵，低 Σ 指 $\Sigma \leqslant 29$）晶界比例来改善材料与晶界有关的多种性能[8-11]. 已有研究[12-15]表明，可以用 GBE 的方法来进一步提高奥氏体不锈钢抗 IGC 和抗 IGSCC 性能. Shimada 等[12]研究表明，将低 ΣCSL 晶界比例提高到 85%（Brandon 标准[16]）以上，304 不锈钢的抗 IGC 性能得到大幅提高，认为 304 不锈钢耐晶间腐蚀性能的提高是由于低 ΣCSL 晶界比例提高后打断了随机晶界连通性所致. Gertsman 和 Bruemmer[17]分析了 304 不锈钢和 600 合金的裂纹扩展路径，指出并不是所有的低 ΣCSL 晶界都对应力腐蚀开裂免疫，只有共格的 $\Sigma3$ 晶界具有这一特殊性能，其他类型的晶界都有可能出现应力腐蚀裂纹. Marrow 等[18]研究了不锈钢中 IGSCC 裂纹在三维空间的扩展，认为低 ΣCSL 晶界相互连接可以形成桥接带（bridging ligaments），这些桥接带可以阻碍 IGSCC 裂纹扩展，但其阻碍能力受到构成此桥接带的晶界类型的影响. 本课题组前期研究[19-21]表明，通过 GBE 处理，可以提高低层错能 fcc 金属材料中低 ΣCSL 晶界比例，并形成大尺寸的"互有 $\Sigma3^n$ 取向关系晶粒的团簇"显微组织.

　　本文研究了晶界网络特征对 304 不锈钢 IGSCC 的影响，并考察了不同低 ΣCSL 晶界比例

---

* 本文合作者：胡长亮、夏爽、李慧、刘廷光、陈文觉. 原发表于《金属学报》，2011，47(7)：939 - 945.

的样品发生 IGSCC 的差异,旨在阐明 304 不锈钢经 GBE 处理后形成的大尺寸的晶粒团簇显微组织以及高比例的 $\Sigma 3^n$ 类型晶界和三叉界角影响 304 不锈钢的抗 IGSCC 性能的机制.

1 实验方法

将实验所用的 304 不锈钢样品经 1 100 ℃,15 min 固溶处理后淬入水中,再用电火花线切割机将样品加工成片状,以便进行冷轧变形及再结晶退火处理. 304 不锈钢样品在固溶处理后分别冷轧 5％和 50％,再分别经 1 100 ℃再结晶退火 5 和 60 min,水淬后得到样品 A 和 B. 试样编号及处理工艺如表 1 所示.本文标记 A 的样品为 GBE 处理的样品.将处理后的样品再进行 650 ℃,12 h 敏化热处理,以便进行应力腐蚀实验.

表 1 样品 A 和 B 的晶界特征分布统计及产生应力腐蚀裂纹的时间
Table 1 Grain boundary character distribution and the results of
SCC tests of specimens A (treated by GBE) and B

Specimen	Cold rolling, ％	1 100 ℃ annealing time, min	Length fraction of grain boundary, ％			Overall low ΣCSL, ％	Mean time of generating SCC crack, h
			Σ1	Σ3	Σ9+Σ27		
A	5	5	4.9	59.2	8.7	73.1	472(408—524)
B	50	60	4.8	35	1.5	42	192(144—220)

在电子背散射衍射(EBSD)测试之前,先将样品用 SiC 砂纸研磨,然后进行电解抛光.所用电解抛光液为 20％HClO₄＋80％CH₃COOH(体积分数),在室温下用 40 V 直流电抛光 60 s.对于出现应力腐蚀裂纹的样品,先用 1200 号的 SiC 砂纸研磨,然后用 0.5 μm 的 SiC 抛光膏进行机械抛光,最后进行上述的电解抛光,获得平整的表面,以备后续的 EBSD,扫描电子显微镜(SEM)和光学显微镜(OM)分析.

应力腐蚀试样采用 C 型环加载方式,试样尺寸按 GB15970.5 规定加工.将 C 型环试样嵌入内开口的模具以施加恒定载荷,模具为相同的 304 不锈钢材料,以避免发生电偶腐蚀.所有的 C 型环样品及模具用电火花线切割机按照编排的程序加工成型,以保证尺寸一致,从而使试样的受力状态一致.应力腐蚀实验介质参考 ASTMG 123 - 2000 标准,将 20％NaCl(质量分数)溶液作为腐蚀溶液,用 H₃PO₄ 调节 pH 值到 2.0,加热至沸腾,溶液沸点为 (112±1)℃.将样品 A 和 B 同时放入配有冷凝回流装置的锥形瓶中进行应力腐蚀开裂(SCC)实验,每隔 4 h 取出试样并用 OM 观察是否有 SCC 裂纹产生.

利用配有 TSL - EBSD 系统的 SEM 对电解抛光后的样品表面进行逐点逐行扫描,得到材料表面各点的取向,通过晶界两侧晶粒的取向差判定晶界类型.由取向成像显微系统(orientation image microscopy, OIM)自动统计不同类型晶界的长度比例. EBSD 测试步长根据晶粒大小不同而改变,在 5—10 μm 之间.本工作采用 Palumbo-Aust 标准[22]:$\Delta\theta_{max}=15°\Sigma^{-5/6}$ 判定晶界类型.

2 实验结果与讨论

2.1 GBE 处理后样品晶界网络显微组织的特征

由于样品 A 和 B 冷轧变形量不同,为了使退火后的平均晶粒尺寸相当,延长样品 B 的退火时间至 60 min. 利用 OIM 系统统计样品 A 和 B 的晶界特征分布（Palumbo-Aust 标准[22]）结果也列于表 1 中.

图 1 给出了样品 A 和 B 晶界类型的 OIM 图. 从图 1a 中可以看出，A 样品经 5% 变形和 1 100 ℃ 退火 5 min 后，形成了大尺寸"互有 $\Sigma 3^n$ 取向关系晶粒的团簇"（以下简称晶粒团簇），如图 1a 中的 C1，C2，C3 和 C4 所示. 在晶粒团簇内部，由于再结晶时多重孪晶的充分发展形成了大量的 $\Sigma 3^n$ 晶界，并且它们相互连接形成了大量 $\Sigma 3 - \Sigma 3 - \Sigma 9$ 和 $\Sigma 3 - \Sigma 9 - \Sigma 27$ 等

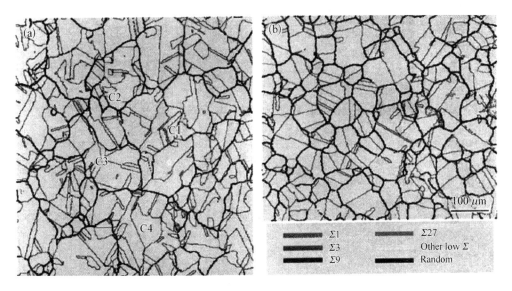

图 1 样品 A 和 B 中不同类型晶界的 OIM 图

Fig. 1 OIM maps of different types of grain boundaries in specimens A（a）and B（b），larger grain-clusters C1，C2，C3，C4 formed in specimen A (the clusters C1，C2，C3 and C4 enclosed by random boundary (R) consist of grains with $\Sigma 3^n$ orientation relationships)

类型三叉界角，而晶粒团簇之间基本为连续的随机晶界. 这样的显微组织含有高比例（73%，长度比例，下同）的低 ΣCSL 晶界，其中 $\Sigma 3$ 晶界占主要部分，约为 60%，$\Sigma 9$ 晶界和 $\Sigma 27$ 晶界的总体比例约为 9%，如表 1 所示. 样品 B 经 50% 变形和 1 100 ℃ 退火 60 min 后，虽然也形成了晶粒团簇，但尺寸较样品 A 中的小得多，如图 1b 和图 2 所示. 低 ΣCSL 晶界比例比较低，只有 42% 左右，如表 1 所示. 样品经过 GBE 处理后，平均晶粒团簇尺寸和低 ΣCSL 晶界的比例明显增大.

2.2 晶界类型对 304 不锈钢 IGSCC 行为的影响

将样品 A 和 B 经 650 ℃ 敏化处理 12 h 后，制成 C 型环样品后加载，浸入经 H_3PO_4 酸化的 NaCl 溶液中加热至沸腾，进行 SCC 实验. 图 3a 是样品 B 浸泡 210 h 后出现 SCC 裂

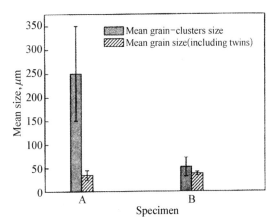

图 2 样品 A 和 B 中平均晶粒尺寸和平均晶粒团簇尺寸统计

Fig. 2 Statistics of mean dimensions of grains and grain-clusters in specimens A and B

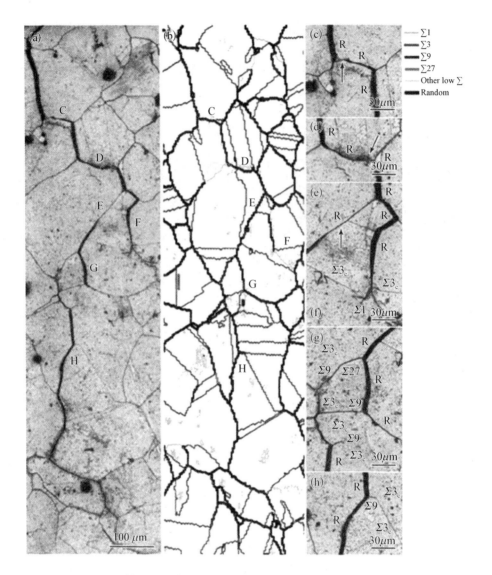

图 3 样品 B 经腐蚀溶液浸泡 210 h 后出现的 SCC 裂纹扩展

Fig. 3 Analyses of the stress corrosion crack propagation in specimen B after stress corrosion for 210 h in acidified boiling 20％NaCl solution （a）metallograph，showing propagation path of crack （b）OIM map of Fig. 3a （c—h）high magnified photos of cracks C—H in Fig. 3a，cracking along random boundary （R）except a few of R shown as arrows in Figs. 3c—e （Σ3_c-coherent twin boundary）

纹扩展的 OM 像. 图 3b 是与图 3a 对应的不同类型晶界的 OIM 图. 图 3c—h 分别对应图 3a 中的 C—H 点. 从图 3a 和图 3c—h 中可以看出,SCC 裂纹大都沿着随机晶界(R)扩展,但也发现同样处于主裂纹扩展路径上的部分 R 晶界未开裂,如图 3c—e 中箭头所示. 研究者[17,23] 在研究晶界类型对应力腐蚀开裂的影响时发现了类似的现象. 在图 3f 中,SCC 裂纹沿 R 晶界向前扩展,但在三叉界角 R‑Σ1‑Σ3 前停止,裂纹被 Σ1 和 Σ3 晶界阻止. 当裂纹扩展至 G 处时,IGSCC 裂纹绕过三叉界角 Σ3‑Σ3‑Σ9 和 Σ3‑Σ9‑Σ27 后继续沿着下方的 R 晶界扩展(图 3g). 在 H 处,SCC 裂纹沿着 R 晶界扩展,但同样处于裂纹扩展路径上的 Σ9 晶界却未开裂,如图 3h 所示. 通过以上分析可以看出,低 ΣCSL 晶界较 R 晶界有更好的抗 IGSCC 能力,SCC 裂纹优先沿着 R 晶界扩展,而几乎所有的 Σ3$_c$(孪晶的共格界面)晶界都未开裂. IGSCC 裂纹扩展会受到三叉界角 Σ3 - Σ3 - Σ9 的阻碍. 其次,在 SCC 裂纹扩展过程中,并不是所有处于主裂纹路径上的 R 晶界都会开裂,如图 3c—e 中均有未开裂的 R 晶界,这是因为 IGSCC 裂纹扩展除了与晶界类型有关,还与晶界面相对于外加应力的方向有关[23-25],这些 R 晶界因处在不利的应力方向上而不能开裂,IGSCC 裂纹只能在三维空间选择其他 R 晶界扩展[23,26]. 因此,样品发生 IGSCC 的难易,一方面取决于样品中 Σ3 - Σ3 - Σ9 三叉界角的多少,另一方面取决于裂纹尖端 R 晶界的晶界面相对于外加应力方向是否有利于开裂.

图 4 是样品 A 经过 436 h 腐蚀实验后观察到的 IGSCC 裂纹扩展的情况. 图 4a 是 OM 像,图 4b 和 c 分别是图 4a 中方框 1 对应的 SEM 二次电子像和晶界类型 OIM 图;图 4d 和 e

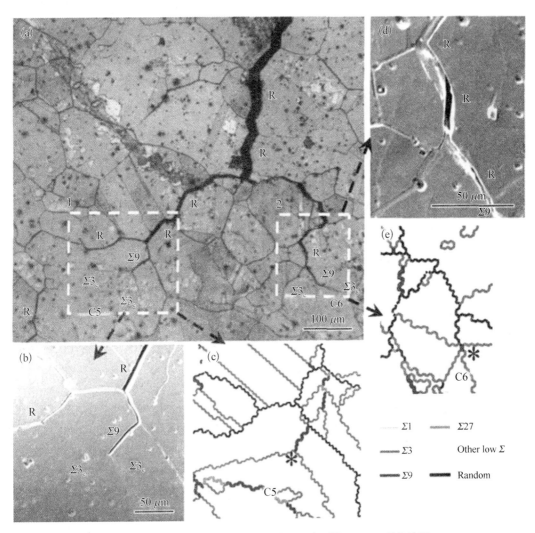

图 4 样品 A 经过 436 h 浸泡实验后观察到的 IGSCC 裂纹扩展

Fig. 4 Analyses of the propagation path of stress corrosion cracks in specimen A stress-corroded for 436 h in acidified boiling 20%NaCl solution （a) optical micrograph；(b) SEM image of area 1, the type of boundary marked；(c) OIM map of area 1, crack propagated along R boundary but stopped at the crossed position of two Σ3$_c$ boundaries, shown as a mark* ；(d) SEM image of area 2；(e) OIM map of area 2, crack propagation stopped also at the crossed position of two Σ3$_c$ boundaries. shown as a mark*

分别是图4a中方框2的SEM二次电子像和晶界类型OIM图.C5和C6是图4a中的2个晶粒团簇.由图4a中可以看出,SCC裂纹大都沿着R晶界扩展,并且裂纹可通过$\Sigma 9$晶界扩展至晶粒团簇内部,如图4b和d中的$\Sigma 9$晶界均发生了开裂.但IGSCC裂纹却最终被前方2个$\Sigma 3_c$晶界阻止而停止在团簇内部,如图4c和e中的*所示,这进一步证明了由2个$\Sigma 3_c$晶界组成的$\Sigma 3$-$\Sigma 3$-$\Sigma 9$三叉界角可以阻止IGSCC裂纹的扩展.

2.3 晶界网络特征对304不锈钢IGSCC行为的影响

图5a是试样A(低ΣCSL晶界比例为73%)应力腐蚀504 h后表面的OM像.可以看出,样品A在腐蚀溶液中较长时间后才出现应力腐蚀裂纹,并且开裂后样品中主裂纹也比较少,应力腐蚀开裂方式为IGSCC和穿晶开裂(TGSCC)的混合型.图5b是试样B(低ΣCSL晶界比例为42%)应力腐蚀196 h后表面的OM像.可以看出,样品中有多条IGSCC主裂纹,与晶界类型OIM图对比后发现,裂纹大都是沿着R晶界扩展,这与图3中的结果一致.

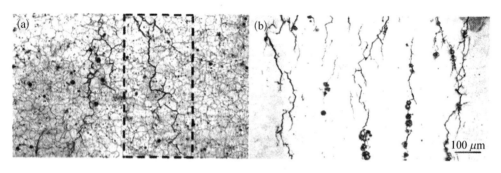

图5 样品A和B发生IGSCC后的表面形貌图

Fig. 5 OM photos of specimens A and B undergone IGSCC in acidified boiling 20%NaCl solution
（a）optical micrograph of specimen A undergone 504 h stress corrosion, a few primary cracks
（b）optical micrograph of specimen B undergone 196 h stress corrosion, more primary cracks

图6a是图5a中方框的OM放大像.图6b,c是与图6a对应的晶界类型OIM图.样品A发生SCC时,最初以TGSCC方式进入材料内部,如图6a中的1所示,在随后的SCC裂纹扩展中,SCC裂纹优先选择晶粒团簇外围的R晶界扩展.当IGSCC裂纹扩展至晶粒团簇内部时,由于$\Sigma 3$-$\Sigma 3$-$\Sigma 9$,$\Sigma 3$-$\Sigma 9$-$\Sigma 27$等$\Sigma 3^n$类型的三叉界角相互连接,裂纹扩展到$\Sigma 3$-$\Sigma 3$-$\Sigma 9$三叉界角时受到阻碍,此时只有通过IGSCC→TGSCC的转变,SCC裂纹才能够继续沿R晶界扩展,如图6a中的3→5→7.SCC裂纹扩展到晶粒团簇内部时,IGSCC裂纹总会遇到三叉界角$\Sigma 3$-$\Sigma 3$-$\Sigma 9$的阻碍,IGSCC裂纹或停止在团簇内部,如图6a中8,9,10,11,12所示,或通过IGSCC→TGSCC转变后继续沿R晶界扩展(图6a中5),这取决裂纹尖端的应力状态及裂纹在三维空间扩展时前方遭遇的晶界类型.

图7a和b分别给出了IGSCC裂纹在GBE处理样品和未经GBE处理样品中扩展的示意图,并以此来说明GBE处理提高样品抗IGSCC性能的机制,由图1可知,GBE样品中形成了大尺寸的形状不规则的晶粒团簇,同时团簇内部所有$\Sigma 3^n$晶界相互连接形成了许多$\Sigma 3$-$\Sigma 3$-$\Sigma 9$,$\Sigma 3$-$\Sigma 9$-$\Sigma 27$等$\Sigma 3^n$类型的三叉界角.当IGSCC裂纹沿团簇外围R晶界扩展时,由于晶粒团簇尺寸较大时呈现出形状的不规则性,使得裂纹尖端R晶界的晶界面相对于外加应力方向不利于开裂的概率增加.同时由于GBE样品中低ΣCSL晶界比例较高,IGSCC裂纹扩展的通道减少.当裂纹扩展到晶粒团簇内部时,大量的相互连接的$\Sigma 3^n$类型三

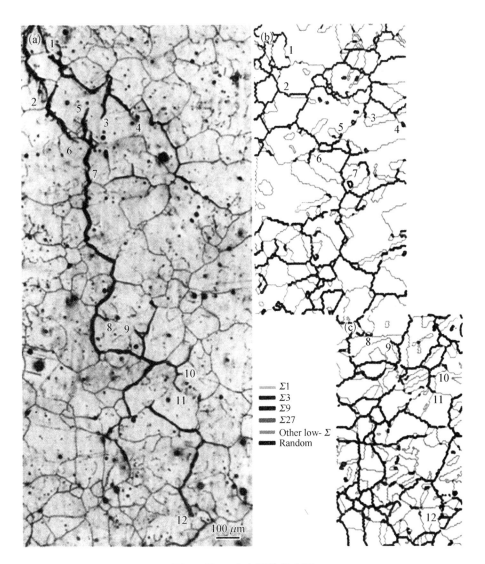

图 6 图 5a 中方框的放大图

Fig. 6 Further analyses of the area outlined by framework in Fig. 5a (a) OM photo (the crack sites of interests were numbered on the micrographs, site 1: the crack propagates into the sample in a transgranular manner; site 2: the crack propagates initially along random grain boundaries then into a grain in a transgranular manner; sites 3 − 6 and 8 − 12: the cracks propagate into grain-clusters and encounter $\Sigma 3 - \Sigma 3 - \Sigma 9$ type triple junctions; sites 7: the cracks propagate along random grain boundaries) (b, c) OIM maps, showing boundaries types of crack propagation

叉界角又阻碍了 IGSCC 裂纹进一步扩展,所以延长了 GBE 样品中 IGSCC 裂纹扩展的时间.当应力增大到一定程度时,迫使 IGSCC 向 TGSCC 转变,并释放了裂纹尖端的应力[6,27],这更加减缓了 IGSCC 裂纹向样品内部扩展,从而提高了样品的抗 IGSCC 性能.而未经 GBE 处理的样品中晶粒团簇尺寸小,低 ΣCSL 晶界比例低,$\Sigma 3^n$ 类型的三叉界角也较少,使得 IGSCC 裂纹扩展时受到的阻碍较小,所以 SCC 在较短的时间内就可沿 R 晶界发生并扩展成多条 IGSCC 裂纹.

因此,GBE 样品中形成的大尺寸的晶粒团簇及内含大量的相连接的 $\Sigma 3^n$ 类型三叉界角,

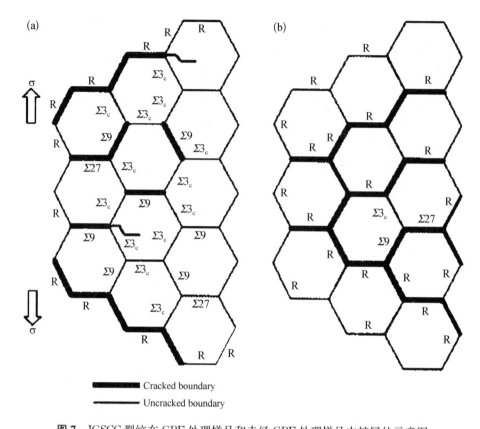

Cracked boundary
Uncracked boundary

图 7 IGSCC 裂纹在 GBE 处理样品和未经 GBE 处理样品中扩展的示意图

Fig. 7 Schematic illustrations of intergranular stress corrosion crack propagation in GBE specimen (a) and non-GBE specimen (b)

阻碍了 IGSCC 裂纹的扩展,并促使 IGSCC 向 TGSCC 转变,从而提高了 304 不锈钢样品的抗 IGSCC 性能

3 结论

GBE 处理可以使 304 不锈钢样品中形成大尺寸的晶粒团簇,团簇内部 $\Sigma 3^n$ 晶界相互连接,形成了大量 $\Sigma 3$-$\Sigma 3$-$\Sigma 9$ 和 $\Sigma 3$-$\Sigma 9$-$\Sigma 27$ 的三叉界角,这样的显微组织中低 ΣCSL 晶界比例较高,超过了 70% (Palumbo-Aust 标准). 未经 GBE 处理的样品在应力腐蚀实验平均 192 h 后就产生多条 IGSCC 主裂纹,而 GBE 处理的样品经应力腐蚀实验平均 472 h 后才出现应力腐蚀裂纹,且裂纹扩展方式为 IGSCC 和 TGSCC 的混合型. GBE 处理后样品中大尺寸的晶粒团簇及团簇内大量的相互连接的 $\Sigma 3^n$ 类型的三叉界角,阻碍了 IGSCC 裂纹的扩展,并迫使 IGSCC 向 TGSCC 转变,从而提高了 304 不锈钢样品的抗 IGSCC 性能.

参 考 文 献

[1] Yang W D. *Nuclear Reactor Materials Science*. Beijing: Atomic Energy Press, 2000: 195.

(杨文斗. 反应堆材料学. 北京: 原子能出版社, 2000: 195)

[2] Trillo E A, Murr L E. *Acta Mater*, 1999: 47: 235.

[3] Zhou Y, Aust K T, Erb U, Palumbo G. *Scr Mater*, 2001;45; 49.

[4] Horn R M, Gordon G M, Ford F P, Cowan R L. *Nucl Eng Des*, 1997; 174: 313.

[5] Saito N, Tsuchiya Y, Kano F, Tanaka N. *Corrosion*, 2000;56: 57.

[6] Watanable T. *Res Mech*, 1984; 11: 47.

[7] Kronberg M L, Wilson F H. *Trans AIME*, 1949; 185: 501.

[8] Randle V. *Acta Mater*, 2004; 52: 4067.

[9] Lehockey E M, Limoges D, Palumbo G, Sklarchuk J, Tomantscher K, Vincze A. *J Power Sour*, 1999; 78: 79.

[10] Thaveeprungsriporn V, Was G S. *Metall Trans*, 1997; 28: 2101.

[11] Lin P, Palumbo G, Erb U. *Scr Metall Mater*, 1995; 33: 1387.

[12] Shimada M, Kokawa H, Wang Z J, Sato Y S, Karibe I. *Acta Mater*, 2002; 50: 2331.

[13] Hu C L, Xia S, Li H, Liu T G, Zhou B X, Chen W J, Wang N. *Corros Sci*, 2011; 53: 1880.

[14] Jin W Z, Yang S, Kokawa H, Wan Z J. *J Mater Sci Technol*, 2007; 23: 785.

[15] Rahimi S, Engelberg D L, Duff J A, Marrow T J. *J Microsc*, 2009; 233: 423.

[16] Brandon D G. *Acta Mater*, 1966; 14: 1479.

[17] Gerstman V Y, Bruemmer S M. *Acta Mater*, 2001; 49: 1589.

[18] Marrow T J, Babout L, Jivkor A P, Wood P, Engelberg D, Stevens N, Withers P J, Newman R C. *J Nucl Mater*, 2006; 352: 62.

[19] Xia S, Zhou B X, Chen W J, Wang W G. *Scr Mater*, 2006;54: 2019.

[20] Xia S, Zhou B X, Chen W J. *J Mater Sci*, 2008; 43: 2990.

[21] Xia S, Zhou B X, Chen W J. *Metall Mater Trans*, 2009;40A: 3016.

[22] Palumbo G, Aust K T, Lehockey E M, Erb U, Lin P. *Scr Mater*, 1998; 38: 1685.

[23] Arafin M A, Szpunar J A. *Corr Sci*, 2009; 51: 119.

[24] Arafin M A, Szpunar J A. *Mater Sci Eng*, 2009; A513 – 514: 254.

[25] Arafin M A, Szpunar J A. *Comput Mater Sci*, 2010; 47: 890.

[26] Pan Y, Adams B L, Olson T, Panayotou N. *Acta Mater*, 1996; 44: 4685.

[27] Alexandreanu B, Was G S. *Scr Mater*, 2006; 54: 1047.

Effect of Grain Boundary Network on the Intergranular Stress Corrosion Cracking of 304 Stainless Steel

Abstract: The grain boundary network in a 304 stainless steel can be controlled by grain boundary engineering (GBE). The total length proportion of $\Sigma 3^n$ coincidence site lattice (CSL) boundaries was increased to more than 70%, and the large size highly twinned grain cluster microstructure formed through the treatment of GBE. Stress corrosion cracking (SCC) susceptibility of 304 stainless steel was evaluated through C-ring specimen tests conducted in acidified boiling 20% NaCl solution. Based on the characterization by SEM, EBSD and OM, it was found that the large grain-clusters associated with many interconnected $\Sigma 3 - \Sigma 3 - \Sigma 9$ and $\Sigma 3 - \Sigma 9 - \Sigma 27$ triple junctions produced by GBE arrest the IGSCC cracks and improve the resistance to IGSCC.

690 合金中晶界网络分布的控制及其对晶间腐蚀性能的影响*

摘　要：利用形变及热处理工艺提高了 690 合金的低 Σ 重位点阵（Coincidence Site Lattice，CSL）晶界比例，通过电子背散射衍射（EBSD）技术表征了由不同类型晶界构成的网络特征，结果表明通过晶界工程处理，能够形成以大尺寸"互有 Σ3^n 取向关系晶粒的团簇"显微组织为特征的晶界网络分布，这种显微组织是再结晶过程中多重孪晶充分发展的结果。通过晶间腐蚀浸泡实验表明通过晶界工程处理的样品抗晶间腐蚀性能较未经过晶界工程处理的样品明显提高。腐蚀后样品的显微形貌表明大尺寸"互有 Σ3^n 取向关系晶粒的团簇"能够阻止晶间腐蚀向样品内部扩展，并且能够保护下层的显微组织。

1　前言

镍基 690 合金作为压水反应堆蒸汽发生器传热管材料，已经应用于许多核电站，虽然取得了较好的效果，但随着核电站设计寿命延长的要求，进一步提高蒸汽发生器可靠性及使用寿命是急需研究的问题。蒸汽发生器传热管道失效的主要原因在不同时期有不同的情况，从近些年统计的结果来看，与晶界有关的应力腐蚀和晶间腐蚀是蒸汽发生器传热管道失效的主要原因[1]。

包括 690 合金在内的绝大部分工程材料都是多晶体，材料的性能与其晶界特性有着非常紧密的联系。利用晶界工程（Grain Boundary Engineering，GBE），通过合适的处理工艺，可以明显提高材料的低 ΣCSL（CSL 是"Coincidence Site Lattice"的缩写，即重位点阵。低 ΣCSL 是指 Σ≤29[2]）晶界比例，控制晶界网络拓扑分布，改善材料与晶界有关的多种性能[3]。

本文利用晶界工程的办法提高了 690 合金中低 ΣCSL 晶界的比例，采用浸泡腐蚀的方法测试了经过晶界工程处理和未经过晶界工程处理的 690 合金样品的耐晶间腐蚀能力，并表征了经腐蚀后样品的显微形貌。

2　实验方法

对 690 合金（60Ni - 30Cr - 10Fe，质量百分数）中进行晶界工程处理。晶界工程处理方法为：将样品冷加工 5%，再抽真空密封在石英管中，在 1 100 ℃下再结晶退火 5 min，然后立即淬入水中，并同时砸破石英管。这样既可以获得高比例的低 ΣCSL 晶界，又起到了固溶处理的作用。为避免晶粒尺寸对腐蚀效果的影响，将一部分样品在 1 100 ℃保温 30 min，使晶粒长大到与晶界工程处理后样品的晶粒尺寸相当。然后将这 2 种样品在 715 ℃下敏化处理 2 h。这样就得到了 2 种样品，即敏化后的晶界工程处理的样品（GBE 样品）和敏化后的普通样品（Non - GBE 样品）。

* 本文合作者：李慧、夏爽、陈文觉、刘廷光、胡长亮。原发表于《中国材料进展》，2011，30（5）：11 - 14.

在浸泡腐蚀前先将样品电解抛光获得干净表面. 使用的电解液是 20% HClO₄＋80% CH₃COOH, 在室温下用 30 V 直流电抛光 30 s. 最后测量样品表面积, 称量样品的原始重量(精确到 0.01 mg). 然后将样品悬挂浸泡在 60% HNO₃＋0.4% HF 水溶液中(体积比)进行晶间腐蚀实验, 每隔 1 d 对样品进行称重并计算腐蚀失重量, 得到腐蚀失重曲线.

在不同腐蚀时间进行取样, 观察腐蚀后样品的显微形貌. 晶界两侧晶粒取向关系利用配备在 CamScan Apollo 300 SEM 中的 HKL EBSD 附件确定, 扫描步长为 2 μm. 数据采集后利用 Channel-5 取向分析系统自动确定晶界类型, 晶界类型的判定采用 Palumbo-Aust 标准[4], 相对于 CSL 模型最大偏差角为 $\Delta\theta_{\max}=15°\Sigma^{-\frac{5}{6}}$.

3 结果与讨论

图 1 给出了 GBE 与 Non-GBE 样品不同类型晶界的 OIM 图. Non-GBE 样品中虽然含有较多退火孪晶(Σ3), 但是多重孪晶界(Σ9, Σ27)很少, 总体低 ΣCSL 晶界比例较低, 为 46%. CBE 样品中含有很多退火孪晶及多重孪晶界, 再结晶过程中多重孪晶得到了充分发

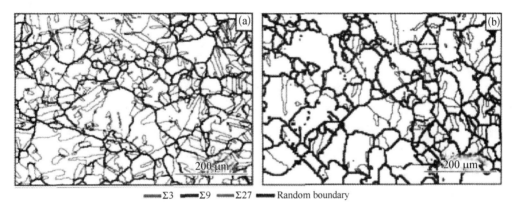

$\blacksquare$ Σ3 $\blacksquare$ Σ9 $\blacksquare$ Σ27 $\blacksquare$ Random boundary

图 1 (a) GBE 样品和(b) Non-GBE 样品不同类型晶界的 OIM 图

Fig. 1 The OIM micrographs of grain boundary characters of (a) GBE specimen and (b) Non-GBE specimen

展, 显微组织中含有很多相互连接的 Σ3-Σ3-Σ9, Σ3-Σ9-Σ27 等 Σ^{3n} $(n=1,2,3\cdots)$ 类型三叉界角, 形成了以数百微米大尺寸"互有 Σ^{3n} 取向关系晶粒的团簇"显微组织为特征的晶界网络分布[5-7], 总的低 ΣCSL 晶界比例接近 75%. 晶粒团簇内任何 2 个晶粒之间, 不论是否相邻均具有 Σ^{3n} 的取向关系, 晶粒团簇之间是由随机晶界构成的网络. 2 种工艺处理后样品的平均晶粒尺寸约为 19 μm(统计时包括孪晶界).

图 2 给出了 GBE 样品与 Non-GBE 样品敏化后经过 550 h 晶间腐蚀浸泡实验后的

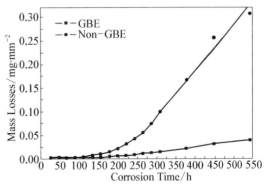

图 2 样品浸泡腐蚀后的腐蚀失重曲线

Fig. 2 The weight losses of specimens immersed in corrosive

腐蚀失重曲线. 可以看出, GBE 样品的腐蚀失重明显小于 Non - GBE 样品, 当腐蚀时间为 550 h 时, 二者腐蚀失重相差约 1 个数量级.

图 3 给出了经过不同时间晶间腐蚀浸泡实验后样品的表面形貌. 当腐蚀 3 d 后样品表面开始有小尺寸的晶粒脱落, 如图 3a, b 中箭头所示. GBE 样品与 Non - GBE 样品未显示出明显区别. 当延长腐蚀时间至 8 d 后, Non - GBE 样品表面晶粒已经明显脱落, 而 GBE 样品表面仍然只有小的晶粒脱落 (如图 3c, d 所示). 当延长腐蚀时间至 14 d, Non - GBE 样品表面层晶粒已经基本脱落, 而 GBE 样品表面仍然只有很少小晶粒脱落, 只是晶界的腐蚀痕迹在一直加深.

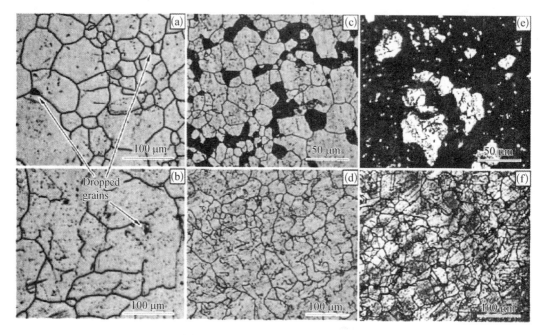

图 3 样品经浸泡晶间腐蚀不同时间后表面形貌

Fig. 3 The surface morphology of specimens immersed in corrosive solution for (a, b) 3 days, (c, d) 8 days, and (e, f) 14 days, (a, c, e) are of GBE specimens and (b, d, f) are of Non - GBE specimens

为了进一步了解 Non - GBE 样品与 GBE 样品的耐晶间腐蚀能力, 对腐蚀后样品的截面进行了观察 (图 4). 可以看出, 即使经过相同时间的腐蚀, 样品内不同晶界的腐蚀深度也有很大差别, Non - GBE 样品表面晶粒很容易脱落; 而 GBE 样品的晶粒很难脱落. 在 Non - GBE 样品中, 孪晶界对晶间腐蚀具有明显的抵抗能力 (图 4c), 而在 GBE 样品中, 由于碳化物更容易在随机晶界处析出[8-9], 产生明显的贫铬区, 腐蚀主要沿晶粒团簇外围的随机晶界扩展, 而晶粒团簇内部的低 ΣCSL 晶界的腐蚀深度要明显浅很多 (图 4d).

通过图 2~4 可以明显看出, CBE 样品抗晶间腐蚀能力明显提高. 一般认为, GBE 处理会打断随机晶界的网络, 从而提高样品的耐晶间腐蚀能力[10-12]. 从图 1 可以看出, 样品的随机晶界网络并未被打断, 而腐蚀后样品表面形貌中也未看到随机晶界网络被打断的迹象 (图 3), 所以 GBE 处理后对样品耐晶间腐蚀能力的影响应该取决于其他因素. 图 5a 给出了 Non -GBE 样品的显微组织特征, 样品中大部分晶界为随机晶界. 经过浸泡实验后, 晶界处腐蚀沿着晶界向样品内部扩展, 如果将表面晶粒周围晶界全部腐蚀, 则这个晶粒就会脱落, 下面的晶界继续受到腐蚀溶液的腐蚀, 随着腐蚀时间的延长晶粒会一层一层脱落 (图 3, 4).

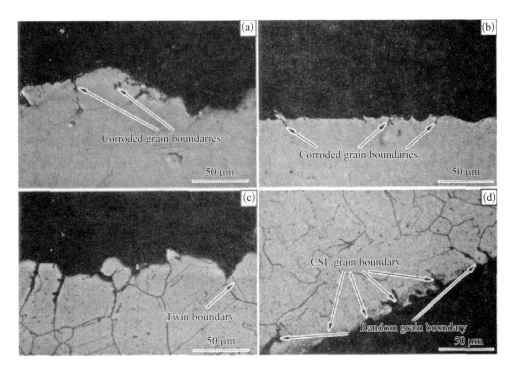

图4 样品经不同时间晶间腐蚀浸泡实验后截面的形貌

Fig. 4 The cross-section morphology of specimens after being immersed in corrosive solution for (a, b) 8 days and (c, d) 24 days, (a, c) are of Non－GBE specimens and (b, d) are of GBE specimens

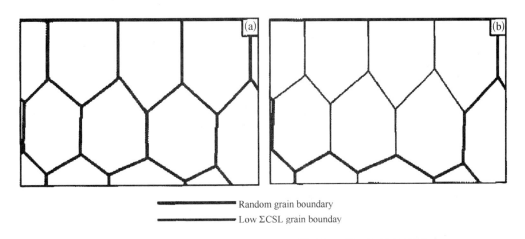

图5 (a) Non－GBE 样品和(b) GBE 样品的显微组织特征简图

Fig. 5 Schematic illustration of the microstructure features of (a) Non－GBE specimen and (b) GBE specimen. The thick lines demonstrate random grain boundary and the thin lines demonstrate low ΣCSL grain boundaries

低 ΣCSL 晶界的耐晶间腐蚀能力较随机晶界强,尤其是共格 Σ3 晶界几乎不会被腐蚀(图3, 4). GBE 样品中含有大量的低 ΣCSL 晶界(如图 5b 中细线所示). 晶间腐蚀将沿着晶粒团簇外围的随机晶界扩展(图 4d),这样,晶粒的脱落变为晶粒团簇的脱落. 在晶粒尺寸相差不多的情况下,造成晶粒团簇脱落所需要腐蚀掉的晶界要比造成晶粒脱落要腐蚀掉的晶界长得多.当晶界腐蚀扩展很深,到达尖端的腐蚀液浓度也会明显降低,会减慢晶间腐蚀的扩展速度.如果晶间腐蚀沿着高阶的 $\Sigma 3^n$ 晶界向晶粒团簇内部扩展时,由于晶粒团簇内部的所有 $\Sigma 3^n$ 类型晶界相互连接构成网络结构,腐蚀扩展最终会遇到 $\Sigma 3 - \Sigma 3 - \Sigma 9$ 三叉界角而被阻止继续扩展[13].所以大尺寸"互有 $\Sigma 3^n$ 取向关系晶粒的团簇"显微组织可以明显降低晶间腐蚀的扩展速度.在晶粒团簇脱落之前,下层组织不会受到腐蚀,因此大尺寸晶粒团簇保护了下层的显微组织.所以晶界工程处理的样品具有更好的抗晶间腐蚀能力.

4 结论

(1) 经过 GBE 处理后,样品中含有高比例的低 ΣCSL 晶界,形成了以大尺寸"互有 $\Sigma 3^n$ 取向关系晶粒的团簇"为特征的晶界网络显微组织.

(2) 经过 GBE 处理后,样品的耐晶间腐蚀能力明显提高,GBE 样品腐蚀失重要比 Non-GBE 样品的腐蚀失重明显降低.

参 考 文 献

[1] Diercks D R, Shack W J, Muscar J. Overview of Steam Generator Tube Degradation and Integrity Issues [J]. *Nuclear Engineering and Design*, 1999, 194: 19-31.

[2] Kronberg M L, Wilson F H. Secondary Recrystallization in Copper [J]. *Transactions of the American Institute of Mining, Metallurgical & Petroleum Engineers*, 1949, 185: 501-514.

[3] Watanabe T. An Approach to Grain Boundary Design for Strong and Ductile Polycrystals [J]. *Res Mechanica*, 1984, 11: 47-84.

[4] Palumbo G, Aust K T, Lehockey E M et al. On a More Restrictive Geometric Criterion for "Special" CSL Grain Boundaries [J]. *Scripta Materialia*, 1998, 38: 1685-1690.

[5] Xia S, Zhou B X, Chen W J, et al. Effect of Strain and Annealing Processes on the Distribution of $\Sigma 3$ Boundaries in a Ni-Based Super Alloy [J]. *Scripta Materialia*, 2006, 54: 2019-2022.

[6] Xia S, Zhou B X, Chen W J. Effect of Single-Step Strain and Annealing on Grain Boundary Character Distribution and Intergranular Corrosion in Alloy 690 [J]. *Journal of Materials Science*, 2008, 43: 2990-3000.

[7] Xia S, Zhou B X, Chen W J. Grain Cluster Microstructure and Grain Boundary Character Distribution in Alloy 690 [J]. *Metallurgical and Materials Transactions A*, 2009, 40: 3016-3030.

[8] Li H, Xia S, Zhou B X, et al. The Dependence of Carbide Morphology on Grain Boundary Character in the Highly Twinned Alloy 690 [J]. *Jounral of Nuclear Materials*, 2010, 399: 108-113.

[9] Li Hui(李慧), Xia Shuang(夏爽), Zhou Bangxin(周邦新), et al. Evolution of Carbide Morphology Precipitated at Grain Boundaries in Ni-Based Alloy 690(镍基 690 合金时效过程中碳化物的形貌演化)[J]. *Acta Metallugica Sinica*(金属学报), 2009, 45: 195-198.

[10] Shimada M, Kokawa H, Wang Z J, et al. Optimization of Grain Boundary Character Distribution for Intergranular Corrosion Resistant 304 Stainless Steel by Twin-Induced Grain Boundary Engineering [J]. *Acta Materialia*, 2002, 50: 2331-2341.

[11] Gertsman V Y, Bruemmer S M. Study of Grain Boundary Character Along Intergranular Stress Sorrosion Crack Paths in Austenitic Alloys [J]. *Acta Materialia*, 2001, 49: 1589-1598.

[12] Lehockey E M, Brennenstuhl A M, Thompson I. On the Relationship Between Grain Boundary Connectivity, Coincident Site Lattice Boundaries, and Intergranular Stess Corrosion Cracking [J]. *Corrosion Science*, 2004, 46: 2383-2404.

[13] Hu C L, Xia S, Li H, et al. Improving the Intergranular Corrosion Resistance of 304 Stainless Steel by Grain Boundary Network Control [J]. *Corrosion Science*, 2011, 53: 1880-1886.

Controlling the Grain Boundary Network to Enhance the Intergranular Corrosion Resistance in Alloy 690

Abstract: The proportion of low Σ coincidence site lattice (CSL) grain boundaries was enhanced by grain boundary engineering (GBE) in Alloy 690. The grain boundary network was characterized by electron backscatter diffraction (EBSD) technique. Formation of large size grain-cluster is the feature of the grain boundary network after GBE treatment. All of the grains in this kind of cluster had $\Sigma 3^n$ mutual misorientations regardless of whether they were adjacent. This microstructure is the result of multiple twinning during recrystallization. The resistance to intergranular corrosion was greatly enhanced by GBE. By observing the morphology of the corroded specimens, it was found that the large size grains-cluster microstructure reduced the penetration of the intergranular corrosion, and protected the inner microstructure.

镍基 690 合金中晶界碳化物析出的研究[*]

摘　要：利用 HRTEM,SEM 和 EBSD 技术研究了固溶处理后镍基 690 合金在 715 ℃时效时晶界处析出碳化物的形貌及与基体的取向关系. 结果表明,碳化物在晶界处于高指数晶体学面的一侧晶粒中析出,并与此侧晶粒具有共格的取向关系. 但是碳化物会向无共格取向关系一侧晶粒生长更快一些,碳化物向两侧晶粒生长速度的不同导致晶界附近 Cr 浓度曲线不对称,并且使晶界附近两侧晶粒的耐腐蚀能力不同.

690 合金是一种含 Cr 量高的镍基合金,目前被广泛用做压水堆核电站的蒸汽发生器传热管材料的制备. 在使用 690 合金之前,晶间应力腐蚀开裂(IGSCC)是蒸汽发生器传热管材料的一个主要失效原因[1],而富 Cr 的碳化物在晶界处析出引起的晶界附近贫 Cr 是产生 IGSCC 的主要原因[2-5]. 为了提高核电的经济性,需要延长核电站关键部件的服役寿命,对蒸汽发生器来说,提高传热管材料的性能就成为关键. 因而,有必要对 690 合金中晶界处碳化物的析出进行更细致的研究.

在 690 合金与奥氏体不锈钢的晶界处析出的碳化物都为富 Cr 的 $M_{23}C_6$[2,4,6,7],它们的析出规律相似. 文献中对 690 合金和奥氏体不锈钢在 500—800 ℃时效时,晶界处析出碳化物的形貌与时效温度[2,8-10]、时效时间[2,9-12]以及晶界类型[7,9,13]等的关系进行了大量的研究. 除了对碳化物的形貌研究外,还利用透射电子显微镜(TEM)和电子背散射衍射技术(EBSD)研究了碳化物与基体之间的取向关系,发现碳化物总是只与一侧晶粒具有共格的取向关系[2,4,7-14]. 但是,仍然不了解碳化物到底容易与哪侧晶粒具有共格的取向关系,这种共格的取向关系对碳化物形貌以及晶界耐腐蚀性能有什么影响也不清楚.

本文利用高分辨透射电子显微镜(HRTEM)、扫描电子显微镜(SEM)和取向成像显微系统(OIM)研究了镍基 690 合金时效热处理后晶界上析出的碳化物与基体的取向关系,讨论了这种取向关系对碳化物形貌和晶界耐腐蚀性能的影响.

1　实验方法

实验所用镍基 690 合金成分(质量分数,%)为：Cr 28.91, Fe 9.45, C 0.025, N 0.008, Ti 0.4, Al 0.34, Si 0.14, Ni 余量. 将 690 合金在 1 100 ℃固溶热处理 15 min,并立即水淬. 然后将固溶处理后的样品冷轧 10%,再经 1 100 ℃再结晶退火 5 min. 按这种工艺方法处理的 690 合金样品,既实现了固溶热处理的目的,同时样品中又会形成大量低 Σ 重位点阵(CSL)晶界[15-17]. 然后将样品在 715 ℃时效热处理 2,5,15,30,50,100 和 200 h,研究晶界碳化物的析出.

HRTEM 的薄膜样品采用电解双喷的办法制备,使用的电解液为 20% HClO₄ + 80%

---

* 本文合作者：李慧、夏爽、彭剑超. 原发表于《金属学报》,2011,47(7)：853 - 858.

C_2H_5OH(体积分数),经液氮冷却至大约-40 ℃用30 V直流进行双喷电解抛光. 在SEM和EBSD分析之前,先将样品电解抛光和电解蚀刻以显示晶界处的碳化物. 使用的电解液是20%$HClO_4$+80% CH_3COOH,经冰水混合物冷却至约0 ℃用30 V直流电解抛光30 s,然后用5 V直流电解蚀刻3—5 s.

利用JEM 2010F HRTEM对样品进行观察与分析,利用CCD相机记录样品的HRTEM像. 使用Digital Micrograph软件(包括快速Fourier变换(FFT)与反Fourier变换(IFFT))对HRTEM像所包含的晶体结构信息进行分析. 用HRTEM配备的能谱仪(EDS)对晶界附近的成分进行分析.

利用配备在Hitachi S570 SEM中的TSL-EBSD附件确定晶界两侧晶粒取向关系,扫描步长为7 μm. 数据采集后利用OIM系统自动确定晶界类型,晶界类型的判定采用Palumbo-Aust标准[18],相对于重位点阵模型[19]最大偏差角为$\Delta\theta_{max}=15°\Sigma^{-5/6}$. 晶粒内特定晶体学面与样品表面的交线(迹线)利用OIM系统分析获得. 利用JSM-6700F SEM对晶界上的碳化物形貌进行观察.

2 实验结果与分析

图1给出了715 ℃时效2 h后晶界处析出的碳化物及其与基体的取向关系. 可以看出,715 ℃时效2 h后在随机晶界处析出的碳化物平行于晶界方向的平均长度约为50 nm,垂直于晶界方向的宽度约为10 nm(图1a).

利用TEM的双倾台使样品旋转倾斜至可以同时观察到晶界两侧的晶粒以及碳化物的晶格条纹像,这样可以利用它们的HRTEM像分析三者之间的取向关系. 通过对HRTEM像进行分析以及对FFT图进行标定后可以看出,图1a中上侧晶粒的$[\bar{2}1\bar{1}]$方向平行于电子束入射方向,如图1c所示;下侧晶粒的[011]方向平行于电子束入射方向,如图1e所示. 经过分析,晶界的取向处在上侧晶粒的$(\bar{5}3\bar{7})$晶面上,同时处在下侧晶粒的$(0\bar{1}1)$晶面上. 晶界碳化物与上侧晶粒具有$(11\bar{1})_{matrix}$ // $(11\bar{1})_{carbide}$共格的取向关系(图1b和c),但与下侧晶粒无共格取向关系(图1d和e). 因此,碳化物与晶界处于高指数晶面的晶粒具有共格的取向关系,而与晶界处于低指数晶面的晶粒无共格的取向关系. 这说明了碳化物是在晶界处于高指数晶面的晶粒中形核,但是,碳化物却向无共格取向关系一侧的晶粒(下侧晶粒)生长得更快(图1a).

受晶界特性的限制,碳化物不能在Σ9晶界[19]上充分生长,但晶界上碳化物的二次枝晶会向与晶界碳化物有共格取向关系一侧晶粒充分生长[13],所以更容易分析这些碳化物与基体的取向关系. 利用EBSD技术可以获得晶粒内任意晶面在样品表面的迹线. 图2给出了715 ℃时效不同时间后Σ9晶界处析出的碳化物形貌,并标出了由OIM系统分析得到的{111}晶面迹线. 通过迹线分析表明,绝大部分的Σ9晶界相对于两侧晶粒的取向为:一侧为{111}面;另一侧为一高指数晶面,这与其他文献[20,21]中统计得到的结果一致. 棒状的碳化物二次枝晶会向着晶界处于高指数晶面的一侧晶粒内生长,而不是向着晶界处于{111}面的一侧晶粒内生长,这些特征在示意图2d中给出. 由于这些棒状碳化物实质是晶界上析出碳化物的二次枝晶,它们与晶界上碳化物有相同的取向[13]. 棒状碳化物枝晶可以向晶界处于高指数晶面的一侧晶粒内生长,说明碳化物与此侧晶粒具有共格取向关系,因此,晶界碳化物是在晶界处于高指数晶面的一侧晶粒中形核的,这与前面利用HRTEM分析得到的结果

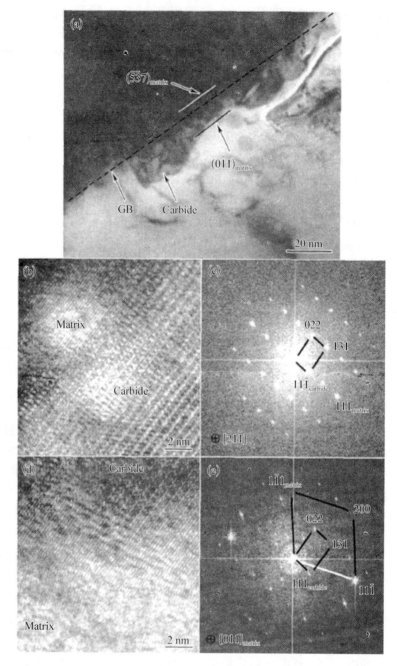

图1 715 ℃时效 2 h 后随机晶界处析出的碳化物与晶界两侧晶粒的取向关系分析

Fig. 1 TEM and HRTEM analyses on $M_{23}C_6$ precipitated along a random boundary in alloy 690 aged at 715 ℃ for 2 h after solid solution, cold rolling and recrystallization annealing （a）TEM image，$(\bar{5}37)_{matrix}$ and $(0\bar{1}1)_{matrix}$ of two grains parallel to the boundary，$M_{23}C_6$ precipitated along the boundary and grew into the lower grain （GB‐grain boundary）；（b）HRTEM image of the upper grain and carbide；（c）indexed FFT pattern of Fig. 1b；（d）HRTEM image of the lower grain and carbide；（e）indexed FFT pattern of Fig. 1d

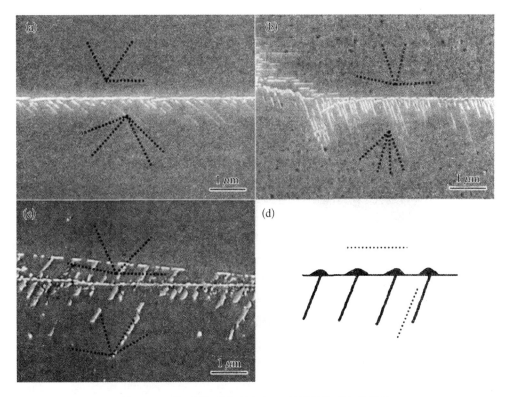

图 2 715 ℃ 时效不同时间后 Σ9 晶界处的碳化物形貌

Fig. 2 Morphologies of second dendrites carbides precipitated at Σ9 grain boundaries in the specimens aged at 715 ℃ for 2 h (a), 30 h (b) and 200 h (c), and the schematic illustration of the morphology features of carbides precipitated at Σ9 grain boundary (d) (the dash lines indicate the intersection between {111} crystal planes and surfaces of the specimens ({111} traces) measured by OIM)

一致.

当碳化物向晶粒内部生长时会涉及基体原子的重新排布,这样就可能在碳化物附近生成局部内应力. 图 3 给出了 715 ℃ 时效 2 h 后随机晶界处析出的碳化物与其附近两侧晶粒的 HRTEM 像. 图 3a 是碳化物与共格取向关系一侧基体的 HRTEM 像,碳化物与基体的晶格条纹像很整齐. 而在与碳化物无取向关系一侧基体内的晶格条纹像比较模糊,如图 3b 中虚线区域所示,说明在此侧晶粒内部存在局部的应力集中,并引起晶格的局部畸变. 对图 3b 中虚线区域进行放大,可以更清楚地观察到这种现象,如图 3c 中虚线区域所示. 在离碳化物稍远的基体处也可以观察到这种情况,如图 3d 中虚线区域所示.

$M_{23}C_6$ 的单胞由 92 个金属原子和 24 个 C 原子组成,占据 27 个基体的单胞体积,在相同体积的基体内有 108 个金属原子[22]. 如果碳化物向基体内生长,单位体积内的金属原子会减少,总的体积会膨胀,这就会在其附近基体中产生压应力,当这种压应力足够大时就会使基体的晶格发生扭曲. 由于碳化物向不具有共格取向关系一侧的晶粒生长得更快(图 1),所以此侧晶粒内部产生的内应力更明显(图 3).

当富 Cr 的碳化物在晶界处析出之后,会在晶界附近形成贫 Cr 区. 图 4 给出了 715 ℃ 时效 15 h 后晶界处的碳化物的形貌及晶界附近的 Cr 浓度分布图. 通过图 4a 的暗场像可以看出,碳化物与图中下侧的晶粒具有共格取向关系,但是向上侧晶粒中生长得更多些. 图 4b 中

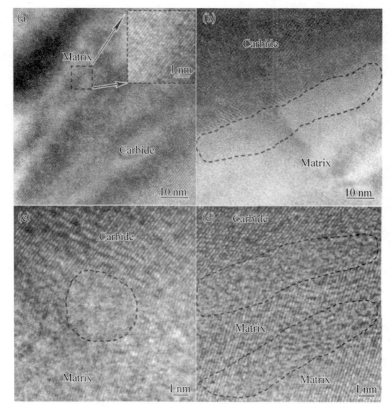

图3 715 ℃时效 2 h 后晶界碳化物附近的 HRTEM 像

Fig. 3 HRTEM images near the carbide precipitated at random grain boundary in the specimen aged at 715 ℃ for 2 h （a）coherent orientation relationship between carbide and matrix existed；（b）no coherent orientation relationship between the two phases，lattice distorsion induced by interior stress；（c，d）high magnified images of lattice distortion area

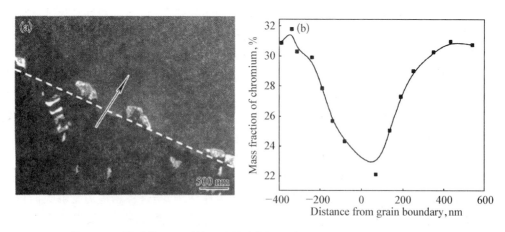

图4 715 ℃时效 15 h 后晶界处的碳化物形貌与晶界附近 Cr 浓度的分布

Fig. 4 Morphology of carbide precipitated at a random grain boundary in the specimen aged at 715 ℃ for 15 h（a）and Cr concentration distribution along arrow in Fig. 4a（b）

的 Cr 浓度曲线明显不对称,文献[23]认为这是由于晶界迁移造成的,但是在 715 ℃时效比较短的时间后晶界不会明显迁移[24]. 对比图 4a 中虚线两侧区域发现,贫 Cr 区在晶界两侧不对称,这主要是由于碳化物向晶界两侧生长速度不同造成的,生长快的一侧,也就是碳化物与基体晶粒不具有共格取向关系一侧,会消耗更多的 Cr 原子.

由于晶界附近 Cr 浓度的不对称,导致晶界附近两侧晶粒的耐腐蚀能力也有差别. 经过深度蚀刻后的形貌如图 5 所示. 对于与碳化物无取向关系一侧的晶粒(上侧晶粒),晶界附近的区域已经全部被腐蚀掉(两条虚线之间),而对于下侧晶粒,仅在碳化物枝晶明显生长的区域被严重腐蚀,如图 5 中箭头所示.

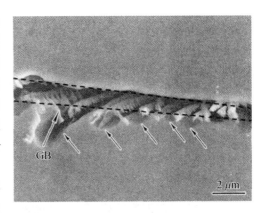

图 5 经过深度蚀刻后晶界附近两侧晶粒的腐蚀形貌

Fig. 5 Morphology of the deeply etched grain boundary (the upper grain corroded more and the lower grain corroded only near carbide dendrites shown by arrows)

3 分析讨论

对碳化物的形貌已进行了大量的研究,但是碳化物到底容易在哪侧晶粒形核仍然不清楚,文献[6,22,23]中认为碳化物在低能的晶界面上形核,如{001}或者{111}面. 如果是这样,孪晶的共格界面将是碳化物形核的最佳位置,因为孪晶的共格界面为{111}面,与 $M_{23}C_6$ 的{111}面有最好的匹配关系. 而实际上在孪晶的共格界面上很少观察到碳化物的析出[4,6,7,13].

当晶界相对于两侧晶粒的取向都处于较高指数时,晶界处原子排布更为混乱且含有更多的自由体积,当晶界相对于其中一侧晶粒的取向处于较高指数时,此侧晶粒在晶界附近将含有更多自由体积. 实验与理论推导[25]证实了碳化物更容易在高空位浓度区域形核. 原子探针层析技术的研究结果表明,C 与 Cr 原子在形成碳化物之前会偏聚在晶界附近一侧的晶粒上,而不是偏聚在晶界核心区,并且容易在"粗糙的"晶界面(晶界面相对此侧晶粒处于高指数面)一侧形成 C-Cr 共偏聚现象[26],之后碳化物会在这个区域形核并与此侧晶粒具有共格取向关系,这类似于毛玻璃表面比光滑玻璃表面更容易附着灰尘一样. 通过 HRTEM 与 OIM 分析,证明碳化物只与晶界处于高指数面一侧的晶粒具有共格取向关系(图 1 和 2). 碳化物析出之后,为了降低系统总的自由能,它与基体的界面会逐渐变得平直,并处于低指数面[22]. 因而经常观察到碳化物与基体的界面为低指数面,导致了碳化物在低指数面上形核的错误判断.

仔细观察在随机晶界处析出碳化物的截面形貌,可以发现具有一些共同点,图 6 给出了碳化物在随机晶界处析出后截面形貌的示意图. 图

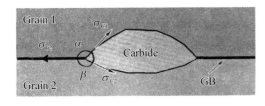

图 6 碳化物在随机晶界处析出后的截面形貌示意图

Fig. 6 Schematic illustration of the cross-section morphology of carbide precipitated at a random grain boundary (the carbide has coherent orientation relationship with grain 2, the grain boundary energy is σ_{12}, the interface energies of carbide and nearby matrix are σ_{C1} and σ_{C2}, $\sigma_{C1} > \sigma_{C2}$)

中碳化物与晶粒 2 具有共格的取向关系,而与晶粒 1 不具有共格的取向关系. 由于碳化物向不具有共格取向关系一侧晶粒生长得更快,因而 $\alpha < \beta$. 假设晶界的界面能为 σ_{12},碳化物与 2 个晶粒基体的界面能分别为 σ_{C1} 和 σ_{C2},由于碳化物与晶粒 1 不具有共格取向关系,因此 $\sigma_{C1} > \sigma_{C2}$. 如果碳化物的析出形貌仅受界面能的控制,3 个界面能在水平与竖直方向的分力应该相等[27],即:$\sigma_{12} = \sigma_{C1} \times \cos(\pi - \alpha) + \sigma_{C2} \times \cos(\pi - \beta)$,水平方向平衡;$\sigma_{C1} \times \sin(\pi - \alpha) = \sigma_{C2} \times \sin(\pi - \beta)$,竖直方向平衡.

但是,实验观察到的形貌明显不满足竖直方向的平衡公式,说明除了界面能的因素还会有其他因素影响碳化物的生长形貌. 一般原子扩散穿越共格界面要比穿越非共格界面困难,所以碳化物向着非共格取向的晶粒中生长更快一些,这侧的相界面也会迁移得更快,这样就会出现在多界面汇聚处界面能之间的不平衡现象(图 6). 由于非共格的相界面迁移快,消耗的 Cr 也会更多,导致此侧的 Cr 浓度更低一些,于是晶界附近的 Cr 浓度曲线表现出明显的不对称,造成了晶界两侧晶粒的耐腐蚀性能差别.

4 结论

碳化物在晶界上析出时,与晶界处于高指数晶面的一侧晶粒具有共格的取向关系,但向非共格取向关系的另一侧晶粒生长得更快,消耗的 Cr 原子也更多,因而造成晶界附近两侧 Cr 浓度分布的不对称. 与碳化物非共格取向关系一侧晶界附近的 Cr 浓度更低一些,耐腐蚀性能也差一些.

参 考 文 献

[1] Diercks D R, Shack W J, Muscar J. *J Nucl Eng Des*, 1999; 194: 19.

[2] Kai J J, Yu G P, Tsai C H, Liu M N, Yao S C. *Metall Trans*, 1989; 20A: 2057.

[3] Stiller K, Nilsson J O, Kjell N. *Metall Trans*, 1996; 27A: 327.

[4] Murr L E, Advani A, Shankar S, Atteridge D G. *Mater Charact*, 1990; 24: 135.

[5] Povich M J. *Corrosion*, 1978; 34: 60.

[6] Li Q, Zhou B X. *Acta Metall Sin*, 2001; 37: 8.
(李强,周邦新. 金属学报,2001; 37: 8)

[7] Lim Y S, Kim J S, Kim H P, Cho H D. *J Nucl Mater*, 2004; 335: 108.

[8] Wilson F G. *J Iron Steel Inst*, 1971; 209: 126.

[9] Trillo E A, Murr L E. *J Mater Sci*, 1998; 33: 1263.

[10] Carolan R A, Faulkner R G. *Acta Metall*, 1988; 36: 257.

[11] Li H, Xia S, Zhou B X, Chen W J, Ni J S. *Acta Metali Sin*, 2009; 45: 195.
(李慧,夏爽,周邦新,陈文觉,倪建森. 金属学报,2009; 45: 195)

[12] Trillo E A, Murr L E. *Acta Mater*, 1999; 47: 235.

[13] Li H, Xia S, Zhou B X, Chen W J, Hu C L. *J Nucl Mater*, 2010; 399: 108.

[14] Hong H U, Rho B S, Nam S W. *Mater Sci Eng*, 2001; A318: 285.

[15] Xia S, Zhou B X, Chen W J, Wang W G. *Scr Mater*, 2006; 54: 2019.

[16] Xia S, Zhou B X, Chen W J. *J Mater Sci*, 2008; 43: 2990.

[17] Xia S, Zhou B X, Chen W J. *Metall Mater Trans*, 2009; 40A: 3016.

[18] Palumbo G, Aust K T, Lehockey E M. *Scr Mater*, 1998; 38: 1685.

[19] Kronberg M L, Wilson F H. *Trans AIME*, 1949; 185: 501.

[20] Randle V, Coleman M, Waterton M. *Metall Mater Trans*, 2011; 42A: 582.

[21] Gokon N, Kajihara M. *Mater Sci Eng*, 2008; A477: 121.

[22] Lewis M H, Hattersley B. *Acta Metall*, 1965; 13: 1159.

[23] Hall E L, Briant C L. *Metall Trans*, 1984; 15A: 793.

[24] Downey S, Kalu P N, Han K. *Mater Sci Eng*, 2008; A480: 96.

[25] Wolft U E. *Trans AIME*, 1968; 242: 814.

[26] Li H, Xia S, Zhou B X, Liu W Q. *Mater Charact*, revised.

[27] McLean D. *Grain Boundaries in Metals*. London: Oxford University Press, 1957: 67.

Study of Carbide Precipitation at Grain Boundary in Nickel Base Alloy 690

Abstract: The morphology and orientation relationship between carbide precipitated at grain boundary and both side matrixes in nickel base alloy 690 aged at 715 ℃ for 2 – 200 h after solution treated were investigated by HRTEM, SEM and EBSD. The results show that the boundary carbide is easy to nucleate in the grain with high indexed crystal plane parallel to boundary, and the carbide has coherent orientation relationship (COR) with this grain. The carbide grows preferentially into the grain without COR between carbide and grain, which leads to lower chromium concentration in this side matrix near boundary. The asymmetry of chromium concentration profile leads to the different corrosion resistance of the two side grains nearby the grain boundary.

原子探针层析方法研究 690 合金晶界偏聚的初步结果[*]

摘　要： 利用脉冲电解抛光的方法制备了含有晶界的三维原子探针（3DAP）针尖样品，用原子探针层析方法（APT）研究了杂质原子在 690 合金晶界附近的偏聚情况. 结果表明：样品经过固溶处理并在 500 ℃～0.5 h 时效后，碳化物仅在部分有利于形核的晶界位错纠结处析出，大小为 5～10 nm. 用三维原子探针可观察到晶界处杂质原子的偏聚现象，C 主要偏聚在晶界附近 1 个原子层内，Si 和 P 偏聚在晶界附近约 2 个原子层的厚度内. 根据这些元素的偏聚数据，讨论了它们发生偏聚时的规律.

Ni 基 690 合金具有优异的耐腐蚀性能，目前被广泛用作压水堆核电站蒸汽发生器的传热管材料. 从上世纪 90 年代投入运行以来，虽然还未见到关于 690 合金传热管发生破损的报道，但运行至 40 年甚至 60 年后，晶间应力腐蚀破裂仍然是最可能发生的失效形式，就像采用 600 合金或 800 合金作为传热管的情况一样[1]. 杂质或溶质原子在晶界附近发生偏聚对金属材料的性能有很大的影响，如晶界的迁移[2]、应力腐蚀开裂[3] 等都与晶界偏聚有关. 因此有必要研究 690 合金中晶界处偏聚的问题.

原子探针层析技术（APT）具有原子尺度的分辨率，可在纳米尺度空间内对原子的位置与种类进行表征，用这种方法来研究晶界处的偏聚及纳米析出相等问题十分有利. APT 技术是目前定量分析纳米尺度范围内不同元素原子分布最微观的技术. 但是，用这种方法分析时的样品必需制备成曲率半径小于 100 nm 的针尖状，由于数据采集有效区域的长度十分有限（～100 nm），只有在距离针尖样品尖端几十纳米附近存在晶界时，才可能收集到晶界处不同元素偏聚的数据.

利用普通电解抛光的方法[4] 制备符合要求样品的概率很低. Seidman 等利用脉冲电解抛光的办法，研究了制备含有晶界针尖样品的方法[5]，并对 Fe(Si) 和 W(Re) 合金中的晶界偏聚[6,7]、Ni 基合金中相界面偏聚[8]、复合材料中金属相和陶瓷相界面偏聚[9] 等问题进行了研究. 结果表明，杂质或溶质原子会偏聚在界面附近 1～2 个原子层范围内，并且偏聚的浓度与界面特性有很大关系. Miller[10,11] 和 Smith[12] 等采用聚焦离子束（FIB）加工方法制备含有晶界的针尖样品，并对钢铁材料晶界偏聚进行了研究，也发现杂质或溶质原子会偏聚在晶界附近，但用 FIB 加工时会引起 Ga 注入样品[13]. Frykholm 等[14] 利用离子减薄制备含有渗碳体析出相的针尖样品，发现离子减薄过程对样品也有损伤.

本研究利用毫秒脉冲电解抛光的方法制备含有晶界的 690 合金针尖样品，利用透射电子显微镜（TEM）观察并筛选出符合 APT 分析要求的样品. 利用选区电子衍射（SAED）方法分析了晶界两侧的晶粒取向关系，用 APT 分析获得了杂质原子在晶界附近发生偏聚后的空间分布情况.

---

* 本文合作者：李慧、夏爽、刘文庆. 原发表于《电子显微学报》，2011，30（3）：206 - 209.

1 实验方法

实验所用 690 合金的成分(质量分数,%)为 Ni 60.52,Cr 28.91,Fe 9.45,C 0.025,N 0.008,Ti 0.4,Al 0.34 ,Si 0.14.将样品真空(真空度约 $5×10^{-3}$ Pa)密封在石英管中,在 1 100 ℃保温 15 min,然后立即淬入水中,并同时砸破石英管进行固溶处理.将固溶处理后的样品冷轧 50%,获得约 0.5 mm 厚的片状样品,样品经过清洗干燥后真空密封在石英管中,在 1 000 ℃再结晶退火 1 min 后立即淬入水中,并同时砸破石英管.淬火后的片状样品在 500 ℃保温 0.5 h 进行时效处理.

样品经过电解抛光后用电子背散射衍射(EBSD)方法确定晶粒大小及取向分布,采用直线截距法统计得到平均晶粒尺寸 $D=3.5~\mu m$(将孪晶包含在内).按照平均概率推测,获得在距离针尖样品尖端 d nm 内含有晶界样品的概率为 d/D,如 d 取 100 nm,这个概率约 3% ,d 取 1 000 nm 则概率约 30%.因此,用这样处理后的样品可比较容易制备符合 APT 分析要求的针尖样品.

利用电火花线切割将片状样品加工成 0.5 mm×0.5 mm×15 mm 棒状样品.采用两次普通电解抛光的办法获得晶界距离样品尖端仅几十纳米的概率很低,但距离为几百纳米的概率会高很多.电解抛光后的针尖状样品安装在改造后的 TEM 样品杆上,利用 JEM-200CX 型 TEM 观察针尖样品,确定针尖的曲率半径,晶界与针尖尖端的距离,及通过 SAED 确定晶界两侧晶粒的取向关系,将那些样品尖端附近几百纳米范围内含有晶界的样品挑选出来进行毫秒脉冲电解抛光,经过这样的精细抛光后,可使晶界距离样品尖端更近.经过多次试验统计,样品经过 20 V,1 ms 的脉冲电解抛光可以使样品尖端减短 75～400 nm.这样,利用毫秒脉冲电解抛光的办法,对已含有晶界的针尖样品进行精细抛光,可以获得适合 APT 分析晶界偏聚的针尖样品.

使用 Oxford nanoScience 公司的三维原子探针(3 DAP)对针尖样品进行分析.3DAP 分析时样品的温度为 70 K,离子化蒸发电压的脉冲分数为 20%.采用 PoSAP 软件分析数据,对不同元素的三维空间分布进行重构.

2 实验结果与讨论

690 合金在 500 ℃时效时需要很长时间晶界处才会明显析出碳化物[15],图 1 给出了 500 ℃时效 0.5 h 后 690 合金晶界附近的 TEM 像.将晶界面旋转至与电子束方向成较大角度时,可清晰地看到晶界处的位错网,仅在一些有利于碳化物形成的晶界位错纠结处观察到有碳化物析出(如图 1 中箭头所示),这些碳化物的大小为 5～10 nm.这样的样品正适合于研究碳化物析出前晶界处元素的偏聚情况.

图 2 给出了 APT 分析晶界偏聚的结果.图

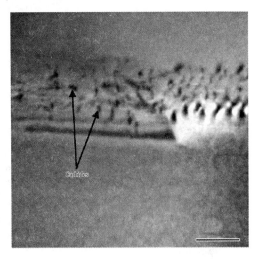

图 1 样品在 500 ℃时效 0.5 h 后的 TEM 像. Bar=100 nm

Fig. 1 TEM image of the specimen aged at 500 ℃ for 0.5 h. Bar=100 nm

2(a，b，c)给出了针尖的 TEM 明、暗场像及对应的 SAED 图. 通过 TEM 明、暗场像可见,在样品尖端附近含有 3 个晶粒 2 个晶界(3 个晶粒分别用 a,b,c 代表,如图 2a 所示),在这两段晶界处都没观察到碳化物的析出,通过对 SAED 图的标定得知这两个晶界均为一般大角度晶界. 在 APT 测试之前需要观察场离子显微镜像(FIM),做 FIM 像主要有如下作用:(1) 可使样品表面吸附的离子脱附(主要是表面碳氢化合物蒸发),使样品的尖端形成光滑半球形;(2) 可获得针尖的曲率半径和晶体取向等信息. FIM 成像前要将样品室抽至很高的真空,再充入少量惰性气体氛($\sim 2 \times 10^{-3}$ Pa),然后对样品外加电场使惰性气体电离成像. 在电场的作用下,样品尖端会承受很大的应力,导致在 FIM 成像时样品针尖处的 a 晶粒脱落,APT 分析开始时所收集的数据即为 a/b 晶粒晶界面上靠近 b 侧晶粒的元素分布情况. 因为 a 晶粒是在高真空下观察 FIM 像时脱落的,断口表面不会产生额外的原子吸附,且 a 晶粒脱落后立即进行 APT 分析,因此所得到的结果应该为 a/b 晶粒晶界面处元素分布的情况.

APT 可以分析的区域有限,仅为针尖样品轴线处 10×10 nm^2 范围内,所以收集到的数据中没能包含 b/c 晶粒晶界处的情况. 图 2d 和图 2e 仅给出了晶界附近 $7 \times 7 \times 6.5$ nm^3 体积内 H,C,Si,P 的空间分布. 由于 Fe,Cr,Ni 为 690 合金的主要组成元素,所以未给出其空间分布. 从图 2(d,f)可见,C 和 H 在晶界处发生了明显的偏聚,并且只偏聚在 0.1 nm 范围内(约 1 个原子层). 690 样品本来不含 H,通过电解抛光制备针尖样品时,H 很容易向样品内部扩散,晶界为其主要的扩散通道,所以 H 更容易富集在晶界上(最高可达 8 at.%),这也可能是导致样品尖端容易在晶界处折断的原因. Si 和 P 在晶界附近也发生偏聚,但是偏聚的范围要比 C 和 H 宽一些,约在 0.5 nm 范围内(约 2 个原子层)(图 2e,2f).

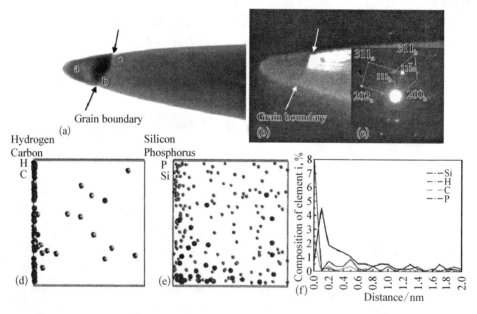

图 2 APT 分析晶界处元素偏聚的结果. a,b：针尖样品的 TEM 明、暗场像；c：与之对应的 SAD 图；a/b 晶界处,d：C、H,e：Si、P 的空间分布图(7 nm×7 nm×6.5 nm)；f：C、H、Si、P 的成分曲线

Fig. 2 The results of needle specimen containing gain boundary analyzed by APT. a,b：TEM bright and dark field images；c：The corresponding select area electron diffraction；Spatial distribution of carbon and hydrogen atoms；d：Silicon and phosphorus atoms；e：at the a/b grain boundary；f：Composition profiles of carbon, hydrogen, phosphorus and silicon

在图 2f 中可见杂质原子在晶界处的含量非常高,尤其是电解抛光引入的 H 含量可以达到 8 at.%,会使材料明显脆化. 考虑到不同种类溶质或杂质原子在基体中的含量不同,可以通过计算它们在晶界处浓度与在基体中浓度的比值(S_{av})来确定其在晶界处偏聚倾向的强弱,实验得到的 S_{av} 数值如表 1 所示. 这几种元素 S_{av} 值的大小顺序为 C>P>Si,另一方面,这几种原子的原子尺寸大小顺序是 Si>P>C,比较这两者之间的关系可见,杂质原子尺寸越小,在晶界处的偏聚程度越强.

表 1 不同杂质元素在晶界处和基体中的含量(S_{av} 代表杂质原子在晶界处浓度与基体中浓度的比值)

Table 1 The concentration of C, Si, P in matrix and grain boundary(S_{av} stands for the ration of concentration of certain element in grain boundary and matrix)

	C	Si	P
At grain boundary (at.%)	2.58	2.04	0.88
In matrix (at.%)	0.043	0.33	0.027
S_{av}	60	6.18	32.59
Atom radius[16] (pm)	67	111	98

3　结论

利用原子探针层析技术可以分析 690 合金中晶界处元素的偏聚行为. 经过固溶处理并在 500 ℃保温 0.5 h 时效后,合金碳化物仅在部分有利于形核的晶界位错纠结处析出,大小为 5~10 nm. 在碳化物还未形成的晶界片段处,C 偏聚在晶界附近 1 个原子层内,Si 和 P 偏聚在晶界附近约 2 个原子层范围内,电解抛光时引入的 H 也明显富集在晶界处. 原子尺寸大小对其向晶界处偏聚的倾向有明显的影响,尺寸越小,偏聚倾向越大.

参 考 文 献

[1] Diercks D R, Shack W J, Muscar J. Overview of steam generator tube degradation and integrity issues [J]. Nucl Engin Design, 1999, 194: 19 – 30.

[2] Kim S G, Park Y B. Grain boundary segregation, solute drag and abnormal grain growth [J]. Acta Mater, 2008, 56: 3739 – 3753.

[3] Stiller K, Nilsson J O, Norring K. The relation between grain boundary chemistry and intergranular stress corrosion cracking in Alloys 600 and 690 [J]. Metall Mater Trans, 1996, A27: 327 – 341.

[4] Seto K, Larson D J, Warren P J, Smith G D W. Grain boundary segregation in boron added interstitial free steels studied by 3 - dimensional atom probe [J]. Scr Mater, 1999, 40: 1029 – 1034.

[5] Krakauer B W, Seidman D N. A system for systematically preparing atom-probe field-ion-microscope specimens for the study of internal interfaces [J]. Rev Sci Instrum, 1990, 3390 – 3398.

[6] Krakauer B W, Seidman D N. Subnanometer scale study of segregation at grain boundaries in an Fe(Si) alloy [J]. Acta Mater, 1998, 46: 6145 – 6161.

[7] Hu J G, Seidman D N. Atomic-scale observations of two-dimensional Re segregation at an internal interface in W (Re) [J]. Phys Rev Lett, 1990, 65: 1615.

[8] Amouyal Y, Mao Z G, Seidman D N. Segregation of tungsten at γ'(L12)/γ(fcc) interfaces in a Ni-based superalloy: An atom-probe tomographic and first-principles study [J]. Appl Phys Lett, 2008, 93: 201905.

[9] Shashkov D A, Muller D A, Seidman D N. Atomic-scale structure and chemistry of ceramic/metal interfaces-II. solute segregation at MgO/Cu(Ag) and CdO/Ag(Au) interfaces [J]. Acta Mater, 1999, 47: 3953 – 3963.

[10] Kelly T F, Larson D J, Miller M K, et al. Three dimensional atom probe investigation of vanadium nitride precipitates and the role of oxygen and boron in rapidly solidified 316 stainless steel [J]. Mater Sci Eng, 1999, A270: 19 – 26.

[11] Pereloma E V, Timokhina I B, Jonas J J, Miller M K. Fine-scale microstructure investigations of warm rolled low-carbon steels with and without Cr, P, and B additions [J]. Acta Mater, 2006, 54: 4539 – 4551.

[12] Maruyama N, Smith G D W, Cerezo A. Interaction of the solute niobium or molybdenum with grain boundaries in α – iron [J]. Mater Sci Eng, 2003, A353: 126 – 132.

[13] Thompson G B, Miller M K, Fraser H L. Some aspects of atom probe specimen preparation and analysis of thin film materials [J]. Ultramicroscopy, 2004, 100: 25 – 34.

[14] Frykholm R, Andren H O. Method for atom probe study of near surface zones in cemented carbides [J]. Ultramicroscopy, 1999, 79: 283 – 286.

[15] Kai J J, Yu G P, Tsai C H, et al. The effect of heat treatment on the chromium depletion, precipitate evolution, and corrosion resistance of INCONEL Alloy 690 [J]. Metall Trans, 1989, 20A: 2057 – 2067.

[16] Clementi E, Raimondi D L, Reinhardt W P. Atomic screening constants from SCF functions [J]. J Chem Phys, 1963, 38: 2686 – 2689.

Preliminary results of grain boundary segregation in alloy 690 by atom probe tomography

Abstract: Needle-like specimens of three dimensional atom probes (3DAP) were prepared by pulse electro-polishing. The segregation of impurity atoms at grain boundaries in Alloy 690 was studied by atom probe tomography (APT). The results show that the carbide only initiated at some grain boundary dislocations after solid solution treatment at 1 100 ℃ and then by aging at 500 ℃ for 0.5 h. At the grain boundary segment where carbide does not precipitate, the segregation of carbon atoms can be detected. The width of carbon atoms segregate at grain boundary is only 1 atomic layer. The widths of silicon and phosphorus atoms segregate at grain boundary are about 2 atomic layers. It can be concluded that the segregation tendency of the impurity atoms depends on their atomic radius.

Cu 对 Zr‐2.5Nb 合金在 500 ℃/10.3 MPa 过热蒸汽中腐蚀行为的影响[*]

摘 要: 添加微量合金元素 Cu 的 Zr‐2.5Nb‐xCu($x=0.2$,0.5,质量分数,%)合金样品,经 β 相水淬、冷轧变形及 580 ℃,50 h 和 620 ℃,2 h 退火处理,在静态高压釜中进行 500 ℃/10.3 MPa 的过热蒸汽腐蚀实验. 利用 SEM 和 TEM 研究了氧化膜截面的显微组织. 结果表明,添加少量 Cu 可以提高 Zr‐2.5Nb 合金的耐腐蚀性能;合金的耐腐蚀性能与氧化膜中的柱状晶的生长及形态有关,添加合金元素 Cu 有利于提高锆合金氧化膜中柱状晶比例,并使柱状晶尺寸增大且排列有序,从而提高锆合金的耐腐蚀性能.

锆合金的热中子吸收截面小,在高温高压水(蒸汽)中具有良好的耐腐蚀性能和力学性能,与 UO_2 具有较好的相容性,而且容易冷加工成型,因此被广泛用做水冷核动力堆的燃料包壳和堆芯结构材料的制备. 锆合金的腐蚀行为主要受氧化膜生长速度的影响,而氧化膜的生长过程是 O^{2-} 经过氧化膜扩散到氧化膜与锆合金基体界面处并与基体反应生成 ZrO_2 的过程. 过去的 40 年,国内外对氧化膜/金属(oxide/metal, O/M)界面处的微观结构以及合金元素对氧化动力学的影响做了大量深入的研究,但对于锆合金的腐蚀机理还没有一个明确的结论. 锆合金氧化膜的相组成主要可分为四方 ZrO_2(t‐ZrO_2)和单斜 ZrO_2(m‐ZrO_2),而按照晶粒形态可以分为柱状晶和等轴晶. 研究[1‐4]表明,锆合金的耐腐蚀性能随着氧化膜中柱状晶以及四方相的含量增加而提高.

添加适当的合金元素可以进一步提高锆合金的耐腐蚀性能,如添加 Nb 的新型锆合金 ZIRLO(Zr‐1Nb‐1Sn‐0.1Fe,质量分数,%,下同)[5]和 M5(Zr‐1Nb‐O)[6],其耐腐蚀性能都比 Zr‐4 合金(Zr‐1.5Sn‐0.2Fe‐0.1Cr)更加优异,并已经成功取代了 Zr‐4 合金而应用于轻水反应堆中. 然而,含 Nb 锆合金的耐腐蚀性能对于 Nb 的含量和热处理制度都非常敏感[7,8],因此,Nb 的添加常常需要复杂的热处理制度. Park 等[9]的研究表明,含 0.05% Cu 的 Zr‐1.1Nb‐0.05Cu 合金可以降低 Zr‐Nb 合金在含 B 和 Li 的 360 ℃水溶液中的腐蚀速率. 文献[10]进一步报道了该合金(HANA‐6)在各种腐蚀环境中都表现出了比 Zr‐4 合金更好的耐腐蚀性能. 但是,Cu 在 Zr‐Nb 合金腐蚀过程中的作用机理至今还不清楚.

本文研究添加不同含量 Cu 的 Zr‐2.5Nb‐xCu($x=0.2$, 0.5)合金,在 500 ℃/10.3 MPa 的过热蒸汽中的腐蚀行为及生成氧化膜的显微组织演化过程,探讨 Cu 影响 Zr‐Nb 合金腐蚀性能的作用机理,为开发新型锆合金提供新的思路.

1 实验方法

以 Zr‐2.5Nb 为对比合金,加入少量 Cu,用非自耗真空电弧炉熔炼成约 60 g 的合金锭,制

* 本文合作者:李强、梁雪、彭剑超、余康、姚美意. 原发表于《金属学报》,2011,47(7):877‐881.

备 Zr‑2.5Nb‑xCu 合金. 为了保证成分均匀,反复熔炼 5 次,并在每次熔炼前将铸锭翻转. 经 ICP‑AES(电感耦合等离子体原子发射光谱)检验,成分符合要求. 合金锭加热至 700 ℃热轧成 3 mm 厚的坯料,用 10%HF+30%HNO$_3$+30%H$_2$SO$_4$+30%H$_2$O(体积分数)混合酸酸洗去除氧化膜后,水洗干燥,经 1 030 ℃,40 min 的 β 相均匀化处理后空冷. 经多次中间退火(580 ℃, 2 h)冷轧至 0.6 mm 厚,切成 10 mm×18 mm×0.6 mm 的片状试样,经酸洗和水洗后分别进行 580 ℃,50 h 和 620 ℃,2 h 再结晶退火处理. 退火后的样品经酸洗和去离子水清洗后,放入静态高压釜中进行 500 ℃/10.3 MPa 的过热蒸汽腐蚀实验,腐蚀增重为 3 个样品的平均值.

用聚焦离子束(focus ion beam, FIB, QUANTA 3D FEG, USA)制备氧化膜截面的透射电镜(TEM)薄膜样品. 用带有 INCA 能谱仪(EDS)的 JEM‑2010F 场发射 TEM 和 JSM‑6700F 场发射扫描电镜(SEM)对锆合金腐蚀前后的合金基体及腐蚀生成氧化膜的显微组织进行观测,为避免对 Cu 的检测造成干扰,在 TEM‑EDS 分析时使用了 Be 制双倾样品台. 高分辨 TEM 图像分析使用设备自带的 Gatan Digital Micrograph 软件处理.

2 实验结果与讨论

2.1 腐蚀增重

图 1 是 Zr‑2.5Nb‑xCu 样品在 500 ℃/10.3 MPa 过热蒸汽中的腐蚀增重曲线. 580 ℃,50 h 退火处理的 Zr‑2.5Nb‑0.2Cu 和 Zr‑2.5Nb‑0.5Cu 样品耐腐蚀性能较好,腐蚀750 h 后增重只有约 200 mg/dm^2;腐蚀性能较差的是 620 ℃,2 h 退火处理的 Zr‑2.5Nb 样品,腐蚀 750 h 后增重达到约 340 mg/dm^2.

总体来说,对于同种合金,580 ℃,50 h 退火样品的耐腐蚀性能明显优于 620 ℃,2 h 退火的样品,这是由于在 620 ℃处理超过了 Zr‑Nb 合金的偏析反应温度,合金中生成部分 β‑Zr 相,使其耐腐蚀性能较差[11]. 值得注意的是,在相同退火条件下,添加 Cu 的 Zr‑2.5Nb‑0.2Cu 和 Zr‑2.5Nb‑0.5Cu 合金样品的耐腐蚀性能较好,而且随着 Cu 含量的增加,中长期的耐腐蚀性能有所提高,这说明在 Zr‑2.5Nb 合金中添加适量合金元素 Cu 对提高耐腐蚀性能有益.

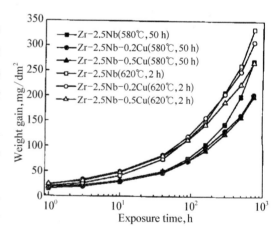

图 1 Zr‑2.5Nb‑xCu 样品在 500 ℃/10.3 MPa 过热蒸汽中的腐蚀增重曲线

Fig. 1 Corrosion behaviors of Zr‑2.5Nb‑xCu alloys annealed at different conditions in 500 ℃/ 10.3 MPa superheated steam

2.2 氧化膜断口的形貌

当多晶材料发生断裂时,裂纹沿着某些晶面或晶界从一个晶粒传播到另一个晶粒. 因此,氧化膜的断口形态不仅反映了断裂特征,同时反映了显微结构特征. 图 2 为 Zr‑2.5Nb‑xCu 样品在 500 ℃/10.3 MPa 过热蒸汽中腐蚀 40 h 的氧化膜的断口形貌,其中插图分别是相应虚框区域的高倍照片. 对于 580 ℃,50 h 退火处理的样品,Zr‑2.5Nb 合金氧化膜的厚度为 2.9 μm(图 2a), Zr‑2.5Nb‑0.2Cu(图 2c)和 Zr‑2.5Nb‑0.5Cu 合金(图 2e)

氧化膜的厚度接近,分别是 2.6 和 2.7 μm,氧化膜中除了极少数微裂纹和孔洞外,并没有观察到大的裂纹,这是腐蚀转折前氧化膜的组织特征. 从高倍率照片可以看出,添加 Cu 的 Zr-2.5Nb-0.2Cu(图 2c)和 Zr-2.5Nb-0.5Cu 合金(图 2e)氧化膜中的柱状晶排列更加整齐紧密,尺寸较大,在柱状晶区域只观察到少量细小的微裂纹,这说明 Cu 的添加有助于氧化膜中柱状晶的生成长大. 对于 620 ℃,2 h 退火处理的样品,Zr-2.5Nb 合金(图 2b)的氧化膜比 Zr-2.5Nb-0.2Cu(图 2d)和 Zr-2.5Nb-0.5Cu 合金(图 2f)更厚,并且断口更加粗糙不平,裂纹也更多,柱状晶区域较少,同时还可以观察到一些台阶,这种台阶显然与氧化膜中原来已经存在的裂纹或易开裂区有关.

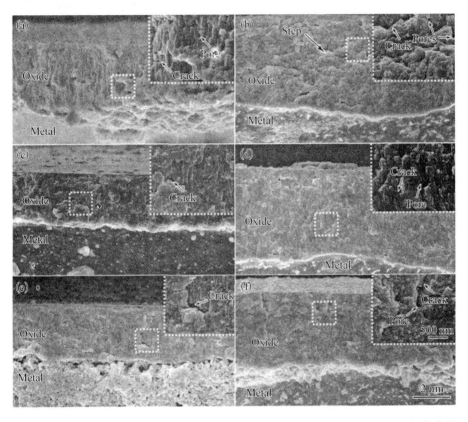

图 2 Zr-2.5Nb-xCu 样品在 500 ℃/10.3 MPa 过热蒸汽中腐蚀 40 h 生成的氧化膜的断口形貌

Fig. 2 Fractographs of oxide layers formed on Zr-2.5Nb-xCu specimens corroded in 500 ℃/10.3 MPa superheated steam for 40 h (a) Zr-2.5Nb, annealed at 580 ℃ for 50 h (b) Zr-2.5Nb, annealed at 620 ℃ for 2 h (c) Zr-2.5Nb-0.2Cu, annealed at 580 ℃ for 50 h (d) Zr-2.5Nb-0.2Cu, annealed at 620 ℃ for 2 h (e) Zr-2.5Nb-0.5Cu, annealed at 580 ℃ for 50 h (f) Zr-2.5Nb-0.5Cu, annealed at 620 ℃ for 2 h

对比图 2a,c,e 和图 2b,d,f 可以看出,在合金成分和腐蚀时间相同的条件下,620 ℃,2 h 退火处理的 Zr-2.5Nb-xCu 样品的氧化膜比 580 ℃,50 h 退火处理的样品更厚,断口表面更加粗糙不平,裂纹和孔洞也明显增多,柱状晶比例明显减少且排列较混乱.

2.3 氧化膜截面的显微组织

图 3 为 Zr-2.5Nb 和 Zr-2.5Nb-0.2Cu 合金在 500 ℃/10.3 MPa 的过热蒸汽中腐蚀

3 h后的氧化膜截面的 TEM 像. 可以在同一视场中同时观察到金属、O/M 界面及氧化膜. 580 ℃,50 h 退火处理的 Zr-2.5Nb 的氧化膜的厚度为 1.4 μm,Zr-2.5Nb-0.2Cu 的氧化膜厚度最小,为 1.2 μm,而 620 ℃,2 h 退火处理的 Zr-2.5Nb-0.2Cu 的氧化膜的厚度最大,约为 1.9 μm. 相同的腐蚀时间内,氧化膜越厚,氧化速率越快,耐腐蚀性能越差,这与前面给出的腐蚀增重曲线(图 1)及 SEM 观测的结果一致.

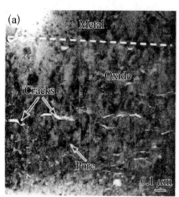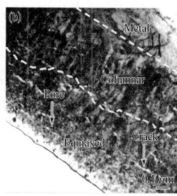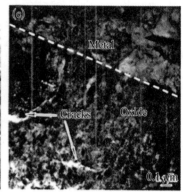

图 3 Zr-2.5Nb 和 Zr-2.5Nb-0.2Cu 样品在 500 ℃/10.3 MPa 过热蒸汽中腐蚀 3 h 的氧化膜截面的 TEM 像

Fig. 3 TEM micrographs of the cross-sections of the oxide films formed on Zr-2.5Nb-xCu specimens after corrosion testing for 3 h in 500 ℃/10.3 MPa superheated steam (a) Zr-2.5Nb specimen annealed at 580 ℃ for 50 h (b) Zr-2.5Nb-0.2Cu specimen annealed at 580 ℃ for 50 h (c) Zr-2.5Nb-0.2Cu specimen annealed at 620 ℃ for 2 h

在氧化膜内侧(靠近 O/M 界面处)及氧化膜中间部位,ZrO_2 主要是柱状晶,而在氧化膜的外侧主要是等轴晶. 所有样品在 O/M 界面附近的柱状晶相对于中间层的柱状晶排列更加紧密,晶粒尺寸也更大. 580 ℃,50 h 退火处理的 Zr-2.5Nb-0.2Cu 样品的氧化膜中柱状晶粒的尺寸和长径比明显较大,宽度约为 50 nm,长度约为 500 nm,长径比约为 10,并且排列比较有序,显得较为致密;580 ℃,50 h 退火处理的 Zr-2.5Nb 样品以及 620 ℃,2 h 退火处理的所有 3 种合金的样品的氧化膜中柱状晶粒的尺寸相近,宽度约为 40 nm,长度约为300 nm,除了尺寸和长径比较小,排列也较无序.

从图 3 中可以观察到氧化膜中不均匀分布的微裂纹和孔洞. 在近氧化膜外侧区域存在的微裂纹和孔洞明显多于靠近 O/M 界面的内侧,微裂纹和孔洞主要存在等轴晶之间,这与氧化膜断面 SEM 观察(图 2)结果一致. 锆合金氧化膜截面样品中微裂纹的数量与耐腐蚀性能之间存在着一定的对应关系,即微裂纹越多,锆合金的耐腐蚀性能越差,如 Zr-2.5Nb 和 620 ℃,2 h 退火处理的 Zr-2.5Nb-0.2Cu 合金样品,氧化膜中观察到的微裂纹较多,这与文献[11-14]报道一致. ZrO_2 基本的破坏方式为晶间断裂,因此,在氧化层内存在压应力的条件下,相对于等轴晶,裂纹不易在排列齐整的柱状晶内萌生和扩展. 在压应力存在的条件下,对压应力进行分解,最大剪切力在倾斜 45°平面内,如果晶粒为柱状晶,则晶界方向均垂直或接近垂直于压应力方向,不易产生微裂纹;而在等轴晶区,则存在部分等轴晶粒的晶界与压应力平面夹角为 45°,有利于微裂纹的产生和扩散. 因此,等轴晶区域观察到的微裂纹要多于柱状晶区域.

本课题组前期工作[12-14]对锆合金腐蚀生成氧化膜的显微组织演变过程及机制进行了大

量论述,认为在腐蚀过程中,锆合金的氧化膜是在氧化膜/金属基体的界面上不断生成的,O^{2-}需要穿过氧化膜后才能到达氧化膜/金属基体的界面. 因此,氧化膜的显微组织及其在腐蚀过程中的演化过程又会直接影响 O^{2-} 或阴离子空位在氧化膜中的扩散,这就会影响氧化膜的生长. 氧化膜生成时由于体积膨胀,在氧化膜内形成压应力,使 ZrO_2 晶体生成时就产生许多缺陷(空位、间隙原子等),稳定了一些亚稳相(立方、四方 ZrO_2 甚至非晶);缺陷在温度、应力和时间的共同作用下,发生扩散、湮没和凝聚,空位被氧化膜中的晶界吸收后形成纳米尺度的孔隙簇,弱化了晶粒之间的结合力;孔隙簇进一步发展成为微裂纹,使氧化膜失去了原有的良好保护性,从而发生了腐蚀速率的转折. 由于氧化膜生成时产生的压应力是无法避免的,所以氧化膜的显微组织在腐蚀过程中发生不断演化也是必然的,如果能延缓这种演化过程就可以提高锆合金的耐腐蚀性能. 在本工作中,合金元素 Cu 的添加显然能够延缓氧化膜的这种演化进程,特别是能够促使柱状晶长大并排列有序,延缓向等轴晶的演变,从而使氧化膜具有更好的保护性,提高锆合金的耐腐蚀性能.

3 结论

(1) 在 Zr-2.5Nb 合金中加入适量 Cu 可以明显提高其在 500 ℃/10.3 MPa 过热蒸汽条件下的耐腐蚀性能.

(2) 在 Zr-2.5Nb 合金中添加适量 Cu 可以提高腐蚀生成的氧化膜中的柱状晶比例,并使柱状晶晶粒尺寸增大,排列整齐,从而提高锆合金的耐腐蚀性能.

参 考 文 献

[1] Garzarolli F, Seidel H, Tricot R, Gros J P. In: Eucken C M, Garde A M, eds., *Zirconium in the Nuclear Industry: Ninth International Symposium*, *ASTM STP 1132*, Philadelphia: American Society for Testing and Materials, 1991: 395.

[2] Takeda K, Anada H. In: Sabol G P, Moan G D, eds., *Zirconium in the Nuclear Industry: Twelfth International Symposium*, *ASTM STP 1354*, West Conshohocken, PA: American Society for Testing and Materials, 2000: 592.

[3] Bouvier P, Godlewski J, Lucazeau G. *J Nucl Mater*, 2002; 300: 118.

[4] Jeong Y H, Kim H G, Kim T H. *J Nucl Mater*, 2003; 317: 1.

[5] Sabol G P. *J ASTM Int*, 2005; 2: 3.

[6] Bossis P, Pêcheur D, Hanifi K, Thomazet J, Blat M. *J ASTM Int*, 2006; 3: 494.

[7] Kim H G, Jeong Y H, Kim T H. *J Nucl Mater*, 2004; 326: 125.

[8] Jeong Y H, Lee K O, Kim H G. *J Nucl Mater*, 2002; 302: 9.

[9] Park J Y, Choi B K, Jeong Y H, Jung Y H. *J Nucl Mater*, 2005; 340: 237.

[10] Park J Y, Choi B K, Jeong Y H, Kim K T, Jung Y H. *Proc of 2005 Water Reactor Fuel Performance Meeting*, Kyoto, Japan, 2005: 188.

[11] Li Q. *PhD Thesis*, Shanghai University, 2008.
   (李强. 上海大学博士论文, 2008)

[12] Zhou B X, Li Q, Yao M Y, Liu W Q, Chu Y L. *Nucl Power Eng*, 2005; 26: 364.
   (周邦新, 李强, 姚美意, 刘文庆, 褚玉良. 核动力工程, 2005; 26: 364)

[13] Zhou B X, Li Q, Liu W Q, Yao M Y, Chu Y L. *Rare Met Mater Eng*, 2006; 35: 1009.
   (周邦新, 李强, 刘文庆, 姚美意, 褚玉良. 稀有金属材料与工程, 2006; 35: 1009)

[14] Yao M Y. *PhD Thesis*, Shanghai University, 2007.
   (姚美意. 上海大学博士论文, 2007)

Effect of Cu-Addition on the Corrosion Behavior of Zr - 2. 5Nb Alloys in 500 ℃/10. 3 MPa Superheated Steam

Abstract: Zr - 2. 5Nb - xCu (x=0. 2, 0. 5, mass fraction, %) specimens were annealed at 580 ℃ for 50 h and 620 ℃ for 2 h after β- quenching and cold rolling, and then corroded in autoclave with 500 ℃/10. 3 MPa superheated steam. SEM and TEM equipped with EDS were employed to investigate their corrosion behaviors through analyzing the microstructures of the fracture surface of the oxide layers or of the cross-section of oxide layers. It was found that the addition of copper can improve the corrosion resistance of Zr - 2. 5Nb alloy, which is closely related to the increases in the fraction and size of columnar grains in the oxide film.

Zr-2.5Nb 合金中 *β*-Nb 相的氧化过程[*]

摘　要： Zr-2.5Nb 合金经 β 相水淬、冷轧变形及 580 ℃,50 h 退火处理,在静态高压釜中进行 500 ℃/10.3 MPa 的过热蒸汽腐蚀实验. HRTEM 观测表明,580 ℃,50 h 退火处理的锆合金析出了颗粒细小(<100 nm)的 β-Nb 第二相,其数量较多,分布较均匀. β-Nb 相的氧化速率比锆合金基体缓慢,在氧化过程中首先生成 NbO_2,呈现晶态和非晶态的混合组织,然后转变成非晶态占主导的氧化物,最后流失到腐蚀介质中.

锆合金作为核反应堆燃料包壳材料已经应用并发展数十年[1,2]. 添加 Nb 可以提高锆合金的耐腐蚀性能,其中以 ZIRLO 和 E635 为代表的 Zr-Sn-Nb-Fe 系列和 M5 合金为代表的 Zr-Nb 系列新包壳材料已进入商用阶段. 大量堆外高压釜的实验[3-7]表明,M5 合金及与其成分相近的 E110 合金在高温高压水中和过热蒸汽中的耐腐蚀性均优于 ZIRLO,E635 和 Zr-4 等合金. 从 Zr-Nb 相图可知,Zr-Nb 合金在室温下由 hcp 结构的 α-Zr 和 bcc 结构的 β-Nb 组成[2]. Zr-Nb 合金的耐腐蚀性能与热处理后的组织状态有关[8,9]：α-Zr 基体中过饱和的 Nb 使合金的耐腐蚀性能变差；而淬火时效处理后生成的 β-Nb 相对改善锆合金耐腐蚀性能有利[3,10-12],Kim 等[13]研究了 Zr-1.5Nb 合金在 360 ℃/18.9 MPa 水中腐蚀后 β-Nb 第二相的氧化过程. 但是,对于氧化膜中 β-Nb 第二相颗粒的氧化过程及其对合金耐腐蚀性能影响的研究并不充分. 为了能够获得数量较多的 β-Nb 第二相,本文选用含 Nb 量较高的 Zr-2.5Nb(质量分数,%)合金,研究其在 500 ℃/10.3 MPa 的过热蒸汽中腐蚀后,氧化膜中 β-Nb 颗粒的氧化过程,并探讨 β-Nb 第二相影响锆合金耐腐蚀性能的机制.

1　实验方法

实验用 Zr-2.5Nb 合金用非自耗真空电弧炉熔炼,钮扣锭重约 60 g,为了保证成分均匀,反复熔化 5 次,并在每次熔化前将铸锭翻转. 经电感耦合等离子体原子发射光谱(ICP-AES)检验,成分符合要求. 合金锭加热至 700 ℃,轧制成约 3 mm 厚的坯料,用 10% HF+30% HNO_3+30% H_2SO_4+30% H_2O(体积分数)混合酸酸洗去除氧化膜后,经水洗及干燥,在 1 030 ℃,40 min 条件下进行 β 相均匀化处理后空冷,经过多次中间退火(580 ℃,2 h)及冷轧至约 0.6 mm 厚,切成 10 mm×18 mm×0.6 mm 大小的片状试样,经酸洗、水洗后在 580 ℃,50 h 条件下进行再结晶退火处理. 退火后样品经酸洗和去离子水清洗后,放入静态高压釜中进行 500 ℃/10.3 MPa 条件下的过热蒸汽腐蚀实验. 样品的热处理均在真空中进行,真空度优于 10^{-3} Pa.

用双喷电解抛光方法制备 Zr-2.5Nb 合金的透射电镜(TEM)样品,电解液为 10% $HClO_4$+90% C_2H_5OH 溶液,冷却至低温(-30 ℃)抛光减薄. 用聚焦离子束(focus ion

---

* 本文合作者：李强、梁雪、彭剑超、刘仁多、余康. 原发表于《金属学报》,2011,47(7)：893-898.

beam，FIB，QUANTA 3D EEG，USA)制备氧化膜截面的 TEM 薄膜样品. 用带有 INCA
能谱仪(EDS)的 JEM－2010F 场发射 TEM 和 JSM－6700F 场发射扫描电镜(SEM)对合金
基体及氧化膜的显微组织进行观测. 高分辨透射电镜(HRTEM)图像分析使用设备自带的
Gatan Digital Micrograph 软件处理.

2　实验结果与讨论

2.1　Zr－2.5Nb 合金中 β－Nb 第二相的显微组织

图 1a 给出了 580 ℃，50 h 退火处理的 Zr－2.5Nb 合金的 TEM 像. 可以看出，颗粒细小
(<100 nm)并呈球状的第二相数量较多，分布也比较均匀. 通过选区电子衍射(SAD)标定
(图 1a 中插图)以及 EDS 分析(图 1b)，确定为 bcc 结构的 β－Nb 相. 值得注意的是，在 β－
Nb 相中有时会检测出少量的 Fe，这来自海绵 Zr 中的杂质元素，由于 Fe 在 α－Zr 中的固溶
度很低，是稳定 β 相的元素，所以易于偏聚在 β－Nb 相中.

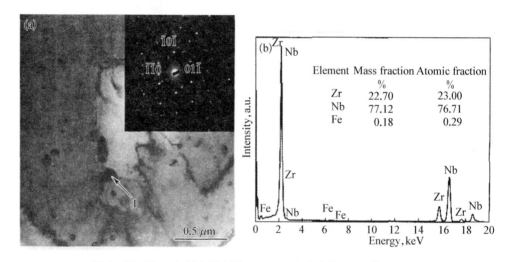

图 1　580 ℃，50 h 退火处理的 Zr－2.5Nb 合金的 TEM 像和 EDS

Fig. 1　TEM micrograph (a) of Zr－2.5Nb specimen annealed at 580 ℃ for 50 h (inset showing
SAD pattern of the paticle 1 with bcc structure) and EDS of particle 1 (b) showing the particle 1
to be rich in Nb

2.2　β－Nb 第二相的氧化过程

图 2 给出了 Zr－2.5Nb 合金在 500 ℃/10.3 MPa 的过热蒸汽中腐蚀 3 h 后，氧化膜/金
属(oxide/metal，O/M)界面区域的 TEM 像. 可以看到嵌在界面处的 2 颗衬度较深的球状
物相. 对其中 P1 处的物相分析表明该类物相属于 bcc 结构，含 Nb 量较高(EDS 图中的 Cu
峰来自 FIB 样品支架)，确定为尚未氧化的 β－Nb 相. 在该 β－Nb 相靠近氧化膜一端出现了
月牙形的亮衬度区，HRTEM 观察显示靠近亮衬度区的 β－Nb 相边缘存在一层非晶态的氧
化物，如图 2b 所示. 可以看出，在 O/M 界面附近的锆合金基体已经氧化，但是嵌在 O/M 界
面上的 β－Nb 相只是在靠近 ZrO₂ 侧发生局部氧化，形成了一层很薄的非晶态氧化物，而在
靠近金属的一侧，依然结合完好，显然 β－Nb 相的抗氧化能力高于锆合金基体. 月牙形空隙

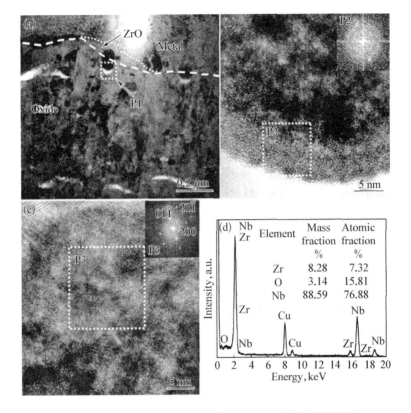

图 2 在 O/M 界面处的 β-Nb 的 HRTEM 分析与成分测量

Fig. 2 HRTEM analyses and EDS measured composition of β-Nb particle at oxide/metal (O/M) interface （a）TEM micrograph near the interface；（b）HRTEM image obtained from area P1 in Fig. 2a (inset showing FFT map of area P2 with amorphous structure)；（c）HRTEM image of β-Nb phase (inset showing FFT map of area P3 with bcc structure)；（d）EDS of β-Nb phase

的形成可能是制样过程造成的，由于 β-Nb 相非晶氧化物的结合力较弱，易于在 FIB 制样时被减薄甚至减穿而出现了空隙. 如在该 β-Nb 第二相颗粒的右侧，另一个 β-Nb 相颗粒的端部虽然出现月牙形亮白衬度（非晶，较薄），但并没有出现空隙.

图 3 给出了图 2a 中 P1 处 β-Nb 相颗粒附近的 Zr 的氧化物的 HRTEM 像. 通过 Fourier 转换及分析，确定是立方晶系的 ZrO（PDF 卡片编号：89-4768），并非通常的 ZrO_2，表明在 O/M 界面处存在 ZrO 过渡层（图 2a）. 由于该过渡层厚度不均匀，并且其电子衍射特性与合金基体很接近，所以不易从衍衬像上直接辨别. 对 β-Nb 相周围的 ZrO 相区进行 EDS 成分分析，没有检测到 Nb，表明 Nb 在 ZrO 中的固溶度低，很难向周围的 ZrO 中扩散，β-Nb 相的氧化应是 O 向 β-Nb 相的扩散过程.

图 4 给出了在氧化膜内部距离 O/M 界面约 650 nm 处含 Nb 相及其周围 Zr 氧化物的 TEM 像. 对含 Nb 相的形态及成分分析确定其为 β-Nb 相的氧化物. 此处 β-Nb 相的氧化物为晶态和非晶态的混合物（图 4b），其中的晶态氧化物，经分析确定为四方结构的 NbO_2（图 4b 插图）. 利用 EDS 分析 β-Nb 相氧化物的元素成分，与未氧化的 β-Nb 相相比，Nb 的相对含量有所增加，而 Zr 减少，表明 β-Nb 相在氧化过程中，Zr 容易向周围扩散流失. 由于

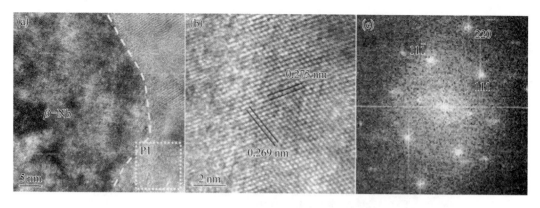

图3 在 O/M 界面处 ZrO 的 HRTEM 像及 FFT 图

Fig. 3 Analyses of ZrO phase at the O/M interface in Fig. 2a （a）HRTEM image；（b）HRTEM image taken from area P1 in Fig. 3a；（c）FFT map obtained from area P1 in Fig. 3a

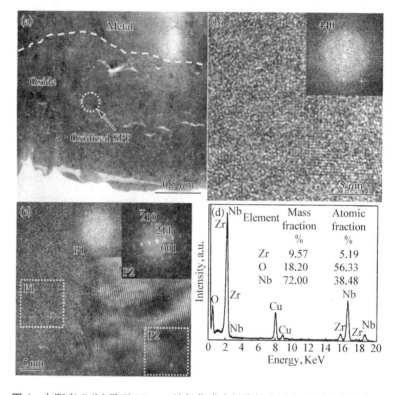

图4 在距离 O/M 界面 650 nm 处氧化膜中氧化的 β- Nb 粒子的显微分析

Fig. 4 Analyses of the oxidized β- Nb phase in the oxide layer at the distance of 650 nm from O/M interface （a）TEM micrograph；（b）HRTEM image of niobium oxide consisted of tetragonal and amorphous structures（inset showing FFT map of tetragonal NbO_2 lattice image）；（c）HRTEM image of zirconium oxide near NbO_2 particle（insets showing FFT maps of amorphous area P1 and monoclinic structure area P2）；（d）EDS of oxidized SPP

晶态和非晶态的氧化物是在纳米尺度上的混合物,因此难以用 EDS 确定晶态和非晶态氧化物在成分上的区别.对其周围 Zr 的氧化物进行分析,确定其为单斜结构的 ZrO_2 (图 4c 中 P2 区及其插图).在 β- Nb 相氧化物周围的 ZrO_2 相中,EDS 没有检测到 Nb.

图 5 给出了远离 O/M 界面的氧化膜的显微照片. 图 5a 中颗粒 1 是含 Nb 的氧化物, 该相呈现完全非晶态, 与周围的 ZrO₂ 的边界清晰, 应该是由原 $\beta - Nb$ 相氧化演变而成. 在其周围的 ZrO₂ 相中同样没有检测到 Nb. 在远离 O/M 界面最外侧的氧化膜中几乎检测不到 Nb, 这与文献[14]报道类似. 图 5b 是腐蚀 40 h 氧化膜的断口的 SEM 像. 可以看出, 明显存在一些与原 $\beta - Nb$ 第二相尺度相近的微小的孔洞, 结合文献[15]中报道的关于氧化膜中微孔洞观测结果, 认为 Nb 氧化物中的 Nb 最终会在腐蚀介质的作用下迁移到腐蚀介质中, 而在原来的位置留下微孔洞.

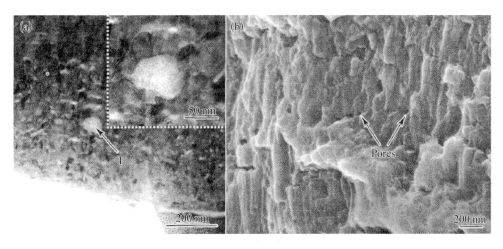

图 5 远离 O/M 界面的氧化膜的显微照片

Fig. 5 Micrographs of the oxide layer far away from O/M interface （a）TEM micrograph of oxide film （particle 1 is a Nb containing oxide and has amorphous structure, see inset） （b）fractograph of oxide film formed after autoclave test for 40 h

图 6 给出了 Zr - 2.5Nb 在 500 ℃/10.3 MPa 的过热蒸汽中腐蚀 3 h 后, 氧化膜截面样品中 $\beta - Nb$ 第二相在氧化过程中转变行为的示意图. $\beta - Nb$ 第二相在氧化前是 bcc 结构, 在靠近 O/M 界面处, 由于 O 供应不足, 因此生成的氧化物为含 O 量较低的晶态 NbO₂ 和含量较高的非晶态氧化物的混合物. 在 ZrO₂ 层的中部, O 供应充足, 因此观察到的 $\beta - Nb$ 相的氧化物主要为 O 含量较高的非晶态氧化物. 这与文献[13]报道的结果基本一致. 而在 ZrO₂ 膜的外层, 没有观察到 $\beta - Nb$ 相的氧化物, 甚至未发现 Nb, 这是由于 Nb 在腐蚀介质的作用下, 流失进入腐蚀液中[14].

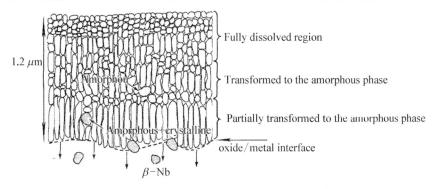

图 6 氧化膜截面中 $\beta - Nb$ 相在氧化过程中转变行为的示意图

Fig. 6 Schematic diagram of the transformation behavior of the $\beta - Nb$ particles in the cross-sectional sample during oxidation

2.3 讨论

Zr - 2.5Nb 合金样品经过 β 相水淬和冷轧变形后,促使 β - Nb 在 580 ℃ 最终热处理时析出[16],生成大量细小且均匀分布的 β - Nb 第二相. 一方面, β - Nb 第二相的析出降低了基体中 Nb 的固溶含量,提高了锆合金的耐腐蚀性能[3,11,12];另一方面, β - Nb 第二相的氧化行为也会对锆合金的耐腐蚀性能产生重要影响. 首先, β - Nb 第二相的氧化进程较缓慢,在 O/M 界面及附近的氧化膜中存在大量细小的 β - Nb 第二相,由于第二相粒子的塑性和韧性远比氧化膜好,可以弛豫界面及附近氧化膜中的部分压应力,同时氧化膜中存在的未完全氧化的第二相粒子会阻止其附近的微裂纹进一步发展[12],从而能够改善锆合金的耐腐蚀性能. 其次, β - Nb 第二相在氧化过程中,Nb 向周围的 ZrO_2 扩散困难,在未接触到腐蚀介质的情况下会保持 Nb 氧化物的独立相的形态,而在接触到腐蚀介质后,又会流失并在氧化膜中形成弥散分布的微孔,这些都会使氧化膜中的应力得到弛豫,延缓氧化膜中裂纹的形成和扩展,改善锆合金的耐腐蚀性能.

3 结论

(1) β - Nb 相在氧化过程中,由于 Nb 在 ZrO_2 中的固溶度小,不会向周围扩散,在氧化初期形成晶态 NbO_2 和非晶态氧化物的混合物,然后大部分转变成非晶态的氧化物;Nb 的非晶氧化物会在腐蚀介质作用下流失形成微孔洞.

(2) β - Nb 第二相的氧化行为会使氧化膜中的应力得到弛豫,延缓氧化膜中微裂纹的形成和扩展,改善锆合金的耐腐蚀性能.

感谢 FEI 公司陈文霞在氧化膜截面 TEM 薄膜样品制备中给予的大力帮助.

参 考 文 献

[1] Ells C E, Evans W. *Can Min Metall Bull*, 1981; 74; 105.

[2] Yang W D. *Reactor Materials Science*. Beijing; Atomic Energy Press, 2006; 19.
(杨文斗. 反应堆材料学. 北京;原子能出版社,2006; 19)

[3] Comstock R J, Schoenberger G, Sable G P. In; Bradley E R, Sabol G P, eds., *Zirconium in the Nuclear Industry*. Philadelphia; American Society for Testing and Materials, 1996; 710.

[4] Ramasubramanian N, Balakrishnan P V. In; Garde A M, Bradley E R, eds., *Zirconium in the Nuclear Industry: Tenth International Symposium*, *ASTM STP 1245*, Philadelphia; American Society for Testing and Materials, 1994; 378.

[5] Anada H, Takeda K. In; Bradley E R, Sabol G P, eds., *Zirconium in the Nuclear Industry: Eleventh International Symposium*, *ASTM STP 1295*, Philadelphia; American Society for Testing and Materials, 1996; 35.

[6] Perkins R A, Busch R A. In; Eucken C M, Garde A M, eds., *Zirconium in the Nuclear Industry: Ninth International Symposium*, *ASTM STP 1132*, Philadelphia; American Society for Testing and Materials, 1991; 595.

[7] Shebaldov P V, Peregud M M, Nikulina A V, Bibilashvili Y K, Lositski A F, Kuz'menko N V, Belov V I, Novoselov A E. In; Sabol G P, Moan G D, eds., *Zirconium in the Nuclear Industry: Twelfth International Symposium*, *ASTM STP 1354*, West Conshohocken, PA; American Society for Testing and Materials, 2000; 545

[8] Kim H G, Jeong Y H, Kim T H. *J Nucl Mater*, 2004; 326; 125.

[9] Jeong Y H, Lee K O, Kim H G. *J Nucl Mater*, 2002; 302; 9.

[10] Koutsky J, Kocik J. *Radiation Damage of Structural Materials*. Amsterdam: Materials Science Monographys, 1994: 315.

[11] Jeong Y H, Kim H G, Kim T H. *J Nucl Mater*, 2003: 317: 1.

[12] Nikulina A V, Markelov V A, Peregud M M, Bibilashvili Y K, Kotrekhov V A, Lositsky A F, Kuzmenko N V, Shevnin Y P, Shamardin V K, Kobylyansky G P, Novoselov A E. In: Bradley E R, Sabol G P, eds. , *Zirconium in the Nuclear Industry: Eleventh International Symposium*, *ASTM STP 1295*, Philadelphia: American Society for Testing and Materials, 1996: 785.

[13] Kim H G, Choi B K, Park J Y, Cho H D, Jeong Y H. *J Alloys Compd*, 2009: 481: 867.

[14] Liu W Q, Zhou B X, Li Q. *Nucl Power Eng*, 2002: 23: 69.
(刘文庆,周邦新,李强. 核动力工程,2002: 23: 69)

[15] Li Q. *PhD Thesis*, Shanghai University, 2008.
(李强. 上海大学博士论文,2008)

[16] Li Q, Liu W Q, Zhou B X. *Rare Met Mater Eng*, 2002: 31: 389.
(李强,刘文庆,周邦新,稀有金属材料与工程,2002;31: 389)

Oxidation Behavior of the β- Nb Phase Precipitated in Zr - 2. 5Nb Alloy

Abstract: The corrosion behavior for Zr - 2. 5Nb specimens heat-treated at 580 ℃ for 50 h after β- quenching and cold rolling has been investigated in 500 ℃/10. 3 MPa superheated steam by autoclave tests. HRTEM equipped with EDS was employed to investigate the matrix microstructure and the oxidation behavior of the β- Nb second phase particles (SPPs). It was found that many β- Nb SPPs with small sizes (<100 nm) randomly precipitated after heat treating at 580 ℃ for 50 h. It was noted that the β- Nb SPPs were more slowly oxidized than the zirconium matrix. The β- Nb SPPs of bcc structure were oxidized to form the mixed structure of amorphous oxide and crystalline NbO_2 at the initial oxidation stage, and then the amorphous phase was changed to the main structure at the middle oxidation stage, finally the niobium oxides were dissolved into the corrosion medium.

添加 Cu 对 M5 合金在 500 ℃过热蒸汽中
耐腐蚀性能的影响*

摘　要：用高压釜腐蚀实验研究了添加 0.05%—0.5% Cu(质量分数)对 M5(Zr-1% Nb)合金在 500 ℃/10.3 MPa 过热蒸汽中耐腐蚀性能的影响；用 TEM 和 SEM 分别观察了合金基体和腐蚀后氧化膜的显微组织. 结果表明：Cu 含量低于 0.2%时，随着 Cu 含量的增加，合金的耐腐蚀性能得到明显改善；继续提高 Cu 含量则对进一步改善合金耐腐蚀性能的作用不明显. Cu 含量不超过 0.2%时，Cu 主要固溶在 α-Zr 基体中；当 Cu 含量高于 0.2%时，Cu 除了固溶在 α-Zr 基体中之外，其余的 Cu 主要以 Zr_2Cu 型第二相析出. 固溶在 α-Zr 基体中的 Cu，在 α-Zr 被氧化后可以延缓氧化膜中空位扩散凝聚形成孔隙和孔隙发展成为微裂纹的过程，增加氧化膜的致密度，从而提高合金的耐腐蚀性能. 可见，对 M5 合金耐腐蚀性能影响起主要作用的是固溶在 α-Zr 中的 Cu，而不是含 Cu 第二相.

　　锆合金是核反应堆中一种重要的结构材料，用作核燃料元件的包壳. 为了提高核电的经济性，需要进一步加深核燃料的燃耗，这样需要延长核燃料的换料周期，增长燃料元件在反应堆内停留的时间，这对锆合金的耐腐蚀性能提出了更高的要求，为此，世界各国都在开发自己的新型锆合金.

　　我国目前还没有自主知识产权的高性能新型锆合金，核电用的锆合金主要是从美、俄、法等国家进口，其中从法国引进的 AFA-3G 燃料组件包壳用的材料是 M5(Zr-1% Nb)(质量分数，下同)合金. 因此，可以尝试在 M5 合金基础上通过添加其他合金元素得到耐腐蚀性能更好的合金. 例如 Park 等[1,2]报道了在 Zr-1.1% Nb 基础上添加 0.05% Cu 发展起来的 HANA-6 合金具有优良的耐腐蚀性能. Hong 等[3]研究发现，Cu 对 Zr-4 合金的耐腐蚀性能也有一定的改善作用. 从这些结果看 Cu 似乎是对提高锆合金耐腐蚀性能有益的添加元素，但是目前有关 Cu 影响锆合金耐腐蚀性能的研究并不系统，其影响机理也不清楚. 因此，本研究在 M5 合金基础上添加不同含量的 Cu，研究其在 500 ℃/10.3 MPa 过热蒸汽中的腐蚀行为，探讨 Cu 影响 M5 合金耐腐蚀性能的规律和机理，为开发具有我国自主知识产权的新型锆合金时合金元素的选择提供依据和指导.

1　实验方法

　　以 M5 为母合金，通过添加不同含量的 Cu 制备成 Zr-1% Nb-xCu ($x=0.05\%$，0.1%，0.2%，0.35%和 0.5%)锆合金，合金熔炼过程中考虑了 Cu 的烧损. 同时用 M5 重熔(M5-remelted)和 M5 原材(M5-received)试样作为对照. 用电感耦合等离子体原子发射光谱(ICPAES)分析得到的合金成分见表 1，可见分析得到的 Cu 含量与名义成分吻合. 合金试样制备过程(图 1)如下：先用真空非自耗电弧炉熔炼成约 60 g 的合金锭，熔炼时通高纯 Ar

---

* 本文合作者：李士炉、姚美意、张欣、耿建桥、彭剑超. 原发表于《金属学报》，2011,47(2)：163-168.

气保护,为保证成分均匀,合金锭每熔炼一次都要翻转,共熔炼 6 次.合金锭经过热压成条块状后在 β 相区(1 030 ℃)进行均匀化处理,再通过热轧、β 相水淬和冷轧等工艺制得 0.6 mm 厚的片状试样,最后进行 580 ℃/50 h 退火处理.试样每次热处理前需要经过酸洗,以除去试样表面的氧化物和杂物等,所用的酸洗液为 30% H_2O(体积分数)＋30% HNO_3＋30% H_2SO_4＋10% HF 混合溶液.制备好的试样经过酸洗和去离子水清洗后放入 500 ℃/10.3 MPa 过热蒸汽中进行腐蚀实验,腐蚀增重由 5 块试样平均得出.

表1 实验用合金成分
Table 1 Chemical compositions of experimental alloys （mass fraction, %）

Alloy	Nb	Cu	Zr
M5+0.05Cu	0.95	0.055	Bal.
M5+0.1Cu	0.98	0.10	Bal.
M5+0.2Cu	0.96	0.20	Bal.
M5+0.35Cu	0.92	0.33	Bal.
M5+0.5Cu	0.97	0.50	Bal.
M5-remelted	0.98	0.010	Bal.
M5-received	0.94	0.010	Bal.

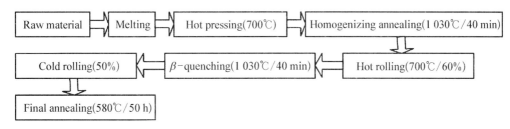

图1 试样制备过程流程图
Fig. 1 Preparation procedure of samples

用 JEM-200CX 型透射电子显微镜(TEM)观察合金试样的显微组织;用 JEM-2010F 型高分辨透射电镜配置的能谱仪(EDS)分析合金中的第二相成分;用 JSM-6700F 型扫描电子显微镜(SEM)观察腐蚀样品氧化膜的断口形貌.TEM 观察用薄试样采用双喷电解抛光制备,电解液为 10%(体积分数)高氯酸酒精溶液,电压 50 V,温度－40 ℃.从腐蚀后的样品上切下一块合金,用混合酸溶去金属基体,然后将氧化膜折断来制备氧化膜断口形貌观察用试样.在进行氧化膜形貌观察前,为了提高图像质量,试样表面蒸镀 Ir.

2 实验结果及分析

2.1 腐蚀增重

图 2 是 M5＋Cu 系列合金在 500 ℃/10.3 MPa 过热蒸汽中腐蚀 750 h 的增重曲线.图 2a 是合金腐蚀增重随腐蚀时间变化的曲线.可以看出,在 M5 合金中添加适量(0.05%—0.5%)的 Cu 后,合金的耐腐蚀性能得到明显改善.图 2b 是 M5＋Cu 系列合金腐蚀 750 h 后腐蚀增重随 Cu 含量变化的曲线.当添加的 Cu 含量小于 0.2% 时,随着 Cu 含量的增加,合金

的腐蚀增重呈线性递减;当添加的 Cu 含量达到 0.2%时,继续提高 Cu 含量则对进一步改善合金耐腐蚀性能的作用不大. 如腐蚀 750 h 时,M5 重熔试样的增重为 358 mg/dm²; M5＋0.2Cu 合金试样的增重只有 226 mg/dm²,比 M5 重熔试样低 37%;而 M5＋0.35Cu 和 M5＋0.5Cu 试样的增重均为 208 mg/dm²,与 M5＋0.2Cu 试样的增重相差不大.

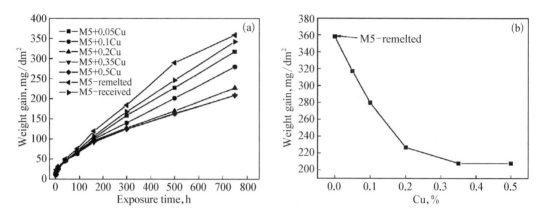

图 2 试样在 500 ℃/10.3 MPa 过热蒸汽中腐蚀后的增重曲线

Fig. 2 Weight gain of the specimens *vs* corrosion time (a) and the Cu content after 750 h exposure (b) in 500 ℃/10.3 MPa superheated steam

2.2 合金的显微组织

图 3 是腐蚀前合金显微组织的 TEM 像. 可见,进行 580 ℃/50 h 退火处理后,M5＋Cu 系列合金均发生了再结晶,晶粒均为等轴晶. 合金中有 2 种尺寸不同的第二相弥散分布在基体中,一种为非常细小的球形颗粒,尺寸在 60 nm 以下;另一种相对较大,在 100—300 nm 之间,形状并不规则,随着 Cu 含量增加,较大第二相的比例增加,且尺寸也增大.

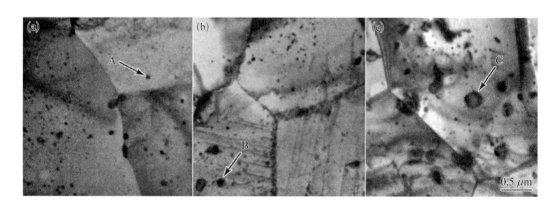

图 3 腐蚀前合金试样的 TEM 像

Fig. 3 TEM micrographs of M5 + 0.1Cu (a), M5 + 0.2Cu (b) and M5 + 0.5Cu (c) before corrosion test

对上述第二相进行 EDS 分析,结果表明: M5＋Cu 系列合金中的细小第二相为 β-Nb 型,其中少量含有 Fe,Cr 或 Cu 元素[1,4];较大的第二相主要是 Zr₂Cu 型,其中部分含有 Nb,Fe 或 Cr 元素[3,5];另外还有 ZrNbFe 型第二相[1,4],尺寸介于前 2 者之间,约为 60—150 nm,

ZrNbFe 型第二相中含有少量 Cr. 第二相中的 Fe 和 Cr 元素是从海绵锆中引进的,Fe 和 Cr 在 α-Zr 中的溶解度很低,最大溶解度分别为约 120（820 ℃）和 200 μg/g（860 ℃）[6],因此多余的 Fe 和 Cr 以第二相形式析出. 图 4 列出了图 3 中 A,B 和 C 所示第二相的能谱图,分别是 β-Nb 型、ZrNbFe 型和 Zr_2Cu 型. 表 2 总结了第二相的种类、数量和尺寸. 从表 2 可以看出,当合金中 Cu 含量较低（≤0.2%）时,析出的第二相以 β-Nb 型为主,也有少量的 ZrNbFe 型,Cu 主要固溶在 α-Zr 基体中;当合金中 Cu 含量较高（>0.2%）时,Zr_2Cu 型第二相析出,并且随着添加的 Cu 含量增加,Zr_2Cu 型第二相增多.

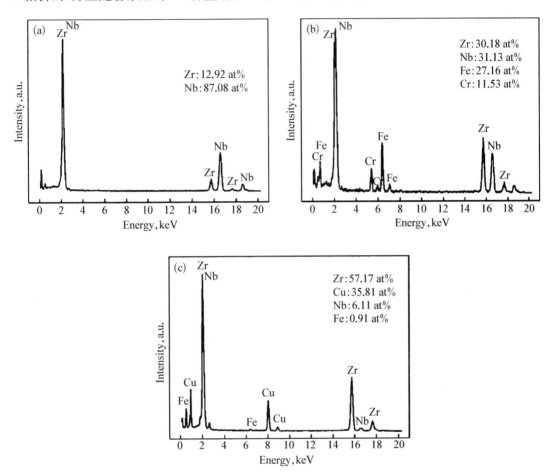

图 4 合金中第二相的 EDS 谱

Fig. 4 EDS patterns of the second phase particles in the alloys (a) β-Nb type, position A in Fig. 3a (b) ZrNbFe type, position B in Fig. 3b (c) Zr_2Cu type, position C in Fig. 3c

表 2 M5+Cu 系列合金中的第二相信息

Table 2 Details of the second phase particles in M5+Cu alloys

Precipitate type	Frequency			Particle size, nm
	M5+0.1Cu	M5+0.2Cu	M5+0.5Cu	
β-Nb	Major	Major	Partly	<60 (major) and 60—150 (minor, contain Fe, Cr or Cu)
ZrNbFe	Minor	Minor	—	60—150 (all, contain Cr)
Zr_2Cu	—	Minor	Partly	100—200 (partly) and 200—300 (partly, contain Fe, Cr or Nb)

2.3 腐蚀样品氧化膜的断口形貌

图 5 是 M5+Cu 系列合金在 500 ℃/10.3 MPa 过热蒸汽中腐蚀 750 h 后的氧化膜断口形貌. 从图 5 可以看出,试样氧化膜厚度随着 Cu 含量的增加而减小,M5 重熔试样氧化膜最厚,约为 24 μm,M5+0.5Cu 试样氧化膜最薄,约为 14 μm,这与腐蚀增重结果是一致的. 随着 Cu 含量的增加,氧化膜断口起伏程度减小,致密度增加. 所有腐蚀样品氧化膜断口都有近似平行金属/氧化膜界面的横向裂纹,随着 Cu 含量的增加,氧化膜中裂纹的数量减少. 此外,样品氧化膜中的 ZrO_2 晶粒主要是柱状晶,即使是靠近氧化膜外表面处仍以柱状晶为主,几乎看不到等轴(接近球形)ZrO_2 晶粒(图 5a1, b1 和 c1). 之前的研究结果[7,8]认为,柱状晶中的缺陷通过扩散凝聚而构成新的晶界,发展成等轴晶. 比较图 5a1、b1 和 c1 可以看出,随着合金中 Cu 含量的减少,氧化膜中柱状晶长度变短,尺寸在几十纳米的微孔隙数量增加. 这些微孔隙的不断增加会发展成为裂纹,也使得柱状晶变短,这也是添加 Cu 元素后 M5 合金耐腐蚀性能提高的原因之一. 关于微孔隙及裂纹的形成过程在本文后面将作详细讨论.

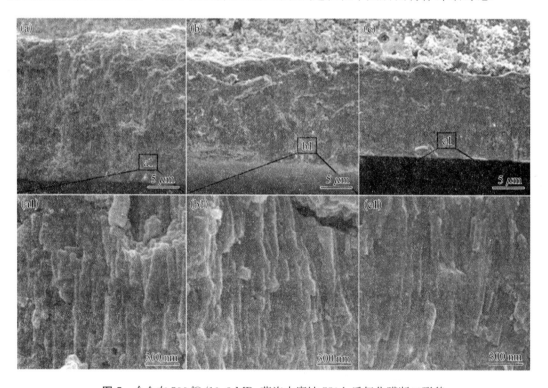

图 5 合金在 500 ℃/10.3 MPa 蒸汽中腐蚀 750 h 后氧化膜断口形貌

Fig. 5 Oxide film fracture micrographs of M5-remelted (a, a1), M5+0.1Cu (b, b1) and M5+0.5Cu (c, c1) alloys corroded in 500 ℃/10.3 MPa superheated steam for 750 h exposure

3 分析讨论

有学者[2,9-11]认为:在 Zr-Nb 或 Zr-Cu 二元合金中,固溶在基体中的 Nb 或 Cu 对提高锆合金的耐腐蚀性能起主要作用,而不是含 Nb 或 Cu 的第二相;当添加的 Nb 或 Cu 含量在固溶度以下时,随着 Nb 或 Cu 含量的增加耐腐蚀性能增加. 由前面的实验结果可知,当

M5＋Cu 系列合金中添加的 Cu 含量小于 0.2％时,Cu 主要固溶在 α-Zr 基体中,含 Cu 第二相很少,随着 Cu 含量的增加,腐蚀增重明显减小,耐腐蚀性能显著提高;当添加的 Cu 含量大于 0.2％时,有大量的 Zr_2Cu 型第二相析出,合金的耐腐蚀性能区别不大,都与 M5＋0.2Cu合金相近,说明 Zr_2Cu 型第二相对合金的耐腐蚀性能影响不明显. 由 Zr-Cu 二元相图可知,580 ℃时 Cu 在 α-Zr 中的固溶度约为 0.15％[12]. 这进一步说明固溶在 α-Zr 基体中的 Cu 含量是影响耐腐蚀性能的关键因素.

Zr 在高温水或水蒸气中的腐蚀是 O^{2-} 通过氧化膜扩散到金属/氧化膜界面处与 Zr 反应形成 ZrO_2 膜或阴离子空位沿反方向扩散而得以不断进行的[13]. 因此,氧化膜的性质和氧化膜显微组织在腐蚀过程中的演化与锆合金的腐蚀行为密切相关. 由于 Zr 氧化成 ZrO_2 时体积发生膨胀,同时又受到金属基体的约束,因此,氧化膜内部会形成很大的压应力,使 ZrO_2 晶体中产生许多缺陷,稳定了一些亚稳相;空位、间隙原子等缺陷在温度、应力和时间的作用下,发生扩散、湮没和凝聚,空位被晶界吸收后形成纳米大小的孔隙簇,弱化了晶粒之间的结合力;孔隙簇进一步发展成为裂纹,环境因素以及合金自身的成分和组织因素必然会对这种演化过程产生影响,从而影响锆合金的耐腐蚀性能[14-16]. 锆合金的耐腐蚀性能与其在腐蚀过程中氧化膜显微组织和晶体结构的演化有关. 这种演化过程是在氧化膜受到压应力情况下发生的,因此裂纹的扩展方向一定是平行于压应力方向,而压应力是平行于金属/氧化膜界面的,这就解释了为何图 5 中的裂纹近似平行于金属/氧化膜界面. 锆合金发生氧化时,固溶在基体中的合金元素及析出的第二相在 α-Zr 基体发生氧化的同时也会逐渐被氧化,进入氧化膜后会影响氧化膜中的内应力和 ZrO_2 的表面自由能,从而影响空位扩散、凝聚成孔隙和孔隙发展成为微裂纹的过程. 固溶在 α-Zr 中的合金元素将比第二相中的合金元素氧化后更容易固溶到 ZrO_2 中,提高了 ZrO_2 的表面自由能,延缓氧化膜中的空位凝聚形成孔隙、孔隙发展成为微裂纹的过程. 氧化膜断口(图 5)形貌观察的结果也证实,随着 M5 合金中 Cu 含量的增加,氧化膜形貌表面和断口起伏程度减小,致密度增加. 这可以合理解释 Cu 含量影响 M5 合金耐腐蚀性能的规律(图 2).

4 结论

(1) M5＋Cu 系列合金的显微组织观察结果表明:当 Cu 含量不超过 0.2％时,析出的第二相主要是 β-Nb 型,尺寸相对较小,Cu 则主要固溶在 α-Zr 基体中;当 Cu 含量高于0.2％时,析出较多的 Zr_2Cu 型第二相,尺寸相对较大. 这说明 580 ℃时 Cu 在 M5 合金 α-Zr 基体中的溶解度不超过 0.2％.

(2) 在 M5 合金成分基础上添加 0.05％—0.5％Cu 可以改善其在 500 ℃/10.3 MPa 过热蒸汽中的耐腐蚀性能:当 Cu 含量在 0.2％以下时,随着 Cu 含量的增加,合金的耐腐蚀性能得到明显改善,继续提高 Cu 含量则对进一步改善合金耐腐蚀性能的作用不明显. 这说明固溶在 α-Zr 中的 Cu 含量对 M5 合金的耐腐蚀性能有很大影响.

(3) M5＋Cu 系列合金腐蚀样品氧化膜断口形貌随着 Cu 含量的增加起伏程度减小,致密度增加. 这说明固溶在 α-Zr 基体中的 Cu 在 α-Zr 被氧化后可以延缓氧化膜中空位扩散凝聚形成孔隙和孔隙发展成为微裂纹的过程,从而提高合金的耐腐蚀性能.

参 考 文 献

[1] Park J Y, Choi B K, Yoo S J, Jeong Y H. *J Nucl Mater*, 2006;359:59.

[2] Park J Y, Yoo S J, Choi B K, Jeong Y H. *J Alloys Compd*, 2007; 437: 274.

[3] Hong H S, Moon J S, Kim S J, Lee K S. *J Nucl Mater*, 2001; 297: 113.

[4] Jeong Y H, Kim H G, Kim T H. *J Nucl Mater*, 2003; 317: 1.

[5] Kim J M, Jeong Y H. *J Nucl Mater*, 1999; 275: 74.

[6] Charquet D, Hahn R, Ortlieb E, Gros J P, Wadier J F. In: Van Swam L F P, Eucken C M eds. , *Zirconium in the Nuclear Industry: 8th International Symposium*, *ASTM STP 1023*, Baltimore: ASTM International, 1988: 405.

[7] Anada H, Takeda K. In: Sabol G P, Bradley E R eds. , *Zirconium in the Nuclear Industry: 11th International Symposium*, *ASTM STP 1295*, Ann Arbor: ASTM International, 1996: 35.

[8] Wadman B, Lai Z, Andrén H O, Nyström A L, Rudling P, Pettersson H. In: Garde A M, Bradley E R eds. , *Zirconium in the Nuclear Industry: 10th International Symposium*, *ASTM STP 1245*, Ann Arbor: ASTM International, 1994: 579.

[9] Jeong Y H, Lee K O, Kim H G. *J Nucl Mater*, 2002; 302: 9.

[10] Kim H G, Jeong Y H, Kim T H. *J Nucl Mater*, 2004; 326: 125.

[11] Jeong Y H, Kim H G, Kim D J. *J Nucl Mater*, 2003; 323: 72.

[12] Abe T, Shimono M, Ode M, Onodera H. *Acta Mater*, 2006; 54: 909.

[13] Yang W D. *Reactor Materials Science*. 2nd Ed. , Beijing: Atomic Energy Press, 2006: 260.

（杨文斗. 反应堆材料学. 第二版. 北京：原子能出版社,2006: 260）

[14] Zhou B X, Li Q, Yao M Y, Liu W Q. *Nucl Power Eng*, 2005;26: 364.

（周邦新,李强,姚美意,刘文庆. 核动力工程,2005; 26: 364）

[15] Zhou B X, Li Q, Liu W Q, Yao M Y, Chu Y L. *Rare Metal Mat Eng*, 2006; 35: 1009.

（周邦新,李强,刘文庆,姚美意,褚于良. 稀有金属材料与工程,2006;35: 1009）

[16] Zhou B X, Li Q, Yao M Y, Liu W Q, Chu Y L. *J ASTM Intl*, 2008; 5: 360.

Effect of Adding Cu on the Corrosion Resistance of M5 Alloy in Super-heated Steam at 500 ℃

Abstract: The effect of Cu content on the corrosion resistance of Zr - 1% Nb - xCu (x=0. 05%—0. 5%, mass fraction) was investigated in superheated steam at 500 ℃ and 10. 3 MPa by autoclave tests. The microstructures of the alloys and oxide films on the corroded specimens were observed by TEM and SEM, respectively. The results showed that when the Cu content was below 0. 2%, the corrosion resistance of the alloys was markedly improved with the increase of Cu content, while further addition of Cu did not lead to a further improvement in the corrosion resistance. When the Cu content was below 0. 2%, the Cu mainly dissolved in the α - Zr matrix. And when the Cu content was more than 0. 2%, part of Cu precipitated as Zr_2Cu second phase particles. When the α - Zr matrix was oxidized, the Cu dissolved in the α - Zr could delay the process that the vacancies in the oxide film diffused and coalesced to form pores, and the pores developed into micro-cracks. Therefore, the corrosion resistance of the alloys was enhanced. It can be concluded that the Cu concentration in the α - Zr matrix, rather than the second phase particles containing Cu, is the main reason that the addition of Cu improves the corrosion resistance of M5 alloy in superheated steam at 500 ℃ and 10. 3 MPa.

690 合金原始晶粒尺寸对晶界工程
处理后晶界网络的影响*

摘 要：利用电子背散射衍射(EBSD)和取向成像显微分析技术(OIM)研究了 690 合金原始晶粒尺寸对晶界工程(GBE)处理后晶界特征分布(GBCD)的影响. 结果表明,原始晶粒尺寸对 GBE 处理提高低 ΣCSL 晶界比例及控制晶界网络分布有明显的影响. 在最终退火工艺相同时,根据不同的原始晶粒尺寸,在 GBE 处理中需要采用不同的冷变形量,才能够获得最佳的晶界网络分布. 可以利用参数"晶粒平均应变量"来表达原始晶粒尺寸和冷变形量共同影响 GBE 处理效果的综合作用.

镍基 690 合金作为压水堆核电站蒸汽发生器传热管材料,已经得到广泛应用,虽然取得了较好的效果,但随着核电站设计寿命延长的要求,进一步提高蒸汽发生器可靠性及使用寿命是需要研究的课题. 蒸汽发生器传热管道失效的主要原因在不同时期有不同的情况,从近些年统计的结果来看,与晶界有关的应力腐蚀开裂和晶间腐蚀仍然是蒸汽发生器传热管道失效的主要原因[1].

从材料的晶界结构及分布的角度考虑如何提高材料性能的研究工作于上世纪 80 年代逐渐受到重视. 日本学者 Watanabe[2] 提出晶界设计及控制的概念,随后发展为晶界工程(GBE)研究领域[3]. Lin 等[4] 对镍基 600 合金进行了加工及退火处理,可以使低 Σ 重位点阵(CSL)晶界比例由 37％ 提高到 71％(长度比例,下同),这使该材料在沸腾 $FeSO_4$ 中浸泡 24 h 的腐蚀实验中腐蚀速率降低 30％—60％. Lehockey 等[5] 将铅酸电池的铅合金电极板的低 ΣCSL 晶界比例提高到 70％ 以上,电池充放电循环次数提高 1—3 倍. 这 2 个成功的应用案例使 GBE 研究被广泛认可. 虽然 GBE 研究已经取得了很大的进展[6-18],但是对一些关键科学问题还未达成共识,尤其是针对不同种类合金提高特殊晶界比例的工艺及其机理问题. 比如,在 304 不锈钢中,通过 5％ 的冷轧及 1 200 K 长时间(72 h)退火,能够将低 ΣCSL 晶界比例提高到 85％ 以上[7];对铅合金[5] 及 600 合金[4] 通过中变形量加工及高温短时间退火,并将这一工艺反复进行多次,能够明显提高低 ΣCSL 晶界比例;本课题组的前期工作[13,19]表明,在 690 合金中通过小变形量冷加工及高温($0.85T_m$,T_m 为熔点)短时间退火能够将低 ΣCSL 晶界比例提高到 70％ 以上. 在不同种类材料中提高低 ΣCSL 晶界比例的工艺不同,一方面是由于材料成分的差别;另一方面,GBE 处理之前材料的原始显微组织差异可能是更重要的因素.

本课题组自主开发的 GBE 技术[19-21]与现行 690 合金管材的生产工艺参数能够很好衔接,并能使 690 合金管材的耐腐蚀性能得到明显提高[20,21]. 为了更好地控制 GBE 处理技术,必须要知道原始显微组织对 GBE 处理之后显微组织的影响. 本文研究了原始晶粒尺寸对 690 合金 GBE 处理之后晶界网络分布的影响.

---

* 本文合作者：刘廷光、夏爽、李慧、陈文觉. 原发表于《金属学报》,2011,47(7)：859-864.

1 实验方法

实验用 690 合金的化学成分(质量分数,%)为:Cr 30.39,Fe 8.88,C 0.023,Al 0.22,Ti 0.26,Si 0.07,Mn 0.23,Cu 0.02,S 0.002,P 0.006,Ni 余量. 对固溶后的片状原始样品进行 50% 的冷轧,然后分别在 1 100 ℃ 真空退火 2 和 120 min,获得了不同晶粒尺寸的样品 SS 和 SL,如表 1 所示. 采用前期工作[19]中得到的 GBE 处理方法(小量冷变形后高温短时间退火). 对 2 种不同晶粒尺寸的样品分别进行 5% 和 10% 的拉伸变形,然后统一在 1 100 ℃ 退火 6 min,得到了不同类型的样品,如表 1 所示. 所有样品均密封在真空石英管中进行热处理,在热处理后快速砸破石英管进行水淬.

表 1 690 合金形变及热处理工艺

Table 1 Thermomechanical treatments of the alloy 690 specimens

Starting material	Pre-processing				GBE		
	Cold rolling reduction ratio in thickness	Annealing time at 1 100 ℃ min	Grain size μm	Specimen	Tensile elongation ratio	Annealing time at 1 100 ℃ min	Specimen
Plate of alloy 690	50%	2	11.4	SS	5%	6	A
	50%	2	11.4	SS	10%	6	B
	50%	120	20.6	SL	5%	6	C
	50%	120	20.6	SL	10%	6	D

利用电解抛光制备符合电子背散射衍射(electron backscatter diffraction,EBSD)分析要求的样品,电解液为 20% $HClO_4$ + 80% CH_3COOH(体积分数),在室温下用 40 V 直流电抛光约 60 s. 利用配备在 CamScan Apollo 300 热场发射枪扫描电子显微镜(SEM)中的 HKL - EBSD 附件对样品表面微区进行分析. 扫描步长为 1—2 μm,扫描区域大小约为 800 μm×500 μm. 测试结果用 HKL - Channel5 软件分析,采用 Palumbo-Aust 标准判定晶界类型[22].

2 实验结果与讨论

图 1 是 2 种不同晶粒尺寸的原始态样品 SS 和 SL 经过 EBSD 测试,再经 HKL - Channel5 软件分析得到的不同类型晶界分布图. 利用 Channel5 软件中的等效圆直径法统计样品 SS 和 SL 的平均晶粒尺寸分别为 11.4 和 20.6 μm(本文中的晶粒尺寸将孪晶统计在内). 样品 SS 和 SL 的低 ΣCSL 晶界比例分别为 46.0% 和 45.7%,其中绝大部分都是 Σ3 晶界(孪晶界),而多重孪晶界(Σ9,Σ27 等)比例很低,如图 2 所示.

2 种不同晶粒尺寸的原始态样品 SS 和 SL 分别经过不同的小形变量拉伸变形及 1 100 ℃ 短时间退火处理,分别得到样品 A,B 和 C,D(表 1). 这些样品的低 ΣCSL 晶界比例也绘入图 2 中,4 种样品的低 ΣCSL 晶界比例较原始样品(SS 和 SL)都有不同程度的提高. 其中样品 A,C 和 D 的低 ΣCSL 晶界比例达到 70% 以上,Σ3 晶界超过 60%,多重孪晶界(Σ9 和 Σ27)占 9% 左右. 而样品 B 的低 ΣCSL 晶界比例为 63.2%,其中 Σ3 晶界为 54.8%,Σ9 和 Σ27 晶界之和为 8.4%. 当原始晶粒尺寸较小时(样品 SS),经过 5% 的拉伸变形在再结晶退

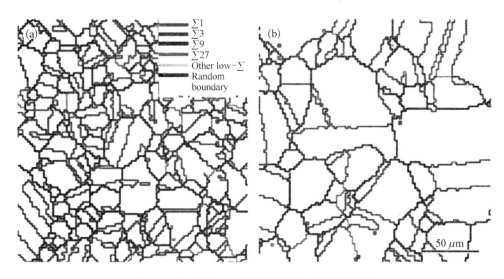

图 1 2种晶粒尺寸的原始态样品的晶界类型图

Fig. 1 EBSD measured distributions of the types of grain boundary of the starting specimens SS (a) and SL (b)

火后能够获得高比例低 ΣCSL 晶界（样品 A）；而经过 10% 的拉伸变形在再结晶退火后低 ΣCSL 晶界比例没有进一步增加反而下降（样品 B）. 当原始晶粒尺寸较大时（样品 SL），经过 5% 和 10% 的拉伸变形及同样的再结晶退火后低 ΣCSL 晶界比例均有提高，且后者相对更高一些. 所以，原始晶粒尺寸对 GBE 处理效果有着显著的影响. 在最终退火工艺相同时，退火前最佳的冷变形量需要随原始晶粒尺寸的不同而变化. 但是对于 GBE 处理后获得的显微组织，不能够只考虑低 ΣCSL 晶界比例的多少，更需要关注晶界网络分布及晶粒尺寸大小.

　　图 3 是样品 A，B，C 和 D 中不同类型晶界的分布图. GBE 处理之后形成大尺寸的"互有 $\Sigma 3^n$ 取向关系晶粒的团簇"显微组织是晶界

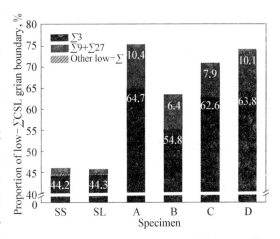

图 2 样品 SS，SL，A，B，C 和 D 的低 ΣCSL 晶界比例直方图

Fig. 2 Histogram of the proportion of low-ΣCSL grain boundary in the specimens SS，SL，A，B，C and D

网络的显著特征[23-25]. 这种晶粒团簇有以下特征：团簇内所有晶粒之间均具有 $\Sigma 3^n$（$n=1$，2，3，…）的取向关系，构成许多相互连接的 $\Sigma 3^n$ 类型的三叉界角（triple junction），如：$\Sigma 3 - \Sigma 3 - \Sigma 9$，$\Sigma 3 - \Sigma 9 - \Sigma 27$ 等；而团簇与团簇之间基本是随机晶界，如图 3a 中的晶粒团簇 L. 这种大尺寸的晶粒团簇显微组织对材料的耐腐蚀性能有着重要影响[20,21,26]. 对图 3a 中的一个晶粒团簇 L 进行详细分析，如图 4 所示. 该晶粒团簇尺寸 D 约为 300 μm，包含 108 个晶粒. HKL - Channel5 软件能够统计出扫描区域内每一个晶粒的平均取向 Euler 角[27,28]，从而可以计算任意 2 晶粒之间的取向关系. 表 2 给出了在该晶粒团簇中随机选取的 8 个晶粒之间的取向关系，结果表明它们之间均具有 $\Sigma 3^n$ 取向关系. 这是因为晶粒团簇是在再结晶时通过

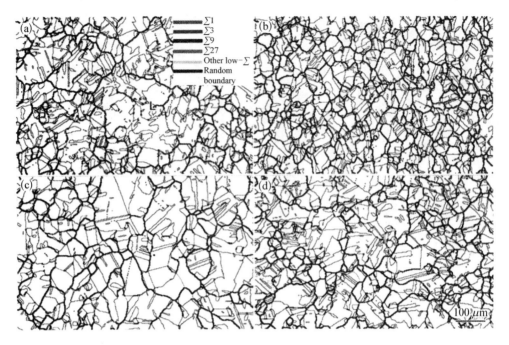

图 3 GBE 处理后得到 4 种样品的不同类型晶界图

Fig. 3 Distributions of the types of grain boundaries in the specimens A (a), B (b), C (c) and D (d) (area L enclosed by random boundary in Fig. 3a is called grain cluster)

多重孪晶(multiple twinning)过程形成的[23,24,29,30],也就是再结晶晶核长大过程中依次形成第一代孪晶、第二代孪晶、第三代孪晶及更高代次的孪晶,形成的这种孪晶链(twin chain)构成一个晶粒团簇.所以晶粒团簇内的晶粒从产生过程来看,它们与晶核的取向关系都符合$\Sigma 3^n$关系,从而可以推出晶粒团簇内任意两晶粒之间符合$\Sigma 3^n$取向关系.除了样品 A 外,其他样品也形成了晶粒团簇显微组织,然而晶粒团簇尺寸、团簇内的晶粒尺寸及$\Sigma 3^n$类型晶界多少等都有差异.

表 2 图 4 所示晶粒团簇 L 内随机选取的 8 个晶粒之间的取向关系(取向关系分别用轴角对的形式和Σ值的形式给出:左下部分用角-轴对形式θ,$[hkl]$表达;右上部分用重位点阵的Σ值及其偏差角$\Sigma / \Delta \theta$表达)

Table 2 Misorientations of eight randomly selected grains within the grain-cluster L in Fig. 4 (the bottom left data are shown in the form of angle-axis pair：θ, $[hkl]$; the top right data are shown in the form of Σ - value of CSL and deviation angle：$\Sigma / \Delta \theta$)

Grain No.	a	b	c	d	e	f	g	h
a		3/0	3/0.1°	3/0.2°	9/0.2°	9/1.0°	3/0.2°	9/0
b	60.0°, [1 1 1]		9/0.5°	9/0.3°	3/0.5°	27a/0.9°	9/0.2°	27a/0
c	59.9°, [1 $\bar{1}$ 1] 39.4°, [0 $\bar{1}$ 1]			9/0.2°	27b/0.5°	3/0	9/0.3°	27b/0
d	59.8°,[$\bar{1}$ 1 1] 38.6°, [0 $\bar{1}$ 1] 39.1°,[$\bar{1}$ 0 $\bar{1}$]				27b/0.5°	27b/0.2°	9/0.1°	3/0.2°
e	38.7°, [0 $\bar{1}$ 1] 59.5°, [$\bar{1}$ 1 1] 34.9°, [0 1 2] 35.9°, [0 $\bar{1}$ 2]					7 291/0.4°	27a/0.2°	81c/0.8°
f	39.9°,[$\bar{1}$ 0 $\bar{1}$] 30.7°, [0 1 $\bar{1}$] 60.0°, [1 $\bar{1}$ $\bar{1}$] 35.2°, [2 0 1] 39.3°,[$\bar{4}$ 1 $\bar{3}$]						27b/0.3°	81b/0.7°
g	59.8°, [$\bar{1}$1$\bar{1}$] 39.1°, [$\bar{1}$ 1 0] 38.6°, [1 0 $\bar{1}$] 38.8°, [$\bar{1}$ 0 1] 31.8°, [0 1 $\bar{1}$] 35.1°,[$\bar{1}$2 0]							27b/0.3°
h	38.9°, [1 0 1] 31.6°, [0 1 $\bar{1}$]35.4°, [$\bar{2}$ 0 $\bar{1}$] 59.8°, [$\bar{1}$$\bar{1}$ 1] 37.6°, [5 $\bar{1}$$\bar{3}$] 53.8°, [2 $\bar{3}$$\bar{2}$] 35.7°, [0 $\bar{2}$ 1]							

利用 HKL－Channel5 软件中的等效圆直径法统计样品 A，B，C 和 D 的平均晶粒团簇尺寸 D 和平均晶粒尺寸 d（等面积圆的直径），结果如图 5 所示. 可见,样品 A，C，D 的平均晶粒团簇尺寸约 $100\ \mu m$,而样品 B 的平均晶粒团簇尺寸却小很多,约为 $50\ \mu m$. 图 5 中的竖线表示晶粒团簇和晶粒尺寸的波动范围,其最高点和最低点的数值相差很大,可见晶粒团簇和晶粒尺寸很不均匀.

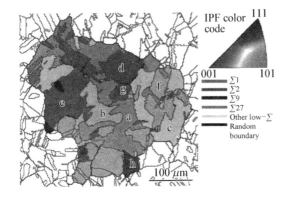

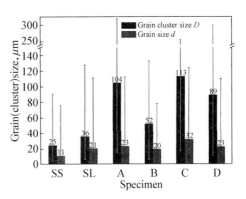

图 4 图 3a 中晶粒团簇 L 中的晶粒取向在反极图（IPF）中的分布

Fig. 4 Distributions of orientations of grains in grain cluster L described by inverse pole figure (IPF) color code

图 5 各个样品的平均晶粒团簇尺寸 D 和平均晶粒尺寸 d

Fig. 5 Mean sizes of grain-clusters, D, and mean sizes of grains in the clusters, d, for the six specimens

从前面的实验结果可以看出,样品 A 和 D 的低 ΣCSL 晶界比例明显高于样品 B,不仅形成了大尺寸的"互有 $\Sigma 3^n$ 取向关系晶粒的团簇"显微组织,而且晶粒团簇内的晶粒尺寸相对较小,内含许多 $\Sigma 3^n$ 类型晶界及相互连接的三叉晶角,得到了很好的晶界网络分布. 而样品 C 的低 ΣCSL 晶界比例虽然也比较高,但是晶粒尺寸较其他样品明显大. 这样的显微组织虽然有较好的抗腐蚀能力,但是由于晶粒尺寸偏大可能会导致其他性能(如室温强度和塑性)明显下降,所以希望通过 GBE 处理之后得到类似样品 A 和 D 的晶界网络分布,即得到大尺寸的晶粒团簇. 而团簇内的晶粒尺寸相对较小,可以用 $(D/d)^2$ 值近似表示晶粒团簇内部包含的平均晶粒数. 这一参数能够反映在显微组织中是否形成了以大尺寸"互有 $\Sigma 3^n$ 取向关系晶粒的团簇"显微组织为特征的晶界网络分布.

比较样品 A 和 D 在 GBE 处理前的原始晶粒尺寸及其最佳的 GBE 处理工艺,可以得出,当增大原始晶粒尺寸后,需要在最终再结晶退火前适当增大冷加工变形量,这样才能得到优化的晶界网络分布. 因此,通过 GBE 处理控制材料的晶界网络分布时,需要综合考虑原始晶粒尺寸和再结晶退火前的冷加工变形量,可以通过"晶粒平均应变量"这一参数来表达. 拉伸变形时,一般用延伸率表示应变量,但延伸率是一个宏观概念,而晶粒平均应变量中的晶粒尺寸是微观概念. 因此,应当寻找一个微观量来表示应变量. 变形后显微组织的最显著特征是形成了位错界面(dislocation boundary),在较小形变量条件下,位错界面主要是小角晶界[31]. 因此,小角晶界密度及其分布能够反映应变程度及其均匀性. 虽然 EBSD 分析结果无法给出单个位错的分布,但能够统计出由位错构成的小角晶界的长度及其分布,进而计算出小角晶界密度(单位面积晶界长度),这一数值可以从微观角度反映应变量的大小. 本文中计算晶粒平均应变量时,用小角晶界密度除以晶粒尺寸,得到晶粒平均应变量. 样品 A，B，C 和 D 退火前的变形状态样品分别记为 A′，B′，C′ 和 D′，它们显微组织中的应变量（小角晶界密度）以及"晶粒平均应变量"参数均给出在表 3 中. 本实验定义取向差小于 15° 的界面为小角晶界.

表 3　样品 A′，B′，C′ 和 D′ 的小角晶界密度以及晶粒平均应变量

Table 3　Densities of low angle grain boundary and the mean strains of grain (ratio of strain to original grain size) of specimens A′，B′，C′ and D′ corresponding to A，B，C and D before annealing

Specimen	Density of low angle grain boundary，μm^{-1}	Mean strain of grain，$10^{-3}\mu m^{-2}$
A′	0.107	9.342
B′	0.354	31.074
C′	0.107	5.176
D′	0.221	10.713

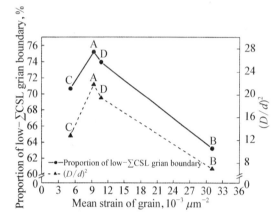

图 6　低 ΣCSL 晶界比例和 $(D/d)^2$ 值与晶粒平均应变量的关系曲线

Fig. 6　Curves of proportion of low-ΣCSL boundary and the value of $(D/d)^2$ *vs* mean strain of grain

样品 A，B，C 和 D 的晶粒平均应变量与低 ΣCSL 晶界比例之间的关系如图 6 中的实线所示，虚线是晶粒平均应变量与 $(D/d)^2$ 值之间的关系. 可以看出，在最终退火工艺相同时，通过 GBE 处理若要获得高比例的低 ΣCSL 晶界和高的 $(D/d)^2$ 值，需要合适的晶粒平均应变量. 若晶粒平均应变量过大（样品 B），再结晶形核密度就高，难以形成大尺寸晶粒团簇的显微组织[23,24]，从而 $(D/d)^2$ 值和低 ΣCSL 晶界比例都比较低（图 3b）；若晶粒平均应变量过小（样品 C），再结晶时形核密度低，虽然形成的晶粒团簇尺寸较大，但由于多重孪晶过程中形成的孪晶尺寸大，使得 $(D/d)^2$ 值较低，这也是不希望得到的显微组织（图 3c）. 所以，通过 GBE 处理要获得高比例的低 ΣCSL 晶界，控制晶界网络分布，得到高的 $(D/d)^2$ 值，在再结晶退火前需要合适的晶粒平均应变量. 在进行 GBE 处理时，除了需要考虑原始晶粒尺寸大小外，其他一些原始显微组织的情况也需要考虑，比如是否存在织构、第二相碳化物的分布状态等，这些原始显微组织对 GBE 处理之后晶界网络分布的影响正在进一步研究之中.

3　结论

原始晶粒尺寸对 690 合金经 GBE 处理后的晶界特征分布及晶界网络有明显的影响. 在最终退火工艺相同时，退火前最佳的冷变形量需要随原始晶粒尺寸的不同而变化，这样才能显著提高低 ΣCSL 晶界比例并形成最佳的晶界网络分布. 可以采用参数"晶粒平均应变量"来表达原始晶粒尺寸和冷加工变形量对 GBE 处理结果的综合作用，若要获得高比例的低 ΣCSL 晶界和高 $(D/d)^2$ 值的晶界网络分布，需要合适的"晶粒平均应变量".

参 考 文 献

[1] Dicrcks D R，Shack W J，Muscar J. *Nucl Eng Des*，1999；194；19.

[2] Watanable T. *Res Mech*, 1984; 11: 47.

[3] Randle V. *The Role of the Coincidence Site Lattice in Grain Boundary Engineering*. Cambridge: Cambridge University Press, 1996: 91.

[4] Lin P, Palumbo G. Erb U, Aust K T. *Scr Metall Mater*, 1995; 33: 1387.

[5] Lehockey E M, Limoges D, Palumbo G, Sklarchuk J, Tomantschger K, Vincze A. *J Power Sources*, 1999; 78: 79.

[6] Randle V. *Acta Mater*, 2004; 52: 4067.

[7] Shimada M, Kokawa H, Wang Z J, Sato Y S, Karibe I. *Acta Mater*, 2002; 50: 2331.

[8] Kumar M, Schwartz A J, King W E. *Acta Mater*, 2002; 50: 2599.

[9] Engelberg D L, Humphreys F J, Marrow T J. *J Microsc*, 2008; 230: 435.

[10] Saylor D M, El-Dasher B S, Adams B L, Rohrer D S. *Metall Mater Trans*, 2004; 35A: 1981.

[11] Gertsman V Y, Bruemmer S M. *Acta Mater*, 2001; 49: 1589.

[12] Marrow T J, Babout L, Jivkov A P, Wood P, Engelberg D, Stevens N, Withers P J, Newman R C. *J Nucl Mater*, 2006; 352: 62.

[13] Xia S, Zhou B X, Chen W J, Wang W G. *Scr Mater*, 2006; 54: 2019.

[14] Wang W G, Zhou B X, Rohrer G S, Guo H, Cai Z X. *Mater Sci Eng*, 2010; A527: 3695.

[15] King W E, Schwartz A J. *Scr Mater*, 1998; 38: 449.

[16] Xia S, Zhou B X, Chen W J, Wang W G. *Acta Metall Sin*, 2006; 42: 129.

（夏爽,周邦新,陈文觉,王卫国. 金属学报,2006; 42: 129）

[17] Zhang Y B, Godfrey A, Liu W, Liu Q. *Acta Metall Sin*, 2009; 45: 1159.

（张玉彬,Godfrey A,刘伟,刘庆. 金属学报,2009; 45: 1159）

[18] Cai Z X, Wang W G, Fang X Y, Guo H. *Acta Metall Sin*, 2010; 46: 769.

（蔡正旭,王卫国,方晓英,郭红. 金属学报,2010; 46: 769）

[19] Xia S, Zhou B X, Chen W J, Yao M Y, Li Q, Liu W Q, Wang J A, Ni J S, Chu Y L, Luo X, Li H. *Chin Pat*, 200710038731. 5, 2007.

（夏爽,周邦新,陈文觉,姚美意,李强,刘文庆,王均安,倪建森,褚于良,罗鑫,李慧. 中国专利,200710038731. 5, 2007）

[20] Xia S, Li H, Liu T G, Zhou B X. *J Nucl Mater*, doi: 10. 1016/j. jnucmat. 2011. 06. 017.

[21] Xia S, Li H, Hu C L, Liu T G, Zhou B X, Chen W J. *Fontevraud 7: Int Symp Contribution of Materials Investigations to Improve the Safety and Performance of LWRs*, Avignon, France: 2010: A025 T06 (CD-ROM).

[22] Palumbo G, Aust K T. *Acta Metall*, 1990; 38: 2343.

[23] Xia S, Zhou B X, Chen W J. *Metall Mater Trans*, 2009; 40: 3016.

[24] Xia S, Zhou B X, Chen W J. *J Mater Sci*, 2008; 43: 2990.

[25] Xia S, Zhou B X, Chen W J, Luo X, Li H. *Mater Sci Forum*, 2010; 638 - 642: 2870.

[26] Hu C L, Xia S, Li H, Liu T G, Zhou B X, Chen W J, Wang N. *Corros Sci*, 2011; 53: 1880.

[27] Humphreys F J, Bate P S, Hurley P J. *J Microsc*, 2001; 201: 50.

[28] Yang P. *Electron Backscattering Diffraction Technique and its Application*. Beijing: Metallurgical Industry Press, 2007: 73.

（杨平. 电子背散射衍射技术及其应用. 北京: 冶金工业出版社,2007: 73）

[29] Berger A, Wilbrandt P J, Ernst F, Klement U, Haasen P. *Prog Mater Sci*, 1988; 32: 1.

[30] Gertsman V Y, Henager C H. *Interface Sci*, 2003; 11: 403.

[31] Doherty R D, Hughes D A, Humphreys F J, Jona J J. Juul J D, Kassner M E, King W E, McNelley T R, McQueen H J, Rollett A D. *Mater Sci Eng*, 1997; A238: 219.

Effect of Original Grain Size on the Boundary Network in
Alloy 690 Treated by Grain Boundary Engineering

Abstract: Effect of original grain size on the grain boundary character distribution (GBCD) in alloy 690 after treatment by grain boundary engineering (GBE) was studied using electron backscatter diffraction (EBSD) and orientation image microscopy (OIM). The proportion of low $-\Sigma$CSL grain boundary and grain boundary network after GBE are obviously influenced by the original grain size. Optimized GBE grain boundary network after the same annealing process can be obtained by altering the cold work degree based on the original grain size. Thus the microstructure after GBE is influenced by the combined effect of original grain size and deformation amount before annealing. Mean strain of grain is used to express the combined effect. A proper value of mean strain of grain is necessary to achieve the desired GBE grain boundary network.

Zr-4合金中氢化物析出长大的透射电镜原位研究[*]

摘 要: 用透射电子显微镜拉伸试样台原位研究了应力、电子束辐照以及第二相对 Zr-4 合金中氢化物析出和长大的影响. 结果表明,在拉应力作用下,裂纹易于沿氢化物扩展,并在裂尖垂直于拉应力方向析出新的氢化物. 氢化物在拉应力诱发下的析出、开裂、再析出……过程,导致了氢致延迟开裂. 在较强的会聚电子束辐照下,Zr-4 合金中的氢化物会分解,新的氢化物会围绕着附近的 Zr(Fe,Cr)$_2$ 第二相粒子析出,新析出的氢化物为面心立方结构的 δ 相.

锆合金因其热中子吸收截面小,并在高温高压水中具有较好的耐腐蚀性能及较高的高温强度,是水冷核反应堆中用作核燃料包壳的一种重要的结构材料. 在反应堆运行中,锆合金包壳受到高温高压水腐蚀时必然会吸收锆水反应生成的一部分氢. 由于氢在 α-Zr 中的固溶度低,过剩的氢将以氢化锆的形态从合金基体中析出. 氢化锆的析出一方面会因自身体积膨胀使锆基体产生畸变而造成应力集中,另一方面氢化锆是一种脆性相,会明显降低锆合金的塑性. 这些都会造成锆合金燃料包壳破损,威胁反应堆的安全,所以锆合金的腐蚀吸氢问题一直倍受关注.

应力、温度等会影响氢在锆合金中的分布,氢会沿应力梯度朝拉应力大的方向或沿温度梯度朝温度低的方向扩散,使得局部氢浓度增高,导致氢化物的析出[1-3]. 姚美意等[4]的研究表明,Zr-4 合金不同热处理后的吸氢能力与 Zr(Fe,Cr)$_2$ 第二相的尺寸和数量密切相关,基于 Zr(Fe,Cr)$_2$ 本身是一种强烈吸氢物质的事实[5],提出了镶嵌在金属/氧化膜界面处未被氧化的 Zr(Fe,Cr)$_2$ 第二相可作为吸氢优先通道的观点. Lelièvre 等[6]对 Zr-4 合金在高温高压水中腐蚀吸氢的研究中,也提出氧化膜中延迟氧化的第二相为氢的扩散提供了短路通道的观点. 上述第二相影响吸氢的观点都是从一些实验数据得出的合理推论,并没有观察到在第二相周围优先形成氢化物的直接实验证据.

在透射电镜(TEM)中,对样品进行原位拉伸或电子束辐照加热,可以直接观察到材料在应力梯度或温度梯度作用下显微组织的变化. 因此,本实验用 TEM 原位观察的方法分别研究了拉应力、电子束辐照加热以及 Zr(Fe,Cr)$_2$ 第二相对 Zr-4 合金中氢化物析出和长大的影响.

1 实验

实验材料选用西北有色金属研究院提供的退火态 Zr-4(Zr-1.5Sn-0.2Fe-0.1Cr,质量分数,%)合金板,厚度 0.6 mm. 先将 Zr-4 合金板冷轧至 0.3 mm,再在 600 ℃的真空炉中退火 2 h,然后经 10% HF+45% HNO$_3$+45% H$_2$O(体积比)混合酸酸洗及去离子水清

---

* 本文合作者:彭剑超、李强、刘仁多、姚美意. 原发表于《稀有金属材料与工程》,2011,40(8): 1377-1381.

洗后置于高压釜中,在 500 ℃,10.3 MPa 的过热水蒸气中腐蚀 3 h,腐蚀后试样的含氢量用美国 LECO 公司的 RH - 600 型定氢分析仪进行测定.

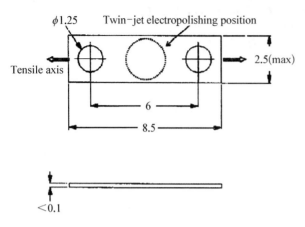

图 1 透射电镜原位拉伸试样简图

Fig. 1 Schematic drawing of the tensile sample for in situ TEM observation

为了获得细小并分布均匀的氢化物,便于 TEM 观察研究,腐蚀后的样品在空气中经 400 ℃,0.5 h 均匀化处理并水淬. 机械磨抛去除氧化层后,用酸洗液化学腐蚀减薄至 70 μm 厚,分别制成 Φ3 mm 圆片和 8.5 mm×2.5 mm×0.1 mm 的标准拉伸样品(图 1). TEM 最终观测的薄样品用电解双喷方法制备,电解液为 20% HClO₄＋80% C₂H₄O₂(体积比),电压 40 V,温度为 20 ℃. 分别用 GATAN 645 单倾拉伸台和 JEOL 双倾台在配有 INCA EDS(OXFORD) 和 CCD(GATAN) 的 JEOL - 2010F 透射电镜中进行原位拉伸及电子束原位辐照观察研究.

原位拉伸试验时,在电解双喷穿孔样品的薄区周围总会产生一些裂纹,在裂纹尖端处会产生应力集中,选择那些与拉伸方向大致垂直的裂纹来观察应力诱发氢化物的析出和长大. 在实验过程中,控制试样的拉伸速度在 1 μm/s 左右,并记录拉伸量和对应的时间. 当裂纹尖端出现大量位错运动时,即停止拉伸并拍照记录. 随后几分钟内可看到裂纹的扩展,待应力松弛,裂纹不再扩展时继续拉伸. 为便于观察比较,整个过程都选取氢化物的同一个衍射斑点获得氢化物中心暗场像.

会聚电子束原位辐照观察时,选择最大的聚光镜光阑并将电子束会聚成约 200 nm 大小的束斑照在氢化物的局部区域,每次 10 min,随后用明暗场像观察氢化物受辐照后的变化情况.

用同样的电子束辐照方法,研究第二相对氢化物析出长大的影响. 先通过 EDS 确定第二相的成分,然后用电子束辐照致使第二相附近的氢化物发生分解,观察在第二相周围新的氢化物析出情况,并拍摄基体、氢化物以及第二相的高分辨像,用 Gatan DigitalMicrograh 软件做快速傅里叶转化(fast fourier transform,FFT)分析.

2 结果与讨论

2.1 拉应力作用下氢化物的开裂、析出和长大

经定氢分析仪分析,腐蚀后试样中的氢含量为 240 μg/g,高于室温时 Zr - 4 合金中氢的最大固溶量(<1 μg/g).

图 2 为 Zr - 4 合金中氢化物在拉应力作用下开裂、析出与长大的过程. 图 2a、2b 是拉伸前裂纹尖端已存在的氢化物的明、暗场像,这应是试样在装入样品台的过程中造成的应力集中产生的,文献[3]报道这种氢化物为体心四方的 γ 相. 从明场像中形成环状的消光条纹可看出,氢化物的存在使得其周围锆基体中应力高度集中.

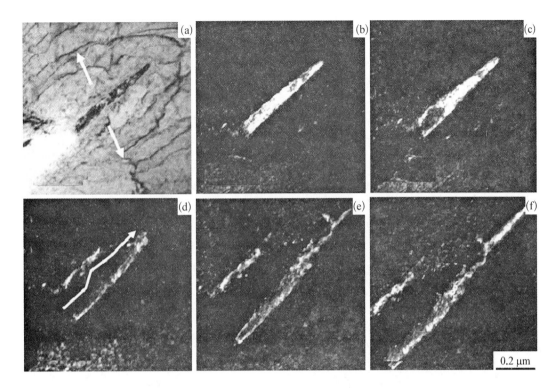

图 2 拉应力作用下氢化物开裂、析出和长大的 TEM 原位观察结果

Fig. 2 In situ TEM observation of hydride precipitating, growing and cracking induced by tensile stress: (a) bright field image and (b) dark field image of a hydride at the cracktip before tensile test, the arrows indicate the direction of the applied stress; (c) the hydride starts cracking after the whole sample tension for 21 μm; (d) the hydride further cracking and (e) new hydride precipitating after tension for 45 μm; (f) hydride growing to 300 nm after a more tension for 0.5 h

拉伸样品时,在裂纹尖端氢化物的周围可看到位错的运动. 当试样整体拉伸位移至 21 μm 时,裂纹开始沿氢化物纵向扩展(图 2c).停止拉伸,裂纹继续沿"Z"字形向前扩展,并贯穿整个氢化物,如图 2d 所示. 当裂纹扩展逐渐靠近锆基体时,扩展速度明显减慢,并逐渐停止.继续拉伸至位移 45 μm,等待 520 s 后,在原开裂的氢化物尖端又析出了新的氢化物(图 2e).新析出的氢化物在拉应力诱发下不断长大,图 2f 是拉伸 30 min 后得到的暗场像,此时氢化物已经长大到长约 300 nm、宽约 40 nm 的针状物. 这种裂纹沿氢化物扩展,并在裂纹尖端应力诱发下再析出新的氢化物,随后裂纹再沿新的氢化物扩展的过程即为氢致延迟开裂.

与文献[1]、[3]报道的相比,本工作观察到应力诱发析出氢化物的孕育期更短(为 6～7 min),长大更加迅速.这可能是本工作所用的拉伸试样含氢量为 240 μg/g,比文献[1]、[3]试样中含氢量高的缘故. 文献[1]、[3]试样中氢含量分别为 100、50 μg/g.

在整个拉伸过程中,开裂后的氢化物一直没有消失,这与文献[1]、[3]中的报道是一致的.说明应力释放后,氢化物的化学势仍然低于基体中固溶氢的化学势,新的氢化物的析出长大所需要的氢来自基体而不是通过开裂的氢化物的分解产生,氢化物相对来说是很稳定的.

2.2 会聚电子束辐照对氢化物的影响

在透射电镜中,利用会聚的强电子束可以对样品进行辐照加热处理,这样可以促进样品中元素的扩散,焊接纳米线或纳米颗粒,还可以使样品的局部区域熔化甚至蒸发[7,8]. 在辐照过程中,由于样品具有良好的导热性,电子束对整个样品的加热作用并不明显[7],而对于会聚的电子束(200~10 nm)所辐照的局部区域,虽然难以直接测得其温度的高低,但根据文献[7,8]的报道,可以估计随束斑直径(对应于电子束电流密度)大小以及辐照时间长短的不同,温度会升高几十到上千度.

用直径为 100~200 nm 的会聚电子束辐照氢化物的方法,观察到了它们的分解现象,图 3 为氢化物经几次辐照后得到的明暗场像. 图 3a 为未辐照氢化物的明场像,图中的圆圈对应每次电子束斑辐照的位置. 经 5 次每次 10 min 的辐照后,与辐照对应处的氢化物发生了分解,图 3b、3c 是辐照后原来氢化物处的明暗场像. 做辐照区域的选区电子衍射时,斑点模糊不清,难以确定该区域的晶体点阵类型,这应该是辐照导致了晶格损伤的缘故. 但通过拍摄氢化物周围基体的中心暗场像对照分析,可以确定氢化物分解后生成了与基体相同的 α-Zr.

停止辐照后,并没有发现氢化物在原来位置重新析出,而在离原氢化物 0.5~1 μm 处出现了如图 3c 暗场像中所看到的许多 10~20 nm 大小的颗粒. 这些颗粒可能是重新析出的氢化物. 同时因为温度较低,氢原子的扩散能力有限,所以氢化物析出时都倾向于形成细小的颗粒.

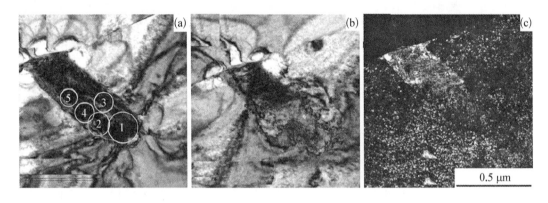

图 3　会聚电子束 5 次辐照后,氢化物部分分解的 TEM 照片

Fig. 3　TEM images of the hydride decomposing, after irradiated for 5 times by converged electron beam: (a) original hydride image; (b) bright field image and (c) dark field image of the partly dissolved hydride

2.3　Zr-4 合金中的第二相对氢化物析出长大的影响

从 2.2 节叙述的结果中已经知道,经电子束辐照后的氢化物发生了分解,并在周围形成了细小的氢化物颗粒,由于驱动力太小无法长大. 如果氢化物旁边存在一种与氢有更大亲和力的物质,情况则有所不同.

图 4a、4b 是 Zr-4 晶粒内的一条氢化物的明暗场像. 在氢化物的左端 A 和中间 B 各有一个 $Zr(Fe,Cr)_2$ 颗粒. 能谱分析表明,两颗粒在成分上存在差异,A 颗粒中 Fe/Cr 比为 1.2,B 颗粒中 Fe/Cr 比为 1.5,其中 B 颗粒中还含有 7% 的氧.

图 5 为用会聚电子束辐照氢化物中部 1 h 后的暗场像. 从图 5a 暗场像中(选取的衍射斑点与图 4b 相同)可以看到,氢化物中间部分已经分解,且在 A 颗粒周围生成了一圈宽约 50 nm 的新相(图 5b 为图 5a 中 A 颗粒区域的放大像),而在 B 颗粒附近则没有类似的发现.

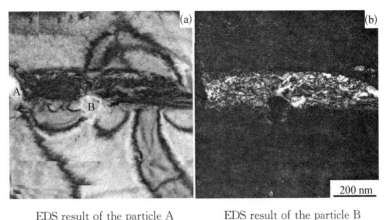

EDS result of the particle A			EDS result of the particle B		
Element	ω/%	at%	Element	ω/%	at%
Cr	20.82	27.07	O	7.38	25.64
Fe	27.06	32.76	Cr	15.05	16.09
Cu	4.70	5.00	Fe	25.02	24.90
Zr	47.43	35.16	Cu	5.06	4.43
			Zr	47.49	28.94

图 4 氢化物的明暗场像以及(a)图中 A、B 两颗粒对应的 EDS 结果

Fig. 4 Bright-field image (a) and dark-field image (b) of hydride and EDS results of the A and B particles in Fig. 4a

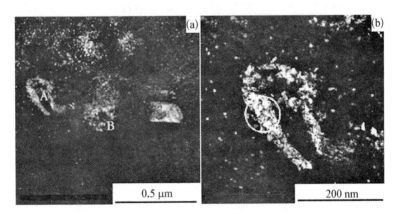

图 5 会聚电子束辐照 1 h 后，Zr（Fe，Cr)₂ 周围氢化物的暗场像

Fig. 5 Dark-field image of hydride around Zr(Fe, Cr)$_2$ particle after irradiating for 1 h by the converged electron beam (a) and the magnification of the particle A in Fig. 5a (b)

　　对图 5a 中的新相进行能谱分析表明，除了 Zr、Cu 两种元素以外，没有别的合金元素．Zr-4合金中没有加入 Cu 元素，Cu 峰来自制造样品台的黄铜材料，同时能谱不能判断 H 元素的存在与否，所以应为 Zr 或者 ZrH$_x$．

　　图 6 是选取图 5b 中圆环所示区域的高分辨像．其中两条点划线之间对应的区域即新析出的相，对其进行快速傅里叶转换分析证明，新相为 δ - ZrH$_2$（fcc），$d_{(111)} = 0.28$ nm，$d_{(200)} = 0.24$ nm．

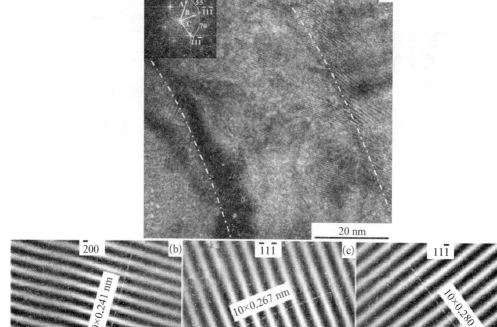

图 6　新析出氢化物的 HRTEM 像,以及基于 FFT 对应的衍射花样与各斑点对应的反 FFT 像

Fig. 6　HRTEM lattice image of the hydride (a), and FFT (inset in Fig. 6a) and interplanar distances based on inverse FFT (b, c, d)

在图 5a 中 B 颗粒周围之所以没有发现类似的氢化物,可能是因为 B 颗粒靠近电子束辐照的位置,温度较高,导致氢原子往低温处扩散的缘故. 此外在图 5a 的上方也出现了如同图 3c 中聚集的颗粒,结合傅里叶转换的高分辨图像分析,这些颗粒与 $Zr(Fe,Cr)_2$ 周围的氢化物具有相同的晶体结构.

上述结果表明,氢化锆因电子束辐照分解后,又会优先围绕着 $Zr(Fe,Cr)_2$ 第二相析出、长大.

3　结论

(1) Zr－4 合金中的氢化物会在拉应力诱发下析出、长大、开裂. 析出过程有孕育期,为 6～7 min,30 min 内氢化物纵向生长了 200 nm.

(2) 在会聚电子束辐照下,Zr－4 合金中的氢化物发生分解,并在原先氢化物周围重新析出 10～20 nm 大小的颗粒状氢化物.

(3) 用会聚电子束辐照氢化物时,分解产生的氢会在 $Zr(Fe,Cr)_2$ 第二相周围优先析出形成新的氢化物,新析出的氢化物为面心立方结构.

参 考 文 献

［1］Cann C D，Sexton E F. *Acta Metall*［J］，1980，28：1215.

［2］Eadie R L，Metzger D R，Leger M. *Scripta Metallurgica et Materialia*［J］，1993，29：335.

［3］Zhou Bangxin(周邦新)，Zheng Sikui(郑斯奎)，Wang Shunxin(汪顺新). *Acta Meta Sinica*(金属学报)［J］，1989，25 (3)：900.

［4］Yao Meiyi(姚美意)，Zhou Bangxin(周邦新)，Li Qiang(李强) *et al. Rare Metal Materials and Engineering*(稀有金属材料与工程)［J］，2007，36(11)：1915.

［5］Shaltiel D，Jacob I，Davidov D. *Less-Commom Metals*［J］，1977，53：117.

［6］Lelièvre G，Tessier C *et al. Alloys and Compounds*［J］，1998，268：308.

［7］Yokota T，Murayama M，Howe J M. *Physical Review Letters*［J］，2003，91(26)：265,504.

［8］Xu Shengyong，Tian Mingliang，Wang Jinguo *et al. Small*［J］，2005，12：1221.

In Situ Investigation of Hydride Precipitation and Growth in Zircaloy‐4 by Transmission Electron Microscopy

Abstract：The effects of stress，electron beam irradiation and second phase particles on zirconium hydride precipitation and growth in Zircaloy‐4 were investigated using in-situ transmission electron microscope observation. Results show that with the tensile stress the cracks are likely to propagate along hydrides and new hydrides are formed at the crack tip along the vertical direction of the applied stress. A process of precipitation，cracking，re-precipitation and so on，induced by tensile stress，causes delayed hydride cracking (DHC). Under the high irradiation of converged electron beam，the hydrides decompose in the Zircaloy‐4，and new hydrides prefer to precipitate around the unoxidized $Zr(Fe, Cr)_2$ particles，and the re-precipitated hydrides are fcc-structure δ phase.

β 相水淬对 Zr‑4 合金在 LiOH 水溶液中耐腐蚀性能的影响*

摘　要：采用 β 相水淬处理后再经 480—600 ℃ 保温 2—200 h 的工艺,研究了 β 相水淬对 Zr‑4 合金在 360 ℃/18.6 MPa 和 0.01 mol/L LiOH 水溶液中腐蚀行为的影响;用 TEM 和 HRSEM 分别观察了合金的显微组织和氧化膜断口形貌. 结果表明：β 相水淬时控制合适的冷却速率,避免残留 β‑Zr 的生成,提高固溶在 α‑Zr 基体中 Fe 和 Cr 含量,Zr‑4 合金也可获得与 Zr‑Sn‑Nb 合金一样优良的耐腐蚀性能. 但当 β 相水淬速率过快时,由于残留 β‑Zr 的存在使 Zr‑4 合金的耐腐蚀性能降低;而随后进行 480—600 ℃ 退火处理,随着退火温度的升高和退火时间的延长,β 相水淬快冷样品的耐腐蚀性能得到明显改善,这主要与残留 β‑Zr 的分解有关.

目前,大多数压水堆采用 Zr‑4 合金作为核燃料的包壳材料,通过优化热加工制度,可以进一步提高 Zr‑4 合金的耐腐蚀性能. 大量的研究[1‑4]表明,再结晶退火态的 Zr‑4 合金在 LiOH 水溶液中腐蚀后,腐蚀转折时间缩短,转折后的腐蚀速率显著增加,而在调整 Zr‑4 合金成分基础上添加 Nb 发展起来的 ZIRLO (Zr‑1.0Sn‑1.0Nb‑0.1Fe,质量分数,%,下同),E635(Zr‑1.2Sn‑1.0Nb‑0.4Fe),以及 N18(Zr‑1.0Sn‑0.35Nb‑0.3Fe‑0.1Cr)和 N36 (Zr‑1.0Sn‑1.0Nb‑0.3Fe)等 Zr‑Sn‑Nb 系合金[4‑8],在 LiOH 水溶液中却表现出优良的耐腐蚀性能,Nb 的添加明显抑制了 LiOH 对 Zr‑4 合金耐腐蚀性能的有害作用. 然而,本课题组前期在研究热处理对 Zr‑4 合金在 360 ℃ LiOH 水溶液中耐腐蚀性能影响时却发现：β 相水淬处理的 Zr‑4 样品在 360 ℃ LiOH 水溶液中也表现出非常优良的耐腐蚀性能,明显优于常规处理的再结晶退火态 Zr‑4 样品,在长达 529 d 的腐蚀实验过程中,腐蚀增重一直与 Zr‑Sn‑Nb 系的 ZIRLO 和 N18 合金相当[9]. 这一规律尚未见类似报道,也与类似热处理的 Zr‑4 样品在 400 ℃ 过热蒸汽中的腐蚀规律不同[10,11]. 合金元素 Fe 和 Cr 添加到 Zr 中后,其中少部分固溶到 α‑Zr 中,大部分以 Zr(Fe, Cr)$_2$ 第二相析出,改变热处理制度对这两者都会产生影响. 因此,经不同热处理的 Zr‑4 合金在不同水化学条件下表现出的腐蚀行为差别,其原因可能与析出的 Zr(Fe, Cr)$_2$ 第二相有关[12‑14],也可能与过饱和固溶在 α‑Zr 中的合金元素含量有关[10,15]. 进一步研究合金元素的过饱和固溶程度和第二相析出程度对 Zr‑4 合金耐腐蚀性能的影响,有助于加深对热处理影响 Zr‑4 合金腐蚀行为机理的认识,也可为了解合金元素如何影响锆合金耐腐蚀性能提供有价值的数据.

1　实验方法

实验所用 0.6 mm 厚的 Zr‑4(Zr‑1.5Sn‑0.2Fe‑0.1Cr)合金板由西北有色金属研究院提供,按常规工艺制备并经过 600 ℃ 再结晶退火. 为了研究 Fe 和 Cr 合金元素在 α‑Zr 中

---

* 本文合作者：沈月锋、姚美意、张欣、李强、赵文金. 原发表于《金属学报》,2011,47(7)：899‑904.

的过饱和固溶程度和第二相析出程度变化,把 Zr‐4 板切成 8 mm 宽的小条,将它们分别真空(约 3×10^{-3} Pa)密封在数支直径 12 mm 的石英管中,再在管式电炉中加热到 1 020 ℃保温 20 min,然后采用两种方法淬入水中:一种是淬入水中后快速敲碎石英管,根据样品从亮红转入暗红(约 600 ℃)的时间,估算最大冷却速率为 500 ℃/s(样品定义为 500 ℃/s‐β‐WQ);另一种是淬入水中后不敲碎石英管,这时冷却速率约为 100 ℃/s(样品定义为 100 ℃/s‐β‐WQ).为减小β相水淬时样品冷却速率的差别,每支石英管中只装 2 个样品.采用不同的冷却方法是为了研究β相淬火时的冷却速率对 Zr‐4 合金耐腐蚀性能的影响.将 500 ℃/s‐β‐WQ 试样切成 20 mm×8 mm 的样品,放入真空石英管管式电阻炉中,分别在 480,540 和 600 ℃保温 2,50 和 200 h,然后将管式电阻炉推离石英管,在石英管外壁淋水冷却,称为空冷(AC).图 1 给出了 Zr‐4 合金的β相淬火和退火工艺.

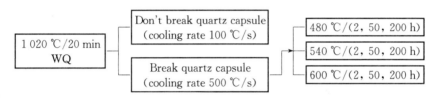

图 1 Zr‐4 合金样品β相淬火及后续的退火工艺

Fig. 1 Procedure of β‐quenching and subsequent annealing for Zr‐4 specimens

将上述热处理样品放入高压釜中进行腐蚀实验,腐蚀条件为 360 ℃/18.6 MPa 的 0.01 mol/L LiOH 水溶液.腐蚀实验前,样品按标准方法用 10% HF+45% HNO_3+45% H_2O(体积分数)混合酸酸洗和去离子水清洗,腐蚀增重是 3—5 个试样增重的平均值.

用 JEM‐200CX 透射电镜(TEM)观察合金样品腐蚀前的显微组织及第二相的大小和分布,薄样品用双喷电解抛光制备,电解液为 10%过氯酸酒精溶液.用 JEM‐2010F 高分辨透射显微镜(HRTEM)及其配备的能谱仪(EDS)分析第二相的晶体结构和化学成分.用 JSM‐6700F 型高分辨扫描电镜(HRSEM)观察腐蚀样品氧化膜的断口形貌,为了提高图像质量,样品表面蒸镀了一层金属 Ir.

2 实验结果与讨论

2.1 β相淬火及退火温度和时间对 Zr‐4 合金耐腐蚀性能的影响

图 2a 是β相淬火速率不同的 Zr‐4 样品在 360 ℃/18.6 MPa 的 0.01 mol/L LiOH 水溶液中腐蚀 340 d 的增重曲线,同时给出了相同条件下 N18 合金的腐蚀增重曲线[9].可以看出,100 ℃/s‐β‐WQ 样品表现出了与 Zr‐Sn‐Nb 系的 N18 合金一样优良的耐腐蚀性能,这也进一步证实了以前得到的β相水淬 Zr‐4 样品表现出优良耐腐蚀性能的实验结果[9].100 ℃/s‐β‐WQ 样品的耐腐蚀性能明显优于 500 ℃/s‐β‐WQ 的样品,前者腐蚀 280 d 的平均增重只有 135 mg/dm²,而后者的腐蚀增重已高达 650 mg/dm²,是前者的 5 倍.这说明β相水淬时冷却速率的差别对 Zr‐4 合金在 360 ℃/LiOH 水溶液中的耐腐蚀性能产生很大影响.

图 2b‐d 是 500 ℃/s‐β‐WQ 样品经 480—600 ℃退火后在 360 ℃/18.6 MPa 的 0.01 mol/L LiOH 水溶液中腐蚀 340 d 的增重曲线.与 500 ℃/s‐β‐WQ 样品相比,随退火

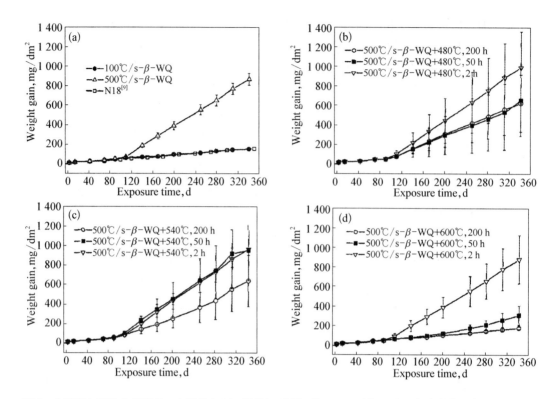

图 2 β相淬火及退火处理 Zr-4 样品在 360 ℃/18.6 MPa 的 0.01 mol/L LiOH 水溶液中的腐蚀增重曲线

Fig. 2 Curves of weight gain *vs* exposure time of Zr-4 specimens during corroding in 0.01 mol/L LiOH aqueous solution at 360 ℃/18.6 MPa （a) curves of two specimens with cooling rate 100 and 500 ℃/s, and curve of N18 specimen also given as comparison （b-d) curves of the 500 ℃/s quenched specimens annealed at different temperatures and times

温度的升高和退火时间的延长耐腐蚀性能得到明显改善,且平行样品腐蚀增重数据的分散性也减小,其中经 600 ℃退火 200 h 后的样品达到与 100 ℃/s-β-WQ 样品一样优良的耐腐蚀性能(图 2a 和 d).平行样品腐蚀增重数据存在很大分散性应该与这些样品退火前来自不同批次的 β 相水淬有关,因为每次水淬的速率很难保证完全相同.这一事实进一步说明 β 相水淬时冷却速率的差别会对耐腐蚀性能产生很大的影响.

2.2 合金的显微组织

图 3a 和 b 是 β 相淬火速率不同的 Zr-4 样品显微组织的 TEM 像.可以看出,与500 ℃/s-β-WQ 样品相比,100 ℃/s-β-WQ 样品中的 α-Zr 板条晶粒较宽,板条晶界处析出的棒状第二相(second phase particals, SPPs)数量较多且尺寸较大,位错密度较低.高倍下可以观察到 500 ℃/s-β-WQ 样品基体中有类似于周邦新等[11]报道的残留 β-Zr 组织,如图 4 所示.选区电子衍射(SAD)分析表明该残留相为 bcc 结构(图 4c),其点阵常数 $a = 0.358$ nm,接近于 PDF 卡片 34-0657 的 β-Zr ($a = 0.355$ nm);该残留相中的 Fe 和 Cr 含量分别是 Zr-4 合金中平均含量的 11 倍和 8 倍(图 4d,其中 Cu 来自 TEM 样品台),且 Sn 含量与合金中的相当,因此可以进一步确定该组织是 β-Zr.

图 3c—k 是 500 ℃/s-β-WQ 样品经 480—600 ℃退火后显微组织的 TEM 像.可以看出,不同样品中 α-Zr 板条的宽度有较大差别,这是由 β 相水淬时冷却速率的差别(每管试样

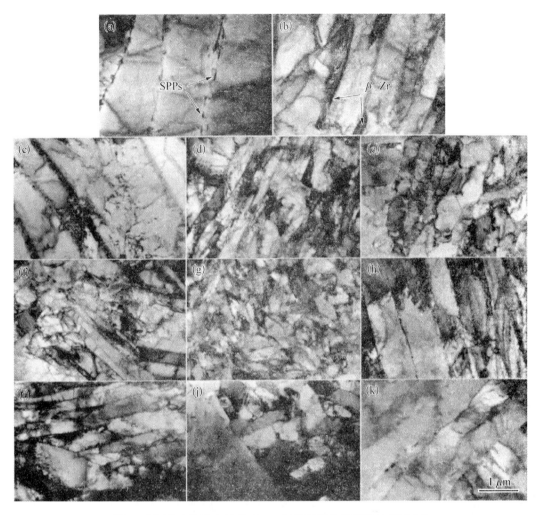

图3 β相淬火及退火处理 Zr-4 样品显微组织形貌的 TEM 像

Fig. 3 TEM micrographs of β-quenched and annealed Zr-4 specimens (a) 100 ℃/s-β-WQ (b) 500 ℃/s-β-WQ (c) 500 ℃/s-β-WQ+480 ℃, 2 h, AC (d) 500 ℃/s-β-WQ+480 ℃, 50 h, AC (e) 500 ℃/s-β-WQ+480 ℃, 200 h, AC (f) 500 ℃/s-β-WQ+540 ℃, 2 h, AC (g) 500 ℃/s-β-WQ+540 ℃, 50 h, AC (h) 500 ℃/s-β-WQ+540 ℃, 200 h, AC (i) 500 ℃/s-β-WQ+600 ℃, 2 h, AC (j) 500 ℃/s-β-WQ+600 ℃, 50 h, AC (k) 500 ℃/s-β-WQ+600 ℃, 200 h, AC

分别水淬)引起的,因为在这样低的温度范围内退火时,不会由于晶界迁移而产生板条宽度的明显变化.这与平行试样腐蚀增重数据存在较大分散性的结果是一致的.随退火温度的升高和退火时间的延长,位错密度减小,再结晶晶粒数量增多且尺寸增大,但即使经 600 ℃,200 h 退火处理的样品也没有发生完全再结晶(图 3k).600 ℃,200 h 退火样品中的第二相尺寸和数量以及位错密度与 100 ℃/s-β-WQ 样品相差不大.

2.3 分析讨论

β相水淬时的冷却速率一方面会影响 α-Zr 基体中过饱和固溶的 Fe 和 Cr 含量,另一方面会影响基体中的位错密度、板条晶粒的宽度及析出的第二相(图 3a 和 b).另外,在 β相淬

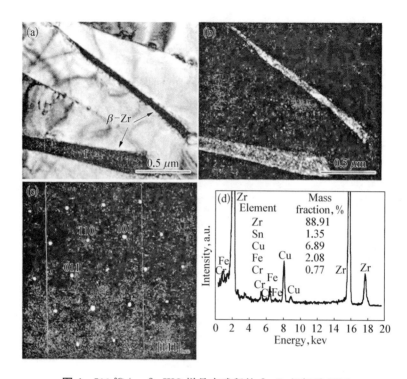

图4 500 ℃/s-β-WQ样品中残留的β-Zr组织及EDS

Fig. 4 TEM bright field image (a), dark field image (b), SAD pattern (c) of β-Zr in 500 ℃/s-β-WQ specimen and EDS of position 1 in Fig. 4a (d)

火的快冷过程中还会将少量的β-Zr保留下来(图4).β相水淬快冷后随着退火温度的升高和退火时间的延长,一方面β-Zr发生分解,α-Zr基体中过饱和固溶的Fe和Cr会以Zr(Fe,Cr)$_2$第二相析出,使基体中过饱和固溶的Fe和Cr含量降低;另一方面,基体中的位错密度也会相应降低(图3c—k).研究[16,17]表明,在含Nb的锆合金中,β-Zr的存在对耐腐蚀性能是非常有害的,β-Zr分解后耐腐蚀性能得到明显改善.周邦新等[11]也曾报道Zr-4合金中存在残留β-Zr时,Zr-4样品在400 ℃过热蒸汽中的耐腐蚀性能明显降低.Garzarolli等[18]也报道当从β相区以大于500 ℃/s的速率冷却时,Zr-4合金在350 ℃去离子水中的耐腐蚀性能降低,本研究中100 ℃/s-β-WQ样品的耐腐蚀性能明显优于500 ℃/s-β-WQ的样品(图2a),表现出与Zr-Sn-Nb系合金一样优良的耐腐蚀性能[9].当β相水淬快冷再经480—600 ℃,2—200 h退火处理后,样品的耐腐蚀性能得到不同程度的改善(图2b—d),这说明残留β-Zr的存在是500 ℃/s-β-WQ样品耐腐蚀差的主要原因.β相水淬快冷后随着退火温度的升高和退火时间的延长耐腐蚀性能得到明显改善,这主要与β-Zr的分解有关.在不存在残留β-Zr时,提高α-Zr基体中的Fe和Cr含量可明显改善Zr-4合金的耐腐蚀性能.

Zr在LiOH水溶液中的腐蚀过程是O^{2-}或OH$^-$通过氧化膜中的晶界或阴离子空位扩散到金属/氧化膜界面处,在金属/氧化膜界面处生长[19-21].由于Zr的P.B.(Pilling-Bedworth)为1.56,锆合金氧化时氧化膜内部会形成很大的压应力,使ZrO$_2$晶体中产生许多缺陷,稳定了一些亚稳相;空位、间隙原子等缺陷在温度、应力和时间的作用下,发生扩散、湮没和凝聚,空位被晶界吸收后形成纳米大小的孔隙簇,弱化了晶粒之间的结合力,孔隙簇进一步发展成为微裂纹,同时亚稳相转变为稳定的单斜 ZrO_2[22-24]. 锆合金氧化时,氧化膜显微组织的这一演化过程是不可避免的,任何影响这一演化过程的因素都会影响锆合金的耐腐蚀性能. 图 5 比较了 500 ℃/s-β-WQ+540 ℃,2 h 和 100 ℃/s-β-WQ 样品腐蚀 340 d 的氧化膜断口形貌. 其腐蚀增重分别为 485 和 156 mg/dm². 可以看出,500 ℃/s-β-WQ+540 ℃,2 h 样品氧化膜中已存在很多平行于金属/氧化膜界面的微裂纹,如图 5a 中箭头所示,并且氧化膜断口的起伏程度很大(图 5b);而 100 ℃/s-β-WQ 样品的氧化膜非常致密,几乎看不到微裂纹(图 5c),只在高倍下看到一些微孔隙,氧化膜断口也比较平整(图 5d),说明提高 α-Zr 基体中的 Fe 和 Cr 含量确实可以延缓 Zr-4 合金腐蚀时氧化膜显微组织的演化过程. 这可能与固溶在 α-Zr 基体中的 Fe 和 Cr 在转入到氧化膜中后可以减小因 Li^+ 或 OH^- 进入氧化膜引起 ZrO_2 表面自由能的降低程度有关[23,24],因为这些固溶的 Fe 和 Cr 比第二相中的 Fe 和 Cr 氧化后更有可能固溶在 ZrO_2 基体中. β-Zr 对锆合金耐腐蚀性能的有害作用可能主要与 β-Zr 和 α-Zr 的腐蚀电位不同而引起的电化学腐蚀有关.

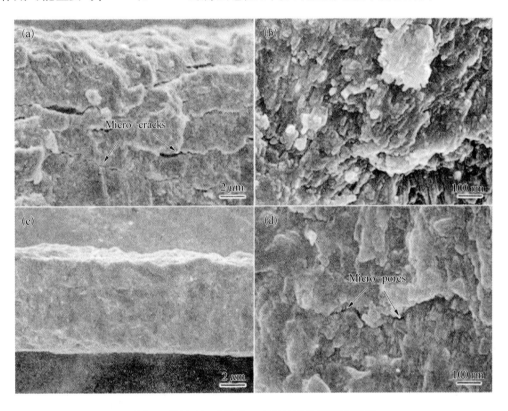

图 5　腐蚀 340 d 的 Zr-4 样品氧化膜断口形貌

Fig. 5　Low (a, c) and high (b, d) magnified fractographs of the oxide films on Zr-4 specimens treated by 500 ℃/s-β-WQ+540 ℃, 2 h (a, b) and 100 ℃/s-β-WQ (c, d) after corrosion in LiOH aqueous solution for 340 d (the fracture surface of the latter is denser and smoother than the former)

从以上的结果可以看出,β 相淬火时的冷却速率对 Zr-4 合金在 LiOH 水溶液中的耐腐蚀性能有非常显著的影响,只要控制合适的冷却速率,可以获得与 Zr-Sn-Nb 系合金一样优良的耐腐蚀性能. 这是一个非常复杂又重要的问题,要彻底了解其中的机理,还有待开展更深入的研究,特别是 Fe 和 Cr 在 ZrO_2 中的存在状态及其对氧化膜显微组织演化过程的影响,以及 β-Zr 影响耐腐蚀性能的作用机理.这不仅对 Zr-4 合金,对于了解整个锆合金的腐蚀机理也会有帮助.

3 结论

(1)当 β 相水淬速率过快时,由于残留 β-Zr 的存在使 Zr-4 合金的耐腐蚀性能降低,这与 β-Zr 对含 Nb 锆合金耐腐蚀性能的影响规律一致.

(2) β 相水淬快冷后再进行 480—600 ℃ 退火处理,随着退火温度的升高和退火时间的延长,耐腐蚀性能得到明显改善,这主要与残留 β-Zr 的分解有关.

(3) β 相水淬时,控制合适的冷却速率,避免残留 β-Zr 的生成,同时提高 α-Zr 基体中 Fe 和 Cr 的过饱和固溶含量后,Zr-4 合金可获得与 Zr-Sn-Nb 合金一样优良的耐腐蚀性能,这说明 α-Zr 基体中固溶的 Fe 和 Cr 合金元素含量是影响耐腐蚀性能的主要因素.

参 考 文 献

[1] Cox B, Wong Y M. In: Eucken C M, Garde A M, eds. , *Zirconium in the Nuclear Industry: 9th International Symposium*, *ASTM STP 1132*, Baltimore: ASTM International, 1991: 643.

[2] Pêcheur D, Godlewski J, Peybernès J, Fayette L, Noé M, Frichet A, Kerrec O. In: Sabol G P, Moan G D, eds. , *Zirconium in the Nuclear Industry: 12th International Symposium*, *ASTM STP 1354*, Philadelphia: ASTM International, 2000: 793.

[3] Zhou B X, Li Q, Huang Q, Miao Z, Zhao W J, Li C. *Nuc Power Eng*, 2000; 21: 43.
(周邦新,李强,黄强,苗志,赵文金,李聪. 核动力工程,2000;21: 43)

[4] Sabol G P, Comstock R J, Weiner R A, Larouere P, Stanutz R N. In: Garde A M, Bradley E R, eds. , *Zirconium in the Nuclear Industry: 10th International Symposium*, *ASTM STP 1245*, Ann Arbor: ASTM International, 1994: 724.

[5] Nikulina A V, Markelvo V A, Peregud M M, Bibilashvili Y K, Kotrekhov V A, Lositsky A F, Kuzmenko N V, Shevnin Y P, Shamardin V K, Kobylyansky G P, Novoselov A E. In: Bradley E R, Sabol G P, eds. , *Zirconium in the Nuclear Industry: 11th International Symposium*, *ASTM STP 1295*, Ann Arbor: ASTM International, 1996: 785.

[6] Zhao W J, Miao Z, Jiang H M, Yu X W, Li W J, Li C, Zhou B X. *J Chin Soc Corros Prot*, 2002; 22: 124.
(赵文金,苗志,蒋宏曼,于晓卫,李卫军,李聪,周邦新. 中国腐蚀与防护学报,2002;22: 124)

[7] Li Z K, Liu J Z, Li P Z, Zhou L, Song Q Z. *Rare Met Mater Eng*, 1999; 28: 101.
(李中奎,刘建章,李佩志,周廉,宋启忠. 稀有金属材料与工程,1999;28: 101)

[8] Li Z K, Liu J Z, Zhu M S, Song Q Z. *Rare Met Mater Eng*, 1996; 25: 43.
(李中奎,刘建章,朱梅生,宋启忠. 稀有金属材料与工程,1996;25: 43)

[9] Yao M Y, Zhou B X, Li Q, Liu W Q, Geng X, Lu Y P. *J Nucl Mater*, 2008; 374: 197.

[10] Yao M Y, Zhou B X, Li Q, Liu W Q, Wang S A, Huang X S. *Rare Met Mater Eng*, 2007; 36: 1915.
(姚美意,周邦新,李强,刘文庆,王树安,黄新树. 稀有金属材料与工程,2007;36: 1915)

[11] Zhou B X, Li Q, Miao Z. *Nucl Power Eng*, 2000; 21: 339.
(周邦新,李强,苗志. 核动力工程,2000;21: 339)

[12] Anada H, Herb B J, Nomoto K, Hagi S, Graham R A, Kuroda T. In: Bradley E R, Sabol G P, eds. , *Zirconium in the Nuclear Industry: 11th International Symposium*, *ASTM STP 1295*, Ann Arbor: ASTM International, 1996: 74.

[13] Foster P, Dougherty J, Burke M G, Bates J F, Worcester S. *J Nucl Mater*, 1990; 173: 164.

[14] Garzarolli G, Steinberg E, Weidinger H G. In: Van Swam L F P, Eucken C M, eds. , *Zirconium in the Nuclear Industry: 8th International Symposium*, *ASTM STP 1023*, Baltimore: ASTM International, 1989: 202.

[15] Zhou B X, Zhao W J, Miao Z, Pan S F, Li C, Jiang Y R. *Chin J Nucl Sci Eng*, 1995；15：242.

(周邦新,赵文金,苗志,潘淑芳,李聪,蒋有荣. 核科学与工程,1995;15:242)

[16] Choo K N, Kang Y H, Pyun S I, Vrbanic V F. *J Nucl Mat*, 1994；209：226.

[17] Jeong Y H, Kim H G, Kim T H. *J Nucl Mater*, 2003；317：1.

[18] Garzarolli F, Pohlmeyer I, Steinberg E, Trapp-pritsching S. *Proc Technical Committee Meeting*, Cadarache：the International Atomic Energy Agency, 1985：66.

[19] Liu J Z. *Structure Nuclear Materials*. Beijing：Chemical Industry Press, 2007：111.

(刘建章. 核结构材料. 北京:化学工业出版社,2007:111)

[20] Yao M Y, Wang J H, Peng J C, Zhou B X, Li Q. *J ASTM Intl*, 2011；8：13.

[21] Liu W Q, Li Q, Zhou B X, Yao M Y. *Rare Met Mater Eng*, 2004, 33：728.

(刘文庆,李强,周邦新,姚美意. 稀有金属材料与工程,2004,33:728)

[22] Zhou B X, Li Q, Yao M Y, Liu W Q, Chu Y L. *Nucl Power Eng*, 2005；26：364.

(周邦新,李强,姚美意,刘文庆,褚于良. 核动力工程,2005;26:364)

[23] Zhou B X, Li Q, Liu W Q, Yao M Y, Chu Y L. *Rare Met Mater Eng*, 2006；35：1009.

(周邦新,李强,刘文庆,姚美意,褚于良. 稀有金属材料与工程,2006;35:1009)

[24] Zhou B X, Li Q, Yao M Y, Liu W Q, Chu Y L. *J ASTM Intl*, 2008；5：360.

Effect of β - Quenching on the Corrosion Resistance of Zr - 4 Alloy in LiOH Aqueous Solution

Abstract：To investigate the effect of β - quenching on the corrosion behavior, Zr - 4 specimens were β - quenched and subsequently annealed at 480—600 ℃ for 2—200 h. The corrosion tests were carried out in 0. 01 mol/L LiOH aqueous solution at 360 ℃/18. 6 MPa. The microstructure of Zr - 4 specimens and fracture surface morphology of oxide films on the corroded specimens were observed by TEM and HRSEM, respectively. The results show that Zr - 4 specimens exhibit excellent corrosion resistance as Zr - Sn - Nb alloys in LiOH aqueous solution after increasing the solid solution content of Fe and Cr in α - Zr matrix by suitable quenching rate from β - phase to avoid the formation of β - Zr. When the cooling rate is too fast, the specimens show worse corrosion resistance due to the formation of β - Zr. After annealing at 480—600 ℃, the better corrosion resistance of Zr - 4 specimens can be obtained with increasing annealing temperture and time due to the decomposition of β - Zr.

Fe/Cr 比对 Zr(Fe，Cr)$_2$ 吸氢性能的影响*

摘　要：参照锆合金中的第二相成分，熔炼 2 种不同 Fe/Cr 比的 Zr(Fe，Cr)$_2$(Fe/Cr=2 和 0.5)和 ZrCr$_2$(Fe/Cr=0)3 种试验合金，对它们进行压力-成分-温度(P-C-T)测试和吸放氢动力学测试. 同时用 XRD 对 360 ℃吸氢前后的样品进行物相分析. 实验结果表明，这 3 种合金在 50，200，360 ℃ 3 个温度下，均有吸氢量随 Fe/Cr 比值的降低而增加的现象. 另外，在 360 ℃，4 MPa 氢气气氛中测试时，这 3 种不同合金的可逆吸放氢量和吸放氢速度随 Fe/Cr 比的降低而增加. XRD 测试表明，在 360 ℃吸氢后并没有氢化物生成.

目前，人们常在锆中添加 Sn、Fe、Cr、Ni、Nb、Cu 等合金元素来提高锆合金的强度和耐腐蚀性能，但这些合金元素在改善锆合金强度及耐腐蚀性能的同时也影响了锆合金腐蚀时的吸氢. 除了 Sn 和 Nb 之外，由于添加的这些合金元素在 α-Zr 中的固溶度很低，大部分会以第二相的形式析出[1]，如 Zr-Sn 系的 Zr-2 合金中存在 Zr$_2$(Fe，Ni)和 Zr(Fe，Cr)$_2$[2]，多余的 Cr 也可以形成 ZrCr$_2$[1]，Zr-4 中存在 Zr(Fe，Cr)$_2$；Zr-Sn-Nb 系的 N18 和 N36 合金中有 Zr-Nb-Fe 和 β-Nb[3,4]；在 Zr-4 中加入 Cu 时会得到 Zr$_2$Cu[5]. 普遍认为 Zr$_2$(Fe，Ni)和 Zr(Fe，Cr)$_2$ 这类第二相对锆合金的吸氢行为有重要影响[6-8]. 因此，单独研究这些第二相的吸放氢性能，可以为探究锆合金的吸氢机制提供一些有价值的参考.

众所周知，Zr(Fe，Cr)$_2$ 和 ZrCr$_2$ 是 AB_2 型储氢合金，已有很多关于它们的储氢性能的研究[9-11]. 根据储氢材料的性能要求，一般只需要研究它们在接近室温下的吸放氢能力，还很少有关于这些合金在反应堆运行的高温条件(如 360 ℃)下吸放氢性能研究的报道. 本实验对不同 Fe/Cr 比的 Zr(Fe，Cr)$_2$ 合金在高温和室温附近的吸放氢性能进行系统研究，包括吸放氢能力，吸放氢动力学以及吸氢前后的物相分析，以便进一步认识锆合金中的第二相在腐蚀吸氢过程中的作用.

1　实验

本实验用非自耗真空电弧炉熔炼 Fe/Cr 比分别为 2，0.5 和 0(原子比，下同)的 Zr(Fe，Cr)$_2$ 3 种金属间化合物(其中 Fe/Cr 比为 0 的合金也就是 ZrCr$_2$). 由于锆合金中的大部分 Zr(Fe，Cr)$_2$ 第二相的 Fe/Cr 比在 0~2 之间[12,13]，同时会随着热处理的不同发生改变[13]，所以研究 Fe/Cr 比分别为 2，0.5 和 0 的 3 种 Zr(Fe，Cr)$_2$ 合金的吸氢性能可以大致表征锆合金中 Zr(Fe，Cr)$_2$ 第二相的吸氢性能. 合金锭经过 5 次熔炼，每熔炼 1 次，翻转 1 次，以确保成分均匀. 再将合金锭研磨成粉，并取粒度在 38.5~55 μm 的粉末在日本 SUZUKI HOKAN. CO. 公司的全自动 Sievert 型 P-C-T 测试仪中进行吸放氢的 P-C-T 曲线和动力学测试. P-C-T 曲线测试温度为 50，200 和 360 ℃，吸放氢动力学测试温度

* 本文合作者：王锦红、姚美意、耿建桥、夏爽、曹潇潇. 原发表于《稀有金属材料与工程》，2011，40(6)：1084-1088.

为 360 ℃,氢压为 4 MPa.测试前样品均在 360 ℃下活化 3 次,每次活化时吸氢 1 h,放氢1 h,确保活化完全.

为了确定所熔炼的合金的结构,把合金锭磨成粉末后进行 XRD 测试.每种样品做完 P-C-T 曲线和吸放氢动力学测试后,均在 360 ℃下吸氢至饱和,并让其在氢气气氛中冷却到室温后取出,然后用 XRD 测定样品吸氢后的晶体结构.

2 结果与讨论

2.1 P-C-T 曲线

图 1a 所示的是 3 种合金在 360 ℃,氢压为 4 MPa 下的活化曲线,图 1b,1c,1d 分别是 Zr(Fe,Cr)$_2$(Fe/Cr=2,0.5)和 ZrCr$_2$ 合金各自在 50,200,360 ℃下的 P-C-T 曲线. 由图 1a 可知,第 2 次活化和第 3 次活化时吸氢量基本相等,这说明经 3 次活化后,3 种合金已基本上活化完全.从图 1b,1c,1d 可以看出,在相同氢压下,温度越高,吸氢量越少,这符合

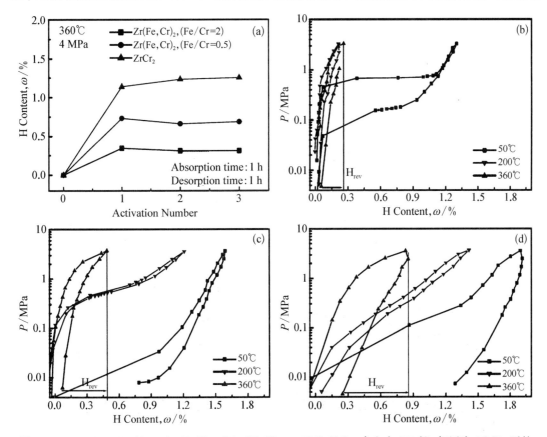

图 1 Zr(Fe,Cr)$_2$(Fe/Cr=2),Zr(Fe,Cr)$_2$(Fe/Cr=0.5)和 ZrCr$_2$ 合金在 360 ℃,氢压为 4 MPa 下的活化曲线(a),Zr(Fe,Cr)$_2$(Fe/Cr=2),Zr(Fe,Cr)$_2$(Fe/Cr=0.5),ZrCr$_2$ 合金分别在 50,200,360 ℃下的 P-C-T 曲线(b,c,d)

Fig. 1 Dependence of the H content on activation number for Zr(Fe,Cr)$_2$(Fe/Cr=2 and 0.5), and ZrCr$_2$ at 360 ℃, 4 MPa H$_2$(a), P-C-T curves at 50, 200 and 360 ℃ of Zr(Fe,Cr)$_2$(Fe/Cr=2) (b), Zr(Fe,Cr)$_2$(Fe/Cr=0.5) (c) and ZrCr$_2$(d), respectively

储氢材料的普遍规律[14]. 从图1b中还能看出,Zr(Fe, Cr)₂(Fe/Cr=2)在50 ℃下,吸氢平台压为0.67 MPa,放氢平台压为0.15 MPa,说明在50 ℃且氢压超过0.67 MPa的状态下吸氢,合金中有氢化物相生成;而当温度为200 ℃和360 ℃,氢压小于4 MPa时,曲线未到达平台压,说明氢只是固溶在合金中,并没有生成氢化物,故吸氢量较低. 要使合金在这2个温度下也能生成氢化物,还需再升高氢压. 从图1c和1d中还能看出,当氢压小于4 MPa时,Zr(Fe, Cr)₂(Fe/Cr=0.5)和ZrCr₂样品在200 ℃和50 ℃下均出现了平台压,说明这2种合金都比Zr(Fe, Cr)₂(Fe/Cr=2)更容易生成氢化物. 而且从图1d的放氢曲线还可以看出,ZrCr₂在50 ℃吸收的氢有很大一部分放不出来. 综合图1b, 1c, 1d可知,在同一氢压下,这3种合金在50,200,360 ℃这3个温度下的吸氢量由小到大的次序都是:Zr(Fe, Cr)₂(Fe/Cr=2)＜Zr(Fe, Cr)₂(Fe/Cr=0.5)＜ZrCr₂,也就是说,合金的吸氢量随Fe/Cr比的降低而增加. Kakiuchl[15]研究Fe/Cr=1和2的Zr(Fe, Cr)₂在室温及300 ℃下的P-C-T曲线时也得到类似的结论.

本课题组的前期研究结果[12]表明,锆合金在腐蚀时的吸氢行为与第二相的种类、尺寸和数量密切相关,认为这可能与第二相粒子自身的可逆吸放氢能力不同有关. 可逆吸放氢能力是指合金可以自由吸收且放出氢的量(即可逆吸放氢量)的大小及可逆吸放氢速率的快慢. 所以需要分析这3种合金在360 ℃,氢压为4 MPa时的可逆吸氢量(H_{rev}). 从图1b, 1c, 1d中可以读出在360 ℃,氢压为4 MPa时Zr(Fe, Cr)₂(Fe/Cr=2),Zr(Fe, Cr)₂(Fe/Cr=0.5)和ZrCr₂的可逆吸氢量依次为:0.16%,0.32%,0.60%(质量分数),即这3种合金在该条件下的可逆吸氢量随Fe/Cr比的降低而增加.

2.2 吸放氢动力学

图2所示的是不同Fe/Cr比的Zr(Fe, Cr)₂粉末样品在360 ℃,4 MPa氢压下的吸放氢动力学曲线(测吸氢动力学曲线时,起始氢压是4 MPa;测放氢动力学曲线时,起始氢压为0 MPa,即在真空环境下测放氢动力学曲线). 从图2可见,不管是吸氢还是放氢,它们都有两个阶段,一开始迅速增加,到一定量后,速度变慢. 在吸氢初期,ZrCr₂和Zr(Fe, Cr)₂(Fe/Cr=0.5)样品在15 s的时间内分别吸了0.65%和0.20%(质量分数)的氢,接近它们最大吸氢量的一半,说明这2种合金在吸氢初期的速度是相当快的;而Zr(Fe, Cr)₂(Fe/Cr=2)样品则

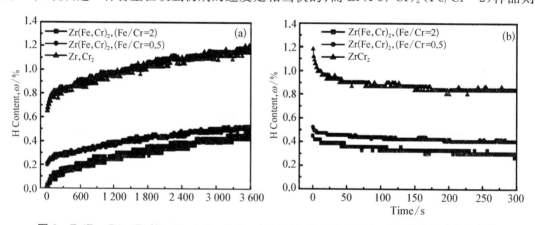

图 2 Zr(Fe, Cr)₂(Fe/Cr=2, 0.5),ZrCr₂ 在 360 ℃,氢压为 4 MPa 时的吸放氢动力学曲线

Fig. 2 Kinetic curves of hydrogen absorbing (a) and desorbing (b) of Zr(Fe, Cr)₂(Fe/Cr=2, 0.5), and ZrCr₂ at 360 ℃, 4 MPa

没有这种短时间内吸氢量迅速增加的现象,说明它吸氢速度相对较慢.因此,在吸氢初期(前 20 s),这 3 种合金的吸氢速度从小到大依次为:$Zr(Fe, Cr)_2(Fe/Cr=2)<Zr(Fe, Cr)_2(Fe/Cr=0.5)<ZrCr_2$.从图 2b 的曲线斜率可知,这 3 种合金的放氢速度从小到大也是依次为:$Zr(Fe, Cr)_2(Fe/Cr=2)<Zr(Fe, Cr)_2(Fe/Cr=0.5)<ZrCr_2$.所以,这 3 种合金的吸放氢速度从小到大依次为 $Zr(Fe, Cr)_2(Fe/Cr=2)<Zr(Fe, Cr)_2(Fe/Cr=0.5)<ZrCr_2$.结合上述可逆吸放氢量的分析结果可知,这 3 种合金的可逆吸放氢能力从小到大也是依次为:$Zr(Fe, Cr)_2(Fe/Cr=2)<Zr(Fe, Cr)_2(Fe/Cr=0.5)<ZrCr_2$,即可逆吸放氢能力随 Fe/Cr 比的降低而增加.

本课题组曾提出一种第二相影响锆合金腐蚀时吸氢行为的模型[12]:在金属/氧化膜界面处 Zr 和 OH⁻ 反应生成的氢可以优先被镶嵌在金属/氧化膜界面处的可逆吸放氢能力强于 Zr 的第二相捕获,它们可作为吸氢的优先通道.根据这一模型以及以上 $Zr(Fe, Cr)_2$ 的可逆吸放氢能力随 Fe/Cr 比的降低而增加的实验结果,就可以解释锆合金在反应堆的服役过程中或在堆外高压釜中腐蚀时,添加合金元素 Cr 会使吸氢量增加这一普遍现象[16,17].

2.3 吸氢前后物相分析

图 3 所示的是 $Zr(Fe, Cr)_2(Fe/Cr=2, 0.5)$,$ZrCr_2$ 在 360 ℃吸氢前后的 XRD 衍射图谱.对照标准 PDF 卡片可以发现,吸氢前的 $Zr(Fe, Cr)_2(Fe/Cr=2)$ 和 $Zr(Fe, Cr)_2(Fe/Cr=0.5)$ 样品的 XRD 图谱都与 $ZrFe_{1.5}Cr_{0.5}$(PDF:42-1289,六方结构,$a=0.500\,7$ nm,$c=0.819\,3$ nm)的谱线接近,但都略微左移,主要衍射峰晶面如图所示.这是因为,$ZrFe_{1.5}Cr_{0.5}$ 的 Fe/Cr=3,大于 2 和 0.5,故谱线都往 θ 值变小的方向偏移,即往左移,且 Fe/Cr 越接近 3,左移程度越小.同样,吸氢前 $ZrCr_2$ 样品的 XRD 图谱与 $ZrCr_2$(PDF:06-0613,六方结构,$a=0.508\,9$ nm,$c=0.827\,9$ nm)谱线接近,但整体左移约 0.3°.表 1 所示的是根据 XRD 图谱计算所得的这 3 种合金的晶格常数,从中可以看出,计算所得的 $Zr(Fe, Cr)_2$(Fe/Cr=2 和 0.5)合金在吸氢前的晶格常数介于上述的标准 $ZrFe_{1.5}Cr_{0.5}$ 和 $ZrCr_2$ 的晶格常数之间,这是因为它们的 Fe/Cr 比介于 3 和 0 之间.锆合金中的 $Zr(Fe, Cr)_2$ 和 $ZrCr_2$ 为六方结构[18-20].以上结果表明,熔炼的合金与锆合金中的 $Zr(Fe, Cr)_2$ 第二相具有相同的结构,并覆盖了可能的多种第二相成分.

比较这 3 种合金吸氢前后的 XRD 衍射图谱,发现它们的峰形基本相同,但吸氢后的 XRD 图谱与吸氢前 XRD 图谱都整体左移了约 3°.这说明 360 ℃吸氢后在随后的冷却过程

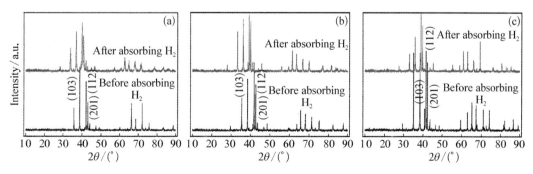

图 3 $Zr(Fe, Cr)_2(Fe/Cr=2)$,$Zr(Fe, Cr)_2(Fe/Cr=0.5)$ 和 $ZrCr_2$ 吸氢前后的 XRD 图谱

Fig. 3 XRD patterns of $Zr(Fe, Cr)_2$ ($Fe/Cr=2, 0.5$) (a, b) and $ZrCr_2$ (c) before and after hydrogen absorbing

中氢又被释放出来,并没有生成新的物相,只是一部分氢固溶在合金中,使得点阵常数变大.这与 P-C-T 曲线上得出的结论是一致的.

表 1 计算得出 3 种合金吸氢前后的晶格常数
Table 1 Lattice constant of alloys before and after hydrogen absorbing (nm)

Alloy	Before absorption	After absorption
$Zr(Fe, Cr)_2$, Fe/Cr=2	a=0.502 4	a=0.527 5
	c=0.819 7	c=0.859 5
$Zr(Fe, Cr)_2$, Fe/Cr=0.5	a=0.505 5	a=0.530 1
	c=0.829 3	c=0.878 6
$ZrCr_2$	a=0.510 9	a=0.543 2
	c=0.828 1	c=0.883 1

3 结论

(1) Fe/Cr 比不同的 3 种 $Zr(Fe, Cr)_2$ 合金在 50, 200 和 360 ℃ 3 个温度下的吸氢量由小到大的次序都为: $Zr(Fe, Cr)_2$(Fe/Cr=2)<$Zr(Fe, Cr)_2$(Fe/Cr=0.5)<$ZrCr_2$,即随着 Fe/Cr 比值的降低,吸氢量增加.

(2) 这 3 种合金在 360 ℃,氢压为 4 MPa 下的可逆吸放氢量和吸放氢速度均按 $Zr(Fe, Cr)_2$(Fe/Cr=2)<$Zr(Fe, Cr)_2$(Fe/Cr=0.5)<$ZrCr_2$(Fe/Cr=0)的顺序变大,即随 Fe/Cr 比值的降低,可逆吸放氢能力增强.这一实验结果可以解释锆合金在反应堆的服役过程中或在堆外高压釜中腐蚀时,添加合金元素 Cr 会使吸氢量增加这一普遍现象.

(3) 3 种合金在 360 ℃ 吸氢后并没有新的物相生成,H 主要以固溶的形式存在于合金中.

参 考 文 献

[1] Charquet D, Hahn R. *Zirconium in the Nuclear Industry: Eighth International Symposium* [C]. West Conshohocken, PA: ASTM International, 1989: 405.

[2] Pettersson K, Bergqvist H. *External Cladding Corrosion in Water Power Reactors* [C]. Vienna: International Atomic Energy Agency, 1985: 53.

[3] Toffolon-Masclet C, Brachet J C, Jago G. *J Nucl Mater* [J], 2002, 305: 224.

[4] Toffolon-Masclet C, Barberis P, Brachet J C et al. *Zirconium in the Nuclear Industry: Fourteenth International Symposium* [C]. West Conshohocken, PA: ASTM International, 2004: 81.

[5] Hong H S, Moon J S. *J Nucl Mater* [J], 2001, 297: 113.

[6] Etoh Y, Schimada S, Yasuda T. *Zirconium in the Nuclear Industry: Eleventh International Symposium* [C]. West Conshohocken, PA: ASTM International, 1996: 825.

[7] Tägstrom P, Limbäck M. *Zirconium in the Nuclear Industry: Thirteenth International Symposium* [C]. West Conshohocken, PA: ASTM International, 2002: 96.

[8] Barberis P, Ahlberg E. *Zirconium in the Nuclear Industry: Thirteenth International Symposium* [C]. West Conshohocken, PA: ASTM International, 2002: 33.

[9] Qian Shenghua. *Windsor University Doctoral Dissertation* [D]. Canada: Windsor University, 1989.

[10] Douglas G Ivey. *Windsor University Doctoral Dissertation* [D]. Canada: Windsor University, 1985.

[11] Adel Esayed. *Windsor University Doctoral Dissertation* [D]. Canada: Windsor University, 1993.

[12] Yao Meiyi(姚美意), Zhou Bangxin(周邦新). *Thesis for Doctorate*(博士论文) [D]. Shanghai: Shanghai University,

2007.

[13] Zhou Bangxin(周邦新), Yang Xiaoling(杨晓林). *Nuclear Power Engineering*(核动力工程)[J], 1997, 18 (6): 511.

[14] Hu ZiLong(胡子龙), *Hydrogen Storage Materials*(贮氢材料)[M]. Beijing: Chemical Industry Press, 2003: 49.

[15] Kakiuchi K, Itagaki N, Furuya T *et al*. *Zirconium in the Nuclear industry: Fourteenth International Symposium* [C]. West Conshohocken, PA: ASTM International, 2004: 349.

[16] Yao Meiyi(姚美意), Zhou Bangxin(周邦新). *Rare Metal Materials and Engineering*(稀有金属材料与工程)[J], 2004, 33(6): 641.

[17] Yao M Y, Zhou B X. *J Nucl Mater*[J], 2006, 350: 195.

[18] Rao P. *General Electric Company Report 76 CRD*[R]. Calif: San Jose, 1977: 183.

[19] Li Zhongkui(李中奎), Liu Jianzhang(刘建章) *et al*. *Rare Metal Materials and Engineering*(稀有金属材料与工程) [J], 1996, 25(5): 43.

[20] Zhou Bangxin(周邦新), Yao Meiyi(姚美意), Li Qiang(李强) *et al*. *Rare Metal Materials and Engineering*(稀有金属材料与工程)[J], 2007, 36(8): 1317.

Effect of Fe/Cr Ratio on the Hydrogen Absorption Performance of $Zr(Fe, Cr)_2$

Abstract: Referring to the compositions of second phase particles (SPPs) in zirconium alloys, three kinds of experimental $Zr(Fe, Cr)_2$ alloy with different Fe/Cr ratio i. e. Fe/Cr=2 and 0. 5 for $Zr(Fe, Cr)_2$ and Fe/Cr=0 for $ZrCr_2$ were prepared. Then the pressure-composition-temperature ($P - C - T$) curves and kinetic curves of hydrogen absorbing and desorbing were obtained. XRD was employed to analyze the structures of the specimens before and after hydrogen absorbing at 360 ℃. Results show that the amount of hydrogen absorption of the three alloys all increases with the decrease of Fe/Cr ratio at 50, 200 and 360 ℃. In addition, the amount and rate of reversible hydrogen absorption and desorption of the three alloys all increase with the decrease of Fe/Cr ratio when tested in 360 ℃, 4 MPa H_2 atmosphere. X-ray diffraction analysis indicates that no hydride is detected after they absorb hydrogen at 360 ℃.

Zr－Sn 系合金在过热蒸汽中的腐蚀吸氢行为*

摘　要：研究第二相大小和种类对 Zr－2 和 Zr－4 合金在 400 ℃，10.3 MPa 过热蒸气中腐蚀吸氢行为的影响. 结果表明：吸氢量的多少和耐腐蚀性能的好坏之间并不一定存在严格的对应关系，而是与第二相大小和种类密切相关. 在相同腐蚀增重下，含粗大第二相样品的吸氢量均大于含细小第二相样品；第二相比较粗大的 Zr－2 样品的吸氢量比 Zr－4 样品大得多；而第二相比较细小的 Zr－2 和 Zr－4 样品的吸氢量差别却很小. 对 $Zr(Fe, Cr)_2$ 和 $Zr_2(Fe, Ni)$ 金属间化合物及纯锆进行的 PCT 和吸放氢动力学测试表明，前两者可自由吸放氢，且吸放氢速度快，而纯锆只能吸氢难以放氢. 据此，Zr－Sn 系合金在 400 ℃，10.3 MPa 过热蒸气中腐蚀吸氢行为可以用提出的"在金属/氧化膜界面处 Zr 和 OH⁻ 反应生成的氢可以优先被镶嵌在金属/氧化膜界面处的可逆吸放氢能力强于 Zr 的 $Zr(Fe, Cr)_2$ 和 $Zr_2(Fe, Ni)$ 第二相捕获，它们可作为吸氢的优先通道"的吸氢模型得到合理解释.

　　核反应堆包壳用锆合金在腐蚀过程中不可避免会吸氢，吸氢后引起的氢脆和氢致延迟开裂是影响锆合金包壳使用寿命的主要因素之一[1]. 随着燃料组件燃耗的进一步提高，对锆合金腐蚀过程中吸氢规律及其机制的研究越来越受到关注. 由于添加到锆合金中的 Fe、Cr、Ni 等合金元素与锆形成的第二相本身是强烈吸氢的物质[2]，所以普遍认为锆合金的吸氢行为与第二相密切相关. 但对于如何影响并没有达成一致认识，归结起来主要有以下两种观点.

　　第一种观点认为，第二相首先影响氧化膜特性（指致密层厚度、孔隙、裂纹等）后才引起吸氢行为的差别（间接影响）. Rudling 等[3,4]认为 Zr－2 中大于某一临界尺寸的第二相粒子会通过影响致密层厚度来影响吸氢行为；Cox 等[5-7]认为，氧化膜中的孔隙、裂纹可以作为吸氢的快速通道，而第二相的尺寸、数量及成分会对这些孔隙和裂纹的几何形状和数量产生影响. 第二种观点认为，氧化膜中暂时未被氧化的第二相提供了氢的短路扩散通道（直接影响）. Hatano 等[8,9]用这一模型解释了在腐蚀初期，相同腐蚀增重下，含粗大第二相的 Zr－2 和 Zr－4 样品的吸氢量比含细小第二相样品多的实验结果. Tägstrom 等[10]也用这一观点解释了他们的实验结果，并认为第二相的数量比第二相的大小对吸氢行为的影响更大；Lelièvre 等[11]用核分析技术研究氢在氧化膜中的分布时，发现在氧化膜中仍与金属接触的未被氧化的第二相周围经常能检测到氢化锆，由此也提出了与上述类似的观点.

　　研究第二相尺寸不同的 Zr－2 和 Zr－4 合金样品腐蚀时的吸氢行为，以及对与这两种合金中第二相成分相同的试验合金 $Zr(Fe, Cr)_2$ 和 $Zr_2(Fe, Ni)$ 做吸放氢性能测试，可以加深对第二相种类、尺寸影响锆合金腐蚀时的吸氢规律及其机制的认识.

---

* 本文合作者：王锦红、姚美意、耿建桥、张欣、张金龙. 原发表于《稀有金属材料与工程》，2011，40（5）：833－838.

1 实验

试验用厚 0.6 mm 的 Zr-2 和 Zr-4 合金板(化学成分见表1)是按常规工艺制备并经过 600 ℃, 1 h 再结晶退火处理. 将其切成尺寸为 25 mm×20 mm 的样品, 其中一部分在真空中 (真空度优于 10^{-3} Pa) 按如图1所示热处理制度进行 810~750 ℃ 循环加热冷却处理 20 次, 目的是使第二相粗化.

表 1 Zr-2 和 Zr-4 合金的名义成分

Table 1 Nominal composition of Zr-2 and Zr-4 alloys (mass fraction, %)

Alloy	Element				
	Sn	Fe	Cr	Ni	Zr
Zr-2	1.5	0.2	0.1	0.05	Balance
Zr-4	1.5	0.2	0.1	—	Balance

将热处理前后的样品一同放入高压釜中进行 400 ℃, 10.3 MPa 过热蒸气腐蚀试验. 腐蚀前, 样品按标准方法用混合酸(体积分数为 10% HF+45% HNO_3+45% H_2O)酸洗和去离子水清洗. 腐蚀增重量由 5 块试样平均值得出.

用美国 LECO 公司的 RH-600 型定氢分析仪测定腐蚀后样品中的氢含量. 由于样品的厚度存在一定的差别, 即使吸氢分数相同, 测量得到的氢含量也会不同, 因而对测得的氢含量数据按 0.57 mm 这一厚度做归一化处理, 具体分析过程及归一化处理方法详见文献[12].

用 JSM-6700 高分辨扫描电镜观察第二相形貌. 用德国 SIS 图像分析软件进行第二相尺寸分布统计, 每种样品中统计的第二相颗粒在 400 颗以上.

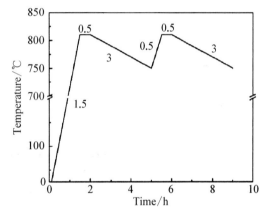

图 1 粗化锆合金样品中第二相的热处理制度

Fig. 1 Heat-treatment for coarsened second phase particles (SPPs) in Zr sample (repeating two times only as shown)

众所周知, Zr-2 和 Zr-4 中第二相的氧化比锆基体慢[13-15], 第二相先以金属态存在于氧化膜中, 然后逐渐被氧化. 因此, 镶嵌在氧化膜/金属界面处的第二相一定是金属态, 所以研究与锆合金中第二相成分相同的试验合金 Zr(Fe, Cr)$_2$ 和 Zr$_2$(Fe, Ni) 的吸放氢特性可以代表锆合金中第二相的吸放氢行为. Zr-2 合金中存在的第二相有 Zr(Fe, Cr)$_2$ 和 Zr$_2$(Fe, Ni), Zr-4 合金中存在的第二相只有 Zr(Fe, Cr)$_2$. EDS 分析结果表明: Zr-4 合金中大部分 Zr(Fe, Cr)$_2$ 第二相的 Fe/Cr 比(原子比, 没有特别说明以下相同)在 1.5~2.1 之间; Zr-2 合金中大部分 Zr(Fe, Cr)$_2$ 第二相的 Fe/Cr 比在 0.5~1.5 之间, Zr$_2$(Fe, Ni) 第二相的 Fe/Ni 比在 1 左右[15]. 本研究用非自耗真空电弧炉熔炼了 Fe/Cr 比分别为 2 和 0.5 的 Zr(Fe, Cr)$_2$ 以及 Fe/Ni 比为 1 的 Zr$_2$(Fe, Ni) 这三种金属间化合物. 熔炼时反复熔炼五次, 每熔炼一次, 翻转一次, 以确保成分均匀. 再将合金锭研磨成粉, 并取粒度在 38.5~55 μm 之间的粉末在日本 SUZUKI HOKAN. CO. 公司的全自动 Sievert 型 PCT(压力-成分-温度)

测试仪中进行 PCT 和吸放氢动力学测试. 测试前样品均在 360 ℃下活化三次,每次活化时吸氢1 h,放氢1 h,确保活化完全.

2 实验结果

2.1 热处理前后第二相尺寸的分布

图 2 是热处理前后 Zr-2、Zr-4 样品中第二相的尺寸分布图,图中同时给出了参与统计的第二相粒子大小的平均值. 热处理前 Zr-2,Zr-4 样品中第二相粒子的平均尺寸分别是 136 和 180 nm,以下分别记为 Zr-2 SPPs fine 和 Zr-4 SPPs fine,表示含细小第二相的样品;热处理后第二相粒子的平均尺寸则增大到 807 和 650 nm,分别记为 Zr-2 SPPs coarse 和 Zr-4 SPPs coarse,表示含粗大第二相的样品. Zr-2 SPPs fine 样品中有 90％以上的第二

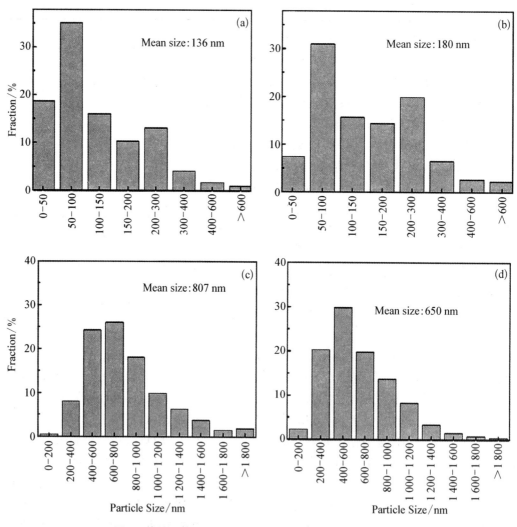

图 2 热处理前后 Zr-2,Zr-4 样品中第二相的尺寸分布

Fig. 2 SPPs size distribution in Zr-2 and Zr-4 specimens before and after heat-treatment: (a) Zr-2 SPPs fine,(b) Zr-4 SPPs fine,(c) Zr-2 SPPs coarse, and (d) Zr-4 SPPs coarse

相粒子的尺寸小于 300 nm(图 2a),而 Zr-2 SPPs coarse 样品中的第二相粒子尺寸则主要分布在 400~1 200 nm 之间(图 2c). Zr-4 SPPs fine 样品中也是有将近 90% 的第二相粒子尺寸小于 300 nm(图 2b), Zr-4 SPPs coarse 样品中的第二相粒子尺寸主要分布在 200~1 200 nm 之间(图 2d). 可见,经过多次循环热处理后样品中的第二相的确长大了.

2.2 腐蚀吸氢行为

图 3 和图 4 分别是第二相大小不同的 Zr-2, Zr-4 样品在 400 ℃, 10.3 MPa 过热蒸气中的腐蚀增重曲线和腐蚀后样品中的吸氢量与腐蚀增重的关系曲线. 从图 3 和图 4 可知,无论 Zr-2 还是 Zr-4 样品,含细小第二相的样品比含粗大第二相样品的耐腐蚀性能更好,同时吸氢量也低;在相同腐蚀增重下,含粗大第二相样品的吸氢量均大于含细小第二相样品. 这与 Hatano 等[8,9]在研究不同大小的第二相对 Zr-2 和 Zr-4 样品腐蚀初期(腐蚀转折前)的吸氢行为影响中得到的结果一致. 这进一步说明第二相大小对 Zr-2 和 Zr-4 合金的吸氢行为确实产生显著影响.

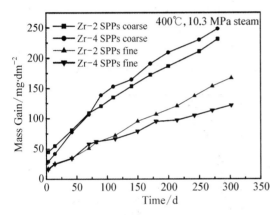

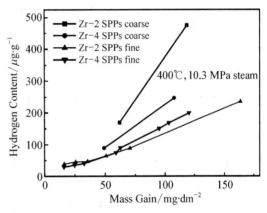

图 3 第二相大小不同的 Zr-2, Zr-4 样品在 400 ℃, 10.3 MPa 过热蒸气中腐蚀时的增重曲线

Fig. 3 Mass gain versus exposure time for Zr-4 and Zr-2 specimens with different SPPs size in 400 ℃, 10.3 MPa super-heated steam

图 4 第二相大小不同的 Zr-2, Zr-4 样品在 400 ℃过热蒸气中腐蚀后样品吸氢量与腐蚀增重的关系曲线

Fig. 4 Hydrogen absorption content versus mass gain for Zr-2 and Zr-4 specimens with different SPPs size in 400 ℃, 10.3 MPa super-heated steam

Zr-2 SPPs coarse 样品的耐腐蚀性能略优于 Zr-4 SPPs coarse 样品,但前者的吸氢量随腐蚀增重的变化却比后者的大很多;Zr-4 SPPs fine 样品的耐腐蚀性能优于 Zr-2 SPPs fine 样品,但两者的吸氢量随腐蚀增重的变化相差不大. 如在腐蚀增重均为 110 mg/dm² 时,Zr-2 SPPs coarse 样品的吸氢量(470 μg/g)是 Zr-4 SPPs coarse 样品吸氢量(248 μg/g)的 1.9 倍;而 Zr-2 SPPs fine 样品的吸氢量(165 μg/g)反而比 Zr-4 SPPs fine 样品(185 μg/g)小 20 μg/g. 由此可见,当第二相比较粗大时,第二相的种类对吸氢行为的影响比第二相细小时更加显著.

从以上实验结果可以看出,样品吸氢量的多少与耐腐蚀性能的好坏之间并不一定存在严格的对应关系,而是与第二相的尺寸和种类密切相关.

2.3 Zr(Fe, Cr)₂, Zr₂(Fe, Ni)及纯锆样品的可逆吸氢量及吸放氢速率比较

图 5 是两种 Zr(Fe, Cr)₂ 样品和 Zr₂(Fe, Ni)及纯 Zr 样品在 360 ℃下的活化曲线. 可以

看出,样品第一次活化时的吸氢量比后二次活化时的大,这是因为第一次活化时所消耗的氢有一部分用来还原合金表面存在的氧化物.第二次活化时的吸氢量与第三次活化时的吸氢量基本相等,这表明,经三次活化后,这些样品已基本活化完全.其中,两种 $Zr(Fe, Cr)_2$ 样品第二次活化时与第一次活化时吸氢量相差不大,这说明 $Zr(Fe, Cr)_2$ 在第一次活化时吸收的氢基本能全部放出,因而在第二次活化时仍可吸收同样量的氢;而 $Zr_2(Fe, Ni)$ 及纯 Zr 第二次活化时与第一次活化时吸氢量相差很大,第二次活化时的吸氢量小于第一次活化时的吸氢量.不同的是,$Zr_2(Fe, Ni)$ 第二次活化时,吸氢量虽比第一次少,但是仍有 0.6%(质量分数,下同)的吸氢量,说明 $Zr_2(Fe, Ni)$ 第一次活化时吸收的一部分氢生成了不可逆的氢化物,因而只有一部分氢可以再次放出;而纯 Zr 第一次活化时的吸氢量是四种材料里最大的,但第二次活化时的吸氢量却大大降低,从 2.1% 降到仅有 0.1%.说明纯 Zr 第一次活化时全部生成了不可逆的氢化物.

图 6 分别是 $Zr(Fe, Cr)_2$($Fe/Cr=0.5$ 和 $Fe/Cr=2$),$Zr_2(Fe, Ni)$ 及纯 Zr 样品在 360 ℃下的吸氢 PCT 曲线.因为这三种样品都先经过了三次活化,所以可以认为这时样品的最大吸氢量基本上等于它们的可逆吸氢量.从图 6 可以看出,这四种样品在 360 ℃,氢压为 3.3 MPa 的条件下,可逆吸放氢量大小依次为:$Zr_2(Fe, Ni) > Zr(Fe, Cr)_2$[$Fe/Cr=0.5$]$> Zr(Fe, Cr)_2$[$Fe/Cr=2$]$>$纯 Zr. Kakiuchi 等[16] 在研究 Fe/Cr 比为 1 和 2 的 $Zr(Fe, Cr)_2$ 在 200 ℃的 PCT 曲线时,也发现随着 Fe/Cr 比的增加,$Zr(Fe, Cr)_2$ 合金的吸氢量降低.

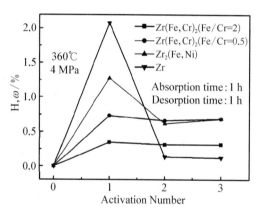 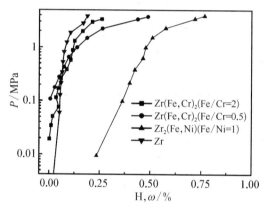

图 5 $Zr(Fe, Cr)_2$($Fe/Cr=2$ 和 $Fe/Cr=0.5$),$Zr_2(Fe, Ni)$ 及纯 Zr 在 360 ℃,氢压为 4 MPa 下的活化曲线

Fig. 5 H absorption value versus activation number for $Zr(Fe, Cr)_2$ ($Fe/Cr=2$ and $Fe/Cr=0.5$), $Zr_2(Fe, Ni)$ and zirconium at 360 ℃, 4 MPa H_2

图 6 $Zr(Fe, Cr)_2$($Fe/Cr=2$ 和 $Fe/Cr=0.5$)和 $Zr_2(Fe, Ni)$($Fe/Ni=1$)在 360 ℃下的吸氢 PCT 曲线

Fig. 6 Absorption PCT curves of $Zr(Fe, Cr)_2$ ($Fe/Cr=2$ and $Fe/Cr=0.5$) and $Zr_2(Fe, Ni)$ at 360 ℃

图 7 是两种 $Zr(Fe, Cr)_2$ 样品和 $Zr_2(Fe, Ni)$ 样品在 360 ℃下的吸放氢动力学曲线.由于纯 Zr 第一次活化时生成了不可逆氢化物,即吸收的氢不能再次放出,故不考虑它的吸放氢速率.由图 7a 可知,$Zr_2(Fe, Ni)$,$Zr(Fe, Cr)_2$($Fe/Cr=0.5$),$Zr(Fe, Cr)_2$($Fe/Cr=2$)样品在氢气气氛中保持 12 s 后吸氢量依次为:0.5%,0.226%,0.032%.说明这三种合金在吸氢初期的吸氢速率按 $Zr_2(Fe, Ni)$($Fe/Ni=1$)$> Zr(Fe, Cr)_2$($Fe/Cr=0.5$)$> Zr(Fe, Cr)_2$($Fe/Cr=2$)的顺序依次变小.由图 7b 可知,开始放氢时,即放氢时间为 0 s 时,$Zr_2(Fe, Ni)$

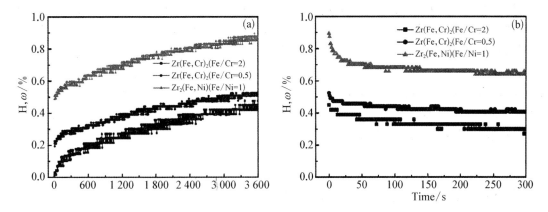

图7 Zr(Fe，Cr)₂(Fe/Cr=2 和 Fe/Cr=0.5)，Zr₂(Fe，Ni)(Fe/Ni=1)及纯 Zr 在 360 ℃下的吸氢和放氢动力学曲线

Fig. 7 H absorption (a) and H desorption (b) kinetics of Zr(Fe，Cr)₂(Fe/Cr=2 and Fe/Cr=0.5), Zr₂(Fe，Ni) and zirconium at 360 ℃

(Fe/Ni=1)，Zr(Fe，Cr)₂(Fe/Cr=0.5)，Zr(Fe，Cr)₂(Fe/Cr=2)内含有的氢依次为：0.895%，0.525%，0.449%；当放氢进行到 12 s 时，上述样品中所含的氢依次为 0.767%，0.474%，0.419%. 也就是说，在前 12 s 内，Zr₂(Fe，Ni)(Fe/Ni=1)，Zr(Fe，Cr)₂(Fe/Cr=0.5)，Zr(Fe，Cr)₂(Fe/Cr=2)放出的氢量依次为：0.128%，0.051%，0.030%. 综合图 7a，7b 可知，无论是在吸氢初期还是放氢初期，这三种样品的吸氢速率和放氢速率均是按 Zr₂(Fe，Ni)(Fe/Ni=1)＞Zr(Fe，Cr)₂(Fe/Cr=0.5)＞Zr(Fe，Cr)₂(Fe/Cr=2)的顺序依次变小. 结合活化曲线可知，在放氢初期，Zr₂(Fe，Ni)和上述两种 Zr(Fe，Cr)₂ 的放氢速度均大于纯锆.

3 讨论

一般来说，提高锆合金的耐腐蚀性能可以减少锆与水反应放出的氢（Zr＋H₂O → ZrO₂＋H₂)，从而间接减少吸氢量. Cox 等[5-7]认为第二相先影响氧化膜的特性，即影响氢通过氧化膜传输的难易程度后再影响吸氢行为. 按照这一观点，那么氢在耐腐蚀性能好的样品氧化膜中的传输会更困难，因而吸氢量也低，但本研究的实验结果并非完全如此. Hatano 等[8]提出氧化膜中暂时未被氧化的第二相可以提供氢的短路扩散通道，用这一观点解释氧化初期是合理的，但对于氧化后期并不适用，因为随着氧化膜的增厚，镶嵌在氧化膜外层中的第二相已经完全被氧化.

本研究热处理对 Zr-4 合金在 400 ℃，10.3 MPa 过热蒸气中腐蚀吸氢行为的影响时，也曾发现吸氢量的多少与耐腐蚀性能好坏之间并不存在严格的对应关系，并提出一种第二相影响吸氢行为的模型[17]（如图 8 所示，图中左边一条线表示金属/氧化膜界面，右边一条线表示氧化膜/腐蚀介质界面，图上的圆球和椭圆球表示第二相）：当锆合金与水反应发生腐蚀时，水得到电子发生 H₂O＋e → H＋OH⁻ 反应，除一部分氢在氧化膜/水介质界面处产生外，OH⁻ 通过氧化膜扩散到金属/氧化膜界面处与锆反应也可以直接生成氧化锆和氢；这样，镶嵌在金属/氧化膜界面处的第二相由于其本身吸氢能力的差别就会对锆合金腐蚀时的吸氢行为产生影响：当第二相的吸氢能力比锆强时，金属/氧化膜界面处的第二相可作为吸氢的优先通道,因而它们的大小和数量会对锆合金腐蚀时的吸氢行为产生明显影响. 本研究对纯锆及与锆合金中第二相成分相同的试验合金做吸放氢性能测试(图5,图6,图7)结果表明,作为吸氢优先通道的第二相作用除了与吸氢能力有关外,还与放氢能力及吸放氢的速率有关,应该说取决于该第二相相对于锆基体可逆吸放氢能力的强弱. 这里,将可逆吸放氢量的大小及吸放氢速率的快慢定义为可逆吸放氢能力. 由此,将第二相影响锆合金腐蚀时吸氢行为的模型进一步描述为"在金属/氧化膜界面处 Zr 和 OH⁻ 反应生成的氢可以优先被镶嵌在金属/氧化膜面处的可逆吸放氢能力强于 Zr 的第二相捕获,它们可作为吸氢的优先通道".

用上述吸氢模型可以合理解释得到: 样品吸氢量的多少与耐腐蚀性能好坏之间并不一定存在严格的对应关系,而是与第二相的大小和种类密切相关.

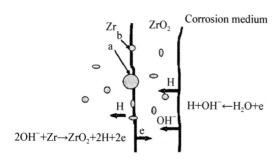

图 8 锆合金腐蚀时的吸氢模型示意图

Fig. 8 Schematic diagram of hydrogen absorption model when zirconium is corroded

Zr-2 合金中存在的第二相有 Zr(Fe, Cr)₂ 和 Zr₂(Fe, Ni), Zr-4 合金中存在的第二相只有 Zr(Fe, Cr)₂[13]. 由于 Zr₂(Fe, Ni) 和 Zr(Fe, Cr)₂ 第二相可逆吸放氢能力强于锆基体(图5,图6,图7),这样金属/氧化膜面处产生的氢就可以优先被镶嵌在金属/氧化膜界面处的第二相(如图8的颗粒 a 和 b)捕获并快速传递到未被氧化的锆基体中. 比较粗大的第二相由于本身尺寸大(如颗粒 a),这样就可以将所捕获的氢传递到更内层的锆基体中,更好地起到吸氢优先通道的作用;比较细小的第二相由于本身尺寸小(颗粒 b),故传递的距离就相应较短,这种吸氢通道的作用就不太显著. 这就可合理解释在相同腐蚀增重下,即理论放氢量相同时,含粗大第二相样品的吸氢量均大于含细小第二相样品(图4). 同时,Zr₂(Fe, Ni) 第二相的可逆吸放氢能力强于 Zr(Fe, Cr)₂ 第二相,它捕获和传输氢的能力更强,所以含粗大第二相的 Zr-2 样品吸氢量多于 Zr-4 样品(图4). 这也说明添加合金元素 Ni 的 Zr-2 合金会明显增加腐蚀时的吸氢量.

4 结论

(1) 第二相大小对 Zr-2 和 Zr-4 合金的腐蚀和吸氢均有很大影响:含细小第二相样品的耐腐蚀性能均优于含粗大第二相的样品,同时在相同腐蚀增重下的吸氢量也更低.

(2) 第二相大小对 Zr-2 合金的腐蚀吸氢影响比 Zr-4 合金的大,这与第二相的成分密切相关.

(3) Zr₂(Fe, Ni) 试验合金的可逆吸放氢能力大于 Zr(Fe, Cr)₂,且都大于纯 Zr.

(4) 提出的"在金属/氧化膜界面处 Zr 和 OH⁻ 反应生成的氢可以优先被镶嵌在金属/氧化膜界面处的可逆吸放氢能力强于锆基体的 Zr(Fe, Cr)₂ 和 Zr₂(Fe, Ni) 第二相捕获,它们可作为吸氢的优先通道"的吸氢模型可以合理解释本研究结果.

参 考 文 献

[1] Lemaignan C, Motta A T. In: Kahn R W eds. Translated by Zhou Bangxin(周邦新). *Nuclear Materials*(核科学)

[M]. Beijing: Science Press, 1999: 3.

[2] Shaltiel D, Jacob I, Davidov D. *J Less-Commom Metals*[J], 1977, 53: 117.

[3] Rudling P, Wikmark G, *J Nucl Mater*[J], 1999, 265: 44.

[4] Rudling P, Wikmark G et al. *Zirconium in Nuclear Industry: Twelfth International Symposium*[C]. Toronto, Canada: ASTM STP 1354, 2000.

[5] Cox B. *J Alloy Compd*[J], 1997, 256: 244.

[6] Cox B. *J Nucl Mater*[J], 1999, 264: 283.

[7] Cox B. Wong Y M. *J Nucl Mater*[J], 1999, 270: 134.

[8] Hatano Y, Isobe K et al. *J Nucl Sci Technol*[J], 1996, 33: 944.

[9] Hatano Y, Hitaka R et al. *J Nucl Mater*[J], 1997, 248: 311.

[10] Tägstrom P, Limbäck M et al. *Thirteenth International Symposium*[C]. Philadelphia: ASTM, 2002: 96.

[11] Lelièvre G, Tessier C et al. *J Alloy Compd*[J], 1998, 268: 308.

[12] Yao Meiyi(姚美意), Zhou Bangxin(周邦新). *Thesis for Doctorate*(博士论文)[D]. Shanghai: Shanghai University, 2007.

[13] Lelièvre G, Tessier C et al. *J Alloy Compd*[J], 1998, 268: 308.

[14] Pecheur D, Lefebvre F, Motta A T et al. *Zirconium in the Nuclear Industry: Tenth International Symposium*[C]. Philadelphia: ASTM, 1994: 687.

[15] Iltis X, Lefebvre F, Lemaignan C. *J Nucl Mater*[J], 1995, 224: 121.

[16] Kakiuchi K, Itagaki N, Furuya T et al. *Journal of ASTM International*[J], 2004, 1(10): 349.

[17] Yao Meiyi(姚美意), Zhou Bangxin(周邦新) et al. *Rare Metal Materials and Engineering*(稀有金属材料与工程)[J], 2007, 36(11): 1915.

Hydrogen Absorption Behavior of Zircaloy Corroded in Super-Heated Steam

Abstract: The hydrogen uptake behavior of Zircaloy-2 and Zircaloy-4 containing different second phase particles (SPPs) in size and category during corrosion test in super-heated steam at 400 ℃/10.3 MPa was investigated. The amount of hydrogen uptake was not always corresponding to the corrosion resistance, while it was closely related to the SPPs size and category. At the same mass gain, the difference is significant in the amount of hydrogen uptake between the Zircaloy-2 and Zircaloy-4 specimens with coarse SPPs, while the difference is negligible in the amount of hydrogen uptake between the Zircaloy-2 specimens SPPs and Zircaloy-4 specimens with fine SPPs. The contribution of SPPs category to the hydrogen uptake is more notable to the specimens with coarse SPPs than specimens with fine SPPs. The tests, containing absorption pressure-composition-temperature (PCT), absorption and desorption kinetics, were proceeded in zirconium, $Zr(Fe, Cr)_2$ and $Zr_2(Fe, Ni)$ intermetallic specimens. Results indicate that $Zr(Fe, Cr)_2$ and $Zr_2(Fe, Ni)$ specimens could absorb and desorb hydrogen easily with high speed, while zirconium could only absorb hydrogen easily but desorb hydrogen difficultly. Based on these results, the hydrogen uptake behavior of Zircaloy corroded in super-heated steam at 400 ℃/10.3 MPa were successfully explained with the model that the hydrogen generating from the reaction of zirconium and OH^- at metal/oxide film interface can be captured by $Zr(Fe, Cr)_2$ and $Zr_2(Fe, Ni)$ SPPs embedding at the metal/oxide interface. These SPPs with higher reversible absorbing and desorbing ability acted as a preferred path for hydrogen uptake.

核电站关键材料中的晶界工程问题[*]

摘　要：简要介绍上海大学"核电站关键材料的基础问题研究"课题组关于 690 合金晶界工程 (grain boundary engineering，GBE)的研究情况.借助上海大学分析测试中心的先进材料分析测试设备,开展了一些有特色的工作.从显微组织的表征与控制的角度,研究材料微观结构对宏观性能的影响.简要介绍 690 合金晶界元素偏聚、晶界碳化物析出、晶界网络分布控制以及对耐腐蚀性能影响等方面的研究结果.

　　我国已经明确提出要积极发展核电,因此,开发性能更加优异的核反应堆用材料是提高核电经济性和保证核反应堆安全运行的关键,而新型高性能核电站关键材料的开发都以其深入系统的基础研究为前提.坚持安全第一是我国核电发展的原则,核电站关键材料的可靠性是核电安全的重要保证.蒸汽发生器传热管用镍基合金以及反应堆内构件用不锈钢都是核电站的关键材料.下面以压水堆型核电站蒸汽发生器传热管的晶界工程研究为例,简要介绍本课题组近几年获得的主要研究成果[1-23].

　　在压水反应堆内,核裂变反应释放出的能量,以热能形式被一回路循环水带出,这些可能带有放射性的高温高压水在一回路内部流动,通过蒸汽发生器传热管将热量传给二回路. Ni 基 690 合金作为压水反应堆的蒸汽发生器传热管材料,因其具有优良的抗腐蚀能力而应用于许多核电站,虽然已取得了较好的效果,但一方面需要进一步延长反应堆的设计寿命;另一方面,提高反应堆运行参数(温度和压力)后,690 合金传热管的破裂失效问题仍需加以关注.蒸汽发生器传热管道失效的主要原因在不同时期有着不同的情况.从近些年统计的结果来看,与晶界有关的应力腐蚀开裂和晶间腐蚀仍然是目前蒸汽发生器传热管道失效的主要原因[24].

　　包括 690 合金在内的绝大部分工程材料都是多晶体,材料的性能与其显微组织及晶界结构特征有着非常紧密的联系.如应力腐蚀开裂、晶间腐蚀、合金及杂质元素的偏聚、蠕变等问题,都会受到晶界结构特征的影响[25].1984 年,Watanabe[26] 提出了"晶界设计(grain boundary design)"的概念,随即在 20 世纪 90 年代形成了"晶界工程(grain boundary engineering，GBE)"这一研究领域.通过特定的形变及热处理工艺,提高低 ΣCSL (coincidence site lattice[27],即重位点阵)晶界比例,从而调整材料的晶界特征分布(指不同类型晶界的比例),以及晶界网络的空间分布来提高材料与晶界相关的性能.在低层错能面心立方金属材料中,晶界工程的研究工作取得了一定的进展[25,28].已有大量的有关 Ni 基合金、奥氏体不锈钢、铅合金、铜合金等材料的晶界工程研究结果报道.Lin 等[29] 对 Ni 基 600 合金进行处理,使得低 ΣCSL 晶界比例由 37% 提高到 71%,这使材料在沸腾硫酸铁浸泡腐蚀实验中,24 h 的腐蚀速率降低了 30%~60%.Palumbo 等[30-31] 对铅酸电池的铅合金电极板进行了中等变形量冷轧及短时间退火处理,将这一工艺反复进行数次后,使低 ΣCSL 晶界

---

[*] 本文合作者：夏爽、李慧、陈文觉、姚美意、李强、刘文庆、王均安、褚于良、彭剑超、张金龙.原发表于《上海大学学报〈自然科学版〉》,2011,17(4)：522-528.

比例提高到70％以上,耐腐蚀性能明显改善,电池充放电循环次数提高了1~3倍. Shimada 等[32]将冷轧5％的304不锈钢在927 ℃退火72 h后,低ΣCSL晶界比例超过了80％,晶间腐蚀速率下降了约75％. 690合金也是一种低层错能面心立方金属材料,可以通过晶界工程来提高其与晶界有关的抗晶间腐蚀、抗应力腐蚀开裂等性能. 我们在近些年的工作中,以将晶界工程技术运用于实际生产中为目标,为改善690合金的耐腐蚀性能,开展了690合金的晶界工程研究,其中涉及晶界元素偏聚[14,17]、晶界相析出[4,9,13-14,18]、晶界网络分布控制[5-8,11,15,19,21]以及晶界工程对耐腐蚀性能的影响[1-3,12,16,20,22-23]等方面. 借助于上海大学分析测试中心的先进材料检测分析设备,以及该中心雄厚的智力资源,本课题组进行了许多有意义的研究工作.

1 利用原子探针层析技术研究晶界偏聚问题

镍基合金及奥氏体不锈钢经过固溶处理后,在热时效时晶界处会析出碳化物,导致晶界附近Cr元素贫化,这是引起材料沿晶界腐蚀及开裂等失效问题的重要原因. 利用先进的原子探针层析(atom probe tomography,APT)技术,能够研究各种原子在晶界处的三维空间分布情况,这是目前最精细和最直接的研究晶界元素偏聚的手段. APT技术采用针状样品,其针尖曲率半径小于100 nm. 一般结构材料的晶粒尺寸大约为几十微米,比符合要求的针尖样品尺度大2个数量级左右,所以晶界出现在样品针尖处的概率很低,而且在测试过程中很容易断针,这又增加了实验研究的难度. 我们通过大量的相关研究工作,提高了针尖部位出现晶界的几率,获得了690合金和304不锈钢晶界偏聚的实验数据[14].

图1给出了500 ℃、时效0.5 h处理后不同原子在690合金晶界附近的三维空间分布图像[14]. 可以明显看出,B,C,Si向690合金的晶界附近偏聚,如图1(a)~图1(c)所示,其中GB为晶界,但观察不到其他元素的原子向晶界处偏聚的现象. 图1(d)和图1(e)给出了垂直于晶界方向统计获得的不同元素原子在690合金晶界附近的成分分布情况,可以看出,C,B,Si明显富集在晶界两侧约3 nm处,其中虚线表示晶界位置. 利用Origin软件对采用APT技术获得的数据进行处理,还可以平行于晶界来统计晶界面上元素偏聚后的分布情况. 图1(f)为Cr原子在晶界两侧各2 nm范围内的平均等浓度分布图,暖色调代表高浓度区域. 可以看出,Cr原子在晶界上发生偏聚后的浓度分布并不均匀,而是呈现周期性的变化规律. Cr原子在晶界面上的高浓度区域(暖色调)明显形成相互平行的条带状结构,这种结构的周期性规律很明显,间距约为7.2 nm,这可能与Ke[33]提出的晶界结构中"好区"与"坏区"的周期分布规律有关.

2 利用高分辨率透射电子显微镜研究晶界碳化物的析出过程

晶界处析出碳化物的形貌及分布对材料性能有很大影响,已有大量有关690合金中晶界上析出碳化物的结构种类、形貌及分布状态的研究报道. 与许多其他Ni基合金及奥氏体不锈钢类似,在690合金的晶界处会析出面心立方结构的$M_{23}C_6$碳化物,该碳化物的晶格常数约为晶粒基体的3倍,通常与晶界一侧的晶粒具有共格取向关系[4]. 然而,碳化物析出的过程却仍然不很清楚,如碳化物析出时与哪一侧的晶粒共格、析出时原子如何重新排列、碳化物形核长大时引起周围应力场的变化等问题. 我们利用高分辨率透射电子显微镜(high-resolution

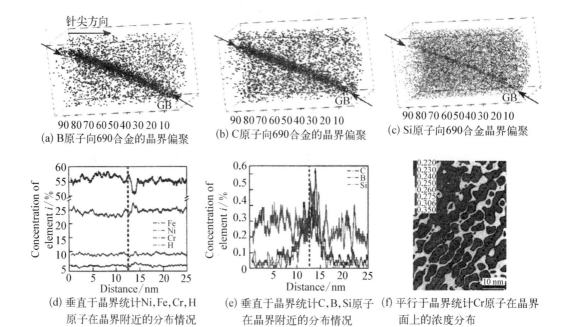

图 1 对固溶处理后的 690 合金进行 500 ℃、时效 0.5 h 处理后,用三维原子探针分析晶界处不同元素原子的三维空间分布图像

Fig. 1 Alloy 690 was aged at 500 ℃ for 0.5 h after solution annealing, three dimensional spatial distribution of different elements obtained by the ATP technique

transmission electron microscopy, HRTEM),研究了不同时效热处理后,晶界处碳化物与基体界面的显微结构,分析了原子排列的晶格条纹像,获得了碳化物析出时的形核规律.

图 2 为 715 ℃、时效 15 h 处理后晶界处析出的碳化物与基体之间原子排列的晶格条纹像[14],其中 γ 为晶粒基体,c 为碳化物.在这些界面弯曲的位置可以看出,在碳化物与基体之间存在与碳化物和基体结构都不同的过渡区域,其宽度不到 30 nm.对高分辨晶格条纹像进行快速傅里叶变换(fast Fourier transform,FFT)后分析标定后的 FFT 图像,结果表明,这种过渡区域具有六方结构,晶格常数 $c = 0.62$ nm,$a = 0.44$ nm,并与基体具有 $\{111\}_\gamma // \{0001\}_{hcp}$,$\{110\}_\gamma // \{10\bar{1}0\}_{hcp}$ 的共格取向关系.这种过渡相的形成与碳化物形成时造成的贫 Cr 区与应变能有关.

3 利用电子背散射衍射技术研究晶界工程处理之后晶界网络分布特征

绝大多数工程应用的材料都是多晶体材料,晶粒之间的界面对材料整体性能有很大影响,因而,"晶界问题"一直受到材料研究者的重视.随着现代分析检测仪器的发展,电子背散射衍射(electron backscattered diffraction,EBSD)技术能够方便、快速地获取及标定晶体样品表面晶粒的取向,以自动判定晶粒之间的取向关系,得出晶界的结构特征.在三维空间,晶界相互连接构成网络,沿晶界腐蚀及开裂的扩展行为,不仅与某个晶界的结构类型有关,更与整个晶界网络的分布特征有关[28].我们通过形变及退火工艺方法,基于再结晶过程中不同类型晶界迁移演化的规律,控制最终形成的晶界网络特征及分布[6-8,22].通过 EBSD 技术可以测定并分析研究二维平面上的晶界网络分布.

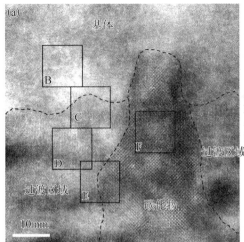

(a) 碳化物与附近晶粒基体的晶格条纹像

(b) (a)图中区域B的FFT像，标定为面心立方结构的晶粒基体区域

(c) (a)图中区域C的FFT像，标定为面心立方结构晶粒基体与六方结构过渡相共存区域

(d) (a)图中区域D的FFT像，标定为六方结构过渡相区域

(e) (a)图中区域E的FFT像，标定为六方结构过渡相与面心立方结构碳化物共存区域

(f) (a)图中区域F的FFT像，标定为面心立方结构碳化物区域

图 2 对固溶处理后的 690 合金进行 715 ℃、时效 15 h 处理后，晶界处析出碳化物与晶粒基体的 HRTEM 图像

Fig. 2 Alloy 690 was aged at 715 ℃ for 15 h after solution annealing, HRTEM images of the carbide precipitate and the grain matrix near the grain boundary

图 3 为经过晶界工程处理（见图 3(a)）和未经过晶界工程处理（见图 3(b)）的 690 合金管材样品中不同类型晶界的取向成像显微（orientation imaging microscopy，OIM）图，其中

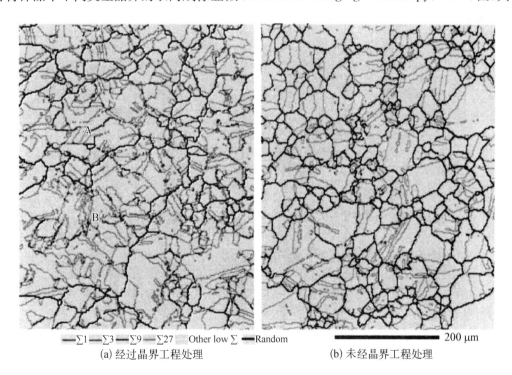

──Σ1 ──Σ3 ──Σ9 ──Σ27 Other low Σ ──Random   200 μm

(a) 经过晶界工程处理   (b) 未经晶界工程处理

图 3 经过晶界工程处理与未经晶界工程处理的 690 合金管材样品中不同类型晶界的 OIM 图

Fig. 3 OIM maps of different types of gain boundaries of the GBE Alloy 690 tube sample and the non-GBE Alloy 690 tube sample

不同颜色的线条代表不同类型的晶界. 可以看出, 晶界工程处理后的样品形成了以大尺寸"互有 $\Sigma 3^n$ 取向关系晶粒的团簇"为特征的晶界网络分布[1], 如图 3(a) 中的"A"和"B"所示. 在晶粒团簇内部, 由于再结晶时多重孪晶的充分发展形成了大量的 $\Sigma 3^n$ 晶界, 并且它们相互连接形成了大量 $\Sigma 3 - \Sigma 3 - \Sigma 9$, $\Sigma 3 - \Sigma 9 - \Sigma 27$ 等类型的三叉界角. 晶粒团簇之间基本为连续的随机晶界, 而未经晶界工程处理的样品就没有这样的晶界网络分布特征. 根据我们在 690 合金中总结的控制低 ΣCSL 晶界比例及晶界网络分布的规律, 通过制定合理的工艺, 在黄铜、304 不锈钢和铅合金中均可以将低 ΣCSL 晶界比例提高到 70% 以上, 并形成以大尺寸"互有 $\Sigma 3^n$ 取向关系晶粒的团簇"为特征的晶界网络分布(见图 4)[5]. 这表明我们针对 690 合金提出的提高低 ΣCSL 晶界比例过程的机理可以普遍适用于多种低层错能面心立方金属材料.

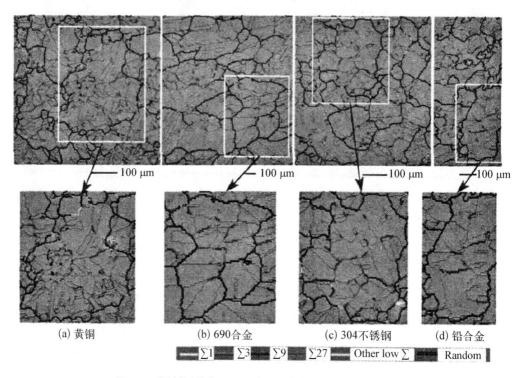

图 4　不同低层错能面心立方金属材料中的晶界网络分布

Fig. 4　Grain boundary networks of various facecentered cubic metallic materials with low stacking fault energy

4　690 合金管材通过晶界工程处理之后耐晶间腐蚀性能明显提高

我们成功地在 690 合金管材中进行了晶界工程处理, 所研发的晶界工程处理工艺能够很好地与现行实际生产所用的生产工艺衔接[1,8,22]. 同时, 运用腐蚀实验评价了晶界工程处理对 690 合金管材耐腐蚀性能的影响, 结果如图 5 所示. 可以看出, 在腐蚀时间相同的情况下, 经过晶界工程处理的管材样品由于晶粒脱落造成的腐蚀失重明显低于未经过晶界工程处理的样品[1,3,7]. 图 6 给出了经过晶界工程处理的管材样品腐蚀后管材截面的显微组织, 其中图 6(d) 中有一个大尺寸"互有 $\Sigma 3^n$ 取向关系晶粒的团簇", 图 6(e) 中的 R 表示随机晶界,

$\Sigma3_c$ 表示共格孪晶界,$\Sigma3_i$ 表示非共格孪晶界. 通过统计分析表明,只有共格 $\Sigma3$ 晶界才具有特殊性能,不被腐蚀,而其他类型晶界均会被腐蚀. 如果在一个三叉界角 $\Sigma3 - \Sigma3 - \Sigma9$ 处,$\Sigma9$ 晶界被腐蚀,那么沿晶界腐蚀的扩展就会被另外 2 条 $\Sigma3$ 晶界所阻挡. 在"互有 $\Sigma3^n$"取向关系晶粒的团簇"内部,$\Sigma3 - \Sigma3 - \Sigma9$ 类型的三叉界角与其他 $\Sigma3^n$ 类型的三叉界角相互连接. 如果腐蚀沿着 $\Sigma27$ 晶界,或者 $\Sigma81$ 晶界,或者更高阶的 $\Sigma3^n$ 晶界向晶粒团簇内扩展时,最终总会遇到 $\Sigma3 - \Sigma3 - \Sigma9$ 三叉界角,从而被阻止继续扩展[1-2]. 所以,大尺寸"互有 $\Sigma3^n$ 取向关系晶粒的团簇"晶界网络能够阻止晶间腐蚀向材料内部扩展.

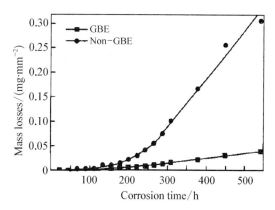

图 5 690 合金管材经过晶界工程处理和未经过晶界工程处理的腐蚀失重曲线

Fig. 5 Weight losses of the GBE Alloy 690 tube sample and non-GBE Alloy 690 tube sample as a function of intergranular corrosion time

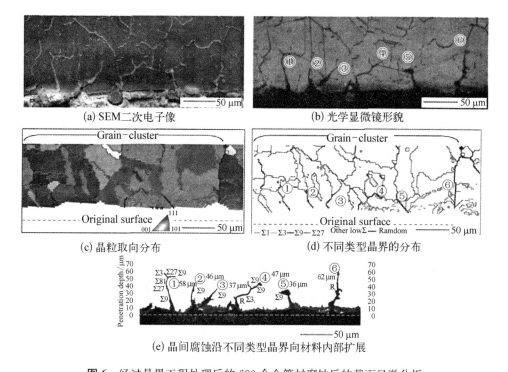

图 6 经过晶界工程处理后的 690 合金管材腐蚀后的截面显微分析

Fig. 6 Microanalysis of the cross-section of the GBE Alloy 690 tube sample undergone intergranular corrosion

5 结束语

上海大学"核电站关键材料的基础问题研究"课题组关于 690 合金晶界工程方面的研究在近几年得到了以下一些主要成果:

（1）利用原子探针层析技术研究了 690 合金晶界偏聚问题,发现在晶界面上偏聚原子的浓度并非均匀分布,而是呈有规律的周期性分布;

（2）利用高分辨率透射电子显微镜研究了 690 合金晶界碳化物在不同晶界上的析出过程及形态演化,指出了碳化物会与哪一侧晶粒存在共格关系及其原因,并发现碳化物与晶粒基体之间存在六方结构的过渡相;

（3）利用电子背散射衍射技术研究经过晶界工程处理之后的晶界网络分布特征,明确了经过晶界工程处理后的样品会形成以大尺寸"互有 $\Sigma 3^n$ 取向关系晶粒的团簇"为特征的晶界网络分布,并阐明了这种团簇形成的原因,指导了"晶界工程"工艺制度的制定;

（4）690 合金管材通过晶界工程处理之后,其耐晶间腐蚀性能明显提高,相互连接的 $\Sigma 3^n$ 类型的三叉界角是阻止沿晶界腐蚀向材料内部扩展的关键因素.

参 考 文 献

［1］XIA S, LI H, LIU T G, et al. Appling grain boundary engineering to Alloy 690 tube for enhancing intergranular corrosion resistance ［J］. Journal of Nuclear Materials, 2011, 416(3)：303－310.

［2］HU C L, XIA S, LI H, et al. Improving the intergranular corrosion resistance of 304 stainless steel by grain boundary network control ［J］. Corrosion Science, 2011, 53(5)：1880－1886.

［3］XIA S, LI H, HU C L, et al. Grain boundary network control and its effect on intergranular corrosion resistance of Alloy 690 ［C/CD］ // Fontevraud 7：Int Symp Contribution of Materials Investigations to Improve the Safety and Performance of LWRs. 2010：A025T06.

［4］LI H, XIA S, ZHOU B X, et al. The dependence of carbide morphology on grain boundary character in the highly twinned Alloy 690 ［J］. Journal of Nuclear Materials, 2010, 399(1)：108－113.

［5］XIA S, ZHOU B X, CHEN W J, et al. Features of highly twinned microstructures produced by GBE in FCC materials ［J］. Materials Science Forum, 2010, 638：2870－2875.

［6］XIA S, ZHOU B X, CHEN W J. Grain cluster microstructure and grain boundary character distribution in Alloy 690 ［J］. Metallurgical and Materials Transactions A, 2009, 40A(12)：3016－3030.

［7］XIA S, ZHOU B X, CHEN W J. Effect of single-step strain and annealing on grain boundary character distribution and intergranular corrosion in Alloy 690 ［J］. Journal of Materials Science, 2008, 43(9)：2990－3000.

［8］XIA S, ZHOU B X, CHEN W J, et al. Effects of strain and annealing processes on the distribution of $\Sigma 3$ boundaries in a Ni-based superalloy ［J］. Scripta Materialia, 2006, 54(12)：2019－2022.

［9］李慧,夏爽,周邦新,等. 690 合金中晶界处碳化物形貌的演化［J］. 金属学报,2009,45(2)：195－198.

［10］夏爽,周邦新,陈文觉,等. 铅合金在高温退火过程中晶界特征分布的演化［J］. 金属学报,2006,42(2)：129－133.

［11］刘廷光,夏爽,李慧,等. 690 合金中原始晶粒尺寸对晶界工程处理后晶界网络的影响［J］. 金属学报,2011,47(7)：859－864.

［12］胡长亮,夏爽,李慧,等. 晶界网络特征对 304 不锈钢晶间应力腐蚀开裂的影响［J］. 金属学报,2011,47(7)：939－945.

［13］李慧,夏爽,周邦新,等. 镍基 690 合金中晶界碳化物析出的研究［J］. 金属学报,2011,47(7)：853－858.

［14］李慧. Ni－Cr－Fe 合金中晶界偏聚与晶界析出的研究［D］. 上海：上海大学,2011：44－79.

［15］夏爽,李慧,周邦新,等. 金属材料中退火孪晶的控制及利用——晶界工程研究［J］. 自然杂志,2010,32(2)：94－100.

［16］罗鑫,夏爽,周邦新,等. 晶界特征分布对 304 不锈钢应力腐蚀开裂的影响［J］. 上海大学学报：自然科学版,2010,16(2)：177－182.

［17］李慧,夏爽,周邦新,等. 原子探针层析方法研究 690 合金晶界偏聚的初步结果［J］. 电子显微学报,2011,30(3)：32－35.

［18］李慧,夏爽,胡长亮,等. 利用 EBSD 技术对 690 合金不同类型晶界处碳化物形貌的研究［J］. 电子显微学报,2010,29(1)：730－735.

［19］夏爽,罗鑫,周邦新,等. 304 不锈钢中"晶粒团簇"显微组织的特征与晶界特征分布的关系［J］. 电子显微学报,2010,

29(1): 678 - 683.

[20] 夏爽,周邦新,陈文觉. 690 合金的晶界特征分布及其对晶间腐蚀的影响[J]. 电子显微学报,2008,27(6): 461 - 468.

[21] 夏爽,周邦新,陈文觉. 形变及热处理对 690 合金晶界特征分布的影响[J]. 稀有金属材料与工程,2008,37(6): 999 - 1003.

[22] 夏爽,周邦新,陈文觉,等. 提高 690 合金材料耐腐蚀性能的工艺方法: 中国,ZL200710038731.5[P]. 2008 - 07 - 08.

[23] 李慧,夏爽,周邦新,等. 690 合金中晶界网络分布的控制及其对晶间腐蚀性能的影响[J]. 中国材料进展,2011,30(5): 11 - 14.

[24] DIERCKS D R, SHACK W J, MUSCAR J. Overview of steam generator tube degradation and integrity issues [J]. Nuclear Engineering and Design, 1999, 194(1): 19 - 30.

[25] RANDLE V. Twinning-related grain boundary engineering [J]. Acta Mater, 2004, 52(14): 4067 - 4081.

[26] WATANABE T. An approach to grain boundary design of strong and ductile polycrystals [J]. Res Mech, 1984, 11(1): 47 - 84.

[27] KRONBERG M L, WILSON F H. Secondary recrystallization in copper [J]. Trans AIME, 1949, 185: 501 - 514.

[28] ROHRER G S. Measuring and interpreting the structure of grain-boundary networks [J]. J Am Ceram Soc, 2011, 94(3): 633 - 646.

[29] LIN P, PALUMBO G, ERB U, et al. Influence of grain boundary character distribution on sensitization and intergranular corrosion of Alloy 600 [J]. Scripta Metall Mater, 1995, 33(9): 1387 - 1392.

[30] LEHOCKEY E M, LIMOGES D, et al. On improving the corrosion and growth resistance of positive Pb-acid battery grids by grain boundary engineering [J]. J Power Sour, 1999, 78(1/2): 79 - 83.

[31] PALUMBO G, ERB U. Enhancing the operation life and performance of lead-acid batteries via grain-boundary engineering [J]. MRS Bulletin, 1999, 24: 27 - 32.

[32] SHIMADA M, KOKAWA H, WANG Z J, et al. Optimization of grain boundary character distribution for intergranular corrosion resistant 304 stainless steel by twin induced grain boundary engineering [J]. Acta Mater, 2002, 50(9): 2331 - 2341.

[33] KE T S. A grain boundary model and the mechanism of viscous intercrystalline slip [J]. Journal of Applied Physics, 1949, 20(3): 274 - 280.

Grain Boundary Engineering Issues in Key Structural Materials Used for Nuclear Power Plants

Abstract: Grain boundary engineering (GBE) issues of Alloy 690 are briefly summarized. The work was carried out by the research group "Fundamental Issues of the Key Structural Materials for Nuclear Power Plants" at Shanghai University. By using advanced material characterization techniques, the microstructures were characterized to establish its relationship with the Properties. This article presents the research findings in grain boundary segregation, carbide precipitation, grain boundary network control and its effect on the intergranular corrosion resistance.

RPV 模拟钢中纳米富 Cu 相的
析出和结构演化研究*

摘　要：提高了 Cu 含量的核反应堆压力容器（RPV）模拟钢经过 880 ℃水淬和 660 ℃调质处理后，在 370 ℃时效 6 000 h，利用 HRTEM，EDS 和原子探针层析（APT）方法研究了纳米富 Cu 相的析出过程和晶体结构演化．观察到 Cu 原子在 α-Fe 基体的{110}晶面上以 3 层为周期发生偏聚，并产生了很大的内应力使晶格发生畸变，这是富 Cu 相析出时的形核过程；随着 Cu 含量的增加和富 Cu 区的扩大，内应力也随着增大，富 Cu 区沿着 α-Fe 基体的{110}晶面发生切变，形成了 ABC/BCA/CAB/ABC 排列的多孪晶 9R 结构；Cu 含量继续增加，富 Cu 相最终转变为 fcc 结构．富 Cu 相的尺寸在 1—8 nm 范围内，数量密度为 0.71×10^{23} m^{-3}．富 Cu 相中还含有 3%—8%（质量分数）的 Ni 和 Mn，并且在相界面上发生偏聚，从而抑制了富 Cu 相的长大．

　　核反应堆压力容器（RPV）是压水堆核电站中不可更换的大型关键部件，由 Mn-Ni-Mo 低合金铁素体钢（A508-Ⅲ钢）制成，在服役工况下，经过长期中子辐照后会引起韧脆转变温度升高，这是影响 RPV 安全运行的关键因素，也是核电站寿命的决定性因素．国外的大量研究[1-5]表明，RPV 钢的辐照脆化效应主要是由于中子辐照损伤以及析出了高数量密度的纳米富 Cu 相．自 20 世纪 60 年代以来，通过热时效对 Fe-Cu 合金中富 Cu 相的析出特点进行了大量研究．最初的研究认为，时效早期析出的纳米富 Cu 相为球形，与铁素体基体保持共格，是 bcc 结构；过时效后富 Cu 相长大．由球形转变为棒状的 fcc 结构，与铁素体基体保持 K-S 位向关系，从而导致 Fe-Cu 合金具有脆性[6-10]．但是，Pizzini 等[11]认为在 Fe-Cu 合金中富 Cu 相的结构从 bcc 向 fcc 转变的过程是比较复杂的，Othen 等[12,13]发现在 Fe-Cu 合金中富 Cu 相由 bcc 结构先转变为 9R 结构，再转变为扭曲的 3R 结构，最后转变为 fcc 结构．然而，对于富 Cu 相在析出时的成核过程以及晶体学信息并不十分清楚．原子探针层析（APT）方法可以获得富 Cu 相析出时富 Cu 团簇的大小和成分变化信息；用高分辨透射电子显微镜（HRTEM）观察可以获得富 Cu 相的晶体学信息，因此，本文采用 APT 和 HRTEM 相结合的方法对富 Cu 相析出时的成核过程和晶体结构演化进行了研究．

　　如果用中子辐照来研究 RPV 钢中富 Cu 相析出，除了辐照费用高昂之外，辐照后样品具有放射性也给实验操作带来诸多不便．因此，本文采用了提高 Cu 含量的 RPV 模拟钢，研究在低温长期时效过程中纳米富 Cu 相的析出和晶体结构的演化特征，这对认识 RPV 钢中富 Cu 相的析出过程以及辐照脆化机理也具有重要意义．

1　实验方法

　　Cu 在 RPV 钢中是一种杂质元素，含量一般低于 0.08%（质量分数，下同），通过热时效使富 Cu 相析出需要相当长的时间，Miller 和 Rusell[4]报道了在 290 ℃经过大约 22 年时效

* 本文合作者：徐刚、楚大峰、蔡琳玲、王伟、彭剑超．原发表于《金属学报》，2011，47(7)：905-911．

后,可以观察到纳米富 Cu 相的析出. 为了加速富 Cu 相的析出,缩短实验时间,本实验采用了提高 Cu 含量至 0.6% 的 RPV 模拟钢,其他成分与 A508-Ⅲ钢保持一致. RPV 模拟钢由真空感应炉冶炼,铸锭重约 40 kg,化学成分见表 1. 钢锭经热锻和热轧制成 4 mm 厚的钢板,将钢板切成 40 mm×30 mm 的小样品在 880 ℃加热 0.5 h 后水淬,然后在 660 ℃加热 10 h 进行调质处理,最后样品在 370 ℃进行 6 000 h 等温时效处理.

表 1 RPV 模拟钢的化学成分

Table 1 Composition of the pressure vessel model steel

Unit	Cu	Ni	Mn	Si	P	C	S	Mo	Fe
Atomic fraction, %	0.53	0.81	1.60	0.77	0.03	1.00	0.011	0.31	Bal.
Mass fraction, %	0.60	0.85	1.58	0.39	0.016	0.22	0.006	0.54	Bal.

用电火花线切割方法从时效处理后的样品中心部位切出长约 20 mm,截面边长约 0.5 mm 的方形细棒,分别用 25%(体积分数)的高氯酸乙酸溶液和 2% 的高氯酸 2-丁氧基乙醇溶液作电解液分两步电解抛光,制备出曲率半径小于 100 nm 的针状样品,APT 样品的制样方法及分析原理详见文献[14]. 用 LEAP 3000HR 型三维原子探针(3DAP)对针尖样品进行分析,样品冷却至 50 K,脉冲频率为 5 kHz,脉冲分数为 20%.

用电火花线切割机从经过时效处理后试样的截面上取下 0.5 mm 厚的薄片,用砂纸将其表面磨平,用 30% HNO_3＋10% HF 的水溶液进行化学减薄至约 120 μm,再用金相砂纸磨薄至约 100 μm,随后用冲样机制得直径为 3 mm 的薄片,再用砂纸磨薄至约 60 μm,最后用 10% $HClO_4$＋90% C_2H_5OH 溶液在 -60 ℃进行双喷电解抛光制备透射电镜(TEM)观察所需的薄样品. 用 H-800 型 TEM 对样品进行一般观察后,再用 JEM-2010F 型场发射 HRTEM 对析出相进行观察,并在 80 kV 下进行能谱(EDS)分析. 所获得的 HRTEM 像用 Gatan 公司的 Digital Micrograph 软件分析晶体结构.

2 实验结果与讨论

2.1 富 Cu 析出相的 TEM 和 HRTEM 分析

RPV 模拟钢为低碳低合金钢,淬火后形成位错密度很高的板条马氏体组织,经过660 ℃高温回火后马氏体分解,位错密度大大降低,但仍然保留着板条形貌,板条内部形成了一些亚晶组织. 在板条间的界面上析出了粒状渗碳体(图 1a),在铁素体基体中还有针状的 Mo_2C 析出(图 1a 中插图),这与 A508-Ⅲ钢在相同热处理条件下的组织类似[15]. 另外,根据 Cu 在铁素体中平衡固溶度的计算公式[16]可知:在 660 和 370 ℃时,Cu 在铁素体中的固溶度(质量分数)分别约为 0.46% 和 0.016%,比模拟钢的 Cu 含量都低,因而调质处理时已经有少量细小的富 Cu 相颗粒析出,呈球状或短棒状(图 1a 中插图). 这些 Cu 颗粒为 fcc 结构,并与基体保持 K-S 位相关系[6]. 在 370 ℃时效 6 000 h 后,板条状的显微组织没有发生明显变化(图 1b),但析出了大量纳米富 Cu 相颗粒(图 1b 中插图).

富 Cu 相析出的早期不仅尺寸小,还与基体保持共格关系,用一般 TEM 观察很难判断,而用 HRTEM 和 EDS 相结合的分析方法,则能够获得更多的信息. 图 2a 是经过 370 ℃时效处理 6 000 h 的样品,当电子束沿着 α-Fe⟨111⟩方向入射时拍摄的晶格条纹像,得到了 {110}

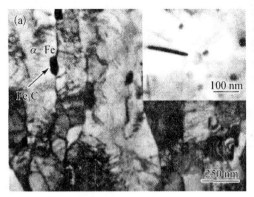
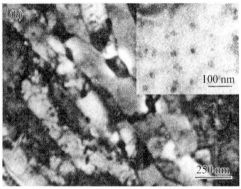

图1 RPV模拟钢在不同热处理条件下的 TEM 像

Fig. 1 TEM micrographs of RPV model steel at different heat treatment conditions (a) tempered at 660 ℃ for 10 h after 880 ℃ water-quenching (modified treatment) (needlelike Mo₂C and Cu-rich particles precipitated in α-Fe matrix shown by the inset); (b) aged at 370 ℃ for 6 000 h after modified treatment (more Cu-rich particles precipitated in matrix shown by the inset)

晶面排列的图像. 照片中存在衬度异常的区域 A, EDS 分析(图 2b)表明 A 区中的 Cu 含量为 3.13%, 高于样品中的平均值, 并含有可觉察的 Ni 和 Mn, 大小约为 3 nm. A 区的 Fourier 变换(FFT)图如图 2a 中插图所示. 经过反 Fourier 变换(IFFT)滤波后得到的图像如图 2c 所示. 从 FFT 图中可以看出, 在 000 和 1̄01 衍射斑点之间出现了强度较弱的斑点(箭头所指), 这是由于 A 区中排列着以 3 层为周期并且衬度较亮的原子面, 如图 2a 和 c 中箭头所示. 在晶格条纹像中产生衬度的差别, 除了因晶格发生了畸变之外, 最大的可能是异种原子在晶面上发生了偏聚. EDS 分析 A 区中的 Cu 含量高于基体中的平均值, 那么偏聚在 {110} 晶面上的这些异种原子就应该是 Cu. 计算模拟的结果[17,18]表明, Cu 原子在 α-Fe 中更容易沿着 ⟨111⟩ 或 ⟨110⟩ 方向迁移. 在 bcc 结构中, 只有 {110} 晶面中同时包含了 ⟨111⟩ 和 ⟨110⟩ 方向, 如

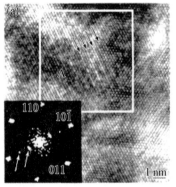
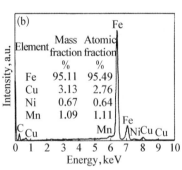
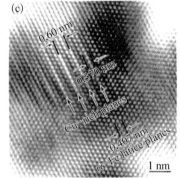

图2 在 370 ℃时效 6 000 h 后富 Cu 相的 HRTEM 像和图中 A 区的 FFT 图, A 区的 EDS 分析及 A 区的 IFFT 图

Fig. 2 HRTEM analyses on Cu-rich precipitate in RPV model steel aged at 370 ℃ for 6 000 h (a) HRTEM image and FFT pattern of region A (weak spots shown by white arrows in FFT pattern corresponding to Cu-rich lattice planes shown by black arrows in HRTEM image); (b) EDS analysis of region A showing more Cu content than that of matrix (see Table 1); (c) IFFT image after filtrating (Cu-rich lattice planes clearly shown by arrows)

果 Cu 原子沿着⟨111⟩和⟨110⟩方向扩散迁移而发生偏聚,那么最终聚集在{110}晶面上的可能性最大. 从 IFFT 图中(图 2c)可以测量晶面间距:α-Fe 基体中两层(10$\bar{1}$)晶面的间距为 0.407 nm,与 α-Fe 的点阵常数非常吻合(PDF number:06-0696);而 Cu 偏聚后衬度较亮的{110}晶面两侧之间的晶面间距为 0.370 nm,晶面间距缩小约 9%. 这表明 Cu 原子在{110}晶面上发生偏聚后产生了较大的应力,导致晶格发生畸变. 由于晶格的畸变,Cu 原子不可能在连续的数层{110}晶面上同时发生偏聚,这或许是观察以以 3 层为周期发生偏聚的原因. 在发生 Cu 原子偏聚的两层晶面之间的间距为 0.60 nm. 在富 Cu 相析出之前,Cu 原子先在 α-Fe 基体的{110}晶面上发生偏聚,与 Al-Cu 合金中 CuAl$_2$ 析出之前,Cu 在 Al 基体的{100}晶面上发生偏聚形成 GP 区的现象[19];与 Al-Zn-Mg 合金中析出 η' 之前 Zn 和 Mg 在 Al 基体的{100}晶面上发生偏聚形成 GPI 区的现象[20];与 Mg-Re-Zn-Zr 合金中析出 β'' 之前 RE,Zn 和 Zr 在基体的(0001)晶面上发生偏聚的现象[21]都十分类似,这或许是第二相析出时成核初期的共同规律.

随着 Cu 原子在 α-Fe 基体{110}晶面上偏聚的数量增加和区域扩大,富 Cu 区中的应力会越来越大,这为富 Cu 区发生晶体结构的演化和相变创造了条件. 图 3a 是另一处富 Cu 相的晶格条纹像,用 EDS 分析(图 3b)得到的 Cu 含量达到 14.29%,由于分析结果受到基体的干扰,富 Cu 相中 Cu 的实际含量会比分析的结果高许多. 该富 Cu 相为椭球形,长轴方向约 13 nm,短轴方向约 9 nm. 从鱼骨状条纹的特点可判断此富 Cu 相为多孪晶的 9R 结构[12]. 将该富 Cu 相分为 A,B 两区,这两区的 FFT 和 IFFT 图分别嵌镶在一张图中,如图 3c 和 d 所

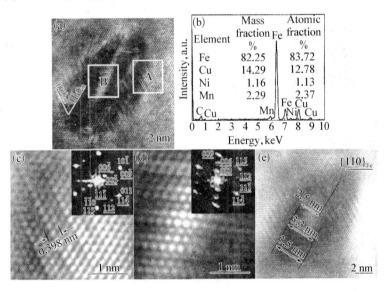

图 3 RPV 模拟钢在 370 ℃时效 6 000 h 后较大尺寸富 Cu 相的 HRTEM 分析

Fig. 3 HRTEM analyses on a larger Cu-rich precipitate in RPV model steel aged at 370 ℃ for 6 000 h (a) HRTEM image of a Cu-rich particle with 9R structure;(b) EDS analysis of the Cu-rich particle;(c) composite FFT pattern (the index numbers of 9R structure were underlined) and IFFT image of region A in Fig. 3a, showing orientation relationship:$(\bar{1}\bar{1}4)_{9R} // (0\ \bar{1}\bar{1})_{\alpha\text{-Fe}}$,$(\bar{1}10)_{9R} // [1\bar{1}1]_{\alpha\text{-Fe}}$;(d) FFT pattern and IFFT image of region B in Fig. 3a, comparison to Fig. 3c showed a twinning relationship between regions A and B with twinning plane;(e) low magnified IFFT image, showing ladder-like twinning plane

示.整个富 Cu 相经过滤波处理后的 IFFT 图如图 3e 所示.从图 3c 中标定后的 A 区 FFT 图可以看出,这时已经形成了按 $ABC/BCA/CAB/ABC$ 排列的 9R 结构,与 α-Fe 基体的取向关系为 $(\bar{1}\bar{1}4)_{9R}\parallel(0\bar{1}\bar{1})_{\alpha\text{-Fe}}$ 和 $[110]_{9R}\parallel[1\bar{1}1]_{\alpha\text{-Fe}}$.从 IFFT 图衬度的变化可以看出,晶体点阵仍然大致保留了以 3 层为周期的特征,不过随着向 α-Fe 基体中延伸,周期性结构逐渐变得模糊并消失,说明这时已经形成 9R 结构的富 Cu 相与 α-Fe 基体的界面之间仍然保持着共格关系.从图 3d 中标定的 B 区 FFT 图可以看出,A 和 B 区同为 9R 结构并互为孪晶关系,孪晶面 $(\bar{1}\bar{1}4)_{9R}$ 和 α-Fe 基体的 $(011)_{Fe}$ 面平行.B 区 FFT 图中的斑点被拉长,说明晶体点阵存在畸变,从 B 区的 IFFT 图像中也能看出晶格发生了扭曲.

从图 3e 可以看出,9R 结构富 Cu 相中孪晶面呈阶梯状分布,阶梯的高度在 0.2—0.8 nm 之间,中部孪晶的宽度在 2.5—3.5 nm 之间.这种结构特征说明了 Cu 原子在 α-Fe 基体的 {110} 晶面上发生偏聚后产生的应力,可以通过在 α-Fe 基体的 {110} 晶面上发生切变,或者说通过已经形成的 9R 结构的晶面上发生切变而得到弛豫.这种通过切变产生孪晶的位错机制还有待进一步分析,但阶梯状的孪晶面说明了这种切变过程正在发展中.

图 4a 是尺寸约为 30 nm 球形富 Cu 相的晶格条纹像.由于富 Cu 相镶嵌在 α-Fe 基体中,因而出现了 morié 条纹,EDS 分析得到的 Cu 含量为 18.73%(图 4b),而团簇中的实际含量还会比分析结果高许多.该富 Cu 区的 FFT 图如图 4a 中插图所示,其标定后的示意图如图 4c 所示.从 FFT 图可以判定此富 Cu 相为 fcc 结构,且与 α-Fe 基体保持 K-S 关系:$(111)_{Cu}\parallel(011)_{\alpha\text{-Fe}}$,$[0\bar{1}1]_{Cu}\parallel[1\bar{1}1]_{\alpha\text{-Fe}}$.利用 HRTEM 和 EDS 研究了萃取复型获得的富 Cu 相,也观察到 Cu 含量达到 80%—90% 后的富 Cu 相是 fcc 结构[22],这是 9R 结构继续演化后的结果.

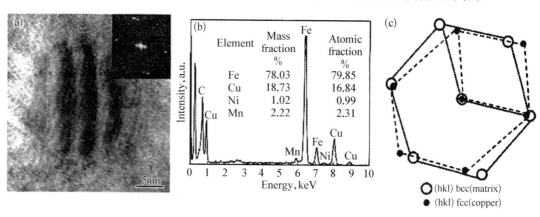

图 4 RPV 模拟钢在 370 ℃时效 6 000 h 后更大尺寸富 Cu 相的 HRTEM 分析

Fig. 4 HRTEM analyses on a larger Cu-rich precipitate in RPV model steel aged at 370 ℃ for 6 000 h (a) HRTEM image and FFT pattern of the Cu-rich particle with fcc structure;(b) EDS analysis of the Cu-rich particle;(c) schematic representation of FFT pattern of Cu-rich particle, showing K-S orientation relationship:$(111)_{Cu}\parallel(011)_{\alpha\text{-Fe}}$,$[0\bar{1}1]_{Cu}\parallel[1\bar{1}1]_{\alpha\text{-Fe}}$

2.2 富 Cu 析出相的 APT 分析

APT 可以精确地测量各种元素的原子在空间的分布以及原子团簇的形成,目前普遍使用 MSEM(maximum separation envelope method)分析 APT 的实验结果来界定纳米团簇的形成.MSEM 是用原子之间的最大距离 d_{max} 来表征某溶质原子聚集的紧密程度,用最小的原子数目 N_{min} 来表征某溶质原子聚集而形成团簇时最少的原子数.通过改变 d_{max} 去检测富 Cu

团簇的析出,可以区分这些团簇在析出过程中所处的先后次序,因为在 d_{max} 大的团簇中 Cu 原子比较松散,应该比 d_{max} 小的团簇处于析出过程的更早期. 设置 $N_{min} = 20$, $d_{max} = 0.4$, 0.5, 0.6 nm 时,APT 分析经过 370 ℃时效处理 6 000 h 后样品中富 Cu 团簇分布的结果如图 5 所示. 为了便于观察富 Cu 团簇的空间分布,MSEM 分析时去除了 Cu 原子的本底,只留下了富 Cu 团簇. 当设定 $d_{max} = 0.4$ nm 时,得到了 19 个富 Cu 团簇,这些富 Cu 团簇应该已经形成了 Cu 的晶体结构,因为 Cu 的晶胞常数 $a = 0.3615$ nm,团簇中 Cu 原子间的距离已经小于 0.4 nm. 当设定 $d_{max} = 0.5$ nm 时,这时仍为 19 个富 Cu 团簇,团簇的分布也没有变化;当设定 $d_{max} = 0.6$ nm 时,得到了 21 个富 Cu 团簇,比设置 $d_{max} = 0.4$ nm 时多了 2 个,在图 5 中用 A_1,A_2 标出了这 2 个富 Cu 团簇的位置. 在设置 d_{max} 较大时得到的富 Cu 团簇中,Cu 原子之间的距离较大,应该是富 Cu 相析出过程的早期阶段,分析这 2 个团簇中 Cu 原子的空间分布,发现它们聚集后呈现一定的规律,图 6 是设置 $d_{max} = 0.6$ nm 时检测到编号为 A_1 团簇的 Cu 原子分布图. 从 2 个相互成 90° 的方向观察,可以看到 Cu 原子聚集在 3 个小面上,面与面之间的距离是 0.6 nm. APT 分析时,原子从样品尖端发生场蒸发并在电场作用下飞向探测器,探测器检测到原子在空间的位置会因为原子的热振动而有偏差,因而 APT 的分析结果不能给出晶体结构的精确信息;另外,探测器的检测效率只有 50%—60%,有一部分 Cu 原子会被漏检. 因此,图 6 中得到 Cu 原子聚集后的分布图应该是富 Cu 相析出初期 Cu 原子在 α-Fe 基体某晶面上发生偏聚后的结果,这与 HRTEM 观察到 Cu 原子在 α-Fe 基体的 {110} 晶面上发生偏聚(图 2)的结果是一致的,Cu 原子偏聚后晶面之间的间距都是 0.6 nm.

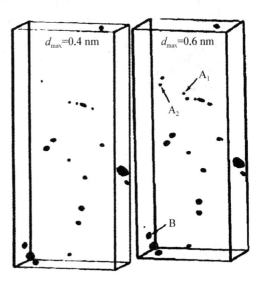

图 5 APT 分析样品在 370 ℃时效 6 000 h 后的富 Cu 相分布图

Fig. 5 Distributions of Cu-rich clusters within an analyzed volume 40 nm×67 nm×144 nm in the specimen aged at 370 ℃ for 6 000 h

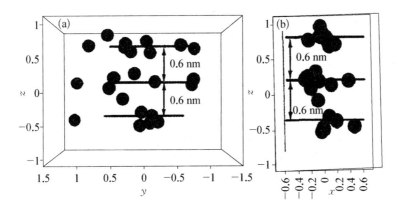

图 6 相互成 90°方向观察图 5 中编号为 A_1 的富 Cu 原子团簇的 Cu 原子分布图

Fig. 6 Distributions of Cu atoms for Cu-rich cluster labeled as A_1 in Fig. 5 within an analyzed volume 1. 5 nm×3 nm×2. 2 nm, observed from two directions, (a) and (b), at 90 degrees to each other

当设置 $d_{max} = 0.6$ nm 和 $N_{min} = 20$ 时,在 40 nm×67 nm×144 nm 的空间中有 21 个富 Cu 团簇析出,通过计算得出富 Cu 团簇的数量密度约为 $0.71×10^{23}$ m^{-3},它们的尺寸分布在 1—8 nm 之间. 富 Cu 相析出后,其中同时还含有大约 3%—8% 的 Ni 和 Mn. 图 7 是图 5 中编号为 B 团簇的 Cu,Ni 和 Mn 原子分布图和成分分布图. 可以看出,Ni 和 Mn 较多的富集在团簇与 α-Fe 基体的界面处. 当富 Cu 相长大时,除了需要 Cu 原子的扩散之外,还需要 Ni 和 Mn 原子向基体中扩散,因而 Ni 和 Mn 富集在相界面上会抑制富 Cu 相的长大.

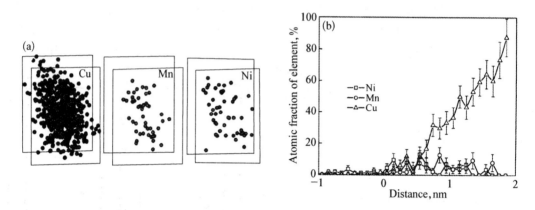

图7　图 5 中编号为 B 的富 Cu 原子团簇中 Cu,Ni 和 Mn 原子分布图和成分图

Fig. 7　Distributions of Cu, Ni and Mn atoms within an analyzed volume 3 nm×4 nm×12 nm for the Cu-rich cluster labeled as B in Fig. 5 (a) and the composition profiles of Cu, Ni and Mn (b)

3　结论

(1) 在析出富 Cu 相的成核过程中,观察到 Cu 原子沿着 α-Fe 基体的 {110} 晶面以 3 层为周期发生偏聚,并产生了较大的内应力,导致晶格发生畸变.

(2) 随着 Cu 原子偏聚量的增加和区域扩大,内应力也增加,通过在 α-Fe 基体的 {110} 晶面上发生切变形成了 ABC/BCA/CAB/ABC 排列的多孪晶 9R 结构,但在富 Cu 相与 α-Fe 基体的界面处晶格失配并不严重.

(3) 富 Cu 相的尺寸在 1—8 nm 之间,数量密度为 $0.71×10^{23}$ m^{-3}. 富 Cu 相中还含有 3%—8%(质量分数) 的 Ni 和 Mn,并偏聚在富 Cu 相和 α-Fe 基体的界面上,从而抑制了富 Cu 相的长大.

参 考 文 献

[1] Toyama T, Nagai Y, Tang Z, Hasegawa M, Almazouzi A, van Walle E, Gerard R. *Acta Mater*, 2007; 55: 6852.

[2] Miller M K, Russell K F, Sokolov M A, Nanstad R K. *J Nucl Mater*, 2007; 361: 248.

[3] Fujii K, Fukuya K, Nakata N, Hono K, Nagai Y, Hasegawa M. *J Nucl Mater*, 2005; 340: 247.

[4] Miller M K, Russell K F. *J Nucl Mater*, 2007; 371: 145.

[5] Miller M K, Nanstad P K, Sokolov M A, Russell K F. *J Nucl Mater*, 2006; 351: 216.

[6] Hornbogen E, Glenn R C. *Trans Metall Soc AIME*, 1960; 218: 1064.

[7] Speich G R, Oriani R A. *Trans Metall Soc AIME*, 1965; 233: 623.

[8] Goodman S R, Brenners S S, Low J R. *Metall Trans*, 1973; 4: 2363.

[9] Buswell J T, English C A, Herherington M G, Phythian W J, Smith G D W, Worral G M. In: Steele L E ed. , *Proc*

14th Int Symp Effects of Radiation on Materials, Vol. 2, Andover, Massachusseetts: The American Society for Testing and Materials, 1990: 127.

[10] Phythian W J, Foreman A J E, English C A, Buswell J T, Hetherington M, Roberts K, Pizzini S. In: Steele L E ed. , *Proc 15th Int Symp Effects of Radiation on Materials*, Nashvile, Tennessee: The American Society for Testing and Materials, 1990: 131.

[11] Pizzini S, Roberts K J, Phythian W J, English C A, Greaves G N. *Philos Mag Lett*, 1990; 61: 223.

[12] Othen P J, Jenkins M L, Smith G D W, Phythian W J. *Philos Mag Lett*, 1991; 64: 383.

[13] Othen P J, Jenkins M L, Smith G D W. *Philos Mag*, 1994; 70A: 1.

[14] Wang W, Zhu J J, Lin M D, Zhou B X, Liu W Q. *J Univ Sci Technol Beijing*, 2010; 1: 39.
（王伟,朱娟娟,林民东,周邦新,刘文庆. 北京科技大学学报,2010;1: 39）

[15] Sheng Z Q, Xiao H, Peng F. *Nucl Power Eng*, 1988; 9: 49.
（盛钟琦,肖红,彭峰. 核动力工程,1988;9: 49）

[16] Yong Q L. *Secondary Phase in Steel*. Beijing: Metallurgical Industry Press, 2006: 127.
（雍岐龙. 钢铁材料中的第二相. 北京: 冶金工业出版社,2006: 127）

[17] Osetsky Y N, Serra A. *Philos Mag*, 1996; 73A: 249.

[18] Lebouar Y. *Acta Mater*, 2001; 49: 2661.

[19] Konno T J, Hiraga K, Kawasaki M. *Scr Mater*, 2001; 44: 2303.

[20] Berg L K, Gjønnes J, Hansen V, Li X Z, Knutson-Wedel M, Waterloo G, Schryvers D, Wallenberg L R. *Acta Mater*, 2001; 49: 3443.

[21] Ping D H, Hono K, Nie J F. *Scr Mater*, 2003; 48: 1017.

[22] Chu D F, Xu G, Wang W, Peng J C, Wang J A, Zhou B X. *Acta Metall Sin*, 2011; 47: 269.
（楚大锋,徐刚,王伟,彭剑超,王均安,周邦新. 金属学报,2011;47: 269）

Investigation on the Precipitation and Structural Evolution of Cu-Rich Nanophase in RPV Model Steel

Abstract: The crystal structure of Cu-rich nanophase in reactor pressure vessel (RPV) model steel was investigated by means of HRTEM, EDS and APT methods. RPV model steel was prepared by vacuum induction furnace melting with higher content of Cu (0. 6%, mass fraction). The ingot about 40 kg in weight was forged and hot rolled to 4 mm in thickness and then cut to specimens of 40 mm×30 mm. Those specimens were further heat treated by 880 ℃ for 0. 5 h water quenching and 660 ℃ for 10 h tempering, and finally aged at 370 ℃ for 6 000 h. It was observed that the Cu atoms segregated on {110} planes of α-Fe matrix in a period every three layers and the distortion of crystal lattice was induced by inner stress produced by the segregation of Cu atoms during the nucleation of the precipitation of Cu-rich nanophase. The inner stress in Cu-rich regions enhanced with the increase of Cu concentration as well as the enlargement in size. It was also observed that the Cu-rich regions underwent a transformation from bcc structure to 9R structure with twins by means of a shear along the {110} plane of α-Fe matrix. Finally, the Cu-rich clusters transformed to fcc structure with further increase of Cu content. The results obtained by APT analysis show that the equivalent diameter and the number density of Cu-rich clusters are about 1—8 nm and 0. 71×10^{23} m^{-3}, respectively. The content of 3%—8% Ni and Mn within the Cu-rich clusters was detected. It is suggested that the growth of Cu-rich clusters was restrained by the segregation of Ni and Mn atoms on the boundaries around the Cu-rich clusters.

Cu 含量对 Zr - 0.80Sn - 0.34Nb - 0.39Fe - 0.10Cr - xCu 合金在 500 ℃ 过热蒸汽中耐腐蚀性能的影响*

摘 要: 采用静态高压釜腐蚀实验研究了 Zr - 0.80Sn - 0.34Nb - 0.39Fe - 0.1 Cr - xCu(x= 0.05—0.5,质量分数,%)合金在 500 ℃,10.3 MPa 过热蒸汽中的耐腐蚀性能,利用 TEM 观察了合金的显微组织.结果表明:添加(0.05—0.5)Cu 对合金在 500 ℃过热蒸汽中的耐腐蚀性能影响不大.当 $x \leq 0.2$ 时,合金中的第二相主要为立方结构的 $Zr(Fe, Cr, Nb)_2$ 和含 Cu 的正交结构的 Zr_3Fe;当 $x > 0.2$ 时,除了 $Zr(Fe, Cr, Nb)_2$ 和含 Cu 的 Zr_3Fe 外,还有四方结构的 Zr_2Cu 析出.$Zr(Fe, Cr, Nb)_2$ 比较细小,而含 Cu 第二相的尺寸较大.即使在添加 0.05Cu 的合金中也有含 Cu 第二相析出,说明 Cu 在该合金 α - Zr 基体中的固溶量很低.因此,添加(0.05—0.5)Cu 对该合金在 500 ℃过热蒸汽中的耐腐蚀性能影响不大的原因可能与固溶在 α - Zr 基体中的 Cu 含量低有关.

锆合金是核反应堆中用作核燃料元件包壳的一种重要结构材料.为了提高核电的经济性,需要进一步加深核燃料的燃耗,这样就要求延长核燃料的换料周期,延长燃料元件在反应堆内停留的时间,因而对燃料元件包壳用材料-锆合金的性能,特别是耐蚀性能提出了更高的要求.核燃料元件在反应堆堆芯中工作时,受到中子辐照、高温高压水的腐蚀和冲刷,腐蚀、氢脆、蠕变、疲劳及辐照损伤等是导致锆合金包壳发生失效的主要原因,其中锆合金包壳的耐水侧腐蚀性能是影响燃料元件使用寿命的最主要因素[1].

合金化是开发高性能锆合金的有效途径.在现有锆合金的基础上,优化合金成分的不同配比或添加其他种类合金元素是开发高性能锆合金的 2 条基本思路,如在 Zr - Sn 和 Zr - Sn -Nb 为基的合金中,降低 Sn 含量到 0.6—0.7(质量分数,%,下同),提高 Fe 含量至 0.4 或(Fe+Cr)含量至 0.6,会进一步提高锆合金的耐腐蚀性能[2-6];韩国学者 Park 等[7]及 Jung 等[8]在 Zr - 1.1Nb 基础上添加 0.05Cu,发展了具有优良耐腐蚀性能的 HANA - 6 合金.可见,Cu 对改善 Zr - Nb 合金耐腐蚀性能是有益的添加元素.

本课题组前期工作[9]系统研究了添加(0.05—0.5)Cu 对 M5(Zr - 1Nb)合金在 500 ℃过热蒸汽中耐腐蚀性能的影响,结果表明 Cu 含量低于 0.2 时,随着 Cu 含量的增加,合金的耐腐蚀性能得到明显的改善;继续提高 Cu 含量对进一步改善合金的耐腐蚀性能作用不大.Hong 等[10]研究了添加 Cu 对 Zr - Sn - Fe - Cr 合金耐腐蚀性能的影响,发现添加 Cu 后合金的耐腐蚀性能也得到一定程度改善.Zr - Sn - Nb 系合金是在综合了 Zr - Sn 和 Zr - Nb 系合金优点后发展起来的新型锆合金,Zr - 0.80Sn - 0.34Nb - 0.39Fe - 0.10Cr(以下简称 S5)是我国自主研发的优化 N18 锆合金.

本文研究了 Cu 含量对 S5 合金在 500 ℃,10.3 MPa 过热蒸汽中耐腐蚀性能的影响,利用 TEM 和 EDS 对析出相进行了分析,探讨 Cu 含量影响 S5 合金耐腐蚀性能的规律和机理.

---

* 本文合作者:姚美意、张宇、李士炉、张欣、周军.原发表于《金属学报》,2011,47(7):872 - 876.

1 实验方法

以 S5 合金为母合金,添加不同含量的 Cu 制备成 Zr - 0.80Sn - 0.34Nb - 0.39Fe - 0.10Cr - xCu (x=0.05—0.5)锆合金(简称为 S5+xCu),合金熔炼过程中考虑了 Cu 的烧损. 同时以重熔的 S5(S5 - remelted)试样作为参照.这些合金样品的熔炼加工制备过程与 M5+ xCu 合金完全相同[9],即在高纯 Ar 气保护下,用真空非自耗电弧炉熔炼成约 60 g 的合金 锭,为保证合金成分均匀,合金锭共熔炼 6 次,每熔炼一次翻转一次合金锭.合金锭经热压成 条块状后进行 1 030 ℃,40 min 的 β 相均匀化处理,再通过热轧、1 030 ℃,40 min 的 β 相水 淬和 50% 冷轧等工艺制得 0.6 mm 厚的片状样品,最后进行 580 ℃,50 h 退火处理.采用这 一制备工艺的目的是为了获得均匀细小弥散分布的第二相,在显微组织优化的前提下比较 Cu 含量的影响[11,12].样品退火加热均在真空中进行,每次热处理前都要采用 30% H_2O+ 30% HNO_3+30% H_2SO_4+10% HF(体积分数)混合酸酸洗的方法去除试样表面的污染和 氧化物;β 相水淬是通过将样品真空封装在石英管中,加热到 1 030 ℃保温 40 min 后立即淬 入水中并快速敲碎石英管来实现的.

电感耦合等离子体原子发射光谱(ICPAES)分析得到的合金成分列在表 1 中,样品中的 Cu 含量与设计成分接近,Sn 含量比原 S5 合金中的略高,这应该是分析误差造成的,而原 S5 合金中的其他合金元素在熔炼过程中可能会发生微量损耗.制备好的样品经过上述混合酸 酸洗和去离子水清洗后放入静态高压釜中进行 500 ℃,10.3 MPa 的过热蒸汽腐蚀实验,腐 蚀增重结果为 5 块试样的平均值.

表 1 实验用合金成分

Table 1 Chemical compositions of experimental alloys measured by ICPAES

(mass fraction %)

Alloy	Sn	Nb	Fe	Cr	Cu	Zr
S5+0.05Cu	0.78	0.35	0.34	0.075	0.058	Bal.
S5+0.1Cu	0.82	0.38	0.35	0.073	0.11	Bal.
S5+0.2Cu	0.82	0.37	0.34	0.073	0.21	Bal.
S5+0.35Cu	0.82	0.35	0.34	0.075	0.35	Bal.
S5+0.5Cu	0.82	0.37	0.37	0.075	0.52	Bal.
S5 - remelted	0.81	0.35	0.33	0.070	0.005 9	Bal.

用 JEM - 200CX 型透射电子显微镜(TEM)观察试样的显微组织,利用 JEM - 2010F 型 高分辨透射电镜配置的能谱仪(EDS)分析合金中第二相成分(由于样品中含有 Cu,所以分 析时使用的 TEM 样品台是 Be 双倾台),并通过选区电子衍射(SAD)分析第二相的晶体结 构.TEM 试样用双喷电解抛光方法制备,电解液为 10%(体积分数)高氯酸酒精溶液,抛光 电压为直流 50 V,温度−40 ℃.

2 实验结果

2.1 腐蚀增重

图 1 是 S5+xCu 合金和 S5 重熔合金在 500 ℃,10.3 MPa 过热蒸汽中腐蚀 750 h 的增

重曲线. 可以看出, S5＋xCu 合金和重熔合金在 500 ℃, 10.3 MPa 过热蒸汽环境中腐蚀时, 都没有产生疖状腐蚀现象, 它们的耐均匀腐蚀性能也没有明显差别. 如腐蚀 750 h 时, S5 重熔试样的腐蚀增重为 399 mg/dm², S5＋0.05Cu, S5＋0.1Cu, S5＋0.2Cu, S5＋0.35Cu 和 S5＋0.5Cu 试样的腐蚀增重分别为 404, 384, 405, 363 和 389 mg/dm². 这说明添加 (0.05—0.5)Cu 对 S5 合金在 500 ℃高压过热蒸汽环境中的耐腐蚀性能没有明显的改善作用, 这与 Cu 对 Zr－Nb 系的 M5 合金耐腐蚀性能影响规律[9]是完全不同的.

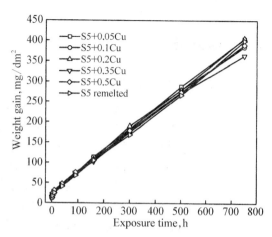

图 1 S5＋xCu 合金在 500 ℃, 10.3 MPa 过热蒸汽中的腐蚀增重曲线

Fig. 1 Weight gains of the S5 ＋ xCu alloys varied with corrosion time in 500 ℃ and 10.3 MPa superheated steam (the samples were heat-treated: β-phase field water qunching at 1 030 ℃ for 40 min and annealed at 580 ℃ for 50 h; (0.05—0.5)Cu addition has little effect on the corrosion resistance of the alloys)

2.2 合金的显微组织

图 2a1, b1 和 c1 分别是 S5＋0.05Cu, S5＋0.2Cu 和 S5＋0.5Cu 合金腐蚀前显微组织的 TEM 像. 可以看出, S5＋xCu 合金经 β 相区水淬、冷轧后在 580 ℃退火处理 50 h 发生了再结晶, 第二相颗粒均匀弥散分布在晶粒内部或晶界上, 同时存在尺寸有一定差别的第二相. 对合金中不同尺寸第二相进行 SAD 和 EDS 分析的结果表明: 小于 60 nm 的第二相为 hcp 结构的 Zr(Fe, Cr, Nb)₂(图 2a2 和 a3), 但也有少量较大尺寸(150—250 nm)的含 Cu 的 Zr(Fe, Cr, Nb)₂; 尺寸为 100—300 nm 的第二相有两类: 一类为含有 Cu 的正交结构 Zr₃Fe(图 2b2 和 b3), 另一类为四方结构的 Zr₂Cu(图 2c2 和 c3), 其中部分含有 Fe. 表 2 总结了第二相的种类、数量和尺寸与 Cu 含量之间的关系. 可以看出, 当合金中的 Cu 含量较低(≤0.2)时, 析出的第二相为 Zr(Fe, Cr, Nb)₂ 和含 Cu 的 Zr₃Fe; 当合金中的 Cu 含量较高(＞0.2)时, 除 Zr(Fe, Cr, Nb)₂ 和 Zr₃Fe 外, 还出现了 Zr₂Cu, 并随着 Cu 含量的增加, Zr₂Cu 数量增多. 从分析结果可以看出: 在 S5 合金中添加 0.05Cu 时, 就已有含 Cu 第二相析出, 这说明 Cu 在 S5 合金 α-Zr 基体中的固溶含量较低.

3 分析讨论

添加(0.05—0.5)Cu 对 Zr—Sn—Nb—Fe—Cr 系的 S5 合金在 500 ℃过热高压蒸汽中耐腐蚀性能的影响不大(图 1), 但对 Zr－Nb 系的 M5 合金在同样环境中的耐腐蚀性能却有显著影响[9]. 这说明合金元素 Cu 对不同体系锆合金耐腐蚀性能影响的规律是不同的. 有学者[13-16]在研究锆合金的腐蚀行为时, 认为起主要作用的是固溶在 α-Zr 基体中的合金元素含量. 本课题组在研究 M5＋xCu 合金的耐腐蚀性能时也认为 Cu 提高合金耐腐蚀性能的原因主要与固溶在 α-Zr 基体中的 Cu 含量有关[9]. 因此, Cu 对 S5 和 M5 合金耐腐蚀性能的影响规律不同很可能与 Cu 在不同体系锆合金基体中的固溶含量不同有关. 一方面, Cu 会与锆合金中的其他合金元素发生交互作用, 以第二相的形式析出, 如 Zr₃Fe 和部分 Zr(Fe, Cr, Nb)₂ 第二相中都含有 Cu(图 2 和表 2), 这必然降低 Cu 在 S5 合金(Zr－Sn－Nb－Fe－Cr

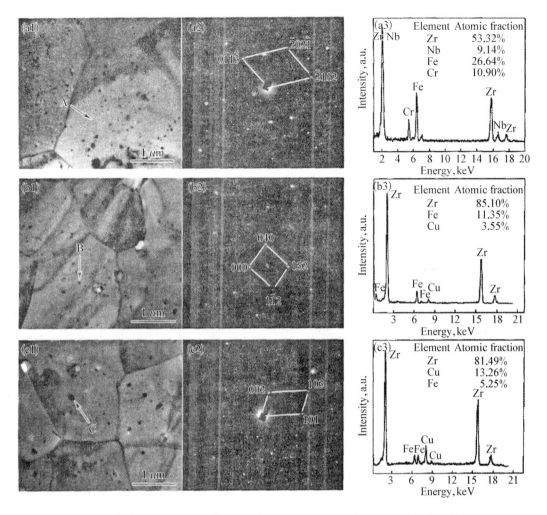

图 2　3 个合金试样腐蚀前的 TEM 像及合金中第二相的 SAD 和 EDS 分析结果

Fig. 2　TEM analyses of three alloys annealed at 580 ℃ for 50 h　(a1) TEM image of alloy S5 + 0.05Cu；(a2) SAD patterns of hcp $Zr(Fe, Cr, Nb)_2$ with size less than 60 nm；(a3) EDS of $Zr(Fe, Cr, Nb)_2$ particle shown by arrow A in Fig. 2a1；(b1) TEM image of alloy S5 + 0.2Cu；(b2) SAD patters of orthogonal $Zr_3Fe(Cu)$；(b3) EDS of $Zr_3Fe(Cu)$ particle shown by arrow B in Fig. 2b1；(c1) TEM image of alloy S5 + 0.5Cu；(c2) SAD patters of tetragonal $Zr_2Cu(Fe)$；(c3) EDS of $Zr_2Cu(Fe)$ particle shown by arrow C in Fig. 2c1

系)α – Zr 基体中的固溶含量. 第二相的 EDS 分析结果也证明了即使 S5 合金中只添加 0.05Cu，也会有含 Cu 的第二相析出(表 2). 另一方面,合金元素 Sn 和 Nb 在 α – Zr 基体中的固溶含量较大,这也可能会影响 Cu 在 α – Zr 基体中的固溶含量. 根据 Zr – Nb 和 Zr – Sn 二元相图可知,580 ℃时,Nb 和 Sn 在 α – Zr 中的最大平衡固溶含量分别约为 0.6 和 1.3,因此,与 M5(Zr – 1Nb)合金比较,S5(Zr – 1Sn – 0.35Nb – 0.4Fe – 0.1Cr)合金在 580 ℃退火时,α – Zr 中固溶的 Sn, Nb 等合金元素的总含量会比 M5 合金高,这也可能使固溶在 α – Zr 基体中的 Cu 含量减少. Hong 等[10]在研究添加 Cu 对 Zr – 4 合金在 360 ℃去离子水中的腐蚀行为影响时发现,当 Sn 含量从 1 增加到 1.5 时,会使同样添加 0.1Cu 的合金中含 Cu 的析出相增多,也使得耐腐蚀性能降低,这说明 Sn 含量对 Cu 在 α – Zr 基体中的固溶含量确实有影响. 因此, Cu 在 S5 合金基体中的固溶含量较小可能就是添加(0.05—0.5)Cu 对 S5 合金耐腐蚀性能影响不大的原因.

表 2 S5+xCu 合金中的第二相信息
Table 2 Details of the precipitates in S5+xCu alloys

Precipitate	Proportion in alloy			Particle size, nm
	S5+0.05Cu	S5+0.2Cu	S5+0.5Cu	
$Zr(Fe, Cr, Nb)_2$	Major	Part	Part	<60 nm (major) and 150—250 nm (minor, containing Cu)
Zr_3Fe	Minor	Part	Part	100—250 nm (all, containing Cu)
Zr_2Cu	—	—	Part	150—300 nm (part, containing Fe)

4 结论

(1) 当 $x \leqslant 0.2$ 时, $Zr-0.80Sn-0.34Nb-0.39Fe-0.10Cr-x$Cu 合金中的第二相主要为密排六方结构的 $Zr(Fe, Cr, Nb)_2$ 和含 Cu 的正交结构 Zr_3Fe; 当 $x > 0.2$ 时, 除了上述两种第二相外, 还有四方结构的 Zr_2Cu 析出. $Zr(Fe, Cr, Nb)_2$ 比较细小, 而含 Cu 第二相的尺寸较大. 即使在添加 0.05Cu 的合金中就有含 Cu 第二相析出, 这说明 Cu 在该合金 α-Zr 基体中的固溶含量很低.

(2) 添加(0.05—0.5)Cu 对 $Zr-0.80Sn-0.34Nb-0.39Fe-0.10Cr$ 合金在 500 ℃过热蒸汽中耐腐蚀性能的影响不大, 这可能与固溶在 α-Zr 基体中的 Cu 含量较低有关.

感谢上海大学微结构重点实验室测试中心的彭剑超老师在合金显微组织观察和分析中给予的帮助与指导.

参 考 文 献

[1] IAEA - TECDOC - 996, IAEA, Vienna, 1998, ISSN 1011 - 4289, January, 1998.

[2] Garzarolli F, Broy Y, Busch R A. In: Bradley E R, Sabol G P, eds., *Zirconium in the Nuclear Industry: 11th International Symposium*, ASTM STP 1295, Garmisch-Partenkirchen, Germany: ASTM International, 1996: 850.

[3] Isobe T, Matsuo Y. In: Eucken C M, Garde A M, eds., *Zirconium in the Nuclear Industry: Ninth International Symposium*, ASTM STP 1132, Philadelphia: ASTM International, 1991: 346.

[4] Takeda K, Anada H. In: Sabol G P, Moan G D, eds., *Zirconium in the Nuclear Industry: Twelfth International Symposium*, ASTM STP 1354, West Conshohocken, PA: ASTM International, 2000: 592.

[5] Broy Y, Garzarolli F, Seibold A, Van Swam L F. In: Sabol G P, Moan G D, eds., *Zirconium in the Nuclear Industry: Twelfth International Symposium*, ASTM STP 1354, West Conshohocken, PA: ASTM International, 2000: 609.

[6] Yueh H K, Kesterson R L, Comstock R J, Shah H H, Colburn D J, Dahlback M, Hallstadius L. *J ASTM Int*, 2005; 2: 330.

[7] Park J Y, Choi B K, Yoo S J, Jeong Y H. *J Nucl Mater*, 2006; 359: 59.

[8] Jung Y H, Lee M H, Kim H G, Park J Y, Jeong Y H. *J Alloys Compd*, 2009; 479: 423.

[9] Li S L, Yao M Y, Zhang X, Geng J Q, Peng J C, Zhou B X. *Acta Metall Sin*, 2011; 47: 163.
（李士炉，姚美意，张欣，耿建桥，彭剑超，周邦新. 金属学报, 2011; 47: 163）

[10] Hong H S, Moon J S, Kim S J, Lee K S. *J Nucl Mater*, 2001; 297: 113.

[11] Mardon J P, Charquet D, Senevat J. In: Sabol G P, Moan G D, eds. , *Zirconium in the Nuclear Industry: Twelfth International Symposium*, *ASTM STP 1354*, West Conshohocken, PA: ASTM International, 2000: 505.

[12] Yao M Y. *PhD Thesis*. Shanghai University, 2007.

(姚美意. 上海大学博士论文, 2007)

[13] Jeong Y H, Lee K O, Kim H G. *J Nucl Mater*, 2002; 302: 9.

[14] Kim H G, Jeong Y H, Kim T H. *J Nucl Mater*, 2004; 326: 125.

[15] Jeong Y H, Kim H G, Kim D J. *J Nucl Mater*, 2003; 323: 72.

[16] Yao M Y, Zhou B X, Li Q, Liu W Q, Geng X, Lu Y P. *J Nucl Mater*, 2008; 374: 197.

Effect of Cu Content on the Corrosion Resistance of Zr – 0. 80Sn – 0. 34Nb – 0. 39Fe – 0. 10Cr – xCu Alloy in Superheated Steam at 500 ℃

Abstract: The effect of Cu content on the corrosion resistance of Zr – 0. 8Sn – 0. 34Nb – 0. 39Fe – 0. 1Cr – xCu alloys (x=0. 05—0. 5, mass fraction, %) was investigated in superheated steam at 500 ℃ and 10. 3 MPa by autoclave tests. The microstructures of the alloys are observed by TEM. The results show that (0. 05—0. 5) Cu addition has little effect on the corrosion resistance of the alloys. When x is below 0. 2, the precipitates Zr (Fe, Cr, Nb)$_2$ with hcp structure and Zr$_3$Fe containing Cu with orthorhombic structure are detected. When x is above 0. 2, besides Zr(Fe, Cr, Nb)$_2$ and Zr$_3$Fe containing Cu, the precipitate of Zr$_2$Cu with tetragonal structure is also detected. Zr(Fe, Cr, Nb)$_2$ precipitates are smaller than the precipitates containing Cu in size. The precipitates containing Cu are found in the alloy even with 0. 05Cu, indicating that the Cu content in α – Zr matrix is very small. Therefore, the reason that the Cu content has little effect on the corrosion resistance of the alloys is maybe related to the lower Cu content in α – Zr matrix.

添加 Nb 对 Zr-4 合金在 500 ℃ 过热蒸汽中耐腐蚀性能的影响*

摘　要： 用高压釜腐蚀实验研究了在 Zr-4 合金成分基础上添加 0.1%—0.3%（质量分数）Nb 的合金在 500 ℃/10.3 MPa 过热蒸汽中的耐腐蚀性能. 用 TEM 和 SEM 分别观察了合金的显微组织和氧化膜断口形貌. 结果表明，合金在 500 ℃/10.3 MPa 过热蒸汽中腐蚀 500 h 均未出现疖状腐蚀，完全抑制了疖状腐蚀的产生，这与 Nb 在 α-Zr 中的固溶量较大有关，固溶在 α-Zr 中的 Nb 能抑制疖状腐蚀斑的形核，提高耐疖状腐蚀性能；合金耐均匀腐蚀性能随着 Nb 含量的增加而降低，这与 Nb 的添加降低了固溶在 α-Zr 中的（Fe+Cr）含量有关，也与 Zr(Fe, Cr, Nb)₂ 第二相的析出有关，这 2 种因素都会加快氧化膜显微组织在腐蚀过程中的演化，促进孔隙和微裂纹的形成.

锆合金是核反应堆中用作核燃料包壳的一种重要结构材料. 在高温高压水或蒸汽中腐蚀时，锆合金表面一般发生均匀腐蚀. 均匀腐蚀的特点为：腐蚀转折前，形成的氧化膜黑色光亮、均匀致密，具有保护性；腐蚀转折后，氧化膜由亮黑色转变为土黄色，当膜厚增加到一定厚度时还会发生剥落现象. 用在沸水堆（BWR）中的 Zr-2 合金包壳（Zr-(1.2—1.7)Sn-(0.07—0.2)Fe-(0.05—0.15)Cr-(0.03—0.08)Ni，质量分数，%，下同）和用在压水堆（PWR）中的 Zr-4 合金包壳（Zr-(1.2—1.7)Sn-(0.18—0.24)Fe-(0.07—0.13)Cr）在富氧水质中还会出现不均匀腐蚀的现象，即疖状腐蚀. 疖状腐蚀的特点为：首先在表面出现白色疖状斑，白斑的剖面呈凸透镜状，随着疖状斑的生长而连成片，此时氧化膜疏松易脱落[1]. 在堆外一般用高压釜在 450—500 ℃ 过热蒸汽中的腐蚀实验来研究和评价锆合金的耐疖状腐蚀性能.

通过改变热加工工艺可以改善 Zr-2 和 Zr-4 合金的耐疖状腐蚀性能[2,3]. 本课题组在前期对经过不同热处理的 Zr-4 合金进行了耐疖状腐蚀性能研究[3]，发现与常规处理的 Zr-4 样品相比，820 ℃，1 h+A.C.（A.C. 是指加热结束后将电炉推离石英管，在石英管外壁淋水冷却的过程）处理的 Zr-4 样品耐疖状腐蚀性能得到了明显改善，在整个腐蚀过程中，片状样品的两个表面上始终没有出现疖状腐蚀斑. 但是片状样品周边由于发生疖状腐蚀而出现氧化膜严重剥落；1 000 ℃，0.5 h+A.C. 处理的 Zr-4 样品的耐疖状腐蚀性能最好，在整个 1 100 h 的腐蚀过程中都没有出现疖状腐蚀斑，样品周边的氧化膜也没有出现剥落；认为这种现象与样品的织构取向以及适当增加的 Fe 和 Cr 在 α-Zr 中的过饱和固溶量有关. 另外，本课题组还研究了几种 Zr-Sn-Nb 锆合金在 500 ℃ 过热蒸汽中的耐疖状腐蚀性能，发现 Nb 含量高于 0.24% 的几种 Zr-Sn-Nb 锆合金在 500 ℃ 过热蒸汽中腐蚀数百小时也未出现疖状腐蚀现象，这可能与 Nb 在 α-Zr 中的固溶度比 Fe 和 Cr 大有关[4,5]. 考虑到 Nb 在 α-Zr 中的固溶度较大，研究了添加微量的 Nb(0.05%—0.1%) 对 Zr-4 合金耐疖状腐蚀性能的影响，发现 Zr-4 合金中添加微量合金元素 Nb 后确实可以改善 Zr-4 合金的耐疖状腐

*　本文合作者：姚美意、李士炉、张欣、彭剑超、赵旭山、沈剑韵. 原发表于《金属学报》，2011，47(7)：865-871.

蚀性能,但添加 0.1% Nb 后还没有完全抑制疖状腐蚀现象[6].本文研究了在 Zr-4 合金基础上添加(0.1%—0.3%)Nb 的合金在 500 ℃过热蒸汽中的耐腐蚀性能,并探讨了 Nb 元素影响锆合金耐腐蚀性能的机制.

1 实验方法

以 Zr-4 为母合金,分别添加 0.1%,0.2%和 0.3%的 Nb,同时以 Zr-4 重熔合金作为实验对照.用真空非自耗电弧炉熔炼成约为 60 g 的合金锭,熔炼时通高纯 Ar 气保护,为保证成分均匀,合金锭每熔炼一次都要翻转,共熔炼 6 次.合金锭经 700 ℃热压成条块状后在 1 030 ℃(β 相)均匀化处理 40 min,然后通过热轧(700 ℃)、β 相水淬(1 030 ℃,40 min)和 50%变形量的冷轧制成 0.7 mm 厚的片状试样,最后进行 580 ℃,50 h+A.C.退火处理.加热退火均在真空中进行.β 相水淬是通过将样品真空封装在石英管中,加热到 1 030 ℃保温 40 min 后立即淬入水中并快速敲碎石英管来实现的.用电感耦合等离子体原子发射光谱(ICP-AES)分析得到的合金成分列在表 1 中,样品中的 Nb 含量与设计值接近,Zr-4 合金中的其他合金元素在熔炼过程中都发生了不同程度的损耗.将上述处理的样品用 30% H_2O+30% HNO_3+30% H_2SO_4+10% HF(体积比)混合酸酸洗和去离子水清洗后,放入高压釜中进行 500 ℃/10.3 MPa 过热蒸汽中腐蚀.用 5 块试样平均值得出腐蚀增重.

表 1　实验用的几种合金成分

Table 1　Chemical compositions of the experimental alloys （mass fraction，%）

Alloy	Sn	Fe	Cr	Nb	Zr
Zr-4-remelted	1.43	0.22	0.098	—	Bal.
Zr-4+0.1Nb	1.33	0.20	0.087	0.10	Bal.
Zr-4+0.2Nb	1.13	0.18	0.079	0.20	Bal.
Zr-4+0.3Nb	1.03	0.16	0.070	0.31	Bal.

用 JEM-2010F 型高分辨透射电镜(TEM)及其配置的能谱仪(EDS)分析合金中第二相的形貌和成分,并通过选区电子衍射(SAD)确认第二相的晶体结构.用 JSM-6700F 型扫描电子显微镜(SEM)观察腐蚀后试样的氧化膜断口形貌.TEM 试样用双喷电解抛光制备,电解液为 10%(体积分数)高氯酸酒精溶液,抛光电压 50 V,温度-40 ℃.氧化膜断口形貌观察用试样通过从腐蚀后的样品上切下一块,用混合酸溶去金属基体,然后将氧化膜折断的方法制备[7].为了提高图像质量,在进行氧化膜断口形貌 SEM 观察前,试样表面蒸镀了一层金属 Ir.

2 实验结果与分析

2.1 腐蚀行为

图 1 和 2 分别是 Zr-4+xNb(x=0.1,0.2,0.3)合金在 500 ℃/10.3 MPa 过热蒸汽中的腐蚀增重随时间的变化曲线和腐蚀后试样表面的形貌.可以看出,腐蚀前期(160 h 前),Zr-4 中添加 Nb 的合金腐蚀速率较慢,腐蚀增重也十分接近,腐蚀 160 h 时的增重均在

110 mg/dm² 左右,而 Zr-4 重熔试样腐蚀速率较快,160 h 时的腐蚀增重达到了 178 mg/dm²,腐蚀增重约为前者的 1.5 倍,试样四周也出现了成片的白色氧化膜,即为疖状腐蚀,但表面仍比较黑亮. 腐蚀后期(160 h 后),Zr-4 重熔试样由于白色氧化膜脱落无法得到正确的增重数据而没有继续腐蚀,Zr-4+xNb 合金耐腐蚀性能较好,腐蚀 500 h 后也没有出现疖状斑,但是 Zr-4+0.3Nb 合金明显比 Zr-4+0.1Nb 和 Zr-4+0.2Nb 合金的腐蚀增重快. 总之,Zr-4 合金中添加一定量的 Nb 元素后,对疖状腐蚀起到了很好的抑制作用,明显改善了耐疖状腐蚀性能,但是添加 Nb 含量不同时对合金的耐均匀腐蚀性能会有不同的影响. 本文作者等[6]研究了微量 Nb(0.05%—0.1%)的添加对 Zr-4 合金耐疖状腐蚀性能的影响,结果发现 Zr-4 合金中添加 0.1% Nb 后就能明显改善耐疖状腐蚀性能,但还不能完全抑制疖状腐蚀现象,试样表面仍有疖状腐蚀斑. 这可能与采用不同的热加工制度有关,本文研究用的 Zr-4+0.1Nb 合金在冷轧和最终热处理之前采用 β 相水淬处理,细化了第二相,同时还可能提高了合金元素在 α-Zr 中的过饱和固溶量. 因而进一步改善了合金的耐疖状腐蚀性能,这与周邦新等[3]的结果是一致的.

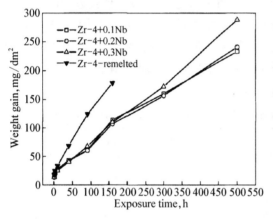

图 1 Zr-4+xNb 合金在 500 ℃/103 MPa 过热蒸汽中腐蚀后的增重曲线

Fig. 1 Curves of weight gain of the Zr-4+xNb alloys during corrosion in 500 ℃/10.3 MPa superheated steam

图 2 Zr-4+xNb 合金在 500 ℃/10.3 MPa 过热蒸汽中腐蚀不同时间后的表面形貌

Fig. 2 Appearances of the Zr-4+xNb alloys corroded in 500 ℃/10.3 MPa superheated steam for different times (a) Zr-4+0.1Nb, 500 h;(b) Zr-4+0.2Nb, 500 h;(c) Zr-4+0.3Nb, 500 h;(d) Zr-4-remelted, 160 h

2.2 合金的显微组织

图 3 是腐蚀前 Zr-4+xNb 合金显微组织的 TEM 像. Zr-4 重熔试样和 Zr-4+0.1Nb 合金基体中第二相尺寸较小(<60 nm);Zr-4+0.3Nb 合金基体中有很多较大的第二相(100—200 nm),合金中第二相多呈条带状分布. 这应该是 β 相水淬时残留的 β-Zr 在随后的冷轧和退火过程中发生分解的结果. 对第二相的能谱(EDS)分析表明:Zr-4+0.1Nb 合金中的第二相成分主要为 Zr-Fe-Cr. Fe/Cr 原子比约为 1.49—1.80,并且随着第二相粒子尺寸增加,Fe/Cr 比值有增大的趋势;Zr-4+0.3Nb 合金中的第二相成分主要为 Zr-Fe-Cr-Nb,其中 Fe/Cr 比值约为 1.85—2.05,同样随着第二相尺寸的增大 Fe/Cr 比值增大(图 4 和表 2). 对上述 2 种成分第二相的选区电子衍射(SAD)分析表明,它们均为 hcp 结构的 Zr

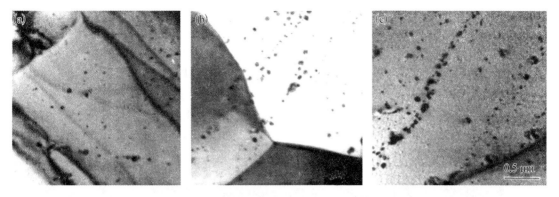

图3 合金显微组织的 TEM 像

Fig. 3 TEM micrographs of Zr‑4‑remelted (a), Zr‑4+0.1Nb (b) and Zr‑4+0.3Nb (c) alloys

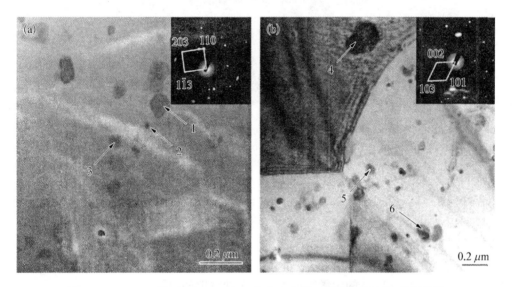

图4 Zr‑4+0.1Nb 和 Zr‑4+0.3Nb 合金中第二相形貌及 SAD 分析结果

Fig. 4 TEM morphologies and SAD patterns of the second phase particles in Zr‑4+0.1Nb (a) and Zr‑4+0.3Nb (b) alloys (SAD patterns indexed by hcp ZrM_2)

$(Fe, Cr)_2$ 型第二相,其中 Nb 元素进入到 $Zr(Fe, Cr)_2$ 中形成 $Zr(Fe, Cr, Nb)_2$ 第二相[8]. 另外,Zr‑4+0.2Nb 合金中的第二相分析结果与 Zr‑4+0.1Nb 合金相似.

表2 图4中第二相的 EDS 结果

Table 2 EDS results of the second phase particles marked in Fig. 4

| Alloy | Precipitate | Atomic fraction of element, % | | | | Formula | Fe/Cr |
		Zr	Fe	Cr	Nb		
Zr‑4+0.1Nb	1	83.87	9.99	6.13	—	$Zr(Fe, Cr)_2$	1.63
	2	86.37	8.15	5.48	—		1.49
	3	85.73	9.17	5.10	—		1.80
Zr‑4+0.3Nb	4	61.10	20.80	10.13	7.98	$Zr(Fe, Cr, Nb)_2$	2.05
	5	70.36	15.13	8.17	6.34		1.85
	6	66.39	18.57	9.33	5.71		1.99

2.3 腐蚀样品氧化膜的断口形貌

图 5 是 Zr-4+xNb 合金在 500 ℃过热蒸汽中腐蚀 160 h 后氧化膜断口形貌. 可以看出,Zr-4 重熔腐蚀样品氧化膜的 ZrO₂ 晶粒均为等轴晶(接近球形),而 Zr-4+0.1Nb 和 Zr-4+0.3Nb 腐蚀样品氧化膜的 ZrO₂ 晶粒以柱状晶为主;Zr-4 重熔氧化膜断口比较疏松,而 Zr-4+0.1Nb 和 Zr-4+0.3Nb 腐蚀样品氧化膜断口相对比较致密,但 Zr-4+0.3Nb 合金的氧化膜中有较多尺寸在 100—200 nm 的微裂纹,如图 5c 中箭头所示,其裂纹尺寸和数量比 Zr-4+0.1 Nb 合金氧化膜中的长且多.

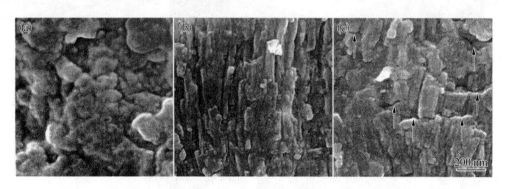

图 5 合金在 500 ℃/10.3 MPa 过热蒸汽中腐蚀 160 h 后氧化膜断口形貌

Fig. 5 SEM images of the fracture surfaces of oxide film on Zr 4 remelted (a), Zr-4+0.1Nb (b) and Zr-4+0.3Nb (c) alloys corroded in 500 ℃/10.3 MPa superheated steam for 160 h (arrows showing micro cracks)

3 分析讨论

3.1 Zr-4 合金中添加 Nb 的含量影响耐均匀腐蚀性能的原因

Zr 在水蒸气中的腐蚀过程是:O^{2-} 通过氧化膜中的晶界或阴离子空位扩散到金属/氧化膜界面处,同时电子从金属表面向反方向运动,使氧化膜在金属/氧化膜界面处生长[9]. 因此,氧化膜的性质以及氧化膜显微组织在腐蚀过程中的演化将直接影响锆合金的腐蚀行为. Zr 氧化成 ZrO₂ 时体积发生膨胀,同时又受到金属基体的约束,因此氧化膜内部会形成很大的压应力,使 ZrO₂ 晶体中产生许多缺陷;空位、间隙原子等缺陷在温度、应力和时间的作用下,发生扩散、湮没和凝聚,空位被晶界吸收后形成纳米大小的孔隙簇,弱化了晶粒之间的结合力;孔隙簇进一步发展成为裂纹,从而影响锆合金的耐腐蚀性能[7,10]. 结合图 5 可知,ZrO₂ 晶粒呈等轴晶的 Zr-4 重熔试样耐均匀腐蚀性能较差,而 ZrO₂ 晶粒呈柱状晶的 Zr-4+(0.1—0.3)Nb 合金耐均匀腐蚀性能较好,这是由于 ZrO₂ 柱状晶中的缺陷通过扩散凝聚而构成新的晶界,发展成等轴晶,因此等轴晶比柱状晶提供给 O^{2-} 扩散的通道更多,合金更容易氧化[11];而在 Zr-4+(0.1—0.3)Nb 合金中,Zr-4+0.3Nb 合金氧化膜有较多微裂纹,提供了 O^{2-} 的扩散通道,耐腐蚀性能下降,这与腐蚀增重结果一致.

合金元素固溶在 α-Zr 中或以第二相析出都会在腐蚀过程中影响氧化膜的性质和结构,从而影响合金的耐腐蚀性能. Kim 等[12]和 Jeong 等[13,14]研究了 Zr-Nb 二元合金的腐蚀行为,发现对锆合金耐腐蚀性能起影响作用的主要是固溶在 α-Zr 中的 Nb 含量,当固溶在 α-Zr 中的 Nb 含量达到平衡固溶度时,氧化膜中柱状晶及四方相的 ZrO₂ 比例最多,合金耐腐蚀性能最好.周邦新等[3]也认为在不同热处理条件下,提高 Fe 和 Cr 在 α-Zr 中的过饱和固溶量,能够调整氧化膜生长的各向异性,从而改善 Zr-4 合金的耐腐蚀性能.从第二相的分析结果可知,Zr-4+0.1Nb 合金的 Nb 元素基本固溶在 α-Zr 中,因此 Zr-4+(0.1—0.3)Nb 合金中固溶在 α-Zr 中的 Nb 含量都不小于 0.1%,这是 Zr-4+(0.1—0.3)Nb 合金耐腐蚀性能较好的原因.但是 Zr-4+0.3Nb 合金的耐均匀腐蚀性能比 Zr-4+0.1Nb 和 Zr-4+0.2Nb 合金差,这可能包含以下两方面的原因.

其一,Nb 的添加改变了第二相和 α-Zr 中 Fe,Cr 含量的分配.图 6 是利用相图计算(CALPHAD)方法得出第二相的 Fe/Cr 比值和 α-Zr 中的 Fe,Cr 含量随 Nb 含量的变化情况.CALPHAD 技术是通过由低元到高元建立热力学模型,进行体系的热力学性质、析出相及中间化合物的稳定性信息预测的一项技术[15].由图 6 可知,Zr-4 合金中随着 Nb 含量的增加,第二相中 Fe/Cr 比值先较快增加,再趋于稳定(图 6a),表 2 中 Zr-4+0.1Nb 和 Zr-4+0.3Nb 合金的能谱分析结果与此符合较好;同时,随着 Nb 含量的增加,固溶在 α-Zr 中的 Fe 含量减少较快,而 Cr 含量缓慢增加(图 6b),因此固溶在 α-Zr 中的 Fe+Cr 含量随 Nb 含量增加而减少.Zr-4+0.3Nb 合金 α-Zr 中的(Fe+Cr)含量比 Zr-4+0.1Nb 和 Zr-4+0.2Nb 合金的低,所以耐腐蚀性能较差,这与文献[16]报道的结果是一致的.另外,Zr-4+0.3Nb 合金在熔炼过程中 Fe 和 Cr 损耗较大(表 1),也可能会影响合金的耐均匀腐蚀性能.

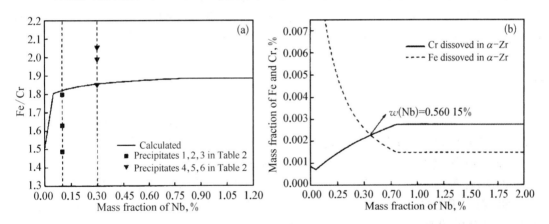

图 6 通过 CALPHAD 计算得出的第二相中 Fe/Cr 比值和 α-Zr 中 Fe,Cr 固溶含量随 Nb 含量的变化曲线

Fig. 6 Variation curves of the atomic ratio of Fe and Cr(Fe/Cr) in the second phase particles (a) and concentrations of Fe and Cr in α-Zr (b) with Nb content calculated by CALPHAD

其二,与 Zr(Fe,Cr,Nb)₂ 第二相析出有关.Zr 发生氧化时的 P.B. 比(金属氧化物与金属体积之比)为 1.56,Nb 氧化时的 P.B. 比为 2.74[17],所以 Nb 氧化后产生的压应力比 Zr 氧化后产生的压应力大得多.当 Nb 固溶在 α-Zr 中时,氧化后产生的压应力均匀分布在氧化膜中;但是当 Nb 以第二相形式存在,并且尺寸较大时,第二相氧化后会产生局部的相当大的压应力.应力的大小对氧化膜的显微组织演化过程产生影响,局部大的压应力容易使氧化膜产生微裂纹,从而加速腐蚀.Zr-4+0.3Nb 合金腐蚀样品氧化膜中可观察到较多的微裂纹(图 5c 中箭头所示)可能就与 Zr(Fe,Cr,Nb)₂ 第二相有关.另外,Zr-4+0.3Nb 合金

中的第二相尺寸较大、分布比较不均匀,也会降低 Zr-4+0.3Nb 合金的耐均匀腐蚀性能,这与 Bangaru[18] 研究 Zr-4 合金第二相影响合金耐腐蚀性能的规律一致.

3.2 Zr-4 合金中添加 Nb 抑制疖状腐蚀的原因

Zhou 等[19] 研究大晶粒的 Zr-4 合金样品在 500 ℃ 过热蒸汽中腐蚀时,观察到氧化膜生长呈明显的各向异性,当晶粒表面的取向分布在 $(10\bar{1}0)$ 至 $(11\bar{2}0)$ 之间时,氧化膜生长较快,腐蚀后期都发展成疖状腐蚀斑,而晶粒表面的取向分布在 (0001) 附近时,氧化膜生长比较慢,在腐蚀后期并没有发展成疖状腐蚀斑.锆合金疖状腐蚀的形成是"成核长大"过程,并与基体的织构取向有关[3,19-21];细晶样品腐蚀时,局部区域的氧化膜可以因晶体的不利取向而生长较快,形成"肿块状"疖状腐蚀斑的"核",由于 P.B. 比为 1.56,肿块状的氧化膜将受到较大的压应力,同时对金属基体会产生张应力并使其发生变形,变形后产生的位错等缺陷成为 O^{2-} 的扩散通道,促进氧化膜生长而发展成为疖状腐蚀;轧制成形的片状样品会形成织构,样品表面取向与氧化膜生长较慢的 (0001) 面比较接近,不易形成疖状腐蚀,而样品四周的截面是氧化膜生长较快的 $(10\bar{1}0)$ 和 $(11\bar{2}0)$ 面,容易生成疖状腐蚀.这就是 Zr-4 重熔试样在腐蚀 160 h 后周边出现严重疖状腐蚀,而表面(轧面)没有出现疖状腐蚀的原因(图 2d).提高固溶在 α-Zr 中的合金元素含量,可以调整晶体氧化时氧化膜生长的各向异性.延缓甚至阻止疖状腐蚀斑的形核,提高耐疖状腐蚀性能. Nb 在 α-Zr 中的固溶度比 Fe,Cr 大得多,因此在 Zr-4 合金中添加 Nb 元素可以获得优良的耐疖状腐蚀性能.

总之,Zr-4+xNb 合金要获得最佳的耐腐蚀性能,添加的 Nb 含量应不少于 0.1%,但又要避免添加过多时会析出 Zr(Fe,Cr,Nb)$_2$ 的第二相.

4 结论

(1) Zr-4 合金中添加(0.1%—0.3%)Nb 后在 500 ℃/10.3 MPa 过热蒸汽中腐蚀500 h 均未出现疖状腐蚀,完全抑制了疖状腐蚀的产生,这与 α-Zr 中固溶了较多的 Nb 有关.固溶在 α-Zr 中的 Nb 能延缓疖状腐蚀斑的形核,提高耐疖状腐蚀性能.

(2) 随着 Nb 含量的增加,Zr-4+(0.1—0.3)Nb 合金的耐均匀腐蚀性能降低,这与 Nb 的添加降低了固溶在 α-Zr 中的(Fe+Cr)含量和 Zr(Fe,Cr,Nb)$_2$ 第二相的析出有关.这两方面都会加快氧化膜显微组织在腐蚀过程中的演化,促进氧化膜中孔隙和微裂纹的形成,从而降低合金的耐腐蚀性能.

参 考 文 献

[1] Liu J Z. *Nuclear Structural Materials*. Beijing: Chemical Industry Press, 2007: 112.
   (刘建章. 核结构材料. 北京: 化学工业出版社, 2007: 112)

[2] Ogata K. In: Van Swam L F P, Eucken C M, eds., *Zirconium in the Nuclear Industry: 8th International Symposium*, ASTM STP 1023, San Diego, CA: ASTM International, 1988: 346.

[3] Zhou B X, Li Q, Yao M Y, Xia S, Liu W Q, Chu Y L. *Rare Met Mater Eng*, 2007; 36: 1129.
   (周邦新, 李强, 姚美意, 夏爽, 刘文庆, 褚于良. 稀有金属材料与工程, 2007; 36: 1129)

[4] Zhou B X, Zhao W J, Miao Z, Huang D C, Jiang Y R. In: Zhou B X, Shi Y K, eds., *Proc Ⅲ - 2 China Nuclear Materials Academic Conference in 1996*, Beijing: Chemical Industry Press, 1997: 183.
   (周邦新, 赵文金, 苗志, 黄德诚, 蒋有荣著; 周邦新, 石永康主编. 96 中国核材料研讨会论文集Ⅲ-2, 北京: 化学工业

出版社,1997:183)

[5] Zhou B X, Yao M Y, Li Q, Xia S, Liu W Q, Chu Y L. *Rare Met Mater Eng*, 2007; 36: 1317.

(周邦新,姚美意,李强,夏爽,刘文庆,褚于良.稀有金属材料与工程,2007;36:1317)

[6] Yao M Y, Zhou B X, Li Q, Xia S, Liu W Q. *Shanghai Met*, 2008; 30: 1.

(姚美意,周邦新,李强,夏爽,刘文庆.上海金属,2008;30:1)

[7] Zhou B X, Li Q, Yao M Y, Liu W Q. *Nucl Power Eng*, 2005; 26: 364.

(周邦新,李强,姚美意,刘文庆.核动力工程,2005;26:364)

[8] Toffolon-Masclet C, Brachet J C, Jago G. *J Nucl Mater*, 2002; 305: 224.

[9] Yang W D. *Reactor Materials Science*. 2nd Ed., Beijing: Atomic Energy Press, 2006: 260.

(杨文斗.反应堆材料学.第二版.北京:原子能出版社,2006:260)

[10] Zhou B X, Li Q, Yao M Y, Liu W Q, Chu Y L. In: Kammenzind B, Limbäck M, eds., *Zirconium in the Nuclear Industry: 15th International Symposium*, *ASTM STP 1505*, Sunriver, Oregon, USA: ASTM International, 2009: 360.

[11] Anada H, Takeda K. In: Bradley E R, Sabol G P. eds., *Zirconium in the Nuclear Industry: 11th International Symposium*, *ASTM STP 1295*, Garmisch-Partenkirchen, Germany: ASTM International, 1996: 35.

[12] Kim H G, Jeong Y H, Kim T H. *J Nucl Mater*, 2004; 326: 125.

[13] Jeong Y H, Kim H G, Kim D J, Choi B K, Kim J H. *J Nucl Mater*, 2003; 323: 72.

[14] Jeong Y H, Kim H G, Kim T H. *J Nucl Mater*, 2003; 317: 1.

[15] Hao S M. *Thermodynamics of Materials*. Beijing: Chemical Industry Press, 2004: 333.

(郝士明.材料热力学.北京:化学工业出版社,2004:333)

[16] Charquet D. *J Nucl Mater*, 2001; 288: 237.

[17] Li T F. *High Temperature Oxidation and Thermal Corrosion of Metals*. Beijing: Chemical Industry Press, 2003: 52.

(李铁藩.金属高温氧化和热腐蚀.北京:化学工业出版社,2003:52)

[18] Bangaru N V. *J Nucl Mater*, 1985; 131: 280.

[19] Zhou B X, Peng J C, Yao M Y, Li Q, Xia S, Du C X, Xu G. *J ASTM Int*, 8(1), Paper ID JAI102951, www. astm. org.

[20] Zhou B X. In: Van Swam L F P, Eucken C M, eds., *Zirconium in the Nuclear Industry: 8th International Symposium*, *ASTM STP 1023*, San Diego, CA: ASTM International, 1988: 360.

[21] Zhou B X, Yao M Y, Xia S, Liu W Q. *J Shanghai Univ* (*Nat Sci Ed*), 2008; 14: 441.

(周邦新,姚美意,李强,夏爽,刘文庆.上海大学学报(自然科学版),2008;14:441)

Effect of Nb on the Corrosion Resistance of Zr-4 Alloy in Superheated Steam at 500 ℃

Abstract: The corrosion resistance of Zr-4+xNb alloys (x=0.1—0.3, mass fraction, %) was investigated in a superheated steam at 500 ℃ and 10.3 MPa by autoclave tests. The microstructure of the alloys and fracture surface morphology of the oxide film formed on the alloys were observed by TEM and SEM, respectively. Results show that no nodular corrosion appears on Zr-4+xNb alloys in 500 ℃/10.3 MPa superheated steam even for 500 h, which is related to the higher Nb concentration dissolved in α-Zr matrix. The Nb dissolved in α-Zr matrix can restrain the nucleation of nodular corrosion, thus improve the nodular corrosion resistance. The uniform corrosion resistance of Zr-4+xNb alloys is lowered with the increase of Nb content. which is related to the decrease of the solid solution concentration of (Fe+Cr) in α-Zr matrix and the precipitation of the second phase particles of Zr(Fe, Cr, Nb)$_2$, and the two aspects will accelerate the microstructural evolution of the oxide film during corrosion process to promote the formation of pores and micro-cracks.

Ni 对 RPV 模拟钢中富 Cu 原子团簇析出的影响[*]

摘　要：用原子探针层析技术和时效模拟方法，研究了不同 Ni 含量并且提高了 Cu 含量的反应堆压力容器(RPV)用模拟钢中富 Cu、富 Ni 和富 Mn 原子团簇的形成. 结果表明，提高钢中的 Ni 含量会促使富 Cu 原子团簇的析出，富 Cu 原子团簇中含有 Ni 和 Mn. 实验检测到富 Ni 的原子团簇，团簇中含有 Cu 和 Mn，富 Ni 原子团簇可以作为富 Cu 原子团簇析出时的形核区. 实验还检测到富 Mn 原子团簇，当 Mn 原子团簇中含有较高的 Ni 时，它也可以成为富 Cu 原子团簇析出时成核的地方. 由于钢中的合金元素 Ni 在形成富 Ni 原子团簇后会成为富 Cu 原子团簇析出时成核区，因而提高 Ni 的含量将促进富 Cu 原子团簇的析出，这是合金元素 Ni 会增加压力容器钢中子辐照脆化敏感性的本质原因.

1　前言

压水堆核电站的压力容器(RPV)一般用 Mn - Ni - Mo 低合金铁素体钢(A508 - Ⅲ)制造，添加合金元素 Ni 可以增加钢的淬透性，提高韧性，降低韧脆转变温度，但是却增加了中子辐照脆化的敏感性，因而对这种合金的研究受到广泛关注. 国际原子能机构为此组织了国际间的合作研究，研究结果已经汇编成册[1]. 目前已经认识到由于中子辐照诱发析出了富 Cu 原子团簇[1-6]和其他 Ni，Mn 等原子团簇[1,4,6-7]是促使 RPV 钢发生辐照脆化的主要原因. 提高钢中 Ni 的含量会促使富 Cu 原子团簇的析出[1,8]，但是其根本原因还不十分清楚.

富 Cu 原子团簇的大小只有几纳米，即使用高分辨率透射电子显微镜观察研究，也难于弄清楚其形核过程，而原子探针层析技术是研究这类问题的非常理想的方法[9-11]. 原子探针层析技术可以对原子的种类和它们在空间的位置逐个进行分析，因而可以探测到只有几个溶质原子发生团聚时的情况，还可以精确测定团簇中不同元素原子的分布和含量[10-11]. 过去采用这种方法分析了压力容器钢的辐照监督试样中富 Cu 原子团簇析出的情况，以及用热时效模拟方法研究富 Cu 原子团簇析出的过程，并且获得了不少的结果，如观察到富 Cu 原子团簇析出时的清晰图像，还能分析只有几纳米大小团簇中的化学成分[1-9,12]. 本研究工作采用热时效模拟的方法研究了 Ni 含量不同并且提高了 Cu 含量的压力容器模拟钢中富 Cu 原子团簇的析出，以及富 Ni 和富 Mn 原子团簇的形成，并根据这些团簇中溶质原子之间距离大小的差别，判断它们在析出和形成过程中的先后次序，分析并研究它们之间成分的变化规律，可以了解溶质原子间的相互作用及其对富 Cu 原子团簇析出的影响.

2　实验方法

参照 A508 - Ⅲ 钢的化学成分，用真空感应电炉冶炼了 Ni 含量分别为 0.79%(质量分

---

* 本文合作者：王均安、刘庆东、刘文庆、王伟、林民东、徐刚、楚大锋. 原发表于《中国材料进展》，2011，30(5)：1 - 6.

数,下同)和1.52％ 2种成分不同的RPV模拟钢,为了能够用热时效处理研究富Cu原子团簇的析出过程,将Cu含量提高到0.5％,样品的化学成分列于表1中.铸锭质量约40 kg,经过热锻和热轧得到4 mm厚的板材.50 mm×40 mm大小的样品经过880 ℃,30 min加热,水淬,再经过660 ℃,10 h回火调质处理,然后在400 ℃时效100~2 000 h,选取富Cu原子团簇刚析出时的样品,以便研究富Cu原子团簇析出的过程,以及钢中其他溶质原子对富Cu原子团簇析出的影响.

表1 实验用RPV模拟钢的化学成分(wt％)

Table 1 Composition of experimental PRV model steels(wt％)

Specimen number	C	Si	Mn	P	S	Mo	Ni	Cu
7	0.21	0.31	1.51	0.025	0.006	0.01	0.79	0.56
8	0.24	0.34	1.58	0.033	0.007	0.02	1.52	0.62

经过时效处理后的样品用电火花线切割法从样品中心截取截面为0.5 mm×0.5 mm的棒状样品,样品长度不小于20 mm,然后在显微镜下进行显微电解抛光,制备曲率半径小于100 nm的针尖状样品.利用Oxford NanoScience公司生产的三维原子探针对样品进行分析,数据采集时样品冷却至50 K,直流脉冲电压频率为5 kHz,脉冲分数为20％,所得数据用PoSAP软件进行分析.

为了能够准确分析原子团簇的存在以及团簇中的成分,应用了最大分离包络法(MSEM, maximum separation envelope method)对获得的数据进行了处理[10],得到了每个团簇中各种原子的数量和团簇的化学成分.这种方法首先需要设置溶质原子之间(如Cu, Ni或Mn原子)的最大距离d_{max},还要设置满足d_{max}条件下最少的溶质原子数量N_{min},在探测到能满足这样条件的溶质原子分布时,就得到了该种溶质原子的团簇.如果固定N_{min}值(如$N_{min}=10$),改变d_{max}值(如$d_{max}=0.4, 0.5, 0.6$ nm等),并使其逐渐增大,这样探测到的几个团簇,由于其中溶质原子的距离不同,也就是溶质原子间的"松散"程度不同,它们应该处于析出过程的不同阶段,溶质原子之间距离稍大的团簇应该比溶质原子之间距离更小的团簇处于析出过程的更早期,比较这些团簇之间的成分差别,可以研究Ni, Mn等合金元素对富Cu原子团簇析出过程的影响,这涉及富Cu原子团簇析出时成核机理方面的问题.

计算团簇数量密度的公式为:$N_v = N_p \times \zeta/(N \times \Omega)$,其中$N_v$为团簇数量密度;$N_p$为分析体积内所检测到的团簇数量,由MSEM分析数据得出;ζ为检测原子场蒸发后的效率参数,取决于设备的探测器,采用0.6;N为所搜集的原子总数;Ω为原子的平均体积.体心立方结构的铁原子Ω为1.178×10^{-2} nm³,由于搜集的所有原子中主要为铁原子,所以这里用铁原子的体积来计算.

3 结果和讨论

根据硬度值和原子探针的分析结果,选取了经100 h和1 000 h时效的2组样品进行对比分析,这组样品能够反映富Cu原子团簇析出早期的情况,也容易观察到Ni对富Cu原子团簇析出时的影响.图1是用原子探针层析技术获得的Cu原子分布图,图中每一个小点是仪器检测得到的一个Cu原子空间位置.由于几个样品的针尖曲率半径不同,收集的原子数量不同,得到的分析体积也有差异.直接观察Cu原子在空间的分布,可以看出2组样品经过

1 000 h 时效后，都析出了富 Cu 原子团簇，团簇非常小，大约只有 1～2 nm. 但是经过 100 h 时效后的样品很难用肉眼观察来判断是否有富 Cu 原子团簇析出，需要用 MSEM 进行数据处理，这正是富 Cu 原子团簇析出的初期，是研究 Ni，Mn 合金元素影响富 Cu 原子团簇析出的最好时机.

在用 MSEM 分析数据时，设置团簇中最少的溶质原子数量为固定值，$N_{min} = 10$，但是设置 d_{max} 值时要根据具体情况在 0.3～1.0 nm 之间变化，这是因为根据 d_{max} 值的大小，也就是团簇中原子的"松散"程度，可以判断团簇在析出过程中所处的不同阶段. 用这种方法分析了 7 ♯ 和 8 ♯ 样品经过 400 ℃ 时效 100 h 和 1 000 h 后的 4 种样品，结果表明 7 ♯ 样品经过 100 h 时效后并没有富 Cu 原子团簇析出，而经过 100 h 时效后的 8 ♯ 样品只有在设置 d_{max} 值大于 0.7 nm 后才能检测出存在富 Cu 原子团簇，这是富 Cu 原子团簇析出的早期阶段，团簇中的 Cu 原子还没有聚集得"十分紧密"，还没有形成由 Cu 原子单独构成的晶体点阵，因为 Cu 的晶胞常数为 0.361 5 nm. 经过 1 000 h 时效后的 7 ♯ 和 8 ♯ 样品在设置 d_{max} 值为 0.3～0.4 nm 时就可以检测到富 Cu 原子团簇的存在，这时已经形成了由 Cu 原子构成的晶体点阵. 分析结果说明，提高钢中的 Ni 含量可以促使富铜原子团簇的析出.

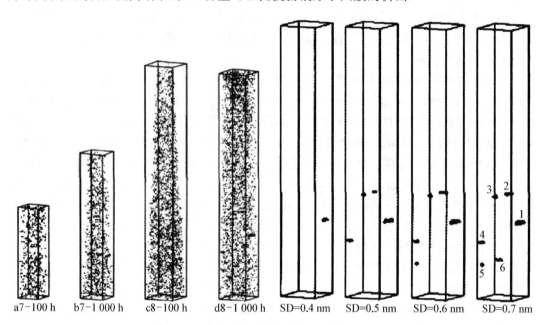

a7-100 h　　b7-1 000 h　　c8-100 h　　d8-1 000 h　　SD=0.4 nm　　SD=0.5 nm　　SD=0.6 nm　　SD=0.7 nm

图 1 7 ♯ 和 8 ♯ 样品经过 400 ℃，100 h 和 1 000 h 时效处理后 Cu 原子的空间分布图. 空间大小为：(a) 11 nm×13 nm×43 nm，(b) 14 nm×15 nm×75 nm，(c) 20 nm×21 nm×147 nm，(d) 18 nm×20 nm×112 nm

Fig. 1 Copper atom maps of 7 ♯ and 8 ♯ specimens aged at 400 ℃, 100 h and at 400 ℃, 1 000 h. The box volume: (a) 11 nm×13 nm×43 nm, (b) 14 nm×15 nm×75 nm, (c) 20 nm×21 nm×147 nm and (d) 18 nm×20 nm×112 nm

图 2 8 ♯ 样品经过 400 ℃，1 000 h 时效处理后，富 Cu 原子团簇的分布图（当设定 $N_{min} = 10$，改变 d_{max} 值(SD)时可以检测到不同数量的富 Cu 原子团簇，依次编号为 1～6）

Fig. 2 Maps of Cu-rich clusters obtained at different d_{max} values with the same $N_{min} = 10$ for 8 ♯ specimen aged at 400 ℃, 1 000 h. The matrix solute atoms have been eliminated based on MSEM

8 ♯ 样品经过 400 ℃，1 000 h 时效后，用 MSEM 分析 Cu 原子团簇的分布，结果如图 2 所示，图中将 Cu 原子的本底去除，只留下了 Cu 原子团簇的分布. 从图中可以看出，当设置

$d_{max}=0.4$ nm 时，只得到 1 个 Cu 原子团簇，当设置 $d_{max}=0.5$ nm 时，得到 4 个 Cu 原子团簇，依次增加 d_{max} 值至 0.7 nm，共得到了 6 个 Cu 原子团簇，将团簇编号 1 至 6，表示它们的先后次序. 图 3 是这 6 个富 Cu 原子团簇中 Cu，Ni 和 Mn 元素的成分分布，这些 Cu 团簇中

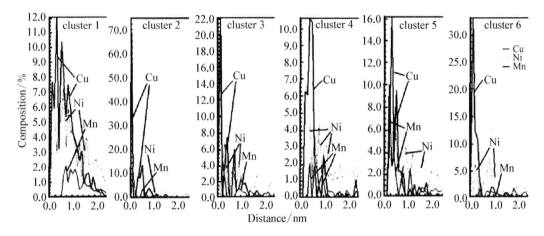

图 3 8#样品经过 400 ℃，1 000 h 时效处理后探测到 6 个富 Cu 原子团簇的 Cu，Ni，Mn 成分分布

Fig. 3 Concentration profiles of Cu, Ni and Mn elements in 6 Cu-rich clusters detected in 8# specimen aged at 400 ℃ for 1 000 h

都含有 Ni 和 Mn，并且常常倾向偏聚在 Cu 团簇和 α-Fe 基体的界面处. 应用同样的方法分析了富 Ni 和富 Mn 原子团簇存在的情况，结果如图 4 所示. 当设置 Ni 原子之间的 $d_{max}=0.35$ nm 时就可探测到 6 个富 Ni 原子团簇（图 4a），而设置 Mn 原子之间的 $d_{max}=0.6$ nm 时才探测到 5 个富 Mn 原子团簇（图 4b），这说明形成富 Ni 原子团簇的倾向比形成富 Mn 原子团簇的更明显. 这些富 Ni 和富 Mn 原子团簇中 Ni，Cu 和 Mn 成分的分布给出在图 5 和图 6 中. 在 Ni 原子间距比较小的富 Ni 原子团簇中（$d_{max}=0.3$ nm），都可以探测到 Cu 原子的存在，这时 Cu 原子的数量还不满足统计富 Cu 原子团簇 $N_{min}=10$ 的最低要求，因而这种含 Cu 的富 Ni 团簇并没有包含在富 Cu 原子团簇的统计中，但是它们更像是富 Cu 原子团簇析出时的成核阶段，这说明富 Ni 原子团簇可以作为 Cu 团簇析出时的成核位置. 从富 Mn 原子团簇的成分分布图（图 6）中可以看出，在富 Mn 原子团簇中 Ni 含量较高时，也可以探测到 Cu 原子的存在. 虽然单一的富 Mn 原子团簇与富 Cu 原子团簇的

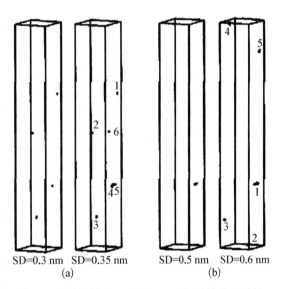

SD=0.3 nm SD=0.35 nm   SD=0.5 nm SD=0.6 nm
(a)   (b)

图 4 8#样品经过 400 ℃，1 000 h 时效处理后的富 Ni 原子团簇分布图（a）和富 Mn 原子团簇分布图（b）（当设定 $N_{min}=10$，$d_{max}=0.3$ nm 时检测到 4 个富 Ni 原子团簇，$d_{max}=0.35$ nm 时检测到 6 个富 Ni 原子团簇；当设定 $d_{max}=0.5$ nm 时检测到 1 个富 Mn 原子团簇，$d_{max}=0.6$ nm 时检测到 5 个富 Mn 原子团簇）

Fig. 4 Maps of Ni-rich and Mn-rich clusters obtained at different d_{max} values with the same $N_{min}=10$ for 8# specimen aged at 400 ℃ for 1 000 h. The matrix solute atoms have been eliminated based on MSEM

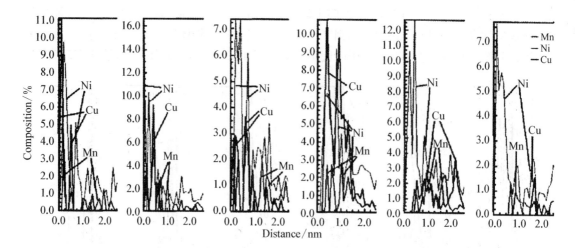

图5 8♯样品经过400℃,1 000 h时效处理后探测到6个富Ni原子团簇的Ni,Cu,Mn成分分布

Fig. 5 Concentration profiles of Ni, Cu and Mn elements in 6 Ni-rich clusters detected in 8♯ specimen aged at 400 ℃ for 1 000 h

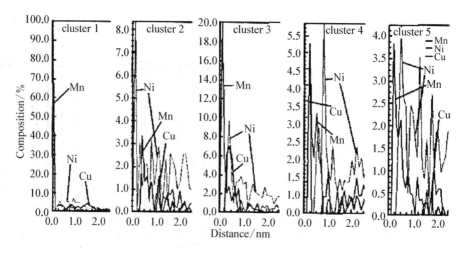

图6 8♯样品经过400℃,1 000 h时效处理后探测到5个富Mn原子团簇的Mn,Ni,Cu成分分布

Fig. 6 Concentration profiles of Mn, Ni and Cu elements in 5 Mn-rich clusters detected in 8♯ specimen aged at 400 ℃ for 1 000 h

析出不一定有直接的关系,但是在Ni含量较高的富Mn原子团簇处也可以作为富Cu原子团簇析出时的成核位置.

　　富Cu,富Ni和富Mn原子团簇的数量密度随d_{max}值的变化如图7所示,模拟钢中1%～2%的Ni和Mn在400 ℃加热时虽然都可以固溶在α-Fe中,但是由于浓度起伏,仍然可以形成富Ni和富Mn的原子团簇,富Ni原子团簇中Ni原子之间的距离比Mn团簇中Mn原子之间的距离更近,形成富Ni团簇的趋势更大.随着时效时间从100 h增加到1 000 h,不仅富Cu原子团簇的数量密度增加,富Ni原子团簇的数量密度增加更加明显,但是富Mn原子团簇的数量密度却有减少的趋势.这种变化趋势不同的原因还有待进一步研究.

　　用同样的方法还分析了经过400 ℃,100 h,1 000 h时效的7♯样品和经过400 ℃,100 h

时效后的 8♯样品,由于篇幅的关系,这些数据在文章中不再列出. 仔细比较这些富 Cu,富 Ni 和富 Mn 团簇中 Cu, Ni, Mn 的成分随 d_{max} 值增大时的变化规律,可以得出这样一些结果:

(1) 在 Cu 原子团簇析出的初期就可以检测到其中含有 Ni,在大多数情况下还有 Mn. Ni(Mn)的富集应该在 Cu 原子发生扩散聚集之前就已经存在,它们不应该是与 Cu 原子一起扩散而发生聚集的结果. 如果是这样,就应该能够检测到富 Ni 和富 Mn 原子团簇的存在,实验结果证明了这一推论.

(2) 在富 Ni 原子团簇中,尤其是在那些 Ni 原子 d_{max} 值更小一些的团簇中,都可以检测到 Cu 原子的存在,这绝不是 Cu 原子随机分布的结果. 我们用嵌入原子势做过理论上

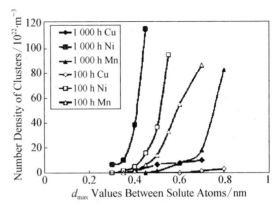

图 7　8♯样品经过 400 ℃, 1 000 h 和 400 ℃, 100 h 时效处理后富 Cu,富 Ni 和富 Mn 原子团簇数量密度随 d_{max} 值不同时的变化

Fig. 7　The variation of Cu-rich, Ni-rich and Mn-rich cluster density with the different d_{max} values for 8♯ specimens aged at 400 ℃ for 1 000 h and at 400 ℃ for 100 h

的估算,当 α-Fe 中存在 1%(原子分数)Ni 时,这时 Cu 原子发生聚集时所需的能量比没有 Ni 时要低[12]. 在富 Ni 原子团簇中能够检测到 Cu 原子,说明了富 Ni 原子团簇本身可以作为富 Cu 原子团簇析出时的成核位置.

(3) 在富 Mn 原子团簇中一般都可以检测到 Ni,在 Ni 含量较高的富 Mn 原子团簇中,还可以检测到 Cu 原子的存在. 富 Mn 原子团簇本身不一定是富 Cu 原子团簇析出时成核的位置,但是当团簇中 Ni 含量高时,富 Mn 原子团簇也可以作为富 Cu 原子团簇析出时成核的位置,因而在富 Cu 原子团簇析出的早期,在团簇中检测到 Ni 原子时,也经常可以检测到 Mn 原子.

(4) 富 Cu 原子团簇析出后,在团簇与 α-Fe 基体的界面处总是含有较高的 Ni 和 Mn,文献中曾讨论过这种现象[1,3,6,8],认为 Ni 和 Mn 的存在降低了 Cu/α-Fe 之间的界面能,因而钢中的合金元素 Ni 可以促使富 Cu 原子的析出[8]. 如果富 Ni 原子团簇可以作为富 Cu 原子析出时的成核位置,而富 Ni 原子团簇中总是含有 Mn,当 Cu 原子扩散进入 Ni 原子团簇中并在 d_{max} 值逐渐减小后,会将 Ni 和 Mn 原子排挤至 Cu 原子团簇的四周,最后形成了富 Cu 原子团簇被富 Ni 和 Mn 层包裹的"壳层"结构,这是必然的事情. "包裹着"Ni 或 Mn 的富 Cu 原子团簇,比单一的富 Cu 原子团簇不易长大,尺寸也更小[3-4],这种现象说明在 Cu 团簇和 α-Fe 基体界面上存在 Ni 和 Mn 时应该是增加了界面能,而不是降低了界面能.

(5) 富 Cu 原子团簇析出时,除了容易在 α-Fe 基体中的界面、位错等能量较高处形核之外,还会在富 Ni 原子团簇处成核. 增加钢中的 Ni 含量后,会增加富 Ni 原子团簇的数量密度,因而会促使富 Cu 原子团簇析出. 这就阐明了合金元素 Ni 会增加 RPV 钢中子辐照脆化敏感性的原因,也许这是最重要的本质原因.

4　结论

采用原子探针层析技术和热处理模拟方法,研究了压力容器模拟钢中合金元素 Ni 对富

Cu 原子团簇析出的影响,得到了如下一些结论:

(1) 压力容器模拟钢中的合金元素 Ni 从 0.8%(质量分数)提高至 1.5%后,会明显促使富 Cu 原子团簇的析出.析出初期的富 Cu 原子团簇中都含有 Ni 和 Mn.

(2) 含有 0.8%~1.5% Ni 的模拟钢在 400 ℃时效处理时,检测到存在富 Ni 的原子团簇,团簇中的 Ni 含量一般在 10%(原子分数)以上.在 Ni 原子之间距离 d_{max} 值比较小的富 Ni 原子团簇中,还可以检测到 Cu 和 Mn.富 Ni 原子团簇可以作为富 Cu 原子团簇析出时的形核位置.

(3) 含有 1.5%(质量分数)Mn 的模拟钢在 400 ℃时效处理时,检测到存在富 Mn 原子团簇,团簇中的 Mn 含量一般在 8%(原子分数)以上.当 Mn 原子团簇中含有较高的 Ni 时,它也可以成为富 Cu 原子团簇析出时成核的位置.

(4) 富 Cu 原子团簇析出时会在富 Ni 的原子团簇中成核,由于 Ni 原子团簇中总是含有 Mn,当 Cu 原子扩散进入 Ni 原子团簇中并在 d_{max} 值减小后,逐渐将 Ni 和 Mn 原子排挤至 Cu 原子团簇的四周,最后形成了富 Cu 原子团簇被富 Ni 和 Mn 层包裹的"壳层"结构.

(5) 由于钢中的合金元素 Ni 会形成富 Ni 原子团簇,Ni 原子团簇又是富 Cu 原子团簇析出时成核的地方,因而增加钢中的 Ni 含量会促使富 Cu 原子团簇的析出.这是合金元素 Ni 会增加 RPV 钢中子辐照脆化敏感性的本质原因.

参 考 文 献

[1] IAEA‐TECDOC‐1441. *Effects of Nickel on Irradiation Embrittlement on Light Water Reactor Pressure Vessel Steels* [R]. Vienna: International Atomic Energy Agency, 2005.

[2] Auger P, Pareige P, Welzel S, *et al*. Synthesis of Atom Probe Experiments on Irradiation-Induced Solute Segregation in French Ferritic Pressure Vessel Steels [J]. *J Nucl Mater*, 2000, 280: 331 – 344.

[3] Miller M K, Wirth B D, Odette G R. Precipitation in Neutron-Irradiated Fe‐Cu and Fe‐Cu‐Mn Model Alloys: a Comparison of APT and SANS Data [J]. *Mater Sci Eng*, 2003, A353: 133 – 139.

[4] Miller M K, Sokolov M A, Nanstad R K, *et al*. APT Characterization of High Nickel RPV Steels [J]. *J Nucl Mater*, 2006, 351: 187 – 196.

[5] Toyama T, Nagai Y, Tang Z, *et al*. Nanostructural Evolution in Surveillance Test Specimens of Commercial Nuclear Reactor Pressure Vessel Studied by Three Dimensional Atom Probe and Positron Annihilation [J]. *Acta Mater*, 2007, 55: 6852 – 6860.

[6] Miller M K, Russel K F, Embrittlement of RPV Steels: an Atom Probe Tomography Perspective [J]. *J Nucl Mat*, 2007, 371: 145 – 160.

[7] Miller M K, Chernobaeva A A, Shtrombakh Y I, *et al*. Evolution of the Nanostructure of VVER‐1000 PRV Materials under Neutron Irradiation and Post Irradiation Annealing [J]. *J Nucl Mat*, 2009, 385: 615 – 622.

[8] Cerezo A, Hirisawa S, Rozdilsky I, *et al*. Combined Atomicscale Modeling and Experimental Studies of Nucleation in the Solid State [J]. *Phil Trans R Soc Lond*, 2003, A361: 463 – 476.

[9] Lozano-Perez S, Sha G, Titchmarsh J M, *et al*. Comparison of the Number Density of Nanosized Cu-Rich Precipitates in Ferritic Alloys Measured Using EELS and EDX Mapping, HREM and 3DAP [J]. *J Mater Sci*, 2006, 41: 2559 – 2565.

[10] Miller M K. *Atom Probe Tomography: Analysis at the Atomic Level* [M]. New York: Kluwer Academic/Plenum Publishers, 2000.

[11] Zhou Bangxin(周邦新), Liu Wenqing(刘文庆). 三维原子探针及其在材料科学研究中的应用[J]. *Materials Science and Technology*(材料科学与工艺), 2007, 15(3): 405 – 408.

[12] Lin Mindong(林民东), Zhu Juanjuan(朱娟娟), Wang wei(王伟), *et al*. 核反应堆压力容器模拟钢中富 Cu 原子团簇的析出与嵌入原子势计算[J]. *Acta Physica Sinica*(物理学报), 2010, 59(2): 1163 – 1168.

Effect of Nickel Alloying Element on the Precipitation of Cu-Rich Clusters in RPV Model Steel

Abstract: The effect of nickel alloying element on the precipitation of Cu-rich clusters in RPV model steel has been investigated by means of atom probe tomography. 40 kg ingots of RPV model steel were prepared by vacuum induction furnace melting with two different Ni contents (0.79 wt. % and 1.52 wt. %) and with higher Cu content (0.5 wt. %). The specimens of 50×40 mm^2, 4 mm in thickness, were heat-treated by 880 ℃—20 min water quenching, 660 ℃—10 h tempering, and 400 ℃ aging, different from 100 h to 2 000 h. The results show that the increase of nickel content in RPV model steels will promote the precipitation of Cu-rich clusters. The contents of Ni and Mn were detected in the Cu-rich clusters at the early stage of the precipitation. Ni-rich clusters containing Cu and Mn were detected, and they could act as the nucleation sites for the precipitation of Cu-rich clusters. Mn-rich clusters could also be detected. Mn-rich clusters were not the nucleation sites for the precipitation of Cu-rich clusters, but the segregation of Cu atoms in those Mn-rich clusters containing much higher Ni could he found. Ni-rich clusters would act as the nucleation sites during the precipitation of Cu-rich clusters; therefore the increase of nickel content in RPV steels could promote the precipitation of Cu-rich clusters. This is the essential reason that the presence of nickel in PRV steel could increase its sensitivity to neutron irradiation embrittlement.

Optimization of N18 Zirconium Alloy for Fuel Cladding of Water Reactors*

Abstract: In order to optimize the microstructure and composition of N18 zirconium alloy (Zr－1Sn－0.35Nb－0.35Fe－0.1Cr, in mass fraction, %), which was developed in China in 1990s, the effect of microstructure and composition variation on the corrosion resistance of the N18 alloy has been investigated. The autoclave corrosion tests were carried out in super heated steam at 400 ℃/10.3 MPa, in deionized water or lithiated water with 0.01 mol/L LiOH at 360 ℃/18.6 MPa. The exposure time lasted for 300—550 days according to the test temperature. The results show that the microstructure with a fine and uniform distribution of second phase particles (SPPs), and the decrease of Sn content from 1% (in mass fraction, the same as follows) to 0.8% are of benefit to improving the corrosion resistance; It is detrimental to the corrosion resistance if no Cr addition. The addition of Nb content with upper limit (0.35%) is beneficial to improving the corrosion resistance. The addition of Cu less than 0.1% shows no remarkable influence upon the corrosion resistance for N18 alloy. Comparing the corrosion resistance of the optimized N18 with other commercial zirconium alloys, such as Zircaloy－4, ZIRLO, E635 and E110, the former shows superior corrosion resistance in all autoclave testing conditions mentioned above. Although the data of the corrosion resistance as fuel cladding for high burn-up has not been obtained yet, it is believed that the optimized N18 alloy is promising for the candidate of fuel cladding materials as high burn-up fuel assemblies. Based on the theory that the microstructural evolution of oxide layer during corrosion process will affect the corrosion resistance of zirconium alloys, the improvement of corrosion resistance of the N18 alloy by obtaining the microstructure with nano-size and uniform distribution of SPPs, and by decreasing the content of Sn and maintaining the content of Cr is discussed.

1 Introduction

Zirconium alloys are used as nuclear fuel cladding in water reactors because of low cross-section for thermal neutron absorption and high corrosion resistance. In order to reduce the cost of nuclear power, it is necessary to increase the burn-up of nuclear fuel. Thus refueling cycle must be extended and the fuel assemblies need to run longer in the reactor core. It is well known that waterside corrosion resistance of the fuel cladding is a key factor to restrict service life of high burn-up fuel assemblies[1]. Therefore, it is necessary to develop new zirconium alloys with advanced performance.

---

* In collaboration with Yao M Y, Li Z K, Wang X M, Zhou J, Long C S, Liu Q, Luan B F. Reprinted from J. Mater. Sci. Technol., 2012, 28(7): 606-613.

There are three categories of zirconium alloys, including Zr – Sn, Zr – Nb and Zr – Sn – Nb alloys now. By adding alloying elements of Fe, Cr, Ni and Cu *etc.* in the three categories[2-12], Zircaloy – 2, Zircaloy – 4, E110, M5, ZIRLO and E635 alloys have been developed for the fuel cladding materials. However, the results of existing studies show that the composition of the applied zirconium alloys is not in the best range, for example reducing the Sn content to 0. 6%—0. 7% (in mass fraction, the same as follows) and increasing the Fe content to 0. 4% or Fe+Cr content to 0. 6% can further improve the corrosion resistance of Zr – Sn and Zr – Sn – Nb alloys[3,13-16]. HANA – 6 alloy, developed by adding 0. 05% Cu in the Zr – 1. 1% Nb alloy, also exhibits excellent corrosion resistance[8]. Therefore, optimizing composition or alloying with other elements based on the existing zirconium alloys, is basic ways to develop new alloys with excellent corrosion resistance. On the other hand, optimizing the alloy microstructure, including the size and distribution of second phase particles (SPPs), can also further improve the corrosion resistance of zirconium alloys[4,11,16].

N18 zirconium alloy was developed by China in 1990s. It is a kind of Zr – Sn – Nb alloys with low Nb content and its composition is Zr – 1Sn – 0. 35Nb – 0. 3Fe – 0. 1Cr (in mass fraction, %). Comparing to the zirconium alloys with high Nb content, N18 alloy has relatively high temperature for α/β phase transformation. The appearance of β – Zr phase can be easily avoided during the thermal mechanical processes, and it is beneficial to obtaining a fine and uniform distribution of SPPs. To optimize the microstructure and composition of N18 zirconium alloy, the effect of microstructure and composition variation on the corrosion resistance of the N18 alloy has been investigated in this paper.

2　Experimental

2. 1　Preparation of the specimen with different microstructure

Sheet specimens in 1. 4 mm thickness provided by Northwest Institute for Non-ferrous Metal Research in China were fabricated from ingots in weight of 500 kg by conventional procedures. To obtain the optimized microstructure, four different procedures were employed to study the relationship between the corrosion resistance and microstructure of N18 alloy (Fig. 1). The procedures for specimen 41 are as follows: (1) Strips of 10 mm in width were cut from N18 plates with 1. 4 mm thickness. (2) The strips were sealed in vacuum quartz capsules and then heated to 1 020 ℃ for 20 min. (3) The capsules were then quenched into water and broken simultaneously (denoted as WQ). (4) The strips were cold rolled from 1. 4 to 0. 6 mm and were cut into specimens with 25 mm×10 mm. In the case of specimens 42, 43 and 44, strips of 25 mm in width were cut from the plates with 1. 4 mm thickness. For specimen 42, they were cold rolled to 0. 6 mm, then heated to 820 ℃ for 2 h in vacuum. For specimens 43 and 44, they were heated to 820 ℃ for 2 h and 700 ℃ for 4 h in vacuum, respectively, then cold rolled to 0. 6 mm. Specimens 42, 43 and 44

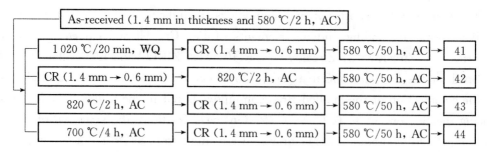

Fig. 1 Procedures and corresponding designation of N18 alloy. Here "AC" refers to air cooling the specimens within the vacuum quartz capsules

were all cut into specimens with a size of 25 mm×20 mm. Finally, all of the specimens were heated to 580 ℃ for 50 h in vacuum for recrystallization and decomposition of β- Zr phase if it exists, as well as for making the solid solution contents of alloying elements in α-Zr matrix at a same level. Because β- Zr phase is very harmful to the corrosion resistance of zirconium alloys[17-21], it can be eliminated by the decomposition after heat treatment at 580 ℃.

2. 2　Composition variation based on N18 alloy

In order to optimize the compositions of N18 alloy, it is necessary to investigate the effect of the composition variation on the corrosion resistance of N18 alloy. The composition variation based on N18 alloy is listed in Table 1. The variation program is illustrated as follows: (1) For N-0 alloy, the lower limit of Nb content (0. 25%) was chosen based on N18 alloy; (2) Comparing with N-0 alloy, N-1 alloy did not contain Cr; (3) For N-2 and N-3 alloys, 0. 05% Cu and 0. 1% Cu were added on the base of the N-1 alloy, respectively; (4) For N-4 alloy, 0. 05% Cu was added based on N-0 alloy; (5) For N-5 alloy, Sn content was decreased to 0. 8% based on N-0 alloy; (6) For N-6 alloy, 0. 05% Cu was added based on N-5 alloy; (7) For N18 alloy, the upper limit (0. 35%) of Nb content was chosen. Each ingot of N-0-N-6 alloys was 20 kg in weight, and N18 alloy ingot was 500 kg in weight. All the ingots were melted and the sheets in 0. 7 mm thickness were fabricated by Northwest Institute for Non-ferrous Metal Research

Table 1　Chemical compositions (mass fraction, %) of adjusting experimental alloys based on the composition of N18 alloy

| Alloy | Element content | | | | | | Comments | |
	Sn	Nb	Fe	Cr	Cu	Zr		
N-0	1. 04	0. 25	0. 33	0. 06	—	Bal.	N18 alloy with lower limit of Nb content	From a 20 kg ingot
N-1	1. 04	0. 25	0. 33	—	—	Bal.	No Cr addition	for each alloy
N-2	1. 04	0. 25	0. 32	—	0. 05	Bal.	No Cr addition, but adding 0. 05 Cu	
N-3	1. 00	0. 25	0. 33	—	0. 1	Bal.	No Cr addition, but adding 0. 1 Cu	
N-4	0. 99	0. 26	0. 34	0. 06	0. 05	Bal.	Adding 0. 05 Cu	
N-5	0. 79	0. 26	0. 33	0. 06	—	Bal.	Decreasing Sn content to 0. 8	
N-6	0. 81	0. 26	0. 35	0. 06	0. 05	Bal.	Decreasing Sn content to 0. 8 and adding 0. 05 Cu	
N18	1. 0	0. 35	0. 35	0. 1	—	Bal.	N18 alloy with upper limit of Nb content	From a 500 kg ingot

in China. Finally, the specimens of 25 mm × 20 mm in size were cut from the plates and were heated to 580 ℃ for 2 h.

2. 3　Corrosion tests

The corrosion tests were carried out in super heated steam at 400 ℃/10. 3 MPa, in deionized water or lithiated water with 0. 01 mol/L LiOH at 360 ℃/18. 6 MPa in static autoclaves according to the ASTM standard test (G 2M - 88) which specifies the use of water with a deaeration practice to reduce dissolved oxygen to low levels. The deaeration was achieved by venting the autoclave several times while the temperature was increased. The exposure time lasted for 300—550 days according to the test temperature. The refreshing period of testing solutions was not more than 30 days. The corrosion behavior was evaluated by weight gain as a function of exposure time. The reported weight gains were a mean value obtained from five specimens. Prior to the corrosion tests, the specimens were cleaned and pickled in a mixed acid solution (30% H_2O+30% HNO_3+30% H_2SO_4+10% HF, in volume fraction), sequentially rinsed in cold tap water, boiling deionized water, and then blow-dried with warm air.

2. 4　Microstructural observation

The microstructure of the specimens was examined by transmission electron microscopy (TEM, JEM - 200CX). The compositions of SPPs were measured by using energy-dispersive spectroscopy (EDS) fitted in a JEM - 2010F transmission electron microscope. Disk specimens, 3 mm in diameter, were mechanically grinded to about 70 μm in thickness, and then thinned by twin-jet electro-polishing technique at a voltage of 40—50 V in a mixture of 10% perchloric acid and 90% ethanol (in volume fraction) at −40 ℃.

3　Results

3. 1　Effect of microstructure on the corrosion resistance of N18 alloy

Fig. 2 shows the TEM micrographs of N18 alloy specimens prepared by different procedures. Full recrystallization structure is obtained for all specimens heat treated by annealing at 580 ℃ for 50 h. The grain size is reduced for the specimens in the sequence of 42≅43>44>41. For specimen 41 (Fig. 2(a)), the fine SPPs distribute uniformly in the matrix and the mean size of SPPs is only about 58 nm with no much difference among them. For specimen 42 (Fig. 2(b)), the SPPs distribute mainly on the grain boundaries, which are obtained by the decomposition of β- Zr. The average size of SPPs in specimen 42 is about 350 nm, which is larger but less than that in specimen 41. For specimen 43 (Fig. 2(c)), the SPPs distribute also on the grain boundaries, which are obtained by the decomposition of deformed β- Zr. For specimen 44, the distribution of SPPs is uniform, but their size is quite different. The big ones are up to 1 μm, but the small ones are less

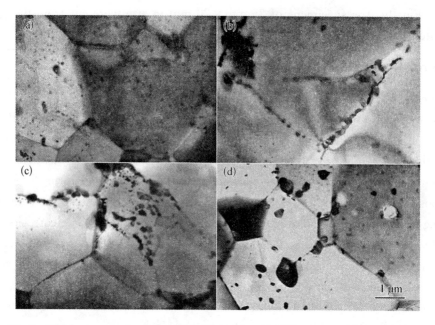

Fig. 2 TEM micrographs of the N18 specimens with different procedures: (a) 41, (b) 42, (c) 43, (d) 44

than 50 nm. The average size is 242 nm (Fig. 2(d)).

The results obtained by EDS analysis show that the composition of SPPs is Zr, Fe, Cr and Nb, which is consistent with the results obtained by Isobe and Matsuo[13] in Zr – 0.5Sn –0.2Fe – 0.1Cr – 0.4Nb alloy. According to the analysis results (Fig. 3), the SPPs can be classified into two groups: one contains lower Nb with a Fe/Cr ratio of 3—4, the other contains higher Nb with a Fe/Cr ratio of 5—7. This result is the same as the report given by Liu *et al.* [22]. All of the SPPs should be Zr(Fe, Cr, Nb)$_2$ metallic compounds and no β – Nb particles are detected[7,23,24]. The difference of the composition of Zr(Fe, Cr, Nb)$_2$ particles could be caused by their formation in different ways. The SPPs with lower contents of Nb are produced by the precipitation from α – Zr matrix, and the SPPs with higher contents of Nb are produced by the decomposition of β – Zr. Because the SPPs are incorporated in α – Zr matrix, it could be understood that the analysis results show the

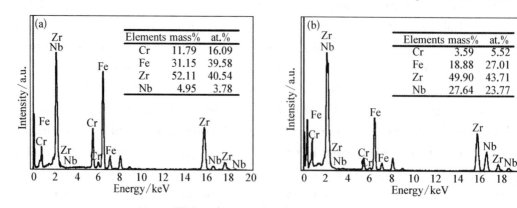

Fig. 3 EDS analysis results of SPPs in N18 specimens

composition of the SPPs with higher contents of Zr than the compounds of Zr(Fe, Cr, Nb)$_2$.

Fig. 4 shows the weight gains vs exposure time for the specimens exposed to super heated steam at 400 ℃/10. 3 MPa and in lithiated water with 0. 01 mol/L LiOH at 360 ℃/ 18. 6 MPa, and the exposure time was 310 and 529 days, respectively. It is obvious that the corrosion resistance of specimens is closely related to size and distribution of the SPPs. The specimen 41 with nano-size and uniform distributed SPPs shows the best corrosion resistance in both super heated steam at 400 ℃ and lithiated water at 360 ℃. The specimen 42 with coarse size and nonuniform distributed SPPs shows the worst corrosion resistance in both super heated steam at 400 ℃ and lithiated water at 360 ℃. The corrosion resistance for the specimens prepared by different procedures varies considerably, and decreases in a same sequence of 41>43>44>42 in both exposure conditions. In order to improve the corrosion resistance for the zirconium alloys containing Nb, it is necessary to obtain a proper microstructure of nano-size and uniformly distributed SPPs, which are the same as the zirconium alloys containing higher Nb of 1%[10-12,18,19,22].

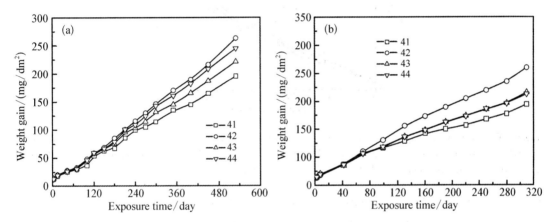

Fig. 4 Weight gains vs exposure time for N18 specimens prepared by different thermal mechanical processes corroded in lithiated water with 0. 01 mol/L LiOH at 360 ℃/18. 6 MPa (a) and in super heated steam at 400 ℃/10. 3 MPa (b)

3. 2　Effect of the composition variation on the corrosion resistance of N18 alloy

Fig. 5 shows the weight gains vs exposure time for the specimens exposed to in lithiated water with 0. 01 mol/L LiOH or deionized water at 360 ℃/18. 6 MPa, and in super heated steam at 400 ℃/10. 3 MPa after varying the composition of alloys based on N18. In comparison with the weight gains of the specimens in the three testing conditions (Fig. 5), some relations between the corrosion resistance and the variation of alloy composition can be deduced: (1) The corrosion resistance can be improved in all three testing conditions by decreasing the content of Sn from 1% to 0. 8%. The weight gains reduce 9%—13% after 400 days exposure. (2) It is detrimental to the corrosion resistance in all three testing conditions if no Cr addition. (3) The addition of Nb content with upper limit (0. 35%) is beneficial to improving the corrosion resistance tested in lithiated water or

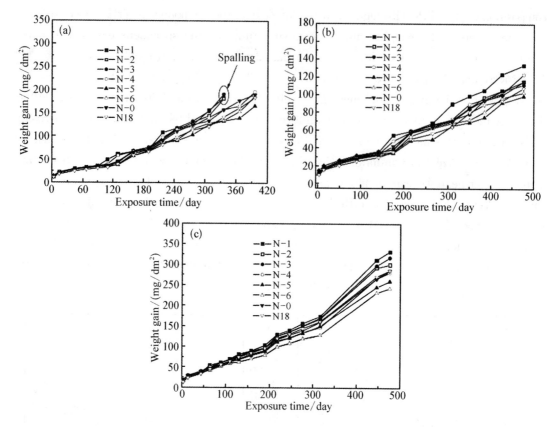

Fig. 5 Weight gains *vs* exposure time for the alloys with the variation of composition based on N18 alloy corroded in lithiated water with 0. 01 mol/L LiOH at 360 ℃/18. 6 MPa (a), in deionized water at 360 ℃/ 18. 6 MPa (b) and in super heated steam at 400 ℃/10. 3 MPa (c)

in deionized water than that with lower limit (0. 25%), but the effect of Nb content between upper and lower limit on the corrosion resistance is not obvious when corrosion tests are carried out in super heated steam at 400 ℃. (4) The addition of Cu less than 0. 1% shows no remarkable influence upon the corrosion resistance for N18 alloy. In previous work[25], it was found that the addition of 0. 05%—0. 5% Cu also has no remarkable influence upon the corrosion resistance for Zr – 0. 80Sn – 0. 34Nb – 0. 39Fe –0. 1Cr – xCu alloy in superheated steam at 500 ℃.

The corrosion resistance of optimized N18 alloy (N – 5 alloy) is compared with that of commercial zirconium alloys, such as Zircaloy – 4, ZIRLO, E110 and E635, as shown in Fig. 6. The data of weight gains of commercial zirconium alloys are taken from literature[10]. In general, the results of corrosion resistance obtained by autoclave tests cannot completely represent the corrosion behavior of zirconium alloys operated as fuel cladding in reactors. For example, the corrosion resistance of ZIRLO and E635 alloys is much better than that of Zircaloy – 4 when they are operated as fuel cladding in reactors or autoclave tested in lithiated water at 360 ℃, but it is not the case when they are tested in autoclave at 400 ℃ in super heated steam. It can be seen from Fig. 6 that the corrosion resistance of optimized N18 alloy (N – 5 alloy) is superior to that of all commercial

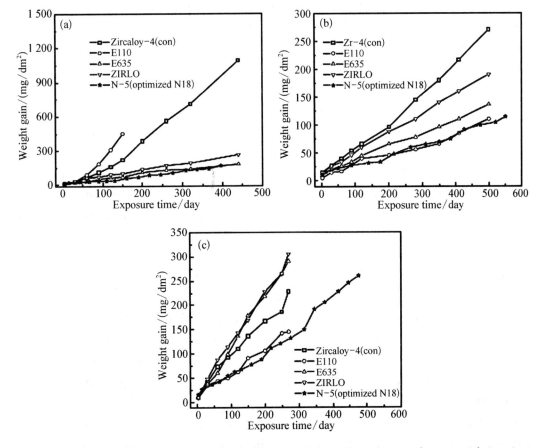

Fig. 6 Weight gains *vs* exposure time for optimized N18 (N−5) alloy and several commercial zirconium alloys (Zircaloy−4, ZIRLO, E635 and E110) tested in autoclave at 360 ℃ in lithiated water with 0.01 mol/L LiOH (a) or in deionized water (b), at 400 ℃ in super heated steam (c). Here, the data of weight gains for the commercial zirconium alloys are taken from literature[10]

zirconium alloys at the three corrosion conditions mentioned in section 2.3. Although the corrosion behavior of the optimized N18 alloy operated as fuel cladding has not been obtained yet, it is believed that the optimized N18 alloy is a promising material for fuel cladding as high burn-up fuel assemblies.

4   Discussion

A new oxide layer is always formed on the oxide/metal interface. So the microstructure of oxide will certainly affect the diffusion of oxygen ions or anion vacancies to result in a different behavior for the growth of oxide layers, a different corrosion resistance in other words. Some previous work[26-28] has pointed out that the defects consisting of vacancies and interstitials are produced during the growth of oxide layers due to the existence of compressive stress. The characteristic of oxide layers with such kind of microstructure is an internal cause for the microstructural evolution during the corrosion process. The diffusion, annihilation and condensation of vacancies and interstitials under the action of stress, temperature and time caused the stress relaxation in oxide layers and phase transformation of meta-stable phases to stable phase. Temperature and time are the external causes for the microstructural evolution during the corrosion process. The vacancies were absorbed by grain boundaries to form the nuclei of pores. The pores grew through the absorption of the vacancies from the grains of oxide to weaken the bonding strength between grains. Based on the formation of pores-clusters, the development of micro-cracks in oxide layers will lead to the loss of protective characteristic, thus the phenomenon of corrosion transition appears. This is an inevitable result caused by the microstructural evolution of oxide. The variation of corrosion kinetics is induced by the microstructural evolution of oxide, so the progress of microstructural evolution is closely related to the corrosion behavior of zirconium alloys. Therefore, to find and control the factors, which could retard the microstructural evolution, will improve the corrosion resistance of zirconium alloys.

The solid solution limitation in $\alpha - Zr$ for most alloying elements used for zirconium alloys, except Sn and Nb, is very small. These alloying elements, such as Fe, Cr, Ni and Cu etc., form metallic compounds with zirconium to precipitate as SPPs in $\alpha - Zr$ matrix. In general, the corrosion resistance of SPPs is better than that of $\alpha - Zr$ matrix. So the SSPs in $\alpha - Zr$ will transform into zirconium oxide at first when the specimens are corroded. Therefore, the SPPs incorporated in oxide layers will play an important role in the microstructural evolution of oxide in two aspects. On the one hand, the SPPs incorporated in oxide with nanometer size provide a large quantity of phase boundaries for the sink of vacancies. In this case, the growth of pores in oxide layers will be slowed down, the microstructural evolution will be retarded in other words, and the corrosion resistance will be enhanced. On the other hand, the oxidation of the SPPs incorporated in oxide layers produce an additional stress in oxide layers to promote the diffusion of the vacancies and enhance the microstructural evolution. In order to improve the corrosion resistance of zirconium alloys, the SPPs in matrix should be in nano-size and uniform distribution, and have a good resistance to oxidation for themselves.

The corrosion resistance of the powder of $Zr(Fe_x, Cr_{1-x})_2$ ($x=1$, 2/3, 1/3) alloys was investigated by autoclave test at 400 °C super heated steam[29]. The results show that the decrease of x value in the alloys, i. e. the increase of Cr content, enhances significantly the corrosion resistance. Takeda and Anada[14] investigated the mechanism of corrosion rate degradation due to the addition of Sn in zirconium alloys. The distribution of Sn in the oxide layers was analyzed by high resolution EDX. It was observed that Sn was enriched at the boundaries of oxide crystallites. Because the solubility of Sn in zirconium oxide is lower than that in $\alpha - Zr$ metal, Sn will be squeezed out from zirconium oxide to segregate at grain boundaries as a metallic state at first during the oxidation process, and then it is turned to Sn oxide in later. An additional stress in oxide layer, produced by the expansion of volume from metal to oxide of Sn (Pilling-Bedworth ratio 1. 31), promotes the diffusion of vacancies to enhance the microstructural evolution of oxide layers.

It is better understood on the basis of the discussion above that the improvement of corrosion resistance for the N18 alloy is due to the microstructure with nano-size and uniform distribution of SPPs, and the decrease in the Sn content as well as maintaining the Cr content.

5 Conclusions

(1) The corrosion resistance of N18 alloy is affected remarkably by the microstructure of SPPs both in the size and distribution. It is necessary to obtain uniform distribution of SPPs in nano-size by adopting appropriate procedures to improve the corrosion resistance of the alloy.

(2) The corrosion resistance can be improved by decreasing the Sn content from 1% to 0.8%. It is detrimental to the corrosion resistance if no addition Cr. The addition of Nb content with upper limit (0.35%) is beneficial to improving the corrosion resistance. The addition of Cu element less than 0.1% shows no remarkable influence upon the corrosion resistance of N18 alloy.

(3) Comparing the corrosion resistance of the optimized N18 with other commercial zirconium alloys, such as Zircaloy‐4, ZIRLO, E635 and E110, the optimized N18 alloy shows superior corrosion resistance in all testing environments during long-term autoclave tests. Therefore, the optimized N18 can be a promising candidate cladding material as high burn-up fuel assemblies.

Acknowledgements

The authors would like to express their thanks to Dr. Qiang Li and Mr. Jianchao Peng of Instrumental Analysis and Research Center of Shanghai University for help in the microstructural analysis. This study is partly supported by the National Natural Science Foundation of China (Nos. 50871064 and 50971084) and Shanghai Leading Academic Discipline Project (No. S30107).

References

[1] IAEA‐TECDOC‐996, *Waterside Corrosion of Zirconium Alloys in Nuclear Power Plants*, IAEA, Vienna, ISSN 1011‐4289, 1998, January.

[2] C. M. Eucken, P. T. Finden, S. Trapp-Pritsching and H. G. Weidinger: *Zirconium in the Nuclear Industry: Eighth International Symposium*, ASTM STP 1023, eds L. F. P. van Swam and C. M. Eucken, West Conshohocken PA, ASTM International, 1989, 113.

[3] F. Garzarolli, Y. Broy and R. A. Busch: *Zirconium in the Nuclear Industry: Eleventh International Symposium*, ASTM STP 1295, eds E. R. Bradley and G. P. Sabol, West Conshohocken PA, ASTM International, 1996, 850.

[4] R. A. Graham, J. P. Tosdale and P. T. Finden: *Zirconium in the Nuclear Industry: Eighth International Symposium*, ASTM STP 1023, eds L. F. P. van Swam and C. M. Eucken, West Conshohocken PA, ASTM International, 1989, 334.

[5] M. Harada, M. Kimpara and K. Abe: *Zirconium in the Nuclear Industry: Ninth International Symposium*, ASTM STP 1132, eds C. M. Eucken and A. M. Garde, West Conshohocken PA, ASTM International,

1991, 368.

[6] B. X. Zhou, W. J. Zhao, Z. Miao, D. C. Huang and Y. R. Jiang: *Biomaterials and Ecomaterials Ⅲ - 2*, *Proceedings of 1996 Chinese Materials Symposium*, eds B. X. Zhou and Y. K. Shi, 1996, November, 17 - 21, Chinese Materials Research Society, Chemical Industry Press, Beijing, China, 1997, 183. (in Chinese)

[7] Z. K. Li, J. Z. Liu, M. S. Zhu and Q. Z. Sun: *Rare Metal Mater. Eng.*, 1996, **25**(5), 43. (in Chinese)

[8] J. Y. Park, B. K. Choi, S. J. Yoo and Y. H. Jeong: *J. Nucl. Mater.*, 2006, **359**, 59.

[9] G. P. Sabol, G. R. Kilp, M. G. Balfour and E. Roberts: *Zirconium in the Nuclear Industry: Eighth International Symposium*, *ASTM STP 1023*, eds L. F. P. van Swam and C. M. Eucken, West Conshohocken PA, ASTM International, 1989, 227.

[10] A. V. Nikulina, V. A. Markelov, M. M. Peregud, Y. K. Bibilashvili, V. A. Kotrekhov, A. F. Lositsky, N. V. Kuzmenko, Y. P. Shevnin, V. K. Shamardin, G. P. Kobylyansky and A. E. Novoselov: *Zirconium in the Nuclear Industry: Eleventh International Symposium*, *ASTM STP 1295*, eds E. R. Bradley and G. P. Sabol, West Conshohocken PA, ASTM International, 1996, 785.

[11] J. P. Mardon, D. Charquet and J. Senevat: *Zirconium in the Nuclear Industry: Twelfth International Symposium*, *ASTM STP 1354*, eds G. P. Sabol and G. D. Moan, West Conshohocken PA, ASTM International, 2000, 505.

[12] R. J. Comstock, G. Schoenberger and G. P. Sabol: *Zirconium in the Nuclear Industry: Eleventh International Symposium*, *ASTM STP 1295*, eds E. R. Bradley and G. P. Sabol, West Conshohocken PA, ASTM International, 1996, 710.

[13] T. Isobe and Y. Matsuo: *Zirconium in the Nuclear Industry: Ninth International Symposium*, *ASTM STP 1132*, eds C. M. Eucken and A. M. Garde, West Conshohocken PA, ASTM International, 1991, 346.

[14] K. Takeda and H. Anada: *Zirconium in the Nuclear Industry: Twelfth International Symposium*, *ASTM STP 1354*, eds G. P. Sabol and G. D. Moan, West Conshohocken PA, ASTM International, 2000, 592.

[15] Y. Broy, F. Garzarolli, A. Seibold and L. E. van Swam: *Zirconium in the Nuclear Industry: Twelfth International Symposium*, *ASTM STP 1354*, ed G. P. Sabol and G. D. Moan, West Conshohocken PA, ASTM International, 2000, 609.

[16] H. K. Yueh, R. L. Kesterson, R. J. Comstock, H. H. Shah, D. J. Colburn, M. Dahlback and L. Hallstadius: *Zirconium in the Nuclear Industry: Fourteenth International Symposium*, *ASTM STP 1467*, eds P. Rudling and B. Kammenzind, West Conshohocken PA, ASTM International, 2004, 330.

[17] K. N. Choo, Y. H. Kang, S. I. Pyun and V. F. Urbanic: *J. Nucl. Mater.*, 1994, **209**, 226.

[18] Y. H. Jeong, K. O. Lee and H. G. Kim: *J. Nucl. Mater.*, 2002, **302**, 9.

[19] Y. H. Jeong, H. G. Kim and T. H. Kim: *J. Nucl. Mater.*, 2003, **317**, 1.

[20] Y. H. Jeong, H. G. Kim and D. J. Kim: *J. Nucl. Mater.*, 2003, **323**, 72.

[21] H. G. Kim, Y. H. Jeong and T. H. Kim: *J. Nucl. Mater.*, 2004, **326**, 125.

[22] W. Q. Liu, Q. Li, B. X. Zhou, Q. S. Yan and M. Y. Yao: *Nucl. Power Eng.*, 2005, **26**(3), 249. (in Chinese)

[23] B. X. Zhou, M. Y. Yao, Q. Li, S. Xia, W. Q. Liu and Y. L. Chu: *Rare Metal Mater. Eng.*, 2007, **36**(8), 1317. (in Chinese)

[24] Y. Z. Liu, W. J. Zhao, Q. Peng and C. L. Sun: *Nucl. Power Eng.*, 2005, **26**, 158. (in Chinese)

[25] M. Y. Yao, Y. Zhang, S. L. Li, X. Zhang, J. Zhou and B. X. Zhou: *Acta. Metall. Sin.*, 2011, **47**(7), 872. (in Chinese)

[26] B. X. Zhou, Q. Li, M. Y. Yao, W. Q. Liu and Y. L. Chu: *Nucl. Power Eng.*, 2005, **26**(4), 364. (in Chinese)

[27] B. X. Zhou, Q. Li, W. Q. Liu, M. Y. Yao and Y. L. Chu: *Rare Metal Mater. Eng.*, 2006, **35**(7), 1009. (in Chinese)

[28] B. X. Zhou, Q. Li, M. Y. Yao, W. Q. Liu and Y. L. Chu: *Zirconium in the Nuclear Industry: Fifteenth International Symposium*, *ASTM STP 1505*, eds B. Kammenzind and M. Limbäck, West Conshohocken PA, ASTM International, 2009, 360.

[29] X. Cao, M. Y. Yao, B. X. Zhou and J. C. Peng: *Acta Metall. Sin.*, 2011, **47**(7), 882. (in Chinese)

C – Cr Segregation at Grain Boundary before the Carbide Nucleation in Alloy 690[*]

Abstract: The grain boundary segregation in Alloy 690 was investigated by atom probe tomography. B, C and Si segregated at the grain boundary. The high concentration regions for each segregation element form a set of straight arrays that are parallel to each other in the grain boundary plane. The concentration fluctuation has a periodicity of about 7 nm in the grain boundary plane. Before the $Cr_{23}C_6$ nucleation at grain boundaries, the C – Cr co-segregate on one side of the grain boundaries while not the exact grain boundary core regions have been detected. The reasons why grain boundary carbides have coherent orientation relationship only with one side of nearby grain which grain boundary is located at high index crystal plane were discussed.

1 Introduction

The precipitation of grain boundary carbides is an important factor that influences the properties related to the grain boundaries of metallic materials, such as intergranular corrosion (IGC) and intergranular stress corrosion cracking (IGSCC) [1, 2]. The precipitation of Cr-rich $M_{23}C_6$ in Alloy 690 and austenitic stainless steels (SS) has been extensively studied for many years [3 – 13].

The precipitation features of grain boundary carbides including crystal structure and morphology in Alloy 690 and SS are well established. Transmission electron microscopy (TEM) studies of grain boundary precipitation have revealed that Cr rich $M_{23}C_6$ is the most common types of the precipitates in these alloys [3, 4, 6, 9]. During the process of precipitation, the grain boundary carbides still have coherent orientation relationship with one of the grains. The morphology evolution of the grain boundary carbides precipitated at various temperatures, time and grain boundary characters had been studied by TEM and scanning electron microscopy (SEM) [3 – 13].

Solute and impurity atoms may segregate at the grain boundaries before the nucleation of carbides. The study of distribution of solute and impurity atoms at the grain boundaries before carbide nucleation is valuable for understanding the nucleation and precipitation behavior of carbides. But, this topic has not been investigated adequately. In this paper, the atom probe tomography (APT) was adopted to directly measure the segregation of solute and impurity atoms at grain boundaries in Alloy 690, and the nucleation of grain boundary carbides was discussed.

* In collaboration with Li Hui, Xia Shuang, Liu Wenqing. Reprinted from Materials Characterization, 2012, 66: 68 – 74.

2　Experimental Procedures

The composition of Alloy 690 used in this experiment is given in Table 1. The Alloy 690 sheets were solution treated at 1 100 ℃ for 15 min in vacuum. After solution treating, the sheets were subsequently quenched into water (WQ). Then a 50% cold reduction and recrystallization annealing at 1 000 ℃ for 1 min was carried out to reduce the grain size. The sizes of specimen sheets are about 0. 5 mm×3 mm×15 mm. In order to control the annealing temperature and time, the small specimens were vacuum sealed less than 5×10^{-3}Pa in long quartz tube about 60 cm in length. Firstly, one end of the quartz tube was heated and achieved the annealing temperature. At the same time, the small specimens were put at another end of the quartz tube. Because the quartz tube is not a good conductor for heat, the small specimens still keep at room temperature. After one end of the quartz tube was heated to 1 000 ℃, the small samples were moved to this end quickly and kept them at 1 000 ℃ for 1 min, and then WQ. After annealing treatment, the specimens were aged at 500 ℃ for 0. 5 h to create grain boundary segregation and early stage of carbide nucleation.

Table 1　The composition of Alloy 690 used in this experiment

	Ni	Cr	Fe	C	N	Si	Ti	Al
Atom fraction	57. 88	31. 09	9. 45	0. 09	0. 03	0. 28	0. 47	0. 70
Weight fraction	60. 73	28. 91	9. 45	0. 02	0. 008	0. 14	0. 40	0. 34

The sheets were electro-polished in a solution of 20% $HClO_4$ +80% CH_3COOH at room temperature with 30 V direct current for 30 s. Electron backscatter diffraction (EBSD) technique was employed to determine the microstructure of the specimens and misorientation of the grains using the TSL laboratory orientation imaging microscopy (OIM) system attached to an SEM. The scanning step is 1 μm. The grain boundary was identified if the misorientation between two points is higher than 2°.

The sheets were cut into small rods with a section of 0. 5 mm×0. 5 mm in 15 mm length using a spark machine. The needle specimens for atom probe tomography (APT) analysis were prepared by standard two steps electro-polishing [14]. JEM – 200 CX transmission electron microscopy (TEM) was used to select the needles sharp enough and containing grain boundary near the apex of the specimens. The needle specimens were analyzed using an Imago Scientific Instruments 3 000 HR local electrode atom probe (LEAP). The specimens were cooled to a temperature of 50 K. The pulse amplitude was kept at 15% of the standing voltage applied to the specimen. The voltage pulsing repetition rate was 200 kHz. The standing voltage on the apex of the specimen was varied automatically in order to maintain an evaporation rate of 0. 2% per voltage pulse. Though the relatively low rate produces a low fracture tendency, these materials have very high fracture tendency when the apex of the specimen contains grain boundary.

3 Results and Discussion

Fig. 1 gives the EBSD analysis of the microstructure of the Alloy 690 specimens annealed at 1 000 ℃ for 1 min. Each line stands for a grain boundary derived from OIM system (Fig. 1a), and the average grain size measured by average intercept method is about 3. 5 μm. Whether the grain is fully recrystallized can be determined by grain average misorientation (GAM) (Fig. 1b). If the GAM is less than 1°, the grain is identified as fully recrystallized in the current paper. About 96. 4% grains have fully recrystallized. So the dislocation density is very low in the microstructure.

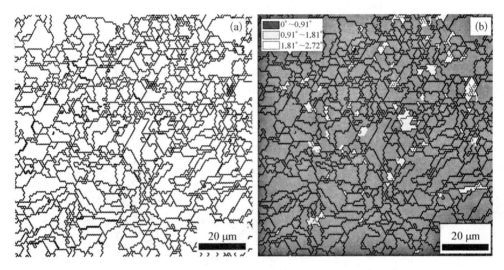

Fig. 1 EBSD analysis of the microstructure of Alloy 690 specimens annealed at 1 000 ℃ for 1 min. (a) OIM map of grain boundaries, (b) map of grain average misorientation

Fig. 2 shows the microstructure of the Alloy 690 specimens aged at 500 ℃ for 0. 5 h. Carbide precipitates are not be observed at grain boundary with few dislocations (Fig. 2a), while the carbides are observed at the dislocation junctions on the grain boundaries with high density dislocation networks (Fig. 2b). The maximum size of the observed carbides is about 7 nm.

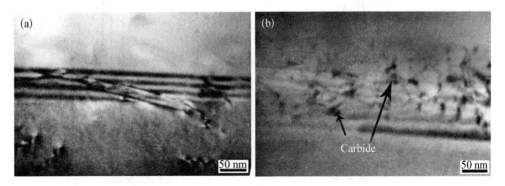

Fig. 2 TEM images of the Alloy 690 specimens aged at 500 ℃ for 0. 5 h

Fig. 3 shows the grain boundary segregation features of the specimen. Carbide precipitates and dislocations are not be observed at grain boundary in the TEM image of the needle specimen which used for APT test (Fig. 3a). And the C atoms are randomly distributed at the grain boundary region indicating that the carbides may not have nucleated yet (Fig. 3c). The atom maps have been oriented so that the views are along (Fig. 3b) and perpendicular to (Fig. 3c) the grain boundary planes, in order to show the atoms segregated at two dimensional grain boundary plane.

One can randomly select two 20 nm×20 nm×25 nm regions across the grain boundary to analyze the grain boundary segregation features of Alloy 690. The distance of the two regions is about 30 nm. Fig. 3d and e gives the depth concentration profiles related to the crossing of the grain boundary. The concentration profiles can give the width of the enriched zone more clearly than the atom maps. Local magnification effect occurs at interfaces when analyzed in the atom probe [15], leading to the spreading of segregated atoms over several nanometers near the grain boundaries (Fig. 3). The local magnification effect depended on the orientation of the grain boundary with respect to the specimen axis in nickel based alloys [16]. By comparing with Fig. 7 in Ref. [16], it can be concluded that the real thickness of the segregation zone should be about 0. 4 nm (a little more than

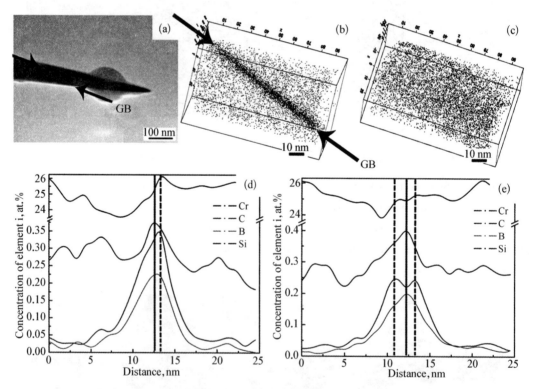

Fig. 3 The grain boundary segregation features in Alloy 690. (a) TEM image of the needle specimen which tested by APT. (b), (c) Three dimensional distribution of C and B atoms at the grain boundaries. The observation directions are along the grain boundary in (b), but perpendicular to the grain boundary in (c). The box sizes are 54 nm×56 nm×98 nm. (d), (e) Depth concentration profiles of Cr, C, B and Si near the grain boundaries. The solid lines stand for the position of grain boundaries, and the dash lines stand for the position of C–Cr co-segregation zones. GB, grain boundary

the atomic spacing between (001) planes of the matrix). It agrees with the traditional knowledge that the solute and impurity atoms segregate at only one to several atomic layers in grain boundaries.

It shows that B, C and Si segregate at the grain boundaries in Alloy 690 (Fig. 3d and e). The other elements were not detected to segregate at grain boundary, and their concentration profiles were not shown in Fig. 3d and e. The Cr atoms cosegregate with the C atoms at some regions at the grain boundary (Fig. 3d), though the carbides had not nucleated at the grain boundary (Fig. 3a and c). The analysis of thermodynamic data about the grain boundary segregation will be reported elsewhere, as this paper concerns with the C and Cr segregation behavior before carbide nucleation at grain boundaries.

Fig. 3 shows that the C and Cr atoms co-segregated at the grain boundaries. Whether stoichiometric $Cr_{23}C_6$ has nucleated at grain boundaries can be analyzed as following. The $Cr_{23}C_6$ should be formed by the excess Cr and C atoms segregated at grain boundaries. Fig. 4 shows the cumulative compositional profile of C and Cr atoms detected across the grain boundary in Fig. 3d. For an interface such as a grain boundary, we would expect to find that the material in the grains on either side of grain boundary shares a common composition. For such a grain boundary, the grain boundary excess for various elements (N_i^{excess}) can be calculated by measuring the distance between the two parallel regions of the profile that represent the concentration of solute or impurity within either grain (Fig. 4a), as shown in other literatures [17,18]. The ratio of $N_{Cr}^{excess}/N_C^{excess}$ at grain boundary is 6.67. The ratio is more than that of stoichiometric of $Cr_{23}C_6$ (23/6, 3.83).

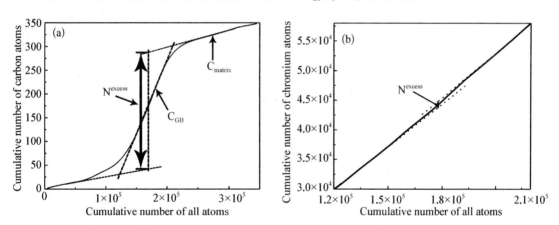

Fig. 4 Cumulative compositional profile showing the number of C (a) and Cr (b) atoms detected across the grain boundary in Fig. 3d

The above analysis shows that the stoichiometry $Cr_{23}C_6$ has not been detected from the C-Cr co-segregation zone. Because one $Cr_{23}C_6$ cell unit needs 24 C atoms [3], and needs many C atoms for nucleation ($\sim$2 000 C atoms per 100 nm^2). The excess density of segregation atoms (Γ) in terms of the excess atoms per square meters can be estimated by: $\Gamma = N_i^{excess}/e_d * A$, where e_d is the atom probe detector efficiency (for the LEAP $e_d \sim 0.38$), and A is the area of the region of interest which the cumulative composition profile was taken (400 nm^2 for the current paper). The density of excess C atoms ($\Gamma_{carbon} \sim 155$ C atoms per 100 nm^2) derived by Fig. 4a is too small for the nucleation of $Cr_{23}C_6$ under the current experiment condition. So the carbide still in the early stage of nucleation. Because the early stage of carbide nucleation does not consume the chromium atoms quickly, the chromium depleted zone near the grain boundary which observed by other literatures [1,4] does not form in the current study (Fig. 3).

It is interesting that the peaks of depth concentration profiles for different elements are at different positions (Fig. 3c and d), though the thickness of the segregation zone is only one to several atomic layers. It needs to define the peaks of the depth concentration profiles for light elements such as B and Si are the grain boundary core regions (indicated by solid lines in Fig. 3c and d), because the carbide nucleation do not affect their segregation behavior [19,20]. One can find that the positions of C – Cr co-segregation zones (the dash line in Fig. 3c) have a little deviation with the grain boundary core regions. The peaks of depth concentration profile for C still has little deviation with the grain boundary core region though the C – Cr co-segregation is not detected in that region (Fig. 3d).

Owing the 3D character of APT reconstruction images, the distribution of segregated species within the plane of grain boundary can be carried out. Fig. 5 gives the 2D concentration distribution of B, C, Si and Cr atoms at grain boundary region. Considering that the grain boundary is not a perfect planar, we select a layer of 4 nm thickness at the grain boundary to analyze the 2D concentration distribution of segregated species at the grain boundary region. In Fig. 5, the warmer colors stand for higher concentration of segregation species. All the maps in Fig. 5 were found to exhibit spatial inhomogeneous in the plane of the grain boundary, e. g. the high concentration regions for each elements form a set of straight arrays that are parallel to each other. The concentration fluctuation of the Cr map is very strong, and has a periodicity of about 7 nm (Fig. 5d). This periodicity is much more than the crystal constants of matrix (~ 0.35 nm) and $M_{23}C_6$ (~ 1.06 nm). So, it may not be induced by the formation of grain boundary carbide or other precipitates. Lemarchand et al. [21] suggested that the periodic fluctuation of concentration is induced by the relaxation of accommodation stress between the two grains. In the "disorder atoms group" grain boundary structure model, Ge [22] considered that the grain boundary region was consisting of numerous ordered regions and disordered regions, the distribution of ordered and disordered region mainly be influenced by the misorientation of the grain boundary. The solute atoms have high tendency to segregate at the disordered region of grain boundary. So, the authors think this periodicity segregation features can give some structure information of the grain boundaries.

Fig. 6 shows the concentration profiles of Cr, C, B and Si along a direction lying on the grain boundary plane to analyze the periodicity of concentration. The statistics direction is showed in Fig. 5d by the arrow, and the statistics crosssection is 4 nm×10 nm. The dash lines show the maxima of concentration for C and Cr appear at the same positions, while the maxima of concentration for B and Si appear at different positions. It

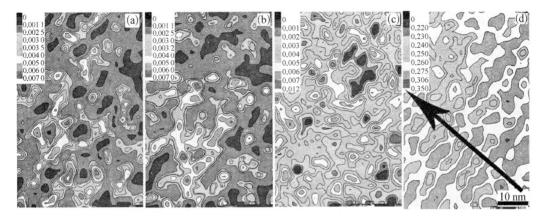

Fig. 5 2D concentration distribution of B (a), C (b), Si (c), Cr (d) at grain boundary regions. The thickness of the slice analyzed is 4 nm

indicates the C and Cr co-segregated at the grain boundary whether observed perpendicular to the grain boundary (Fig. 3) or observed along the grain boundary (Fig. 6). In general, the Cr atoms should homogenously solute in the matrix, while not segregate at grain boundary, and the segregated atoms should randomly distribute at grain boundary region. The formation of these segregation features will be discussed in the following sections.

Fig. 7 schematically illustrates the reason why C atoms segregate at different positions with B and Si atoms in the grain boundaries. Fig. 7a is an ideal illustration of the grain boundary geometry. Normally, after proper

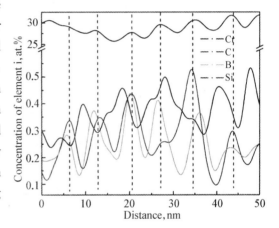

Fig. 6 The concentration profiles of Cr, C, B and Si along a direction lying on the grain boundary plane. The statistics direction is shown in Fig. 5d by the arrow, and the statistics cross-section is 4 nm×10 nm

thermal treating, the solute or impurity atoms would segregate at the grain boundaries randomly as shown in Fig. 7b. But, the C and Cr atoms have strong interaction, because they will form $Cr_{23}C_6$. The Cr atoms will migrate to near the grain boundary by the attraction of C atoms, and then the C atoms will migrate to this side of the grain boundary. If the C atoms segregate at the position indicated by the arrows in Fig. 7c, the interaction between C and Cr atoms will be stronger. The C and Cr atoms have high tendency to segregate at these regions nearby the lower grain. Because the high indexed grain boundary plane contains more free volume like the regions indicated by the arrows in Fig. 7c, the C and Cr cosegregate on the lower part of the grain boundary (Fig. 7c). The area which contains more free volume should have a periodicity that depends on the misorientation between the two grains. This periodicity leads to the periodic variation of segregation

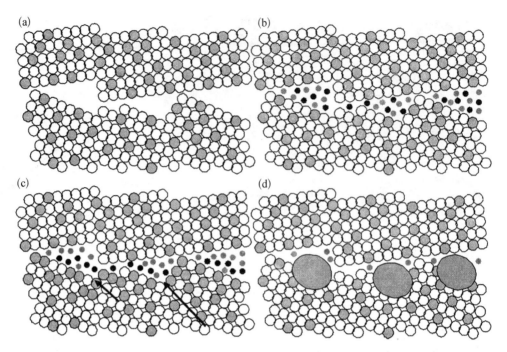

Fig. 7 Schematic illustration of the grain boundary segregation process. (a) Grain boundary without segregation atoms, (b) the atoms of B, Si and C randomly segregated at grain boundary, (c) the atoms of C and Cr co-segregated on the lower part of the grain boundary, (d) the carbides nucleated on the lower grain nearby the grain boundary. The large white spots stand for Fe and Ni atoms, the large gray spots stand for Cr atoms, the small gray spots stand for B and Si atoms, the small black spots stand for C atoms, and the large particles stand for carbide

density for segregated species in the grain boundary.

Then the C and Cr will further enrich at this region, the carbides will nucleate and precipitate to have a coherent orientation relationship with the nearby grain which grain boundary plane lying on a high index crystal plane of the lower grain (Fig. 7d). It may be one of the reasons why all of the observed grain boundary carbides have coherent orientation relationship only with the nearby grain which grain boundary plane lying on a high index crystal plane, while they are not randomly oriented. The effect of this typical crystal orientation relationship feature between carbides and nearby grains on the intergranular corrosion resistance has been established by the HRTEM and SEM analysis [23].

4 Conclusion

This paper aimed to clarify some aspects of grain boundary segregation before the carbide nucleation in nickel based Alloy 690. The approach used was the APT analysis, which gives access to nanometer 3D chemical information at the grain boundary. Based on the results and discussion above, the following conclusion can be derived.

The three dimensional distribution of atomic species at grain boundary was investigated. The B, C, Si and Cr segregate at grain boundaries. The high concentration regions for each segregation elements form a set of straight arrays that are parallel to each other. The concentration fluctuation of Cr has a periodicity of about 7 nm in the grain boundary plane. These periodic segregation features are helpful for revealing the structure of grain boundaries. The C and Cr co-segregate on one side of grain boundaries while not the exact grain boundary core regions before the carbides nucleated at grain boundaries. The number of Cr atoms in C – Cr co-segregation zone is more than that the amount demanding for $Cr_{23}C_6$ nucleation. But the nucleation of $Cr_{23}C_6$ carbides does not occur, because the density of C atoms is too small.

Acknowledgments

This work was supported by the National Basic Research Program of China (No. 2011CB610502), the National Natural Science Foundation of China (No. 50974148), Shanghai Leading Academic Discipline Project (No. S30107), and Innovation Fund of Shanghai University.

References

[1] Was GS. Grain boundary chemistry and intergranular fracture in austenitic nickel — base alloys — a review. Corrosion 1990;46: 319 – 30.

[2] Stiller K, Nilsson JO, Norring K. Structure, chemistry, and stress corrosion cracking of grain boundaries in alloys 600 and 690. Metall Trans 1996;27A: 327 – 41.

[3] Lewis MH, Hattersley B. Precipitation of $M_{23}C_6$ in austenitic steels. Acta Metall 1965;13: 1159 – 68.

[4] Kai JJ, Yu GP, Tsai CH, Liu MN, Yao SC. The effect of heat treatment on the chromium depletion, precipitate evolution, and corrosion resistance of Inconel alloy 690. Metall Trans 1989; 20A: 2057 – 67.

[5] Angeliu TM, Was GS. Behavior of grain boundary chemistry and precipitates upon thermal treatment of controlled purity Alloy 690. Metall Trans 1990; 21A: 2097 – 107.

[6] Trillo EA, Murr LE. A TEM investigation of $M_{23}C_6$ carbide precipitation behavior on varying grain boundary misorientations in 304 stainless steels. J Mater Sci 1998; 33: 1263 – 71.

[7] Trillo EA, Murr LE. Effects of carbon content, deformation, and interfacial energetics on carbide precipitation and corrosion sensitization in 304 stainless steels. Acta Mater 1998; 47: 235 – 45.

[8] Thuvander M, Stiller K. Microstructure of boron containing high purity Alloy 690. Mater Sci Eng 2000; A281: 96 – 103.

[9] Li Q, Zhou BX. A study of microstructure of Alloy 690. Acta Metall Sin (Chin Lett) 2001;37: 8 – 12.

[10] Kurban M, Erb U, Aust KT. A grain boundary characterization study of boron segregation and carbide precipitation in alloy 304 austenitic stainless steel. Scr Mater 2006; 54: 1053 – 8.

[11] Lim YS, Kim JS, Kim HP, Cho HD. The effect of grain boundary misorientation on the intergranular $M_{23}C_6$ carbide precipitation in thermally treated Alloy 690. J Nucl Mater 2004; 335: 108 – 14.

[12] Li H, Xia S, Zhou BX, Chen WJ, Ni JS. Evolution of carbide morphology precipitated at grain boundaries in Ni-based Alloy 690. Acta Metall Sin (Chin Lett) 2009; 45: 195 – 8.

[13] Li H, Xia S, Zhou BX, Chen WJ, Hu CL. The dependence of carbide morphology on grain boundary character in the highly twinned Alloy 690. J Nucl Mater 2010; 399: 108 – 13.

[14] Miller MK. Atom probe tomography: analysis at the atomic level. 1st ed. New York: Kluwer Academic/Pienum Publishers; 1999. p. 33.

[15] Vurpillot F, Cerezo A, Blavette D, Larson DJ. Modeling image distortions in 3DAP. Microsc Microanal 2004; 10: 384 – 90.

[16] Blavette D, Duval P, Letellier L, Guttmann M. Atomic-scale APFIM and TEM investigation of grain boundary microchemistry in Astroloy nickel base superalloys. Acta Mater 1996; 44: 4995 – 5005.

[17] Krakauer BW, Seidman DN. Absolute atomic-scale measurements of the Gibbsian interfacial excess of solute at internal interfaces. Phys Rev 1993; 48B: 6724 – 7.

[18] Hudson D, Smith GDW. Initial observation of grain boundary solute segregation in a zirconium alloy (ZIRLO) by three-dimensional atom probe. Scr Mater 2009; 61: 411 – 4.

[19] Thuvander M, Stiller K, Blavette D, Menand A. Grain boundary precipitation and segregation in Ni – 16Cr – 9 Fe model materials. Appl Surf Sci 1996; 94 – 95: 343 – 50.

[20] Thuvander M, Miller MK, Stiller K. Grain boundary segregation during heat treatment at 600 ℃ in a model alloy 600. Mater Sci Eng 1999; A270: 38 – 43.

[21] Lemarchand D, Cadel E, Chambreland S, Blavette D. Investigation of grain-boundary structure-segregation relationship in an N18 nickel-based superalloy. Philos Mag 2002; 82A: 1651 – 69.

[22] Ge TS. A grain boundary model and the mechanism of viscous intercrystalline slip. J Appl Phys 1949; 20: 274 – 80.

[23] Li H, Xia S, Zhou BX, Peng JC. Study of carbide precipitation at grain boundaries in nickel based Alloy 690. Acta Metall Sin (Chin Lett) 2011; 47: 853 – 8.

核反应堆压力容器模拟钢中
纳米富 Cu 相的变形特征[*]

摘　要：提高了 Cu 含量的核反应堆压力容器(reactor pressure vessel, RPV)模拟钢样品,经过 880 ℃水淬、660 ℃调质处理和 400 ℃ 1 000～4 000 h 的等温时效处理,观察到纳米富 Cu 相的析出;随后进行 20％～30％冷轧变形,采用萃取复型(extraction replica, ER)和高分辨透射电镜(high resolution transmission electron microscopy, HRTEM)的方法研究纳米富 Cu 相的变形特征. 研究结果表明,镶嵌在 α - Fe 基体中的纳米富 Cu 相,在冷轧变形时的变形机制较为复杂,存在多种变形方式. 当纳米富 Cu 相的晶体处于有利取向时,可以跟随基体一起发生滑移变形,表现为"软"颗粒的特性;当晶体处于不利取向时,会发生孪生变形,甚至诱发马氏体相变,有时生成"轮毂辐条"状的孪晶结构,大大提高了纳米富 Cu 相继续变形时的抗力,表现为"硬"颗粒的特征,因而析出纳米富 Cu 相会产生明显的强化作用.

核反应堆压力容器(RPV)是装载核燃料元件、支撑堆内构件和容纳一回路冷却剂并维持其压力的大型重要部件. RPV 长期在高温、高压和中子辐照下运行,它的完整性对于核反应堆及整个核电站的安全和寿命至关重要,其中脆性破坏对反应堆的安全威胁最大. 目前,国内外广泛采用 Mn - Mo - Ni 低合金铁素体钢(A508 -Ⅲ)制造 RPV. 大量研究[1-5]表明,RPV 在工作温度(288 ℃)下经过中子长期辐照后,钢中的杂质元素 Cu 会以富 Cu 纳米相析出,这是引起 RPV 钢韧脆转变温度升高的主要原因. 随着检测技术的不断进步,人们对于这种纳米富 Cu 相析出过程的认识也逐渐深入. Othen 等[6-7]用高分辨透射电镜(HRTEM)研究了 Fe - 1. 30％ Cu 和 Fe - 1. 28％ Cu - 1. 43％ Ni 合金中 4～30 nm 范围内富 Cu 析出相的晶体结构. 结果表明,当析出相尺寸大于 4 nm 时,其结构由 bcc 转变为 9R;当尺寸大于 18 nm 时,则转变为更稳定的 3R 结构. 3R 结构是一种畸变的 fcc 结构,与基体的取向接近 K - S 关系,并随着析出相的进一步长大,可最终转变为 fcc 结构.

这些富 Cu 析出相的尺寸仅有几到几十纳米,研究其在基体变形时的行为较为困难,但研究这些富 Cu 析出相,对于认识纳米富 Cu 相析出后导致钢的强化以及对韧脆转变温度的影响十分必要. Kimihiro 等[8]曾用电镜原位拉伸的方法研究了 Fe - 1％ Cu 合金中纳米富 Cu 相的大小和数量密度对变形的影响,观察到富 Cu 相阻碍了位错运动并迫使位错弓出产生强化作用,但是并未研究富 Cu 相自身的变形问题. 萃取复型(ER)是一种将第二相颗粒从基体中分离出来进行研究的方法. 第二相颗粒在复型上的分布基本保持了它们在基体中的状态,在分析其晶体结构和成分时,可以排除基体的干扰. 因此,用萃取复型方法对 RPV 钢中的纳米富 Cu 相进行研究,可以得到它们的大小、形状、成分、结构和分布状态等信息. 如果将变形后的 RPV 钢中的富 Cu 相萃取出来,则可以研究纳米富 Cu 相经过变形后的状态. 因此,本工作采用 HRTEM 和 ER 相结合的方法,研究了经热时效和冷轧变形提高了 Cu 含量的 RPV 模拟钢中的富 Cu 相,希望获得更多关于纳米富 Cu 相在 RPV 钢变形时的信息.

* 本文合作者：蔡琳玲、徐刚、冯柳、王均安、彭剑超. 原发表于《上海大学学报(自然科学版)》,2012,18(3)：311 - 316.

1 实验材料和方法

本实验所用的材料是在 A508-Ⅲ 钢成分的基础上提高了 Cu 含量的 RPV 模拟钢,这是为了时效时容易观察到富 Cu 相的析出过程[9]. 该模拟钢由真空感应炉冶炼,铸锭质量约40 kg,其化学成分如表 1 所示. 钢锭经过热锻和热轧得到 4 mm 厚的板材,最后切成 30 mm×30 mm 的小样品. 将这些样品加热到 880 ℃保温 0.5 h 后水淬,再加热到 660 ℃保温 10 h 进行调质处理,最后将样品在 400 ℃分别进行 1 000, 2 000 和 4 000 h 的时效处理. 为了研究纳米富 Cu 析出相在 α-Fe 基体变形时的行为,时效后的部分样品还进行了 20%或 30%的冷轧变形.

表 1 实验用压力容器模拟钢的化学成分

Table 1 Composition of the pressure vessel model steel %

	Cu	Ni	Mn	Si	P	C	S	Mo	Fe
Atomic fraction	0.55	1.45	1.60	0.67	0.062	1.09	0.013	0.011	Balance
Mass fraction	0.62	1.52	1.58	0.34	0.033	0.24	0.007	0.020	Balance

用 HRTEM 研究薄样品可以得到纳米富 Cu 相的晶体学信息,但是由于纳米相镶嵌在 α-Fe 基体中,分析晶体结构时会受到基体的干扰. 因此,通过 ER 方法把第二相从基体中分离出来进行单独研究,可以获得更准确的信息. 析出了富 Cu 相的 α-Fe 样品置于硝酸酒精溶液中,Fe 和 Cu 之间将形成微电池. 由于 Fe 的电极电位低于 Cu,所以 α-Fe 作为阳极会被腐蚀,而富 Cu 相作为阴极则受到保护. 根据这个原理,本实验采用 4%硝酸酒精溶液作为腐蚀剂,将纳米富 Cu 相从 α-Fe 基体中萃取出来,ER 样品的制备步骤如文献[10]所述. 为了使用能谱仪(energy dispersive spectrometer, EDS)分析析出相中的 Cu 质量分数,ER 的碳膜用 Mo 网捞取. 使用 JEM-2010F 透射电镜进行观察,获得的 HRTEM 晶格条纹像采用 Gatan 公司的 Digital Micrograph 软件进行 Fourier 变换(Fourier transform, FFT)和反 Fourier 变换(inverse FFT, IFFT)分析,研究富 Cu 相的晶体结构和变形情况.

2 实验结果与讨论

2.1 富 Cu 相的形貌及分布

RPV 模拟钢在经过调质热处理后,淬火时得到的板条马氏体分解,位错密度大幅降低. 在随后的时效过程中,碳化物会在晶界处析出,基体中也会析出很多弥散分布的富 Cu 相,大小仅有几到几十纳米,如图 1(a)和图 1(b)所示. 从图 1(b)中可以看到,很多富 Cu 相缀饰在位错上,这是因为富 Cu 相析出时容易在位错和界面处形核的缘故,这与文献[11-12]中报道的结果类似. 图 1(c)为 ER 方法得到的 TEM 照片,其中黑箭头所示为碳化物,白箭头所示为富 Cu 相,与图 1(b)薄样品的照片相比,富 Cu 相的分布状况非常一致.

2.2 纳米富 Cu 相的变形行为

在 400 ℃分别时效 1 000, 2 000 和 4 000 h 及 20%～30%变形的 6 个样品中,用 ER 方

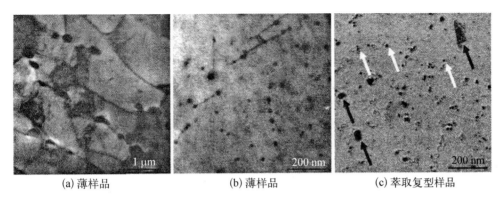

(a) 薄样品　　　　　　　(b) 薄样品　　　　　　　(c) 萃取复型样品

图 1　样品经 400 ℃时效 1 000 h 后的 TEM 图像

Fig. 1　TEM micrographs of the specimen aged at 400 ℃ for 1 000 h

法共获得了 160 个富 Cu 相,其中 54 个拍摄到了清晰的晶格条纹像. 从这些富 Cu 相中可以观察到,30 个晶体的晶格像发生了变化,说明它们发生了变形. 那些晶格像没有发生明显变化的富 Cu 相,可能是由于与基体之间的取向关系不利于变形,在基体发生变形时位错绕过了富 Cu 相;或者是在有利于变形时,当位错切过富 Cu 相发生滑移时,位错并没有滞留在晶体内. 一般纯 Cu 的变形方式为滑移,只有在低温条件下或高速变形时才能观察到孪生[13-14]. 在纳米晶的金属如纳米金线中,也曾观察到以孪生的方式发生变形[15]. 由于纳米大小的富 Cu 相是镶嵌在 α-Fe 基体中,当基体变形时,它们既要与基体协调一致地发生变形,又会受到基体的约束,因此,它们的变形机制也较为复杂. 下面选取几个典型情况进行分析研究.

图 2 为从时效 2 000 h 并经过 30%冷轧变形样品的萃取复型中观察到的一个富 Cu 相,大小约为 8 nm. 利用 HRTEM 晶格条纹像获得的 FFT 图,标定晶面指数后镶嵌在图 2(a)中,可判定为 fcc 结构,点阵常数 $a=0.356$ nm. 从图 2(b)的 IFFT 图中可以看出,富 Cu 相中的晶格产生了畸变,图中画出的各个圆圈中心都有一个位错,表明该富 Cu 相在 α-Fe 基体变形时曾经沿($1\bar{1}1$)和($\bar{1}11$)晶面发生过滑移变形,这时富 Cu 相在 α-Fe 基体中处于有利的取向,可以跟随基体一同发生滑移变形,表现出"软"的特性. 该富 Cu 相的 EDS 分析结果如图 2(c)所示,Cu 的原子数分数为 52.35,其余主要为 Fe 和 Si. 在 EDS 谱线中可以清晰地看到 Mo 和 C 的谱线,这是因为富 Cu 相是用碳膜萃取,并由 Mo 网支撑的,计算富 Cu 相的成分时,C 和 Mo 的谱线不应考虑在内.

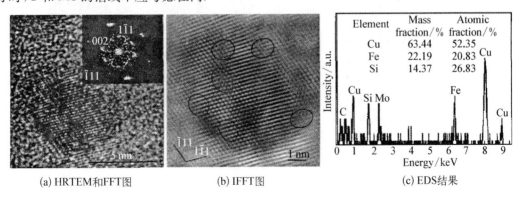

(a) HRTEM和FFT图　　　　　(b) IFFT图　　　　　(c) EDS结果

图 2　从 400 ℃时效 2 000 h 和 30%变形样品萃取复型上得到的一个富 Cu 相的分析结果

Fig. 2　Analytic results of a Cu-rich precipitate in the specimen deformed by 30% after aging at 400 ℃ for 2 000 h

图 3 为从时效 4 000 h 并经过 30％冷轧变形样品的萃取复型上观察到的一个富 Cu 相，大小约为 7 nm. 从图 3(a)中可以看出，该富 Cu 相中的晶格条纹比较"杂乱"，表明晶体已经发生过变形. 在经过滤波处理后的 IFFT 图中(见图 3(b))，可以根据晶格条纹的分布和走向，用 A，B，C，D，E 5 条线将该富 Cu 相的晶体分为 1，2，3，4，5 这 5 个区域，并对每个相邻部分的晶体取向关系进行分析. 分析标定后的 FFT 图表明，该富 Cu 相为 fcc 结构，晶格常数 $a=0.37$ nm，与纯 Cu 非常接近. EDS 分析结果(见图 3(c))显示，Cu 的原子数分数为63.34％，其余主要为 Fe 每两个相邻区标定后的 FFT 图以及该相邻区进一步放大后的 IFFT 图如图 3(d)～图 3(h)所示，每个图中分别标出了与图 3(b)中位置相同的 A～E5 条线和相同位置关系的 1～5 区域. 分析标定后的 FFT 图，发现 A～E5 条线中每条线两侧的晶体取向都为孪晶取向关系，也就是说 1～5 这几个区域依次相互之间都为孪晶关系. 以图

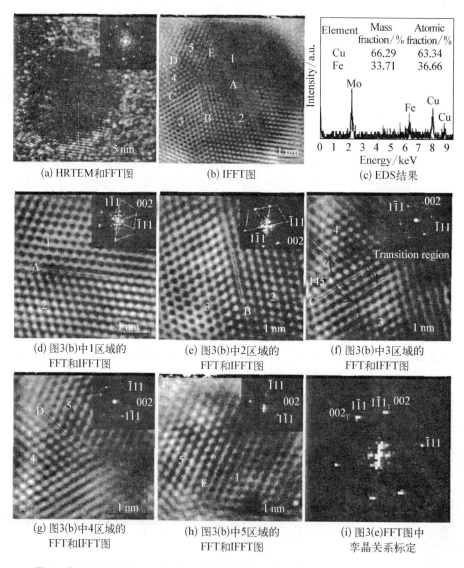

(a) HRTEM和FFT图　　(b) IFFT图　　(c) EDS结果

(d) 图3(b)中1区域的　　(e) 图3(b)中2区域的　　(f) 图3(b)中3区域的
　　FFT和IFFT图　　　　　FFT和IFFT图　　　　　FFT和IFFT图

(g) 图3(b)中4区域的　　(h) 图3(b)中5区域的　　(i) 图3(e)FFT图中
　　FFT和IFFT图　　　　　FFT和IFFT图　　　　　孪晶关系标定

图 3 从 400 ℃时效 4 000 h 及 30％变形样品的萃取复型上得到的一个富 Cu 相

Fig. 3 Analytic results of a Cu-rich precipitate in the specimen deformed by 30％ after aging at 400 ℃ for 4 000 h

3(e)中的 FFT 图为例,在这组对称的斑点中,用 002—002$_T$ 和 1$\bar{1}$1—1$\bar{1}$1$_T$ 标出了互为孪晶取向关系的两组斑点(见图 3(i)). (111)晶面与孪晶面之间的夹角应为 70°32′,则图 3(b)中5 组孪晶夹角之和应为 352°40′,这与一个圆周 360°还相差 7°20′. 仔细观察发现,A~E 这 5条孪晶界附近的晶面排列并不十分规则,往往包含了几层原子的"过渡区". 以图 3(f)为例,过渡区包含了 4 层原子,且测得孪晶面"C"两侧镜面对称的(111)晶面之间的夹角约为 145°,并不是 141°4′,正是由于这种"过渡区"中原子面的错动才逐渐补偿了 7°20′ 的差别. 纳米大小的富 Cu 相在 α-Fe 基体发生变形时形成了这种"轮毂辐条"状的孪晶结构,这是一种非常特殊的变形方式. 这时富 Cu 相在 α-Fe 基体中处于不利的取向,不容易跟随基体一同发生变形,此时的富 Cu 相表现为"硬"的特性,尤其是在发生了这种"轮毂辐条"状的孪生以后,该富 Cu 相更不容易再发生变形. 平均晶粒大小为 400 nm 的纯 Cu 样品,当其中的孪晶层厚度由 100 nm 减薄至 20 nm 后,屈服强度将由大约 500 MPa 增加至 900 MPa[16]. 纳米孪晶结构的纯 Cu 具有非常高的强度,何况本实验观察到的纳米孪晶层厚度只有 2~5 nm. 因此,镶嵌在 α-Fe 基体中的纳米富 Cu 相也可以成为"硬"颗粒,起到强化作用.

图 4 为从时效 1 000 h 并经过 20% 冷轧变形样品的萃取复型中观察到的一个富 Cu 相,大小约为 6 nm. 标定晶面指数后的 FFT 图镶嵌在图 4(a)的 HRTEM 晶格条纹像中,图4(b)为图 4(a)的 IFFT 图,图 4(c)为该富 Cu 相的 EDS 分析结果,其中 Cu 的原子数分数为

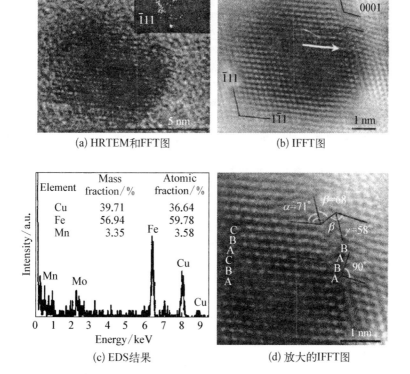

(a) HRTEM和FFT图　　(b) IFFT图

(c) EDS结果　　(d) 放大的IFFT图

图 4 从 400 ℃时效 1 000 h 及 20%变形样品的萃取复型上得到的一个富 Cu 相

Fig. 4 Analytic results of a Cu-rich precipitate in the specimen deformed by 20% after aging at 400 ℃ for 1 000 h

36.64%，Fe 的原子数分数为 59.78%，Mn 的原子数分数为 3.58%．从标定的 FFT 图中可判定该富 Cu 相为 fcc 结构．从图 4(b) 的 IFFT 图中可以看出，该富 Cu 相内部原子排列比较整齐，但在右侧边缘处的原子排列有些错动，如箭头所指处．该区域进一步放大后的结果如图 4(d) 所示．可以看出，在箭头所指 $(\bar{1}11)$ 晶面处通过三层原子面的堆垛层错使得晶体结构从左边 ABCABC 排列的 fcc 结构转变为右边 ABAB 排列的 hcp 结构，晶面夹角从 fcc 的 71° 逐渐转变为 hcp 的 58°，测得 hcp 的点阵常数 $a = 0.25$ nm，$c = 0.40$ nm，并与 fcc 保持着 $(\bar{1}11)_{fcc} // (01\bar{1}1)_{hcp}$，$[110]_{fcc} // [11\bar{2}0]_{hcp}$ 的取向关系．这与高锰钢 (fcc) 中应变诱发 ε 相 (hcp) 马氏体相变时的取向关系 $(111)_{fcc} // (0001)_{hcp}$，$[1\bar{1}0]_{fcc} // [11\bar{2}0]_{hcp}$ 有一些不同[17]，但表明当纳米富 Cu 相处于不利取向时，甚至会因为变形而诱发马氏体相变．

3 结束语

镶嵌在 α-Fe 基体中的纳米富 Cu 相因受到基体的约束，使其在基体变形时的变形特征十分复杂．观察到有多种变形方式：当纳米富 Cu 相的晶体处于有利取向时，可以跟随基体一起发生滑移变形，表现为"软"颗粒的特性；当处于不利取向时，会发生孪生变形，甚至诱发马氏体相变，有时会生成"轮毂辐条"状的孪晶结构，大大提高了纳米富 Cu 相继续变形时的抗力，表现为"硬"颗粒的特征，因而析出纳米富 Cu 相会产生明显的强化作用．

参 考 文 献

[1] TOYAMA T，NAGAI Y，TANG Z，et al. Nanostructural evolution in surveillance test specimens of a commercial nuclear reactor pressure vessel studied by three-dimensional atom probe and positron annihilation [J]. Acta Mater，2007，55：6852 - 6860.

[2] MILLER M K，RUSSELL K F，SOKOLOV M A，et al. APT characterization of irradiated high nickel RPV steels [J]. J Nucl Mater，2007，361：248.

[3] FUJII K，FUKUYA K，NAKATA N，et al. Hardening and microstructural evolution in A533B steels under high-dose electron irradiation [J]. J Nucl Mater，2005，340：247 - 258.

[4] MILLER M K，RUSSELL K F. Embrittlement of RPV steels：an atom probe tomography perspective [J]. J Nucl Mater，2007，371：145.

[5] MILLER M K，NANSTAD P K，SOKOLOV M A，et al. The effects of irradiation，annealing and reirradiation on RPV steel [J]. J Nucl Mater，2006，351：216 - 222.

[6] OTHEN P J，JENKINS M L，SMITH G D W. High-resolution electron microscopy studies of the structure of Cu precipitates in α-Fe [J]. Philos Mag A，1994，70(1)：1.

[7] OTHEN P J，JENKINS M L，SMITH G D W，et al. Transmission electron microscopy investigations of the structure of copper precipitates in thermally-aged Fe - Cu and Fe - Cu - Ni [J]. Philos Mag Lett，1991，64(6)：383 - 391.

[8] KIMIHIRO N，NOBUYASU N，HIDEKI M. Quantitative analysis of the dependence of hardening on copper precipitate diameter and density in Fe - Cu alloys [J]. J Nucl Mater，2007，367：392 - 398.

[9] 朱娟娟，王伟，林民东，等. 用三维原子探针研究压力容器模拟钢中富铜原子团簇的析出[J]. 上海大学学报：自然科学版，2008，14(5)：526.

[10] 楚大锋，徐刚，王伟，等. APT 和萃取复型研究压力容器模拟钢中富 Cu 团簇的析出[J]. 金属学报，2011，47(3)：269 - 274.

[11] THOMPSON S W，KRAUSS G. Copper precipitation during continuous cooling and isothermal aging of A710 - type steels [J]. Metall Mater Trans A，1996，27A：1573 - 1588.

[12] VARUQHESE R，HOWELL P R. Application of transmission electron microscopy to the study of a low-carbon

steel: HSLA - 100 [J]. ASTM Spec Tech Publ, 1993, 1165: 199 - 211.

[13] CHRISTIAN J W, MAHAJAN S. Deformation twinning [J]. Prog Mater Sci, 1995, 39(1/2): 4.

[14] MEYERS M A, VÖHRINGER O, LUBARDA V A. The onset of twinning in metals: a constitutive description [J]. Acta Mater, 2001, 49: 4025 - 4039.

[15] YANG L, JUN S, JIAN Y H, et al. Fracture of sub - 20 nm ultrathin gold nanowires [J]. Adv Funct Mater, 2011, 21(20): 3982 - 3989.

[16] SHEN Y F, LU L, LU Q H, et al. Tensile properties of copper with nano-scale twins [J]. Scripta Mater, 2005, 52 (10): 989 - 994.

[17] OLSON G B, COHEN M. A general mechanism of martensitic nucleation (I): general concepts and fcc → hcp transformation [J]. Metall Trans A, 1976, 7(12): 1897 - 1904.

Deformation Characterization of Nano-Scale Cu-Rich Precipitates in Reactor Pressure Vessel Model Steel

Abstract: The nano-scale Cu-rich precipitates in reactor pressure vessel (RPV) model steel after water quenching, tempering and thermal aging at 400 ℃ has been investigated by extraction replica (ER) and high resolution transmission electron microscopy (HRTEM). Samples were cold-rolled by 20%~30% after aging to study the deformation characterization of Cu-rich precipitates. The results show that the deformation characterization of the nanometer Cu-rich precipitates is complicated due to the restriction of α - Fe matrix. Different deformation mechanisms are detected. When the Cu-rich precipitates are located in a favorable orientation, they will be deformed by slipping together with the deformation of α - Fe matrix to show the particles of "soft" characteristic. When the Cu-rich precipitates are located in an unfavorable orientation, they will be deformed by twinning to form the spoke-like twins within a Cu-rich precipitate, even a martensitic phase transformation induced by strain. In this case, the Cu-rich precipitates show the particles of "hard" characteristic. Therefore the precipitation hardening occurs after the formation of Cu-rich precipitates in α - Fe matrix.

Cu 在 Zr－2.5Nb－0.5Cu 合金及其腐蚀生成氧化膜中的存在形式*

摘　要：用电子显微镜研究了合金元素 Cu 在 Zr－2.5Nb－0.5Cu 合金及其腐蚀生成氧化膜中的存在形式.合金基体中的元素 Cu 主要以四方结构的 $CuZr_2$ 第二相形式存在,$CuZr_2$ 相中会富集杂质元素 Fe.合金经过 550 ℃,25 MPa 超临界水条件深度腐蚀后,伴随着合金基体发生的 H 致 β 相变,$CuZr_2$ 第二相也发生了相消溶、扩散及凝聚长大过程.在腐蚀生成的氧化膜中,合金元素 Cu 以非晶氧化物的形式存在于氧化锆的晶界.

锆合金是水冷核反应堆中燃料元件的包壳材料.为降低核电成本,需要增加核燃料的燃耗,延长燃料组件的换料周期,因而需要开发性能更好的锆合金.在已添加的合金元素中,Sn、Nb、Fe、Cr 等元素开发较早,有大量的研究报道,对其作用了解较多,而 Cu 元素在近几年才开始受到关注,所以对其了解有限.Cu 元素在核用锆合金中的添加应用始于 20 世纪 60 年代开发的 Zr－2.5Nb－0.5Cu 合金[1],在已作为 CANDU 堆压力管材料的 Zr－2.5Nb 合金中添加少量 Cu 元素,是为了获得强度更高的材料,用于制造隔离压力管与排管容器的夹紧盘簧(garter spring).因其接触介质是起隔热作用的 CO_2,要求的是力学性能,所以其耐腐蚀性能没有受到关注.在核用锆合金发展的数十年过程中,只有零星的关于含 Cu 锆合金报道.20 世纪 80 年代,L. Castaldelli 报道了添加 Cu 能提高 Fe－Cu－Zr 合金耐腐蚀性能[2].后来 Aylin Yilmazbayhan 等报道了在 360 ℃水中,Zr－2.5Nb－0.5Cu 合金的耐腐蚀性能优于 Zr－4、ZIRLO 和 Zr－2.5Nb[3].韩国从 1997 年起开始进行 HANA 系列新型锆合金的研发工作,在此期间开始关注添加 Cu 元素的重要作用,对于 Zr－4 及含微量 Cu 的 Zr－Cu 二元合金,适量的 Cu 有利于提高耐腐蚀性能等.特别是公布了含合金元素 Cu 的新型锆合金 HANA－3 和 HANA－6 表现出了优异的耐腐蚀性能[4-6].Zr－2.5Nb－0.5Cu 合金在 550 ℃,25 MPa 超临界水中也表现出较好的耐腐蚀性能[7].虽然添加适量 Cu 元素对锆合金耐腐蚀性能的提高作用明显,但到目前为止的研究工作不多,特别是缺乏金属基体中合金元素 Cu 在腐蚀氧化时的变化过程,在氧化膜中的存在形式和对氧化膜结构变化的影响等重要研究数据,因此对于添加适量 Cu 元素后能提高锆合金耐腐蚀性能的作用机制认识非常有限,有必要对此进行研究.

本工作利用高分辨透射电镜等先进手段研究了合金元素 Cu 在 Zr－2.5Nb－0.5Cu 合金及其经过 550 ℃,25 MPa 超临界水腐蚀后,在合金基体和生成氧化膜中的存在形式.

1　实验

Zr－2.5Nb－0.5Cu 合金板材来自西北有色金属研究院.将尺寸约为 10 mm×20 mm,

---

* 本文合作者：李强、梁雪、姚美意、彭剑超.原发表于《稀有金属材料与工程》,2012,14(1)：92－95.

厚 0.6 mm 的锆合金片状样品,经酸洗(10% HF+30% HNO₃+30% H₂SO₄+30% H₂O 体积比的混合酸)与水洗后,放置在石英管中,在真空(<5×10⁻³ Pa)中进行 580 ℃, 5 h 的退火热处理,然后将加热炉推离石英管,用水浇淋石英管进行冷却.

将热处理后的样品经过标准方法酸洗和水洗,在静态高压釜中进行 550 ℃, 25 MPa 超临界水腐蚀试验,定期降温取出样品称重.

用带有 INCA 能谱仪(EDS)的 JEM-2010F 场发射透射电镜及 JSM-6700F 场发射扫描电镜,对锆合金腐蚀前后的合金基体及腐蚀生成氧化膜的显微组织进行观测研究,为避免对检测 Cu 元素的干扰,透射电镜 EDS 分析时使用了铍双倾样品台.

合金基体 TEM 样品的制备过程:将片状样品机械磨抛及酸洗减薄至约 0.08 mm 厚,冲出 Φ3 mm 圆片,用双喷电解抛光的方法制备 TEM 样品,所用电解液为 10% HClO₄+90% C₂H₅OH,减薄电压为 45 V,温度低于−30 ℃.

腐蚀后氧化膜 TEM 样品的制备过程:机械磨抛去除一个平面的氧化膜及大部分合金基体,制成 Φ3 mm 圆片,先在圆片金属一侧中心位置酸蚀出 Φ0.6 mm 的小面积浅坑,然后再同心酸蚀出 Φ1.5 mm 的大面积坑,直至中心露出氧化膜后进行离子减薄.

对于腐蚀前后合金基体,机械磨抛后酸洗,酸洗时用棉签轻轻擦抹表面,然后用水冲洗,同时用棉签轻轻擦抹表面,再经超声波水洗后晾干. 对于腐蚀后氧化膜断面,机械磨抛去除端面氧化膜,酸洗法去除部分端面的合金基体,露出一截氧化膜,用水冲洗及烘干后用导电胶固定在 SEM 样品台上,折断露出的氧化膜,将要观察的断面镀 Au 导电层.

2 结果和讨论

2.1 Zr-2.5Nb-0.5Cu 合金中 Cu 的主要存在形式

由于得到的 Zr-2.5Nb-0.5Cu 合金样品数量较少,所以本工作中只对出厂态样品进行了 580 ℃, 5 h 退火处理,处理后的显微组织见图 1 及图 2a. 等轴的 α-Zr 晶粒尺度较大,合金元素 Cu 主要以四方结构的 CuZr₂ 相(a=0.32 nm, c=1.12 nm)第二相形式析出,颗粒尺度较大,粒径多数超过 200 nm,远大于 β-Nb 第二相. 合金样品中 β-Nb 第二相的分布不均匀,明显呈现团簇或条带分布,这与来样曾在 β 淬火后的后续热加工或冷加工后的退火温度进入了 α+β 双相区有关.

EDS 分析表明,Fe 元素会富集在 CuZr₂ 相中,如区域 A1 的 Fe 含量(质量分数)约为 0.46%,区域 A2 的为 0.32%,Fe 在 CuZr₂ 相中的分布并不是均匀的,Fe 是来自海绵锆中的杂质元素,在 β-Nb 第二相也有 Fe 的富集. TEM 样品中观测到的 CuZr₂ 相有着核壳结构(图 1b),内核的部分区域富集 Fe,而在如 A3 的外层区域,EDS 检测不到 Fe,但是这种现象尚不能确定是否为制样引起,因为在 TEM 观察时发现,相对于锆合金基体来说,CuZr₂ 相在电解双喷制样时更易于被消溶,如在图 1c, 1d 中看到的原 CuZr₂ 相已经消溶,留下了 Zr 的非晶+微晶氧化物,这种现象普遍存在.

2.2 Zr-2.5Nb-0.5Cu 合金在 550 ℃, 25 MPa 超临界水条件下腐蚀后的合金基体中 Cu 的变化

图 2 为 Zr-2.5Nb-0.5Cu 合金样品腐蚀前及在 550 ℃, 25 MPa 超临界水条件下腐蚀

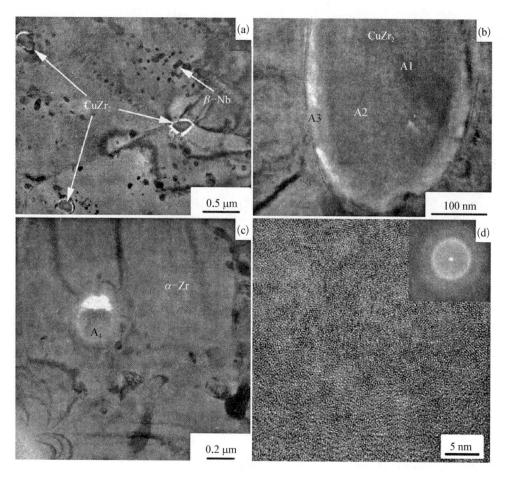

图1 Zr‑2.5Nb‑0.5Cu合金样品的 TEM 照片,(图 1d 为 1c 中 A4 区域的 HRTEM 及其傅里叶转换像)

Fig. 1 TEM images of matrix for Zr‑2.5Nb‑0.5Cu specimen (a, b, c) and HRTEM images (1d) showing the region marked as A4 in Fig. 1c, and the patterns obtained by FFTs

后的合金基体 SEM 照片. 可以看出,经 340 h 腐蚀后,合金基体显微组织的变化并不明显,特别是 $CuZr_2$ 和 β‑Nb 第二相基本保持了腐蚀前的大小和分布. 长期腐蚀后(1 150 h)的样品,合金基体发生了与 Zr‑2.5Nb 样品基本一样的 H 致 β 相变[8]. EDS 及选区电子衍射研究表明 Cu 元素仍然以 $CuZr_2$ 形式存在,此时 $CuZr_2$ 第二相颗粒明显长大,腐蚀 1 150 h 的 $CuZr_2$ 颗粒尺寸与腐蚀前相比长大了近 1 倍,同时数量减少,表明在合金基体发生相变的同时,$CuZr_2$ 相也发生了相消溶、扩散及凝聚过程.

2.3 Zr‑2.5Nb‑0.5Cu 合金在 550 ℃, 25 MPa 超临界水中腐蚀后,合金元素 Cu 在氧化膜中的存在形式

图 3a 为 Zr‑2.5Nb‑0.5Cu 合金样品在 550 ℃, 25 MPa 超临界水中腐蚀 40 h 时生成氧化膜断面的显微组织. 可以看到在氧化锆晶界处存在衬度亮白的区域,由于 Cu 的二次电子产额相对较高,这可能是合金元素 Cu 在氧化锆晶界上发生偏聚的结果,但因分辨率的限制,用 SEM 及其 EDS 不能得到直接的证明.

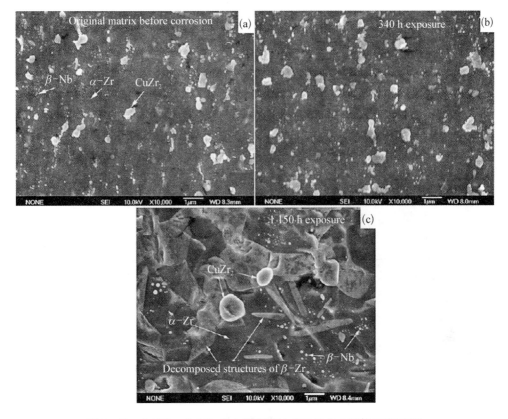

图 2 Zr‑2.5Nb‑0.5Cu 合金样品腐蚀前后的合金基体 SEM 照片

Fig. 2 SEM images showing the matrix microstructure of Zr‑2.5Nb‑0.5Cu specimen before (a) and after (b, c) corrosion testing in supercritical-water at 550 ℃, 25 MPa

图 3b,3c 为腐蚀 16 h 生成氧化膜的 TEM 照片. 值得注意的是许多氧化锆晶粒间出现亮白的衬度区域,这与图 3a 中 SEM 观测到的氧化锆晶界存在亮白衬度区域现象吻合. 用 EDS 对这些区域进行了研究,较窄的亮白衬度区难以检测出其他合金元素的存在(图 3c,P3 区),这可能与元素相对含量少有关,但在一些与之相连的较大面积的亮白衬度区(图 3c,P1 区),可以检测出合金元素 Cu 和 Fe 的存在,HRTEM 研究表明这些区域呈现出非晶形态,这表明合金基体中的 CuZr₂ 相氧化后,Cu 及原富集在 CuZr₂ 相中的 Fe 元素会存在于氧化锆晶界. CuZr₂ 相在合金基体中是尺度较大的第二相,但在合金腐蚀生成的氧化膜中极难找到合金元素 Cu 的富集存在,这表明 CuZr₂ 在合金腐蚀过程中更容易被氧化,并且氧化后合金元素 Cu 会沿着氧化锆的晶界扩散并分布. 合金元素 Cu 在晶界上的偏聚(沿晶界的扩散),必然会改善氧化锆晶界的特性,从而对锆合金的耐腐蚀性能产生影响. 关于合金元素 Cu 对锆合金耐腐蚀性能的影响,将另作详细论述.

3 结论

(1) Zr‑2.5Nb‑0.5Cu 合金样品经 580 ℃,5 h 处理后,合金元素 Cu 主要以四方结构的 CuZr₂($a=0.32$ nm,$c=1.12$ nm)第二相形式存在,该相颗粒尺度较大,粒径多数超过 200 nm,远大于同时存在的 β‑Nb 第二相;杂质元素 Fe 会富集在 CuZr₂ 和 β‑Nb 相中.

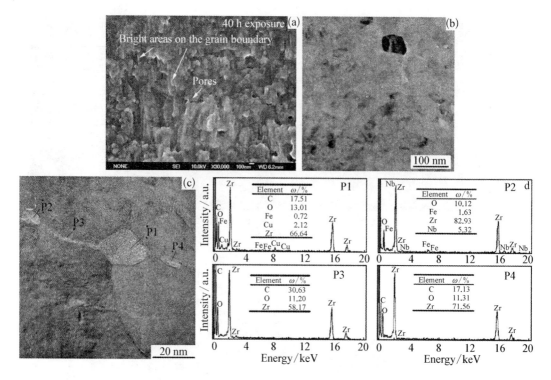

图 3 Zr‐2.5Nb‐0.5Cu 合金样品在 550 ℃，25 MPa 超临界水中腐蚀 16 h 生成氧化膜的 SEM、TEM 照片及 EDS 分析结果（a：SEM；b，c：TEM）

Fig. 3 Images of the oxide films on Zr‐2.5Nb‐0.5Cu specimens corroded in supercritical-water at 550 ℃，25 MPa during 16 h：（a）fracture surface morphology（SEM）；（b，c）TEM images，（d）EDS analysis results for different micro areas in Fig. 3c

（2）Zr‐2.5Nb‐0.5Cu 合金在 550 ℃，25 MPa 超临界水条件下长期腐蚀后（1 150 h），原合金基体中的 $CuZr_2$ 相也发生了相消溶、扩散及凝聚过程,颗粒尺寸明显增大,同时数量减少.

（3）Zr‐2.5Nb‐0.5Cu 合金在 550 ℃，25 MPa 超临界水条件下腐蚀时,$CuZr_2$ 相比合金基体更易被氧化,并且氧化后合金元素 Cu 会沿着氧化锆的晶界扩散并分布.

参 考 文 献

［1］Chakravartty J K，Dey G K，Banerjee S *et al*. *J Nucl Mater*［J］，1995，218：247.

［2］Castaldelli L，Fizzotti C，Lunde L. *Eighth International Symposium on Zirconium in the Nuclear Industry*［C］. US：ASTM，1982，754：105.

［3］Aylin Yilmazbayhan，Motta Arthur T，Comstock Robert J *et al*. *J Nucl Mater*［J］，2004，324：6.

［4］Yong Hwan Jeong，Jeong-Yong Park，Jong-Hyuk Baek *et al*. *Technical Meeting on Behavior of High Corrosion Resistance Zr-based Alloys*［C］. Buenos Aires，Argentina：The National Library，2005.

［5］Hyun Seon Hong，Moon Jae Sik，Kim Seon Jin *et al*. *J Nucl Mater*［J］，2001，297：113.

［6］Jeong-Yong Park，Choi Byung-Kwon，Yoo Seung Jo *et al*. *J Nucl Mater*［J］，2006，359：59.

［7］Li Q，Zhou Bangxin，Yao Meiyi *et al*. *15th International Symposium on Zirconium in the Nuclear Industry*［C］. Sunriver，Oregon，USA，2007.

［8］Li Qiang(李强)，Zhou Bangxin（周邦新），Yao Meiyi(姚美意) *et al*. *Rare Metal Materials and Engineering*（稀有金属材料与工程）［J］，2008，37(10)：1815.

Existing Form of Cu in Zr – 2. 5Nb – 0. 5Cu Alloys and the Oxide Films during Corrosion Testing in SCW at 550 ℃/25 MPa

Abstract: Scanning electron microscopy (SEM) and high resolution transmission electron microscopy (HRTEM), equipped with an energy dispersive X-ray spectroscopy (EDS) were employed for examining the existing form of Cu in Zr – 2. 5Nb – 0. 5Cu alloys and the oxide films during corrosion testing in SCW (supercritical water) at 550 ℃/25 MPa. In Zr – 2. 5Nb – 0. 5Cu alloy, Cu exists mainly in the form of tetragonal second phase $CuZr_2$, rich in impurity element Fe. The $CuZr_2$ are transferred and agglomerated with the β phase transformation of the alloy induced by hydrogen during the corrosion testing. In the form of amorphous cuprous oxide, Cu segregates on the grain boundaries of zirconia in the oxide films during corrosion testing.

镍基 690 合金晶界成分演化规律的研究*

摘 要: 采用原子探针层析(APT)技术研究晶界碳化物析出之前杂质或溶质原子在 690 合金晶界处的偏聚规律,并采用配备在高分辨透射电子显微镜(HRTEM)上的能谱(EDS)设备研究晶界碳化物析出之后 690 合金晶界处的成分演化规律. 结果表明:在晶界碳化物析出之前,C、B、Si 等原子偏聚在晶界处,同时 C 与 Cr 共偏聚在晶界处,晶界富 Cr;碳化物析出之后,晶界处 Cr 浓度越来越低,在 715 ℃时效 15 h 后晶界处 Cr 浓度达到最低值,然后随着时效时间的延长 Cr 浓度逐步提高;当碳化物基本覆盖整个晶界后晶界富 Cr;除了晶界处 Cr 贫化程度外,晶界处贫 Cr 区内的 Cr 浓度梯度也是影响晶界耐腐蚀能力的主要原因.

1 前言

690 合金是一种高 Cr 的镍基合金,由于其优异的耐腐蚀性能及综合力学性能,被广泛应用为压水堆核电厂蒸汽发生器传热管材料. 核反应堆蒸汽发生器传热管材料损坏的原因在不同时期略有区别,但是晶间应力腐蚀破裂一直是这类材料失效的重要原因之一[1],而产生晶间应力腐蚀破裂的主要原因就是晶界上碳化物的析出及其引起的晶界附近 Cr 的贫化.

在实际应用中,为了获得较好的综合性能,一般先对 690 合金进行固溶处理,然后再在 715 ℃时效热处理一定时间. 通过时效热处理可以控制晶界上碳化物的析出,调整晶界附近贫 Cr 区的成分,这样可以显著提高其耐晶间腐蚀性能[2]. 因此,690 合金时效后晶界碳化物的析出问题受到了大量的关注,如时效后晶界碳化物析出形貌的演化规律[3,4],不同类型晶界处碳化物析出形貌及演化规律的差异[5,6],时效过程中晶界附近贫 Cr 区内 Cr 浓度的演化规律[4,7,8]等. 但是,晶界碳化物析出之前晶界处的成分分布情况,晶界碳化物析出之后晶界贫 Cr 区内成分与晶界耐腐蚀能力的对应关系仍然研究得很少.

本文利用原子探针层析(APT)技术研究晶界碳化物析出之前各元素在 690 合金晶界处的空间分布情况,采用配备在高分辨透射电子显微镜(HRTEM)上的能谱(EDS)设备测定碳化物析出之后 690 合金晶界处的成分演化规律,通过浸泡晶间腐蚀的方法考验不同样品的耐腐蚀能力,基于实验结果分析时效过程中 690 合金晶界处的成分演化规律及其对晶界耐腐蚀能力的影响.

2 实验方法

实验所用 690 合金成分为:Cr(28.91%,质量分数,以下同)、Fe(9.45%)、C(0.025%)、N(0.008%)、Ti(0.4%)、Al(0.34%)、Si(0.14%)、Ni 余量. 将所有片状 690 合金样品密

---

* 本文合作者:李慧、夏爽、刘文庆、彭剑超. 原发表于《核动力工程》,2012,33(S2):65 - 69.

封在真空(真空度优于 5×10^{-3} Pa)石英管中,在 1 100 ℃保温 15 min,然后立即淬入水中,并同时砸破石英管进行固溶处理.

为了方便 APT 测试分析,将一部分样品冷轧 50%后再在 1 000 ℃再结晶处理 0.5 min,获得完全再结晶的样品[9];然后将样品在 500 ℃时效 0.5 h,获得晶界碳化物尚未析出的样品.将片状的 690 合金样品用电火花线切割成 0.5 mm×0.5 mm×15 mm 的棒状样品,然后利用标准的两次抛光的方法制备成三维原子探针(3DAP)针尖样品.利用 JEM 200CX TEM 筛选合适的针尖样品,选择针尖曲率半径小于 100 nm 且尖端含有晶界的样品进行分析.即使样品晶粒尺寸很小,获得可用于 3DAP 分析的样品的概率仍然很小(约 3%~5%).3DAP 分析采用 Imago Scientific 公司生产的 3 000 HR 型局部电极原子探针(LEAP).分析温度为 50 K,脉冲分数为 15%,控制蒸发速率使每一千个电脉冲能收集到 2 个离子.由于含有晶界的样品极易在电场力的作用下而折断,3DAP 分析成功的概率也很小(~5%).采用 IVAS 软件对获得的数据进行重构得到各元素的三维空间分布图与成分信息.

固溶后的样品在 715 ℃时效 0.5~200 h 后获得晶界碳化物析出程度不同的样品,利用双喷电解抛光的方法制备透射电镜(TEM)薄膜样品.双喷液为 20% $HClO_4$ + 80% CH_3COOH(体积比),电压为 30 V 直流.利用配备在 JEM 2010F HRTEM 的 EDS 设备分析晶界附近的成分分布,电子束加速电压为 200 kV.

将经过 715 ℃时效处理后的片状样品仔细抛光获得干净表面,利用千分尺测量并计算其表面积(精确到 1 μm^2),利用电子天平测量起始重量(精确到 0.1 mg).将不同时效处理后的样品室温浸泡在 65% HNO_3+0.4% HF 的水溶液中进行浸泡晶间腐蚀实验.每隔 1 d 将样品取出烘干并称重,获得腐蚀失重曲线.

3 结果与讨论

图 1 给出了通过 APT 技术获得的固溶处理后 690 合金在 500 ℃时效 0.5 h 后元素在晶界附近偏聚的三维空间分布图.图 1a 和图 1b 分别为沿着和垂直于晶界方向(空间尺寸为 54 nm×56 nm×98 nm),可以明显看到偏聚的原子分布在一个二维的晶界面上.由于 C 原子均匀地偏聚在晶界面上,没有团聚在一块,说明碳化物尚未析出(图 1b).通过软件选取一个垂直于晶界面方向的 20 nm×20 nm×25 nm 的区域(图 1b 中的 A 区域)分析各元素在晶界附近的成分分布情况(图 2).可以明显看出,Ni 和 Fe 原子在晶界处贫化,Cr 原子富集在晶界处(图 2a);Ti 和 Al 原子在晶界附近的分布没有明显变化(图 2b);C、B、Si 原子明显偏聚在晶界处(图 2c).通过 APT 技术分析结果还可看出,即使碳化物尚未在晶界处析出,C 与 Cr 原子已经共偏聚在晶界处,形成 C-Cr 共偏聚区,这是以往实验方法无法观察到的现象.

当碳化物析出之后,由于碳化物的析出消耗了大量的 Cr 原子,Cr 原子扩散很慢来不及补充,就会在晶界附近形成贫 Cr 区.贫

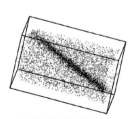

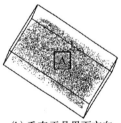

(a) 沿着晶界面方向   (b) 垂直于晶界面方向

图 1 固溶处理后 690 合金在 500 ℃时效 0.5 h 后元素在晶界附近偏聚的三维空间分布

Fig. 1 Grain Boundary Segregation Features of Alloy 690 Aged at 500 ℃ for 0.5 h after Solution Treatment

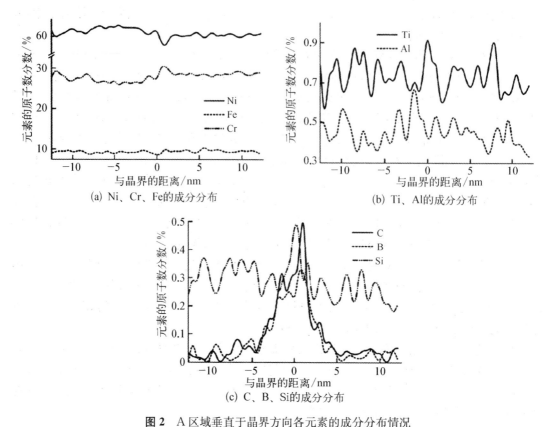

(a) Ni、Cr、Fe的成分分布

(b) Ti、Al的成分分布

(c) C、B、Si的成分分布

图2 A区域垂直于晶界方向各元素的成分分布情况

Fig. 2 Concentration of Each Element along Grain Boundary of Region A

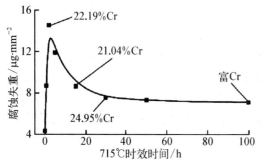

图3 不同热处理的690合金样品的晶间腐蚀失重量以及晶界处的最低Cr浓度

Fig. 3 Weight Losses and Grain Boundary Chromium Concentration of Alloy 690 after Different Heat Treatment

Cr区的形成会明显降低晶界的耐腐蚀性能. 图3给出了715 ℃时效不同时间后样品在室温浸泡晶间腐蚀276 h后的腐蚀失重量及其对应的晶界处最低Cr浓度;图4给出了晶间腐蚀后样品表面的扫描式电子显微镜(SEM)图像. 由图3可以看出,在715 ℃时效2 h的样品的腐蚀失重量最大,并且此样品表面已有部分小晶粒脱落(图4a),表明此样品腐蚀很严重,但是Cr贫化程度却不是最严重的. 当时效时间达到15 h后,晶间处最低Cr浓度降低到21.04%(图3),但是样品的耐腐蚀能力已经提高,只在部分晶界看到腐蚀裂纹(图4b). 随着时效时间的延长,样品的耐腐蚀能力逐步提高,同时晶界处Cr浓度也逐步提高,当时效时间超过30 h后样品的腐蚀失重量已经很小并且不再明显变化(图3),样品表面也只能看到晶间腐蚀沟痕而看不到腐蚀裂纹(图4c和图4d). 从图3可以明显看出,样品耐晶间腐蚀能力的演化规律与样品晶界处Cr贫化程度的演化规律并不相同,这与一般认为的Cr贫化程度直接影响晶界耐腐蚀能力的观点不尽相同[4].

由于690合金基体的Cr浓度高达30%,所以图3中给出的690合金在不同时效热处理

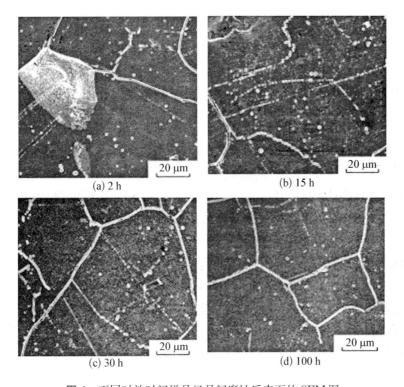

图 4　不同时效时间样品经晶间腐蚀后表面的 SEM 图

Fig. 4　SEM Micrograph of Different Specimens after Different Aging Treatments.

后晶界处最低的 Cr 浓度都没有低于一般认为的能引起晶间腐蚀的 Cr 浓度阈值（12%），即使没有低于 Cr 浓度阈值. 在本实验的腐蚀环境下，690 合金仍然会出现晶间腐蚀问题，说明除了 Cr 贫化程度外还有其他因素影响 690 合金的耐晶间腐蚀性能. 如果同时考虑到晶界处贫 Cr 区的深度与宽度，则可以发现两者都会对晶界的耐腐蚀性能有影响. 图 5 给出了在 715 ℃时效不同时间后样品晶界处贫 Cr 区内 Cr 浓度梯度的演化规律，可以看出贫 Cr 区内 Cr 浓度梯度的演化规律与样品的晶间腐蚀失重的演化规律一致. 这就表明，在高 Cr 含量的镍基 690 合金中晶界处贫 Cr 区内 Cr 浓度梯度是影响其耐腐蚀能力的一个重要原因.

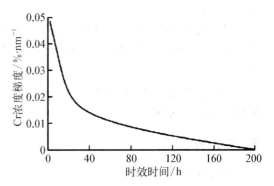

图 5　690 合金样品在 715 ℃时效不同时间后晶界处贫 Cr 区内的 Cr 浓度梯度

Fig. 5　Chromium Gradient of Chromium Depletion Zone at Grain Boundary in Alloy 690 Aged at 715 ℃

以往的 EDS 研究结果表明，晶界附近成分的演化规律为：时效初期，晶界处 Cr 浓度逐渐降低，达到最低值后随着时效时间的延长而 Cr 浓度逐步恢复，材料的耐晶间腐蚀能力提高（图 6a）[4, 8].

如果同时考虑 APT 与 EDS 的研究结果，可以获得更为细致的晶界成分演化规律. 首先，在时效初期，杂质或溶质原子偏聚在晶界处，随着晶界处 C 偏聚量的升高，Cr 原子也会

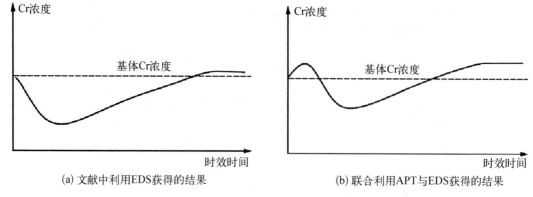

图 6　晶界处 Cr 浓度的演化规律

Fig. 6　Evolution of Chromium Concentration at Grain Boundary

共偏聚在晶界处,此时晶界富 Cr;进一步时效后,碳化物在晶界处析出,消耗大量的 Cr,晶界处开始贫 Cr,随着时效时间的延长,晶界处 Cr 的贫化程度越来越严重;继续延长时效时间后,碳化物长大已经基本完成,晶界处 Cr 贫化程度降低,样品的耐晶间腐蚀能力提高;当碳化物基本覆盖整个晶界后,晶界处又开始富 Cr,但此时的大尺寸的晶界碳化物不利于样品的耐晶间腐蚀能力(图 6b).

4　结论

(1) 利用 APT 技术可以获得晶界处不同元素的三维空间分布图. 在碳化物析出之前 C、B、Si 等原子偏聚在 690 合金的晶界处,C 与 Cr 共偏聚在晶界处.

(2) 碳化物析出后,晶界处 Cr 浓度随着时效时间的延长而降低,在 715 ℃时效 15 h 后晶界处的 Cr 贫化程度最严重,随后晶界处 Cr 浓度逐渐升高,碳化物覆盖整个晶界后晶界处富 Cr.

(3) 除了晶界处 Cr 贫化程度外,晶界处贫 Cr 区内 Cr 浓度梯度也是影响晶界耐晶间腐蚀能力的重要原因.

参 考 文 献

[1] Diercks D R,Shack W J,Muscar J. Overview of Steam Generator Tube Degradation and Integrity Issues [J]. Nuclear Engineering and Design,1999,194:19 - 30.

[2] 邱绍宇,苏兴万,文燕,等. 热处理对 690 合金腐蚀性能影响的实验研究[J]. 核动力工程,1995,16:336 - 341.

[3] 李慧,夏爽,周邦新,等. 镍基 690 合金时效过程中晶界碳化物的形貌演化[J]. 金属学报,2009,45:195 - 198.

[4] Kai J J,Yu G P,Tsai C H,et al. The Effects of Heat Treatment on the Chromium Depletion,Precipitate Evolution,and Corrosion Resistance of INCONEL Alloy 690 [J]. Metallurgical Transactions,1989,20A:2057 - 2067.

[5] Li H,Xia S,Zhou B X,et al. The Dependence of Carbide Morphology on Grain Boundary Character in the Highly Twinned Alloy 690 [J]. Journal of Nuclear Materials,2010,399:108 - 113.

[6] Lim Y S,Kim J S,Kim H P,et al. The Effect of Grain Boundary Misorientation on the Intergranular $M_{23}C_6$ Carbide Precipitation in Thermally Treated Alloy 690 [J]. Journal Nuclear Materials,2004,335:108 - 114.

[7] 李慧,夏爽,周邦新,等. 镍基 690 合金中晶界碳化物析出的研究[J]. 金属学报,2011,47:853 - 858.

[8] 李强,周邦新. 690 合金的显微组织研究[J]. 金属学报,2001,37:8 - 12.

[9] Li H, Xia S, Zhou B X, et al. C - Cr Segregation at GrainBoundary before the Carbide Nucleation in Alloy 690 [J]. Materials Characterization, 2012, 66: 68 - 74.

Study on Evolution of Grain Boundary Composition in Nickel Based Alloy 690

Abstract: The solute and impurity atoms segregation at grain boundaries before carbide precipitation in Alloy 690 were examined by atom probe tomography. The evolution of grain boundary chemistry after carbide precipitation in Alloy 690 was examined by X-Ray energy dispersive spectrum adapted at high resolution transmission electron microscopy. C, B and Si atoms segregation, C and Cr co-segregation, and Cr enrichment at grain boundaries were observed before carbide precipitation at the grain boundaries. The chromium depletion zone formed at the grain boundaries after carbide precipitation. The chromium concentration at grain boundaries decreased with the aging time prolonging. The chromium concentration at grain boundaries reaches the minimum value after aging at 715 ℃ for 15 h. And then, the chromium concentration at grain boundaries increased with the aging time prolonging. After the carbides were fully covering the grain boundaries, the enrichment of chromium at grain boundaries can be observed. The chromium depletion zone, especially the chromium content gradient in the chromium depletion zone was the main factor that influenced the corrosion resistance of the grain boundary in alloys with high chromium content.

添加 2 wt% Cu 对 Zr-4 合金显微结构和耐腐蚀性能的影响*

摘　要：研究了在 Zr-4 合金中添加 2 wt% Cu 的合金显微组织及其在 500 ℃ 和 10.3 MPa 过热蒸汽中的耐腐蚀性能. 结果表明,该合金经过热轧、冷轧以及经 2 h、580 ℃ 真空退火处理后,得到以 α-Zr 为基体的显微组织,合金中主要存在四方结构的 Zr_2Cu 和密排六方结构的 $Zr(Fe, Cr, Cu)_2$ 第二相,Zr_2Cu 相有长度 1～4 μm,厚度约 1 μm 的片状和直径 300～500 nm 的球形两种形态,并且都会富集一些 Fe 元素. 在 10.3 MPa、500 ℃ 过热水蒸气中,添加 2 wt% Cu 的 Zr-4 合金不发生疖状腐蚀,表明 Cu 是改善锆合金耐疖状腐蚀性能的有益元素.

随着经济迅速发展,环保理念的提升,人们对洁净能源的需求日益迫切. 其中核动力反应堆正朝着提高燃料燃耗、热效率以及安全可靠性的方向发展,因此对燃料元件包壳用锆合金提出了更高的要求[1]. 通常提高锆合金耐腐蚀性能的主要途径是改进合金成分[2]. Cu 元素在核用锆合金中的添加及应用始于 20 世纪 60 年代开发的 Zr-2.5 Nb-0.5Cu 合金[3-4]. 一些研究者发现添加 Cu 元素可以有效提高锆合金的耐腐蚀性能,且在含 Nb 锆合金和 Zr-4 合金中作用效果不同[5-7].

由于以往的研究基本都是关注添加较低含量 Cu 元素(≤0.5 wt%)的锆合金,虽然更切合核用锆合金的实际工程需求,但较难得到含 Cu 相的更多信息. 本文以 Zr-4 合金为母材,添加较高含量(2 wt%)的 Cu,研究该合金的显微组织及在 500 ℃/10.3 MPa 过热蒸汽中的耐腐蚀性能,探索 Cu 元素影响锆合金耐腐蚀性能的机制.

1 实验材料及方法

实验 Zr-4 合金来自西北有色金属研究院,Cu 为市售分析纯电解铜(99.9 wt%). 使用真空非自耗电弧炉将原材料熔炼成合金锭,熔炼过程中使用高纯氩气保护. 为保证成分均匀,每次熔炼后均翻转一次,共熔炼 5 次. 使用电感耦合等离子体原子发射光谱分析熔炼后的合金成分如表 1 所示,为描述方便,标称为 Zr-4-2Cu 合金. 作为对比的 Zr-4 合金成分也列于表中,试验中 Sn 元素略有挥发损耗. 试验合金经过 700 ℃ 热压成型后,再经热轧(700 ℃)及冷轧获得 0.7 mm 的片状试样,最后进行 580 ℃/2 h 退火处理. 最终热处理及入高压釜腐蚀前均需酸洗,以除去试样表面的氧化膜和杂质,所用酸洗液为 45% H_2O+45% HNO_3+10% HF(体积比)混合溶液.

样品经 3 次去离子水煮沸清洗后,在 500 ℃/10.3 MPa 的过热蒸汽中进行静态高压腐蚀试验,定期取出试样称重,腐蚀增重为 4 个试样的平均值.

使用 D/MAX-2200 型 X 射线衍射仪对合金进行物相分析,测试参数为：铜靶 Ka,电

---

* 本文合作者：李强、余康、刘仁多、梁雪、姚美意. 原发表于《上海金属》,2012,34(4)：1-6.

表 1　实验合金的成分(质量分数,%)

Table 1　Chemical composition of the modified Zry-4 alloys(wt%)

合　金	Cu	Fe	Cr	Sn	Zr
Zr-4-2Cu	1.98	0.21	0.12	1.12	余量
Zr-4	—	0.22	0.13	1.3	余量

压 40 kV,电流 40 mA.使用 JSM-6700 型扫描电子显微镜(SEM)对合金形貌及氧化膜断口进行观察.采用带 INCA 能谱仪(EDS)的 JEM-2010F 高分辨透射电子显微镜(HRTEM)观察合金中第二相并分析其成分.合金 TEM 试样采用双喷电解抛光方法制备,电解液为 80% CH₃COOH+20% HClO₃(体积比)混合液.氧化膜断口 SEM 样品制备方法为酸洗溶去试样的部分金属基体,折断露出的氧化膜后蒸镀 Ir 导电层.

2　实验结果

2.1　合金 XRD 分析

图 1 为实验合金 X 射线衍射图.其中密排六方结构的 α-Zr 与四方结构 Zr₂Cu 所对应的衍射峰比较明显,但其他相的衍射峰较难检测到,说明该合金中以 α-Zr 相和 Zr₂Cu 相为主.

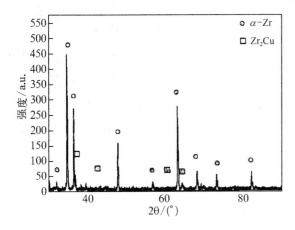

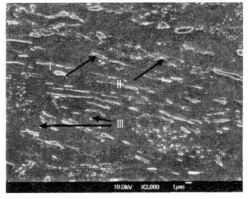

图 1　Zr-4-2Cu 合金经 580 ℃/2 h 退火后的 X 射线衍射图

Fig. 1　XRD spectra of the Zr-4-2Cu alloys after 580 ℃/2 h annealing

图 2　Zr-4-2Cu 合金经 580 ℃/2 h 退火后 SEM 照片

Fig. 2　TEM image of Zr-4-2Cu alloys after 580 ℃/2 h annealing

α-Zr 基体的衍射峰与标准 α-Zr 衍射峰相比略微左移,这可能是由于基体中固溶了少量的 Fe、Cr、Sn,使得 α-Zr 基体的晶格常数稍有变大,至今仍未有有关室温中 Cu 在 α-Zr 中的固溶度的实验数据,但是 Hong H S 等[8]人根据溶解度方程计算出室温下 Cu 在 α-Zr 中的固溶度约为 $5×10^{-47}$ wt%,远小于 Fe、Cr、Sn 的固溶度[9],所以 Cu 元素主要存在于 Zr₂Cu 相中.

2.2　合金 SEM 及 TEM 分析

图 2 是合金经 580 ℃/2 h 退火后的 SEM 照片.可以看出 Zr-4-2Cu 合金中的第二相

主要有三种形态：其一为尺寸较大的片状第二相（Ⅰ型），长度约 1～4 μm，厚度约 1 μm；其二为直径约 300～500 nm 的球形颗粒（Ⅱ型）；其三为直径约 100～200 nm 的细小球形颗粒（Ⅲ型）.

图 3 为 Zr-4-2Cu 合金中长度约 1.5 μm、宽度约 300 nm 的第二相（Ⅰ型）的明暗场像及其衍射花样，可以看出其中心位于 α-Zr 晶界，两端分别处于两个 α-Zr 晶粒内，本身由三个晶粒构成. 对其做明暗场像，并通过 SAD 标定以及 EDS 分析（见表 2），确定三个晶粒均为四方结构的、富集 Fe 元素的 Zr_2Cu 相.

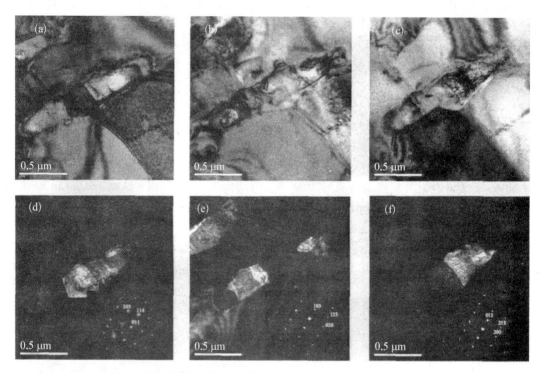

图 3 合金条状第二相中三个晶粒的明暗场像及其电子衍射花样

Fig. 3 Bright field image, dark field image and corresponding SAD pattern of lamellar precipitate

表 2 相应区域的 EDS 测试结果

Table 2 EDS test results of corresponding region

区 域	元素含量/(wt%)		
	Zr	Cu	Fe
1	75.26	21.16	1.43
2	75.61	20.86	1.37
3	74.56	20.88	1.34

这种形态第二相的形成是由于 Cu 在 α-Zr 中的固溶度极低，而添加的 Cu 含量较多，熔炼冷凝后会析出大量的 Zr_2Cu，特别容易在 α-Zr 晶粒的晶界处呈较大尺寸片状析出. 经过热压、热轧、冷轧等形变过程，这种尺寸较大的第二相，首先会在应力作用下也参与形变，导致位错增值，并且在晶粒的内部不同区域滑移系开动的情况不同，逐步造成内部各区域"分割"形成各个"胞块"，形成亚晶. 从而形成这种片状第二相内分为若干个亚晶的情况.

图 4 为 Zr-4-2Cu 合金中直径约为 400 nm 球形颗粒（Ⅱ型）第二相的 TEM 照片，分析

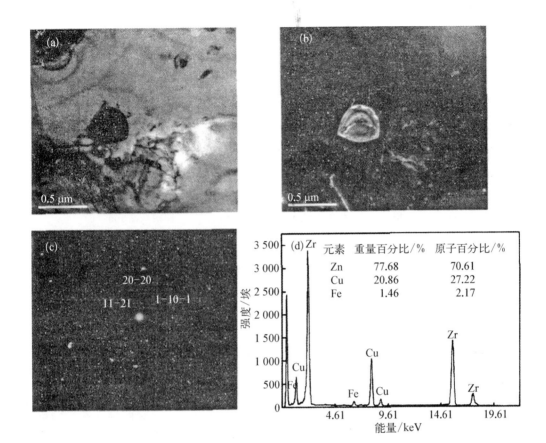

图 4 Zr-4-2Cu 合金样品中 Zr₂Cu 第二相的 TEM 图片 （a）明场像；（b）暗场像；（c）衍射花样；（d）Zr₂Cu 第二相 EDS 谱图

Fig. 4 TEM image of Zr-4-2Cu precipitates in Zr-4-2Cu alloy （a）bright field image；（b）dark field image；（c）corresponding SAD pattern；（d）EDS spectrum

确定也为四方晶体结构的 Zr_2Cu 相.

在 Zr_2Cu 相中检测出含量较多的 Fe 元素(约 1.5 wt%)，远高于母材 Zr-4 合金中 Fe 元素含量，而 EDS 分析并未检测到 Cr、Sn 等元素，说明 Fe 更易于富集在 Zr_2Cu 第二相中. 这可能是由于合金元素在发生扩散聚集时，Fe 原子扩散速度较快[10]. Fe 在 Zr_2Cu 相中的富集，必然会影响原 Zr-4 合金中合金元素的存在状态，如 α-Zr 中的合金元素固溶含量，第二相数量、大小及元素构成等，从而影响试验合金的耐腐蚀性能.

图 5 是 Zr-4-2Cu 合金中尺寸约为 100~200 nm 的细小颗粒状第二相(Ⅲ型)的 TEM 照片和 EDS 谱图. 可以确定图中第二相为含 Cu 的 $Zr(Fe, Cr)_2$，其中 Fe/Cr 比约为 0.88. Fe 和 Cr 在 α-Zr 中的固溶度很低，Fe 为 120 $\mu g/g$，Cr 为 200 $\mu g/g$[10]，合金中的 Fe 和 Cr 大部分以 $Zr(Fe, Cr)_2$ 形式析出. EDS 分析的结果并不完全符合 $Zr(Fe, Cr)_2$ 第二相中的含量，这是由于微区分析时的电子束有一定的穿透深度和扩散范围，而第二相尺寸较小，因此分析时会受到周围 Zr 基体的影响，但 Fe/Cr 值是相对准确的.

$Zr(Fe, Cr)_2$ 的结构取决于 Fe/Cr 比，当<0.2 或>0.9 时与 $ZrFe_2$ 结构一致，为面心立方结构，其间则与 $ZrCr_2$ 一致为密排六方结构. 根据电子衍射花样标定，可以确定这种第二相为密排六方结构，符合能谱结果中 Fe/Cr 比.

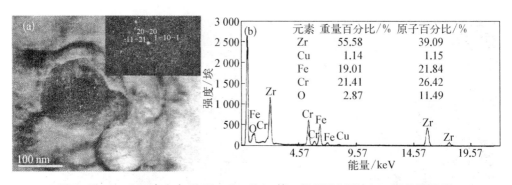

图 5　Zr-4-2Cu 合金中 Zr(Fe，Cr，Cu)₂ 第二相 TEM 图片(a)；EDS 谱图(b)

Fig. 5　TEM image of Zr(Fe, Cr, Cu)₂ precipitates (a)；EDS spectrum (b)

值得注意的是,Zr(Fe，Cr)₂ 第二相中存在少量的 Cu 元素(约 1 wt%),低于合金的名义成分,但远高于 Cu 在锆基体中的固溶度.这是由于高添加量的 Cu 元素除了大量以 Zr₂Cu 第二相形式析出外,少量会在 Zr(Fe，Cr)₂ 第二相中取代一部分 Fe 或 Cr 原子,而形成 Zr(Fe，Cr，Cu)₂第二相.

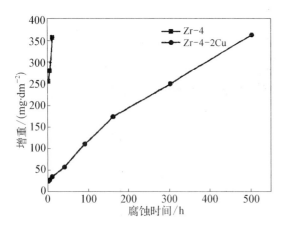

图 6　样品在 500 ℃/10.3 MPa 过热蒸汽水中的腐蚀增重曲线

Fig. 6　Corrosion behaviors of the modified Zry-4 alloys in the steam of 500 ℃/10.3 MPa

2.3　腐蚀行为

图 6 给出了锆合金样品在 500 ℃/10.3 MPa 过热蒸汽中的腐蚀增重曲线.可以看出添加 Cu 的 Zr-4 合金耐腐蚀性能明显优于 Zr-4 合金,Zr-4 合金试样的腐蚀增重非常迅速.图 7 为试验合金腐蚀后的表面形貌,从图中可以看出,Zr-4 合金试样腐蚀 4 h 后已发生较为严重的疖状腐蚀,原来黑色致密的氧化膜已经大部分变白(图 7(a)).腐蚀时间超过 50 h 后,氧化膜已全部变白,并且样品边缘的氧化物发生脱落,无法准确称重(图 7(b));而 Zr-4-2Cu 试样即使腐蚀 500 h 后,依旧为具有保护性的黑亮色氧化膜,未发生疖状腐蚀(图 7(d)),表现出较好的耐腐蚀

性能.显然,Cu 的添加有益于 Zr-4 合金在 500 ℃/10.3 MPa 过热蒸汽中的耐腐蚀性能,特别是能抑制疖状腐蚀的发生.

图 8 为试验合金腐蚀后氧化膜断口的 SEM 照片,可以看出 Zr-4 合金腐蚀 4 h 后已经产生严重的疖状腐蚀,生成凸透镜状、结构疏松的疖状腐蚀区,最大厚度达到 20 μm.Zr-4-2Cu合金未产生疖状腐蚀,腐蚀 250 h 后生成的氧化膜厚度 8 μm 左右,氧化膜表面及氧化膜/金属(O/M)界面较平整,结构比较致密,只有少量尺寸约 2 μm 左右的横向裂纹.从其局部放大图中看到,氧化膜以柱状晶为主,并且形态较完整,排列紧密.这与在 Zr-2.5 Nb 合金中添加少量 Cu 的相关研究结果[11]类似,即添加 Cu 有利于提高锆合金氧化膜中柱状晶比例,并使柱状晶尺寸增大且排列有序,从而提高锆合金在 500 ℃/10.3 MPa 过热蒸汽中的耐腐蚀性能.

图 7 样品腐蚀后的表面形貌 （a）Zr-4 腐蚀 4 h；（b）Zr-4 腐蚀 50 h；（c）Zr-4-2Cu 腐蚀 300 h；（d）Zr-4-2Cu 腐蚀 500 h

Fig. 7 Surface topography of samples after the corrosion （a）Zr-4 alloy, 4 h；（b）Zr-4 alloy, 50 h；（c）Zr-4-2Cu alloy, 300 h；（d）Zr-4-2Cu alloy, 500 h

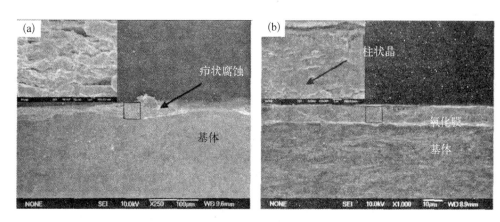

图 8 试验样品腐蚀生成氧化膜的断口形貌照片 （a）Zr-4 腐蚀 4 h；（b）Zr-4-2Cu 腐蚀 250 h

Fig. 8 Fractographs of oxide layers formed on samples after the corrosion （a）Zr-4 alloy, corrosion for 4 h；（b）Zr-4-2Cu alloy, corrosion for 250 h

3 分析讨论

　　锆合金疖状腐蚀的腐蚀机理和影响因素非常复杂,目前还未形成统一的认识. Taylor[12] 认为疖状腐蚀首先发生在合金元素含量低的区域,在给定的氧化气氛中,应有一个临界最低合金浓度,如果在合金的氧化界面的每一点,合金元素浓度不低于该临界值,就可以阻止疖状腐蚀的发生. 另外还有研究认为第二相粒子尺寸大于 175 nm 时可能会发生疖状腐蚀. 李强[13]发现疖状腐蚀易在腐蚀初期微孔、微裂纹少,比较致密的氧化膜区域发生,这种氧化膜会局部破裂,易产生贯穿裂纹,这些贯穿裂纹会成为提供充足氧的直接供氧通道,从而产生疖状腐蚀. 本实验在 Zr-4 合金中加入了 2wt% 的 Cu,虽然合金元素总量较高,但因 Cu 在 α-Zr 中的固溶含量极低,会以较大尺寸的第二相形式存在,并且 Zr-4 中原有的（α-Zr 和 Zr(Fe, Cr)$_2$ 相中)元素 Fe 易于富集在 Zr$_2$Cu 相中,导致合金内部元素微观分布不均匀,因而不利于耐疖状腐蚀. Zr-4-2Cu 合金未发生疖状腐蚀,显然与合金元素 Cu 有关. 在研究

Zr-2.5Nb-0.5Cu 合金中发现,Zr_2Cu 相易被氧化,氧化后的 Cu 元素会聚集到氧化锆的晶界上,并沿着晶界扩散,可以抑制氧化膜中微孔洞的生成,同时抑制并延迟平行微裂纹的产生[13]. Cu 的这种易氧化、沿晶界扩散的特性,可以消减合金元素分布不均匀的不利影响,也可以弛豫氧化膜的内应力,使氧化膜能够保持均匀、完整,避免疖状腐蚀的发生.

4 结论

(1) Zr-4-2Cu 合金中主要存在四方结构的 Zr_2Cu 和密排六方结构的 Zr (Fe, Cr, Cu)$_2$ 第二相;Zr_2Cu 相有长度约 1~4 μm、厚度约 1 μm 的片状和直径约 300~500 nm 的球形两种形态,并且都会富集一定量的 Fe 元素.

(2) Zr-4-2Cu 合金腐蚀后生成的氧化膜以柱状晶为主,并且形态较完整,排列紧密.

(3) 添加 2 wt% Cu 的 Zr-4 合金在 500 ℃/10.3 MPa 过热蒸汽中表现出较好的耐腐蚀性能,不发生疖状腐蚀,表明 Cu 是改善锆合金耐疖状腐蚀的有益元素.

参 考 文 献

[1] Liu Wenqing, Li Qiang, Zhou Bang xin, et al. Effect of heat treatment on the microstructure and corrosion resistance of a Zr-Sn-Nb-Fe-Cr alloy[J]. J Nucl Mater, 2005, 341(2-3): 97-102.

[2] 李强,周邦新,姚美意,等.锆合金在 550 ℃,25 MPa 超临界水中的腐蚀行为[J].稀有金属材料与工程,2007,36(8): 1358-1360.

[3] 卡恩 R W,哈森 P,克雷默 E J.核应用中的锆合金[M].材料科学与技术丛书,核材料,10B 卷,科学出版社.1999: 1-45.

[4] Chakravartty J K. Characterization of hot deformation behaviour of Zr-2.5Nb-0.5Cu using processing maps[J]. J. Nucl. Mater., 1995, 218(2): 247-255.

[5] Castaldelli L, Fizzotti C. Long-term test results of promising new zirconium alloys[C]//Zirconium in the Nuclear Industry: Fifth International Symposium, ASTM STP 754, 1982: 105-126.

[6] Yilmazbayhan A, Motta A T. Structure of zirconium alloy oxides formed in pure water studied with synchrotron radiation and optical microscopy: relation to corrosion rate[J]. J. Nucl. Mater., 2004, 324(1): 6-22.

[7] Hong H S, Moon J S. Investigation on the oxidation characteristics of copper-added modified Zircaloy-4 alloys in pressurized water at 360 ℃ [J]. J. Nucl. Mater., 2001, 297(2): 113-119.

[8] Hong H S, Kim H S, Kim S J, et al. Effects of copper addition on the tensile properties and microstructures of modified Zircaloy-4[J]. J. Nucl. Mater., 2000, 280(2): 230-234.

[9] Ch Arquetd, Hahnr, Ortlibe, et al. Solubility limits and formation of intermetallic precipitates in Zr-Sn-Fe-Cr alloys [C]//van SWAMLFP, EUCKEN C M. Zirconium in the Nuclear Industry, Eighth International Symposium, ASTM STP 1023. Philadelphia, PA: ASTM, 1989: 405-422.

[10] 周邦新,苗志,李聪. Zr(Fe, Cr)$_2$ 金属间化合物在 500 ℃ 过热蒸汽中的腐蚀研究[J].核动力工程,1997,18(1): 53-60.

[11] 李强,梁雪,彭剑超,等. Cu 对 Zr-2.5Nb 合金在 500 ℃/10.3 MPa 过热蒸汽中腐蚀行为的影响[J].金属学报,2011, 47(07): 877-881.

[12] Taylor D F. An oxide-semiconductance model of nodular corrosion and its application to Zirconium alloy development [J]. J. Nucl. Mater, 1991, 184(1): 65-77.

[13] 李强.锆合金在 550 ℃/25 MPa 超临界水中腐蚀行为的研究[D].上海:上海大学,2008.

Effect of 2 wt% Copper Addition on Microstructure and the Corrosion Behaviors of Zircaloy-4 Alloy

Abstract: The effects of the 2 wt% copper addition on microstructures and the corrosion behaviors of Zircaloy-

4 alloys in the 500 ℃ superheated steam of 10.3 MPa were investigated. The main phase was α - Zr after hot rolled, cold rolled and subsequent vacuum annealing treatment (580 ℃ for 2 h). Furthermore, tetragonal-Zr_2Cu and hexagonal-Zr (Fe, Cr, Cu)$_2$ precipitates were observed by TEM. Enrichment of Fe element was found in lamellar (about 1～4 μm and 1 μm in length and thickness, respectively) and spherical (300～500 nm in diameter)Zr_2Cu phases. Nodular corrosion was not observed in Zr - 4 - 2Cu after 500 h corrosion in 500 ℃ superheated steam of 10.3 MPa, indicating that Cu was beneficial to improving the nodular corrosion resistance of Zircaloy alloy in the environment stated above.

N18 锆合金疖状腐蚀问题研究[*]

摘　要：N18 锆合金样品经过不同条件的热处理后，用高压釜在 500 ℃/10.3 MPa 过热蒸汽中进行腐蚀实验，研究其耐疖状腐蚀性能. 结果表明：样品经过 780 ℃的 $\alpha+\beta$ 双相及 700 ℃的 α 相长时间的热处理后易出现疖状腐蚀，其原因是出现了合金元素，特别是 Nb 元素的贫化区. 采用大变形量加工及低温退火处理并不能有效消除已形成的合金元素贫化区，只有重新进行充分的 β 相高温固溶处理才能消除.

1　引言

反应堆运行时，锆合金作为核燃料元件包壳材料会受到高温高压水的腐蚀，生成氧化膜. 由 Zr-2 和 Zr-4 合金的研究结果可知，在加氢除氧的压水堆工况下，一般发生的是均匀腐蚀，形成较致密的黑色氧化膜；而在沸水堆的工况条件下，由于含氧量较高，有时会出现局部不均匀的腐蚀斑，即疖状腐蚀. 发生疖状腐蚀时，先在黑色氧化膜上生成白色斑点，直径 0.1~0.5 mm，截面呈凸透镜状，深度约为直径的 1/5[1]. 随着疖状斑的增多和长大，最终会连成一片，形成疏松且容易剥落的白色氧化膜[2,3]. 这种情况的发生一方面会造成包壳管的过早破损，另一方面形成的疏松氧化层易脱落，会引起回路中部件的磨损[4]. 因此，研究锆合金疖状腐蚀的问题对于提高核反应堆的安全可靠性具有重要意义. 实验室通常采用 450~500 ℃过热蒸汽腐蚀试验来研究和评价锆合金的耐疖状腐蚀性能[5].

N18 合金是我国自主研发的新锆合金，在 360 ℃/LiOH 水溶液工况条件下表现出了优良的耐腐蚀性能[6,7]；在 400~550 ℃高温高压水或蒸汽条件下，也表现出了很好的耐疖状腐蚀性能[8-10]. 已有的研究结果显示，在不同相区加热处理的 N18 合金，无论是快冷还是慢冷，在 500 ℃/10.3 MPa 过热蒸汽中都表现出良好的耐疖状腐蚀性能[8,9]. 但是，最近的一些实验研究表明，在某些加工及热处理条件下，N18 合金也会出现疖状腐蚀，因此，有必要对此进行深入研究.

2　实验方法

研究用 N18 锆合金由西北有色金属研究院提供，板材厚度为 2 mm，化学成分列于表 1. 为了研究 N18 锆合金出现疖状腐蚀的问题，采用了图 1 中所列出的实验加工及热处理过程. 其中，过程 1 的主要特征是在 $\alpha+\beta$ 两相区实施 780 ℃长时间保温，希望出现易发生疖状腐蚀的合金显微组织；过程 2 是在较高温度（700 ℃）的 α 相区长时间保温、$\alpha+\beta$ 两相区不同时间以及高温（1 020 ℃）β 相区短时间处理后，进行超过 50% 的轧制变形，最后进行统一的低温

---

* 本文合作者：李强、黄昌军、杨艳平、徐龙、梁雪. 原发表于《核动力工程》，2012，33(S2)：22-27,38.

长时间退火处理,考察不同相区及不同保温时间对疖状腐蚀的影响;过程 3 是尝试通过高温 β 相区热处理及轧制变形等处理,改善 780 ℃长时间保温处理所形成的易发生疖状腐蚀的不良显微组织. 真空热处理在石英管式真空(优于 10^{-3} Pa)炉中进行,保温结束后石英管外淋水冷却. 在真空热处理及放入静态高压釜腐蚀前均用混合酸进行标准酸洗(体积分数为 10% HF＋45% HNO_3＋45% H_2O). 腐蚀条件为 500 ℃/10.3 MPa 的过热蒸汽.

表 1 N18 合金的化学成分
Table 1 Chemical Composition of N18 Alloy

元　素	Zr	Sn	Nb	Fe	Cr
质量分数/%	余量	0.98	0.25	0.35	0.08

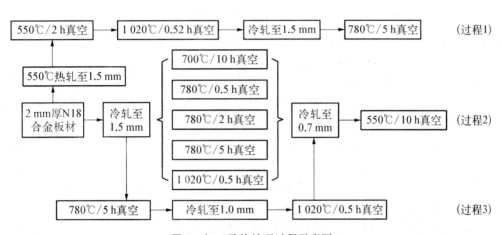

图 1 加工及热处理过程示意图
Fig. 1 Processing and Heat Treatment Diagram

用带有 INCA(牛津仪器公司的特定型号)能谱仪(EDS)的 JEM‐2010F 透射电子显微镜(TEM)观察腐蚀前基体显微组织;用 VHX‐100 光学显微镜(OM)观察腐蚀生成氧化膜表面;用 SU1510、JSM‐6700F 和 Apollo300 扫描电子显微镜(SEM)对腐蚀前、后的合金基体及腐蚀生成氧化膜的显微组织进行分析. 合金基体试样经机械研磨及酸洗减薄至厚度小于 0.1 mm,冲切成直径 3 mm 的小圆片,用双喷电解抛光[电解液:10%(容积)$HClO_4$＋90%(容积)C_2H_5OH,−30 ℃]的方法获得 TEM 可观测薄区. 合金试样经酸洗及水洗后即可用于 SEM 观测;腐蚀后的试样用混合酸将断面局部基体溶去,露出氧化膜并将其折断,制备出氧化膜断口样品. 为了提高成像质量,观察前将氧化膜断面或表面样品进行了喷金处理.

3　实验结果与讨论

3.1　过程 1 处理合金显微组织及腐蚀结果

图 2 是过程 1 处理的 N18 合金样品显微组织. 可以看出样品已完全再结晶;在晶界和三晶交界区域有大量的析出相,而晶粒内析出相很少. 在两个晶粒之间的晶界处,析出相沿晶界呈条片状,轮廓较平滑(Ⅰ型,图 2b),而在多晶交汇处的析出相较复杂,既有Ⅰ型,也有形状复杂的Ⅱ型(图 2c). 经选区电子衍射(SAD)及 EDS 分析,确定Ⅰ型析出相为体心立方

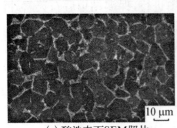 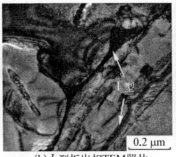 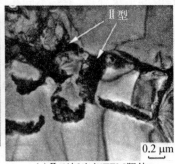

(a)酸洗表面SEM照片　　　(b)Ⅰ型析出相TEM照片　　　(c)Ⅱ型析出相TEM照片

图 2　780 ℃/5 h 处理后 N18 合金显微组织

Fig. 2　Micrographs of N18 Specimens after 780 ℃/5 h Annealing

(bcc)结构的 β-Zr,Cr 质量分数在 0.3%左右,Fe 元素质量分数小于 2.0%,Nb 元素质量分数大于 10.0%;Ⅱ型析出相为六方结构的 Zr(Fe,Cr,Nb)$_2$ 相,Fe 质量分数 14.43%~32.38%,Cr 质量分数为 4.04%~8.89%,Nb 质量分数小于 10.0%. 两种析出相都富含Nb、Fe、Cr 合金元素,远高于合金的平均含量. 在析出相附近及远离晶界的晶粒内部作EDS 检测,均很难探测出 Fe 和 Nb 元素. 显然,当试样加热到 780 ℃时,处于 $\alpha+\beta$ 两相区,Fe、Nb 元素大量扩散进入 β-Zr 中,造成 α-Zr 相晶粒内部包括以固溶及第二相形式存在的 Nb、Fe 元素严重贫化.

Nb、Fe 作为可以稳定 β 相的元素,在退火温度与冷却速度相同条件下,其在 β-Zr 中的固溶含量对 β-Zr 是否发生分解起着重要的作用. 从 Zr-Nb 和 Zr-Fe 二元相图可以知道,Fe、Nb 元素都能使 β 相区向较低温度扩展,Nb 元素质量分数在 0.6%~18.8%之间时,稍高于 610 ℃就会形成 β-Zr,而 Fe-Zr 合金在低于 730 ℃时都不会形成 β-Zr. 因此,当冷却速率较快时,含 Nb 量较高的 β-Zr 相更容易被稳定至室温. 轧制变形处理的 N18 合金在 780 ℃保温时,发生回复与再结晶,早期处理形成的析出相也会发生重溶,Nb、Fe、Cr 等合金元素会发生迁移,在晶界,特别是三晶交汇处形成富含合金元素 Nb、Fe、Cr 的 β-Zr 相. 在实验条件的降温过程中,部分 β-Zr 相分解形成 α-Zr 相和含 Nb 量相对较少的 Zr(Fe,Cr,Nb)$_2$ 相;而在 β-Zr 相分解过程中的合金元素迁移,使得部分 β-Zr 相中的 Nb 含量较高,得以稳定到室温,形成观察到的合金显微组织. 这种合金组织最明显的特征就是合金元素偏聚到晶界,造成晶粒内合金元素贫化,特别是 Nb、Fe 元素的贫化.

图 3 为腐蚀后试样氧化膜表面及断口形貌. 从图 3a 试样表面 SEM 照片可以看出,经500 ℃/10.3 MPa 过热蒸汽腐蚀 3 h 后的 N18 试样表面出现了大量的疖状腐蚀斑. 图 3b 为较高倍率的 SEM 照片,对比图 2a 的合金基体 SEM 照片,可以看到氧化膜表面保留了合金基体表面晶粒的基本形貌特征. 异常增厚的氧化膜疖状凸起都是发生在合金 α-Zr 相晶粒中部的合金元素贫化区,而第二相富集的晶界区域的氧化膜较薄;疖状凸起的表面都会出现裂纹. 图 3c 为腐蚀生成氧化膜断面的 SEM 照片,包含一个表面开裂的疖状斑,疖状斑呈凸透镜状,厚度约为 4 μm,厚度与直径之比约为 1/3,表面裂纹为垂直裂纹,但并未贯穿整个氧化膜;疖状斑的生长暂时被限制在一个晶粒内,结构还较致密,应为疖状腐蚀的初期阶段.

3.2　过程 2 和过程 3 处理合金显微组织及腐蚀结果

图 4 和图 5 是过程 2 和过程 3 处理后的部分 N18 合金样品显微组织. 从 TEM 图中可

(a) 低倍表面形貌　　　　(b) 较高倍率表面形貌　　　　(c) 断口形貌

图 3　腐蚀 3 h 后样品氧化膜表面及断口 SEM 照片

Fig. 3　Surface and Fracture Morphology of Oxide Film Corroded in 500 ℃/10. 3 MPa Superheated Steam for 3 h Exposure

以看出,由于都经过了变形及相同的低温 580 ℃退火处理,所有试样都发生了再结晶,形成等轴的 α-Zr 晶粒,粒径尺寸基本相同,都在 2～10 μm. 结合 SEM 图可以了解第二相的情况,在 700 ℃进行 10 h 处理的样品中第二相分布相对均匀,但尺寸较大,在 100～500 nm,经 SAD 及 TEM-EDS 分析,为含少量 Cr 的密排六方结构(hcp)的 Zr(Nb, Fe)$_2$ 相,Nb 质量分数达到 7%～20%. 经过 780 ℃处理过的样品,析出相既有含少量 Cr 的 Zr(Nb, Fe)$_2$ 相,Nb 质量分数超过 10%,也有含少量 Nb 的 Zr(Fe, Cr)$_2$ 相,并且都出现了析出相在原 780 ℃处理下形成的 α 相晶界附近位置聚集分布的现象,表明后续的超过 50%的变形及低温 α 相区退火处理没有明显改变类似于过程 1 的富含合金元素的第二相的偏聚情况. 随着在 780 ℃处理时间的延长,析出相的聚集度增加,即分布的不均匀性增加. 在 1 020 ℃进行 0.5 h 处理的试样显微组织见图 4e,析出的主要是 hcp 结构 Zr(Nb, Fe, Cr)$_2$ 第二相,估计 Nb 质量分数在 1.5%左右[11],呈条带分布. 图 4f 和图 5d 是在 780 ℃进行 5 h 处理,再经 1 020 ℃、0.5 h 固溶处理的样品显微组织,标识为 780 ℃/1 020 ℃,可以看到重新进行的高温固溶处理

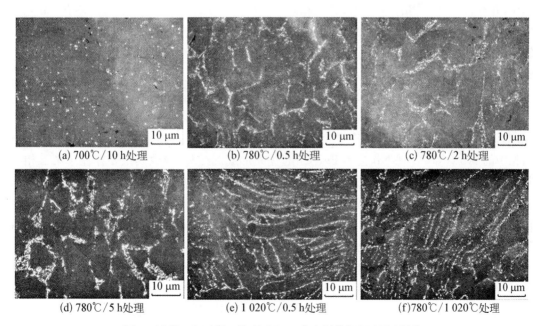

(a) 700℃/10 h处理　　　　(b) 780℃/0.5 h处理　　　　(c) 780℃/2 h处理

(d) 780℃/5 h处理　　　　(e) 1 020℃/0.5 h处理　　　　(f)780℃/1 020℃处理

图 4　过程 2 和过程 3 处理后 N18 合金显微组织 SEM 照片

Fig. 4　SEM Micrographs of N18 Specimens after Processing 2 and Processing 3

(a) 700℃/10 h处理   (b) 780℃/0.5 h处理

(c) 780℃/2 h处理   (d) 780℃/1 020℃处理

图 5 过程 2 和过程 3 处理后部分 N18 合金显微组织 TEM 照片

Fig. 5 TEM Micrographs of N18 Specimens after Processing 2 and Processing 3

可以改善合金元素的偏聚情况. 当锆合金加热到 1 020 ℃时,处于 β 单相区,合金元素完全固溶,冷却时,部分 β-Zr 通过贝氏体转变成板条状 α-Zr. β 共析体元素 Fe 和 Cr 被相变前沿排斥,并在板条状 α 相的晶界上以第二相的形式析出. N18 除 Fe、Cr 元素外,还有 Nb 元素,当富集的合金元素达到可以稳定 β 相的浓度及温度条件,将形成亚稳态的 β-Zr 相. 进一步轧制变形及最终再结晶退火处理,β-Zr 分解成 α-Zr 和富合金元素的第二相,形成所观察到的显微组织.

    N18 合金中 Nb 质量分数只有 0.25%. 在 Zr-Nb 合金中,Nb 在 α-Zr 中的固溶度随退火温度升高而提高,500 ℃下 Nb 在 α-Zr 中的固溶度可达 0.6%,一般的理解 N18 合金中的 Nb 应该全部固溶在 α-Zr 中. 李中奎用 SEM 波谱仪研究了 N18 (NZ2)合金中 α-Zr 与析出相中的 Nb 含量,发现 Nb 会和 Fe、Cr 起优先形成沉淀相,导致其主要存在于第二相中,并出现其在 α-Zr 中固溶量随退火温度升高而降低的异常现象[11]. 本实验中除了 1 020 ℃/变形/580 ℃处理的样品中析出相的 Nb 含量稍低,其他都富含 Nb 元素,并且第二相的数量较多,这必然导致 α-Zr 中 Nb 含量的严重贫化,影响其发挥耐腐蚀的有益作用. 本实验中在 700 ℃和 780 ℃较长时间的保温,会使 Nb 等合金元素的扩散偏聚析出更充分. 从图 4 和图 5 中可以看到,实验样品合金基体中都存在无析出相的区域(SEM 照片中无亮点的区域),应该是合金元素贫化区. 其中,780 ℃/5 h 处理的大尺度贫化区较多,而 780 ℃/0.5 h 处理的较少(虽然析出相偏聚,但非偏聚区域的析出相也较多);其他条件处理的介于二者之间,其中具有可比性的 780 ℃/1 020 ℃比 1 020 ℃/0.5 h 处理的大块无析出相区多,表明前者析出相的分布不如后者均匀.

图 6 是过程 2 和过程 3 处理后的 N18 合金试样经 500 ℃/10.3 MPa 过热蒸汽腐蚀 3 h 后部分样品氧化膜表面的 OM 照片. 经过 780 ℃/5 h 处理的样品表面出现了大量的疖状腐蚀斑,这与过程 1 处理的样品类似;随着 780 ℃ 保温时间的减少以及处理温度的降低,合金耐疖状腐蚀性能明显得到改善. 其中,700 ℃/10 h、780 ℃/2 h 和 780 ℃/1 020 ℃ 都出现了少量疖状腐蚀,而 780 ℃/0.5 h 和 1 020 ℃/0.5 h 几乎未观察到疖状腐蚀斑. 780 ℃/1 020 ℃ 处理的样品出现少量疖状腐蚀的原因可能是固溶处理时间不够,致使固溶不均匀,影响了合金元素分布的均匀性. 可以看出疖状腐蚀斑的出现与数量与合金基体中大块合金元素贫化区有关. 结合对过程 1 处理样品的研究,显然疖状腐蚀易在大块合金元素贫化区发生,这与 Taylor 报道的对 Zr‐2 或 Zr‐4 疖状腐蚀的研究结果类似[12].

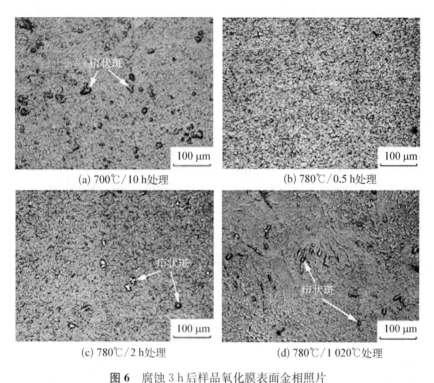

(a) 700℃/10 h处理 (b) 780℃/0.5 h处理

(c) 780℃/2 h处理 (d) 780℃/1 020℃处理

图 6 腐蚀 3 h 后样品氧化膜表面金相照片

Fig. 6 Surface Morphology of Oxide Film Corroded for 3 h Exposure

Jeong 等在研究不含 Fe、Cr 合金元素的 Zr‐xNb [x=0~0.6%(质量)]时发现,采用正常加工工艺,当 Nb 质量分数低于 0.2% 时会出现疖状腐蚀[13]. 姚美意等报道在 Zr‐4 合金中添加 Nb 元素,采用正常加工工艺,当添加的 Nb 质量分数<0.1% 时,便能明显改善 Zr‐4 合金的耐疖状腐蚀性能,但还不能完全抑制;当添加量为 0.1%~0.3%(质量分数)时,长时间腐蚀均未出现疖状腐蚀[14,15]. 这些研究表明,α‐Zr 中合金元素含量低不利于锆合金耐疖状腐蚀,Nb 元素的添加可以提高锆合金的耐疖状腐蚀性能,但也要达到一定的含量. 但是,即使合金中 Nb 含量不低,若处理不当,一方面使 α‐Zr 中固溶的有益合金元素严重贫化,另一方面富含合金元素的析出相分布不均匀,将出现大到一定的体积的合金元素贫化区,容易发生疖状腐蚀. 文献[8,9]报道 N18 合金不发生疖状腐蚀,可能是在 $\alpha+\beta$ 两相区处理时间不够长,尚未产生足够的合金元素贫化区.

N18 合金在 $\alpha+\beta$ 两相区热处理时,在晶界处形成的 β‐Zr,Nb、Fe、Cr 等合金元素会

迁移至 β-Zr 中,在 α-Zr 晶粒内出现合金元素贫化区,特别是 Nb 的贫化;在 α 相区温度上限附近热处理时,若保温时间较长,固溶在 α-Zr 中得 Nb 元素会充分扩散偏析到析出相中,也会形成 Nb 贫化区.采用后续大变形量加工及低温退火处理并不能有效消除合金元素贫化区,只有重新进行高温充分的固溶处理才能消除.

4 结论

(1) N18 合金经高温热处理会出现合金元素贫化区,特别是 Nb 的贫化区,这是发生疖状腐蚀的主要原因.

(2) N18 合金在 $\alpha+\beta$ 两相区热处理时,在晶界处形成 β-Zr,Nb、Fe、Cr 等合金元素会迁移至 β-Zr 中,并会在 α-Zr 晶粒内出现合金元素贫化区;在较高温度的 α 相区热处理,Nb 元素会扩散偏析到析出相中,也会形成 Nb 贫化区.

(3) 已出现合金元素贫化区的 N18 合金,很难通过后续变形及低温 α 相区退火处理消除,只有重新进行充分的 β 相高温固溶处理才能消除.

参 考 文 献

[1] Ogata K, Mishima Y, Okubo T, et al. Zirconium in the Nuclear Industry[C]. ASTM STP 1023, American Society for Testing and Materials, Philadelphia, 1989: 291.

[2] Cheng B, Adamson R B. Zirconium in the Nuclear Industry[C], ASTM STP 939, American Society for Testing and Materials, Philadelphia, 1987: 387.

[3] Zhou B X. Zirconium in the Nuclear Industry[C]. ASTM STP 1023, American Society for Testing and Materials, Philadelphia, 1989: 360.

[4] 杨文斗. 反应堆材料学[M]. 北京:原子能出版社,2000.

[5] Johnson A B, Horton R M. Zirconium in the Nuclear Industry[C]. Philadelphia: ASTM, 1977: 295.

[6] 刘文庆,王泽明,刘庆东,等. Zr-Sn-Nb-Fe 合金显微组织及耐腐蚀性能研究[J]. 原子能科学技术,2009, 43(7): 630.

[7] 张欣,姚美意,李士炉. 加工工艺对 N18 锆合金在 360 ℃/18.6 MPa LiOH 水溶液中的腐蚀行为的影响[J]. 金属学报,2011,47(9): 1112.

[8] 赵文金,苗志,蒋宏曼,等. Zr-Sn-Nb 合金的腐蚀行为研究[J]. 中国腐蚀与防护学报,2002,22(2): 124.

[9] 周邦新,姚美意,李强,等. Zr-Sn-Nb 合金耐疖状腐蚀性能的研究[J]. 稀有金属材料与工程,2007,36: 1317.

[10] 李强,周邦新,姚美意,等. 锆合金在 550 ℃/25 MPa 超临界水中的腐蚀行为[J]. 稀有金属材料与工程,2007, 36: 1358.

[11] 李中奎,周廉,张建军,等. Zr-Sn-Nb-Fe 合金中铌的存在方式及其与热处理的关系[J]. 稀有金属材料与工程, 2004,33(12): 1362-1364.

[12] Taylor D F. An Oxide-semiconductance Model of Nodular Corrosion and its Application to Zirconium Alloy Development [J]. J Nucl Mater. 1991,184: 65.

[13] Jeong Y H, Kim H G, Kim D J. Influence of Nb Concentration in the α-Matrix on the Corrosion Behavior of Zr-xNb Binary Alloys[J]. J Nucl Mater,2003,323: 72.

[14] 姚美意,周邦新,李强,等. 微量 Nb 的添加对 Zr24 合金耐疖状腐蚀性能的影响[J]. 上海金属,2008.30(6): 1.

[15] 姚美意,李士炉,张欣,等. 添加 Nb 对 Zr-4 合金在 500 ℃过热蒸汽中的耐腐蚀性能的影响[J]. 金属学报, 2011, 47(7): 865.

Nodular Corrosion Investigation of N18 Zirconium Alloys

Abstract: In the present work, the nodular corrosion properties in the steam (500 ℃/10.3 MPa) of N18 zirconium alloys after different heat treatment were investigated. The nodular corrosion of N18 alloys was liable to occur after $\alpha + \beta$ binary phases at 780 ℃ and α uniary phase treatment at 700 ℃. Microstructure observation found that the dilution of alloying element (especially Nb) at high temperature contributed the deterioration of nodular resistance. Furthermore, the formed alloying elements depleted zone could not be eliminated effectively by large deformation or low-temperature annealing treatment. A sufficient solution treatment at β phases has been proved as an effective way to eliminate the alloying elements depleted zone.

锆合金在 LiOH 水溶液中腐蚀的各向异性研究*

摘　要：选用了具有相同织构的 Zr-4，N18 和 ZIRLO 锆合金片状样品，利用高压釜在 360 ℃，18.6 MPa 的 0.01 mol/L LiOH 水溶液中进行了 280 d 腐蚀实验，采用 EBSD 和 SEM 研究了织构及合金成分对锆合金耐腐蚀性能的影响。结果表明，腐蚀 280 d 后，Zr-4 样品表现出明显的腐蚀各向异性特征，在织构因子较大的轧面（S_N 面）上氧化膜较厚，耐腐蚀性能差，而在织构因子较小的垂直于轧向的截面（S_R 面）和垂直于横向的截面（S_T 面）上氧化膜较薄，耐腐蚀性能好。添加合金元素 Nb 的 N18 和 ZIRLO 样品氧化膜生长的各向异性受到抑制，在 S_N，S_R 和 S_T 3 个不同面上氧化膜的厚度相同，耐腐蚀性能比 Zr-4 样品的 S_N 面优良。但是，如果只以样品的 S_R 和 S_T 面进行比较，氧化膜的生长速率会随 Nb 含量的增加而增大，耐腐蚀性能变差。从改善合金的耐腐蚀性能考虑，Nb 的添加量不应该高于 0.3%（质量分数）。

锆合金的热中子吸收截面小，并具有良好的高温力学性能，因而被广泛用于制备压水堆核电站中核燃料元件的包壳材料。燃料元件在反应堆中运行时，锆合金包壳与高温高压水接触而发生腐蚀，同时还会吸氢，这是影响包壳使用寿命的主要因素。另外，在一回路水中添加 LiOH 调节 pH 值后还会加剧包壳的腐蚀。为了提高核电的经济性，需要加深核燃料的燃耗和延长燃料组件的换料周期，这对锆合金的耐腐蚀性能提出了更高的要求，因而性能更加优良的锆合金仍在不断研究开发中。

目前，人们对锆合金的腐蚀问题已经进行了广泛研究，并探讨了腐蚀机理[1-4]。由于 Zr 的晶体结构是密排六方（$c/a=1.59$），其氧化各向异性比较显著。Kim 等[5] 的研究表明，纯 Zr 单晶在 360 ℃纯 H_2O 中腐蚀时，$(11\bar{2}0)$ 晶面上氧化膜的生长速率明显比（0001）晶面上的快。Zhou 等[6] 利用 Zr-4 合金大晶粒样品（晶粒尺寸在 0.2—0.8 mm 之间）逐个研究了晶粒上的氧化膜厚度与晶粒表面取向的关系，结果表明，Zr-4 合金在 500 ℃过热蒸汽和 360 ℃，0.01 mol/L 的 LiOH 水溶液中腐蚀时，氧化膜生长的各向异性非常显著，但是在这 2 种不同腐蚀条件下，氧化膜的厚度与锆合金晶粒表面的取向关系却截然不同，这是一个值得深入研究的问题。

锆合金在加工成材的过程中会产生织构，这将影响锆合金的耐腐蚀性能。Charquet 等[7] 和 Wang 等[8] 的研究都表明，Zr-4 合金在 500 ℃过热蒸汽中腐蚀时的增重会随着样品的织构因子 f_N 增大而减小。虽然对 Zr-4 合金腐蚀的各向异性已经有了一定的了解，但是改变合金成分后会有什么影响尚不清楚。本研究选用了 3 种成分不同的锆合金，利用小晶粒有织构的片状样品，从样品 3 个不同面上研究了织构及合金成分对耐腐蚀性能的影响，试图进一步了解锆合金腐蚀时的各向异性特征及其与合金成分的关系。

1　实验方法

研究用的样品是 Zr-4，N18 和 ZIRLO 合金板材，厚度为 2 mm，其化学成分如表 1 所

* 本文合作者：孙国成、姚美意、谢世敬、李强。原发表于《金属学报》，2012，48（9）：1103-1108。

示. Zr-4 合金的 Sn 含量和 ZIRLO 合金的 Nb 含量比其标准值略低, 但是这并不会影响研究织构及成分与耐腐蚀性能的关系. 用电火花线切割机将板材加工成尺寸为 24 mm × 15 mm 的样品, 用水磨砂纸磨掉切割面上的损伤层. 样品经过 30% H_2O + 30% HNO_3 + 30% H_2SO_4 + 10% HF (体积分数, 下同) 的混合酸酸洗后在 580 ℃ 真空退火 10 h. 在腐蚀前, 样品再用上述混合酸酸洗, 并在去离子沸水中煮 3 次, 每次更换去离子水后煮 8 min. 最后用静态高压釜在 360 ℃, 18.6 MPa 的 0.01 mol/L LiOH 水溶液中进行腐蚀实验, 每隔一定的时间停釜取样, 测量样品 3 个不同面上的氧化膜厚度, 这 3 个面分别是样品的轧面 (S_N 面), 垂直于轧向的截面 (S_R 面) 和垂直于横向的截面 (S_T 面). 用手工锯在腐蚀后的样品上截取一小片, 然后浸入上述酸液中溶去部分金属, 留下样品边沿处的氧化膜, 经过清洗干燥后折断氧化膜, 用 JSM-6700F 扫描电镜 (SEM) 观察氧化膜的断口形貌并测量氧化膜的厚度, 若不折断氧化膜, 可用 SEM 观察氧化膜的内表面形貌. 再截取一小片用于观察氧化膜的外表面形貌. 未经腐蚀的 3 种样品, 经过电解抛光后 (抛光液为 20% $HClO_4$ + 80% CH_3COOH, 抛光电压为 25 V), 用装有电子背散射衍射 (EBSD) 系统的 Apollo 300 SEM 分析样品的织构.

表 1 实验所用锆合金的化学成分

Table 1 Chemical compositions of zirconium alloys (mass fraction, %)

Specimen	Sn	Nb	Fe	Cr	Zr
Zr-4	1.17	—	0.23	0.11	Bal.
N18	0.93	0.25	0.32	0.08	Bal.
ZIRLO	0.88	0.85	0.12	0.01	Bal.

2 实验结果与讨论

2.1 样品的织构

图 1 是 ZIRLO 样品经过 580 ℃ 退火 10 h 后的 (0001) 极图和 S_N 面法向 (ND), S_R 面法向 (RD), S_T 面法向 (TD) 的反极图. Zr-4 和 N18 样品的织构与 ZIRLO 样品的非常相似, 都形成了 (0001) 极沿 S_N 面法向, 并向 S_T 面法向倾斜分布的织构. 表 2 列出了这 3 种样品 S_N, S_R 和 S_T 面法向上的织构因子 f_N, f_R 和 f_T, 该因子表示 (0001) 极在这 3 个不同方向上的分布比例[9]. 这 3 种样品的 f_N 都比较大, 说明 (0001) 极沿 S_N 面法向分布的比例较大 (图 1a 和 b). 这 3 种样品的 f_R 和 f_T 都比较小, 说明 (0001) 极沿 S_R 和 S_T 面法向分布的比例较小, 而 (11$\bar{2}$0) 和 (01$\bar{1}$0) 极沿 S_R 和 S_T 面法向分布的比例较大 (图 1c 和 d).

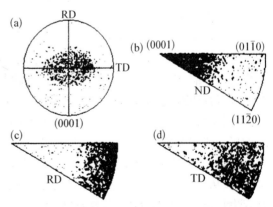

图 1 ZIRLO 样品经 580 ℃ 退火 10 h 后的 (0001) 极图和样品轧面法向 (ND), 样品轧向 (RD), 样品横向 (TD) 的反极图

Fig. 1 (0001) pole figure (a), inverse pole figure of the normal direction (ND) to rolling surface (b), inverse pole figure of rolling direction (RD) (c) and inverse pole figure of transversal direction (TD) (d) of ZIRLO specimen annealed at 580 ℃ for 10 h

表 2 Zr-4，N18 和 ZIRLO 样品 ND，RD 和 TD 的织构因子
Table 2 Texture factors of ND, RD and TD for Zr-4, N18 and ZIRLO specimens

Sepecimen	f_N	f_R	f_T
Zr-4	0.763	0.116	0.121
N18	0.718	0.076	0.206
ZIRLO	0.800	0.049	0.151

Note: texture factors f_N, f_R, f_T describe the effective number of crystallites with the basal pole aligned along ND, RD and TD, respectively

2.2 样品 S_N，S_R 和 S_T 面上氧化膜的生长规律

图 2 是 3 种样品的 S_N，S_R 和 S_T 面上氧化膜厚度随腐蚀时间的变化曲线. 可见,Zr-4 样品氧化膜生长的各向异性特征非常明显,在腐蚀 90—100 d 时,S_N 面上氧化膜的生长速率发生明显变化,氧化膜厚度随腐蚀时间的变化曲线发生转折,转折之后氧化膜生长急剧加快,表现出较差的耐腐蚀性能. 在 S_R 和 S_T 面上氧化膜的生长速率直到腐蚀 280 d 时都没有发生明显的变化,表现出较好的耐腐蚀性能,并且这 2 个面上氧化膜厚度随腐蚀时间的变化非常一致,在腐蚀 160 d 时氧化膜的厚度只有 3.8 μm,而这时在 S_N 面上氧化膜的厚度已经达到 19.1 μm. 腐蚀 280 d 时,在 $S_R(S_T)$ 和 S_N 面上氧化膜厚度的差别进一步增大,分别是 6.7 μm 和 104 μm,相差 15 倍. 这与用 Zr-4 合金大晶粒样品研究每个晶粒表面上氧化膜的厚度与其晶体取向关系的结果完全一致,大晶粒样品在 LiOH 水溶液中腐蚀 340 d 时,晶粒表面的取向在(0001)面附近时,氧化膜的厚度要比晶粒表面取向在(10$\bar{1}$0)至(11$\bar{2}$0)附近的厚 6—8 倍[6]. 一般薄片或管状的 Zr-4 样品在 LiOH 水溶液中腐蚀时,腐蚀增重随时间的变化曲线都存在转折现象[10,11],在转折以后,腐蚀增重迅速增加,这种变化规律与 S_N 面上氧化膜厚度随腐蚀时间的变化规律完全相同,但是与 $S_R(S_T)$ 面上氧化膜厚度随腐蚀时间的变化规律不同,这是因为薄片或管状样品腐蚀时,生成氧化膜的绝大部分表面都是 S_N 面,而不是 S_R 或 S_T 面的缘故.

添加合金元素 Nb 后的 Zr-Sn-Nb 系合金,在 LiOH 水溶液中腐蚀时表现出与 Zr-Sn 系的 Zr-4 合金完全不同的腐蚀特征. 无论是 N18 还是 ZIRLO 样品,氧化膜生长的各向异性完全被抑制,在 S_N 面上氧化膜的生长规律与 S_R 和 S_T 面上的完全一样,氧化膜生长速率在腐蚀过程中都没有明显的变化. 在腐蚀 280 d 后,3 个面上氧化膜的厚度并没有明显的差别. 由图 2 可以计算得出,Zr-4,N18 和 ZIRLO 样品在腐蚀 130—280 d 时 $S_R(S_T)$ 面上氧化膜的生长速率分别为 0.021 $\mu m/d$、0.024 $\mu m/d$ 和 0.036 $\mu m/d$,随 Nb 含量的增加反而增大. 研究[12-14]表明,含 Nb 的 Zr-Sn-Nb 合金在 LiOH 水溶液中的耐腐蚀性能比 Zr-4 合金优良. 从本工作的实验结果可以看出,这主要是由于添加合金元素 Nb 抑制了 S_N 面上氧化膜生长速率的增加. 如果只比较样品 S_R 和 S_T 面上氧化膜的生长速率,那么添加 Nb 后反而会使氧化膜的生长速率增加,氧化膜增厚. 在目前已经得到工业应用的 3 大系列锆合金(Zr-Sn 系,Zr-Sn-Nb 系,Zr-Nb 系)中,如以堆外高压釜腐蚀实验的结果来评价,Zr-Sn-Nb 系锆合金在 360 ℃ 的 LiOH 水溶液中的耐腐蚀性能均优于 Zr-Sn 系的 Zr-4 合金;但是,在 400 ℃ 过热蒸汽中的腐蚀实验结果又会因为 Nb 含量的不同而有差异,添加 1% Nb(质量分数,下同)的 ZIRLO 和 E635 合金(均为 Zr-Sn-Nb 合金)在 400 ℃ 过热蒸汽中的耐腐蚀性能都比 Zr-4 合金差,只有添加 0.3% Nb 的 N18 合金才与 Zr-4 合金相当[15]. 结合

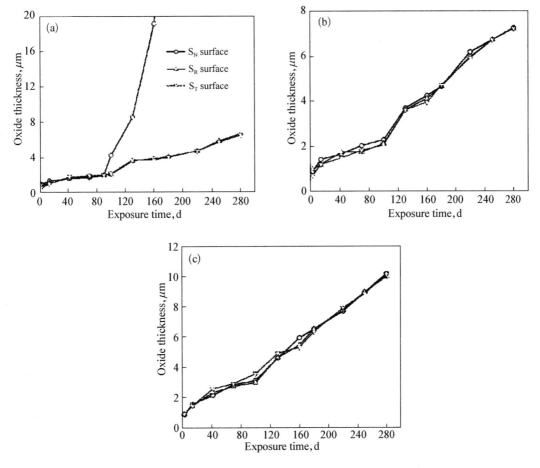

图 2 Zr-4，N18 和 ZIRLO 样品在 360 ℃，18.6 MPa，0.01 mol/L 的 LiOH 水溶液中腐蚀时，其轧面（S_N 面），垂直于轧向的截面（S_R 面）和垂直于横向的截面（S_T 面）上氧化膜厚度随时间的变化

Fig. 2 Oxide thickness on the rolling surface (S_N surface), the surface perpendicular to rolling direction (S_R surface) and the surface perpendicular to transversal direction (S_T surface) as a function of exposure time for specimens corroded in lithiated water with 0.01 mol/L LiOH at 360 ℃，18.6 MPa for specimens Zr-4 (a)，N18 (b) and ZIRLO(c)

本工作的研究结果可以看出，为改善 Zr-Sn-Nb 系锆合金在 LiOH 水溶液中的耐腐蚀性能，Nb 的添加量不应该高于 0.3%，这样既能抑制 S_N 面上氧化膜生长的各向异性特征，又能使氧化膜的生长速率与 Zr-4 合金样品 S_R 和 S_T 面上的相当. 可见，目前某些已经商业应用的 Zr-Sn-Nb 系合金中添加 1% Nb 并不是最佳的选择.

2.3 氧化膜的形貌特征

图 3 是 Zr-4 样品腐蚀 180 d 时 S_N 和 S_R 面上氧化膜的断口形貌以及内外表面的形貌. 可见，S_N 面上的氧化膜厚度已经达到 40.6 μm，从断口中可以观察到很多微裂纹，内表面凹凸不平，周邦新等[16]曾经讨论过这种凹凸不平的内表面与氧化膜中出现微裂纹的必然联系. 从外表面看，晶粒也比 S_R 面上的粗大；而 S_R 面上的氧化膜较薄，只有 4.1 μm，内表面比较平整，从外表面看，晶粒比较细小. 由于金属 Zr 氧化生成 ZrO_2 体积发生膨胀，P.B. 比

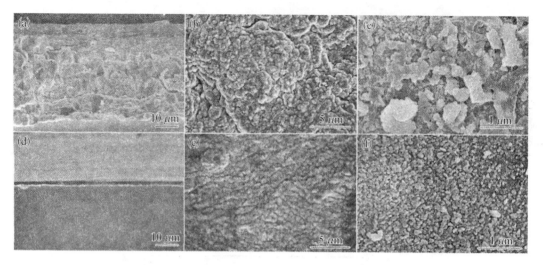

图 3 Zr-4 样品在腐蚀 180 d 时 S_N 和 S_R 面上氧化膜断口形貌和内外表面形貌

Fig. 3 Fracture surface (a, d), inner surface (b, e) and surface (c, f) morphologies of oxide films formed on S_N(a—c) and S_R(d—f) surfaces of Zr-4 specimen after 180 d exposure

(Pilling-Bedworth ratio)是 1.56,并且氧化过程发生在氧化膜/金属界面处,生成 ZrO_2 后会受金属基体的约束,因此氧化膜中存在较大的压应力,在这种条件下生成的 ZrO_2 晶体中会产生许多缺陷. 周邦新等[17]的研究表明,氧化膜中的晶体缺陷在应力、温度的作用下,在腐蚀过程中发生扩散,并在 ZrO_2 晶界上凝聚形成孔隙,并逐步发展为微裂纹. 杜晨曦等[18]的研究表明,在 Zr-4 单晶(大晶粒)表面上氧化初期生成的纳米晶 ZrO_2 都具有明显的"纤维"织构,但是这种织构的集中程度又与金属晶面的取向有关,取向接近(0001)晶面上生成氧化膜的织构比取向接近(01$\bar{1}$0)和(11$\bar{2}$0)晶面上的更分散一些,氧化膜中也包含了更多的小角度晶界及晶体缺陷. 因此,S_N 面上初始生成氧化膜中的缺陷会比 S_R 和 S_T 面多,在腐蚀过程中也容易形成孔隙和微裂纹. 另外,Zhou 等[3]的研究表明,在 LiOH 水溶液中腐蚀时,Li^+ 渗入氧化膜后吸附在孔隙和微裂纹的壁上,会降低 ZrO_2 的表面自由能,促使孔隙和微裂纹的发展. 耿建桥等[19]也观察到 Zr-4 样品在 LiOH 水溶液中腐蚀时,氧化膜中的压应力随氧化膜厚度增加会急剧减小,这说明样品在这种情况下腐蚀时会加速氧化膜中空位等缺陷的扩散凝聚,因而也促使了氧化膜内应力的弛豫. 锆合金的氧化过程由阴离子在氧化膜中的扩散进程所控制[20],S_N 面上生成的氧化膜中有较多的孔隙和微裂纹,提供了 O^{2-} 的扩散通道,从而加剧了 S_N 面上氧化膜的生长,使 S_N 和 S_R(S_T)面上氧化膜厚度在腐蚀后期出现了较大差别.

图 4 是 N18 和 ZIRLO 样品腐蚀 180 d 时 S_R 面上氧化膜的断口形貌. 可见,N18 样品 S_R 面上的氧化膜厚 4.2 μm,断口上观察到较少的微裂纹,而 ZIRLO 样品 S_R 面上氧化膜厚 6.5 μm,断口上可观察到较多的微裂纹,氧化膜也显得比较疏松. 刘文庆等[21]的研究表明,Nb 的氧化物在 LiOH 水溶液中有一定的溶解度. ZIRLO 合金含 Nb 量比较高,会形成一些纳米大小的 β-Nb 第二相,而 N18 合金含 Nb 量比较低,一部分 Nb 会固溶在 α-Zr 中,其余的 Nb 会进入 $Zr(Fe, Cr)_2$ 第二相中,形成 $Zr(Fe, Cr, Nb)_2$ 第二相,不会单独形成 β-Nb 第二相[22]. 当纳米大小的 β-Nb 第二相进入氧化膜最终被氧化后,部分 Nb 的氧化物会溶解到腐蚀介质中,在氧化膜中留下孔隙,增加了阴离子的扩散通道,加剧了锆合金的腐蚀,从而会降低高 Nb 锆合金的耐腐蚀性能.

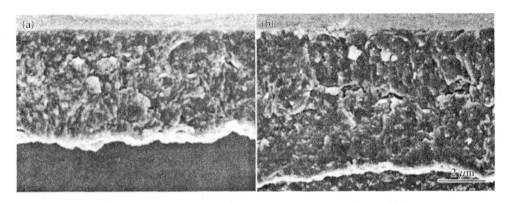

图 4 N18 和 ZIRLO 样品在腐蚀 180 d 时 S_R 面上氧化膜断口形貌

Fig. 4 Fracture surface morphologies of oxide films formed on S_R surface after 180 d exposure for specimens N18 （a) and ZIRLO (b)

在 Zr - Sn 合金中添加 Nb 后能够抑制氧化膜的各向异性生长,改善样品 S_N 面的耐腐蚀性能,因而在 LiOH 水溶液中腐蚀时,N18 和 ZIRLO 合金的耐腐蚀性能明显优于 Zr - 4 合金. 但是,如果只比较 S_R 和 S_T 面的耐腐蚀性能,添加合金元素 Nb 后反而会使耐腐蚀性能变坏. 从氧化膜显微组织在腐蚀过程中的演化会影响耐腐蚀性能的角度考虑,Zhou 等[3] 曾讨论过含 Nb 的 ZIRLO 合金可以抑制 Li^+ 渗入氧化膜后降低 ZrO_2 表面自由能的作用,因而在 LiOH 水溶液中腐蚀时的耐腐蚀性能优于 Zr - 4 合金. 但是要彻底了解为什么添加合金元素 Nb 后会对 S_N 面和 S_R（S_T）面的耐腐性能产生完全不同的影响,还需要进行深入的研究.

3　结论

（1）Zr - 4 样品在 LiOH 水溶液中腐蚀时,表现出明显的各向异性特征. 在织构因子最大的 S_N 面上,随着腐蚀时间的延长,氧化膜的生长速率发生明显变化,转折后生长速率急剧增加,耐腐蚀性能很差. 但是在织构因子较小的 S_R 和 S_T 面上,氧化膜的生长速率没有明显变化,具有良好的耐腐蚀性能.

（2）在 N18 和 ZIRLO 合金中添加合金元素 Nb 后,完全抑制了氧化膜生长的各向异性,织构因子不同的 S_N, S_R 和 S_T 面上,氧化膜的生长速率几乎完全相同. 因而 N18 和 ZIRLO 合金在 LiOH 水溶液中腐蚀时比 Zr - 4 合金的耐腐蚀性能优良. 但是,如果只对样品的 S_R 和 S_T 面进行比较,添加 Nb 后的 N18 和 ZIRLO 合金比 Zr - 4 合金的耐腐蚀性能差,氧化膜的生长速率会随 Nb 含量的增加而增大.

（3）从改善锆合金的耐腐蚀性能考虑,在 Zr - Sn - Nb 系的合金中,Nb 的添加量不应该高于 0.3%（质量分数）.

参 考 文 献

[1] Garzarolli F, Seidel H, Tricot R, Gros J P. In: Eucken C M,Garde A M eds. , *Zirconium in the Nuclear Industry: 9th International Symposium*, *ASTM STP 1132*, Baltimore: ASTM International, 1991: 395.

[2] Yilmazbayhan A, Breval E, Motta A T, Comstock R J. *J Nucl Mater*, 2006;349: 265.

［3］ Zhou B X, Li Q, Yao M Y, Liu W Q, Chu Y L. In: Kammenzind B, Limbäck M eds. , *Zirconium in the Nuclear Industry: 15th International Symposium*, *ASTM STP 1505*, Baltimore: ASTM International, 2009: 360.

［4］ Park J Y, Choi B K, Yoo S J, Jeong Y H. In: Kammenzind B, Limbäck M eds. , *Zirconium in the Nuclear Industry: 15th International Symposium*, *ASTM STP 1505*, Baltimore: ASTM International, 2009: 471.

［5］ Kim H G, Kim T H, Jeong Y H. *J Nucl Mater*, 2002;306: 44.

［6］ Zhou B X, Peng J C, Yao M Y, Li Q, Xia S, Du C X, Xu G. In: Limbäck M, Barbéris P eds. , *Zirconium in the Nuclear Industry: 16th International Symposium*, *ASTM STP 1529*, Bridgeport: ASTM International, 2011: 620.

［7］ Charquet D, Tricot R, Wadier J E. In: Van Swam L F P, Eucken C M eds. , *Zirconium in the Nuclear Industry: 8th International Symposium*, *ASTM STP 1023*, Baltimore: ASTM International, 1989: 374.

［8］ Wang C T, Eucken C M, Graham R A. In: Eucken C M, Garde A M eds. , *Zirconium in the Nuclear Industry: 9th International Symposium*, *ASTM STP 1132*, Baltimore: ASTM International, 1991: 319.

［9］ Kearns J J. *Report No. WAPD-TM-472*, *Report of Westinghouse Electric Corporation*, *Bettis Atomic Power Laboratory*, Pittsburgh, PA, 1965: 2.

［10］ Han J H, Rheem K S. *J Nucl Mater*, 1994;217: 197.

［11］ Pecheur D, Godlewski J, Billot P, Thomazet J. In: Sabol G P, Bradley E R eds. , *Zirconium in the Nuclear Industry: 11th International Symposium*, *ASTM STP 1295*, Ann Arbor: ASTM International, 1996: 94.

［12］ Sabol G P, Kilp G R, Balfour M G, Roberts E. In: Van Swam L F P, Eucken C M eds. , *Zirconium in the Nuclear Industry: 8th International Symposium*, *ASTM STP 1023*, Baltimore: ASTM International, 1989: 227.

［13］ Nikulina A V, Markelov V A, Peregud M M, Bibilashvili Y K, Kotrekhov V A, Lositsky A F, Kuzmenko N V, Shevnin Y P, Shamardin V K, Kobylyansky G P, Novoselov A E. In: Sabol G P, Bradley E R eds. , *Zirconium in the Nuclear Industry: 11th International Symposium*, *ASTM STP 1295*, Ann Arbor: ASTM International, 1996: 785.

［14］ Zhou B X, Li Q, Huang Q, Miao Z, Zhao W J, Li C. *Nuci Power Eng*, 2000;21: 439.
（周邦新,李强,黄强,苗志,赵文金,李聪. 核动力工程,2000;21: 439）

［15］ Zhou B X, Yao M Y, Li Z K, Wang X M, Zhou J, Long C S, Liu Q, Luan B F. *J Mater Sci Technol*, 2012; 28: 606.

［16］ Zhou B X, Li Q, Yao M Y, Liu W Q, Chu Y L. *Nucl Power Eng*, 2005;26: 364.
（周邦新,李强,姚美意,刘文庆,褚于良. 核动力工程,2005;26: 364）

［17］ Zhou B X, Li Q, Yao M Y, Liu W Q, Chu Y L. *Corros Prot*, 2009;30: 589.
（周邦新,李强,姚美意,刘文庆,褚于良. 腐蚀与防护,2009;30: 589）

［18］ Du C X, Peng J C, Li H, Zhou B X. *Acta Metall Sin*, 2011;47: 887.
（杜晨曦,彭剑超,李慧,周邦新. 金属学报,2011;47: 887）

［19］ Geng J Q, Zhou B X, Yao M Y, Wang J H, Zhang X, Li S L, Du C X. *J Shanghai Univ*（*Nat Sci Ed*）, 2011;17: 293.
（耿建桥,周邦新,姚美意,王锦红,张欣,李士炉,杜晨曦. 上海大学学报(自然科学版),2011;17: 293）

［20］ Cox B. *J Nucl Mater*, 2005;336: 331.

［21］ Liu W Q, Zhou B X, Li Q. *Nucl Power Eng*, 2002;23: 68.
（刘文庆,周邦新,李强. 核动力工程,2002;23: 68）

［22］ Zhou B X, Yao M Y, Li Q, Xia S, Liu W Q, Chu Y L. *Rare Met Mater Eng*, 2007;36: 1317.
（周邦新,姚美意,李强,夏爽,刘文庆,褚于良. 稀有金属材料与工程,2007;36: 1317）

Study of Anisotropic Behavior for Zirconium Alloys Corroded in Lithiated Water

Abstract: Zirconium alloys of a hexagonal close-packed crystal structure have prominent anisotropic characteristic in comparison with metals of a cubic crystal structure and a strong texture is produced in sheet or tubular materials during the fabrication process. The anisotropic characteristic is bound to be reflected on the corrosion behavior of zirconium alloys. In order to investigate the effect of texture and compositions on the anisotropic growth of oxide layer formed on zirconium alloys and clarify the mechanism of improving corrosion resistance by adding Nb in zirconium alloys, Zr－4, N18 and ZIRLO zirconium alloys with different contents of Nb were adopted as the experimental materials. All the plate specimens of zirconium alloys 2 mm in thickness have a similar texture. Corrosion tests were carried out in a static autoclave at 360 ℃,18. 6 MPa in lithiated water with 0. 01 mol/L LiOH. The results show that the anisotropic growth of oxide layer on different surfaces of the specimens was only observed for Zr－4 specimen but not for N18 and ZIRLO specimens. The thickness of oxide layer develops much faster on the rolling surface (S_N surface) than that on the surface perpendicular to the rolling direction (S_R surface) and the surface perpendicular to the transversal direction (S_T surface) for Zr－4 specimen after 90—100 d exposure, and the corrosion resistance on the S_R and S_T surfaces was much better than that on the S_N surface. However, for N18 and ZIRLO specimens the anisotropic growth of oxide layer was restrained by the addition of Nb, and the oxide thickness on these three different surfaces was the same after 280 d exposure. Therefore the corrosion resistance of N18 and ZIRLO sheet or tubular specimens was superior to Zr－4 corroded in lithiated water, because the oxide layers grew mainly on the S_N surface of the specimens. If making a comparison among Zr－4, N18 and ZIRLO specimens about the growth rate of oxide layers only on the S_R and S_T surfaces, it is shown that the growth rate of oxide layers increased with the increase of Nb content in these alloys. From a point of view for the improving corrosion resistance, the addition of Nb no more than 0. 3% (mass fraction) is recommended.

Zr - 0.7Sn - 0.35Nb - 0.3Fe - xGe 合金在高温高压 LiOH 水溶液中耐腐蚀性能的研究*

摘　要: 利用静态高压釜腐蚀实验研究了 Zr - 0.7Sn - 0.35Nb - 0.3Fe - xGe(x=0.05, 0.1, 0.2,质量分数,%)系列合金在 360 ℃/18.6 MPa/0.01 mol/L LiOH 水溶液中的耐腐蚀性能;利用 TEM 和 SEM 分别观察了合金基体的显微组织和氧化膜的显微组织. 结果表明: Ge 可以显著改善 Zr - 0.7Sn - 0.35Nb - 0.3Fe 合金在高温高压 LiOH 水溶液中的耐腐蚀性能,当 Ge 含量为 0.1%时,合金的耐腐蚀性能最佳. 在 Zr - 0.7Sn - 0.35Nb - 0.3Fe - xGe 系列合金中发现尺寸较小的 hcp 结构的 Zr(Fe, Cr, Nb)$_2$ 型、Zr(Fe, Cr, Nb, Ge)$_2$ 型第二相和尺寸较大的四方结构的 Zr$_3$Ge 型第二相. 腐蚀 220 d 的 Zr - 0.7Sn - 0.35Nb - 0.3Fe - 0.1Ge 合金氧化膜致密,厚度较薄,几乎没有微孔隙和微裂纹,ZrO$_2$ 柱状晶较多. 这说明添加适量的 Ge 不仅可以有效延缓氧化膜中空位扩散凝聚形成微孔隙和微孔隙发展形成微裂纹的过程,还可以延迟 ZrO$_2$ 由柱状晶向等轴晶的演化,从而改善合金的耐腐蚀性能.

　　Zr 具有优异的核性能,它的热中子吸收截面只有 0.18 barn,并与 UO$_2$ 的相容性好,尤其具有良好的力学性能及耐高温水腐蚀性能,因此在水冷核反应堆中锆合金被广泛用作燃料棒的包壳材料和燃料组件中的结构材料[1]. 为了提高核电经济性,需要降低核燃料的循环成本、加深核燃料燃耗,这就需要延长换料周期,因而对锆合金的耐水侧腐蚀性能提出了更高的要求[2]. 由于 Zr - 4 合金已经不能满足高燃耗燃料组件和延长换料周期的要求,因此许多国家在 Zr - Sn 系锆合金基础上,通过降低 Sn 含量,并加入 Nb, Fe 和 Cr 等合金元素后,开发了 ZIRLO, HANA - 4, NDA 和 E635 等 Zr - Sn - Nb 系新型锆合金[1,3-5].

　　合金化是改善锆合金耐腐蚀性能的一种有效方法. 研究表明[1,6-9]: 固溶在 α - Zr 中的 Sn 可以抵消 N 的有害作用,改善 Zr 的耐腐蚀性能;Nb 对 Zr 有强化作用,并能减少 Zr 的吸 H 量;适量的 Fe 和 Cr 可以提高 Zr 的力学性能;Nb, Fe 和 Cr 等元素形成的第二相粒子会不同程度地改善锆合金的耐腐蚀性能. 可见,选择合适的合金添加元素是优化锆合金成分和提高锆合金耐腐蚀性能的研究重点. 根据 Wagner 氧化膜成长理论和 Hauffe 原子价规律[1,6],位于元素周期表第 IVA 族的 Ge 元素可以增加锆合金氧化膜中的电子浓度,减少阴离子空位,从而抑制氧离子扩散,降低锆合金的腐蚀速率. 尤其是 Ge 具有良好的半导体性质,可以降低氧化膜中的空位迁移[10]. 本课题组曾研究了 Ge 对 Zr - 4 合金耐腐蚀性能的影响,结果表明,Ge 可以显著改善 Zr - 4 合金在 360 ℃/18.6 MPa/0.01 mol/L LiOH 水溶液中的耐腐蚀性能. 但是目前有关 Ge 影响锆合金耐腐蚀性能的研究并不系统,其影响机理也不清楚. 因此,本文主要研究添加不同含量 Ge 对 Zr - 0.7Sn - 0.35Nb - 0.3Fe 合金在 360 ℃/18.6 MPa/0.01 mol/L LiOH 水溶液中耐腐蚀性能的影响,并探讨其影响机理.

---

* 本文合作者:谢兴飞、张金龙、朱莉、姚美意、彭剑超. 原发表于《金属学报》,2012,48(12): 1487 - 1494.

1 实验材料及方法

以 Zr-0.7Sn-0.35Nb-0.3Fe(质量分数,%,下同)合金为母合金,添加不同含量的 Ge 制成 Zr-0.7Sn-0.35Nb-0.3Fe-xGe(x=0.05, 0.1, 0.2)系列锆合金.合金试样制备过程如图 1 所示.先利用非自耗真空电弧炉熔炼成约 65 g 的合金锭,熔炼时通入高纯 Ar,合金锭总共翻转熔炼 6 次.合金锭经 700 ℃ 热压成条块状后在 1 030 ℃ 进行均匀化热处理 40 min,再经过热轧、1 030 ℃/40 min 空冷、多次冷轧和 580 ℃ 中间退火制成 20 mm×15 mm× 0.7 mm 的腐蚀实验用片状样品.最终进行 580 ℃/5 h 再结晶退火.未添加 Ge 的 Zr0.7Sn-0.35Nb-0.3Fe 合金经过相同工艺处理后制成对比样品.将制备好的上述样品经过 30% H_2O+30% H_2SO_4+30% HNO_3+10% HF(体积分数)混合酸酸洗和去离子水清洗后放入静态高压釜中,在 360 ℃/18.6 MPa/0.01 mol/L LiOH 水溶液中进行腐蚀实验.入釜之前的合金表面平整(表面粗糙度 R_a 值约为 0.265 μm).腐蚀样品定期取出称重,绘制腐蚀增重曲线,腐蚀增重是 3—5 片平行样品的平均值.

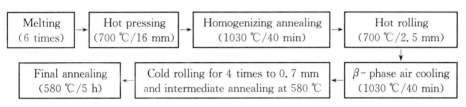

图 1 合金试样制备流程图

Fig. 1 Preparation procedures of samples

用 JEM-200CX 型透射电子显微镜(TEM)观察合金的显微组织:用配置了能谱仪(EDS)的 JEM-2010F 型高分辨透射电子显微镜(HRTEM)观察分析合金中第二相的形貌与成分,并通过选区电子衍射(SAED)确定第二相的晶体结构.为了减少统计误差,每种第二相都选取 10 个以上粒子进行分析.用双喷电解抛光法制备 TEM 观察用薄试样,电解液选用 10% $HClO_4$+90% C_2H_5OH(体积分数)混合溶液.

用 JSM-6700F 型扫描电子显微镜(SEM)观察氧化膜断口、外表面及内表面微观形貌.按照文献[11]中描述的方法制备 SEM 观察用样品.其中,氧化膜内表面形貌观察样品的具体制备过程为:先用低速金刚石切割机切下 4 mm×4 mm 的样品.用金相砂纸打磨去掉一面氧化膜,在透明胶带上穿出直径约为 2 mm 的圆孔并粘在金属一侧,用 30% H_2O+30% H_2SO_4+30% HNO_3+10% HF(体积分数)混合酸溶掉圆孔中的金属基体,当露出氧化膜后迅速用水充分清洗,然后吹干.为了避免电荷积累影响成像质量,样品表面蒸镀了一层金属 Ir.

用 HELIOS-600I 型聚焦离子束(focus ion beam, FIB)制备 TEM 观察用的氧化膜横截面薄试样,文献[12]中介绍了 FIB 制备样品的具体过程.

2 实验结果与分析

2.1 腐蚀增重

图 2 是 Zr-0.7Sn-0.35Nb-0.3Fe-xGe 系列合金在 360 ℃/18.6 MPa/0.01 mol/L

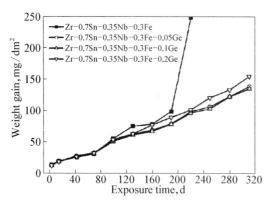

图2 Zr-0.7Sn-0.35Nb-0.3Fe-xGe 系列合金在 360 ℃/18.6 MPa/0.01 mol/L LiOH 水溶液中腐蚀 310 d 的增重曲线

Fig. 2 Weight gain curves of Zr-0.7Sn-0.35Nb-0.3Fe-xGe alloys corroded in lithiated water with 0.01 mol/L LiOH at 360 ℃/18.6 MPa for 310 d

LiOH 水溶液中腐蚀 310 d 的增重曲线. 可以看出,Zr-0.7Sn-0.35Nb-0.3Fe 合金腐蚀 220 d 的平均增重高达 247.7 mg/dm²,而 x 为 0.05,0.1 和 0.2 的合金腐蚀 310 d 时的平均增重分别为 138.1,134.6,和 154.2 mg/dm². 可见,在 Zr-0.7Sn-0.35Nb-0.3Fe 合金中添加适量 (0.05%—0.2%) 的 Ge 后,在 360 ℃/18.6 MPa/0.01 mol/L LiOH 水溶液中耐腐蚀性能得到明显改善,其中 Ge 含量为 0.1% 时,合金的耐腐蚀性能最佳;Ge 含量提高到 0.2% 时,合金的腐蚀增重有所增加,耐腐蚀性能有所下降.

2.2 合金的显微组织

图 3 是 Zr-0.7Sn-0.35Nb-0.3Fe-xGe 系列合金显微组织的 TEM 像. 表 1 统计了不同合金中第二相的化学成分、尺寸和类型. 可以看出,3 种合金的第二相分布在 α-Zr 等轴晶的晶粒内部和晶界上. 根据 EDS 和 SAED 分析结果可以确定,在 3 种合金中,都存在呈球状的 Zr(Fe, Cr, Nb)₂ 和 Zr(Fe, Cr, Nb, Ge)₂ 第二相,Fe/Nb 原子比在 2.3—3.6 之间. 其中,Zr(Fe, Cr, Nb)₂型第二相尺寸约为 50 nm(图 3 中箭头 1,2,3 所示),这与文献[9,13]报道的

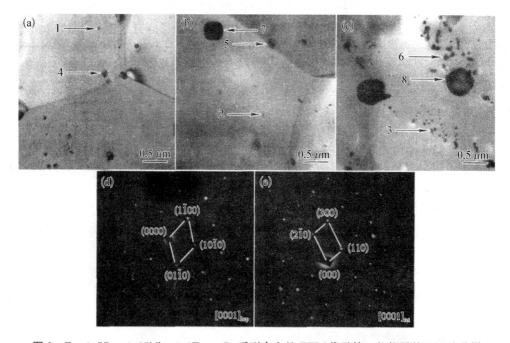

图3 Zr-0.7Sn-0.35Nb-0.3Fe-xGe 系列合金的 TEM 像及第二相粒子的 SAED 花样

Fig. 3 TEM images (a—c) of Zr-0.7Sn-0.35Nb-0.3Fe-xGe alloys and SAED patterns (d, e) of second phase particles (SPPs) (a) Zr-0.7Sn-0.35Nb-0.3Fe-0.05Ge;(b) Zr-0.7Sn-0.35Nb-0.3Fe-0.1Ge;(c) Zr-0.7Sn-0.35Nb-0.3Fe-0.2Ge;(d) SAED pattern of SPP 4 in Fig. 3a, indicated Zr(Fe, Cr, Nb, Ge)₂;(e) SAED pattern of SPP 7 in Fig. 3b, indicated Zr₃Ge

研究结果相一致. Zr(Fe，Cr，Nb，Ge)₂型第二相尺寸约为 100 nm(图 3 中箭头 4,5,6 所示),这两种第二相均为 hcp 结构(图 3d). Zr-0.7Sn-0.35Nb-0.3Fe-xGe 系列合金中并没有特意添加 Cr,而 EDS 分析结果显示第二相中存在少量 Cr,这是来自海绵 Zr 中的杂质元素 Cr,因为 Cr 在 α-Zr 中的最大固溶度仅为 200 μg/g[14],多余的 Cr 会以第二相析出.

表 1 图 3 中 Zr-0.7Sn-0.35Nb-0.3Fe-xGe 系列合金的第二相信息

Table 1 Details of SPPs in Zr-0.7Sn-0.35Nb-0.3Fe-xGe alloys in Fig. 3

| Arrow | Chemical composition of SPPs（atomic fraction，%） | | | | | Fe/Nb | Zr/Ge | Type and size |
	Zr	Fe	Nb	Ge	Cr			
1	79.88	13.83	4.62	—	1.67	2.99	—	Zr(Fe，Cr，Nb)₂
2	74.00	18.30	5.52	—	2.18	3.31	—	hcp, about 50 nm
3	85.85	10.20	3.10	—	0.85	3.29	—	
4	88.13	6.78	1.90	2.77	0.42	3.56	31.81	Zr(Fe，Cr，Nb，Ge)₂
5	74.79	14.78	5.49	2.28	2.66	2.69	32.80	hcp, about 100 nm
6	87.41	6.29	2.63	3.03	0.64	2.39	28.84	
7	76.84	0.35	—	22.34	0.47	—	3.43	Zr₃Ge, TET
8	79.29	0.26	—	19.95	0.50	—	3.97	400—500 nm

在 Zr-0.7Sn-0.35Nb-0.3Fe-0.1Ge 和 Zr-0.7Sn-0.35Nb-0.3Fe-0.2Ge 合金中,除了 Zr(Fe，Cr，Nb)₂ 和 Zr(Fe，Cr，Nb，Ge)₂ 两种第二相外,还存在尺寸为 300—500 nm 的较大第二相(图 3 中箭头 7 和 8 所示),此种第二相主要含有 Zr 和 Ge 元素,Zr/Ge 原子比略大于 3,这是由于电子束有一定的穿透深度和扩散范围,受到 Zr 基体影响的缘故.根据 Zr-Ge 二元相图与 SAED 结果(图 3e)确定这种第二相为四方结构的 Zr₃Ge 型第二相,其标准点阵常数 $a=1.108$ nm,$c=0.548$ nm.以上结果说明,580 ℃时 Ge 在 α-Zr 中的固溶度不超过 0.05%,过量的 Ge 会从基体中析出,先与 Zr，Nb，Fe 和 Cr 形成 Zr(Fe，Cr，Nb，Ge)₂ 型第二相,当 Ge 含量提高到 0.1%时,Ge 还会与 Zr 单独形成 Zr₃Ge 型第二相.对比图 3b 和 3c 可知,当 Ge 含量提高到 0.2%时,Zr₃Ge 型第二相的数量会增多.含 Ge 第二相尺寸较大,这可能是因为 Zr₃Ge 第二相粒子与基体之间的界面能较低,促进第二相粒子长大的缘故.

2.3　氧化膜的外表面形貌

图 4 为 Zr-0.7Sn-0.35Nb-0.3Fe 和 Zr-0.7Sn-0.35Nb-0.3Fe-0.1Ge 合金腐蚀 220 d 的氧化膜外表面形貌(SEM 像).由图可见,Zr-0.7Sn-0.35Nb-0.3Fe 合金表面的局部区域已经出现较长的微裂纹(图 4a 箭头所示),此时,氧化膜逐渐失去保护作用.O²⁻ 或 OH⁻ 可以通过微裂纹、微孔隙到达氧化膜/金属界面与金属基体加速氧化反应.然而,Zr-0.7Sn-0.35Nb-0.3Fe-0.1Ge 合金腐蚀 220 d 的氧化膜外表面晶粒间无明显微裂纹,合金具有良好的耐腐蚀性能(图 4b).

2.4　氧化膜的断口形貌

图 5 是 Zr-0.7Sn-0.35Nb-0.3Fe-xGe 系列合金在 360 ℃/18.6 MPa/0.01 mol/L LiOH 水溶液中腐蚀 220 d 的氧化膜断口形貌(SEM 像).图 5a 是 Zr-0.7Sn-0.35Nb-0.3Fe 合金的氧化膜断口形貌,氧化膜平均厚度约为 14.2 μm,氧化膜内部存在大量的平行于氧化膜/金属界面的裂纹,而且部分裂纹较宽.在靠近氧化膜内表面处几乎全为 ZrO₂ 等轴

图 4 不同合金在 360 ℃/18.6 MPa/0.01 mol/L LiOH 水溶液中腐蚀 220 d 氧化膜外表面的形貌

Fig. 4 Surface morphologies of the oxide films on Zr - 0.7Sn - 0.35Nb - 0.3Fe (a) and Zr - 0.7Sn - 0.35Nb - 0.3Fe - 0.1Ge (b) alloys corroded in lithiated water with 0.01 mol/L LiOH at 360 ℃/18.6 MPa for 220 d

晶,很难发现柱状晶,断口高低起伏明显,存在许多"台阶"状区域(图 5c 所示),部分 ZrO_2 晶粒间存在微裂纹.

图 5b 是 Zr - 0.7Sn - 0.35Nb - 0.3Fe - 0.1Ge 合金的氧化膜断口形貌,可以看到,氧化膜平均厚度约为 4.2 μm,与 Zr - 0.7Sn - 0.35Nb - 0.3Fe 合金相比(图 5a 和 5c),含 Ge 合金

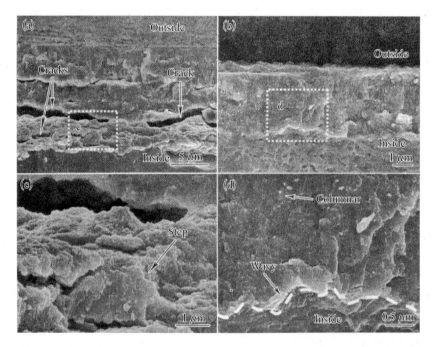

图 5 不同合金在 360 ℃/18.6 MPa/0.01 mol/L LiOH 水溶液中腐蚀 220 d 氧化膜断口的形貌

Fig. 5 Low(a, b) and high (c, d) magnified fracture surface morphologies of the oxide films on Zr - 0.7Sn - 0.35Nb - 0.3Fe (a, c) and Zr - 0.7Sn - 0.35Nb - 0.3Fe - 0.1Ge (b, d) alloys corroded in lithiated water with 0.01 mol/L LiOH at 360 ℃/18.6 MPa for 220 d

的氧化膜较薄，ZrO₂ 晶粒排列比较紧密，界面平整，在靠近氧化膜内表面处观察到较多 ZrO₂ 柱状晶(图 5d). 氧化膜的生长是由柱状晶向等轴晶演化的过程，柱状晶中的缺陷通过扩散凝聚形成新的晶界，从而柱状晶逐渐发展成等轴晶[15,16]. 由此可见，Ge 可以有效延迟氧化膜中柱状晶向等轴晶演化的过程. 此外，发现两种样品的氧化膜/金属界面处起伏不平，呈波浪型(图 5d 虚线所示)，这种现象形成的原因将会在后文作详细分析.

2.5 氧化膜的内表面形貌

图 6 是 Zr‐0.7Sn‐0.35Nb‐0.3Fe‐xGe 系列合金在 360 ℃/18.6 MPa/0.01 mol/L LiOH 水溶液中腐蚀 220 d 的氧化膜内表面形貌(SEM 像). 图 6a 和 6c 是 Zr‐0.7Sn‐0.35Nb‐0.3Fe合金的氧化膜内表面形貌，可以看到，内表面不仅凹凸不平，而且出现大量微裂纹，这是因为氧化膜是在氧化膜/金属界面处不断生成，锆合金氧化时体积膨胀，金属对氧化膜产生束缚作用，在氧化膜/金属界面处会存在很大的压应力. 随着腐蚀时间延长，当应力超过金属的束缚作用时，就会在应力集中区域产生微孔隙. 在应力持续作用下，这些微孔隙扩展连结可以形成微裂纹. 在局部区域放大的图 6c 中，可以清晰地观察到微裂纹延伸扩展到氧化膜内部. 由此判断，氧化膜已经比较疏松.

图 6b 和 6d 所示 Zr‐0.7Sn‐0.35Nb‐0.3Fe‐0.1Ge 合金的氧化膜内表面比较致密，没有发现微裂纹. 说明 Ge 的加入抑制了微裂纹的形成. 在图 6b 中还可以清晰地发现菜花状突起. 图 6a 中凹凸不平的形貌与图 6b 和 6d 中菜花状突起都验证了图 5 所示的氧化膜/金

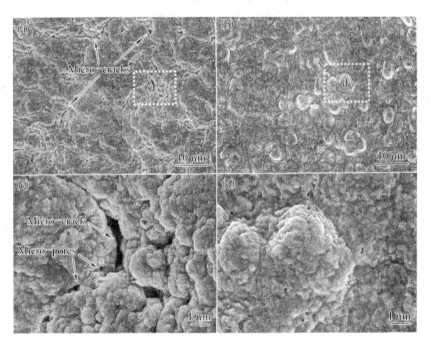

图 6 不同合金在 360 ℃/18.6 MPa/0.01 mol/L LiOH 水溶液中腐蚀 220 d 氧化膜内表面的形貌

Fig. 6 Low (a, b) and high (c, d) magnified inner surface morphologies of the oxide films on Zr‐0.7Sn‐0.35Nb‐0.3Fe (a, c) and Zr‐0.7Sn‐0.35Nb‐0.3Fe‐0.1Ge (b, d) alloys corroded in lithiated water with 0.01 mol/L LiOH at 360 ℃/18.6 MPa for 220 d

属界面起伏不平,这可能与 α-Zr 的晶体取向有关[17],也可能是因为第二相粒子在氧化膜/金属界面与 ZrO_2 形成相界面[18],降低了附近合金基体的氧化速率,第二相分布不均匀导致的界面氧化不均匀.

2.6 氧化膜横截面的显微组织

用 FIB 制备 Zr-0.7Sn-0.35Nb-0.3Fe-0.2Ge 合金腐蚀 70 d 的氧化膜横截面薄试样,在 TEM 的扫描透射(STEM)模式下观察到氧化膜内部存在许多尺寸较小的球状第二相(图 7a),根据 EDS 结果确定是发生部分氧化的 Zr(Fe, Cr, Nb, Ge)$_2$ 型第二相(图 7b).通过 TEM 观察和 EDS 分析可知,氧化膜内部存在发生部分氧化的 Zr$_3$Ge 型第二相(图 7c 和 7d),可见 Zr(Fe, Cr, Nb, Ge)$_2$ 型和 Zr$_3$Ge 型第二相的氧化速率都比 α-Zr 基体慢.同时,在 Zr$_3$Ge 型第二相靠近氧化膜外表面一侧,出现了平行于氧化膜/金属界面的微裂纹(图 7c). TEM 暗场像显示在氧化膜内部存在大量柱状晶(图 7e).特别说明,图 7b 和 7d 的 EDS 谱中出现的 Cu 峰是透射电镜样品台的结构材料引起的.

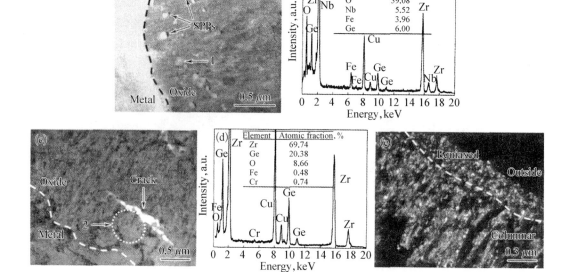

图 7 Zr-0.7Sn-0.35Nb-0.3Fe-0.2Ge 合金在 360 ℃/18.6 MPa/0.01 mol/L LiOH 水溶液中腐蚀 70 d 的氧化膜横截面的形貌

Fig. 7 Cross-sectional morphologies of oxide film on Zr-0.7Sn-0.35Nb-0.3Fe-0.2Ge alloy corroded in lithiated water with 0.01 mol/L LiOH at 360 ℃/18.6 MPa for 70 d　(a) STEM image　(b) EDS of SPP 1 in Fig. 7a　(c) TEM image　(d) EDS of SPP 2 in Fig. 7c　(e) dark field image

3 分析讨论

从上述实验结果可以看出,添加的 Ge 可以明显改善 Zr-0.7Sn-0.35Nb-0.3Fe-xGe 系列合金在 360 ℃/18.6 MPa/0.01 mol/L LiOH 水溶液中的耐腐蚀性能.添加的 Ge 一部分会固溶在 α-Zr 基体中,多余的 Ge 会以 Zr(Fe, Cr, Nb, Ge)$_2$ 和 Zr$_3$Ge 型第二相形式析

出,这两方面因素会共同影响氧化膜显微组织和 ZrO_2 晶体结构演化,从而影响锆合金的耐腐蚀性能.

周邦新等[11,19]从氧化膜显微组织结构演化的角度阐述了锆合金的腐蚀机理:因为 Zr 氧化生成 ZrO_2 时的 P. B. 比(金属氧化物与金属体积之比)为 1.56,氧化过程中生成的氧化膜体积膨胀,同时受到金属基体约束,氧化膜内部将会产生很大的压应力,在这种条件下生成的 ZrO_2 中会形成空位、间隙原子等各种缺陷;在温度和压力作用下,这些缺陷发生扩散、湮没和凝聚,空位被晶界吸收后形成孔隙簇,孔隙簇进一步发展成为微裂纹,降低了氧化膜的保护作用.

Ge 可以改善 Zr-0.7Sn-0.35Nb-0.3Fe 合金在 LiOH 水溶液中的耐腐蚀性能的可能原因如下:

首先,固溶在 α-Zr 中的 Ge 比第二相中的 Ge 更容易固溶进入 ZrO_2 中,增加 ZrO_2 中的电子浓度,减少阴离子空位,从而抑制 O^{2-} 和 OH^- 在 ZrO_2 中扩散,降低了锆合金的腐蚀速率[1,6].LiOH 水溶液中的 Li^+ 渗入氧化膜中会降低 ZrO_2 的表面自由能[20],固溶在 α-Zr 中的 Ge 可能会减少 ZrO_2 表面自由能降低的程度,延缓空位凝聚形成微孔隙、微孔隙形成微裂纹的过程,从而改善合金耐腐蚀性能.另外,根据公式[21]:P. B. 比 $=(M \cdot d_o)/(n \cdot d_{ox} \cdot A)$ (式中,M 为氧化物分子量,d_o 为金属密度,n 为氧化物分子中金属原子数目,d_{ox} 为氧化物密度,A 为金属原子量),计算得到 Ge 的 P. B. 比为 1.23.而 Zr,Nb,Fe 和 Cr 的 P. B. 比[21]分别为 1.56,2.74,1.77 和 2.02.固溶在 ZrO_2 中的 Ge 可能会减少氧化膜的膨胀程度,降低压应力,延缓形成微裂纹的过程.但是,在 Zr-0.7Sn-0.35Nb-0.3Fe 合金的氧化膜中会产生更大的压应力,随着腐蚀时间的延长,当应力集中达到一定程度后,会产生微裂纹来释放应力,所以,在 Zr-0.7Sn-0.35Nb-0.3Fe 合金腐蚀 220 d 的氧化膜中,存在较多的微裂纹(图 4a 和 5a).

其次,Zr(Fe,Cr,Nb,Ge)$_2$ 和 Zr_3Ge 型第二相会对改善锆合金耐腐蚀性能起到关键作用.在尺寸较小的 Zr(Fe,Cr,Nb,Ge)$_2$ 型第二相中,Ge 取代了 Nb,Fe 或 Cr 的点阵位置,从而减少了 Zr(Fe,Cr,Nb,Ge)$_2$ 型第二相在氧化过程中体积膨胀的程度,降低了局部附加应力.另外,许多研究表明,第二相氧化后与 ZrO_2 基体形成的界面可以作为空位的尾闾,延缓空位通过扩散在氧化膜中凝聚形成孔隙的过程[20].同时,纳米大小分布均匀的第二相可以延迟微裂纹的形成,从而改善合金的耐腐蚀性能[7,9,22-24].但是,Ge 含量较高的 Zr_3Ge 型第二相尺寸较大,被包裹进入氧化膜中发生氧化时,在其周围极易产生局部附加应力,引起应力集中,产生微裂纹等缺陷(图 7c),促使氧化膜中的空位扩散,加速显微组织演化,破坏氧化膜的完整性,从而削弱氧化膜的保护作用.所以,含有较多 Zr_3Ge 型第二相的 Zr-0.7Sn-0.35Nb-0.3Fe-0.2Ge 合金的耐腐蚀性能有所下降(图 2).

再次,锆合金的耐腐蚀性能还与氧化膜的显微组织和 ZrO_2 晶体结构的演化有重要关系.有学者认为,ZrO_2 柱状晶中的缺陷凝聚形成新晶界,从而柱状晶演化成为等轴晶,大量柱状晶的存在可以显著降低锆合金的腐蚀速率,改善锆合金的耐腐蚀性能[15,16,25].本文研究发现,含 Ge 合金的耐腐蚀性能比 Zr-0.7Sn-0.35Nb-0.3Fe 合金优良,对应的氧化膜内部的柱状晶也更多(图 5 和图 7e),认为这是由于加 Ge 可以有效地延缓 ZrO_2 由柱状晶向等轴晶的演化过程的缘故,从而保留了更多的柱状晶.

4 结论

(1) $Zr-0.7Sn-0.35Nb-0.3Fe-xGe$ 系列合金的显微组织观察结果表明：当 Ge 含量在 0.05%—0.2% 时，都会析出 hcp 结构的 $Zr(Fe, Cr, Nb)_2$ 型和 $Zr(Fe, Cr, Nb, Ge)_2$ 型第二相. 这说明 580 ℃时 Ge 在 $\alpha-Zr$ 中的固溶度不超过 0.05%. 当 Ge 含量达到或超过 0.1%时，还会析出四方结构的 Zr_3Ge 型第二相，尺寸较大.

(2) 在 $Zr-0.7Sn-0.35Nb-0.3Fe$ 合金基础上添加 0.05%—0.2% Ge 可以明显改善合金在 360 ℃/18.6 MPa/0.01 mol/L LiOH 水溶液中的耐腐蚀性能. 当 Ge 含量达到 0.1%时，合金耐腐蚀性能最佳. 继续提高 Ge 含量时，合金耐腐蚀性能有所下降. 说明 Zr_3Ge 型第二相对改善合金耐腐蚀性能是不利的.

(3) 在 360 ℃/18.6 MPa/0.01 mol/L LiOH 水溶液中腐蚀 220 d 的 $Zr-0.7Sn-0.35Nb-0.3Fe-0.1Ge$ 合金氧化膜致密，厚度较薄，几乎没有微裂纹，ZrO_2 柱状晶较多.

参 考 文 献

[1] Liu J Z. *Structure Nuclear Materials*. Beijing：Chemical Industry Press，2007：19.
（刘建章. 核结构材料. 北京：化学工业出版社，2007：19）

[2] Zhao W J，Zhou B X，Miao Z，Peng Q，Jiang Y R，Jiang H M，Pang H. *Atom Energ Sci Technol*，2005；39(suppl)：1.
（赵文金，周邦新，苗志，彭倩，蒋有荣，蒋宏曼，庞华. 原子能科学技术，2005；39(增刊)：1）

[3] Nikulina A V，Markelov V A. In：Bradley E R，Sabol G P eds.，*Zirconium in the Nuclear Industry: 11th International Symposium*，*ASTM STP 1295*，Garmisch Partenkirchen Germany：ASTM International，1996：785.

[4] Sabol G P，Comstock R J. In：Garde A M，Bradley E R eds.，*Zirconium in the Nuclear Industry: 10th International Symposium*，*ASTM STP 1245*，Baltimore, MD：ASTM International，1994：724.

[5] Jung Y I，Lee M H，Kim H G，Park J Y，Jeong Y H. *J Alloys Compd*，2009；479：423.

[6] Yang W D. *Reactor Materials Science*. 2nd Ed.，Beijing：Atomic Energy Press，2006：260.
（杨文斗. 反应堆材料学. 第二版，北京：原子能出版社，2006：260）

[7] Liu W Q，Zhu X Y，Wang X J，Li Q，Yao M Y，Zhou B X. *Atom Energ Sci Technol*，2010；44：1477.
（刘文庆，朱晓勇，王晓娇，李强，姚美意，周邦新. 原子能科学技术，2010；44：1477）

[8] Kim J M，Jeong Y H，Kim I S. *J Nucl Mater*，2000；280：235.

[9] Liu W Q，Li Q，Zhou B X，Yao M Y. *Nucl Power Eng*，2003；24(1)：33.
（刘文庆，李强，周邦新，姚美意. 核动力工程，2003；24(1)：33）

[10] Wang J K. *Modern Ge Metallurgy*. Beijing：Metallurgy Industry Press，2005：58.
（王吉坤. 现代锗冶金，北京：冶金工业出版社，2005：58）

[11] Zhou B X，Li Q，Yao M Y，Liu W Q，Chu Y L. In：Kammenzind B，Limback M eds.，*Zirconium in，the Nuclear Industry: 15th International Symposium*，*ASTM STP 1505*，West Conshohochen：American Society for Testing and Materials，2009：360.

[12] Schaffer M，Schaffer B，Ramasse Q. *Ultramicroscopy*，2012；114：62.

[13] Zhou B X，Yao M Y，Li Q，Xia S，Liu W Q，Chu Y L. *Rare Met Mater Eng*，2007；36：1317.
（周邦新，姚美意，李强，夏爽，刘文庆，褚于良. 稀有金属材料与工程，2007；36：1317）

[14] Charquet D，Hahn R，Ortlib E. In：Van Swam L F P，Eucken C M eds.，*Zirconium in the Nuclear Industry: 8th International Symposium*，*ASTM STP 1023*，Philadelphia：ASTM International，1989：405.

[15] Anada H，Takeda K. In：Sabol G P，Bradley E R eds.，*Zircorium in the Nuclear Industry: 11th International Symposium*，*ASTM STP 1295*，Ann Arbor：ASTM International，1996：35.

[16] Wadman B，Lai Z，Andren H O，Nystrom A L，Rudling P，Pettersson H. In：Garde A M，Bradley E R eds.，*Zirconium in the Nuclear Industry: 10th International Symposium ASTM STP 1245*，Ann Arbor：ASTM

International，1994：579.

[17] Zhou B X，Peng J C，Yao M Y，Li Q，Xia S，Du C X，Xu G. In：Limback M，Barbéris P eds.，*Zirconium in the Nuclear Industry: 16th International Symposium*，*ASTM STP 1529*，Bridgeport：ASTM International，2011：620.

[18] Zhang X，Yao M Y，Li S L，Zhou B X. *Acta Metall Sin*，2011；47：1112.
（张欣，姚美意，李士炉，周邦新. 金属学报，2011；47：1112）

[19] Zhou B X，Li Q，Liu W Q，Yao M Y，Chu Y L. *Rare Met Mater Eng*，2006；35：1009.
（周邦新，李强，刘文庆，姚美意，褚于良. 稀有金属材料与工程，2006；35：1009）

[20] Zhou B X，Li Q，Yao M Y，Liu W Q，Chu Y L. *Corros Prot*，2009；30：589.
（周邦新，李强，姚美意，刘文庆，褚于良. 腐蚀与防护，2009；30：589）

[21] Li T F. *Metal High Temperature Oxidation and Thermal Corrosion*. Beijing：Chemical Industry Press，2003：51.
（李铁藩. 金属高温氧化和热处理. 北京：化学工业出版社，2003：51）

[22] Yang X L，Zhou B X，Jiang Y R，Li C. *Nucl Power Eng*，1994；15(1)：79.
（杨晓林，周邦新，蒋有荣，李聪. 核动力工程，1994；15(1)：79）

[23] Toffolon-Masclet C，Brachet J C，Jago G. *J Nucl Mater*，2002；305：224.

[24] Cao X X，Yao M Y，Peng J C，Zhou B X. *Acta Metall Sin*，2011；47：882.
（曹潇潇，姚美意，彭剑超，周邦新. 金属学报，2011；47：882）

[25] Yilmazbayhan A，Breval E. *J Nucl Mater*，2006；349：265.

Study on the Corrosion Resistance of Zr‑0.7Sn‑0.35Nb‑0.3Fe‑xGe Alloy in Lithiated Water at High Temperature Under High Pressure

Abstract：The corrosion resistance of Zr‑0.7Sn‑0.35Nb‑0.3Fe‑xGe (x=0.05, 0.1, 0.2, mass fraction, %) alloys was investigated in lithiated water with 0.01 mol/L LiOH at 360 ℃/18.6 MPa by autoclave tests. The microstructures of the alloys and oxide films on the corroded specimens were observed by TEM and SEM. The results show that the corrosion resistance of the Zr‑0.7Sn‑0.35Nb‑0.3Fe alloys in lithiated water at high temperature under high pressure is markedly improved by Ge addition. The alloy with 0.1%Ge shows the best corrosion resistance. In Zr‑0.7Sn‑0.35Nb‑0.3Fe‑xGe alloys, there exists fine Zr(Fe, Cr, Nb)$_2$ and Zr(Fe, Cr, Nb, Ge)$_2$ second phase particles (SPPs) with a close-packed hexagonal crystal structure (hcp) and coarse Zr$_3$Ge SPPs with a tetragonal crystal structure (TET). The oxide films formed on the Zr‑0.7Sn‑0.35Nb‑0.3Fe‑0.1Ge alloys corroded for 220 d are compact and thin. The micro-pores and micro-cracks are hardly detected and many ZrO$_2$ columnar grains exist in the oxide films formed on the Zr‑0.7Sn‑0.35Nb‑0.3Fe‑0.1Ge alloys. This indicates that the suitable amount of Ge could not only delay the process that the vacancies diffuse to form micro-pores and micro-pores develop to form micro-cracks, but also could retard the evolution from ZrO$_2$ columnar grains to ZrO$_2$ equiaxed grains.

富 Cu 团簇的析出对 RPV 模拟钢韧-脆转变温度的影响*

摘 要：将 Cu 含量高于实际核反应堆压力容器(RPV)钢的模拟钢在 880 ℃水淬后,在 660 ℃进行调质处理,然后在 370 ℃时效不同时间,采用 TEM,原子探针层析法(APT)和冲击实验对其进行研究.结果表明,时效 1 150 h 后,富 Cu 团簇的析出仍处于形核阶段,对韧-脆转变温度(DBTT)没有明显的影响;时效 3 000 h 后,试样中析出了平均尺寸为 1.5 nm 的富 Cu 团簇,主要分布在位错线上,数量密度达到 4.2×10^{22} m^{-3},DBTT 由调质处理后的 -100 ℃升高至 -60 ℃;时效 13 200 h 后,富 Cu 团簇略有长大,平均尺寸达到 2.4 nm,团簇的数量密度与时效 3 000 h 的试样处于相同数量级,DBTT 升高至 -45 ℃.采用热时效方法使富 Cu 团簇析出后,DBTT 只提高了 55 ℃,没有中子辐照引起的那样显著,这不仅是因为富 Cu 团簇的数量密度低,基体中没有中子辐照产生的晶体缺陷也是重要的原因.

核反应堆压力容器(RPV)装载核燃料组件及堆内其他构件,支撑并引导控制棒,维持着一回路中高温高压冷却水的压力,并对放射性极强的堆芯具有辐射屏蔽作用,是确保压水堆核电站运行安全,并且是不可更换的大型关键部件,它的服役寿命决定着核电站的运行寿命.当前,RPV 大多采用 Mn-Ni-Mo 低合金铁素体 A508-III 钢制造,在服役工况下(290 ℃)经过长期中子辐照后会引起辐照脆化,材料的韧-脆转变温度(DBTT)会升高,严重影响核反应堆的运行安全.大量研究[1-8]表明,RPV 钢的辐照脆化效应主要是由于中子辐照损伤产生晶体缺陷,并诱发高数量密度富 Cu 团簇的析出所造成的.据报道[4],RPV 钢的焊缝在经过高注量(5×10^{23} n/m^2)中子辐照后,会析出高数量密度(1×10^{23} m^{-3})的富 Cu 团簇,其 DBTT(T_{41J})比未辐照时升高 169 ℃.采用中子辐照实验研究 RPV 钢中富 Cu 团簇的析出操作不便,且成本较高.此外,由于辐照所产生的各种晶体缺陷相互干扰,不容易明确区分富 Cu 团簇析出对材料 DBTT 的影响.采用低温热时效的办法也可以使过饱和固溶的 Cu 以富 Cu 团簇形式析出,同时,不会引起其他晶体缺陷的变化,这就为单独研究富 Cu 团簇的析出对 RPV 钢 DBTT 的影响创造了条件.Cu 在 RPV 钢中是一种杂质元素,含量应该控制在 0.08%(质量分数)以下,研究[9]表明,在 290 ℃大约需要时效 24 年才能观察到富 Cu 团簇析出.研究[10-19]表明,适当提高 RPV 钢中 Cu 的含量,通过热时效也会使大量富 Cu 团簇析出,用这种方法可以方便地研究析出富 Cu 团簇中的成分和晶体结构的变化.材料的脆性受冶金、化学成分、显微组织以及析出相等多种因素的影响,成分相同的 RPV 模拟钢在相同的冶炼条件下,其脆性主要是受显微组织和析出相的影响.目前,尚未见富 Cu 团簇析出对材料 DBTT 影响的研究报道.

采用提高 Cu 含量的 RPV 模拟钢,用透射电镜(TEM),原子探针层析法(atom probe tomography,APT)和冲击实验等研究在低温长期时效过程中富 Cu 团簇的析出对 RPV 模拟钢 DBTT 的影响.

---

* 本文合作者:徐刚、蔡琳玲、冯柳、王均安、张海生.原发表于《金属学报》,2012,48(6):753-758.

1 实验方法

实验用 RPV 模拟钢中的 Cu 含量相对于商用 A508－III 钢有较大程度提高,其他合金成分与 A508－III 钢相同,具体成分(质量分数,%)为:Cu 0.6,Ni 0.85,Mn 1.58,Mo 0.54,Si 0.39,P 0.016,C 0.22,S 0.006,Fe 余量.RPV 模拟钢采用真空感应炉冶炼,铸锭重约 40 kg,钢锭经热锻和热轧制成 7 mm 厚的钢板后在 880 ℃保温 0.5 h 后水淬,然后在 660 ℃回火 10 h 进行调质处理.将热处理后的钢板用电火花线切割方法沿轧制方向切割尺寸为 7 mm×12 mm×60 mm 的样品,在 370 ℃进行不同时间的时效处理后,按照 ASTM E23 标准打磨制备成尺寸为 5 mm×10 mm×55 mm 的 V 型缺口 Charpy 冲击试样.

用 TINIUS OLSEN 84 型摆锤示波冲击试验机进行冲击实验,最大冲击能量为 406 J,冲击速度为 5.47 m/s.在不同温度下进行冲击实验后,得到每个试样的冲击断裂吸收能,并采用双曲正切函数拟合得到冲击断裂吸收能-温度曲线.在工程应用中,都采用截面为 10 mm×10 mm 的全尺寸试样进行冲击实验,选取断裂吸收能量为 41 J 处对应的温度定义为 DBTT 值.由于本文中采用的是小截面试样,不能采用通常全尺寸试样的方法来确定 DBTT 值,因而将调质态试样的 DBTT 曲线中的上平台与下平台断裂吸收能之间的中值对应的温度定为 DBTT 值.

用电火花线切割方法在时效后试样垂直轧制方向的截面上切割 0.5 mm 厚的薄片,将表面用砂纸磨平后,用组成为 30% HNO_3＋10% HF(体积分数)的溶液化学减薄至约 120 μm 后,用冲样机制得直径为 3 mm 的薄片,再用细砂纸磨薄至约 60 μm 厚,最后用双喷电解减薄方法制得 TEM 试样,减薄温度为－60 ℃,电解液为 10%高氯酸酒精溶液.用 H－800 型 TEM 对试样进行观察.

用电火花线切割方法从冲击试样的中心部位切出长约 20 mm,截面边长约 0.5 mm 的方形细棒,分别用体积分数为 25%的高氯酸乙酸溶液和 2%的高氯酸 2-丁氧基乙醇溶液作为电解液分 2 步进行电解抛光,制备出曲率半径小于 100 nm 的针状试样,具体的 APT 试样制备方法及分析原理见文献[20].用 LEAP 3000HR 型三维原子探针(3DAP)对针尖试样进行分析,实验温度为－223 ℃,脉冲频率为 2 MHz,脉冲分数为 20%.

APT 方法是在分析逐个原子种类的基础上构建纳米空间中不同元素原子的分布,可以精确测定纳米空间中合金的化学成分,但是不能给出完整的晶体结构信息.用 APT 方法研究 RPV 钢的显微组织时,如果观察到 Cu 析出后,文献中习惯称为"富 Cu 团簇(Cu-rich clusters)",从严格的意义来说,应该是"纳米富 Cu 相(Cu-rich nano-phases)",为了与大多数文献的表述一致,本文仍采用"富 Cu 团簇"的习惯用法.

2 实验结果与讨论

2.1 TEM 显微组织观察

图 1 示出了 RPV 模拟钢在 660 ℃回火处理 10 h 及继续在 370 ℃时效不同时间后的 TEM 像.由图 1a 可见,RPV 模拟钢淬火后形成了具有高位错密度的板条马氏体组织,再经过 660 ℃高温回火 10 h 后马氏体发生分解,但仍然保留着板条形貌,在板条间的界面上有颗

粒状渗碳体析出,晶粒内部的位错密度大大降低,同时形成了一些亚晶组织;在铁素体基体中有针状的 Mo₂C 析出和少量球状或短棒状富 Cu 颗粒析出,如图 1a 中插图所示,这是由于模拟钢的 Cu 含量高于 660 ℃时 Cu 在铁素体中的平衡固溶度,过饱和的 Cu 在铁素体基体中析出所致[21]. 由图 1b—d 可见,试样继续在 370 ℃时效 1 150 h,3 000 h 和 13 200 h 后,板条状的组织没有发生明显变化;由于 370 ℃时 Cu 在铁素体中的平衡固溶度比 660 ℃时的低,试样在 370 ℃时效时,过饱和的 Cu 会析出,但由于调质处理后降低了铁素体中固溶的 Cu 浓度,试样的时效温度也较低,Cu 原子在铁素体基体中的扩散速度较慢,时效 1 150 h 后,用 TEM 并没有观察到富 Cu 析出相,而只有少量调质处理时形成的球状或短棒状的富 Cu 颗粒,如图 1b 中插图所示;当试样在 370 ℃时效 3 000 h 后,铁素体基体中有弥散细小的纳米富 Cu 颗粒析出,并且大多数都在位错线上,如图 1c 中插图所示;与时效 3 000 h 试样相比,时效 13 200 h 的试样中富 Cu 颗粒的数量没有明显变化,但尺寸略有增加,如图 1d 中插图所示. 通过 HRTEM,EDS 和 APT 分析发现,试样在时效过程析出的颗粒都是富 Cu 相,没有其他合金碳化物析出[18,19,22].

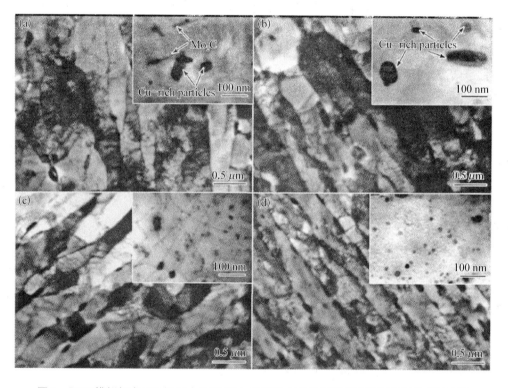

图 1 RPV 模拟钢在 660 ℃回火处理 10 h 及继续在 370 ℃时效不同时间后的 TEM 像

Fig. 1 TEM images of reactor pressure vessel (RPV) simulated steel specimens after different heat treatments （a）tempered at 660 ℃ for 10 h after 880 ℃ water-quenched （modified treatment）(inset show the needle-like Mo₂C and Cu-rich particles precipitated in α-Fe matrix); （b）continue aged at 370 ℃ for 1 150 h after modified treatment （inset show a few Cu-rich particles precipitated in α-Fe matrix); （c）continue aged at 370 ℃ for 3 000 h after modified treatment （inset show small Cu-rich particles precipitated in α-Fe matrix); （d）continue aged at 370 ℃ for 13 200 h after modified treatment （compared to the specimen aged at 370 ℃ for 3 000 h after modified treatment, equivalent but a little larger Cu-rich particles precipitated in matrix shown by the inset）

2.2 APT 分析

APT 可以精确地表征材料中各种元素的原子在纳米空间中的分布,判断纳米团簇的析出一般采用最大间隔包络法(maximum separation envelope method, MSEM),这种方法需要设置2 个基本参数:一个是团簇中某溶质原子之间的最大距离 d_{max},取值范围可在 0.3—0.7 nm 之间;另外一个是某溶质原子在满足 d_{max} 时最少的原子数目 N_{min}. 溶质原子若满足这 2 个基本参数所设定的值,则团簇即被确定. 若 d_{max} 较小时就能检测到团簇,则表示形成该团簇的溶质原子已经形成了自身的晶体结构,这种团簇已经处于析出后的成长阶段;若团簇是在 d_{max} 较大时才能检测到,那么这种团簇中的溶质原子还处于比较"松散"的状态,即为团簇析出时的成核阶段. 经过淬火和调质处理后的试样在 370 ℃时效不同时间后富 Cu 团簇析出的分析结果在文献[22]中已有详细叙述,本文只引用其中的主要结果,以方便对冲击实验结果进行分析.

在时效 1 150 h 的试样中,设定 $N_{min}=10$ 和 $d_{max}=0.5$ nm 时,在分析的体积中没有检测到富 Cu 团簇;而设定 $N_{min}=10$ 和 $d_{max}=0.6$ nm 时,可检测到 2 个富 Cu 团簇,表明此时团簇中的 Cu 原子还比较松散,这种富 Cu 团簇正处于析出时的形核阶段. 当设定 $N_{min}=10$ 和 $d_{max}=0.5$ nm 时,在时效 3 000 h 和 13 200 h 的试样中都能检测到富 Cu 团簇. 图 2 给出了在 370 ℃时效 3 000 h 试样中富

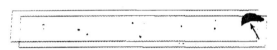

图 2 APT 分析试样在 370 ℃时效 3 000 h 后析出的富 Cu 团簇分布

Fig. 2 Distributions of Cu-rich clusters within an analyzed volume 57 nm×57 nm×440 nm in the specimen aged at 370 ℃ for 3 000 h (arrow denotes the largest Cu-rich cluster precipitated during modified treatment)

Cu 团簇的分布. 为了便于观察,图中已经将基体中的 Cu 原子剔除,只留下了富 Cu 团簇. 可见,在所分析的 57 nm×57 nm×440 nm 空间中,共获得了 25 个富 Cu 团簇,其中有一个尺寸较大的团簇(箭头所指),这应该是在 660 ℃调质处理过程中析出的富 Cu 相,在统计 370 ℃时效过程中富 Cu 团簇的析出数据时应该排除在外. 表 1 中列出了试样在 370 ℃时效不同时间后测量得到的富 Cu 团簇的平均等效直径(D)和数量密度(N_v),团簇中 Cu 的平均含量(C_p)以及 Cu 在基体中的浓度(C_m). 可见,当时效从 3 000 h 延长至 13 200 h,团簇的数量密度并没有数量级的变化,只是富 Cu 团簇的平均等效直径略有增大,表明富 Cu 团簇的长大速率比较慢.

表 1 时效不同时间的 RPV 模拟钢试样中富 Cu 团簇的 D, N_v, C_p 和 C_m

Table 1 D, N_v, C_p and C_m of RPV simulated steel specimens aged at 370 ℃ for different time

Aging time, h	D, nm	N_v, 10^{22} m^{-3}	C_p, %	C_m, %
0	—	—	—	0.27±0.02
1 150	—	—	—	0.25±0.02
3 000	1.5	4.2	45	0.18±0.02
13 200	2.4	4.3	55	0.15±0.02

Note:D—average equivalent diameter of Cu-rich clusters, N_v—number density of Cu-rich cluster, C_p—average Cu content (atomic fraction) in Cu-rich clusters, C_m—average Cu content (atomic fraction) in matrix

2.3 冲击实验结果与分析

图 3 示出了 RPV 调质态模拟钢试样经 370 ℃时效不同时间后的冲击断裂吸收能与温度的关系,可从各曲线获得试样的 DBTT. 由图 3a 可知,调质态试样冲击断裂吸收能-温度

曲线的上平台与下平台能量值之间的中值为 36 J,对应温度为−100 ℃,即可以冲击断裂吸收能为 36 J 的标准判断各试样的 DBTT,调质态试样的 DBTT 为−100 ℃. 由图 3b—d 可以获得,调质态 PRV 模拟钢继续在 370 ℃时效 1 150 h,3 000 h 和 13 200 h 后,对应于冲击断裂吸收能为 36 J 的 DBTT 分别为−95 ℃、−60 ℃和−45 ℃. 由于 RPV 模拟钢试样经过淬火和 660 ℃调质处理,再在 370 ℃时效不同时间后对晶粒组织不会有明显的影响(图 1),因而其 DBTT 的变化就只与析出的富 Cu 团簇有关. 根据析出强化理论,富 Cu 团簇通过晶格失配强化[24],化学强化[25]、模量差异强化[26]和位错与析出粒子的交互作用强化[27, 28]等 4 种方式实现析出强化,其中位错与析出粒子的交互作用是最主要的强化方式. 位错在滑移面上开动所需的力,也就是金属发生塑性变形时的流变应力会随温度的降低而升高,当温度降低至流变应力大于断裂应力时,材料就发生了脆性断裂,此温度即为 DBTT. 纳米尺度的富 Cu 团簇析出后,会使断裂应力有所增加,但是对位错在滑移面上开动所需的力会有更大的影响,纳米颗粒对位错的运动产生钉轧作用,使流变应力随温度降低而升高更加明显,因而富 Cu 团簇析出会导致 DBTT 的升高.

图 3 所示结果说明,当试样在 370 ℃时效 1 150 h 后,富 Cu 团簇处于析出过程的成核阶段,这时对 DBTT 的影响并不明显;只有在时效 3 000 h 富 Cu 团簇析出后,并且团簇数量密度达到 10^{22} m^{-3}时,DBTT 才会明显升高,由−100 ℃升高至−60 ℃;继续延长时效时间至 13 200 h,由于团簇数量密度没有发生数量级的变化,DBTT 继续升高的幅度并不太大,只升

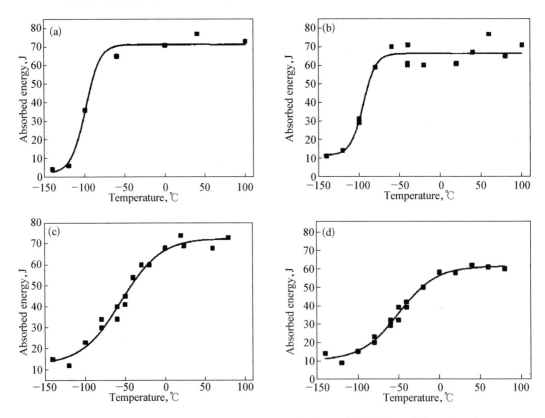

图 3 调质态 RPV 模拟钢试样在 370 ℃时效不同时间后冲击断裂吸收能与温度的关系

Fig. 3 Relationships between absorbed energy and temperature of impact tests of the RPV simulated steel specimens after modified treatment (a) and continue aged at 370 ℃ for 1 150 h (b), 3 000 h (c) and 13 200 h (d)

高至$-45\ ℃$. 中子辐照注量达到$3×10^{23}\ n/m^2$时,RPV 钢中富 Cu 团簇的数量密度可以达到$10^{24}\ m^{-3}$,其 DBTT 能达到$100\ ℃$以上,比未辐照试样的 DBTT 高$150\ ℃$以上[23],而 RPV 模拟钢时效后的 DBTT 比未时效试样的 DBTT 升高幅度较小(约$55\ ℃$),这主要是由于经过时效的 RPV 模拟钢中富 Cu 团簇的数量密度较低,而且基体也未受到中子辐照损伤的缘故.

3 结论

调质处理后的 Cu 含量为0.6%的 RPV 模拟钢在$370\ ℃$时效$1\ 150\ h$,富 Cu 团簇正处于析出的形核阶段,对韧-脆转变温度(DBTT)没有明显的影响;经过$3\ 000\ h$时效后,大部分富 Cu 团簇在位错线上析出,团簇的平均尺寸为$1.5\ nm$,数量密度达到$4×10^{22}\ m^{-3}$,PRV 模拟钢的 DBTT 由调质态的$-100\ ℃$升高至$-60\ ℃$;继续延长时效时间至$13\ 200\ h$,富 Cu 团簇略有长大,平均尺寸达到$2.4\ nm$,但其数量密度并没有明显变化,DBTT 升高至$-45\ ℃$,只比调质态的试样提高了$55\ ℃$.

参 考 文 献

[1] Toyama T, Nagai Y, Tang Z, Hasegawa M, Almazouzi A, van Walle E, Gerard R. *Acta Mater*, 2007;55: 6852.

[2] Auger P, Pareige P, Welzel S, van Duysen J C. *J Nucl Mater*, 2000;280: 331.

[3] Miller M K, Russell K F, Sokolov M A, Nanstad R K. *J Nucl Mater*, 2007;361: 248.

[4] Miller M K, Russell K F, Sokolov M A, Nanstad R K. *J Nucl Mater*, 2003;320: 177.

[5] Miller M K, Nanstad P K, Sokolov M A, Russell K F. *JNucl Mater*, 2006;351: 216.

[6] Phythian W J, English C A. *J Nucl Mater*, 1993;205: 162.

[7] Carter R G, Soneda N, Dohi K, Hyde J M, English C A, Server W L. *J Nucl Mater*, 2001;298: 211.

[8] Miller M K, Chernobaeva A A, Shtrombakh Y I, Russell K F, Nanstad R K, Erak D Y, Zabusov O O. *J Nucl Mater*, 2009;385: 615.

[9] Miller M K, Russell K F. *J Nucl Mater*, 2007;371: 145.

[10] Hornbogen E, Glenn, R C. *Trans Metall Soc AIME*, 1960;218: 1064.

[11] Zhang R Q, Hong X F, Peng Q. *Nucl Power Eng*, 2010;31(1): 4.
　　(张瑞谦,洪晓峰,彭倩,核动力工程. 2010;31(1): 4)

[12] Goodman S R, Brenners S S, Low J R. *Metall Trans*, 1973;4: 2363.

[13] Pizzini S, Roberts K J, Phythian W J, English C A, Greaves G N. *Philos Mag Lett*, 1990;61: 223.

[14] Othen P J, Jenkins M L, Smith G D W, Phythian W J. *Philos Mag Lett*, 1991;64: 383.

[15] Speich G R, Oriani R A. *Trans Metall Soc AIME*, 1965;233: 623.

[16] Othen P J, Jenkins M L, Smith G D W. *Philos Mag*, 1994;A70: 1.

[17] Styman P D, Hyde J M, Wilford K, Morley A, Smith G D W. *Prog Nucl Energ*, 2011, doi: 10.1016/j. pnucene. 2011.10.010.

[18] Chu D F, Xu G, Wang W, Peng J C, Wang J A, Zhou B X. *Acta Metall Sin*, 2011;47: 269.
　　(楚大锋,徐刚,王伟,彭剑超,王均安,周邦新. 金属学报,2011;47: 269)

[19] Xu G, Cai L L, Feng L, Zhou B X, Wang W, Peng J C. *Acta Metall Sin*, 2011;47: 906.
　　(徐刚,蔡琳玲,冯柳,周邦新,王伟,彭剑超. 金属学报,2011;47: 906)

[20] Miller M K. *Atom Probe Tomography: Analysis at the Atomic Level*. New York: Kliwer Academic/Plenum Publishers, 2000: 25.

[21] Yong Q L. *Secondary Phase in Steel*. Beijing: Metallurgical Industry Press, 2006: 127.
　　(雍岐龙. 钢铁材料中的第二相. 北京: 冶金工业出版社,2006: 127)

[22] Xu G, Cai L L, Feng L, Zhou B X, Liu W Q, Wang J A. *Acta Metall Sin*, 2012;48: 407.

(徐刚,蔡琳玲,冯柳,周邦新,刘文庆,王均安,金属学报,2012;48: 407)

[23] Buswell J T, English C A, Herherington M G, Phythian W J, Smith G D W, Worral G M. In: Steele L E ed. , *Proc 14th Int Symp Effects of Radiation on Materials*, Vol. 2, Andover, Massachusseetts: The American Society for Testing and Materials, 1990: 127.

[24] Brown L M, Ham R K. In: Kelly A, Nicholson R B, eds. , *Strengthening Methods in Crystals*. London: Applied Science Publishers;1965: 9.

[25] Friedel J. *Disloctions*. London: Pergamon Press;1964: 1.

[26] Russell K C, Brown L M. *Acta Metall*, 1972;20: 969.

[27] Harry T, Bacon D J. *Acta Mater*, 2002;50: 195.

[28] Harry T, Bacon D J. *Acta Mater*, 2002;50: 209.

Effect of the Precipitation of Cu‑Rich Clus‑ ters on the Dbtt of RPV Simulated Steel

Abstract: Reactor pressure vessels (RPVs) are usually made of low alloy ferritic steels, among which A508‑III steel is a typical one. The long-term neutron irradiation can induce the embrittlement of RPV steels, and the embrittlement may lead to a reduction of the RPV service life. Generally, this behavior of the embrittlement is well established and is typically assessed by the increase in the ductile-to-brittle transition temperature (DBTT) of the RPV steels. For many years, extensive studies have revealed that irradiation-induced ultrafine Cu-rich clusters (CRCs) play an important role and CRCs with high number density cause hardening and embrittlement of the RPV steels. In order to investigate the effect of the precipitation of CRCs on DBTT of RPV steels by thermal aging, it is necessary to increase the Cu content in RPV steel. A 40 kg ingot of RPV simulated steel based on the composition of A508‑III steel with higher Cu content (0.6% in mass fraction) was prepared by vacuum induction melting, and it was forged and hot rolled to a plate with 7 mm in thickness. Specimens with a dimension of 7 mm×12 mm×60 mm were cut from the hot-rolled plate. The heat treatment routes of the specimens consists of a soaking at 880 ℃ for 0.5 h, a water quenching, a tempering at 660 ℃ for 10 h, and a final aging at 370 ℃ for various times. The effect of the precipitation of CRCs on the DBTT of the RPV simulated steel was investigated by Charpy impact tests, as well as the microstructure analysis was carried out by TEM and atom probe tomography (APT). According to ASME 23 standard, Charpy‑V specimens with a dimension of 5 mm×10 mm×55 mm were prepared and tested by TINIUS OLSEN 84 impact test machine. The TEM analysis shows that CRCs precipitate on dislocations in the specimen aged at 370 ℃ for 3 000 h, and the clusters become a little coarsened when the aging time is extended to 13 200 h. For the specimens aged for 1 150 h, CRCs were on the stage of the nucleation assessed by TEM as well as APT analysis, and they did not have an effect on the DBTT of the RPV simulated steel. For the specimens aged for 3 000 h, CRCs precipitated with an average equivalent diameter of 1.5 nm and a number density of 4.2×10^{22} m^{-3}, and it results in the increase of the DBTT from -100 ℃ to -60 ℃. For the specimens aged for 13 200 h, CRCs slightly coarsened to 2.4 nm of the average equivalent diameter, while the number density is similar to that of the specimens aged for 3 000 h. In this case the DBTT rose to -45 ℃. Therefore, the present work shows the precipitation of CRCs induced by thermal aging reveals a smaller impact on the DBTT than that by neutron irradiation. From the thermal aging aspect, the much lower number density of CRCs and the absence of the defects induced by neutron irradiation in the matrix could account for this phenomenon.

利用 APT 对 RPV 模拟钢中
富 Cu 原子团簇析出的研究*

摘　要：提高了 Cu 含量的核反应堆压力容器(RPV)模拟钢经过 880 ℃水淬和 660 ℃调质处理，在 370 ℃时效不同时间后，利用原子探针层析技术(APT)进行分析. 结果表明：样品经过 1 150 h 时效后，富 Cu 团簇正处于析出过程的形核阶段；经过 3 000 h 和 13 200 h 时效后析出了富 Cu 团簇，团簇的平均等效直径分别为 1.5 nm 和 2.4 nm，团簇中 Cu 的平均浓度分别为 45% 和 55%(原子分数)，团簇的数量密度约为 4.2×10^{22} m^{-3}；样品经过 13 200 h 时效后，α-Fe 基体中的 Cu 含量为 (0.15 ± 0.02)%，仍然高于 Cu 在 α-Fe 中平衡固溶度的理论计算值，说明这时富 Cu 团簇的析出过程还没有达到平衡. 对渗碳体的分析结果表明，Ni，Si 和 P 偏聚在渗碳体和 α-Fe 基体的相界面附近，Mn，Mo 和 S 富集在渗碳体中；并没有观察到 Cu 在相界面上偏聚的现象.

核反应堆压力容器(RPV)装载核燃料组件并承载着一回路的高温高压冷却水，是压水堆核电站中不可更换的大型关键部件，也是保障核电站运行安全的重要屏障之一，因而它的服役寿命决定着核电站的运行寿命. RPV 是由 Mn-Ni-Mo 低合金铁素体钢(A508-III 钢)制成，Cu 是钢中的一种杂质元素，含量应控制在 0.08%(质量分数)以下，在运行工况下，经过长期中子辐照后会引起 RPV 钢的韧脆转变温度(DBTT)升高，这是决定 RPV 服役寿命的重要因素，也是影响核电站安全运行的关键问题. 研究[1-9]表明，RPV 钢的辐照脆化效应主要是由于中子辐照损伤产生晶体缺陷，并诱发析出高数量密度纳米富 Cu 析出物所致. 由于原子探针层析技术(APT)对这种富 Cu 析出物进行分析时，只能获得成分和尺寸的信息，不能获得晶体结构的信息，并且 APT 的分辨率极高，可以检测到十来个原子发生团聚时的情况，这时析出相只处于形核初期，大小也只有 1 nm 左右，因此，在 APT 分析时，人们习惯将这种富 Cu 析出物称为富 Cu 团簇(Cu-rich cluster). 用透射电镜分析时，容易分辨的析出物往往已长大至数纳米，而且可以通过电子衍射或拍摄高分辨晶格条纹像来确定其晶体结构，因而称为富 Cu 相. 如果用中子辐照来研究 RPV 钢中富 Cu 团簇析出，辐照费用高昂，实验操作也很不便. 采用低温热时效的办法也可以使过饱和固溶的 Cu 以纳米富 Cu 相形式析出，有文献报道，在 290 ℃时效大约 24 年也可以观察到纳米富 Cu 团簇的析出[10]，但在中子辐照下这种过程将会大大加快. 研究[11-17]表明，适当提高钢中 Cu 的含量，通过热时效也会使大量富 Cu 相析出. 因此，研究者就以提高了 Cu 含量的 RPV 模拟钢为研究对象，采用低温长期时效的办法并利用高分辨透射电镜(HRTEM)对富 Cu 相的析出过程和结构演化规律进行了研究[18,19]，但是，配备了能谱仪(EDS)的 HRTEM 并不能对富 Cu 相在析出过程中的成分变化进行精确分析，因为分析纳米相的化学成分时会受到基体的干扰. APT 可以获得材料中各种原子的空间分布信息，能够对时效过程中富 Cu 团簇的大小和成分变化规律进行分析，而且还可以研究合金元素和杂质元素在界面上的偏聚情况，这些问题对进一步了解富 Cu 团簇的析出过程及其对 PRV 钢的辐照脆化机理具有重要意义.

＊ 本文合作者：徐刚、蔡琳玲、冯柳、刘文庆、王均安. 原发表于《金属学报》，2012，48(4)：407-413.

1 实验方法

实验所用 RPV 模拟钢是在商用 A508 - III 钢基础上将 Cu 含量提高至 0.6%(质量分数),其他成分保持不变.RPV 模拟钢由真空感应炉冶炼,铸锭重约 40 kg,化学成分见表 1.钢锭经热锻和热轧制成 4 mm 厚的钢板,利用线切割机切成 40 mm×30 mm×4 mm 的小样品,在 880 ℃加热 0.5 h 后水淬,然后在 660 ℃加热 10 h 进行调质处理,最后在 370 ℃进行 0,1 150 h、3 000 h 和 13 200 h 的时效处理.

表 1 反应堆压力容器(RPV)模拟钢的化学成分

Table 1 Chemical compositions of the reactor pressure vessel (RPV) model steel

Composition	Cu	Ni	Mn	Si	P	C	S	Mo	Fe
Atomic fraction, %	0.53	0.81	1.60	0.77	0.03	1.00	0.011	0.31	Bal.
Mass fraction, %	0.60	0.85	1.58	0.39	0.016	0.22	0.006	0.54	Bal.

用电火花线切割方法从样品的中心部位切出长约 20 mm,截面边长约 0.5 mm 的方形细棒,分别用体积分数为 25%的高氯酸乙酸和 2%的高氯酸 2 - 丁氧基乙醇作电解液分两步抛光细棒,制备出曲率半径小于 100 nm 的针状样品,用于 APT 的实验分析,具体的样品制备方法及分析原理可参阅文献[20].用 LEAP 3000HR 型三维原子探针(3DAP)对针尖样品进行分析,样品冷却至−223 ℃,脉冲频率为 2 000 kHz,脉冲分数为 20%.

APT 可以表征材料中各种元素的原子在纳米空间中的分布,判断纳米团簇的析出一般采用 MSEM (maximum separation envelope method)方法,这种方法需要设置 2 个基本参数:一个参数是团簇中某溶质原子之间的最大距离 d_{max},取值范围可在 0.3—0.7 nm 之间,当 d_{max} 满足 0.3—0.4 nm 时,则表示团簇中的溶质原子已经形成了自己的晶体结构,如果 d_{max} 只能满足 0.6—0.7 nm 时,则表示团簇中的溶质原子还比较"松散",该团簇处于析出过程的初期;另外一个参数是某溶质原子在满足某一 d_{max} 时最少的原子数目 N_{min}.若满足这两个基本参数所设定的值,则团簇即被确定.团簇数量密度 N_v 用以下公式计算:

$$N_v = \frac{N_p \zeta}{N_a \Omega} \tag{1}$$

式中,N_p 是分析体积内所检测到的团簇数量;检测参数 ζ 为 0.6;N_a 为所收集的原子总数;Ω 是原子的平均体积,对于 bcc 结构的 Fe 为 1.178×10^{-2} nm^{-3},由于收集的所有原子中主要为 Fe 原子,所以这里以 Fe 原子的体积来计算.假定富 Cu 团簇为球形;每个团簇的等效直径 D_p 用以下公式得出:

$$D_p = \sqrt[3]{\frac{3n_p \Omega}{\pi \zeta}} \tag{2}$$

式中,n_p 为满足 d_{max} 和 N_{min} 设定值时单个富 Cu 团簇体积中所检测到的原子数.

2 实验结果与讨论

Starink 和 Zahra[21,22]给出了 Cu 在 α - Fe 中固溶度随温度变化的理论公式

$$c_{Cu}(T) = c_0 \exp(-\Delta H_{sol}/N\kappa_B T) \tag{3}$$

式中，$c_{Cu}(T)$ 表示 Cu 在 α - Fe 中固溶度；c_0 为常数；ΔH_{sol} 为富 Cu 团簇的形成焓；N 为 Avogadro 常数，约为 6.02×10^{23} mol^{-1}；κ_B 为 Boltzmann 常数，约为 1.38×10^{-23} J/K；T 为热力学温度. Miller 等[23]通过热时效 Fe - Cu 合金测得 c_0 约为 3.1，ΔH_{sol} 约为 4 kJ/mol，由此可以得出不同温度下 Cu 在 α - Fe 中的固溶度. 图 1 为淬火态 RPV 模拟钢经 660 ℃保温 10 h 调质处理后的 Cu，Ni 和 Mn 原子的空间分布图. 可以看出，各元素在铁素体基体中的分布比较均匀，没有观察到富 Cu 团簇析出. 在基体中选取适当大小的体积，根据该体积内采集到的所有不同元素原子之间的比例，得出基体中 Cu 的平均浓度为(0.27 ± 0.02)%（原子分数，下同），低于材料中的平均 Cu 含量（表 1），与式（3）所计算出的理论值 0.29%基本吻合，这是由于 RPV 模拟钢在调质处理时，铁素体基体中过饱和的 Cu 已经析出形成尺寸较大的短棒状富 Cu 相，其为 fcc 结构[7]. 这种富 Cu 相在基体中的数量密度较低，在图 1 的分析体积中并未截获.

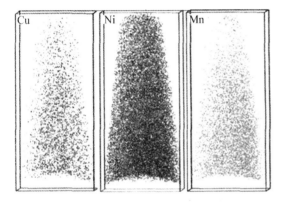

图 1 调质态 RPV 模拟钢样品中的 Cu，Ni 和 Mn 原子分布图

Fig. 1 Distributions of Cu, Ni and Mn atoms within an analyzed volume 86 nm×84 nm×186 nm in the RPV model steel specimen tempered at 660 ℃ for 10 h after water quenching

图 2 调质态 RPV 模拟钢样品在 370 ℃时效 1 150 h 后的 Cu，C 和 Mn 原子分布图

Fig. 2 Distributions of Cu, C and Mn atoms within an analyzed volume 82 nm×80 nm×174 nm in the RPV model steel specimen aged at 370 ℃ for 1 150 h (The chemical composition profile of the interface between cementite and the ferritic matrix was analyzed within the pipe along the direction shown by the arrow in the distribution map of C atoms)

图 2 为调质态 RPV 模拟钢样品经 370 ℃时效 1 150 h 后的 Cu，C 和 Mn 原子的空间分布图. 可以看出，Cu 原子在基体中分布比较均匀，其平均浓度约为(0.25 ± 0.02)%. 直接观察 Cu 原子的分布图无法判断是否有富 Cu 团簇析出，当设 $N_{min} = 10$，$d_{max} = 0.5$ nm 时，在分析的体积中没有检测到富 Cu 团簇，但是当设 $N_{min} = 10$，$d_{max} = 0.6$ nm 时，可以检测到 2 个富 Cu 团簇，这表明样品在 370 ℃时效 1 150 h 后，富 Cu 团簇正处于析出过程的形核初期，这种 Cu 原子比较"松散"的富 Cu 团簇的数量密度也很低. 从 C 和 Mn 原子分布图中可以看出，C 和 Mn 在相同区域有明显的富集，由于样品在调质处理时会形成渗碳体，由此认为 C 和 Mn 富集区应为渗碳体. 渗碳体与铁素体基体之间的界面是弯曲的，为了分析界面处各元素成分分布的准确性，只对与界面尽可能垂直的圆管内的元素进行分析，圆管的位置如图 2

中 C 原子的分布图中所示,圆管的直径约为 8 nm,长度约为 30 nm,管子的轴向基本垂直于渗碳体与铁素体基体的界面.

图 3 为渗碳体和铁素体界面附近几种元素的成分分布图,渗碳体和铁素体基体中各元素的平均浓度列于表 2 中.本文将图 3a 中 Fe 元素浓度由低到高变化的中值处定义为渗碳体与铁素体基体的界面,用虚线在图 3a,b 和 c 中标出.从表 2 中的数据可知,渗碳体中 C 的浓度约是基体中的 90 倍,而 Mn 在渗碳体中的浓度约是铁素体基体中的 14 倍.由于 Mo 与 C 有较强的亲和力,因而在渗碳体中也会有 Mo 原子聚集(图 3b),而 Mn 和 Mo 对 S 也有较好的亲和力,因而在渗碳体中存在 Mo 和 Mn 时,S 原子也会向这一区域偏聚,渗碳体中 Mo 和 S 的浓度比基体中的分别高 10 倍和 95 倍,这表明含有 Mo 和 Mn 的渗碳体形成后,有可能降低杂质元素 S 在钢中的有害作用.从图 3b 中还可以看出,Ni 和 Si 的浓度在界面附近明显高于它们分别在渗碳体和铁素体中的浓度,这表明 Ni 和 Si 会偏聚在渗碳体和铁素体的界面处,这种偏聚可能会阻止渗碳体的继续长大.从图 3c 可以看出,P 在界面处也有明显的偏聚,与 Ni 和 Si 相似;Cu 在渗碳体中的浓度((0.10±0.04)%)比较低,在界面处并未观察到 Cu 偏聚的现象.

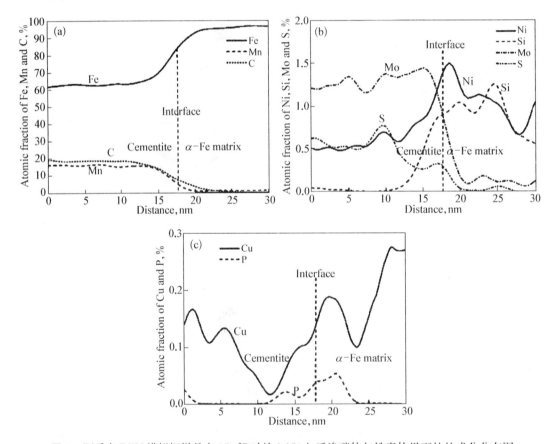

图 3 调质态 RPV 模拟钢样品在 370 ℃时效 1 150 h 后渗碳体与铁素体界面处的成分分布图

Fig. 3 Composition profiles of Fe, Mn, C (a), Ni, Si, Mo, S (b) and Cu, P (c) at the interface between the cementite and the ferritic matrix in the RPV model steel specimen aged at 370 ℃ for 1 150 h

图 4 是调质态 RPV 模拟钢样品经 370 ℃时效 3 000 h 后的 Cu, Ni 和 Mn 原子的空间分布图.直接从 Cu 原子的分布图中已经可以看出有较多的富 Cu 团簇析出,当设 $N_{min}=10$,

表2 调质态 RPV 模拟钢样品在 370 ℃时效 1 150 h后渗碳体和铁素体基体中各元素的平均含量

Table 2 Average contents of different elements in the cementite and the ferritic matrix for the
RPV model steel specimen aged at 370 ℃ after 1 150 h　　　　　（atomic fraction，%）

Position	Cu	Ni	Mn	Si	P	C	S	Mo	Fe
Cementite	0.10± 0.04	0.56± 0.04	15.7± 0.5	0.02± 0.01	0.002± 0.000 3	18.3± 0.3	0.57± 0.06	1.3± 0.1	63.4± 0.5
α-Fe Matrix	0.25± 0.03	0.82± 0.25	1.14± 0.08	0.68± 0.04	0.006± 0.004	0.21± 0.05	0.006± 0.002	0.12± 0.03	96.7± 0.6

d_{max}=0.5 nm，在大小为 57 nm×57 nm×440 nm 的分析体积中，用 MSEM 法检测出有 25
个富 Cu 团簇，其中有一个尺寸较大的团簇（图 4 中箭头所示），等效直径约为 26 nm，这应该
是在 660 ℃调质处理过程中析出的富 Cu 团簇，而不是在 370 ℃时效过程中形成的. 由于本
文主要是研究 370 ℃时效过程中富 Cu 团簇的析出特征，故在统计分析时将其剔除，只对余
下的 24 个团簇进行统计分析，由此得出富 Cu 团簇的数量密度为 4.2×10^{22} m^{-3}. 这一数值
与商用 RPV 钢在 290 ℃时效约 24 年后析出富 Cu 团簇的数量密度值相当[5,10]，但是比同样
成分的 RPV 模拟钢经淬火后直接在 400 ℃时效得到的富 Cu 团簇数量密度低 1 个数量
级[24]，比商用 RPV 钢经中子辐照至 3×10^{23} n/m^2 注量后的大约低 2 个数量级[25]. 这种数量
密度的差别与材料中晶体缺陷的数量有关，因为富 Cu 原子团簇析出时容易在位错及界面等
晶体缺陷处形核. 淬火后形成高位错密度的板条马氏体，中子辐照时引起大量的晶体缺陷及
位错是富 Cu 原子团簇数量密度增加的直接原因.

图4 调质态 RPV 模拟钢样品在 370 ℃时效
3 000 h后的 Cu，Ni 和 Mn 原子分布图

Fig. 4 Distributions of Cu，Ni and Mn atoms
within an analyzed volume 57 nm×57 nm×440 nm
in the RPV model steel specimen aged at 370 ℃ for
3 000 h（The largest Cu-rich cluster shown by the
arrow was rejected during the data analysis as it
should be formed during tempering at 660 ℃ for
10 h after water quenching）

图5 时效 3 000 h 的 RPV 模拟钢样品中团簇中平
均 Cu 含量与富 Cu 团簇的等效直径的关系

Fig. 5 Average Cu content in individual Cu-rich
cluster as a function of equivalent cluster diameter in
the RPV model steel specimen aged at 370 ℃ for
3 000 h

　　图 5 为调质态 RPV 模拟钢样品时效 3 000 h 后所形成的富 Cu 团簇中 Cu 的平均含量
与团簇等效直径之间的关系. 可以看出，这时析出的富 Cu 团簇尺寸较小，团簇的等效直径分

布在 0.7—2.8 nm 之间,平均等效直径大约为 1.5 nm,团簇中的平均 Cu 浓度约为 45%. 从图中可以看出,随着等效直径的增大,团簇中 Cu 的平均含量反而有所下降,这是由于统计方法造成的结果. 当团簇的等效直径较小时,富 Cu 团簇中 Fe 原子数量较少,从而提高了 Cu 的平均含量;当团簇的等效直径较大时,在团簇的中心 Cu 原子的聚集程度比较紧密,但是距离中心稍远的区域,Cu 原子的分布比较松散,在这一区域中含有较多的 Fe 原子,这样使得团簇中 Cu 的平均含量又会下降. 图 6a 是一个等效直径为 1.46 nm 的富 Cu 团簇中的 Fe,Cu,Ni 和 Mn 原子分布图,团簇以外的各种原子在图中都没有显示. 从团簇中各元素的浓度分布图(图 6b)可以看出,团簇的中心 Cu 原子的浓度约为 40%,团簇中仍然存在较多的 Fe 原子,其浓度在 60% 左右,同时在团簇内部还聚集有少量 Ni 和 Mn 原子. 除 Cu 团簇外,Cu 在基体中的平均浓度为 $(0.18\pm0.02)\%$,比调质处理后未时效样品中的 Cu 浓度低,这是由于富 Cu 团簇析出后导致的结果.

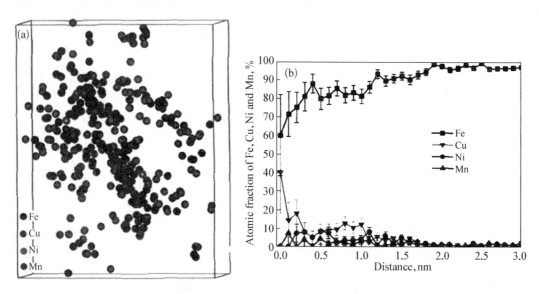

图 6　时效 3 000 h 样品中一个富 Cu 原子团簇的 Fe,Cu,Ni 和 Mn 分布图和成分分布图

Fig. 6　Distributions of Fe, Cu, Ni and Mn atoms within an analyzed volume 3 nm×3 nm×4 nm for a Cu-rich cluster in the RPV model steel specimen aged at 370 ℃ for 3 000 h (a) and the composition profiles of Fe, Cu, Ni and Mn (b)

　　图 7 是调质态 RPV 模拟钢样品经 370 ℃时效 13 200 h 后的 Cu,Ni 和 Mn 原子的空间分布图. 从图中可以明显看出有富 Cu 团簇析出,当设 $N_{min}=10$,$d_{max}=0.5$ nm,在大小为 67 nm×64 nm×99 nm 的分析体积中,用 MSEM 法检测出有 8 个富 Cu 团簇,富 Cu 团簇的数量密度约为 4.3×10^{22} m^{-3},与时效 3 000 h 的样品中富 Cu 团簇的数量密度相当,并没有显著的增加. 图 8 为这 8 个富 Cu 团簇中 Cu 的平均含量与团簇等效直径之间的关系. 与图 5 的情况一样,由于统计分析方法的原因,造成了团簇中 Cu 的平均含量随着等效直径从 1 nm 增大到 4 nm 时反而有所下降的结果. 不过当等效直径进一步增大到 11 nm 时,团簇中的 Cu 含量有了明显增加. 从图 8 中还可以看出,样品经过 370 ℃时效 13 200 h 后,只有少数富 Cu 团簇发生了明显长大,大部分富 Cu 团簇长大较慢,它们平均等效直径约为 2.4 nm,比时效 3 000 h 样品中富 Cu 团簇的平均等效直径(1.5 nm)略有增加;团簇中 Cu 的平均浓度约为 55%,比时效 3 000 h 样品团簇中的 Cu 平均浓度高,随着团簇长大,团簇中 Cu 的含量也在

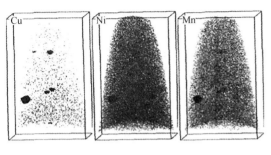

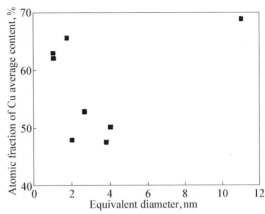

图7 调质态 RPV 模拟钢样品在 370 ℃ 时效 13 200 h后的 Cu，Ni 和 Mn 原子分布图

Fig. 7 Distributions of Cu，Ni and Mn atoms within an analyzed volume 67 nm×64 nm×99 nm in the RPV model steel specimen aged at 370 ℃ for 13 200 h

图8 时效 13 200 h 的 RPV 模拟钢样品中团簇中平均 Cu 含量与富 Cu 团簇的等效直径的关系

Fig. 8 Average Cu content in individual Cu-rich cluster as a function of equivalent cluster diameter in the RPV model steel specimen aged at 370 ℃ for 13 200 h

提高. 图 9a 是等效直径为 2.5 nm 团簇中的 Fe，Cu，Ni 和 Mn 原子分布图. 从团簇中 Fe，Cu，Ni，Mn 各元素的浓度分布图（图 9b）可以看出，团簇的中心几乎全部为 Cu 原子，在团簇内还有 Fe，Ni 和 Mn 原子聚集，尤其是在富 Cu 团簇与 α-Fe 基体的界面处，还有 Mn 和 Ni 原子的偏聚现象. 除 Cu 团簇外，Cu 在基体中的平均浓度为 $(0.15\pm0.02)\%$，比时效 3 000 h样品基体中的 Cu 浓度更低，但是仍然比 370 ℃ 时 Cu 在 α-Fe 中平衡固溶度的理论

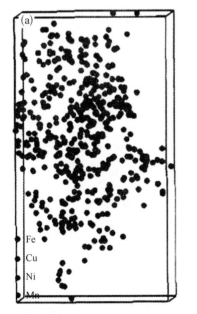

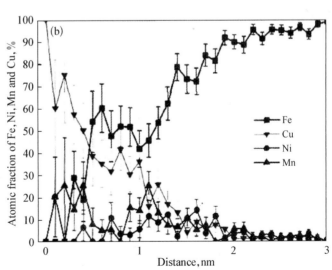

图9 时效 13 200 h 样品中富 Cu 原子团簇的 Fe，Cu，Ni 和 Mn 原子分布图和成分分布图

Fig. 9 Distributions of Fe，Cu，Ni and Mn atoms within an analyzed volume 3 nm×4 nm×6 nm for a Cu-rich cluster in the RPV model steel specimen aged at 370 ℃ for 13 200 h（a）and the composition profiles of Fe，Cu，Ni and Mn（b）

计算值(约 0.004%)高,说明这时富 Cu 团簇的析出过程还没有达到平衡.

3 结论

(1) 调质态 RPV 模拟钢样品经过 370 ℃时效 1 150 h 后,富 Cu 团簇正处于析出的形核过程中;时效 3 000 h 后,样品中才析出了富 Cu 团簇,其数量密度约为 4.2×10^{22} m^{-3},团簇的等效直径在 0.7—2.8 nm 之间,平均等效直径约为 1.5 nm,团簇中 Cu 的平均浓度约为 45%;时效 13 200 h 后,样品中富 Cu 团簇的数量密度没有明显变化,但团簇的平均等效直径增大至 2.4 nm,团簇中 Cu 的平均浓度增加至 55%. 这时 α-Fe 基体中的 Cu 含量为(0.15 ± 0.02)%,仍然高于 Cu 在 α-Fe 中平衡固溶度的理论计算值,说明富 Cu 团簇的析出过程还没有达到平衡.

(2) 对 370 ℃时效 1 150 h 样品中的渗碳体分析结果表明,Ni, Si 和 P 会偏聚在铁素体和渗碳体的相界面附近,Mn, Mo 和 S 富集在渗碳体中,但没有观察到 Cu 在相界面上偏聚的现象.

参 考 文 献

[1] Toyama T, Nagai Y, Tang Z, Hasegawa M, Almazouzi A, van Walle E, Gerard R. *Acta Mater*, 2007;55: 6852.

[2] Auger P, Pareige P, Welzel S, Van Duysen J C. *J Nucl Mater*, 2000;280: 331.

[3] Miller M K, Russell K F, Sokolov M A, Nanstad R K. *J Nucl Mater*, 2007;361: 248.

[4] Miller M K, Russell K F, Sokolov M A, Nanstad R K. *J Nucl Mater*, 2003;320: 177.

[5] Miller M K, Nanstad P K, Sokolov M A, Russell K F. *J Nucl Mater*, 2006;351: 216.

[6] Phythian W J, English C A. *J Nucl Mater*, 1993;205: 162.

[7] Fukuya K, Ohno K, Nakata H, Dumbill S, Hyde J M. *J Nucl Mater*, 2003;312: 163.

[8] Carter R G, Soneda N, Dohi K, Hyde J M, English C A, Server W L. *J Nucl Mater*, 2001;298: 211.

[9] Miller M K, Chernobaeva A A, Shtrombakh Y I, Russell K F, Nanstad R K, Erak D Y, Zabusov O O. *J Nucl Mater*, 2009;385: 615.

[10] Miller M K, Russell K F. *J Nucl Mater*, 2007;371: 145.

[11] Hornbogen E, Glenn R C. *Trans Metall Soc AIME*, 1960;218: 1064.

[12] Speich G R, Oriani R A. *Trans Metall Soc AIME*, 1965;233: 623.

[13] Goodman S R, Brenners S S, Low J R. *Metall Trans*, 1973;4: 2363.

[14] Pizzini S, Roberts K J, Phythian W J, English C A, Greaves G N. *Philos Mag Lett*, 1990;61: 223.

[15] Othen P J, Jenkins M L, Smith G D W, Phythian W J. *Philos Mag Lett*, 1991;64: 383.

[16] Othen P J, Jenkins M L, Smith G D W. *Philos Mag*, 1994;70A: 1.

[17] Styman P D, Hyde J M, Wilford K, Morley A, Smith G D W. *Prog Nucl Energ*, 2011, doi: 10.1016/j. pnucene. 2011.10.01.

[18] Chu D F, Xu G, Wang W, Peng J C, Wang J A, Zhou B X. *Acta Metall Sin*, 2011;47: 269.
    (楚大锋,徐刚,王伟,彭剑超,王均安,周邦新. 金属学报,2011;47: 269)

[19] Xu G, Chu D F, Cai L L, Zhou B X, Wang W, Peng J C. *Acta Metall Sin*, 2011;47: 905.
    (徐刚,楚大锋,蔡琳玲,周邦新,王伟,彭剑超. 金属学报,2011;47: 905)

[20] Miller M K. *Atom Probe Tomography: Analysis at the Atomic Level*. New York: Kliwer Academic/Plenum Publishers, 2000: 25.

[21] Starink M J, Zahra A M. *Philos Mag*, 1998;77: 187.

[22] Starink M J, Zahra A M. *Thermochim Acta*, 1997;292: 159.

[23] Miller M K, Russell K F, Pareige P, Starink M J, Thomson R C. *Mater Sci Eng*, 1998;A250: 49.

[24] Zhu J J, Wang W, Lin M D, Liu W Q, Wang J A, Zhou B X. *J Shanghai Univ (Nat Sci)*, 2008;5: 525.
(朱娟娟,王伟,林民东,刘文庆,王均安,周邦新. 上海大学学报(自然科学版),2008;5: 525)
[25] Buswell J T, English C A, Herherington M G, Phythian W J, Smith G D W, Worral G M. In: Steele L E ed. , *Proc 14th Int Symp Effects of Radiation on Materials*, Vol. 2, Andover, Massachusseetts: The American Society for Testing and Materials, 1990: 127.

Study on the Precipitation of Cu − Rich Clusters in the RPV Model Steel by APT

Abstract: Reactor pressure vessel (RPV) is nonreplaceable component for the pressurized water reactor (PWR) in the nuclear power plants. RPVs are usually made of low alloy ferritic steels and A508 − III steel is one type of these materials. After long-term service under the neutron irradiation, the ductile-to-brittle transition temperature (DBTT) of the RPV steel, which is the main parameter used to measure the degree of the embrittlement, will shift towards higher temperature. This phenomenon is termed irradiation-induced embrittlement, and it is a main factor to affect the operation safety and the lifetime of nuclear power plants. It is realized that the irradiation-induced embrittlement is mainly attributed to the precipitation of Cu-rich nanophases with a high number density. The precipitation process of Cu-rich nanophases can be well characterized by an atom probe tomography (APT) analysis for their size, composition and number density, and the Cu-rich nanophases obtained by the APT analysis are usually termed Cu-rich clusters. It is worthwhile to investigate the precipitation process of Cu-rich clusters by thermal aging for better understanding the mechanism of embrittlement. In order to accelerate the precipitation of Cu-rich clusters, experiment was performed by a RPV model steel containing higher Cu content than commercially available A508 − III steel. RPV model steel was prepared by vacuum induction melting with higher content of Cu (0. 6%, mass fraction). The specimens of the RPV model steel were tempered at 660 ℃ for 10 h followed by air cooling after water quenching from 880 ℃, and then they were isothermally aged at 370 ℃ for different time. The precipitation process of Cu-rich clusters is investigated by APT analysis. The results show that the Cu-rich clusters are on the stage of the nucleation when the specimens were aged at 370 ℃ for 1 150 h. After specimens were aged for 3 000 and 13 200 h, the average equivalent diameter of the Cu-rich clusters increases from 1. 5 nm to 2. 4 nm, and the average Cu content in the Cu-rich clusters vary from 45% to 55% (atomic fraction). The number density of the Cu-rich clusters in both types of the specimens is at the order of 10^{22} m^{-3}. The Cu concentration in the ferritic matrix is $(0.15 \pm 0.02)\%$ for the specimen aged at 370 ℃ for 13 200 h, which is still higher than the limitation of Cu solubility in the ferritic matrix at 370 ℃. It means that the precipitation process of Cu-rich clusters does not reach the equilibrium state. The analysis results also show that Ni, Si, P atoms, but not Cu atoms, segregate near the interface between the cementite and the ferritic matrix, and Mn, Mo, S atoms are enriched in the cementite.

利用 APT 对 RPV 模拟钢中界面上原子偏聚特征的研究*

摘　要： 核反应堆压力容器(RPV)模拟钢样品经过 660 ℃调质处理和 370 ℃时效 3 000 h 后，用原子探针层析法研究了晶界和相界面上原子偏聚的特征. 结果表明，Ni，Mn，Si，C，P 和 Mo 在晶界处均有不同程度的偏聚，偏聚倾向由强到弱依次为：C，P，Mo，Si，Mn 和 Ni. Cu 在晶界处会出现贫化现象. Si 在晶界上的偏聚程度与晶界的特性有关. 在这几种元素中，C 在晶界上偏聚的宽度最大，如以成分分布图中浓度峰的半高宽来比较，C 的偏聚宽度是 Mn，Ni 和 Mo 的 1.5 倍. 在富 Cu 相与 α-Fe 的相界面处，Ni 和 Mn 有明显的偏聚，而 C，P，Mo 和 Si 倾向偏聚在相界面的 α-Fe 一侧，且偏聚的程度比晶界处的低.

　　工程中应用的金属材料绝大多数都是多晶材料，不仅材料的成分体系和相组织结构会影响材料的性能，材料中的晶界和相界面对材料的性能也有重要的影响. 晶界上的化学特性（如：第二相析出和杂质原子或溶质原子的偏聚）会对材料的强度[1]、韧脆性[2]、蠕变[3]、疲劳[4,5]、腐蚀[6,7]和应力腐蚀[8,9]等性能产生重要的影响. 溶质原子或杂质原子在相界面上的偏聚会影响相界面的能量、相界面的迁移和相组织的稳定性，因而对材料的高温性能也会产生重要的作用[10,11]. 虽然使用 Auger 能谱仪或 X 射线能谱仪可对界面的偏聚进行研究[12,13]，但是结果不够精细. 原子探针层析法（APT）是在逐个分析原子种类的基础上构建不同原子在空间的三维分布，能够得出不同原子在界面上的偏聚浓度以及相互之间的关系. 因此，可借助 APT 对界面的偏聚问题进行细致的研究. 不过应用 APT 进行分析时，需要将样品制成曲率半径小于 100 nm 的针尖状样品，而且还要使被分析的界面处于距离针尖端部数十纳米以内的位置，这是该种实验技术困难之处. 近几年来，用这种方法研究了锆合金[14]、铝合金[15]、低碳马氏体高强钢[11,16]、316 不锈钢以及 690 合金[17,18]中的晶界偏聚问题，为了解晶界的化学特性积累了有意义的数据.

　　目前，大多数压水堆核电站的反应堆压力容器（RPV）都用 Mn-Ni-Mo 低合金铁素体钢（A508-III）制造，RPV 在运行工况下，长期经受中子辐照会导致材料的韧脆转变温度（DBTT）升高，影响反应堆的运行安全，并成为制约核电站服役寿命的主要因素. 这种辐照脆化主要是由于材料在辐照过程中引起高数量密度的富 Cu 团簇析出以及 P 等有害杂质在晶界偏聚所致[19,20]. 本工作采用 APT 研究了提高 Cu 含量后的 RPV 模拟钢中晶界偏聚以及富 Cu 相析出后相界面处的偏聚特征，这对于了解 RPV 钢的辐照脆化、析出强化机理和其热稳定性等具有重要意义.

1　实验方法

　　实验所用 RPV 模拟钢是在商用 A508-III 钢基础上将 Cu 含量由 0.08%（质量分数）以

---

* 本文合作者：徐刚、蔡琳玲、冯柳、刘文庆、王均安. 原发表于《金属学报》，2012，48(7)：789-796.

下提高至 0.6%,其他成分保持不变. RPV 模拟钢由真空感应炉冶炼,铸锭重约 40 kg,化学成分见表 1. 钢锭经热锻和热轧制成 4 mm 厚的钢板,切割成 4 mm×40 mm×30 mm 的样品,在 880 ℃加热 0.5 h 后水淬,然后在 660 ℃加热 10 h 进行调质处理,最后在 370 ℃等温时效 3 000 h. 制定该热处理制度的目的主要是为了研究 RPV 模拟钢中富 Cu 团簇的析出问题.

表 1 反应堆压力容器(RPV)模拟钢的化学成分

Table 1 Chemical compositions of the reactor pressure vessel (RPV) model steel

Composition	Cu	Mn	Ni	Si	P	C	S	Mo	Fe
Atomic fraction,%	0.53	1.60	0.81	0.77	0.03	1.00	0.011	0.31	Bal.
Mass fraction,%	0.60	1.58	0.85	0.39	0.016	0.22	0.006	0.54	Bal.

用电火花线切割方法从样品的中心部位切出长约 20 mm,截面边长约 0.5 mm 的方形细棒,分别用体积分数为 25% 的高氯酸乙酸和 2% 的高氯酸 2-丁氧基乙醇作电解液分两步抛光细棒,制备出曲率半径小于 100 nm 的针状样品,用于 APT 的实验分析,具体的样品制备方法及分析原理可参阅文献[21]. 用 LEAP 3000HR 型三维原子探针对针尖样品进行分析,样品冷却至 -223 ℃,脉冲频率为 2 000 kHz,脉冲分数为 20%.

2 结果与讨论

2.1 界面偏聚特征的观察

图 1 是样品经 APT 分析后得到的 C 原子空间分布图. 从图中可以看出 C 原子在图 1a 中所示的虚线附近发生了偏聚,从而将分析区域分为 3 部分,分别用数字 1,2 和 3 表示. 将 C

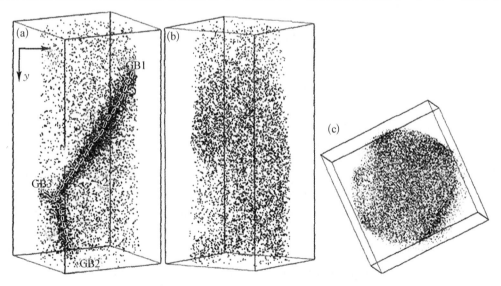

图 1 相互成 90°方向观察 C 原子在分析体积内的分布图

Fig. 1 Distributions of C atoms within an analyzed volume of 57 nm×56 nm×118 nm (a), Fig. 1a rotated around y direction for 90° (b) and Fig. 1a rotated around x direction for 90° (c) (GB1, GB2 and GB3 represent grain boundary 1, grain boundary 2 and grain boundary 3, respectively; 1, 2 and 3 represent grain 1, grain 2 and grain 3, respectively)

原子的三维空间分布图沿 y 轴旋转 90°后(图 1b)则未见明显的 C 原子偏聚现象,这表明 C 原子不是在一维的缺陷处偏聚,而是在二维的界面上偏聚,由此可以判定区域 1 和 2 及 2 和 3 为不同的晶粒,相互之间的界面为晶界;同样,沿 x 轴旋转 90°后(图 1c),可判定区域 1 和 3 也为 2 个不同的晶粒.这 3 个晶粒之间的晶界分别用 GB1,GB2 和 GB3 表示.

图 2 为分析体积中(57 nm×56 nm×118 nm)各元素的原子空间分布图.从图中可以看出,有 1 个富 Cu 相处于这 3 个晶粒的相交处,等效直径约为 26 nm.根据 Cu 在 α-Fe 中平衡固溶度公式[22]可知:在 660 ℃加热时,Cu 在 α-Fe 中的固溶度约为 0.46%(质量分数),比 RPV 模拟钢中的 Cu 含量低,因而,淬火后过饱和固溶的 Cu 在调质处理时会在位错或晶界处析出,为 fcc 结构[23].已有的研究[24]表明,调质态 RPV 模拟钢在 370 ℃时效 6 000 h 时,析出富 Cu 团簇的等效直径不超过 8 nm,由此可以确定这种较大的富 Cu 相应该是在 660 ℃调质处理过程中析出的,而不是在 370 ℃时效过程中形成的.从图中还可以看出,在富 Cu 相中还有 Ni 和 Mn 富集,其他元素则不明显.观察 C,Ni,Mn,P,Mo 和 Si 原子在空间的分布,可以看出它们在 3 个晶界处都有不同程度的偏聚,为了弄清楚各元素的原子在晶界或富 Cu 相与 α-Fe 基体的相界上的偏聚情况,需要对界面附近的成分进行定量分析.

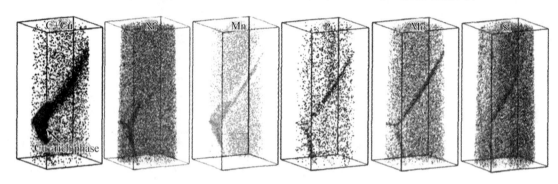

图 2 分析体积中 C,Cu,Ni,Mn,P,Mo 和 Si 原子的分布图

Fig. 2 Distributions of C, Cu, Ni, Mn, P, Mo and Si atoms within an analyzed volume of 57 nm×56 nm× 118 nm

2.2 晶界偏聚的成分分析

以界面的法向为 z 轴随机选取一个长方体区域进行晶界附近成分的分析,以便了解不同原子在晶界上的偏聚情况.在 GB1,GB2 和 GB3 处所选取的长方体位置及沿 z 轴分析获得的成分分布如图 3-5 所示.在成分分布图中,以 C 原子偏聚浓度最高处作为晶界的位置,在图中用虚线标出.从成分分布图中可以看出,不同原子在晶界附近几个纳米的区域内发生偏聚,比一般认为的晶界厚度要宽.这种差异主要有 2 方面的原因:在原子尺度上晶界不是一个平面,在分析的范围内晶界也不会很平整,因而沿 z 轴分析时晶界就比实际的宽,但是这种影响不会太大;APT 分析过程中,当原子从样品尖端发生场蒸发离开样品到达探测器时,因原子的热振动横向会发生偏移,这样,当界面与针状样品轴向不垂直时,测量得到杂质原子在界面上偏聚的宽度会出现放大效应[25].Blavette 等[26]研究发现这种放大效应与针状样品轴向和界面的夹角有关,当界面与样品的轴向夹角为 0°时,放大效应最大;当界面与样品轴向夹角为 90°时,放大效应最小,可以认为此时测定界面偏聚的宽度为真实的宽度;当界面与样品的轴向夹角为锐角时,放大效应介于二者之间.本实验中 GB1,GB2 和 GB3 晶界

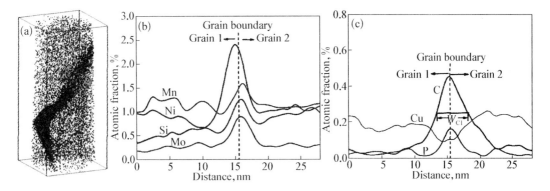

图 3　C 和 Cu 的原子分布图及 GB1 附近的成分分布图

Fig. 3　C and Cu atom map (a) and the composition profiles of elements Mn，Ni，Mo and Si (b) as well as Cu，P and C (c) on the grain boundary GB1 within an analyzed volume of 14 nm×17 nm×29 nm (W_{C1} is the width of the composition profiles at the half intensity for C atoms at GB1)

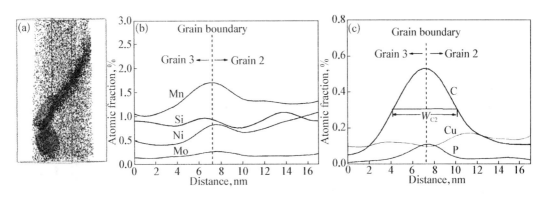

图 4　C 和 Cu 的原子分布图及 GB2 附近的成分分布图

Fig. 4　C and Cu atom map (a) and the composition profiles of elements Mn，Ni，Mo and Si (b) as well as Cu，P and C (c) on the grain boundary GB2 within an analyzed volume of 13 nm×13 nm×17 nm (W_{C2} is the width of the composition profiles at the half intensity for C atoms at GB2)

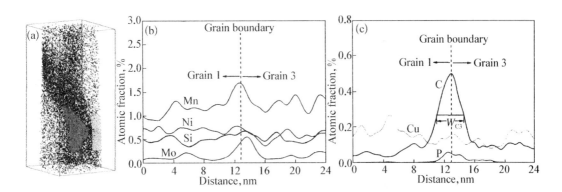

图 5　C 和 Cu 的原子分布图及 GB3 附近的成分分布图

Fig. 5　C and Cu atom map (a) and the composition profiles of elements Mn，Ni，Mo and Si (b) as well as Cu，P and C (c) on the grain boundary GB3 within an analyzed volume of 8 nm×8 nm×24 nm (W_{C3} is the width of the composition profiles at the half intensity for C atoms at GB3)

与样品轴向之间的夹角分别约为 $35°,15°$ 和 $75°$. 同一种原子在 3 个不同晶界上的偏聚宽度也存在差异,这应该是晶界与样品轴向的夹角不同产生的放大效应. 以 C 原子为例,其在 GB1,GB2 和 GB3 晶界上偏聚峰的半高宽 W_{C1},W_{C2} 和 W_{C3} 分别为 4.9,6.2 和 3.8 nm. 这与上面讨论的放大效应完全吻合,GB2 晶界与样品轴向的夹角最小,放大效应最大,C 原子偏聚浓度峰的半高宽也最宽. 比较不同原子在 3 个晶界处的成分分布图后可以看出,不同原子在不同晶界处的偏聚宽度存在差异,其中 C 原子的偏聚宽度最宽,其偏聚浓度峰的半高宽约为 Ni,Mn 和 Mo 原子的 1.5 倍,这应该与 C 的原子半径比较小有关.

从图 3—5 中不同原子在晶界附近的浓度分布曲线可以看出,同一种原子在同一个晶粒的不同区域中的浓度存在一定的差别,这是因为分析的体积比较小,在不同位置的小体积内,原子的浓度存在涨落起伏. 为了便于分析,取各晶粒中每种原子的平均浓度与其相应的原子在 3 个晶粒中的总平均浓度进行比较,结果如表 2 所示. 比较表 1 和表 2 中的成分差别可以看出,各种原子在 3 个晶粒中的总平均浓度比块体样品的化学成分低,这主要是由于第二相析出以及晶界偏聚都会不同程度地消耗合金元素或杂质元素所致. 各种原子在晶界处偏聚的最高浓度如表 3 所示,评判它们的偏聚倾向可以用原子偏聚的最高浓度与 3 个晶粒中相应原子总平均浓度的比值 Γ 来表示,$\Gamma>1$ 时,表明该原子在晶界处发生了偏聚,Γ 越大则偏聚倾向越大;$\Gamma<1$ 时,表明该原子在晶界处贫化. 图 6 为 3 个晶界处各种原子偏聚倾向的直方图. 从图 3—6 可以看出,C,P 和 Mo 原子在 3 个晶界处都有较强的偏聚倾向. Faulkner 和 Suzuki 等[27,28]用 Auger 能谱仪研究了 C 和 P 在晶界处的偏聚,认为它们在晶界上偏聚时存在竞争现象:在同一个晶界上,当 C 原子的偏聚浓度较大时,P 原子的偏聚倾向下降,而当 P 原子的偏聚浓度较大时,C 原子的偏聚倾向会下降. 但是从表 3 中 C 和 P 在晶界处偏聚的最高浓度来看,这种竞争现象并不明显. Ni 和 Mn 原子在晶界的偏聚倾向较弱. 以往的研究[16,29,30]表明,不管是低温长期时效还是中温回火,Si 原子在铁素体或马氏体钢的晶界处会贫化,而经过辐照后 Si 原子才会在晶界处偏聚. 本实验观察到 Si 原子在 GB1 处存在明显的偏聚现象,但是在 GB2 和 GB3 处并不明显,这表明 Si 原子在晶界处发生偏聚时具有晶界选择性,这可能与晶界的结构类型有关,或者说与晶界两侧的晶粒取向关系有关,也就是与晶界的能量有关. 当某一晶界的特性有利于 Si 原子发生偏聚时,Si 从两侧晶粒中向晶界扩散的速度又可能受到晶粒与晶界之间相对取向关系的影响,在取向有利侧的晶粒中会有更多的 Si 原子扩散至晶界上,从而导致该侧晶粒中 Si 的贫化. 在 GB1 晶界上存在 Si 原子明显偏聚的现象,而在晶界两侧的 GB1 和 GB2 中 Si 的浓度存在明显差别(图 3b),GB1 中的 Si 可能更容易向晶界扩散发生偏聚,因而造成晶粒中 Si 的贫化. 从图 6 中还可以看出,Cu 原子在晶界处发生贫化,这可能是因为当富 Cu 相在晶界处析出时,Cu 原子会沿晶界扩散,速率

表 2 晶界附近各晶粒中不同元素的平均含量以及 3 个晶粒中总的平均含量

Table 2 Average concentrations of different elements within grain 1, grain 2 and grain 3 and the average concentration of different elements on all of the three grains

(atomic fraction, %)

Position	Cu	Mn	Ni	Si	P	C	Mo
Grain 1	0.18±0.04	1.22±0.10	0.75±0.17	0.51±0.10	0.03±0.015	0.04±0.008	0.15±0.06
Grain 2	0.17±0.05	1.31±0.05	0.94±0.11	1.02±0.05	0.02±0.004	0.09±0.01	0.24±0.03
Grain 3	0.14±0.03	1.29±0.06	0.61±0.11	0.69±0.13	0.03±0.013	0.07±0.02	0.14±0.02
Average	0.16±0.02	1.27±0.15	0.77±0.16	0.74±0.22	0.023±0.007	0.07±0.04	0.18±0.06

表3　3个晶界处不同元素的原子发生偏聚后的最高浓度

Table 3　Maximum segregation concentrations of the different atoms at three grain boundaries

(atomic fraction, %)

Position	Cu	Mn	Ni	Si	P	C	Mo
GB1	0.09	1.59	1.26	2.4	0.17	0.45	0.90
GB2	0.10	1.71	0.84	0.97	0.11	0.53	0.28
GB3	0.14	1.69	0.77	0.67	0.06	0.49	0.55

较快,同时,基体中的 C,P,Si,Ni,Mn 和 Mo 原子会向晶界偏聚,从而阻碍晶粒中的 Cu 原子向晶界扩散,因而造成 Cu 原子在晶界处的贫化现象. 从总体上说,各元素在晶界处偏聚倾向由强到弱依次为:C,P,Mo,Si,Mn 和 Ni.

2.3　相界面处原子偏聚的成分分析

富 Cu 相与晶粒 1,晶粒 2 和晶粒 3 的界面分别用 PI1,PI2 和 PI3 表示,在 PI1,PI2 和 PI3 处所选取的长方体位置及沿 z 轴分析获得的成分分布图如图 7—9 所示. 设定 Cu 和 Fe 原子随距离变化的浓度曲线的相交处为相界面,在成分分布图中用虚线标出. 从图 7—9 的相界面附近成分分布图可以看出,Ni 和 Mn 原子在相界面处发生明显偏聚,其他元素的原子在各相界面处有不同程度的偏聚,并倾向偏聚在相界面的 α-Fe 一侧.

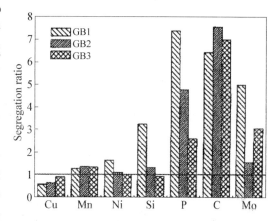

图 6　各元素原子在 3 个晶界处偏聚后的最高浓度与 3 个晶粒中相对应元素总平均浓度的比例关系对比图

Fig. 6　Comparison of the ratio of the maximum segregation concentration for different elements at three grain boundaries from the average content on all of the three grains

富 Cu 相中主要的合金元素是 Fe,其平均浓度约为 3.5%(原子分数);Ni 在富 Cu 相中的平均浓度约为 1.5%(原子分数),比基体中的平均浓度高约 1 倍;而 Mn,Si,C,P 和 Mo 的浓度都比基体中的低. 从 Cu-Fe 二元相图可知,Cu 中不可能固溶如此多的 Fe,本课题组的研究[31]结果表明,即使纳米富 Cu 相中含有更多的 Fe,Ni 和 Mn 等元素,它们仍然是固溶体,这应该是纳米尺度效应的结果.

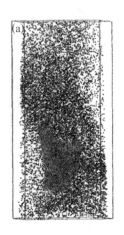
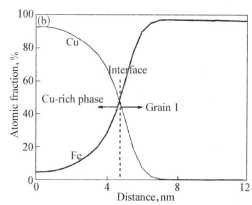

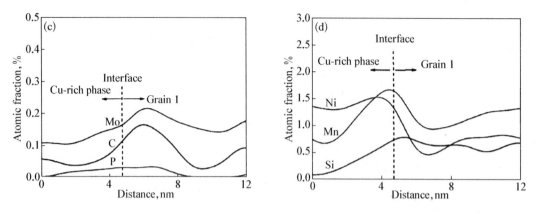

图 7 C 和 Cu 的原子分布图及富 Cu 相与晶粒 1 界面附近的成分分布图

Fig. 7 C and Cu atom map (a) and composition profiles of elements Fe and Cu (b), Mo, P and C (c) as well as Ni, Mn and Si (d) at the interface between Cu-rich phase and grain 1 within an analyzed volume of 7 nm×7 nm×12 nm

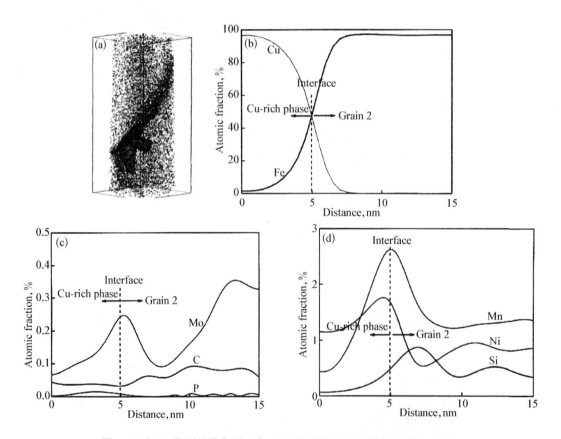

图 8 C 和 Cu 的原子分布图及富 Cu 相与晶粒 2 界面附近的成分分布图

Fig. 8 C and Cu atom map (a) and composition profiles of elements Fe and Cu (b), Mo, P and C (c) as well as Ni, Mn and Si (d) at the interface between Cu-rich phase and grain 2 within an analyzed volume of 7 nm×7 nm×15 nm

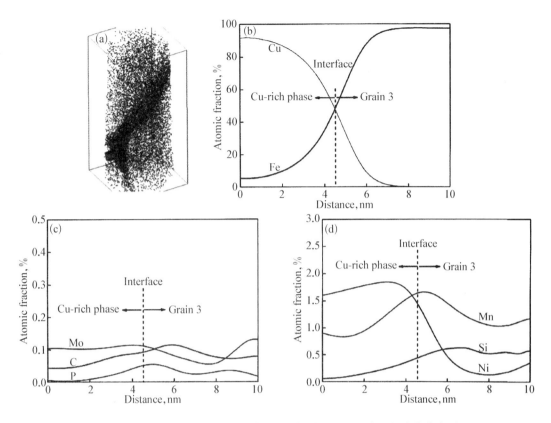

图 9 C 和 Cu 的原子分布图及富 Cu 相与晶粒 3 界面附近的成分分布图

Fig. 9 C and Cu atom map (a) and composition profiles of elements Fe and Cu (b), Mo, P and C (c) as well as Ni, Mn and Si (d) at the interface between Cu-rich phase and grain 3 within an analyzed volume of 6 nm×6 nm×10 nm

表 4 富 Cu 相与各晶粒的界面处几种元素的最高含量

Table 4 Maximum concentrations of the elements at the interfaces among the Cu-rich phase and the different grains (atomic fraction，%)

Position	Cu	Mn	Ni	Si	P	C	Mo	Fe
PI1	—	1.67	1.51	0.78	0.034	0.16	0.21	—
PI2	—	2.63	1.76	0.87	0.013	0.06	0.24	—
PI3	—	1.65	1.84	0.62	0.055	0.11	0.11	—

Note：PI1，PI2 and PI3 represent the interfaces between the Cu-rich phase and the grain 1，grain 2 and grain 3，respectively

　　相界面 PI1，PI2 和 PI3 处不同原子的最高浓度值统计结果见表 4,结合表 2 中各原子在基体中的平均浓度值可以得出相界面处各种原子偏聚程度的直方图(图 10).从图中可以看出,Ni 和 Mn 在相界面处的偏聚程度较高,并高于其在晶界处的偏聚程度,这应该与偏聚后可以降低相界面能有关[24]. Mo, P, C 和 Si 原子在相界面处的偏聚程度远低于其在晶界处的偏聚程度,甚至有的相界面处还会发生贫化,这可能是由于相界面处的能量比晶界处的高,不利于某些原子的偏聚所导致.

3 结论

（1）Ni，Mn，Si，C，P 和 Mo 原子在晶界处有不同程度的偏聚，其中，C 在晶界上偏聚的宽度最宽，如以成分分布图中峰的半高宽比较，C 原子的偏聚宽度是 Mn，Ni，Mo 原子的 1.5 倍；各元素在晶界处偏聚倾向由强到弱依次为：C，P，Mo，Si，Mn 和 Ni；Cu 在晶界处会出现贫化现象.

（2）Si 原子在晶界上的偏聚与晶界特性有关.

（3）在富 Cu 相与 α‑Fe 的相界面处，Ni 和 Mn 原子有明显的偏聚；而 C，P，Mo 和 Si 原子倾向偏聚在相界面的 α‑Fe 一侧，且偏聚的程度比晶界处的低.

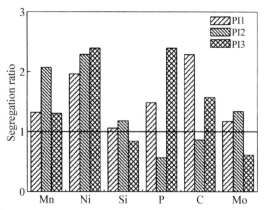

图 10 富 Cu 相与各晶粒的界面处各元素原子的偏聚浓度与基体中相对应原子浓度的比例关系对比图

Fig. 10 Comparison of the segregation ratio at the interfaces of Cu‑rich phase and α‑Fe grains from the average composition of the matrix

参 考 文 献

[1] Takaki S, Furjioka M, Aihara S, Nagataki Y, Yamashita T, Sano N, Adachi Y, Nomura M, Yaguchi H. *Mater Trans*, 2004;45: 2239.

[2] Garcia-Mazario M, Lancha A M, Hernández-Mayoral M. *J Nucl Mater*, 2007;360: 293.

[3] Laha K, Kyono J, Kishimoto S, Shinya N. *Scr Mater*, 2005;52: 675.

[4] Bowen P, Hippsley C A, Knott J F. *Acta Metall*, 1984;32: 637.

[5] Bulloch J H. *Int J Pres Ves Pip*, 1988;33: 197.

[6] Wang K, Xu T D, Shao C, Yang C. *J Iron Steel Res Int*, 2011;18: 61.

[7] Wei W, Grabke H J. *Corros Sci*, 1986;26: 223.

[8] Atrens A, Wang J Q, Stiller K, Andren H O. *Corros Sci*, 2006;48: 79.

[9] Heo N H, Jung Y C, Lee J K, Kim K T. *Scr Mater*, 2008;59: 1200.

[10] Lemarchand D, Cadel E, Chambreland S, Blavette D. *Philos Mag*, 2002;82A: 1651.

[11] Kolli R P, Seidman D N. *Acta Mater*, 2008;56: 2073.

[12] Wu J, Song S H, Weng L Q, Xi T H, Yuan Z X. *Mater Charact*, 2008;59: 261.

[13] Khalid F A. *Scr Mater*, 2001;44: 797.

[14] Hudson D, Smith G D W. *Scr Mater*, 2009;61: 411.

[15] Sha G, Yao L, Liao X Z, Ringer S P, Duan Z C, Langdon T G. *Ultramicroscopy*, 2011;111: 500.

[16] Isheim D, Kolli R P, Fine M E, Seidman D N. *Scr Mater*, 2006;55: 35.

[17] Etienne A, Radiguet B, Cunningham N J, Odette G R, Valiev R, Pareige P. *Ultramicroscopy*, 2011;111: 659.

[18] Li H, Xia S, Zhou B X, Liu W Q. *Mater Charact*, 2012;66: 68.

[19] Toyama T, Nagai Y, Tang Z, Hasegawa M, Almazouzi A, van Walle E, Gerard R. *Acta Mater*, 2007;55: 6852.

[20] Bischler P J E, Wild R K. In: Gelles D S, Nanstad R K, Kumar A S, Little E A eds. , *Effects of Radiation on Materials: 17th International Symposium*, *ASTM STP 1270*, West Conshohocken, PA: American Society for Testing and Materials, 1996: 260.

[21] Miller M K. *Atom Probe Tomography: Analysis at the Atomic Level*. New York: Kliwer Academic/Plenum Publishers, 2000: 25.

[22] Yong Q L. *Secondary Phase in Steel*. Beijing: Metallurgical Industry Press, 2006: 127.

（雍其龙. 钢铁材料中的第二相. 北京：冶金工业出版社，2006：127）

[23] Hornbogen E, Glenn R C. *Trans Metall Soc AIME*, 1960；218：1064.

[24] Xu G, Chu D F, Cai L L, Zhou B X, Wang W, Peng J C. *Acta Metall Sin*, 2011；7：905.

（徐刚，楚大锋，蔡琳玲，周邦新，王伟，彭剑超. 金属学报，2011；47：905）

[25] Vurpillot F, Cerezo A, Blavette D, Larson D J. *Microsc Microanal*, 2004；10：384.

[26] Blavette D, Duval P, Letellier L, Guttmann M. *Acta Mater*, 1996；44：4995.

[27] Faulkner R G, Jones R B, Zheng L, Flewett P E J. *Philos Mag*, 2005；85：2065.

[28] Suzuki S, Obata M, Abiko K, Kimura H. *Scr Metall*, 1983；17：1325.

[29] Cerezo A, Clifton P H, Lozano-Perez S, Panayi P, Sha G, Smith G D W. *Microsc Microanal*, 2007；13：408.

[30] Jiao Z, Was G S. *Acta Mater*, 2011；59：4467.

[31] Chu D F, Xu G, Wang W, Peng J C, Wang J A, Zhou B X. *Acta Metall Sin*, 2011；47：269.

（楚大锋，徐刚，王伟，彭剑超，王均安，周邦新. 金属学报，2011；47：269）

Segregation of Atoms on the Interfaces in the RPV Model Steel Studied by APT

Abstract：The segregation of impurity or solute atoms to grain boundaries as well as phase interfaces can either improve or degrade the chemical, physical and mechanical properties of alloys. This phenomenon has been studied widely for iron based alloys, and the analysis method by an atom probe tomography (APT) is a powerful tool for better understanding this problem. The resulting composition changes of grain boundaries and phase interfaces, as well as the precipitation of Cu-rich nanophases, are frequently associated with the phenomenon of embrittlement in ferritic reactor pressure vessel (RPV) steels. The present work was carried out to study the segregation of impurity or solute atoms to grain boundaries as well as phase interfaces in a RPV model steel with higher content of Cu (0.53%, atomic fraction) than commercially available one. The RPV model steel was prepared by vacuum induction melting. The specimens were further heat treated by water quenching at 880 ℃ for 30 min and tempering at 660 ℃ for 10 h, and finally aged at 370 ℃ for 3 000 h. The results show that the segregation amount of Ni, Mn, Si, C, P and Mo atoms on grain boundaries are varied. The sequence of segregation tendency for different atoms from strong to weak is C, P, Mo, Si, Mn and Ni, whilst Cu atoms were clearly depleted at the grain boundaries. Si atoms also segregate to the grain boundaries, but it depends on the characteristic of the grain boundaries. The C segregation range at grain boundaries is the widest. According to the width of the composition profiles at the half intensity for different atoms at the grain boundaries, the segregation range of C atoms is 1.5 times wider than that of Mn, Ni and Mo atoms. Furthermore, Ni and Mn atoms evidently segregate to the interfaces between the Cu-rich phase and the α-Fe matrix, while C, P, Mo, Si atoms prefer to segregate towards the α-Fe matrix near the interfaces, but their segregation amount at the interfaces of Cu-rich phase and the α-Fe matrix is less than that at the grain boundaries.

添加 Bi 对 Zr-4 合金在 400 ℃/10.3 MPa 过热蒸汽中耐腐蚀性能的影响*

摘　要： 在 Zr-4 合金基础上添加 0.1%—0.5% Bi(质量分数)制备成 Zr-4+xBi 合金,用高压釜腐蚀实验研究了 Bi 含量对 Zr-4+xBi 合金在 400 ℃/10.3 MPa 过热蒸汽中耐腐蚀性能的影响;用 TEM, EDS 和 SEM 观察了合金和合金腐蚀后氧化膜的显微组织. 结果表明：随着 Bi 含量的增加,Zr-4+xBi 合金中第二相的尺寸和形状变化不大,但数量增多,并出现了不同成分的第二相,包括 Zr(Fe, Cr)$_2$, Zr-Fe-Cr-Bi, Zr-Fe-Sn-Bi 和 Zr-Fe-Cr-Sn-Bi. 在 Zr-4+0.1Bi 合金中检测到了含 Bi 的第二相,这说明 580 ℃时 Bi 在 Zr-4+xBi 合金 α-Zr 基体中的固溶度小于 0.1%. 另外,适量 Bi 的添加促进了原先固溶在 α-Zr 基体中 Sn 的析出. 与 Zr-4 合金相比,在 Zr-4 中添加 0.1%—0.5% Bi 后合金的耐腐蚀性能反而下降,并随着 Bi 含量的增加耐腐蚀性能恶化趋势越显著,这说明 Zr-4 合金中添加 Bi 并不能改善合金的耐腐蚀性能,反而产生有害的影响,这应该与含 Bi 第二相和同时含有 Bi, Sn 第二相的析出有关.

锆合金具有较低的热中子吸收截面(0.18 barn),对核燃料具有良好的相容性,在 300—400 ℃的高温高压水和过热蒸汽中具有较好的耐腐蚀性能和适中的力学性能,因此是压水堆中用作核燃料元件包壳的一种重要结构材料. 为降低核电成本,需要加深核燃料的燃耗,延长换料周期,这就需要开发高燃耗的燃料组件,要求高性能的锆合金包壳材料作为支撑. 核燃料元件在反应堆堆芯中工作时,锆合金包壳处在中子辐照和高温高压冷却水高速冲刷等极端苛刻的工况环境中,因受到高温高压水的腐蚀生成 ZrO$_2$ 膜,使锆合金包壳的有效厚度减薄,影响锆合金的使用寿命. 因此,提高包壳的耐水侧腐蚀性能是开发高性能锆合金主要关注的问题之一.

对现有工程应用锆合金进行成分调整(包括现有合金成分的优化和添加其他合金元素)是研制性能优异锆合金的基本方法. Zr-4(Zr-1.5Sn-0.2Fe-0.1Cr,质量分数,%,下同)合金是用作压水堆中的第一代包壳材料,在 Zr-4 合金成分基础上添加 Nb 发展了 ZIRLO(Zr-1Sn-1Nb-0.1Fe)[1,2],E635(Zr-1.2Sn-1Nb-0.4Fe)[3] 和 N18(Zr-1Sn-0.35Nb-0.3Fe-0.1Cr)[4] 等 Zr-Sn-Nb 系合金,其中 ZIRLO 和 E635 合金已得到了工程应用. 韩国在 Zr-1Nb 合金中添加少量的 Cu 发展了 HANA-6(Zr-1.1Nb-0.05Cu)合金[5]. Hong 等[6] 研究发现,添加 Cu 对 Zr-4 合金的耐腐蚀性能有一定的改善作用. 基于 Cu 对改善 Zr-4 和 Zr-1Nb 合金耐腐蚀性能的有益作用,本课题组系统研究了添加不同含量的 Cu 对 Zr-Sn-Nb-Fe-Cr 和 Zr-1Nb 合金在不同腐蚀条件下耐腐蚀性能的影响,结果发现其影响规律是不同的,这主要取决于合金体系和腐蚀条件[7,8]. 关于添加其他合金元素对合金耐蚀性的影响,初步的研究结果[9] 表明,添加适量的合金元素 Bi 对 Zr-1Nb 合金的耐腐蚀性能有改善作用. 另外,从热中子吸收截面这一特性考虑,Bi 也是一种合适的合金添加元素. 因此,本课题组系统开展了添加不同含量 Bi 对 Zr-4, Zr-Sn-Nb-Fe-Cr 和 Zr-1Nb 合金在多

---

* 本文合作者：姚美意、邹玲红、谢兴飞、张金龙、彭剑超. 原发表于《金属学报》,2012,48(9)：1097-1102.

种腐蚀条件下耐腐蚀性能的影响研究,探究其影响规律与锆合金体系和腐蚀条件之间的关系,这有助于深入认识合金元素影响锆合金在不同腐蚀条件下的腐蚀机理.本工作主要研究了添加不同含量 Bi 对 Zr-4 合金在 400 ℃/10.3 MPa 过热蒸汽中耐腐蚀性能的影响.

1 实验材料及方法

以 Zr-4 为母合金,在其成分基础上添加不同含量 Bi(熔炼时考虑了 Bi 的挥发损耗),采用非自耗真空电弧炉熔炼成不同 Bi 含量的 Zr-4+xBi 锆合金(x=0, 0.1, 0.3, 0.5),其中 Bi 含量为 0 的合金即为重熔的 Zr-4(定义为 Zr-4 remelted),这是为了在相同的制备工艺下比较 Bi 含量对 Zr-4 合金耐腐蚀性能的影响.合金锭经热压、β 相均匀化处理、热轧、β 相水淬、冷轧和最终 580 ℃/50 h 退火处理制备成腐蚀实验用的片状样品,采用 580 ℃/50 h 最终退火的目的是使 α-Zr 基体中固溶的合金元素尽可能达到平衡,具体制备工艺见文献[8].用 ICPAES(电感耦合等离子体原子发射光谱)分析得到的合金成分列在表 1 中,与 Zr-4 母合金相比,Sn 含量有不同程度的挥发损耗,Bi 含量与设计值吻合.将上述制备的样品用混合酸(体积比为 45% H$_2$O+45% HNO$_3$+10% HF)酸洗和去离子水清洗后,放入高压釜中进行 400 ℃/10.3 MPa 过热蒸汽的腐蚀实验,用 5 块试样增重平均值得出腐蚀增重.

表 1 实验用 Zr-4+xBi 合金成分
Table 1 Chemical compositions of Zr-4+xBi alloys (mass fraction, %)

Alloy	Sn	Fe	Cr	Bi	Zr
Zr-4 remelted	1.43	0.22	0.10	—	Bal.
Zr-4+0.1Bi	1.33	0.23	0.10	0.10	Bal.
Zr-4+0.3Bi	1.32	0.23	0.11	0.29	Bal.
Zr-4+0.5Bi	1.34	0.23	0.11	0.51	Bal.

用 JEM-200CX 透射电镜(TEM)观察合金的显微组织.用 JEM-2010F 型高分辨透射电镜及其配置的能谱仪(EDS)观察分析合金中第二相的形貌和成分.用 JSM-6700F 型扫描电子显微镜(SEM)观察腐蚀样品氧化膜的断口形貌.TEM 试样采用双喷电解抛光方法制备,电解液为 10%(体积分数)高氯酸酒精溶液,抛光电压为 50 V,温度为 -40 ℃.氧化膜断口形貌观察用试样的制备是从腐蚀样品切下一小块,用混合酸腐蚀掉金属基体,然后用镊子将氧化膜折断[10],在试样表面蒸镀一层金属 Ir 以提高 SEM 图像质量.

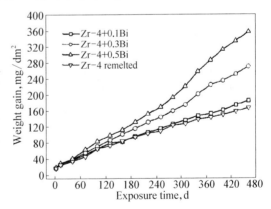

图 1 Zr-4+xBi 合金在 400 ℃/10.3 MPa 过热蒸汽中的腐蚀增重曲线

Fig. 1 Weight gain of the Zr-4+xBi alloys *vs* exposure time in superheated steam at 400 ℃/10.3 MPa

2 实验结果

2.1 腐蚀增重

图 1 给出了 Zr-4+xBi(x=0, 0.1, 0.3,

0.5)合金在 400 ℃/10.3 MPa 过热蒸汽中腐蚀后的增重曲线. 与重熔的 Zr-4 相比,添加 0.1%—0.5% Bi 后合金的耐腐蚀性能反而下降,并且随 Bi 含量增加耐腐蚀性能恶化程度愈明显,如 Zr-4+0.1Bi 合金腐蚀 462 d 的增重为 185.7 mg/dm²,略高于 Zr-4 重熔样品 (167.9 mg/dm²),而 Zr-4+0.3Bi 和 Zr-4+0.5Bi 合金腐蚀 462 d 的增重分别高达 270.8 和 357.9 mg/dm²,分别是 Zr-4 重熔样品的 1.6 倍和 2.1 倍. 由此可见,在 Zr-Sn 系的 Zr-4 合金中添加少量的 Bi 并不能改善合金在 400 ℃/10.3 MPa 过热蒸汽中的耐腐蚀性能,反而使耐腐蚀性能恶化.

2.2 合金的显微组织

图 2 是腐蚀前 Zr-4+xBi 系列合金显微组织的 TEM 像. 从图 2 可以看出绝大多数第二相非常细小(小于 60 nm),呈球形,但也有少量大于 100 nm 的第二相,形状不规则;Bi 含量的增加对第二相的尺寸影响不大,但使第二相的数量增多;合金中第二相多呈条带状分布,这应该是 β 相水淬时残留的 β-Zr 在随后的冷轧和退火过程中发生分解的结果.

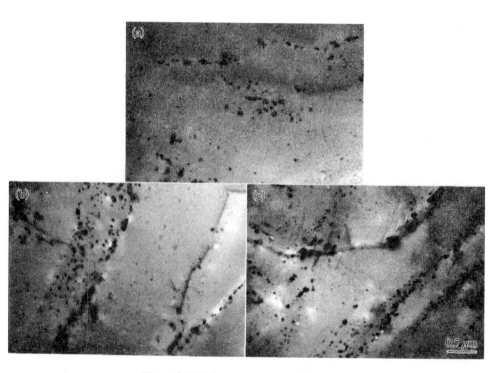

图 2 腐蚀前 Zr-4+xBi 合金的 TEM 像

Fig. 2 TEM images of Zr-4+0.1Bi (a), Zr-4+0.3Bi (b) and Zr-4+0.5Bi (c) alloys before corrosion tests

表 2 和表 3 分别总结了 Zr-4+xBi 系列合金中由 EDS 分析得到的典型第二相的成分和各种第二相出现的几率等信息. 从表 2 和表 3 可以看出: (1) Zr-4+0.1Bi 合金中的大部分第二相成分为 Zr-Fe-Cr,Fe/Cr 原子比在 1.7—2.0,这应该是 Zr(Fe, Cr)₂[7];另外,还发现了极少量的 Bi/Sn 比为 0.25 的 Zr-Fe-Sn-Bi 第二相,这说明 580 ℃时 Bi 在 Zr-4+xBi 合金 α-Zr 基体中的固溶度较小(<0.1%). (2) Zr-4+0.3Bi 合金中的部分第二相成分为 Zr-Fe-Cr,Fe/Cr 原子比在 1.6—1.9,这应该是 Zr(Fe, Cr)₂[7];部分为 Zr-Fe-Cr-

Sn‐Bi,Fe/Cr 原子比为 3.4—3.8,Bi/Sn 原子比约为 0.5.（3）Zr‐4+0.5Bi 合金中的第二相都含有 Bi,主要有 3 种成分的第二相,一种为 Zr‐Fe‐Cr‐Bi,Bi 含量较低,Fe/Cr 原子比为 0.9—1.3,这可能是 Zr(Fe,Cr,Bi)$_2$,一部分 Bi 取代了 Fe 和 Cr;一种是 Zr‐Fe‐Sn‐Bi,Bi/Sn 比 0.8—1.0;还有一种是 Zr‐Fe‐Cr‐Sn‐Bi,Fe/Cr 原子比为 1.6—2.2,Bi/Sn 原子比为 1.0—1.3.可见随着 Bi 含量的增加,Zr‐4+xBi 合金中第二相的成分和种类都会发生变化.大量的研究结果表明 Zr‐4 合金中的第二相为 Zr(Fe,Cr)$_2$,并不存在含 Sn 的第二相[11-13],但很明显 Bi 的添加促进了 Sn 以第二相形式析出,降低了原先固溶在 α‐Zr 基体中的 Sn 含量.

表 2　Zr‐4+xBi 合金中典型第二相的成分

Table 2　Compositions of the typical second phase particles in Zr‐4+xBi alloys

(atomic fraction,%)

Precipitate	Sn	Fe	Cr	Bi	Zr
Zr‐Fe‐Cr	—	29.18	14.79	—	55.64
Zr‐Fe‐Sn‐Bi	5.62	2.71	—	1.40	90.26
Zr‐Fe‐Sn‐Bi	3.50	3.94	—	3.49	89.06
Zr‐Fe‐Cr‐Bi	0.55	1.02	0.98	0.51	96.94
Zr‐Fe‐Cr‐Sn‐Bi	2.54	11.31	6.04	2.54	77.14

表 3　Zr‐4+xBi 合金中各种第二相信息

Table 3　Details of the precipitates in Zr‐4+xBi alloys

Alloy	Zr‐Fe‐Cr		Zr‐Fe‐Cr‐Bi		Zr‐Fe‐Sn‐Bi		Zr‐Fe‐Cr‐Sn‐Bi		
	Frequency	Fe/Cr	Frequency	Fe/Cr	Frequency	Bi/Sn	Frequency	Fe/Cr	Bi/Sn
Zr‐4+0.1Bi	Major	1.7—2.0	—	—	Minor	0.25	—	—	—
Zr‐4+0.3Bi	Partly	1.6—1.9	—	—	—	—	Partly	3.4—3.8	0.5
Zr‐4+0.5Bi	—	—	Partly	0.9—1.3	Partly	0.8—1.0	Partly	1.6—2.2	1.0—1.3

Note：Fe/Cr and Bi/Sn‐atomic ratio

3　分析讨论

　　大量的研究结果[14-17]表明,在 Zr‐Sn 系和 Zr‐Sn‐Nb 系合金中适当降低 Sn 含量均可以改善锆合金的耐腐蚀性能,以上文献中研究的 Sn 含量主要以固溶在 α‐Zr 基体中的形式存在.这说明适当降低固溶在 α‐Zr 基体中的 Sn 含量可以改善合金的耐腐蚀性能.因此,从添加 Bi 降低了固溶在 α‐Zr 基体中 Sn 含量的角度应该是有利于改善合金耐腐蚀性能的.但本研究发现 Bi 的添加会降低合金的耐腐蚀性能,并随着 Bi 含量的增加进一步降低(图1).显微组织观察结果表明,Bi 在 α‐Zr 基体中的固溶度较小,主要以含 Bi 第二相析出(表 2 和表 3),这说明含 Bi 第二相的析出是降低合金耐腐蚀性能的主要原因.

　　Zr 在高温高压水或者过热水蒸气中发生腐蚀,是由于 O^{2-} 或 OH$^-$ 通过氧化膜中的阴离子空位或者晶界扩散到金属/氧化膜界面处,与 Zr 反应生成 ZrO$_2$,同时电子沿反方向扩散而不断进行的[18,19].因此,氧化膜的性质与显微组织将会在 Zr 发生腐蚀氧化的过程中产生影响.Cox[20]很早就注意到锆合金腐蚀转折后氧化膜中的裂纹和孔隙,并用汞压孔隙仪测定了氧化膜中的孔隙.对于其是如何产生的,周邦新等[10,21]认为：由于 Zr 的 P.B.比(金属氧化物与金属的体积比)为1.56,因此Zr在发生氧化时体积会发生膨胀,同时受到金属基体的约束,使得氧化膜的内部产生压应力,从而使ZrO₂晶体内产生许多缺陷. 这些缺陷在时间、应力和温度的作用下,发生扩散、湮没和凝聚,空位被晶界吸收后形成纳米大小的孔隙簇,孔隙簇进一步发展成微裂纹,引起腐蚀加速. 这从氧化膜的断口形貌上可以得到证实. 图3是Zr-4+xBi合金在400℃/10.3 MPa过热蒸汽中腐蚀462 d的氧化膜断口形貌. 图中可以观察到微孔隙、微裂纹以及柱状晶演变为等轴晶的区域(图3中的方框所示);另外,还可以看到氧化膜断口并不平整. 凹凸起伏程度较大,这是氧化膜内部存在微孔隙、微裂纹等缺陷在折断后表现出的结果.

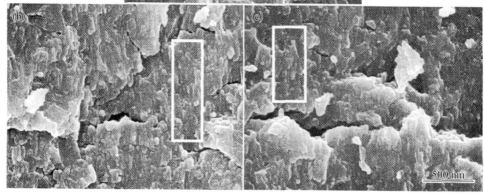

图3 Zr-4+xBi合金在400℃/10.3 MPa过热蒸汽中腐蚀462 d的氧化膜断口形貌

Fig. 3 Fracture surface morphology of the oxide film formed on Zr-4+0.1Bi (a), Zr-4+0.3Bi (b) and Zr-4+0.5Bi (c) alloys corroded in 400℃/10.3 MPa superheated steam for 462 d exposure

在Zr发生氧化时,合金中的第二相和固溶在α-Zr中的合金元素也会发生氧化,影响氧化膜的性质和结构,最终影响合金的耐腐蚀性能. Zr的P.B.比为1.56,Fe生成FeO的P.B.比为1.77,Cr的P.B.比为2.02,而Bi和Sn的P.B.比分别为2.27和1.32[18],可见,Bi的P.B.比最大. 第二相中的Bi发生氧化时将在局部产生更大的附加应力,使氧化膜中容易出现孔隙和微裂纹,对氧化膜的显微组织结构演化产生影响,最终加速腐蚀. 同时Bi的添加促进了原本固溶在α-Zr基体中的一部分Sn以第二相形式析出,由于第二相分布的局部性,与固溶在α-Zr基体中的Sn影响相比,第二相中的Sn氧化后对腐蚀形成的氧化膜影响也更具局部性. Sn的P.B.比大于1,这也必然会在局部产生附加应力,促进氧化膜中孔隙和微裂纹的形成,从而加速腐蚀.

另外,由于 Fe 和 Cr 在 α-Zr 中的最大固溶度仅为 120 和 200 $\mu g/g$[22],因此绝大部分 Fe 和 Cr 存在于第二相中. 本课题组前期研究[23,24]发现,适当提高固溶在 α-Zr 基体中的合金元素 Fe 和 Cr 的含量,能显著提高 Zr-4 合金在 LiOH 水溶液中的耐腐蚀性能. 在 Zr-4 合金中添加 Bi 后,可能会促进 Fe 和 Cr 的进一步析出,使固溶在 α-Zr 基体中的 Fe 和 Cr 的含量降低,从而也可能导致合金耐腐蚀性能下降.

4 结论

(1) Bi 含量对 Zr-4+xBi 合金中的第二相尺寸和形状影响不大,但对其数量和成分有很大影响:随着 Bi 含量的增加,第二相数量增多,出现了包括 Zr(Fe, Cr)$_2$, Zr-Fe-Sn-Bi, Zr-Fe-Cr-Sn-Bi 和 Zr-Fe-Cr-Bi 不同成分的第二相. 适量 Bi 的添加促进了原先固溶在 α-Zr 基体中的 Sn 以第二相形式析出.

(2) 在 Zr-4+0.1Bi 合金中检测到了含 Bi 的第二相,这说明 580℃时 Bi 在 Zr-4+xBi 合金 α-Zr 基体中的固溶度小于 0.1%.

(3) 在 Zr-4 中添加 0.1%—0.5% Bi 后的合金在 400 ℃/10.3 MPa 过热蒸汽中的耐腐蚀性能均不如 Zr-4 合金,Bi 含量越高耐腐蚀性能越差,说明添加 Bi 会对 Zr-4 合金的耐腐蚀性能产生有害影响,这应该与合金中含 Bi 第二相和同时含有 Bi, Sn 第二相的析出有关.

参 考 文 献

[1] Sabol G P. In: Rudling P, Kammenzind B eds. , *Zirconium in the Nuclear Industry: Fourteenth Internatianal Symposium*, *ASTM STP 1467*, Stockholm: ASTM International, 2004: 3.

[2] Sabol G P, Comstock R J, Weiner R A. In: Garde A M, Bradley E R eds. , *Zirconium in the Nuclear Industry: Tenth International Symposium*, *ASTM STP 1245*, Baltimore, MD: ASTM International, 1994: 724.

[3] Nikulina A V, Markelov V A, Peregud M M. In: Bradley E R, Sabol G P eds. , *Zirconium in the Nuclear Industry: Eleventh International Symposium*, *ASTM STP 1295*, Garmisch-Partenkirchen, Germany: ASTM International, 1996: 785.

[4] Zhao W J, Miao Z, Jiang H M, Yu X W, Li W J, Li C, Zhou B X. *J Chin Soc Corros Prot*, 2002;22: 124.
(赵文金,苗志,蒋宏曼,于晓卫,李卫军,李聪,周邦新. 中国腐蚀与防护学报,2002;22:124)

[5] Park J-Y, Yoo S J, Choi B-K, Jeong Y H. *J Nucl Mater*, 2008;373: 343.

[6] Hong H S, Moon J S, Kim S J, Lee K S. *J Nucl Mater*, 2001;297: 113.

[7] Yao M Y, Li S L, Zhang X, Peng J C, Zhou B X, Zhao X S, Shen J Y. *Acta Metall Sin*, 2011;47: 865.
(姚美意,李士炉,张欣,彭剑超,周邦新,赵旭山,沈剑韵. 金属学报,2011;47: 865)

[8] Li S L, Yao M Y, Zhang X, Geng J Q, Peng J C, Zhou B X. *Acta Metall Sin*, 2011;47: 163.
(李士炉,姚美意,张欣,耿建桥,彭剑超,周邦新. 金属学报,2011;47: 163)

[9] Li P Z, Li Z K. *Rare Met Mater Eng*, 1998;27: 356.
(李佩志,李中奎. 稀有金属材料与工程,1998;27: 356)

[10] Zhou B X, Li Q, Yao M Y, Liu W Q. *Nucl Power Eng*, 2005;26(4): 364.
(周邦新,李强,姚美意,刘文庆. 核动力工程,2005;26(4): 364)

[11] Li C, Li P, Zhou B X, Zhao W J, Peng Q, Ying S H, Shen B L. *Nucl Power Eng*, 2002;23(4): 20.
(李聪,李蓓,周邦新,赵文金,彭倩,应诗浩,沈保罗. 核动力工程,2002;23(4): 20)

[12] Foster J P, Dougherty J, Burke M G. *J Nucl Mater*, 1990;173: 164.

[13] Charquet D. In: Sabol G P, Moan G D eds. , *Zirconium in the Nuclear Industry: Twelfth International Symposium*, *ASTM STP 1354*, West Conshohocken: ASTM International, 2000: 3.

[14] Eucken C M, Finden P T. In: Van Swam L F P, Eucken C M eds. , *Zirconium in the Nuclear Industry: Eighth*

International Symposium, *ASTM STP 1023*, Philadelphia: ASTM International, 1989: 113.

[15] Takeda K, Anada H. In: Sabol G P, Moan G D eds. , *Zirconium in the Nuclear Industry: Twelfth International Symposium*, *ASTM STP 1354*, West Conshohocken: ASTM International, 2000: 592.

[16] Garzarolli F, Broy Y, Busch R A. In: Bradley E R, Sabol G P Eds. , *Zirconium in the Nuclear Industry: Eleventh International Symposium*, *ASTM STP 1295*, Garmisch-Partenkirchen, Germany: ASTM International, 1996: 850.

[17] Graham R A, Tosdale J P, Finden P T L F P. In: Van Swam L F P, Eucken C M eds. , *Zirconium in the Nuclear Industry: Eighth International Symposium*, *ASTM STP 1023*, Philadelphia: ASTM International, 1989: 334.

[18] Yang W D. *Reactor Materials Science*. 2nd Ed. , Beijing: Atomic Energy Press, 2006: 260.
(杨文斗. 反应堆材料学. 第二版, 北京: 原子能出版社, 2006: 260)

[19] Zhou B X, Li Q, Liu W Q, Yao M Y, Chu Y L. *Rare Met Mater Eng*, 2006; 35: 1009.
(周邦新, 李强, 刘文庆, 姚美意, 褚于良. 稀有金属材料与工程, 2006; 35: 1009)

[20] Cox B. *J Nucl Mater*, 1969; 29: 50.

[21] Zhou B X, Li Q, Yao M Y, Liu W Q, Chu Y L. *J ASTM Int*, 2008; 5: 360.

[22] Charquet D, Hanh R, Ortlib E. In: Van Swam L F P, Eucken C M eds. , *Zirconium in the Nuclear Industry*, *Eighth International Symposium*, *ASTM STP 1023*, Philadelphia: ASTM International, 1989: 405.

[23] Yao M Y, Zhou B X, Li Q, Liu W Q, Yu W J, Chu Y L. *J Nucl Mater*, 2008; 374: 197.

[24] Shen Y F, Yao M Y, Zhang X, Li Q, Zhou B X, Zhao W J. *Acta Metall Sin*, 2011; 47: 899.
(沈月锋, 姚美意, 张欣, 李强, 周邦新, 赵文金. 金属学报, 2011; 47: 899)

Effect of Bi Addition on the Corrosion Resistance of Zr-4 in Superheated Steam at 400 ℃/10. 3 MPa

Abstract: The effect of Bi contents on the corrosion resistance of Zr-4+xBi (x=0. 1%—0. 5%, mass fraction) alloys, which were prepared by adding Bi to Zr-4, was investigated in superheated steam at 400 ℃ and 10. 3 MPa by autoclave tests. The microstructures of the alloys and fracture surface morphology of the oxide film formed on the alloys were observed by TEM, EDS and SEM. The results show that with the increase of Bi content, the second phase particles (SPPs) are almost the same in size and shape, but increase in amount and vary in composition, including Zr(Fe, Cr)$_2$, Zr-Fe-Cr-Bi, Zr-Fe-Sn-Bi and Zr-Fe-Cr-Sn-Bi. Even in the Zr-4+0. 1Bi alloy, Bi-containing SPPs were detected. This indicates that the solid solubility of Bi in α-Zr matrix of Zr-4+xBi alloys is less than 0. 1% at 580 ℃. Moreover, the addition of Bi promotes the precipitation of Sn which originally dissolved in the α-Zr matrix of Zr-4. Compared with Zr-4, the addition of Bi makes the corrosion resistance worse, and it becomes more obvious with the increase of Bi content. This illustrates that the addition of Bi can not improve the corrosion resistance, on the contrary, it brings a harmful influence. This may be related to the precipitation of the Bi-containing and Bi-Sn-containing SPPs.

退火温度对 Zr-0.85Sn-0.16Nb-0.38Fe-0.18Cr 合金耐腐蚀性能的影响[*]

摘　要：为了研究退火温度对 Zr-0.85Sn-0.16Nb-0.38Fe-0.18Cr 合金耐腐蚀性能的影响，在 740～820 ℃温度范围内改变冷轧前后的退火温度制备样品，通过静态高压釜腐蚀实验研究样品在 360 ℃/18.6 MPa 去离子水、400 ℃/10.3 MPa 过热蒸汽和 500 ℃/10.3 MPa 过热蒸汽中的腐蚀行为；用透射电镜（TEM）和能谱仪（EDS）研究合金的显微组织，包括第二相的尺寸、成分与种类. 研究结果表明，这种锆合金中的第二相为含少量 Nb 的密排六方结构的 Zr(Fe, Cr, Nb)₂，提高退火温度使第二相尺寸增大，第二相中的 Nb 含量降低；在 740～800 ℃温度范围内改变退火温度，样品在上述中性水质中的耐腐蚀性能与常规工艺处理的相当，且都优于 Zr-4 合金，表明该合金的耐腐蚀性能对退火温度并不敏感.

1　引言

　　锆合金是核反应堆中一种重要的结构材料，用作核燃料包壳，在高温高压水中工作. 随着核动力反应堆技术朝着加深燃料燃耗、提高反应堆热效率以及安全可靠性的方向发展，对燃料元件包壳用锆合金的性能提出了更高的要求，为此，许多国家都在研究开发新型锆合金[1]. 目前国际上开发的锆合金主要有 3 个系列：Zr-Sn、Zr-Nb 和 Zr-Sn-Nb，已经应用的有 Zr-2、Zr-4、E110、M5、ZIRLO、E635 等锆合金，具有应用前景的有 N18、N36 和 HANA 等锆合金. 添加合金元素及优化合金成分的不同配比是开发高性能锆合金的基本方法. N18 锆合金是我国自主研制的新型 Zr-Sn-Nb 系合金，具有良好的耐腐蚀性能. 周邦新等[2]的研究表明适当降低 N18 合金中的 Sn 含量还可以进一步提高其耐腐蚀性能，而去除 Cr 后对 N18 合金的耐均匀腐蚀性能是有害的. 常规加工工艺制备的 Zr-4 合金在 500 ℃过热蒸汽中表现出严重的疖状腐蚀，而 Nb 元素对改善锆合金耐疖状腐蚀性能有重要的作用[3-6].

　　另外，Zr-4 合金的耐疖状腐蚀性能和耐均匀腐蚀性能均会受到中间退火和最终退火温度以及退火时间的影响，其影响规律可以用累积退火参数 A 来表达：要获得好的耐疖状腐蚀性能要求 $A \leqslant 10^{-18}$ h，而要获得好的耐均匀腐蚀性能则要求 2×10^{-18} h $\leqslant A \leqslant 5 \times 10^{-17}$ h[7-12]. 但在含 Nb 锆合金的研究中却发现耐腐蚀性能与 A 值之间并没有很好的对应关系[13,14]，这是因为含 Nb 锆合金的显微组织比 Zr-4 复杂，所以不能用 A 判断退火温度和时间对含 Nb 锆合金耐腐蚀性能的影响. 本研究发现中间退火温度对含 0.35% Nb 的 N18 合金耐腐蚀性能也会产生较大影响[2,4,15]. 已有的研究表明热处理温度对低 Nb 锆合金耐腐蚀性能的影响远低于对高 Nb 锆合金的影响[16-19].

　　基于以上的研究结果，采用模拟加工过程中的中间退火方法对在 740～820 ℃范围内制

---

* 本文合作者：姚美意、张伟鹏、周军、李强. 原发表于《核动力工程》，2012，33(6)：88-92.

备的 Zr-0.85Sn-0.16Nb-0.38Fe-0.18Cr 样品在多种中性水中的腐蚀行为进行研究,为加工过程中选择合适的退火温度提供依据.

2 实验方法

2.1 热处理工艺

实验用 1.4 mm 厚的 Zr-0.85Sn-0.16Nb-0.38Fe-0.18Cr 锆合金由西北有色金属研究院熔炼和加工制备,铸锭重量为 20 kg. 为了研究在 740~820 ℃范围内改变中间退火温度或最终退火温度对合金耐腐蚀性能的影响,采用 8 种退火处理工艺制备样品,编号为 7B、7C、7D、7E 和 7F 的样品分别先进行 820 ℃/2 h、800 ℃/2 h、780 ℃/2 h、760 ℃/2 h、740 ℃/2 h 的中间退火,再从 1.4 mm 冷轧到 0.7 mm,然后切成 20 mm×20 mm 大小的试样,经混合酸(体积分数为 10% HF+45% HNO_3+45% H_2O)酸洗后进行 580 ℃/5 h 的最终退火处理;编号为 7H、7J 和 7K 的样品则先从 1.4 mm 厚的板材直接冷轧到 0.7 mm,然后也切成 20 mm×20 mm 大小的试样,用相同的混合酸酸洗后再分别进行 800 ℃/2 h、780 ℃/2 h 和 580 ℃/2 h 的最终退火处理,其中 7K 样品为对比的样品,是按常规工艺处理的.

2.2 腐蚀试验

为了比较 Zr-0.85Sn-0.16Nb-0.38Fe-0.18Cr 经不同温度中间退火或最终退火制备的样品与 Zr-4 样品耐腐蚀性能的差别,将其一起放入静态高压釜中进行腐蚀试验,腐蚀条件分别为:360 ℃/18.6 MPa 去离子水、400 ℃/10.3 MPa 过热蒸汽和 500 ℃/10.3 MPa 过热蒸汽. 腐蚀试验前,所有样品均按标准方法酸洗(酸洗液同上)和去离子水清洗,腐蚀增重由 5 个试样的平均值得出.

2.3 显微组织观察

用 JEM-200CX 透射电镜(TEM)观察合金样品的显微组织及第二相的大小和分布,薄样品用双喷电解抛光制备,电解液为体积分数 10% 高氯酸+90% 乙醇的混合溶液. 用 JEM-2010F 高分辨 TEM 配置的能谱仪(EDS)分析第二相的成分,每种样品分析 4 个以上的第二相颗粒. 根据耐腐蚀性能的差异,本文重点观察了 7B 和 7K 样品的显微组织.

3 实验结果与讨论

3.1 合金的显微组织

图 1 是 7B 和 7K 样品显微组织的 TEM 像. 从图 1 可以看出,2 种样品经过 580 ℃退火处理之后都发生了完全再结晶,第二相分布在晶粒内部或晶界上,尺寸有一定差别. α 相上限温区处理的 7B 样品中的第二相明显较 α 相下限温区处理的 7K 样品中的少,这是因为在 820 ℃加热时 α-Zr 基体中固溶的合金元素含量较高,原先析出的一部分第二相发生了溶解,而在最终 580 ℃退火时还没有充分析出.

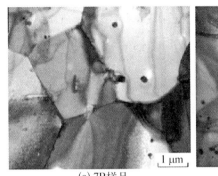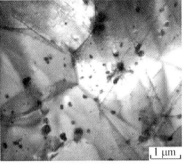

(a) 7B样品   (b) 7K样品

图 1 样品显微组织的 TEM 像

Fig. 1 TEM micrographs of specimens

表 1 和表 2 分别给出了 7K 和 7B 样品中用于能谱分析的第二相尺寸、成分分析结果和 Fe/Cr 比值等. 由表 1 可以看出: ① 7K 样品中的第二相由 Zr‑Fe‑Cr‑Nb 4 种元素组成, 其中 Nb 含量在 2%~5.8%(原子数分数)之间, 这说明 7K 样品 α‑Zr 基体中固溶的 Nb 含量肯定低于 0.16%, 因为添加的合金元素 Nb 有一部分已经进入到第二相; ② 第二相中的 Fe/Cr 比值在 1.8~2.2 之间, 与 Zr‑0.85Sn‑0.16Nb‑0.38Fe‑0.18Cr 平均 Fe/Cr 比值 2.06 接近. 在 EDS 分析时, 由于电子束有一定的穿透深度和扩散范围, 所以不可避免会包含 α‑Zr 基体, 这样 EDS 分析只能是半定量的, 并不能由 EDS 的结果判断是哪种第二相. 结合文献[6]中的 EDS 和选区电子衍射(SAD)分析结果可知, 7K 样品中只含有 1 种第二相, 即密排六方结构 $Zr(Fe, Cr, Nb)_2$ 第二相.

表 1 7K 样品中第二相的 EDS 分析结果及第二相尺寸

Table 1 EDS results and size of second phase particles in specimen 7K

第二相编号	第二相尺寸/nm	第二相化学成分(原子数分数 %)				Fe/Cr
		Zr	Fe	Cr	Nb	
1	172	44.03	34.37	15.84	5.76	2.17
2	118	53.59	28.70	14.47	3.24	1.98
3	164	71.09	17.42	9.48	2.01	1.84
4	45	45.24	33.46	16.92	4.38	1.98

表 2 7B 样品中第二相的 EDS 分析结果及第二相尺寸

Table 2 EDS results and size of second phase particles in specimen 7B

第二相编号	第二相尺寸/nm	第二相化学成分(原子数分数 %)				Fe/Cr
		Zr	Fe	Cr	Nb	
1	340	54.16	28.24	16.07	1.53	1.75
2	207	67.93	19.87	10.91	1.29	1.82
3	146	82.37	10.83	5.91	0.89	1.83
4	329	57.92	26.00	14.56	1.52	1.79
5	535	51.70	29.69	17.17	1.45	1.73

由表 2 可以看出: ① 7B 样品中的第二相也是由 Zr‑Fe‑Cr‑Nb 4 种元素组成, 其中 Nb 含量在 0.8%~1.6%(原子数分数)之间, 比 7K 样品第二相中的 Nb 含量低(表 1), 说明

7B 样品 α-Zr 基体中固溶的 Nb 含量比 7K 中的高;② 7B 样品中第二相的 Fe/Cr 比值为 1.7~1.9. 同样,结合文献[6]中的 EDS 和 SAD 分析结果可知,7B 合金基体中也只有 1 种第二相,即密排六方结构的 Zr(Fe, Cr, Nb)$_2$ 第二相.

3.2 腐蚀行为

3.2.1 360 ℃/18.6 MPa 去离子水中的腐蚀行为

图 2 给出了 4 种典型样品在 360 ℃/18.6 MPa 去离子水中腐蚀 330 d 的增重曲线,其他样品的增重曲线介于 7K 和 7H 样品之间. 由图 2 可知:除了 7B 样品(腐蚀 330 d 的增重为 76 mg/dm^2)的耐腐蚀性能略低于 Zr-4 合金(腐蚀 330 d 的增重为 75 mg/dm^2)外,其他样品的耐腐蚀性能都略优于 Zr-4 合金,腐蚀 330 d 的增重在 67~72 mg/dm^2 范围内. 这表明本文设计的合金在 740~800 ℃ 范围内改变中间退火温度或最终退火温度对其在 360 ℃/18.6 MPa 去离子水中的耐腐蚀性能影响不大,并优于 Zr-4 合金.

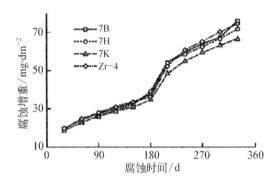

图 2　几种典型样品在 360 ℃/18.6 MPa 去离子水中的腐蚀增重曲线

Fig. 2　Weight gains versus exposure time for typical specimens in deionized water at 360 ℃/18.6 MPa

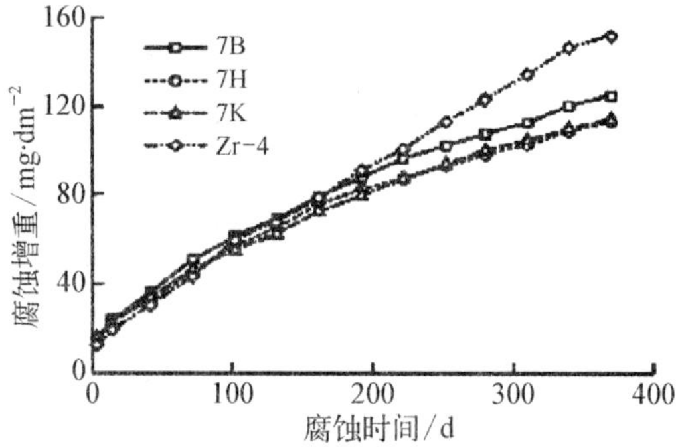

图 3　几种典型样品在 400 ℃/10.3 MPa 过热蒸汽中的腐蚀增重曲线

Fig. 3　Weight gains versus exposure time for typical specimens in super-heated steam at 400 ℃/10.3 MPa

3.2.2 400 ℃/10.3 MPa 过热蒸汽中的腐蚀行为

图 3 给出了 4 种典型样品在 400 ℃/10.3 MPa 过热蒸汽中腐蚀 370 d 的增重曲线,其余样品的腐蚀增重曲线介于 7B 和 7D 样品之间. 由图 3 可以看出,与 Zr-4 合金(腐蚀 370 d 的增重为 153 mg/dm^2)相比,不管是常规工艺处理的 7K 样品(腐蚀 370 d 的增重为 115 mg/dm^2),还是在 740~820 ℃ 温度范围内改变中间退火温度或最终退火温度处理的其他样品(腐蚀 370 d 的增重为 113~126 mg/dm^2 范围内),在 400 ℃ 过热蒸汽中的耐腐蚀性能都明显优于 Zr-4 合金,这说明本文设计的合金在 400 ℃ 过热蒸汽中也具有优良的耐腐蚀性能,并且对退火温度不敏感.

3.2.3 500 ℃/10.3 MPa 过热蒸汽中的腐蚀行为

图 4 给出了 4 种典型样品在 500 ℃/10.3 MPa 过热蒸汽中腐蚀 40~360 h 的增重曲线,其余样品的腐蚀增重曲线介于 7B 和 7C 样品之间. 由图 4 可以看出,Zr-4 合金在腐蚀 40 h 时就出现明显的疖状腐蚀斑,氧化膜发生脱落而无法继续腐蚀实验,而本文设计的合金不管采用哪种温度进行退火处理,样品在 500 ℃/10.3 MPa 过热蒸汽中腐蚀时都没有产生疖状腐蚀,耐腐蚀性能明显优于 Zr-4 合金. 与常规处理的 7K 样品(腐蚀 360 h 的增重为 201 mg/

dm²)相比,改变退火温度的样品中除了 7B 样品腐蚀 360 h 的增重为 219 mg/dm²,耐腐蚀性能稍差之外,其他样品(腐蚀 360 h 的增重在 188～201 mg/dm² 范围内)的耐腐蚀性能与 7K 样品的相差不大. 这说明退火温度对这种锆合金在 500 ℃ 过热蒸汽中的耐腐蚀性能影响也不大.

从不同温度退火制备的 Zr - 0.85Sn - 0.16Nb - 0.38Fe - 0.18Cr 样品在 3 种水化学条件下的腐蚀行为(图 2～图 4)可知,与常规工艺处理的 7K 样品相比,该合金在最终冷轧前后,在 740～800 ℃ 范围内改变退火温度后仍表现出优良的耐腐蚀性能,并且耐腐蚀性能均优于 Zr - 4 合金,少量 Nb 的添加确实抑制

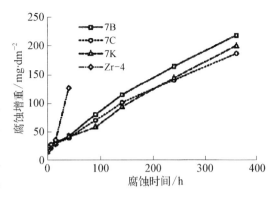

图 4　几种典型样品在 500 ℃/10.3 MPa 过热蒸汽中的腐蚀增重曲线

Fig. 4　Weight gains versus exposure time of typical specimens in super heated steam at 500 ℃/10.3 MPa

了疖状腐蚀,与本研究前期的研究结果[6]一致. 另外,由于添加的 Nb 含量较低,并且有一部分 Nb 还会进入第二相,因而与 Zr - 4 合金相比不会明显降低 α/β 的相变温度;而与 N18 合金相比则具有更高的 α/β 相变温度,这样在较高温度退火也不会出现 β - Zr. 另一方面,当退火温度升高后,虽然第二相的长大会使耐腐蚀性能变差,但同时又适当提高了 α - Zr 基体中固溶的合金元素(Fe、Cr 和 Nb)含量,这对耐腐蚀性能又会起到有利的作用[2,8,20],Zr - 0.85Sn - 0.16Nb - 0.38Fe - 0.18Cr 中的 Nb 含量比 N18 合金中的低,这样 α - Zr 基体中固溶的 Nb 含量随退火温度的变化没有 N18 合金的大. 因而该合金的耐腐蚀性能对退火温度并不像 N18 合金那样敏感,与文献中报道的低 Nb 锆合金的耐腐蚀性能对热处理温度不敏感的结果一致[16-19]. 可见,本文设计的合金是一种既耐疖状腐蚀,又具有优良耐均匀腐蚀性能并且对中间退火温度不敏感的锆合金.

4　结论

(1) Zr - 0.85Sn - 0.16Nb - 0.38Fe - 0.18Cr 合金中的第二相为含少量 Nb 的密排六方结构的 Zr(Fe,Cr,Nb)₂,提高冷轧退火前的热处理温度使第二相尺寸增大,第二相中的 Nb 含量降低,这说明固溶在 α - Zr 基体中的 Nb 含量相应增加.

(2) Zr - 0.85Sn - 0.16Nb - 0.38Fe - 0.18Cr 合金在 360 ℃/18.6 MPa 去离子水、400 ℃/10.3 MPa 过热蒸汽和 500 ℃/10.3 MPa 过热蒸汽 3 种中性水质中的耐腐蚀性能对 740～800 ℃ 范围内的退火温度不敏感,均表现出了比 Zr - 4 合金优良的耐腐蚀性能,少量 Nb 的添加明显抑制了疖状腐蚀的产生.

参 考 文 献

[1] 赵文金,周邦新,苗志,等. 我国高性能锆合金的发展[J]. 原子能科学技术,2005,39(Suppl.):2 - 9.

[2] Zhou Bangxin, Yao Meiyi, Li Zhongkui, et al. Optimization of N18 Zirconium Alloy for Fuel Cladding of Water Reactors[J]. J Mater Sci Technol, 2012,28(7):606 - 613.

[3] 周邦新,赵文金,苗志,等. 新锆合金的研究[C]. 96 中国核材料研讨会,生物及环境材料,Ⅲ - 2. 北京:化学工业出版社. 1997:183 - 186.

［4］周邦新,姚美意,李强,等. Zr－Sn－Nb 合金耐疖状腐蚀性能的研究［J］. 稀有金属材料与工程,2007,36(8)：1317－1321.

［5］姚美意,周邦新,李强,等. 微量 Nb 的添加对 Zr－4 合金耐疖状腐蚀性能的影响［J］. 上海金属,2008；30(6)：1－3.

［6］姚美意,李士炉,张欣,等. 添加 Nb 对 Zr－4 合金在 500 ℃ 过热蒸汽中耐腐蚀性能的影响［J］. 金属学报,2011,47(07)：865－871.

［7］Steinberg E, Weidinger H G, Schaa A. Analytical Approachs and Experimental Verification to Describe the Influence of Cold Work and Heat Treatment on the Mechanical Properties of Zircaloy Cladding tubes［R］. Zirconium in the Nuclear Industry：Sixth International Symposium, ASTM STP 824,1984, pp. 106－122.

［8］周邦新,李强,姚美意,等. 热处理影响 Zr－4 合金耐疖状腐蚀性能的机制［J］. 稀有金属材料与工程,2007,36(7)：1129－1134.

［9］Garzarolli G, Steinberg E, Weiginger H G. Microstructureand Corrosion Studies for PWR and BWR Zircaloy Cladding［R］. Zirconium in the Nuclear Industry：Eighth International Symposium, ASTM STP 1023, 1989, pp. 202－212.

［10］Thorvaldsson T, Andersson T, Wilson A, et al. orrelation between 400 ℃ Steam Corrosion Behavior, Heat Treatment and Microstructure of Zircaloy－4 Tubing ［C］. Zirconium in the Nuclear Industry：Eighth International Symposium, ASTM STP 1023, 1989, pp. 128－140.

［11］Franklin D G, Lang P M. Zirconium-Alloy Corrosion：aReview Based on an International Atomic Energy Agency (IAEA) Meeting［R］. Zirconium in the Nuclear Industry：Ninth International Symposium, ASTM STP 1132, 1991, pp. 3－32.

［12］Foster P, Dougherty J, Burke M G, et al. Influence of inal Recrystallization Heat Treatment on Zircaloy－4 Strip Corrosion［J］. J Nucl Mat, 1990,173(2)：164－178.

［13］Isobe T, Matsuo Y. Development of Highly Corrosion esistant Zirconium-Based Alloys［R］. Zirconium in the Nuclear Industry：Ninth International Symposium, ASTM STP 1132,1991, pp. 346－367.

［14］Kim J M, Jeong Y H, Jung Y H. Correlation of HeatTreatment and Corrosion Behavior of Zr－Nb－Sn－Fe－Cu Alloys［J］. J Mater Proc Tech, 2000, 104：145－149.

［15］张欣,姚美意,周邦新,等. 加工工艺对 N18 锆合金 360 ℃ LiOH 水溶液中腐蚀行为的影响［J］. 金属学报,2011,47(09)：1112－1116.

［16］Choo K N, Kang Y H, Pyun S I, et al. Effect of omposition and Heat Treatment on the Microstructure and Corrosion Behavior of Zr－Nb Alloys ［J］. J Nucl Mater, 1994,209：226－235.

［17］Jeong Y H, Lee K O, Kim H G. Correlation between icrostructure and Corrosion Behavior of Zr－Nb Binary ［J］. J Nucl Mater,2002,302：9－19.

［18］Kim H G, Jeong Y H, Kim T H. Effect of Isothermal-Annealing on the Corrosion Behavior of Zr－xNb alloys［J］. J Nucl Mater,2004,326：125－131.

［19］Jeong Y H, Kim H G, Kim D J. Influence of Nb oncentration in the α－Matrix on the Corrosion Behavior of Zr－xNb Binary alloys［J］. J Nucl Mater, 2003,323：72－80.

［20］Yao M Y, Zhou B X, Li Q, et al. A Superior Corrosion ehavior of Zircaloy－4 in Lithiated Water at 360 ℃/18.6 MPa by β－Quenching［J］. J Nucl Mater, 2008,374(1－2)：197－203.

Effect of Annealing Temperatures on Corrosion Resistance of a Zr－0.85Sn－0.16Nb－0.38Fe－0.18Cr Alloy

Abstract：To investigate the effect of annealing temperatures on the corrosion resistance of a Zr－0.85Sn－0.16Nb－0.38Fe－0.18Cr alloy, different annealing temperatures (740～820 ℃) before and after the final cold rolling were employed to prepare the specimens for corrosion tests. The specimens were corroded in deionized water at 360 ℃/18.6 MPa, in super heated steam at 400 ℃/10.3 MPa and at 500 ℃/10.3 MPa, respectively by autoclave test. The microstructure including the size and composition of second phase particles (SPPs) was examined by TEM and EDS. Results show that the SPPs in the Zr－0.85Sn－0.16Nb－0.38Fe－

0. 18Cr alloy are Zr(Fe, Cr, Nb)$_2$ with a hcp structure; With the increasing of annealing temperatures, the size of the SPPs increases and the Nb content in the SPPs decreases; Under the three test conditions, the specimens annealed at different temperatures between 740 ℃ and 800 ℃ are all with corrosion resistance behavior as good as the specimen prepared by conventional procedures, and their corrosion resistance is superior to Zircaloy – 4. This indicates that the corrosion resistance of this alloy is insensitive to the annealing temperatures.

The Effect of Final Annealing after β – Quenching on the Corrosion Resistance of Zircaloy – 4 in Lithiated Water with 0.04 M LiOH*

Abstract: To further understand the effect of heat treatments on the corrosion resistance of Zircaloy – 4, some specimens were treated by β – quenching at 1 020 ℃ for 20 min and then annealing at 480 – 600 ℃ for 2 – 200 h. The specimens were corroded in lithiated water with 0.04 M LiOH at 360 ℃/18.6 MPa. The microstructures of the specimens and the oxide films were observed by TEM and SEM. The results show that the β – quenched specimen with the cooling rate of 100 ℃/s behaves the best corrosion resistance. After annealing at 480 – 600 ℃, the corrosion resistance gets worse with increasing annealing temperature. The annealing time has little effect on the corrosion resistance for the specimens annealed at 480 and 540 ℃ after β – quenching. However, the corrosion resistance of the specimens annealed at 600 ℃ becomes worse with the increase of annealing time. The grain size of the oxide on the specimen with high Fe and Cr concentrations in α – Zr is coarser than that on the specimen with low Fe and Cr concentrations in α – Zr, and the pores and cracks are fewer in the oxide film on the former specimen than those on the latter specimen. This indicates that a suitable increase in Fe and Cr concentrations in α – Zr matrix is beneficial to retarding the microstructural evolution of oxide, thereby improving the corrosion resistance.

1 Introduction

Zircaloy – 4 has been used as nuclear fuel cladding in pressurized water reactors (PWRs) for its good corrosion resistance. In order to meet the demand for higher burn-up and longer refueling period in-reactor, instead of Zircaloy – 4, several advanced zirconium alloys, such as ZIRLO and M5 have been developed. In addition, the corrosion resistance of zirconium alloys can be also improved by optimizing the microstructure of alloys[1-4]. By replacing either final or intermediate annealing with a β – quenching process for Zircaloy – 4, the corrosion resistance can be further improved[1,3,5]. In our previous work[2], it was surprisingly found that Zircaloy – 4 treated by β – quenching exhibited similar excellent corrosion resistance to Zr – Sn – Nb alloys in lithiated water with 0.01 M LiOH, which was attributed to the increase in concentrations of Fe and Cr in α – Zr matrix. Graham et al. [6], Wadman and Andrén[7] and Zhou et al. [5] also thought the concentrations of alloying elements in α – Zr matrix have a great effect on corrosion resistance. Other researchers[3,8,9]

---

* In collaboration with Yao M Y, Shen Y F, Li Q, Peng J C, Zhang J L. Reprinted from Journal of Nuclear Materials, 2013, 435: 63 – 70.

owed the better corrosion resistance to the second phase particles (SPPs). In fact, the average size and area fraction of SPPs, and the concentrations of alloying elements in α-Zr matrix are varied in a systematic manner during heat treatments at different temperatures. In order to further understand the mechanism about the effect of heat-treatments on the corrosion behavior, Zircaloy-4 specimens were treated by β-quenching with different cooling rates and then annealing at 480-600 ℃ for 2-200 h. The corrosion results in lithiated water with 0.01 M LiOH at 360 ℃ and 18.6 MPa have been reported in a previous paper[10]. It was found that the specimens with a slow cooling rate behaved similar good corrosion resistance to N18 alloy, a Zr-Sn-Nb alloy with a lower content of Nb, which is accordant with our previous result[2]. While the specimens with a fast cooling rate got worse corrosion resistance due to the existence of β-Zr. In this work, the corrosion behavior of Zircaloy-4 specimens, treated by β-quenching with a slow cooling rate and subsequent annealing at 480-600 ℃ for 2-200 h, is investigated in lithiated water with 0.04 M LiOH. This investigation is also a part of the systematic study to examine the corrosion behavior in different water chemistries for the Zircaloy-4 specimens treated by β-quenching with different cooling rates and subsequent annealing at different temperatures for different time periods.

2　Experimental procedures

2.1　Heat treatment procedures

As-received Zircaloy-4 plates with 4 mm thickness were fabricated by conventional procedures[11]. The plates were rolled to 1.4 mm at 700 ℃, annealed at 580 ℃ for 2 h, and then cold rolled to 0.7 mm. Some strips with 8 mm in width were cut from the Zircaloy-4 plates. In order to obtain the same cooling rate from β quenching, only two strips were sealed in each vacuum quartz capsule at about 3×10^{-3} Pa. All of the specimens in quartz capsules were treated at 1 020 ℃ for 20 min and then quenched into the water immediately. Here, the quartz capsules were not broken. The cooling rate is about 100 ℃/s determined by counting the period from the bright red to dark red (~600 ℃) of the specimens within the vacuum quartz capsule after quenching into water (defined as 100 ℃/s-β-WQ). Such a cooling rate is slower comparing to the case that the specimens quenched into the water and broken the quartz capsules at the same time. The specimens with 25 mm in length were cut from the strips and annealed at 480-600 ℃ for 2-200 h in a vacuum furnace at about 3×10^{-3} Pa (Fig. 1). The purpose of such annealing is to change the concentrations of Fe and Cr in α-Zr matrix. Of course, the size and area fraction of SPPs were also changed in a systematic manner.

2.2　Characterization of microstructural of the specimens

The microstructures were observed using a JEM-200CX transmission electron microscope

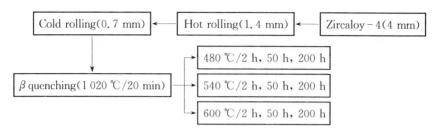

Fig. 1 Preparation procedures of Zircaloy – 4 specimens with different heat treatments

(TEM) and a JEM – 2010F high resolution transmission electron microscope (HRTEM) with EDS. The area fraction and the mean diameter of SPPs were calculated using an image analyzer. The specimens for the TEM observation were mechanically ground to about 70 μm in thickness, then were punched out 3 mm disks in diameter, and were finally thinned by twin-jet electro-polishing technique in a solution of 10% perchloric acid and 90% ethanol (in volume) at -40 ℃ with a DC voltage of 40 – 50 V.

2.3 Corrosion testing

The corrosion tests were performed with a static autoclave in lithiated water with 0.04 M LiOH at 360 ℃ and 18.6 MPa. The corrosion behavior was evaluated by weight gain as a function of exposure time. The reported weight gain is a mean value obtained from 3 – 5 specimens. Prior to the corrosion tests, the specimens were cleaned and pickled in a mixed acid (10% HF, 45% HNO_3 and 45% H_2O, in volume), sequentially rinsed in cold tap water, boiling deionized water and then blow-dried with warm air.

2.4 Characterization of fracture surfaces and outer surfaces of oxide films

JSM – 6700F high resolution scanning electron microscope (HRSEM) was used to characterize the morphology of fracture surfaces and outer surfaces of oxide films formed on the specimens after corrosion. The specimen preparation for fracture surface observation has been described in references [2,12] in detail. A piece of 5 mm×5 mm specimen was directly cut from the corroded specimen for the observation of the outer surface. In order to avoid charging by bombardment of electron beams in the microscope, a thin layer of Pt was deposited by sputtering on the oxide samples.

3　Results

3.1　Characteristics of microstructures and precipitates of the specimens with different heat treatments

The microstructures of β – quenched and annealed Zircaloy – 4 specimens are shown in Fig. 2. It is clear that the SPPs in the β – quenched specimen are very fine and disperse uniformly (Fig. 2a). Compared to the β – quenched specimen, there is no obvious change in the size and distribution of SPPs among the annealed specimens, except the specimens

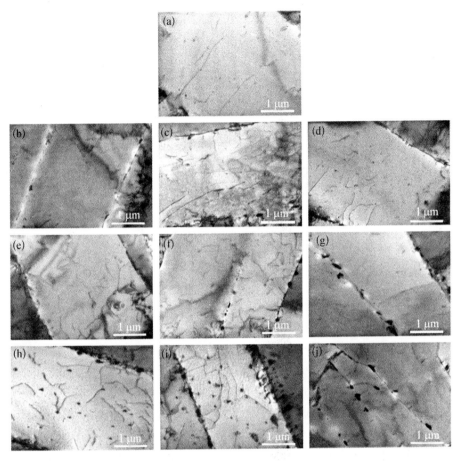

Fig. 2 TEM micrographs of $\beta-$ quenched and annealed Zircaloy-4 specimens: (a) 100 ℃/s-β-WQ; (b) 100 ℃/s-β-WQ+480 ℃/2 h; (c) 100 ℃/s-β-WQ+540 ℃/2 h; (d) 100 ℃/s-β-WQ+600 ℃/2 h; (e) 100 ℃/s-β-WQ+480 ℃/50 h; (f) 100 ℃/s-β-WQ+540 ℃/50 h; (g) 100 ℃/s-β-WQ+600 ℃/50 h; (h) 100 ℃/s-β-WQ+480 ℃/200 h; (i) 100 ℃/s-β-WQ+540 ℃/200 h; (j) 100 ℃/s-β-WQ+600 ℃/200 h

annealed at 480 ℃ and 540 ℃ for 200 h, and 600 ℃ for 50 h and 200 h (Fig. 2b – j). It is obvious that the SPPs distribution in 100 ℃/s-β-WQ and 100 ℃/s-β-WQ + 600 ℃/2 h specimens is more homogeneous than that in other specimens. It is noted that the SPPs at the boundaries of α – Zr lath grains are usually appear as rod-like and are larger than those in the α – Zr lath grains. These SPPs may form during $\beta-\alpha$ phase transformation and grow up during annealing. The SPPs are Zr(Fe, Cr)$_2$ of a close-packed hexagonal structure (hcp) according to the results of SAD and EDS analyses (Fig. 3). The Cu peaks in EDS of Fig. 3b comes from the specimen holder in TEM. The width of lath grains is similar among the quenched and annealed specimens. The mean value is about 3 μm in width. This means it is easy to keep the cooling rate at a same level by such a $\beta-$ quenching manner, while it is difficult in the case of broking the quartz capsule[10]. This also further confirms that annealing at 480 – 600 ℃ for 2 – 200 h should not result in boundary migration of the lath grains. Only local recrystallization has taken place even in the specimen 100 ℃/s-β-WQ+

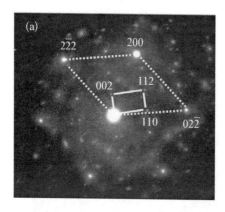 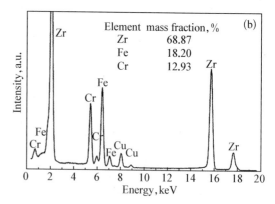

Fig. 3 SAD patterns (a) and EDS result, (b) of the SPPs in the Zircaloy – 4 specimen. The solid lines refer to the $Zr(Fe, Cr)_2$ SPPs and the dot lines refer to the α – Zr in Fig. 3a

600 ℃/200 h. In addition, many dislocations are found in the grains resulting from the β-quenching stress or the deformation during the preparation of TEM specimens.

The mean diameter and area fraction of the observed SPPs are shown in Table 1. For reducing the calculation errors, the TEM micrographs taken from different fields in the same specimen were selected to analyze the SPPs characteristics. The parameters such as the mean diameter and area fraction were determined from the analysis on more than 150 precipitates. It is clear that in specimen 100 ℃/s-β-WQ, the SPPs are the finest in size (35 nm) and are the lowest in area fraction (0.42%). In the annealed specimens at 480 – 600 ℃ for 2 h, the mean diameter (38 – 39 nm) of SPPs is close to that in specimen 100 ℃/s-β-WQ, but the area fraction of SPPs has a little increase from 0.5% to 0.7% with increasing annealing temperature. In the annealed specimens at 480 – 600 ℃ for 200 h, as the annealing temperature increases, the mean diameter of SPPs increases from 48 nm to 64 nm, and the area fraction of SPPs also increases from 0.78% to 1.58%. Although the area fraction of SPPs is very small in these specimens, the difference in the area fraction of SPPs can indirectly reflect the difference of Fe and Cr concentrations in α – Zr matrix. It is obvious that the concentrations of Fe and Cr in α – Zr matrix are highest in specimen 100 ℃/s-β-WQ, while they are the lowest in specimen 100 ℃/s-β-WQ + 600 ℃/200 h. This meets the thermo-dynamics and dynamics laws of SPPs precipitation.

Table 1 The mean diameter and area fraction of second phase particles in Zircaloy – 4 with different heat treatments.

Specimen	Particle size (nm)	Area fraction (%)
100 ℃/s-β-WQ	35	0.42
100 ℃/s-β-WQ+480 ℃/2 h	38	0.50
100 ℃/s-β-WQ+480 ℃/200 h	48	0.78
100 ℃/s-β-WQ+540 ℃/2 h	39	0.69
100 ℃/s-β-WQ+540 ℃/200 h	55	0.88
100 ℃/s-β-WQ+600 ℃/2 h	39	0.70
100 ℃/s-β-WQ+600 ℃/200 h	64	1.58

3.2 Corrosion behavior

The corrosion results of β- quenched and annealed Zircaloy - 4 specimens are shown in Fig. 4. Corrosion rate increases dramatically after the corrosion transition time. For most of Zircaloy - 4 specimens with different heat treatments, the corrosion transition happens at 70 days. For the specimens annealed at 600 ℃ for 50 h and 200 h, the time of corrosion transition is 60 days. Obviously, the corrosion resistance of specimen 100 ℃/s-β-WQ is the best among the specimens with different heat treatments. For the specimens annealed at 480 ℃ and 540 ℃ for different time periods respectively, there is little difference in weight gain. It means that annealing time has little influence on the corrosion resistance when the annealing temperature is below 540 ℃. The corrosion resistance of the specimens annealed at 600 ℃ decreases with the increase of annealing time. Moreover, it is obvious that the corrosion resistance of the annealed specimens decreases with increasing annealing temperature for the same annealing time.

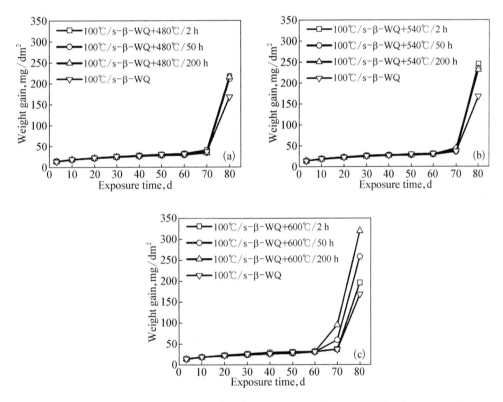

Fig. 4 Weight gain vs. exposure time for β- quenched and annealed Zircaloy - 4 specimens corroded in lithiated water with 0. 04 M LiOH at 360 ℃/18. 6 MPa for 80 days

3.3 Characterization of oxides by HRSEM

It is well known that the oxidation of zirconium alloys takes place at the oxide/metal interface. So the corrosion behavior of zirconium alloys is closely associated with the oxide characteristics, such as shape and size of oxide grains, pores and cracks in the oxide. In this study, the oxide films at different stages of oxide growth (10 days, 30 days and 80 days) on specimens 100 ℃/s-β-WQ+600 ℃/2 h and 100 ℃/s-β-WQ+600 ℃/200 h, which exhibit a large difference in corrosion resistance at the later stage of corrosion, were selected to observe using HRSEM. The information of the specimens used for HRSEM observation of oxide films is listed in Table 2.

Table 2 Information of the specimens used for HRSEM observation of oxide films.

Sample	Weight gain (mg/dm^2)		
	10 d	30 d	80 d
100 ℃/s-β-WQ+600 ℃/2 h	19	27	196
100 ℃/s-β-WQ+600 ℃/200 h	19	25	319

3.3.1 Outer surface morphology of oxide films

Fig. 5 shows the outer surface morphologies of the oxide films formed on specimens 100 ℃/s-β-WQ+600 ℃/2 h and 100 ℃/s-β-WQ+600 ℃/200 h corroded in lithiated water for 10 days exposure. Many whisker-like grains stretching out from a large number of protruded round oxide can be observed in the two specimens (indicated by arrows in Fig. 5a and b), but the size of whisker-like grains in the two specimens is different. In the oxide film on specimen 100 ℃/s-β-WQ+600 ℃/2 h (Fig. 5a), the whiskers are long (100 – 300 nm in length) and fine (~30 nm in diameter). In the oxide film on specimen 100 ℃/s-β-WQ+600 ℃/200 h (Fig. 5b), the whiskers are short (<100 nm in length) and coarse (~60 nm in diameter). The pores in the oxide film on specimen 100 ℃/s-β-WQ+600 ℃/2 h are hardly detected (Fig. 5a). However, there are numerous pores at the boundaries of ZrO$_2$ grains on specimen 100 ℃/s-β-WQ+600 ℃/200 h, as indicated by arrows in Fig. 5b. This illustrates that the microstructure of the oxide film at the early stage of oxide growth, has exhibited an obvious difference even for the specimens having the same weight gain (19 mg/dm^2). The difference lays the foundation for different corrosion behavior at the later stage.

Fig. 6 shows the outer surface morphologies of the oxide films formed on specimens 100 ℃/s-β-WQ+600 ℃/2 h and 100 ℃/s-β-WQ+600 ℃/200 h corroded in lithiated water

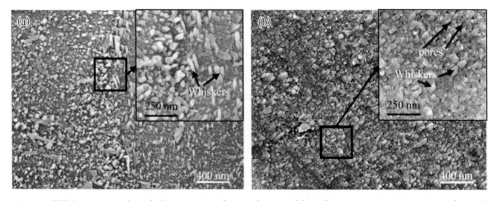

Fig. 5 SEM micrographs of the outer surfaces of oxide films formed on specimens 100 ℃/s-β-WQ+600 ℃/2 h (a), and 100 ℃/s-β-WQ+600 ℃/200 h (b), with equal weight gain corroded in lithiated water with 0.04 M LiOH for 10 days

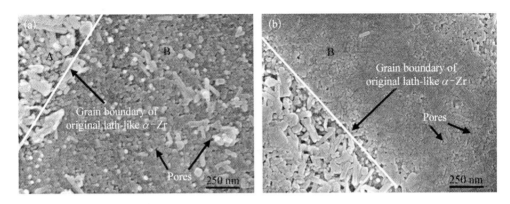

Fig. 6 SEM micrographs of the outer surfaces of oxide films formed on specimens 100 ℃/s-β-WQ+600 ℃/2 h (a), and 100 ℃/s-β-WQ+600 ℃/200 h (b), corroded in lithiated water with 0.04 M LiOH for 80 days exposure

for 80 days. Here, the weight gains of the two specimens show a great difference (196 mg/dm^2 and 319 mg/dm^2) at the post-transition stage. Compared to the oxide morphologies at the pre-transition stage, only a few whisker-like grains are observed. The oxide morphologies are different from grain to grain of the α - Zr matrix, as shown in A-area and B-area in Fig. 6a and b. It is considered that the growth rate and the morphology of an oxide depend greatly on the crystal orientation of the α - Zr matrix[13,14]. A-area in Fig. 6a and b consists of bar-type and granular-type oxides. The bar-type oxide looks like the laid whisker. B-area oxide in Fig. 6a and b shows the fine and round morphology with numerous pores at the boundaries of round oxide. It is obvious that the grain size in the oxide film on specimen 100 ℃/s-β-WQ+600 ℃/200 h is smaller than that on specimen 100 ℃/s-β-WQ+600 ℃/2 h. This indicates that the oxide film on specimen 100 ℃/s-β-WQ+600 ℃/200 h can provide more grain boundaries to promote the diffusion of O^{2-} or OH$^-$ through the oxide layer. Hence, specimen 100 ℃/s-β-WQ+600 ℃/200 h exhibits a fast corrosion rate (Fig. 4c).

3.3.2 Fracture surface morphology of oxide films

Fig. 7 shows the fracture surface morphologies of the oxide films formed on specimens 100 ℃/s-β-WQ+600 ℃/2 h and 100 ℃/s-β-WQ+600 ℃/200 h corroded in lithiated water for 30 days. At the early stage of oxide growth with 30 days exposure, the oxide films on both specimens 100 ℃/s-β-WQ+600 ℃/2 h and 100 ℃/s-β-WQ+600 ℃/200 h are compact and crack-free. The oxide in the whole thickness is composed of grains with equiaxed structure, but the trace of evolution from columnar to equiaxed grains is clear, as indicated by the dot lines in Fig. 7a and b. The equiaxed structure was also observed in β - quenching Zircaly - 4[2] and was reported in a reference[13]. It is also observed that the grain size in the oxide film on specimen 100 ℃/s-β-WQ+600 ℃/200 h (Fig. 7b) is smaller than that on specimen 100 ℃/s-β-WQ+600 ℃/2 h (Fig. 7a). The result is consistent with the observed results from the outer surface of oxide films (Fig. 6). This further implies that the oxide film formed at the early stage can provide more grain

Fig. 7 SEM micrographs of the fracture surface of oxide films formed on specimens 100 ℃/s-β-WQ+600 ℃/2 h (a), and 100 ℃/s-β-WQ+600 ℃/200 h (b), corroded in lithiated water with 0. 04 M LiOH for 30 days

boundaries to promote the diffusion of O^{2-} or OH^- through the oxide layer to accelerate the corrosion at the later stage.

Fig. 8 shows the fracture surface morphologies of the oxide films formed on specimens 100 ℃/s-β-WQ+600 ℃/2 h and 100 ℃/s-β-WQ+600 ℃/200 h corroded in lithiated water for 80 days. At the later stage of oxide growth with 80 days exposure, cracks are observed in the oxide film (indicated by white arrows in Fig. 8a and b). The oxide film on specimen 100 ℃/s-β-WQ+600 ℃/2 h is compact and no pore is observed even near the outer surface (Fig. 8c). However, the oxide film on specimen 100 ℃/s-β-WQ+600 ℃/200 h is loose, numerous pores and some cracks can be observed even in the middle of oxide layer, and the formation process of cracks from pores is evident (Fig. 8d). The equiaxed grains are observed to be dominant for the two specimens, but the trace of evolution from columnar to equiaxed grains is still clear on specimen 100 ℃/s-β-WQ+600 ℃/2 h (indicated by the dot lines in Fig. 8c). It is also observed that the grain size in the oxide film on specimen 100 ℃/s-β-WQ+600 ℃/200 h (Fig. 8c) is smaller than that on specimen 100 ℃/s-β-WQ+600 ℃/2 h (Fig. 8d). The result is consistent with the observed results from the oxide film of 30 days exposure (Fig. 7).

4 Discussion

It is well-known that a new oxide layer is formed on the oxide/metal interface. The microstructure such as shape and size of oxide grains, and pores and cracks in oxides, will certainly affect the diffusion of O^{2-} or OH^- ions or anion vacancies to result in a different growth behavior of oxide. In other words, the corrosion behavior of zirconium alloys is closely associated with the oxide characteristics. The Pilling-Bedworth ratio of Zr is 1. 56, so a compressive stress exits inevitably in the oxide film on zirconium alloys. Based on the above fact, Zhou et al. [12,15,16] put forward a mechanism from the microstructural evolution of the oxide film. The main viewpoints are described as follows: the defects (vacancies and

Fig. 8 Low (a and b) and high (c and d) magnified SEM micrographs of the fracture surfaces of oxide films formed on specimens 100 ℃/s-β-WQ+600 ℃/2 h (a and c), and 100 ℃/s-β-WQ+ 600 ℃/200 h (b and d), corroded in lithiated water with 0.04 M LiOH for 80 days

interstitials) in the oxide film during its formation will be produced and the metastable phases (amorphous, cubic- and tetragonal - ZrO_2) will be stabilized due to the existence of compressive stress and defects in the oxide film; the diffusion, annihilation and condensation of vacancies and interstitials under the action of stress, temperature and time cause the stress relaxation and phase transformation from metastable phases to stable monoclinic - ZrO_2 (m - ZrO_2); vacancies will be absorbed by grain boundaries to form pores; the condensation of pores will form micro-cracks and degrade the protective characteristic of the oxide film. The pores at grain boundaries and the formation process from pores to cracks are clearly visible (Figs. 5 - 8).

When Zircaloy - 4 corroded in lithiated water, micro-electrochemical cells are consisted of SPPs and α - Zr at the Zr/ZrO_2 interface. α - Zr acts as an anode for its lower potential to be oxidized prior to SPPs[9]. The oxidation of SPPs takes place after they are incorporated into the oxide film. The SPPs incorporated in oxide layers will affect the microstructural evolution of oxide in two aspects[17-19]. On the one hand, the SPPs with nanometer size and homogeneous distribution provide a large quantity of phase boundaries for the sink of vacancies in oxides. In this case, the formation of pores and cracks in oxide layers is slowed down. In other words, the microstructural evolution will be retarded and the corrosion resistance will be enhanced. On the other hand, the oxidation of SPPs incorporated in oxide layers produces an additional stress in oxide layers to promote the diffusion of vacancies to form pores and cracks, but may stabilize the surrounding cubic- or tetragonal - ZrO_2. The observation of oxide film has shown that the pores and cracks in the oxide on specimen 100 ℃/s-WQ+600 ℃/200 h are more than those on specimen 100 ℃/s-β-WQ+ 600 ℃/2 h (Figs. 5, 6 and 8). This agrees with the corrosion results that the latter specimen exhibits superior corrosion resistance to the former specimen (Fig. 4c and Table 2). However, it is noted that the area fraction of $Zr(Fe, Cr)_2$ SPPs is lower than 2% in this study (Fig. 2 and Table 1). It is difficult to understand that the $Zr(Fe, Cr)_2$ SPPs can produce such a great effect on the corrosion behavior (Fig. 4).

It is well understood that the effect of Fe and Cr dissolved in the $\alpha - Zr$ matrix on the oxide characteristics is more homogeneous and unitary than SPPs due to the local distribution of SPPs. As mentioned in the introduction, the average size and area fraction of SPPs and the concentrations of alloying elements in $\alpha - Zr$ matrix are varied in a systematic manner during heat treatments at different temperatures from 480 ℃ to 600 ℃. We tried to measure the concentrations of Fe and Cr in $\alpha - Zr$ matrix using atom probe tomography (LEAP - 3000HR) and wave dispersion spectrometer (WDS), but we failed to obtain valuable results because the sample easily fractured in the analysis of atom probe tomography and the concentrations of Fe and Cr in $\alpha - Zr$ matrix were lower than the limited resolution of WDS. We analyzed the average size and area fraction of observed SPPs to assume the Fe and Cr concentrations in α-matrix with the annealing condition in an indirect manner. It is reasonable to conclude that the Fe and Cr concentrations in $\alpha - Zr$ matrix decrease with the increase of SPPs area fraction (Table 1). Moreover, on the base of previous studies[7,20], it can be inferred that Fe and Cr are supersaturated in $\alpha - Zr$ matrix for specimen 100 ℃/s-β-WQ. According to the results of Wadman and Andrén[7] and Hood and Schultz[21], it is difficult to measure the diffusion coefficient of Fe (D_{Fe}) in Zr - Sn alloys when temperature is below 500 ℃. This means that the diffusion of Fe and Cr is very slow at lower than 500 ℃. Fe and Cr are still supersaturated in $\alpha - Zr$ matrix of specimens 100 ℃/s-β-WQ + 480 ℃. From the equation $D = D_0 \exp(-Q/kT)$, it is understood that the diffusion coefficient increases with increasing temperature. It is reasonable to deduce that the concentrations of Fe and Cr in $\alpha - Zr$ matrix decrease with increasing annealing temperature for the specimens annealed at 480, 540 and 600 ℃ for the same annealing time.

Concerning the oxygen diffusion in the oxide, Cox and Pemsler[22] suggested that O^{2-} passed through the oxide layer into the metal by non-lattice diffusion via the routes, such as grain boundaries and pores in the oxide. The observation on the oxide film shows that the grain size in the oxide film on specimen 100 ℃/s-β-WQ+600 ℃/200 h is smaller than that on specimen 100 ℃/s-β-WQ+600 ℃/2 h, on both the outer surface and the fracture surface of the oxide films (Figs. 6 and 8). This implies that the Fe and Cr dissolved in $\alpha - Zr$ would affect the nucleation of ZrO_2. It is also observed that the pores and cracks are more in the oxide film on specimen 100 ℃/s-β-WQ+600 ℃/200 h than those on specimen

100 ℃/s-β-WQ+600 ℃/2 h (Figs. 5 – 8). Fine grains can provide more grain boundaries and more pores and cracks can also provide more routes to promote the diffusion of O^{2-} or OH^- through the oxide layer. In our previous work[2], it was also observed that the cracks in the oxide film on specimen βQ Zircaloy – 4 (treated by β – quenching) was much fewer than those on specimen RA Zircaloy – 4 (treated by recrystallization annealing), and the development of pores to cracks on the former specimen was noticeably slower than that on the latter specimen. In investigating the relationship between the corrosion behavior and microstructure evolution of oxide films on zirconium alloys tested in lithiated water, Zhou et al. [12,15,16] also found that the pore clusters and micro-cracks along grain boundaries were more easily formed in the oxide film on Zircaloy – 4 than ZIRLO, and the cracks parallel to the oxide/metal interface were also more easily produced in Zircaloy – 4 than that in ZIRLO. Moreover, the higher concentrations of Fe and Cr in α – Zr also make it possible to stabilize the cubic- or tetragonal – ZrO_2, as observed by Pecheur et. al. [17]. This further illustrates that a suitable increase in concentrations of alloying elements in α – Zr matrix is beneficial to retarding the microstructural evolution of oxide, thereby improving the corrosion resistance. This is possibly related to the effect of Fe and Cr dissolved in α – Zr on the surface free energy of ZrO_2 after the oxidation of α – Zr.

It was reported that when Zircaloy – 4 specimens were corroded in lithiated water, Li^+ was easier to incorporate in the oxide film for its small ionic radius. The incorporation of Li^+ into oxide films and absorption of these ions on the wall of pores will reduce the surface free energy of ZrO_2 during corrosion tests in lithiated water. As a result, the diffusion of vacancies and the formation of pores and micro-cracks were accelerated to degrade the corrosion resistance[12,16]. The whiskers are longer in the oxide film on specimen 100 ℃/s-β-WQ+600 ℃/2 h with high Fe and Cr concentrations in α – Zr than those on specimen 100 ℃/s-β-WQ+600 ℃/200 h with low Fe and Cr concentrations in α –Zr (Fig. 5). The whisker prepared by scraping method from the outer surface of oxide film was observed by HRTEM to be m – ZrO_2 containing a small amount of Fe and Cr (Fig. 9 and Table 3). It is well understood that Fe and Cr dissolved in the α – Zr matrix are more possible to dissolve in ZrO_2 grains than those in $Zr(Fe, Cr)_2$ SPPs. Based on the growth mechanism of whiskers that the grain nuclei having higher surface energy preferentially grow to whisker morphology[23], it is suggested that higher concentrations of Fe and Cr in α – Zr result in a higher surface free energy of ZrO_2 to lower the effect of Li^+ incorporated in the oxide, and thereby retard the formation of pores to improve the corrosion resistance. Moreover, Jeong et. al. [13] thought that the corrosion in aqueous LiOH solution was accelerated owing to the easy substitution of Li^+ for Zr^{4+} in the oxide layer due to the similar ionic radio ($Li^+ = 76$ pm and $Zr^{4+} = 72$ pm). More grain boundaries and pores result in more Li^+ incorporated into the oxide films to accelerate the corrosion (Fig. 4). Based on the above discussion, the concentrations of Fe and Cr in α – Zr matrix are possibly the main factors to influence the corrosion resistance of the quenched and annealed Zircaloy – 4 specimens.

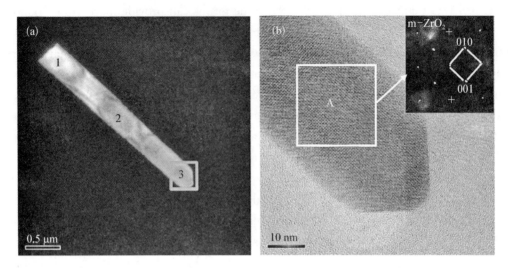

Fig. 9 Dark field image (a), and HRTEM lattice image (b), in the square of Fig. 9a. about the whisker prepared by scraping method from the outer surface of oxide film. The patterns obtained by FFT from A-area are also incorporated in Fig. 9b and are indicated $m - ZrO_2$. EDS analysis shows that the whisker is $m - ZrO_2$ with a small amount of Fe and Cr

Table 3 EDS results of positions 1, 2 and 3 in Fig. 9a (mass faction, %).

No.	Zr	Fe	Cr	O
1	76.53±0.84	0.53±0.22	1.22±0.23	21.72
2	78.54±0.77	—	1.15±0.22	20.20
3	75.66±0.91	—	1.09±0.24	23.25

5 Conclusions

(1) $\beta-$ quenched Zircaloy $-$ 4 specimens with the cooling rate of about 100 ℃/s behave the best corrosion resistance by autoclave tests in lithiated water with 0.04 M LiOH at 360 ℃/18.6 MPa. The corrosion resistance degrades with increasing the annealing temperature from 480 ℃ to 600 ℃ after $\beta-$ quenching. The corrosion resistance of specimens annealed at 600 ℃ after $\beta-$quenching decreases with increasing annealing time. Annealing time has little effect on the corrosion resistance for the specimens annealed at 480 and 540 ℃.

(2) Whisker-like grains are observed on the outer surface of the oxide film in pre-transition region. HRTEM observation shows that the whisker-like grain is $m - ZrO_2$ with a small amount of Fe and Cr.

(3) The observation on both the outer surface and fracture surface of the oxide films shows that the grain of the oxide on the specimen with high Fe and Cr concentrations in $\alpha -$ Zr is coarser than that on the specimen with low Fe and Cr concentrations in $\alpha - $ Zr. Pores and cracks are fewer in the oxide film on the specimen with high Fe and Cr concentrations in $\alpha - $ Zr. This indicates that a suitable increase in Fe and Cr concentration in $\alpha - $ Zr matrix is beneficial to retarding the microstructural evolution of oxide, thereby improving the

corrosion resistance.

Acknowledgements

The authors would like to express their thanks to Mr. Yuliang Chu and Mr. Weijun Yu of Instrumental Analysis and Research Center of Shanghai University for help in the microstructural analysis. This study is partly supported by National Natural Science Foundation of China (No. 50971084) and National Advanced Pressurized Water Reactor Project of China (No. 2011ZX06004 - 023).

References

[1] T. Andersson, G. Vesterlund, in: Zirconium in the Nuclear Industry: Fifth International Symposium, ASTM STP 754,1982, pp. 75 - 95.

[2] M. Y. Yao, B. X. Zhou, Q. Li, et al. , J. Nucl. Mater. 374(2008) 197 - 203.

[3] N. V. Bangaru, R. A. Busch, J. H. Schemel, in: Zirconium in the Nuclear Industry: Seventh International Symposium, ASTM STP 939,1987, pp. 341 - 363.

[4] H. K. Yueh, R. L. Kesterson, R. J. Comstock, et. al. , in: Zirconium in the Nuclear Industry: Fourteenth International Symposium, ASTM STP 1467,2005, pp. 330 - 346.

[5] B. X. Zhou, W. J. Zhao, Z. Miao, et al. , Chin. J. Nucl. Sci. Eng. 15(1995) 242 - 246 (in Chinese).

[6] R. A. Graham, J. P. Tosdale, E. T. Finden, in: Zirconium in the Nuclear Industry: Eighth International Symposium, ASTM STP 1023,1989, pp. 334 - 345.

[7] B. Wadman, H. -O. Andrén, in: Zirconium in the Nuclear Industry: Eighth International Symposium, ASTM STP 1023,1989, pp. 423 - 434.

[8] G. Garzarolli, E. Steinberg, H. G. Weidinger, in: Zirconium in the Nuclear industry: Eighth International Symposium, ASTM STP 1023,1989, pp. 202 - 212.

[9] H. G. Weidinger, H. Ruhmann, G. Cheliotis, et. al. , in: Zirconium in the Nuclear Industry: Ninth International Symposium, ASTM SIP 1132,1991, pp. 499 - 535.

[10] Y. F. Shen, M. Y. Yao, X. Zhang, et al. , Acta. Metall. Sin. 47(2011) 899 - 904 (in Chinese).

[11] C. M. Eucken, P. T. Finden, et al. , in: Zirconium in the Nuclear Industry: Eighth International Symposium, ASTM STP 1023,1989, pp. 113 - 127.

[12] B. X. Zhou, Q. Li, M. Y. Yao, et. al. , in: Zirconium in the Nuclear Industry: Fifteenth International Symposium, ASTM STP 1505,2009, pp. 360 - 383.

[13] Y. H. Jeong, J. H. Baek, S. J. Kim, et al. , J. Nucl. Mater. 270(1999) 323 - 333.

[14] B. X. Zhou, J. C. Peng, M. Y. Yao, et al. , J ASTM Intl, Paper ID JAI102951, ⟨http: //www. astm. org⟩, 8(1) (2011) 620 - 648.

[15] B. X. Zhou, Q. Li, M. Y. Yao, et al. , Nucl. Power Eng. 26(2005) 364 - 371 (in Chinese).

[16] B. X. Zhou, Q. Li, W. Q. Liu, M. Y. Yao, et al. , Rare Metal Mat. Eng. 35(2006) 1009 - 1016 (in Chinese).

[17] D. Pêcheur, F. Lefebvre, A. T. Mottal, J. Nucl. Mater. 189(1992) 318 - 332.

[18] J. Godlewski, in: Zirconium in the Nuclear Industry: Tenth International Symposium, ASTM STP 1245,1994, pp. 663 - 686.

[19] B. X. Zhou, M. Y. Yao, Z. K. Li, et al. , J. Mater. Sci. Tech. 28(2012) 606 - 613.

[20] C. Li, B. X. Zhou, W. J. Zhou, et al. , J. Nucl. Mater. 304(2002) 134 - 138.

[21] G. M. Hood, R. J. Schultz, in: Zirconium in the Nuclear Industry: Eighth International Symposium, ASTM STP 1023,1989, pp. 435 - 450.

[22] B. Cox, J. P. Pemsler, J. Nucl. Mater. 28 (1968) 73.

[23] W. Li, Inorganic whisker, Chemistry and applied chemistry Press, Beijing, China, 2005, pp. 2 - 15 (in Chinese).

添加 Bi 对 Zr-1Nb 合金在 360℃和 18.6 MPa 去离子水中耐腐蚀性能的影响*

摘　要：利用高压釜腐蚀实验研究了 Zr-1Nb-xBi(x=0.05—0.3,质量分数,%)合金在 360℃和 18.6 MPa 去离子水中的耐腐蚀性能.结果表明,在 Zr-1Nb 合金的基础上添加 Bi 能明显改善其耐腐蚀性能,且随着 Bi 含量的增加,合金的耐腐蚀性能进一步提高.合金显微组织的 TEM 观察和 EDS 分析表明,合金中存在 ZrNbFe 型和 β-Nb 第二相,Bi 含量对第二相的种类、尺寸和数量没有明显的影响;0.3%的 Bi 可全部固溶在 α-Zr 基体中,且不影响 Nb 的固溶含量.氧化膜断口和内表面形貌的 SEM 观察表明,固溶在 α-Zr 基体中的 Bi 能够明显延缓氧化膜显微组织的演化,包括孔隙发展成为微裂纹的过程和柱状晶向等轴晶的转变.

由于锆合金具有低的热中子吸收截面,良好的耐腐蚀性能以及力学性能,因此被广泛用于核反应堆中燃料元件的包壳材料[1-3].法国产 M5(Zr-1Nb-0.16O)合金是 Zr,Nb,O 三元合金,是 Framatome 公司开发并已应用于法国第三代燃料组件 AFA-3G 的燃料棒包壳材料,具有良好的耐腐蚀性能、抗蠕变性能、抗辐照增长和低吸氢性能[4].为提高核电的经济性,需要加深核燃料的燃耗,延长换料周期,这就对反应堆包壳材料锆合金的耐腐蚀性能提出了更高的要求.

合金化是有效提高锆合金耐腐蚀性能的主要方法之一,基于 Cu 对改善 Zr-4 和 Zr-1Nb 合金耐腐蚀性能的有益作用[5,6],本课题组曾系统研究了添加不同含量的 Cu 对 Zr-Sn-Nb-Fe-Cr 和 Zr-1Nb 合金在不同腐蚀条件下耐腐蚀性能的影响,结果发现其影响规律不同,这主要取决于合金体系和腐蚀条件[7,8].Bi 作为一种热中子吸收截面低(0.082×10^{-24} cm^2),在 α-Zr 中的溶解度较大,且满足空位扩散理论[9]的元素,可以考虑作为一种合金添加元素.李佩志等[10]的研究表明,在 Zr-1Nb 合金基础上添加少量 Bi 能够提高合金的耐腐蚀性能,然而并没有系统分析 Bi 含量对显微组织及其耐腐蚀性能的影响.基于 Bi 的有益作用,本课题组开展了添加不同含量 Bi 对 Zr-1Nb,Zr-4 和 Zr-Sn-Nb-Fe-Cr 合金在多种腐蚀条件下耐腐蚀性能的影响研究,探究其影响规律与锆合金体系和腐蚀条件之间的关系.文献[11]报道了添加不同含量 Bi 对 Zr-4 合金显微组织和耐腐蚀性能的影响,结果表明,适量 Bi 的添加促进了 Zr-4 合金中原先固溶在 α-Zr 基体中的 Sn 以含 Bi 和 Sn 的第二相的形式析出;但在 Zr-4 合金中添加(0.1%—0.5%)Bi 后在 400℃和 10.3 MPa 过热蒸汽中的耐腐蚀性能反而下降,并随着 Bi 含量的增加,耐腐蚀性能下降趋势越显著.Zr-1Nb 作为一种不含 Sn 的合金,添加不同含量 Bi 对其显微组织和耐腐蚀性能的影响尚未报道.本工作研究了添加不同含量 Bi 对 Zr-1Nb 合金显微组织以及在 360℃和 18.6 MPa 去离子水中的腐蚀行为的影响.

---

* 本文合作者：朱莉、姚美意、孙国成、陈文觉、张金龙.原发表于《金属学报》,2013,49(1)：51-57.

1 实验方法

以 Zr-1Nb 合金为母合金,在此基础上添加不同含量的 Bi,制备成 Zr-1Nb-xBi($x=$ 0.05,0.1,0.2,0.3,质量分数,%)合金,配料过程中考虑了 Bi 元素 10% 的损耗. 腐蚀试样制备过程为:先用真空非自耗电弧炉熔炼成 65 g 的合金锭,熔炼时通高纯 Ar 气保护,为保证成分均匀,合金锭每熔炼一次都要翻转,共熔炼 6 次,熔炼好的合金按 700 ℃ 热压、1 030 ℃,40 min 的 β 相均匀化处理、700 ℃ 热轧、1 030 ℃,40 min 的 β 相空冷、四道次冷轧和中间退火、最终 580 ℃,5 h 退火的顺序制成 20 mm×15 mm×0.7 mm 的腐蚀实验片状样品. 样品退火加热均在真空中进行,每次热处理前都要采用 30% H_2O+30% HNO_3+30% H_2SO_4+10% HF(体积分数)混合酸酸洗的方法去除试样表面的污染及氧化物;β 相均匀化处理和 β 相空冷均是将样品放在真空度约为 $5×10^{-3}$ Pa 的石英管内,在管式电炉中加热到 1 030 ℃ 保温 40 min,然后将电炉推离石英管浇水冷却. 为了在相同的制备工艺下比较 Bi 含量对 Zr-1Nb 合金耐腐蚀性能的影响,将未添加 Bi 的 Zr-1Nb 合金进行了重熔处理,定义为 Zr-1Nb-remelted. 用电感耦合等离子体原子发射光谱(ICPAES)分析得到的合金成分列在表 1 中,结果显示合金中的 Bi 含量与设计值接近. 制备好的样品经过酸洗和去离子水清洗后放入高压釜中进行 360 ℃ 和 18.6 MPa 去离子水的腐蚀实验. 腐蚀结果用 5 个平行试样的平均增重表征.

表 1 实验用合金的化学成分

Table 1 Chemical compositions of experimental alloys (mass fraction,%)

Alloy	Nb	Bi	Zr
Zr-1Nb-remelted	0.97	$<3×10^{-4}$	Bal.
Zr-1Nb-0.05Bi	1.01	0.04	Bal.
Zr-1Nb-0.1Bi	1.01	0.09	Bal.
Zr-1Nb-0.2Bi	1.01	0.19	Bal.
Zr-1Nb-0.3Bi	1.02	0.29	Bal.

用 10%(体积分数)高氯酸酒精溶液作为电解液,在电压 50 V,温度−40 ℃ 下采用双喷电解抛光的方法制备透射电子显微镜(TEM)观察的薄试样. 用 JEM-200CX 型 TEM 观察合金试样的显微组织,利用 JEM-2010F 型高分辨透射电镜(HRTEM)及其配置的能谱仪(EDS)观察和分析合金中第二相的形貌和成分,并通过选区电子衍射(SAD)确认第二相的晶体结构. 用 JSM-6700F 型扫描电子显微镜(SEM)观察氧化膜的断口和内表面形貌,观察用的腐蚀样品制备方法如文献[12]中所述. 在进行氧化膜形貌观察前,为提高图像质量,试样表面蒸镀 Pt.

2 实验结果

2.1 腐蚀行为

图 1 是 Zr-1Nb-xBi 合金在 360 ℃ 和 18.6 MPa 去离子水中腐蚀 290 d 后的腐蚀增重曲线. 可以看出,在 Zr-1Nb 合金的基础上添加 Bi 可以明显改善其耐腐蚀性能,且随着 Bi 含量

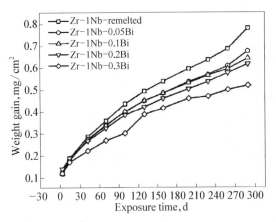

图 1 Zr‑1Nb‑xBi 合金在 360 ℃和 18.6 MPa 去离子水中腐蚀 290 d 后的增重曲线

Fig. 1 Weight gain of the Zr‑1Nb‑xBi alloys *vs* exposure time in deionized water at 360 ℃ and 18.6 MPa

的增加,合金的耐腐蚀性能进一步提高,如腐蚀 290 d 时 Zr‑1Nb 中添加 0.1%和 0.3%的 Bi 样品的平均腐蚀增重分别为 0.64 和 0.52 mg/cm²,比 Zr‑1Nb 重熔样品(0.78 mg/cm²)分别降低 17.9%和 33.3%.这与李佩志等[10]报道的在 Zr‑1Nb 合金中添加少量 Bi 能改善合金耐腐蚀性能的结果相一致.

2.2 合金显微组织

图 2 为 Zr‑1Nb‑xBi 合金显微组织的 TEM 像.可以看出,4 种合金中的第二相均呈条带状分布,这一方面可能是因为 β 相空冷时冷却速率较慢,一部分第二相在 β 相空冷冷却过程中沿 α‑Zr 板条晶界处析出引起的,另一方面也可能是因为 β 相空冷后残留的 β‑Zr 在随后的冷轧和 580 ℃(低于偏析温度 610 ℃)退火过程中发生分解而形成的.第二相按形状和大小可分为两种:一种为平均尺寸在 50 nm 左右的球状细小第二相,如图 2 中箭头 1—4 所示;另一种为平均尺寸约为 100 nm 的椭球状第二相,如图 2 中箭头 5—8 所示.

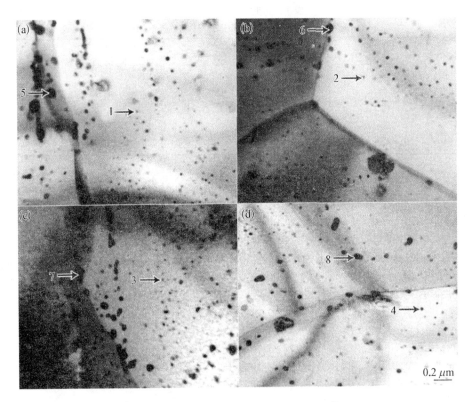

图 2 Zr‑1Nb‑xBi 系列合金的 TEM 像

Fig. 2 TEM images of the Zr‑1Nb‑xBi alloys with $x=0.05$ (a), $x=0.1$ (b), $x=0.2$ (c) and $x=0.3$ (d) (Arrows 1—4 indicate β‑Nb; arrows 5—8 indicate ZrNbFe type)

EDS 分析表明：Zr‑1Nb‑xBi 合金中细小球状第二相由 Zr，Nb，Fe 和 Cr 4 种元素组成，其中 Fe 和 Cr 含量很低，Fe/Cr 比在 0.18—0.77 之间（图 3a），SAD 分析证实这种第二相为 bcc 结构 β‑Nb（图 3d 中实框所示），这与李士炉等[8]报道的 Zr‑1Nb‑xCu 合金中细小第二相的种类一致. 图 3d 中虚框所示为 α‑Zr 基体. 椭球状第二相也是由 Zr，Nb，Fe 和 Cr 4 种元素组成，其中 Cr 含量比较低，但 Fe 含量明显高于细小球状第二相，且 Fe/Cr 比也明显高于后者，在 2.0—4.76 之间（图 3b）. 结合文献[13—15]中的报道，这种椭球状第二相应该是 ZrNbFe 型. Zr‑1Nb‑xBi 合金中并没有特意添加 Fe 和 Cr，因此，这 2 种第二相中的 Fe 和 Cr 均来自海绵 Zr 中的杂质. 另外，对 Zr‑1Nb‑xBi 合金的基体进行 EDS 分析（图 3c），结果表明，添加的 0.3% Bi 可全部固溶在 α‑Zr 中，这说明 580 ℃时 Bi 在 Zr‑1Nb 合金 α‑Zr 基体中的固溶度不小于 0.3%；同时基体中 Nb 含量（0.59%）接近该温度下 Nb 在 α‑Zr 中的最大固溶度 0.6%，表明 Zr‑1Nb 合金中添加 Bi 对 Nb 在 α‑Zr 中的固溶含量没有明显的影响. 需指出的是，图 3 中 EDS 的 Cu 峰为 TEM 样品台引起的.

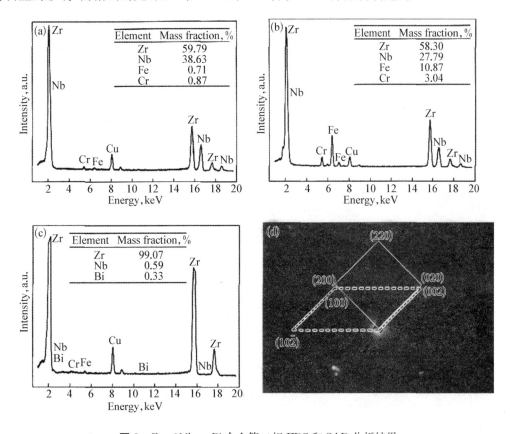

图 3 Zr‑1Nb‑xBi 合金第二相 EDS 和 SAD 分析结果

Fig. 3 EDS and SAD patterns of the second phase particles (SPPs) in the Zr‑1Nb‑xBi alloys (a) β‑Nb type, position 2 in Fig. 2b;(b) ZrNbFe type, position 5 in Fig. 2a;(c) α‑Zr matrix; (d) SAD patterns of SPP 2 in Fig. 2b (Dotted parallelogram indicate the hexagonal α‑Zr matrix, solid parallelogram the bcc β‑Nb)

2.3 氧化膜断口形貌

图 4 为 Zr‑1Nb‑xBi 合金在 360 ℃和 18.6 MPa 去离子水中腐蚀 200 d 后的氧化膜断

口形貌. 可以看出, 腐蚀样品氧化膜厚度随着 Bi 含量的增加而减小, Zr-1Nb 重熔试样的氧化膜厚度约为 4.4 μm, Zr-1Nb-0.1Bi 合金的氧化膜厚度约为 3.7 μm, 而 Zr-1Nb-0.3Bi 合金的氧化膜厚度约为 3 μm, 这与腐蚀增重结果相符. 从 Zr-1Nb 重熔样品的断口形貌中可以观察到平行于氧化膜/金属基体界面的微裂纹, 部分氧化膜厚的区域还存在垂直于氧化膜/金属界面的裂纹(图 4a2 和 a3 中箭头所示), Zr-1Nb-0.1Bi 样品的断口上有少量裂纹(图 4b2 和 b3 中箭头所示), 而 Zr-1Nb-0.3Bi 样品的断口上几乎观察不到微裂纹. 此外, 3 种样品的氧化膜都是由 ZrO_2 柱状晶和等轴晶组成, 能够明显看到柱状晶与等轴晶的分界线(图 4a2, b2 和 c2 中虚线所示), 靠近外表面的是明显的等轴晶区域. 通过统计得出: Zr-1Nb 重熔样品氧化膜中的等轴晶占整个氧化膜厚度的 21% 左右, 并且柱状晶中出现向等轴晶转变的痕迹(图 4a3 中放大框所示), Zr-1Nb-0.1Bi 合金氧化膜中等轴晶的比例约为 18%, 而 Zr-1Nb-0.3Bi 合金氧化膜中等轴晶的比例约为 16%, 且柱状晶的长度大于 Zr-1Nb 重熔样品的柱状晶长度. 研究[16,17]指出, 氧化膜中等轴晶的出现是由于柱状晶中的缺陷通过扩散凝聚形成新的晶界, 从而形成等轴晶. 因此可以看出, 添加 Bi 延缓了氧化膜中柱状晶向等轴晶转化的过程, 从而提高了其耐腐蚀性能.

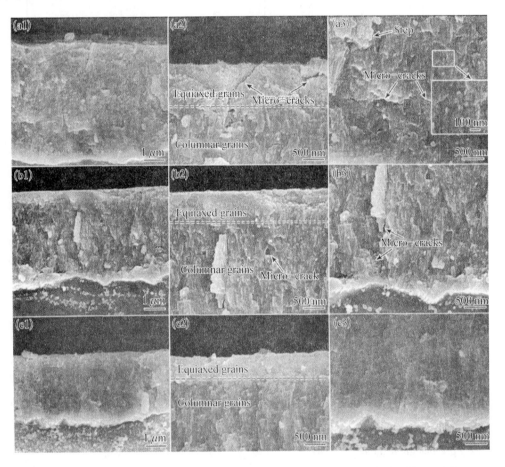

图 4　不同合金在 360 ℃和 18.6 MPa 去离子水中腐蚀 200 d 后的氧化膜断口形貌

Fig. 4　Fracture surface morphologies of the oxide film formed on Zr-1Nb-remelted (a1, a2, a3), Zr-1Nb-0.1Bi (b1, b2, b3) and Zr-1Nb-0.3Bi (c1, c2, c3) alloys corroded in deionized water at 360 ℃ and 18.6 MPa for 200 d exposure

同时,通过比较图 4a3，b3 和 c3 可以看出,Zr‐1Nb‐remelted 样品的氧化膜断口起伏不平,比较粗糙,存在一些明显的"台阶"和微裂纹,说明这些氧化膜在被折断之前内部就存在一些弱的结合面,因此折断时裂纹沿着这些弱的结合面发生扩展,形成了起伏不平的断口形貌;而随着合金中 Bi 含量的增加,氧化膜断口越来越平整,同时更加致密.

2.4 氧化膜内表面形貌

图 5 为 Zr‐1Nb‐xBi 合金在 360 ℃和 18.6 MPa 去离子水中腐蚀 200 d 后的氧化膜内表面形貌.Zr‐1Nb 重熔样品的氧化膜内表面凹凸不平(图 5a1),说明不同 α‐Zr 晶粒间氧化生长速率不均匀;高倍下可观察到尺寸为 700 nm×400 nm 的孔隙(图 5a2),这应该是未被氧化的第二相被酸蚀刻后掉落造成的,这也反映出第二相的氧化比 α‐Zr 基体慢;随着合

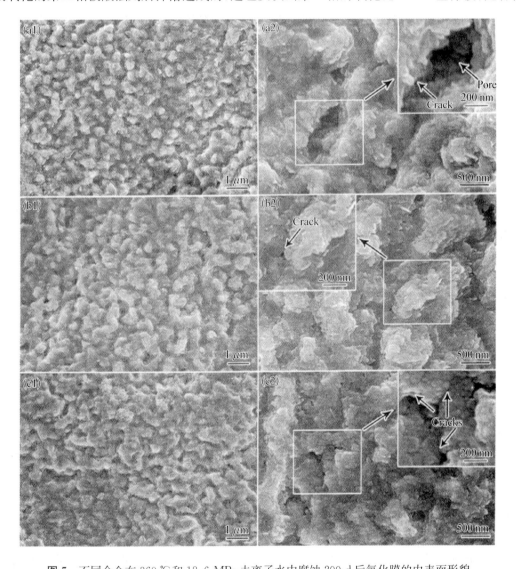

图 5　不同合金在 360 ℃和 18.6 MPa 去离子水中腐蚀 200 d 后氧化膜的内表面形貌

Fig. 5　Inner-surface micrographs of the oxide film of Zr‐1Nb‐remelted (a1, a2), Zr‐1Nb‐0.1Bi (b1, b2) and Zr‐1Nb‐0.3Bi (c1, c2) alloys corroded in demonized water at 360 ℃ and 18.6 MPa

金中 Bi 含量的增加,样品氧化膜内表面逐渐趋于平整(图 5b1 和 c1),这说明添加 Bi 可以减弱不同 α-Zr 晶粒间氧化生长速率的差别. 高倍下 3 种样品都能观察到少量微裂纹,这可能是制样过程中的蚀刻使氧化膜中压应力得到释放造成的.

3 分析与讨论

文献[11]报道了添加适量 Bi 的确能促进 Zr-4 合金中原先固溶在 α-Zr 基体中的 Sn 以含 Bi 和 Sn 第二相的形式析出,同时合金在 400 ℃ 和 10.3 MPa 过热蒸汽中的耐腐蚀性能随着 Bi 的添加反而下降,并且随着 Bi 含量的增加耐腐蚀性能下降趋势越显著. 而本工作中,在 Zr-1Nb 合金中添加的 Bi 全部固溶在 α-Zr 基体中,同时 Bi 的添加能改善 Zr-1Nb 合金在 360 ℃ 和 18.6 MPa 去离子水中的耐腐蚀性能,并随着 Bi 添加量的增加,合金耐腐蚀性能进一步提高. 这说明 Bi 对 Zr-4 和 Zr-1Nb 合金在不同水化学条件下耐腐蚀性能的影响规律和机理不同,这可能与 Bi 在合金中的存在形式有关. 初步的结果已表明,Bi 以固溶的形式存在能改善合金的耐腐蚀性能,而以第二相形式存在则会恶化合金的耐腐蚀性能.

对于锆合金的氧化过程,周邦新等[18-22]提出:由于 Zr 的 P. B. (Pilling-Bedworth)比为 1.56,在氧化过程中发生体积膨胀,同时受到金属基体的约束,在氧化膜内部必然存在很大的压应力;在压应力的作用下氧化膜中的空位和间隙原子发生扩散、凝聚从而形成点、线、面或体的各种缺陷,这些缺陷在时间、应力和温度的作用下,发生扩散、湮没和凝聚,空位被晶界吸收后形成纳米大小的孔隙簇,孔隙簇进一步发展成微裂纹,从而加快 O^{2-} 或 OH^- 的扩散,降低其耐腐蚀性能. 在氧化膜的生长过程中,这种压应力是无法避免的,因此,氧化膜的显微组织在腐蚀过程中不断发生演化也是必然的,如果通过改变合金的化学成分能够延缓这种演化过程,那么锆合金的耐腐蚀性能也会得到提高. 从图 4 和图 5 可以看出,Zr-1Nb 合金中添加 Bi 后,氧化膜较未添加 Bi 的合金致密,并且起伏较小,说明固溶在 α-Zr 基体中的 Bi 明显延缓了氧化膜中显微组织的演化过程,包括氧化膜中的孔隙发展成微裂纹的过程和柱状晶向等轴晶的转变,进而提高了合金的耐腐蚀性能.

周邦新等[19,20]用扫描探针显微镜(SPM)研究了 Zr-4 和一种 Zr-Sn-Nb 合金在 360 ℃ 和 18.6 MPa,0.04 mol/L LiOH 水溶液中腐蚀 14 d 后氧化膜表面的晶粒形貌,结果表明,Zr-Sn-Nb 合金氧化膜表面 ZrO_2 的晶粒起伏比 Zr-4 合金小,认为与 Zr-Sn-Nb 合金中固溶的 Nb 可减少因 Li^+ 和 OH^- 渗入引起的 ZrO_2 表面自由能降低的程度,使氧化膜中的空位不容易扩散凝聚形成微孔隙和晶界微裂纹,从而改善合金的耐腐蚀性能. 固溶在 α-Zr 中的 Bi 也可能有类似的作用,从氧化膜的断口形貌(图 4)上可以看出,添加 Bi 后确实减缓了孔隙和微裂纹的形成过程. 同时,由于 Zr 是 hcp 金属,因而锆合金具有明显的各向异性,这一特征必然反映到锆合金不同取向晶粒间的氧化速率的差别上[23],从而在氧化膜内表面体现出凹凸不平的形貌. 当 α-Zr 中固溶 Bi 后可能减小不同取向晶粒氧化生长的各向异性,即不同取向晶粒间的氧化速率趋于一致,从而得到较为平整的氧化膜内表面(图 5). 另外,锆合金在高温高压水中的腐蚀也是一个电化学过程[24],氧化时 Zr/ZrO_2 界面上的第二相与 α-Zr 基体组成许多微电池,α-Zr 基体作为阳极优先氧化[25]. 当 α-Zr 中固溶 Bi 后会改变基体的电位,这可能减小了其与第二相的电势差,从而减缓电化学腐蚀进程,提高合金的耐腐蚀性能.

4 结论

(1) Zr-1Nb-xBi 合金中包含 2 种第二相: 一种是含少量 Cr 的 Fe/Cr 比在 2.0—4.76 之间的椭球状 ZrNbFe 型第二相;一种是含少量 Fe 和 Cr, Fe/Cr 比在 0.18—0.77 之间的细小球状 β-Nb 第二相. 析出的第二相不含 Bi,添加的 0.3% Bi 全部固溶在 α-Zr 基体中,适量 Bi 的添加不影响 α-Zr 基体中 Nb 的固溶含量.

(2) 在 360 ℃ 和 18.6 MPa 去离子水中腐蚀 200 d 的 Zr-1Nb-xBi 合金氧化膜由柱状晶和等轴晶组成,随着 Bi 含量的增加,柱状晶比例增大,同时断口和内表面都趋于平整.

(3) 在 Zr-1Nb 合金基础上添加 (0.05%—0.3%) Bi 可以改善合金在 360 ℃ 和 18.6 MPa 去离子水中的耐腐蚀性能,且随着 Bi 含量的增加,合金耐腐蚀性能进一步提高.

参 考 文 献

[1] Motta A T, Yilmazbayhan A, Gomes da Silva M J, Comstock R J, Was G S, Busby J T, Gartner E, Peng Q J, Jeong Y H, Park J Y. *J Nucl Mater*, 2007;371: 61.

[2] Cox B. *J Nucl Mater*, 2005;336: 331.

[3] Billot P, Yagnik S, Ramasubramanian N, Peybernes J, Pêcheur D. In: Moan G D, Rudling P eds., *Zirconium in the Nuclear Industry: 13th International Symposium*, *ASTM STP 1423*, Annecy: ASTM International, 2002: 169.

[4] Mardon J P, Charquet D, Senevat J. In: Sabol G P, Moan G D eds., *Zirconium in the Nuclear Industry: 12th International Symposium*, *ASTM STP 1354*, Toronto: ASTM International, 2000: 505.

[5] Park J Y, Yoo S J, Choi B K, Jeong Y H. *J Nucl Mater*, 2008;374: 343.

[6] Hong H S, Moon J S, Kim S J, Lee K S. *J Nucl Mater*, 2001;297: 113.

[7] Yao M Y, Li S L, Zhang X, Peng J C, Zhou B X, Zhao X S, Shen J Y. *Acta Metall Sin*, 2011;47: 865.
(姚美意,李士炉,张欣,彭剑超,周邦新,赵旭山,沈剑韵. 金属学报,2011;47: 865)

[8] Li S L, Yao M Y, Zhang X, Geng J Q, Peng J C, Zhou B X. *Acta Metall Sin*, 2011;47: 163.
(李士炉,姚美意,张欣,耿建桥,彭剑超,周邦新. 金属学报,2011;47: 163)

[9] Chen H M, Ma C L, Bai X D. *The Corrosion and Protection of Nuclear Materials*. Beijing: Atomic Energy Press, 1984: 1.
(陈鹤鸣,马春来,白新德. 核反应堆材料腐蚀及其防护. 北京: 原子能出版社,1984: 1)

[10] Li P Z, Li Z K, Xue X Y, Liu J Z. *Rare Met Mater Eng*, 1998;27: 356.
(李佩志,李中奎,薛祥义,刘建章. 稀有金属材料与工程,1998;27: 356)

[11] Yao M Y, Zou L H, Xie X F, Zhang J L, Peng J C, Zhou B X. *Acta Metall Sin*, 2012;48: 1098.
(姚美意,邹玲红,谢兴飞,张金龙,彭剑超,周邦新. 金属学报,2012;48: 1098)

[12] Yao M Y, Zhou B X, Li Q, Liu W Q, Geng X, Lu Y P. *J Nucl Mater*, 2008;374: 197.

[13] Kim Y S, Kim S K, Bang J G, Jung Y H. *J Nucl Mater*,2000;279: 335.

[14] Comstock R J, Schoenberger G, Sabol G P. In: Bradley E R, Sabol G P eds.,*Zirconium in the Nuclear Industry: 11th International Symposium*, *ASTM STP 1295*, Garmisch-Partenkirchen: ASTM International, 1996: 710.

[15] Park J Y, Choi B K, Yoo S J, Jeong Y H. *J Nucl Mater*, 2006;359: 59.

[16] Anada H, Takeda K. In: Bradley E R, Sabol G P eds., *Zirconium in the Nuclear Industry: 11th International Symposium*,*ASTM STP 1295*, Garmisch-Partenkirchen: ASTM International, 1996: 35.

[17] Wadman B, Lai Z, Andrén H O, Nyström A L, Rudling P,Pettersson H I. In: Garde A M, Bradley E R eds., *Zirconium in the Nuclear Industry: 10th International Symposium*, *ASTM STP 1245*, Baltimore M D: ASTM International, 1994: 579.

[18] Zhou B X, Li Q, Yao M Y, Liu W Q. *Nucl Power Eng*, 2005;26: 364.

（周邦新,李强,姚美意,刘文庆. 核动力工程,2005;26:364）

[19] Zhou B X, Li Q, Liu W Q, Yao M Y, Chu Y L. *Rare Met Mater Eng*, 2006;35:1009.

（周邦新,李强,刘文庆,姚美意,褚于良. 稀有金属材料工程,2006;35:1009）

[20] Zhou B X, Li Q, Yao M Y, Liu W Q, Chu Y L. In: Kammenzind B, Limback M eds., *Zirconium in the Nuclear Industry: 15th International Symposium*, *ASTM STP 1505*, Sunriver Oregon: ASTM International, 2008:371.

[21] Zhou B X, Li Q, Huang Q, Miao Z, Zhao W J, Li C. *Nucl Power Eng*, 2000;21:339.

（周邦新,李强,黄强,苗志,赵文金,李聪. 核动力工程,2000;21:339）

[22] Zhou B X, Li Q, Yao M Y, Liu W Q, Chu Y L. *Corros Prot*, 2009;30:589.

（周邦新,李强,姚美意,刘文庆,褚于良. 腐蚀与防护,2009;30:589）

[23] Zhou B X, Peng J C, Yao M Y, Li Q, Xia S, Du C X, Xu G. In: Limback M, Barberis P eds., *Zirconium in the Nuclear Industry: 16th International Symposium*, *ASTM STP 1529*, Chengdu: ASTM International, 2010:620

[24] Zhong X Y, Yang B, Li M C, Yao M Y, Zhou B X, Shen J N. *Rare Met Mater Eng*, 2010;39:2167.

（钟祥玉,杨波,李谋成,姚美意,周邦新,沈嘉年. 稀有金属材料工程,2010;39:2167）

[25] Weidinger H G, Ruhmann H, Cheliotis G, Maguire M, Yau T L. In: Eucken C M, Garde A M eds., *Zirconium in the Nuclear Industry: 9th International Symposium*, *ASTM STP 1132*, Kobe: ASTM International, 1991:499.

Effect of Bi Addition on the Corrosion Resistance of Zr – 1Nb Alloy in Deionized Water at 360 ℃ and 18.6 MPa

Abstract: The effect of Bi contents on the corrosion resistance of Zr – 1Nb – xBi ($x = 0.05$—0.3, mass fraction, %) was investigated in deionized water at 360 ℃ and 18.6 MPa by autoclave tests. The results show that the corrosion resistance of Zr – 1Nb alloy can be improved by adding Bi, and the more the Bi content is, the better the corrosion resistance is. TEM and EDS analyses on the microstructures of the alloys show that there are two types of second phase particles (SPPs), including ZrNbFe and β – Nb. The Bi contents have little effect on the type, size and amount of SPPs, 0.3% Bi can be completely dissolved in α – Zr matrix and has no influence on the solution content of Nb in α – Zr matrix. From the fracture and inner surface morphology of oxide films observed by SEM, it can be seen that the Bi dissolved in the α – Zr could noticeably slow down the microstructural evolution of oxide film, including the propagation of micro-cracks and the transformation from columnar grains to equiaxed grains in the oxide film.

Crystal Structure Evolution of the Cu-Rich Nano Precipitates from bcc to 9R in Reactor Pressure Vessel Model Steel*

Abstract: The crystal structure evolution of the Cu-rich nano precipitates from bcc to 9R during thermal aging was studied in nuclear reactor pressure vessel (RPV) model steels. The specimens, contained higher copper and nickel contents than commercially available one, were heated at 890 ℃ for 0.5 h and then water quenched followed by tempering at 660 ℃ for 10 h and aging at 400 ℃ for 1 000 h. It was observed that bcc and 9R orthogonal structure, as well as 9R orthogonal and 9R monoclinic structure, coexist in a single Cu-rich nano precipitate. Further analyses pointed out that Cu-rich nano precipitates of bcc structure were not stable, it may preferentially transform to 9R orthogonal structure and then to 9R monoclinic structure. This results showed that the crystal structure evolution of the Cu-rich nano precipitates was complex.

1 Introduction

Cu-rich nano precipitates usually lead to the increase of the ductile-brittle transition temperature (DBTT) of reactor pressure vessel (RPV) steels, the degradation of mechanical properties and the limitation of the service life of the nuclear reactors[1-3]. However, Cu-rich nano precipitates can be used as the strengthening phase in high strength low carbon steels such as natural gas pipe steel and ship structure steels, etc. [4-6] Therefore, the precipitation process in steels has attracted much attention since early 1960s[7]. Over the last fifty years, lots of work has been done focused on the precipitation in iron-copper (Fe–Cu) alloys or low alloy ferritic steels by thermal aging experiments and computer simulations[8-13]. The results indicate that Cu-rich nano precipitates undergo a complex precipitation process and their different structures are relative to the dimension. Coherent Cu-rich precipitates of body centred cubic (bcc) structure appear firstly as Cu is supersaturated in α–Fe matrix. When growing into a critical size, Cu-rich nano precipitates then transform into a 9R twin structure with the stacking sequence of ABC/BCA/CAB and its orientation relationship with α–Fe matrix is $(\bar{1}\bar{1}4)_{9R}//(011)_{Fe}$, $[\bar{1}\bar{1}0]_{9R}//[11\bar{1}]_{Fe}$. The critical size of the transformation from bcc to 9R structure is proportional to the aging temperature[14-18]. The crystal structure evolution from bcc to 9R of Cu-rich nano precipitates described above has been confirmed by experiments and computer simulations,

* In collaboration with Feng Liu, Peng Jianchao, Wang Junan. Reprinted from Acta Metallurgical Sinica (English Letters), 2013,26(6): 707 – 712.

and also been accepted broadly, yet another different results have been reported in the last two years. Recently, Wang et al. [19] reported that, not only the monoclinic 9R structure but also 2H close packed structure and stacking faults are exist in one individual Cu-rich precipitate. Heo et al. [20] found that in many 9R Cu-rich precipitates, besides the common twin-related structures there do exist some other untwined structures. These imply that there may have some unknown structures evolution process which needs to be clarified.

In this paper, the more detail crystal structure evolution from bcc to 9R of Cu-rich nano precipitates in RPV model steels is investigated by high resolution transmission electron microscopy (HRTEM) and energy dispersive spectrum (EDS). The results are significant for understanding the precipitation nature as well as the structure evolution process of Cu-rich precipitates, and thus it is helpful for the study on irradiation embrittlement mechanism of RPV steels.

2　Experimental

RPVs are usually made of manganese-nickel-molybdenum (Mn – Ni – Mo) low alloy ferritic steel. Cu is the residual element and its content is generally lower than 0. 08 wt. % (the chemical compositions are referred to mass fraction unless special noted in this paper), and alloy element Ni content is in the range of 0. 4%– 1. 0%. In order to let the Cu-rich nano precipitates appear in a shorter time during the thermal aging treatment, the contents of Cu and Ni are increased, and other elements have the same level with those of the commercial available RPV steels. The experimental material is named RPV model steel in this paper, and the chemical composition is given in Table 1. The ingot of the RPV model steel, about 40 kg in weight, was prepared by vacuum induction melting, and then forged and hot rolled into a plate with 6 mm in thickness. The specimens with the size of 35 mm × 40 mm were cut from the plate and heated at 890 ℃ for 0. 5 h, then water quenched followed by tempering at 660 ℃ for 10 h and air cooled. Finally, the specimens were isothermally aged at 400 ℃ for 1 000 h.

Table 1　Chemical composition of the RPV model steel

Content	Cu	Ni	Mn	Si	P	C	S	Mo	Fe
Atomic fraction (at. %)	0. 45	1. 37	1. 16	0. 36	0. 029	1. 02	0. 014	0. 006	Balanced
Mass fraction (wt. %)	0. 52	1. 46	1. 15	0. 18	0. 016	0. 22	0. 008	0. 01	Balanced

HRTEM specimens were prepared by punching discs of 3 mm in diameter from the 0. 1 mm thin sheet of the aged samples and then mechanically ground to a thickness of about 50 μm. Finally the specimens were electro-polished at −40 ℃ in a electrolyte of 10% perchloric acid and 90% alcohol (volume fraction). The microstructural observation of the Cu-rich nano precipitates by HRTEM was carried out on a JEM – 2010F operating at 200 kV. Oxford-INCA EDS was used to analyze the chemical composition of the precipitates. Since the Cu-rich nano precipitates were about 5 – 15 nm in size, digital micrograph software provided by Gatan Company was employed as an assistance to analyze the high resolution electron micrographs and to study the crystal structure.

3   Results and Discussion

Figure 1(a) is the distribution of Cu-rich nano precipitates (the gray dots and one of them is marked by white circle) observed by scanning transmission electron microscopy (STEM). The mean size of these precipitates is about 6. 5 nm. EDS analysis of a Cu-rich nano precipitate taken from Fig. 1(a) is shown in Fig. 1(b), which indicates the precipitate contains Cu. But the Cu content measured by EDS is lower than the actual one, this is caused by the interference of matrix due to the small size of Cu-rich nano precipitates.

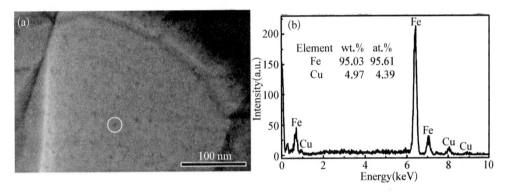

Fig. 1   (a) STEM micrograph of Cu-rich precipitates in RPV model steel after aging at 400 ℃ for 1 000 h; (b) EDS analysis of a Cu-rich nano precipitate marked with the white circle taken from (a)

HRTEM is adopted to investigate the crystal structure and the crystal structure evolution of Cu-rich nano precipitates during thermal aging. The specimens are tilted to let the electron beam direction closely parallel to a $\langle 111 \rangle$ direction of one grain in the α - Fe matrix. 25 Cu-rich nano precipitates with clear high resolution lattice image have been observed, and 16 precipitates of them are 9R structures determined by the fast Fourier transform (FFT) and inverse fast Fourier transform (IFFT) analysis. In order to investigate the crystal structure evolution of Cu-rich nano precipitates, some micrographs with special lattice images are chosen and the analysis results are described as follows.

A Cu-rich nano precipitate of spherical shape is shown in Fig. 2(a), its size is about 5. 5 nm and its Cu content is around 5% shown in Fig. 2(c). The FFT pattern of region A in Fig. 2(a) with the crystal-plane indices is embedded in the top right corner of Fig. 2(a), which indicates this region is a bcc structure. While the spots of FFT pattern are enlonged in different extents, this means that distortion occurs in the crystal lattice. The measured interplanar spacing is 0. 197 nm for (011) and (101), 0. 184 nm for (110). It is reduced by 2. 8% and 9. 2% respectively compared with the standard value of α - Fe (110) (PDF number: 06 - 0696). However the interplanar spacings of {110} in the region apart from the Cu-rich region are all 0. 201 nm, which is close to the standard value. So, Cu

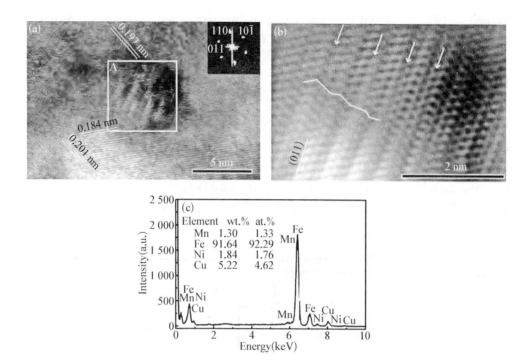

Fig. 2 (a) HRTEM micrograph taken along the ⟨111⟩ direction and FFT pattern of a Cu-rich precipitate; (b) IFFT pattern of region A in Fig. 2 (a); (c) EDS analysis of the Cu-rich precipitate

atoms segregate to the crystal plane of (110) in α-Fe matrix can result in the interplanar spacing reduction, which has ever been observed in our previous study[21]. The IFFT pattern of region A in Fig. 2(a) is shown in Fig. 2(b). The segregation of Cu atoms yields stress concentration in this region and results in the lattice distortion. The tiny "staggers" appeared frequently among the (011) crystal planes are observed, some of them are marked by the white arrows. Many tiny "kinks" are also observed on (110) and (101) crystal planes and a group of "kinks" are labeled with the white lines. Such Cu-rich region should be the initial stage of the Cu-rich nano precipitates. With more segregation of Cu atoms, the distortion will be increased continuously, which can make the crystal structure transform further.

Another Cu-rich nano precipitate in the same specimen shown in Fig. 3 (a) has significant "herring-bone fringes". The Cu-rich nano precipitate is spindle in shape, and the long axis and short axis is about 14 nm and 6.5 nm respectively. EDS analysis results in Fig. 3(b) indicate that Cu content is about 20%, Ni and Mn atoms are included besides Fe in the Cu-rich nano precipitate.

In order to analyze conveniently, the Cu-rich nano precipitate is divided into two parts marked with A and B, which is separated by the white lines. FFT has been conducted in region A and B and shown in Fig. 3(c) and Fig. 3(d). Typical 9R structure in region A is deduced from the crystal-plane indices indexed in Fig. 3(c). In a similar way, bcc structure in region B is deduced from Fig. 3(d). Figure 3(e) shows the IFFT pattern in Fig. 3(a).

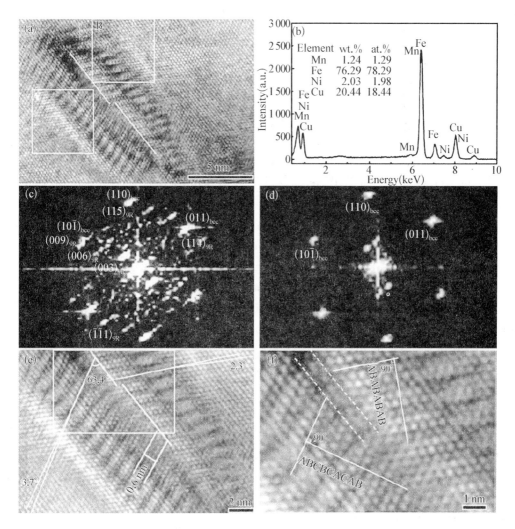

Fig. 3 (a) HRTEM micrograph of a Cu-rich precipitate taken along the ⟨111⟩ direction;
(b) EDS analysis of the Cu-rich precipitate; (c) FFT pattern of region A in (a); (d) FFT pattern
of region B in (a); (e) IFFT pattern of (a); (f) enlarged pattern of the rectangle region in (e)

From the pattern, it is observed that the contrast of the (009) crystal plane changes signifcantly every period of three-atomic layers. The spacing of the period is 0. 6 nm. The reason is that Cu atoms segregate to the {110} crystal plane in α – Fe matrix in the periods every three layers at the nucleation stage of Cu-rich precipitates, which has been reported by Xu et al. using the analysis of atom probe tomography (APT) and HRTEM[21]. Compared with that in region A, the atoms in region B, bcc structure, arrange regularly and the arrangement mode is different obviously from that in region A. The rectangle region in Fig. 3(e) which covers parts of region A and region B as marked in Fig. 3(a) is enlarged so as to observe the atomic arrangement in the Cu-rich nano precipitate clearly, the enlarged pattern is shown in Fig. 3 (f). The stacking sequence of the close packed crystal plane in region A is ABC/BCA/CAB, the angle between (009) crystal plane and [001] crystal direction is 90°, which indicates that this region is an orthogonal 9R structure. The orientation relationship with α - Fe is $(\bar{1}\bar{1}4)_{9R}//(0\bar{1}\bar{1})_{Fe}$, $[\bar{1}10]_{9R}//$ $[1\bar{1}1]_{Fe}$, which is consistent with the results reported in the previous literature [22]. In addition, the angle between (009) and $(\bar{1}\bar{1}4)$ crystal plane is 63. 4°, (009) and the adjacent $(10\bar{1})$ crystal plane in the matrix is 3. 7°. This can produce a certain elastic strain field, so the structure is not stable. The stacking sequence of the close packed plane in region B is AB/AB/AB/AB, the angle between (110) crystal plane and [110] crystal direction is 90°, which further demonstrates that this region is a bcc structure. The (110) crystal plane and the adjacent (110) crystal plane in the matrix is not parallel, and there exists an angle of 2. 3°. This shows that the lattice have a distortion when coherent bcc Cu-rich nano precipitates nucleate. A transition zone, about 3 atom layers in width, between region A and B is observed in Fig. 3(f) marked with two parallel dotted lines. In this zone, the atomic arrangement is disordered, which is caused by the different crystal structures on both sides.

The above analysis results indicate that the Cu-rich nano precipitates cannot be considered as a 9R twin structure directly according to the "herring-bone fringes", which is different from the former works[23], but contains both bcc and 9R structure sometimes. This structure can be deduced to a transitional state when bcc transform to 9R twin structure. However this structure is not stable, it would transform to a more stable 9R twin structure.

Figure 4(a) shows the HRTEM micrograph of one more Cu-rich nano precipitate in the same specimen. The Cu-rich nano precipitate is ellipsoid in shape, and the length of long axis and short axis is about 10. 4 nm and 6. 3 nm respectively. EDS analysis result indicates that the Cu content in this precipitate is around 17% Cu shown in Fig. 4(b). Figure 4(c) is the FFT pattern of region A in Fig. 4(a), this is a 9R multiple-twin structure. IFFT pattern of region A is shown in Fig. 4(d). It can be seen that the (009) atomic planes in the twin segments of region B and the one near to the matrix in region C are parallel to the {110} crystal plane of the matrix, the angle between two adjacent (009) atomic planes is 120°. IFFT patterns of region B and region C in Fig. 4(a) are shown in Fig. 4(e) and Fig. 4(f). In the three twin segments, the stacking sequences of the close packed planes along the [001] direction are all ABC/BCA/CAB/ABC, which are all typical 9R structure. Whereas there are some differences between them, the angle between [100] crystal direction and (009) crystal plane in region B is 90°, while those of the two twin segments in region C are both about 86°. It is shown that one orthogonal 9R structure and two monoclinic 9R structures coexist in one single Cu-rich nano precipitate.

The different structures of Cu-rich nano precipitates as shown in Fig. 2, Fig. 3 and Fig. 4 reveal in somewhat the possible process of structural evolution of such precipitates from the initial stage of nucleation to the more stable 9R structure. In the initial stage, internal stresses introduced by the segregation of Cu atoms lead to the lattice distortion, so the tiny "staggers" and "kinks" among crystal planes appear as shown in Fig. 2(b). The bcc structure of Cu-rich nano precipitates at this moment is coherent with the α - Fe

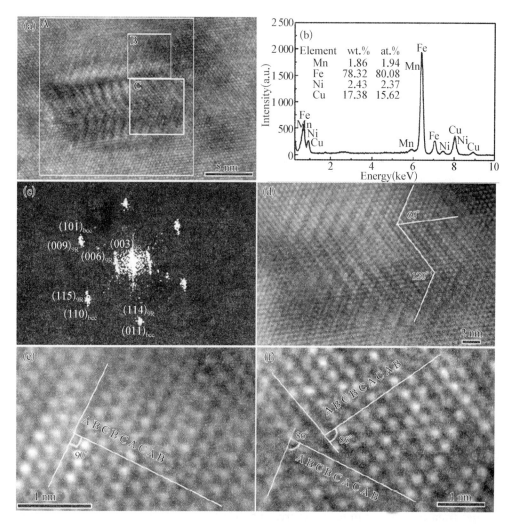

Fig. 4 (a) HRTEM micrograph of a Cu-rich precipitate taken along the ⟨111⟩ direction;
(b) EDS analysis of the Cu-rich precipitate; (c) FFT pattern of region A in (a); (d) IFFT pattern
of region A in (a); (e) IFFT pattern of region B in (a); (f) IFFT pattern of region C in (a)

matrix. When it is growing by the segregation of more Cu atoms, the bcc structure would transform into 9R structure preferentially in the high strain energy region which is located around the interface between Cu-rich nano precipitate and α - Fe matrix. Such transformation does not occur in the core of the precipitates and the regions far from the high strain energy locations. So bcc and 9R structures coexisting in one single Cu-rich nano precipitate can be revealed as in Fig. 3. This state should be the transitional process from bcc structure to 9R twin structure. The precipitate in Fig. 4 is a multiple-twin structure containing both orthogonal 9R and monoclinic 9R structures. It was reported that only two twin-related segments exist in the precipitates of small size, generally in 5 nm or less, further twining segments would take place in a larger precipitates, so three or more twin segments could be observed[24-26]. The three twin segments of the Cu-rich nano precipitate do not have the same structure, one of them is orthogonal, the other two are monocilinc.

Therefore such Cu-rich nano precipitate is considered to be the transitional state from two twin-related segments to multiple-twin segments. Combined with the fact of which Cu-rich nano precipitate having bcc and orthogonal 9R structure coexist as shown in Fig. 3, it is inferred that the 9R orthogonal structure may appear at first and then transform into a 9R monocilinc structure. The analysis results of the precipitates in Fig. 3 and Fig. 4 demonstrate that (009) crystal planes in 9R structure would have a orientation adjustment during the evolution process of the initial 9R twins to the later multiple twin structure. This makes the angle between (009) and ($\bar{1}\bar{1}4$) decrease from about 63.7° to 60°. Monzen et al.[27] also reported that (009) crystal plane rotated to align more closely with $\{110\}_{bcc}$ in the matrix during electron irradiation or annealing in Fe – Cu alloys, the angle decreased from 64.5° to 61°. Concerning the Fig. 2(a) and Fig. 2(b) in the reference paper[27], we measured the angle between (009) crystal plane and [001] crystal diretion and found it is 90° for 9R orthogonal but 86.4° for 9R monoclinic. It indicates that 9R monocilinic structure is a more stable one. These results suggest that the crystal structure evolution of Cu-rich nano precipitates is complex.

4 Conclusion

The specimens of RPV model steel were tempered at 660 ℃ for 10 h followed air cooling after heat treatment at 890 ℃ for 0.5 h and water quenching, then they were isothermally aged at 400 ℃ for 1 000 h. The crystal structure evolution of Cu-rich nano precipitates was studied by HRTEM. It is observed that bcc and 9R orthogonal structure or 9R orthogonal and 9R monoclinic structure coexist in one single Cu-rich nano precipitate. It is revealed that bcc structure may transform to 9R orthogonal structure at first, and then to a monoclinic structure. The crystal structure evolution of Cu-rich nano precipitates is complex.

Acknowledgements

This work was financially supported by the National Basic Research Program of China (No. 2011CB610503), the National Natural Science Foundation of China (No. 50931003) and Ministry of Major Subject of Shanghai (No. S30107).

References

[1] P. D. Styman, J. M. Hyde, K. Wilford and A. Morley, Prog. Nucl. Energy **57**(2012) 86.

[2] T. Toyama, Y. Nagai, Z. Tang and M. Hasegawa, Acta Mater. **55**(2007) 6852.

[3] M. K. Miller, R. K. Nanstad, M. A. Sokolov and K. F. Russell, J. Nucl. Mater. **351**(2006) 216.

[4] R. P. Kolli, R. M. Wojes, S. Zaucha and D. N. Seidman, Int. J. Mater. Res. **99**(2008) 513.

[5] R. P. Kolli and D. N. Seidman, Acta Mater. **56**(2008) 2073.

[6] C. Zhang, M. Enomoto, T. Yamashita and N. Sano, Metall. Mater. Trans. A **35**(2004) 1263.

[7] E. Hornbogen and R. C. Glen, Trans. Metall. Soc. AIME. **218**(1960) 1061.

[8] D. Molnar, R. Mukherjee, A. Choudhury and A. Mora, Acta Mater. **60**(2012) 6961.

［9］P. Grammatikopoulos, D. J. Bacon and Y. N. Osetsky, Model. Simul. Mater. Sci. Eng. **19**(2011) 1.

［10］J. B. Yang, T. Yamashita, N. Sano and M. Enomoto, Mater. Sci. Eng. A. **487**(2008) 128.

［11］A. Deschamps, M. Militzer and W. J. Poole, ISIJ Int. **41**(2001) 196.

［12］Y. Kamada, S. Takahashi, H. Kikuchi and S. Kobayashi, J. Mater. Sci. **44**(2009) 949.

［13］A. Ghosh, B. Mishra, S. Das and S. Chatterjee, Metall. Mater. Trans. A **36**(2005) 703.

［14］A. Ghosh and S. Chatterjee, Mater. Charct. **55**(2005) 298.

［15］H. Nakamichi, K. Yamada and K. Sato, J. Microsc. **242**(2011) 55.

［16］V. N. Urtseva, D. A. Mirzaevb, I. L. Yakovlevac and N. A. Tereshchenkoc, Phys. Met. Metall. **110**(2010) 346.

［17］J. J. Blackstock and G. J. Ackland, Philos. Mag. A **81**(2001) 2127.

［18］S. L. Perez, M. L. Jenkins and J. M. Titchmarsh, Philos. Mag. Lett. **86**(2006) 367.

［19］W. Wang, B. X. Zhou, G. Xu and D. F. Chu, Mater. Charact. **62**(2011) 438.

［20］Y. U. Heo, Y. K. Kimb, J. S. Kim and J. K. Kim, Acta Mater. **61**(2013) 519.

［21］G. Xu, D. F. Chu, L. L. Cai, B. X. Zhou and W. Wang, Acta Metall. Sin. **47**(2011) 905. (in Chinese)

［22］P. J. Othen, M. L. Jenkins, G. D. W. Smith and W. J. Phythian, Philos. Mag. Lett. **64**(1991) 383.

［23］T. H. Lee, Y. O. Kim and S. J. Kim, Philos. Mag. A **87**(2007) 209.

［24］Y. L. Bouar, Acta Mater. **49**(2001) 2661.

［25］H. R. Habibi-bajguir and M. L. Jenkins, Philos. Mag. Lett. **73**(1996) 155.

［26］P. J. Othen, M. L. Jenkins and G. D. W. Smith, Philos. Mag. A **70**(1994) 1.

［27］R. Monzen, M. L. Jenkins and A. P. Sutton, Philos. Mag. A **80**(2000) 711.

The Growth Mechanism of Grain Boundary Carbide in Alloy 690 *

Abstract: The growth mechanism of grain boundary $M_{23}C_6$ carbides in nickel base Alloy 690 after aging at 715 ℃ was investigated by high resolution transmission electron microscopy. The grain boundary carbides have coherent orientation relationship with only one side of the matrix. The incoherent phase interface between $M_{23}C6$ and matrix was curved, and did not lie on any specific crystal plane. The $M_{23}C_6$ carbide transforms from the matrix phase directly at the incoherent interface. The flat coherent phase interface generally lies on low index crystal planes, such as $\{011\}$ and $\{111\}$ planes. The $M_{23}C_6$ carbide transforms from a transition phase found at curved coherent phase interface. The transition phase has a complex hexagonal crystal structure, and has coherent orientation relationship with matrix and $M_{23}C_6$: $\{111\}_{matrix}//\{0001\}_{transition}//\{111\}_{carbide}$, $\langle11\bar{2}\rangle_{matrix}//\langle2\bar{1}10\rangle_{transition}//\langle11\bar{2}\rangle_{carbide}$. The crystal lattice constants of transition phase are $c_{transition} = \sqrt{3}\times a_{matrix}$ and $a_{transition} = \sqrt{6}/2 \times a_{matrix}$. Based on the experimental results, the growth mechanism of $M_{23}C_6$ and the formation mechanism of transition phase are discussed.

1 Introduction

The carbides of $M_{23}C_6$ generally precipitate at grain boundaries in Alloy 690 and austenitic stainless steels during thermal aging. Because precipitation of Cr rich $M_{23}C_6$ affects the corrosion resistance of the alloys, the precipitation of $M_{23}C_6$ carbide has been extensively studied in Alloy 690 and austenitic stainless steels [1-8].

The grain boundary $M_{23}C_6$ carbides generally have the orientation relationship of $(100)_m//(100)_c$ and $[001]_m//[001]_c$ (m: matrix, c: carbide) with the matrix only on one side near the grain boundaries, but not with the matrix on the other side [1-4,6-8]. The atoms near the interface of carbide and matrix will redistribute during the growth of $M_{23}C_6$, so that the Cr atoms migrate into the carbide, while the Fe and Ni atoms migrate into the matrix [9,10]. However, the formation of $M_{23}C_6$ carbides may not be a simple phase transformation from the matrix. Beckitt et al. [11] suggested that $M_{23}C_6$ carbides were formed by the migrating of dislocations. Singhal et al. [12] observed that the stacking faults emanated after the precipitation of carbide at the grain boundary. The carbides grew across the stacking faults, and then the precipitation process was repeated. Some literature suggested that the $M_{23}C_6$ was formed by the decomposition of M_6C, or formed by the phase transformation of M_7C_3 [1,13]. However, more detailed research is needed to clarify the

* In collaboration with Hui Li, Shuang Xia, Jianchao Peng. Reprinted from Materials Characterization, 2013, 81: 1-6.

growth mechanism of the grain boundary carbide.

In this work, high resolution transmission electron microscopy (HRTEM) was used to characterize the transformation process of the crystal structure from the matrix to the $M_{23}C_6$ carbides at the interface in detail. The fast Fourier transformation (FFT) and inverse (FFT) (IFFT) were used to analyze the HRTEM images. Here, we report an observation of a transition phase with complex hexagonal crystal structure formed near the $M_{23}C_6$ in Alloy 690. It is expected that the current results can provide additional information for a comprehensive understanding of the growth mechanism of $M_{23}C_6$ in Alloy 690 and austenitic stainless steels.

2 Experimental Procedures

The nominal chemical composition of the investigated Alloy 690 is Cr-28. 91, Fe-9.45, C-0. 025, N-0. 008, Ti-0. 4, Al-0. 34, Si-0. 14, Ni-balance in wt. %. The specimens were vacuum sealed under 5×10^{-3} Pa in quartz capsules and homogeneous solution treated at 1 100 ℃ for 15 min, then quenched with water simultaneous with the breaking of the quartz capsules (WQ). This homogeneous solution treatment can fully dissolve the grain boundary carbides [14]. The aging treatments were carried out at 715 ℃ for 2-100 h for a significant growth of the $M_{23}C_6$ at grain boundaries.

HRTEM foils of the specimens are prepared by twin jet electro-polishing. Twin jet elecro-polishing was carried out with 30 V direct current in a solution of 20% $HCLO_4$ + 80% C_2H_5OH (volume fraction) at −40 ℃. HRTEM observations were conducted with a JEM 2010 F TEM operated at 200 kV to obtain the high resolution lattice images of carbide and matrix. DigitalMicrograph™ was used for processing the high resolution lattice images recorded by a CCD camera. The FFT and IFFT, which are included in DigitalMicrograph™, can reveal the periodic contents in regions of interest of the images. The size of the mask applied for obtaining IFFT images is 3. It is useful for image analysis to obtain the information about the crystal structure.

3 Experimental Results

The Cr rich $M_{23}C_6$ carbides precipitate at grain boundaries easily after proper thermal treating. The interface between the carbides and the matrix can be classified into coherent interface and incoherent interface, since the grain boundary $M_{23}C_6$ carbides generally have coherent orientation relationship with only one nearby grain [1-4,6-8].

Fig. 1 gives the HRTEM images of the $M_{23}C_6$ precipitated at the grain boundary and the nearby matrix. The high resolution lattice images of $M_{23}C_6$ are quite clear, while the lattice images of the matrix are distinct, which indicates the incoherent phase interfaces between $M_{23}C_6$ and the matrix as indicated by dashed lines. It can also be easily identified that the phase interface is incoherent in Fig. 1d, by comparing the high resolution images of

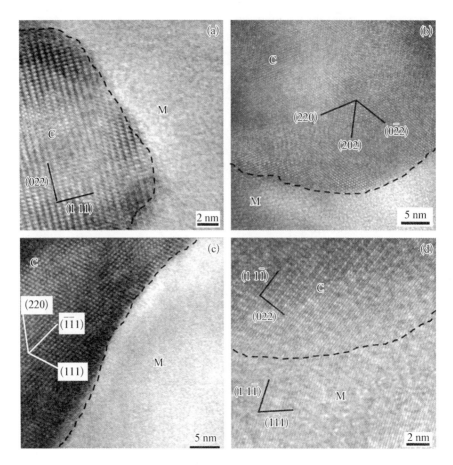

Fig. 1   The HRTEM images of $M_{23}C_6$ precipitated at the grain boundary. The nearby matrix does not have coherent orientation relationship with the particular side of the carbide. The specimens were aged at 715 ℃ for (a) 2 h, (b) 15 h, (c) 30 h, and (d) 2 h. C, carbide, M, matrix

the $M_{23}C_6$ and the matrix. It can be seen that all the incoherent interfaces are curved and do not lie on any specific crystal planes of the matrix.

Fig. 2 gives the HRTEM images of grain boundary $M_{23}C_6$ carbides which have coherent orientation relationship with the nearby matrix. The phase interfaces denoted by the dashed lines in Fig. 2 can be categorized into flat interface and curved interface. The flat interfaces generally lie on the low indexed crystal planes (Fig. 2a, c and e), e. g. {011} and {111}, which are similar to the results of the literature [15,16].

Transition regions can be observed between the carbides and the matrix by different incident beam directions in the specimens aged at 715 ℃ for different time (Fig. 2b, d and f). The high resolution lattice images of transition regions are different from that of the $M_{23}C_6$ and the matrix.

The FFT and IFFT were carried out at regions R1, R2 and R3 in Fig. 2b, c and e, respectively, to enhance the periodicity of HRTEM images and analyze the crystal structure of these regions (Fig. 3). It can be observed that the interplanar spacing is larger

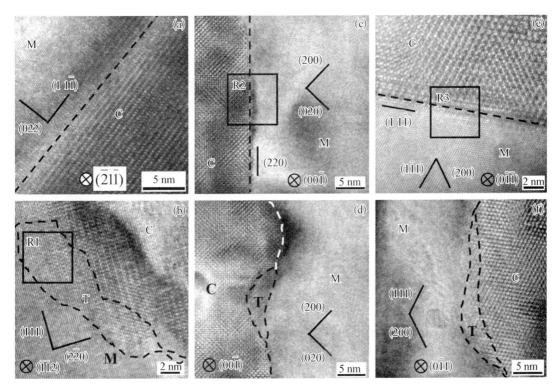

Fig. 2 The HRTEM images of $M_{23}C_6$ precipitated at the grain boundary. One side of the adjacent matrix has coherent orientation relationship with the carbide. The specimens were aged at 715 ℃ for (a), (b) 2 h, (c), (d) 15 h, (e), and (f) 30 h. C, carbide, M, matrix, T, transition regions

at every third layer or {111} planes in the transition region, as indicated by the arrows in Fig. 3a and c. The interplanar spacing is also larger at every third layer of {022} planes in the matrix near the flat interface of matrix and carbide, as indicated by the arrows in Fig. 3b.

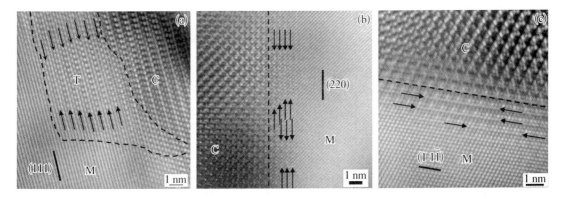

Fig. 3 The IFFT images of $M_{23}C_6$ and nearby matrix as indicated by R1 in Fig. 2b (a), R2 in Fig. 2c (b), R3 in Fig. 2e (c). C, carbide, M, matrix, T, Transition

Fig. 4 gives the FFT and IFFT analysis of an $M_{23}C_6$ carbide with irregular shape and its nearby matrix for a comprehensive understanding of the formation mechanism of the transition regions. The carbide has coherent orientation relationship with the matrix. The largest transition region, about 30 nm in width, can be observed, when the phase interface is curved, as shown in Fig. 4a. FFT and IFFT were carried out in regions b, c, d, and e and shown in Fig. 4 (b, b-1, c, c-1, d, d-1, e and e-1), respectively.

Fig. 4b and b-1 show the face centered cubic structure of the matrix in region b of Fig. 4a. The indexed FFT image shows that region c in Fig. 4a has partial matrix structure and partial hexagonal crystal structure (Fig. 4c-1). No clear interface can be identified between the matrix and the transition region (Fig. 4c). If the analysis region is moved closer to the $M_{23}C_6$ during the FFT operation, the indexed FFT image shows that region d in Fig. 4a has partial hexagonal crystal structure and partial $M_{23}C_6$ cubic structure (Fig. 4d-1). It is also difficult to define a clear interface between the transition region and the $M_{23}C_6$ in Fig. 4d. After moving the analysis region to region e in Fig. 4a, the region is entirely $M_{23}C_6$ cubic structure (Fig. 4e and e-1).

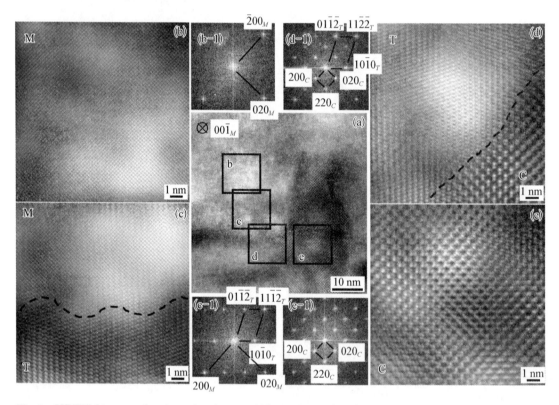

Fig. 4 HRTEM images of an irregular shape carbide precipitated at the grain boundary in the specimen aged at 715 °C for 15 h. (a) HRTEM image of the carbide with irregular shape and the nearby matrix. IFFT (b) and indexed FFT (b-1) images of region b in Fig. 4a, IFFT (c) and indexed FFT (c-1) images of region c in Fig. 4a, IFFT (d) and indexed FFT (d-1) images of region d in Fig. 4a, IFFT (e) and indexed FFT (e-1) images of region e in Fig. 4a. M, matrix, T, transition region, C, carbide

Fig. 4 shows that the transition region between the matrix and the carbide has complex hexagonal crystal structure. The crystal structure of the transition phase does not match with that of any Ni–Cr–Fe phases or carbide phases already known from the literature. The appearance of the transition regions between the matrix and the $M_{23}C_6$ revealed the precipitation process of the $M_{23}C_6$. The transition phase has the following coherent orientation relationship with the matrix and the $M_{23}C_6$:

$\{111\}_{matrix} // \{0001\}_{transition} // \{111\}_{carbide}$,

and $\langle 11\bar{2} \rangle_{matrix} // \langle 2\bar{1}10 \rangle_{transition} // \langle 11\bar{2} \rangle_{carbide}$.

The lattice constants of the complex hexagonal structure for the transition phase

are:
$$c_{transition} = \sqrt{3} \times a_{matrix} = 0.62 \text{ nm}$$
$$a_{transition} = \frac{\sqrt{6}}{2} \times a_{matrix} = 0.44 \text{ nm}$$

4　Discussion

The experimental results show that a transition phase formed at the coherent interface between $M_{23}C_6$ and matrix. Sleeswyk et al. [13]assumed that the $M_{23}C_6$ may transform from an hcp transition phase, which is neither carbide nor metallic compound. But, they did not detect any transition phase during the growth of $M_{23}C_6$ by TEM. The current study provides the direct evidence of the transition phase formed during the growth of $M_{23}C_6$.

The transition phase may be formed by the Cr atoms enriched at one $\{111\}$ atomic layer in every third $\{111\}$ atomic layer. The stacking sequence of $\{111\}$ planes is ABCABCABC for face centered cubic structure, where A, B and C stand for different $\{111\}$ layers. The distance between each AB, BC and CA is $\sqrt{3}/3 \times a_{matrix}$ (left part of Fig. 5a). The Cr atom is a little larger than Ni and Fe atoms. If Cr atoms enrich at one $\{111\}$ atomic layer in every third $\{111\}$ atomic layer, for example, enrich at layer C, the thicknesses of layers A and B will reduce a little due to the depletion of Cr atoms, while the thickness of layer C will increase a little due to the enrichment of Cr atoms. The three layers should be renamed as A', B', and X, respectively. Then the stacking sequence is $A'B'XA'B'XA'B'X$. The distance between each $B'X$ and XA' layers are similar and both are a little larger than that of $A'B'$ (the right part of Fig. 5a). This stacking sequence is the same as that of complex hexagonal crystal structure. The crystal lattice constant c of the transition phase is the distance between two A' layers, $\sqrt{3} \times a_{matrix}$, as shown in Fig. 5a.

The atom arrangements of the (111) plane of matrix and the (0001) plane of transition phase are shown in Fig. 5b. In Fig. 5b, the arrows are used to denote the enrichment of Cr atoms at one $\{111\}$ atomic layer in every third $\{111\}$ atomic layer. The gray dots are used to denote the equivalent atoms after the Cr enriching at one $\{111\}$ atomic layers in every third $\{111\}$ atomic layers. Hence, it can be concluded that if Cr atoms enrich at one $\{111\}$ atomic layer in every third $\{111\}$ atomic layer, the atom arrangement of (111) plane of the matrix would change to (0001) plane of transition phase (Fig. 5b). And it can be concluded that the other crystal lattice constant of the transition phase is $a_{transition} = \sqrt{6}/2 \times a_{matrix}$.

Furthermore, the crystal lattice constant of $M_{23}C_6$ is three times that of the matrix, and $\{111\}$ plane is the best match plane between $M_{23}C_6$ and matrix [1]. So, Cr atoms enriching at one $\{111\}$ atomic layer in every third $\{111\}$ atomic layer are very beneficial for

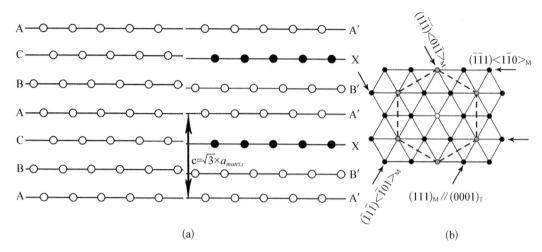

Fig. 5 The schematic illustration of stacking sequence of (111) planes of the matrix and (0001) planes of the transition phase (a), and the atom arrangement of (111) plane of matrix and (0001) plane of the transition phase (b). The arrows point out the Cr enriching at which layers. The gray dots show the equivalent atoms in (0001) plane of the transition phase

the growth of $M_{23}C_6$. Based on the above discussion, we believe that the transition phase is a result of Cr atoms enriching at one {111} atomic layer in every third {111} atomic layer.

The transition phase is more evident at curved coherent phase interfaces than at flat coherent phase interfaces. One $M_{23}C_6$ unit cell contains 92 metal atoms and 24 C atoms, and occupies 27 units of the matrix which contains 108 metal atoms in the same volume [1]. When the carbide grows into the matrix after nucleation at grain boundary, the volume expansion will induce a compressive stress in the matrix [17]. This stress may act as driving force for Cr atoms substituting for Ni and Fe atoms and segregating at one {111} atomic layer in every third {111} atomic layer. The matrix near curved interface has more complex stress state than that near the flat interface, so the transition phase is more easily formed near the curved interface.

The transition phase cannot be observed at the incoherent interface between $M_{23}C_6$ and matrix (Fig. 1). The transition phase formed at coherent phase interface should have coherent orientation relationship with both matrix and $M_{23}C_6$, such as $\{111\}_{matrix}//$ $\{0001\}_{transition}//\{111\}_{carbide}$, and $\langle11\bar{2}\rangle_{matrix}//\langle2\bar{1}10\rangle_{transition}//\langle11\bar{2}\rangle_{carbide}$. But for the incoherent interface, the transition phase can only have coherent orientation relationship with one of the matrices or $M_{23}C_6$, which is not promoting the growth of $M_{23}C_6$. Furthermore, Cr atoms can easily substitute Ni and Fe atoms at the incoherent phase interface and promote the growth of $M_{23}C_6$, due to the assistant flux of vacancies and high interface energy at incoherent phase interface [18]. So the transition phase did not form at incoherent phase interface.

In general, the migrating rate of incoherent interface is faster than that of coherent interface, and the migrating rate of a curved interface is faster than that of a flat interface during the growth of $M_{23}C_6$. This is consistent with the current results. The incoherent phase interface has the fastest migrating rate as stated above. Furthermore, the growth of carbide consumes more Cr atoms near the incoherent phase interface than near the coherent phase interface. So, the Cr depletes more seriously near the grain boundary in the matrix which is incoherent with the carbide [17,19]. Less active energy is needed for the carbide to transform from the transition phase than to transform from the matrix directly. So the migrating rate of a curved interface is much faster than that of a flat interface.

5   Conclusions

Based on the HRTEM observations and discussion of the growth mechanism of grain boundary $M_{23}C_6$ carbide after aging treatment, the following conclusions can be derived.

(1) The incoherent phase interface between $M_{23}C_6$ and matrix is curved, and does not lie on any specific crystal plane. The $M_{23}C_6$ transforms directly from the matrix phase at the incoherent interface.

(2) The coherent phase interface between $M_{23}C_6$ and matrix can be classified into flat interface and curved interface. The flat coherent interface generally lies on low index crystal planes, such as (011) and (111) planes. The $M_{23}C_6$ transforms from a transition phase at coherent phase interface, and not directly from the matrix phase. The transition phase is more evident at curved coherent interface than at flat coherent interface.

(3) The transition phase has a complex hexagonal crystal structure, and has coherent orientation relationship with matrix and $M_{23}C_6$: $\{111\}_{matrix}//\{0001\}_{transition}//\{111\}_{carbide}$, and $\langle11\bar{2}\rangle_{matrix}//\langle2\bar{1}10\rangle_{transition}//\langle11\bar{2}\rangle_{carbide}$. The crystal lattice constants of the transition phase are $c_{transition}=\sqrt{3}\times a_{matrix}$ and $a_{transition}=\sqrt{6}/2\times a_{matrix}$.

Acknowledgments

This work was financially supported by the National Basic Research Program of China (No. 2011CB610502), the National Natural Science Foundation of China (No. 50974148), and the Innovation Fund of Shanghai University.

References

[1] Lewis MH, Hattersley B. Precipitation of $M_{23}C_6$ in austenitic steels. Acta Metall 1965; 13: 1159–68.

[2] Kai JJ, Yu GP, Tsai CH, Liu MN, Yao S C. The effects of heat treatment on the chromium depletion, precipitate evolution, and corrosion resistance of INCONEL Alloy 690. Metall Trans A 1989; 20A: 2057–67.

[3] Angeliu TM, Was GS. Behavior of grain boundary chemistry and precipitates upon thermal treatment of controlled purity Alloy 690. Metall Trans A 1990; 21A: 2097–107.

[4] Trillo EA, Murr LE. A TEM investigation of $M_{23}C_6$ carbide precipitation behaviour on varying grain boundary misorientations in 304 stainless steels. J Mater Sci 1998; 33: 1263–71.

[5] Trillo EA, Murr LE. Effects of carbon content, deformation, and interfacial energetics on carbide precipitation and corrosion sensitization in 304 stainless steel. Acta Mater 1998; 47: 235–45.

[6] Lim YS, Kim JS, Kim HP, Cho HD. The effect of grain boundary misorientation on the intergranular $M_{23}C_6$ carbide precipitation in thermally treated Alloy 690. J Nucl Mater 2004; 335: 108–14.

[7] Li H, Xia S, Zhou BX, Chen WJ, Ni JS. Evolution of carbide morphology precipitated at grain boundaries in Ni-based Alloy 690. Acta Metall Sin (Chin Lett) 2009; 45; 195-8.

[8] Li H, Xia S, Zhou BX, Chen WJ, Hu CL. The dependence of carbide morphology on grain boundary character in the highly twinned Alloy 690. J Nucl Mater 2010; 399; 108-13.

[9] Tytko D, Choi PP, Klower J, Kostka A, Inden G, Raabe D. Microstructural evolution of a Ni-based superalloy (617B) at 700 ℃ studied by electron microscopy and atom probe tomography. Acta Mater 2012; 60; 1731-40.

[10] Was GS, Tischner HH, Latanision RM. The influence of thermal treatment on the chemistry and structure of grain boundaries in Inconel 600. Metall Trans A 1981; 12A; 1397-408.

[11] Beckitt FR, Clark BR. The shape and mechanism of formation of $M_{23}C_6$ carbide in austenite. Acta Metall 1967; 15; 113-29.

[12] Singhal LK, Martin JW. The growth of $M_{23}C_6$ carbide on incoherent twin boundaries in austenite. Acta Metall 1967; 15; 1603-10.

[13] Sleeswyk AW, Helle JN, von Rosenstel AP. Nucleation of the thermal F. C. C. → H. C. P. transformation. Philos Mag 1964; 9; 891-6.

[14] Jiao SY, Zhang MC, Zheng L, Dong JX. Investigation of carbide precipitation process and chromium depletion during thermal treatment of Alloy 690. Metall Mater Trans A 2010; 41A; 26-42.

[15] Hong HU, Rho BS, Nam SW. Correlation of the $M_{23}C_6$ precipitation morphology with grain boundary characteristics in austenitic stainless steel. Mater Sci Eng A 2001; A318; 285-92.

[16] Hong HU, Nam SW. The occurrence of grain boundary serration and its effect on the $M_{23}C_6$ carbide characteristics in an AISI 316 stainless steel. Mater Sci Eng A 2002; A332; 255-61.

[17] Li H, Xia S, Zhou BX, Peng JC. Study of carbide precipitation at grain boundary in nickel base Alloy 690. Acta Metall Sin (Chin Lett) 2011; 47; 853-8.

[18] Wolff UE. Orientation and morphology of $M_{23}C_6$ precipitated in high-nickel austenite. Trans Metall Soc AIME 1968; 242; 814-23.

[19] Kaneko K, Fukunaga T, Yamada K, Nakada N, Kikuchi M, Saghi Z, et al. Formation of $M_{23}C_6$ -type precipitates and chromium-depleted zones in austenite stainless steel. Scr Mater 2011; 65; 509-12.

Atomic Scale Study of Grain Boundary Segregation before Carbide Nucleation in Ni – Cr – Fe Alloys[*]

Abstract: Three dimensional chemical information concerning grain boundary segregation before carbide nucleation was characterized by atom probe tomography in two Ni – Cr – Fe alloys which were aged at 500 ℃ for 0. 5 h after homogenizing treatment. B, C and Si atoms segregation at grain boundary in Alloy 690 was observed. B, C, N and P atoms segregation at grain boundary in 304 austenitic stainless steel was observed. C atoms co-segregation with Cr atoms at the grain boundaries both in Alloy 690 and 304 austenitic stainless steel was found, and its effect on the carbide nucleation was discussed. The amount of each segregated element at grain boundaries in the two Ni – Cr – Fe alloys were analyzed quantitatively. Comparison of the grain boundary segregation features of the two Ni – Cr – Fe alloys were carried out based on the experimental results.

1 Introduction

Ni – Cr – Fe alloys are known to preserve their excellent mechanical and corrosion properties even at elevated temperature. These alloys, e. g. 304 austenitic stainless steel (304 SS), Inconel Alloy 600, Incoloy Alloy 800 and Inconel Alloy 690, are therefore used as steam generator tubes in pressurized water reactor nuclear power plant. However, despite the expected good performance of the alloys, numerous environmentally induced failures, especially intergranular attack (IGA), of the alloys have been reported [1,2]. One of the most important microstructure factors that induced IGA is grain boundary segregation and grain boundary precipitation. This problem leads to many investigations on the grain boundary chemistry.

The precipitaion of Cr rich $M_{23}C_6$ carbide in Alloy 690 and austenitic stainless steels has been extensively studied for many years [3 – 13]. The precipitation features of grain boundary carbides including crystal structure and morphology in Alloy 690 and austenitic stainless steels are well established. Transmission electron microscopy (TEM) studies of grain boundary precipitation have revealed that Cr rich $M_{23}C_6$ is the most common types of the precipitates in these alloys [3,4,6,9]. During the process of precipitation, the grain boundary carbides still have coherent orientation relationship with one of the nearby grains. The morphology evolution of the grain boundary carbides precipitated at various temperatures, time and grain boundary characters had been studied by TEM and scanning

* In collaboration with Hui Li, Shuang Xia, Wenqing Liu, Tingguang Liu. Reprinted from Journal of Nuclear Materials, 2013, 439: 57 – 64.

electron microscopy (SEM) [3 - 13].

During carbide precipitation, the solute and impurity atoms segregate at grain boundaries. Thuvander et al. [14] reported the amount of C atoms segregation at grain boundary was correlated with the grain boundary Cr concentration during the carbide precipitation process in Alloy 600. Solute and impurity atoms are expected to segregate at the grain boundaries before the nucleation of carbides. The study of distribution of solute and impurity atoms at the grain boundaries before carbide nucleation is valuable for understanding the nucleation and precipitation behavior of carbides. However, this topic has not been investigated adequately.

In this paper, the atom probe tomography (APT) was adopted to directly measure the three dimensional distribution of atoms at grain boundaries before carbides precipitation in Alloy 690 and 304 SS. It aims to clarify the grain boundary segregation features before carbide precipitation and the difference features of grain boundary segregation in Ni based and Fe based Ni – Cr – Fe alloys.

2   Experimental procedures

The chemical compositions of as-received Alloy 690 and 304 SS used in this experiment are given in Table 1. The sample sheets were homogenized at 1 100 ℃ for 15 min in vacuum. After solution treating, the sheets were subsequently quenched into water (WQ). Then a 50% cold reduction and recrystallization annealing at 1 000 ℃ for 1 min was carried out to reduce the grain size. The detailed annealing processes are described in elsewhere [15]. After annealing treatment, the specimens were aged at 500 ℃ for 0. 5 h to create grain boundary segregation and early stage of carbide nucleation.

Table 1   The chemical compositions of as-received Alloy 690 and 304 SS in this experiment (at. %)

	Ni	Cr	Fe	Ti	Al	Mn	Si	C	N	B	P
Alloy 690	57. 89	31. 09	9. 45	0. 47	0. 70	—	0. 28	0. 09	0. 03	—	—
304 SS	9. 70	20. 45	66. 89	—	—	1. 13	1. 59	0. 19	—	0. 02	0. 03

The sheets were cut into small rods with a scale of 0. 5 mm×0. 5 mm in 15 mm length using a spark machine. The needle specimens for APT analysis were prepared by standard two steps electro-polishing [16]. JEM – 200 CX TEM was used to select the sharp enough needle specimens with grain boundary near the apex. The needle specimens were analyzed using a CAMECA 3 000 HR local electrode atom probe (LEAP). The specimens were cooled to a temperature of 50 K. The pulse amplitude was kept at 15% of the standing voltage applied to the specimen. The voltage pulsing repetition rate was 200 kHz. The standing voltage on the apex of the specimen was varied automatically in order to maintain an evaporation rate of 2 ions in every 1 000 pulses. Though the relatively low evaporation rate produces a low fracture tendency, these materials have very high fracture tendency when the apex of the specimen contains grain boundary. High background vacuum levels were kept, less than 9.24×10^{-9} Pa, due to the low evaporation rates of this study. Hence, the background noise levels seen in the mass spectra were low, $\sim 10^{-4}$ of the height of the largest signal in the mass spectra (see Fig. S1). There is overlap in the spectrum between $_{28}Si^{2+}$ and $_{14}N^{+}$ at 14 amu. If there is $_{14}N^{2+}$ peak the mass spectrum, the concentration of Si can be estimated by the isotopic abundance of the $_{29}Si^{2+}$ and $_{30}Si^{2+}$ peaks, the concentration of N can be estimated by the $_{14}N^{2+}$ and partial $_{14}N^{+}$ peaks. Else, there is not $_{14}N^{2+}$ peak the mass spectrum, it is estimated that the N can not be observed in this specimen.

3 Results and discussion

3.1 Grain boundary segregation features in Alloy 690

Grain boundary segregation of many elements in Alloy 690 aged at 500 ℃ for 0.5 h was characterized using APT. After this heat treatment, carbides only initially nucleated at the dislocation junctions on the grain boundary, most of grain boundaries do not contain carbides [15]. Fig. 1 gives the three dimensional atom distribution maps at grain boundary in Alloy 690. The atom maps have been oriented so that the views are along (Fig. 1a－h) and perpendicular (Fig. 1i) to the grain boundary planes, in order to show the atoms segregating at two dimensional grain boundary plane. The carbide still does not nucleate at

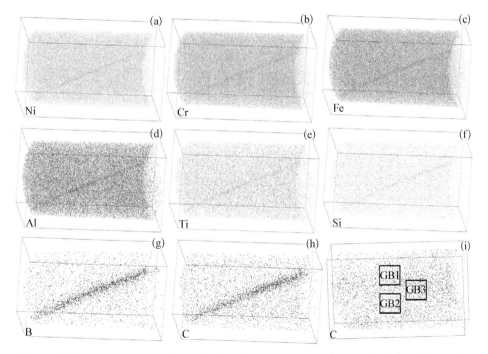

Fig. 1 APT atom maps showing grain boundary segregation features in Alloy 690 aged at 500 ℃ for 0.5 h. The views of (a)－(h) are parallel to the grain boundary, while the view is perpendicular to the grain boundary in (i). The dimensions of the boxes are 54 nm×56 nm× 98 nm. GB: grain boundary. GB1, GB2 and GB3 are different regions in this grain boundary

grain boundary because of the C atoms distributing at grain boundary plane randomly (Fig. 1 h and i). Enrichment of B, C and Si atoms are evident from the atom maps (Fig. 1 e−g). Trajectory aberrations occur at the interfaces in the atom probe [17], leading to an apparent increase in the total atom density in that region (Fig. 1). The average composition in Alloy 690 characterized by APT is shown in Table 2. The measured composition is a little different from the mean composition.

Table 2　The average compositions in Alloy 690 and 304 SS analyzed by APT (at. %)

	Ni	Cr	Fe	Mn	Ti	Al	Si	B	C	P	N
Alloy 690	60.20± 0.04	28.72± 0.02	9.59± 0.01	—	0.400± 0.002	0.771± 0.003	0.251± 0.002	0.026 9± 0.000 6	0.042 5± 0.000 8	—	—
304 SS	9.08± 0.01	20.81± 0.01	67.25± 0.02	1.074± 0.002	—	—	1.631± 0.003	0.022 6± 0.000 3	0.109 7± 0.000 7	0.021 6± 0.000 3	0.008 6± 0.000 1

Because the area of grain boundary obtained in the current study is comparatively large ($>3 000$ nm^2), three randomly selected boxes (20 nm×20 nm×25 nm in size, GBs 1, 2 and 3 in Fig. 1i) across the grain boundary were analyzed. They are designated as GB 1♯, GB 2♯, and GB 3♯, respectively. The distance between each box is about 30 nm. This method can improve the statistical significance and can be used to compare the segregation variations in the grain boundary.

Segregation variations at different regions in the same grain boundary in Alloy 690 can be seen from the depth compositional profiles across the grain boundary in Fig. 2. It shows that B, C and Si atoms segregated in a layer about 6 nm in width at grain boundary. The segregation layers seem to be much wider than the real results, ~ several atoms layers (less than 1 nm) [18]. It is mainly introduced by the following respects: firstly, the grain boundary planes are not perfect planar though in only 20 nm×20 nm scale, so the grain boundaries seems to be wider than the real thickness, however, this is not the main reason. Secondly, the trajectory aberrations occur during the APT analyzing the interface, which leads to that the segregants spread at wider region [17].

The concentration of Ni changes from 56.7±0.3 at. % (depletion) at GB 1♯ to 64.3±0.3 at. % (enrichment) at GB 2♯, and neither depletion nor enrichment at GB 3♯. Similar variations were also observed in the case of Cr and Fe (Fig. 2a, d and g), for instance, the enrichment of Cr is observed at GBs 1♯ and 2♯, but not at GB 3♯. Ti and Al homogenously distribute at matrix and grain boundary (Fig. 2b, e and h). All three regions show enrichment of B, C and Si, but the concentrations are highest at GB 1♯ compared to GBs 2♯ and 3♯, though in the same grain boundary (Fig. 2c, f and i). If the C concentration is higher, the Cr atoms will be attracted by the C atoms and segregate at grain boundary (GBs 1♯ and 2♯). The N atoms are not observed in the atom maps, because N atoms may be exhausted by TiN precipitated in the matrix [19].

3.2　Grain boundary segregation features in 304 SS

Grain boundary segregation of many elements in 304 SS aged at 500 ℃ for 0.5 h is also

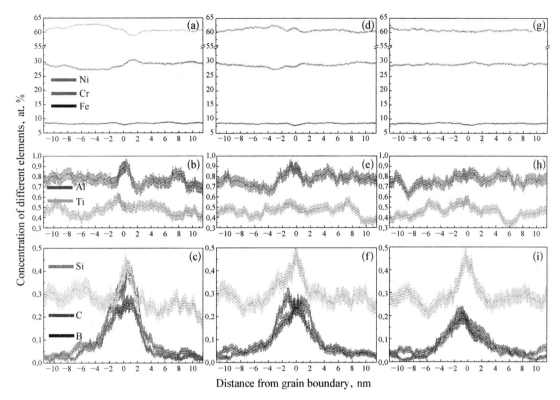

Fig. 2 The depth compositional profiles of Ni, Cr, Fe, Ti, Al, B, C and Si near the grain boundary in Alloy 690 aged at 500 ℃ for 0.5 h. (a‐c) GB1 in Fig. 1, (d‐f) GB2 in Fig. 1, (g‐i) GB3 in Fig. 1. The cross‐section of the sampling box is 20 nm×20 nm, the depth of the sampling box is 2 nm, the moving step of the sampling box is 0.1 nm

evident from APT analysis (Fig. 3). The atom maps have been oriented so that the views are parallel to (Fig. 3a‐i) or perpendicular to (Fig. 3j) the grain boundary planes, in order to show the atoms segregating at two dimensional grain boundary plane. The carbide still does not nucleate at grain boundary because of the C atoms distributing at grain boundary plane randomly. Depletion of Fe (Fig. 3c) and enrichment of B, C, N and P is evident from the individual atom maps (Fig. 3f‐i). The average composition in 304 SS characterized by APT is shown in Table 2.

The area of grain boundary obtained in 304 SS specimen in the current study is comparatively large (>2 500 nm²), three randomly selected boxes (20 nm×20 nm×25 nm in size, GBs 4, 5 and 6 in Fig. 3) across the grain boundary are analyzed. They are designated as GB 4♯, GB 5♯, GB 6♯, respectively. The distance between each box is about 30 nm.

Segregation variations at different regions in the same grain boundary in 304 SS can be seen from the 1D depth profiles across the grain boundary in Fig. 4. It shows that B, C, P and N atoms segregated in a layer about 5 nm in width at grain boundary. The segregation layers seem to be much wider than the real results, ～several atoms layers (less than 1 nm) [18], due to the trajectory aberrations [17].

Similar to the case of Alloy 690, segregation variations at different regions in the same grain boundary are evident in 304 SS (Fig. 4). But some grain boundary segregation

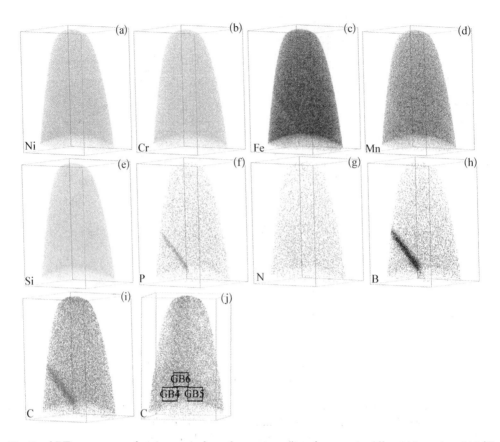

Fig. 3 APT atom maps showing grain boundary segregation features in Alloy 304 aged at 500 ℃ for 0.5 h. The views of (a)-(i) are parallel to the grain boundary, and the view is perpendicular to the grain boundary in (j). The dimensions of the boxes are 103 nm×105 nm×160 nm. GB: grain boundary. GB4, GB5 and GB6 are different regions in this grain boundary

features in 304 SS are different from that in Alloy 690. The Ni and Fe atoms deplete at grain boundary in all the three regions, due to the enrichment of other segregated atoms (Fig. 4a, d and g). Si atoms do not segregate at grain boundary in 304 SS, and Mn atoms deplete at grain boundary in 304 SS a little (Fig. 4b, e and h). Further more, if the C concentration is higher, the Cr atoms will be attracted by the C atoms and segregate at grain boundary (Fig. 4).

The peaks of concentration profiles for different elements are a little different (Figs. 2 and 4). Such behavior may originate from C and Cr attraction, or other effect as illustrated by Ref. [20]. There may be some other effects induce the variation of concentration profiles, such as artefacts associated to the basic phenomenon of atom probe analysis, e.g. the different evaporation field of different elements, also may originate from chromatic effects, local magnification effects, preferential retention or evaporation for certain species.

Fig. 5 schematically illustrates the carbide nucleation and growth process. At the beginning of aging, trace elements such as C and B atoms segregate at grain boundary, as shown in the dashed rectangular region in Fig. 5a. And then, Cr atoms migrate to and enrich near the grain boundary, due to the attraction of C atoms, as shown in the dashed

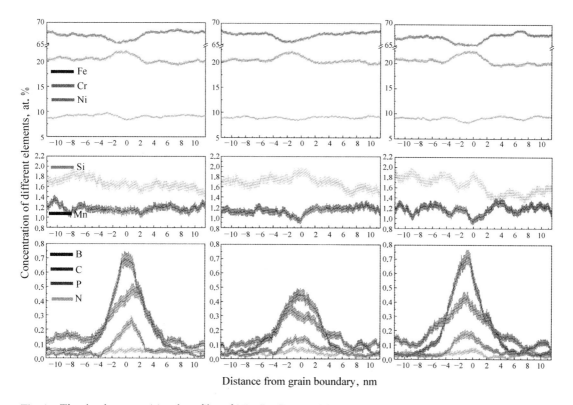

Fig. 4 The depth compositional profiles of Ni, Cr, Fe, Si, Mn, B, C and Si near the grain boundary in 304 SS aged at 500 ℃ for 0.5 h. (a-c) GB4 in Fig. 3, (d-f) GB5 in Fig. 3, (g-i) GB6 in Fig. 3. The cross-section of the sampling box is 20 nm×20 nm, the depth of the sampling box is 2 nm, the moving step of the sampling box is 0.1 nm

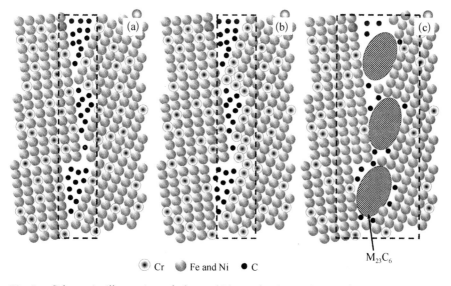

Fig. 5 Schematic illustration of the carbide nucleation and growth process, (a) C atoms randomly segregate at grain boundary, (b) C and Cr atoms co-segregate on the right side of grain boundary and (d) the carbide precipitated on the grain boundary and Cr depletion zone formed near the carbide

rectangular region in Fig. 5b. It leads to the C and Cr co-segregate at one side of grain boundary which lies on high indexed plane, while not the grain boundary core region[15]. This phenomenon can not be observed by other analyzing method, such as HRTEM or X-ray energy dispersive spectrum (EDS). If the grain boundary contains enough C and Cr, the carbide will nucleate and growth at the grain boundary, and have coherent orientation relationship with one side of matrix. At the same time, the grain boundary lies on the high index crystal plane of this side matrix [21]. During the growth of carbide, Cr depletion zone forms near the carbide, the atoms of trace elements continue to segregate at grain boundary and phase boundary [22], as shown in the dashed rectangular region in Fig. 5c. The evolution of Cr depletion zone after carbide precipitation at grain boundary was well established by EDS [4,5,9,14]. However, the current study shows that Cr enriched at grain boundary before carbide nucleation by APT. Hence, the Cr concentration evolution at grain boundary is fully established by the combination of the results from APT and EDS, and it should be enrichment, depletion, healing at grain boundary.

3.3   The quantitative analysis of the grain boundary segregation in Ni – Cr – Fe alloys

3.3.1   Analysis method

Despite of the depth concentration profiles, the relative proportions of the various elements (C_i) present at different regions across the grain boundary can be determined by the cumulative compositional profiles [23]. Several profiles should be taken at different regions on the grain boundary and their results are averaged in order to improve the significance of the measurement.

For a grain boundary, the number of excess atoms detected at the grain boundary, N_i^{excess}, can be found by the cumulative compositional profiles [23,24]. Gibbs interface excess (Γ) is the excess amount of solute or impurity in terms of atoms per square meter. The value of Γ can be made using:

$$\Gamma_i = \frac{N_i^{excess}}{e_d \times A} \tag{1}$$

e_d is the atom probe detector efficiency, $e_d \sim 0.38$ for LEAP. A is the surface area of the region of interest over which the cumulative composition profile was taken. In the current study, $A = 400 \text{ nm}^2$ as illustrated above. The calculated value for the Γ can be related to knowing the properties of the material [25].

The concentration of element i at grain boundary (C_i^{gb}) will be higher than that at matrix (C_i), if the atoms of element i segregate at grain boundary. But direct comparison of the C_i^{gb} value of different elements can not give useful information, because of different bulk concentration of the different elements. So, the enrichment factor in the grain boundary of element i, S_{av}, should be defined and derived from:

$$S_{av} = \frac{C_i^{gb}}{C_i} \tag{2}$$

The comparison of S_{av} for different elements can give the grain boundary segregation tendency of different elements.

The free energy of segregation, $\triangle G_i$, can be expressed by the Guttmann theory for a multicomponent system [26]:

$$\frac{C_i^{GB}}{C_0^{GB} - \sum C_i^{GB}} = \frac{C_i^{matrix}}{1 - \sum C_i^{matrix}} e^{-\frac{\triangle G_i}{RT}} \tag{3}$$

In Eq. (3), the interaction of different segregation atoms is not considered. C_0^{GB} is the fraction of the grain boundary region available for segregated atoms at saturation, and generally set as 1. C_i^{GB} is the actual concentration of element i at grain boundary, C_i^{matrix} is the concentration of element i at matrix. $\triangle G_i$ is the free energy of segregation per mole of segregated element i. $\triangle G_i > 0$ means the increase of free energy and depletion tendency of this element, $\triangle G_i < 0$ means the decrease of free energy and high segregation tendency of this element. T is the experimental temperature, and $T = 773$ K.

3.3.2 Experimental and analysis results

Fig. 6 gives the cumulative compositional profiles of C, B and Si detected across the three different regions of the grain boundary in Alloy 690, and cumulative compositional profiles of C, B, P and N detected across the three different regions of the grain boundary in 304 SS.

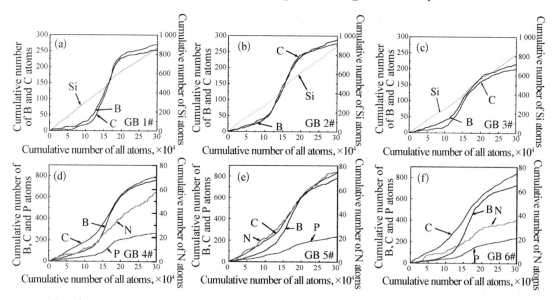

Fig. 6 Cumulative compositional profiles of C, B and Si detected across the GB 1# (a), GB 2# (b), GB 3# (c) in Alloy 690, and C, B, P and N detected across the GB 4# (d), GB 5# (e), GB 6# (f) in 304 SS

The values of Γ, S_{av}, and $\triangle G_i$ of different elements segregated at grain boundaries in Alloy 690 and 304 SS were calculated using the method showed in the above section. They are showed in Fig. 7. Three profiles were taken at different regions on the grain boundary and their results averaged in order to improve the significance of the measurement for all the parameters.

Benhadad et al. [27] assumed that an alloy with 0.033 wt. % C contains sufficient C

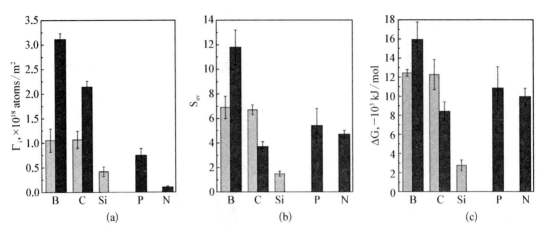

Fig. 7 The values of Γ (a), S_{av} (b) and ΔG_i (c) of different elements segregated at grain boundaries in Alloy 690 and 304 SS. Light gray: Alloy 690, dark gray: 304 SS

to compete with B for grain boundary sites. The levels of bulk B concentration are similar in Alloy 690 and 304 SS (0. 026 9± 0. 000 6 at. % and 0. 022 6± 0. 000 3 at. %), and the level of B segregation in 304 SS ($\Gamma = 3. 11 \pm 0. 12 \times 10^{18}$ atoms/m²) is higher than in Alloy 690 ($\Gamma = 1. 05 \pm 0. 24 \times 10^{18}$ atoms/m²), although increased C segregation is observed at grain boundary in 304 SS than in Alloy 690 (a doubling of $\Gamma = 1. 07 \pm 0. 18 \times 10^{18}$ atoms/m² to $\Gamma = 2. 14 \pm 0. 12 \times 10^{18}$ atoms/m²). It indicates that B has higher tendency to segregate at grain boundary in 304 SS than in Alloy 690.

If the interaction of different segregation atoms is not considered, the total reduction of grain boundary free energy due to segregation is 27. 489 kJ/mol for Alloy 690 and 45. 207 kJ/mol for 304 SS. These results suggest that the grain boundary composition change due to segregation may influences the properties of grain boundary much more in 304 SS than in Alloy 690 [20].

3. 4   The comparison of grain boundary segregation features in Fe – based and Ni – Based Ni – Cr – Fe alloys

Based on the results and discussion above, the grain boundary segregation features in these two kinds of Ni – Cr – Fe alloys, Fe – based 304 SS and Ni – based Alloy 690 can be summarized as follows:

(1) The impurity and solute atoms segregate inhomogeneously in the same grain boundary both in 304 SS and Alloy 690.

(2) The grain boundary segregation tendencies (S_{av}) are B (11. 8±1. 4)>P (5. 4± 1. 4)>N (4. 7±0. 3)>C (3. 7±0. 4) in 304 SS, and B (6. 9±0. 9)> C (6. 7±0. 4)> Si (1. 5±0. 2) in Alloy 690.

(3) Cr atoms may co-segregate with C atoms at grain boundaries before carbide nucleation at the grain boundaries both in 304 SS and Alloy 690.

(4) Ni atoms generally deplete at grain boundary both in 304 SS and Alloy 690. The literature shows that the Ni atoms may co-segregate with P atoms at grain boundaries

[28], but the P atoms segregation do not leads to Ni segregation in the current study.

(5) In the current study, Fe atoms may segregate or deplete at grain boundary in Alloy 690. But Fe atoms generally deplete at grain boundary in 304 SS.

(6) B atoms have the strongest grain boundary segregation tendency both in 304 SS and Alloy 690. The grain boundary segregation tendency and Gibbs free energy of B in 304 SS is higher than in Alloy 690.

(7) C atoms are easy to segregate at grain boundaries both in 304 SS and Alloy 690. The grain boundary segregation tendency and Gibbs free energy of C in Alloy 690 is higher than in 304 SS, due to the higher bulk C concentration and the site competition of P atoms which segregate at grain boundary[29,30]. It is imply that the segregation tendency is influenced by the bulk concentration of the segregates.

(8) Si atoms slightly segregate at grain boundaries in Alloy 690, but do not segregate at grain boundaries in 304 SS.

(9) N and P atoms segregate at grain boundary in 304 SS, and their segregation Gibbs free energy are similar. N atoms may be exhausted by the TiN precipitated in the matrix and can not be observed in the grain boundary of Alloy 690 [19].

(10) Mn atoms deplete at grain boundary in 304 SS. This phenomenon is similar to that of proton irradiation induced segregation in 304 SS [32].

(11) B, C, N, P segregation Gibbs energies are similar both in 304 SS and Alloy 690.

4   Conclusion

This paper uses the APT analysis, which gives access to the nanometer three dimensional chemical distribution information at the grain boundary, aims to clarify some aspects of grain boundary segregation before the carbide nucleation in two kinds of Ni-Cr-Fe alloys, Ni based Alloy 690 and Fe based 304 SS. Based on the results and discussion above, the following conclusion can be derived:

(1) B and C atoms segregate at grain boundary both in Alloy 690 and 304 SS, P and N segregate at grain boundary in 304 SS. Si atoms segregate at grain boundary in Alloy 690, but do not segregate at grain boundary in 304 SS. Cr enriches at grain boundary both in Alloy 690 and 304 SS, although carbide does not nucleate. Ni and Fe may segregate, deplete or homogeneously distribute at grain boundary in Alloy 690, but they deplete at grain boundary in 304 SS.

(2) C and Cr atoms co-segregate at grain boundaries before carbide nucleation in Alloy 690 and 304 SS. Combination with other results in literatures, the evolution of Cr concentration at grain boundary should be enrichment at grain boundary before carbide nucleation, depletion at grain boundary after carbide precipitation, and healing after obvious growth of carbide.

(3) After aging treatment at 500 ℃ for 0.5 h, the total reduction of grain boundary free energy due to segregation is 27.489 kJ/mol for Alloy 690 and 45.207 kJ/mol for 304.

Acknowledgements

This work was supported by the National Basic Research Program of China (No. 2011CB610502), the National Natural Science Foundation of China (No. 50974148), Innovation Fund of Shanghai University.

Appendix A. Supplementary data

Supplementary data associated with this article can be found, in the online version, at http://dx. doi. org/10. 1016/j. jnucmat. 2013. 03. 067.

References

[1] G. S. Was, Corrosion 46 (1990) 319 – 330.

[2] K. Stiller, J. O. Nilsson. K. Norring, Metall. Trans. A 27A (1996) 327 – 341.

[3] M. H. Lewis, B. Hattersley, Acta Metall. 13 (1965) 1159 – 1168.

[4] J. J. Kai, G. P. Yu, C. H. Tsai, M. N. Liu, S. C. Yao, Metall. Trans. A 20A (1989) 2057 – 2067.

[5] T. M. Angeliu, G. S. Was, Metall. Trans. A 21A (1990) 2097 – 2107.

[6] E. A. Trillo, L. E. Murr, J. Mater. Sci. 33 (1998) 1263 – 1271.

[7] E. A. Trillo, L. E. Murr, Acta Mater. 47 (1998) 235 – 245.

[8] M. Thuvander. K. Stiller, Mater. Sci. Eng. A A281 (2000) 96 – 103.

[9] Q. Li, B. X. Zhou, Acta Metall. Sin. (Chin. Lett.) 37 (2001) 8 – 12.

[10] M. Kurban, U. Erb, K. T. Aust, Scr. Mater. 54 (2006) 1053 – 1058.

[11] Y. S. Lim, J. S. Kim, H. P. Kim, H. D. Cho. J. Nucl. Mater. 335 (2004) 108 – 114.

[12] H. Li, S. Xia, B. X. Zhou, W. J. Chen, J. S. Ni, Acta Metall. Sin. (Chin. Lett.) 45 (2009) 195 – 198.

[13] H. Li, S. Xia, B. X. Zhou, W. J. Chen, C. L Hu, J. Nucl. Mater. 399 (2010) 108 – 113.

[14] M. Thuvander, M. K. Miller, K. Stiller, Mater. Sci. Eng. A A270 (1999) 38 – 43.

[15] H. Li, S. Xia, B. X. Zhou, W. Q. Liu, Mater. Charact. 66 (2012) 68 – 74.

[16] M. K. Miller, Atom Probe Tomography: Analysis at the Atomic Level, first ed. , Kluwer Academic/Pienum Publishers, New York, 1999. p. 33.

[17] F. Vurpillot, A. Cerezo, D. Blavette, D. J. Larson, Microsc. Microanal. 10 (2004) 384 – 390.

[18] D. A. Muller, M. J. Mills. Mater. Sci. Eng. A260 (1999) 12 – 28.

[19] S. Xia, On the grain boundary character distribution and its evolution mechanism in Alloy 690, Ph. D thesis, Shanghai University, 2007.

[20] S. Bail, M. J. Olszta, S. M. Bruemmer, D. N. Seidman, Scripta. Mater. 66 (2012) 809 – 812.

[21] H. Li, S. Xia, B. X. Zhou, J. C. Peng, Acta Metall. Sin. (Chin. Lett.) 47 (2011) 853 – 858.

[22] L. Karlsson, H. Norden, J. Phy. Coll. C9 (1984) 391 – 396.

[23] B. W. Krakauer, D. N. Seidman, Phys. Rev. B 48B (1993) 6724 – 6727.

[24] D. Hudson, G. D. W. Smith, Scripta. Mater. 61 (2009) 411 – 414.

[25] J. Takahashi, K. Kawakami, K. Ushioda, S. Takaki, N. Nakata, T. Tsuchiyama, Scripta. Mater. 66 (2012) 207 –210.

[26] M. Guttmann, D. Mclean, Grain boundary Segregation in multicomponent systems, in: W. C. Johnson, J. M. Blakely (Eds.), Interfacial segregation, ASM, Metals Park, OH, 1979, p. 261.

[27] S. Benhadad, N. Richards, M. C. Chaturvedi, Metall. Mater. Trans. A A33 (2002) 2005 – 2017.

[28] P. Lejcek, S. Hofmann, Acta Metall. Mater. 39 (1991) 2469 – 2476.

[29] P. Lejcek, A. V. Krajnikov, Y. N. Ivashchenko, M. Militzer, J. Adamek, Surf. Sci. 280 (1993) 325 – 334.

[30] Z. Jiao, G. S. Was, Acta Mater. 59 (2011) 1220 – 1238.

Effect of Initial Grain Sizes on the Grain Boundary Network during Grain Boundary Engineering in Alloy 690 *

Abstract: Grain boundary engineering (GBE) has been carried out in nickel-based Alloy 690 with different initial grain sizes. The microstructure evolution during GBE–processing is characterized using electron backscatter diffraction to study the initial grain size effects on the grain boundary network (GBN). The microstructures of the partially recrystallized samples revealed that the GBE-processing is a strain-recrystallization process, during which each grain-cluster is formed by "multiple twinning" starting from a single recrystallization nucleus. Taking into consideration the coincidence site lattices (CSLs) and $\sum$, which is defined as the reciprocal density of coincidence sites, a high proportion of low- $\sum$ CSL grain boundaries (GBs) and large grain-clusters are found to be the features of GBE-processed GBN. The initial grain size has a combined effect on the low- $\sum$ CSL GBs proportion. A large initial grain size reduces the number of recrystallization nuclei that form, increasing the cluster size, but decreasing twin boundary density. On the other hand, smaller initial grain sizes increase the density of twin boundary after recrystallization, while decreasing grain-cluster size. Neither the grain-cluster size nor the twin boundary density is the sole factor influencing the proportion of low- $\sum$ CSL GBs. The ratio of the grain cluster size over the grain size governs the proportion of low- $\sum$ CSL GBs.

1 Introduction

The grain boundary network (GBN) plays a noticeable role in the grain-boundary-related properties in polycrystalline materials. The concept of "grain boundary engineering" (GBE),[1,2] means "grain boundary design and control"that aims at improving the grain-boundary-related properties. Coincidence site lattices (CSLs) play an important role in GBE, with the notation $\sum$ denoting the reciprocal density of coincidence sites; the improvement in grain-boundary-related properties are sought to be effected by enhancing the proportion of the so-called low- $\sum$ CSL grain boundaries (GBs)[3] ($\sum$ -value $\leq$ 29). GBE has been successfully applied in many low stacking fault energy (SFE) face centered cubic (FCC) metallic materials, such as nickel-based alloys,[4-9] copper alloys,[10-13] lead

* In collaboration with Tingguang Liu, Shuang Xia, Hui Li, Qin Bai. Reprinted from Journal of Materials Research Society, 2013, 28(9): 1165 – 1176.

alloys[14,15] and stainless steels.[16-20]

Two types of thermomechanical treatment are reported for GBE-processing: iterative strain annealing[4,5,8,10-14] and one-step strain annealing.[6,7,9,15-20] Although different processing schedules are applied during GBE-processing, they have the same principal purposes of increasing the proportion of low-$\sum$ CSL GBs, increasing the proportion of triple junctions with $\sum 3^n$ boundaries[8,12] and disrupting the connectivity of the random boundary network[4,10,11,21] or forming large grain-clusters.[7,22,23] Grain-cluster is a domain of twin-related grains[24,25]: all of the boundaries within the cluster have $\sum 3^n$ misorientations, whereas the outer boundaries of the cluster are crystallographically random.[23,26]

Regarding the GBN evolution during GBE-processing, the "$\sum 3$ regeneration model" proposed by Randle,[10,11] and "boundary decomposition mechanisms" proposed by Kumar[4] and "twinning emission" proposed by Shimada and Kokawa[16] could interpret the formation of high proportion of $\sum 3$ boundaries and the disrupted connectivity of the random boundary network after GBE. The key point is that a relatively low-strain deformation is chosen for inducing local grain boundary migration upon heating rather than the full recrystallization route.[6] Our previous works[7,22,23] showed that the GBE microstructures are featured by the formation of large grain-cluster and large amount of inner connected $\sum 3^n$ type triple junctions. All the grains maintain $\sum 3^n$ mutual misorientations in a grain-cluster regardless of whether they are adjacently positioned or not. It is believed that[7,22] the grain-cluster is formed by "multiple-twinning"[25-31] starting from a single nucleus during the recrystallization process.

Prestrain and annealing are two factors of GBE-processing schedule. Many published articles studied the influence of prestrain and annealing on the resulting GBN. However, high proportion of low-$\sum$ CSL GBs could be formed after GBE with different processing schedules. The reason may be the different initial microstructures. The microstructure after deformation and annealing is influenced by its initial microstructure,[32-35] such as grain size, texture and carbide precipitation at GBs. The present work aims to study the influence of initial grain sizes on the GBN after GBE in nickel (Ni)-based Alloy 690, which is usually used as the steam generator tube material in pressurized water nuclear reactors. Comparison between GBE microstructures with different initial grain sizes was made. Analysis of partially recrystallized microstructure using electron backscatter diffraction-based orientation imaging microscopy (EBSD-OIM) technique gives an indication of the GBN evolution during GBE. Influence of initial grain sizes on GBE microstructures can be understood following the GBN evolution characterization.

2 Experiment

A solution-annealed Alloy 690, with the measured chemical composition (in wt%):

30. 39 Cr, 8. 88 Fe, 0. 023 C, 0. 002 S, 0. 006 P, 0. 07 Si, 0. 23 Mn, 0. 22 Al, 0. 26 Ti, 0. 02 Cu, and the balance Ni, was used as the experimental material in this work. The Alloy 690 sample with a shape of 1. 12 × 8 × 80 mm was single pass cold-rolled with a thickness reduction of 30%. The deformed strips were vacuum-sealed in quartz capsules and annealed at 1 100 ℃ for 1 min or 5 min or at 1 150 ℃ for 120 min, respectively, then quenched in water. Thereafter, the quartz capsules (WQ) were broken open simultaneously to obtain the desired starting state specimens of the cold rolled and annealed Ni-based Alloy 690 strips with different grain sizes, which have been designated as specimens S, M and L correspondingly. After the GBE-processing, the strips were scheduled to be characterized by low-strain deformation prior to short-duration annealing at adequately high temperature, this part of the data were adopted from our previous works. [7,22,23,35] The specimens were deformed by tensile deformations at room temperature. Specimens S were deformed with tensile strains of 3%, 5%, and 10% respectively. Specimens M were deformed with tensile strains of 5%, 8%, and 10% respectively. Specimens L were deformed with tensile strains of 5%, 8%, 10%, 13%, and 17% respectively. These tensile-strained strips were vacuum-sealed in quartz capsules and subsequently annealed at 1 100 ℃ for 5 min, then quenched in water. Thereafter the quartz capsules were broken open simultaneously so that the specimens S1, S2, S3, M1, M2, M3, L1, L2, L3, L4, and L5 were obtained correspondingly. These thermomechanical processing procedures are displayed in Table Ⅰ. The specimens are classified into S-series, M-series and L-series according to their initial grain sizes. In this work, all types of boundaries are included as defining individual grains, and thus the twins are regarded as grains. The partially recrystallized specimens S2p (S-5% tensile strained—1 100 ℃/60 s), M1p (M-5% tensile strained—1 100 ℃/120 s) and L3p (L-10% tensile strained—1 100 ℃/ 60 s) were prepared to study the GBN evolution during recrystallization.

Table Ⅰ　The thermomechanical processing procedure of specimens

Preprocessing			GBE		
Cold rolling (%)	Annealing	Sample ID	Tensile strain (%)	Annealing	Sample ID
30	1 100 ℃×60s	S	3	1 100 ℃×5 min	S1
30	1 100 ℃×60s	S	5	1 100 ℃×5 min	S2
30	1 100 ℃×60s	S	10	1 100 ℃×5 min	S3
30	1 100 ℃×300 s	M	5	1 100 ℃×5 min	M1
30	1 100 ℃×300 s	M	8	1 100 ℃×5 min	M2
30	1 100 ℃×300 s	M	10	1 100 ℃×5 min	M3
30	1 150 ℃×2 h	L	5	1 100 ℃×5 min	L1
30	1 150 ℃×2 h	L	8	1 100 ℃×5 min	L2
30	1 150 ℃×2 h	L	10	1 100 ℃×5 min	L3
30	1 150 ℃×2 h	L	13	1 100 ℃×5 min	L4
30	1 150 ℃×2 h	L	17	1 100 ℃×5 min	L5

The specimens for electron backscatter diffraction (EBSD) experiments were mechanically polished, and subsequently electropolished in an electrolyte containing 20%

$HClO_4$—80% CH_3COOH at room temperature with 30 V direct current for 90 s. The specimen surface crystallographic orientation data were obtained using the HKL-Technology EBSD system (Oxford Instruments, Cambridge, UK), which was attached to a CamScan Apollo 300 scanning electron microscope (SEM; Obducat CamScan Ltd., Cambridge, UK). The EBSD mapping area was 300×300 or 800×500 μm with step size of 1 or 2 μm according to the grain size. Two maps were measured for each specimen. Grain boundary characters were defined by coincidence site lattice (CSL) model according to the Palumbo-Aust's criterion ($\Delta\theta_{max} = 15° \sum^{[-5/6]}$). [36]

3 Results and discussion

3.1 GBNs after GBE with different initial grain sizes

Figure 1, the GBN maps with boundary types, shows the microstructures of three starting state samples with different grain sizes, designated as S (6.6 μm), M (17.1 μm), and L (31.3 μm), respectively. The proportions of low- $\sum$ CSL GBs are 53.6%, 51.6%, and 64.5% by length for samples S, M and L, as shown in Fig. 4. Sample L has the highest $\sum$ 3 proportion and the highest total low- $\sum$ CSL GBs proportion. The sum of $\sum$ 9 and $\sum$ 27 proportion is higher in sample S than that in the other two. However, for all these three microstructures the $\sum$ 3 boundaries appear in the shape of straight single line or parallel line pairs, either grain spanning or terminated within a single grain.

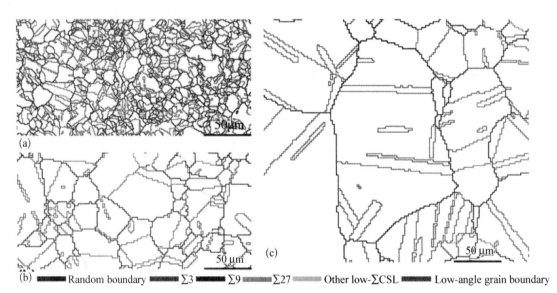

(a) (b) (c) Random boundary ▬ $\sum$3 ▬ $\sum$9 ▬ $\sum$27 ▬ Other low-$\sum$CSL ▬ Low-angle grain boundary

Fig. 1 The GBNs of starting state samples: (a) sample S, (b) sample M, and (c) sample L. Random boundaries, $\sum$ 3, $\sum$ 9, $\sum$ 27, and other low- $\sum$ CSL boundaries, and low-angle grain boundaries ($\sum$1) are black, red, blue, green, yellow and gray respectively. (The same below)

Figure 2 shows the GBNs of samples after GBE-processing, which is a thermomechanical treatment featured by low-strain deformation followed by high-temperature annealing.

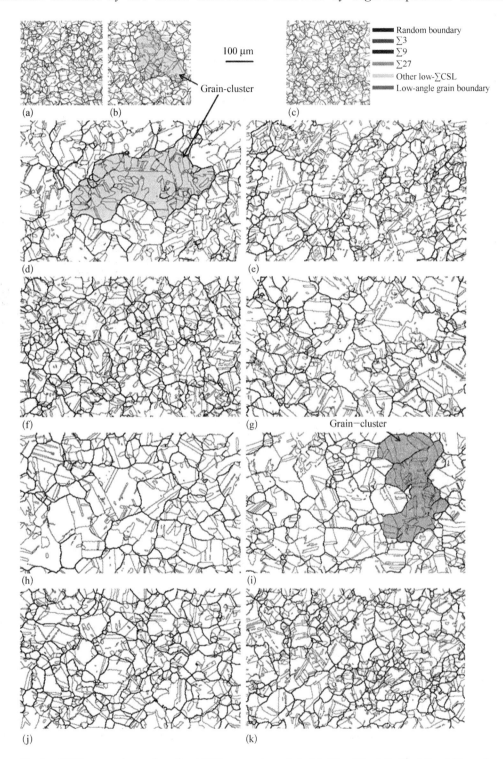

Fig. 2 GBN maps of samples after GBE-processing: (a)-(c) S1 - S3, (d)-(f) M1 - M3, and (g)-(k) L1 - L5. Some grain-clusters encircled by random boundaries are highlighted by gray and in figures (b), (d), (i)

There are obvious differences in the grain sizes between S2 and M1, L3. The average grain size and distribution for all the samples are shown in Fig. 3. The area with unique orientation was regarded as a grain in the current work, and therefore the annealing twins are included for the purpose of grain size calculation. The average grain size of GBE-processed samples in S-series is obviously smaller than those in M-series and L-series. The grain sizes of GBE-treated samples in M-series and L-series are at the same level. The average grain sizes after GBE-processing are similar among the S-series samples and among the M-series samples respectively regardless of the extent of deformation. But the grain size distributions for all the samples have quite wide fluctuations.

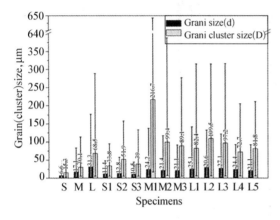

Fig. 3 The mean sizes of grain and grain-cluster of all specimens

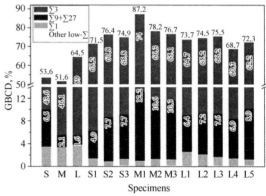

Fig. 4 The grain boundary character distribution (GBCD) of all specimens: $\sum 1$, $\sum 3$, $\sum 9 + \sum 27$ and other low- $\sum$ CSL GBs

The grain boundary character distributions (GBCDs) of all the samples are shown in Fig. 4, including $\sum 1$ (low-angle boundaries, misorientation angle: $2 - 15°$), $\sum 3$, the sum of $\sum 9$ and $\sum 27$, and other low- $\sum$ CSL GBs. There is an evident increasing of the proportion of low- $\sum$ CSL GBs in the samples after strain deformation and annealing processing as compared with the starting samples. The proportion of low- $\sum$ CSL GBs in sample M1 is near 90%. In S-series, the proportions of low- $\sum$ CSL GBs after GBE are around 75% despite the different strains before annealing. The proportions of low- $\sum$ CSL GBs in the M-series decrease obviously with increase of the strain deformations before annealing. No obvious differences are found in the proportion of low- $\sum$ CSL GBs for the samples of L-series except L4. Nearly all of the low- $\sum$ CSL GBs are of $\sum 3^n$ type, and about 85% of the special boundaries are $\sum 3$ boundaries. The ($\sum 9 + \sum 27$) fraction is nearly in direct proportion to the $\sum 3$ boundary fraction for all the samples after GBE-processing.

The marked characteristic of the GBN with high proportion of low- $\sum$ CSL GBs after GBE is the formation of large-size gain-clusters.[23,24,26] As shown in Fig. 2, some grain-clusters highlighted by gray background are encircled by random high-angle GBs. The mean size of grain-clusters and the size spread for each sample are included in Fig. 3. In the size statistics, the border clusters in the OIM map are included because the clusters are very large and only a small number of clusters are wholly included in the EBSD-investigated areas, especially for M-series and L-series. Figures 2 and 3 show that the grain-cluster size is quite inhomogeneous. An extremely large cluster is found in sample M1. The circle equivalent diameter of the cluster is over 600 μm. The mean cluster size of the GBE-treated samples in S-series is less than those in M-series and L-series. The low- $\sum$ CSL GBs fraction is determined not only by the grain-cluster size but also by the grain size.

In general, a high proportion of low- $\sum$ CSL GBs is expected for the GBE-processed material. Large grain-cluster associated with large amount of inner connected $\sum 3^n$ type triple junctions is believed to be the reason for the improved corrosion resistance of GBE-processed material.[7,18,37] Small grain size is believed to be beneficial for the mechanical properties. Therefore, the proportion of low- $\sum$ CSL GBs, the grain-cluster size and the grain size are three most significant measures of the GBN. The three parameters determine the characteristics of GBN in the GBE-processed specimens. The plots of the three measures versus the initial grain size are shown in Fig. 5, including low- $\sum$ CSL GBs proportion, grain size (d) and grain-cluster size (D). Two lines are plotted in each map for the samples with different strain deformation amounts during GBE-processing, which are designated as the "series of 5% tensile deformation in GBE" (red lines) and as the "series of 10% tensile deformation in GBE" (black lines) respectively. Because it is observed that only the sample L1 is not fully recrystallized, the dashed lines are used to link between "M1" and "L1" instead of solid lines. Figure 5(a) shows that the low- $\sum$ CSL GBs percentage changes dramatically with the initial grain sizes for the samples of"series of 5% tensile deformation in GBE,"but changed slightly for the samples of "series of 10% tensile deformation in GBE. " The grain sizes after GBE-processing increase with the increasing of the initial grain sizes for both of the two series samples as shown in Fig. 5(b). Regarding the dependence of grain-cluster size on the initial grain sizes, the 5% tensile deformation series sample "M1" shows a dramatic increase as compared with that of "S2"; however, in the case of the 10% tensile deformation series, the increase in grain-cluster size is not so large. Figure 5(d) shows the ratio of the grain-cluster size over the grain size, (D/d), as a function of initial grain sizes. In this figure, both of the two series samples show a very similar changing trend with respect to that of Fig. 5 (a). This establishes that the parameter (D/d) and low- $\sum$ CSL GBs proportion show a consistent dependence on the initial grain size.

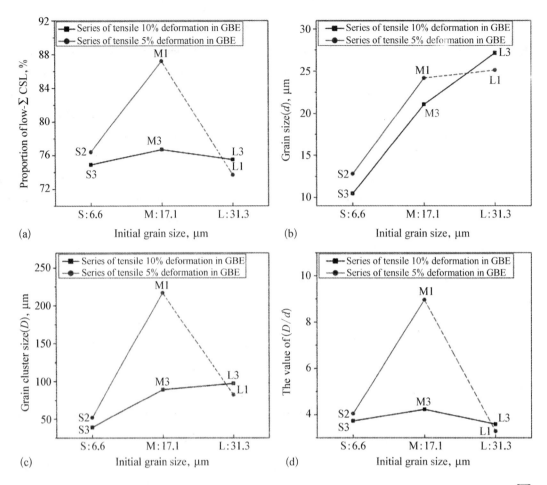

Fig. 5 The dependence of GBE microstructures on the initial grain sizes: (a) the proportion of low-$\sum$ CSL GBs, (b) the mean grain size, (c) the mean grain-cluster size, and (d) the ratio of grain-cluster size over grain size, (D/d)

3.2 Analysis of GBN evolution during GBE using partially recrystallized microstructure

The initial grain size has a significant influence on the GBN after GBE-processing, as shown in Fig. 5. The GBCD data on Fig. 4 revealed that the single-step GBE-processing can increase the proportion of low-$\sum$ CSL GBs to more than 75% irrespective of the initial grain size. However, a microstructure with medium grain size is required for achieving the highest low-$\sum$ CSL GBs proportion. The grain size and grain-cluster size data (Fig. 3) revealed that the GBN after GBE is quite different for the different series samples with different initial grain sizes. The mean sizes of grains and grain-clusters after GBE tend to increase with increase in the initial grain sizes.

Partially recrystallized specimens were prepared to study the microstructural evolution at the early stage of annealing. Specimen S was annealed 60 s at 1 100 ℃ after 5% tensile deformation, and thus the partially recrystallized specimen S2p was obtained. The crystallographic features of the partially recrystallized microstructure in sample S2p was characterized with different methods by HKL-Channel 5 software, as shown in Fig. 6. Figure 6(a) shows the GBN with different grain boundary types. The gray lines are low-angle boundaries, which is formed during deformation and recovery annealing. The low-angle GB density can be used to distinguish between the recrystallized and the nonrecrystallized regions in the partially recrystallized sample. In Fig. 6(a), the middle region is nearly free of low-angle GBs, but there is a larger amount of such low-angle GBs in the surroundings. This can be used to distinguish the areas between recrystallized and nonrecrystallized microstructures.

Figure 6(b) shows the GBN distribution in which the background is shaded according to the local orientation gradient[38,39] (or local misorientation in HKL/Channel 5 software), which is a parameter describing the deformation amount. The orientation imaging microscopy (OIM) map is a grid map in which each grid point corresponds to the crystallographic orientation of that point in the sample. The local orientation gradient is

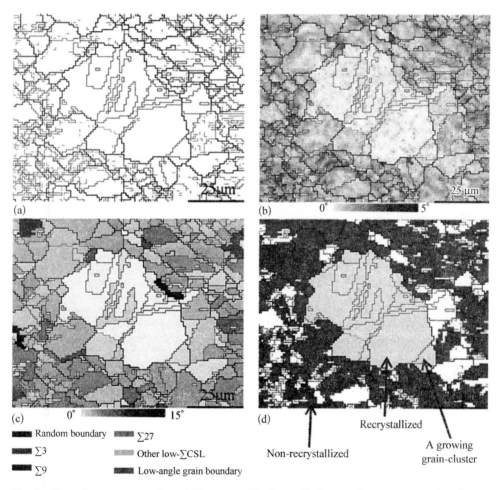

Fig. 6 The microstructure of partially recrystallized sample S2p is characterized with different methods by HKL/Channel 5: (a) GBN distribution map, (b) GBN distribution map in which the background is shaded according to the local orientation gradient, (c) GBN distribution map in which the background is shaded according to the grain mean misorientation, and (d) recrystallized fraction component map

the average misorientation between the analyzed grid point and its adjacent points, which indicates the amount of the local strains. The local orientation gradient is close to zero in the fully recrystallized grains. The recrystallized region (light gray) is surrounded by the nonrecrystallized darker gray region in Fig. 6(b).

In HKL/Channel 5 software, grain mean misorientation is another parameter describing the deformation amount. Grain mean misorientation gives a quantification measurement of the grain's deformation.[38,39] The background of Fig. 6(c) is shaded according to this parameter. The light gray region is recrystallized and surrounded by the nonrecrystallized darker gray region in Fig. 6(c). A critical value of 0.85° for the grain-mean-misorientation was set to distinguish the recrystallized and nonrecrystallized regions in Fig. 6(d). Nonrecrystallized regions were colored purple. The white and gray regions are recrystallized.

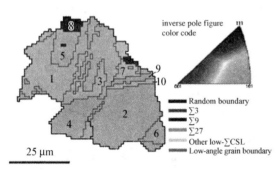

Fig. 7 A growing grain-cluster in the partially recrystallized sample S2p as shown in Fig. 6, contrasted by color of inverse pole figure. The grain labels are according to Table Ⅱ

From the analysis above, it is believed that the specimen S2p is partially recrystallized. As shown in Fig. 6, there are recrystallized and nonrecrystallized regions, which are separated by random GBs. A grain-cluster formed in the recrystallized region, that is, the gray region in Fig. 6(d). All GBs are of $\sum 3$, $\sum 9$ or $\sum 27$ types in the cluster, that is, all the inner neighboring grains have $\sum 3^n$ misorientations. Ten grains are selected in the grain-cluster, as shown in Fig. 7. They have six different orientations, and the mutual misorientations between any two orientations are of $\sum 3^n$, as listed in Table Ⅱ. Therefore, it can be known that all grains inside of the grain-cluster maintain $\sum 3^n$ mutual misorientations regardless of whether they are adjacent.

Table Ⅱ Mutual misorientations of the ten grains selected from Fig. 7. (The bottom left data are rotation angle-axis pair: θ, $[hkl]$; the top right data are the closest CSL $\sum$-value and deviation: $\sum/\Delta\theta$)

	A	B	C	D	E	F
A	...	$\sum 9/0.4°$	$\sum 3/0°$	$\sum 27a/0.3°$	$\sum 3/0.1°$	$\sum 9/0.3°$
B	38.5°,[−1 0 1]	...	$\sum 3/0.4°$	$\sum 27b/0.6°$	$\sum 3/0.2°$	$\sum 9/0.6°$
C	60.0°,[−1 −1 1]	59.6°,[−1 1 1]	...	$\sum 9/0.2°$	$\sum 9/0.1°$	$\sum 3/0°$
D	59.9°[1 1 1]	36.0°[0 1 2]	38.7°[1 0 −1]	...	$\sum 81c/0°$	$\sum 27a/0.8°$
E	31.9°,[0 −1 1]	59.8°,[1 −1 1]	39.0°,[0 −1 −1]	38.4°,[3 −5 −1]	...	$\sum 27b/0.7°$
F	38.6°,[−1 0 1]	39.5°,[1 0 1]	60.0°,[1 −1 1]	32.4°,[1 0 −1]	34.7°,[−1 0 −2]	...

Note: Grains are grouped by grain average orientation. Orientation relationship is given in the form: rotation angle and axis in the bottom left of the table, the closest CSL $\sum$-value/deviation in the top right.
A: 1; B: 2, 9; C: 3, 5, 7, 10; D: 4; E: 6; F: 8

The M-5% tensile deformation — 1 100 ℃/120 s annealing specimen M1p is also characterized with different methods by HKL/Channel 5 software. Orientation distribution map, local misorientation map, grain mean misorientation map and recrystallized fraction component map are shown in Fig. 8. Following the analysis above, it is believed that the sample M1p is also a partially recrystallized sample. A large grain-cluster formed in the recrystallized region, as shown in Fig. 9. Twenty big grains are randomly selected for analysis of the mutual misorientations, which are listed in Table Ⅲ. The same conclusion can be drawn that any two grains in the grain-cluster have $\sum 3^n$ misorientations regardless

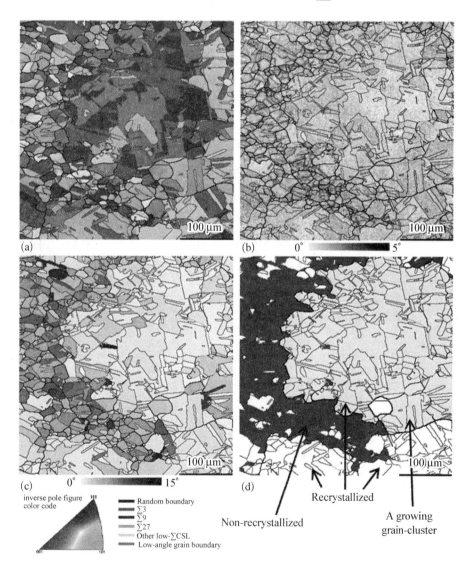

Fig. 8 The microstructure of partially recrystallized sample M1p is characterized with different methods by HKL/Channel 5: (a) grain orientation map contrasted by color of inverse pole figure, (b) GBN distribution map in which the background is shaded according to the local orientation gradient, (c) GBN distribution map in which the background is shaded according to the grain mean misorientation, and (d) recrystallized fraction component map

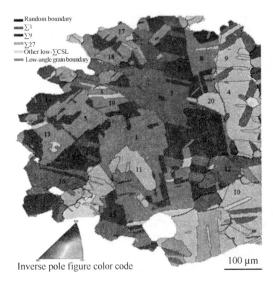

Random boundary
Σ3
Σ9
Σ27
Other low-ΣCSL
Low-angle grain boundary

Inverse pole figure color code   100 μm

Fig. 9 A growing grain-cluster in the partially recrystallized sample M1p as shown in Fig. 8, contrasted by color of inverse pole figure. The grain numbers are according to Table Ⅲ.

of whether they are adjacent.

The L-10% tensile deformation —1 100 ℃/ 60 s annealing specimen L3p is characterized with different methods by HKL/Channel 5 software, as shown in Fig. 10. It is also a partially recrystallized sample. A small grain-cluster is found in the recrystallized region. All GBs are $\sum 3$, $\sum 9$ or $\sum 27$ misorientation inside of the grain-cluster.

As shown in Figs. 6,8, and 10, it is clear that the samples S2p, M1p, and L3p are partially recrystallized samples. On the other hand, GBE microstructures are featured by high proportion of low-$\sum$CSL GBs and large grain-clusters in the fully recrystallized samples S2, M1, and L3. Therefore, the GBE-processing should be a recrystallization process. In the partially recrystallized samples, grain-cluster formed in the recrystallized regions that are surrounded by deformed

Table Ⅲ Mutual misorientations of the twenty grains selected from Fig. 9

	A	B	C	D	E	F	G	H	I
A	...	$\sum 3/0.2°$	$\sum 9/0.5°$	$\sum 9/0.3°$	$\sum 27b/0.7°$	$\sum 3/0.2°$	$\sum 9/0.3°$	$\sum 3/0.3°$	$\sum 9/0.1°$
B	59.8°, [1 1 1]	...	$\sum 3/0.2°$	$\sum 3/0.5°$	$\sum 81c/0.6°$	$\sum 9/1.0°$	$\sum 27a/0.6°$	$\sum 9/0.3°$	$\sum 27b/0.9°$
C	38.4°, [0 −1 1]	59.8°, [1 −1 1]	...	$\sum 9/0.3°$	$\sum 243e$ 49.7°,[6 5 5]	$\sum 27a/1.6°$	$\sum 81c/0.3°$	$\sum 27b/0.2°$	$\sum 81b/0.7°$
D	38.6° [−1 0 1]	59.5° [1 1 1]	39.2° [−101]	...	$\sum 243g/0°$	$\sum 27b/0.4°$	$\sum 81c/0.3°$	$\sum 27b/0.5°$	$\sum 81c/0.3°$
E	34.7°, [1 0 −2]	39.0°, [5 3 −1]	49.7°, [111]	31.6°, [−1 −1 −4]	...	$\sum 81b/0.1°$	$\sum 3/0°$	$\sum 9/0.2°$	$\sum 243a$ 43.1°,[955]
F	59.8°, [−1 1 1]	37.9°, [1 0 −1]	33.0v, [−101]	35.8°, [0 −1 2]	54.4°, [−2 3 −2]	...	$\sum 27b/0.2°$	$\sum 9/0.6°$	$\sum 27a/0.5°$
G	39.2°, [0 1 −1]	31.0°, [0 1 1]	38.7°, [−3 −1 −5]	38.1°, [−3 1 −5]	60.0°, [1 −1 1]	35.6°, [−2 0 1]	...	$\sum 3/0.7°$	$\sum 81a/0°$
H	59.7°, [1 −1 −1]	38.6°, [1 0 1]	35.6°, [−1 2 0]	35.9°, [0 1 −2]	38.7°, [0 −1 −1]	39.5°, [−1 0 1]	59.3°, [1 −1 −1]	...	$\sum 27b/0.5°$
I	39.0°, [0 −1 −1]	34.5°, [2 0 −1]	55.2°, [−3 2 2]	39.1°, [5 −1 −3]	43.1°, [−1 1 −2]	31.1°, [0 −1 1]	38.9°, [4 1 −1]	35.9°, [−2 0 1]	...

Note: Grains are grouped by grain average orientation. Orientation relationship is given in the form: rotation angle and axis in the bottom left of the table, the closest CSL $\sum$-value/deviation in the top right.

A: 1, 8, 19; B: 2, 6, 7, 14, 15; C: 5, 13, 17, 18; D: 3, 16; E: 4; F: 10, 20; G: 9; H: 11; I: 12

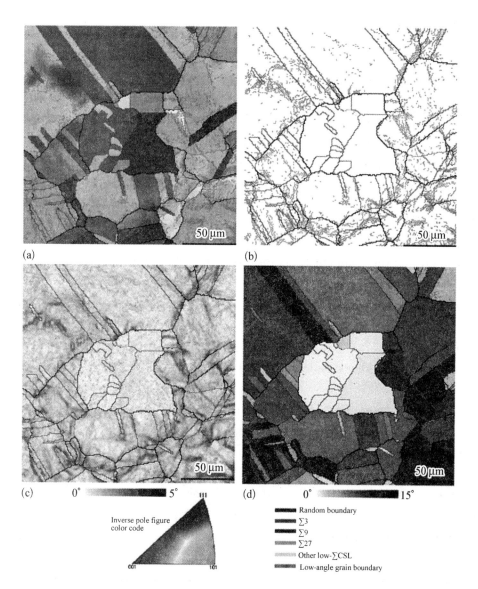

(a)

(b)

(c)   0° ▭▭▭ 5°   ⟨111⟩

Inverse pole figure
color code

⟨001⟩   ⟨101⟩

(d)   0° ▭▭▭ 15°

▬▬ Random boundary
▭▭ Σ3
▬▬ Σ9
▭▭ Σ27
▭▭ Other low-ΣCSL
▬▬ Low-angle grain boundary

Fig. 10 The microstructure of partially recrystallized sample L3p is characterized with different methods by HKL/Channel 5: (a) grain orientation map contrasted by color of inverse pole figure, (b) GBN distribution map, (c) GBN distribution map in which the background is shaded according to the local orientation gradient, and (d) GBN distribution map in which the background is shaded according to the grain mean misorientation

microstructures. All GBs have $\sum 3^n$ misorientations inside of these grain-clusters, no matter what the cluster sizes are. It is suggested that the grain-clusters would grow during further annealing by the migration of the GBs between the cluster and the deformed matrix. Therefore, the GBE process is a recrystallization process with the grain-cluster growth starting from a single nucleus. During the nucleus growth, the first, the second and the higher order generation twins were produced sequentially to form a twin-chain. This is the "multiple twinning" process[25,28-31]. Therefore, any grain in a grain-cluster has a

$\sum 3^n$ misorientation with its mother orientation, that is, the recrystallization nucleus. Therefore a further inference can be got that any two grains in a cluster have $\sum 3^n$ misorientation regardless of whether they are adjacent.

3.3 Influence of the initial grain sizes on the GBE microstructures

The initial grain size influences both the grain-cluster size and the grain size. It is evident that a grain boundary network with larger size grain-clusters and smaller size grains are expected to have higher proportion of low- $\sum$ CSL GBs. The ratio of the grain-cluster size over the grain size, (D/d), determined the proportions of low- $\sum$ CSL GBs.

In the GBE microstructures, the random GBs are formed by the impingement of the grain-clusters, and the grain-cluster is formed by multiple twinning starting from a single recrystallization nucleus, as can be seen from the partially recrystallized microstructure. The lower the nucleus density is, the larger the grain-clusters are. The nucleation density decreases with the increasing of original grain size because GBs are the preferential nucleation sites. Hence, the average size of grain-cluster increases with the increasing of initial grain sizes for the samples of "series of 10% tensile deformation in GBE" in Fig. 5(c). However, in the case of the samples of "series of 5% tensile deformation in GBE," the grain-cluster size first increases and then decreases. This is because the 5% tensile deformation is too low for the large grained sample L1 to induce completed recrystallization during annealing at $1\,100\,°C/5$ min. So, except the unrecrystallized sample L1, the grain-cluster size increased with the increasing of the initial grain sizes for both series samples. On the other hand, a large initial grain size reduces the twinning frequency because the disorientation of the moving boundary is changing less frequently during recrystallization. The grain size, which can partly represent the size of twins, therefore increases with the increasing of the initial grain sizes.

Although both the grain-cluster size and the grain size increase with the increases of the initial grain sizes in the fully recrystallized samples, the proportions of low- $\sum$ CSL GBs do not monotonically increase with the initial grain sizes. For example, the proportions of low- $\sum$ CSL GBs of the "M3" are slightly higher than that of "S3" and "L3," those of which are belonging to the "series of 10% tensile deformation in GBE," as shown in Fig. 5(a). This can be attributed to the fact that the initial grain size has a combined effect on the low- $\sum$ CSL GBs proportion, as shown in Fig. 11. Smaller initial grain sizes induce higher density of nucleation sites. The final GBN would have smaller grain-clusters and higher density of random boundaries, which is a disadvantageous factor for the formation of high proportion of low- $\sum$ CSL GBs. On the other hand, twinning is easier to occur when the misorientation of the front boundary between the growing grain and the deformed microstructure frequently change in a smaller grained initial microstructure. [40-44] The sizes of twins are therefore smaller, in other words, higher twin density defined by twin boundary amount per unite area. This is an advantageous factor for the formation of high proportion of low-$\sum$ CSL GBs. Because of this combined effect, neither grain-cluster size nor the twin boundary density is the solely sufficient condition for determining the low-$\sum$ CSL GBs proportion. The low-$\sum$ CSL GBs proportion is proportional to the ratio of the grain-cluster size over the grain size, (D/d), as shown in Fig. 5(d).

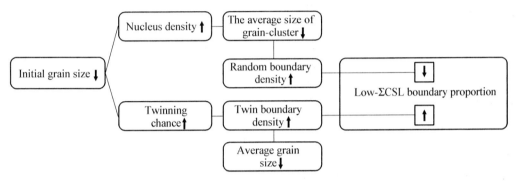

Fig. 11   Effects of initial grain sizes on GBE microstructures: grain-cluster size, grain size and low-$\sum$ CSL GBs proportion

The sample series with smaller deformation amount (5%) has higher low-$\sum$ CSL GBs, larger grain-cluster sizes and larger grain sizes than that of the 10% strained samples, the only exception being the unrecrystallized sample L1. This indicated that the smaller deformation used in the GBE-processing results in higher proportion of low-$\sum$ CSL GBs. But, for a material being considered for critical use in certain real and rigorous industrial applications, one cannot only consider the proportion of low-$\sum$ CSL GBs but also should consider the grain size, because the too large grain size will decrease the mechanical properties. Therefore, when using the GBE technique, cares must be taken to make a compromise between the proportion of low-$\sum$ CSL GBs and other microstructure parameters, such as grain size.

4   Conclusions

In Alloy 690, GBE-processing, which is featured by low-strain deformation, followed by short-duration annealing at high temperature, can increase the proportion of low-$\sum$ CSL GBs to more than 75%. The initial grain size has a significant influence on the GBN resulting from GBE-processing. The partially recrystallized microstructures showed that the GBE-processing is a recrystallization process, in which each grain-cluster was formed by "multiple twinning" starting from a single recrystallization nucleus. The large grain-clusters and high proportion of low- $\sum$ CSL GBs are the features of GBN after GBE-processing. The initial grain size has a combined influence on the low- $\sum$ CSL GBs proportion. Smaller initial grain sizes induce higher twin boundary density, but smaller grain-cluster and higher random GB density, and vice versa. In the GBE-processed microstructure, neither grain-cluster size nor the twin boundary density is the solely sufficient condition for determining the low- $\sum$ CSL GBs proportion. The ratio of the grain-cluster size over the grain size (D/d) governs the proportion of low- $\sum$ CSL GBs. However, when applying the GBE technique to materials involved in manufacture for real and critical industrial or other applications, a compromise between the proportion of low- $\sum$ CSL GBs and grain size should be considered.

Acknowledgments

This work was jointly supported by National Science Foundation of China (NSFC) (Grant No. 50974148), Major State Basic Research Development Program of China (Grant No. 2011CB610502), the fund from Jiuli Hi-Tech Metals Co., Ltd., China, and Shanghai Leading Academic Discipline Project (Grant No. S30107).

References

[1] T. Watanabe: Approach to grain boundary design for strong and ductile polycrystals. *Res. Mech.* 11,47 (1984).

[2] V. Randle: *The Role of the Coincidence Site Lattice in Grain Boundary Engineering* (Cambridge University Press, Cambridge, England, 1996), p. 91.

[3] P. Lin, G. Palumbo, U. Erb, and K. T. Aust: Influence of grain boundary character distribution on sensitization and intergranular corrosion of Alloy 600. *Scr. Metall. Mater.* 33,1387 (1995).

[4] M. Kumar, A. J. Schwartz, and W. E. King: Microstructural evolution during grain boundary of low to medium stacking fault energy fcc materials. *Acta Mater.* 50,2599 (2002).

[5] M. Kumar, W. E. King, and A. J. Schwartz: Modifications to the microstructural topology in f. c. c. materials through thermomechanical processing. *Acta Mater.* 48, 2081 (2000).

[6] S. L. Lee and N. L. Richards: The effect of single-step low strain and annealing of nickel on grain boundary character. *Mater. Sci. Eng.*, A 390,81 (2005).

[7] S. Xia, B. X. Zhou, and W. J. Chen: Effect of single-step strain and annealing on grain boundary character distribution and intergranular corrosion in Alloy 690. *J. Mater. Sci.* 43,2990 (2008).

[8] C. A. Schuh, M. Kumar, and W. E. King: Analysis of gain boundary networks and their evolution during grain boundary engineering. *Acta Mater.* 51,687 (2003).

[9] L. Tan, K. Sridharan, T. R. Allen, R. K. Nanstad, and D. A. McClintock: Microstructure tailoring for property improvements by grain boundary engineering. *J. Nucl. Mater.* 374, 270 (2008).

[10] V. Randle: Mechanism of twinning-induced grain boundary engineering in low stacking-fault energy materials. *Acta Mater.* 47,4187 (1999).

[11] V. Randle and M. Coleman: A study of low-strain and medium-strain grain boundary engineering. *Acta Mater.* 57, 3410 (2009).

[12] G. S. Rohrer, V. Randle, C-S. Kim, and Y. Hu: Changes in the five-parameter grain boundary character distribution in α-brass brought about by iterative thermomechanical processing. *Acta Mater.* 54,4489 (2006).

[13] V. Randle and H. Davies: Evolution of microstructure and properties in alpha-brass after iterative processing.

Metall. Mater. Trans. A 33, 1853 (2002).

[14] W. G. Wang and H. Guo: Effects of thermo-mechanical iterations on the grain boundary character distribution of Pb-Ca-Sn-Al alloy. *Mater. Sci. Eng.*, A 445 – 446, 155 (2007).

[15] S. Xia, B. X. Zhou, and W. J. Chen: Evolution of grain boundary character distributions in Pb Alloy during high temperature annealing. *Acta Metall. Sin.* 42, 129 (2006).

[16] M. Shimada, H. Kokawa, Z. J. Wang, Y. S. Sato, and I. Karibe: Optimization of grain boundary character distribution for intergranular corrosion resistant 304 stainless steel by twin-induced grain boundary engineering. *Acta Mater.* 50, 2331 (2002).

[17] D. L. Engelberg, F. J. Humphreys, and T. J. Marrow: The influence of low-strain thermo-mechanical processing on grain boundary network characteristics in type 304 austenitic stainless steel. *J. Microsc.* 230, 435 (2008).

[18] C. L. Hu, S. Xia, H. Li, T. G. Liu, B. X. Zhou, W. J. Chen, and N. Wang: Improving the intergranular corrosion resistance of 304 stainless steel by grain boundary network control. *Corros. Sci.* 53, 1880 (2011).

[19] M. Sekine, N. Sakaguchi, M. Endo, H. Kinoshita, S. Watanabe, H. Kokawa, S. Yamashita, Y. Yano, and M. Kawai: Grain boundary engineering of austenitic steel PNC316 for use in nuclear reactors. *J. Nucl. Mater.* 414, 232 (2011).

[20] M. Michiuchi, H. Kokawa, Z. J. Wang, Y. S. Sato, and K. Sakai: Twin-induced grain boundary engineering for 316 austenitic stainless steel. *Acta Mater.* 54, 5179 (2006).

[21] C. A. Schuh, M. Kumar, and W. E. King: Analysis of grain boundary networks and their evolution during grain boundary engineering. *Acta Mater.* 51, 687 (2003).

[22] S. Xia, B. X. Zhou, W. J. Chen, and W. G. Wang: Effects of strain and annealing processes on the distribution of $\sum 3$ boundaries in a Ni-based superalloy. *Scr. Mater.* 54, 2019 (2006).

[23] S. Xia, B. X. Zhou, and W. J. Chen: Grain cluster microstructure and grain boundary character distribution in Alloy 690. *Metall. Mater. Trans. A* 40, 3016 (2009).

[24] B. W. Reed, M. Kumar, R. W. Minich, and R. E. Rudd: Fracture roughness scaling and its correlation with grain boundary network structure. *Acta Mater.* 56, 3278 (2008).

[25] B. W. Reed and M. Kumar: Mathematical methods for analyzing highly-twinned grain boundary networks. *Scr. Mater.* 54, 1029 (2006).

[26] V. Y. Gertsman and C. H. Henager: Grain boundary junctions in microstructure generated by multiple-twinning. *Interface Sci.* 11, 403 (2003).

[27] G. Gottstein: Annealing texture development by multiple-twinning in f. c. c. crystals. *Acta Mater.* 32, 1117 (1984).

[28] P. Haasen: How are new orientations generated during primary recrystallization. *Metall. Mater. Trans.* B 24, 225 (1993).

[29] V. Y. Gertsman and K. Tangri: Computer simulation study of grain boundary and triple junction distributions in microstructures formed by multiple-twinning. *Acta Metall. Mater.* 3, 2317 (1995).

[30] C. Cayron: Multiple-twinning in cubic crystals: Geometric/algebraic study and its application for the identification of the $\sum 3^n$ grain boundaries. *Acta Crystallogr.*, Sect. A 63, 11 (2007).

[31] C. Cayron: Quantification of multiple twinning in face centred cubic materials. *Acta Mater.* 59, 252 (2011).

[32] K. Matsumoto, T. Shibayanagi, and Y. Umakoshi: Effect of grain size on grain orientations and grain boundary character distribution in recrystallized Al – 0.3mass% Mg alloy. *Scr. Metall. Mater.* 33, 1321 (1995).

[33] F. Caleyo, F. Cruz, T. Baudin, and R. Penelle: Texture and grain size dependence of grain boundary character distribution in recrystallized Fe – 50%Ni. *Scr. Mater.* 41, 847 (1999).

[34] W. G. Wang, B. X. Zhou, and H. Guo: Relations of initial microstructure with grain boundary character distributions in a cold rolled and annealed lead alloy. *Mater. Sci. Forum* 638 – 642, 2864 (2010).

[35] T. G. Liu, S. Xia, H. Li, B. X. Zhou, and W. J. Chen: Effect of original grain size on the boundary network in Alloy 690 treated by grain boundary engineering. *Acta Metall. Sin.* 47, 859 (2011).

[36] G. Palumbo and K. T. Aust: Structure-dependence of intergranular corrosion in high purity nickel. *Acta Metall.* 38, 2343 (1990).

[37] S. Xia, H. Li, T. G. Liu, and B. X. Zhou: Appling grain boundary engineering to alloy 690 tube for enhancing intergranular corrosion resistance. *J. Nucl. Mater.* 416, 303 (2011).

[38] F. J. Humphreys, P. S. Bate, and P. J. Hurley: Orientation averaging of electron backscattered diffraction data. *J. Microsc.* 201, 50 (2001).

[39] D. P. Field, P. B. Trivedi, S. I. Wright, and M. Kumar: Analysis of local orientation gradients in deformed single crystals. *Ultramicroscopy* 103, 33 (2005).

[40] R. L. Fullman and J. C. Fisher: Formation of annealing twins during grain growth. *J. Appl. Phys.* 22, 1350 (1951).

[41] W. Charnock and J. Nutting: The factors determining the frequency of occurrence of annealing twins. *Metal Sci.* 1, 78 (1967).

[42] C. S. Pande, M. A. Imam, and B. B. Rath: Study of annealing twins in fcc metals and alloys. *Metall. Mater. Trans.* A 21, 2891 (1990).

[43] B. B. Rath, M. A. Imam, and C. S. Pande: Nucleation and growth of twin interfaces in fcc metals and alloys. *Mater. Phys. Mech.* 1, 61 (2000).

[44] D. P. Field, L. T. Bradford, M. M. Nowell, and T. M. Lillo: The role of annealing twins during recrystallization of Cu. *Acta Mater.* 55, 4233 (2007).

形变及热处理对白铜 B10 合金晶界
特征分布的影响[*]

摘　要：研究晶界工程处理过程中的冷轧变形量和再结晶退火对白铜 B10 合金晶界特征分布的影响，采用电子背散射衍射(EBSD)技术表征分析晶界网络的变化. 结果表明：白铜 B10 合金经冷轧 7％后在 800 ℃退火 10 min 可使低 Σ CSL(Coincidence site lattice，$\Sigma \leqslant 29$)晶界比例提高到 75％以上，同时形成尺寸较大的"互有 $\Sigma 3^n$ 取向关系晶粒的团簇"显微组织. 当变形量小于 7％时，经 800 ℃退火后没有完全再结晶；当变形量大于 7％时，低 Σ CSL 晶界比例和平均晶粒团簇的尺寸随冷轧变形量的增加而下降.

　　绝大多数工程应用金属材料都是多晶体材料. 晶界相对于晶粒内部来说结构有序性差，具有更大的自由体积，更高的自由能，因此，晶界对材料的多种性能都有很大影响. KRONBERG 等[1]于 1949 年提出局部原子回旋再结晶成核的模型，从定向成核的观点来说明再结晶织构与加工织构取向间的关系，这种取向关系可以构成特殊的重位点阵(CSL)晶界. 重位点阵晶界常用 Σn CSL 晶界表示，其中 n 表示两个晶粒点阵构成的超点阵中有 $1/n$ 的点阵位置相互重合. 低 Σ CSL 晶界具有特殊的结构和性能，如抗晶界偏聚[2]、抗晶间腐蚀[3]、抗晶间应力腐蚀开裂[4]、抗蠕变[5]等.

　　1984 年，WATANABE[6]提出"晶界设计"(Grain boundary design)的概念，其目的就是控制金属的晶界特征分布，增加低 Σ CSL 晶界的比例，从而改善其与晶界相关的性能. 这一概念随后被 PALUMBO 等[3,7]发展成为"晶界工程"(Grain boundary engineering，GBE)的研究领域，即通过适当的冷加工变形，并控制再结晶过程中热处理工艺参数，达到调整晶界特征分布的目的，从而提高材料与晶界有关的性能. 晶界工程技术能够应用于多种低层错能面心立方结构的金属材料，如奥氏体不锈钢[8]、镍及其合金[8]、铅及其合金[7]、铜及其合金[9]等.

　　白铜是以镍为主要添加元素的铜基合金，具有低层错能面心立方结构，铜镍之间可无限固溶，形成连续固溶体[10]. 白铜 BFe10 - 1 - 1(简称白铜 B10 合金)，由于其具有优良的导电性、导热性、耐腐蚀性和较好的加工性能、中等以上的强度等[11]，因此，作为换热器冷凝管被广泛用于火力发电、核电、造船、海水淡化和海洋工程等行业[12]. 在火电行业，白铜冷凝管的腐蚀问题一直没有得到彻底解决. 作为发电机组的重要构件，其腐蚀泄漏是影响发电机组稳定运行的一大问题. 它不仅会带来换管损失和停机损失，而且换管过程还可能会进一步污染水质，加速剩余白铜冷凝管的进一步腐蚀[13].

　　本文作者借助 EBSD 技术研究形变及热处理工艺对白铜合金晶界特征分布的影响，得出控制白铜合金晶界特征分布的工艺，为通过晶界工程技术提高白铜 B10 合金的耐腐蚀性能提供可能的处理方法.

---

* 本文合作者：茹祥坤、刘廷光、夏爽、马爱利、郑玉贵. 原发表于《中国有色金属学报》，2013，23(8)：2176 - 2181.

1 实验

本实验用材料为白铜 B10 合金,化学成分见表 1. 首先对原材料进行 50％的冷轧变形,并在 800 ℃下保温 10 min 后水淬,获得本实验的始态样品. 然后对始态样品分别进行 3％、5％、7％、10％、20％和 50％的冷轧变形,在 800 ℃保温 10 min 后进行水淬,得到样品 S1、S2、S3、S4、S5 和 S6. 样品制备工艺如表 2 所示.

表 1 白铜 B10 合金的化学成分
Table 1 Composition of investigated cupronickel B10 alloy (mass fraction,％)

Ni	Fe	Mn	C	Pb	S	P	Zn	Cu
10.4	1.73	0.68	0.027 - 0.014	<0.001	0.004	0.003	<0.01	Bal.

表 2 样品的制备工艺
Table 2 Thermal-mechanical treatments of specimens

Pre-treatment	Cold rolling/％	Annealing	Sample
50％ cold rolling+(800 ℃,10min)+WQ	3	(800 ℃,10min)+WQ	S1
	5		S2
	7		S3
	10		S4
	20		S5
	50		S6

样品经过金相砂纸预磨后进行电解抛光,制备出适合 EBSD 检测的样品. 电解液成分(体积分数):25％ H_3PO_4+25％ C_2H_5OH+50％ H_2O,抛光电压为直流 26 V,时间约为 120 s. 在相同电解液中进行电解蚀刻后进行金相观察,蚀刻电压为直流 6 V,时间约为 10 s,用于金相观察. 采用 KEYENCE-VHX 数码显微镜对样品表面进行金相照片的拍摄. 采用配备在 CamScan Apollo 300 型热场发射枪扫描电子显微镜上的 Oxford/HKL-EBSD 系统对样品表面选定微区进行 EBSD 测试,并采用 Channel 5 数据处理软件进行取向分析. 采用 Brandon 标准($\Delta\theta_{max} = 15°\Sigma^{-1/2}$)[14]对 CSL 晶界类型进行判定,不同类型的晶界百分比例均为晶界长度百分比例.

2 结果与分析

图 1 所示为白铜 B10 合金始态样品的金相照片、不同类型晶界图、样品表面取向分布(IPF)图和(001)/(110)极图. 利用 Channel 5 软件对始态样品的晶界特征分布进行统计,低 Σ CSL 的比例为 53.2％,其中 Σ3 为 46.6％, Σ9+ Σ27 为 3.5％,其他低 Σ CSL 晶界比例为 3.1％. 利用等效圆直径法统计该样品的平均晶粒尺寸,如果将孪晶算为晶粒,那么晶粒平均尺寸为 11.45 μm;如果不将孪晶算为晶粒,那么晶粒平均尺寸为 26.01 μm. 从(001)极图和(110)极图来看,材料有弱的(110)[001]织构,15°偏差范围内的织构含量为 9.8％.

经过不同工艺的形变及退火处理得到样品 S1～S6,并进行 EBSD 测定. 图 2(a)～(f)分

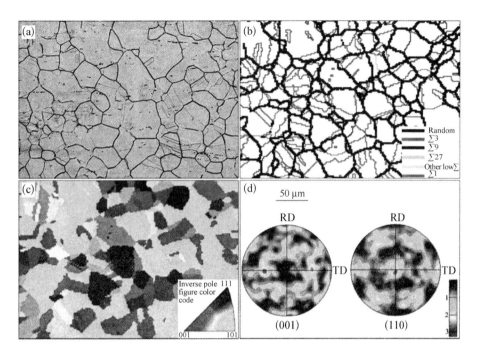

图 1 白铜 B10 合金始态样品的金相照片、不同类型晶界图、表面取向分布(IPF)图和 (001)/(110)极图

Fig. 1 Optical metallographs of starling-state specimen of cupronickel BIO alloy (a), OIM map of different types of gain boundaries (b), surface orientation distribution (IPF) map (c), (001) and (110) pole figures (d) (In Fig. 1(b), black lines denote random boundaries, red lines denote $\Sigma 3$ boundaries, blue lines denote $\Sigma 9$ boundaries, green lines denote $\Sigma 27$ boundaries, yellow lines denote other low Σ boundaries, and gray lines denote $\Sigma 1$ boundaries)

别为样品 S1～S6 的不同类型晶界图. 由图 2 可知,样品 S1～S6 的晶界网络有明显差异,样品 S2～S5 含有较高比例的孪晶界,构成大尺寸的晶粒团簇,尤其是样品 S3,而样品 S1 和 S6 的孪晶界含量明显较低.

出现高比例的孪晶及其相关界面和大尺寸晶粒团簇是 GBE 处理后显微组织的重要特征[15-16],图 2(c)中的阴影区域 M 就是一个晶粒团簇. 团簇内部的晶界基本上全是 $\Sigma 3$、$\Sigma 9$ 和 $\Sigma 27$ 类型晶界,边界全是随机晶界. 对晶粒团簇 M 进行分析,如图 3 所示. 该晶粒团簇的等效圆直径约为 $300\ \mu m$,包含 88 个晶粒,从中随机选取 8 个较大晶粒,对它们互相之间的取向关系进行分析[17],如表 3 所列. 由此可以得出,晶粒团簇内任意两个晶粒之间互有 $\Sigma 3^n$ 取向关系,无论它们是否相邻,这种互有 $\Sigma 3^n (n=1,2,3,\cdots)$ 取向关系的晶粒在团簇内部构成了大量的 $\Sigma 3^n$ 类型的三叉界角(Triple junction)[15],如 $\Sigma 3 - \Sigma 3 - \Sigma 9$ 和 $\Sigma 3 - \Sigma 9 - \Sigma 27$ 等. 这种大尺寸的晶粒团簇显微组织是 GBE 处理提高材料的耐腐蚀性能的原因[18]. 下面对样品 S1 - S6 的晶界网络特征进行统计,包括晶界特征分布、平均晶粒尺寸和晶粒团簇平均尺寸.

图 4(a)所示为样品 S1～S6 不同类型晶界比例. 可见,S2～S5 样品中形成了高比例的低 Σ CSL 晶界,明显高于始态样品的低 Σ CSL 晶界比例,尤其是样品 S3 的低 Σ CSL 晶界比例接近 80%,达到了明显的 GBE 处理效果;而样品 S1 和 S6 的低 Σ CSL 晶界比例与始态样品

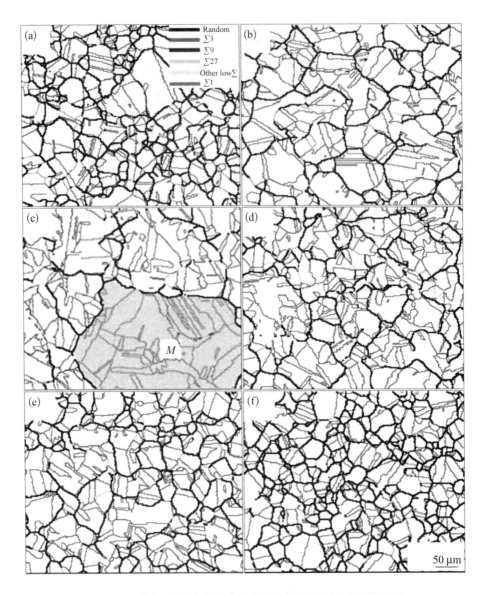

图 2　经不同工艺形变及退火处理后各样品的不同类型晶界图

Fig. 2　Grain boundary networks of specimens after thermal-mechanical treatments (Black lines denote random boundaries，red lines denote $\Sigma 3$ boundaries，blue lines denote $\Sigma 9$ boundaries，green lines denote $\Sigma 27$ boundaries，yellow lines denote other low Σ boundaries and gray lines denote $\Sigma 1$ boundaries)：(a) S1；(b) S2；(c) S3；(d) S4；(e) S5；(f) S6

基本相当，没有达到 GBE 处理的效果. 图 4(b)所示为各样品的平均晶粒尺寸和晶粒团簇平均尺寸，随 GBE 处理过程中的形变量变化趋势与低 Σ CSL 晶界比例的变化趋势相似. 与始态样品相比，晶粒尺寸和晶粒团簇尺寸都有显著增加，尤其是晶粒团簇尺寸，其中样品 S3 中的晶粒团簇尺寸平均约为 80 μm.

　　GBE 处理后样品的晶界网络中，含有的大尺寸晶粒团簇是通过再结晶过程形成的[15]，退火时再结晶晶核长大过程中发生多重孪晶(Multiple twinning)[19]，依次形成了一代孪晶、二代孪晶、三代孪晶和更高代次的孪晶，构成孪晶链(Twin chain)[15]，从而构成一个晶粒团簇. 整个晶粒团簇由一个再结晶晶核通过多重孪晶形成,从晶粒团簇内部的晶粒产生过程可以看出,它们都与初始的再结晶晶核符合 $\Sigma 3^n$ 的取向关系,从而晶粒团簇内部任意两个晶粒之间都具有 $\Sigma 3^n$ 的取向关系.

晶粒团簇长大过程中发生多重孪晶,形成 $\Sigma 3^n (n=1,2,3,\cdots)$ 类型晶界,这是 GBE 处理后产生高比例低 Σ CSL 晶界的主要原因. 因此,形成大尺寸的"互有 $\Sigma 3^n (n=1,2,3,\cdots)$ 取向关系晶粒的团簇"后,样品的低 Σ CSL 晶界比例才会明显提高[16]. GBE 处理过程中,形变量越大的样品在随后的退火过

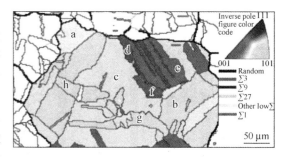

图 3 图 2(c)中晶粒团簇 M 中的晶粒取向在反极图(IPF)中的分布

Fig. 3 Orientations distributions of grains in grain-cluster M described by inverse pole figure (IPF) color code in Fig. 2(c)

表 3 图 3 所示晶粒团簇 M 内随机选取的 8 个晶粒之间的取向关系

Table 3 Misorientations of eight randomly selected grains within grain-cluster M in Fig. 3

Grain	Misorientation							
	a	b	c	d	e	f	g	h
a		3/0.1°	9/0.3°	81b/0.2°	27b/0.1°	27a/0.5°	27b/0.1°	9/0.1°
b	59.9°,$[11\bar{1}]$		3/0.3°	27b/0.1°	9/0.1°	9/0.5°	9/0.2°	3/0.4°
c	38.6°,$[011]$	59.7°,$[\bar{1}11]$		9/0.7°	3/0.2°	3/0.3°	3/0.7°	9/0.1°
d	54.3°,$[322]$	35.5°,$[\bar{1}\bar{2}0]$	39.6°,$[\bar{1}01]$		3/0.1°	27a/0.1°	27b/0.4°	81a/0.4°
e	35.5°,$[02\bar{1}]$	38.8°,$[0\bar{1}1]$	59.8°,$[\bar{1}11]$	59.9°,$[\bar{1}11]$		9/0.1°	9/0.6°	27b/0.3°
f	31.1°,$[101]$	39.4°,$[0\bar{1}1]$	59.7°,$[1\bar{1}1]$	31.7°,$[0\bar{1}1]$	38.8°,$[01\bar{1}]$		9/0.3°	27b/0.3°
g	35.5°,$[0\bar{2}1]$	39.1°,$[\bar{1}01]$	59.3°,$[\bar{1}11]$	35.8°,$[02\bar{1}]$	38.3°,$[011]$	39.2°,$[01\bar{1}]$		27a/0°
h	39.0°,$[011]$	59.6°,$[1\bar{1}1]$	38.8°,$[01\bar{1}]$	39.3°,$[\bar{4}11]$	35.1°,$[0\bar{2}1]$	35.7°,$[\bar{2}01]$	31.6°,$[101]$	

Note: θ denotes misorientation angle; $[hkl]$ denotes miller index of misorientation rotation axis; Σ denotes reciprocal density of coinciding sites; $\triangle\theta$ denotes deviation of experimentally measured misorientation from exact CSL misorientation; a and b from 27a, 27b and 81a, 81b mean two kinds of CSL with same lattice density and different misorientation.

程中再结晶形核密度就越高,可供晶核长大的潜在空间就越小,晶粒团簇尺寸就越小,不利于形成高比例的低 Σ CSL 晶界. 因此,随 GBE 处理过程中的形变量增大,处理后样品 S3～S6 的低 Σ CSL 晶界比例降低. 而样品 S1 和 S2 的低 Σ CSL 晶界比例反而比样品 S3 的低,原因是 3%与 5%的变形量太小,没有产生足够的形变储能,在随后的退火过程中没有发生再结晶或者没有完全再结晶. 图 4(a)中样品S1～S3, $\Sigma 1$ 晶界(小角晶界,取向差 2°～15°之间)比例随着冷轧变形量的增加而降低,说明样品 S1 和 S2 没有发生或没有完成再结晶. 因此,GBE 处理过程中需要合适的形变量,才能在随后的退火过程中形成高比例的低 Σ CSL 晶界.

$\Sigma 9$ 和 $\Sigma 27$ 晶界是晶粒团簇长大过程中发生多重孪晶现象时[19]形成的,因此, $\Sigma 9$ 和 $\Sigma 27$ 晶界比例之和,与 $\Sigma 3$ 晶界比例变化趋势相似,如图 4(a)所示. 低 Σ CSL 晶界中基本上全是 $\Sigma 3^n$ 晶界,其中, $\Sigma 3$ 晶界占绝大多数,因此,总体低 Σ CSL 晶界比例也与样品的 $\Sigma 3$ 晶界比例变化趋势相同,在样品 S3(冷轧变形量 7%)处出现最大值,达到 76.81%.

从图 4(b)中可以看出,平均晶粒尺寸和晶粒团簇平均尺寸同样随着冷轧压下量从 3%逐渐增加,先增大,后减小,在样品 S3(冷轧压下量 7%)处出现最大值,分别为 16.20 μm 和

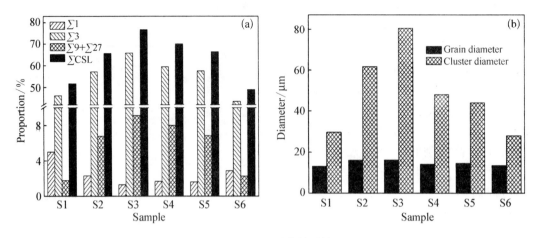

图4 晶界特征分布统计结果

Fig. 4 Statistics of grain boundary character distribution：（a）Proportion of different types of grain boundaries；（b）Mean size of grains and grain-clusters in each sample

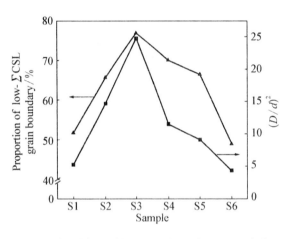

图5 各个样品的低 Σ CSL 晶界比例和 $(D/d)^2$ 值

Fig. 5 Proportion of low- Σ CSL boundary and value of $(D/d)^2$ of each sample

80.49 μm. 这说明，7% 冷轧变形是本实验条件下 GBE 处理时的合适变形量. 样品的低 Σ CSL 晶界比例与样品的晶粒团簇平均尺寸 D 和平均晶粒尺寸 d 之比的平方 $(D/d)^2$ 成正相关关系[16]，如图 5 所示. 因此，虽然平均晶粒尺寸和晶粒团簇平均尺寸的变化趋势相同，均在 S3 处达到最大值，但是平均晶粒尺寸变化幅度比晶粒团簇平均尺寸的变化幅度小得多，样品 S3 的晶粒团簇平均尺寸明显比其他样品的大，使得样品 S3 具有最高比例的低 Σ CSL 晶界和最大的晶粒团簇平均尺寸，形变量过大和过小都不利于获得好的 GBE 处理效果.

3 结论

（1）固溶处理后的白铜 B10 合金，经过冷轧变形 7% 和 800 ℃ 保温 10 min 退火处理，可以将低 Σ CSL 晶界比例提高到 75% 以上，同时形成大尺寸的"互有 $\Sigma 3^n$ 取向关系晶粒的团簇"，并获得了最佳的"形变-热处理"工艺，为提高白铜 B10 合金的耐蚀性提供可能的处理方法.

（2）当冷轧变形量小于 7% 时，因为没有足够的形变储能，样品经 800 ℃ 退火后没有完全再结晶. 当冷轧变形量大于 7% 时，样品在随后的退火过程中能够发生再结晶，随着形变量增加，再结晶形核密度增加，可供晶核长大的潜在空间就越小，因而，冷轧变形量大于 7% 时，样品退火后的平均晶粒团簇尺寸和 $\Sigma 3^n$ 晶界比例都会随着退火前变形量的增加而明显降低,致使低 Σ CSL 晶界比例也明显降低.

参考文献

[1] KRONBERG M L, WILSON F H. Secondary recrystallization in copper[J]. Trans AIME, 1949, 185(1): 501 - 514.

[2] KURBAN M, ERB U, AUST K T. A grain boundary characterization study of boron segregation and carbide precipitation in alloy 304 austenitic stainless steel[J]. Scr Mater. 2006, 54(6): 1053 - 1058.

[3] LIN P, PALUMBO G ERB U, AUST K T. Influence of grain boundary character distribution on sensitization and intergranular corrosion of alloy 600[J]. Scr Metall Mater, 1995, 33(9): 1387 - 1392.

[4] WATANABE T. TSUREKAWA S, KOBAYASHI S, YAMAURA S I. Structure-dependent grain boundary deformation and fracture at high temperatures[J]. Mater Sci Eng A, 2005, 420/411: 140 - 147.

[5] ALEXANDREANU B. WAS G S. The role of stress in the efficacy of coincident site lattice boundaries in improving creep and stress corrosion cracking [J]. Scr Mater, 2006, 54(6): 1047 - 1052.

[6] WATANABE T. Approach to grain boundary design for strong and ductile polycrystals[J]. Res Mechanica: International Journal of Structural Mechanics and Materials Science, 1984, 11(1): 47 - 84.

[7] PALUMBO G, ERB U. Enhancing the operating life and performance of lead-acid batteries via grain-boundary engineering[J]. MRS Bulletin, 1999, 24(11): 27 - 32.

[8] WEST E A, WAS G S. IGSCC of grain boundary engineered 316L and 690 in supercritical water[J]. Journal of Nuclear Materials, 2009, 392(2): 264 - 271.

[9] RANDLE V, OWEN G. Mechanisms of grain boundary engineering[J]. Acta Materialia, 2006, 54(7): 1777 - 1783.

[10]《重有色金属材料加工手册》编写组. 重有色金属材料加工手册第 1 分册[M]. 北京: 冶金工业出版社, 1979.
Edition Group of Heavy Non-ferrous Metal Materials Processing Manual. Book one of heavy non-ferrous metal materials processing manual[M]. Beijing: Metallurgical Industry Press, 1979.

[11] 李文生, 王智平, 路阳, 徐建林. 杨德寿. 高强铜基合金材料的研究与应用现状[J]. 有色金属. 2002, 54(2): 30 - 34.
LI Wen-sheng. WANG Zhi-ping, LU Yang, XU Jian-lin, YANG De-shou. Review of high-strength copper-based alloy on application and research[J]. Nonferrous Metals, 2002, 54(2): 30 - 34.

[12] 陈平, 付亚波. 铜镍合金冷凝管产品市场分析[J]. 上海有色金属. 2007, 28(4): 191 - 195.
CHEN Ping, FU Ya-bo. Market analysis of condensation tubes of Cu-Ni alloy[J]. Shanghai Nonferrous Metals, 2007, 28(4): 191 - 195.

[13] 邓楚平, 黄伯云. 李卫. 潘志勇, 李宏英. 不同服役条件下冷凝器白铜管的腐蚀特性[J]. 中国有色金属学报, 2005, 15(11): 1692 - 1698.
DENG Chu-ping, HUANG Bai-yun. LI Wei. PAN Zhi-yong, LI Hong-ying. Corrosion characteristics of white copper condenser tubes under different serving conditions[J]. The Chinese Journal of Nonferrous Metals. 2005, 15(11): 1692 - 1698.

[14] LEHOCKEYEM, BRENNENSTUHL A M, THOMPSON I. On the relationship between grain boundary connectivity, coincident site lattice boundaries and intergranular stress corrosion cracking [J]. Corrosion Science, 2004, 46(10): 2383 - 2404.

[15] XIA Shuang, ZHOU Bang-xin, CHEN Wen-jue. Grain cluster microstructure and grain boundary character distribution in alloy 690[J]. Metallurgical and Materials Transactions A. 2009, 40(12): 3016 - 3030.

[16] 刘廷光, 夏爽, 李慧, 周邦新, 陈文觉. 690 合金原始晶粒尺寸对晶界工程处理后晶界网络的影响[J]. 金属学报, 2011, 47(7): 859 - 864.
LIU Ting-guang, XIA Shuang, LI Hui, ZHOU Bang-xin, CHEN Wen-jue. Effect of original grain size on the boundary network in alloy 690 treated by grain boundary engineering[J]. Acta Metallurgica Sinica, 2011, 47(7): 859 - 864.

[17] HUMPHREYS F J, BATE P S, HURLEY P J. Orientation averaging of electron backscattered diffraction data[J]. Journal of Microscopy, 2001, 201(1): 50 - 58.

[18] XIA Shuang, LI Hui, LIU Ting-guang, ZHOU Bang-xin. Appling grain boundary engineering to Alloy 690 tube for enhancing intergranular corrosion resistance[J]. Journal of Nuclear Materials, 2011,416(3): 303 – 310.

[19] GERTSMAN V Y, HENAGER C H. Grain boundary junctions in microstructure generated by multiple twinning [J]. Interface Science, 2003, 11(4): 403 – 415.

Effect of deformation and heat-treatment on grain boundary distribution character of cupronickel B10 alloy

Abstract: The effects of cold rolling deformation and annealing on the grain boundary character distribution (GBCD) during grain boundary engineering (GBE) treatment were investigated by electron backscatter diffraction (EBSD) in cupronickel B10 alloy. The results show that the proportion of low-Σ CSL (Coincidence site lattice, $\Sigma \leqslant 29$) grain boundaries increase to more than 75% by 7% cold rolling and subsequent annealing at 800 ℃. In this case, the grain boundary network (GBN) is featured by the formation of highly twinned large size grain-clusters produced by multiple twinning during recrystallization. When the cold rolling deformation amount is less than 7%, the 800 ℃ annealing can not induce perfect recrystallization. The perfect recrystallization occurs when the deformation amount is more than 7%, and the proportion of low-Σ CSL grain boundaries and the average size of grain-clusters decrease with the increase of the cold rolling reduction ratio.

利用晶界工程技术优化 H68 黄铜中的晶界网络*

摘　要：利用电子背散射衍射(EBSD)和取向成像(OIM)技术研究了形变量及退火时间对 H68 黄铜晶界网络的影响. 结果显示,形变量对处理后样品的晶界特征分布及晶粒尺寸和晶粒团簇尺寸都有显著影响,而退火时间(10 min~3 h)所产生的影响不明显;其中经 5% 冷轧及在 550 ℃ 下退火不同时间都能够显著提高 H68 黄铜的低 $\sum$ CSL 晶界比例到 80% 以上,晶界网络中形成了大尺寸的互有 $\sum 3^n(n= 1,2,3\cdots\cdots)$ 取向关系晶粒的团簇.

多晶体材料中晶界原子排列不规则[1],使晶界成为位错滑移障碍和快速扩散通道,对材料的力学性能、腐蚀性能和物理性能都有显著影响. 通过控制晶界的含量进行细晶强化,或制备纳米晶材料和单晶材料,能够改变材料性能;通过控制晶界结构也能改变材料性能. 1984 年 Watanabe 提出"晶界设计与控制"的构想[2],继而在 90 年代形成"晶界工程"(Grain Boundary Engineering, GBE)研究领域[1]. 主要是在中低层错能面心立方金属中,通过合适的形变和热处理工艺提高材料中特殊结构晶界(一般指 $\sum \leqslant 29$ 的低 $\sum$ CSL 晶界,CSL 是重位点阵 coincidence site lattice 的缩写)的比例,从而调整多晶体晶界网络(grain boundary network),能够显著改善材料与晶界有关的性能,比如抗晶界偏聚、抗晶间腐蚀[3-4]、抗晶间应力腐蚀开裂[5]、抗蠕变等性能.

20 世纪 90 年代中后期,GBE 技术在镍基 600 合金[3]和铅酸电池电极板[6]中的应用促进了 GBE 领域的研究,继而在镍基合金[7-8]、铜合金[9-11]、铅合金[12]和奥氏体不锈钢[4,13-14]等材料中进行了 GBE 技术研究,低 $\sum$ CSL 晶界比例被提高到 70% 以上. 并且对特殊结构晶界的特征与性质、GBE 技术的工艺方法、GBE 处理过程中晶界网络的演化和 GBE 技术改善材料性能的机理进行了大量研究. 常用的 GBE 处理工艺有单次应变-退火和多次循环应变-退火.

GBE 处理后形成大量的以 $\sum 3$ (孪晶界)为主的低 $\sum$ CSL 晶界,Kumar 等[7,9-10]认为,这种晶界网络中的随机晶界连通性被打断,从而抑制了沿随机晶界扩展的晶间腐蚀,是 GBE 处理提高材料耐晶间腐蚀的原因;本课题组研究结果表明[15-16],采用 Palumbo-Aust 判据定义晶界类型[17]的晶界网络中,随机晶界连通性并没有被打断,而是形成了大尺寸"互有 $\sum 3^n$ 取向关系晶粒的团簇"显微组织(以下简称晶粒团簇),是 GBE 处理提高材料耐晶间腐蚀能力的原因. 晶粒团簇[17]是一个孪晶相关区域,团簇内所有晶粒能够通过孪晶链相连,区域内的所有晶界有 $\sum 3^n$ 取向差,而外围晶界一般为随机晶界.

H68 型黄铜为单相 α 相,具有良好的力学性能和可加工性,是黄铜中应用最为广泛的一个品种,但在氨气环境中易产生腐蚀开裂. H68 黄铜属于低层错能面心立方结构,GBE 技术有潜力应用在 H68 黄铜中,从而在不改变 H68 黄铜化学成分的基础上改善其耐晶间腐蚀性

---

* 本文合作者：杨辉辉、刘廷光、夏爽、李慧、白琴. 原发表于《上海金属》,2013,35(5):9－13,62.

能.本工作研究了经不同冷轧变形量及不同工艺退火处理后 H68 黄铜的晶界网络特征,探索适合 H68 黄铜的 GBE 处理方法.

1 实验

本实验用 H68 型黄铜的化学成分(质量分数,%)为:Cu 68.36,Zn 31.64,为单相 α-黄铜.首先对 H68 黄铜原料进行 38% 冷轧变形及在 600 ℃ 下真空退火处理 15 min,获得组织均匀的材料,作为本实验用的始态样品 SS,厚度为 0.5 mm. 对始态样品进行不同压下量冷轧变形:3%,5%,7%,9%,12% 和 16%,再在 550 ℃ 下分别真空退火 10 min、1 h 和 3 h,从而研究形变量及退火工艺对 H68 黄铜晶界网络的影响,得到适合于 H68 黄铜的 GBE 处理方法.处理后的样品编号形式为:SSx-time,x 为压下量(%),time 为 550 ℃ 下真空退火时间,如 SS3-10 min 就是 3% 形变量在 550 ℃ 退火 10 min 的样品编号.

把样品制备出适合进行电子背散射衍射(EBSD)测试的表面.首先对加工成合适形状的样品在金相砂纸上机械磨光(1♯→3♯→5♯),再进行电解抛光,电解液为 $25\% H_3PO_4 + 25\% C_2H_5OH + 50\% H_2O$(体积分数),在室温下用 13 V 直流电抛光约 2 min.利用配备在热场发射枪扫描电子显微镜(CamScan Apollo 300)上的 EBSD 仪器(Oxford Instrument/HKL-Channel 5)对样品表面微区进行取向信息采集.扫描步长为 2 μm,扫描区域为 500 μm× 400 μm.统计晶界特征分布时,采用 Palumbo-Aust 标准判定晶界类型[17].

2 结果与讨论

2.1 始态样品显微组织

始态样品 SS 经 EBSD 测试后,利用 Channel 5 软件分析得到的取向成像图如图 1 所示,利用标准反极图颜色码表示不同晶粒取向.可以看出,始态样品没有显著织构,含有一定数量的孪晶,主要是角孪晶(grain-corner twin),其次是半平行孪晶(incomplete parallel-sided twin)和完全平行孪晶(complete parallel-sided twin),也有孪晶环(occluded twin)[19],晶粒团簇内的孪晶个数比较少,孪晶链比较简单.利用 Channel 5 软件分析样品的平均晶粒尺寸为 18.2 μm,本文中统计晶粒尺寸时包含孪晶;另外,根据晶粒团簇的定义-孪晶相关区域,区域都是以孪晶界为主的 $\sum 3^n$ 类型晶界,因此忽略孪晶界及其相关晶界($\sum 9$ 和 $\sum 27$)后的尺寸为晶粒团簇平均尺寸,求出样品 SS 的晶粒团簇平均尺寸为 45.0 μm.样品的晶界特征分布为:低 $\sum$ CSL 晶界比例为 60.5%,$\sum 3$ 晶界56.5%,$\sum 9$ 晶界

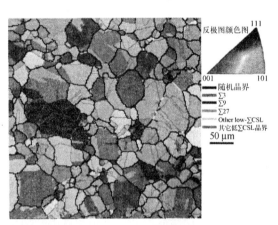

反极图颜色图 111

001   101

随机晶界
$\sum 3$
$\sum 9$
$\sum 27$
Other low-$\sum$CSL
其它低$\sum$CSL晶界

50 μm

图 1 利用 EBSD - OIM 技术测得的始态样品 SS 的显微组织

Fig. 1 OIM map of starting state sample SS characterized by EBSD

2.5%,$\sum 27$ 晶界 0.5%.统计晶界特征分布时均为长度比例,包含取向差大于 $2°$ 的所有晶界,其中取向差在 $2°{\sim}15°$ 之间的为 $\sum 1$ 晶界.可以看出,经一般冷轧退火处理后的 H68 黄铜中含有较多的退火孪晶,这是低层错能面心立方金属材料的一般特征.

2.2 不同工艺 GBE 处理后的晶界网络

始态显微组织如图 1 所示的 H68 黄铜,经不同压下量冷轧及 550 ℃下不同时间真空退火处理后的显微组织如图 2 所示,包含不同类型晶界分布.可以看出,经不同处理工艺得到的样品的晶粒尺寸及孪晶界含量都有显著差异,样品 SS3－1 h(3%形变量 550 ℃退火 1 h)和样品 SS3－3 h 的晶粒尺寸明显较大,样品 SS3－1 h,SS3－3 h,SS5－10 min,SS5－1 h、SS5－3 h 和 SS7－3 h 的孪晶界比例明显较高,随机晶界含量较少,构成大尺寸"互有 $\sum 3^n$取向关系晶粒的团簇"显微组织,图中的阴影区域是晶粒团簇显微组织,团簇内部都是孪晶相关晶界($\sum 3^n$),主要是孪晶界($\sum 3$).下面从晶界特征分布、晶粒尺寸和晶粒团簇尺寸三个方面,对这些样品的晶界网络进行比较分析.

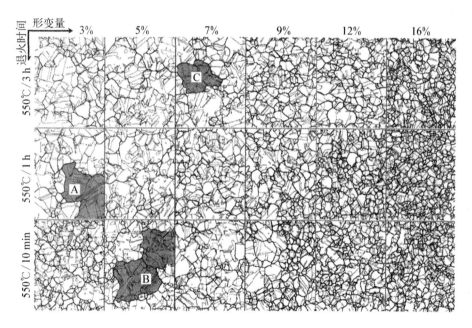

图 2 始态样品 SS 经不同工艺"冷轧-退火"处理后的晶界网络,包含不同类型晶界分布
Fig. 2 Grain boundary networks with grain boundary types of the samples after different "cold rolling annealing" processing on starting state sample SS

图 3 是始态样品 SS 及经不同工艺"形变-退火"处理后样品的低 $\sum$ CSL 晶界比例.可以看出,始态样品 SS 经不同工艺处理后的低 $\sum$ CSL 晶界比例有很大差异.首先,形变量对处理后样品的低 $\sum$ CSL 晶界比例的影响,经 5%冷轧及 550 ℃下不同时间退火后的低 $\sum$ CSL晶界比例最高,都超过 80%,随着冷轧压下量的进一步增加,退火后的低 $\sum$ CSL晶界比例逐步降低;经 3%冷轧及 550 ℃下不同时间退火后的低 $\sum$ CSL 晶界比例稍低于

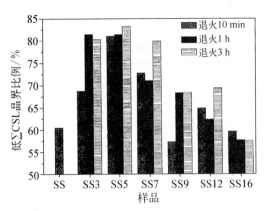

图 3 始态样品 SS 及经不同工艺处理后的低 $\sum$ CSL 晶界比例

Fig. 3 Low-$\sum$ CSL boundary proportions of samples after different processing on starting state sample SS

5% 冷轧及退火后的比例, 尤其是短时间 (10 min) 退火后的样品. 其次, 退火时间对退火后样品的低 $\sum$ CSL 晶界比例的影响, 经 3% 和 9% 冷轧及 10 min 退火处理后样品的低 $\sum$ CSL 晶界比例明显低于更长时间退火处理后样品的比例, 经 7% 冷轧及 3 h 退火处理后样品的低 $\sum$ CSL 晶界比例明显高于短时间退火处理后样品的比例, 经 5% 冷轧及不同时间退火处理后样品的低 $\sum$ CSL 晶界比例基本相同, 其他变形量样品的退火时间对退火后的低 $\sum$ CSL 晶界比例也没有很大影响. 因此, 对于如图 1 所示显微组织的 H68 样品, 5% 冷轧变形及 550 ℃ 退火处理是比较理想的 GBE 处理方法, 能够把样品的低 $\sum$ CSL 晶界比例提高到 80% 以上,

且受退火时间的影响较小; 而冷轧变形量在 3%～7% 范围内, 都能够通过合适的退火工艺把样品的低 $\sum$ CSL 晶界比例提高到约 80%.

样品 SS5‑3 h 和 SS16‑10 min 的晶界特征分布如图 4 所示, 无论低 $\sum$ CSL 晶界比例较高的样品 SS5‑3 h, 还是比例较低样品 SS16‑10 min, 其低 $\sum$ CSL 晶界中都主要是 $\sum$ 3 晶界, 占低 $\sum$ CSL 晶界的 85% 左右, 其次是孪晶相关晶界 $\sum$ 9 和 $\sum$ 27, 还有少量的 $\sum$ 1 晶界, 其他低 $\sum$ CSL 晶界含量很少, 占总晶界长度的比例都不足 1%. 这种晶界特征分布与 $\sum$ 3^n 晶界的"多重孪晶"形成机制有关. 多重孪晶过程中, 新形成的晶界多为 $\sum$ 3 晶界, 孪晶相遇形成其他类型的 $\sum$ CSL 晶界, 从而影响了晶界的特征分布.

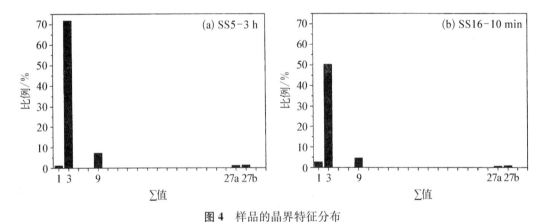

图 4 样品的晶界特征分布

Fig. 4 Grain boundary character distributions of sample

始态样品 SS 及经不同工艺"形变‑退火"处理后样品的晶粒尺寸如图 5(a) 所示. 冷轧压下量在 3% 至 16% 范围内, 随着形变量的增加, 550 ℃ 下不同时间退火后的晶粒尺寸都逐渐

降低,退火时间对晶粒尺寸的影响较小,只有冷轧 3%-退火 550 ℃/10 min 的样品的晶粒尺寸明显小于更长时间退火的样品,这一原因在后文有进一步叙述.与始态样品相比,经 3%～9%冷轧变形及退火后样品的晶粒尺寸都明显变大,除了样品 SS3 - 10 min,尤其是低 $\sum$ CSL晶界比例较高的样品 SS3 - 1 h,SS3 - 3 h,SS5 - 10 min,SS5 - 1 h,SS5 - 3 h 和 SS7 - 3 h,其晶粒尺寸比原始晶粒尺寸明显变大.因此,GBE 处理获得了高比例的低 $\sum$ CSL 晶界,虽然晶粒尺寸明显变大,但这可以通过控制始态样品的晶粒尺寸来调整.

从图 2 可以看出,具有较好的 GBE 处理效果样品的显著特征是,其晶界网络中含有高比例的低 $\sum$ CSL 晶界,随机晶界比例小于 30%;另一个显著特征是,这些少量的随机晶界的分布并不是随机的,而是彼此相连构成晶粒团簇显微组织,如图 2 中的灰色背景区域是选取的三个晶粒团簇.始态样品 SS 及经不同工艺"形变-退火"处理后样品的晶粒团簇平均尺寸如 5(b)所示,可见,具有高比例低 $\sum$ CSL 晶界的样品 SS3 - 1 h,SS3 - 3 h,SS5 - 10 min、SS5 - 1 h,SS5 - 3 h 和 SS7 - 3 h 中都形成了大尺寸的晶粒团簇,平均尺寸超过 $100~\mu m$,与始态样品的晶粒团簇尺寸相比明显变大,而低 $\sum$ CSL 晶界比例较低样品的晶粒团簇尺寸较小.由于晶间腐蚀主要沿随机晶界扩展,形成大尺寸晶粒团簇是 GBE 处理提高材料耐腐蚀能力的关键.这种区别于一般晶界网络的"大尺寸晶粒团簇结构"的形成,反映了 GBE 处理过程中必然有区别于一般再结晶过程的晶界网络演化规律,这是 GBE 研究领域的重要内容。

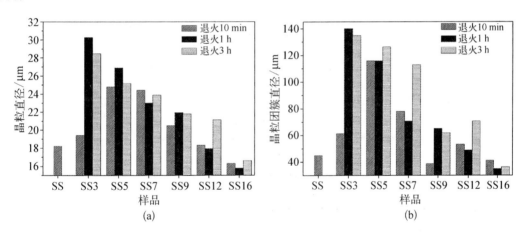

图 5 始态样品 SS 及经不同工艺处理后的平均晶粒尺寸(a)及平均晶粒团簇尺寸(b)

Fig. 5 The mean grain sizes (a) of the samples and the mean sizes of grain - cluster (b) after different processing on the starting state sample SS

GBE 处理过程中,由于采用小形变量变形,退火时再结晶形核密度较低,晶核有足够的潜在空间长大,最终形成大尺寸"互有 $\Sigma 3^n$ 关系晶粒的团簇"显微组织,而晶粒尺寸大小主要受再结晶晶界迁移时的孪晶形成几率及迁移速率影响。再结晶前沿晶界迁移时,孪晶形成几率与形变量成正比,形变量越大,再结晶前沿晶界的晶格不规则程度越高,孪晶形成几率越高,因此,在一定范围内(本实验中 5%～16%),形变量越小,退火后的晶粒尺寸越大;另外,晶界迁移速率也与形变储能成正比,形变量越小时晶界迁移速率越慢,当形变储能足够小时,以至于晶界迁移过程受阻,需要通过形成孪晶改变前沿晶界取向差,形成迁移性较高的晶界,因此,当形变量很小时,退火后的晶粒尺寸未必很大,比如样品 SS3 – 10min 的晶粒尺寸就小于样品 SS5 – 10min 的。

3 结论

H68 黄铜经不同冷轧变形量及在 550 ℃下退火不同时间后的晶界网络显示,冷轧变形量对 GBE 处理后样品的晶界网络有显著影响,而退火温度的影响较小;经 3% ～7% 冷轧及合适时间退火后的样品中形成了高比例的低 ΣCSL 晶界,构成大尺寸"互有 Σ3^n 取向关系晶粒的团簇"显微组织,尤其是 5% 冷轧及退火后的样品,实现了很好的 GBE 处理效果。

参 考 文 献

[1] Randle V. The role of the coincidence site lattice in grain boundary engineering[M]. Cambridge:Cambridge University Press,1996.

[2] Watanabe T. Approach to grain boundary design for strong and ductile polycrystals[J]. Res Mechanica,1984,11:47 – 84.

[3] Lin P,Palumbo G,Erb U,et al. Influence of grain boundary character distribution on sensitization and intergranular corrosion of alloy 600[J]. Scripta Metall. Mater. 1995,33(9):1387 – 1392.

[4] Shimada M,Kokawa H,Wang Z J,et al. Optimization of grain boundary character distribution for intergranular corrosionresistant 304 stainless steel by twin-induced grain boundary engineering[J]. Acta Mater. 2002,50(9):2331 – 2341.

[5] Aust KT,Erb U,Palumbo G. Interface Control for Resistance to Intergranular Cracking[J]. Mater. Sci. Eng. A,1994, 176(1 – 2):329 – 334.

[6] Lehockey E M,Limoges D,Palumbo G,et al. On improving the corrosion and growth resistance of positive Pb-acid battery grids by grain boundary engineering[J]. J. of Power Sources,1999,78(1 – 2):79 – 83.

[7] Kumar M,King W E,Schwartz A J. Modifications to the microstructural topology in fce materials through thermomechanical processing[J]. Acta Mater. 2000,48(9):2081 – 2091.

[8] Tan L,Sridharan K,Allen T R,et al. Microstructure tailoring for property improvements by grain boundary engineering[J]. J. Nucl. Mater. 2008,374(1 – 2):270 – 280.

[9] Randle V. Mechanism of twinning – induced grain boundary engineering in low stacking – fault energy materials[J]. Acta Mater. 1999(15 – 16),47:4187 – 4196.

[10] Randle V,Coleman M. A study of low-strain and medium-strain grain boundary engineering[J]. Acta Mater. 2009, 57(11):3410 – 3421.

[11] 姜英,王卫国,郭红. 同步改善黄铜 H68 晶界腐蚀行为和力学性能[J]. 中国有色金属学报,2011,21(2):377 – 383.

[12] 夏爽,周邦新,陈文觉,等. 高温退火过程中铅合金晶界特征分布的演化[J]. 金属学报,2006,42:129 – 133.

[13] Engelberg D L,Humphreys F J,Marrow T J. The influence of low-strain thermo-mechanical processing on grain boundary network characteristics in type 304 austenitic stainless steel [J]. J. Microsc. 2008,230(3):435 – 444.

[14] Michiuchi M,Kokawa H,Wang Z J,et al. Twin-induced grain boundary engineering for 316 austenitic stainless steel[J]. Acta Mater. 2006,54(19):5179 – 5184.

[15] Xia S,Zhou B X,Chen W J,et al. Effects of strain and annealing processes on the distribution of $\sum$ 3 boundaries in a Ni-based superalloy [J]. Scr. Mater. 2006,54(12):2019 – 2022.

[16] Xia S,Zhou B X,Chen W J. Grain cluster microstructure and grain boundary character distribution in alloy 690 [J], Metall. Mater. Trans. 2009,40A (12):3016 – 3030.

[17] Palumbo G,Aust K T. Structure-dependence of intergranular corrosion in high purity nickel [J]. Acta Metall. 1990,38(11):2343 – 2352.

[18] Reed B W, Kumar M. Mathematical methods for analyzing highly-twinned grain boundary networks [J]. Scr. Mater. 2006, 54(6): 1029 – 1033.

[19] Mahajan S, Pande C S, Imam M A, et al. Formation of annealing twins in f. c. c. crystals. Acta Mater. 1997, 45 (6): 2633 – 2638.

Optimizing the Grain Boundary Network of H68 Brass by Using Grain Boundary Engineering

Abstract: The electron backscatter diffraction (EBSD) and orientation imaging microscopy (OIM) technique was used to analyze the effects of deformation amount and annealing time on the grain boundary network in H68 brass. The result shown that the deformation amount had an important effect on the grain boundary character distribution, the grain sizes and the grain-cluster after annealing, but the annealing time in the range of 10 min and 3 h had little effect. 5% cold rolling reduction and annealing at 550 ℃ for 10 min ～3 h was a good grain boundary engineering (GBE) procedure. The low- $\sum$ CSL grain boundary proportion increased to more than 80%, and large grain-clusters ware formed in the samples. Any two grains in grain-cluster had $\sum 3^n$ mutual mis-orientation regardless of whether they were adjacent or not.

Zr−1Nb−0.7Sn−0.03Fe−xGe 合金在 360 ℃ LiOH 水溶液中耐腐蚀性能的研究*

摘　要： 对添加微量合金元素 Ge 的 Zr−1Nb−0.7Sn−0.03Fe−xGe (x=0,0.05,0.1,0.2,质量分数,%)合金在 360 ℃,18.6 MPa 和 0.01 mol/L LiOH 水溶液中进行静态高压釜腐蚀实验. 利用 TEM 和 SEM 研究了合金和氧化膜的显微组织. 结果表明：添加适量 Ge 可以显著提高 Zr−1Nb−0.7Sn−0.03Fe 合金在 360 ℃,18.6 MPa 和 0.01 mol/L LiOH 水溶液中的耐腐蚀性能；在 Zr−1Nb−0.7Sn−0.03Fe−xGe 合金中,除了存在 bcc 结构的 β−Nb 型第二相和四方结构的 Zr−Nb−Fe−Cr 第二相,还存在四方结构的 Zr−Nb−Fe−Cr−Ge 和四方结构的 Zr_3Ge 型第二相；这些第二相的氧化速率比 α−Zr 基体慢. 腐蚀 190 d 后,Zr−1Nb−0.7Sn−0.03Fe−0.1Ge 合金氧化膜中微裂纹较少,并且存在较多的 ZrO_2 柱状晶；添加 Ge 既可以有效延缓氧化膜中的缺陷形成微孔隙和微裂纹的过程,又可以延迟 ZrO_2 柱状晶向等轴晶的演化,因而可以提高合金的耐腐蚀性能.

　　锆合金是重要的压水堆用核燃料包壳材料,作为反应堆运行时的第一道安全屏障,锆合金包壳会与一回路中的高温高压水发生腐蚀反应,这是影响燃料组件寿命的主要因素,也是影响核电经济性的重要原因. 堆外高压釜腐蚀实验的结果表明,Zn−Sn−Nb 系锆合金在高温高压 LiOH 水溶液中的耐腐蚀性能明显优于 Zr−4 合金(Zr−1.2Sn−0.2Fe−0.1Cr)[1−6]. 随着燃料组件向高燃耗的方向发展,许多国家开发了 ZIRLO (Zr−1Nb−1Sn−0.1Fe)[1,2], E635(Zr−1.2Sn−1Nb−0.4Fe)[3],HANA−4 (Zr−1.5Nb−0.4Sn−0.2Fe−0.1Cr)[4,5], N18 (Zr−1Sn−0.35Nb−0.3Fe−0.1Cr)[6], N36(Zr−1Sn−1Nb−0.3Fe)[6] 和 NDA(Zr−1Sn−0.28Fe−0.16Cr−0.1Nb−0.01Ni)[7] 等 Zr−Sn−Nb 系新型锆合金来提高燃料包壳的耐腐蚀性能. 在 Zr−Sn−Nb 合金中,大体上又可以分为高 Nb(≥1%)和低 Nb(约 0.3%)两大类. 锆合金的耐腐蚀性能与化学成分、第二相性质、氧化膜组织结构演化等因素密切相关. 研究[6−10]表明：合金化是改善锆合金耐腐蚀性能的有效途径,添加适量的 Sn,Nb,Fe 和 Cr 等元素可以不同程度地提高锆合金的耐腐蚀性能. 大多数合金元素在 α−Zr 中的固溶度较低,多余的合金元素会以第二相的形式析出,目前还不清楚究竟是固溶在 α−Zr 中的合金元素还是第二相改善了锆合金的耐腐蚀性能,或是两者兼有. 近年来,选择添加元素来优化锆合金成分已经成为改善锆合金耐腐蚀性能的研究热点. 根据 Wagner 氧化膜成长理论[11]和 Hauffe 原子价规律[12],位于元素周期表第ⅣA 族的 Ge 可以增加锆合金氧化膜中的电子浓度,减少阴离子空位,从而抑制 O^{2-} 扩散,降低锆合金的腐蚀速率. 本课题组前期工作[13]研究了 Ge 对 Zr−4 和 Zr−0.7Sn−0.35Nb−0.3Fe 合金(低 Nb 的 Zr−Sn−Nb 合金)耐腐蚀性能的影响,在这 2 种合金中存在不同类型的含 Ge 第二相,腐蚀实验结果表明,Ge 可以显著改善这 2 种合金在 360 ℃,18.6 MPa 和 0.01 mol/L LiOH 水溶液中的耐腐蚀性能,当 Ge 含量达到 0.1%时,合金耐腐蚀性能最佳. 本课题组前期研究发现,在 Zr−0.7Sn−0.35Nb−

* 本文合作者：张金龙、谢兴飞、姚美意、彭剑超、梁雪. 原发表于《金属学报》,2013,(49)4：443−450.

0.3Fe. 合金中存在 Zr(Fe，Cr，Nb)₂ 型第二相[13]，在 Zr－1Nb－1Sn－0.4Fe 合金中则存在 Zr－Nb－Fe 和 β－Nb 型第二相[8]，可见，不同成分的 Zr－Sn－Nb 合金会在第二相的种类和成分等方面存在差异，所以 Ge 对不同成分 Zr－Sn－Nb 合金耐腐蚀性能的影响机理可能会有所不同. 因此，本工作研究添加不同含量 Ge 对 Zr－1Nb－0.7Sn－0.03Fe 合金(高 Nb 的 Zr－Sn－Nb 合金)在 360 ℃，18.6 MPa 和 0.01 mol/L LiOH 水溶液中耐腐蚀性能的影响，进一步认识 Ge 改善 Zr－Sn－Nb 合金耐腐蚀性能的机理，为开发具有我国自主知识产权的新型 Zr－Sn－Nb 合金提供科学依据.

1 实验方法

以 Zr－1Nb－0.7Sn－0.03Fe(质量分数，%，下同)合金为母合金，在其中添加不同含量 Ge，用非自耗真空电弧炉熔炼制成 Zr－1Nb－0.7Sn－0.03Fe－xGe(x=0.05，0.1，0.2)合金锭，为保证成分均匀，合金锭总共翻转熔炼 6 次. 合金锭经 700 ℃ 热压成条块状后，在 1 030 ℃ 真空加热 40 min 后空冷，再经过热轧(约 700 ℃)，1 030 ℃ 真空加热 40 min 后空冷，多次冷轧和 580 ℃ 中间退火，制成 20 mm×15 mm×0.7 mm 的腐蚀实验用片状样品，最后在 580 ℃ 加热 5 h 进行再结晶退火. 未添加 Ge 的 Zr－1Nb－0.7Sn－0.03Fe 合金经过相同工艺处理后制成对比样品. 将制备好的上述样品经过 30%H_2O+30%H_2SO_4+30%HNO_3+10%HF(体积分数)混合酸酸洗和去离子水清洗后，放入静态高压釜中，在 360 ℃，18.6 MPa 和 0.01 mol/L LiOH 水溶液中进行腐蚀实验. 腐蚀样品定期取出称重，腐蚀增重是 3—5 片平行样品的平均值.

利用 JEM－200CX 型透射电子显微镜(TEM)观察合金的显微组织；利用配置了 INCA 能谱仪(EDS)的 JEM－2010F 场发射高分辨透射电子显微镜(HRTEM)观察分析合金中第二相的形貌与成分，并通过选区电子衍射(SAD)确定第二相的晶体结构；利用自带的 Gatan Digital Micrograph 软件分析处理 TEM 像. 为了减少统计误差，每种第二相都选取 10 个以上粒子进行分析. 用常规的双喷电解抛光法制备 TEM 观察用薄试样，电解液选用 10%(体积分数)高氯酸酒精溶液.

利用 JSM－6700F 冷场发射扫描电子显微镜(SEM)观察腐蚀生成氧化膜的断口和内表面形貌，样品的制备方法如文献[14]中描述，在样品表面蒸镀了一层金属 Ir 来提高成像质量.

利用 HELIOS－600I 型聚焦离子束(focus ion beam，FIB)制备氧化膜截面的 TEM 观察用薄样品. 离子束分辨率为 5 nm，加速电压为 2—30 kV. 为了避免离子束损伤氧化膜，最初在氧化膜外表面沉积了一层厚度约为 1 μm 的金属 Pt.

2 实验结果与分析

2.1 腐蚀增重

图 1 是 Zr－1Nb－0.7Sn－0.03Fe－xGe(x=0，0.05，0.1，0.2)合金在 360 ℃，18.6 MPa 和 0.01 mol/L LiOH 水溶液中的腐蚀增重曲线. Zr－1Nb－0.7Sn－0.03Fe 合金在腐蚀 70 d 后曲线发生转折，随后腐蚀增重急剧升高，腐蚀 130 d 的平均增重高达 6.21 mg/dm². 但是，

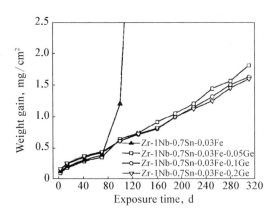

图 1 Zr‑1Nb‑0.7Sn‑0.03Fe‑xGe(x=0, 0.05, 0.1, 0.2) 合金在 360 ℃, 18.6 MPa 和 0.01 mol/L LiOH 水溶液中的腐蚀增重曲线

Fig. 1 Weight gain curves of Zr‑1Nb‑0.7Sn‑0.03Fe‑xGe (x=0, 0.05, 0.1, 0.2) alloys corroded in lithiated water with 0.01 mol/L LiOH at 360 ℃ and 18.6 MPa

腐蚀 310 d 时, x 为 0.05, 0.1 和 0.2 的合金的平均增重分别为 1.81, 1.63 和 1.60 mg/dm². 可见, (0.05%—0.2%)Ge 的添加可以显著改善合金在 360 ℃, 18.6 MPa 和 0.01 mol/L LiOH 水溶液中的耐腐蚀性能.

2.2 合金显微组织

图 2a—c 是 Zr‑1Nb‑0.7Sn‑0.03Fe‑xGe(x=0.05, 0.1, 0.2) 合金显微组织的 TEM 像. 根据 EDS 和 SAD 分析结果可以确定, 在 3 种合金中, 都存在 β‑Nb, Zr‑Nb‑Fe‑Cr 和 Zr‑Nb‑Fe‑Cr‑Ge 3 种第二相 (SPPs). 其中, β‑Nb 型第二相为 bcc 结构 (图 2d 和 g), 尺寸在 50 nm 以下 (图 2a 中箭头 1 所示). Zr‑Nb‑Fe‑Cr 和 Zr‑Nb‑Fe‑Cr‑Ge 均属于四方 (tet) 结构 (图 2e 和 h) 的 Zr₂Fe 型第二相, Nb/Fe 原子比在 1.7—2.2 之间. 以上 3 种第二相均呈球状.

尺寸为 50—100 nm (图 2a 中箭头 2 所示), Nb/Fe 原子比在 1.7—2.2 之间. 以上 3 种第二相均呈球状.

在 Zr‑1Nb‑0.7Sn‑0.03Fe‑0.1Ge 和 Zr‑1Nb‑0.7Sn‑0.03Fe‑0.2Ge 合金中, 除了 β‑Nb, Zr‑Nb‑Fe‑Cr 和 Zr‑Nb‑Fe‑Cr‑Ge 3 种第二相外, 还存在尺寸为 300—500 nm 较大的第二相 (图 2b 中箭头 3 和图 2c 中箭头 4 所示). 根据 Zr‑Ge 二元相图、SAD (图 2f) 和 EDS (图 2i) 分析结果, 确定这种第二相为 tet 结构的 Zr₃Ge 型第二相. 表 1 统计了不同合金中第二相的尺寸、类型和数量. 以上结果说明, 580 ℃ 加热时, Ge 在 α‑Zr 中的最大固溶含量不超过 0.05%, 过量的 Ge 会从金属基体中析出, 先与 Zr, Nb, Fe 形成 Zr‑Nb‑Fe‑Cr‑Ge 第二相, 当 Ge 含量提高到 0.1% 时, Ge 还会与 Zr 形成 Zr₃Ge 型第二相. 对比图 2b 和 c 可知, 当 Ge 含量提高到 0.2% 时, Zr‑Nb‑Fe‑Cr‑Ge 和 Zr₃Ge 型第二相的数量增多. 需要说明的是, Zr‑1Nb‑0.7Sn‑0.03Fe‑xGe 合金中并没有特意添加 Cr, 而 EDS 中出现的 Cr 峰是来自海绵 Zr 中的杂质 Cr, 因为 Cr 在 α‑Zr 中的最大固溶度仅为 200 μg/g[15], 所以多余的 Cr 会以第二相析出. 另外, Cu 峰是由 TEM 样品台的结构材料引起的.

2.3 氧化膜的断口形貌

图 3 是 Zr‑1Nb‑0.7Sn‑0.03Fe 和 Zr‑1Nb‑0.7Sn‑0.03Fe‑0.1Ge 合金在 360 ℃, 18.6 MPa 和 0.01 mol/L LiOH 水溶液中腐蚀后的氧化膜断口形貌. 可以看出, Zr‑1Nb‑0.7Sn‑0.03Fe 合金腐蚀 130 d 的氧化膜平均厚度约为 42.3 μm, 氧化膜内部存在大量平行于或垂直于氧化膜/金属界面的微裂纹 (图 3a 和 b), 这些微裂纹为 O²⁻ 或 OH⁻ 在氧化膜中的扩散提供了通道, 加速了氧化膜的生长. 在中间层还观察到大量等轴晶, 几乎观察不到柱状晶, 氧化膜十分疏松 (图 3c). 然而, Zr‑1Nb‑0.7Sn‑0.03Fe‑0.1Ge 合金腐蚀 190 d 的氧化膜平均厚度仅约为 6.4 μm, 此时氧化膜较为致密, 只是发现少量微孔隙和平行于氧化膜/金属界面微裂纹 (图 3d 和 e). 氧化膜外层是柱状晶与等轴晶共存的区域, 能够清晰地观察到部分柱状晶 (图 3f).

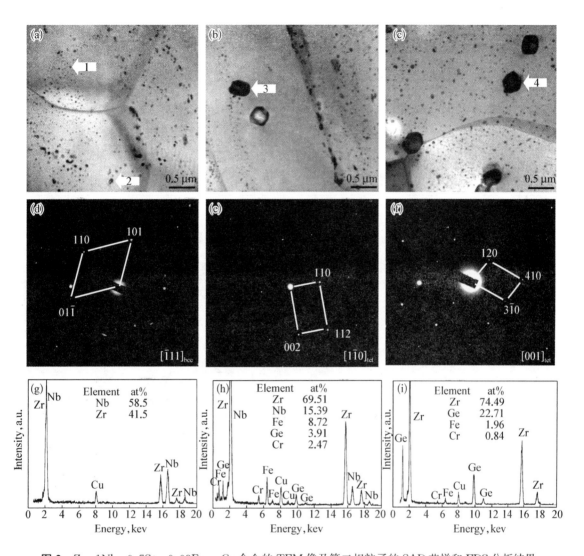

图 2　Zr‐1Nb‐0.7Sn‐0.03Fe‐xGe 合金的 TEM 像及第二相粒子的 SAD 花样和 EDS 分析结果

Fig. 2　TEM images of Zr‐1Nb‐0.7Sn‐0.03Fe‐xGe alloys (a—c) and corresponding SAD patterns (d—f) and EDS analysis results (g—i) of second phase particles (β‐Nb, Zr‐Nb‐Fe‐Cr‐Ge and Zr$_3$Ge) (a) TEM image of Zr‐1Nb‐0.7Sn‐0.03Fe‐0.05Ge；(b) TEM image of Zr‐1Nb‐0.7Sn‐0.03Fe‐0.1Ge；(c) TEM image of Zr‐1Nb‐0.7Sn‐0.03Fe‐0.2Ge；(d) SAD patterns of β‐Nb as pointed by arrow 1 in Fig. 2a；(e) SAD patterns of Zr‐Nb‐Fe‐Cr‐Ge as pointed by arrow 2 in Fig. 2a；(f) SAD patterns of Zr$_3$Ge as pointed by arrow 3 in Fig. 2b；(g) EDS spectra of β‐Nb as pointed by arrow 1 in Fig. 2a；(h) EDS spectra of Zr‐Nb‐Fe‐Cr‐Ge as pointed by arrow 2 in Fig. 2a；(i) EDS spectra of Zr$_3$Ge as pointed by arrow 3 in Fig. 2b

2.4　氧化膜的内表面形貌

图 4 是 Zr‐1Nb‐0.7Sn‐0.03Fe 和 Zr‐1Nb‐0.7Sn‐0.03Fe‐0.1Ge 合金在 360 ℃，18.6 MPa 和 0.01 mol/L LiOH 水溶液中腐蚀后氧化膜的内表面形貌. 可以看出，Zr‐1Nb‐0.7Sn‐0.03Fe 合金腐蚀 130 d 的氧化膜内表面不仅凹凸不平，而且出现大量微裂纹（图 4a 和 b）. 然而，Zr‐1Nb‐0.7Sn‐0.03Fe‐0.1Ge 合金腐蚀 190 d 的氧化膜内表面还比较致密，未观察到微裂纹（图 4c 和 d）. 这说明添加的 Ge 可以有效地延缓氧化膜/金属界面处微裂

表 1 图 2 中 Zr‑1Nb‑0.7Sn‑0.03Fe‑xGe 合金的第二相信息

Table 1 Details of second phase particles in Zr‑1Nb‑0.7Sn‑0.03Fe‑xGe alloys in Fig. 2

Alloy	β‑Nb	Zr‑Nb‑Fe‑Cr	Zr‑Nb‑Fe‑Cr‑Ge	Zr$_3$Ge
Zr‑1Nb‑0.7Sn‑0.03Fe‑0.05Ge	Major, < 50 nm, bcc	Partly, 50—100 nm, tet	Minor, 50—100 nm, tet	—
Zr‑1Nb‑0.7Sn‑0.03Fe‑0.1Ge	Major	Partly	Partly	Minor, 300—500 nm, tet
Zr‑1Nb‑0.7Sn‑0.03Fe‑0.2Ge	Partly	Partly	Major	Partly

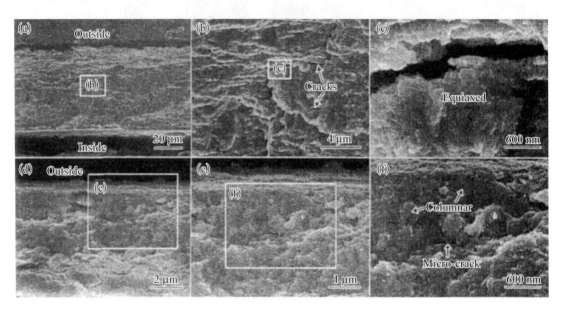

图 3 Zr‑1Nb‑0.7Sn‑0.03Fe 和 Zr‑1Nb‑0.7Sn‑0.03Fe‑0.1Ge 合金在 360 ℃, 18.6 MPa 和 0.01 mol/L LiOH 水溶液中腐蚀后的氧化膜断口形貌

Fig. 3 Low (a, d) and high (b, c, e, f) magnified fracture surface morphologies for the oxide films on Zr‑1Nb‑0.7Sn‑0.03Fe alloy corroded for 130 d (a—c) and Zr‑1Nb‑0.7Sn‑0.03Fe‑0.1Ge alloy corroded for 190 d (d—f) in lithiated water with 0.01 mol/L LiOH at 360 ℃ and 18.6 MPa

纹的形成过程.

另外, Zr‑1Nb‑0.7Sn‑0.03Fe‑0.1Ge 合金腐蚀 70 d 后, 在其氧化膜内表面只观察到 200—500 nm 尺寸范围内的凹凸起伏(图 4e 和 f), 随着腐蚀时间延长, 在腐蚀 190 d 的样品内表面观察到尺寸为 3—5 μm 的菜花状突起, 高低起伏程度较大(图 4c 和 d). 这 2 种形貌是氧化膜生长不均匀引起的, 文献[16]用孔隙簇与腐蚀过程的关系, 解释了氧化膜内表面出现菜花状突起的可能原因: 氧化膜中的孔隙簇能够以自身为圆心向周围提供 O^{2-} 的方式使氧化膜向金属基体推进, 这样, 随着氧化过程的进行, 在氧化膜内表面会形成尺寸较大的菜花状突起.

2.5 氧化膜横截面的显微组织

TEM 观察发现, 用 FIB 制备 Zr‑1Nb‑0.7Sn‑0.03Fe‑0.2Ge 合金腐蚀 70 d 的氧化膜横截面薄试样, 在靠近氧化膜/金属界面的区域内存在大量柱状晶(图 5a), 而在靠近氧化

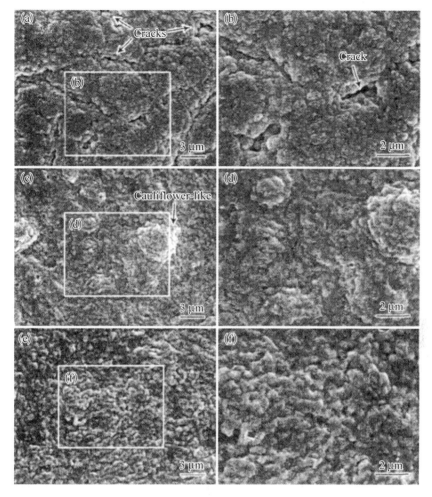

图 4 Zr‑1Nb‑0.7Sn‑0.03Fe 和 Zr‑1Nb‑0.7Sn‑0.03Fe‑0.1Ge 合金在 360 ℃,18.6 MPa 和 0.01 mol/L LiOH 水溶液中腐蚀后氧化膜的内表面形貌

Fig. 4 Low (a, c, e) and high (b, d, f) magnified inner surface morphologies of the oxide films on Zr‑1Nb‑0.7Sn‑0.03Fe alloy corroded for 130 d (a, b) and Zr‑1Nb‑0.7Sn‑0.03Fe‑0.1Ge alloy corroded for 190 d (c, d) and 70 d (e, f) in lithiated water with 0.01 mol/L LiOH at 360 ℃ and 18.6 MPa

膜外表面的区域内主要存在尺寸为 20—40 nm 的等轴晶,在等轴晶区域内存在较多微孔隙(图 5b). 由于锆合金的氧化总是发生在氧化膜/金属界面处,这表明氧化膜的显微组织在生长过程中是由柱状晶向等轴晶不断演化的过程,这与许多研究结果[17,18]相一致. 同时,ZrO_2 的晶粒形态与氧化膜的保护特性密切相关,大量等轴晶存在的区域会表现出较差的保护性能.

在 TEM 的扫描透射(STEM)模式下观察发现,氧化膜内部存在许多发生部分氧化的 β‑Nb‑O,Zr‑Nb‑Fe‑Cr‑O 和 Zr‑Nb‑Fe‑Cr‑Ge‑O 第二相(图 5c),在氧化膜/金属界面(图 5 虚线所示)处存在发生部分氧化的 Zr_3Ge‑O 第二相(图 5d),这表明第二相的氧化速率比 α‑Zr 基体慢. 根据 EDS 分析结果,部分氧化的 Zr‑Nb‑Fe‑Cr‑O 中 Nb/Fe 原子比在 0.6—0.9 之间,这比合金中 Zr‑Nb‑Fe‑Cr 第二相的 Nb/Fe 原子比降低了 47%—59%. 这表明 Zr‑1Nb‑0.7Sn‑0.03Fe‑xGe 合金中的 Zr‑Nb‑Fe‑Cr 第二相氧化时,Nb

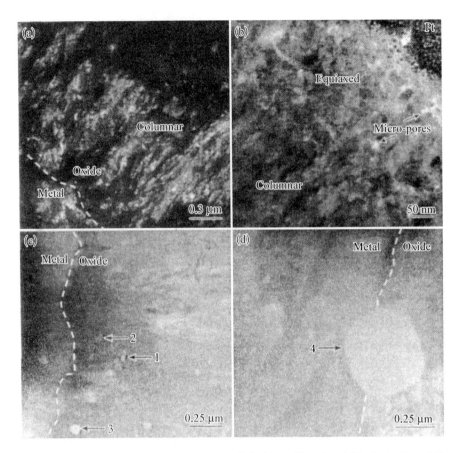

图 5 Zr‑1Nb‑0.7Sn‑0.03Fe‑0.2Ge 合金在 360 ℃, 18.6 MPa 和 0.01 mol/L LiOH 水溶液中腐蚀 70 d 的氧化膜横截面的形貌

Fig. 5 Cross-sectional morphologies of the oxide film on Zr‑1Nb‑0.7Sn‑0.03Fe‑0.2Ge alloy corroded in lithiated water with 0.01 mol/L LiOH at 360 ℃ and 18.6 MPa for 70 d (Dotted lines indicate the oxide/metal interfaces) (a) dark field image; (b) TEM image; (c, d) STEM image

会向周围的 ZrO_2 中扩散,并且 Nb 的扩散速度比 Fe 快. 然而,部分氧化的 Zr‑Nb‑Fe‑Cr‑Ge‑O 第二相中的 Nb/Fe 原子比在 1.9—2.4 之间,与合金中 Zr‑Nb‑Fe‑Cr‑Ge 第二相中的 Nb/Fe 原子比相当,表明 Ge 可能在一定程度上会延缓第二相中的 Nb 向周围的 ZrO_2 中扩散,表 2 统计了发生部分氧化时的第二相信息. 由于 TEM 电子束进入样品后有一定的扩散范围和穿透深度,会受到 ZrO_2 基体的影响,所以 Zr 和 O 的含量会高一些.

2.6 分析与讨论

锆合金的氧化是发生在氧化膜/金属界面处,O^{2-} 或 OH^- 需要穿过氧化膜后才能到达氧化膜/金属界面与 Zr 发生反应. 因此,氧化膜的显微组织及其演化过程会直接影响 O^{2-} 或阴离子空位在氧化膜中的扩散. 周邦新等[14,19]从氧化膜显微组织演化的角度阐述了锆合金的腐蚀机理:因为 Zr 氧化生成 ZrO_2 时的 P. B. 比(金属氧化物与金属体积之比)为 1.56,氧化过程中生成的氧化膜体积膨胀,同时受到金属基体约束,使得氧化膜内部产生很大的压应力,在这种条件下生成的 ZrO_2 会形成许多缺陷;在时间、温度和压力的作用下,这些缺陷发生扩散、湮没和凝聚,空位被晶界吸收后形成孔隙簇,孔隙簇进一步发展成为微裂纹,降低了氧化膜的保护作用,因而引起腐蚀加速.

表 2 图 5 中 Zr-1Nb-0.7Sn-0.03Fe-0.2Ge 合金发生部分氧化时的第二相信息

Table 2 Details of partly oxidized second phase particles in Zr-1Nb-0.7Sn-0.03Fe-0.2Ge alloys in Fig. 5 (atomic fraction,%)

Arrow	Type	Zr	Nb	Fe	Cr	Ge	O	Nb/Fe
1	β-Nb-O	27.92	30.70	—	—	—	41.38	—
2	Zr-Nb-Fe-Cr-O	43.04	10.58	14.09	1.77	—	30.52	0.75
3	Zr-Nb-Fe-Cr-Ge-O	56.84	3.07	1.35	0.67	3.97	34.10	2.27
4	Zr_3Ge-O	67.57	—	—	—	24.56	7.87	—

添加合金元素 Ge 的 Zr-1Nb-0.7Sn-0.03Fe-xGe 合金,在 360 ℃,18.6 MPa 和 0.01 mol/L LiOH 水溶液中表现出较好的耐腐蚀性能.在锆合金氧化时,固溶在 α-Zr 中的合金元素和第二相都会发生氧化.Ge 的 P.B. 比为 1.23[13],而 Zr,Nb,Fe 和 Cr 的 P.B. 比分别为 1.56,2.74,1.77 和 2.02.所以,固溶在 α-Zr 中的 Ge 可能会减少 Zr 氧化生成 ZrO_2 时的体积膨胀程度,降低氧化膜中的压应力.同时,Ge 可以取代 Zr-Nb-Fe-Cr-Ge 中 Zr,Nb,Fe 或 Cr 的点阵位置,从而可能会降低 Zr-Nb-Fe-Cr 第二相在氧化过程中的体积膨胀程度,最终减少氧化膜中在 Zr-Nb-Fe-Cr 第二相周围产生的附加应力.目前很多研究[14,20-23]认为氧化膜中产生的巨大压应力可以加速空位扩散凝聚,形成微孔隙和微孔隙发展形成微裂纹的过程.那么,固溶在 α-Zr 中的 Ge 和 Zr-Nb-Fe-Cr-Ge 第二相都可以有效地延缓上述过程,显著提高合金的耐腐蚀性能.

与 Zr-1Nb-0.7Sn-0.03Fe 合金比较,在相同的腐蚀情况下,Zr-1Nb-0.7Sn-0.03Fe-0.1Ge 合金的氧化膜厚度较薄,微裂纹较少,并且存在较多垂直于氧化膜/金属界面的柱状晶.柱状晶中的缺陷通过扩散凝聚形成新的晶界,从而柱状晶逐渐发展成等轴晶.许多研究[17,24,25]表明,氧化膜中存在的 ZrO_2 柱状晶与锆合金良好的耐腐蚀性能密切相关,这与腐蚀增重的结果相一致(图 1).可见,Ge 还可以延缓氧化膜中 ZrO_2 柱状晶向等轴晶的演化过程.另外,文献[14,19]用扫描探针显微镜(SPM)研究了 Zr-0.92Sn-1.03Nb-0.12Fe 合金在 360 ℃,18.6 MPa 和 0.04 mol/L LiOH 水溶液中腐蚀 14 d 后的氧化膜表面晶粒的起伏形貌,结果表明,LiOH 水溶液中的 Li^+ 或 OH^- 会渗入氧化膜中,进入空位或吸附在孔隙壁上,降低 ZrO_2 表面自由能,促进了空位通过扩散凝聚形成孔隙簇和晶界微裂纹的过程.他们认为 Nb 可以减少表面自由能降低的程度,延缓氧化膜中孔隙和微裂纹的发展过程,那么,固溶在 α-Zr 中的 Ge 可能也会有类似的作用.

在尺寸较大的 Zr_3Ge 周围极易产生局部附加应力,引起应力集中,形成微裂纹等缺陷,不利于改善合金的耐腐蚀性能.但是,合金的显微组织观察结果表明,Zr-1Nb-0.7Sn-0.03Fe-0.2Ge 合金中析出的 Zr-Nb-Fe-Cr-Ge 第二相的数量最多,Zr_3Ge 第二相的数量很少,Zr-Nb-Fe-Cr-Ge 第二相有利于改善合金的耐腐蚀性能.Ge 对合金耐腐蚀性能的综合改善作用超过了 Zr_3Ge 的不利影响,所以,Zr-1Nb-0.7Sn-0.03Fe-0.2Ge 合金的耐腐蚀性能在 Zr-1Nb-0.7Sn-0.03Fe-xGe 合金中是最佳的.

3 结论

(1) 在 Zr-1Nb-0.7Sn-0.03Fe-xGe 合金中,当 Ge 含量为 0.05%—0.2% 时,会析

出 bcc 结构的 β - Nb,属于 tet 结构 Zr_2Fe 型的 Zr - Nb - Fe - Cr 和 Zr - Nb - Fe - Cr - Ge 3 种第二相. 当 Ge 含量达到或超过 0.1% 时,还会析出 tet 结构的 Zr_3Ge 型第二相. 以上这些第二相的氧化速率都比 α - Zr 基体慢.

(2) 在 Zr - 1Nb - 0.7Sn - 0.03Fe 合金基础上添加的(0.05%—0.2%)Ge 可以明显改善合金在 360 ℃,18.6 MPa 和 0.01 mol/L LiOH 水溶液中的耐腐蚀性能. 腐蚀 190 d 后的 Zr - 1Nb - 0.7Sn - 0.03Fe - 0.1Ge 合金氧化膜比较致密,ZrO_2 柱状晶较多. Ge 既可以延迟 ZrO_2 柱状晶向等轴晶的演化,又能够延缓微裂纹的形成过程.

参 考 文 献

[1] Comstock R J, Schoenberger G, Sable G P. In: Bradley E R, Sabol G P eds. , *Zirconium in the Nuclear Industry: 11th International Symposium*, *ASTM STP 1295*, Ann Arbor: ASTM International, 1996: 710.

[2] Sabol G P, Kilp G R, Balfour M G, Roberts E. In: Van Swam L F P, Eucken C M eds. , *Zirconium in the Nuclear Industry: 8th International Symposium*, *ASTM STP 1023*, Baltimore: ASTM International, *1989*: 227.

[3] Nikulina A V, Markelov V A, Peregud M M, Bibilashvili Y K, Kotrekhov V A, Lositsky A F, Kuzmenko N V, Shevnin Y P, Shamardin V K, Kobylyansky G P, Novoselov A E. In: Sabol G P, Bradley E R eds. , *Zirconium in the Nuclear Industry: 11th International Symposium*, *ASTM STP 1295*, Ann Arbor: ASTM International, 1996: 785.

[4] Jung Y I, Lee M H, Kim H G, Park J Y, Jeong Y H. *J Alloy Compd*, 2009: 479: 423.

[5] Jeong Y H, Park S Y, Lee M H, Choi B K, Baek J H, Park J Y, Kim J H, Kim H G. *J Nucl Sci Technol*, 2006: 43: 977.

[6] Yang W D. *Reactor Materials Science*. 2nd Ed. , Beijing: Atomic Energy Press, 2006: 260.
(杨文斗. 反应堆材料学. 第二版,北京: 原子能出版社,2006: 260)

[7] Liu J Z. *Structure Nuclear Materials*. Beijing: Chemical Industry Press, 2007: 19.
(刘建章. 核结构材料. 北京: 化学工业出版社,2007: 19)

[8] Liu W Q, Zhu X Y, Wang X J, Li Q, Yao M Y, Zhou B X. *Atom Energ Sci Technol*, 2010: 44: 1477.
(刘文庆,朱晓勇,王晓娇,李强,姚美意,周邦新. 原子能科学技术,2010: 44: 1477)

[9] Kim J M, Jeong Y H, Kim I S. *J Nucl Mater*, 2000: 280: 235.

[10] Liu W Q, Li Q, Zhou B X, Yao M Y. *Nucl Power Eng*, 2003: 24: 33.
(刘文庆,李强,周邦新,姚美意. 核动力工程,2003: 24: 33)

[11] Wagner C J. *J Chem Phy*, 1950: 18: 62.

[12] Hauffe K. *Reactionen in und an Fasten Stoffen*. Berlin: Springer, 1966: 1.

[13] Xie X F, Zhang J L, Zhu L, Yao M Y, Zhou B X, Peng J C. *Acta Metall Sin*, 2012: 48: 1487.
(谢兴飞,张金龙,朱莉,姚美意,周邦新,彭剑超. 金属学报,2012: 48: 1487)

[14] Zhou B X, Li Q, Yao M Y, Liu W Q, Chu Y L. In: Kammenzind B, Limback M eds. , *Zirconium in the Nuclear Industry: 15th International Symposium*, *ASTM STP 1505*, West Conshohochen: American Society for Testing and Materials, 2009: 360.

[15] Charquet D, Hahn R, Ortlib E. In: Van Swam L F P, Eucken C M eds. , *Zirconium in the Nuclear Industry: 8th International Symposium*, *ASTM STP 1023*, Philadelphia: ASTM International, 1989: 405.

[16] Zhou B X, Li Q, Yao M Y, Liu W Q, Chu Y L. *Nucl Power Eng*, 2005: 26: 364.
(周邦新,李强,姚美意,刘文庆,褚于良. 核动力工程,2005: 26: 364)

[17] Anada H, Takeda K. In: Sabol G P, Bradley E R eds. , *Zirconium in the Nuclear Industry: 11th International Symposium*, *ASTM STP 1295*, Ann Arbor: ASTM International, 1996: 35.

[18] Wadman B, Lai Z, Andren H O, Nystrom A L, Rudling P, Pettersson H. In: Garde A M, Bradley E R eds. , *Zirconiumin in the Nuclear Industry: 10th International Symposium ASTM STP 1245*, Ann Arbor: ASTM International, 1994: 579.

[19] Zhou B X, Li Q, Liu W Q, Yao M Y, Chu Y L. *Rare Met Mater Eng*, 2006: 35: 1009.

（周邦新，李强，刘文庆，姚美意，褚于良. 稀有金属材料工程，2006；35：1009）

[20] Park J Y, Yoo S J, Choi B K, Jeong Y H. *J Nucl Mater*, 2007; 437: 274.

[21] Kim H G, Choi B K, Park J Y. *J Alloy Compd*, 2009; 481: 867.

[22] Garzarolli F, Seidel H, Tricot R, Gros J P. In: Eucken C M, Garde A M eds., *Zirconiumin in the Nuclear Industry: 9th International Symposium ASTM STP 1132*, Baltimore: ASTM International. 1991: 395.

[23] Park J Y, Choi B K, Jeong Y H, Jung Y H. *J Nucl Mater*, 2005; 340: 237.

[24] Liu Y Z, Park J Y, Kim H G, Jeong Y H. *Mater Chem Phys*, 2010; 122: 408.

[25] Yilmazbayhan A, Breval E, Motta A T, Comstock R J. *J Nucl Mater*, 2006; 349: 265.

Study on the Corrosion Resistance of Zr - 1Nb - 0.7Sn - 0.03Fe - xGe Alloy in Lithiated Water at 360 ℃

Abstract: Zirconium alloys have low thermal neutron absorption cross-section, good corrosion resistance and adequate mechanical properties. They have been successfully developed as fuel cladding materials in pressurized water reactors. It's well known that the corrosion resistance of Zr - Sn - Nb alloys is significantly superior to that of Zircaloy - 4 alloy when corroded in lithiated water. The corrosion resistance of zirconium alloys is controlled by their chemical compositions, characteristics of second phase particles (SPPs) and microstructure evolution of the oxide in them. The corrosion tests of Zr - 1Nb - 0.7Sn - 0.03Fe - xGe (x=0, 0.05, 0.1, 0.2, mass fraction, %) alloys were investigated by means of an autoclave test in lithiated water with 0.01 mol/L LiOH at 360 ℃ under a pressure of 18.6 MPa. The microstructures of the alloys and oxide films on the corroded specimens were examed by using TEM and SEM. The sample for the oxide microstructure observation was prepared by a HELIOS - 600I focused ion beam. The results reveal that the corrosion resistance of Zr - 1Nb - 0.7Sn - 0.03Fe - xGe (x=0.05, 0.1, 0.2) alloys was remarkably superior to that of Zr - 1Nb - 0.7Sn - 0.03Fe alloy. The corrosion resistance of Zr - 1Nb - 0.7Sn - 0.03Fe alloys is markedly improved by the addition of (0.05%—0.2%)Ge. In addition to Zr - Nb - Fe - Cr SPPs with the tetragonal crystal structure (tet) and β - Nb SPPs with the bcc crystal structure, the Zr - Nb - Fe - Cr - Ge SPPs with the tet structure and Zr_3Ge SPPs with the tet structure were detected out in Zr - 1Nb - 0.7Sn -0.03 Fe - xGe alloys. The oxidation of SPPs was found to be slower than that of α - Zr matrix. There exist a few micro-cracks and more ZrO_2 columnar grains in the oxide film formed on Zr - 1Nb - 0.7Sn - 0.03Fe - 0.1Ge alloys corroded for 190 d. However, more micro-cracks and ZrO_2 equiaxed grains appear in the oxide film formed on Zr - 1Nb - 0.7Sn - 0.03Fe alloys corroded for 130 d. Because the P. B. ratio of Ge is smaller than those of Zr, Nb, Fe and Cr, it is likely that the volume expansion of the oxide on Zr - 1Nb - 0.7Sn - 0.03 Fe - xGe (x=0.05, 0.1, 0.2) alloys is smaller than that on Zr - 1Nb - 0.7Sn - 0.03Fe alloy, and the compressive stress can be reduced and the micro-cracks can be effectively decreased in the oxide on Zr - 1Nb - 0.7Sn -0.03 Fe - xGe (x=0.05, 0.1, 0.2) alloys. The addition of Ge can not only delay the developing process of the defects in oxide films to form micro-pores and micro-cracks, but also retard the microstructural evolution from columnar grains to equiaxed grains. Therefore, it is concluded that the addition of Ge can improve the corrosion resistance of alloy.

Ge 含量对 Zr－4 合金在 LiOH 水溶液中耐腐蚀性能的影响*

摘　要：采用堆外高压釜腐蚀试验研究添加 0.1% Ge 和 0.5%Ge(质量分数)对 Zr－4 合金在 360 ℃、0.01 mol/L LiOH 水溶液中耐腐蚀性能的影响,用 TEM 和 EDS 研究合金的显微组织和第二相成分. 结果表明：与 Zr－4 合金相比,添加 Ge 后合金的耐腐蚀性能显著提高,但随着 Ge 添加量的增加,对耐腐蚀性能的提高作用减弱. 当 Ge 含量为 0.1% 时,析出密排六方结构的 Zr(Fe,Cr)$_2$ 和 Zr(Fe,Cr,Ge)$_2$ 型第二相；当 Ge 含量达到 0.5% 时,还会析出 Zr－Sn－Fe－Cr－Ge 和尺寸较大的四方结构 Zr$_3$Ge 型第二相,添加 Ge 会使 α－Zr 基体中固溶的 Sn 含量降低. 固溶在 α－Zr 中的 Ge 和尺寸较小的 Zr(Fe,Cr)$_2$、Zr(Fe,Cr,Ge)$_2$ 型第二相可以提高 Zr－4＋xGe 合金的耐腐蚀性能；析出 Zr－Sn－Fe－Cr－Ge 和尺寸较大的 Zr$_3$Ge 第二相后会对耐腐蚀性能产生不利的影响.

锆合金因其热中子吸收截面小,并且在高温高压水中具有良好的耐腐蚀性能和较高的强度,已被用作核电站水冷动力堆核燃料元件的包壳材料和堆芯的其他结构材料. 随着人们希望进一步提高核电的经济性,需要提高核燃料的燃耗,对燃料元件包壳用锆合金提出了更高要求[1-2]. 在现有锆合金基础上调整合金元素的不同配比或添加其他种类合金元素,通过适当的热处理工艺可以获得耐腐蚀性能更加优良的锆合金,同时也可提高合金的强度.

由于 Zr－4 合金已不能满足高燃耗的要求,许多国家都进行了改善 Zr－4 合金耐腐蚀性能的研究[3]. 锆合金中添加的合金元素主要有 Sn、Nb、Fe 和 Cr 等. Sn 在锆合金中能起到固溶强化的作用,同时能有效降低锆合金中 N 对合金耐腐蚀性能的有害影响. TAKEDA 等[4] 研究表明：Zr－xSn－0.19Fe－0.1Cr 合金的耐腐蚀性能随 Sn 含量的降低而提高,将 Sn 含量降到 0.65% 后可进一步改善合金的耐腐蚀性能. Nb 的热中子吸收截面小,可消除 C、Al 和 Ti 等杂质对锆合金耐腐蚀性能的危害. 研究发现[5],锆合金的耐腐蚀性能与 Nb 在 α－Zr 中的固溶量有关. 在锆合金中 Fe 和 Cr 也是重要的合金元素. 有学者认为固溶在 α－Zr 基体中的 Fe 和 Cr 含量是影响耐腐蚀性能的主要因素[6-7]；也有学者认为合金中 Zr(Fe,Cr)$_2$ 等第二相粒子的尺寸、分布和体积分数才是影响耐腐蚀性能的主要因素[8-9]. Fe 和 Cr 对锆合金耐腐蚀性能的影响与 Fe＋Cr 含量和 Fe 与 Cr 的摩尔比以及水化学条件有关. GARZAROLLI 等[10] 发现在 Fe 和 Cr 含量适度的情况下,提高 Fe 与 Cr 的比值,Zr－4 合金在 350 ℃、0.01 mol/L LiOH 水溶液中的耐腐蚀性能提高. 很多学者还研究了 Cu、Ni、V、Bi、Mo、Ta 和 S 等合金元素对锆合金耐腐蚀性能的影响[11-13].

刘建章[14] 认为,Ge 的原子半径与锆的原子半径相差悬殊,不适合作为锆合金的添加元素,近十多年来未见有添加 Ge 元素对锆合金耐腐蚀性能影响的文献报道. 根据 Wagner 氧化膜生长理论和 Hauffe 原子价规律[14-15] 可知,位于元素周期表第 ⅣA 族的 Ge 元素可以增加锆合金氧化膜中的电子浓度,减少阴离子空位,从而抑制氧离子扩散,降低锆合金的腐蚀

---

* 本文合作者：张金龙、谢兴飞、姚美意、彭剑超、李强. 原发表于《中国有色金属学报》,2013,23(6)：1542－1548.

速率.但目前未见有关 Ge 对锆合金耐腐蚀性能影响的详细研究报道,其机理也不清楚.为此,本文作者旨在研究 Ge 对 Zr-4 合金耐腐蚀性能的影响,并初步探讨影响腐蚀行为的机理,为研究新型锆合金提供理论依据.

1 实验

以 Zr-4 为母合金,分别添加 0.1% 和 0.5%(质量分数,下同)的 Ge 制备成 Zr-4+xGe 锆合金样品,并以 Zr-4 重熔试样作为对照.具体制备工艺如下:用 SW-Ⅱ 型非自耗真空电弧炉将配好成分的原料熔炼成约 65 g 的合金锭,熔炼在高纯 Ar 作为保护气体的条件下进行,为保证合金成分的均匀性,合金锭翻转熔炼共 6 次.将合金锭在 700 ℃ 预热后反复热压制成条块状坯料,样品放入石英管真空热处理炉中在 β 相区均匀化处理(1 030 ℃ 保温 40 min),700 ℃ 热轧至 1.4 mm,β 相水淬(在管式电炉中将封装在真空石英管中的样品加热到 1 030 ℃ 保温 40 min,然后淬入水中并快速敲碎石英管),再多道次冷轧至 0.7 mm,最终退火处理采用 580 ℃ 保温 50 h 后空冷.将上述样品用 45%H$_2$O+45%HNO$_3$+10%HF(体积分数)混合酸酸洗和去离子水清洗后,放入静态高压釜中进行 360 ℃、18.6 MPa、0.01 mol/L LiOH 水溶液中腐蚀.

用 JEM-200CX 透射电镜观察腐蚀前合金的显微组织;用带有 INCA 能谱仪(EDS)的 JEM-2010F 场发射透射电镜观察分析合金中第二相的形貌和成分,所有 EDS 分析采用无标样法;通过选区电子衍射(SAD)确认第二相的晶体结构.TEM 样品制备过程如下:将片状样品机械磨抛或化学酸洗减薄至约厚度 0.07 mm,用专用模具冲出 d3 mm 圆片,去除毛边后用双喷电解抛光的方法制备 TEM 样品,电解双喷仪为国产 MTP-1A,所用电解液为 10%HClO$_4$+90%C$_2$H$_5$OH(体积分数),电解抛光时的直流电压为 45 V,温度低于 −30 ℃,用液氮直接注入电解液中降温.用 JSM-6700F 型扫描电子显微镜(SEM)观察腐蚀后样品的氧化膜断口形貌.断口形貌观察用样品用混合酸溶去腐蚀样品的金属基体,然后将氧化膜折断的方法制备[16].为了提高图像质量,在进行氧化膜断口形貌观察前,样品表面蒸镀了一层金属 Ir.

2 结果与分析

2.1 腐蚀质量增加

图 1 所示为 Zr-4+xGe 合金在 360 ℃、0.01 mol/L LiOH 水溶液中的腐蚀质量增加随时间的变化曲线.从图 1 可以看出:Zr-4 重熔样品腐蚀 130 d 时的质量增加达到了 204 mg/dm^2,而添加 0.1% Ge 和 0.5% Ge 的合金样品腐蚀 310 d 时的质量增加分别仅为 113 mg/dm^2 和 125 mg/dm^2.说明在 360 ℃、18.6 MPa、0.01 mol/L LiOH 水溶液中添加 Ge 的锆合金样品的腐蚀速率均比 Zr-4 重熔

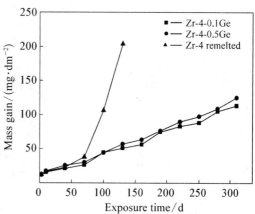

图 1 Zr-4+xGe 合金在 360 ℃、0.01 mol/L LiOH 水溶液中腐蚀质量增加随时间的变化曲线

Fig. 1 Changing curves of corrosion mass gain of Zr-4+xGe alloys with exposure time in lithiated solution with 0.01 mol/L LiOH at 360 ℃

样品的明显变慢,但添加 0.5%Ge 的样品比添加 0.1%Ge 的样品腐蚀速率略有增加.

2.2 合金的显微组织

图 2 所示为腐蚀前 Zr-4+xGe 合金显微组织的 TEM 像. 从图 2 可以看出:Zr-4 重熔样品中晶粒内有尺寸较小且数量较多的第二相(见图 2(a)),为密排六方结构的 Zr(Fe,Cr)$_2$ 型第二相[17];Zr-4+xGe 合金中均有不同程度的呈条带状分布的第二相,这应该是 β 相水淬时残留的 β-Zr 在随后的冷轧和退火过程中发生分解的结果. Zr-4+xGe 合金中晶内和晶界均有较多的第二相;结合第二相的 EDS 分析(见表 1)和 SAD 花样标定(见图 2(d))表明:Zr-4+xGe 合金中尺寸较小的第二相(粒径 50~100 nm)为密排六方结构的 Zr(Fe,Cr)$_2$;尺寸稍大些的第二相(100~200 nm)为密排六方结构的 Zr(Fe,Cr,Ge)$_2$. Zr-4+xGe 合金基体中均未检测到 Ge,说明 Ge 在 α-Zr 基体中的固溶含量很小,主要进入第二相中. Ge 元素进入 Zr(Fe,Cr)$_2$ 中形成 Zr(Fe,Cr,Ge)$_2$ 第二相. Zr-4+0.5Ge 合金中的第二相除了 Zr(Fe,Cr)$_2$ 外,还有尺寸稍大的第二相(100~200 nm)Zr-Sn-Fe-Cr-Ge 和 Ge 含量较高且尺寸较大(400~600 nm)的 Zr-Ge 化合物相(见图 2(c)). 这种 Zr-Ge 化合物中 Zr 与 Ge 的摩尔比略大于 3,通过 SAD 花样标定并结合 Zr-Ge 二元相图确定这种化合物为四方结构的 Zr$_3$Ge 相(见图 2(e)). 对 Zr-4+xGe 合金的能谱分析还表明(见表 1):Zr-4+xGe 合金的 Zr(Fe,Cr)$_2$ 第二相和 Zr-4+0.1Ge 合金的 Zr(Fe,Cr,Ge)$_2$ 第二相中均不含 Sn(少量 Sn 是由于电子束斑大而包含了合金基体中的 Sn);而 Zr-4+0.5Ge 合金的 Zr-Sn-Fe-Cr-Ge 第二相中的 Sn 含量高达 11.37%~12.12%,说明 Ge 含量高时,Ge 会促使

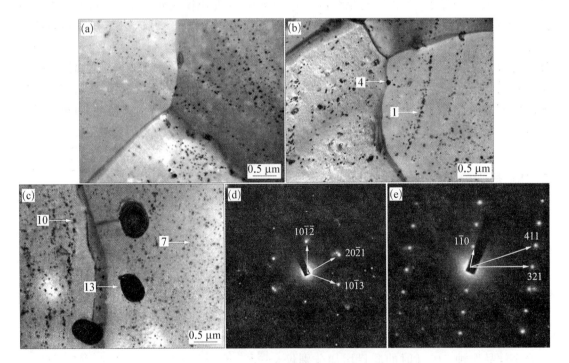

图 2 合金显微组织的 TEM 像及第二相粒子的 SAD 像

Fig. 2 TEM images of alloys and SAD patterns of second phase particles(SPPs): (a) Remelted Zr-4; (b) Zr-4+0.1Ge; (c) Zr-4+0.5Ge; (d) SAD pattern of SPP-4 in Fig. 2(b), indicated as Zr(Fe,Cr, Ge)$_2$; (e) SAD pattern of SPP-13 in Fig. 2(c), indicated as Zr$_3$Ge

表 1   第二相的 EDS 能谱分析结果

Table 1   EDS analysis results for SPPs

Alloy	SPPs No.	Mass fraction/%					Type of SPPs
		Zr	Fe	Cr	Sn	Ge	
Zr-4+0.1Ge	1	79.92	11.77	6.93	1.37	—	Zr(Fe,Cr)₂ HCP
	2	66.87	21.24	11.90	—	—	
	3	60.21	25.16	14.63	—	—	
	4	93.74	0.30	0.26	1.28	4.41	Zr(Fe,Cr,Ge)₂ HCP
	5	86.32	7.10	4.07	2.14	0.36	
	6	86.62	6.81	3.12	0.91	2.54	
Zr-4+0.5Ge	7	76.08	14.38	8.32	1.22	—	Zr(Fe,Cr)₂ HCP
	8	61.55	26.42	12.03	—	—	
	9	80.15	13.23	6.62	—	—	
	10	82.60	3.99	0.32	11.94	1.16	Zr-Sn-Fe-Gr-Ge
	11	81.72	5.33	0.79	11.37	0.78	
	12	80.37	5.39	0.89	12.12	1.22	
	13	80.69	0.39	0.47	0.41	18.03	Zr₃Ge TET
	14	82.83	0.34	0.38	0.78	15.67	
	15	80.97	0.49	0.39	0.68	17.48	

原来固溶在 α-Zr 基体中的 Sn 析出,并进入 Zr-Sn-Fe-Cr-Ge 第二相中.

2.3   氧化膜的外表面形貌

图 3 所示为 Zr-4+0.1Ge 和 Zr-4+0.5Ge 合金腐蚀 250 d 的氧化膜外表面 SEM 像. 由图 3 可见,Zr-4+0.1Ge 合金腐蚀 250 d 的氧化膜外表面晶粒间无明显微裂纹,合金具有良好的耐腐蚀性能(见图 3(a)). Zr-4+0.5Ge 合金表面的局部区域虽然出现较小的微空隙(见图 3(b)),但是整个氧化膜的外表面晶粒间无明显微裂纹,合金仍然具有良好的耐腐蚀性能.

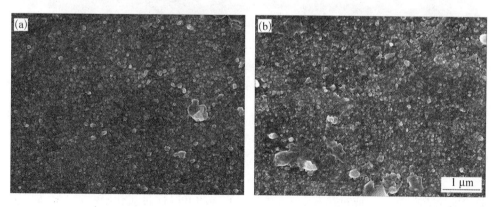

图 3   Zr-4+xGe 合金在 360 ℃、18.6 MPa、0.01 mol/L LiOH 水溶液中腐蚀 250 d 后氧化膜外表面的形貌

Fig. 3   Surface morphologies of oxide films on Zr-4+0.1Ge (a) and Zr-4+0.5Ge (b) alloys corroded in lithiated water with 0.01 mol/L LiOH at 360 ℃ and 18.6 MPa for 250 d

2.4 氧化膜的断口形貌

图 4 所示为 Zr‑4＋xGe 合金在 360 ℃、18.6 MPa、0.01 mol LiOH 水溶液中腐蚀后的氧化膜断口形貌. 图 4(a)所示为 Zr‑4 样品腐蚀 190 d 的氧化膜断口形貌,氧化膜厚度为 20 μm,氧化膜内部存在较多的平行于氧化膜/金属界面的裂纹,而且部分裂纹较宽. 在靠近氧化膜内表面处几乎全为 ZrO_2 等轴晶,很难发现柱状晶,断口高低起伏明显,存在许多较大的微裂纹(见图 4(a_1)). 图 4(b)所示为 Zr‑4＋0.5Ge 样品腐蚀 250 d 的氧化膜断口形貌,可以看到,氧化膜厚度为 6.2 μm,与 Zr‑4 样品比较(见图 4(a)、4(a_1)),含 Ge 样品的氧化膜较薄,界面平整,在靠近氧化膜内表面处发现较多 ZrO_2 柱状晶(见图 4(b_1)). 因此,随着 Ge 添加 Zr‑4＋xGe 氧化膜厚度变薄、氧化膜中柱状晶向等轴晶演化延后.

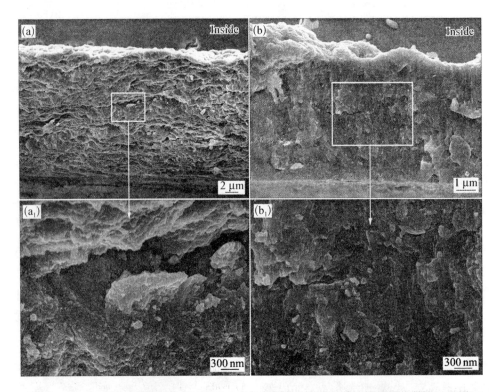

图 4 Zr‑4＋xGe 合金在 360 ℃、0.01 mol/L LiOH 水溶液中腐蚀后的氧化膜断口形貌

Fig. 4 Fracture surface morphologies of oxide films on remelted Zr‑4 alloy (a, a_1) corroded for 190 d and Zr‑4‑0.5Ge alloy (b, b_1) corroded for 250 d in lithiated water with 0.01 mol/L LiOH at 360 ℃

3 讨论

锆合金在高温高压水或过热蒸汽中腐蚀是 O^{2-} 或 OH^- 通过氧化膜扩散到金属/氧化膜界面处与 Zr 反应形成 ZrO_2 膜或阴离子空位沿反方向扩散而得以不断进行的,所以氧化膜的性质和氧化膜的显微组织在腐蚀过程中的演化与锆合金的腐蚀行为密切相关. Zr 氧化成 ZrO_2 时 P. B. 比为 1.56,体积发生膨胀,同时又受到金属基体的约束,因此氧化膜内部会形

成很大的压应力,在这种条件下生成的 ZrO_2 中会形成空位和间隙原子等各种缺陷;在温度和压力作用下,这些缺陷发生扩散、湮没和凝聚,空位被晶界吸收后形成孔隙簇,孔隙簇进一步发展成为微裂纹,降低了氧化膜的保护作用[18-20].

均匀弥散分布的第二相氧化后与 ZrO_2 基体之间形成的界面可以作为空位的尾闾,延缓空位通过扩散在氧化锆晶界上凝聚形成孔隙的过程[21],因而纳米大小分布均匀的 $Zr(Fe,Cr)_2$ 第二相可以延缓氧化膜显微组织的演化,有利于改善合金的耐腐蚀性能.虽然 Zr-4 和 Zr-4+xGe 合金中的第二相都呈现均匀弥散分布,但是,在 LiOH 水溶液中 Zr-4+xGe 合金的耐腐蚀性能比 Zr-4 合金的好得多.其原因可能如下:一方面,在 LiOH 水溶液中腐蚀时,由于 Li^+ 进入氧化膜中被吸附在孔隙壁上会降低 ZrO_2 的表面自由能.当 Zr-4+xGe 合金中 Ge 含量较低时,由于固溶在 ZrO_2 中少量的 Ge 可能会减小 ZrO_2 表面自由能降低的程度,从而提高合金的耐腐蚀性能.另一方面,当添加的 Ge 含量较低时,合金中的第二相为尺寸较小且较均匀分布的 $Zr(Fe,Cr)_2$ 和 $Zr(Fe,Cr,Ge)_2$ 型第二相,根据公式[22]:

$$P.B. = (Md_o)/(nd_{ox}A) \qquad (1)$$

式中:M 为氧化物的相对分子质量;d_o 为金属密度;n 为氧化物分子中金属原子数目;d_{ox} 为氧化物密度;A 为金属的相对原子质量,计算得到 Ge 的 P.B. 比为 1.23. P.B. 比是 Pilling-Bed worth ration,即金属氧化物体积与金属体积之比.而 Zr、Fe 和 Cr 的 P.B. 比分别为 1.56,1.77 和 2.02[22].在尺寸较小的 $Zr(Fe,Cr,Ge)_2$ 型第二相中,Ge 取代了 Fe 或 Cr 的点阵位置,从而降低了 $Zr(Fe,Cr,Ge)_2$ 型第二相在氧化过程中体积膨胀的程度,减小了局部附加应力.因此,添加少量的 Ge 可提高 Zr-4+xGe 合金在 LiOH 水溶液中耐腐蚀性能.

当 Ge 含量增加到 0.5% 时,合金中还会析出尺寸较大(400~600 nm)的四方结构 Zr_3Ge 型第二相,当这种第二相进入氧化膜中发生氧化时,在其周围易产生局部附加应力,引起应力集中,容易产生微裂纹等缺陷,促进氧化膜显微组织结构的演化,降低合金的耐腐蚀性能.另外,Ge 的添加促进了原本固溶在 α-Zr 基体中的一部分 Sn 以第二相析出,由于第二相分布的局部性,因此与固溶在 α-Zr 基体中的 Sn 影响相比,第二相中的 Sn 氧化后对腐蚀形成的氧化膜影响也具有局部性.Sn 的 P.B. 比为 1.32,大于 1[22],这也必然会在局部区域产生附加应力,促进氧化膜中孔隙和微裂纹的形成,从而加速腐蚀.因此,在 LiOH 水溶液中,Zr-4+xGe 合金耐腐蚀性能随 Ge 含量进一步增加而略微降低.

4 结论

(1) 在 Zr-4+xGe 合金中,当 Ge 含量为 0.1% 时,会析出密排六方结构的 $Zr(Fe,Cr)_2$ 和 $Zr(Fe,Cr,Ge)_2$ 型第二相;当 Ge 含量达到 0.5% 时,还会析出尺寸稍大的 Zr-Sn-Fe-Cr-Ge 和尺寸较大的四方结构 Zr_3Ge 型第二相.添加 0.5%Ge 会促使合金中原来固溶在 α-Zr 中的 Sn 析出,并进入 Zr-Sn-Fe-Cr-Ge 第二相中.

(2) 与 Zr-4 合金相比,添加 0.1%~0.5%Ge 能改善 Zr-4+xGe 合金在 360 ℃、0.01 mol/L LiOH 水溶液中的耐腐蚀性能,但 Ge 含量较高时,因析出 Zr-Sn-Fe-Cr-Ge 和尺寸较大的 Zr_3Ge 型第二相,又会对耐腐蚀性能产生不利的影响.

参 考 文 献

[1] 周邦新,李强,姚美意,夏爽,刘文庆,褚于良.热处理影响 Zr-4 合金疖状腐蚀性能的机制[J].稀有金属材料与工

程,2007,36(7):1129-1134.

ZHOU Bang-xin, LI Qiang, YAO Mei-yi, XIA Shuang, LIU Wen-qing, ZHU Yu-liang. Effect mechanism of heat treatments on nodular corrosion resistance of Zircaloy-4[J]. Rare Metal Materials and Engineering, 2007,36(7): 1129-1134.

[2] 李士炉,姚美意,张欣,耿建桥,彭剑超,周邦新. 添加 Cu 对 M5 合金在 500 ℃过热蒸汽中耐腐蚀性能的影响[J]. 金属学报,2011,47(2):163-168.

LI Shi-lu, YAO Mei-yi, ZHANG Xin, GENG Jian-qiao, PENG Jian-chao, ZHOU Bang-xin. Effect of adding Cu on the corrosion resistance of M5 alloy in superheated steam at 500 ℃[J]. Acta Metallurgica Sinica, 2011, 47(2): 163-168.

[3] SABOL G P, COMSTOCK R J, WEINER R A, LAROUERE E, STANUTZ R N. In-reactor corrosion performance of ZIRLO and Zircaloy-4[C]// Zirconium in the Nuclear Industry: The 10th International Symposium. Philadelphia: ASTM STP 1245, 1994: 724-744.

[4] TAKEDA K, ANADA H. Mechanism of corrosion rate degradation due to tin[C]// Zirconium in the Nuclear Industry: The 12th International Symposium. Philadelphia: ASTM STP 1354, 2000: 592-608.

[5] JEONG Y H, KIM H G, KIM D J, CHOI B K, KIM J H. Influence of Nb concentration in the α-matrix on the corrosion behavior of Zr-xNb binary alloys[J]. Journal of Nuclear Materials, 2003,323: 72-80.

[6] 姚美意,周邦新,李强,刘文庆,虞伟均,褚于良. 热处理对 Zr-4 合金在 360 ℃ LiOH 水溶液中腐蚀行为的影响[J]. 稀有金属材料与工程,2007,36(11):1920-1923.

YAO Mei-yi, ZHOU Bang-xin, LI Qiang, LIU Wen-qing, YU Wei-jun, ZHU Yu-liang. Effect of heat treatments on the corrosion resistance of Zircaloy-4 in LiOH aqueous solution at 360 ℃[J]. Rare Metal Materials and Engineering, 2007, 36(11):1920-1923.

[7] 沈月锋,姚美意,张欣,李强,周邦新,赵文金. β-相水淬对 Zr-4 合金在 LiOH 水溶液中腐蚀性能的影响[J]. 金属学报,2011,47(7):899-904.

SHEN Yue-feng, YAO Mei-yi, ZHANG Xin, LI Qiang, ZHOU Bang-xin, ZHAO Wen-jin. Effect of β-quenching on the corrosion resistance of Zr-4 alloy in LiOH aqueous solution[J]. Acta Metallurgica Sinica, 2011,47(7): 899-904.

[8] ANADA H, HERB B J, NOMOTO K, HAGI S, GRAHAM R A, KURODA T. Effect of annealing temperature on corrosion behavior and ZrO₂ microstructure of Zircaloy-4 cladding tube[C]// Zirconium in the Nuclear Industry: The 11th International Symposium. Philadelphia: ASTM STP 1295, 1996: 74-93.

[9] GARZAROLLI F, STEINBERG E, WEIDINGER H G. Microstructure and corrosion studies for optimized PWR and BWR Zircaloy cladding [C]// Zirconium in the Nuclear Industry: The 8th International Symposium. Philadelphia: ASTM STP 1023, 1989: 202-212.

[10] GARZAROLLI F, BROY Y, BUSCH R A. Comparison of the long-time corrosion behavior of certain Zr alloys in PWR, BWR, and laboratory[C]// Zirconium in the nuclear industry: The 11th International Symposium. Philadelphia: ASTM STP 1295, 1996: 850-864.

[11] HONG H S, MOON J S, KIM S J, LEE K S. Investigation on the oxidation characteristics of copper-added modified Zircaloy-4 alloys in pressurized water at 360 ℃[J]. Journal of Nuclear Materials, 2001, 297: 113-119.

[12] ISOBE T, MATSUO Y. Development of highly corrosion resistant zirconium-based alloys[C]// Zirconium in the Nuclear Industry: The 9th International Symposium. Philadelphia: ASTM STP 1132,1991: 346-367.

[13] 李佩志,李中奎,薛祥义,刘建章. 合金元素对 Zr-Nb 合金耐蚀性能的影响[J]. 稀有金属材料与工程,1998, 27(6): 356-359. LI Pei-zhi, LI Zhong-kui, XUE Xiang-yi, LIU Jian-zhang. Influence of alloying elements on the corrosion resistance of Zr-Nb alloys [J]. Rare Metal Materials and Engineering, 1998, 27(6): 356-359.

[14] 刘建章. 核结构材料[M]. 北京:化学工业出版社,2007:19-22.

LIU Jian-zhang. Structure nuclear materials[M]. Beijing: Chemical Industry Press, 2007: 19-22.

[15] 杨文斗. 反应堆材料学[M]. 2 版. 北京:原子能出版社,2006:260-264.

YANG Wen-dou. Reactor materials science[M]. 2nd ed. Beijing: Atomic Energy Press, 2006: 260-264.

[16] YAO Mei-yi, ZHOU Bang-xin, LI Qiang, LIU Wen-qing, GENG Xun, LU Yan-ping. A superior corrosion behavior

of Zircaloy - 4 in lithiated water at 360 ℃/18. 6 MPa by β-quenching[J]. Journal of Nuclear Materials, 2008, 374: 197 - 203.

[17] BROY Y, GARZAROLLI F, SEIBOLD A, VAN-SWAM L E. Influence of transition elements Fe, Cr and V on long-time corrosion in PWRs[C]// Zirconium in the Nuclear Industry: The 12th International Symposium. Philadelphia: ASTM STP 1354, 2000: 609 - 622.

[18] ZHOU Bang-xin, LI Qiang, YAO Mei-yi, LIU Wen-qing, ZHU Yu-liang. Effect of water chemistry and composition on microstructural evolution of oxide on Zr alloys[C]// Zirconium in the Nuclear Industry: The 15th International Symposium. Philadelphia: ASTM STP 1505, 2008: 360 - 380.

[19] 周邦新,李强,姚美意,刘文庆,褚于良. 锆-4合金在高压釜中腐蚀时氧化膜显微组织的演化[J]. 核动力工程,2005, 26(4): 364 - 371.
    ZHOU Bang-xin, LI Qiang, YAO Mei-yi, LIU Wen-qing, ZHU Yu-liang. Microstructure evolution of oxide films formed on Zircaloy - 4 during autoclave tests[J]. Nuclear Power Engineering, 2005,26(4): 364 - 371.

[20] 周邦新,李强,刘文庆,姚美意,褚于良. 水化学及合金成分对锆合金氧化膜显微组织演化的影响[J]. 稀有金属材料与工程,2006,35(7): 1009 - 1016.
    ZHOU Bang-xin, LI Qiang, LIU Wen-qing, YAO Mei-yi, ZHU Yu-liang. The effects of water chemistry and composition on the microstructure evolution of oxide films on zirconium alloys during autoclave tests[J]. Rare Metal Materials and Engineering, 2006,35(7): 1009 - 1016.

[21] 周邦新,李强,姚美意,刘文庆,褚于良. Zr-4合金氧化膜的显微组织研究[J]. 腐蚀与防护,2009,30(9): 589 - 594.
    ZHOU Bang-xin, LI Qiang, YAO Mei-yi, LIU Wen-qing, ZHU Yu-liang. Microstructure of oxide films formed on Zircaloy - 4[J]. Corrosion & Protection, 2009,30(9): 589 - 594.

[22] 李铁藩. 金属高温氧化和热腐蚀[M]. 北京：化学工业出版社,2003: 51 - 56.
    LI Tie-fan. Metal high temperature oxidation and thermal corrosion[M]. Beijing: Chemical Industry Press, 2003: 51 - 56.

Effect of Ge addition on corrosion performance of Zr - 4 alloy in lithiated solution

Abstract: In order to investigate the effect of Ge addition on the corrosion resistance of zircaloy - 4 (Zr - 4) alloy, the corrosion performance of Zr - 4 + xGe (x=0. 1%, 0. 5%, mass fraction) was studied in lithiated solution with 0. 01 mol/L LiOH at 360 ℃ by out reactor autoclave testing. The microstructures of the alloys and the compositions of the second phase particles (SPPs) were investigated by TEM and EDS. The results show that compared with Zr - 4 alloy, the corrosion resistances of Zr - 4 + xGe alloys containing 0. 1% and 0. 5% Ge are markedly improved, while the corrosion resistances decrease slightly with the increase of Ge addition. The addition of Ge decreases the solid solution content of Sn in the α - Zr matrix. When the Ge content is 0. 1%, the second phase particles precipitate as $Zr(Fe,Cr)_2$ and $Zr(Fe,Cr,Ge)_2$ in Zr - 4 + xGe alloy. When the Ge content reaches 0. 5%, a part of Ge precipitates as Zr - Sn - Fe - Cr - Ge or Zr_3Ge SPPs. The solid solution of Ge in the α - Zr matrix and small SPPs of $Zr(Fe,Cr)_2$ and $Zr(Fe,Cr,Ge)_2$ are beneficial to the corrosion resistance of Zr - 4 + xGe alloy, while the corrosion resistance decreases due to the precipitation of Zr - Sn - Fe - Cr - Ge and SPPs of Zr_3Ge with large size.

Zr - 0.8Sn - 0.35Nb - 0.4Fe - 0.1Cr - xBi 合金在 400 ℃过热蒸汽中的腐蚀行为*

摘　要： 采用静态高压釜腐蚀试验研究了 Zr - 0.8Sn - 0.35Nb - 0.4Fe - 0.1Cr - xBi(x=0.1, 0.3 和 0.5，质量分数，%)合金在 400 ℃/10.3 MPa 过热蒸汽中的腐蚀行为；并用 TEM、EDS 和 SEM 观察分析了合金和腐蚀后氧化膜的显微组织。结果表明，在 Zr - 0.8Sn - 0.35Nb - 0.4Fe - 0.1Cr 合金中添加 0.1%，0.3% 和 0.5% 的铋对其在 400 ℃过热蒸汽中的耐腐蚀性能都有较大改善作用，但随着铋含量的增加，其改善作用减弱；在 Zr - 0.8Sn - 0.35Nb - 0.4Fe - 0.1Cr 合金中添加 0.1% 铋后，合金中只有 Zr(Fe,Cr,Nb)₂ 一种第二相；添加 0.3% 铋后，有 Zr(Fe,Cr,Nb)₂ 和 Zr - Fe - Sn - Bi 两种第二相析出；添加 0.5% 铋后，有 Zr(Fe,Cr,Nb)₂，Zr - Fe - Sn - Bi 和 Zr - Fe - Cr - Nb - Sn - Bi 三种第二相析出；Zr - 0.8Sn - 0.35Nb - 0.4Fe - 0.1Cr 合金中添加适量铋会促进原来固溶的锡以第二相析出。以上结果说明 Zr - 0.8Sn - 0.35Nb - 0.4Fe - 0.1Cr - xBi 合金在 580 ℃时 α - Zr 基体中可固溶不少于 0.1% 的铋，这对改善合金的耐腐蚀性能是有利的，但含铋和锡第二相的析出则使合金的耐腐蚀性能下降。

合金化是开发高性能锆合金的重要途径之一，目前国际上开发的锆合金主要有 Zr - Sn，Zr - Nb 和 Zr - Sn - Nb 三大系列。在现有这些锆合金的基础上调整合金成分的不同配比或添加其他合金元素是改善锆合金耐腐蚀性能的有效手段。

本课题组研究发现通过 β 相水淬提高固溶在 α - Zr 基体中的合金元素铁和铬的含量后，能显著提高 Zr - 4 合金在 LiOH 水溶液中的耐腐蚀性能[1]。添加不同含量铜(0.05%～0.5%)对 M5 和 Zr - Sn - Nb - Fe - Cr 合金在 500 ℃过热蒸汽中耐蚀性的影响规律是不同的；随着 Cu 含量的增加，M5+xCu 合金的耐蚀性得到明显改善[2]，而铜含量对 Zr - Sn - Nb - Fe - Cr - xCu 合金耐腐蚀性能的影响不大[3]。铋元素在 α - Zr 中固溶度相对较大，根据 Zr - Bi 二元相图，在 580 ℃时大约可达到 6%(质量分数)，另外铋的热中子吸收截面很小(0.034 b)，符合包壳材料热中子吸收截面低的要求。铋很可能是一种合适的合金添加元素[4]。

因此，本课题组系统开展了添加铋对 Zr - Sn，Zr - Nb 和 Zr - Sn - Nb 三大系列锆合金在多种水化学条件下耐腐蚀性能的影响研究。本工作主要研究添加铋对 Zr - 0.8Sn - 0.35Nb - 0.4Fe - 0.1Cr(以下简称 S5)合金在 400 ℃/10.3 MPa 过热蒸汽中腐蚀行为的影响，探讨铋与锆合金中其他合金元素的相互作用及其对锆合金耐腐蚀性能的影响机制。

1　试验

以 S5 为母合金，添加适量的铋，采用真空非自耗电弧炉熔炼成 S5+xBi(x=0.1,0.3 和 0.5)合金锭，合金熔炼过程考虑了铋的烧损，并以 S5 重熔合金(定义为 S5 - remelted)作为对照。为便于轧制加工，先将熔炼好的合金锭在 700 ℃下热压成长条状，然后在 1 030 ℃下保

---

* 本文合作者：张伟鹏、姚美意、朱莉、张金龙、李强。原发表于《腐蚀与防护》，2013,34(6):463 - 467,481.

温 40 min 进行均匀化处理,再通过热轧、β 相水淬及冷轧等工艺制备成 20 mm×8 mm× 0.6 mm 的腐蚀试验用片状样品,最后进行 580 ℃/50 h 的再结晶退火. 具体制备工艺见文献[2-3]. 在每次热处理前都用体积分数为 10%HF＋30%HNO₃＋30%H₂SO₄＋30%H₂O 的混合酸进行酸洗. 用电感耦合等离子体原子发射光谱(ICP-AES)分析得到的合金成分见表 1. 分析结果显示样品中的铋含量与设计成分较吻合,其他合金元素有少量挥发,但偏差不大. 制备好的样品经过上述混合酸酸洗和去离子水清洗后放入静态高压釜中进行 400 ℃/10.3 MPa 的过热蒸汽腐蚀试验,腐蚀增重结果为 5 块试样的平均值.

表 1　试验用锆合金化学成分　　　　　　　　　　　　　　　　　　　　　%

合金编号	Sn	Nb	Fe	Cr	Bi	Zr
S5 重熔合金	0.76	0.35	0.37	0.075	—	余量
S5＋0.1Bi	0.74	0.33	0.37	0.12	0.07	余量
S5＋0.3Bi	0.78	0.33	0.36	0.11	0.28	余量
S5＋0.5Bi	0.75	0.32	0.37	0.081	0.53	余量

采用 JEM-200CX 型透射电子显微镜(TEM)观察合金试样的显微组织;JEM-2010F 型高分辨透射电子显微镜及其配置的能谱仪(EDS)分析合金中第二相的成分;JSM-6700F 型扫描电子显微镜(SEM)观察腐蚀样品氧化膜的断口形貌. TEM 样品采用双喷电解抛光法制得;氧化膜断口形貌 SEM 观察用样品是先用混合酸将金属基体溶去一部分,再将露出的氧化膜掰断制得. 在进行氧化膜形貌观察前,为提高图像质量,在试样表面蒸镀一层金.

2　结果与讨论

2.1　腐蚀增重

图 1 是 S5＋xBi 系列合金在 400 ℃/10.3 MPa 过热蒸汽中的腐蚀增重曲线. 由图 1 可见,S5 重熔合金的耐蚀性最差,S5＋0.1Bi 合金最好,S5＋0.3Bi 合金和 S5＋0.5Bi 合金略次之. 腐蚀 462 d 时,S5＋0.1Bi 合金的平均增重只有 177.3 mg·dm⁻²,比 S5 重熔合金降低 24%;S5＋0.3Bi 和 S5＋0.5Bi 合金试样的平均增重分别为 191.9 mg·dm⁻² 和 196.8 mg·dm⁻². 与 S5 合金相比,在 S5 合金中添加 0.1%～0.5%铋后时合金的耐蚀性都有明显提高,但随着铋含量的增加,合金的耐蚀性又会变差. 这说明要改善 S5 合金的耐蚀性,不宜添加过多的铋.

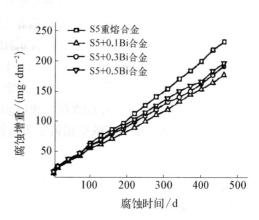

图 1　S5＋xBi 系列合金在 400 ℃/10.3 MPa 过热蒸汽中的腐蚀增重曲线

2.2　合金的显微组织

图 2 为 S5＋xBi 合金腐蚀前显微组织的 TEM 像. 经过 β 相区水淬、冷轧及最终 580 ℃/50 h 退火后,S5＋xBi 系列合金均发生了再结晶. 合金中的第二相呈条带状分布,这可能是

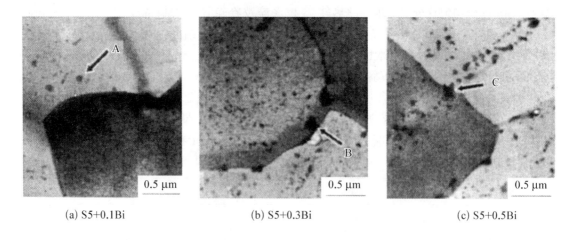

(a) S5+0.1Bi   (b) S5+0.3Bi   (c) S5+0.5Bi

图 2   腐蚀前 S5+xBi 合金显微组织的 TEM 像

因为合金在 β 相水淬后残留的 β-Zr 在随后的冷轧和退火过程中发生分解而形成,也可能是在 β 相水淬冷却过程中已有一部分第二相沿板条晶界析出的结果.合金中的第二相在晶粒内部及晶界处均有分布.

    图 3 为 S5+xBi 系列合金中具有代表性的第二相 EDS 谱,谱图中的铜峰来自 TEM 样品台.表 2 为 S5+xBi 系列合金中的各种第二相信息.由图 2 和表 2 可见:① S5+0.1Bi 合金中的第二相数量较少,多数呈球状,尺寸在 30~60 nm 之间,还有一部分为椭球状及不规则形状,尺寸在 100~150 nm 之间,见图 2(a);EDS 分析表明,大小不同的第二相都由锆、铌、铁和铬四种元素组成,Fe/Cr 比值在 2.3~2.8 之间,结合文献[5]可知,合金中只含有一种第二相,即密排六方结构 Zr(Fe,Cr,Nb)$_2$ 第二相.② S5+0.3Bi 合金中的第二相明显比 S5+0.1Bi 合金中的多,较小的第二相(30~60 nm)呈球状,大多分布于晶内,较大第二相(100~150 mn)呈不规则状,多数分布在晶界处(图 2b).EDS 分析表明,其中一部分第二相含锆、铌、铁和铬四种元素,Fe/Cr 比值在 2.5~3 之间,为密排六方结构 Zr(Fe,Cr,Nb)$_2$ 第二相,与 S5+0.1Bi 合金中的 Zr(Fe,Cr,Nb)$_2$ 第二相成分差别不大;另一部分第二相含锆、铁、锡和铋四种元素,Bi/Sn 比值在 0.7~1 之间,尺寸从 30 nm 到 150 nm 不等,其中尺寸在 100~150 nm 的第二相多数分布在晶界处.③ S5+0.5Bi 合金中第二相的分布均匀程度比 S5+0.3Bi 合金的差,其中较大第二相(100~150 nm)的数量比例明显比 S5+0.3Bi 合金中的高.EDS 分析表明有三种类型第二为密排六方结构的 Zr(Fe,Cr,Nb)$_2$ 第二相,成分与上

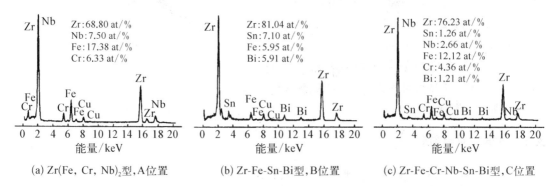

(a) Zr(Fe, Cr, Nb)$_2$型,A位置   (b) Zr-Fe-Sn-Bi型,B位置   (c) Zr-Fe-Cr-Nb-Sn-Bi型,C位置

图 3   S5+xBi 系列合金中典型第二相的 EDS 谱

表 2　S5+xBi 系列合金中各类第二相信息

合金编号	Zr－Fe－Cr－Nb		Zr－Fe－Sn－Bi		Zr－Fe－Cr－Nb－Sn－Bi		
	概率	Fe/Cr	概率	Bi/Sn	概率	Fe/Cr	Bi/Sn
S5+0.1Bi	所有	2.3～2.8	—	—	—	—	—
S5+0.3Bi	部分	2.5～3	部分	0.7～1	—	—	—
S5+0.5Bi	部分	2～2.7	部分	1.0～2.1	部分	2.5～3.2	1.0～1.5

述两种合金中的 Zr(Fe,Cr,Nb)$_2$ 第二相类似;一种为 Zr－Fe－Sn－Bi 第二相,Bi/Sn 比值在 1～2.1 之间;另外还析出一种含锆、铁、铬、铌、锡和铋六种元素的第二相,Fe/Cr 比值在 2.5～3.2 之间,Bi/Sn 比值在 1～1.5 之间. 以上结果说明,580 ℃时在 S5+xBi 合金 α－Zr 基体中至少可以固溶 0.1%的铋,比 Zr－4+xBi 合金基体中铋的固溶度大,这可能主要与 S5 合金中的锡含量比 Zr－4 合金中低有关;同时适量铋的添加促进了原来固溶的锡以第二相析出,这与 Zr－4 合金中添加铋的结果一致.

2.3　腐蚀样品氧化膜断口形貌

图 4 是 S5+0.1Bi 合金和 S5+0.5Bi 合金在 400 ℃/10.3 MPa 过热蒸汽中腐蚀 72 d 的氧化膜断口形貌. 由图 4(a)可见,S5+0.1Bi 试样氧化膜厚度约为 2 μm,氧化膜比较致密,断口的起伏程度较小,ZrO$_2$ 晶粒主要是柱状晶,没有明显的等轴晶区域,但在高倍下可以看到柱状晶转化为等轴晶的痕迹,如图 4(a)中的圆圈区域所示. 从图 4(b)可见,S5+0.5Bi 试样氧化膜厚度约为 3 μm,断口呈台阶状,起伏程度较大,在靠近金属基体的区域有明显的微裂纹,近似平行于金属/氧化膜界面,在靠近氧化膜外表面有明显的等轴晶区域,如图 4(b)中虚线所示. 在高倍下也可看到柱状晶转化为等轴晶的痕迹,如图 4(b)中圆圈区域所示. 氧化膜中等轴晶是由于柱状晶中的缺陷通过扩散凝聚而构成新的晶界发展而成[6-7]. 通过对比发现,在腐蚀初期(72 d),S5+0.5Bi 合金与 S5+0.1Bi 合金腐蚀增重相差不大的情况下,其氧化膜显微组织已有明显区别. S5+0.1Bi 合金氧化膜更致密,没有明显的微裂纹,在进一步腐蚀过程中 O^{2-} 和 OH$^-$ 不容易扩散到金属/氧化膜界面处,从而延缓合金的腐蚀,表现出更佳的耐腐蚀性能.

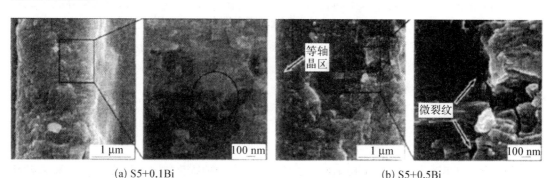

(a) S5+0.1Bi　　　　　　　　　　(b) S5+0.5Bi

图 4　合金在 400 ℃/10.3 MPa 过热蒸汽中腐蚀 72 d 后氧化膜断口形貌

有学者[1-4,8-10]认为 α－Zr 基体中固溶的合金元素含量是影响锆合金耐腐蚀的主要因素,也有学者[11-13]认为第二相是影响锆合金耐腐蚀性能的主要因素. 笔者认为这两方面都会对锆合金的耐腐蚀性能产生影响,只是对于不同成分的锆合金在不同水化学条件下腐蚀时,影响其耐腐蚀性能的主导因素可能不同. 本工作发现在 S5 合金中添加 0.1%～0.5% 的铋对其在 400 ℃/10.3 MPa 过热蒸汽中的耐腐蚀性能有较大的改善作用，但随着铋含量的增加改善作用减弱，即耐腐蚀性能随着铋含量的增加而降低（见图 1）. 显微组织观察表明，580 ℃ 时在 S5+xBi 合金 α-Zr 中可以固溶不小于 0.1% 的铋，这说明这部分固溶的铋可改善合金的耐腐蚀性能. 但随着铋质量分数增加到 0.3% 以后，合金中的一部分铋会以第二相形式析出，同时还会促进 S5 合金中原来固溶的锡以第二相析出（表 2），且随铋含量的增加，含铋和锡第二相的比例增加. 这说明含铋和锡第二相的析出会降低合金的耐腐蚀性能，这与 Zr-4 中添加铋的腐蚀规律一致.

锆合金的腐蚀过程是 O^{2-} 或 OH^- 通过氧化膜中的晶界或阴离子空位扩散到金属/氧化膜界面处，在金属/氧化膜界面处生成 ZrO_2 的过程[14-15]. 因此氧化膜的性质（如孔隙和微裂纹）会对锆合金的耐腐蚀性能产生影响. 周邦新等[16-17]认为：由于锆的 PB 比（金属氧化物与金属的体积比）为 1.56，因此锆在发生氧化时体积会发生膨胀，同时受到金属基体的约束，这使得氧化膜的内部产生压应力，从而使 ZrO_2 晶体内产生许多缺陷. 这些缺陷在应力和温度的作用下，发生扩散、湮没和凝聚，空位被晶界吸收后形成纳米大小的孔隙簇，孔隙簇进一步发展成微裂纹，引起腐蚀加速. 任何延缓这一过程的因素都会改善合金的耐腐蚀性能，反之则会降低合金的耐腐蚀性能. 在锆发生氧化时，合金中的第二相和固溶在 α-Zr 中的合金元素也会发生氧化，影响氧化膜的性质和结构，最终影响合金的耐腐蚀性能. 固溶在 α-Zr 基体中的铋比第二相中的铋更可能固溶在 ZrO_2 基体中，这可能会减少因 O^{2-} 和 OH^- 渗入引起的 ZrO_2 表面自由能降低的程度，使氧化膜中的空位不容易扩散凝聚形成微孔隙和晶界微裂纹，从而改善合金的耐腐蚀性能[17]. 铋的添加促进了原来固溶在 α-Zr 基体中的一部分锡以第二相析出. 由于第二相的影响范围主要在其周边区域，因此与固溶在 α-Zr 基体中的锡相比，第二相中的锡氧化后对腐蚀形成的氧化膜影响也更具局部性. 锡的 PB 比大于 1，这导致局部产生附加应力，促进氧化膜中孔隙和微裂纹的形成，从而加速腐蚀. 含铋和锡第二相的析出会抵消固溶铋对锆合金耐腐蚀性能的改善作用，随着铋含量的增加，含铋和锡第二相的数量增多，这种抵消作用增强，从而 S5+xBi 合金的耐腐蚀性能随着铋含量的增加而变差.

另外，锆合金在高温高压水中的腐蚀也是一个电化学过程，腐蚀时 Zr/ZrO_2 界面上的第二相与 α-Zr 基体组成许多微电池，α-Zr 基体作为阳极优先氧化[18]. 一方面当 α-Zr 基体中固溶铋后会改变其电位，另一方面由于铋的添加改变了第二相的成分和种类，不同第二相的电位是有差异的. 这样，整个体系的电化学行为也随之改变，从而影响锆合金的耐腐蚀性能，这是需要进一步研究的问题.

3 结论

（1）Zr-0.8Sn-0.35Nb-0.4Fe-0.1Cr-xBi 合金中第二相的种类与铋含量密切相关. 在 Zr-0.8Sn-0.35Nb-0.4Fe-0.1Cr 合金中添加 0.1%Bi 后，合金中只有 Zr(Fe,Cr,Nb)₂ 一种第二相析出；添加 0.3% 铋后，有 Zr(Fe,Cr,Nb)₂ 和 Zr-Fe-Sn-Bi 两种第二相析出；添加 0.5% 铋后，有 Zr(Fe,Cr,Nb)₂，Zr-Fe-Sn-Bi 和 Zr-Fe-Cr-Nb-Sn-Bi 三种第二相析出. 适量铋的添加促进了原来固溶在 α-Zr 基体中的锡以第二相形式析出.

（2）在 Zr-0.8Sn-0.35Nb-0.4Fe-0.1Cr 合金中添加的 0.1% 铋可全部固溶在 α-Zr

基体中,而添加 0.3% 铋后,部分铋以第二相形式析出,这说明 580 ℃时铋在 Zr - 0.8Sn - 0.35Nb -0.4Fe - 0.1Cr - xBi 合金中的固溶度不低于 0.1%.

（3）与 Zr - 0.8Sn - 0.35Nb - 0.4Fe - 0.1Cr 合金相比,添加 0.1%～0.5% 的铋可改善合金在 400 ℃过热蒸汽中的耐腐蚀性能,但随铋含量的增多,改善作用减弱,这与含铋和 Sn 第二相的析出有关.这说明 Zr - 0.8Sn - 0.35Nb - 0.4Fe - 0.1Cr - xBi 合金要获得好的耐腐蚀性能,添加的铋含量不宜过高.

参 考 文 献

[1] Yao M Y, Zhou B X, Li Q, et al. A superior corrosion behavior of zircaloy - 4 in lithiated water at 360 ℃/18.6 MPa by β-quenching [J]. J Nucl Mater, 2008, 374(2):197 - 203.

[2] 李士炉,姚美意,张欣,等.添加 Cu 对 M5 合金在 500 ℃过热蒸汽中耐腐蚀性能的影响[J].金属学报,2011,47(2):163 - 168.

[3] 姚美意,张宇,李士炉,等.Cu 含量对 Zr - 0.80Sn - 0.34Nb - 0.39Fe - 0.10Cr - xCu 合金在 500 ℃过热蒸汽中耐腐蚀性能的影响[J].金属学报,2011,47(7):872 - 876.

[4] 李佩志,李中奎,薛祥义,等.合金元素对 Zr - Nb 合金耐蚀性能的影响[J].稀有金属材料与工程,1998,27(6):356 - 359.

[5] 姚美意,李士炉,张欣,等.添加 Nb 对 Zr - 4 合金在 500 ℃过热蒸汽中耐腐蚀性能的影响[J].金属学报,2011,47(7):865 - 871.

[6] Anada H, Takeda K. Microstructure of oxides on zircaloy - 4, 1.0 Nb zircaloy - 4, and zircaloy - 2 formed in 10.3 - MPa steam at 673 K [C]// Zirconium in the Nuclear industry: 11th International Symposium, ASTM STP 1295, Bradley E R and Sabol G P Eds. Garmisch-Partenkirchen: ASTM International, 1996:35 - 54.

[7] Wadman B, Lai Z, Andren H O. et al. Microstructure of oxide layers formed during autoclave testing of zirconium alloys[C]// Zirconium in the Nuclear Industry: 10th International Symposium, ASTM STP 1245, Garde A M, Bradley E R Eds. Baltimore, MD, PA: ASTM International, 1994:579 - 598.

[8] Jeong Y H, Lee K O, Kim H G. Correlation between microstructure and corrosion behavior of Zr - Nb binary alloy [J]. J Nucl Mater, 2002, 302(1):9 - 19.

[9] Kim H G, Jeong Y H, Kim T H. Effect of isothermal annealing on the corrosion behavior of Zr-xNb alloys [J]. J Nucl Mater, 2004, 326(1):125 - 131.

[10] Jeong Y H, Kim H G, Kim D J, et al. Influence of Nb concentration in the α-matrix on the corrosion behavior Zr-xNb binary alloys [J]. J Nucl Mater, 2003, 323(1):72 - 80.

[11] Comstock R J, Schoenberger G, Sable G P. Influence of processing variables and alloy chemistry on the corrosion behavior of ZIRLO nuclear fuel cladding [C]// Zirconium in the Nuclear Industry: Eleventh International Symposium, ASTM STP 1295, Bradley E R and Sabol G P Eds. Garmisch-Partenkirchen, Germany, PA: ASTM International, 1996:710 - 725.

[12] Rudling P, Wikmark G A. Unified model of zircaloy BWR corrosion and hydriding mechanisms [J]. J Nucl Mater, 1999, 265 (1):44 - 59.

[13] Rudling P, Wikmark G. Impact of second phase particles on BWR Zr - 2 corrosion and hydriding performance [C]// Zirconium in Nuclear Industry: Twelfth International Symposium, ASTM STP 1354, Sabol G P, Moan G D Eds. Toronto: ASTM International, 2000:678 - 706.

[14] 刘建章.核结构材料[M].北京:化学工业出版社,2007.

[15] Yao M Y, Wang J H, Peng J C, et al. Study on the role of second phase particles in hydrogen uptake behavior of zirconium alloys [C]//Zirconium in the Nuclear Industry: Sixteenth International Symposium, ASTM STP 1529. Chengdu: ASTM International, 2011:466 - 495.

[16] 周邦新,李强,姚美意,等.锆 - 4 合金在高压釜中腐蚀时氧化膜显微组织的演化[J].核动力工程,2005,26(4):364 - 371.

[17] Zhou B X, Li Q, Yao M Y, et al. Effect of water chemistry and composition on microstructural evolution of oxide on

Zr alloys [C]// Zirconium in the nuclear industry: 15th International Symposium, ASTM STP 1505. Oregon: ASTM International, 2009: 360 - 383.

[18] Weidinger H G, Ruhmann H, Cheliotis G, et al. Corrosion-electrochemical properties of zirconium intermetallics [C]// Zirconium in the Nuclear Industry: 9th International Symposium, ASTM STP 1132, Japan: ASTM International, 1991: 499 - 535.

Corrosion Behavior of Zr - 0.8Sn - 0.35Nb - 0.4Fe - 0.1Cr - xBi Alloys in Superheated Steam at 400 ℃ /10.3 MPa

Abstract: Autoclave test was employed to investigate the effect of Bi contents on the corrosion resistance of Zr -0.8Sn - 0.35Nb - 0.4Fe - 0.1Cr - xBi (x=0.1%—0.5%, mass fraction) alloys in superheated steam at 400 ℃ and 10.3 MPa. The microstructures of the alloys and oxide films on the corroded specimens were observed by TEM and SEM, respectively. The results show that (0.1%—0.5%) Bi addition can improve the corrosion resistance of the Zr - 0.8Sn - 0.35Nb - 0.4Fe - 0.1Cr alloy, but the degree of the improvement effect decreases with the increase of Bi contents. Only Zr(Fe,Cr,Nb)$_2$ second phase particles (SPPs) were detected in the alloy with 0.1% Bi; Zr(Fe, Cr, Nb)$_2$ and Zr - Fe - Sn - Bi SPPs were detected in the alloy with 0.3% Bi. Zr(Fe,Cr,Nb)$_2$, Zi - Fe - Sn - Bi and Zr - Fe - Cr - Nb - Sn - Bi SPPs were detected in the alloy with 0.5% Bi. This implies that the Bi addition promotes the precipitation of Sn as the SPPs and the solid solubility of Bi in α - Zr matrix of Zr - 0.8Sn - 0.35Nb - 0.4Fe - 0.1Cr - xBi alloys is not less than 0.1% at 580 ℃. From the above results, it can be concluded that the Bi as solid solution in α - Zr matrix can improve the corrosion resistance, however, the precipitation of SPPs containing Bi and Sn is deleterious to the corrosion resistance.

Zr − 0.80Sn − 0.4Nb − 0.4Fe − 0.10Cr − xCu 合金 在 400 ℃ 过热蒸气中的耐腐蚀性能*

摘　要： 采用静态高压釜腐蚀试验研究了 Zr − 0.80Sn − 0.4Nb − 0.4Fe − 0.10Cr − xCu(x=0.05～0.5,质量分数,％)合金在 400 ℃,10.3 MPa 过热蒸气中的耐腐蚀性能,用 TEM 和 SEM 分别观察了合金的显微组织和氧化膜的断口形貌. 结果表明:当 Cu 含量不超过 0.2％时,合金中析出的第二相主要是尺寸较小的 Zr(Fe,Cr,Nb)$_2$ 型和少量尺寸相对较大的含 Cu 的 Zr$_3$Fe 型;当 Cu 含量超过 0.2％时,合金中析出了 Zr$_2$Cu 型第二相,随着 Cu 含量的增加,Zr$_2$Cu 型第二相尺寸增大,数量增多;在添加 0.05％Cu 的合金中就有含 Cu 第二相的析出,说明 Zr − 0.80Sn − 0.4Nb − 0.4Fe − 0.10Cr − xCu 合金 α − Zr 基体中固溶的 Cu 含量很低. 当 Cu 含量不超过 0.35％时,合金的耐腐蚀性能基本没有差别;但是当 Cu 含量达到 0.5％时,由于合金中析出了尺寸较大、数量较多的 Zr$_2$Cu 型第二相,致使合金的耐腐蚀性能变差.

　　锆合金具有热中子吸收截面小,耐腐蚀性能和力学性能优良等特点而被广泛用作核动力反应堆燃料元件包壳以及堆内的某些结构构件. 为了提高核电的经济性,核动力反应堆技术正朝着加深燃料燃耗,延长换料周期的方向发展,这对锆合金的性能提出了更高的要求,尤其是耐腐蚀性能. 传统的 Zr − 4 合金即使在优化合金成分和改进热加工工艺后,仍不能满足高燃耗及延长换料周期的要求. 因此,自 20 世纪 80 年代以来,世界各主要核电国家相继推出了多种新型锆合金,其中主要有以美国的 ZIRLO 等合金为代表的 Zr − Sn − Nb 系列和以法国的 M5 等合金为代表的 Zr − Nb 系列. 我国在 20 世纪 90 年代也开发研究了同属 Zr − Sn − Nb 系的 N18 和 N36 新型锆合金[1].

　　近几年来,Park 等[2] 报道了在 Zr − 1.1Nb(质量分数,％)基础上添加 0.05％Cu 发展起来的 HANA − 6 合金具有更加优良的耐腐蚀性能. Hong 等[3] 研究发现合金元素 Cu 对 Zr − 4 合金耐腐蚀性能也有一定的改善作用. 从这些结果看,Cu 似乎对提高锆合金耐腐蚀性能是有益的添加元素,但是目前有关 Cu 影响锆合金耐腐蚀性能的研究并不系统,其影响机理也不清楚. 为此,本研究在 S5 合金(Zr − 0.80Sn − 0.4Nb − 0.4Fe − 0.10Cr,我国自主研发的优化 N18 锆合金,属 Zr − Sn − Nb 系)和 Zr − 1Nb 合金(Zr − Nb 系)基础上添加不同含量的 Cu,系统研究了合金元素 Cu 对 2 种不同体系锆合金耐腐蚀性能的影响规律. 研究结果表明,Cu 对 S5 合金和 Zr − 1Nb 合金在 500 ℃,10.3 MPa 过热蒸气中耐腐蚀性能的影响规律是不同的[4,5];Cu 对 Zr − 1Nb 合金 400 和 500 ℃,10.3 MPa 过热蒸气中耐腐蚀性能的影响规律也不完全相同[4]. 这说明 Cu 对锆合金腐蚀性能的影响与合金成分和腐蚀条件密切相关.

　　本工作是系统研究 Cu 对锆合金耐腐蚀性能影响规律的一部分,主要研究了 Cu 含量对 S5 合金在 400 ℃,10.3 MPa 过热蒸气中耐腐蚀性能的影响,为进一步了解合金元素 Cu 对不同体系锆合金耐腐蚀性能的影响提供实验和理论依据.

---

* 本文合作者：张欣、姚美意、李中奎、周军、李强. 原发表于《稀有金属材料与工程》,2013,42(6)：1210 − 1214.

1 实验

以 S5 为母合金,添加不同含量的 Cu 制备出 Zr-0.80Sn-0.4Nb-0.4Fe-0.10Cr-xCu(x=0.05, 0.1, 0.2, 0.35 和 0.5)合金(简称为 S5+xCu),同时以 S5 重熔(S5-remelted)试样作为对照. 这些合金与文献[5]中报道的合金来自同一批次熔炼,其熔炼加工制备过程与文献[5]基本相同,即在高纯 Ar 气保护下,用真空非自耗电弧炉熔炼成约 60 g 的合金锭,为保证合金成分均匀,合金锭共熔炼 6 次,每熔炼一次都要翻转. 合金锭经热压成条块状后进行 1 030 ℃,40 min 的 β 相均匀化处理,再通过热轧、1 030 ℃,40 mm 的 β 相水淬和 50%冷轧等工艺制得 0.6 mm 厚的片状样品,最后进行 580 ℃,10 h 退火处理. 采用这一制备工艺的目的是为了获得均匀细小弥散分布的第二相,在显微组织优化的前提下比较 Cu 含量对耐腐蚀性能的影响. 样品加热退火均在真空中进行,每次热处理前都要采用混合酸(体积比为 30%H$_2$O+30%HNO$_3$+30%H$_2$SO$_4$+10%HF)酸洗去除试样表面的污染及氧化物. β 相水淬是通过将样品真空封装在石英管中,加热到 1 030 ℃保温 40 min 后立即淬入水中并快速敲碎石英管来实现. 将制备好的样品经过上述混合酸酸洗和去离子水清洗后放入静态高压釜中进行 400 ℃,10.3 MPa 的过热蒸气腐蚀试验,腐蚀增重为 5 块试样的平均值.

用电感耦合等离子体原子发射光谱(ICPAES)分析得到的合金成分列于表 1,与文献[5]中给出的成分完全相同. 用 JEM-2010F 型高分辨透射电镜(HRTEM)观察合金试样的显微组织,用能谱(EDS)仪分析合金中第二相成分,由于合金中含有 Cu,所以分析时使用铍样品台. TEM 试样是用双喷电解抛光制备,电解液为 10%(体积分数)高氯酸乙醇溶液,抛光电压为直流 50 V,温度-40 ℃. 样品腐蚀后切下一块,用混合酸溶去金属基体,然后将氧化膜折断,用 JSM-6700F 型扫描电子显微镜(SEM)观察氧化膜的断口形貌. 在观察氧化膜形貌前,为了提高图像质量,试样表面蒸镀了一层 Ir.

表 1 试验用几种合金成分

Table 1 Chemical compositions of several experimental alloys (mass fraction,%)

Alloy	Sn	Nb	Fe	Cr	Cu	Zr
S5+0.05Cu	0.78	0.35	0.34	0.075	0.058	Bal.
S5+0.1Cu	0.82	0.38	0.35	0.073	0.11	Bal.
S5+0.2Cu	0.82	0.37	0.34	0.073	0.21	Bal.
S5+0.35Cu	0.82	0.35	0.34	0.075	0.35	Bal.
S5+0.5Cu	0.82	0.37	0.37	0.075	0.52	Bal.
S5-remelted	0.81	0.35	0.33	0.070	0.006	Bal.

2 实验结果

2.1 合金的显微组织

图 1 是腐蚀前几种合金样品的 TEM 照片. 从图 1 中可以看出,S5+xCu 系列合金经 580 ℃保温 10 h 后,均发生了再结晶,尺寸不同的第二相弥散分布在晶粒内部或晶界上. 按

照第二相粒子的尺寸大致可以分为 2 类,一类尺寸较小,在 60 nm 以下;另一类尺寸相对较大,在 100~350 nm 之间,且随着合金中 Cu 含量的增加,尺寸较小的第 1 类第二相逐渐减少,而第 2 类第二相则有所增加. 对数十个上述两类第二相分别进行了 EDS 分析,图 2 为图 1 中所示的第二相 A, B 和 C 的 EDS 能谱. 文献[5]已对相同合金中第二相的成分和晶体结构进行了 EDS 和 SAD 分析,结合文献[5]和本研究的 EDS 分析结果可知,小于 60 nm 的第 1 类第二相是 Zr(Fe,Cr,Nb)₂ 型;而尺寸较大的第 2 类第二相根据其成分又可分为 2 种,分别是含少量 Cu 的 Zr₃Fe 型和含少量 Fe 的 Zr₂Cu 型. 表 2 列出几种类型的第二相在 S5+ xCu 系列合金中的信息. 从表 2 可以看出,当合金中的 Cu 含量≤0.2% 时,析出的主要是 Zr(Fe,Cr,Nb)₂ 和少量含 Cu 的 Zr₃Fe 型第二相,Zr₃Fe 型第二相会随着 Cu 含量的增加而增

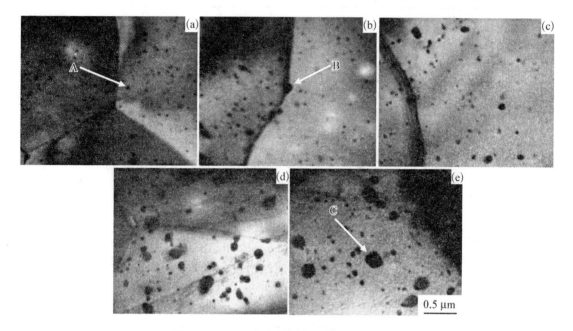

图 1 S5+xCu 合金试样腐蚀前的 TEM 照片

Fig. 1 TEM images of the specimens before corrosion test: (a) S5+0.05Cu, (b) S5+0.1Cu, (c) S5+0.2Cu, (d) S5+0.35Cu, and (e) S5+0.5Cu

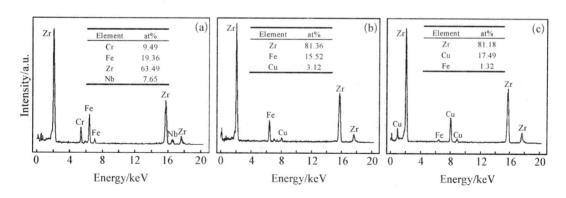

图 2 合金中几种第二相的 EDS 能谱

Fig. 2 EDS spectra of the second phase particles (SPPs) in the alloys: (a) Zr(Fe,Cr,Nb)₂, position A in Fig. 1a; (b) Zr₃Fe, position B in Fig. 1b; (c) Zr₂Cu, position C in Fig. 1c

多，这说明 Cu 可以促进 Zr_3Fe 型第二相的析出；当合金中 Cu 含量＞0.2％时，析出的第二相主要是 $Zr(Fe,Cr,Nb)_2$ 型（部分含 Cu 第二相的尺寸较不含 Cu 的大）、含少量 Cu 的 Zr_3Fe 型和含少量 Fe 的 Zr_2Cu 型，随着 Cu 含量的增加，Zr_2Cu 型第二相数量增多并伴随着尺寸的增大. 从分析结果可以看出，在 S5 合金中添加 0.05％Cu 时，就已有含 Cu 第二相析出，这说明 Cu 在 Zr－Sn－Nb 合金中 α－Zr 基体中的固溶含量较 Zr－Nb 合金中的低[4,5].

表 2　S5＋xCu 系列合金中的第二相
Table 2　Details of the SPPs in S5＋xCu alloys

SPPs type	Frequency					Particle size/nm
	S5+0.05Cu	S5+0.1Cu	S5+0.2Cu	S5+0.35Cu	S5+0.5Cu	
$Zr(Fe,Cr,Nb)_2$	Major	Major	Partly	Partly	Partly	＜60 (major) and 150~250 (minor, contain Cu)
Zr_3Fe	Minor	Minor	Partly	Partly	Partly	100~250 (all, contain Cu)
Zr_2Cu	—	—	—	Partly	Partly	150~300 (partly, contain Fe)

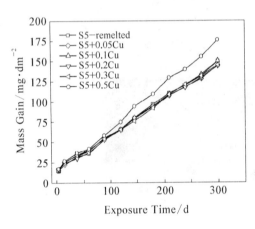

图 3　S5＋xCu 系列合金在 400 ℃, 10.3 MPa 过热蒸气中的腐蚀增重曲线

Fig. 3　Curves of mass gain vs exposure time of the S5＋xCu specimens in 400 ℃, 10.3 MPa superheated steam

2.2　腐蚀增重

图 3 是 S5＋xCu 系列合金在 400 ℃, 10.3 MPa 过热蒸气中腐蚀 300 d 的增重曲线. 从图 3 可以看出，当 S5 合金中添加的 Cu 含量≤0.35％时，合金在腐蚀 300 d 后的增重基本上没有差别. 如腐蚀 300 d 后 S5 重熔样品的增重为 149.9 mg/dm²，S5＋0.1Cu 和 S5＋0.35Cu 样品的腐蚀增重分别为 150.5 和 145.3 mg/dm². 这说明当 S5 合金中的 Cu 含量≤0.35％时，Cu 对 S5 合金在 400 ℃, 10.3 MPa 过热蒸气中的耐腐蚀性能影响不大. 继续提高 S5 合金中的 Cu 含量至 0.5％时，合金的耐腐蚀性变差，S5＋0.5Cu 样品腐蚀 300 d 后增重达到了 175.9 mg/dm². 这说明添加过量的 Cu 会使 S5 合金的耐腐蚀性能变差.

2.3　腐蚀样品氧化膜的断口形貌

图 4 是 S5＋xCu 系列合金在 400 ℃, 10.3 MPa 过热蒸气中腐蚀 270 d 的氧化膜断口形貌. 从图 4 中可以看出，S5 重熔和 S5＋0.35Cu 腐蚀样品的氧化膜厚度差不多，均为 8 μm 左右，而 S5＋0.5Cu 腐蚀样品的氧化膜最厚，约为 10 μm，这与腐蚀增重结果一致. 所有氧化膜断口中都有较多的近似平行于金属/氧化膜界面的横向裂纹（如图 4a，4b，4c 中的箭头所示），这些裂纹可能是样品制备过程中造成的，但也能说明在观察到裂纹的地方在制样前其本身是弱的结合面或已经存在微裂纹. 在离氧化膜表面不远处，S5 重熔和 S5＋0.35Cu 腐蚀样品的氧化膜中以柱状晶为主，且较为致密；而 S5＋0.5Cu 腐蚀样品的氧化膜中已经有大部分柱状晶演化为等轴晶，且有一些微孔隙存在. 这种断口形貌的差别与合金的耐腐蚀性能存在一定的对应关系[6].

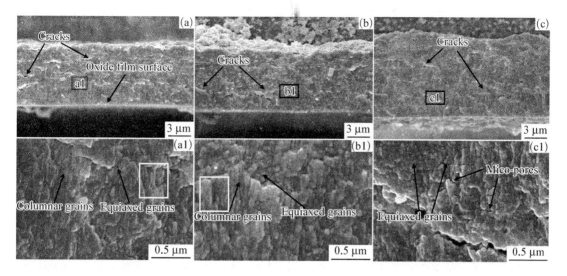

图 4　S5＋xCu 合金样品在 400 ℃，10.3 MPa 过热蒸气中腐蚀 270 d 的氧化膜断口形貌

Fig. 4　SEM images of the fracture surfaces of oxide film on the specimens of S5 - remelted (a, a1), S5＋ 0.35Cu (b, b1) and S5＋0.5Cu (c, c1) corroded in 400 ℃, 10.3 MPa superheated steam for 270 d exposure

3　分析讨论

有学者[7,8]认为 α-Zr 基体中固溶的合金元素含量是影响锆合金耐腐蚀性能的主要因素，也有学者[9-11]认为第二相是影响锆合金耐腐蚀性能的主要因素．应该说这二方面都会对锆合金的耐腐蚀性能产生影响，只是对于不同成分的锆合金在不同腐蚀条件下，影响其耐腐蚀性能的主导因素可能不同．本研究发现在 S5 合金中添加 0.05％～0.35％的 Cu 对其在 400 ℃，10.3 MPa 过热蒸气中的耐腐蚀性能影响不大，但当添加的 Cu 含量增加到 0.5％时，合金的耐腐蚀性能反而下降（图 3）．本工作在研究 Zr-1Nb＋xCu 合金在 500 ℃，10.3 MPa 过热蒸气中的耐腐蚀性能时，发现固溶在 α-Zr 中的 Cu 含量是提高 Zr-1Nb 合金耐腐蚀性能的主要因素[4]．显微组织观察表明，在 S5 合金中添加 0.05％的 Cu 时就会有含 Cu 的第二相析出（表 2），这说明 Cu 在 S5 合金 α-Zr 基体中的固溶含量很低，所以添加适量的 Cu 对 S5 合金耐腐蚀性能的影响不大．当 S5 合金中添加的 Cu 含量进一步增加时，随着 Cu 含量的增加，含 Cu 第二相的数量增多并伴随着尺寸的增大[5]（图 1），说明这些含 Cu 第二相的析出会降低合金的耐腐蚀性能（图 3）．

Zr 在水蒸气中的腐蚀过程是 O^{2-} 或 OH^- 通过氧化膜扩散到金属/氧化膜界面处与 Zr 反应形成 ZrO_2[12]．因此，氧化膜的性质直接影响锆合金的腐蚀行为．合金元素固溶在 α-Zr 中或以第二相析出都会影响氧化膜的性质与结构，从而影响合金的耐腐蚀性能[7]．有研究[6,13]认为氧化膜生长时，是先生长出柱状晶，随着氧化膜显微组织结构的演化，逐渐转变为等轴晶，这在图 4 中也能看出柱状晶转变为等轴晶的痕迹（如图 4a1 和 4b1 中的白框所示）．不同锆合金氧化膜中柱状晶和等轴晶的比例能够反映出氧化膜显微组织结构演化的快慢，也反映了锆合金耐腐蚀性能的差异．也就是说，耐腐蚀性能好的锆合金，柱状晶向等轴晶演化的过程比耐腐蚀性能差的锆合金慢，因而柱状晶的比例高．在 S5＋0.5Cu 样品氧化膜中有较多的等轴晶（图 4c1），这说明其氧化膜显微组织结构演化较快．Zr 发生氧化时的 P. B.

比(金属氧化物与金属体积之比)为 1.56,Cu 氧化时的 P.B. 比为 1.72[14],所以 Cu 氧化后产生的压应力比 Zr 氧化后产生的压应力大. 当 Cu 固溶在 α-Zr 中时,氧化后产生的压应力均匀分布在氧化膜中;但是当 Cu 以第二相形式存在,并且数量较多,尺寸较大时,第二相氧化后在局部会产生相当大的压应力,引起局部应力梯度的变化. 应力梯度的大小必然对氧化膜显微组织结构的演化过程产生影响,应力梯度大的局部区域更容易使氧化膜中空位、间隙原子等缺陷发生扩散和凝聚,从而加速了氧化膜显微组织结构的演化,如柱状晶向等轴晶的转化,孔隙和微裂纹的增多,这使 O^{2-} 或 OH^- 通过氧化膜扩散的通道增多,从而使合金的耐腐蚀性能变差[15]. 因此,这些数量较多、尺寸较大的含 Cu 第二相可能是导致 S5+0.5Cu 合金在 400 ℃,10.3 MPa 过热蒸气中耐腐蚀性能变差的主要原因.

4 结论

(1) 当 Zr-0.80Sn-0.4Nb-0.4Fe-0.10Cr-xCu(S5)合金中添加 0.05% Cu 时就有含 Cu 第二相的析出,这说明 α-Zr 基体中固溶的 Cu 含量很低;当 S5 合金中添加的 Cu 含量不超过 0.2% 时,析出的第二相主要是尺寸较小的 $Zr(Fe,Cr,Nb)_2$ 型和尺寸相对较大的含 Cu 的 Zr_3Fe 型;当添加的 Cu 含量超过 0.2% 时,除上述 2 种第二相外,还析出了含少量 Fe 的 Zr_2Cu,随着 Cu 含量的增加,Zr_2Cu 型第二相尺寸增大,数量增多.

(2) 当 S5 合金中添加的 Cu 含量不超过 0.35% 时,S5+xCu 合金的耐腐蚀性能基本没有差别;但是当 S5 合金中添加的 Cu 含量达到 0.5% 时,由于合金中析出了数量较多、尺寸较大的含 Cu 第二相,致使合金的耐腐蚀性能变差.

参 考 文 献

[1] Zhao Wenjin(赵文金),Zhou Bangxin(周邦新),Miao Zhi(苗志)et al. Atomic Energy Science and Technology (原子能科学技术)[J],2005,39(Suppl):1.

[2] Park J Y, Choi B K, Yoo S J et al. J Nucl Mater [J], 2006,359:59.

[3] Hong H S, Moon J S, Kim S J et al. J Nucl Mater[J], 2001,297:113.

[4] Li Shilu(李士炉),Yao Meiyi(姚美意),Zhang Xin(张欣)et al. Acta Metallurgica Sinica (金属学报)[J],2011,47(2):163.

[5] Yao Meiyi (姚美意),Zhang Yu(张宇),Li Shilu(李士炉)et al. Acta Metallurgica Sinica (金属学报)[J],2011,47(7):872.

[6] Zhou Bangxin(周邦新),Li Qiang(李强),Liu Wenqing (刘文庆)et al. Rare Metal Materials and Engineering (稀有金属材料与工程)[J],2006,35(7):1009.

[7] Jeong Y H, Lee K O, Kim H G. J Nucl Mater[J], 2002,302:9.

[8] Kim H G, Jeong Y H, Kim T H. J Nucl Mater[J], 2004,326:125.

[9] Comstock R J, Schoenberger G, Sable G P. In:Bradley E R, Sabol G P eds. Zirconium in the Nuclear Industry: 11th International Symposium[C]. Garmisch Partenkirchen:ASTM International, 1996:710.

[10] Rudling P, Wikmark G A. J Nucl Mater [J], 1999,265:44.

[11] Rudling P, Wikmark G et al. In:Sabol G P, Moan G D eds. Zirconium in the Nuclear Industry: 12th International Symposium[C]. Toronto:ASTM International, 2000:678.

[12] Yang Wendou(杨文斗). Reactor Materials Science (反应堆材料学)[M]. Beijing:Atomic Energy Press, 2006:260.

[13] Wadman B, Lai Z, Andrén H O et al. Zirconium in the Nuclear Industry: 10th International Symposium[C]. Baltimore:ASTM International, 1994:579.

[14] Li Tiefan(李铁藩). *High Temperature Oxidation and Thermal Corrosion of Metals* (金属高温氧化和热腐蚀)[M]. Beijing: Chemical Industry Press, 2003: 52.

[15] Zhou B X, Li Q, Yao M Y *et al*. In: Kammenzind B, Limbäck M eds. *Zirconium in the Nuclear Industry: 15th International Symposium*[C]. Baltimore: ASTM International, 2009: 360.

Corrosion Resistance of Zr - 0. 80Sn - 0. 4Nb - 0. 4Fe - 0. 10Cr - *x*Cu Alloys in Super-Heated Steam at 400 ℃

Abstract: The effect of Cu content on the corrosion resistance of Zr - 0. 80Sn - 0. 4Nb - 0. 4Fe - 0. 10Cr - *x*Cu alloys ($x=0.05\sim0.5$, mass fraction, %) was investigated in superheated steam at 400 ℃ and 10. 3 MPa by autoclave tests. The microstructure of the alloys and the fracture surface of oxide films on the corroded specimens were observed by TEM and SEM, respectively. The results show that when the addition of Cu is below 0. 2%, the second phase particles (SPPs) are mainly $Zr(Fe,Cr,Nb)_2$ in smaller size, and a few Zr_3Fe with Cu. When the addition of Cu is above 0. 2%, the SPPs of Zr_2Cu are precipitated. The Zr_2Cu particles become larger in size and more in amount with the increase of Cu content. The precipitates containing Cu are found in the alloy even with 0. 05% Cu, which indicates that the Cu concentration in α - Zr matrix is very small. Cu addition of 0. 05%~0. 35% has little effect on the corrosion resistance of the alloys, but when the addition of Cu reaches 0. 5%, the corrosion resistance of the alloy decreases, which is the worst one among the alloys.

附：周邦新及其科研团队发表论文清单

（截至 2013 年）

1958 年

［1］颜鸣皋,周邦新.冷轧铜板再结晶结构的形成[J].物理学报.1958,14(2):121-135.

1959 年

［2］周邦新,王维敏,陈能宽.铁硅合金中立方织构的形成[J].Φ.M.M.(俄文期刊).1959,8(6):885-891.

1960 年

［3］周邦新,王维敏,陈能宽.铁硅合金中立方织构的形成[J].物理学报.1960,16(3):155-159.

［4］王维敏,周邦新,陈能宽.铁硅合金(110)[001]单晶体的形变和再结晶[J].物理学报.1960,16(5):263-271.

1963 年

［5］周邦新,颜鸣皋.纯铜薄带厚度对冷轧及再结晶织构的影响[J].金属学报.1963,6(2):163-175.

［6］周邦新,颜鸣皋.磷对冷轧纯铜再结晶的影响[J].物理学报.1963,19(10):633-648.

［7］周邦新.钼单晶体的范性变形[J].物理学报.1963,19(5):285-295.

［8］周邦新.钼单晶体的冷轧及再结晶织构[J].物理学报.1963,19(5):297-305.

1964 年

［9］陈能宽,周邦新.金属的回复与再结晶[C]//晶体缺陷与强度(下册).北京:科学出版社,1964:55-107.

［10］周邦新.铁硅单晶体的冷轧及再结晶织构[J].金属学报.1964,7(4):423-436.

1965 年

［11］周邦新,刘起秀.α铀冷轧及再结晶织构[J].原子能科学技术.1965(2):138-147.

［12］周邦新,陈能宽.热轧铁硅单晶体再结晶的研究[J].金属学报.1965,8(2):244-252.

［13］周邦新.铁硅合金中(110)[001]和(100)[001]织构的形成[J].金属学报.1965,8(3):380-393.

［14］周邦新,刘起秀.钨和铌单晶体的冷轧及再结晶织构[J].金属学报.1965,8(3):340-345.

［15］周邦新,孔令枢.用普通光照明观察α铀晶粒组织的金相技术[J].原子能科学技术.1965(3):249-255.

［16］周邦新,刘起秀.铀板的再结晶[J].原子能科学技术.1965(8):734-739.

1976 年

［17］周邦新,吴国安.12CrMoV 钢管焊接缺陷的研究[J].金属材料研究.1976(4):475-477.

1980 年

［18］白延祖(白点研究小组的化名,小组负责人周邦新).锆-2 合金管材焊接后在过热蒸汽中的不均匀腐

蚀[J]. 核动力工程. 1980(4):28 - 42.

[19] 周邦新. 制备透射电子显微镜金属薄膜样品的自动控制装置[J]. 物理. 1980,9(5):411 - 413.

1981 年

[20] 周邦新,孔令枢. α铀假单晶体的制备[J]. 原子能科学技术. 1981(5):598 - 601.

[21] 周邦新,刘起秀. Al 和 Cu 单晶体拉伸形变后的再结晶[J]. 金属学报. 1981,17(4): 363 - 373.

1982 年

[22] 周邦新. 锆-2 合金在过热蒸汽中氧化转折机理[C]//1980 核材料会议文集. 北京:原子能出版社,1982:87 -98.

1983 年

[23] 周邦新. 3%Si - Fe 合金屈服前的微应变研究[J]. 金属学报. 1983,19(1):A31 - A39.

[24] 周邦新,盛钟琦. 锆-2 合金与 18/8 奥氏体不锈钢冶金结合层的研究[J]. 核科学与工程. 1983,3(2):153 - 160.

1984 年

[25] 周邦新. 电子显微术在研究 UO_2 燃料芯块中的应用[C]//燃料元件及分析. 1984:1 - 5.

[26] 周邦新,盛钟琦,李卫军. 18/8 奥氏体与 26/1 铁素体不锈钢爆炸焊结合层的电镜研究[J]. 电子显微学报. 1984,3(4):85.

1985 年

[27] 周邦新,赵文金. U - 7.5%Nb - 2.5%Zr 合金的金相及成分均匀性研究[C]//1985 年核材料会议文集(《核科学与工程》增刊). 1985:36 - 40.

1986 年

[28] Zhou Bangxin, Sheng Zhongqi. The Second Phase Particles in Zircaloy - 2 and Their Effects on Corrosion Behavior in Superheated Steam[C]//Progress in Metal Physical Metallurgy, Proceedings of the First Sino-Japanese Symposium on Metal Physics and Physical Metallurgy, Eds by R. R. Hasigutietal. 1986:163 - 170.

1987 年

[29] 周邦新. 钢材闪光焊时灰斑缺陷的形成机理[J]. 理化检验-物理分册. 1987,23(2):20 - 23.

[30] 周邦新,赵文金,黄德诚. 钛和 18/8 不锈钢冶金结合层的研究[J]. 金属科学与工艺. 1987,6(4):26 - 32.

1988 年

[31] 周邦新,马继梅,杨敏华,张琴娣. Zr - 4 板材拉伸性能的研究[J]. 核动力工程. 1988,9(4):64 - 68.

[32] 周邦新,张琴娣,杨敏华. 锆-4 合金渗氢方法的研究[J]. 核动力工程. 1988,9(1):43 - 48.

[33] 周邦新,郑斯奎,汪顺新. 真空电子束焊接对锆-2 合金熔区中成分、组织及腐蚀性能的影响[J]. 核科学与工程. 1988,8(2):130 - 137.

1989 年

[34] Zhou Bangxin. Electron Microscopy Study of Oxide Films Formed on Zircaloy - 2 in Superheated Steam

[C]//Zirconium in the Nuclear Industry: Eighth International Symposium. 1989:360 - 373.

[35] Zhou Bangxin. TEM Study of Bonding Layers in Dissimilar Alloy Explosive Bonding Joints[J]. Acta Metallurgica Sinica (English Letters). 1989,2(4):237 - 243.

[36] 毛合简,周邦新. 国际 RERTR 计划发展现状[J]. 核动力工程. 1989,10(4):71 - 75,46.

[37] 周邦新,郑斯奎,汪顺新. Zr - 2 合金中应力及应变诱发氢化锆析出过程的电子显微镜原位研究[J]. 金属学报. 1989,25(3):A190 - A195.

[38] 周邦新,钱天林. 锆-4 合金氧化膜的结构研究[C]//1989 核材料会议文集(《核科学与工程》增刊). 1989:356 - 360.

[39] 周邦新. 水冷动力堆燃料元件包壳的水侧腐蚀[J]. 核动力工程. 1989,10(6):73.

[40] 周邦新,盛钟琦,彭峰. 异种金属爆炸焊结合层的电子显微镜的研究[J]. 金属学报. 1989,25(1):A7 - A12.

1990 年

[41] 周邦新. (110)〔110〕Fe - Si 单晶体的冷轧和再结晶[J]. 金属学报. 1990,26(5):A340 - A345.

[42] 周邦新,蒋有荣. 锆-2 合金在 500—800 ℃空气中氧化过程的研究[J]. 核动力工程. 1990,11(3):41 - 47.

[43] 周邦新. 锆-2 合金在过热蒸汽中形成氧化膜的电子显微镜研究[J]. 中国腐蚀与防护学报. 1990,10(3):197 -206.

[44] Zhou Bangxin, Jiang Yourong. Oxidation of Zircaloy - 2 in Air from 500 ℃ to 800 ℃[C]// Proceedings of International Symposium on High Temperature Corrosion and Protection. 1990:121 - 124.

1991 年

[45] ZhouBangxin. Cold-rolling and recrystallization of (110)〔110〕iron-silicon single crystals[J]. Acta Metallurgica Sinica (English Letters). 1991,4(2):103 - 107.

[46] 赵文金,周邦新. Zr - 4 合金中第二相的研究[J]. 核动力工程. 1991,12(5):67 - 72,76.

[47] 周邦新,李卫军,递忠信,杨晓林. 316 及含钛或铌 316 不锈钢的显微组织研究[J]. 核科学与工程(增刊)(快堆专集,Vol.1). 1991,11(3):97 - 108.

[48] 周邦新. 核工业中的锆合金及其发展[J]. 核科学与工程. 1991,11(2):28 - 39.

[49] 周邦新,李卫军,杨晓林,递忠信. 模拟裂变物对 316 不锈钢晶界浸蚀的研究[J]. 核科学与工程. 1991,11(3):109 - 115.

[50] Zhou Bangxin, Qian Tianlin. A Study of Microstucture of Oxide Film Formed on Zircaloy - 4[C]// Proceedings of International Symposium on High Temperature Corrosion and Protection. 1991:125 - 128.

[51] Zhou Bangxin, Jiang Yourong. A Study of Stress in Oxide Film Formed on Zircaloy - 2 Tubes[C]// Proceedings of International Symposium on High Temperature Corrosion and Protection. 1991:121 - 124.

1992 年

[52] 周邦新,蒋有荣. Zr - 4 管中氢化物分布的应力再取向研究[J]. 核动力工程. 1992,13(5):66 - 69.

[53] 钱天林,周邦新. 锆-4 合金中第二相大小与腐蚀性能关系的研究[C]//核材料会议文集(1991 年). 成都:四川科学技术出版社,1992:31 - 35.

[54] 周邦新,递忠信. LT24 铝合金腐蚀过程的原位观察[C]//核材料会议文集(1991 年). 成都:四川科学技术出版社,1992:188 - 193.

[55] 周邦新,李卫军,递忠信,李聪. 模拟裂变产物对三种不锈钢晶界浸蚀的研究[C]//核材料会议文集(1991 年). 成都:四川科学技术出版社,1992:70 - 74.

[56] 周邦新,张琴娣,杨敏华,李卫军,黄健庆.模拟裂变产物沿 316(Ti)不锈钢晶界浸蚀后对力学性能的影响[C]//核材料会议文集(1991 年).成都:四川科学技术出版社,1992:75 - 79.

[57] 周邦新,赵文金,潘淑芳,苗志,李聪,蒋有荣.时效处理对锆-4 合金微观组织及腐蚀性能的影响[C]//核材料会议文集(1991 年).成都:四川科学技术出版社,1992:27 - 30.

1993 年

[58] 蒋有荣,周邦新,杨敏华.Zr - 4 板中氢化物应力再取向的研究[J].核动力工程.1993,14(4):368 - 373,380.

[59] 周邦新,李聪,黄德诚.Zr(Fe,Cr)$_2$ 金属间化合物的氧化[J].核动力工程.1993,14(2):149 - 153,190.

[60] 周邦新.锆合金中的疖状腐蚀问题[J].核科学与工程.1993,13(1):51 - 58.

1994 年

[61] 李聪,周邦新.Zr - 4 合金氧化膜(<100 nm)的电镜研究[J].核动力工程.1994,15(2):152 - 157.

[62] 王卫国,周邦新.锆合金板织构的控制[J].核动力工程.1994,15(2):158 - 163.

[63] 杨晓林,周邦新,蒋有荣,李聪.Zr - 4 合金中第二相 Zr(Fe,Cr)$_2$ 的电化学分离[J].核动力工程.1994,15(1):79 - 83,96.

[64] 周邦新,蒋有荣.Cu - Al 爆炸焊结合层的透射电镜研究[J].金属学报.1994,30(3):B104 - B108.

[65] 周海波,张志毅,周邦新.热处理对 U$_3$Si$_2$ - Al 燃料板包壳显微组织及其厚度测量的影响[J].核动力工程.1994,15(3):248 - 253.

[66] Jiang Yourong, Zhou Bangxin. A Study of Stress Reorientation of Hydrides in Zircaloy(锆合金中氢化物应力再取向的研究)[C]// 中国核科技报告.1994:984 - 999.

1995 年

[67] 周邦新,赵文金,苗志,潘淑芳,李聪,蒋有荣.改善锆-4 合金耐腐蚀性能的研究[J].核科学与工程.1995,15(3): 242 - 249.

1996 年

[68] 李聪,周邦新.锆-4 合金氧化膜中 Zr(Fe,Cr)$_2$ 第二相粒子的 HREM 观察[J].腐蚀科学与防护技术.1996,8(3):242 - 246.

[69] Wang Weiguo, Zhou Bangxin. On the Mechanism of Textures Formation in Zircaloy - 4 Plate Rolled at Elevated Temperature in α+β Dual-Phase Region[J]. Textures of Materials ICOTOM - 11. 1996(2): 639 -644.

[70] 王卫国,周邦新.轧制温度对 Zr - 4 合金板织构的影响[J].核动力工程.1996,17(3):255 - 261.

[71] Zhou Bangxin, Li Cong, Miao Zhi, Dai Jiyan. Corrosion Behavior of Zr(Fe,Cr)$_2$ Metallic Compounds in Superheated Steam[C]// 中国核科技报告.1996 (s1):33 - 36.

[72] Zhou Bangxin, Yang Xiaolin. The Effect of Heat Treatments on the Structure and Composition of Second Phase Particles in Zircaloy - 4 [C]// 中国核科技报告.1996:901 - 910.

[73] Zhou Bangxin, Zhao Wenjin, Miao Zhi, Pan Shufang, Li Cong, Jiang Yourong. The Effect of Treatments on the Corrosion Behavior of Zircaloy - 4[C]// 中国核科技报告.1996:911 - 925.

[74] Zhou Hairong, Zhou Bangxin. TEM Study of Microstructure in Explosive Welded Joints Between Zircaloy - 4 and Stainless Steel[C]// 中国核科技报告.1996 (s1):31 - 32.

1997 年

[75] 赵文金,苗志,蒋有荣,蒋宏曼,周邦新.预生膜氧化处理对 Zr - 4 包壳疖状腐蚀的影响[C]//1996 中国

材料研讨会文集(第3卷：生物及环境材料,第2分册：环境材料).北京：化学工业出版社,1997：162-166.

[76] 周邦新,苗志,李聪.Zr(Fe,Cr)$_2$金属间化合物在500 ℃过热蒸汽中的腐蚀研究[J].核动力工程.1997,18(1):53-60.

[77] 周邦新.Zr-Sn-Nb系合金的显微组织研究[C]// 1996中国材料研讨会文集(第3卷：生物及环境材料,第2分册：环境材料).北京：化学工业出版社,1997:187-191.

[78] 周邦新.改善锆合金耐腐蚀性能的概述[J].金属热处理学报.1997,18(3):8-15.

[79] 周邦新.核工业中的有色金属[C]//中国有色金属材料发展现状及迈入21世纪对策(第一次学术研讨会文集).1997:1-14.

[80] 周邦新,杨晓林.热处理对Zr-4合金中第二相结构和成分的影响[J].核动力工程.1997,18(6):511-516.

[81] 周邦新,赵文金,苗志,黄德程,蒋有荣.新锆合金的研究[C]//1996中国材料研讨会文集(第3卷：生物及环境材料,第2分册：环境材料).北京：化学工业出版社,1997:183-186.

[82] 周海蓉,周邦新.Zr-4/1Cr18Ni9Ti爆炸焊结合层的显微组织研究[J].核动力工程.1997,18(1):61-64.

1998 年

[83] 王卫国,周邦新,郑忠民.Fe-Cr-Al-Si合金阻尼性能研究[J].金属学报.1998,34(10):1039-1042.

[84] 王卫国,周邦新.高温轧制Zr-4合金板织构的形成机制[J].核动力工程.1998,19(1):37-42.

[85] 赵文金,苗志,蒋宏曼,李聪,于晓卫,周邦新.加工工艺对Zr-4管抗疖状腐蚀的影响[J].核动力工程.1998,19(5):462-467.

[86] 周邦新,赵文金,蒋有荣,夏邦杰,李京,张立新.Zr-4中合金元素的表面偏聚[J].核动力工程.1998,19(6):506-508.

[87] 周邦新.水化学对燃料元件包壳腐蚀行为的影响[J].核动力工程.1998,19(4):354-359,364.

1999 年

[88] 赵文金,苗志,蒋宏曼,李聪,于晓卫,周邦新.表面处理对锆-4合金抗疖状腐蚀性能的影响[J].稀有金属.1999,23(6):459-460.

2000 年

[89] Wang Weiguo, Zhou Bangxin. Magneto-elastic Interchange in Ferromagnetic Alloys[J]. Materials and Design. 2000,21(6):541-545.

[90] 李强,周邦新.快淬Nd-Fe-B粉末的TEM样品制备[J].稀有金属材料与工程.2000,29(4):283-284.

[91] 王卫国,周邦新.铁磁合金中的磁弹转换[J].金属学报.2000,36(1):81-86.

[92] 周邦新,李强,苗志.β相水淬对锆-4合金耐腐蚀性能的影响[J].核动力工程.2000,21(4):339-343,352.

[93] 周邦新.铁硅合金中形成立方织构的有关问题[J].宝钢技术.2000(5):52-58.

[94] 周邦新,李强,黄强,苗志,赵文金,李聪.水化学对锆合金耐腐蚀性能影响的研究[J].核动力工程.2000,21(5):439-447,472.

[95] 赵文金,蒋宏曼,李聪,周邦新.Zr-4合金基体中Fe,Cr含量的间接测量[C]//中国核科技报告.2000:595-606.

[96] 赵文金,周邦新,苗志,李聪,蒋宏曼,于晓卫,蒋有荣,黄强,苟渊,黄德诚. Development of New Zirconium Alloys for PWR Fuel Rod Claddings(压水堆燃料包壳新锆合金的发展)[C]//中国核科技报

告.2000:608-618.

2001 年

[97] 李强,周邦新.690 合金的显微组织研究[J].金属学报.2001,37(1):8-12.

[98] 刘文庆,周邦新.Zr-4 合金在 LiOH 水溶液中腐蚀机理的概述[J].核动力工程.2001,22(1):65-69.

[99] 刘文庆,李强,周邦新.锆锡合金腐蚀转折机理的讨论[J].稀有金属材料与工程.2001,30(2):81-84.

[100] 姚美意,周邦新.镁合金耐蚀表面处理的研究进展[J].材料保护.2001,34(10):19-21.

2002 年

[101] Li Cong,Zhou Bangxin,Zhao Wenjin,Li Pei,Peng Qian. Determination of Fe and Cr content in α-Zr solid solution of Zircaloy-4 with different heat-treated states[J]. Journal of Nuclear Materials. 2002, 304(2-3):134-138.

[102] 李聪,李蓓,周邦新,赵文金,彭倩,应诗浩,沈保罗.Zr-4 合金 α-Zr 固溶体中的 Fe、Cr 含量分析[J].核动力工程.2002,23(4):20-24.

[103] 李强,刘文庆,周邦新.变形及热处理对 Zr-Sn-Nb 合金中 β-Zr 分解的影响[J].稀有金属材料与工程.2002,31(5):389-392.

[104] 刘文庆,周邦新,李强.Zr-Nb 合金在 LiOH 水溶液中耐腐蚀性的研究[J].核动力工程.2002,23(1):68-70.

[105] 王卫国,周邦新,刘曙光,李卫军.高温退火 Fe-Cr 基减振合金的阻尼性能与晶界析出相[J].材料工程.2002(9):3-6,25.

2003 年

[106] Wang Weiguo, Zhou Bangxin. The local internal stress in ferromagnetic alloys: SB theory and the local internal stress source[J]. Materials and Design. 2003,24(3):163-167.

[107] Zhou Bangxin, Yao Meiyi, Miao Zhi, Li Qiang, Liu Wenqing. The Cracking Induced by Oxidation-Hydriding in Welding Joints of Zircaloy-4 Plates[J]. Journal of Shanghai University (English Edition). 2003,7(1):18-20.

[108] 刘文庆,周邦新,李强.ZIRLO 合金和 Zr-4 合金在 LiOH 水溶液中耐腐蚀性能的研究[J].核动力工程.2003,24(3):215-218,252.

[109] 刘文庆,李强,周邦新,姚美意.显微组织对 ZIRLO 锆合金耐腐蚀性的影响[J].核动力工程.2003,24(1):33-36.

[110] 刘文庆,李强,姚美意,周邦新.显微组织对 Zr-Sn-Nb 合金耐腐蚀性能的影响[J].原子能科学技术.2003(s1):140-144,156.

[111] 周邦新,李强,姚美意,刘文庆.锆-4 合金氧化膜中的晶粒形貌观察[J].稀有金属材料与工程.2003,32(6):417-419.

[112] 周邦新,李强,姚美意,刘文庆.扫描探针显微镜在锆合金氧化膜显微组织研究中的应用[J].原子能科学技术.2003(s1):153-156.

[113] 周邦新,李强,苗志,喻应华,姚美意.真空电子束焊接对锆合金耐腐蚀性能的影响[J].核动力工程.2003,24(3):236-240.

2004 年

[114] Liu Wenqing, Zhou Bangxin, Li Qiang, Yao Meiyi. Degradation of corrosion resistance of Zircaloy-4 in LiOH aqueous solution[J]. Rare Metals. 2004,23(3):286-288.

[115] 刘文庆,李强,周邦新,姚美意.LiOH 水溶液加速 Zr-4 合金腐蚀的研究[J].稀有金属材料与工程.

2004,33(7):728-730.

[116] 刘文庆,李强,周邦新,姚美意. 水化学对 Zr-4 合金氧化膜/基体界面处压应力的影响[J]. 稀有金属材料与工程. 2004,33(10):1112-1115.

[117] 刘文庆,陈文觉,李强,周邦新,姚美意. 水化学对 Zr-4 合金氧化膜形貌的影响[J]. 核动力工程. 2004,25(1):517-521.

[118] 马淳安,甘永平,褚有群,黄辉,成旦红,周邦新. Electro-oxidation behavior of tungsten carbide electrode in different electrolytes[J]. 中国有色金属学会会刊：英文版. 2004,14(1):11-14.

[119] 马淳安,黄烨,褚有群,成旦红,周邦新,徐志花,毛信表. 硝基甲烷在碳化钨电极上的电催化还原[J]. 浙江工业大学学报. 2004,32(2):119-122.

[120] 姚美意,周邦新,李强,刘文庆,苗志,喻应华. 合金成分对锆合金焊接区腐蚀时吸氢性能的影响[J]. 稀有金属材料与工程. 2004,33(6):641-645.

[121] 姚美意,李强,周邦新,苗志,喻应华,刘文庆. 热处理对含 Nb 锆合金焊接试样显微组织和耐腐蚀性能的影响[J]. 核动力工程. 2004,25(2):147-151.

[122] 周邦新,刘文庆,李强,姚美意. LiOH 水溶液提高 Zr-4 合金腐蚀速率的机理[J]. 材料研究学报. 2004,18(3):225-231.

2005 年

[123] Liu Wenqing, Zhou Bangxin, Li Qiang, Yao Meiyi. Detrimental role of LiOH on the oxide film formed on Zircaloy-4[J]. Corrosion Science. 2005,47(7):1855-1860.

[124] Liu Wenqing, Li Qiang, Zhou Bangxin, Yan Qingsong, Yao Meiyi. Effect of heat treatment on the microstructure and corrosion resistance of a Zr-Sn-Nb-Fe-Cr alloy[J]. Journal Nuclear Materials. 2005,341(2-3):97-102.

[125] 陈刚,杨小玲,高谈英,倪建森,徐晖,周邦新. 钕铁硼粘结磁体阴极电泳处理技术的研究[J]. 磁性材料及器件. 2005,36(3):33-35.

[126] 刘文庆,周邦新,李强,姚美意. 不同介质对 Zr-4 合金氧化膜结构的影响[J]. 稀有金属材料与工程. 2005,34(4):562-564.

[127] 刘文庆,李强,周邦新,严青松,姚美意. 热处理制度对 N18 新锆合金耐腐蚀性能的影响[J]. 核动力工程. 2005,26(3):249-253,287.

[128] 刘文庆,严青松,李强,周邦新,姚美意. 显微组织对 Zr-1.0Nb 合金耐腐蚀性能的影响[J]. 电子显微学报. 2005,24(4):307-307.

[129] 马淳安,张维民,李国华,郑遗凡,周邦新,成旦红. 介孔结构空心球状 WC 粉体催化剂的制备与表征[J]. 化学学报. 2005,63(12):1151-1154.

[130] Wang Junan, Zhou Bangxin, Li Qiang, Zhu Yuliang, Sun Huande. AlN+MnS inclusions in oriented electrical steels[J]. Trans. Nonferrous Met. Soc. China. 2005,15(2):460-463.

[131] 王卫国,冯柳,张欣,夏爽,周邦新. 冷轧变形 Pb-Ca 基合金在退火过程中的晶界特征分布[J]. 中国体视学与图像分析. 2005,10(4):215-217.

[132] Wang Zhanyong, Xu Hui, Ni Jiansen, Zhou Bangxin. Texture evolution in nanocomposite $Nd_2Fe_{14}B/\alpha$-Fe magnets prepared by direct melt spinning[J]. Journal of Rare Earths. 2005,23(3):298-301.

[133] 王占勇,周邦新,徐晖,倪建森. 快淬双相纳米复合稀土永磁材料的晶化研究[J]. 稀有金属材料与工程. 2005,34(1):1-6.

[134] 姚美意,周邦新,王均安. 电压对镁合金微弧氧化膜组织及耐蚀性的影响[J]. 材料保护. 2005,38(6):7-10.

[135] 赵文金,周邦新,苗志,彭倩,蒋有荣,蒋宏曼,庞华. 我国高性能锆合金的发展[J]. 原子能科学技术. 2005(s1):2-9.

[136] 周邦新,李强,姚美意,刘文庆,褚于良. 锆-4 合金在高压釜中腐蚀时氧化膜显微组织的演化[J]. 核动力工程. 2005,26(4):364 – 371.

[137] 周邦新. 三维原子探针——从探测逐个原子来研究材料的分析仪器[J]. 自然杂志. 2005,27(3):125 – 129.

2006 年

[138] Wang Zhanyong, Xu Hui, Ni Jiansen, Li Qiang, Zhou Bangxin. Effect of high magnetic field on the crystallization of $Nd_2 Fe_{14} B/\alpha$ - Fe nanocomposite magnets[J]. Rare Metals. 2006,25(4):337 – 341.

[139] Wang Zhanyong, Liu Wenqing, Li Qiang, Zhou Bangxin, Xu Hui, Ni Jiansen. Effect of Nb addition on the microstructure and magnetic properties $Nd_2 Fe_{14} B/\alpha$-Fe nanocomposite magnets[C]//19th International Vacuum Nanoelecronics Conference and 50th International Field Emission Symposium. 2006:81.

[140] Xia S, Zhou B X, Chen W J, Wang W G. Effects of Strain and Annealing Processes on the Distribution of Σ3 Boundaries in a Ni-based Superalloy[J]. Scripta Materialia. 2006,54(12):2019 – 2022.

[141] Yao M Y, Zhou B X, Li Q, Liu W Q, Chu Y L. The Effect of Alloying Modifications on Hydrogen Uptake of Zirconium-alloy Welding Specimens During Corrosion Tests[J]. Journal of Nuclear Materials. 2006,350:195 – 201.

[142] 陈刚,杨小玲,高谈英,倪建森,徐晖,周邦新. Cathode electrophoretic technology for bonded NdFeB permanent magnet[J]. 中国有色金属学会会刊：英文版. 2006,16(A02):93 – 96.

[143] 侯雪玲,周邦新,徐晖,倪建森,孔俊峰,张少杰. GdAl 磁制冷工质材料的磁热效应[J]. 稀有金属材料与工程. 2006,35(5):749 – 751.

[144] 黄照华,倪健森,徐晖,王占勇,周邦新. Preparation technology and magnetic properties of Nd9.5Fe77B6Co5Zr2.5 nanocomposite magnets[J]. 中国有色金属学会会刊：英文版. 2006,16(A02):97 – 99.

[145] 刘文庆,雷鸣,耿迅,李强,周邦新. 显微组织对 Zr – Sn – Nb – Fe 锆合金耐腐蚀性能的影响[J]. 材料热处理学报. 2006,27(6):47 – 51.

[146] 王卫国,周邦新,冯柳,张欣,夏爽. 冷轧变形 Pb – Ca – Sn – Al 合金在回复和再结晶过程中的晶界特征分布[J]. 金属学报. 2006,42(7):715 – 721.

[147] 王占勇,周邦新,倪建森,徐晖. Zr 对 $Nd_2 Fe_{14} B/\alpha$ – Fe 快淬纳米复合永磁材料的影响[J]. 河北工业大学学报. 2006,35(5):13 – 16.

[148] 王占勇,周邦新,倪建森,徐晖. 三维原子探针技术在纳米复合永磁材料中的应用[J]. 科学通报. 2006,51(12):1487 – 1488.

[149] 夏爽,周邦新,陈文觉,王卫国. 高温退火过程中铅合金晶界特征分布的演化[J]. 金属学报. 2006,42(2):129 –133.

[150] 姚美意,周邦新,李强,刘文庆,褚于良. 研究合金元素对锆合金耐腐蚀性能影响的单片试样法[J]. 稀有金属材料与工程. 2006,35(10):1651 – 1655.

[151] 周邦新,李强,刘文庆,姚美意,褚于良. 水化学及合金成分对锆合金腐蚀时氧化膜显微组织演化的影响[J]. 稀有金属材料与工程. 2006,35(7):1009 – 1016.

[152] 周邦新,姚美意,苗志,李强,刘文庆. 氧化-氢化引起的锆合金焊接件开裂问题[J]. 核动力工程. 2006, 27(1):34 – 36.

2007 年

[153] 方晓英,王卫国,郭红,张欣,周邦新. 304 不锈钢冷轧退火 $\Sigma3^n$ 特殊晶界分布研究[J]. 金属学报.

2007,43(12):1239 - 1244.

[154] 方晓英,王卫国,周邦新.金属材料晶界特征分布(GBCD)优化研究进展[J].稀有金属材料与工程.
2007,36(8):1500 - 1504.

[155] 侯雪玲,李士涛,周邦新,倪建森,徐辉,张少杰.添加元素 Sn 对 $Gd_5Si_2Ge_2$ 合金磁热效应的影响[J].
稀有金属材料与工程.2007,36(9):1605 - 1607.

[156] 雷鸣,刘文庆,严青松,李强,姚美意,周邦新.变形及热处理对 Zr - Sn - Nb 新锆合金第二相粒子的影
响[J].稀有金属材料与工程.2007,36(3):467 - 470.

[157] 李强,周邦新,姚美意,刘文庆,褚于良.锆合金在 550 ℃/25 MPa 超临界水中的腐蚀行为[J].稀有金
属材料与工程.2007,36(8):1358 - 1361.

[158] 刘文庆,刘庆冬,李聪,方淑芳,周邦新.铌钒氮微合金钢中碳氮化合物研究[J].材料热处理学报.
2007,28(s1):9 - 13.

[159] 刘文庆,雷鸣,耿迅,李强,姚美意,周邦新.热加工对 Zr - Sn - Nb 锆合金显微组织和耐腐蚀性能的影
响[J].原子能科学技术.2007,41(6):711 - 715.

[160] 刘文庆,周邦新,刘庆冬.温度和脉冲频率对三维原子探针测试结果的影响[J].真空科学与技术学
报.2007,27(增刊):53 - 56.

[161] 严青松,刘文庆,李强,姚美意,周邦新.热处理制度对 Zr - Sn - Nb 新锆合金耐腐蚀性能的影响[J].
稀有金属材料与工程.2007,36(1):104 - 107.

[162] 姚美意,周邦新,李强,刘文庆,王树安,黄新树.第二相对 Zr - 4 合金在 400 ℃过热蒸汽中腐蚀吸氢行
为的影响[J].稀有金属材料与工程.2007,36(11):1915 - 1919.

[163] 姚美意,周邦新,李强,刘文庆,虞伟均,褚于良.热处理对 Zr - 4 合金在 360 ℃ LiOH 水溶液中腐蚀行
为的影响[J].稀有金属材料与工程.2007,36(11):1920 - 1923.

[164] 张海兵,王均安,杨光,周邦新,余海峰.Cr11 铁素体不锈钢退火过程中织构变化[J].上海金属.2007,
29(4):17 - 21.

[165] 周邦新,姚美意,李强,夏爽,刘文庆,褚于良.Zr - Sn - Nb 合金耐疖状腐蚀性能的研究[J].稀有金属
材料与工程.2007,36(8):1317 - 1321.

[166] 周邦新,李强,姚美意,夏爽,刘文庆,褚于良.热处理影响 Zr - 4 合金耐疖状腐蚀性能的机制[J].稀有
金属材料与工程.2007,36(7):1129 - 1134.

[167] 周邦新,刘文庆.三维原子探针及其在材料科学研究中的应用[J].材料科学与工艺.2007,15(3):
405 - 408.

2008 年

[168] Yao M Y, Zhou B X, Li Q, Liu W Q, Geng X, Lu Y P. A Superior Corrosion Behavior of Zircaloy -
4 in Lithiated Water at 360 ℃/18. 6 MPa by β - Quenching[J]. Journal of Nuclear Materials. 2008,
374 (1 - 2):197 - 203.

[169] Xia Shuang, Zhou Bangxin, Chen Wenjue. Effect of Single-Step Strain and Annealing on Grain
Boundary Character Distribution and Intergranular Corrosion in Alloy 690[J]. Journal of Materials
Science. 2008,43:2990 - 3000.

[170] Wang Weiguo, Yin Fuxing, Guo Hong, Li He, Zhou Bangxin. Effects of recovery treatment after
large strain on the grain boundary character distributions of subsequently cold rolled and annealed Pb -
Ca - Sn - Al alloy[J]. Materials Science and Engineering A - Structural Materials Properties
Microstructure and Processing. 2008,491(1 - 2):199 - 206.

[171] Zhou B X, Li Q, Yao M Y, Liu W Q, Chu Y L. Effect of Water Chemistry and Composition on
Microstructural Evolution of Oxide on Zr Alloys[J]. Journal of ASTM International, ASTM
symposium on zirconium in the nuclear industry:15th international symposium on 24 - 28 June 2007 in

Sunriver OR. 2008,5(2):1-21.

[172] 曾奇锋,周邦新,姚美意,彭剑超,夏爽.添加合金元素 Cu 和 Mn 对锆合金中第二相的影响[J].上海大学学报:自然科学版.2008,14(5):531-536.

[173] 陈刚,杨小玲,徐晖,倪建森,周邦新.温压钕铁硼粘结磁体制备技术的研究[J].稀有金属材料与工程.2008,37(2):308-311.

[174] 李强,周邦新,姚美意,刘文庆,褚于良.Zr-2.5Nb 合金在 550 ℃/25 MPa 超临界水中腐蚀时的氢致 α/β 相变[J].稀有金属材料与工程.2008,37(10):1815-1818.

[175] 刘庆冬,褚于良,王泽民,刘文庆,周邦新.Nb-V 微合金钢中渗碳体周围元素分布的三维原子探针表征[J].金属学报.2008,44(11):1281-1285.

[176] 刘庆冬,刘文庆,王泽民,周邦新.Nb-V 微合金钢中碳化物析出的三维原子探针表征[J].金属学报.2008,44(7):786-790.

[177] 刘文庆,王泽民,刘庆冬,李强,周邦新,姚美意.变形热处理对 Zr-1.0Nb 合金耐腐蚀性能的影响[J].材料科学与工艺.2008,16(3):400-402.

[178] 刘文庆,褚于良,王泽民,李聪,方淑芳,周邦新.铌-钒微合金钢中碳氮化合物的析出特点[J].理化检验:物理分册.2008,44(10):540-543.

[179] 刘文庆,刘庆冬,周邦新.三维原子探针对 Ni₄Mo 合金的研究[J].稀有金属材料与工程.2008,37(10):1719-1722.

[180] 刘文庆,刘庆冬,李聪,方淑芳,周邦新.三维原子探针对微合金钢中 G.P.区的观测[J].江苏大学学报:自然科学版.2008,29(2):131-133.

[181] 鲁艳萍,姚美意,周邦新.热处理对 N36 锆合金腐蚀与吸氢性能的影响[J].上海大学学报:自然科学版.2008,14(2):194-199.

[182] 王均安,贺英,邱振伟,周邦新,Frantisek Kovac.小形变量轧制下电工钢中立方织构的形成[J].上海大学学报:自然科学版.2008,14(5):461-466.

[183] 夏爽,周邦新,陈文觉.690 合金的晶界特征分布及其对晶间腐蚀的影响[J].电子显微学报.2008,27(6),461-468.

[184] 夏爽,周邦新,陈文觉.形变及热处理对 690 合金晶界特征分布的影响[J].稀有金属材料与工程.2008,37(6),999-1003.

[185] 姚美意,周邦新,李强,夏爽,刘文庆.微量 Nb 的添加对 Zr-4 合金耐疖状腐蚀性能的影响[J].上海金属.2008,30(6):1-3.

[186] 周邦新,姚美意,李强,夏爽,刘文庆.Zr-4 合金薄板的织构与耐疖状腐蚀性能的关系[J].上海大学学报:自然科学版.2008,14(5):441-445.

[187] 朱娟娟,王伟,林民东,刘文庆,王均安,周邦新.用三维原子探针研究压力容器模拟钢中富铜原子团簇的析出[J].上海大学学报:自然科学版.2008,14(5):525-530.

2009 年

[188] Xia Shuang, Zhou Bangxin, Chen Wenjue. Grain Cluster Microstructure and Grain Boundary Character Distribution in Alloy 690[J]. Metallurgical and Materials Transactions A. 2009,40A:3016-3030.

[189] 黄照华,李强,张士岩,周邦新.快淬 NdFeB 磁粉磁性能不均匀性问题的研究[J].稀有金属材料与工程.2009,38(2):247-250.

[190] 李慧,夏爽,周邦新,倪建森,陈文觉.镍基 690 合金时效过程中晶界碳化物的形貌演化[J].金属学报.2009,45(2):195-198.

[191] 李慧,夏爽,周邦新,陈文觉.晶界类型及时效处理对 690 合金耐晶间腐蚀性能的影响[C]// 中国核科学技术进展报告(第一卷):核材料分卷.2009:50-55.

[192] 刘庆冬,刘文庆,王泽民,周邦新.回火马氏体中合金碳化物的 3D 原子探针表征(Ⅰ.形核)[J].金属学报.2009,45(11):1281-1287.

[193] 刘庆冬,彭剑超,刘文庆,周邦新.回火马氏体中合金碳化物的 3D 原子探针表征(Ⅱ.长大)[J].金属学报.2009,45(11):1288-1296.

[194] 刘庆冬,褚于良,彭剑超,刘文庆,周邦新.回火马氏体中合金碳化物的 3D 原子探针表征(Ⅲ.粗化)[J].金属学报.2009,45(11):1297-1302.

[195] 刘文庆,王泽民,刘庆冬,李强,姚美意 周邦新.Zr-Sn-Nb-Fe 合金显微组织及耐腐蚀性能研究[J].原子能科学技术.2009,43(7):630-635

[196] 王坤,陈文觉,夏爽,周邦新.高温退火过程中 316 不锈钢晶界特征分布的演化[J].上海金属.2009,31(5):13-18.

[197] 夏爽,周邦新,陈文觉,李慧.690 合金中低 $\sum$ CSL 晶界分布规律的研究.现代化工、冶金与材料技术前沿[C]//中国工程院化工、冶金与材料工程学部第七届学术会议论文集.2009:1669-1672.

[198] 张欣,姚美意,周邦新,耿建桥,王锦红,李士炉.热处理对 N18 锆合金在 360℃LiOH 水溶液中腐蚀行为的影响[C]//中国核科学技术进展报告(第一卷):核材料分卷.2009:226-231.

[199] 周邦新,李强,姚美意,刘文庆,褚于良.Zr-4 合金氧化膜的显微组织研究[J].腐蚀与防护.2009,30(9):589-594,610.

2010 年

[200] Xia Shuang, Zhou Bangxin, Chen Wenjue, Luo Xin, Li Hui. Features of Highly Twinned Microstructures Produced by GBE in FCC Materials[J]. Materials Science Forum. 2010, 638-642: 2870-2875.

[201] 方晓英,王卫国,Rohrer G S,周邦新.冷轧退火后铁素体不锈钢的晶界面分布[J].金属学报.2010,46(4):404-410.

[202] 侯雪玲,张鹏,胡星浩,徐晖,倪健森,周邦新.$Gd_{0.95}Nb_{0.05}$ 合金磁热效应的研究[J].稀有金属材料与工程.2010,39(1):126-128.

[203] 侯雪玲,胡星浩,汪学真,曾智,徐晖,周邦新.低磁场下获得巨磁热效应的 GdSiGeZn 合金[J].上海大学学报:自然科学版.2010,16(1):35-37.

[204] 李慧,夏爽,胡长亮,周邦新,陈文觉.利用 EBSD 技术对 690 合金不同类型晶界处碳化物形貌的研究[J].电子显微学报.2010,29(1):730-735.

[205] 李强,刘仁多,周邦新,姚美意.变形及热处理影响 Zr-4 合金显微组织的研究[J].上海金属.2010,32(6):5-8.

[206] 林民东,朱娟娟,王伟,周邦新,刘文庆,徐刚.核反应堆压力容器模拟钢中富 Cu 原子团簇的析出与嵌入原子势计算[J].物理学报.2010,59(2):1163-1168.

[207] 刘文庆,朱晓勇,王晓姣,李强,姚美意,周邦新.Nb 元素和 Fe 元素对锆合金耐腐蚀性能的影响[J].原子能科学技术.2010,44(12):1477-1481.

[208] 刘文庆,朱晓勇,王晓姣,钟柳明,周邦新.铌钒微合金钢中碳化物的析出过程[J].材料科学与工艺.2010,18(2):164-167.

[209] 罗鑫,夏爽,李慧,周邦新,陈文觉.晶界特征分布对 304 不锈钢应力腐蚀开裂的影响[J].上海大学学报:自然科学版.2010,16(2):177-182.

[210] 王均安,王辉,周邦新.低温退火对铁硅合金中立方织构形成的影响[J].电子显微学报.2010,9(5):468-474.

[211] 王伟,朱娟娟,林民东,周邦新,刘文庆.核反应堆压力容器模拟钢中富 Cu 纳米团簇析出早期阶段的研究[J].北京科技大学学报.2010,32(1):39-43.

[212] 夏爽,罗鑫,周邦新,陈文觉,李慧.304 不锈钢中"晶粒团簇"显微组织的特征与晶界特征分布的关系

[J].电子显微学报.2010,29(1):678-683.

[213] 夏爽,李慧,周邦新,陈文觉.金属材料中退火孪晶的控制及利用——晶界工程研究[J].自然杂志.
2010,32(2):94-100.

[214] 杨波,李谋成,姚美意,周邦新,沈嘉年.高温高压水环境中锆合金腐蚀的原位阻抗谱特征[J].金属学
报.2010,46(8):946-950.

[215] 钟祥玉,杨波,李谋成,姚美意,周邦新,沈嘉年.Zr-4合金表面氧化膜的电化学阻抗谱特征[J].稀有
金属材料与工程.2010,39(12):2165-2168.

2011 年

[216] 张欣,姚美意,李士炉,周邦新.加工工艺对 N18 锆合金在 360 ℃/18.6 MPa LiOH 水溶液中腐蚀行
为的影响[J].金属学报.2011,47(09):1112-1116.

[217] Zhou B X, Peng J C, Yao M Y, Li Q, Xia S, Du C X, Xu G. Study of the Initial Stage and
Anisotropic Growth of Oxide Layers Formed on Zircaloy-4[J]. Journal of ASTM International. 2011,
8(1):620-648

[218] Yao M Y, Wang J H, Peng J C, Zhou B X, Li Q. Study on the Role of Second Phase Particles in
Hydrogen Uptake Behavior of Zirconium Alloys[J]. Journal of ASTM International. 2011,8(2):466-
495.

[219] 曹潇潇,姚美意,彭剑超,周邦新.$Zr(Fe_x,C_{rl-x})_2$合金在 400 ℃过热蒸汽中腐蚀行为研究[J].金属学
报.2011,47(7):882-886.

[220] 楚大锋,徐刚,王伟,彭剑超,王均安,周邦新.APT 和萃取复型研究压力容器模拟钢中富 Cu 团簇的
析出[J].金属学报.2011,47(3):269-274.

[221] 杜晨曦,彭剑超,李慧,周邦新.Zr-4 合金腐蚀初期氧化膜的显微组织研究[J].金属学报.2011,47
(7):887-892.

[222] 耿建桥,周邦新,姚美意,王锦红,张欣,李士炉,杜晨曦.水化学和腐蚀温度对锆合金氧化膜中压应力
的影响[J].上海大学学报:自然科学版.2011,17(3):293-296.

[223] 胡长亮,夏爽,李慧,刘廷光,周邦新,陈文觉.晶界网络特征对 304 不锈钢晶间应力腐蚀开裂的影响
[J].金属学报.2011,47(7):939-945.

[224] 李慧,夏爽,周邦新,陈文觉,刘廷光,胡长亮.690 合金中晶界网络分布的控制及其对晶间腐蚀性能的
影响[J].中国材料进展.2011,30(5):11-14.

[225] 李慧,夏爽,周邦新,彭剑超.镍基 690 合金中晶界碳化物析出的研究[J].金属学报.2011,47
(7):853-858.

[226] 李慧,夏爽,周邦新,刘文庆.原子探针层析方法研究 690 合金晶界偏聚的初步结果[J].电子显微学
报.2011,30(3):206-209.

[227] 李慧,夏爽,周邦新,刘文庆.原子探针层析技术对 Ni-Cr-Fe 合金晶界偏聚的研究[C]//2011 中国
材料研讨会论文摘要集.2011:701.

[228] 李强,梁雪,彭剑超,余康,姚美意,周邦新.Cu 对 Zr-2.5Nb 合金在 500 ℃/10.3 MPa 过热蒸汽中腐
蚀行为的影响[J].金属学报.2011,47(7):877-881.

[229] 李强,梁雪,彭剑超,刘仁多,余康,周邦新.Zr-2.5Nb 合金中 β-Nb 相的氧化过程[J].金属学报.
2011,47(7):893-898.

[230] 李士炉,姚美意,张欣,耿建桥,彭剑超,周邦新.添加 Cu 对 M5 合金在 500 ℃过热蒸汽中耐腐蚀性能
的影响[J].金属学报.2011,47(2):163-168.

[231] 刘廷光,夏爽,李慧,周邦新,陈文觉.690 合金原始晶粒尺寸对晶界工程处理后晶界网络的影响[J].
金属学报.2011,47(7):859-864.

[232] 刘文庆,钟柳明,彭剑超,刘仁多,姚美意,周邦新.锆合金中第二相的研究[J].稀有金属材料与工程.

2011,40(07):1216 - 1219.

[233] 彭剑超,李强,刘仁多,姚美意,周邦新. Zr - 4 合金中氢化物析出长大的透射电镜原位研究[J]. 稀有金属材料与工程.2011,40(8):1377 - 1381.

[234] 沈月锋,姚美意,张欣,李强,周邦新,赵文金. β 相水淬对 Zr - 4 合金在 LiOH 水溶液中耐腐蚀性能的影响[J]. 金属学报.2011,47(7):899 - 904.

[235] 王锦红,姚美意,周邦新,耿建桥,夏爽,曹潇潇. Fe/Cr 比对 Zr(Fe,Cr)₂ 吸氢性能的影响[J]. 稀有金属材料与工程.2011, 40(6):1084 - 1088.

[236] 王锦红,姚美意,周邦新,耿建桥,张欣,张金龙. Zr - Sn 系合金在过热蒸汽中的腐蚀吸氢行为[J]. 稀有金属材料与工程.2011, 40(5):833 - 838.

[237] 夏爽,李慧,周邦新,陈文觉,姚美意,李强,刘文庆,王均安,褚于良,彭剑超,张金龙. 核电站关键材料中的晶界工程问题[J]. 上海大学学报:自然科学版.2011,17(4):522 - 528.

[238] 徐刚,楚大锋,蔡琳玲,周邦新,王伟,彭剑超. RPV 模拟钢中纳米富 Cu 相的析出和结构演化研究[J]. 金属学报.2011,47(7):905 - 911.

[239] 姚美意,张宇,李士炉,张欣,周军,周邦新. Cu 含量对 Zr - 0.80Sn - 0.34Nb - 0.39Fe - 0.10Cr - xCu 合金在 500 ℃ 过热蒸汽中耐腐蚀性能的影响[J]. 金属学报.2011,47(7):872 - 876.

[240] 姚美意,李士炉,张欣,彭剑超,周邦新,赵旭山,沈剑韵. 添加 Nb 对 Zr - 4 合金在 500 ℃ 过热蒸汽中耐腐蚀性能的影响[J]. 金属学报.2011,47(7):865 - 871.

[241] 周邦新,王均安,刘庆东,刘文庆,王伟,林民东,徐刚,楚大锋. Ni 对 RPV 模拟钢中富 Cu 原子团簇析出的影响[J]. 中国材料进展.2011,30(5):1 - 6.

2012 年

[242] Zhou B X, Yao M Y, Li Z K, Wang X M, Zhou J, Long C S, Liu Q, Luan B F. Optimization of N18 Zirconium Alloy for Fuel Cladding of Water Reactors[J]. J. Mater. Sci. Technol. 2012, 28(7):606 - 613.

[243] Fang Xiaoying, Wang Weigou, Guo Hong, Qin Congxiang, Zhou Bangxin. Evolutions of texture and grain boundary plane distributions a ferritic stainless steel[J]. Journal of Central South University. 2012,19:3363 - 3368.

[244] Li Hui, Xia Shuang, Zhou Bangxin, Liu Wenqing. C - Cr Segregation at Grain Boundary before the Carbide Nucleation in Alloy 690[J]. Materials Characterization. 2012,66:68 - 74.

[245] 蔡琳玲,徐刚,冯柳,王均安,彭剑超,周邦新. 核反应堆压力容器模拟钢中纳米富 Cu 相的变形特征[J]. 上海大学学报:自然科学版.2012,18(3):311 - 316.

[246] 李强,梁雪,周邦新,姚美意,彭剑超. Cu 在 Zr - 2.5Nb - 0.5Cu 合金及其腐蚀生成氧化膜中的存在形式[J]. 稀有金属材料与工程.2012,41(1):92 - 95.

[247] 李慧,夏爽,刘文庆,彭剑超,周邦新. 镍基 690 合金晶界成分演化规律的研究[J]. 核动力工程.2012,33(s2):65 - 69.

[248] 李强,余康,刘仁多,梁雪,姚美意,周邦新. 添加 2 wt% Cu 对 Zr - 4 合金显微结构和耐腐蚀性能的影响[J]. 上海金属.2012,34(4):1 - 6.

[249] 李强,黄昌军,杨艳平,徐龙,梁雪,周邦新. N18 锆合金疖状腐蚀问题研究[J]. 核动力工程.2012,33(s2):22 - 27,38.

[250] 孙国成,周邦新,姚美意,谢世敬,李强. 锆合金在 LiOH 水溶液中腐蚀的各向异性研究[J]. 金属学报.2012,48(9):1103 - 1108.

[251] 谢兴飞,张金龙,朱莉,姚美意,周邦新,彭剑超. Zr - 0.7Sn - 0.35Nb - 0.3Fe - xGe 合金在高温高压 LiOH 水溶液中耐腐蚀性能的研究[J]. 金属学报.2012,48(12):1487 - 1494.

[252] 徐刚,蔡琳玲,冯柳,周邦新,王均安,张海生. 富 Cu 团簇的析出对 RPV 模拟钢韧-脆转变温度的影响[J]. 金属学报.2012,48(6):753 - 758.

[253] 徐刚,蔡琳玲,冯柳,周邦新,刘文庆,王均安. 利用 APT 对 RPV 模拟钢中富 Cu 原子团簇析出的研究[J]. 金属学报. 2012,48(4): 407 - 413.

[254] 徐刚,蔡琳玲,冯柳,周邦新,刘文庆,王均安. 利用 APT 对 RPV 模拟钢中界面上原子偏聚特征的研究[J]. 金属学报. 2012,48(7): 789 - 796.

[255] 姚美意,邹玲红,谢兴飞,张金龙,彭剑超,周邦新. 添加 Bi 对 Zr - 4 合金在 400 ℃/10.3 MPa 过热蒸汽中耐腐蚀性能的影响[J]. 金属学报. 2012,48(9): 1097 - 1102.

[256] 姚美意,张伟鹏,周军,周邦新,李强. 退火温度对 Zr - 0.85Sn - 0.16Nb - 0.38Fe - 0.18Cr 合金耐腐蚀性能的影响[J]. 核动力工程. 2012,33(6): 88 - 92.

2013 年

[257] Feng Liu, Zhou Bangxin, Peng Jianchao, Wang Junan. Crystal Structure Evolution of the Cu-Rich Nano Precipitates from bcc to 9R in Reactor Pressure Vessel Model Steel[J]. Acta Metallurgica Sinica (English Letters). 2013,26(6):707 - 712.

[258] Li Hui, Xia Shuang, Zhou Bangxin, Peng Jianchao. The growth mechanism of grain boundary carbide in Alloy 690[J]. Materials Characterization. 2013,81:1 - 6.

[259] Li Hui, Xia Shuang, Liu Wenqing, Liu Tingguang, Zhou Bangxin. Atomic scale study of grain boundary segregation before carbide nucleation in Ni - Cr - Fe Alloys[J]. Journal of Nuclear Materials. 2013,439(1 - 3):57 - 64.

[260] Liu Tingguang, Xia Shuang, Li Hui, Zhou Bangxin, Bai Qin, Su Cheng, Cai Zhigang. Effect of initial grain sizes on the grain boundary network during grain boundary engineering in Alloy 690[J]. Journal of Materials Research. 2013, 28(9):1165 - 1176.

[261] Yao M Y, Shen Y F, Li Q, Peng J C, Zhou B X, Zhang J L. The Effect of Final Annealing after β - Quenching on the Corrosion Resistance of Zircaloy - 4 in Lithiated Water with 0.04 M LiOH[J]. Journal of Nuclear Materials. 2013,435: 63 - 70.

[262] 贾向南,王均安,蔡琳玲,徐刚,王晓娇,刘文庆,周邦新. RPV 模拟钢热时效过程中碳化物与基体界面元素的偏聚[J]. 上海大学学报: 自然科学版. 2013,19(1):54 - 60.

[263] 李强,黄昌军,梁雪,彭剑超,周邦新. N18 锆合金的疖状腐蚀[J]. 腐蚀与防护. 2013,34(8):655 - 658.

[264] 茹祥坤,刘廷光,夏爽,周邦新,马爱利,郑玉贵. 形变及热处理对白铜 BIO 合金晶界特征分布的影响[J]. 中国有色金属学报. 2013,23(8):2176 - 2181.

[265] 杨辉辉,刘廷光,夏爽,李慧,周邦新,白琴. 利用晶界工程技术优化 H68 黄铜中的晶界网络[J]. 上海金属. 2013,35(5):9 - 13,62.

[266] 张金龙,谢兴飞,姚美意,周邦新,彭剑超,梁雪. Zr - 1Nb - 0.7Sn - 0.03Fe - xGe 合金在 360 ℃ LiOH 水溶液中耐腐蚀性能的研究[J]. 金属学报. 2013,49(4):443 - 450.

[267] 张金龙,谢兴飞,姚美意,周邦新,彭剑超,李强. Ge 含量对 Zr - 4 合金在 LiOH 水溶液中耐腐蚀性能的影响[J]. 中国有色金属学报. 2013,(6):1542 - 1548.

[268] 张伟鹏,姚美意,朱莉,张金龙,周邦新,李强. Zr - 0.8Sn - 0.35Nb - 0.4Fe - 0.1Cr - xBi 合金在 400 ℃过热蒸汽中的腐蚀行为[J]. 腐蚀与防护. 2013,(6):463 - 467,481.

[269] 张欣,姚美意,李中奎,周军,李强,周邦新. Zr - 0.80Sn - 0.4Nb - 0.4Fe - 0.10Cr - xCu 合金在 400 ℃过热蒸气中的耐腐蚀性能[J]. 稀有金属材料与工程. 2013,42(6):1210 - 1214.

[270] 朱莉,姚美意,孙国成,陈文觉,张金龙,周邦新. 添加 Bi 对 Zr - 1Nb 合金在 360 ℃和 18.6MPa 去离子水中耐腐蚀性能的影响[J]. 金属学报. 2013,49(1): 51 - 57.

后　记

　　作为上海大学文库最重要的组成部分——"上海大学院士文库"的建设将集中展示院士在上海大学学科建设、教学科研中的领军作用,体现院士们的学术风采、人格魅力,从而能够起到激励后学的作用。

　　目前已经收集出版了我校部分院士的著作、文集(文选),如:钱伟长院士的《钱伟长文选》(五卷)、《钱伟长学术论文集》(四卷)、《弹性板壳的内禀理论》(钱伟长博士学位论文)、《格林函数和变分法在电磁场和电磁波计算中的应用》、《教育与教育问题的思考》等,徐匡迪院士的《徐匡迪文集(钢铁冶金卷)》(AB卷)、《我的学术生涯》,以及刘高联院士的《刘高联文选》(上下卷)、刘元方院士的《刘元方文集》、傅家谟院士的《傅家谟文集》等。

　　上海大学图书馆敢为人先,率先在开发利用图书馆海量数据库方面进行大胆尝试,更是独立立项"上海大学院士文库",这无疑为图书馆服务社会尤其是服务学校开了个好头,是具有功在当代的意义。在建设"上海大学院士文库"的同时,上海大学图书馆和上海大学材料学院、上海大学出版社多次论证,并征求了周院士本人的意见,决定出版《周邦新文选》,作为上海大学新组建20周年校庆和周邦新院士80华诞的献礼。

　　本"文选"收录周邦新院士部分论文近200篇,总计200多万字。为尽可能真实再现周院士半个多世纪来的科研风貌,按论文发表的时间顺序编排;为尽可能保持论文发表时的原貌,内容原则上编辑不作改动;为便于读者更全面地了解周院士科研工作的全貌,书后附有周院士(科研团队)的论文目录清单。

　　此书在编辑过程中,上海大学图书馆、上海大学材料学院和上海大学出版社的多位老师付出了很多辛勤的工作和努力;学校各有关部门、各级领导给予了大力支持,在此一并表示感谢!

《周邦新文选》编辑小组

2014年4月15日